py	pyridine
PyBroP®	bromotripyrrolidinophosphonium hexafluorophosphate
RCM	ring-closing metathesis
Red-Al®	sodium bis(2-methoxyethoxy) aluminum hydride
ROM	ring-opening metathesis
ROMP	ring-opening metathesis polymerization
r. s. m.	recovered starting material
SEM	2-(trimethylsilyl)ethoxymethyl
SET	single electron transfer
TBAF	tetra-*n*-butylammonium fluoride
TBAI	tetra-*n*-butylammonium iodide
TBDPS	*tert*-butyldiphenylsilyl
TBS	*tert*-butyldimethylsilyl
TEMPO	2,2,6,6-tetramethyl-1-piperidinyloxy, free radical
Teoc	2-(trimethylsilyl)ethoxycarbonyl
TES	triethylsilyl
Tf	trifluoromethanesulfonate
Tfa	trifluoroacetamide
TFA	trifluoroacetic acid
TFAA	trifluoroacetic anhydride
Thexyl	1,1,2-trimethylpropyl
THF	tetrahydrofuran
THP	tetrahydropyranyl
TIPB	1,3,5-triisopropylbenzene
TIPS	triisopropylsilyl
TMEDA	*N*,*N*,*N'N'*-tetramethylethylenediamine
TMGA	tetramethylguanidinium azide
TMP	tetramethylpiperidide
TMS	trimethylsilyl
TMSE	2-(trimethylsilyl)ethyl
TMSEE	(trimethylsilyl)ethynyl ether
tol	tolyl
TPAP	tetra-*n*-propylammonium perruthenate
TPS	triphenylsilyl
Tr	trityl, triphenylmethyl
Trisyl	2,4,6-triisopropylbenzenesulfonyl
Troc	2,2,2-trichloroethoxycarbonyl
TS	transition state
Ts	4-toluenesulfonyl
TTBP	tri-*tert*-butylpyrimidine

Further Reading from Wiley-VCH

Nicolaou, K. C. / Sorensen, E. J.
Classics in Total Synthesis – Targets, Strategies, Methods
1996. ISBN 3-527-29284-5 (Hardcover)
1996. ISBN 3-527-29231-4 (Softcover)

Nicolaou, K. C. / Hanko, R. / Hartwig, W. (Eds.)
Handbook of Combinatorial Chemistry –
Drugs, Catalysts, Materials, 2 Vols.
2002. ISBN 3-527-30509-2

Reichardt, C.
Solvents and Solvent Effects in Organic Chemistry, 3. Ed.
2002. ISBN 3-527-30618-8

Wasserscheid, P. / Welton, T. (Eds.)
Ionic Liquids in Synthesis
2002. ISBN 3-527-30515-7

Classics in Total Synthesis II

K. C. Nicolaou and
S. A. Snyder

More Targets, Strategies, Methods

With a Foreword by
E. J. Corey

WILEY-VCH GmbH & Co. KGaA

Prof. Dr. K. C. Nicolaou

Department of Chemistry
The Scripps Research Institute
10550 North Torrey Pines Road
La Jolla, CA 92037
USA

and

Department of Chemistry and Biochemistry
University of California, San Diego
9500 Gilman Drive
La Jolla, CA 92093
USA

Scott A. Snyder

Department of Chemistry
The Scripps Research Institute
10550 North Torrey Pines Road
La Jolla, CA 92037
USA

Library of Congress Card No.: applied for

A catalogue record for this book is available from the British Library.

Bibliographic information published by Die Deutsche Bibliothek
Die Deutsche Bibliothek lists this publication in the Deutsche Nationalbibliografie; detailed bibliographic data is available in the Internet at http://dnb.ddb.de

Printed in the Federal Republic of Germany.
Printed on acid-free paper.

Typesetting: Hagedorn Kommunikation, Viernheim
Printing and Bookbinding: aprinta Druck GmbH & Co. KG, Wemding

ISBN Hardcover: 3-527-30685-4
Softcover: 3-527-30684-6

Foreword

The appearance of *Classics in Total Synthesis* by K. C. Nicolaou and E. J. Sorensen in 1996 brought great benefit to the practitioners and students of chemical synthesis as applied to complex naturally occurring molecules. It also stands as a milestone to mark the closing of the twentieth century, which encompassed both the birth of the field and its growth to one of the most intellectually fulfilling and practically important areas of modern science. Now, with the advent of the twenty-first century, we are fortunate to have *Classics in Total Synthesis II* by K. C. Nicolaou and Scott A. Snyder which, following in the tradition of *Classics I*, brings a detailed and authoritative exposition of more than thirty syntheses that have been published during the decade 1993-2003. The accounts of these syntheses are presented with considerable attention to background, strategic ideas, new methodology and unusual chemistry. They are engaging, interesting and crystal clear. One cannot help but be impressed with the care and attention to detail that the authors have taken. Their work also pays tribute to the field of synthesis, as did *Classics I*, because it conveys the dynamism and excitement of today's synthetic chemistry.

Given the enormous advances in the synthesis of complex naturally occurring molecules, especially since the post World War II era, and the ever greater sophistication and complexity of the field, it is tempting to speculate that we may be reaching some kind of scientific or intellectual plateau in which the "golden era" of discovery is behind us. How many challenging and worthy synthetic targets remain to be discovered? How many truly powerful and general new strategies and synthetic reactions remain to be discovered? Is there a prospect for the development of entirely new ways of planning or executing synthesis? In my judgment the opportunities for new developments and discoveries are so vast that today's synthesis is best regarded as a youngster with a brilliant future. The careful study of *Classics I* and *II* will help to show the way to the new synthesis of the future. Thanks are due to Nicolaou and Snyder for their wonderful role in helping to advance synthesis to ever higher levels and newer ideas.

E. J. Corey
Harvard University
7 May 2003

Preface

Although organic synthesis played a pivotal role in many of the most exciting and important scientific breakthroughs of the 20th century, its part in the expected discoveries and inventions of the 21st century is certain to be even more profound. Yet, for this field to fulfill its future obligations, the present generation of synthetic chemists must successfully address two distinct, but equally challenging tasks: advancing chemical synthesis to new heights, and training the next generation of synthetic organic chemists on whose shoulders will lie the responsibility of sharpening this art and science even further. It was precisely with these objectives in mind that *Classics in Total Synthesis* was launched in 1996; its widely acknowledged relevance to these goals today is reflected in the fact that educators, students, and researchers throughout the world embrace it as an educational tool and a source of inspiration.

Nevertheless, because the past decade has witnessed such an unprecedented flourishing in the field of total synthesis, we felt a second volume was needed to assist the original compilation in accomplishing the objectives laid out above. What you are currently holding in your hands is this new book, a treatise that is not meant to be an update of *Classics I*, but rather a unique textbook that both complements and augments the former through its presentation of over thirty innovative and instructive total syntheses accomplished since 1992. Our expectation is that by employing the general pedagogical format of *Classics I* in conjunction with special mini-review sections, which cover important new synthetic reactions, emerging trends, and additional syntheses of the chapter molecule, this text will become an equally valuable tool as its predecessor.

It goes without saying, of course, that bringing *Classics in Total Synthesis II* to fruition required the assistance of many talented individuals. First and foremost, we would like to thank Dr. Tamsyn Montagnon for her thoughtful suggestions and critical comments that improved the manuscript during all of its evolutionary stages. Janise L. Petrey was invaluable for her careful editing of the entire text which ensured its smooth reading and timely completion. We are also grateful to William E. Brenzovich, David Y.-K. Chen, Scott T. Harrison, and Casey J. N. Mathison for their proof-reading of the text in the final stages of production, and to Dr. Helen J. Mitchell and Professors Phil S. Baran, Thomas J. Katz, Erik J. Sorensen, and Georgios E. Vassilikogiannakis for their thoughtful input which enhanced the quality of certain chapters. Some of the specialized graphics, including the cover design, are the product of Robbyn M. Echon's artistic talents. We thank all of our collaborators at Wiley-VCH, especially Dr. Elke Maase, Dr. José Oliveira, and Dr. Peter Gölitz for their hard work and diligence throughout the course of this project. Finally, we owe much to our immediate families (Georgette, Colette, Alex, Chris, and P. J. Nicolaou, and Cathy Warren) for their support, encouragement, and patience during the many months that this text has occupied our hearts and minds.

In closing, we wish to dedicate this book to our mentors who embedded within us the passion for synthesis. If this text can educate and inspire more young scientists in the same way that they have, then we will regard this book as a great success and an immense source of personal satisfaction.

La Jolla
April 2003

K. C. Nicolaou
Scott A. Snyder

About the Authors

K. C. Nicolaou is Professor of Chemistry at the University of California, San Diego and is Chairman of the Department of Chemistry and holds the Aline W. and L. Skaggs Professorship of Chemical Biology and the Darlene Shiley Chair in Chemistry at The Scripps Research Institute. His impact on chemistry, biology and medicine flows from his works in chemical synthesis and chemical biology described in hundreds of publications and 55 patents. For his contributions to research and education, he was elected a Member of the National Academy of Sciences, USA, a Fellow of the American Academy of Arts and Sciences, and a Foreign Member of the Academy of Athens, Greece, and received numerous prizes, awards and honors.

Scott A. Snyder received his B. A. degree in Chemistry, summa cum laude, from Williams College in 1999, and then began graduate studies at The Scripps Research Institute under the guidance of Prof. K. C. Nicolaou. He is the recipient of a Barry M. Goldwater Fellowship in Science and Engineering, a National Science Foundation Predoctoral Fellowship, and graduate fellowships from Pfizer, Inc. and Bristol-Myers Squibb.

Contents Overview

Table of Contents

Chapter 1

Introduction: Perspectives in Total Synthesis

Chapter 2

Isochrysohermidin

D. L. Boger (1993)

Chapter 3

Swinholide A

I. Paterson (1994)

Chapter 4

Dynemicin A

A. G. Myers (1995); S. J. Danishefsky (1995)

Chapter 5

Ecteinascidin 743

E. J. Corey (1996)

Chapter 6

Resiniferatoxin

P. A. Wender (1997)

Chapter 7

Epothilones A and B

K. C. Nicolaou (1997)

Chapter 8

Manzamine A

J. D. Winkler (1998); S. F. Martin (1999)

Chapter 9

Vancomycin

K. C. Nicolaou (1998); D. A. Evans (1998)

Chapter 10

Everninomicin 13,384-1

K. C. Nicolaou (1999)

Chapter 11

Bisorbicillinoids

K. C. Nicolaou (1999)

Chapter 12

Aspidophytine

E. J. Corey (1999)

Chapter 13

CP-Molecules

K. C. Nicolaou (1999)

Chapter 14

Colombiasin A

K. C. Nicolaou (2001)

Chapter 15

Quinine

G. Stork (2001)

Chapter 16

Longithorone A

M. D. Shair (2002)

Chapter 17

(−)-FR182877

E. J. Sorensen (2002)

Chapter 18

Vinblastine

T. Fukuyama (2002)

Chapter 19

Quadrigemine C and Psycholeine

L. E. Overman (2002)

Chapter 20

Diazonamide A

K. C. Nicolaou (2002; 2003)

Chapter 21

Plicamine

S. V. Ley (2002)

Chapter 22

Okaramine N

E. J. Corey (2003)

1

Introduction: Perspectives in Total Synthesis

"What I cannot create, I do not understand."
Richard P. Feynman
(1965 Nobel Laureate in Physics)

Because the field of natural product total synthesis is so rich in history, broad in scope, and deep in impact, it would literally take hundreds of thick volumes just to catalog its many accomplishments. Amazingly, however, only the eight words of the quotation above are needed to capture its essence. Indeed, as anyone who has ever performed research in target-oriented synthesis can attest, the only way to truly understand a compound chemically, be it a complex secondary metabolite or a designed molecule, is to construct it in the laboratory. No matter how many three-dimensional models of a compound are built, the number of times its connectivities are drawn on paper, or the length of time spent mentally contemplating its structural intricacies, these actions alone can never reveal all the secrets of a molecule. It is the physical act of creation, the process of synthesis, that holds the key to unlocking some of these closely guarded and often highly valuable pieces of information.

After all, every molecule, just like a person, is the unique sum of its individual parts. Its physical characteristics, its personality if you will, emerge from the specific combination of its atoms, bonds, and stereocenters; to remove or alter just one of these items would be essentially the same as a close friend transforming into a complete stranger. This feature, more than any other, is what ensures that the field of synthesis will always have vitality, because it

"If a definitive history of twentieth century science is ever written, one of the highlights may well be a chapter on the chemical synthesis of complex molecules, especially the total synthesis of naturally occurring substances." E. J. Corey

© The Nobel Foundation

"The unique challenge which chemical synthesis provides for the creative imagination and the skilled hands ensures that it will endure as long as men write books, paint pictures, and fashion things which are beautiful, or practical, or both."
R. B. Woodward

means that each target molecule offers the practitioner different opportunities for discovery and invention. Attempts to fashion a novel molecular motif may serve to uncover a previously unknown mode of chemical reactivity, thereby leading to a new synthetic method. Challenges confronted in generating a structure with existing strategies and tactics might inspire the development of entirely new synthetic approaches. The possibilities are essentially limitless, restricted only by the tools at our disposal, the compounds that we seek to construct, and our own creativity.

Accordingly, the primary goal of the *Classics in Total Synthesis* series is to provide you with a clear understanding of our present synthetic capabilities in the hope that this information will not only inspire you, but also equip you with the skills needed to extend the frontiers of chemical synthesis in new directions. Our method to achieve this objective resides in analyzing several of the most creative total syntheses that have been accomplished in the laboratory, discussing in detail the strategies and methodologies required for their completion. We presented 36 of these syntheses in the first volume.[1] In this book, you will find over 30 more in 21 chapters, all achieved since 1992. If we are successful in our task, then you will take away the knowledge that was gleaned from these campaigns as well as experienced some of the drama, frustration, and excitement that was part of their execution.

However, before you begin to feast your eyes on the syntheses that comprise the main body of this text, we hope that you will let the next few pages in this introductory chapter whet your appetite. Apart from introducing the molecules whose stories will soon be familiar to you, we also want to highlight some of the major themes that have guided research endeavors in the field of total synthesis in recent years. Our comments are organized under the headings of targets, strategies, and methods, the subtitle for the *Classics* series and the three essential components that must be interwoven in order to complete a total synthesis.

1.1 Targets

"The structure known, but not yet accessible by synthesis, is to the chemist what the unclimbed mountain, the unchartered sea, the untilled field, the unreached planet, are to other men."
R. B. Woodward

Any research program in total synthesis begins, of course, with the selection of a natural product target. Unfortunately, because Nature is such a skillful and imaginative chemist, there are almost too many interesting choices to narrow the field down to just one. To illustrate this point, spend at least a few minutes perusing the small collection of structures shown in Figure 1. Their diversity at the molecular level is truly stunning, particularly if you consider that Nature uses just six atoms (H, C, O, N, S, and Cl) to stitch together their intricate and completely disparate architectures. Some are rich with stereochemical complexity, others are decorated with multiple rings, and many possess domains that are not found in any other secondary metabolite.

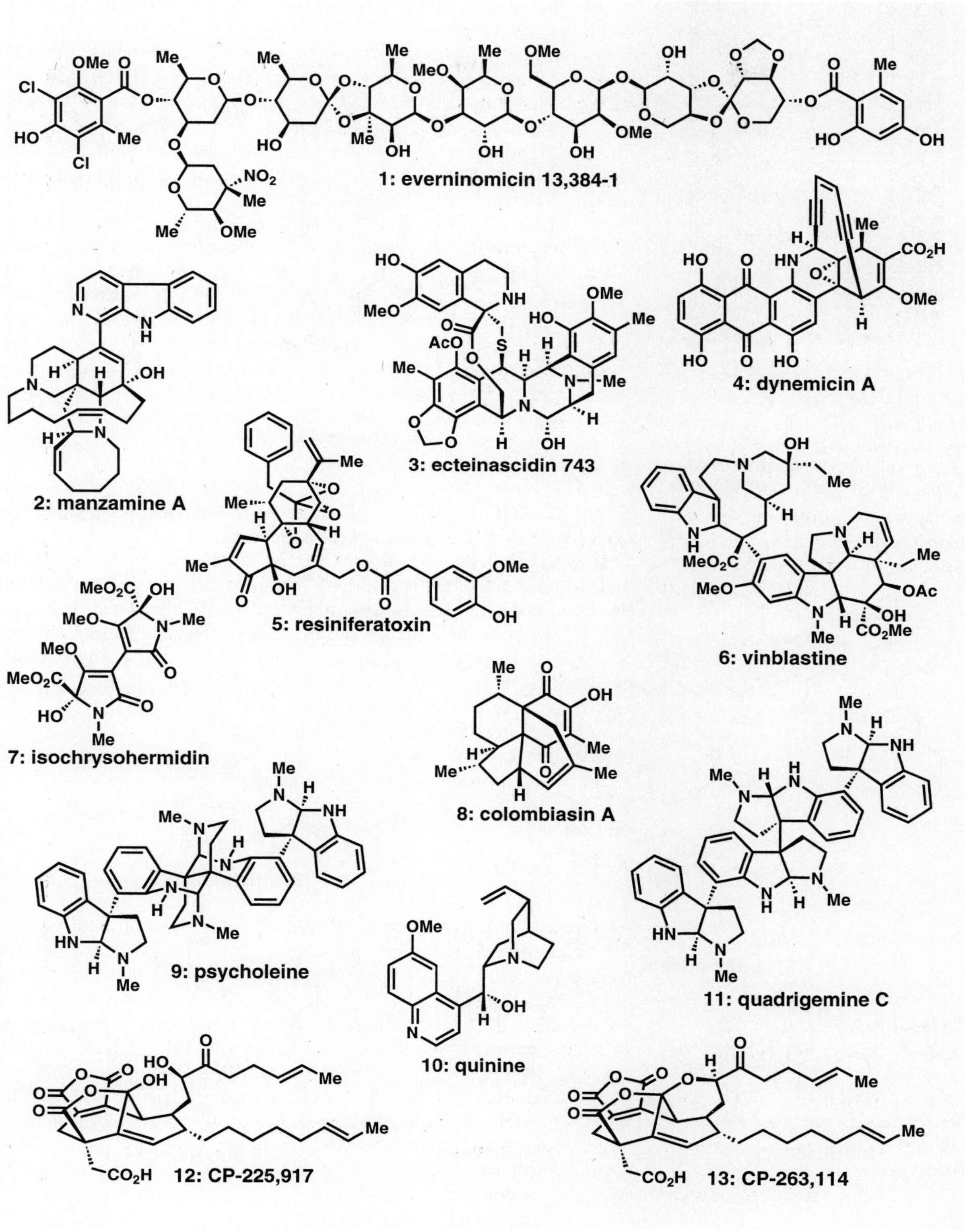

1: everninomicin 13,384-1

2: manzamine A

3: ecteinascidin 743

4: dynemicin A

5: resiniferatoxin

6: vinblastine

7: isochrysohermidin

8: colombiasin A

9: psycholeine

10: quinine

11: quadrigemine C

12: CP-225,917

13: CP-263,114

Figure 1. Selected structures of complex natural products.

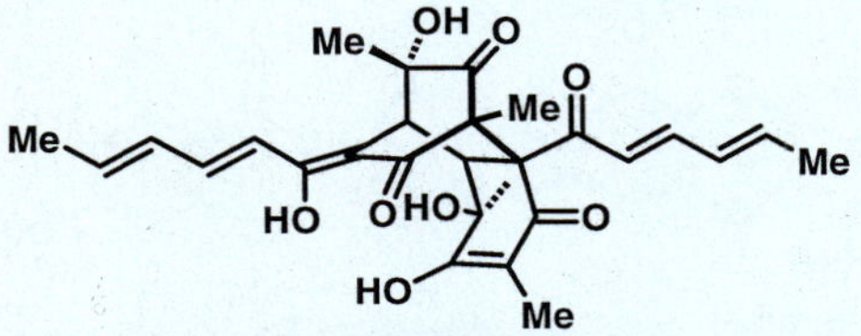

14: bisorbicillinol

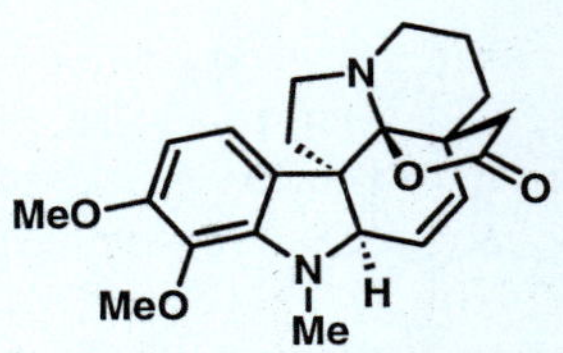

15: aspidophytine

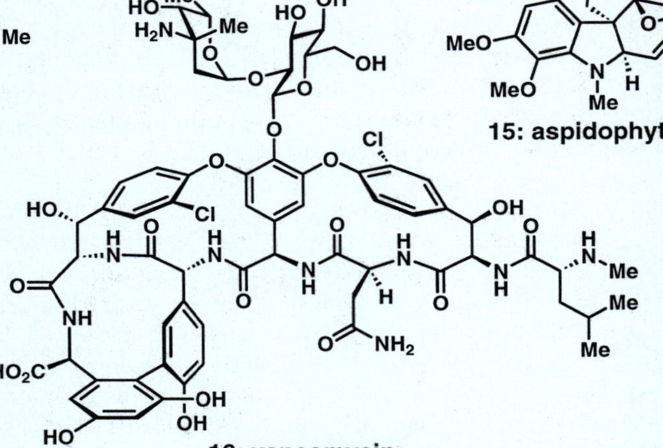

16: vancomycin

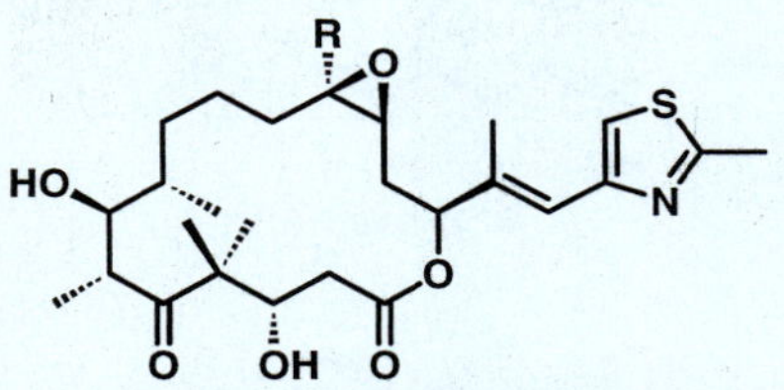

17: R = H, epothilone A
18: R = Me, epothilone B

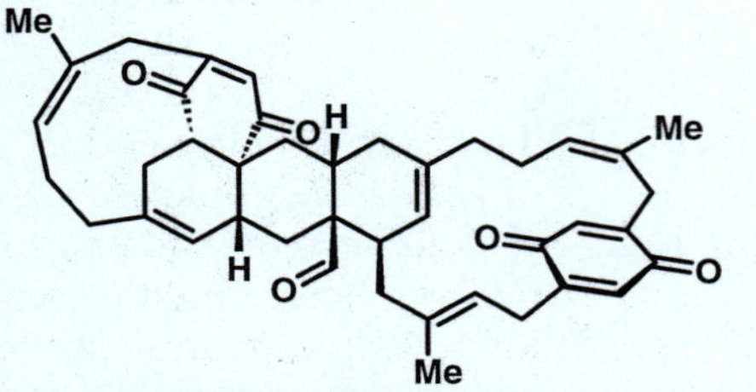

19: longithorone A

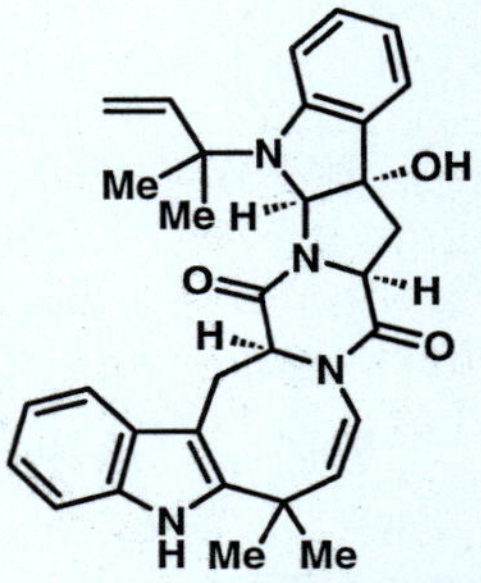

20: okaramine N

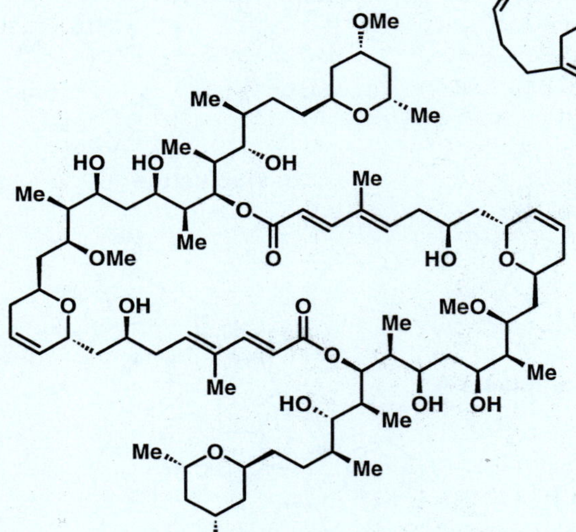

21: swinholide A

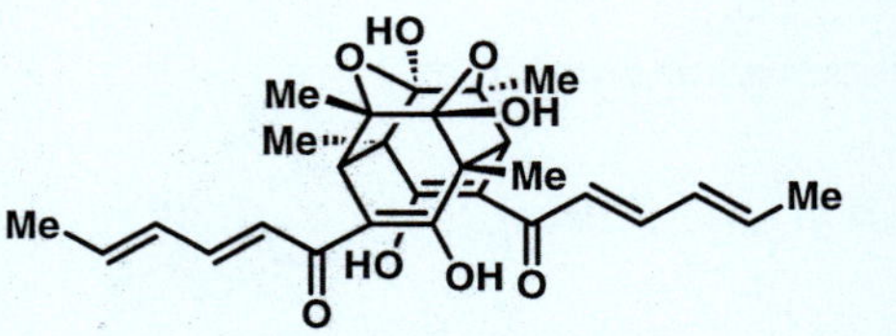

22: trichodimerol

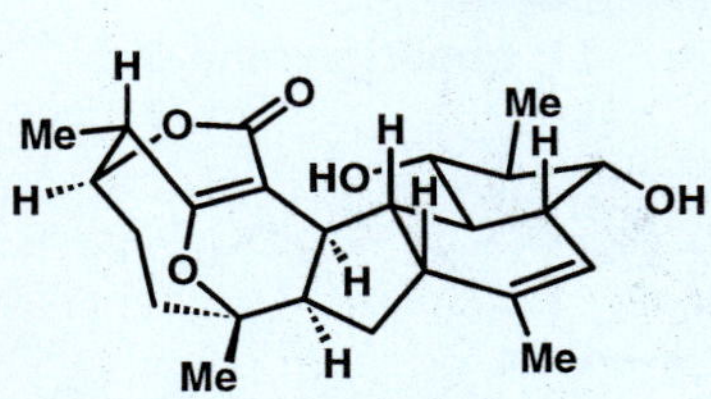

23: FR182877

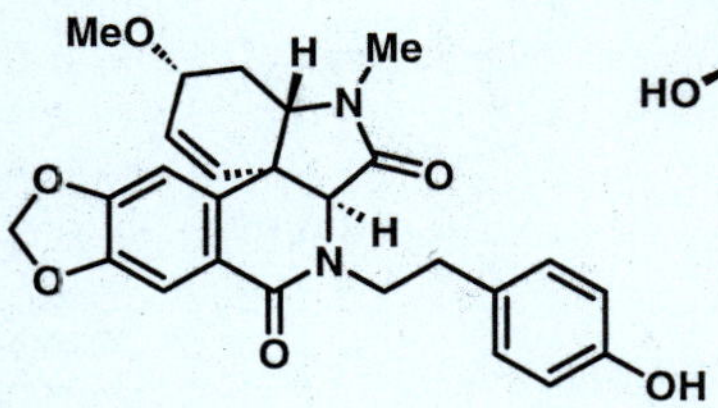

24: plicamine

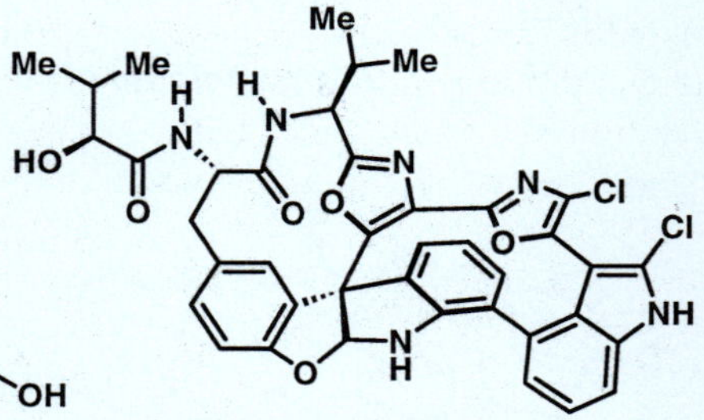

25: diazonamide A

Figure 1. Continued.

By what rationale, then, should one select a target among the thousands of potential choices? For the past half century, the chemical community has attempted to answer this question by choosing those compounds that seem to provide the highest level of synthetic challenge. Indeed, as all of the syntheses in *Classics I* reveal, this motivation has long succeeded in advancing the frontiers of chemistry. Not only do these natural products typically inspire novel and creative synthetic approaches, but they also provide one of the best forums in which to identify synthetic inadequacies worthy of further methodological development.

Importantly, however, throughout the 1990s and to the present day, these criteria have expanded, with natural products increasingly being selected for chemical synthesis with additional objectives in mind. For example, with an ever-growing desire to use natural products as tools to solve problems in chemical biology, synthetic chemists are more typically devoting their attention to those challenging targets that will also provide additional research opportunities in this field. There are many ways that such programs can take shape. For instance, researchers can employ their route to a bioactive natural product in order to synthesize structural analogs of it, with the final goal of identifying the specific structural motifs responsible for its potency. These same studies might also be executed in attempts to discover compounds with improved biological activity profiles or different modes of action than the parent natural product, in the hopes of providing new and/or improved pharmaceutical agents. On other occasions, a total synthesis might be undertaken to provide sufficient supplies of a scarce natural product in order to follow up on interesting preliminary biological findings. All these themes, as well as many others not related to biology or medicine, will be brought to light in the upcoming pages as we present the stories of the compounds shown in Figure 1.

"It is in the course of attack of the most difficult problems, without consideration of eventual applications, that new fundamental knowledge is most certainly garnered."
R. Robinson

1.2 Strategies

As a prelude to undertaking any of these research endeavors, one must initially devote significant time and mental energy in the pursuit of synthetic strategies capable of responding to the challenges presented by a given molecular architecture. For over 40 years, retrosynthetic analysis has served as the primary tool for this task, affording the practitioner with the means to disconnect a given target both logically and systematically based either on known or completely novel reactions. This approach has long led to creative strategies, and, as evidenced by the syntheses in this text, its power is alive and well.

"There is an unparalleled opportunity for the application of chemical synthesis to biological and medical problems at a fundamental level."
E. J. Corey

With that said, however, the past few years have witnessed a renaissance in the use of biosynthetic considerations to guide retrosynthetic planning. The thought being if Nature is the master chemical artisan, then why not attempt to mimic her levels of efficiency in

© TSRI

"The complexity in the behavior of organic molecules is such that the first execution of any chemical synthesis based on a design is always a venture into the uncertain, an experiment run to find out whether and under what conditions the elements of the design do indeed correspond to reality."
A. Eschenmoser

the laboratory by enlisting the same reactions that she employs to prepare these compounds. Of course, this theme is not new, but it seems certain that it will be more wholeheartedly adopted as we move further into the 21st century, driven by both an improved understanding of biosynthetic pathways and the economic need to enhance the efficiency and environmental friendliness of our synthetic sequences. Many syntheses in this book reflect this concept, and, to set the stage for their discussion, we have delineated the beautiful total synthesis of glabrescrol (**29**) by Corey and Xiong in Scheme 1.[2] As you can see, the key biomimetic step is an acid-catalyzed conversion of polyepoxide **26** into **27**, a process that concurrently generate four of the five tetrahydrofuran rings of the final target (**29**). Indeed, the ease and selectivity of this transformation make it quite difficult to envision the existence of a more expedient or effective method to synthesize these domains.

Apart from using clues provided by Nature to generate efficient and novel strategies, synthetic chemists have also been attempting to accomplish the same goal in single-pot cascade sequences by combining known reactions that would normally be executed independently. The example in Scheme 2, from the Paquette group, serves to explain this alternate approach to strategy design.[3] In this instance, the addition of two different nucleophiles to the carbo-

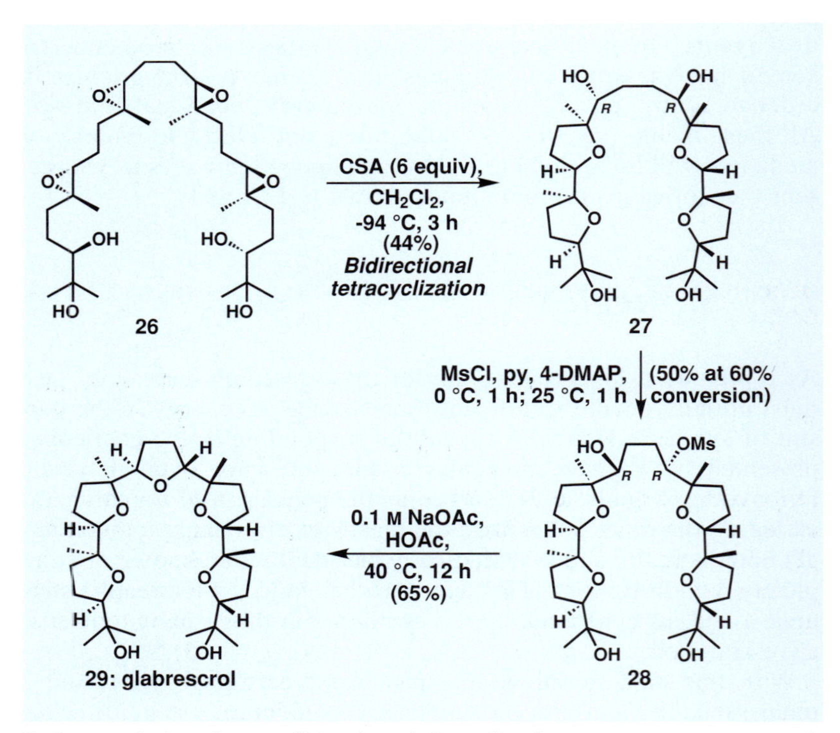

Scheme 1. An elegant "biomimetic" cyclization sequence as part of Corey and Xiong's expedient total synthesis of glabrescol (**29**).

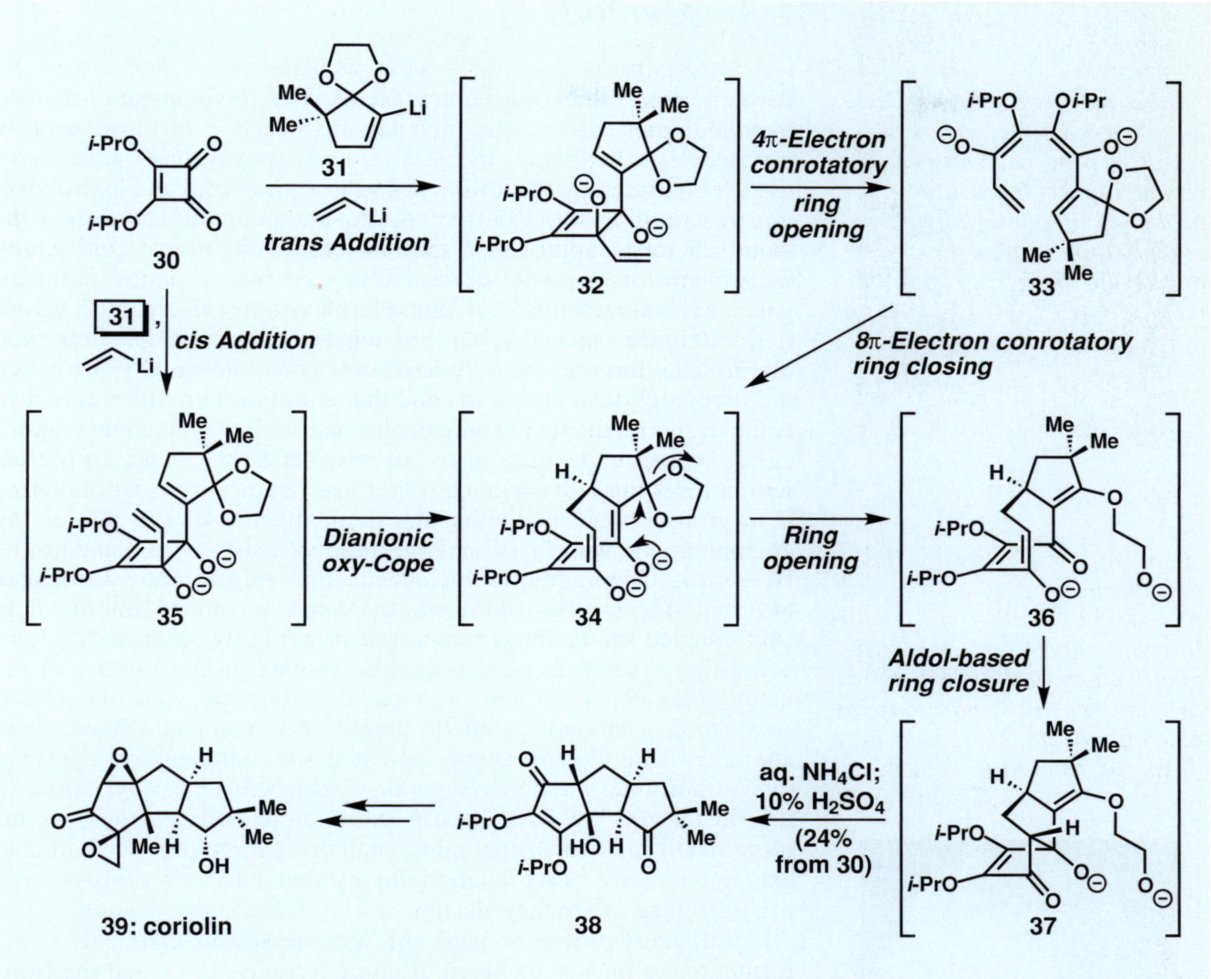

Scheme 2. A highly effective cascade sequence developed by the Paquette group for the synthesis of complex natural products such as coriolin (**39**).

nyl groups of squarate ester **30** enabled entrance into a complicated series of pathways that ultimately delivered a tricyclic product (**38**). The dramatic increase in molecular complexity in just a single step is truly remarkable, and, by having reached this advanced staging area so quickly, the Paquette group was able to complete total syntheses of several natural products, such as coriolin (**39**), in just a few more steps. Many additional variations on this general theme will be encountered in the coming chapters. If experience is any guide, we have only just begun to uncover the possibilities for these types of reaction cascades.

"In chemistry, one's ideas, however beautiful, logical, elegant, imaginative that they may be in their own right, are simply without value unless they are actually applicable to the one physical environment we have—in short, they are only good if they work!"

R. B. Woodward

1.3 Methods

"Many new synthetic processes have been discovered as a result of a perceived need in connection with specific problems involving novel or complicated structures and a deliberate search for suitable methodology." E. J. Corey

Perhaps more than any other factor, the development of new methodologies drives the manner in which total synthesis is practiced. For example, the discovery of transformations such as the Diels–Alder cycloaddition in 1928 or the Wittig and hydroboration reactions in the 1950s catalyzed dramatic shifts in how synthetic chemists approached their science and, more specifically, the process of synthetic design. Over the course of the past few years, several methods have come into existence that can be classified within this special group, the most notable, perhaps, being the olefin metathesis reaction. Indeed, ever since the early 1990s when the first catalysts were developed that could initiate this reaction in contexts pertinent to the practical construction of complex molecules, synthetic chemists have not regarded the presence of olefins within rings in quite the same way. For example, consider the Nicolaou group's total syntheses of two members (**46** and **47**) of the coleophomone family of natural products shown in Scheme 3.[4] Here, ring-closing olefin metathesis on regioisomeric substrates (**41** and **42**) synthesized from the same advanced intermediate (**40**) enabled the facile generation of their highly strained 11-membered ring systems in a stereospecific manner. In the absence of the methodological developments needed to turn this reaction into a sharp tool, a modern synthetic practitioner would not likely have attempted ring closure at the indicated site, contemplating instead more traditional methods (several of which, by the way, failed in this instance). We will discuss this and related reactions in far more detail in several upcoming chapters. Moreover, we shall also encounter many other methodologies that have similarly altered the trajectory of strategy design.

Significantly, however, gaps and weaknesses in "classical" transformations offer just as much of an opportunity to extend the frontiers of chemical synthesis as entirely new reactions. For instance, while the Diels–Alder reaction unquestionably provides a powerful means to create highly functionalized six-membered rings, there are still many scenarios in which this transformation cannot be induced to proceed asymmetrically. One of these instances is the construction of six-membered rings of the type found in the natural product (+)-ambruticin (**56**, Scheme 4). Inspired by this problem, the Jacobsen group at Harvard University developed a novel chromium catalyst (**50**) that brilliantly overcame this weakness in Diels–Alder chemistry, delivering both ring systems of **56** with excellent enantioselectivity.[5] Almost every total synthesis in this book will similarly illustrate the development of at least one new synthetic method to access a specific element of the target molecule, whether based on an initial design or in response to challenges encountered during the synthesis.

"It is interesting to observe how an acorn of hypothesis can become a tree of knowledge." Sir D. H. R. Barton

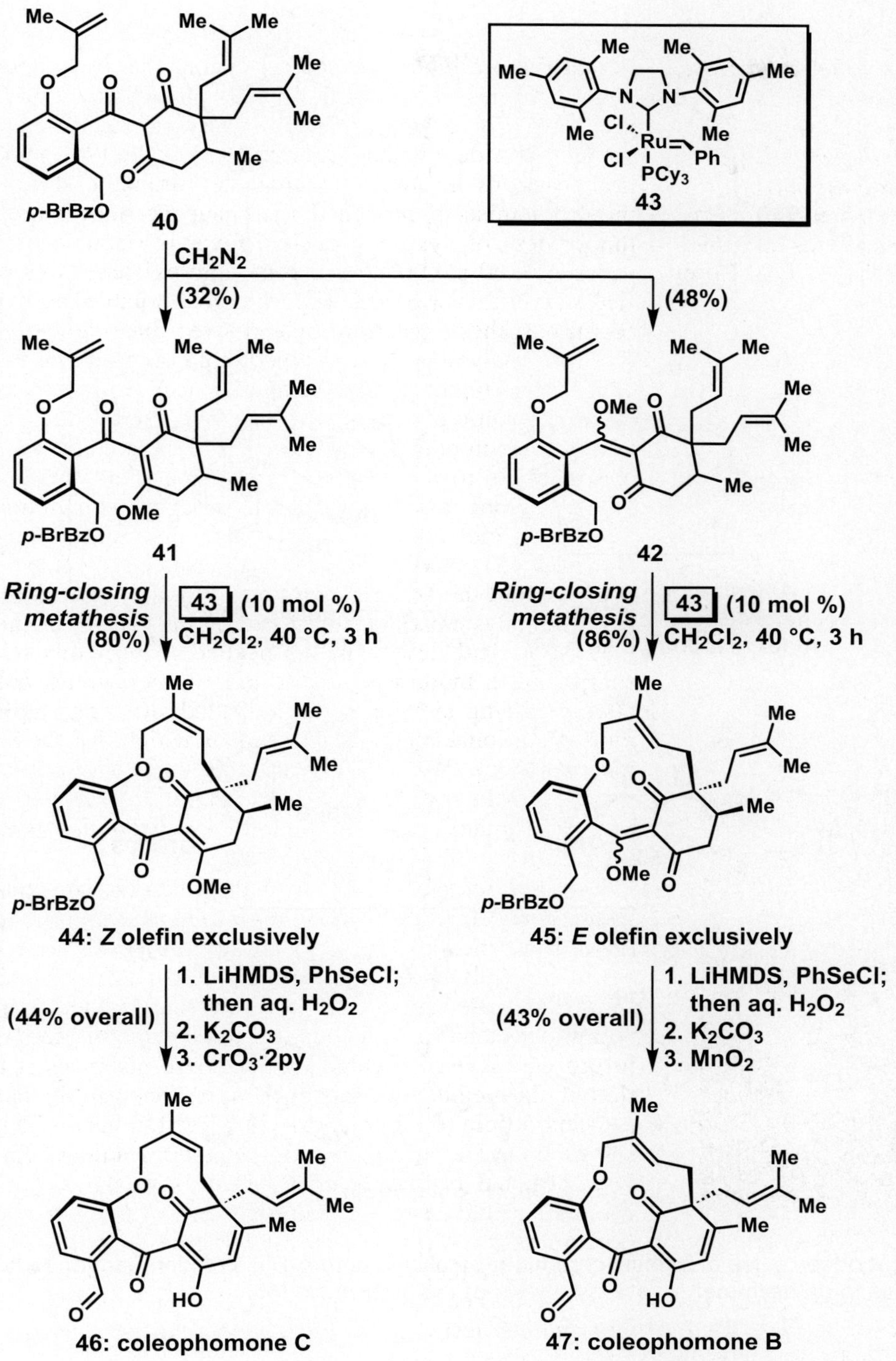

Scheme 3. New directions in organic synthesis: ring-closing metathesis as a key step in Nicolaou's total synthesis of coleophomones C and B (**46** and **47**).

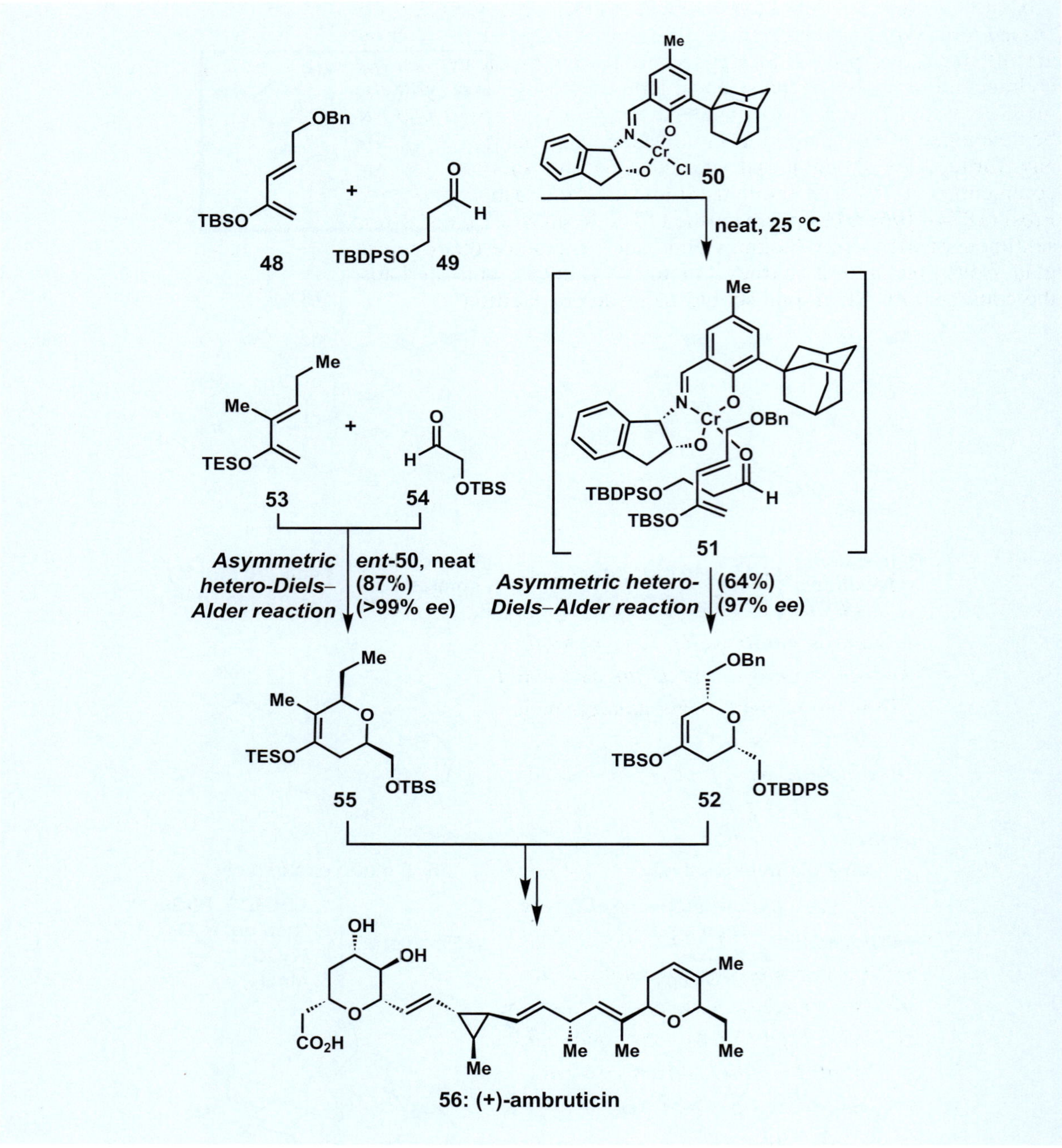

Scheme 4. Jacobsen's use of a novel chromium-catalyzed hetero-Diels–Alder reaction to fashion the two pyran systems in an asymmetric total synthesis of (+)-ambruticin (**56**).

By now, you will surely have realized that although we have discussed total synthesis under three different headings, these features are all intricately linked. Strategies are always predicated on the architectural domains of the selected target molecule, and synthetic methods will either help to dictate those designs or will have to be developed to put them in motion. Rather than belabor this point any further, we thought that we should close this part of our commentary with some thoughts from the American poet Robert Frost (1874–1963). His poem entitled "The Road Not Taken" offers sentiments which echo the knowledge and experience that one can gain from unchartered journeys in total synthesis, and, perhaps, the courage with which one should face such expeditions.

The Road Not Taken
(1915)

Two roads diverged in a yellow wood,
And sorry I could not travel both
And be one traveler, long I stood
And looked down one as far as I could
To where it bent in the undergrowth;
Then took the other, as just as fair,
And having perhaps the better claim,
Because it was grassy and wanted wear;
Though as for that the passing there
Had worn them really about the same,
And both that morning equally lay
In leaves no step had trodden black.
Oh, I kept the first for another day!
Yet knowing how way leads on to way,
I doubted if I should ever come back.
I shall be telling this with a sigh
Somewhere ages and ages hence:
Two roads diverged in a wood, and I –
I took the one less traveled by,
And that has made all the difference.[6]

1.4 Classics in Total Synthesis II

© The Nobel Foundation

"The progress of science today is not so much determined by brilliant achievements of individual workers, but rather by the planned collaboration of many observers." E. Fischer

Although writing this book was certainly far from trivial, any individual challenge encountered during its execution paled in comparison to the initial agony that we faced in selecting the total syntheses that would comprise its chapters. Indeed, even though we restricted the candidate pool to only those works that have been completed since 1992, we knew from the outset that we would have to exclude some very elegant and instructive syntheses. Our final decisions (the molecules shown in Figure 1) were based primarily on the dual concerns of presenting a diverse set of molecular architectures and highlighting as many reactions and synthetic strategies as possible, particularly those that were not covered within the pages of *Classics I*. Several of these syntheses come from our research group, and, in some cases, we have selected these over others because our familiarity with them provided an opportunity to share the details of their execution in a far more elaborate and intimate fashion.

As you progress through the next few chapters, you will find that the pedagogical framework is essentially the same as in the first volume, meaning that every total synthesis is analyzed retrosynthetically and then discussed as a forward synthesis. However, there are some subtle and important differences. First, at the start of each chapter we have placed a concept box to indicate the major themes that will be touched upon during our discussion of that molecule. We mean for this additional feature to help organize your thoughts as you read and to enable a more expedient search of important concepts and reactions within the text. Second, we have striven to condense our presentation of each retrosynthetic analysis, focusing more attention on the overarching strategies and tactics that led to the final synthetic design rather than a specific discussion of the successful synthesis in reverse. However, as in the first book, this presentation is based only on the final strategy for the sake of simplicity and pedagogy, which may mean that we have inadvertently taken liberty with the actual thoughts of the group that accomplished the synthesis. In all instances we have tried to base our discussion on the design considerations mentioned in the original literature; we apologize in advance for any misconstrued presentations in these sections.

"Synthesis remains a dynamic and central area of chemistry. There are many new principles, strategies, and methods of synthesis waiting to be discovered." E. J. Corey

The final change is the most obvious one, and that involves the presentation of multiple syntheses of the same molecule within a single chapter. In some cases, we have provided a complete story for every successful synthesis in order to explore their different elements fully. In others, we have attempted to accomplish the same goal by adding a short section at the end of the chapter highlighting only the key steps of these additional syntheses, focusing on their unique solutions for certain structural motifs compared with those of the synthesis presented in the main text. Our expectation is that the discussion of additional syntheses in these two formats will

"Courage is the beginning of action, but chance is the master of the end."
Democritos, 5th Century B. C.

markedly increase the educational value of this text and will demonstrate the virtuosity that can be achieved with the current repertoire of synthetic transformations in the hands of true masters of the art.

So, with these comments in mind as well as some perspective on the symbiotic relationship between target selection and the resultant strategies and methods marshaled to synthesize them, we will now set you free to explore the syntheses found in the remainder of the book. Any can be read as an individual offering, and in those instances in which an important topic is covered in more depth elsewhere, we will direct you to that chapter. In closing, we truly hope that you will enjoy reading about these syntheses as much as we enjoyed writing about them, and that our selections will provide a path of learning that you will enjoy travelling.

References

Bluebox Artwork: © Ekdotike Athenon S.A., *The Olympic Games: Ancient Greece*

1. K.C. Nicolaou, E.J. Sorensen, *Classics in Total Synthesis*, Weinheim, Wiley-VCH, **1996**, p. 798.
2. Z. Xiong, E.J. Corey, *J. Am. Chem. Soc.* **2000**, *122*, 9328.
3. L.A. Paquette, F. Geng, *J. Am. Chem. Soc.* **2002**, *124*, 9199. For a review of this general cascade sequence, see: L.A. Paquette, *Eur. J. Org. Chem.* **1998**, 1709.
4. K.C. Nicolaou, G. Vassilikogiannakis, T. Montagnon, *Angew. Chem.* **2002**, *114*, 3410; *Angew. Chem. Int. Ed.* **2002**, *41*, 3276.
5. P. Liu, E.N. Jacobsen, *J. Am. Chem. Soc.* **2001**, *123*, 10772. For an earlier use of this catalyst system, see: A.G. Dossetter, T.F. Jamison, E.N. Jacobsen, *Angew. Chem.* **1999**, *111*, 2549; *Angew. Chem. Int. Ed.* **1999**, *38*, 2398.
6. From *The Poetry of Robert Frost* by Robert Frost, edited by Edward Connery Lathem. Copyright 1916, 1923, 1928, 1930, 1934, 1939, 1947, 1949, © 1969 by Holt Rinehart and Winston, Inc. Copyright 1936, 1942, 1944, 1945, 1947, 1948, 1951, 1953, 1954, © 1956, 1958, 1959, 1961, 1962 by Robert Frost. Copyright © 1962, 1967, 1970 by Leslie Frost Ballantine.

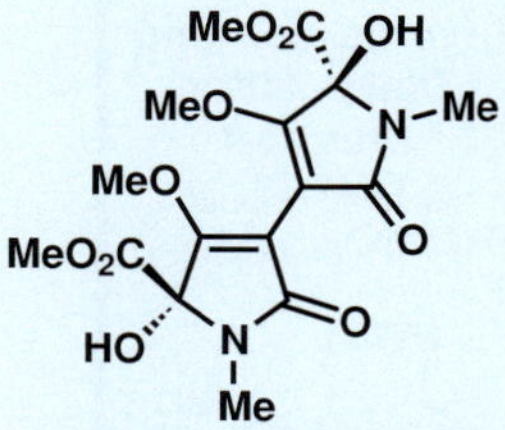

1: isochrysohermidin

2

D. L. Boger (1993)

Isochrysohermidin

2.1 Introduction

Among all the transformations in our current synthetic repertoire, few chemists would question that the Diels–Alder reaction is one of the most valuable and versatile tools to effect the construction of complex molecules, particularly natural products.[1] Indeed, using only the work described in the two volumes of *Classics* as a measure of the overall utility of the Diels–Alder cycloaddition, no other reaction can come close to boasting a similar number of creative solutions to the challenging synthetic puzzles provided by Nature's library of diverse molecular architectures.

Up to this point, however, our discussions have focused almost exclusively on examples of [4+2] cycloadditions derived from $HOMO_{diene}/LUMO_{dienophile}$-controlled processes, reactions of the type first identified by Otto Diels and Kurt Alder in 1928.[2] While such normal-demand Diels–Alder reactions account for the vast majority of reported applications of the [4+2] cycloaddition in organic synthesis, they certainly do not encompass the entire range of possibilities for this type of pericyclic process. For example, unactivated (or neutral) diene and dienophile partners, typified by butadiene and ethylene, can sometimes successfully engage in Diels–Alder fusions, as illustrated in Scheme 1. Generally, though, this type of [4+2] cycloaddition has not found widespread utility in organic synthesis, primarily due to the difficulty in achieving sufficient conversion into product based on the significantly greater magnitude of the HOMO/LUMO energy gap relative to that typically observed in the normal-demand Diels–Alder scenario (ΔE_2 versus ΔE_1). In stark contrast to this rather lethargic combination, especially favorable energy-level separations for [4+2] cyclo-

Key concepts:

- **Inverse-electron-demand Diels–Alder reactions**
- **Azadiene chemistry**
- **Tandem reactions**
- **Reductive ring contractions**
- **Singlet oxygen chemistry**

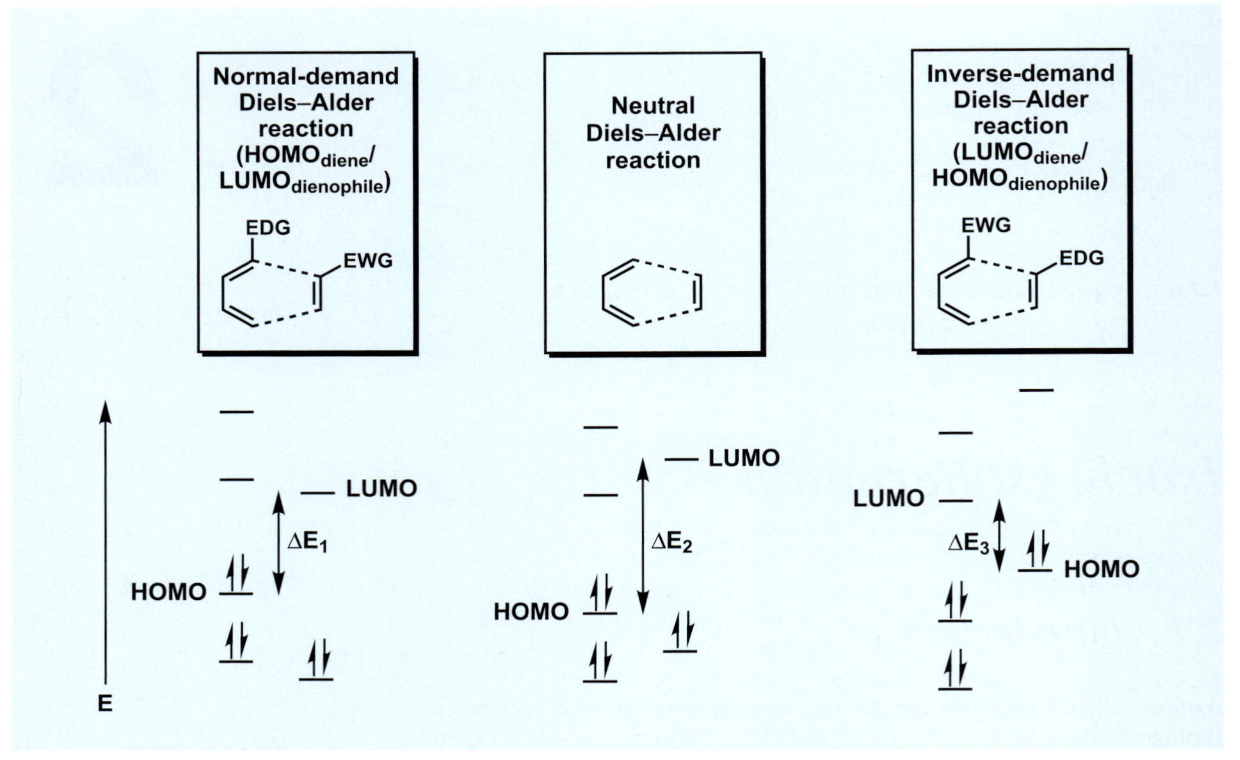

Scheme 1. Classification of Diels–Alder reactions and their corresponding energy diagrams. EDG = electron-donating group, EWG = electron-withdrawing group.

addition can often be reached upon reversal of the electronic properties of the typical Diels–Alder diene/dienophile partners in a $LUMO_{diene}/HOMO_{dienophile}$ paradigm, with cyclization sometimes achieved under milder conditions than the traditional normal-demand reaction (ΔE_3 versus ΔE_1). Indeed, this variant of the Diels–Alder reaction is quite useful and over the course of the past two decades has been the focus of increasing attention, resulting in the elucidation of numerous complementary diene/dienophile partners capable of engaging fruitfully in inverse-electron-demand Diels–Alder chemistry.[3] As such, these achievements have encouraged the application of the developed technology to total synthesis endeavors, often with spectacular results.

In this chapter, we will examine one such total synthesis in detail, namely that of the heteroatom-rich natural product isochrysohermidin (**1**) as achieved by Dale Boger and co-worker Carmen Baldino at The Scripps Research Institute in 1993, which featured an insightful double inverse-electron-demand Diels–Alder reaction in combination with several unique and instructive synthetic transformations.[4] Before engaging in an analysis of this synthesis, however, we felt that it would be prudent to partake in a sojourn through some instructive examples of inverse-electron-demand Diels–Alder chemistry within the vast realm of total synthesis. While our treatment is far from comprehensive, we aspire merely to provide a suitable sampling of the power of this reaction to enable a fuller appreciation of the work at hand, since this class of Diels–Alder reactions constitutes uncharted territory in the *Classics* series. Should your interest be piqued, we encourage you to turn to several excellent review articles to explore this field further.[3]

2.1.1 Azadiene-Based Inverse-Electron-Demand Diels–Alder Reactions

Among systems that are appropriate for inverse-electron-demand Diels–Alder chemistry, azadienes (particularly as part of heteroaromatic systems) are ideally suited as 4π-electron components with electron-rich, strained, or sometimes even simple olefinic dienophile partners. In fact, the vast majority of Diels–Alder reactions in the inverse-electron-demand paradigm incorporate one or more electronegative heteroatoms within the diene portion to create sufficient electron deficiency, as all-carbon-based analogues have proven to be relatively recalcitrant participants in $LUMO_{diene}$-controlled Diels–Alder processes, even when numerous electron-withdrawing groups have been attached. Overall, while several researchers have firmly established the fertility of azadiene-based [4+2] cycloaddition reactions in the landscape of total synthesis, the method has been especially championed and mastered by the Boger group; as such, we will limit ourselves here to expounding upon some of their seminal achievements.

Within the collection of conceivable azadiene systems, it is usually quite challenging to employ simple α,β-unsaturated imine systems as the 4π-electron component in any type of Diels–Alder reaction, since competitive imine addition or tautomerization often precludes successful cycloaddition. Moreover, even if Diels–Alder cyclization occurs, the enamine product often proves unstable and/or difficult to manipulate. As such, the Boger group reasoned that these challenges could potentially be addressed by incorporating a bulky electron-withdrawing substituent on the terminal nitrogen atom, a modification that was anticipated to facilitate the desired Diels–Alder cycloaddition by: 1) preferentially decelerating 1,2-imine addition through effective steric shielding, 2) enhancing the electron-deficient nature of the azadiene system and thus increasing the rate of the reaction by lowering the $LUMO_{diene}/HOMO_{dienophile}$ energy gap (ΔE_3 in Scheme 1), and 3) conveying stability to the final product by affording a deactivated enamine system. Indeed, numerous studies along these lines have culminated in the realization that several *N*-sulfonyl, *N*-phosphoryl, and *N*-acyl addends enable such inverse-electron-demand Diels–Alder reactions to proceed.[5]

The first disclosed route to the antitumor agent streptonigrone (**6**, Scheme 2) serves as an excellent paragon of such Diels–Alder tech-

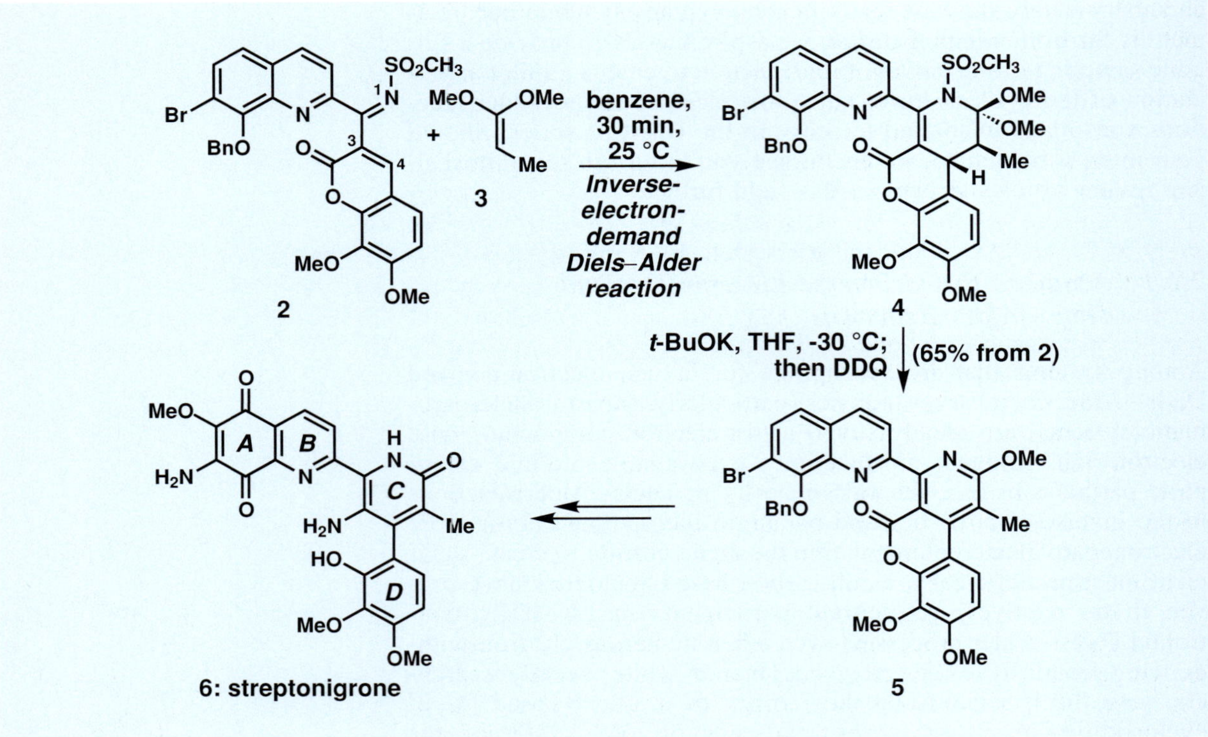

Scheme 2. Boger's total synthesis of streptonigrone (**6**) featuring an inverse-electron-demand Diels–Alder reaction of an *N*-sulfonyl-1-aza-1,3-butadiene (**2**).

nology.[6] With a fully substituted pyridone ring as the central structural lynchpin of this natural product (ring C), it was envisioned that reacting an electron-deficient *N*-sulfonyl-1-aza-1,3-butadiene system (**2**) with a suitable, electron-rich dienophile such as **3** would initiate an inverse-electron-demand Diels–Alder cycloaddition, fashioning a system that could eventually be elaborated into this challenging motif. As expected, upon simple dissolution of these reactants in benzene with subsequent stirring at ambient temperature for 30 minutes, a facile regioselective Diels–Alder reaction provided intermediate adduct **4**. Several features of this remarkable transformation merit further analysis. First, although the neighboring lactone system appended to the diene moiety in **2** facilitates the Diels–Alder reaction by rigidifying the diene system into an *s-cis* conformation, an equally important feature is the complementary nature of the electron-withdrawing ability of this functionality to that of the *N*-sulfonyl system from a frontier-molecular-orbital perspective, thereby leading to acceleration of the reaction by lowering the $LUMO_{diene}/HOMO_{dienophile}$ energy gap (a further decrease in ΔE_3 of Scheme 1).[7,8] Proof for this statement resides in the fact that while in the absence of such electron-withdrawing substituents related *N*-sulfonyl-1-aza-1,3-butadiene systems readily participate in Diels–Alder fusions at temperatures typically ranging from 60 to 100 °C, here the electron-deficient character was accentuated to the degree that the reaction could be carried out at room temperature. Perhaps the more intriguing feature of this Diels–Alder reaction, however, was the exclusive regio- and diastereoselectivity of the operation, a result that can be attributed to two mutually reinforcing occurrences which stabilize the *endo* transition state (as drawn in the column figure): 1) a secondary orbital interaction between the diene C-2 substituent and the dienophile OMe group, and 2) an anomeric-type effect which results from the *trans* periplanar arrangement of the lone pair of electrons on the nitrogen atom and the electron rich σ-bond of the enol ether. The first feature is noticeably absent in the corresponding *exo* transition state, while neither can be achieved in any all-carbon-based Diels–Alder reaction. As such, this type of inverse-electron-demand Diels–Alder reaction proceeds with a level of stereoselectivity that is far higher than that typically observed in normal [4+2] processes. As a final comment on the value of this synthetic approach, the Diels–Alder product **4** could then be smoothly converted into a fully substituted pyridine ring system (**5**) upon sequential treatment with *t*-BuOK and DDQ.

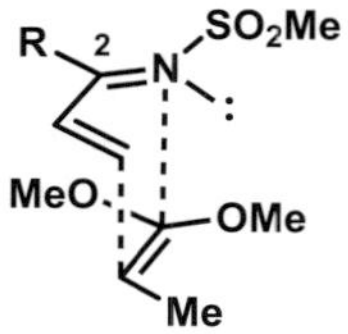

Although the above example verifies that isolated azadienes can be employed with ease in inverse-electron-demand Diels–Alder processes, the reaction can be further facilitated by embedding this partner within heteroaromatic rings, mainly due to the superior stability of an azadiene when it is part of such a system. Indeed, the first demonstrated example of an inverse-electron-demand Diels–Alder reaction utilized such a diene.[9] Since this initial discovery, subsequent efforts have established the general rule of thumb that the ease with which heterocycle-based azadienes participate in such [4+2] cycloadditions increases upon incorporation of additional nitrogen atoms in the aromatic ring, irrespective of whether it is part of the diene or not; similarly, appending electron-withdrawing groups to any such heteroaromatic ring accentuates the electron deficiency of the azadiene system and thus also accelerates the reaction.

Among the numerous total syntheses that have benefited from such Diels–Alder technology, Boger's route to the potent antitumor antibiotic (+)-CC-1065 (**15**, Scheme 3) stands paramount due to the development of an iterative inverse-electron-demand Diels–Alder approach to forge its identical central and right-hand indoline subunits.[10] In the opening maneuver of this synthesis, 1,2,4,5-tetrazine-3,6-dicarboxylate (**8**) smoothly engaged dienophile **7** (which was still sufficiently electron-rich for this reaction despite the incorporation of one electron-withdrawing substituent) in a Diels–Alder reaction to afford intermediate **9**, a transient species which was immediately converted into **10** through an entropically favored retro-Diels–Alder event expelling nitrogen. Although longer-lived, **10** did not persist either, eventually aromatizing to **11** through the loss of methanol in 70 % overall yield for the entire cascade

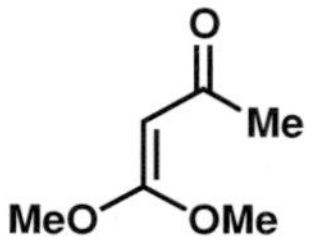

R 2 SO_2Me N MeO OMe Me

O Me MeO OMe
7

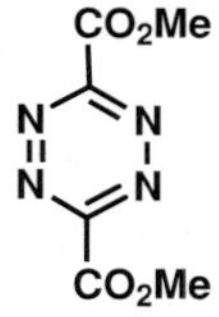

CO_2Me N N N N CO_2Me
8

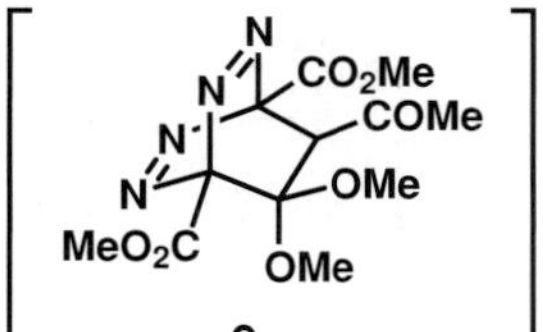

N N CO_2Me COMe N N OMe MeO_2C OMe
9

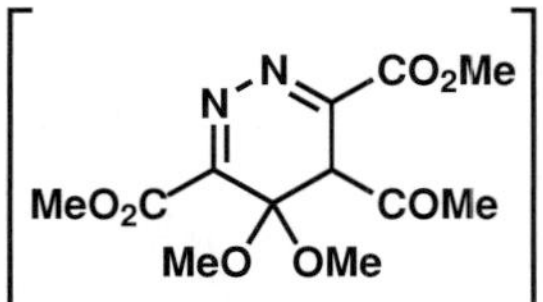

N N CO_2Me MeO_2C COMe MeO OMe
10

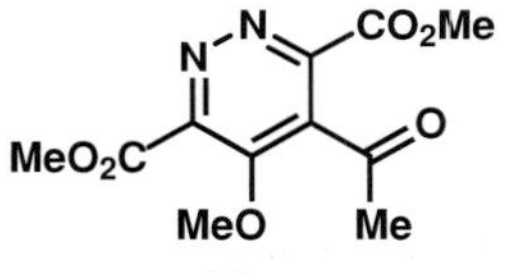

N N CO_2Me O MeO_2C MeO Me
11

sequence. With **11** in hand, this key intermediate was then elaborated into **12** in anticipation of a second, this time intramolecular, inverse-electron-demand Diels–Alder reaction. While such 1,2-diazine systems are typically reluctant diene participants in intermolecular Diels–Alder fusions, the entropic assistance provided by appropriately tethering the dienophile for an intramolecular cycloaddition is sufficient in most instances to override such sluggishness. Indeed, upon heating **12** in 1,3,5-triisopropylbenzene in a sealed vessel for a prolonged period (16–18 hours) at 230 °C, the formation of indoline system **14** was achieved in 87 % yield via intermediate **13**. Overall, as might be anticipated, the scope of this latter type of intramolecular Diels–Alder reaction is highly dependent on both the length of the tether and the nature of any heteroatoms included within, and, to date, solely alkynyl and allenyl dienophiles have proven competent as 2π-electron partners in such reactions.[11] In this particular context, only a nitrogen-based

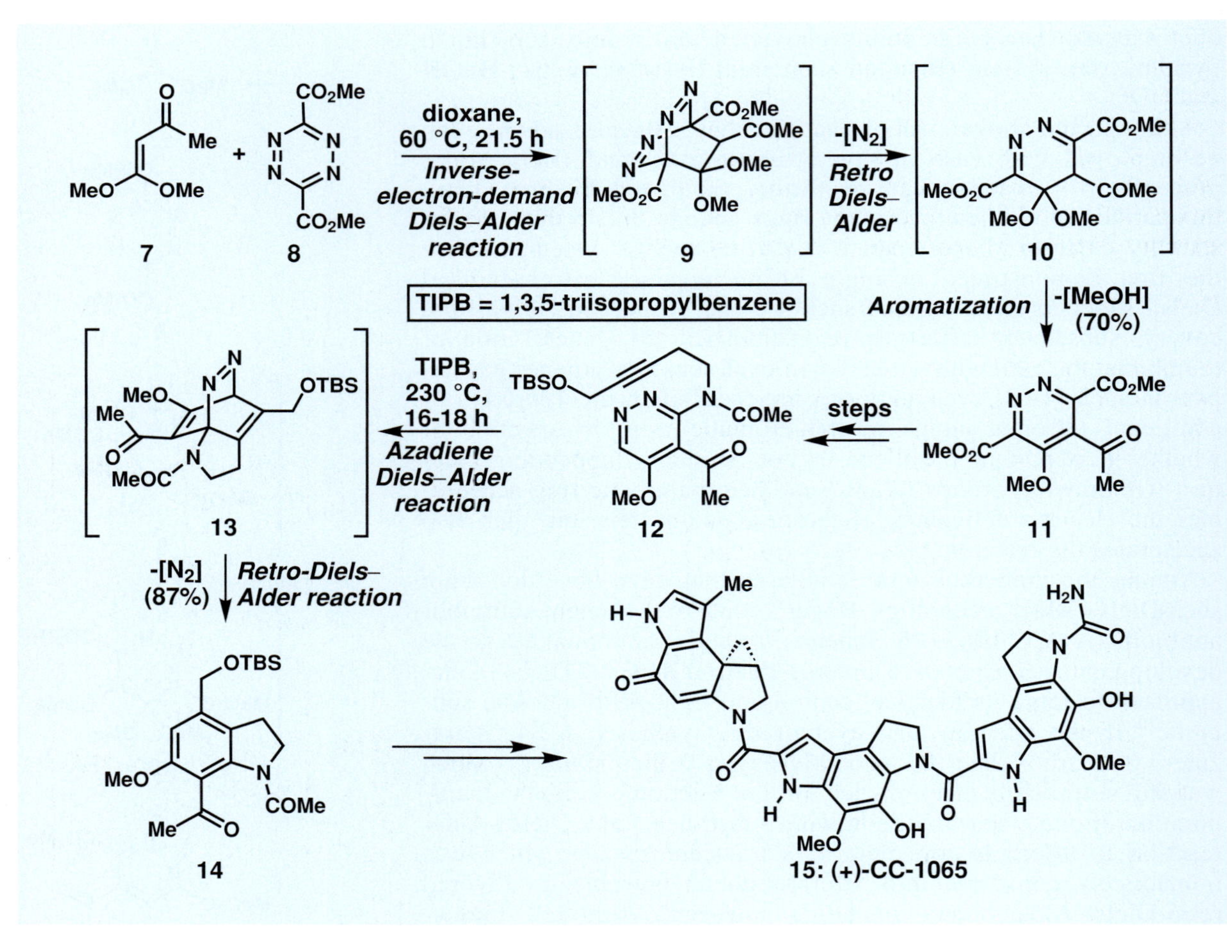

Scheme 3. Use of 1,2,4,5-tetrazine **8** in an inverse-electron-demand Diels–Alder reaction followed by an intramolecular 1,2-diazine/alkyne cycloaddition in Boger's total synthesis of (+)-CC-1065 (**15**).

linker with an electron-withdrawing group directly attached afforded **14** in synthetically useful yields. While one must concede that the successfully identified conditions for this final Diels–Alder reaction are relatively vigorous, the overall approach constitutes a rather novel strategy by which to fashion such substituted ring systems. Indeed, at this juncture the connection between **14** and **8** is far from obvious, as all traces of the four nitrogen atoms in the starting diene have been erased.

As a final example of inverse-electron-demand Diels–Alder chemistry, we turn to a total synthesis of pyrimidoblamic acid (**23**, Scheme 4), the heteroaromatic chromophore of the clinically effective antitumor agent bleomycin A_2.[12] The primary synthetic challenge posed by this target is the creation of a fully substituted pyrimidine ring system. Once again, azadiene Diels–Alder chemistry provided a unique and effective solution for fashioning such a highly substituted aromatic system.[13] In Boger's elegant and unique formulation, 2,4,6-tris(ethoxycarbonyl)-1,3,5-triazine (**16**) smoothly engaged in a [4+2] cycloaddition with 1,1-diamino-1-propene (**18**), a dienophile generated *in situ* through a thermally induced tautomerization of propioamidine **17**, to provide the tetrasubstituted pyrimidine **22** in 85 % yield. The reaction presumably followed the mechanistic course shown in Scheme 4, with the slow step in the overall cascade being the retro-Diels–Alder expulsion of ethyl cyanoformate from **21**. Key to the success of this Diels–Alder event was the use of a polar, aprotic solvent (DMF) and proper thermal activation (90 °C), as too low a

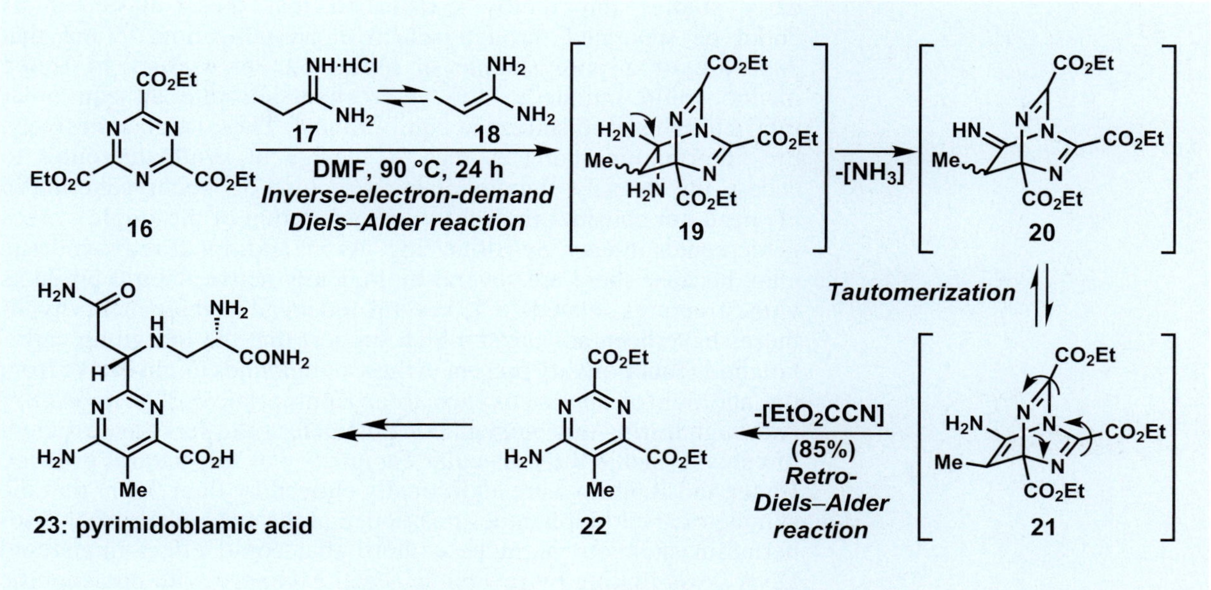

Scheme 4. A 1,3,5-triazine/amidine Diels–Alder reaction as part of Boger's synthesis of pyrimidoblamic acid (**23**), a key intermediate in the total synthesis of bleomycin A_2.

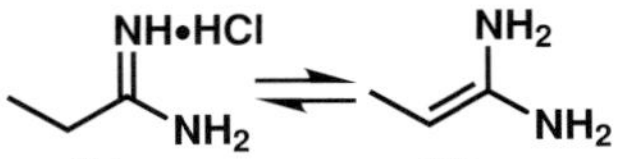

temperature slowed the reaction considerably, particularly in regards to the tautomerization of **17** to **18**, while at higher temperatures numerous side reactions occurred (especially involving ethyl cyanoformate). Moreover, utilizing the HCl salt of **17** was most likely critical in achieving superior conversion into **22**, as the resultant weakly acidic media could reasonably be expected to catalyze both the tautomerization of **17** to **18** as well as the subsequent retro-Diels–Alder process (**21**→**22**).

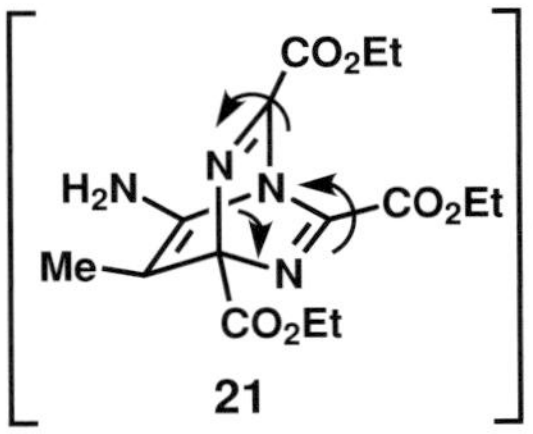

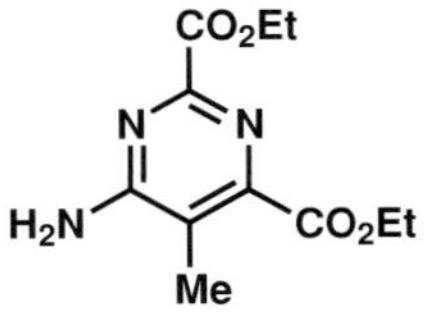

Overall, while this excursion has been brief, the above examples should provide a preliminary indication of the power of azadiene-based inverse-electron-demand Diels–Alder reactions in the context of total synthesis, particularly for the synthesis of highly substituted heterocyclic systems such as pyridines, diazines, and pyrimidines. Arguably, few if any other methods enable such facile, diverse, and consistent constructions of these challenging aromatic systems. With this background in place, we are now prepared to analyze the total synthesis of isochrysohermidin (**1**) by Boger and Baldino.

2.2 *Retrosynthetic Analysis and Strategy*

In 1986, the unique 2-oxo-3-pyrroline dimer isochrysohermidin (**1**, Scheme 5) was isolated from the leaves and stems of the plant *Mercuialis leiocarpia* by Masui and co-workers, and, although **1** is drawn as the D,L-diastereomer in Scheme 5, the *meso* form of **1** also constitutes a secondary metabolite obtained from the same biological source.[14] In addition to structural characterization, these early studies importantly established that these diastereomers could be separated through selective crystallization,[14b] and that exposure of a pure sample of either D,L- or *meso*-**1** to acidic media would initiate a slow interconversion into an equimolar ratio of both stereoisomers at equilibrium.[4b] Thus, taken cohesively, this information indicates that the design of synthetic routes to access the densely functionalized structure of isochrysohermidin (**1**) need not consider the controlled generation of the single stereogenic center in each pyrroline ring. As an additional retrosynthetic clue, because there are several biologically active natural products with structures related to **1**, several independent biogenetic hypotheses have been advanced which suggest that the intriguing carbinolamide functionality present in these compounds might derive from the addition of singlet oxygen to an appropriate 3,3′-bispyrrole.[15] Although testing this biosynthetic postulate could serve as sufficient impetus to tackle the molecular complexity of this natural product, Boger and Baldino were additionally enticed by their belief that the symmetrical carbinolamides positioned at the ends of the isochrysohermidin skeleton might be a motif that could effect interstrand DNA cross-linking by reversible acetal exchange with nucleophilic sites in duplex DNA. If true, this target could serve as a useful tool with which to probe chemical biology. In addition, they viewed this natural product as an excellent platform upon which to explore inverse-electron-demand Diels–Alder chemistry.

Thus, in an initial retrosynthetic simplification, it was envisioned that isochrysohermidin (**1**) could be formed in a single step from pyrrole dimer **25** upon the addition of singlet oxygen in a Diels–Alder fashion, as delineated in Scheme 5. Success in this endeavor would provide an intermediate endoperoxide adduct (**24**), which was anticipated to decompose immediately upon formation to the targeted natural product (**1**) through a fragmentation sequence encouraged by the loss of carbon dioxide from the strategically positioned carboxylic acid groups. Whereas pyrroles bearing such electron-withdrawing groups at C-2 are known to react sluggishly with singlet oxygen,[16] separate model studies on pyrrole monomers of the type present in **25** verified that such an addition was possible

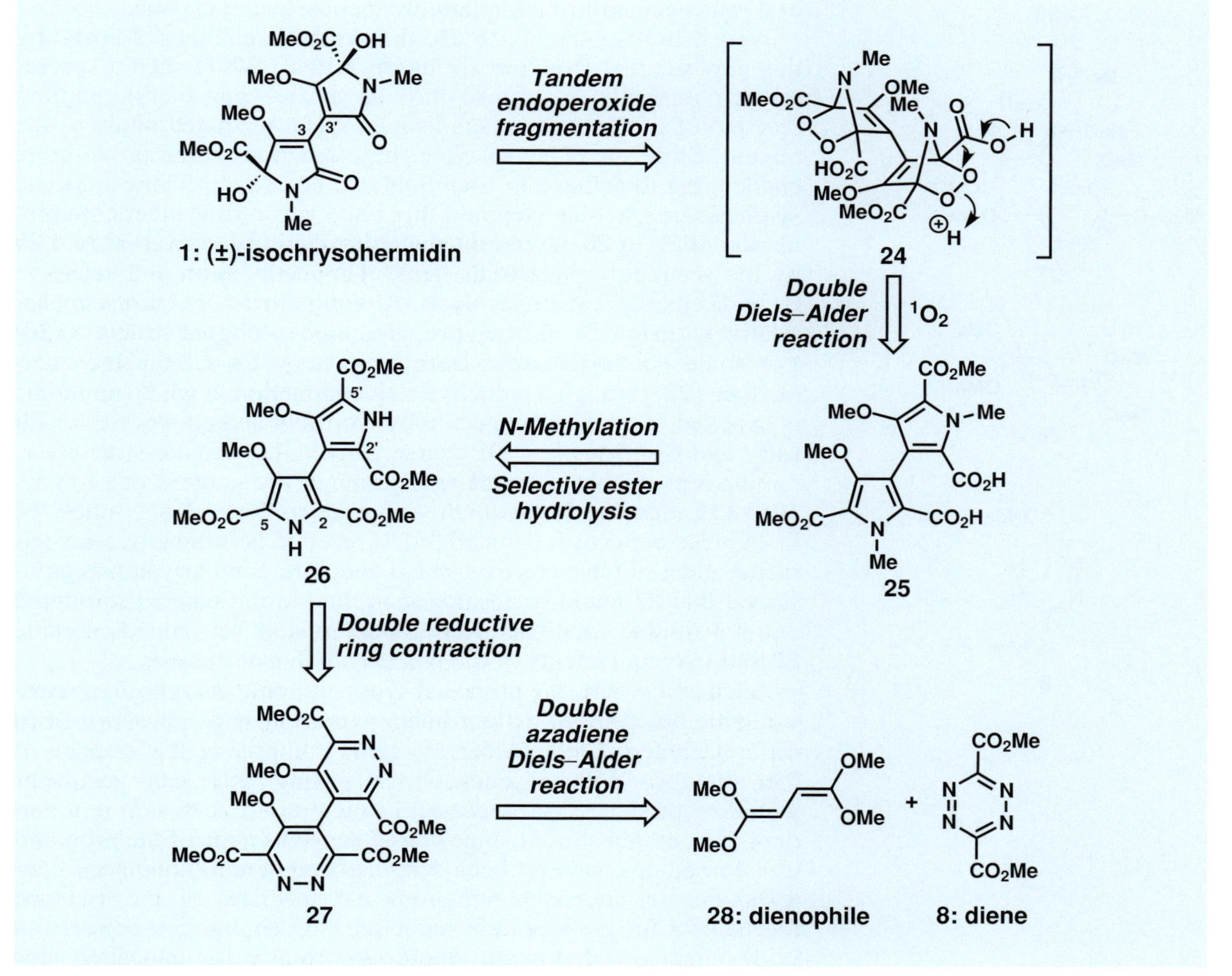

Scheme 5. Retrosynthetic analysis of isochrysohermidin (**1**).

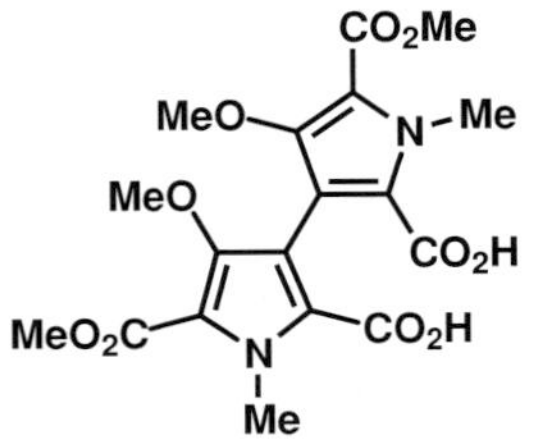

and that oxidative decarboxylation could indeed furnish the carbinolamide moiety of the natural product.[17] Yet, the present context would require this set of conversions to succeed in tandem fashion on two pyrroles, certainly a more challenging proposition than that of the model systems since the risk of undesired side reactions is significantly greater. Based on these preliminary successes and the fact that there are numerous methods to generate singlet oxygen *in situ*, however, it seemed reasonable to assume that this critical sequence converting **25** into **1** could be achieved upon sufficient screening of reaction conditions. At the outset, though, it was anticipated that this reaction would likely also require significant optimization, as subtle features such as solvent and temperature dramatically affect reaction yields within a single identified method of $^1O_2$ generation. One should also be mindful that this sequence was not designed to afford diastereomerically pure **1**, since both **25** and $^1O_2$ are achiral, but, as mentioned earlier, forming a mixture of **1** was deemed to be a relatively inconsequential issue.

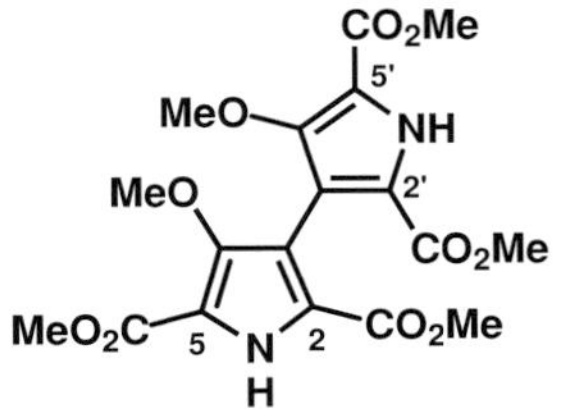

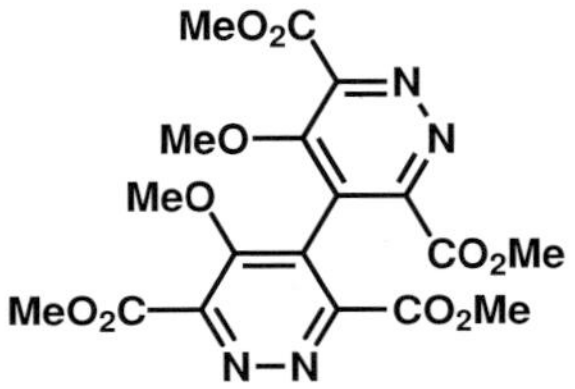

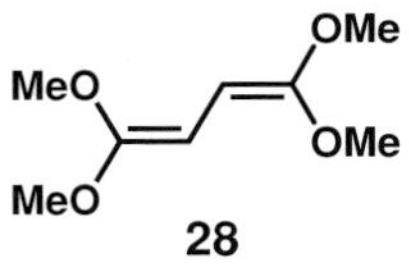

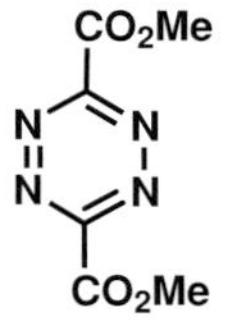

Even with disassembly to **25**, the synthetic challenges posed by this new subgoal structure are hardly trivial. While such a species might potentially be formed through a late-stage biaryl-coupling reaction of pyrrole monomers using a metal-mediated union of the Suzuki, Stille, or Ullmann type, these reactions often prove quite challenging to achieve in such hindered contexts.[18] Thus, a critical synthetic insight side-stepping this issue was retrosynthetic manipulation of **25** to **26**, an intermediate that could be converted into **25** by the seemingly pedestrian steps of *N*-methylation and selective C-2/C-2′ methyl ester hydrolysis. Although these alterations appear relatively insignificant, they provided a new subgoal structure (**26**) that could potentially arise from a precursor bis-1,2-diazine intermediate (**27**) through a reductive ring contraction in which ammonia is expelled, a reaction pioneered by Kornfeld and co-workers at Eli Lilly and Company.[19,20] Of course, such a 1,2-diazine structure is reminiscent of that encountered earlier in the context of (+)-CC-1065 (**15**, Scheme 3) in which such a heterocycle was synthesized in an inverse-electron-demand [4+2] reaction between a 1,2,4,5-tetrazine and a suitable electron-rich dienophile. Similarly, it was envisioned that **27** could be fashioned in the present context through a one-pot double azadiene Diels–Alder fusion between dienophile **28** and two equivalents of the electron-deficient diene **8**.

Taken as a whole, the proposed synthetic route amounts to a retrosynthetic blueprint of extraordinary symmetry and appeal with two different hetero-Diels–Alder reactions comprising the essence of the strategy. Without question, achieving this total synthesis would be predicated on successful execution of each step in a tandem manner, a matter of some consequence as none of the proposed transformations had ever been demonstrated in non-monomeric contexts. For instance, the numerous intermediates of the assumed mechanism for the Kornfeld reductive ring contraction conversion (vide infra) could provide ample opportunity for undesired side reactions in the projected isochrysohermidin synthesis. Thus, conversions such as this one would likely require that both halves of the molecule react in the same manner simultaneously in order to limit the likelihood of precious material travelling down unproductive pathways.

2.3 *Total Synthesis*

Synthetic efforts towards isochrysohermidin (**1**) commenced with exploration of the critical double inverse-electron-demand Diels–Alder union of 1,1,4,4-tetramethoxy-1,3-butadiene (**28**) with tetrazine **8**, as shown in Scheme 6.[4] After considerable experimentation, it was found that the desired sequence based on a double [4+2] cycloaddition reaction to produce the 1,2-diazine dimer **27** could be effected in 65 % yield upon treatment of **28** with

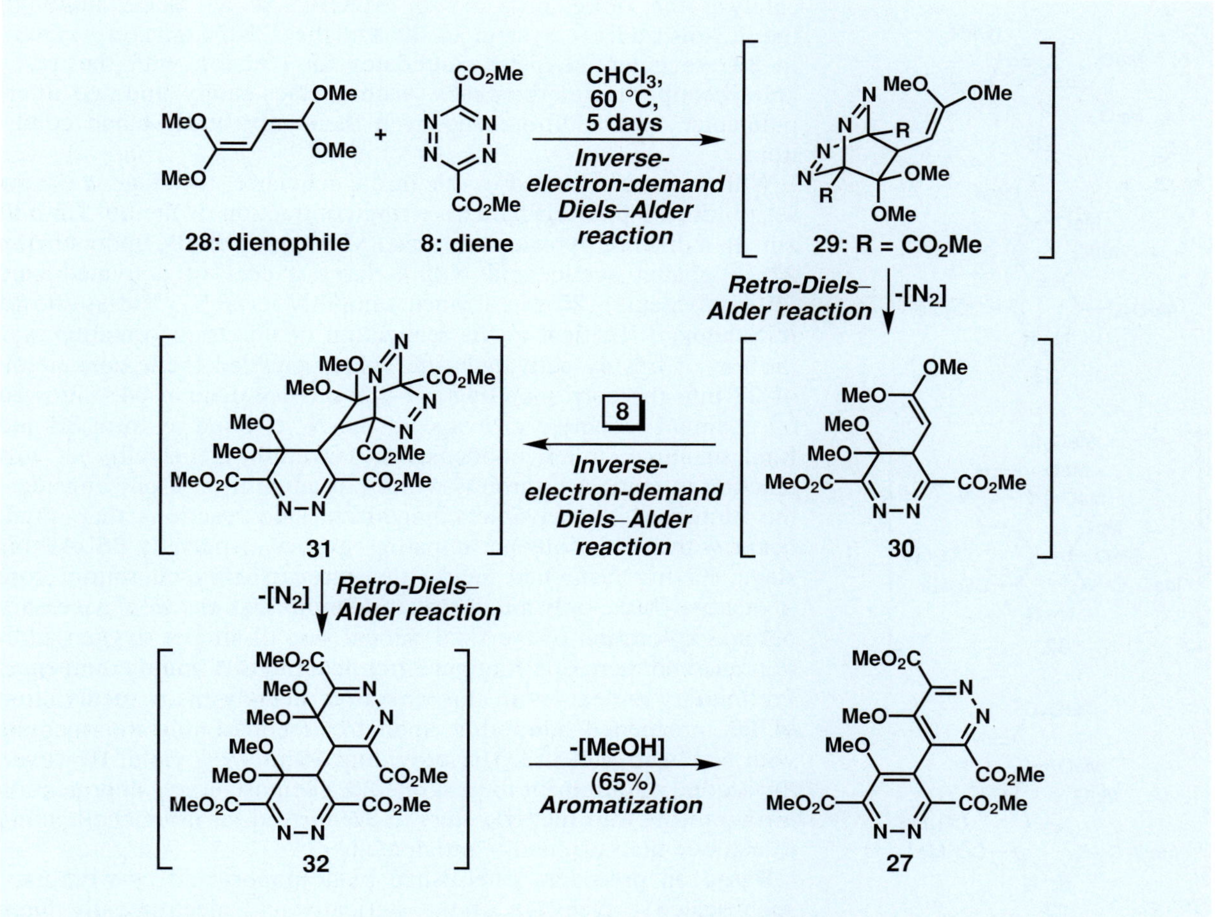

Scheme 6. Initial double azadiene Diels–Alder reaction/retro-Diels–Alder process leading to key biaryl intermediate **27**.

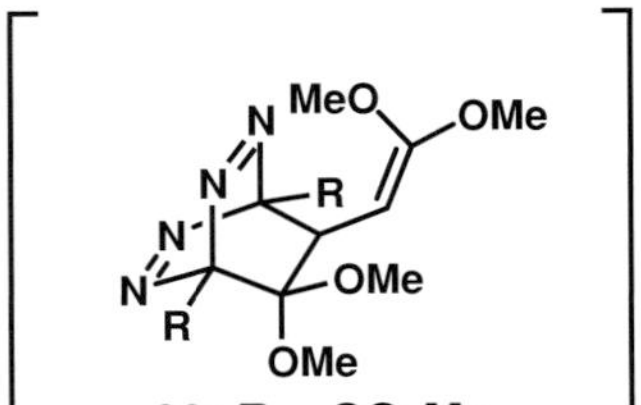

4.5 equivalents of **8** in $CHCl_3$ at 60 °C over the course of 5 days. Detailed studies of this process confirmed that the first Diels–Alder reaction leading to **29** was complete within 30 minutes in a variety of solvents, and was attended by the spontaneous loss of nitrogen, furnishing **30**. In contrast, the second cycloaddition was far more sluggish, most probably the result of increased steric hindrance around the free double bond in **30** relative to **28**, or the decreased nucleophilic character of this intermediate and/or its aromatized variant (**33**, see column figure). Overall, however, the most challenging feature of this reaction was effecting eventual aromatization to **27**, as **32** was isolated in substantial quantities from several experiments and even precipitated during the course of the reaction in nonpolar solvents such as benzene. Fortunately, it was discovered that this final event in the sequence was promoted upon the addition of 4-Å molecular sieves to sequester the formed MeOH. The use of $CHCl_3$ as solvent also improved the reaction's yield since it could provide trace amounts of HCl upon heating to catalyze the obligatory loss of MeOH. As an aside, although the 2,3-bisazadiene system in **30** and the 1,2-diazine ring system in **33** are potential diene candidates for reaction with the available dienophilic intermediates, such species rarely undergo intermolecular cycloaddition, and even then only under harsh conditions.

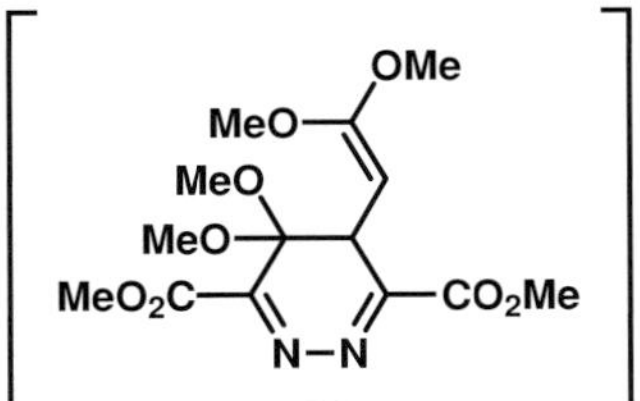

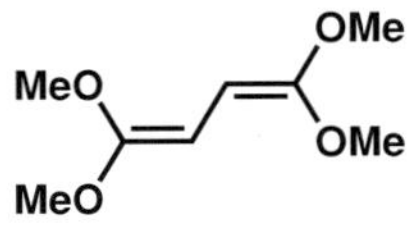

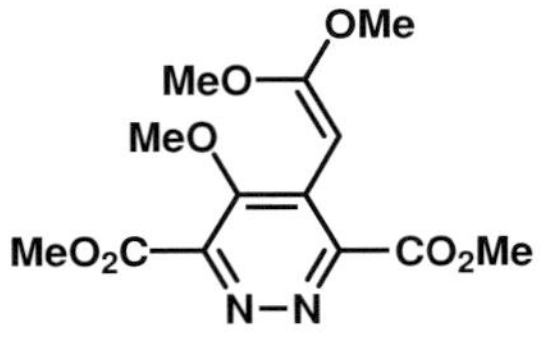

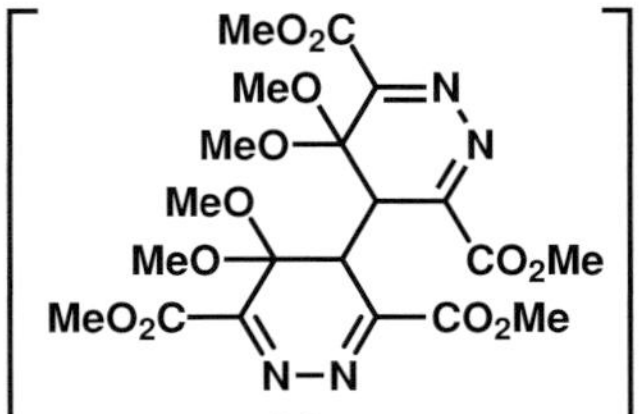

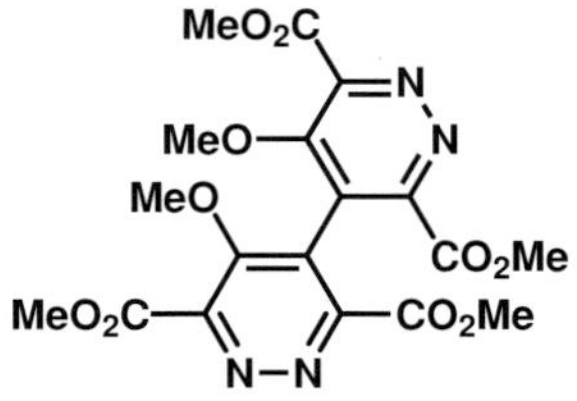

With success achieved in this initial sequence, the stage was now set to test the planned reductive ring contraction of the bis-1,2-diazine to a dimeric pyrrole derivative. Most gratifyingly, upon stirring **27** in glacial acetic acid with a large excess of activated zinc (40 equivalents), **26** was formed smoothly in 68 % yield as shown in Scheme 7. Critical to the realization of this transformation was the use of freshly activated zinc, which enabled facile conversion of **27** into the corresponding 1,4-dihydro-1,2-diazine **34**, followed by eventual reductive cleavage of the N–N bond to form **35** and final enamine–imine condensation. Overall, the fidelity of this reaction in tandem fashion is worthy of admiration if one considers the numerous conceivable competitive side reactions that might occur with the various participating species, especially **35**. At this stage, having fashioned nearly the entire isochrysohermidin core in just two steps, only a few structural modifications were necessary before exploration of the final critical step of singlet oxygen addition and endoperoxide fragmentation leading to **1** could commence. Preliminary endeavors in this direction, namely bis-*N*-methylation of **26**, proceeded admirably upon treatment of this intermediate with NaH and MeI in DMF, providing **36** in 98 % yield. However, the second requirement to generate **25**, chemoselective deprotection of two of the four methyl esters in **36**, proved far more challenging to achieve than originally anticipated.

Based on precedent established by Rapoport and co-workers,[21] such selective cleavage of the sterically and electronically more accessible C-2 methoxycarbonyl group could be achieved on pyrrole monomers (with a H atom at C-3 instead of the C-3/C-3′ biaryl

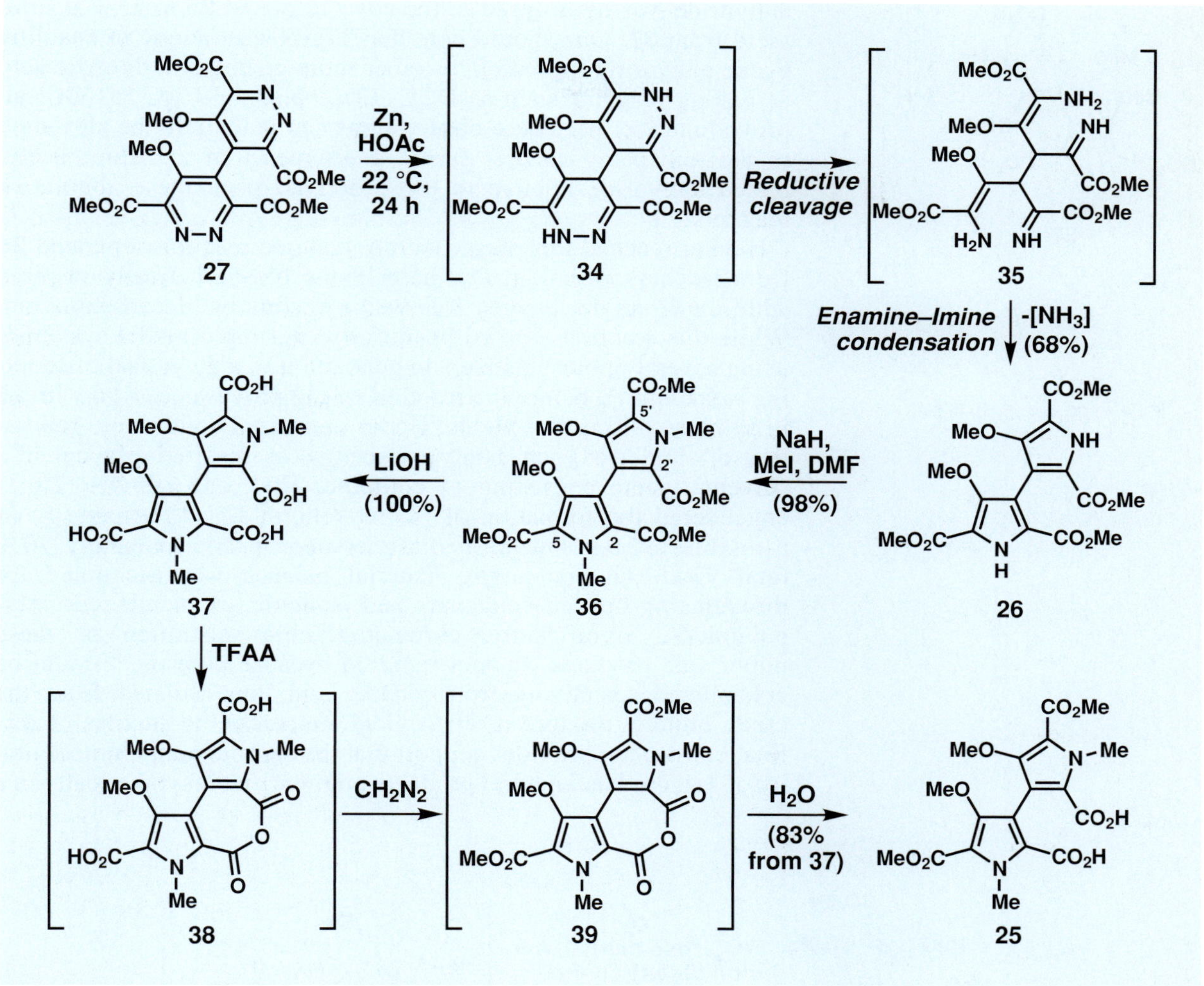

Scheme 7. Synthesis of the advanced key intermediate **25**.

linkage). Boger and Baldino had similarly verified this finding in their own model studies.[17] However, upon inclusion of the pyrrole in a biaryl setting such as **36**, the differential levels of steric accessibility are eroded significantly relative to these simple systems, and extensive efforts to controllably differentiate the C-2/C-2′ and C-5/C-5′ methyl esters in this context failed. Although these results were a minor setback, success was eventually realized upon reversion to a more indirect, but highly insightful, sequence. First, all four esters in **36** were exhaustively hydrolyzed with LiOH, and selective anhydride formation was then achieved upon the addition of TFAA, affording **38**. Both the remaining free acids were then converted into their corresponding methyl esters *in situ* through exposure to diazomethane, and, upon aqueous work-up, the cyclic anhydride was hydrolyzed to the coveted diacid **25** in 83 % overall yield from **37**. One should note that TFAA was unique in enabling cyclic anhydride formation, as other more common carboxylic acid activating reagents such as DCC, EDC, or 2,4,6-$Cl_3C_6H_2COCl$ all proved ineffective. These observations can potentially be attributed to the noticeably smaller steric requirements of a trifluoroacetyl mixed anhydride relative to those derived from these alternative reagents.

Having reached this stage, merely a single reaction separated **25** from isochrysohermidin (**1**), namely the biosynthetically inspired addition of singlet oxygen followed by oxidative decarboxylation. While this reaction worked beautifully on simple model systems[17] using several photosensitizers to generate $^1O_2$, a great deal of scouting was required before the reaction was to prove successful with **25** in synthetically useful yields. Under optimized conditions, generation of singlet oxygen using rose bengal as sensitizing agent in a solvent system consisting of collidine, H_2O, and *i*-PrOH (2:6:1) engendered the formation of isochrysohermidin (**1**, Scheme 8) as a mixture of D,L and *meso* diastereomers in approximately 70 % total yield. The remaining material balance was accounted for through ring-opened congeners and isomeric six-membered carbinolamides. Significantly, chromatographic separation of these minor side products was not required because pure D,L-**1** could be crystallized selectively from EtOAc and thus isolated from the crude product mixture in 40 % yield. As such, the success of this final conversion provides support that the carbinolamide functionality of **1** and related natural products might arise biosynthetically in a

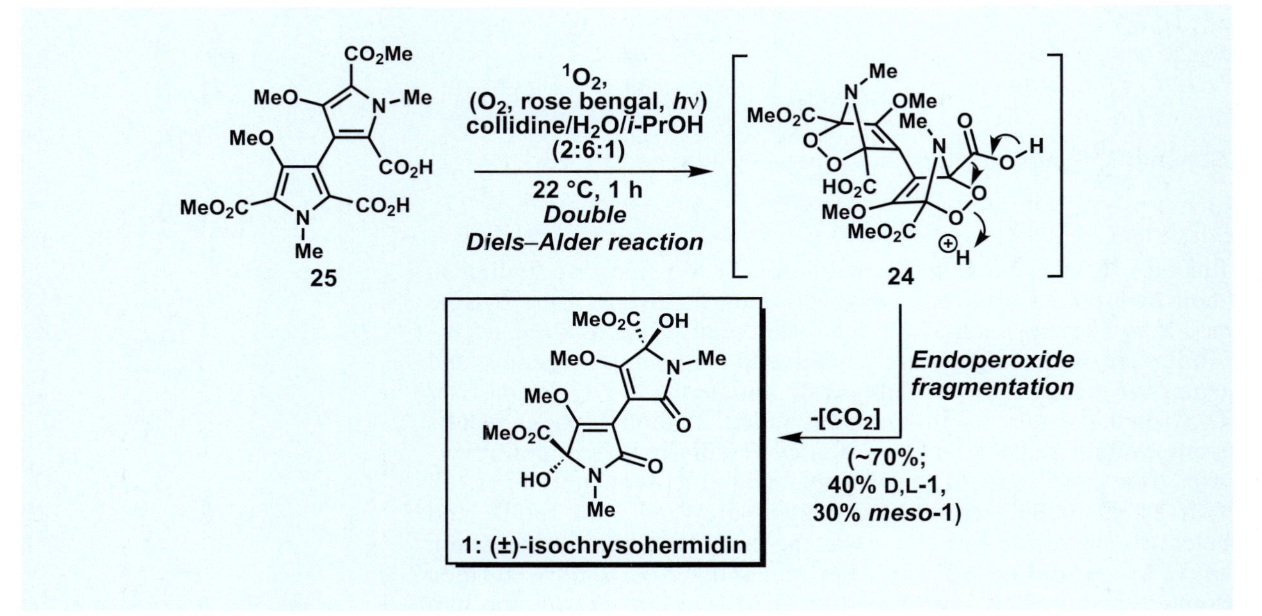

Scheme 8. Completion of the synthesis of isochrysohermidin (**1**).

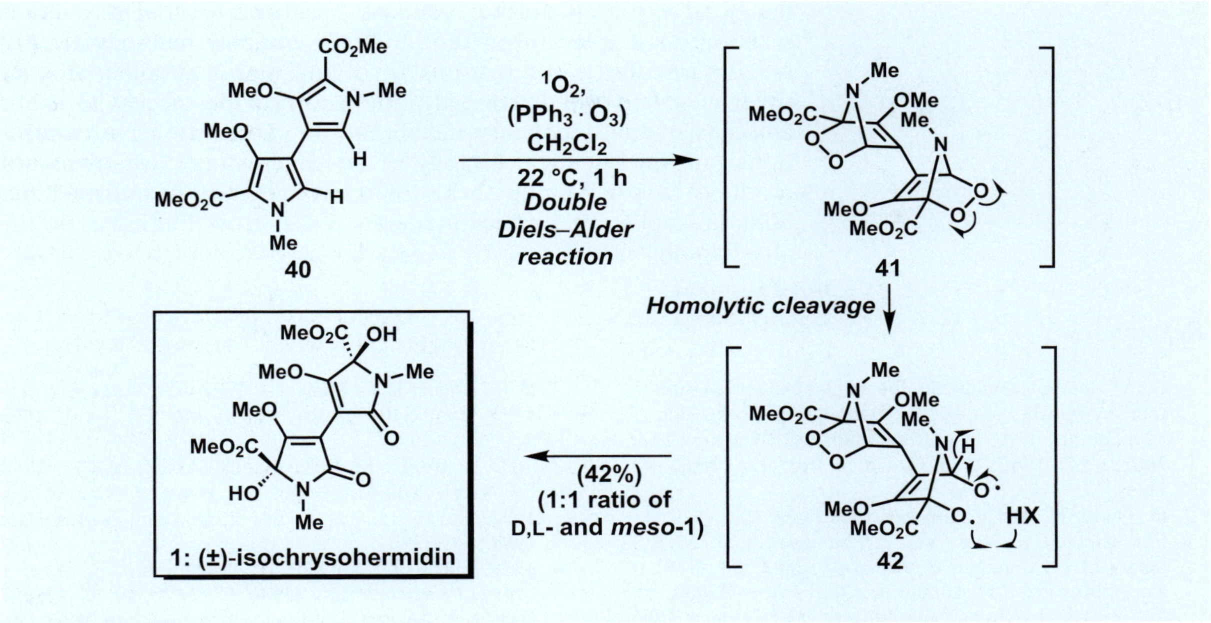

Scheme 9. Wasserman's alternative method of synthesizing isochrysohermidin (**1**) from bispyrrole **40** using singlet oxygen.

similar manner. As it turns out, however, the presence of the C-2/C-2′ carboxylic acid groups in **25** was not a prerequisite to obtain isochrysohermidin (**1**) in this final endeavor. As illustrated by the complementary work of H. H. Wasserman towards this target, exposure of the C-2/C-2′ H-substituted analogue of **25** (**40**, Scheme 9) to singlet oxygen (generated through thermal decomposition of a $PPh_3{\cdot}O_3$ complex formed *in situ*[22]) similarly led to a mixture of D,L- and *meso*-**1** in 42 % yield.[4b,16a,b] Thus, the facility of this key transformation on a related substrate offers further validation of the biosynthetic formation of the carbinolamides in **1** through such a process since the transformation would appear to possess a certain degree of substrate generality.

2.4 Conclusion

Boger and Baldino's total synthesis of isochrysohermidin (**1**) by a sequence encompassing only eight operations, including two different hetero-Diels–Alder reactions, amounts to a beautiful and concise synthetic enterprise. While each tandem operation was impressive in its own right, the power of the inverse-electron-demand Diels–Alder approach for the initial preparation of a highly substituted bis-1,2-diazine ring system, which could then be transformed into a bispyrrole intermediate, is especially striking. Accordingly, this total synthesis further validates azadiene-based Diels–Alder reactions as a powerful method to forge complex heterocyclic systems, particularly in situations involving biaryl systems. Beyond sheer elegance, the developed route also provides access to useful amounts of this secondary metabolite for more extensive explorations into the chemical biology of this interesting class of natural products. Studies along these lines have already established that both D,L- and *meso*-**1** effect interstrand DNA cross-linking as originally hypothesized.[4a]

References

1. For a general review on the Diels–Alder reaction in total synthesis, see: K.C. Nicolaou, S.A. Snyder, T. Montagnon, G.E. Vassilikogiannakis, *Angew. Chem.* **2002**, *114*, 1742; *Angew. Chem. Int. Ed.* **2002**, *41*, 1668.
2. O. Diels, K. Alder, *Annalen* **1928**, *460*, 98.
3. For selected reviews, see: a) S. Jayakumar, M.P.S. Ishar, M.P. Mahajan, *Tetrahedron* **2002**, *58*, 379; b) M. Behforouz, M. Ahmadian, *Tetrahedron* **2000**, *56*, 5259; c) D.L. Boger, *Chemtracts: Org. Chem.* **1996**, *9*, 149; d) D.L. Boger, *J. Heterocycl. Chem.* **1996**, *33*, 1519; e) D.L. Boger in *Comprehensive Organic Synthesis, Vol. 5* (Ed.: B.M. Trost), Pergamon, Oxford, **1991**, pp. 451–512; f) D.L. Boger, *Chem. Rev.* **1986**, *86*, 781; g) S.M. Weinreb, *Acc. Chem. Res.* **1985**, *18*, 16; h) D.L. Boger, *Tetrahedron* **1983**, *39*, 2869.
4. a) D.L. Boger, C.M. Baldino, *J. Am. Chem. Soc.* **1993**, *115*, 11418; b) H.H. Wasserman, R.W. DeSimone, D.L. Boger, C.M. Baldino, *J. Am. Chem. Soc.* **1993**, *115*, 8457.
5. D.L. Boger, W.L. Corbett, T.T. Curran, A.M. Kasper, *J. Am. Chem. Soc.* **1991**, *113*, 1713.
6. a) D.L. Boger, K.C. Cassidy, S. Nakahara, *J. Am. Chem. Soc.* **1993**, *115*, 10733; b) D.L. Boger, S. Nakahara, *J. Org. Chem.* **1991**, *56*, 880.
7. I. Fleming, *Frontier Orbitals and Organic Chemical Reactions*, John Wiley and Sons, Chichester, **1978**, p. 249.
8. For the synthesis of *N*-sulfonyl-azadiene systems, see ref. 5 and D.L. Boger, W.L. Corbett, *J. Org. Chem.* **1992**, *57*, 4777.
9. a) R.A. Carboni, R.V. Lindsey, *J. Am. Chem. Soc.* **1959**, *81*, 4342; b) J. Sauer, H. Wiest, *Angew. Chem.* **1962**, *74*, 353; *Angew. Chem. Int. Ed. Engl.* **1962**, *1*, 269.
10. a) D.L. Boger, R.S. Coleman, *J. Am. Chem. Soc.* **1988**, *110*, 4796; b) D.L. Boger, R.S. Coleman, *J. Am. Chem. Soc.* **1988**, *110*, 1321; c) D.L. Boger, R.S. Coleman, *J. Am. Chem. Soc.* **1987**, *109*, 2717; d) D.L. Boger, R.S. Coleman, *J. Org. Chem.* **1986**, *51*, 3250; e) D.L. Boger, R.S. Coleman, *J. Org. Chem.* **1984**, *49*, 2240.
11. D.L. Boger, S.M. Sakya, *J. Org. Chem.* **1988**, *53*, 1415.
12. D.L. Boger, R.F. Menezes, T. Honda, *Angew. Chem.* **1993**, *105*, 310; *Angew. Chem Int. Ed. Engl.* **1993**, *32*, 273.
13. a) D.L. Boger, M.J. Kochanny, *J. Org. Chem.* **1994**, *59*, 4950; b) D.L. Boger, Q. Dang, *J. Org. Chem.* **1992**, *57*, 1631; c) D.L. Boger, Q. Dang, *Tetrahedron* **1988**, *44*, 3379.
14. a) Y. Masui, C. Kawabe, K. Mastumoto, K. Abe, T. Miwa, *Phytochemistry* **1986**, *25*, 1470; b) K. Abe, T. Okada, Y. Masui, T. Miwa, *Phytochemistry* **1989**, *28*, 960.
15. a) G.A. Swan, *J. Chem. Soc., Perkin, Trans 1* **1985**, 1757; b) A.R. Forrester, *Experientia* **1984**, *40*, 688; c) G.A. Swan, *Experientia* **1984**, *40*, 687.
16. a) H.H. Wasserman, V.M. Rotello, R. Frechette, R.W. DeSimone, J.U. Yoo, C.M. Baldino, *Tetrahedron* **1997**, *53*, 8731; b) H.H. Wasserman, R. Frechette, V.M. Rotello, G. Schulte, *Tetrahedron Lett.* **1991**, *32*, 7571; c) D.A. Lightner, G.S. Bisacchi, R.D. Norris, *J. Am. Chem. Soc.* **1976**, *98*, 802; d) J.W. Scheeren, R.W. Aben, *Tetrahedron Lett.* **1974**, *15*, 1019; e) H.H. Wasserman, J.R. Scheffer, J.L. Cooper, *J. Am. Chem. Soc.* **1972**, *94*, 4991; f) C.S. Foote, S. Wexler, W. Ando, R. Higgins, *J. Am. Chem. Soc.* **1968**, *90*, 975; g) C.S. Foote, S. Wexler, W. Ando, *Tetrahedron Lett.* **1965**, *6*, 4111.
17. D.L. Boger, C.M. Baldino, *J. Org. Chem.* **1991**, *56*, 6942.
18. For general commentary on this subject, see: *Metal-catalyzed Cross-coupling Reactions* (Eds.: F. Diederich, P.J. Stang), Wiley-VCH, Weinheim, **1998**, p. 517.
19. N.J. Bach, E.C. Kornfeld, N.D. Jones, M.O. Chaney, D.E. Dorman, J.W. Paschal, J.A. Clemens, E.B. Smalstig, *J. Med. Chem.* **1980**, *23*, 481.
20. For a representative example, see: a) D.L. Boger, M. Patel, *J. Org. Chem.* **1988**, *53*, 1405; b) D.L. Boger, M. Patel, *Tetrahedron Lett.* **1987**, *28*, 2499.
21. a) H. Rapoport, J. Bordner, *J. Org. Chem.* **1964**, *29*, 2727; b) H. Rapoport, C.D. Willson, *J. Am. Chem. Soc.* **1962**, *84*, 630; c) H. Rapoport, K.G. Holden, *J. Am. Chem. Soc.* **1962**, *84*, 635.
22. R.W. Murray, M.L. Kaplan, *J. Am. Chem. Soc.* **1969**, *91*, 5358.

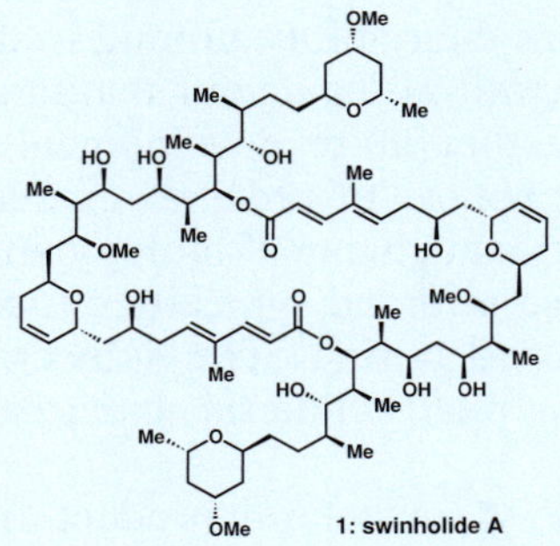

1: swinholide A

3

I. Paterson (1994)

Swinholide A

3.1 Introduction

Because all cancers derive from the uncontrolled replication of abnormal cells, the scientific and medical communities have long used chemical agents capable of disrupting the cell-division process as a standard element of cancer-treatment regimes. The rationale being that if cancerous cells lose their ability to reproduce, then they will be unable to invade neighboring tissues and spread their deleterious effects to other parts of the body. Thankfully, this theory works quite well in practice, as you know if you have read the chapters on Taxol™, vinblastine, and the epothilones in the *Classics* series. Its therapeutic potential, however, has only barely been tapped. Not only have just a small portion of the likely dozens of biomolecules involved in a process as complicated as mitosis been identified, but only a few of the targets that are known are currently being pursued pharmaceutically. For example, although the three compounds just mentioned exert their cytotoxic action through slightly different cellular effects, they all attack the same protein (tubulin). Accordingly, the advancement of cancer chemotherapy in the coming decades will depend crucially on both our ability to obtain a more complete understanding of the mitotic process as well as our skill in identifying small molecules that can selectively target every relevant player in this cycle.

Natural products are certain to play a pivotal role in both these tasks.[1] For example, researchers have long known that the protein actin, one of the two major cytoskeletal components of eukaryotic cells (the other being the aforementioned tubulin), determines both cell shape and motility. Only in more recent times has its non-trivial part in the process of cell division been uncovered, a result

Key concepts:

- **Boron-mediated aldol reactions**
- **Paterson aldol reaction**
- **Mukaiyama aldol reaction**
- **Sakurai reaction**

2: hemiswinholide A

3: preswinholide A

4: misakinolide A

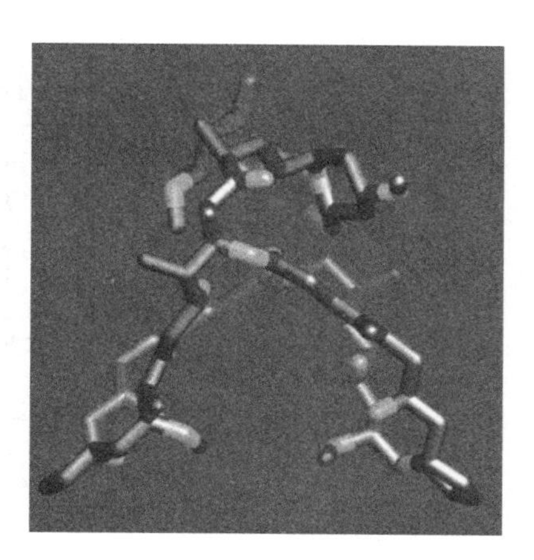

due solely to the discovery of a series of highly cytotoxic secondary metabolites whose activity derives from their ability to bind to this biomolecule.[2] As such, this protein as well as the agents that target it provide a useful starting point for the future development of entirely novel classes of chemotherapeutics. In order to evaluate this potential more fully through additional chemical biology studies, enriched material supplies of these rare and typically structurally complex natural products are needed first. The subject of this chapter is a research program in total synthesis undertaken with just that objective in mind.

Our main protagonist is swinholide A, a natural product first reported in 1985 by S. Carmely and Y. Kashman from Tel-Aviv University as one of several antifungal isolates obtained from *Theonella swinhoei*, a sponge that resides in the Red Sea.[3] As shown in the adjoining column figure, their original structural assignment for this intriguing agent was **2**, a designation indicating that swinholide A possessed a monomeric, 22-membered macrolide ring decorated with 15 stereocenters, a trisubstituted tetrahydropyran ring, a disubstituted dihydropyran system, and two isomerizable double bonds. This assignment, however, was later proven to be in error by the Kitagawa group at Osaka University who rediscovered this natural product from an Okinawan sponge and determined that its actual structure was instead a dimeric and C_2-symmetric version of the original (i. e. **1**).[4] Equally important as establishing the correct molecular architecture of this natural product was the Kitagawa group's demonstration that swinholide A (**1**) possessed strong cytotoxic activity against both the KB and L 1210 tumor-cell lines with actin as its biological target. Intriguingly, although several other natural products with similar structures, such as preswinholide A (**3**)[5] and the 40-membered dilactone misakinolide A (**4**),[6] have since been isolated and fully characterized from *Theonella swinhoei* and its sponge relatives, only swinholide A (**1**) is highly cytotoxic. Thus, this information suggests that the imposing 44-membered ring within **1** is at least partially responsible for its cytotoxic potency since this motif is the only structural feature that is not shared by some portion of these other natural products. As shown in the adjacent column, this outcome is potentially reflected by the unique "twisted saddle" conformation of swinholide A, an orientation that directs its nonpolar residues towards the exterior and points its highly polar oxygen motifs into the inner cavity, poised to sequester actin.

Based on the amalgamation of such a fascinating mode of biological action, a formidable and unique structure, and a need for additional material supplies, swinholide A (**1**) issued a clarion call that had to be answered by the chemical community with a total synthesis. While many would attempt to answer that summons, the first to succeed was Professor Ian Paterson and his group at Cambridge University, who prepared the first laboratory samples of swinholide A (**1**) in 1994.[7] In this chapter, we shall recount the details of their highly inventive route to this formidable target, one in which the strength of numerous asymmetric transformations, most notably the aldol reaction, served to establish much of its stereochemical complexity. Before we embark on a description of this instructive campaign, a brief foray into general aspects of the asymmetric aldol reaction is required in order to set the stage for the most challenging elements of the Paterson group's design towards **1**. Our goal is not to discuss this highly important reaction in a comprehensive manner or to delve into all its various intricacies, as several review articles have already accomplished these objectives quite effectively.[8,9] Rather, we desire to paint a broad picture of the strengths and weaknesses of various aldol protocols to establish stereochemical relationships, focusing our discussion exclusively on the reactions of boron enolates.

3.1.1 Boron-Mediated Asymmetric Aldol Reactions

Among all the possible enolates that one could form from a ketone, boron enolates would appear to provide the greatest potential for accomplishing high levels of aldol asymmetry on the basis of their physical properties. Not only are these intermediates homogeneous and uncomplicated in terms of aggregation state (which is not true of lithium enolates, for example), but they provide access to tight and organized transition states because B–O bonds are relatively short and far less nucleophilic than their metallated counterparts. As such, chiral information that has been encoded into the reactants should, in principle, be faithfully expressed in the aldol products. In practice, in order to accomplish any level of aldol stereoselectivity, one must first be able to convert a given ketone into its corresponding *E* and/or *Z* enolate at will and with complete stereospecificity.[9] Although for many years this simply stated objective eluded the synthetic community, the technology to address this issue presently exists due primarily to the efforts of the Mukaiyama, Evans, Masamune, Heathcock, and Paterson groups.[10]

As shown in Figure 1 in the neighboring column, boron enolates (**6**) can be fashioned under kinetic control merely by treating a ketone (**5**) with a dialkyl boron reagent and a tertiary amine base. Which enolate regio- and stereochemistry ultimately results, however, is governed by several factors, including the steric requirements of the substrate and the boron ligands, the leaving group on the boron reagent (we will consider only a triflate or a chloride here), and the size of the amine base. Based on the broad nature and number of these features, you might expect that these outcomes are hard to foresee. In fact, they are quite predictable. The general rule of thumb is simply that the use of dialkyl boron chloride reactants with large ligands (such as a cyclohexyl ring) in combination with small bases (such as Et_3N) affords *E* enolates (**10**), while the joint power of dialkyl boron triflates bearing small alkyl ligands (such as *n*-butyl) and sterically demanding amine bases (such as

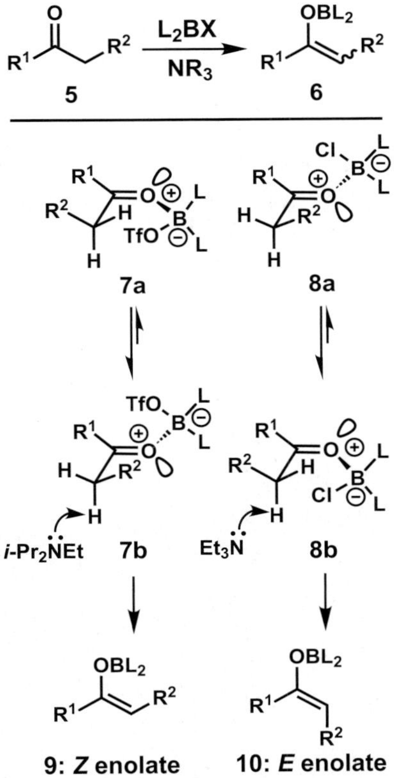

Figure 1. Rationale for kinetic deprotonation leading to *Z* or *E* enolates.

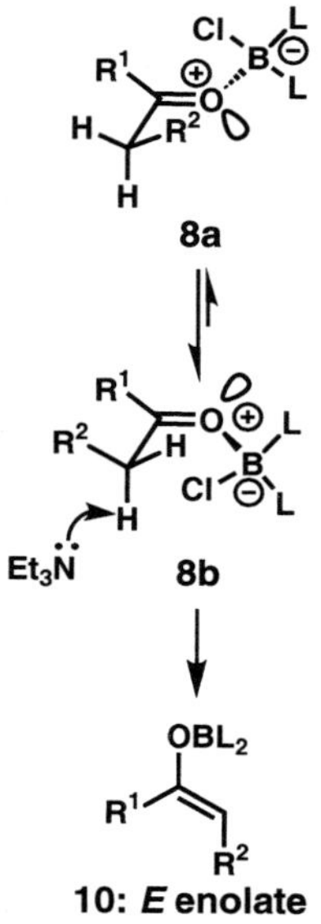

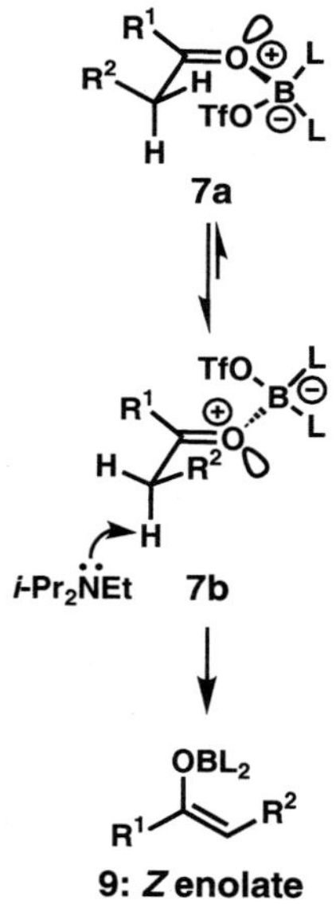

10: *E* enolate

9: *Z* enolate

i-Pr_2NEt) provides *Z* enolates (**9**).* In order to understand why these results occur, we will have to delve into some physical organic chemistry, advancing arguments based both on steric and electronic effects.[11]

Considering first those reactions involving dialkyl boron chlorides, all evidence points to the fact that these reagents preferentially complex the lone pair of the ketone oxygen atom *cis* to the alkyl group that can best stabilize a negative charge, meaning simply the less-substituted alkyl group (Me>Et>*i*-Pr). As such, this orientation is superior on both steric and electronic grounds, and thus its selective formation should be enhanced by larger ligands on the boron species. Once chelated, the chlorine atom then serves to electronically activate the proton nearest to it for selective abstraction by the amine base. In picture form, these comments translate into an arrangement such as **8b** being favored over the alternative **8a**, with the complex poised over the eventual site of deprotonation. Abstraction of the activated proton *cis* to the boron chelate then leads to the regio- and stereoselective formation of an *E* boron enolate (**10**). Accordingly, since that proton is slightly sterically shielded over its available alternatives, the use of small amine bases ensures that only this proton is excised (i. e. that deprotonation is controlled by electronic activation, and not steric accessibility).

For reactions involving dialkyl boron triflates, slightly different arguments come into play because steric factors alone rather than any electronic effects, such as the ability of the alkyl group to stabilize a negative charge, dictate the initial direction of complexation. Thus, for the generic ketone (**5**) shown in Figure 1, intermediate **7b** would be favored over **7a** since it points the bulk of the triflate away from the substrate. This orientation is favored by small ligands on the boron reagent since it enhances the impact of the triflate in determining the direction of complexation. Intriguingly, though, the side of complexation does not correlate to the site of deprotonation as was the case above. Indeed, the proton that is removed is the one periplaner to the carbonyl in the conformation of **7b** which minimizes the steric interaction between R^1 and R^2, thereby leading to a *Z* enolate (**9**). As such, the use of relatively large amine bases like *i*-Pr_2NEt ensures that only the most accessible proton in transition state **7b** is removed.[11]

Having discussed a qualitative picture for how configurationally pure enolates can be synthesized (with a mnemonic for the trends discussed above in the boxed sections of Scheme 1), we can finally discuss their reaction with aldehydes in aldol reactions. In theory, up to four different reaction products can be generated in any given aldol reaction since two new stereocenters can be created by the process. In reality, because boron enolates react in tight tran-

* Within the context of this chapter, the oxygen–boron substituent always has a higher priority than the R^1 group of structures **9** and **10** in assigning *E* and *Z* enolate titles. This assignment pattern is reflected in the vast majority of the primary chemical literature.

sition states, at most only two of these four products will typically be formed. Which of these adducts is produced can be predicted accurately by the Zimmerman–Traxler model, a picture for aldol reactions shown in Scheme 1 in which boron enolates react with aldehydes through chair-like transition states.[12] For example, if a *Z* boron enolate (**9**) is added to an aldehyde, then two different reacting conformations can be adopted (**11** and **12**), leading either to 1,2-*syn* (**13**) or 1,2-*anti* (**14**) aldol products, respectively. However, since the latter transition state (**12**) results in severe 1,3-diaxial interactions as shown, **11** constitutes the more favorable reaction pathway. Accordingly, 1,2-*syn* products (**13**) are observed almost exclusively in aldol reactions involving *Z* boron enolates. Similar arguments serve to explain the selective formation of 1,2-*anti* products (**17**) in aldol events employing *E* enolates (**10**). Thus, relative stereocontrol in aldol reactions results merely through the initial formation of a stereochemically pure enolate. In order to accomplish

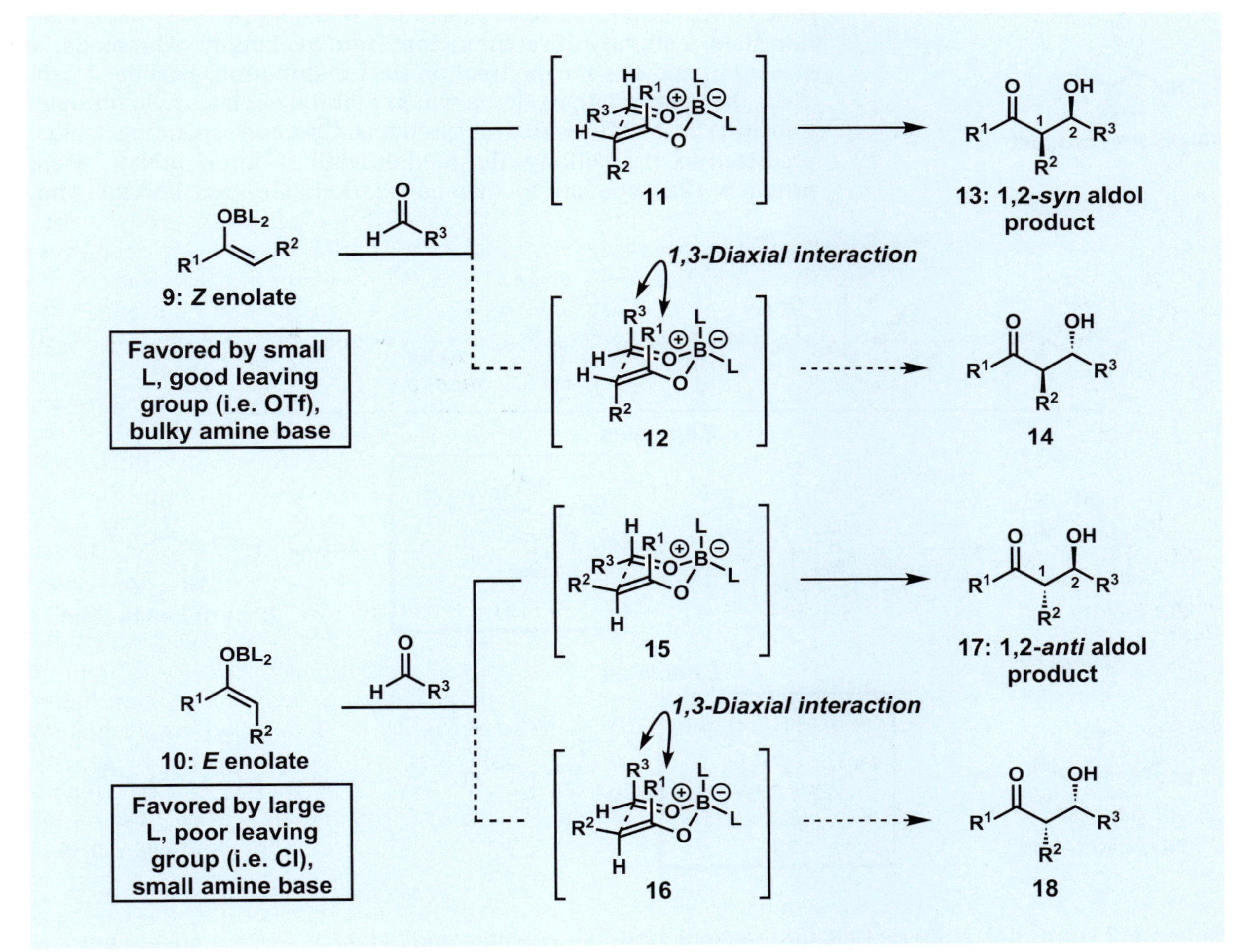

Scheme 1. Zimmerman–Traxler models for aldol stereoselectivity with boron enolates.

absolute stereocontrol (i. e. form only one aldol product instead of the two possible 1,2-*syn* or 1,2-*anti* adducts), however, one must find techniques that can discriminate between the π faces of both the aldehyde and the boron enolate components within transition states **11** and **15**. The rest of this section will focus on the general substrate- and reagent-based methods that enable this highly useful form of stereoselectivity to be reached.

The simplest and most common method for accomplishing asymmetric induction in any type of reaction is to utilize a chiral element within one of the reactants to direct its interaction with another. For example, when an aldehyde bearing an α-stereogenic center is attacked by a nucleophile, the approach of that reagent is dictated by the steric demands of the groups situated at that adjacent stereocenter. As shown in the neighboring column, we are simply talking about the standard Felkin–Ahn model for stereoselectivity.[13] In the aldol reaction, the use of chiral aldehydes can similarly provide access to homochiral products. However, since a cyclic transition state is active in these reactions rather than an open, acyclic transition state, a slightly different picture from the Felkin–Ahn model is needed to account for the final product distribution. Scheme 2 provides the appropriate model in which *E* boron enolates react through transition state **23** to provide products (**24**) corresponding to that predicted by the Felkin–Ahn model, while *Z* boron enolates yield products (**22**) counter to that convention (i. e. anti-Felkin–Ahn)

Felkin–Ahn product

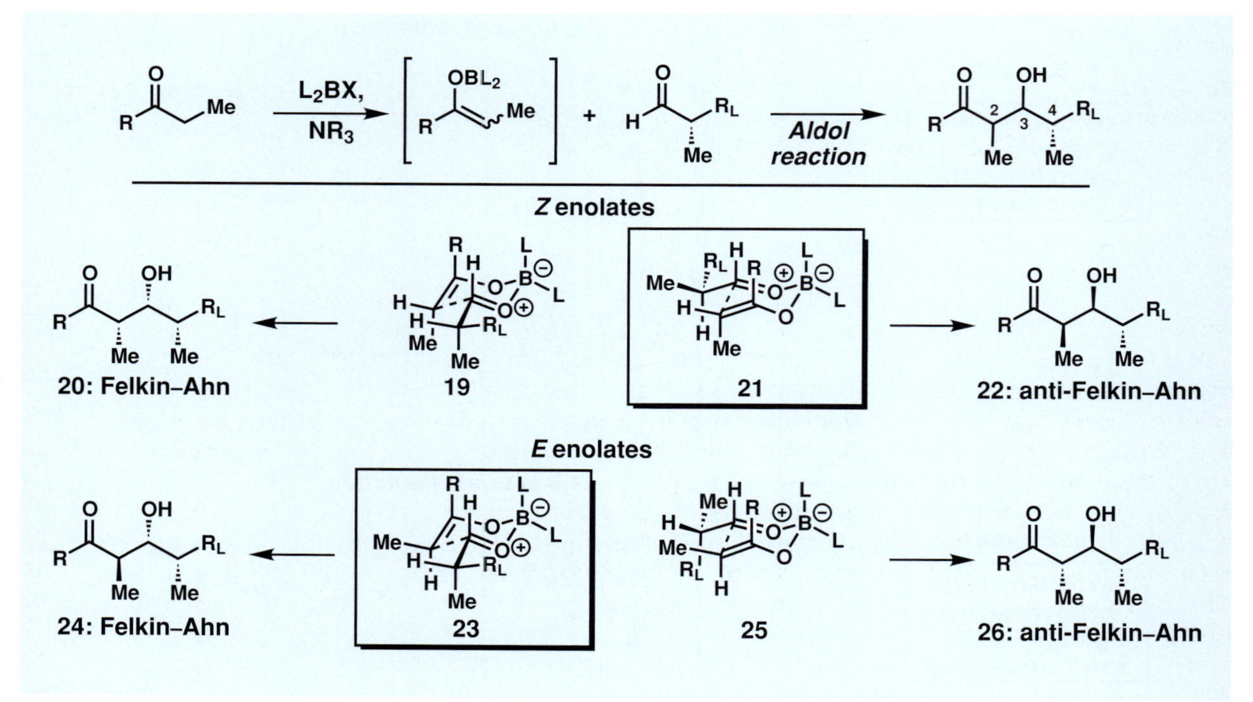

Scheme 2. Transition state models for reactions between enolates and aldehydes with a stereogenic center at the α-position (valid as long as R_L>Me).

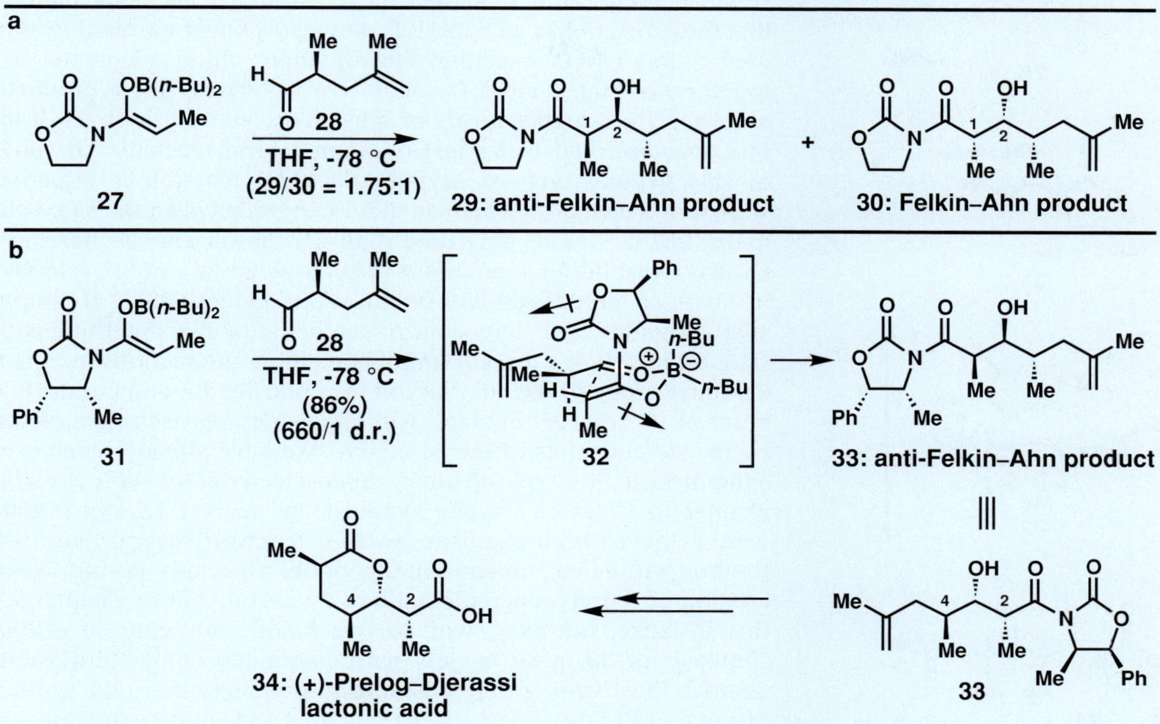

Scheme 3. While asymmetric induction using chiral aldehydes to achieve selectivity usually provides modest results (a), incorporation of an auxiliary provides excellent induction (b) as demonstrated in Evans' total synthesis of the (+)-Prelog–Djerassi lactonic acid (**34**). (1982)[14]

through transition state **21**. In each case, the stereogenic center of the aldehyde dictates the enolate approach, with the alternate facial presentation disfavored due to steric penalties arising from the *syn*-pentane interaction in **19** and the eclipsed *gauche* interaction in **25**.

In practice, aldehydes bearing an adjacent stereogenic center, particularly one devoid of a bulky group, typically provide only modest aldol stereoselectivity. For example, consider the reaction of aldehyde **28** with *Z* boron enolate **27** as shown in Scheme 3a.[14] According to the models just discussed, one would expect this reaction to provide only 1,2-*syn* products with anti-Felkin–Ahn stereoselectivity. Indeed, that conjecture proved to be true for the most part as exclusively 1,2-*syn* aldol adducts resulted with **29**, a compound whose stereotriad reflected anti-Felkin–Ahn selectivity, constituting the predominant product. However, a fair amount of the alternate 1,2-*syn* Felkin–Ahn adduct (**30**) was also observed, such that the final ratio of **29** to **30** was a disappointing 1.75:1.

In order to circumvent this general lack of aldol selectivity with chiral aldehydes, one typically needs to employ a chiral enolate as well, bearing a homochiral element that is either a permanent stereocenter or a removable chiral auxiliary. The expectation is that these reactions can benefit from what is known as double asymmetric induction, meaning that if their chiral elements work together, or stated more formally, are "matched" for a productive aldol reaction, higher levels of diastereoselectivity will result than that accomplished using just one homochiral reactant. The power of this general type of asymmetric aldol reaction is effectively shown in Scheme 3b where an aldol merger between the same aldehyde (**28**) as before, this time with a *Z* boron enolate bearing an Evans' oxazolidinone auxiliary (**31**), now led to a 660:1 selectivity in favor of the desired anti-Felkin–Ahn adduct (**33**).[14] Its impressively high level of stereocontrol can be rationalized with transition state **32** in which a drive to minimize dipole interactions orients the carbonyl group of the auxiliary in the opposite direction to the C–O bond of its adjacent enolate, with aldehyde approach then dictated by the stereocenters of the auxiliary.[9] We have already seen several examples of this type of aldol stereoselectivity in the cytovaricin chapter of *Classics I* where syntheses of several 1,2-*syn* products were achieved with exquisite levels of control; several more illustrations will be encountered in this book, especially during our discussion of the glycopeptide antibiotic vancomycin in Chapter 9. In this instance, however, with this technique allowing an efficient synthesis of **33** in 86 % yield with essentially complete diastereocontrol, the Evans group was able to complete the total synthesis of the (+)-Prelog–Djerassi lactonic acid (**34**) quite expeditiously.

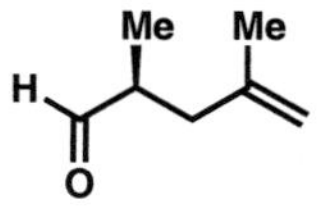

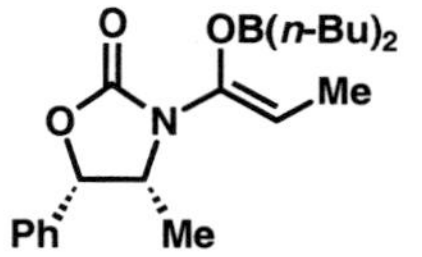

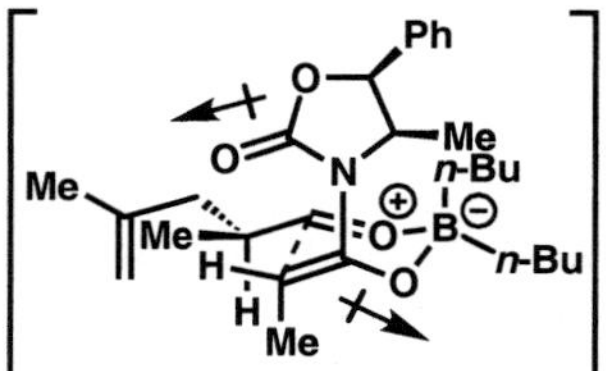

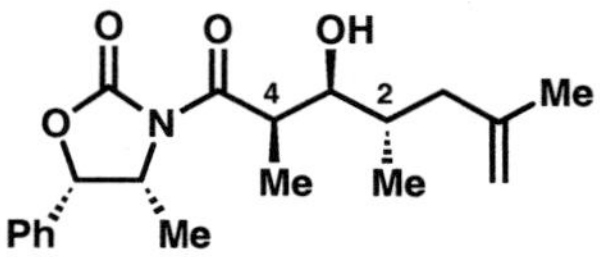

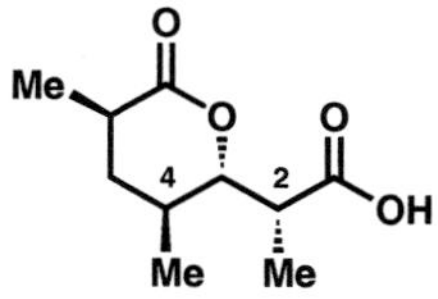

Despite its formidable power, the one drawback of the Evans auxiliary approach is that it does not typically enable the facile synthesis of 1,2-*anti* aldol products, irrespective of the type of aldehyde (chiral or achiral) employed, because the bulk of the auxiliary precludes the direct formation of *E* boron enolates. As such, other metals, and hence open transition states, are required to accomplish such stereoselectivity. Fortunately, there are other substrate-based techniques that can come to the rescue and provide these products with excellent diastereoselectivity using boron enolates. One of the best of these methodologies emanates from the Paterson group in which stereocenters of the type present in ketone **35** (Scheme 4) can orchestrate the union of their corresponding *E* boron enolates with almost any aldehyde to afford a single 1,2-*anti* aldol product.[15] For example, the aldol reaction of *E* boron enolate **36** with achiral aldehyde **37** gave rise to **39** in 86 % yield and with a diastereomeric ratio better than 99:1. The feature responsible for the high level of aldol asymmetry in this case is a desire to minimize steric interactions as well as the opportunity to acquire secondary stabilization through a hydrogen bond between the carbonyl of the benzoate and the neighboring aldehyde proton as shown in transition state **38**. The true power and ingenuity of this asymmetric aldol technique, however, is the fact that the stereogenic center on the enolate is actually an auxiliary. Indeed, as indicated in part b of the same scheme, the benzoylated oxygen atom of a reaction product (i. e.

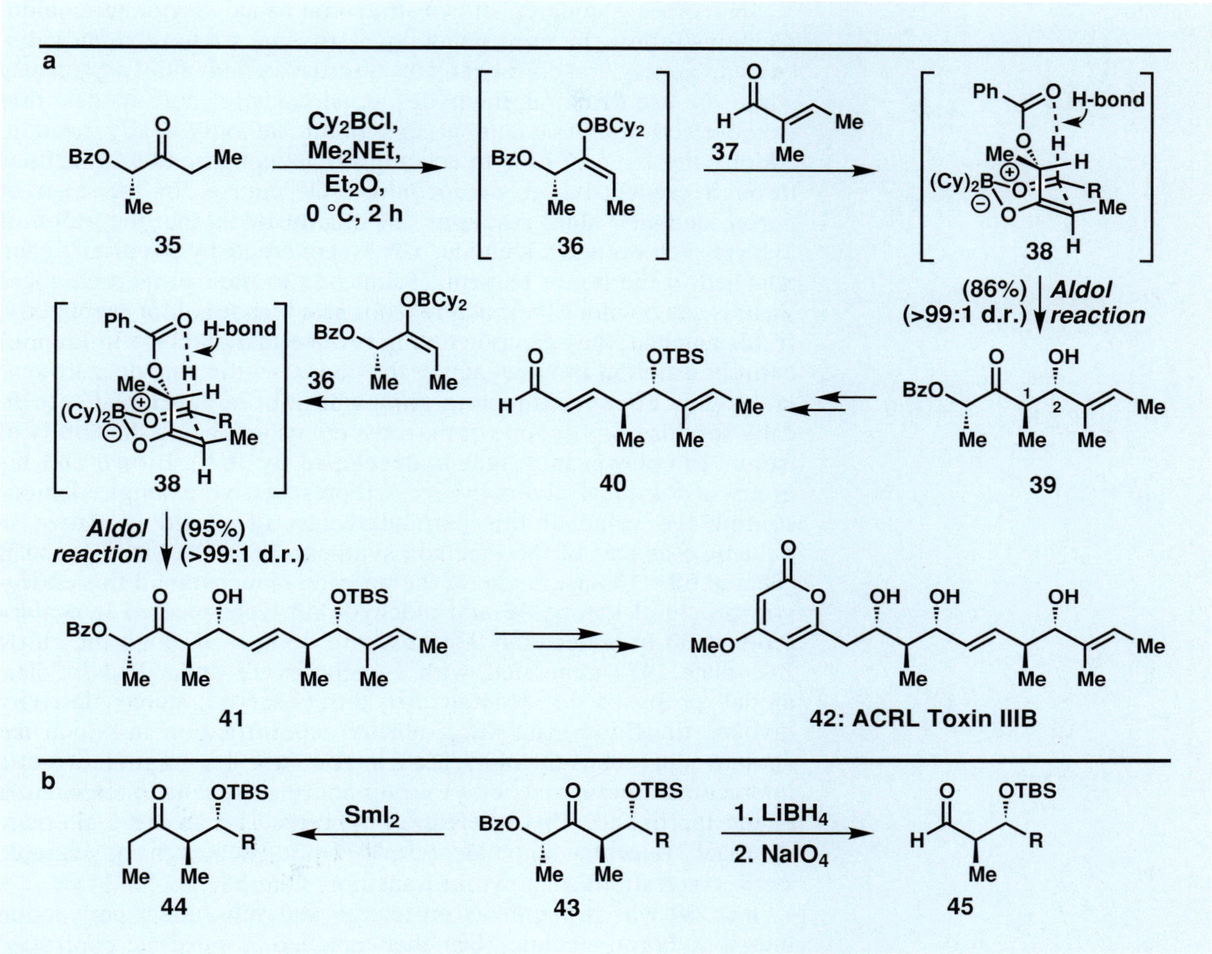

Scheme 4. (a) Chiral auxiliaries in the form of α-oxygenated substrates to accomplish 1,2-*anti* aldol stereoselectivity in Paterson's total synthesis of ACRL Toxin IIIB (**42**) and (b) methods to remove the directing group. (1994)[16]

43) can either be excised completely using SmI_2 to afford a compound such as **44**, or converted into an aldehyde over two steps to provide a substrate for a subsequent aldol reaction.[15] In the Paterson group's total synthesis of ACRL Toxin IIIB (**42**),[16] the latter technique is exactly how **39** was manipulated to ultimately afford **40**, a material which then underwent a second 1,2-*anti*-selective asymmetric aldol reaction with boron enolate **36** to produce **41** in excellent yield (95 %) and diastereoselectivity (>99:1 d. r.). Overall, these two highly successful 1,2-*anti* aldol reactions served to complete most of the stereochemical complexity of the final target (**42**). We shall encounter another highly effective method from the Paterson group to accomplish 1,2-*anti* selectivity using propionate-derived chiral ketones in our discussion of swinholide A (vide infra).

Most other examples of substrate-controlled asymmetric aldol reactions follow the same general patterns that we have already discussed, namely that chiral enolates afford excellent aldol asymmetry while the use of chiral aldehydes alone leads to more modest diastereoselectivity. As such, at this point we only really need to address the issue of how to accomplish asymmetric aldol reactions through reagent-based stereocontrol. Of course, in the case of boron-mediated aldol reactions, all this means is that the union of a given ketone and aldehyde pair is governed by a chiral ligand attached to the boron reagent. If matched to their properties, these ligands can enhance their natural substrate bias for aldol asymmetry. If mismatched, they can potentially reverse that outcome in favor of entirely different products. Over the course of the past decade, several highly effective directing groups have been developed specifically for this purpose, one of the most competent being the family of isopinocampheyl (Ipc) ligands developed by H. C. Brown and his group at Purdue University.[17,18] A representative example demonstrating the value of this particular class of ligand is shown in Scheme 5 as part of the Paterson synthesis of ebelactones A and B (**62** and **63**).[19] Thus, in one of the opening maneuvers of this endeavor, prochiral ketone **46** and aldehyde **48** were merged into aldol product **50** in better than 97 % *ds* with 86 % *ee* through the chiral *Z* enolate (**47**) generated with (−)-Ipc_2BOTf and *i*-Pr_2NEt. The model proposed to account for the observed stereoselectivity invokes transition state **49**, a reactive conformation in which the enolate approaches the aldehyde from the face that minimizes steric interactions between its ethyl group and the indicated stereocenter of the Ipc ligand. This orientation is favored because the alternate approach trajectory onto aldehyde **48** would encounter considerable steric congestion as shown in transition state **51**.

Once **50** was in hand, its protection and subsequent conversion into a *Z* boron enolate (**52**) then enabled a substrate-controlled union with **48** in a second asymmetric aldol reaction, which ultimately gave rise to **54** in 83 % overall yield from **50** and, more importantly, with a d. r. of 95:5. As such, this reaction once again serves to indicate the high levels of substrate-based stereocontrol that can be achieved using chiral enolates. To install the final stereocenters of the target, the Paterson group then converted **54** into aldehyde **55**, hoping that its merger with the *E* boron enolates **56a** and **56b** could afford access to aldol products (**58** and **60**) which could then be advanced to the ebelactones (**61** and **62**). Although both of these aldol products possessed an anti-Felkin–Ahn stereotriad, an unfavorable arrangement based on the model discussed in Scheme 2, aldol stereoselectivity using only chiral aldehydes are usually poor, so the expectation was that both Felkin–Ahn and anti-Felkin–Ahn products would result. Indeed, this hypothesis proved correct as the reaction of **55** with *E* boron enolate **56a** afforded an almost equimolar ratio of the two 1,2-*anti* aldol products (**57** and **58**), while its reaction with *E* enolate **56b** led to increased selectivity for the Felkin–Ahn product (**59**) in approxi-

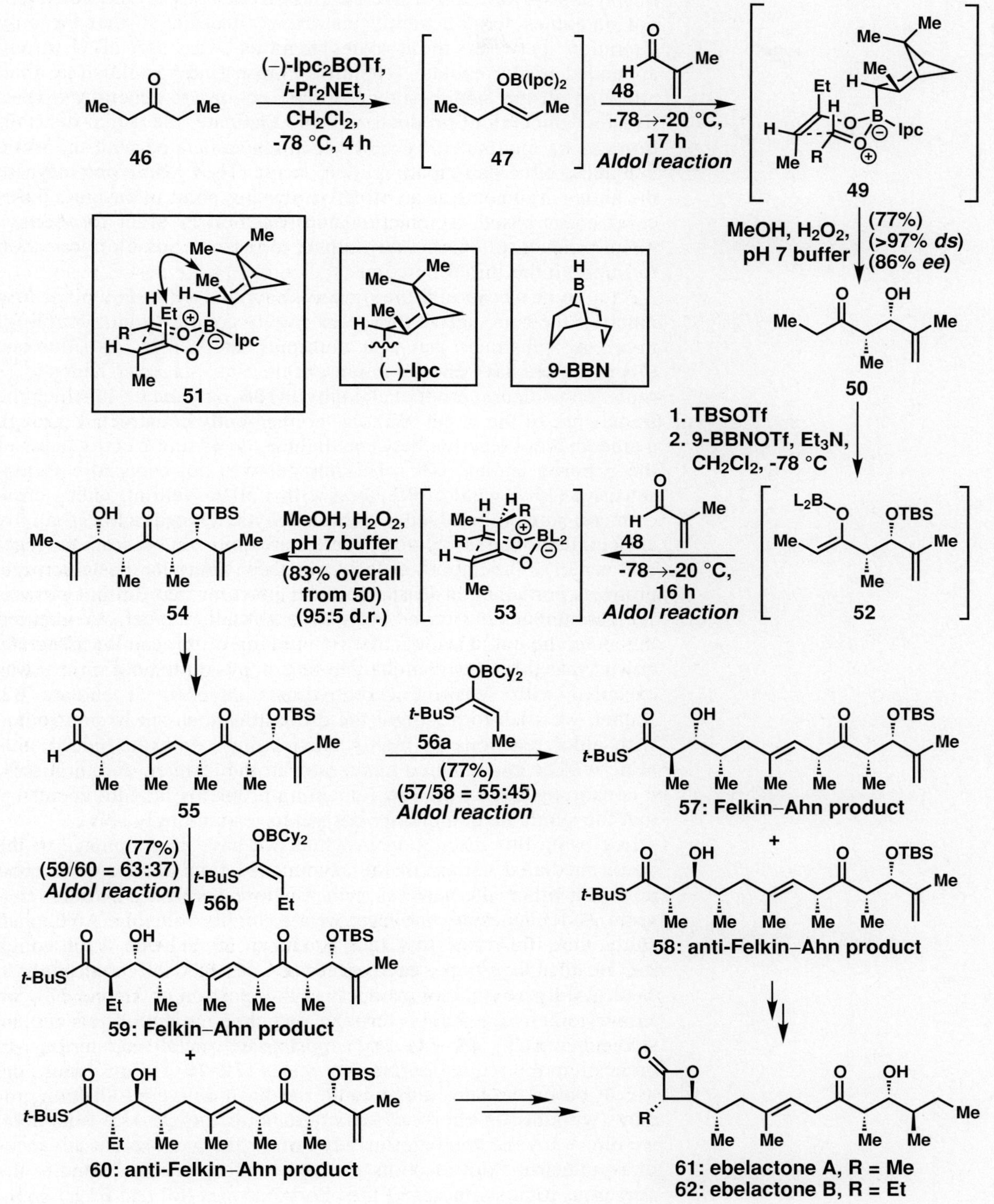

Scheme 5. Use of three different approaches for aldol stereoselectivity in Paterson's total synthesis of ebelactones A and B (**61** and **62**). (1990)[19]

mately a 2:1 ratio with **60**. While a slightly disappointing finale to this synthesis in terms of overall diastereoselectivity, their differential outcomes serve a highly instructive function in that the only alteration between these two reactions was the alkyl group appended to the enolate. Therefore, these findings illustrate that, although all the models shown in this section are generally predictive for which aldol products will predominate, the actual distribution can be modulated by subtle features associated with the steric demands of the participating components. Thus, while one may use the above arguments as an effective starting point to design a particular boron-based asymmetric aldol reaction in order to obtain a given product, often one may expect to perform some optimization to improve the initial outcome.

As a point of caution, the models described up to this point lose much of their predictive powers in highly complex aldol couplings involving substrates that bear multiple stereocenters. To illustrate this point, we have chosen to present the final stages of Kinoshita's route to the natural product elaiophylin (**66**, Scheme 6), in which the framework of the target (**65**) was convergently constructed through a tandem aldol reaction between dialdehyde **64** and 2 equivalents of the *Z* boron enolate **63**.[20] This merger was not especially stereoselective (13 % yield of **65** along with a 50 % yield of other stereoisomers) since it involved a chiral aldehyde reactant and essentially an achiral enolate (as its stereocenters are quite remote from the reaction center). The important feature is these alternate diastereomeric products contained at least one aldol junction that did not possess 1,2-*syn* stereochemistry as one would normally expect. We mention this outcome not to impugn the fine quality of this total synthesis in any way, as this tandem aldol coupling step is quite bold since it was executed with sensitive carbohydrates attached to enolate **63**. Rather, we wish to highlight the difficulties that can be confronted with aldol reactions in highly challenging contexts. Indeed, this issue will be encountered again later in the chapter, and, arguably, it constitutes one of the few remaining problems in aldol chemistry that the synthetic community has yet to resoundingly solve.

In closing this discussion in which we have paid homage to the boron-mediated variant of the asymmetric aldol reaction, we must note that other alternatives, such as those involving lithium, titanium, and magnesium enolates, are also highly valuable. An elegant study that illustrates this fact is shown in Scheme 7 in which the Heathcock group stereoselectively obtained all four possible α-alkyl-β-hydroxy aldol products in the reaction of ketone **67** with isobutyraldehyde simply through judicious selection of enolate stereochemistry (**68**–**71**) and appropriate metal counterions to orchestrate the requisite transition states (**72**–**75**).[21] For certain, the use of boron enolates alone could not have delivered all four products without some form of substrate modification. As a final note, we direct anyone wishing to study further the most recent advances in asymmetric aldol reaction to several review articles[8,9] and to the numerous total syntheses of the phorboxazoles (**80** and **81**),[22] spon-

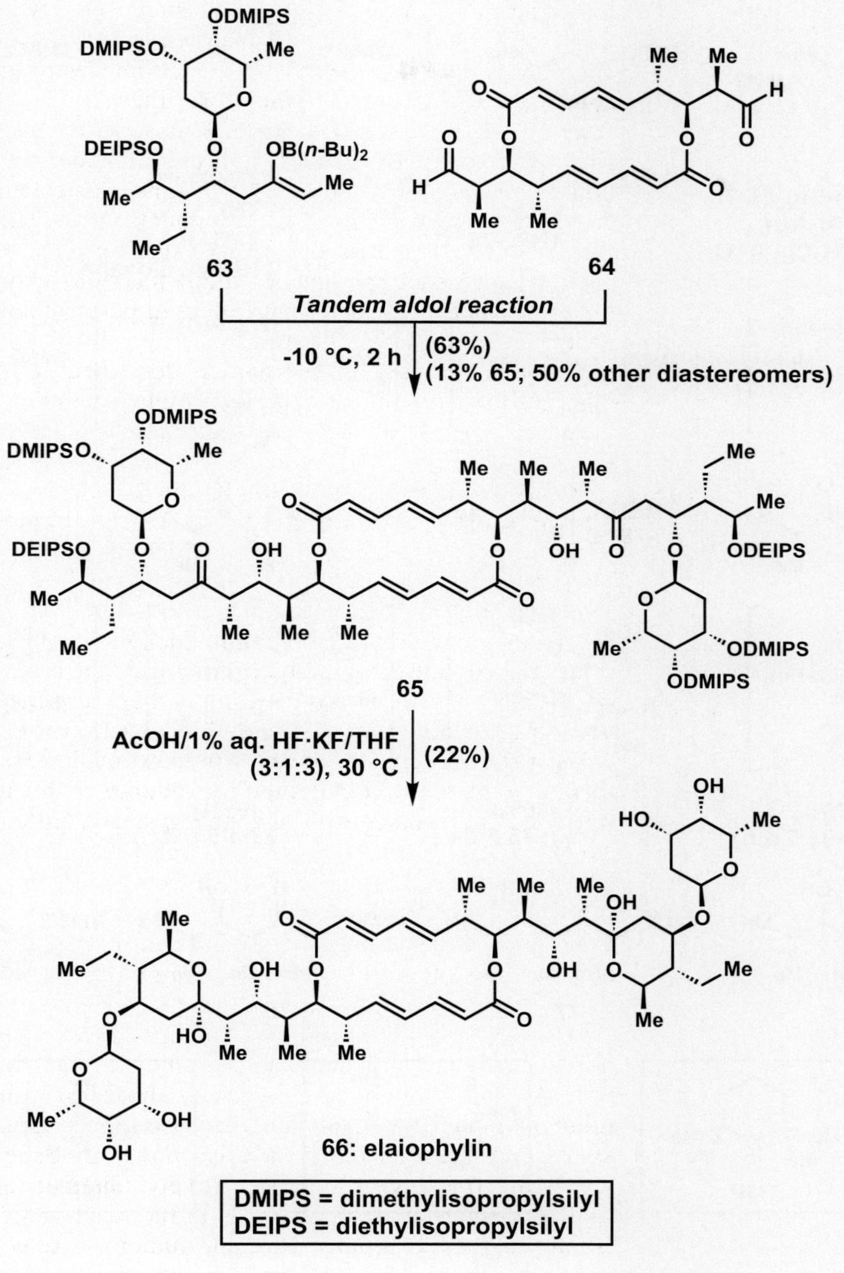

63

64

Tandem aldol reaction

-10 °C, 2 h (63%)
(13% 65; 50% other diastereomers)

65

AcOH/1% aq. HF·KF/THF (3:1:3), 30 °C (22%)

66: elaiophylin

DMIPS = dimethylisopropylsilyl
DEIPS = diethylisopropylsilyl

Scheme 6. The challenges in accomplishing diastereoselective aldol couplings during the late stages of a total synthesis as shown by Kinoshita's route to elaiophylin (**66**). (1986)[20]

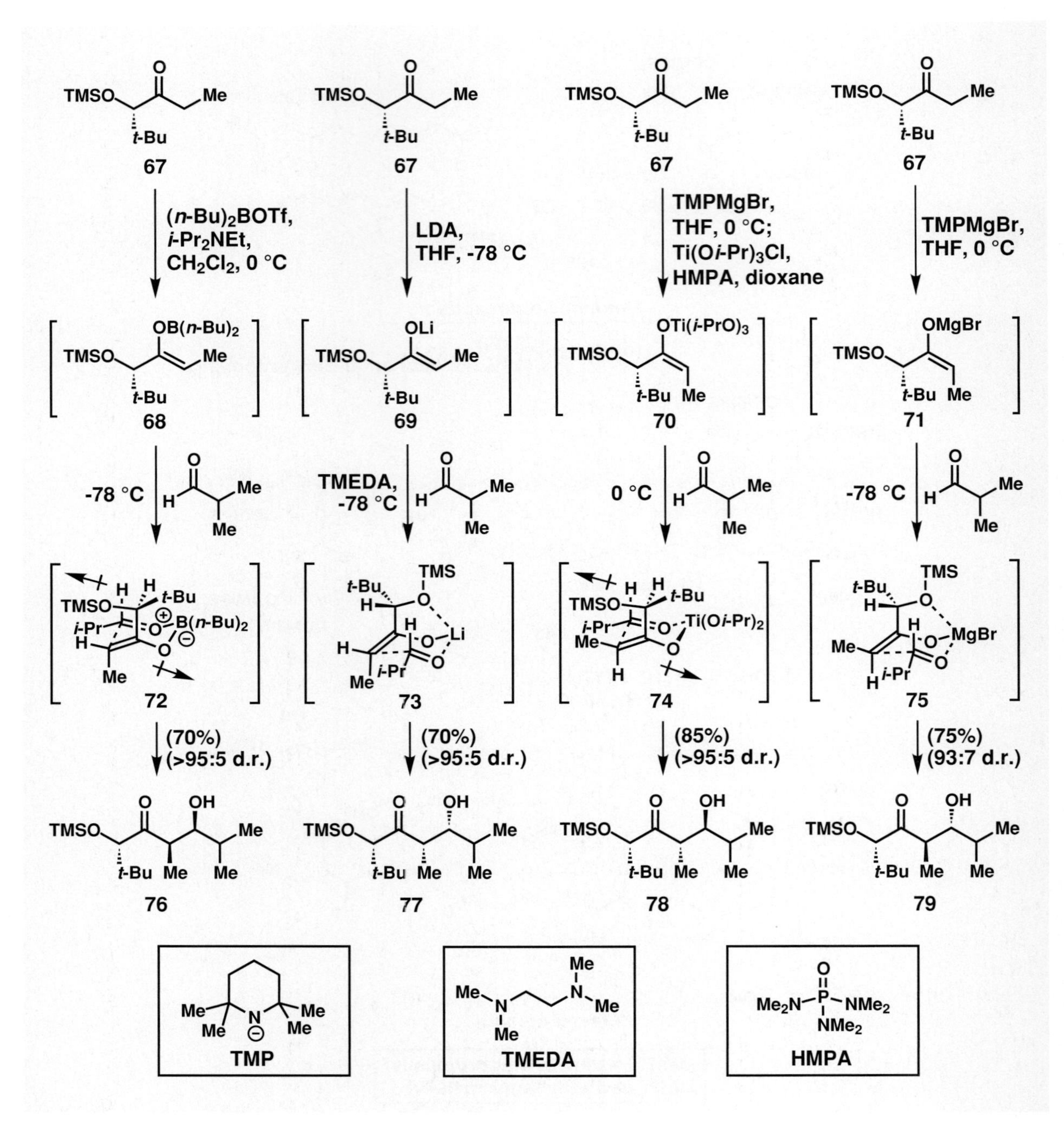

Scheme 7. Generation of all four possible stereoisomeric α-alkyl-β-hydroxy aldol products from the same chiral ketone starting material. (Heathcock, 1991)[21]

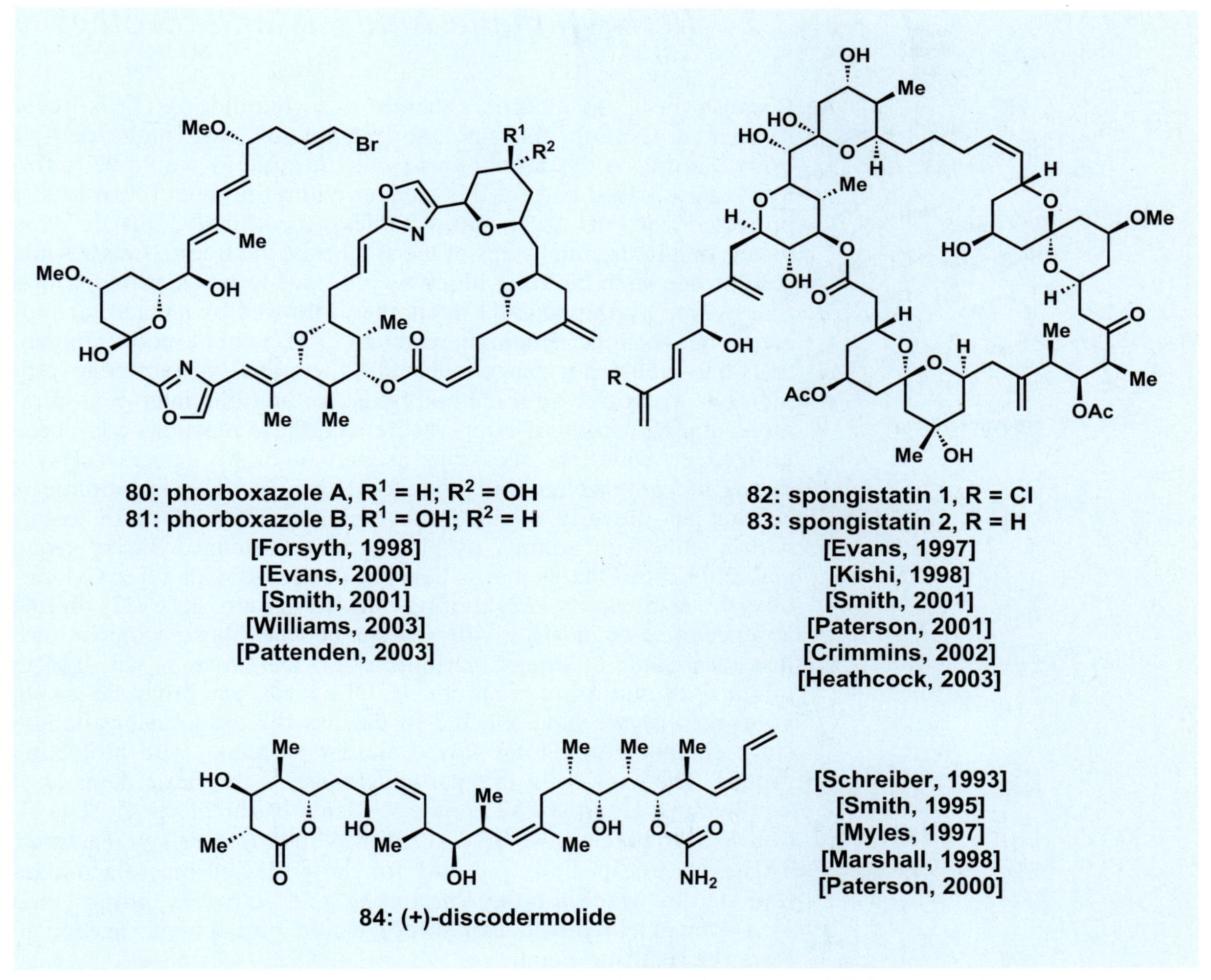

Scheme 8. Selected polyketide natural products synthesized between 1993 and 2003.

gistatins (**82** and **83**),[23] and discodermolide (**84**)[24] as listed in Scheme 8. These three molecules in particular have inspired many elegant solutions for their complex motifs, including a wide variety of different aldol solutions to create their complicated stereochemical elements, as well as some non-aldol techniques to establish the same architectural domains. For certain, they provide an excellent reflection of the current state-of-the-art methods in asymmetric aldol chemistry and a qualitative measure of the strengths and weaknesses of several different aldol techniques. With this scheme concluding our brief foray into the key elements of asymmetric aldol chemistry, we are now ready to embark on the main theme of this chapter, the total synthesis of swinholide A (**1**).[7]

3.2 Retrosynthetic Analysis and Strategy

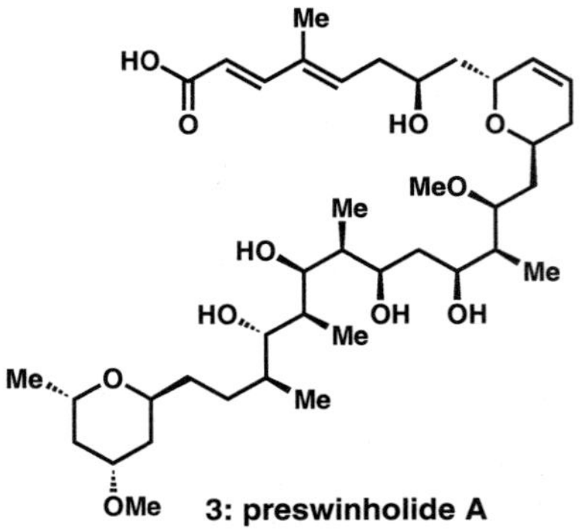

3: preswinholide A

Because the C_2-symmetric structure of swinholide A (**1**) is likely formed in Nature through the merger of two molecules of preswinholide A (**3**), an obvious synthetic strategy would be to follow Nature's lead and set this smaller natural product (or at least a protected version) as an initial target. Accordingly, this decision means that in the final steps of the synthesis, the free C-1 carboxylic acid of one such building block would have to be esterified with a free hydroxy group at C-21 in another, followed by a macrolactonization between the remaining acid and C-21 alcohol motifs. In general, one would not expect such tasks to pose any particular problems as many powerful methods exist for both the inter- and intramolecular formation of esters. Moreover, these reactions have been utilized on countless occasions as part of highly successful syntheses of complex natural products. However, with swinholide A Nature has cleverly clouded the typical facility of these events with a veil of uncertainty by placing an additional hydroxy group at C-23, a motif that is just as likely to condense with a free C-1 carboxylic acid as its neighboring alcohol group at C-21. In the absence of some means to differentiate between these two positions, it is reasonable to expect that both esterification steps will lead to mixtures of regioisomeric products. Of course, you might be asking yourself why we have elected to discuss this issue at any length since chemists have long solved similar problems with protecting groups. The answer, in this particular case, is the steric demand of a protecting group at C-23 could sufficiently shield the C-21 position so as to preclude, or at least severely retard, these key reactions. This issue is especially pressing for the projected macrolactonization step in which a bulky carboxylic acid activating group (such as a Yamaguchi mixed anhydride) would probably be needed to form the requisite bond.

It was with these considerations in mind that the Paterson group elected to pursue a strategy towards swinholide A (**1**) in which they would attempt to make these esterification reactions site-selective on intermediates with both hydroxy groups at C-21 and C-23 unprotected. Thus, as shown in Scheme 9, they set **85** as their critical target, projecting that hydrolysis of its C-1 methyl ester would provide one of the requisite coupling partners, and selective removal of the cyclic di-*t*-butylsilyl group guarding the alcohol groups at C-21 and C-23 would afford the other. Hopefully, once these preparatory events were executed on **85**, the appropriate selection of reaction conditions, combined with a bit of serendipity in the form of substrate bias, could then enable their initial union and subsequent macrolactonization to proceed with some measure of regioselectivity. Although such conjecture could only be probed empirically, the Paterson group could take comfort in the fact that if product mixtures did result, then it was likely that, at least for the initial coupling, the undesired material could be recycled

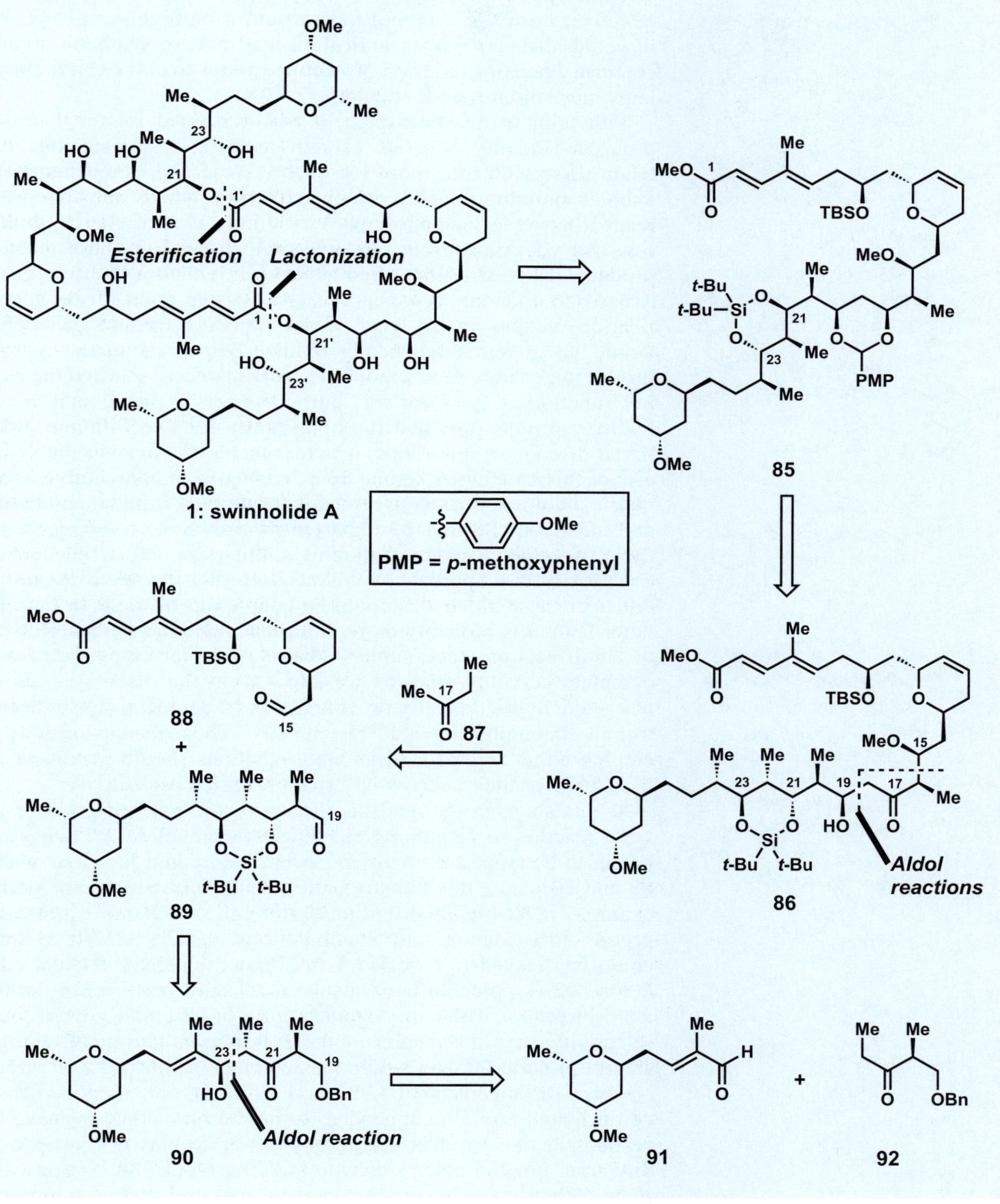

Scheme 9. Paterson's retrosynthetic analysis of swinholide A (**1**): initial stages.

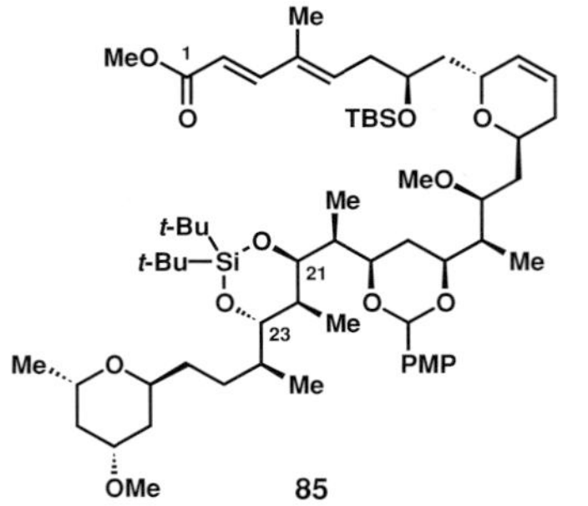

through hydrolysis of the newly formed ester. If the sequence was effective, however, then not only would it be highly efficient, but it would also serve as an indication that modern synthetic organic chemistry perhaps relies on protecting groups to solve certain problems more often than it should.

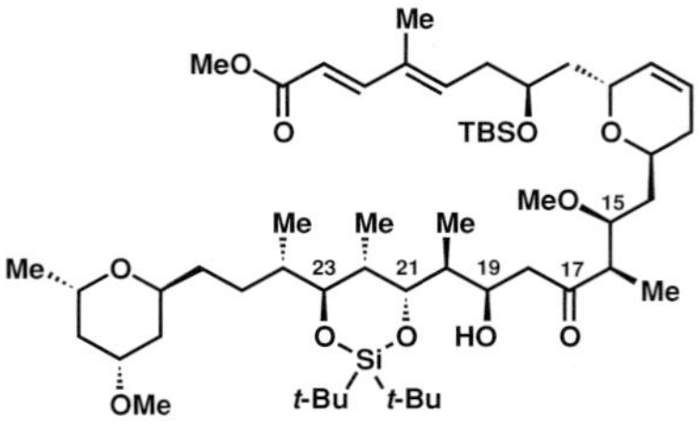

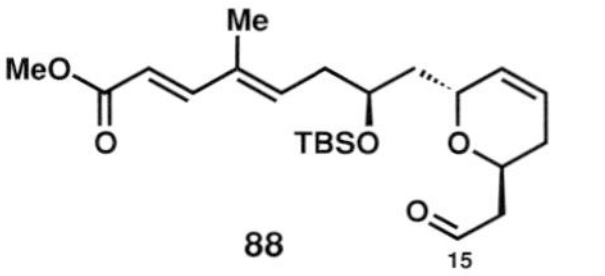

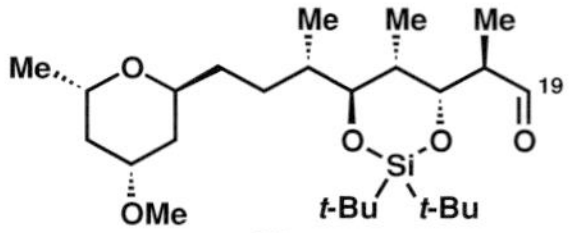

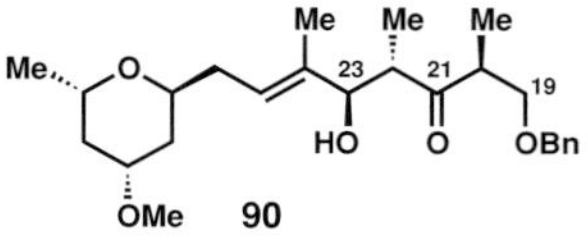

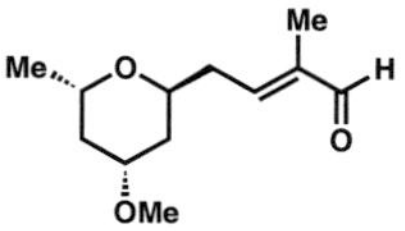

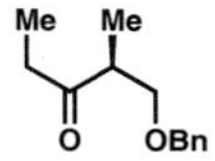

With faith in this bold plan, it now remained for the Paterson group to identify a series of retrosynthetic disconnections that could dissect **85** into more reasonably sized and stereochemically simpler building blocks, an objective for which substrate- and reagent-based asymmetric aldol reactions would play a crucial role. As you know from our preceding discussion, aldol reaction products are readily identified as those in which a hydroxy group exists β to a ketone. As such, a useful starting point for the identification of sites in **85** where aldol transforms could be applied would be to retrosynthetically oxidize one of its many hydroxy groups to generate such a motif. These researchers selected the alcohol function at C-17 for this purpose since its location β to two hydroxy groups provided the opportunity for two different aldol-based disconnections. Thus, it is reasonable to envision the synthesis of this new target ketone **86** as arising from consecutive asymmetric aldol reactions between enolates derived from butanone (**87**) and aldehydes **88** and **89**. Apart from providing convergence, as these two aldehydes are of roughly similar size and stereochemical complexity, this approach would also afford a flexible route in that either of these aldehydes could be coupled to **87** first. In fact, the latter feature is probably more important since the stereoselectivity of aldol reactions on complex substrates cannot be predicted with complete certainty, and, as we know from the discussion above, chiral aldehydes typically provide the worst degree of substrate control in accomplishing aldol asymmetry. Thus, the opportunity to employ either order for these aldol reactions should maximize the likelihood of their success in terms of diastereoselectivity.

At this stage in the analysis, however, the potential role of the aldol reaction in fashioning elements of swinholide A (**1**) has just begun to be tapped as there are several additional locations within **88** and **89** where this transformation could form stereocenters. For example, if **89** was modified to **90** through some simple protecting group and oxidation state manipulations, then its C22–C23 bond could be cleaved to reveal α,β-unsaturated aldehyde **91** and ethyl ketone **92** as potential participants in an aldol reaction. In the forward direction, if the stereogenic center in **92** could govern facial selectivity, then the merger of the *E* boron enolate of **92** with **91** should provide **90** as a single diastereomer with a 1,2-*anti* stereochemical arrangement at C-22 and C-23. If not, then perhaps a chiral ligand could be appended to that boron enolate to make the union selective by overturning any substrate bias in the opposite direction. For the other aldehyde building block (**88**, Scheme 10), if its carbon chain were shortened by one unit of unsaturation to afford **93** through the indicated Horner–Wadsworth–Emmons transform, then its C-7 stereocenter could potentially arise through

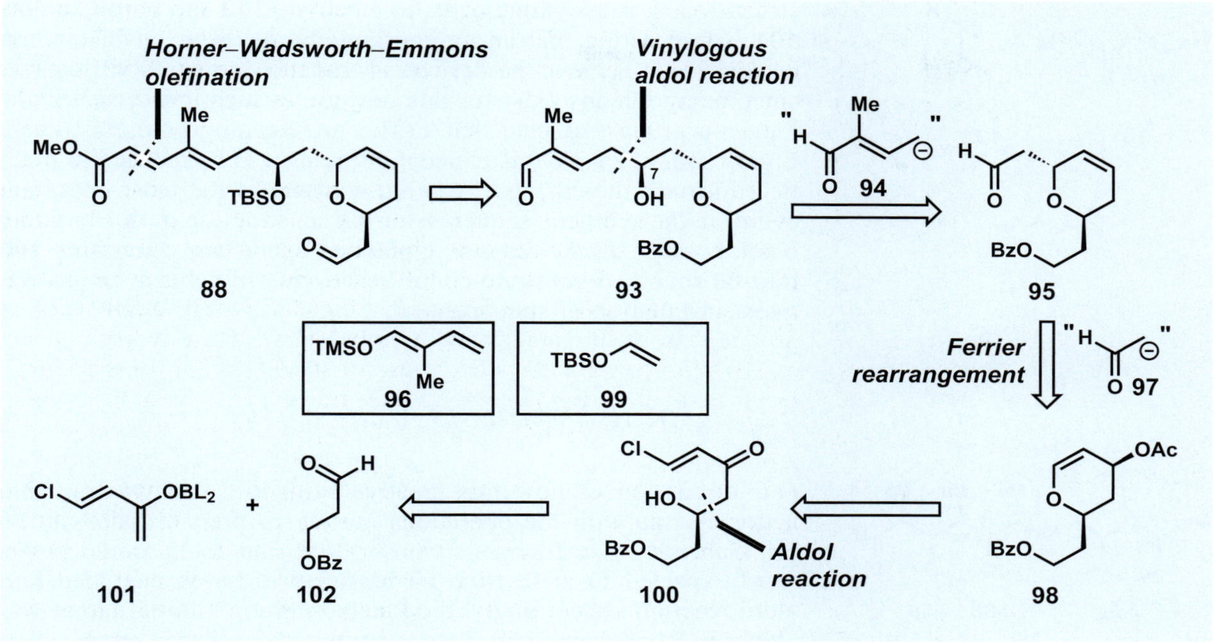

Scheme 10. Paterson's retrosynthetic analysis of swinholide A (**1**): final stages.

a vinylogous aldol reaction using a synthetic equivalent of **94**, such as the anion derived from cleavage of the TMS group in silyl enol ether **96**. Since this version of the aldol reaction, more commonly known as the Mukaiyama aldol reaction,[25] does not enlist the direct participation of boron to organize the reactants, it proceeds through an open (i. e. acyclic) transition state. Accordingly, for this aldol reaction to prove effective, the addition of "**94**" would have to occur with Felkin–Ahn selectivity to the *Si* face of the aldehyde group in **95**. This requirement did not seem too daunting as molecular models revealed that the substrate was likely biased in the desired direction by the presence of the pyran ring. Moreover, chelation of the aldehyde and the adjacent oxygen atom in this ring with a cation (e. g. titanium) should serve to enhance this predisposition by affording a temporary complex capable of shielding the *Re* face of the aldehyde from the incoming nucleophile.

Before we conclude this retrosynthetic analysis, however, we should mention that there was one more aldol-based transform that the Paterson group thought could account for one of the remaining two stereocenters in **95**. Thus, if the stereochemical complexity of **95** was reduced to that of **98**, anticipating that a Ferrier rearrangement[26] involving an acyl anion equivalent in the form of acetate **97** could accomplish this conversion during the synthesis (i. e. **99**), then subsequent opening of its pyran ring would provide a substrate (**100**) amenable to an aldol disconnection. Indeed, as shown in Scheme 10, cleavage of the indicated bond within this new subgoal structure led, retrosynthetically, to aldehyde **102** and boron enolate **101**,[27] forecasting that an appropriate chiral ligand on that boron enolate could achieve the desired asymmetric induction. Of course, since many ligands exist for this purpose as mentioned earlier, the Paterson group expected that sufficient screening of the available library should enable this element of the plan to be reduced to practice. In truth, though, its execution was one of the most important events in the synthetic sequence for the construction of this building block because all the remaining planned operations converting **100** into **88** involved substrate-controlled events to achieve the subsequent installation of stereocenters.

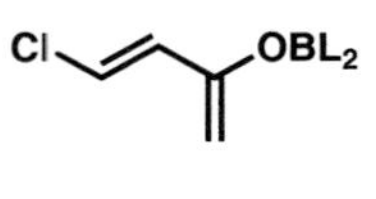

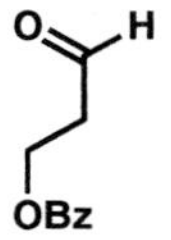

3.3 Total Synthesis

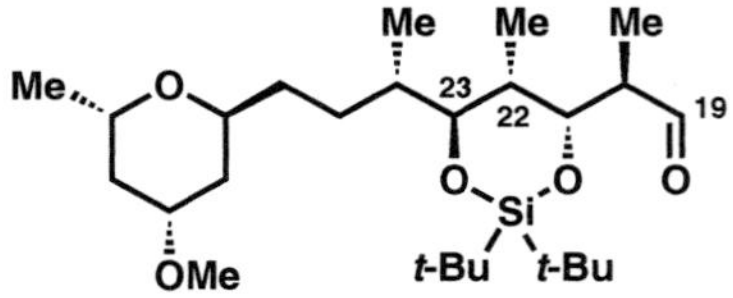

Our discussion of how this creative synthetic plan was put into action begins with the operations needed to prepare aldehyde **89** (cf. Scheme 9), a fragment whose C-22 and C-23 stereocenters were expected to arise from an asymmetric boron-mediated *anti* aldol reaction. Accordingly, the Paterson group's initial target was aldehyde **91** (Scheme 11), a precursor whose critical synthetic challenge constituted the generation of a pyran ring bearing three stereocenters, two of which are axially disposed in its most stable chair conformation. As we shall see, this task did not prove overly difficult to accomplish once an appropriate chiral starting material was identified.

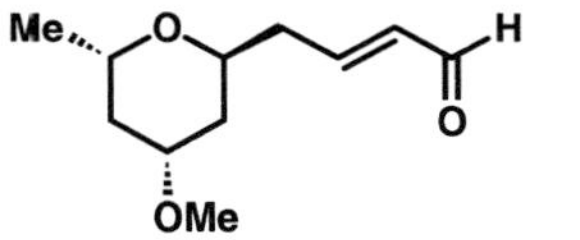

Starting from the known racemic alcohol **103**, the Paterson group projected that if they could stereoselectively epoxidize the allylic olefin of this compound and effect a kinetic resolution of the allylic alcohol in the same step to provide **104**, they would have succeeded in establishing two of the three stereocenters needed for **91**. As will be discussed in far more detail in Chapter 6, this objective can generally be accomplished with a Sharpless asymmetric epoxidation protocol employing a reagent combination matched to react faster with the desired enantiomer of the allylic alcohol starting material. In this case, the use of $Ti(O\textit{i}\text{-}Pr)_4$ and *t*-BuOOH in conjunction with (+)-DIPT in CH_2Cl_2 at −20 °C performed this task by providing the desired substrate with excellent enantioselectivity (96 % *ee* at 43 % conversion).[28] Seeking now to complete the pyran ring, the newly installed epoxide was selectively opened with hydride as supplied by Red-Al® to afford **105** (with the hydroxy group directing the site of attack).[29] Upon ozonolytic cleavage of the terminal olefin of **105** in methanol, tetrahydropyranyl acetal **106** was then formed in 67 % yield for these two steps. As such, only one additional stereocenter remained to be installed onto this template, a task that the Paterson group expected could be accomplished through a stereocontrolled alkylation of an intermediate oxonium species derived from **106** (such as **108**). Thankfully, following the conversion of **106** into **107** through a simple methylation, the subsequent

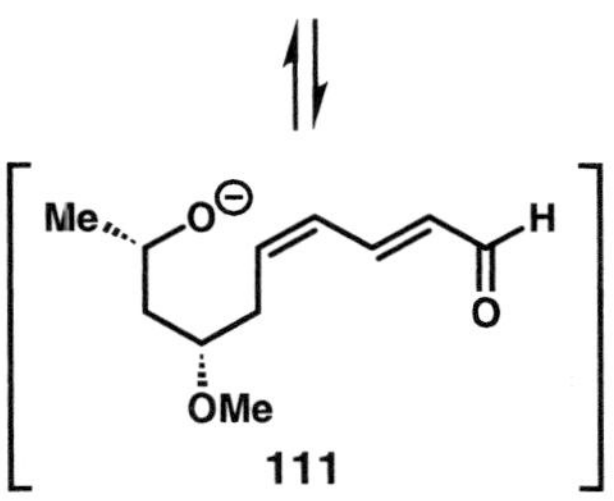

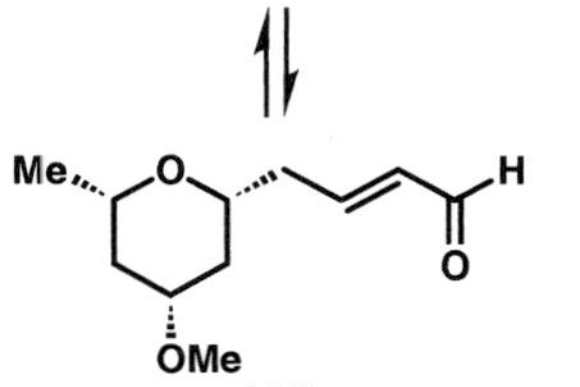

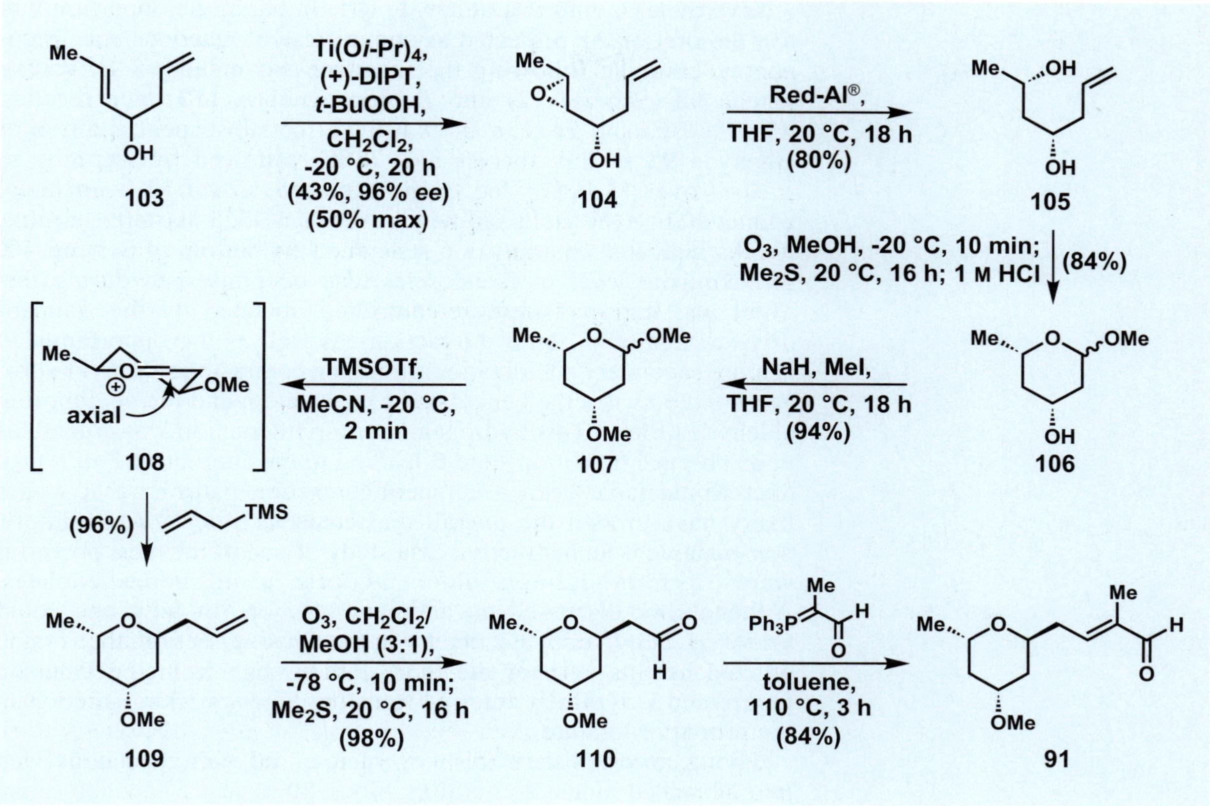

Scheme 11. Synthesis of building block **91**.

exposure of this intermediate to allyltrimethylsilane and TMSOTf in MeCN at −20 °C gave rise to **109** in 96 % yield with complete axial delivery of the required allyl tether. With all the chiral information now encoded in the substrate, only a conventional two-step ozonolysis/Wittig-olefination sequence was required to complete the target (**91**). These events occurred in a combined yield of 82 %.

Before moving on, two features of these final operations are worthy of commentary. First, it is reasonable to envision that the addition of a vinyl nucleophile (**94**) derived from a reagent such as **96** (see column figure) could have provided **91** directly upon its addition to oxonium species **108**. In fact, the Paterson group was able to accomplish this union following extensive experimental exploration.[7] Unfortunately, the operation did not prove amenable to large-scale synthesis, and thus was abandoned in favor of the shown sequence. Second, the terminating Wittig olefination step actually provided a small amount of **112** (the C-27 epimer of **91**) in a ratio of 1:6 with **91**. This outcome can be rationalized through the equilibration sequence shown in the column figure on page 50, a chemical event that occurred either following successful olefination as indicated, or prior to reaction with aldehyde **110**.

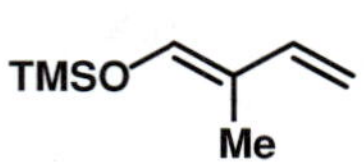

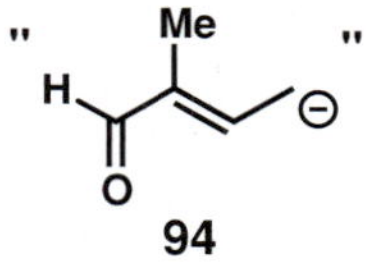

Nevertheless, with this aldehyde (**91**) in hand, the opportunity to test the first of the projected asymmetric aldol reactions was imminent. Pleasingly, following the selective conversion of the known ketone **92** (Scheme 12) into *E* boron enolate **113** upon reaction with Cy_2BCl and Et_3N in Et_2O at 0 °C, the subsequent addition of aldehyde **91** to this species at −78 °C, followed by warming to −20 °C over 14 hours, led to the expected 1,2-*anti*-2,4-*anti* aldol adduct **90** in 84 % yield and with better than 97 % diastereoselectivity. As indicated by transition state **A** at the bottom of Scheme 12, the exquisite level of facial selectivity accomplished during this event was entirely substrate-controlled, dictated by the standard drive to minimize steric interactions as well as the opportunity to acquire secondary stabilization through a contrasteric formyl hydrogen bond between the benzylated oxygen atom and the neighboring aldehyde proton. This hydrogen-bonding interaction was crucial, as in its absence transition state **B** leading to the alternate 1,2-*anti* product would have been a competing reaction pathway that would likely have eroded the overall diastereoselectivity.[30] Accordingly, this example is an instructive case study of one of the most powerful ways to form 1,2-*anti* aldol products using boron enolates. Although we discussed an alternate solution for how one could achieve 1,2-*anti* aldol asymmetry in the opening section, the version utilized here is perhaps the more famous one from the Paterson group, and is typically referred to as the Paterson aldol reaction in the primary literature.

Having accomplished this key merger, **90** was then converted into advanced aldehyde building block **89** in just a few additional steps. First, the newly installed alcohol function at C-23 was enlisted to effect a stereoselective reduction of its β-disposed C-21 ketone neighbor upon its complexation with $Me_4NBH(OAc)_3$ in a 1:1 mixture of AcOH and MeCN.[31] With this step affording a 1,3-*anti* diol intermediate, this compound was then captured as a cyclic *t*-butylsilylene derivative upon reaction with di-*t*-butylsilylbistriflate in the presence of 2,6-lutidine to afford **114** in 78 % yield for the two steps. Seeking now to establish the remaining stereocenter of the building block, the C-24 position of swinholide A (**1**), a substrate-controlled asymmetric hydroboration using thexylborane in THF, served to create the desired configuration at this site by delivering a hydrogen atom to the appropriate face.[32] Of course, since this hydroboration also installed a superfluous hydroxy group at C-25 in the product (**115**), additional synthetic operations were required to excise this motif. Fortunately, this issue posed no inherent challenges as the standard two-step Barton–McCombie deoxygenation protocol admirably served to remove this unwanted oxygen function, providing **116** in 87 % yield.[33] Interestingly, this product was contaminated with a minuscule amount of material bearing a stereocenter (either the C-29 or C-31 center, unassigned) epimeric to the arrangement shown for **116**, presumably due to 1,5-hydrogen abstraction caused by the intermediate radical shown in the inset box in Scheme 12. Never-

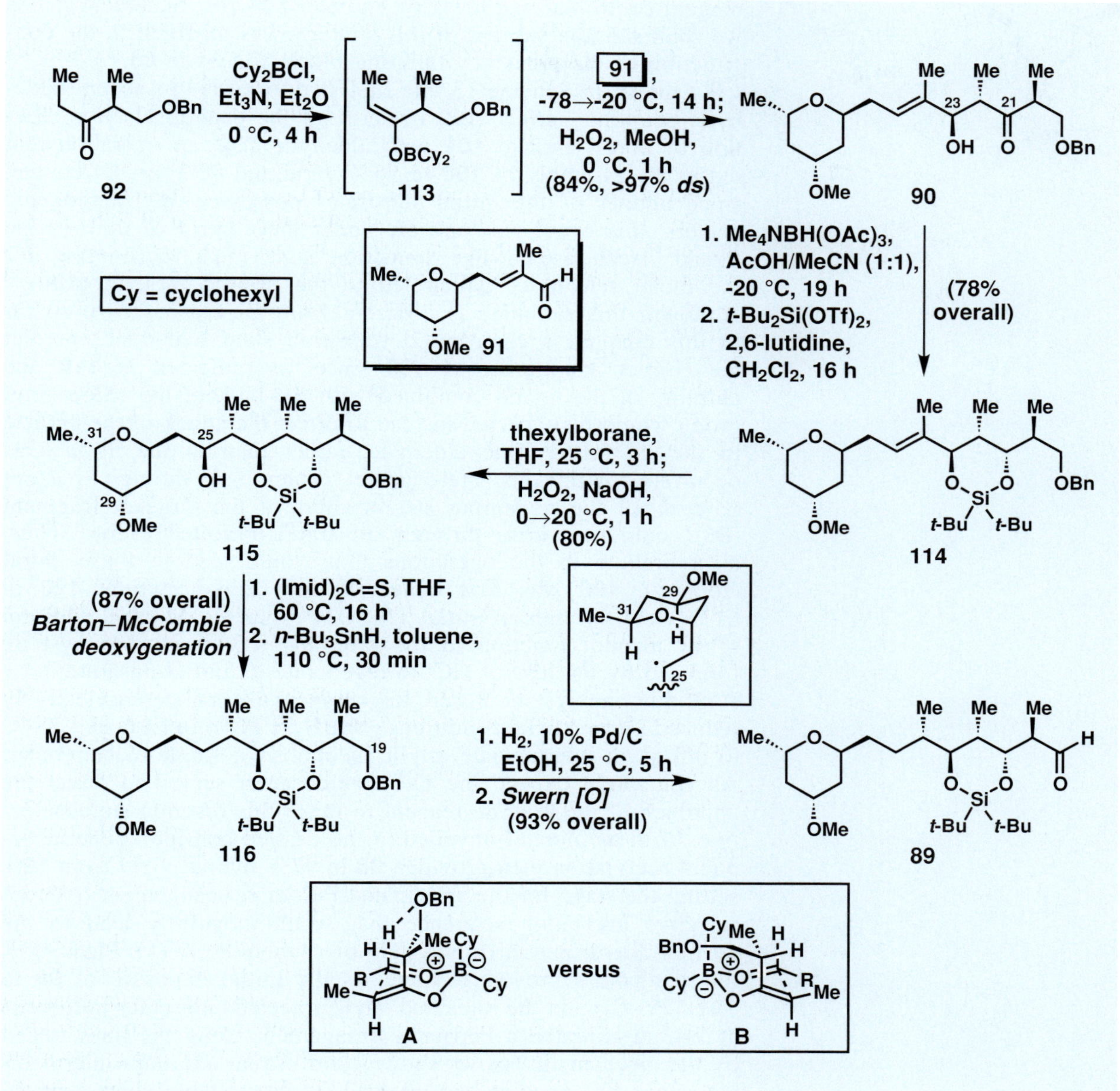

Scheme 12. Completion of key building block **89**.

theless, with an efficient route to **116** identified, the synthesis of aldehyde **89** was completed in 93 % yield by removing the benzyl ether guarding the hydroxy group at C-19 under standard conditions (H_2, 10 % Pd/C) and then employing a conventional Swern protocol. Before continuing on, however, it is interesting to note that all the stereocenters within this fragment (**89**) were generated through substrate-controlled processes following the initial

Me
O
OH
104

reagent-controlled synthesis of epoxide **104** (cf. Scheme 11). As we shall see, the success of this strategy was mirrored in the construction of the other key building block, aldehyde **88**.

As shown in Scheme 13, the first step towards this second aldehyde coupling partner (**88**) began with the enantioselective reaction of boron enolate **101** with aldehyde **102**, an operation that led to the assembly of **100** in 56 % yield and 80 % *ee*. Of course, since neither of the initial substrates possessed any stereogenic centers, this event was entirely under the control of the (+)-Ipc ligand[17] with a boat-like transition state (**118**) accounting for the facial selectivity which led to the desired product (**100**).[30] Although this rationale is different from all those shown so far in this chapter, a chair-based transition state would not predict the correct enantiomer of **100** since, as indicated in **119**, the chirality of the ligand combined with the bulk of the chlorovinyl group would be expected instead to direct the attack of the enolate to the *Re* face of the aldehyde, rather than to the *Si* face as occurred. With this stereogenic center set through reagent control, all the remaining stereocenters of the targeted fragment (**88**) could now arise through substrate-controlled events. Thus, as a prelude to the operations that would lead to these chiral elements, **100** was first converted into dihydropyrone **120** in 61 % yield through an initial TMSOTf-promoted Michael addition of its alcohol function to the proximal α,β-unsaturated system, followed by the loss of HCl to regenerate its initial unsaturation.* With this ring system closed, the ketone was then stereoselectively reduced under Luche conditions ($NaBH_4$, $CeCl_3$) in Et_2O at −78 °C to afford an intermediate allylic alcohol as a single diastereomer. As you might expect, the C-14 stereocenter served to direct the approach of the hydride reagent to make this reaction stereoselective. With an alcohol unveiled in the desired configuration, subsequent acetylation then provided **98** in 97 % overall yield from **120**, setting the stage for the anticipated Ferrier rearrangement[26]/stereoselective alkylation sequence that would hopefully lead to the desired dihydropyran ring system of swinholide A (**1**). Pleasingly, this conjecture proved accurate, as the initial exposure of **98** to $Ti(Oi\text{-}Pr)_2Cl_2$ and the indicated silyl-protected enol ether in toluene at −42 °C initiated a Ferrier rearrangement along the lines traced by the mechanistic arrows shown to afford an oxonium intermediate. Once this species had adopted its most stable conformation, subsequent formation of a titanium chelate to the alkylating agent (i. e. **121**) then enabled the delivery of the needed functional group through a chair-like transition state to afford **95** as a single stereoisomer in 85 % yield.

---

* Although the yield for this transformation was modest (61 %), the result is unrepresentative of its general effectiveness; the Paterson group has demonstrated on several other substrates that it more typically constitutes a high-yielding method to access such systems.[27]

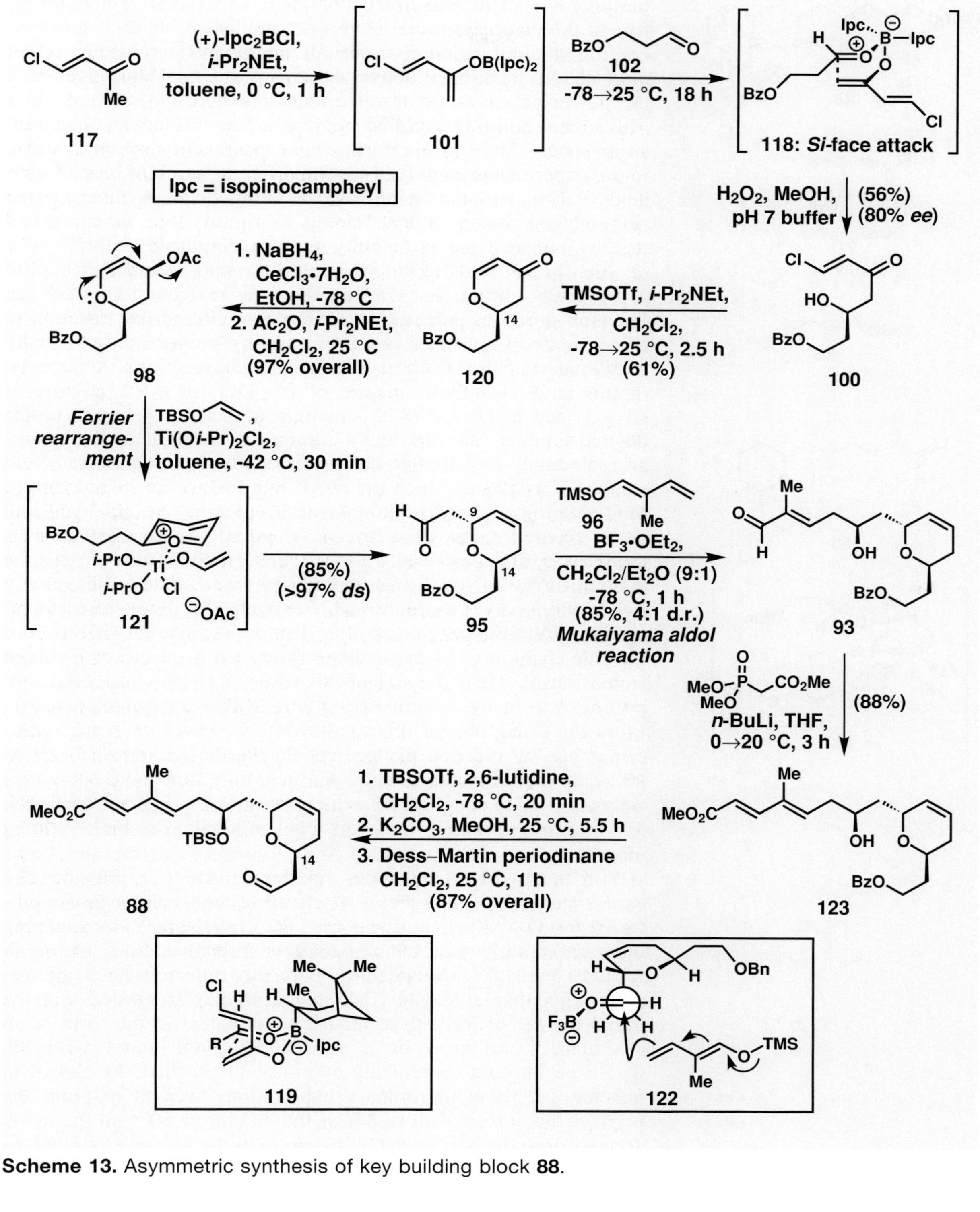

Ipc = isopinocampheyl

Scheme 13. Asymmetric synthesis of key building block **88**.

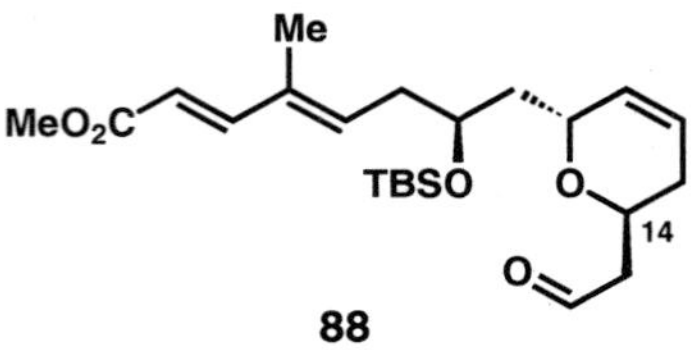

88

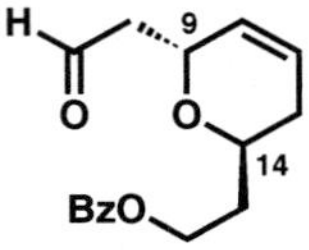

95

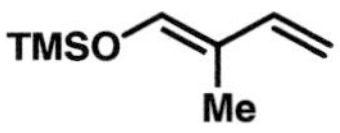

96

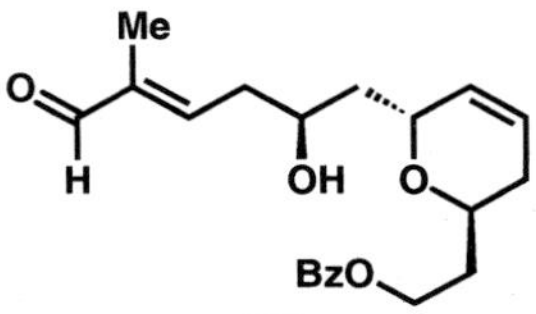

93

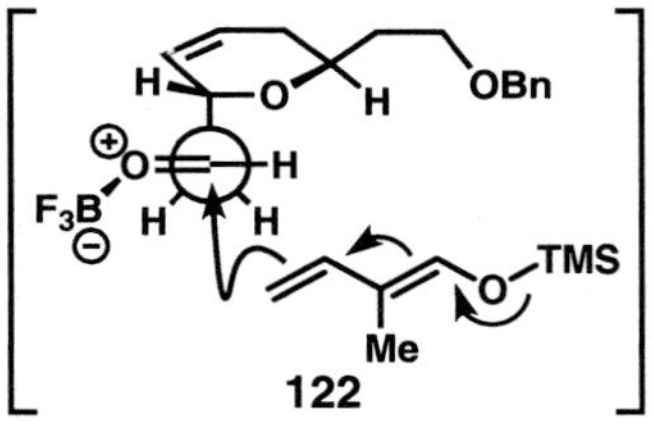

122

With this critical operation accomplished, the desired aldehyde building block (**88**) was nearly complete, with the only main objective of the projected route not yet accomplished being a substrate-controlled aldol reaction to install one more stereogenic center onto **95**. As mentioned above, achieving this goal did not rest on the power of a boron-controlled aldol reaction, but, instead, on a Mukaiyama addition using **96** and proceeding through an open transition state.[25] Thus, in an effort to make this reaction stereoselective, initial experiments employed several different titanium-based Lewis acid catalysts with the intention of chelating the aldehyde and pyran ring oxygen atoms of **95**, hoping to rigidify the substrate and thereby enhance the probability for stereocontrol. Unfortunately, all such Lewis acids led to substrate decomposition rather than to a productive union. As a result, the only real possibility for the Paterson group to proceed forward was to attempt the reaction in the absence of substrate chelation, a risky proposition as usually only modest control is observed in such aldol reactions. Amazingly, in this case a catalytic amount of $BF_3{\cdot}OEt_2$ in a 9:1 mixture of CH_2Cl_2 and Et_2O at −78 °C smoothly afforded a 4:1 ratio of the desired product (**93**) and its C-7 epimer in 85 % yield. The exquisite selectivity for this step can be rationalized through a transition state such as **122** in which the overriding feature for its outcome is minimization of its dipole moment by placing the aldehyde and pyran oxygen atoms diametrically opposed to one another. This model was first proposed by the Evans group to account for the similar formation of 1,3-*anti* products from related substrates.[34] Yet, while it serves as a reasonable explanation, dipole effects cannot constitute the only controlling feature because modification of the side chain at C-14 to a smaller group led to far lower levels of stereocontrol. Thus, the size of this tether must play at least a supporting role in the asymmetry of this Mukaiyama aldol reaction. From the standpoint of this total synthesis, however, it was gratifying that the reaction did proceed in the desired sense to afford **93** as the predominant product. As such, only four more operations were required to complete **88**, which, as shown in Scheme 13, speak for themselves based on their conventional and high-yielding nature.

With the two key aldehydes finally in hand (i. e. **88** and **89**), explorations could now begin at effecting their merger into a protected form of preswinholide A (i. e. **85**, cf. Scheme 9) by attaching them sequentially to a butanone spacer through aldol reactions or related chemistry. Although in theory this objective could be met by using either aldehyde first, model studies suggested that the initial merger of **89** with a butanone equivalent in the form of an allylsilane,[35] followed by a subsequent aldol union with **88**, would be the more promising approach in practice. As shown in Scheme 14, these preliminary explorations seemed to point the way for the correct path to **85**, as the merger of **89** with the anion derived from the loss of the TMS group in the indicated allylsilane reagent gave rise to **124** in an impressive yield of 94 % with 95 %

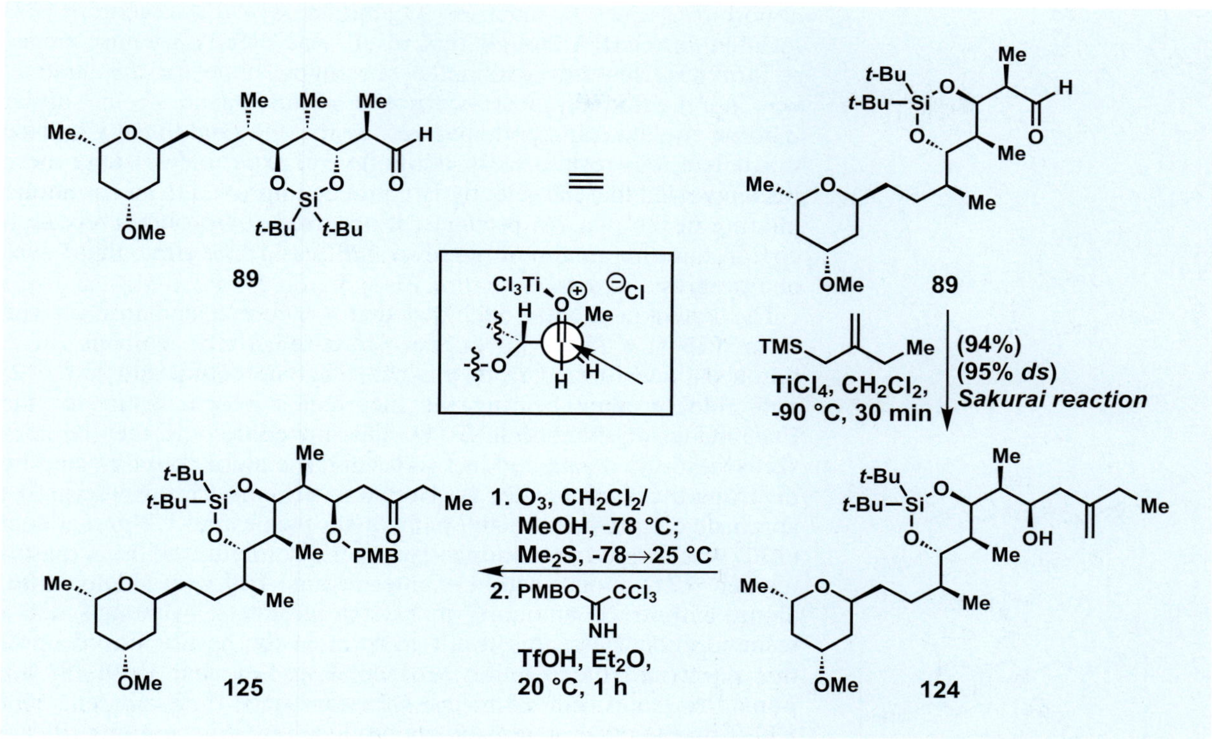

Scheme 14. First strategy to complete preswinholide A (**3**): preparation of intermediate **125**.

diastereoselectivity.* Of course, the high levels of asymmetry observed in this titanium-mediated reaction, more commonly known as the Sakurai reaction,[36] were not overly surprising since the product was predicted by the standard Felkin–Ahn addition model as shown in the inset box in Scheme 14. As we shall see, while this addition followed conventional wisdom regarding transition states, the second merger using an aldol reaction to complete **85**, unfortunately, would not.

After **124** had been converted into **125** (Scheme 14) through an ozonolysis reaction and an alcohol protection step, the Paterson group then transformed this ketone into a *Z* boron enolate (**126**, see column figure) expecting that its subsequent reaction with **88** would afford the 1,2-*syn*-disposed aldol product **128**. Indeed, this reaction did provide primarily *syn* products as one would anticipate based on the Zimmerman–Traxler models shown in Scheme 1. Disappointingly, the desired *syn* product (**128**) was the minor adduct, as

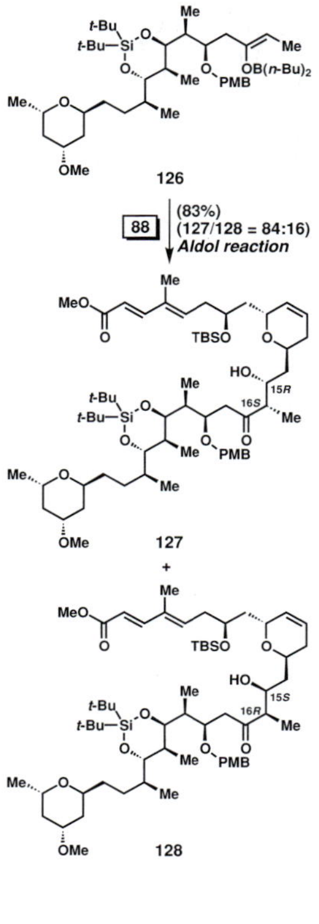

* Initial investigations sought to add boron enolates of butanone directly to **89**; however, as often occurs with chiral aldehydes, diastereoselectivity was poor. Since the attachment of chiral ligands to the boron enolate failed to change this outcome, the Paterson group turned to reactions that proceeded through acyclic stereocontrol.

almost five times as much of its alternate *syn* diastereomer (**127**) resulted instead. Although this result was clearly a consequence of structural features within the substrates, hope for the strategy was not completely lost because a chiral ligand on a similar *Z* boron enolate could perhaps override that bias and thereby reverse this deleterious result. Sadly, while several experiments along these lines revealed that the selectivity could be improved to an equimolar mixture of the two *syn* products, it occurred at a prohibitively high cost in that the maximum yield of **128** was 18 %. How, then, does one progress forward?

The Paterson group concluded that if *Z* boron enolates derived from **125** (i. e. **126**) gave poor stereoselectivity, perhaps an *E* boron enolate formed from this same ketone could lead to a 1,2-*anti* aldol product bearing the incorrect stereochemistry for the C-15 center of swinholide A (**1**). This material could then be converted into the desired adduct following the aldol coupling step by inverting that stereocenter. As shown in Scheme 15, this alternative approach did provide a better pathway as the desired 1,2-*anti* adduct (**130**) was formed predominantly in a 1.5:1 mixture with its diastereomer (**131**) upon reaction of intermediate **129** with aldehyde **88**, along with trace amounts of 1,2-*syn* products. Although not a resounding success, this result constituted the best achieved up to that point and the reaction proceeded in excellent yield (83 %). Thus, the group decided to press forward once they had separated **130** from the other stereoisomers produced by this coupling. Pleasingly, **130** could be advanced smoothly to key target **85** (see Scheme 16) over five steps, with the key transformations being two substrate-controlled reductions of ketone intermediates to create properly formatted C-15 and C-17 stereocenters, thereby correcting the original aldol outcome. Within this effective sequence, however, there is one operation worthy of some additional commentary, and that is the formation of the *p*-methoxyphenyl acetal in **132** by exposing the initial alcohol product from catecholborane reduction of **130** to DDQ under anhydrous conditions. Although a relatively unconventional technique to create these protecting groups, it is used on occasion in the synthesis of complex molecules (another example will be encountered in Chapter 13 on the CP-molecules). As shown in the neighboring column figures, the transformation begins with the formation of a benzylic cation as accomplished by the formation of a charge-transfer complex between the reagent and the substrate. This event activates the latter for the shown cyclization, which completes the assembly of the acetal.[37]

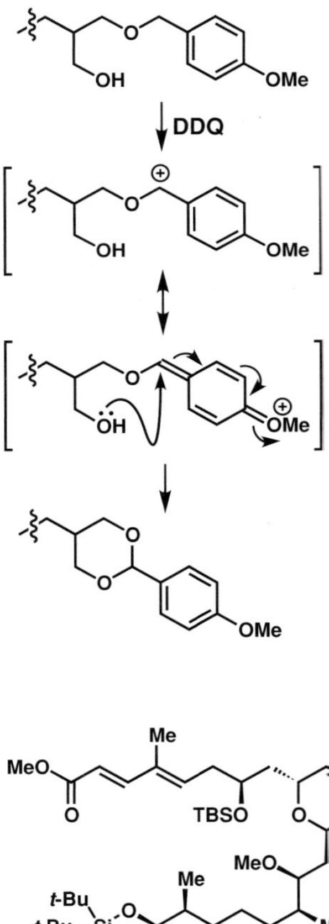

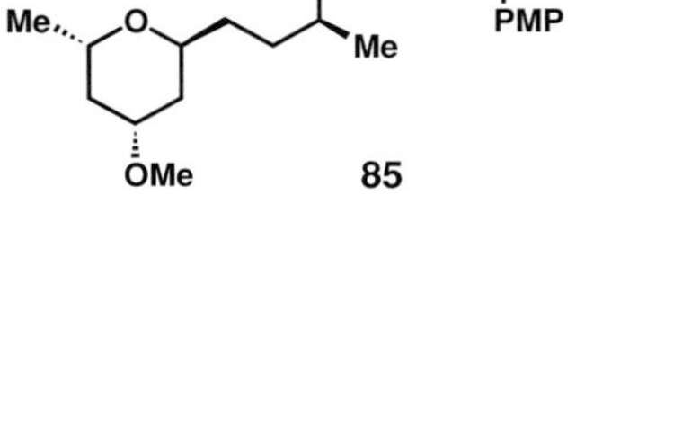

At this point, with a protected form (**85**) of preswinholide A in hand, the opportunity existed to examine the final steps envisioned to complete the total synthesis of swinholide A (**1**). While enticing, the Paterson group elected to pursue two other objectives before tackling these final operations. First, to prove the stereochemical outcome of all the preceding steps, they converted **85** into preswinholide A (**3**) using the three operations defined in Scheme 16, and, in the process, intersected two synthetic conjugates

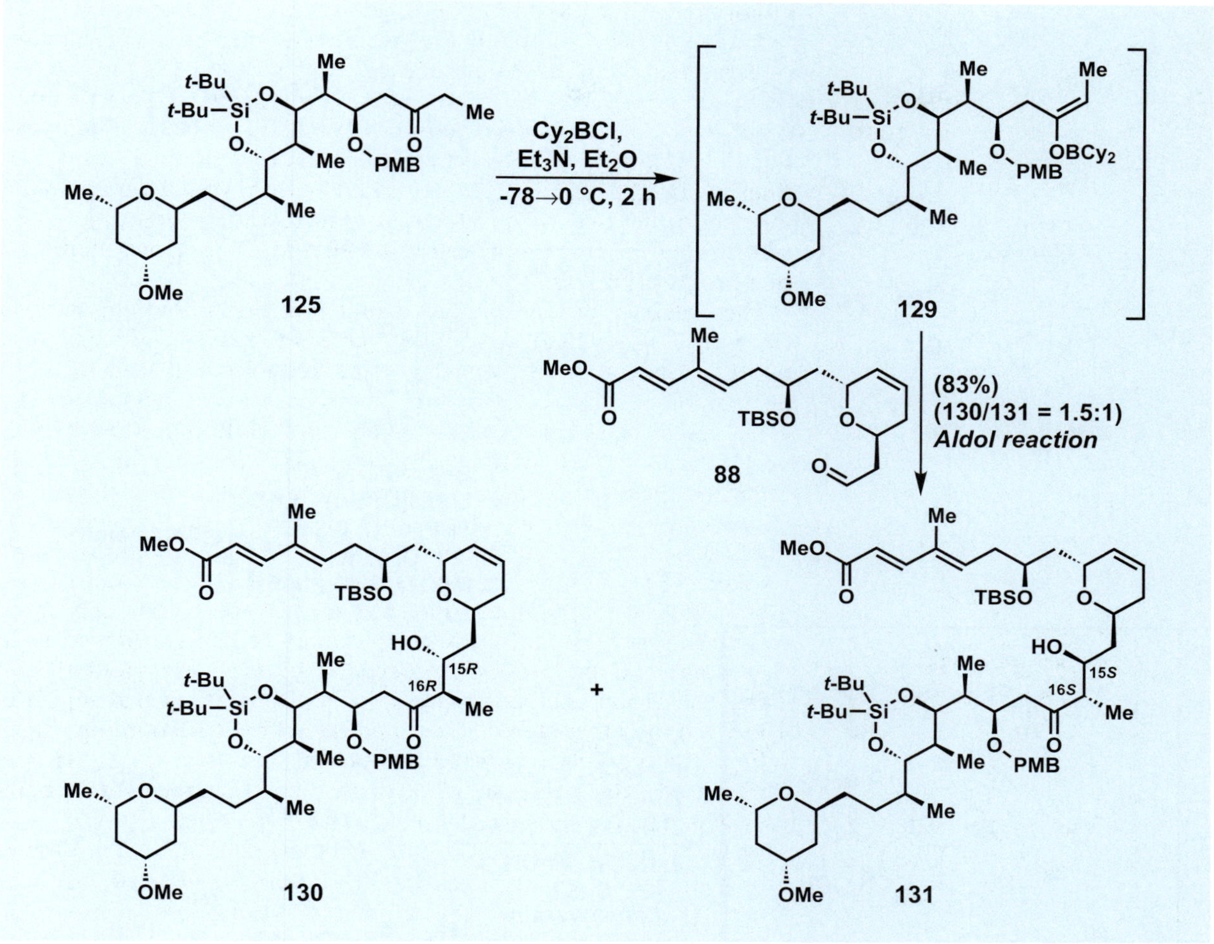

Scheme 15. First approach towards preswinholide A (**3**): final aldol coupling.

of preswinholide A (**3**) generated by the Kitagawa group from the naturally derived material. With the data sets for these two derivatives as well as preswinholide A in harmony, these researchers then decided to direct their attention towards improving the material throughput of their sequence to **85**. As you can readily ascertain by our discussion thus far, the main problem with all approaches developed towards this key fragment centered on an inability to use **88** effectively in an asymmetric aldol reaction. Accordingly, the Paterson group expected that a reasonable starting point to find an improved route would be to abandon this particular method of coupling. More specifically, since **89** appeared highly suitable for participating in diastereoselective additions under Felkin–Ahn control (cf. Scheme 14), perhaps an ideal sequence would be to attach the four-carbon unit to **88** first through a different series of reactions, and then to merge **89** with this advanced fragment through a Mukaiyama-type aldol reaction.

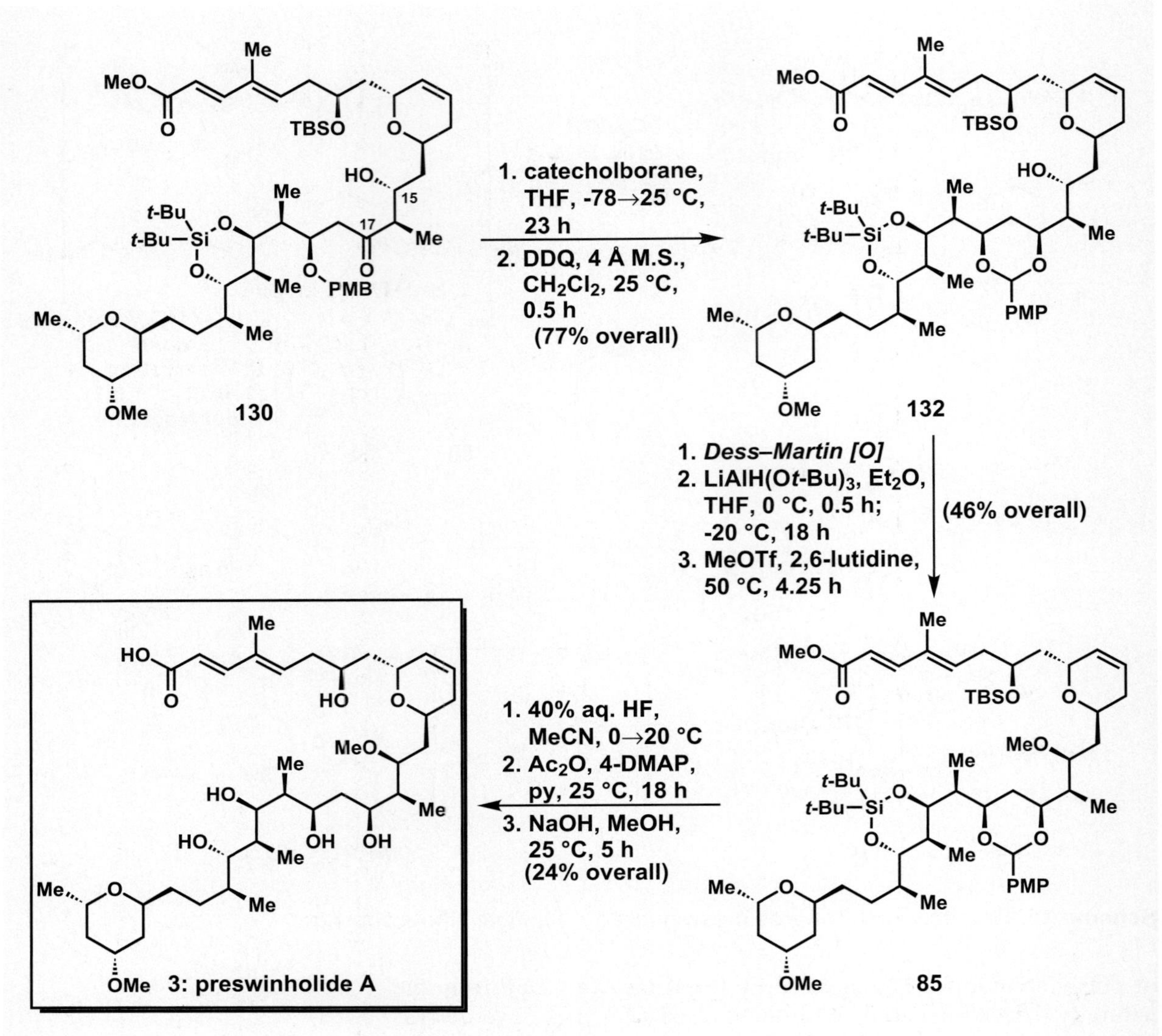

Scheme 16. Completion of preswinholide A (**3**).

As shown in Scheme 17, this alternate approach to **85** worked beautifully. Thus, adopting an alternate asymmetric reaction as a means to attach the butanone-like spacer to **88**, this aldehyde was exposed to the indicated (+)-Ipc-derived crotylboration reagent[38] in an event that afforded **133** with complete diastereoselectivity. Subsequent methylation of the newly formed alcohol followed by a Wacker oxidation step then served to create methyl ketone **134**. With these steps completing the synthesis of the required four-carbon spacer in 44 % overall yield, the substrate was then treated with LiHMDS and TMSCl to generate a silyl enol ether (**135**) in

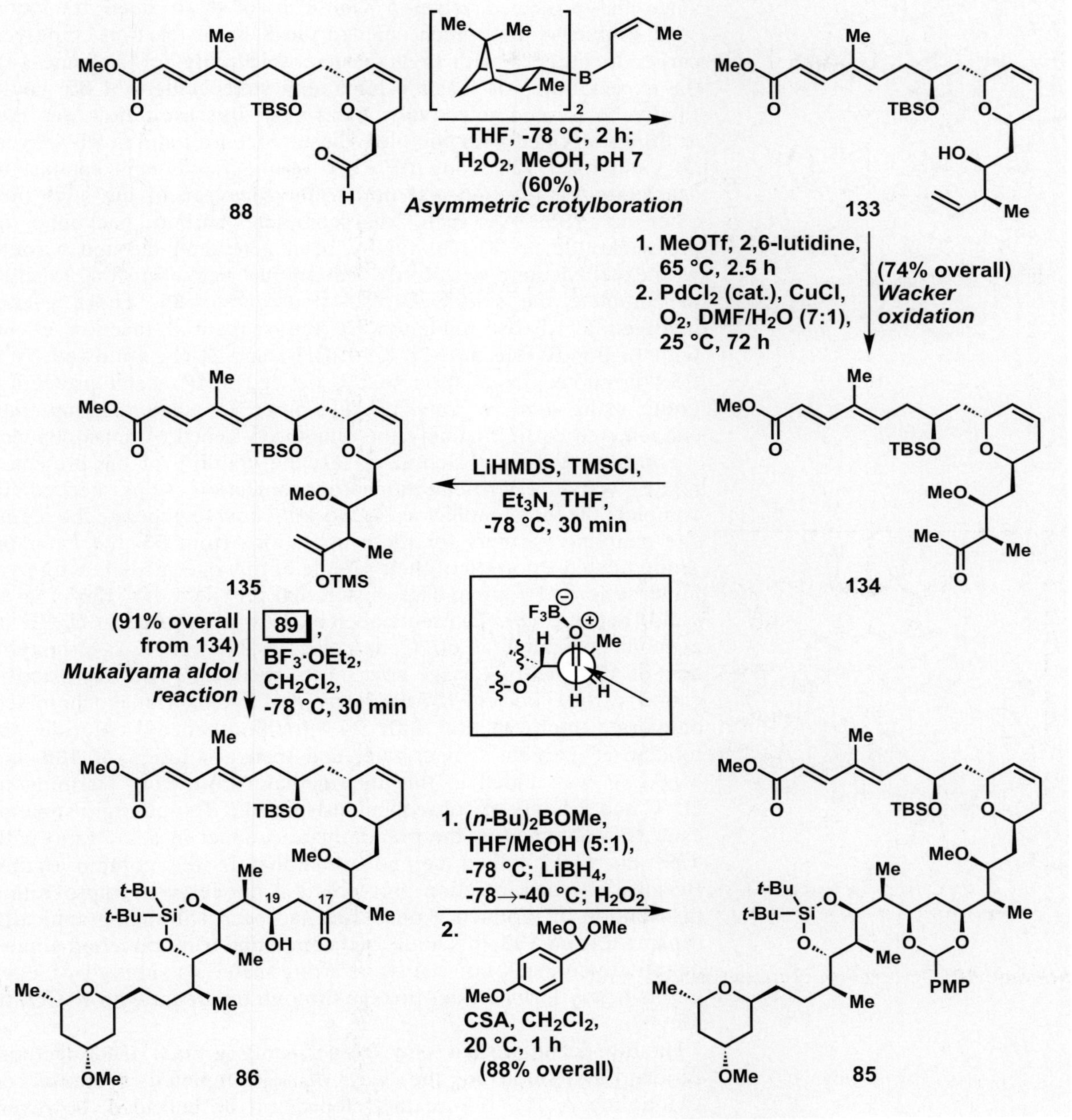

Scheme 17. Second approach to merge the main building blocks (88 and 89).

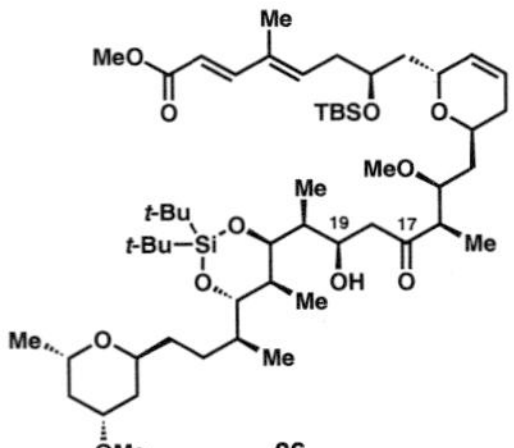

preparation for a Mukaiyama aldol reaction with **89**. Of course, since the projected reaction would involve an open transition state similar to that which enabled the Sakurai reaction employed earlier to proceed with high diastereoselectivity (cf. Scheme 14), the expectation was that the remaining stereocenters of **85** would finally be formed selectively. Pleasingly, this aspiration was met as **86** was formed with complete diastereocontrol and in 91 % overall yield from **134** using $BF_3{\cdot}OEt_2$ as the Lewis acid catalyst to accelerate the reaction and initiate the cleavage of the silyl enol ether in **135**. As such, the complete carbon backbone of preswinholide A (**3**) had finally been generated through a route with excellent stereoselectivity, leaving just a few finishing touches to complete the synthesis of key precursor **85**. Those events involved a 1,3-*syn* reduction by the sequential reaction of **86** with $(n\text{-Bu})_2BOMe$ at −78 °C, $LiBH_4$, and H_2O_2, followed by a CSA-promoted installation of the desired PMP acetal protecting group using a more conventional source for such a group than encountered earlier, namely the dimethoxy acetal of anisaldehyde.

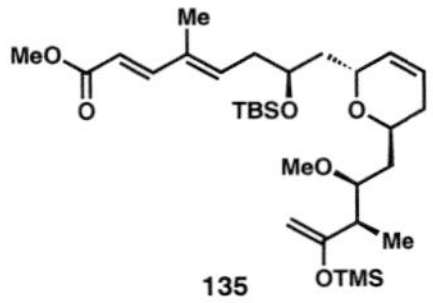

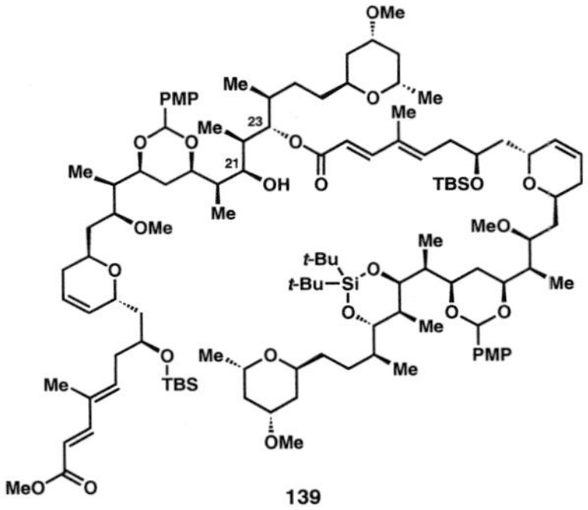

The time had finally come to test the viability of the projected regioselective esterification/macrolactonization steps needed to complete the target molecule (**1**). Seeking now to generate the requisite coupling partners for these operations from **85**, the Paterson group treated a portion of their sample of this intermediate with pyridine-buffered HF•py in THF, a step that provided diol **136** in 94 % yield (Scheme 18). The remainder of **85** was exposed to NaOH in aqueous methanol at 60 °C to generate the free C-1 carboxylic acid of **137**. With the stage now set to attempt the first key esterification, carboxylic acid **137** was converted into its Yamaguchi mixed anhydride upon reaction with 2,4,6-trichlorobenzoyl chloride[39] in toluene at ambient temperature, and then a solution of **136** and 4-DMAP was added to this intermediate. Following warming to 60 °C and 3 hours of subsequent stirring, the desired regioisomeric ester was obtained as the predominant product in a 2:1 ratio with **139**, the C-23 linked compound shown in the column figure. Because alternate reaction protocols did not serve to improve this distribution of products,* once **139** had been chromatographically separated from **138**, this undesired compound was converted almost quantitatively back into the two starting materials simply by cleaving its newly formed ester linkage through treatment with K_2CO_3 in MeOH.

Having accomplished one merger, only a final macrolactonization now stood in the way of a completed synthesis of swinholide A (**1**). Before that step could be executed, however, the protecting groups guarding oxygen functions at C-1 and C-21′ in **138** had to be removed. Although such operations are usually cursory events, the complexity of swinholide A (**1**)

* In fact, some protocols using modified coupling reagents, such as DCC, led to a complete reversal in selectivity, favoring the C-23-linked product **139**.

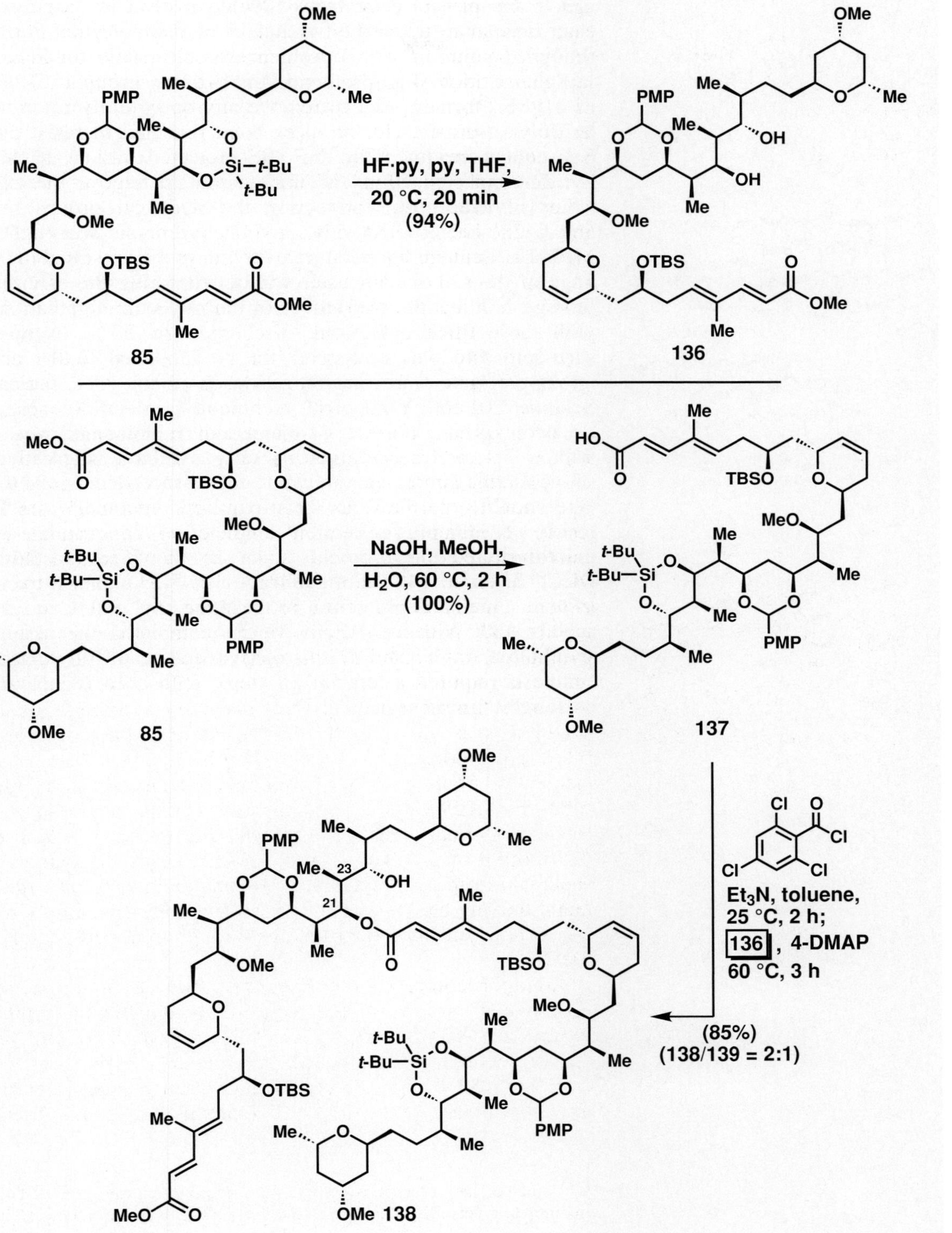

Scheme 18. Initial coupling of monomeric preswinholide A building blocks.

rendered them quite challenging. For example, all attempts to cleave the methyl ester from **138** led instead to cleavage of the ester bond that linked the two halves of the molecule. Fortunately, though, a solution to this problem was ultimately found as shown in Scheme 19 by first protecting the hydroxy group at C-23 of **138** as a TBS ether in order to remove any opportunity for it to direct hydrolytic reagents to the ester bond that had to resist cleavage. Subsequent treatment of the TBS-protected variant of **138** with pyridine-buffered HF•py in THF served to remove the cyclic di-*t*-butylsilylene group protecting the hydroxyl groups at C-21′ and C-23′ chemoselectively. A final hydrolysis with $Ba(OH)_2$ in MeOH at ambient temperature over 4 days then resulted in the cleavage of the C-1 methyl ester without rupturing the essential ester linkage holding the two halves of the molecule together. As such, with these three operations discharged in 53 % overall yield, seco-acid **140** was now ready to be subjected to the action of an appropriate macrolactonization protocol. As indicated in Scheme 20, the Yamaguchi technique[39] splendidly rose to the occasion as the correct 44-membered regioisomer was formed with a 5:1 preference over its only available alternative (**141**, see column figure) in an event that proceeded in 84 % yield. One should note, however, this exquisite outcome was solely the result of judicious selection of reaction conditions, as other macrolactonization protocols such as the Keck method using DCC[40] afforded **141** (with cyclization at C-23′) almost exclusively instead. Finally, deprotection of the three silyl ethers and the PMP acetals with aqueous HF in MeCN completed the assembly of swinholide A (**1**). Overall, this concise and beautifully orchestrated synthesis required a total of 38 steps, with only 28 operations in its longest linear sequence.

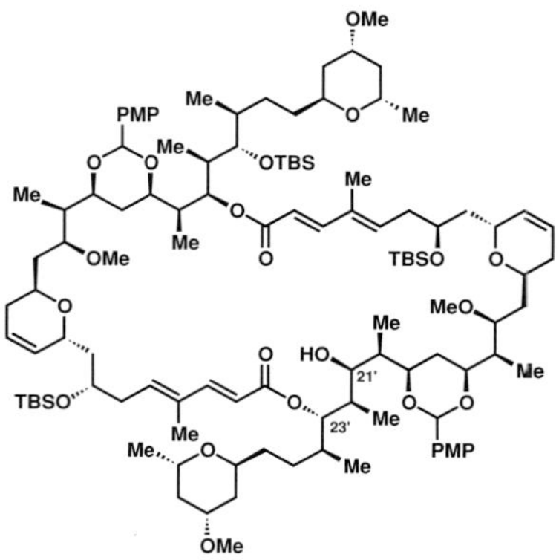

OMe PMP Me Me OTBS Me Me OMe Me O TBSO MeO Me HO 21′ Me TBSO 23′ Me Me Me PMP Me OMe

141

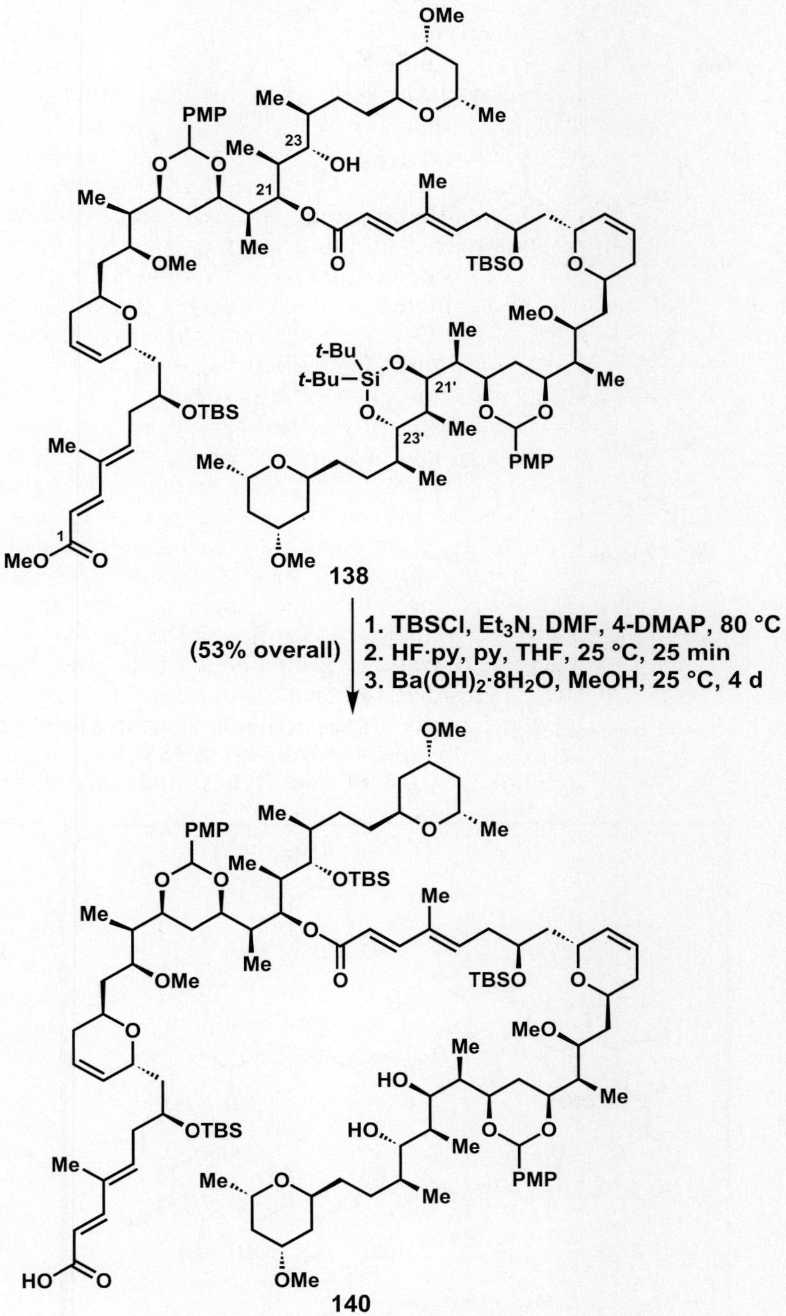

138

1. TBSCl, Et_3N, DMF, 4-DMAP, 80 °C
2. HF·py, py, THF, 25 °C, 25 min
3. $Ba(OH)_2·8H_2O$, MeOH, 25 °C, 4 d

(53% overall)

140

Scheme 19. Preparation of macrolactonization precursor **140**.

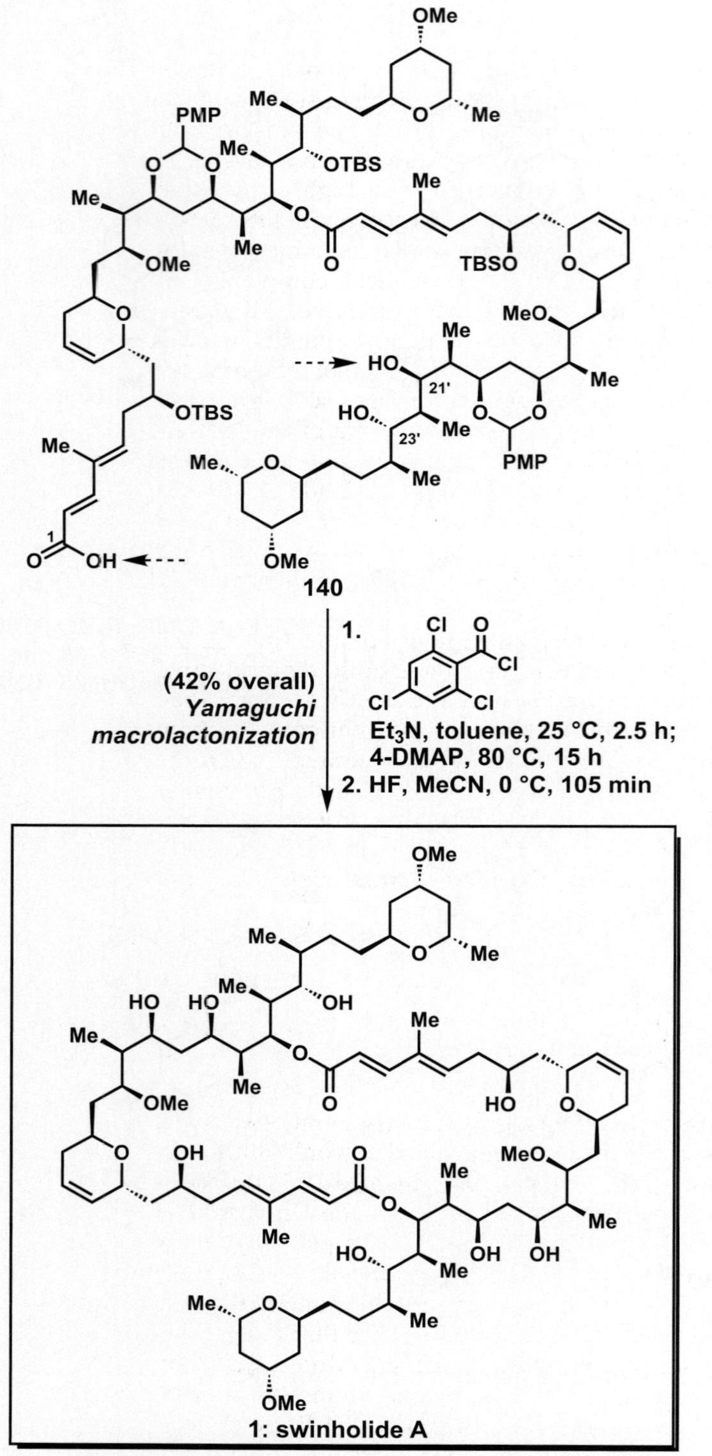

Scheme 20. Final stages and completion of Paterson's total synthesis of swinholide A (1).

3.4 Conclusion

Although many of the syntheses presented in this book effectively illustrate the value of the aldol reaction to generate stereocenters, none affords a comparable demonstration of the true virtuosity of this transformation as does the Paterson synthesis of swinholide A (**1**). Indeed, the success of this convergent and highly inventive synthetic sequence relied on the execution of five completely different substrate- and reagent-based variants of the asymmetric aldol reaction to establish much of the stereochemical complexity of the target molecule. As such, these events effectively illustrate the benefits of this reaction for the construction of complex natural products and the presently advanced state of aldol methodology. By the same token, however, this synthesis also underscores the fact that there are still gaps in our current understanding of this powerful transformation, as the stereochemical outcomes of some of the final aldol mergers seeking to generate preswinholide A (**3**) did not follow established trends. Only through continued exploration of the aldol reaction in such complex contexts can we ever hope to acquire the knowledge needed to render these couplings equally predictable.

Finally, we cannot end this chapter without commenting on the concluding stages of the synthesis in which both regioselective esterification and macrolactonization were efficiently accomplished on unprotected intermediates. Although conventional wisdom always dictates against the execution of such a strategy when the potential for side reactions is high, the success of this particular sequence indicates that there are cases in which protecting groups can be avoided without any loss in productivity. The more we dare to explore such sequences, the more our efficiency in constructing complex molecules will improve.

3.5 Nicolaou's Total Synthesis of Swinholide A (1)

While the Paterson group was the first to synthesize swinholide A (**1**), their route did not constitute the only sequence that would ultimately prove capable of accessing this compound. Indeed, in 1996 the Nicolaou group at The Scripps Research Institute and the University of California, San Diego reported an equally effective pathway that is summarized in Schemes 21–24.[41] As one might expect, this alternate route to **1** was based on a strategy in which monomeric preswinholide A-type building blocks were merged through a late-stage esterification/macrolactonization sequence. Moreover, many of the stereocenters on these pieces arose from asymmetric aldol reactions. Apart from these broad similarities with the Paterson synthesis, all other elements of the synthetic plan were wholly unique.

Striving for convergence, the Nicolaou group elected to generate their key dimerization precursors through the union of two functionalized carbon chains, just as the Paterson group had accomplished during their synthesis. However, rather than rely on an aldol reaction to merge these fragments and concomitantly generate stereocenters, these researchers instead projected that a dithiane coupling[42] between **152** (see Scheme 21) and an appropriate electrophilic acceptor such as **157** (Scheme 22) could accomplish the same objective without encoding any new chiral information. Although this latter feature might appear inconsequential, as we already know from

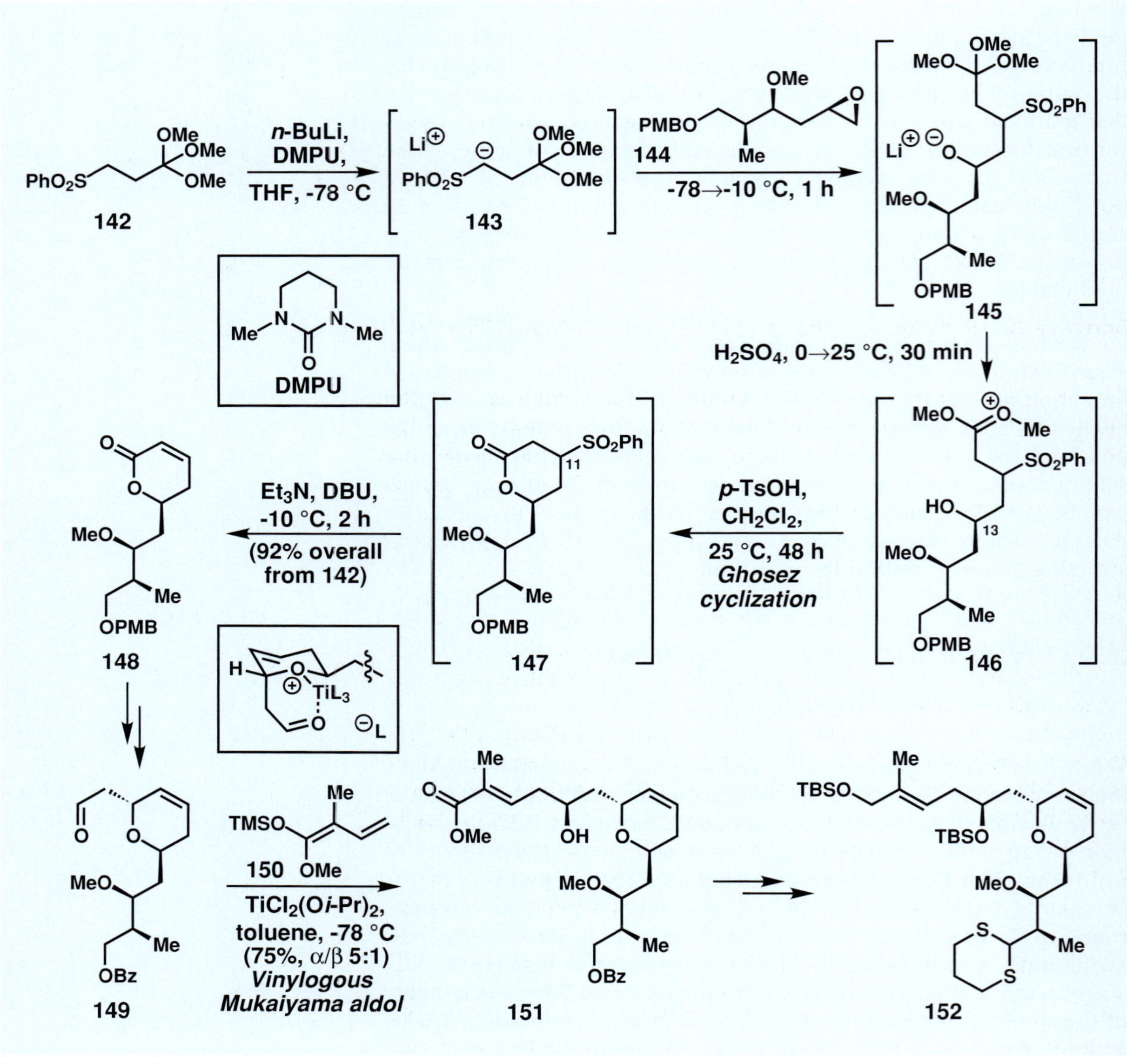

Scheme 21. Nicolaou's synthesis of advanced intermediate **152**.

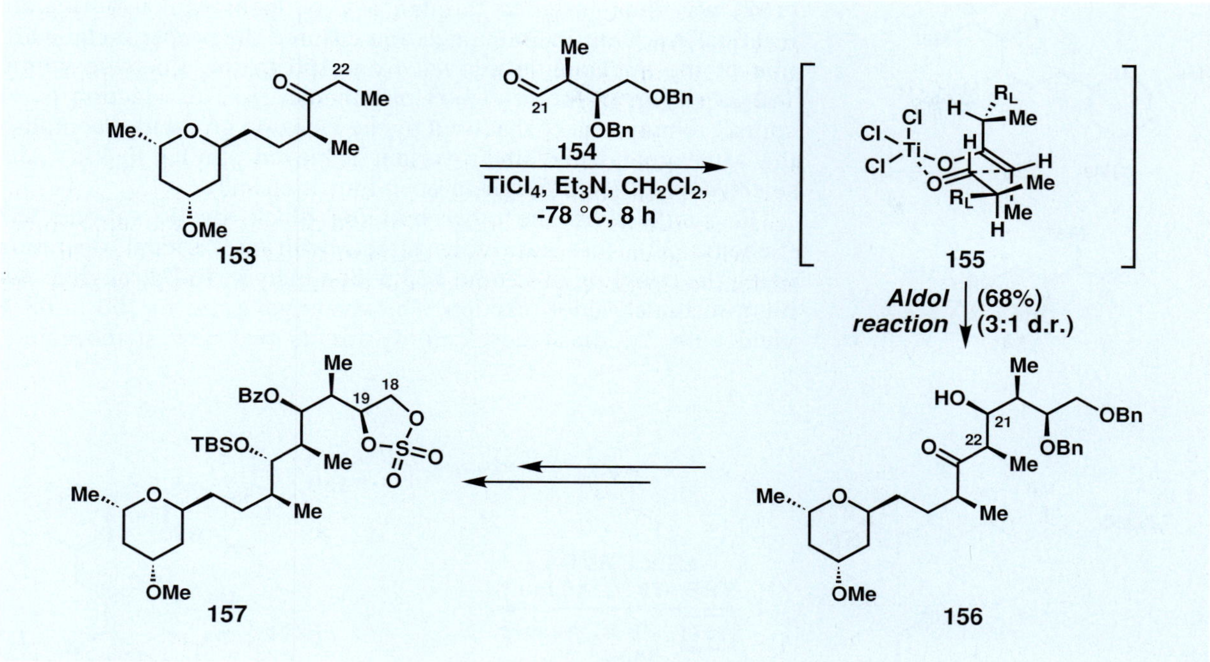

Scheme 22. Nicolaou's synthesis of advanced building block 157.

the discussion above, it proved quite difficult in some cases for the Paterson group to effect such unions with efficient stereocontrol for newly formed stereogenic centers. Accordingly, by adopting this alternate strategy the Nicolaou group avoided such potential complications in this case, since only its viability, was in question not its stereochemical outcome.

Schemes 21 and 22 outline the key features of the synthetic routes that ultimately provided both of these key precursors (**152** and **157**) as single stereoisomers. For the eventual dithiane nucleophile (**152**, Scheme 21), the Nicolaou group set as its initial objective the homochiral assembly of its dihydropyran ring, a goal that was accomplished through an inventive use of the Ghosez cyclization reaction.[43] In the event, treatment of phenylsulfone **142** with *n*-BuLi in the presence of DMPU in THF at −78 °C afforded an intermediate nucleophile (**143**) capable of smoothly attacking the terminal position of epoxide **144** once the latter was introduced into the reaction flask. With this event giving rise to **145**, the reaction was quenched with aqueous H_2SO_4 in order to protonate the resultant alkoxide at C-13 and convert the strategically positioned orthoester into an nucleophilic acceptor, operations that activated the substrate for a subsequent *p*-TsOH-initiated ring closure which provided **147**. Finally, the addition of Et_3N and DBU to the reaction medium induced the phenylsulfone within **147** to eliminate, completing the assembly of **148** in 92 % overall yield from **142**. Having efficiently orchestrated this sequence, the remaining stereochemical complexity of the target

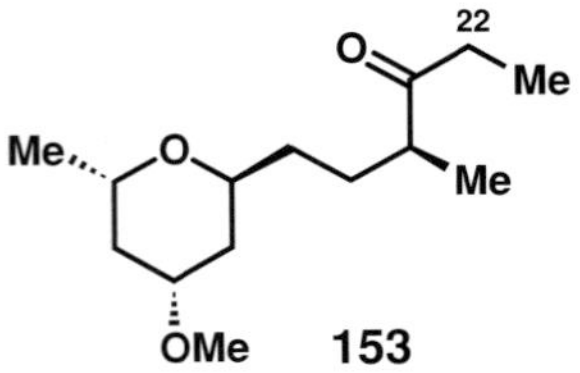

(**152**) was then installed through a vinylogous Mukaiyama aldol reaction in which titanium chelation ensured the proper facial addition of the nucleophile derived from **150** to the aldehyde within **149** as shown in the inset box in Scheme 21. This reaction is, of course, reminiscent of that used by the Paterson group to accomplish the same goal where their variant achieved similar Felkin–Ahn selectivity, except through an open transition state.

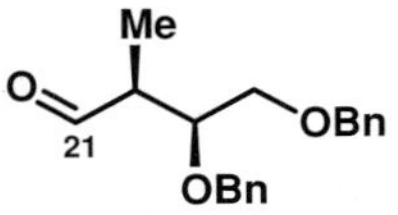

The synthesis of the other building block, cyclic sulfone **157** (Scheme 22), was far more direct, with the critical operation being the coupling of ketone **153** with aldehyde **154** through a titanium-mediated aldol reaction. This event gave rise to **156** in 68 % yield with 3:1 diastereoselectivity for its two new stereocenters,

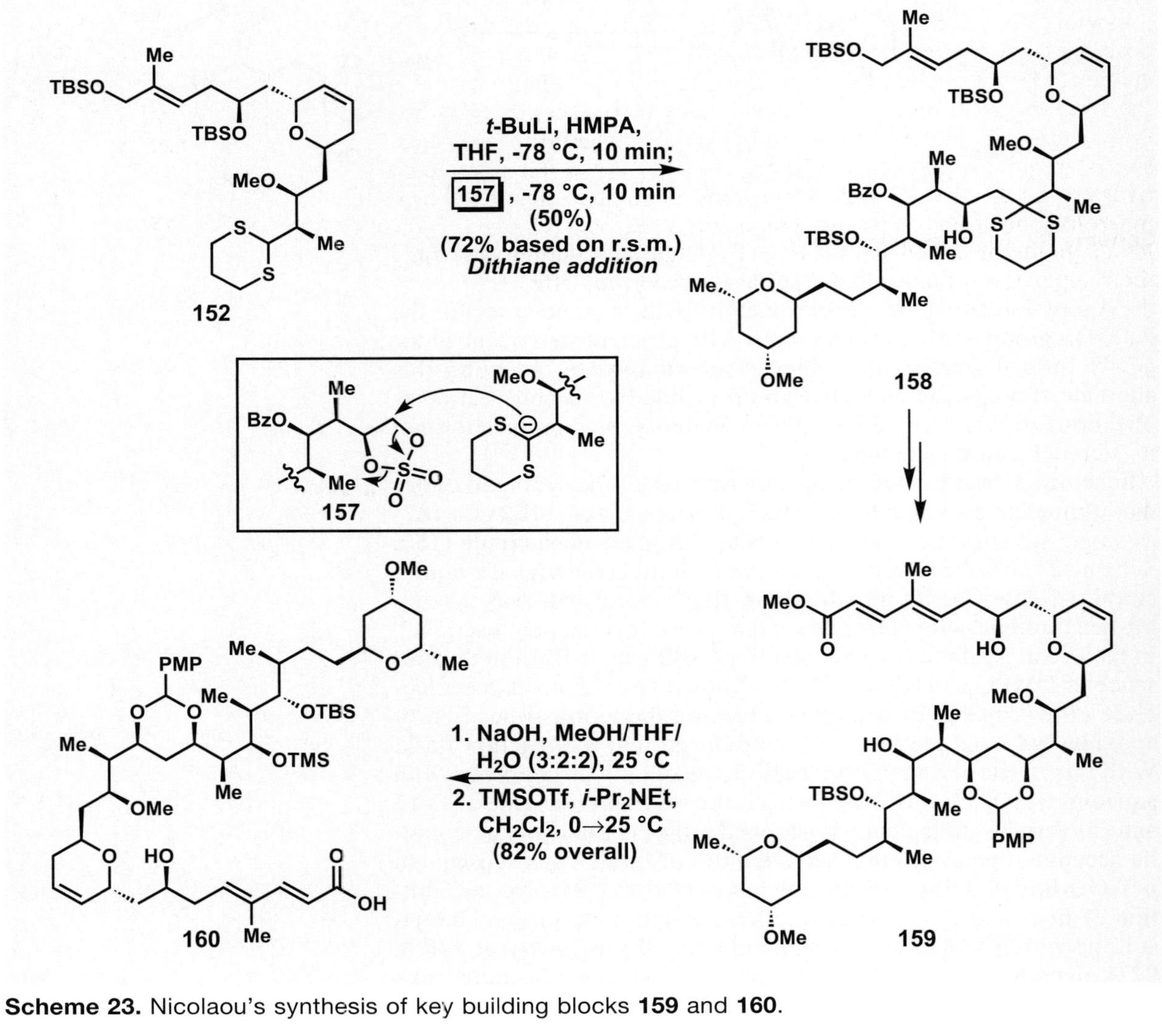

Scheme 23. Nicolaou's synthesis of key building blocks **159** and **160**.

an outcome that can be rationalized by the shown transition state (**155**) in which the bulky substituents on the stereocenters adjacent to the aldehyde and the *Z* enolate are diametrically opposed to minimize steric interactions. This aldol reaction represents an example of a matched case between two chiral substrates leading to double asymmetric induction.

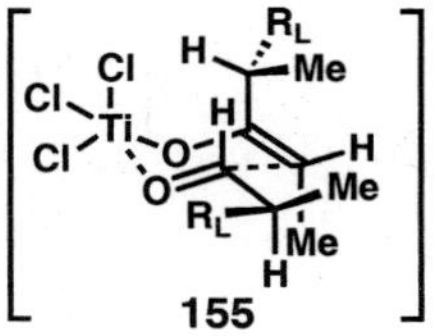

Once **156** had been converted into **157**, its merger with the dithiane fragment (**152**) to provide **158** was accomplished in 50 % yield as shown in Scheme 23, with the attack of the lithiated species derived from **152** occurring exclusively at the C-18 position of **157**. Accordingly, this transformation illustrates that one can view the five-membered cyclic sulfate within **157** as the synthetic equivalent of an epoxide. In fact, the cyclic sulfate was superior to this alternate electrophile, as attempts to use an epoxide in this dithiane addition step met with complete failure. From **158**, the two monomeric preswinholide A building blocks **159** and **160** bearing differentiable protecting groups on the alcohol functions at C-21 and C-23 were synthesized in short order, and then merged and elaborated into the final target through the sequence shown in Scheme 24. Since these steps mirror those of Paterson synthesis, no additional discussion is required, except to note that the bulk of a TBS-protected hydroxy group at C-23 did not prevent either the esterification or macrolactonization steps as conjectured earlier. However, the lower yields observed for these last operations highlight not only their superfluous nature, but also their deleterious effects.

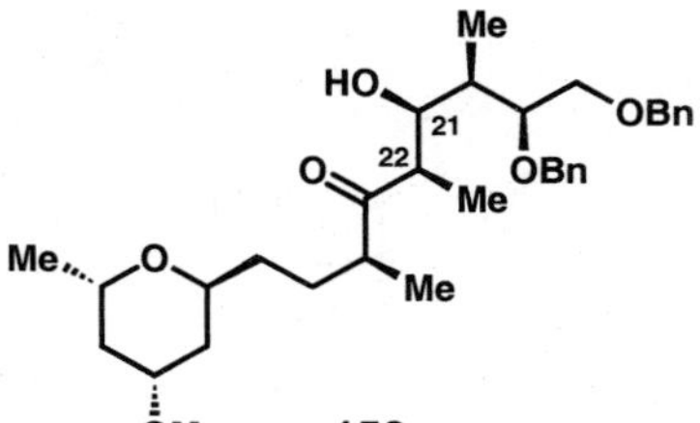

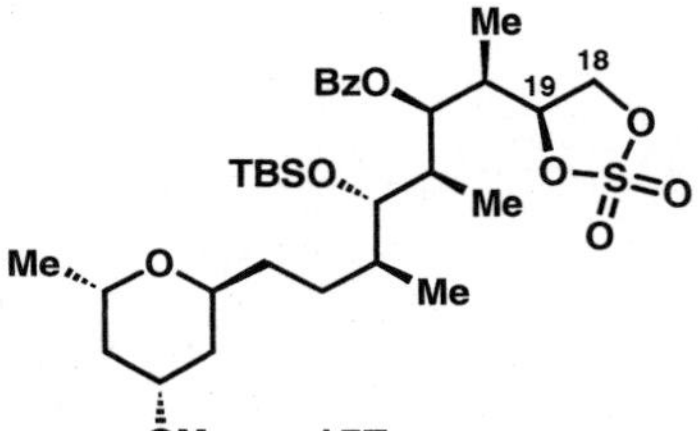

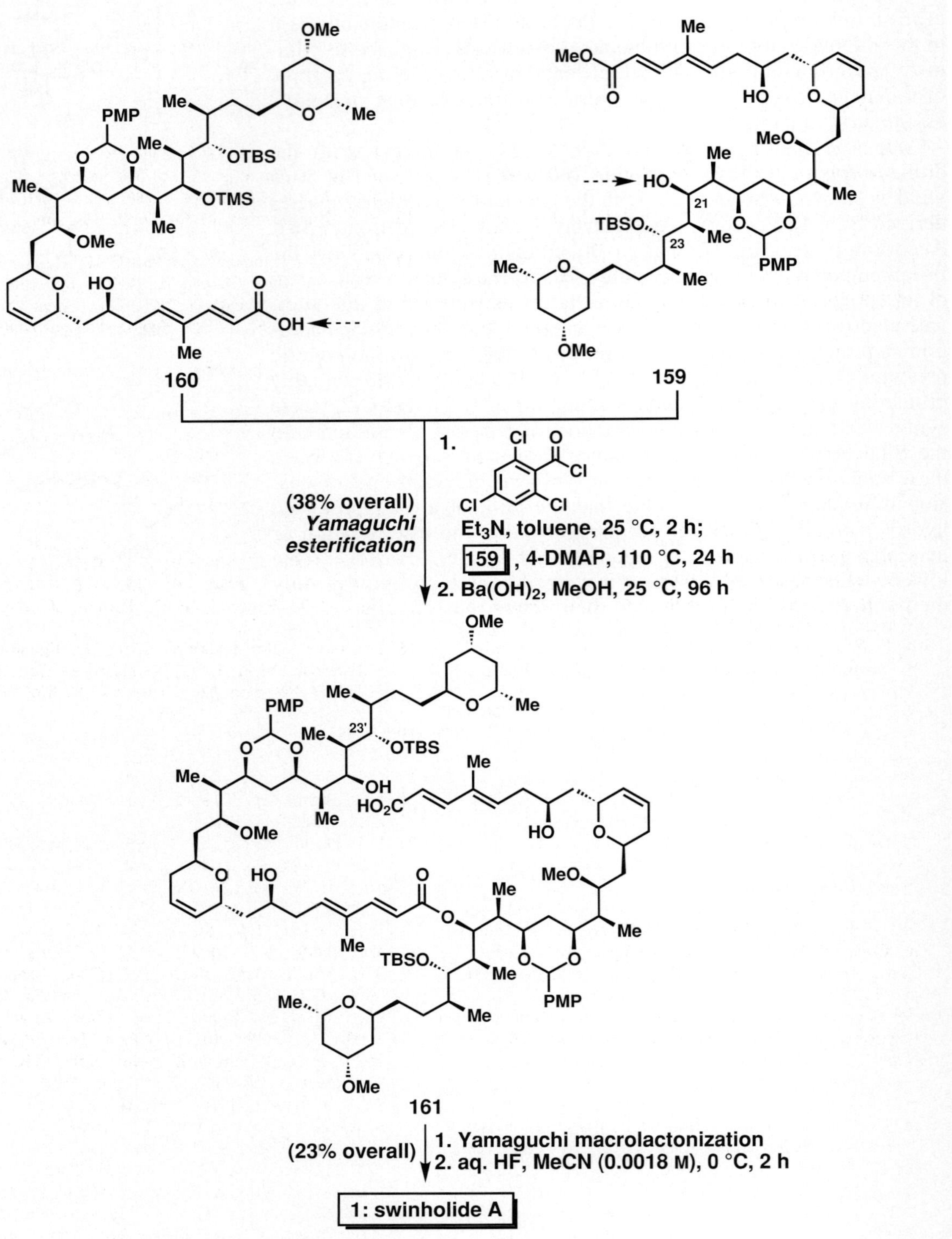

Scheme 24. Final stages and completion of Nicolaou's total synthesis of swinholide A (**1**).

References

1. For a general review on cancer therapies using natural products, see: J. Mann, *Nature Reviews: Cancer* **2002**, *2*, 143.
2. For an excellent review on actin-binding macrolides, see: K.-S. Yeung, I. Paterson, *Angew. Chem.* **2002**, *114*, 4826; *Angew. Chem. Int. Ed.* **2002**, *41*, 4632.
3. S. Carmely, Y. Kashman, *Tetrahedron Lett.* **1985**, *26*, 511.
4. a) M. Kobayashi, J. Tanaka, T. Katori, M. Matsuura, I. Kitagawa, *Tetrahedron Lett.* **1989**, *30*, 2963; b) I. Kitagawa, M. Kobayashi, T. Katori, M. Yamashita, J. Tanaka, M. Doi, T. Ishida, *J. Am. Chem. Soc.* **1990**, *112*, 3710; c) M. Kobayashi, J. Tanaka, T. Katori, M. Matsuura, M. Yamashita, I. Kitagawa, *Chem. Pharm. Bull.* **1990**, *38*, 2409; d) M. Kobayashi, J. Tanaka, T. Katori, I. Kitagawa, *Chem. Pharm. Bull.* **1990**, *38*, 2960; e) M. Doi, T. Ishida, M. Kobayashi, I. Kitagawa, *J. Org. Chem.* **1991**, *56*, 3629.
5. J.S. Todd, K.A. Alvi, P. Crews, *Tetrahedron Lett.* **1992**, *33*, 441.
6. J. Tanaka, T. Higa, M. Kobayashi, I. Kitagawa, *Chem. Pharm. Bull.* **1990**, *38*, 2967.
7. For the full account of this synthesis, see: a) I. Paterson, J.G. Cumming, R.A. Ward, S. Lamboley, *Tetrahedron* **1995**, *51*, 9393; b) I. Paterson, J.D. Smith, R.A. Ward, *Tetrahedron* **1995**, *51*, 9413; c) I. Paterson, R.A. Ward, J.D. Smith, J.G. Cumming, K.-S. Yeung, *Tetrahedron* **1995**, *51*, 9437; d) I. Paterson, K.-S. Yeung, R.A. Ward, J.D. Smith, J.G. Cumming, S. Lamboley, *Tetrahedron* **1995**, *51*, 9467. For preliminary communications, see: e) I. Paterson, K.-S. Yeung, R.A. Ward, J.G. Cumming, J.D. Smith, *J. Am. Chem. Soc.* **1994**, *116*, 9391; f) I. Paterson, J.D. Smith, R.A. Ward, J.G. Cumming, *J. Am. Chem. Soc.* **1994**, *116*, 2615.
8. For recent reviews of the aldol reaction, see: a) C.H. Heathcock in *Comprehensive Organic Synthesis, Vol. 2* (Eds.: B.M. Trost, I. Fleming), Pergamon, Oxford, **1991**, pp. 133–179; b) C.H. Heathcock in *Comprehensive Organic Synthesis, Vol. 2* (Eds.: B.M. Trost, I. Fleming), Pergamon, Oxford, **1991**, pp. 181–238; c) B.M. Kim, S.F. Williams, S. Masamune in *Comprehensive Organic Synthesis, Vol. 2* (Eds.: B.M. Trost, I. Fleming), Pergamon, Oxford, **1991**, pp. 239–275.
9. For a review specifically on boron-mediated aldol reactions, see: C.J. Cowden, I. Paterson in *Organic Reactions, Vol. 51* (Eds.: L.A. Paquette et al.), John Wiley & Sons, New York, **1997**, pp. 1–200.
10. For selected examples of early papers in this field, see: a) T. Mukaiyama, K. Inomata, M. Muraki, *J. Am. Chem. Soc.* **1973**, *95*, 967; b) M. Hirama, S. Masamune, *Tetrahedron Lett.* **1979**, *20*, 2225; c) D.E. Van Horn, S. Masamune, *Tetrahedron Lett.* **1979**, *20*, 2229; d) D.A. Evans, E. Vogel, J.V. Nelson, *J. Am. Chem. Soc.* **1979**, *101*, 6120; e) T. Inoue, T. Mukaiyama, *Bull. Chem. Soc. Jpn.* **1980**, *53*, 174; f) S. Masamune, W. Choy, F.A.J. Kerdesky, B. Imperiali, *J. Am. Chem. Soc.* **1981**, *103*, 1566; g) D.A. Evans, J.V. Nelson, E. Vogel, T.R. Taber, *J. Am. Chem. Soc.* **1981**, *103*, 3099.
11. J.M. Goodman, I. Paterson, *Tetrahedron Lett.* **1992**, *33*, 7223. For a related model, see: E.J. Corey, S.S. Kim, *J. Am. Chem. Soc.* **1990**, *112*, 4976.
12. H.E. Zimmerman, M.D. Traxler, *J. Am. Chem. Soc.* **1957**, *79*, 1920.
13. For leading papers, see: a) C.H. Heathcock, L.A. Flippin, *J. Am. Chem. Soc.* **1983**, *105*, 1667; b) E.P. Lodge, C.H. Heathcock, *J. Am. Chem. Soc.* **1987**, *109*, 2819; c) C.H. Heathcock, *Aldrichim. Acta* **1990**, *23*, 99.
14. D.A. Evans, J. Bartroli, *Tetrahedron Lett.* **1982**, *23*, 807.
15. a) I. Paterson, D.J. Wallace, S.M. Velázquez, *Tetrahedron Lett.* **1994**, *35*, 9083; b) I. Paterson, D.J. Wallace, *Tetrahedron Lett.* **1994**, *35*, 9087.
16. a) I. Paterson, D.J. Wallace, *Tetrahedron Lett.* **1994**, *35*, 9477; b) I. Paterson, D.J. Wallace, C.J. Cowden, *Synthesis* **1998**, 639.
17. a) H.C. Brown, M.C. Desai, P.K. Jadhav, *J. Org. Chem.* **1982**, *47*, 5065; b) H.C. Brown, B. Singaram, *J. Org. Chem.* **1984**, *49*, 945; c) P.K. Jadhav, K.S. Bhat, P.T. Perumal, H.C. Brown, *J. Org. Chem.* **1986**, *51*, 432.
18. For other aldol reactions using Ipc ligands, see: a) I. Paterson, M.A. Lister, *Tetrahedron Lett.* **1988**, *29*, 585; b) I. Paterson, M.A. Lister, C.K. McClure, *Tetrahedron Lett.* **1986**, *27*, 4787.
19. a) I. Paterson, A.N. Hulme, *J. Org. Chem.* **1995**, *60*, 3288; b) I. Paterson, A.N. Hulme, *Tetrahedron Lett.* **1990**, *31*, 7513.
20. K. Toshima, K. Tatsuta, M. Kinoshita, *Tetrahedron Lett.* **1986**, *27*, 4741.
21. N.A. Van Draanen, S. Arseniyadis, M.T. Crimmins, C.H. Heathcock, *J. Org. Chem.* **1991**, *56*, 2499.
22. a) C.J. Forsyth, F. Ahmed, R.D. Cink, C.S. Lee, *J. Am. Chem. Soc.* **1998**, *120*, 5597; b) D.A. Evans, D.M. Fitch, T.E. Smith, V.J. Cee, *J. Am. Chem. Soc.* **2000**, *122*, 10033; c) D.A. Evans, V.J. Cee, T.E. Smith, D.M. Fitch, P.S. Cho, *Angew. Chem.* **2000**, *112*, 2633; *Angew. Chem. Int. Ed.* **2000**, *39*, 2533; d) D.A. Evans, D.M. Fitch, *Angew. Chem.* **2000**, *112*, 2636; *Angew. Chem. Int. Ed.* **2000**, *39*, 2536; e) A.B. Smith, K.P. Minbiole, P.R. Verhoest, M. Schelhaas, *J. Am. Chem. Soc.* **2001**, *123*, 10942; f) A.B. Smith, P.R. Verhoest, K.P. Minbiole, M. Schelhaas, *J. Am. Chem. Soc.* **2001**, *123*, 4834; g) M.A. González, G. Pattenden, *Angew. Chem.* **2003**, *115*, 1293; *Angew. Chem. Int. Ed.* **2003**, *42*, 1255; h) D.R. Williams, A.A. Kiryanov, U. Emde, M.P. Clark, M.A. Berliner, J.T. Reeves, *Angew. Chem.* **2003**, *115*, 1296; *Angew. Chem. Int. Ed.* **2003**, *42*, 1258.

23. a) D. A. Evans, B. W. Trotter, P. J. Coleman, B. Côté, L. C. Dias, H. A. Rajapakse, A. N. Tyler, *Tetrahedron* **1999**, *55*, 8671; b) D. A. Evans, P. J. Coleman, L. C. Dias, *Angew. Chem.* **1997**, *109*, 2951; *Angew. Chem. Int. Ed. Engl.* **1997**, *36*, 2738; c) D. A. Evans, B. W. Trotter, B. Côté, P. J. Coleman, *Angew. Chem.* **1997**, *109*, 2954; *Angew. Chem. Int. Ed. Engl.* **1997**, *36*, 2741; d) D. A. Evans, B. W. Trotter, B. Côté, P. J. Coleman, L. C. Dias, A. N. Tyler, *Angew. Chem.* **1997**, *109*, 2957; *Angew. Chem. Int. Ed. Engl.* **1997**, *36*, 2744; e) J. Guo, K. J. Duffy, K. L. Stevens, P. I. Dalko, R. M. Roth, M. M. Hayward, Y. Kishi, *Angew. Chem.* **1998**, *110*, 198; *Angew. Chem. Int. Ed.* **1998**, *37*, 187; f) M. M. Hayward, R. M. Roth, K. J. Duffy, P. I. Dalko, K. L. Stevens, J. Guo, Y. Kishi, *Angew. Chem.* **1998**, *110*, 202; *Angew. Chem. Int. Ed.* **1998**, *37*, 192; g) A. B. Smith, V. A. Doughty, Q. Lin, L. Zhuang, M. D. McBriar, A. M. Boldi, W. H. Moser, N. Murase, K. Nakayama, M. Sobukawa, *Angew. Chem.* **2001**, *113*, 197; *Angew. Chem. Int. Ed.* **2001**, *40*, 191; h) A. B. Smith, Q. Lin, V. A. Doughty, L. Zhuang, M. D. McBriar, J. K. Kerns, C. S. Brook, N. Murase, K. Nakayama, *Angew. Chem.* **2001**, *113*, 202; *Angew. Chem. Int. Ed.* **2001**, *40*, 196; i) I. Paterson, D. Y.-K. Chen, M. J. Coster, J. L. Aceña, J. Bach, K. R. Gibson, L. E. Keown, R. M. Oballa, T. Trieselmann, D. J. Wallace, A. P. Hodgson, R. D. Norcross, *Angew. Chem.* **2001**, *113*, 4179; *Angew. Chem. Int. Ed.* **2001**, *40*, 4055; j) M. T. Crimmins, J. D. Katz, D. G. Washburn, S. P. Allwein, L. F. McAtee, *J. Am. Chem. Soc.* **2002**, *124*, 5661.
24. a) D. T. Hung, J. B. Nerenberg, S. L. Schreiber, *J. Am. Chem. Soc.* **1996**, *118*, 11054; b) J. B. Nerenberg, D. T. Hung, P. K. Somers, S. L. Schreiber, *J. Am. Chem. Soc.* **1993**, *115*, 12621; c) A. B. Smith, T. J. Beauchamp, M. J. LaMarche, M. D. Kaufman, Y. Qiu, H. Arimoto, D. R. Jones, K. Kobayashi, *J. Am. Chem. Soc.* **2000**, *122*, 8654; d) A. B. Smith, Y. Qiu, D. R. Jones, K. Kobayashi, *J. Am. Chem. Soc.* **1995**, *117*, 12011; e) S. S. Harried, G. Yang, M. A. Strawn, D. C. Myles, *J. Org. Chem.* **1997**, *62*, 6098; f) J. A. Marshall, B. A. Johns, *J. Org. Chem.* **1998**, *63*, 7885; g) I. Paterson, O. Delgado, G. J. Florence, I. Lyothier, J. P. Scott, N. Sereinig, *Org. Lett.* **2003**, *5*, 35; h) I. Paterson, G. J. Florence, K. Gerlach, J. P. Scott, *Angew. Chem.* **2000**, *112*, 385; *Angew. Chem. Int. Ed.* **2000**, *39*, 377.
25. T. Mukaiyama, K. Banno, K. Narasaka, *J. Am. Chem. Soc.* **1974**, *96*, 7503.
26. R. J. Ferrier, *J. Chem. Soc.* **1964**, 5443.
27. I. Paterson, S. Osborne, *Tetrahedron Lett.* **1990**, *31*, 2213.
28. a) Y. Gao, R. M. Hanson, J. M. Klunder, S. Y. Ko, H. Masamune, K. B. Sharpless, *J. Am. Chem. Soc.* **1987**, *109*, 5765; b) W. R. Roush, A. D. Palkowitz, *J. Org. Chem.* **1989**, *54*, 3009; c) W. R. Roush, R. J. Brown, *J. Org. Chem.* **1983**, *48*, 5093.
29. J. M. Finan, Y. Kishi, *Tetrahedron Lett.* **1982**, *23*, 2719.
30. I. Paterson, J. M. Goodman, M. Isaka, *Tetrahedron Lett.* **1989**, *30*, 7121.
31. D. A. Evans, K. T. Chapman, E. M. Carreira, *J. Am. Chem. Soc.* **1988**, *110*, 3560. For an example in a total synthesis using this combination, see: D. A. Evans, J. S. Clark, R. Metternich, V. J. Novack, G. S. Sheppard, *J. Am. Chem. Soc.* **1990**, *112*, 866.
32. W. C. Still, J. C. Barrish, *J. Am. Chem. Soc.* **1983**, *105*, 2487.
33. a) D. H. R. Barton, S. W. McCombie, *J. Chem. Soc., Perkin Trans 1* **1975**, 1574; b) D. H. R. Barton, D. Crich, A. Löbberding, S. Z. Zard, *Tetrahedron* **1986**, *42*, 2329.
34. a) D. A. Evans, J. L. Duffy, M. J. Dart, *Tetrahedron Lett.* **1994**, *35*, 8537; b) D. A. Evans, M. J. Dart, J. L. Duffy, *Tetrahedron Lett.* **1994**, *35*, 8541.
35. B. A. Narayanan, W. H. Bunnelle, *Tetrahedron Lett.* **1987**, *28*, 6261.
36. For an early review on this reaction, see: D. Schinzer, *Synthesis* **1988**, 263.
37. Y. Oikawa, T. Yoshioka, O. Yonemitsu, *Tetrahedron Lett.* **1982**, *23*, 889.
38. a) H. C. Brown, K. S. Bhat, R. S. Randad, *J. Org. Chem.* **1989**, *54*, 1570; b) H. C. Brown, K. S. Bhat, *J. Am. Chem. Soc.* **1986**, *108*, 5919.
39. J. Inanaga, K. Hirata, H. Saeki, T. Katsuki, M. Yamaguchi, *Bull. Chem. Soc. Jpn.* **1979**, *52*, 1989.
40. E. P. Boden, G. E. Keck, *J. Org. Chem.* **1985**, *50*, 2394.
41. a) K. C. Nicolaou, K. Ajito, A. P. Patron, H. Khatuya, P. K. Richter, P. Bertinato, *J. Am. Chem. Soc.* **1996**, *118*, 3059; b) K. C. Nicolaou, A. P. Patron, K. Ajito, P. K. Richter, H. Khatuya, P. Bertinato, R. A. Miller, M. J. Tomaszewski, *Chem. Eur. J.* **1996**, *2*, 847.
42. D. Seebach, E. J. Corey, *J. Org. Chem.* **1975**, *40*, 231.
43. J. C. Carretero, L. Ghosez, *Tetrahedron Lett.* **1988**, *29*, 2059.

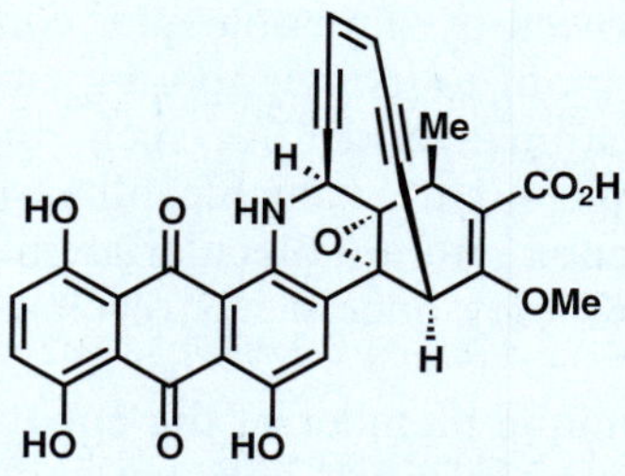

1: dynemicin A

4

A. G. Myers (1995)
S. J. Danishefsky (1995)

Dynemicin A

Key concepts:

- **Enediyne chemistry**
- **Molecular design and chemical biology**
- **Novel Diels–Alder strategies**
- **Substrate-based stereocontrol**

4.1 Introduction

Although the molecular connectivities of certain natural products often serve to humble synthetic chemists by their sheer complexity, one should not assume that Nature assembles such compounds merely to show off its synthetic abilities. Rather, most secondary metabolites exist for some specific biochemical purpose, and, more often than not, their presence has conferred some form of evolutionary advantage to the producing organism that has enabled it to survive. This concept has important ramifications, because if we can determine both how and why Nature has evolved these molecules to achieve certain objectives, then we can likely unravel many of the mysteries of chemical biology and gain the insight needed to impart function to molecules of our own imagining.[1]

Over the course of the past decade, one of the most important families of natural products that has helped to guide scientists in such directions is the cyclic enediynes, typified by the neocarzinostatin (**2**) and kedarcidin (**3**) chromophores and calicheamicin γ_1^I (**4**), a secondary metabolite presented in Chapter 30 of *Classics I*.[2] All of these compounds are potent antitumor agents with a shared mode of activity, namely the cleavage of single and double stranded DNA through a highly damaging diradical generated upon Bergman cycloaromatization of their enediyne motif.[3] Intriguingly, while this feature is uniform, Nature has endowed each natural product with a uniquely engineered triggering device to set this critical event into motion. For example, neocarzinostatin (**2**) only undergoes the Bergman reaction after it has been selectively attacked by a sulfur-based nucleophile at C-12, an event that leads to a rearrangement of the 9-membered enediyne ring into an intermediate

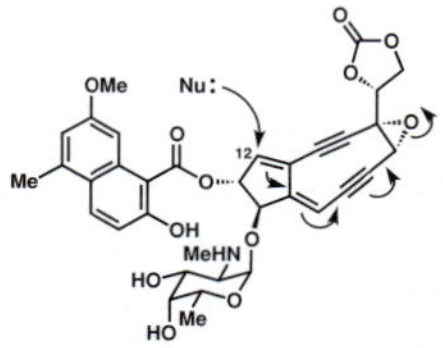

2: neocarzinostatin chromophore

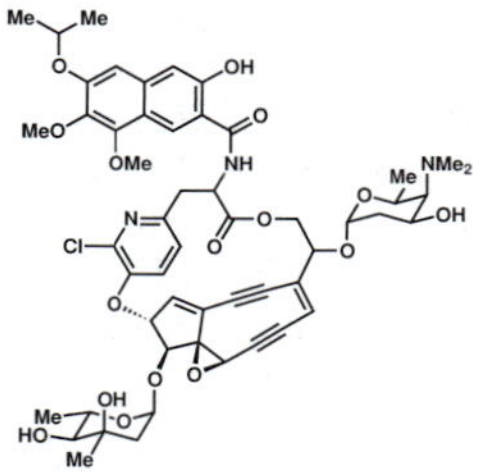

3: kedarcidin chromophore

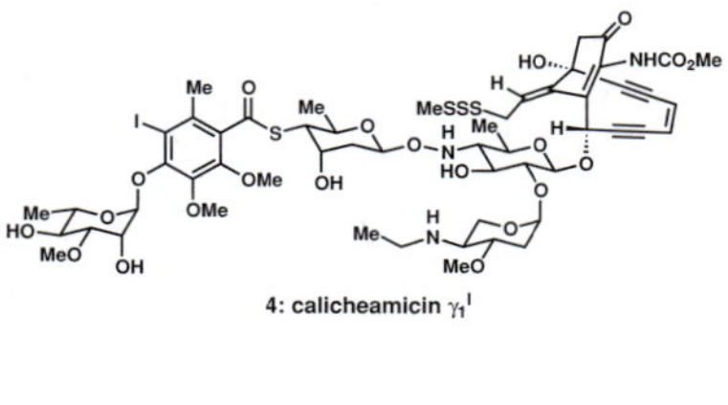

4: calicheamicin γ_1^I

cumulene.[4] In contrast, calicheamicin γ_1^I (**4**) requires initial cleavage of its trisulfide to reveal a thiol, followed by intramolecular conjugate addition of this newly unveiled motif to its neighboring enone in order to initiate the same cycloaromatization. As such, these divergent activation pathways provide highly valuable blueprints as to how latent reactivity can be loaded into a molecular architecture, hoping that it will be discharged only once it has reached its specific molecular target.

The subject of this chapter is an unique member of the enediyne family, dynemicin A (**1**), which Nature has gifted with yet another ingenious method to spark the Bergman cycloaromatization reaction that confers its DNA-damaging activity. Dynemicin A was first isolated in the mid-1980s from the fermentation broth of *Micromonospora chernisa*, a bacterial colony found in a soil sample from the Gujarat State in India, and later characterized by scientists at the Bristol-Myers Pharmaceutical Company in Japan.[5] Of course, a cursory examination of its molecular structure quickly reveals the basis for its membership into the enediyne class as it possesses a strained ten-membered ring containing a 1,5-diyne-3-ene system. Despite sharing this requisite characteristic, however, dynemicin A (**1**) can reasonably be considered the "black sheep" of this family in that it is the only one which does not possess any carbohydrate units, a feature that confers upon it two major differences from its relatives. First, while the other enediynes possess a fair degree of sequence selectivity in their interactions with DNA, dynemicin A is far less discriminating.[6] Second, while all of the other enediynes are positively charged at physiological pH since at least one of their sugars possesses an amine, the absence of this group and the presence of a carboxylic acid in **1** renders it negatively charged. Moreover, because **1** is the only enediyne that possesses an anthraquinone chromophore, it is also the only member of the family that is brilliantly colored: violet in the solid state and deep blue in solution.

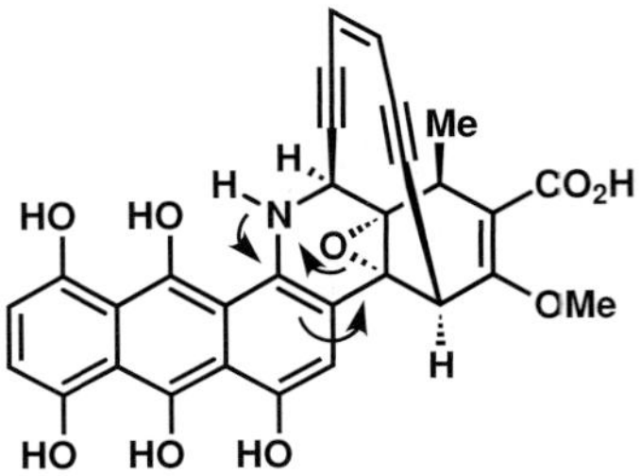

Apart from its aesthetic properties, the anthraquinone unit is also believed to be the critical motif responsible for the preliminary events that convert dynemicin A (**1**) from a pro-drug into a highly active agent. As first proposed by Professor Martin Semmelhack at Princeton University, this portion is thought to be the initial contact point between this natural product and DNA as its planarity should facilitate its smooth intercalation into the minor groove.[7,8] Once there, the anthraquinone ring is reduced, presumably by a biomolecule such as NADPH, leading to an activated substrate (**5**, Scheme 1) possessing phenols with electron pairs that are now free to assist in the intramolecular opening of its strategically positioned epoxide. If this event occurs, then a highly reactive quinone methide (**6**) results which can subsequently be intercepted by a nucleophile such as the free amine in a guanine residue from the adjacent DNA sequence to provide **7**. Alternatively, one could also invoke nitrogen participation within **5** to achieve this same outcome by initially forming **10** (see column figures), the tautomeric

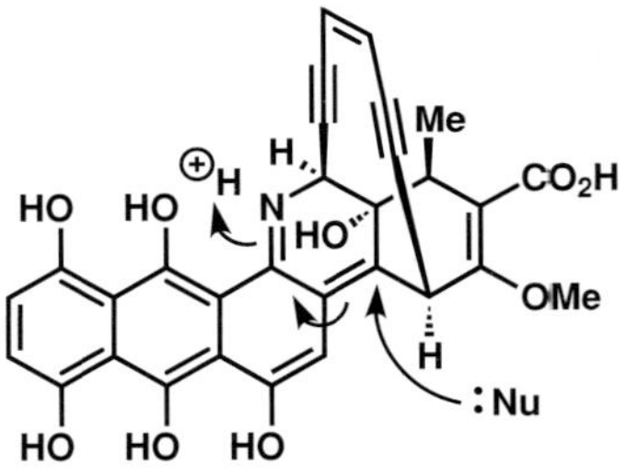

5

Epoxide opening

10

Alkylation

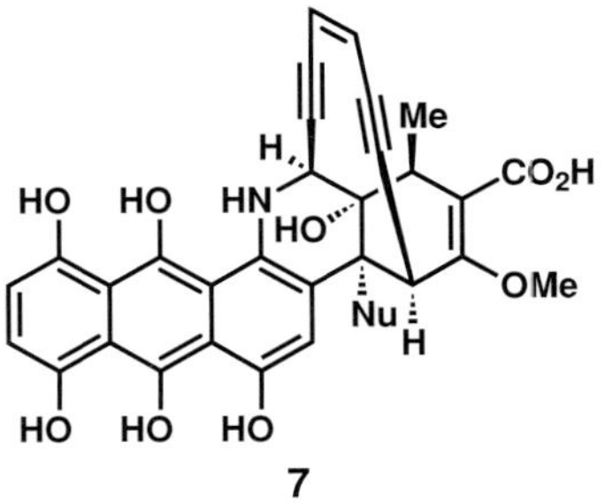

7

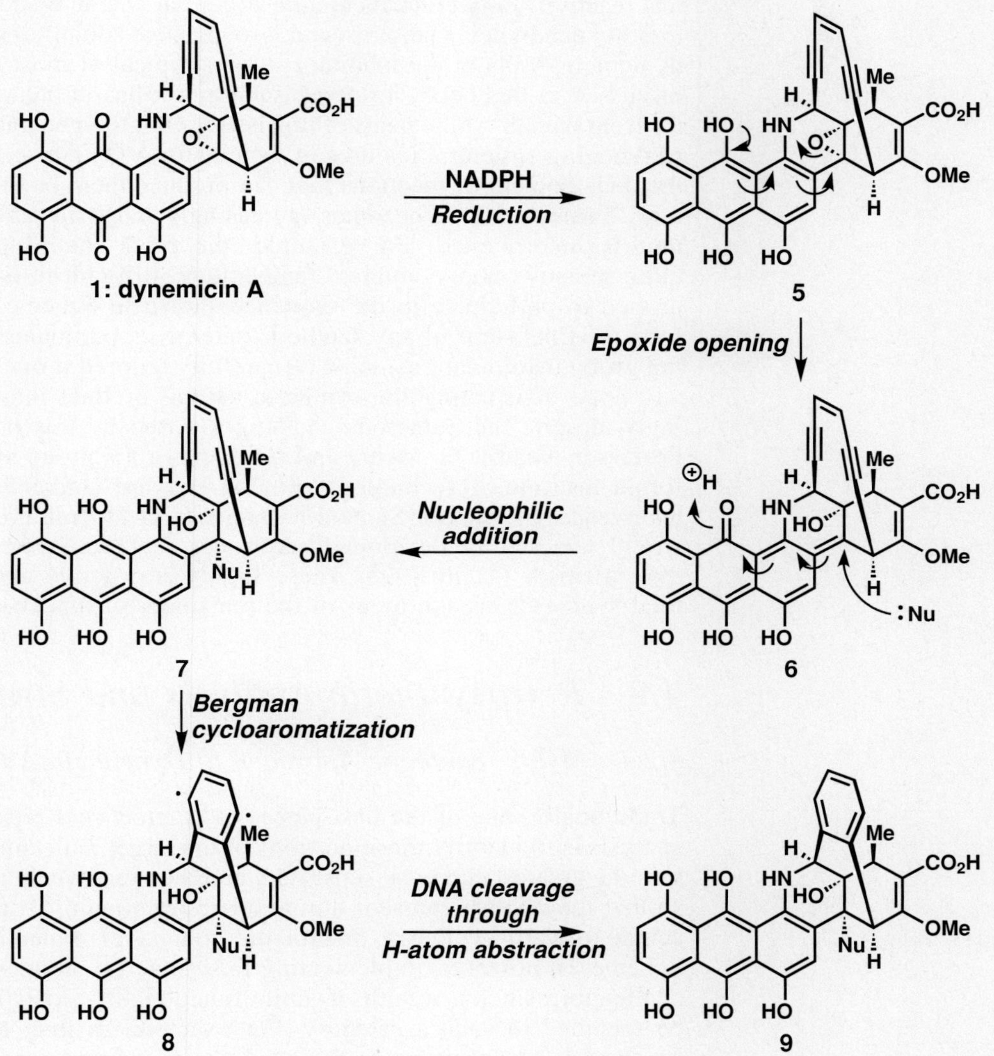

Scheme 1. Proposed mechanism for the biological activity of dynemicin A.

form of **6**, in which the aromaticity of only one ring has been broken. Irrespective of which of these pathways to **7** is accurate, the important feature is that once the nucleophile has added, the conformational disposition of the original enediyne motif in **1** has been greatly restructured such that the distance between the termini of the transannular alkyne functions is now significantly shorter. As a result, Bergman cyclization of **7** to 1,4-diradical intermediate **8** proceeds quite readily, leading to DNA damage through the abstraction of hydrogen atoms.

Due to the combination of this intriguing mechanism of biochemical reactivity, as well as its unique structural characteristics relative to other enediynes, many research groups have sought to synthesize dynemicin A (**1**) in the laboratory. As is typical of most molecules discussed in this book, however, such a task has ultimately proven far from trivial. At the heart of this issue is not the generation of the challenging structural features of dynemicin A (**1**) per se, but rather the development of reactions that can produce them on an architectural framework that becomes increasingly more unstable as each unit is incorporated. For example, the final natural product is quite sensitive above ambient temperature, difficult to isolate, and primed to participate in the sequence shown in Scheme 1. Therefore, the final steps of any synthetic enterprise, particularly protecting group manipulations, must be carefully tailored if one is to have any hope of isolating the synthetic natural product intact. Amazingly, despite the numerous challenges posed by this target, both Professor Andrew G. Myers and members of his group at the California Institute of Technology (now at Harvard University)[9] and a team lead by Professor Samuel J. Danishefsky at Columbia University[10] successfully developed sequences that delivered synthetic dynemicin A (**1**) in 1995. These highly instructive and creative total syntheses are the focus of the remainder of this chapter.

4.2 *Retrosynthetic Analysis and Strategy*

4.2.1 *Myers' Synthetic Approach to Dynemicin A (1)*

Traditionally, one of the best places to commence a retrosynthetic analysis is to identify those portions of the target molecule that possess the greatest chemical sensitivity and then remove them early on so that they will be present during a bare minimum of steps in the actual forward synthesis. Within the context of dynemicin A (**1**, Scheme 2), however, implementing this powerful strategy is rather challenging since a wealth of labile functionality exists that could be included in such a category. For example, as long as there is an epoxide in proximity to the enediyne motif, the propensity for nonproductive Bergman cyclization should be relatively high. Moreover, even if this activating ring is removed, the enediyne itself could still participate in the same event under the right circumstances. Equally problematic, the anthraquinone system on the other side of the molecule could likely prove to be just as capricious, since it is well documented that such motifs often instigate challenges due to their redox chemistry, even during the most "routine" of transformations.

Thus, in order to select between these various options and formulate a viable retrosynthetic analysis, the Myers group experimented with several model systems and carefully considered published studies toward related targets, such as calicheamicin γ_1^I (**4**), in order to get a qualitative feel for just how sensitive these different

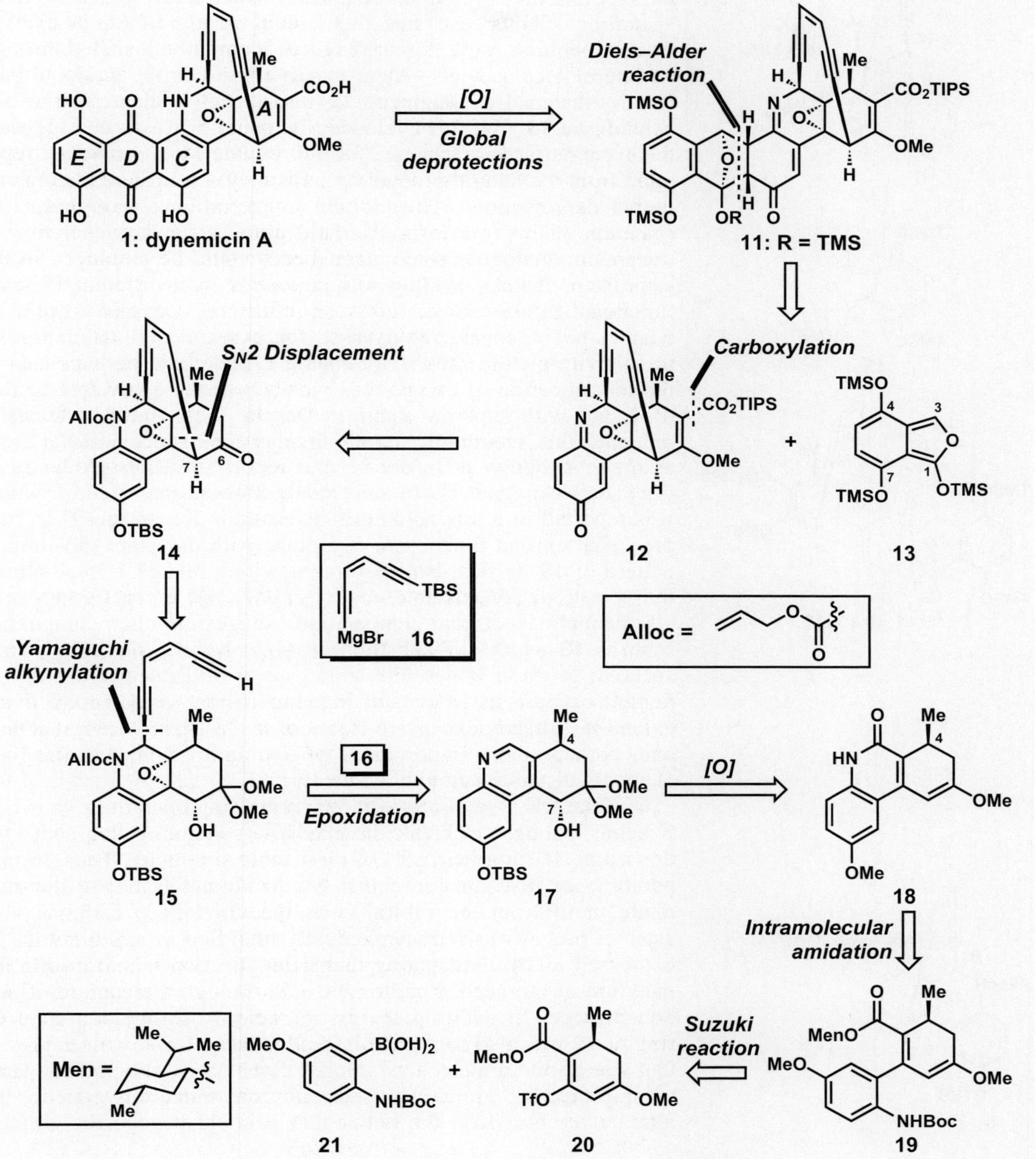

Scheme 2. Myers' retrosynthetic analysis of dynemicin A (**1**).

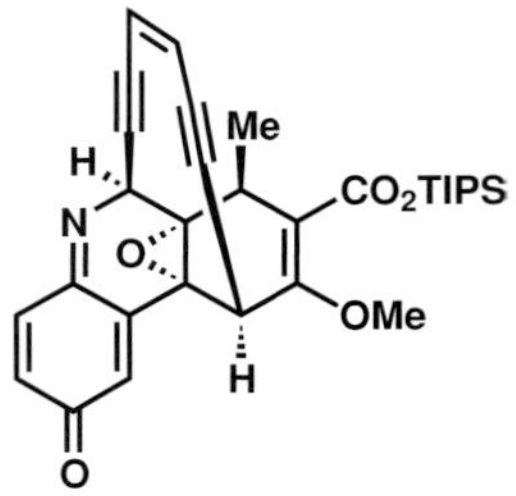

motifs might be. Information from both channels quickly pointed to the fact that the anthraquinone portion would likely cause the most "handling" challenges. Thus, this system was the first to be excised from dynemicin A (**1**), forecasting that it could be installed through an intermolecular Diels–Alder reaction between the imidoquinone form of the enediyne segment (**12**) and a reactive diene such as isobenzofuran **13**.[11a] If this cycloaddition could prove successful, then the event defined in Scheme 2 would provide **11**, a compound separated from the natural product by only a two-electron oxidation and global deprotections.* In addition to providing convergence, the selection of this transform also held promise for the generation of dynemicin analogues since other dienes could be employed in the same step, thereby leading to compounds with different DE-ring functionalization. Access to such structural variants would, of course, be of considerable value for experimental validation of the activity picture shown in Scheme 1, and could perhaps lead to the identification of compounds with a potency equivalent to that of **1**, but with superior stability. Despite these enticing features, reducing this seemingly simple strategy to practice was far from assured. Although there are several reports of Diels–Alder reactions between isobenzofurans and quinones, the first of which was reported in a total synthesis by Andrew Kende in 1977,[11b] no precedent existed for the use of a diene with the exact substitution pattern of **13**. In fact, isobenzofurans with a related 1,3,4,7-substitution pattern are unstable above −20 °C and unreactive towards all dienophiles at that temperature, suggesting that compounds such as **13** with 1,4,7-substitution might behave similarly.[12] Irrespective of these issues, the strategy was still deemed worthy of pursuit because its successful implementation would productively extend the already expansive frontiers of the Diels–Alder reaction, and, perhaps more importantly, would rapidly complete the final elements of the dynemicin architecture.[13]

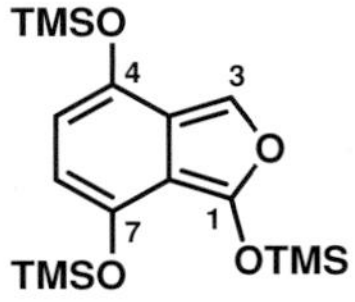

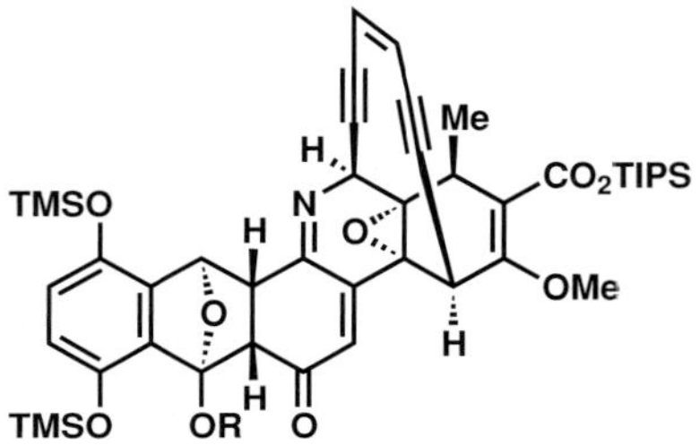

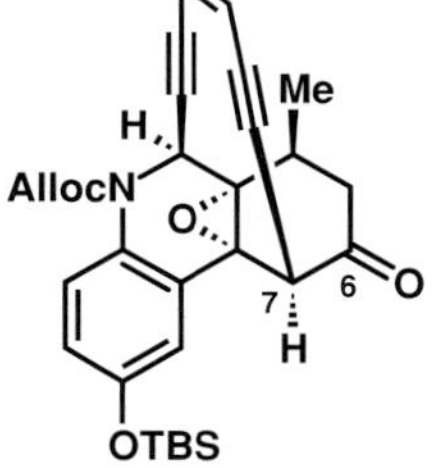

With the DE system removed, the remaining operations shown in Scheme 2 sought to break the enediyne-containing fragment (**12**) down into starting materials of reasonable simplicity. Thus, in preparation for disconnections that would ultimately excise the enediyne motif from the central core, the vinylogous carbonic acid ester in ring A was retrosynthetically modified to a C-6 ketone as expressed in **14**, anticipating that its enolization would enable the insertion of the needed carboxylic acid through the capture of carbon dioxide. In addition to this alteration, the imidoquinone C-ring of **12** was also converted into its reduced form, with the resultant free amine protected as an allyl carbamate (Alloc). This particular protecting group was selected from numerous conceivable alternatives based on the belief that it could engage the amine's

* One should note that although the Diels–Alder product (**11**) is drawn with both facial- and regioselectivity for the addition of the isobenzofuran diene, no such control was required as all stereochemical information in this segment of the molecule would be erased upon completion of the symmetrical anthraquinone within **1**.

lone pair of electrons by resonance, thus preventing the epoxide lysis cascade that would lead to immediate Bergman cyclization (i. e. **5**→**10**→**7**→**8**). Although this electron-withdrawing property is inherent to all carbamates, the group selected also had to possess the capability for facile removal since the same deliterious cascade could be induced during the deprotection step if the conditions were either too strongly acidic or basic. Thus, because allyl carbamates are readily and chemoselectively excised under neutral conditions with palladium, this group appeared to be the ideal choice to meet this criterion (at least on paper).

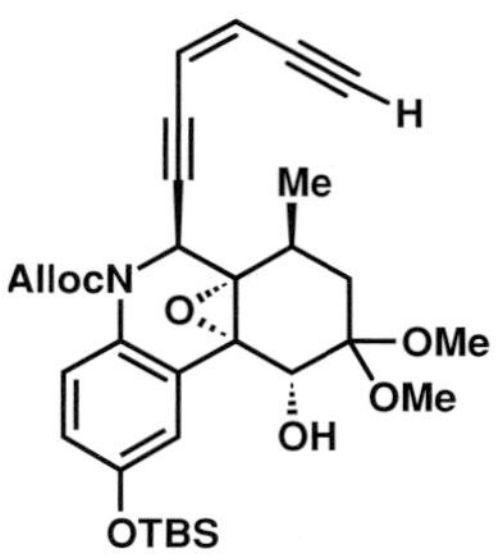

Having implemented these minor, but crucial, adjustments, the C-7 arm of the 10-membered ring was then detached from **14**, which led back to **15** as a new subgoal structure, forecasting that this C–C bond could be forged during the actual synthesis by the attack of a metallated acetylide onto a suitable group at this position. Potential choices in this regard could be a leaving group formed from an alcohol, such as a mesylate or tosylate, or, more indirectly, a ketone, whose employment would then require a subsequent deoxygenation step to remove the tertiary alcohol that would result from successful ring closure. Next, following removal of the endocyclic epoxide to a precursor olefin and subsequent retro-aromatization of the B-ring to form a quinoline system, the remaining bond connecting the enediyne fragment to the dynemicin core was then removed as alkynyl Grignard reagent **16**, ultimately revealing **17**. The adoption of this sequence was based on precedent established by Yamaguchi, who demonstrated that alkynyl Grignard reagents can add smoothly adjacent to the nitrogen atom in pyridine rings following their activation as *N*-acylpyridinium intermediates.[14] Accordingly, it was anticipated that the same 1,2-selectivity would arise here once **17** had been converted into an appropriate *N*-acylquinolium. While this is a reasonable assumption, the larger issue surrounding the viability of this strategy was the stereoselectivity of the addition because it required the approach of the Grignard nucleophile (**16**) *syn* to the adjacent methyl group at C-4. With little question, this proposition was a tenuous one based on the typical role played by such groups in directing the approach of external reagents. However, since there is a wealth of Lewis basic functionality within **17**, perhaps some form of chelation could modify the typical conformation of the A-ring to bias the addition of **16** in the desired sense by eroding the impact of the methyl group at C-4. Equally conceivable, if the same type of complex could be formed on the α-face of the substrate and was of suitable size, then it could serve as a blocking group capable of overriding any steric influence of the methyl group on the opposite side of the molecule. As such, there was a reasonable measure of hope that this critical process could be implemented during the forward synthesis, although conventional wisdom would seem to dictate otherwise.

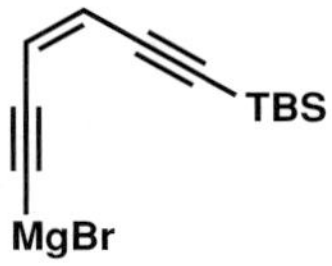

Having removed the enediyne portion of the molecule, only a few modifications remained to complete this retrosynthetic analysis.

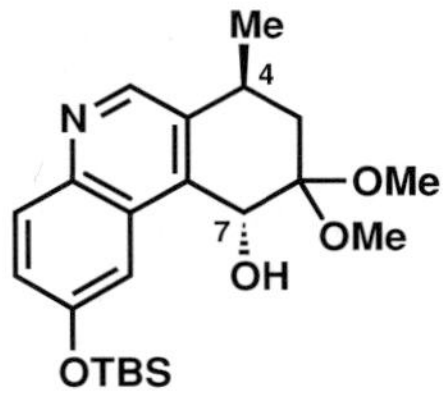

H
Me
AllocN
O
OMe
OMe
OH
OTBS
15

TBS
MgBr
16

Me
4
N
OMe
7
OMe
OH
OTBS
17

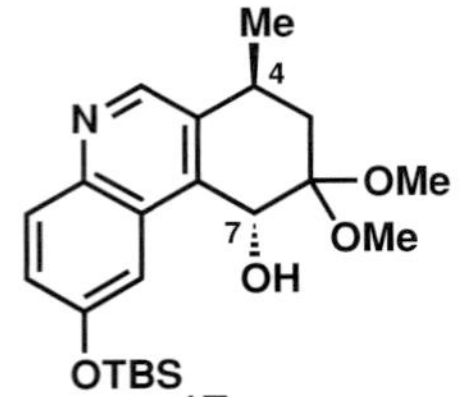

First, the alcohol group at C-7 within **17** was disconnected to an enol ether, now counting on the methyl group at C-4 within **18** to direct an epoxidation reaction that would establish this stereogenic center during the synthesis. It was envisioned that the quinoline ring of **17** could be derived from the oxidized form (**18**). Although this latter modification might seem slightly odd, as quinoline ring systems are relatively common entities and could certainly constitute viable starting materials, its implementation enabled the identification of a highly convergent approach for the construction of this projected target. As shown, if the newly installed amide linkage within **18** was fragmented, then the resultant bicyclic system could be broken in half to fragments of commensurate size through one of many metal-mediated coupling transforms such as a Suzuki reaction between boronic acid **21** and vinyl triflate **20**. As such, the task of synthesizing dynemicin A (**1**) has been reduced to the assembly of two relatively simple building blocks. Critical for the viability of the entire strategy, however, was finding a means to access stereochemically pure **20** since its stereogenic center was projected to play a leading role in orchestrating the installation of the other stereocenters of dynemicin A. Although we will leave a discussion of how this problem might be handled for the actual synthesis, we should note at this juncture that the chiral menthol moiety planted in **20** was expected to serve as an auxiliary to help accomplish this objective.

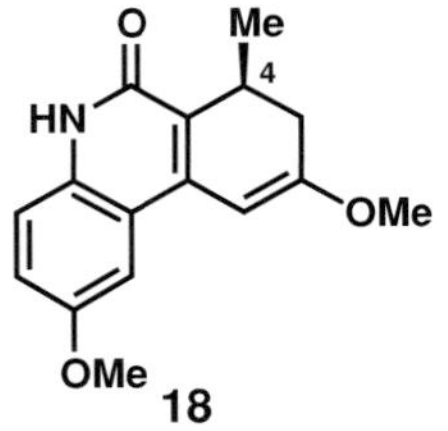

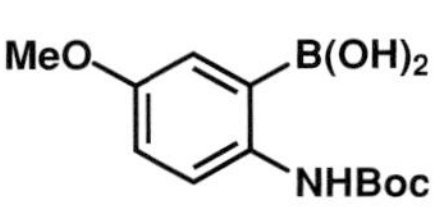

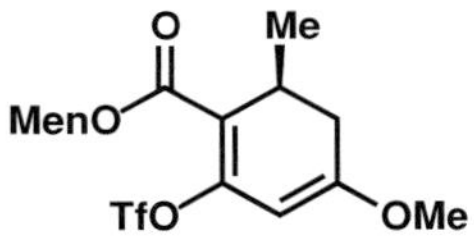

4.2.2 Danishefsky's Synthetic Approach to Dynemicin A (1)

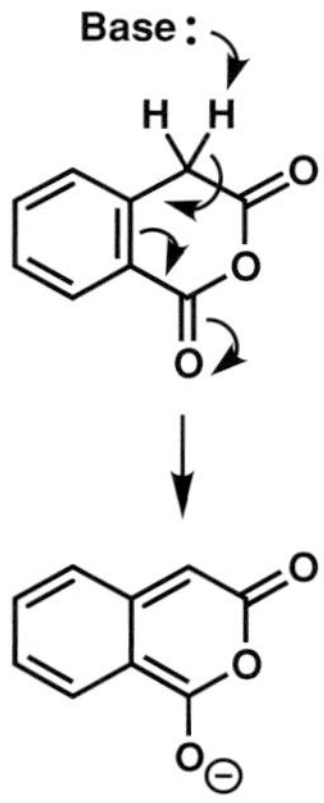

In assessing plausible synthetic sequences that could access dynemicin A (**1**, Scheme 3), the Danishefsky group came to similar conclusions as the Myers group regarding the stability of the most nefarious elements of this target, and disconnected the anthraquinone system first from a modified form of the target molecule (**22**) through an intermolecular Diels–Alder reaction.[10] While the same general type of dienophile was anticipated for this event as in the Myers analysis, namely imidoquinone **23**, the diene projected in this instance was **24**, a member of the class of Diels–Alder participants accessible by deprotonating the methylene unit of an appropriately formatted homophthalic anhydride (see column figure). Apart from their demonstrated ability to readily engage a wealth of dienophiles such as quinones, as first shown by Tamura,[15] these 4π-electron components are highly valuable due to their predisposition to lose a molecule of CO_2 following successful cycloaddition, ultimately affording a ketone product. Here, that benefit translated into a Diels–Alder adduct (**22**) separated from the complete anthraquinone system of **1** merely by deprotections and an oxidation reaction. As with the Diels–Alder reaction mentioned in the preceding section, the only uncertainty involved with this enticing approach was the lack of any precedent for the successful use of a diene with the exact substitution pattern

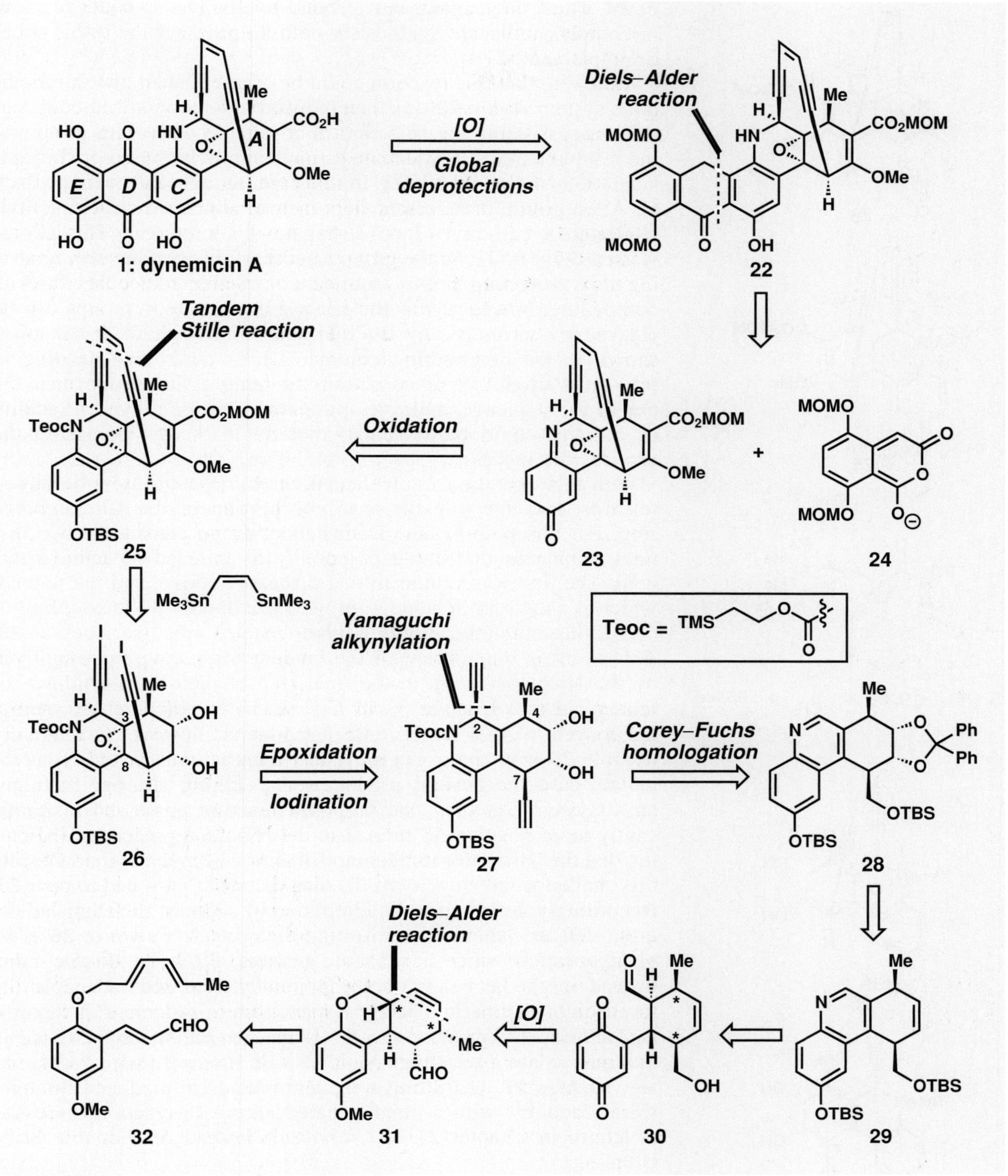

Scheme 3. Danishefsky's retrosynthetic analysis of dynemicin A (**1**).

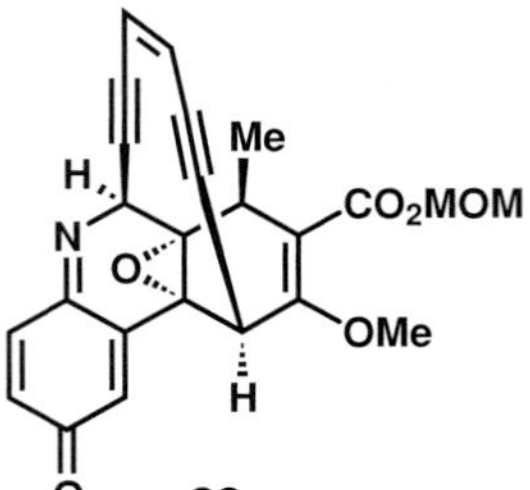

of **24**. Thus, once again, new ground for the Diels–Alder reaction in complex molecule synthesis would be paved if the event could be implemented.

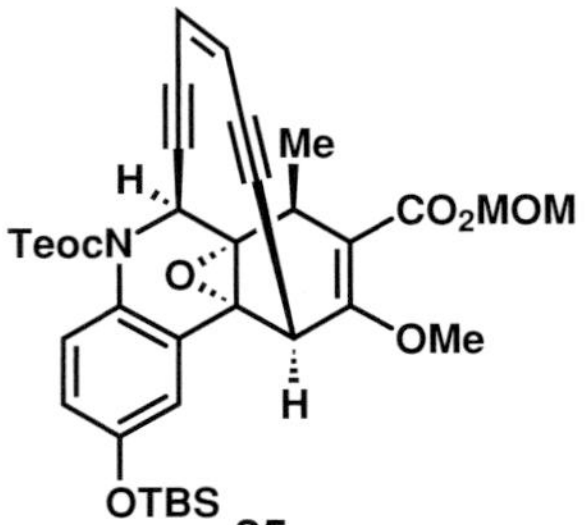

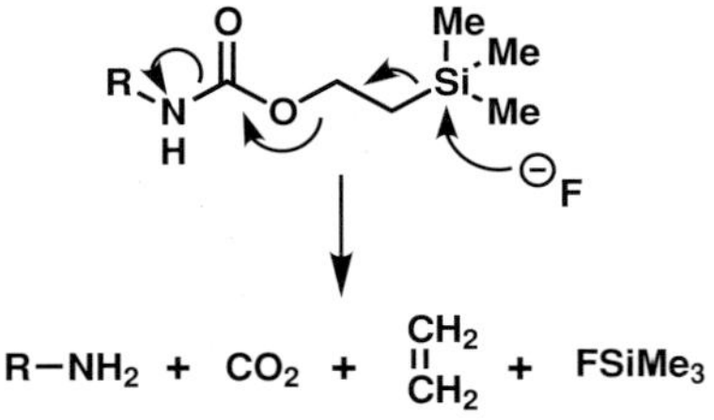

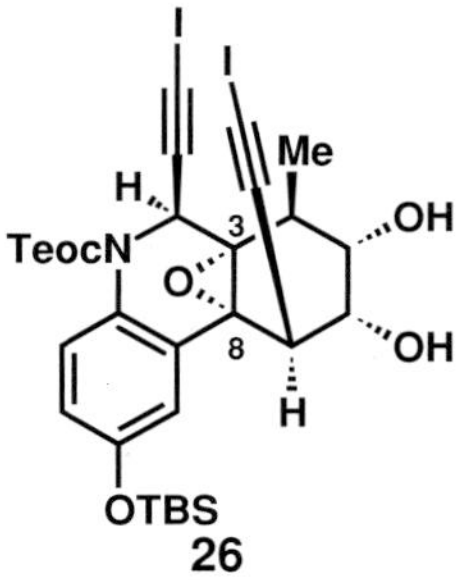

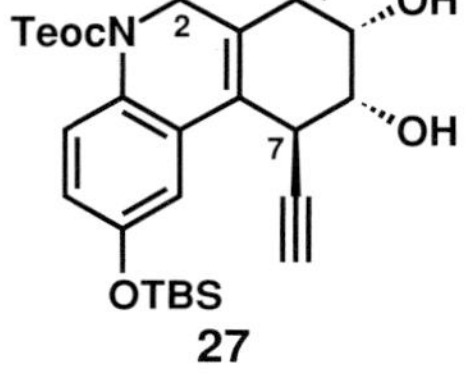

Assuming that this reaction could be accomplished, the imidoquinone system within **23** was then reduced to its aromatized congener (**25**), necessitating the selection of a suitable protecting group for the resultant amine to attenuate its nucleophilicity based on the considerations discussed earlier. In this case, though, rather than attach an Alloc group, these researchers instead appended a 2-(trimethylsilyl)ethoxy carbonyl (Teoc) group to this subtarget. This choice was partially based on the group's successful experiences in removing this protecting group from other unstable molecules.[16] More compelling, however, was the knowledge that Teoc groups can be cleaved by virtually any fluoride source through the mechanism shown in the neighboring column. Thus, rather than needing to rely on a single set of conditions to remove this group near the end of the sequence, multiple options with varying levels of acidity or basicity could be screened, thereby markedly increasing the chances for success.

With this operation implemented, attention could now be turned towards disconnecting the enediyne portion of the 10-membered ring. For this purpose, the Danishefsky group considered two distinct sequences that could potentially be enlisted to achieve this goal. The first was similar to that discussed above, and the second, which is shown in Scheme 3, projected a double Stille coupling of *cis*-1,2-distannylethene with a dihalogenated precursor such as **26**. This event is reminiscent of the tandem Stille coupling employed by the Nicolaou group in the final step of their total synthesis of rapamycin (see Chapter 31 in *Classics I*) to complete the central macrocycle of that target.[17] In that context, however, a relatively flexible 29-membered macrocyclic ring was constructed, whereas in this case the desired product was a highly strained medium-sized system. As such, the proposed reaction possessed a significantly more positive ΔS term than the Nicolaou precedent, indicating that the barrier for its accomplishment was much higher. Despite this challenge, the novelty of the idea dictated that it had to be tried. Accordingly, to enhance the likelihood of success, the Danishefsky group left the juncture for inserting the epoxide shown in **26** as an open question, since a substrate bearing a C3–C8 double bond instead might be better conformationally disposed to enable the insertion of the olefinic bridge. Apart from this element of flexibility, additional expressions of this same general strategy existed in that interpolative couplings could also be attempted with intermediates such as **27**, projecting a tandem palladium-mediated Sonogashira reaction[18] with a dihalogenated alkene (a reaction discussed at length in Chapter 31 of *Classics I*) instead of a double Stille coupling.

Unfortunately, synthesizing the test compounds for any of these reactions also constituted a formidable endeavor, since it meant that a route had to be identified that could provide substrates with three groups *syn* to each other (the methyl group at C-4 and both the alkyne functions at C-2 and C-7). This problem is much the same as that encountered during the Myers analysis of their planned insertion of the entire enediyne motif through a nucleophilic addition; in this case, the Danishefsky group projected a similar solution, but with a creative twist. Since both the C-4 and C-7 centers within either **26** or **27** are in the same ring, one could reasonably anticipate that attempts to excise either of these chiral elements individually would not prove particularly fruitful in facilitating the installation of the others. Thus, based on this conjecture, a more productive entry might be to find a way to insert the alkyne group stereoselectively at C-2 in the presence of the other two stereogenic centers. Of course, a Yamaguchi-type acetylide addition would be the ideal reaction to append such a group onto a suitable substrate. As we already know from the discussion above, the requisite approach trajectory of such a nucleophile may be impossible to accomplish, particularly with a precursor such as **28** in which there are *two* groups blocking that side of the molecule instead of just one! Yet, since it had already been forecasted that the vinylogous carboxylic acid in the A-ring of **1** could be fashioned from a diol precursor, if that diol was formed on the α-face as drawn in **27**, then perhaps this motif could be used to counterbalance the effect of the C-4 and C-7 centers in directing acetylide approach. For example, if these alcohols were engaged with a large protecting group, such as the diphenyl acetal shown in **28**, then the steric bulk of this entire motif could potentially favor approach from the β-face.

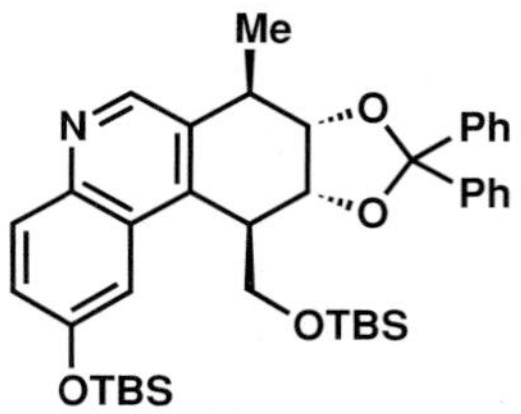

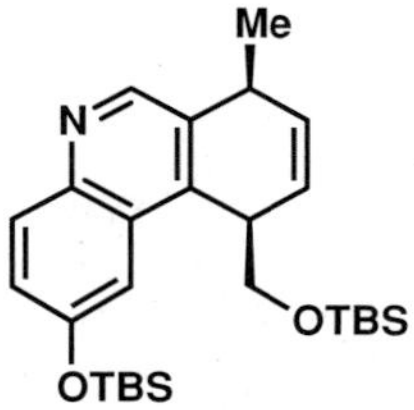

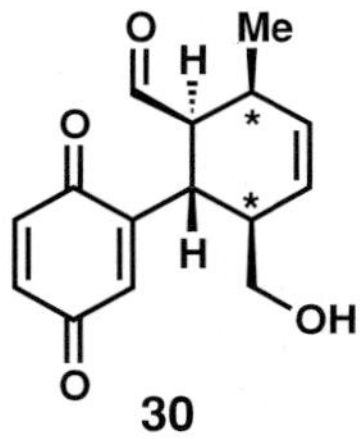

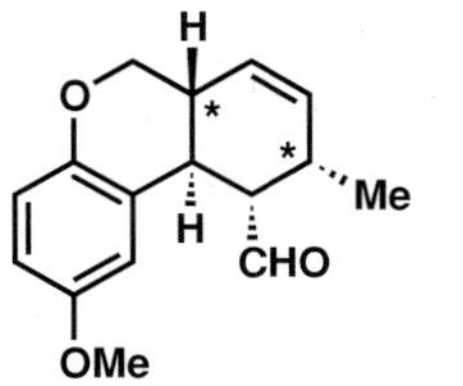

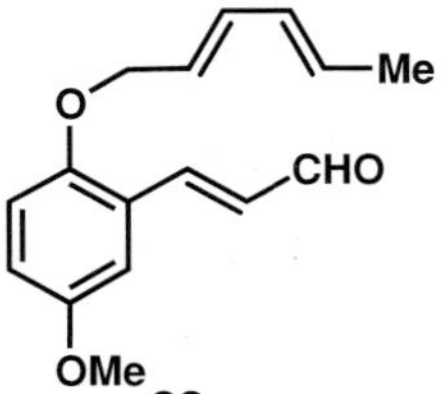

Assuming that this Yamaguchi alkynylation could indeed be brought about, the requisite diol-based directing group in **28** could arise from a facially selective dihydroxylation of a precursor alkene (**29**) governed by the resident stereogenic centers. With the identification of this subgoal structure, a means to generate the remaining stereochemical elements then became obvious since they all are part of a six-membered ring bearing an element of unsaturation, thus soliciting the possibility of a Diels–Alder reaction for their installation. However, while the recognition of this transform was of critical importance for the final elements of this strategy, several additional modifications of **29** were required before this powerful disconnection could be applied (as it is virtually impossible to prepare a diene system directly from a quinoline). Accordingly, the B-ring portion of this aromatic ring was disconnected to quinone aldehyde **30**, anticipating that amine capture of its aldehyde could then enable ring closure onto the adjacent ketone of the quinone during the synthesis.[19] While effecting this structural modification may seem far from obvious, its projection was critical in that it enabled the identification of **31** as a material that could constitute a viable precursor for this new goal structure (**30**), a target that was finally amenable to a Diels–Alder-based retrosynthetic disconnection. Thus, retrosynthetic cleavage of the six-membered ring within **31** at the indicated junctions led to **32** as a suitable substrate from which to start synthetic investigations towards dynemicin A (**1**). One should note, however, that while **32** should willingly participate in an *endo* selective union to establish the two critical centers in **31** marked with an asterisk, they could only be set in a relative sense since **32** is achiral.

4.3 Total Synthesis

4.3.1 Myers' Total Synthesis of Dynemicin A (1)

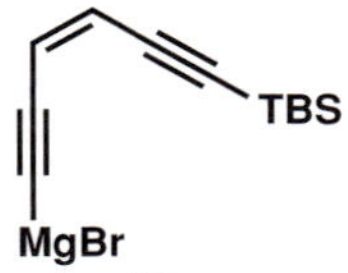

We begin our discussion of this total synthesis with those steps that were required to access the precursor for the enediyne Grignard reagent **16**, namely **36** (Scheme 4).[9] As alluded to in both Chapters 30 and 31 of *Classics I*, such building blocks are typically generated by using the Sonogashira variant[18] of the Stephens–Castro reaction,[20] or, as more commonly and simply named, the Sonogashira reaction, to forge the bonds between their central olefin and alkyne arms. Indeed, this pattern was replicated here with the sequential use of this transformation starting with the union of **33** and TMS-acetylene promoted by catalytic CuI and $Pd(PPh_3)_2Cl_2$, followed by reaction of the resultant product (**34**) with TBS-acetylene in the presence of catalytic CuI and $Pd(PPh_3)_4$. With these operations affording **35** in 61 % overall yield, final treatment with K_2CO_3 in MeOH led to the desired substrate (**36**) by selective excision of the TMS protecting group. As such, with only one of the two alkyne termini unblocked, the conversion of **36** into the eventual Grignard nucleophile (**16**) was now anticipated to be relatively simple. Before we can discuss that transformation in more detail, however, we must first present the synthesis of its projected coupling partner, quinoline **17**.

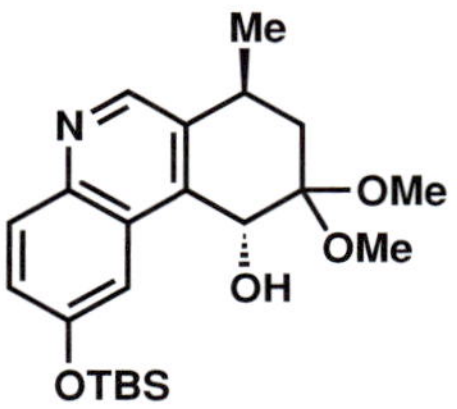

As mentioned during the planning stages, it was anticipated that most of the complexity of this fragment could arise from the initial union of arylboronic acid **21** and vinyl triflate **20** (see Scheme 2). Since the first of these preliminary targets was one step away from a known compound,[9] the Myers group only needed to concern themselves with developing a sequence to synthesize the second (**20**). Based on the pattern of functionality within this building block, particularly its lone stereocenter, they anticipated that an effective starting point would be to synthesize a cyclohexane-1,3-

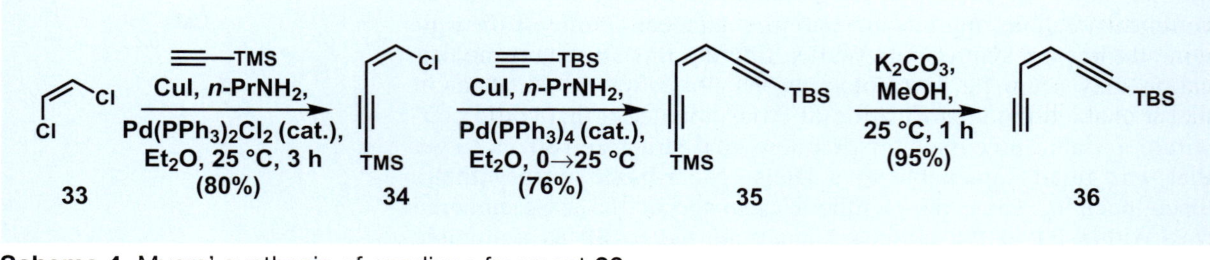

Scheme 4. Myers' synthesis of enediyne fragment **36**.

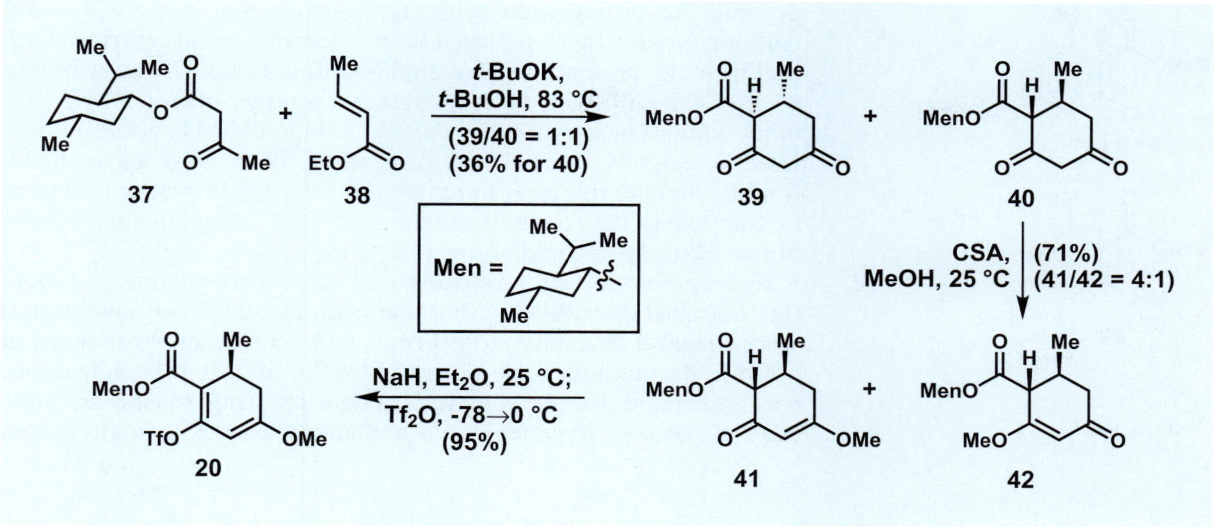

Scheme 5. Myers' synthesis of intermediate **20**.

dione system such as **40** (Scheme 5) since it could likely be elaborated to **20** through sequential, functionalizing enolizations. Equally enticing was the prospect that this initial target (**40**) could be formed in a single step through a one-pot double condensation by merging a β-ketoester monoanion (derived from a material such as **37**) in a Michael reaction with an α,β-unsaturated ester (such as **38**) followed by a Dieckmann cyclization, a transformation first described in 1899 by von Schilling and Vorlünder.[21] While simply stated, the challenge resided in finding a method to make this powerful transformation proceed with some form of stereoselectivity, an objective that had not been previously demonstrated. Accordingly, the Myers group reasoned that if they appended a chiral auxiliary such as a menthyl group to the somewhat remote carboxylic acid position of the requisite β-ketoester starting material (**37**), then perhaps its reaction with **38** would proceed in a biased manner. Unfortunately, initial experiments using *t*-BuOK in refluxing *t*-BuOH revealed that this conjecture was not to be the case, as both **40** and its diastereomeric partner **39** were generated in equimolar ratios. The desired product (**40**) could, however, be selectively crystallized from this mixture in 36% yield. Thus, the menthyl auxiliary served merely as a resolving agent in this process, facilitating the isolation of the desired product.* Although not quite as selective as originally anticipated, the ease of this reaction (which could provide hundreds of grams of **40**) was more than sufficient to enable the synthesis to

* One should note that several other chiral auxiliaries were appended to the β-ketoester starting material and some did lead to modest diastereoselectivity (~1.5:1). However, none of these products crystallized as readily, leaving the initially developed chemistry with **37** as the most practical route to **40**.

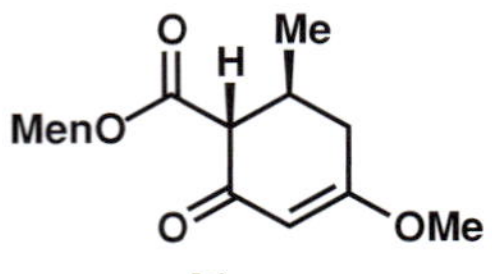

proceed. Accordingly, the sterically more accessible of its two enol tautomers was then preferentially captured upon stirring with MeOH in the presence of 10-camphorsulfonic acid (CSA), affording **41** in a 4:1 ratio with its regioisomeric product (**42**) in 71 % combined yield. Once again, the desired compound (**41**) could be isolated selectively through crystallization. With pure **41** in hand, final treatment with NaH to remove its only labile proton, followed by trapping of the resultant enolate with Tf_2O, completed the assembly of **20** in 95 % yield.

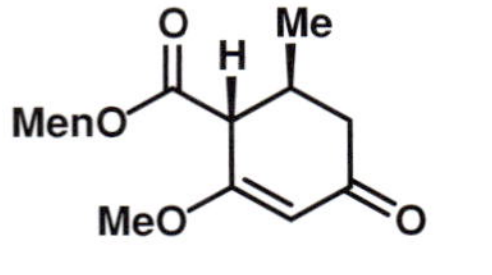

Seeking now to test the Suzuki reaction which was meant to join this intermediate (**20**) with boronic acid **21**, these two compounds were stirred with catalytic $Pd(PPh_3)_4$ and Na_2CO_3 in 1,4-dioxane at 100 °C. As predicted, the desired bicyclic adduct (**19**, Scheme 6) was generated in 90 % yield.[22] The final ring of the tricyclic ABC domain of dynemicin A was then formed in a single opera-

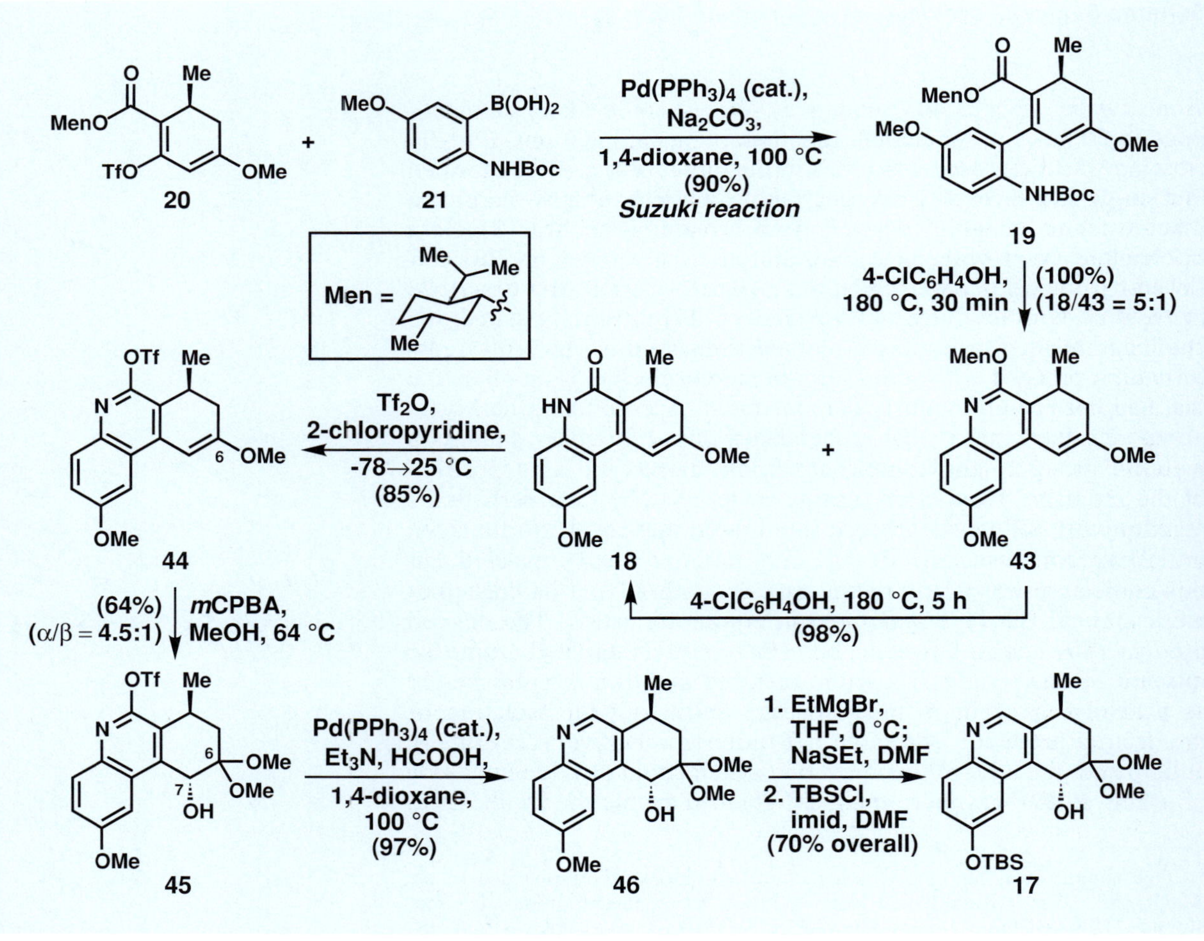

Scheme 6. Myers' synthesis of intermediate **17**.

tion by heating this compound in 4-chlorophenol at 180 °C, conditions that accomplished both deprotection of the *t*-butylcarbamate (Boc) group as well as intramolecular amidation to form **18**.[23] The facility of this reaction is particularly remarkable, considering that strongly acidic conditions (such as TFA in CH_2Cl_2) are typically required to cleave Boc groups. In this case, though, the 4-chlorophenol solvent accomplished the same task since its phenol proton behaved as a Brönsted acid, leading to an unusually mild set of reaction conditions that ensured that the acid-labile enol ether functional group within the product survived. Despite its mildness, the reaction also gave rise to a minor by-product in the form of menthyl quinolyl ether **43**, a compound presumably formed through a condensation reaction between the deprotected amine and the carbonyl group. Fortunately, this material did not have to be discarded, as 5 hours of additional heating in refluxing 4-chlorophenol funneled it into the desired product (**18**) almost quantitatively.

With the B-ring installed, the format of **18** now needed to be adjusted to that of **17** in order to prepare the ground for the eventual merger of the enediyne bridge. Thus, the initial operations focused on generating the required quinoline ring system, starting with the conversion of **18** into **44** as effected in 85 % yield with the addition of 2 equivalents of Tf_2O and 4 equivalents of 2-chloropyridine in CH_2Cl_2 at −78 °C followed by slow warming to ambient temperature. Although this transformation is relatively routine, one should take note that it was accomplished without using 2,6-di-*t*-butylpyridine, the relatively expensive base typically employed to accomplish such reactions. Thus, the conclusion that can be drawn from this insightful protocol is that 2-chloropyridine possesses a similar activity profile to this bulkier base (weak basicity and inactivity towards triflic anhydride) and might constitute a cheaper alternative for it in a host of other common transformations.

Pressing forward, **44** was then subjected to the action of *m*CPBA in MeOH at elevated temperature, hoping that its lone stereocenter could guide the approach of this epoxidation reagent to the desired face and that the reaction conditions would lead to concurrent epoxide opening to afford a dimethoxy ketal at C-6. As expected, this conjecture proved to be correct, with **45** formed as the major product, along with only a minor amount of the alternative, β-disposed C-7 alcohol isomer. Significantly, one should appreciate that the epoxide opening only proceeded as it did due to the ability of the methoxy group at C-6 in the starting material to promote facile epoxide opening. In contrast, had MeOH alone effected epoxide lysis through external S_N2 attack, it undoubtedly would have attacked the C-7 position exclusively because this site is sterically less encumbered and, perhaps more importantly, more activated since it is adjacent to an aromatic system. However, since the substrate had been carefully designed, the reaction proceeded smoothly in the desired manner, enabling the synthesis of **17** to then be completed by the initial removal of the triflate from the aromatic system of **45** using catalytic $Pd(PPh_3)_4$ in 1,4-dioxane at 100 °C with formic acid as a hydride source,[24] followed by exchange of the phenolic methyl ether for a TBS ether over two steps. This final protecting group exchange was deemed to be critical for the viability of the terminating steps of the projected synthesis since the harsh conditions typically required for methyl ether deprotection were expected to be too severe for the survival of an advanced dynemicin substrate (vide infra). In this case, the transformation was readily accomplished on this, far simpler, intermediate.

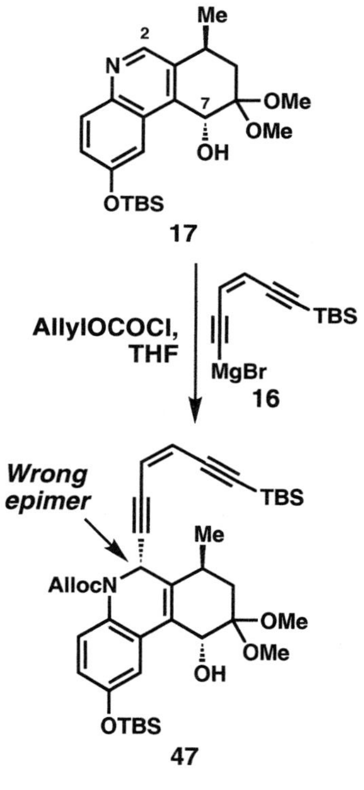

Having reached this staging area, forays began in the hopes of installing the enediyne bridge onto the existing scaffold, starting with the stereo- and regioselective merger of **16** (formed by heating **36** with EtMgBr at 50 °C in THF) with **17** through a Yamaguchi-type[14] acetylide addition. Unfortunately, initial probes quickly revealed that although the reactant did add at the desired position (C-2), its approach occurred almost exclusively from the undesired face as shown in the column figures to afford **47**. As mentioned during the planning stages, one potential tactic to overcome this seemingly inherent, and certainly expected, bias was to chelate the numerous Lewis basic sites within **17**, such as the alcohol at C-7 and its adjacent methyl ethers. Gratifyingly, this strategy proved capable of directing **16** to the desired β-face. In the event, **17** (see Scheme 7) was initially stirred with EtMgBr in THF at 0 °C, presumably leading to the conformation of the A-ring shown in **48** in which the magnesium counterion of the intermediate alkoxide engaged the oxygen atom of either one or both methoxy groups. Following cooling of the reaction mixture to −78 °C, both allyl chloroformate and **16** were then added, leading initially to *N*-acylquinolium species **48**. With the substrate appropriately activated and the methyl group in a pseudoequatorial position, **16** then approached the least hindered face (*cis* to the methyl group at C-4) along the typical Burgi–Dunitz trajectory, leading to the formation of **49** in 89 % yield with better than 25:1 β/α selectivity. As evidence for the role of chelation in this event, when the alcohol group at C-7 in the starting material was protected as a silyl ether, application of the same reaction conditions led to the predominant approach of **16** from the undesired α-face.

With this critical reaction beautifully orchestrated, all efforts could now be directed towards closing the final juncture of the 10-membered enediyne ring, a requirement whose implementation was projected to involve the addition of an acetylide ion to a suitable group at C-7. However, since this operation would ultimately lead to the excision of the alcohol function at C-7, it was decided to use this motif productively for one additional operation before attempting its displacement, namely directing the formation of the central C3–C8 endocyclic epoxide. Thus, **49** was treated with *m*CPBA in buffered CH_2Cl_2 at 0 °C, giving rise to **50** in 88 % yield with complete facial selectivity for the drawn product. Next, treatment of **50** with TBAF in THF at 0 °C, followed by silyl reprotection of the phenolic position under standard conditions (TBSCl,

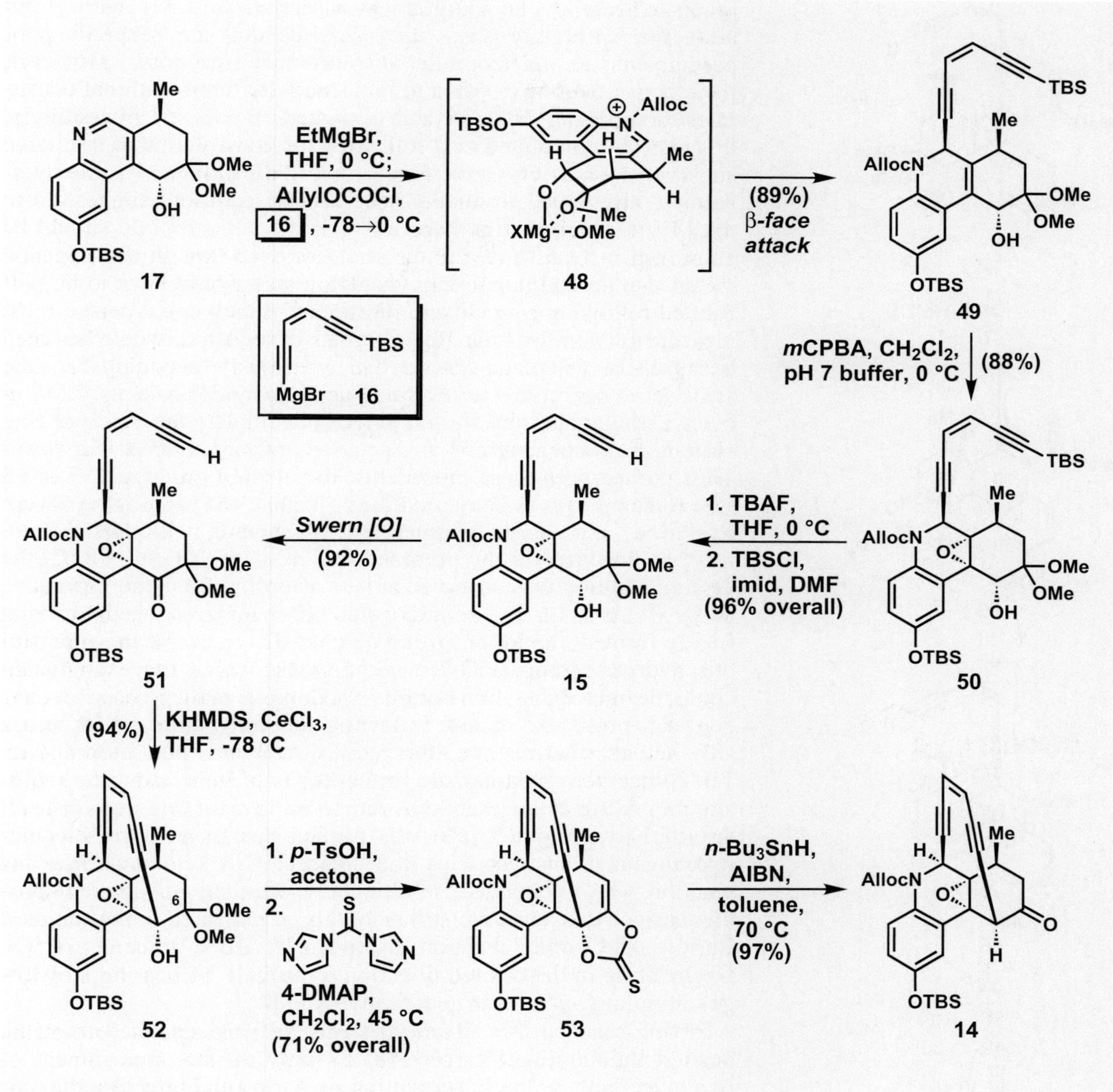

Scheme 7. Myers' elaboration of **17** into **14**.

imid, DMF) provided **15** in which the free acetylene could now be converted into a nucleophile.

The main question at this advanced stage was what group at C-7 within **15** would enable the final ring closure to occur. Initial operations logically sought to generate a leaving group at this position such as the mesylate in **54** (see column figure on next page). However, upon exposure of this compound to base in order to form the requisite acetylide, no addition was observed. This outcome was not altogether surprising given that acetylide ions are generally poor participants in nucleophilic displacement reactions. However, there is one major exception to this reactivity trend, and that constitutes the opening of Lewis acid activated epoxides.[25] Accordingly, this idea was attempted next following the construction of a suitable substrate such as **56** (see column figure). Unfortunately, this intermediate also failed to undergo the desired addition, suggesting to the Myers group that perhaps a C-7 ketone electrophile should be employed instead to rescue the strategy, even though this decision meant that an additional deoxygenation step would have to be performed following ring closure. Fortunately, they could derive some measure of comfort that this alternative sequence would succeed because the Nicolaou group had previously accomplished the same set of operations in several dynemicin model systems.[26] Moreover, a similar addition was employed to complete the enediyne ring system of calicheamicin γ_1^{I} as discussed in Chapter 30 of *Classics I*. Thus, armed with these precedents, the alcohol group at C-7 in **15** (see Scheme 7) was converted into a ketone (**51**) through a Swern oxidation, and, upon treatment of this adduct with KHDMS to form an acetylide in the presence of $CeCl_3$ in THF at $-78\,°C$, the desired addition was observed at last, affording **52** in an impressive 94 % yield! With the long-coveted 10-membered enediyne ring finally formed, the Myers group now needed to excise the superfluous hydroxy group at C-7, a requirement which they anticipated could be met using the Barton–McCombie radical-based deoxygenation protocol.[27] Thus, following the conversion of **52** into a C-6 ketone intermediate, this new compound was treated with 1,1′-thiocarbonyldiimidazole in the hopes of generating the requisite thiohydroxamate ester substrate to implement this step; instead, an alternative product (**53**) was formed due to an intramolecular rearrangement instigated by the adjacent C-6 ketone. While this reaction was unintended, it ultimately afforded no inherent problems since this substrate still smoothly participated in the standard Barton–McCombie reaction (n-Bu_3SnH, AIBN, toluene, 70 °C), fragmenting in the desired direction to provide **14** bearing a hydrogen atom at C-7 in near quantitative yield.

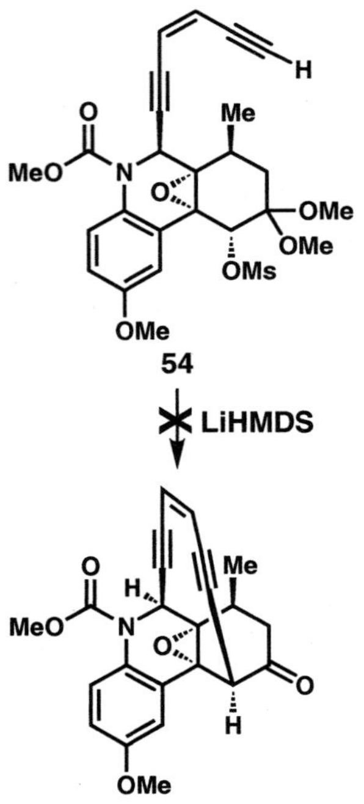

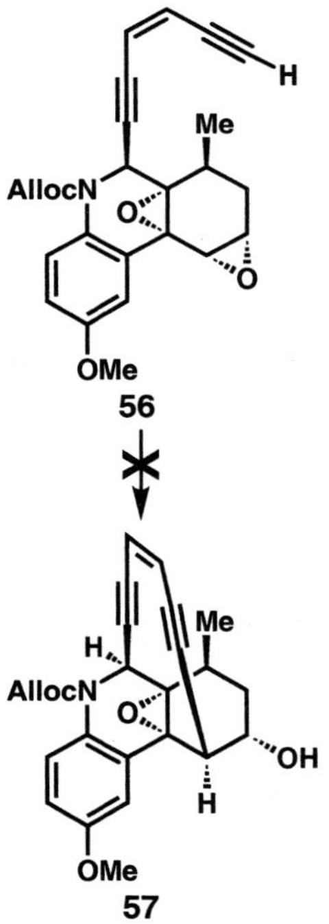

Having reached this advanced stage, **14** was separated from the desired imidoquinone target (**12**) by only the accomplishment of two objectives: properly formatting its A-ring and effecting the oxidation of its C-ring to an imidoquinone without inducing the cascade reaction that would lead to Bergman cycloaromatization of the enediyne bridge. The former of these requirements was attempted first, and, in fact, proved to be one of the most challenging transformations to effect. The difficulty resided in installing the carboxylic acid group at C-5, a requirement that projected the selective formation of an enolate from the C-6 ketone followed by capture of the resultant anion with carbon dioxide. At first, a host of bases failed in this purpose, leading only to extensive decomposition. Fortunately, examination of enough conditions eventually led to a glimmer of hope as a protocol developed by Rathke and co-workers ($MgBr_2$ and Et_3N in CH_3CN at ambient temperature under an atmosphere of CO_2) gave rise to traces of enol acid **58**.[28] Subsequent optimization of this reaction was accomplished by converting the relatively unstable **58** directly into its more resilient enol ether congener (**59**) through sequential reaction with *t*-BuOK and methyl triflate, ultimately enabling a reproducible yield of 54 % for this key sequence.

With the A-ring appropriately decorated through a set of reactions that could advance sufficient amounts of material, it was now time to effect the final conversions needed to reach imidoquinone **12**. Thus, in order to set the stage for the oxidation event that would eventually accomplish this objective, the phenolic TBS ether within **59** was deprotected upon exposure to $3HF{\cdot}Et_3N$ in CH_3CN at ambient temperature and the free carboxylic acid was protected as a TIPS ester to afford **60** in 63 % overall yield. Having left the ABC tricycle fully protected except for its phenol, subsequent treatment with iodosylbenzene in MeOH[29] then effected its smooth oxidation to **61** in 89 % yield, leading to a quinone ring engaged as a hemiaminal since the Alloc group had not yet been deprotected. Thus, this result might lead one to wonder why the Alloc group had not been removed first from **60** before attempting this oxida-

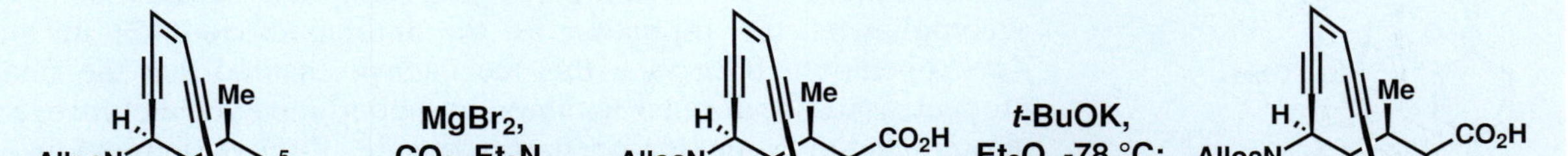

Scheme 8. Myers' synthesis of advanced intermediate **12**.

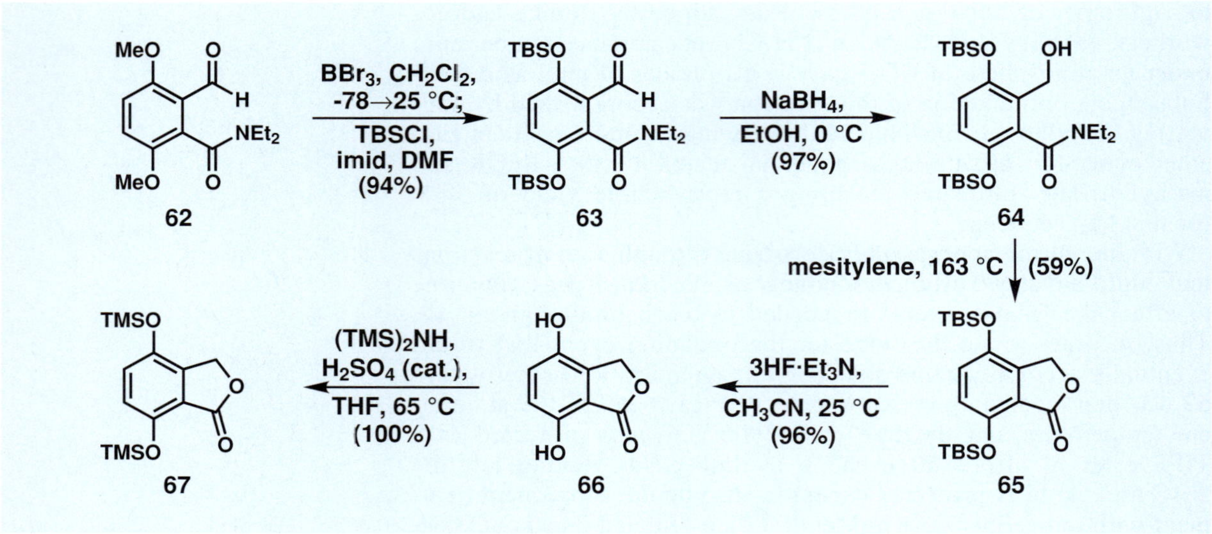

Scheme 9. Myers' synthesis of key intermediate **67**.

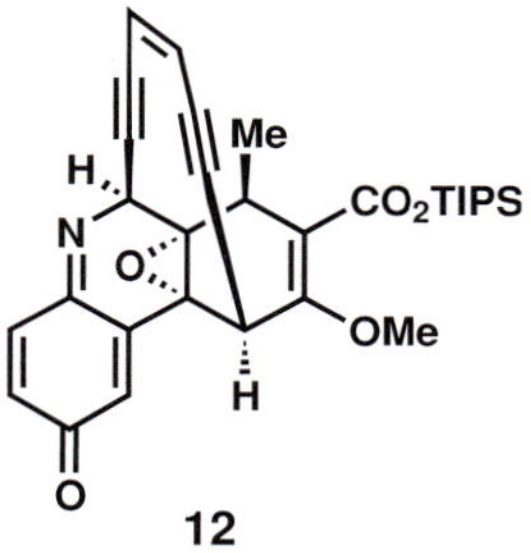

tion, as the final imidoquinone target (**12**) should have been formed directly. The answer is that by having installed an extraneous, but strategically positioned, methoxy group onto **61**, subsequent Alloc deprotection was expected to force the resultant free amine to commit its lone pair of electrons in displacing this group, thereby accomplishing the formation of the imidoquinone (**12**) in the same operation. If correct, this mechanism ensured that the final product would be formed without any opportunity for the nitrogen to initiate the epoxide-opening cascade discussed previously (**5**→**10**→**7**). Indeed, this insightful design proceeded exactly as envisioned, as **12** was generated in 78 % yield from **61** following standard Alloc cleavage with *n*-Bu_3SnH and catalytic $Pd(PPh_3)_2Cl_2$.

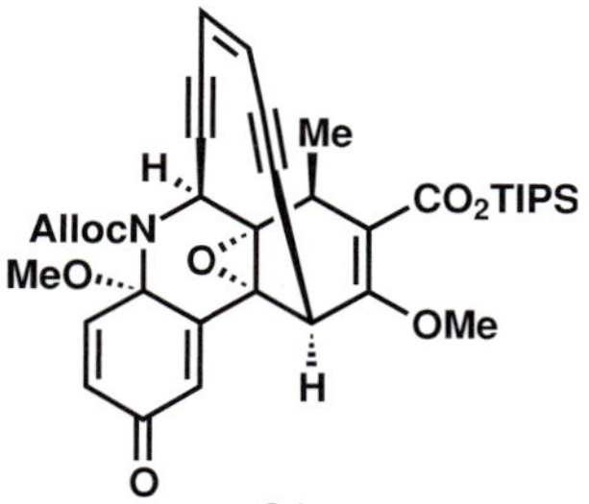

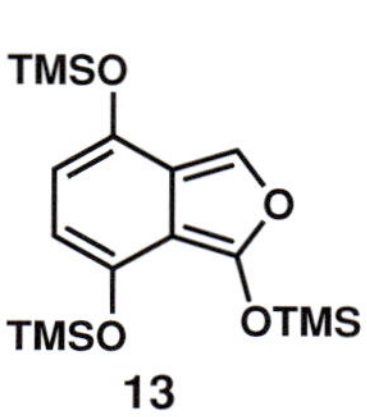

At this point, all that remained to complete the targeted natural product (**1**) was the attachment of the anthraquinone system, which, as discussed above, projected a Diels–Alder reaction with an isobenzofuran (i. e. **13**). Thus, a synthesis of its precursor, phthalide **67**, was developed as shown in Scheme 9.[30] Since most of these transformations are relatively routine, we will not describe them in any detail here, except to note that the ability to incorporate several different protecting groups for the phenol functions at different stages of the synthesis (cf. **62**, **63**, **67**) was a critical element for the completion of dynemicin A (**1**). As we shall see, only **67** ultimately proved capable of both participating in the desired Diels–Alder reaction and being elaborated to the complete DE anthraquinone system.

Thus, pressing forward with the final stages of the total synthesis, **67** (Scheme 10) was smoothly converted into the required isobenzofuran diene system (**13**) upon sequential treatment with KHMDS in THF at −78 °C and $TMSCl{\cdot}Et_3N$. Since this diene was not espe-

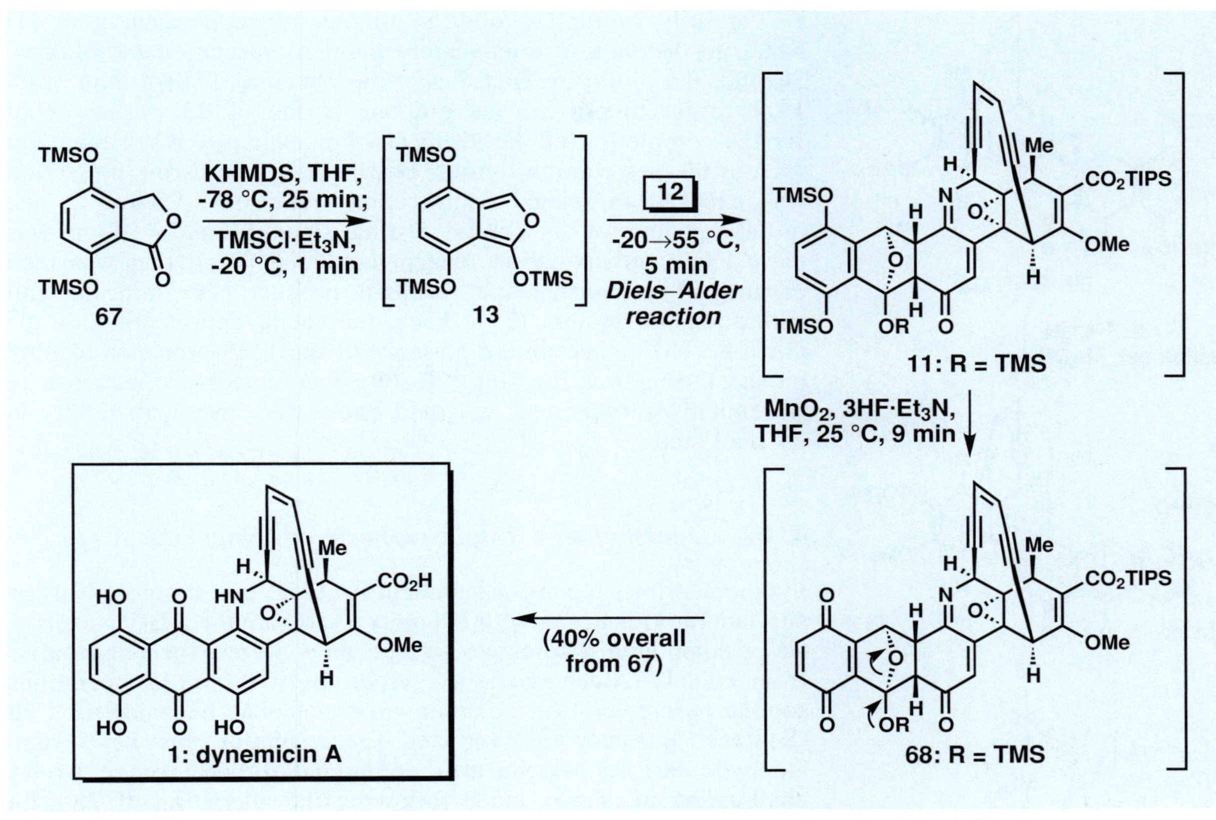

Scheme 10. Final stages and completion of Myers' total synthesis of dynemicin A (**1**).

cially stable, its Diels–Alder partner, imidoquinone **12**, was added directly to the reaction solution at −20 °C; subsequent warming to 55 °C over the course of 5 minutes enabled the desired [4+2] reaction leading to **11** to proceed. Because this product (**11**) also proved difficult to isolate, it was treated in the same reaction vessel with an oxidant and a silyl cleaving agent to induce the remaining operations needed to complete **1**. Numerous protocols had to be explored to find conditions capable of efficiently accomplishing all these objectives, but eventually it was discovered that MnO_2 and $3HF{\cdot}Et_3N$ in THF were the ideal combination,[31] leading first to the excision of the phenolic TMS groups in **11** to generate an E-ring quinone (i. e. **68**), followed by cleavage of the remaining TMS group, rearrangement, and oxidation to afford the first synthetic sample of dynemicin A (**1**) in 40 % overall yield from **67**.

Overall, the stereoselective route to dynemicin A developed by the Myers group was highly creative, and, impressively, very concise as only 24 steps were involved in its longest linear sequence. Before concluding this section, however, two features of the final one-pot sequence shown in Scheme 10 are worthy of further commentary. First, the final reaction time converting **11** into **1** had to

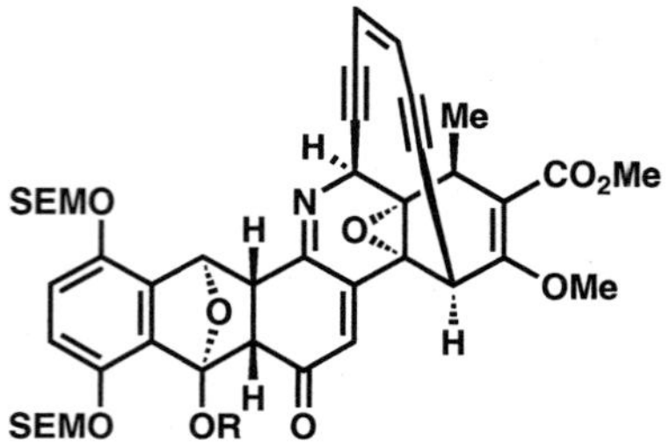

SEMO, H, N, O, Me, CO_2Me, OMe, H, H, O, SEMO, OR, O

69: R = TMS

3HF·Et_3N, silica gel, MeCN

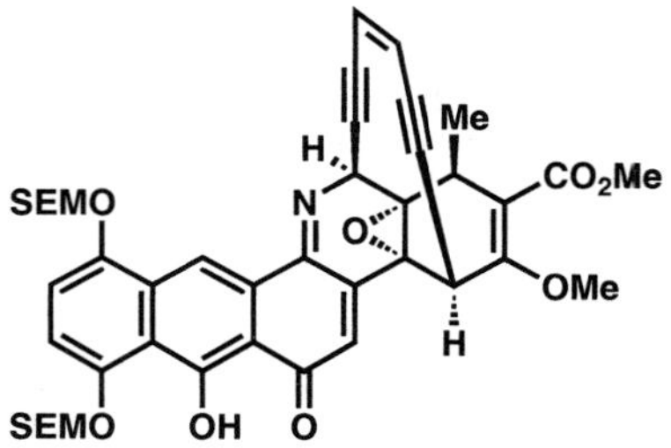

70

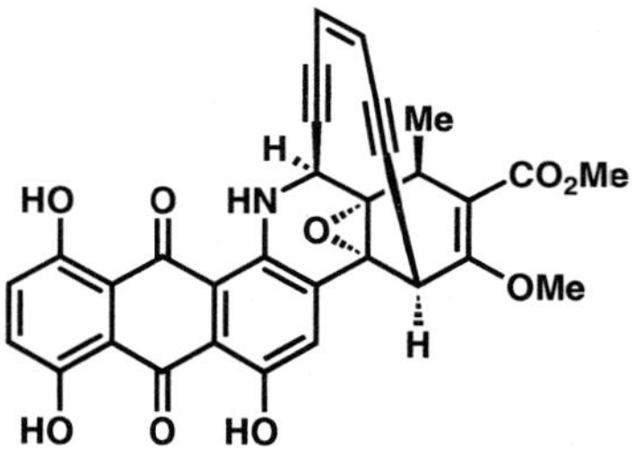

71

be carefully limited to only 9 minutes since dynemicin A (**1**) began to decompose immediately upon its formation in solution. Second, the ability to first cleave the phenolic TMS groups from **11** in order to provide the quinone E-ring of **68** was essential for the completion of the synthesis. For example, if a compound such as **69** (see column figures) bearing alternate E-ring protection was exposed to related deprotection conditions, leading to the initial opening of its central D-ring since its TMS group was more labile, no oxidation reagent could be found that was then capable of converting the resultant product (**70**) into the full anthraquinone system (**71**). Thus, the facile deprotection of the phenolic TMS ethers in the presence of the TMS-protected alcohol on the D-ring was the single feature that saved this sequence to dynemicin A; otherwise, it would likely have met with failure in its final step.

4.3.2 Danishefsky's Total Synthesis of Dynemicin A (1)

Synthetic efforts towards dynemicin A (**1**) in the Columbia University laboratories began with attempts to prepare the ABC system of **28**, a compound whose two stereocenters were expected to arise from a Diels–Alder reaction.[10] Accordingly, the initial operations sought to convert the commercially available benzaldehyde **72** (Scheme 11) into the projected precursor for this key event, aldehyde **32**. As shown, this compound did not prove overly challenging to access, since following the alkylation of **72** with sorbyl bromide using K_2CO_3 in acetone, a subsequent Horner–Wadsworth–Emmons olefination provided the ester analogue (**73**) of this target in 91 % yield. From here, only a standard two-step reduction/oxidation sequence was needed to complete **32**. With the stage now set to examine the critical [4+2] cycloaddition, **32** was then dissolved in CH_2Cl_2 and, following its activation with a Lewis acid ($ZnCl_2$), converted slowly into the desired Diels–Alder product (**31**) in 60 % yield. As one would expect, this event proceeded with complete *endo* selectivity since the dienophile has a conjugating substituent. Interestingly, the same reaction could also be conducted simply through thermal activation by heating **32** in refluxing benzene, leading to a superior yield of **31**, but, as is often typical of Diels–Alder reactions conducted without Lewis acids, reduced *endo*/*exo* selectivity (3:1). Accordingly, the Lewis acid catalyzed procedure was adopted to advance material into the next phases of the program since it provided the desired product more cleanly.

Having generated the two critical stereocenters of the initial target (i. e. **28**), the next set of operations sought to effect the conversion of this Diels–Alder cycloadduct (**31**) into a quinoline-bearing intermediate such as **75**. Although the connection between **31** and **75** may not seem clear, in fact, all that was required to meet this goal was the sequential opening and closing of the B-ring within **31**.

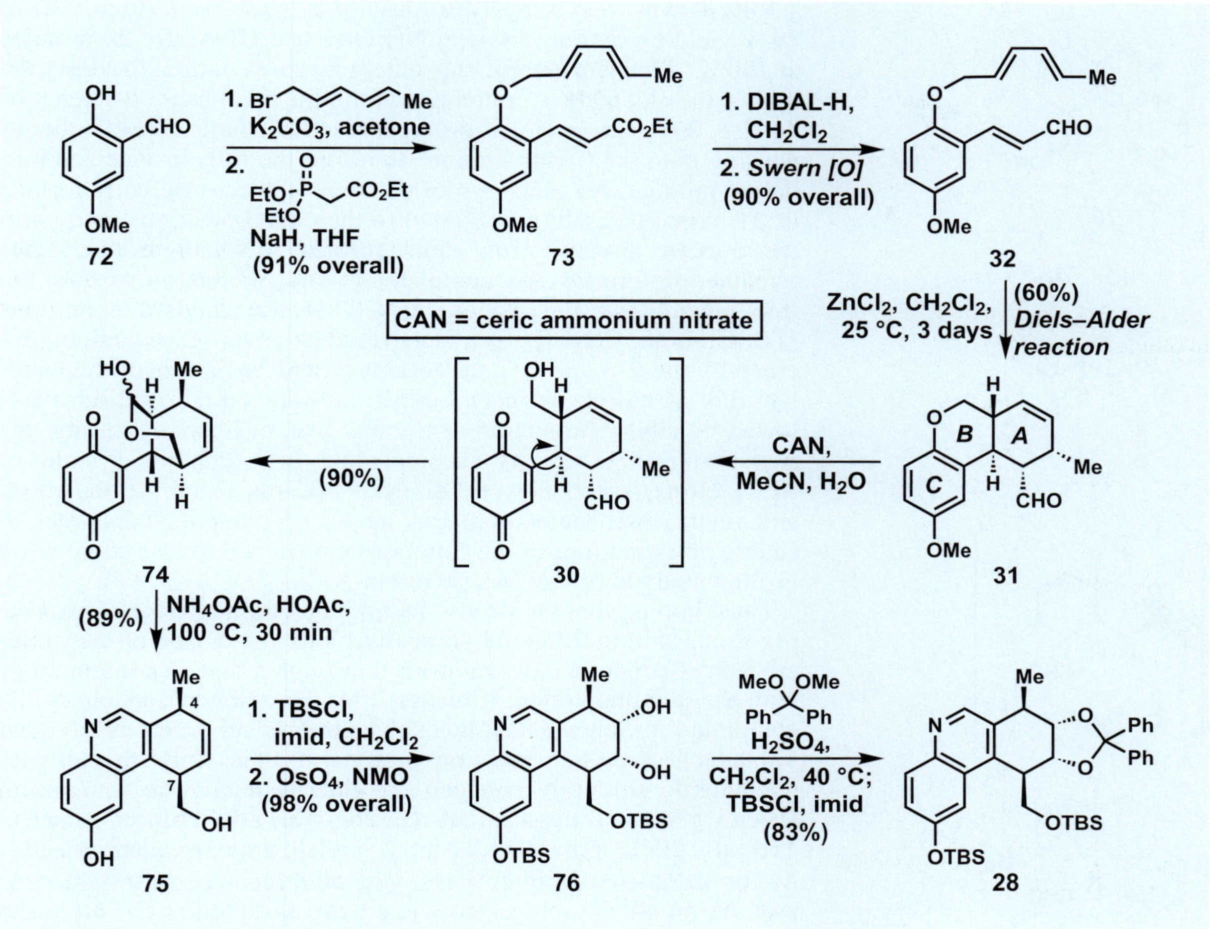

Scheme 11. Danishefsky's synthesis of intermediate **28**.

Thus, in the first operation, **31** was treated with ceric ammonium nitrate (CAN) in aqueous MeCN to effect the oxidative conversion of its C-ring into a quinone, an event attended by the concomitant scission of its B-ring to afford a hydroxymethyl function in the product (**30**). However, since a residual aldehyde from the Diels–Alder reaction was present on the same side of the A-ring as this newly generated alcohol, the isolated product was not **30**, but instead was lactol **74**. Although this latter ring closure was not critical for the operations ultimately needed to reach **28**, it did serve to protect the oxidized substrate (**30**) from the action of excess CAN. For example, only decomposition resulted when the *exo* Diels–Alder variant of **31** was subjected to the same conditions, presumably because the hydroxymethyl function that was produced in this case could not form a lactol since its aldehyde partner was on the opposite side of the molecule.

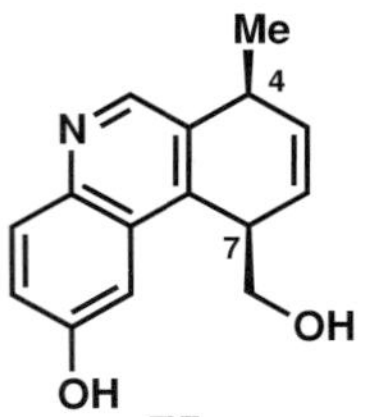

Ring B was then reclosed to provide a quinoline system (**75**) in 89 % yield by treating **74** with NH_4OAc and HOAc for 30 minutes at 100 °C. Mechanistically, this outcome can be rationalized as capture of the aldehyde by ammonia within the open lactol congener of **74** (i. e. **30**) as an imine, affording a reactive handle capable of condensing onto the C-ring quinone to afford the fully aromatized quinoline product. As such, two of the stereocenters originally formed in **31** have successfully migrated to their final positions, and, with the needed aromatic ring now formed, only three operations remained before **28** was complete: 1) silyl protection of both the free phenol and the alcohol at C-7 under standard conditions (TBSCl, imid, CH_2Cl_2), 2) a facially selective dihydroxylation governed by the C-4 and C-7 stereocenters, and 3) capture of the resultant diol as a diphenyl acetal under strongly acidic conditions followed by silyl reprotection of the free phenol liberated during this step. With these events accomplished in a combined yield of 81 %, efforts could now be directed towards installing the third, and final, stereocenter onto this advanced compound in order to enable investigations of the couplings envisioned for the completion of the enediyne bridge in dynemicin A.

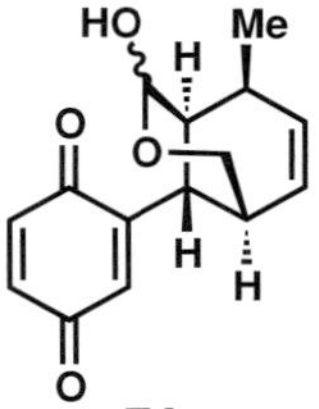

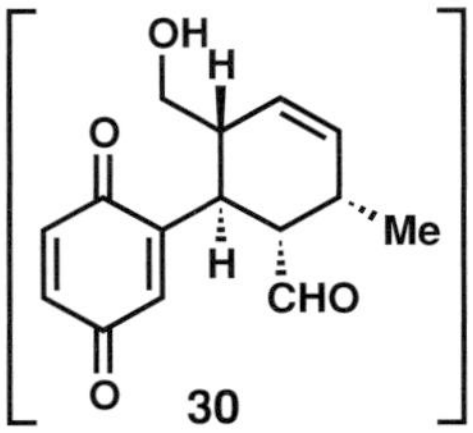

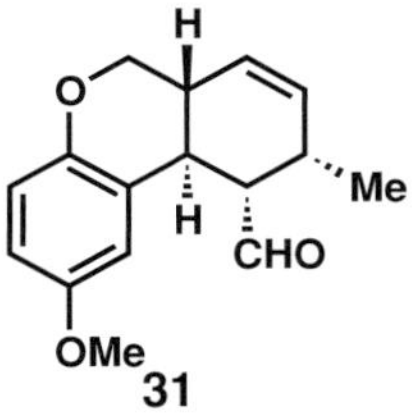

Thus, hoping that the newly incorporated diphenyl acetal protecting group within **28** could sterically block the α-face of the molecule, an effect potentially reinforced through π-stacking interactions with the quinone system (illustrated by the redrawn version of this compound in Scheme 12), this substrate was subjected to a typical Yamaguchi acetylide addition protocol.[14] Pleasingly, the alkynyl nucleophile smoothly engaged the intermediate *N*-acylquinolium species generated upon initial reaction with allyl chloroformate in THF at −20 °C, providing **77** in 80 % yield with complete selectivity for the desired product. Thus, with all three necessary stereocenters set, an alkyne moiety now had to be installed at C-7 from the resident protected alcohol to set the stage for the bridging couplings at the heart of the approach. Accordingly, the TBS group appended to the primary alcohol at this position in **77** was chemoselectively excised upon carefully monitored exposure to HCl in THF in the presence of two other silyl groups. Such selective removal in the presence of what one might anticipate would be a more labile phenolic silyl group was due to the use of acidic conditions rather than fluoride, as the phenol oxygen is less Lewis basic due to resonance. The resultant deprotected alcohol was then oxidized under Swern conditions to afford **78** in 86 % overall yield. With an aldehyde unveiled off C-7, a subsequent Corey–Fuchs homologation[32] completed the alkyne in 62 % overall yield. A few minor protecting group manipulations then converted **79** into **80**, a substrate amenable for attempts to insert the olefinic linker that would finalize the 1,5-diyne-3-ene motif of dynemicin A (**1**).

Unfortunately, initial probes of Sonogashira-type reactions[18] on substrates related to **80** led to no material in which the enediyne bridge had been fashioned, suggesting that perhaps the terminal alkynes possessed too poor a conformational disposition for the

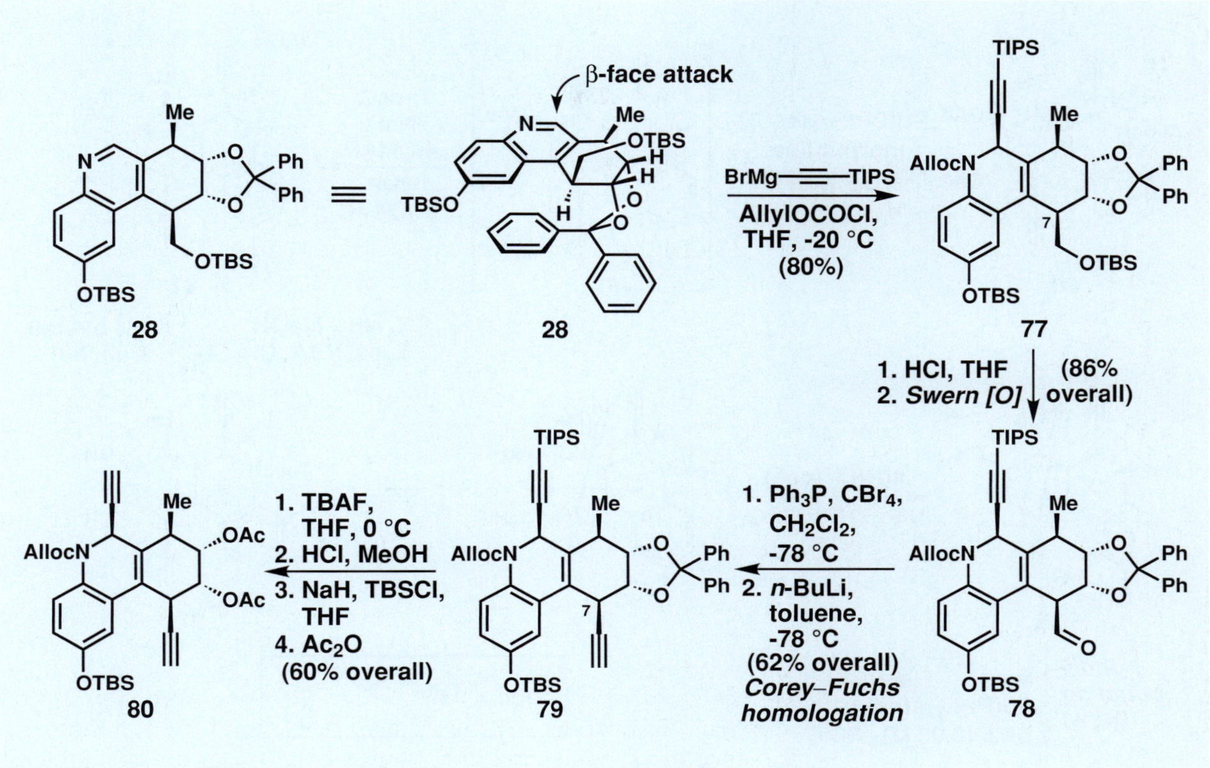

Scheme 12. Danishefsky's elaboration of **28** into diacetylene **80**.

event to succeed. But, as mentioned above, it had been forecasted that the presence of the C3–C8 epoxide would provide a series of additional test substrates that might possess a different, and potentially more favorable, conformation. Indeed, molecular modeling suggested that such a compound might constitute a superior substrate for ring closure, as the formation of a constrained ring at this position appeared to bring the termini of the two alkyne groups closer together. Thus, following the exchange of the Alloc group for Teoc protection (to facilitate the final stages of the synthesis) and cleavage of both acetate groups with NH_3 in MeOH, the central C3–C8 double bond was then epoxidized with *m*CPBA in CH_2Cl_2 as shown in Scheme 13. Unfortunately, once again **27** and related substrates failed to participate in a tandem Sonagashira reaction, suggesting recourse to a double Stille coupling reaction as the only remaining strategy to bring this compelling design to fruition. Remarkably, this strategy worked. Following the conversion of **27** into **26** with *N*-iodosuccinamide (NIS) in the presence of catalytic $AgNO_3$, reaction with *cis*-1,2-distannylethene as promoted by $Pd(PPh_3)_4$ in DMF at 75 °C led to the desired product (**83**) in 81 % yield! Two features of this amazing reaction that completed the quintessential structural motif of dynemicin A (**1**) are worthy

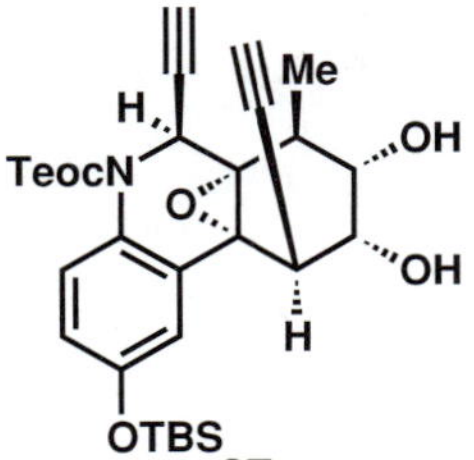

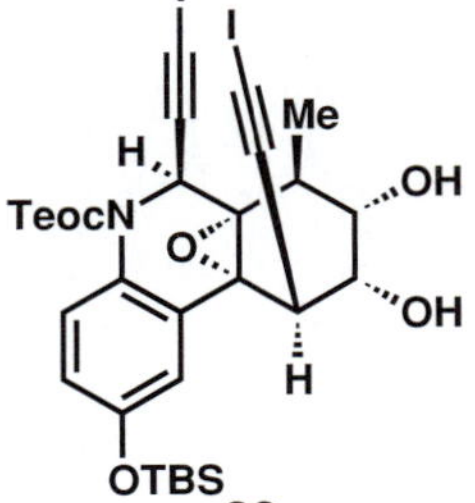

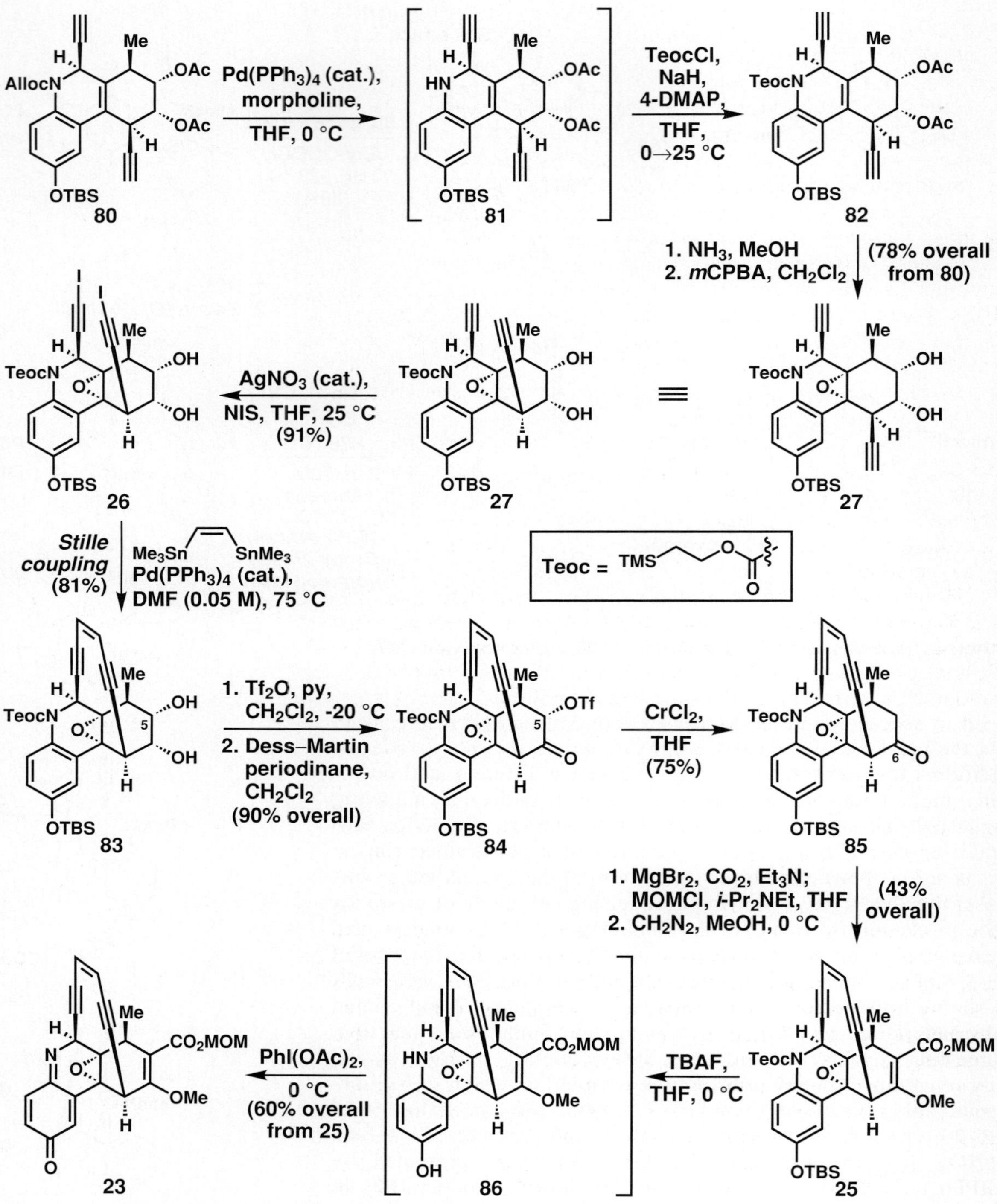

Scheme 13. Danishefsky's synthesis of advanced intermediate **23**.

of note. First, the reaction had to be conducted at a relatively low dilution (0.05 M) to minimize intermolecular dimerization, and, second, in the absence of the central epoxide motif (i. e. a C3–C8 double bond instead), Stille coupling failed to proceed in the desired direction. Thus, this latter finding serves as an excellent example of how subtle conformational effects can play a deciding role in the success or failure of a given synthetic operation.

Having now reached this advanced staging area after much toil, the completion of the targeted imidoquinone (**23**) was nearly within reach, needing only the conversion of the A-ring within **83** to that of the target molecule and the successful oxidation of its C-ring without triggering Bergman cycloaromatization. The first of these requirements was accomplished along similar lines as in the Myers synthesis. Accordingly, following selective protection of the more accessible alcohol function at C-5 as a triflate group with Tf_2O and catalytic amounts of pyridine in CH_2Cl_2 at −20 °C, the remaining free alcohol was then oxidized to a ketone (**84**) in 90 % overall yield using Dess–Martin periodinane. A final reduction with $CrCl_2$ then excised the triflate,[33] leading to **85** in which the needed carboxylic acid could be inserted by enolization and subsequent capture with CO_2. Once again, Rathke's conditions[28] proved to be the only means capable of accomplishing this task, and following treatment of the crude carboxylic acid with MOMCl to trap it as a methoxymethyl ester, a final reaction with diazomethane (CH_2N_2) in MeOH then converted the enol into its methyl ether derivative (**25**) in 43 % overall yield. At this point, the Teoc group appended to the amine and the phenolic TBS group had to be cleaved, and then the resultant product had to be quickly and mildly converted into an imidoquinone (i. e. **23**). Under optimized conditions, these requirements were met in 60 % yield by treatment of **25** with TBAF in THF at 0 °C to afford the transient **86**, followed by immediate reaction with $PhI(OAc)_2$ at 0 °C.

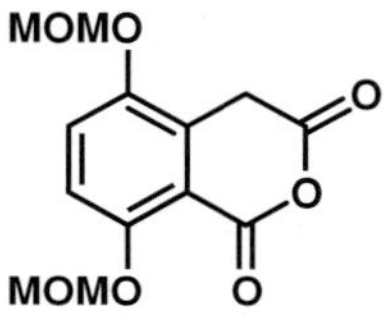

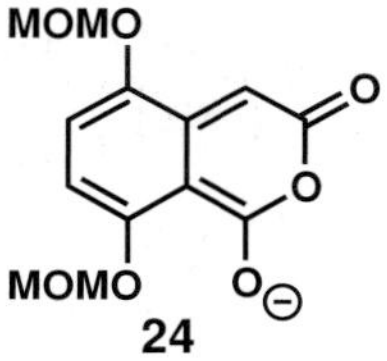

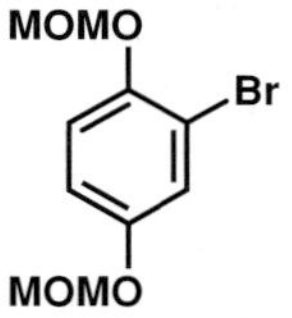

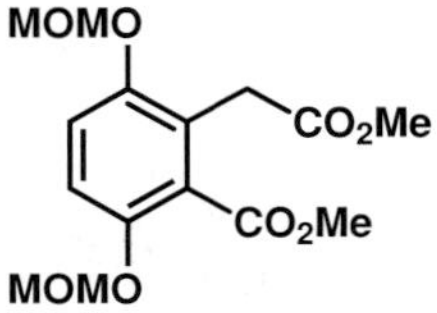

With the complete imidoquinone building block finally in hand, the Danishefsky group had only to identify a route to the Tamura-type homophthalic anhydride[15] in order to create the diene system (cf. Scheme 3) that would hopefully permit the incorporation of the still-missing anthraquinone. The target for this endeavor was **95** (see Scheme 14), which upon deprotonation was expected to afford the reactive 4π-component needed for the projected Diels–Alder reaction (i. e. **24**). Although numerous sequences exist to synthesize such relatively simple fragments, the Danishefsky group elected to develop a unique sequence that ultimately accomplished its construction in a highly creative and concise manner. Because the key precursor to a cyclic anhydride is typically a dicarboxylic acid such as **93**, these operations commenced with a very powerful, but relatively under-utilized transformation that proved capable of converting **87** directly into a diester analogue of that target, **92**, in a single step.[34] In the event, treatment of **87** with lithium tetramethylpiperidide (LiTMP) in THF at −78 °C presumably led to the initial formation of benzyne intermediate **88**,

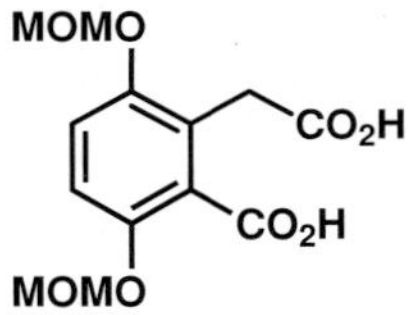

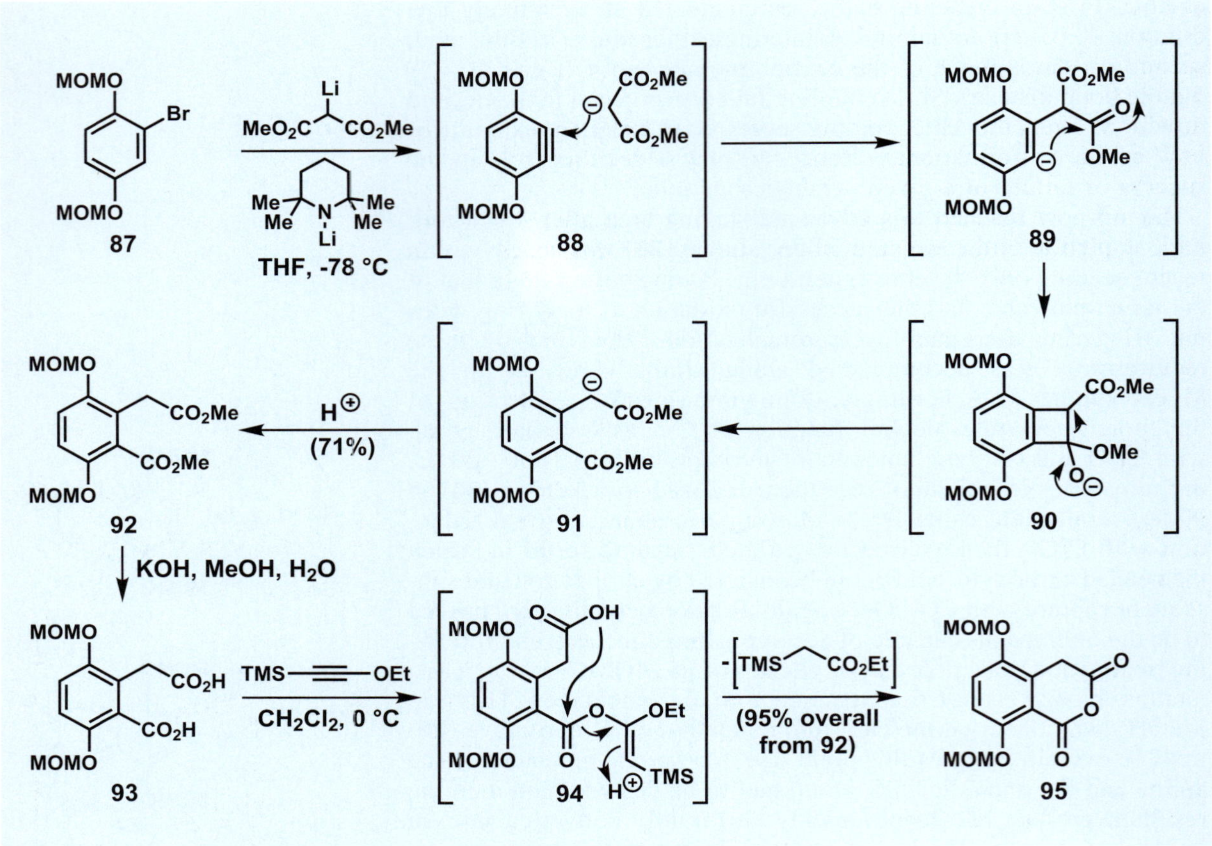

Scheme 14. Danishefsky's synthesis of homophthalic anhydride **95**.

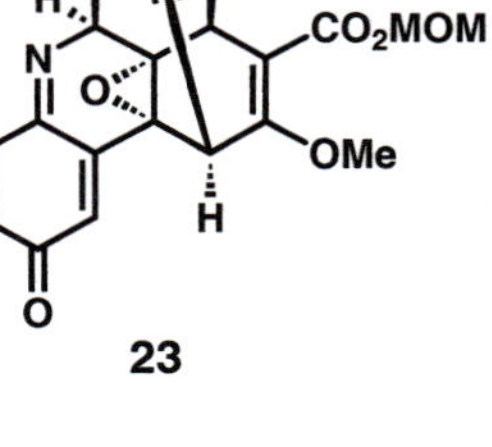

which was then engaged by the anion of dimethyl malonate (also present in solution) to initiate a series of conversions that concluded with the formation of **92** in 71 % yield. One should note, however, that the mechanism shown in this scheme has been presented only to provide a reasonable picture as to how this reaction might proceed; no experiments seeking to verify its accuracy have yet been performed. Irrespective of this mechanistic uncertainty, it was clear that with **92** in hand only two operations remained to complete **95**. First, both methyl esters were saponified using KOH in aqueous methanol, and the resultant dicarboxylic acid product (**93**) underwent cyclodehydration in near quantitative yield in the presence of ethyl 2-(trimethysilyl)ethynyl ether (TMSEE)[35] in CH_2Cl_2 at 0 °C.

Having accomplished the synthesis of both coupling partners, the final steps of the projected total synthesis of dynemicin A (**1**) were explored immediately, starting, of course, with the merger of the DE subunit with the enediyne-containing tricycle **23**. Thus, as shown in Scheme 15, the homophthalic anhydride (**95**) was treated with

LiHMDS in THF at 0 °C for 35 minutes, leading to a bright yellow solution which presumably reflected the successful formation of diene **24**. Following this event, the imidoquinone dienophile (**23**) was then added to the flask. After 35 minutes of additional stirring at 0 °C, the reaction mixture was treated with $PhI(OCOCF_3)_2$, hoping that if Diels–Alder cycloaddition and expulsion of CO_2 had proceeded as planned to generate **22**, this unstable intermediate would then be successfully converted into anthracenol **97**. Admirably, these conditions accomplished that goal, leading to the assem-

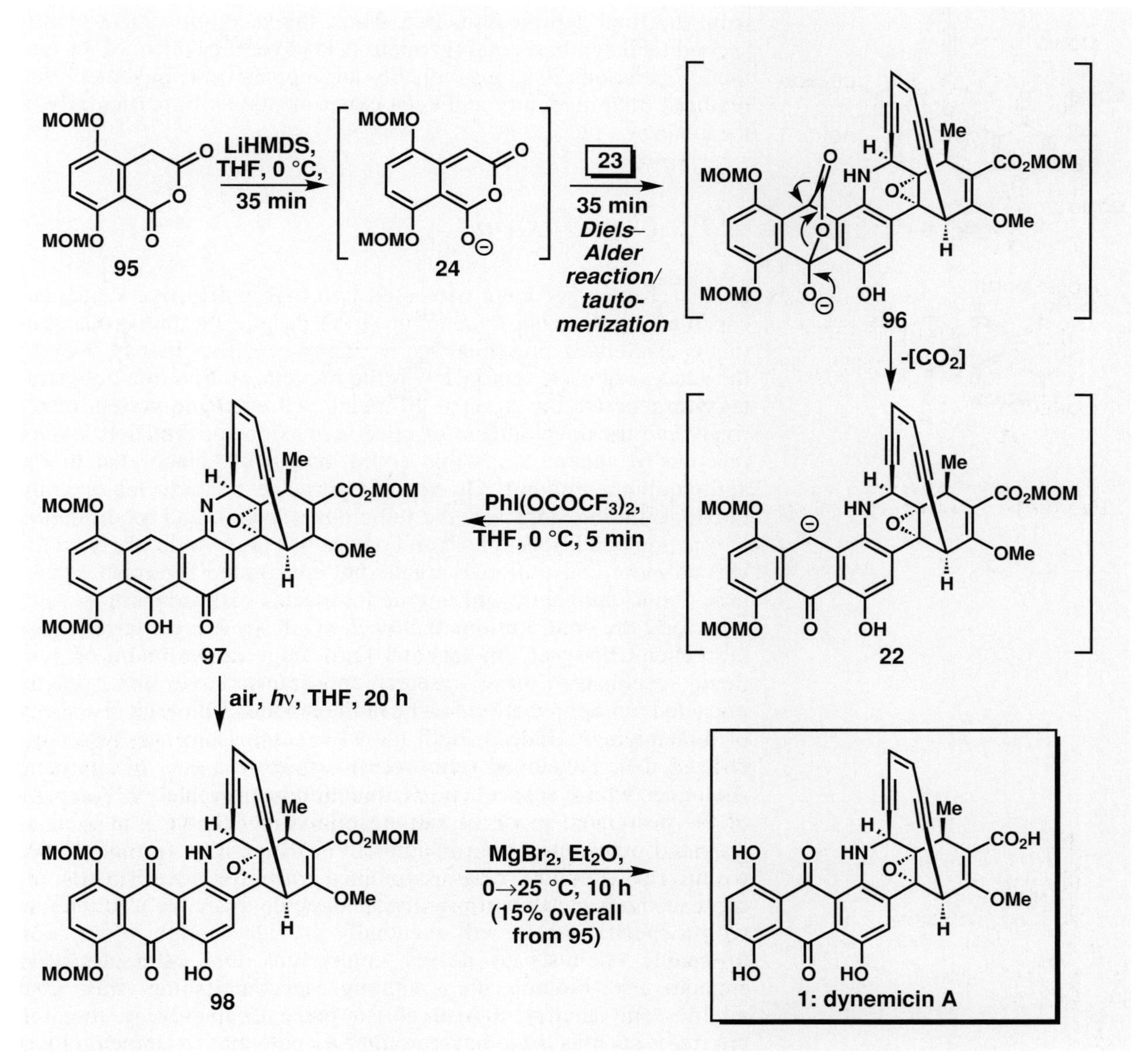

Scheme 15. Final stages and completion of Danishefsky's total synthesis of dynemicin A (**1**).

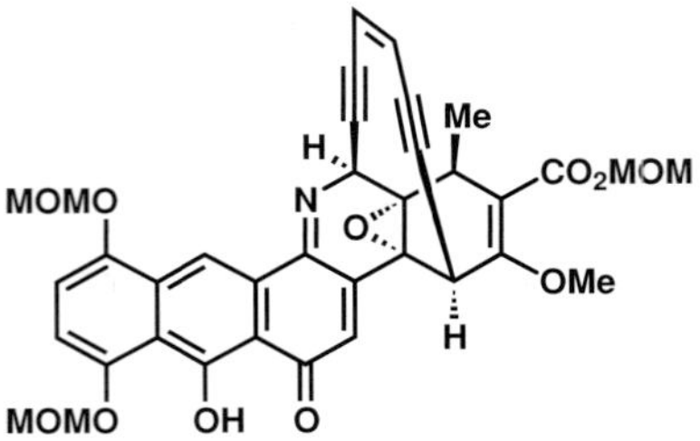

bly of the anticipated product (**97**) and thus leaving only an oxidation and global deprotection to complete dynemicin A (**1**). Following a fair amount of reaction scouting, these terminating events were ultimately accomplished by first exposing **97** to daylight under aerobic conditions, followed by treatment of the resultant anthraquinone (**98**) with $MgBr_2$ in Et_2O.[36] Overall, synthetic dynemicin A (**1**) was obtained in a combined yield of 15 % for the four operations that started from **95**. While this numerical outcome might seem low, it was not due to any inefficiency in the reactions themselves, but rather reflects the instability of several of the intermediates as well as the difficulty in purifying and isolating **1** from the final deprotection step. Thus, the accomplishment of this second total synthesis of dynemicin A (**1**) over the course of 33 synthetic operations is unquestionably an impressive achievement that required both ingenuity and keen experimental skill, particularly in its finale.

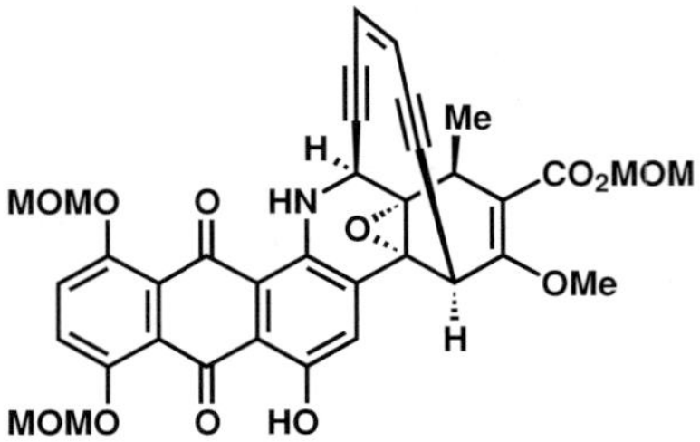

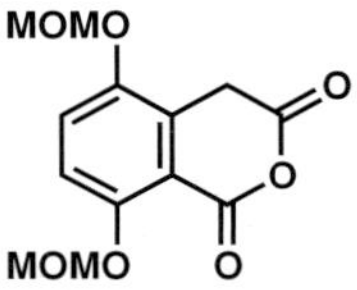

4.4 Conclusion

In this chapter, we have witnessed two highly effective sequences capable of delivering dynemicin A (**1**) despite the numerous synthetic challenges presented by its many sensitive motifs. Key to the success of each approach was the implementation of a bold strategy to generate the strained 10-membered enediyne system effectively and the development of novel extensions of the Diels–Alder reaction to append rings that could be quickly elaborated to the anthraquinone subunit. Of equal importance to these major complexity-building steps was the judicious selection and modification of reaction conditions for several operations that would likely be trivial on more conventional targets, but constituted formidable tasks here due to the subtle and unique intricacies of dynemicin A.

Beyond the contributions that each synthesis has conferred upon the general body of fundamental knowledge in the realm of synthetic chemistry, these research programs have also greatly impacted our appreciation for the unique mode of biological activity of dynemicin A. Indeed, both the Myers and Danishefsky groups utilized their developed sequences to generate a host of synthetic analogues whose subsequent examination has revealed the veracity of the postulated mode of action defined in Scheme 1 as well as provided other interesting insights.[9,10,37] Hopefully, further studies within the enediyne class, combined with the identification of other methods capable of priming a molecule to enable it to interact with a specific target, will eventually provide enough information to enable chemists to design compounds that can selectively engage any biomolecule. Although accomplishing this goal might seem fanciful in light of our present capabilities, research programs such as these have brought its potential attainment closer to reality.

4.5 Schreiber's Synthesis of Tri-O-Methyl Dynemicin A Methyl Ester

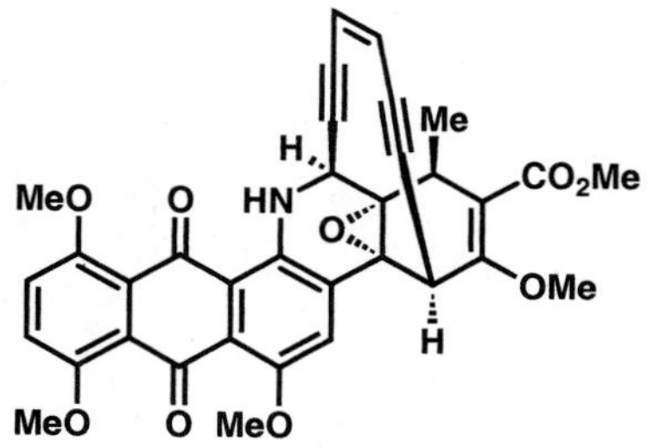

Although the Myers and Danishefsky groups were the first to complete total syntheses of dynemicin A (**1**), the core architecture of the entire natural product had, in fact, been completed several months earlier by Professor Stuart L. Schreiber and his group at Yale University (now at Harvard University).[38] All that separated the advanced structure synthesized by these researchers (**107**, Scheme 16) from the final target were three phenolic methyl ethers, which, unfortunately, proved resilient to cleavage. As such, the developed route cannot constitute a total synthesis in the strict sense of the definition. However, based on our discussion above, the accomplishment of even this advanced compound is certainly laudable, and, for this reason, this work is included here because it provides an alternative and especially creative solution for the construction of the key architectural elements of dynemicin A.

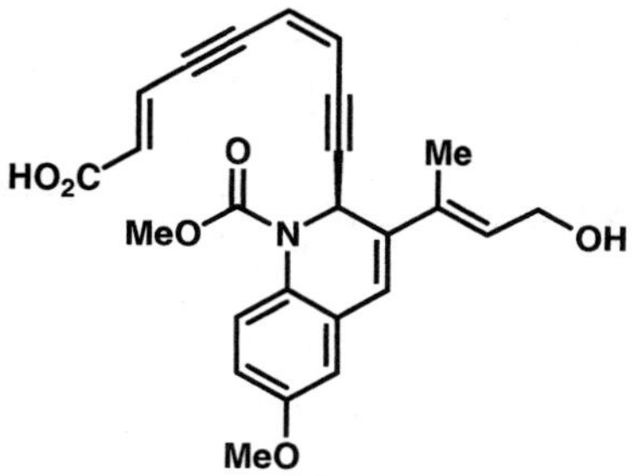

The critical transformation that these researchers sought to accomplish was the formation of the 10-membered enediyne ring and the A-ring in a single step, rather than the appendage of the enediyne motif onto an already established tricyclic system as both the Myers and Danishefsky groups had done. Thus, they forecasted that an initial macrolactonization between the free acid and alcohol motifs in **99** could lead to **100**, an intermediate programmed for a transannular Diels–Alder reaction that could provide **101** in the same step. As shown, this strategy was indeed reduced to practice with the formation of **101** in 50 % yield following the exposure of **99** to the standard Yamaguchi macrolactonization conditions (2,4,6-trichlorobenzoyl chloride, Et_3N, PyBroP®) at ambient temperature over 13 hours. The remarkable facility of this amazing sequence is likely due to the inferred excellent proximity and alignment of the diene in relation to the macrocyclic dienophile, enforced by the unique structural characteristics of the enediyne system, combined with a low energy of activation. Significantly, in the absence of the macrolactone, **99** itself could not be induced to participate in an intramolecular [4+2] cycloaddition even though it possesses both a diene and dienophile. Accordingly, transannular activation, accomplished by initial macrocycle formation, clearly represented the crucial facet for the success of this conversion.[39]

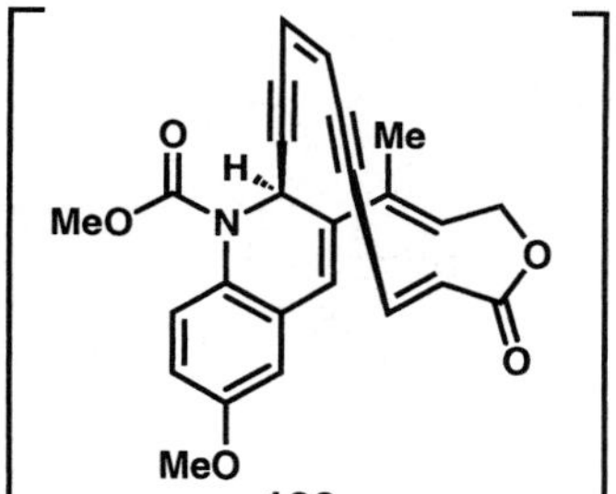

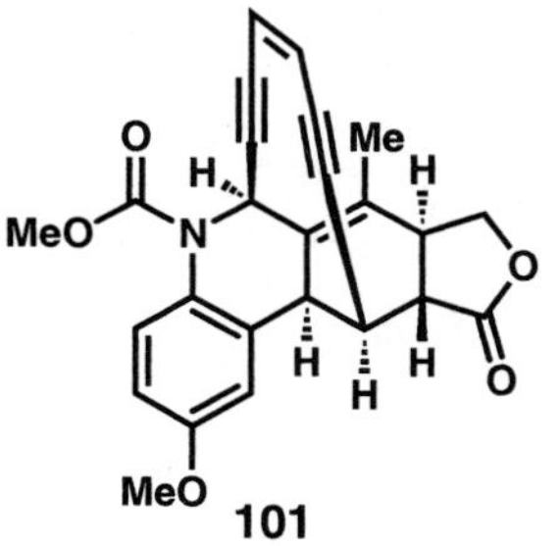

With this advanced structure (**101**) in hand, subsequent adjustment of its A-ring functionality to **102** enabled the anthraquinone portion to be attached not through a Diels–Alder reaction, but instead through an interesting Friedel–Crafts alkylation using **103** and activated with AgOTf. Methyl capture by the enol function within this product using K_2CO_3 and MeI in acetone then completed the assembly of **104** in 57 % yield, enabling subsequent conversion into **107** over the course of several steps. For certain, the Schreiber

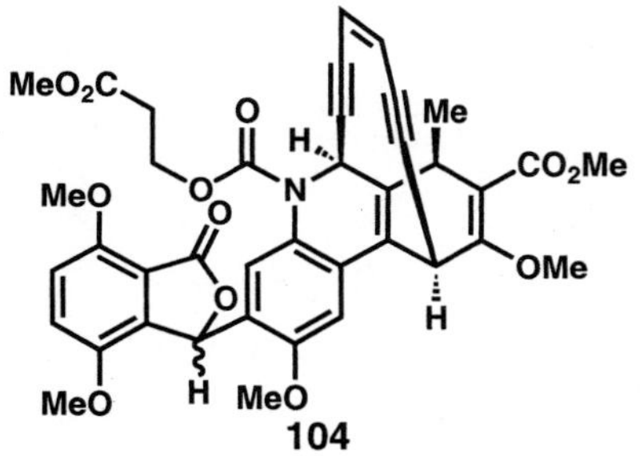

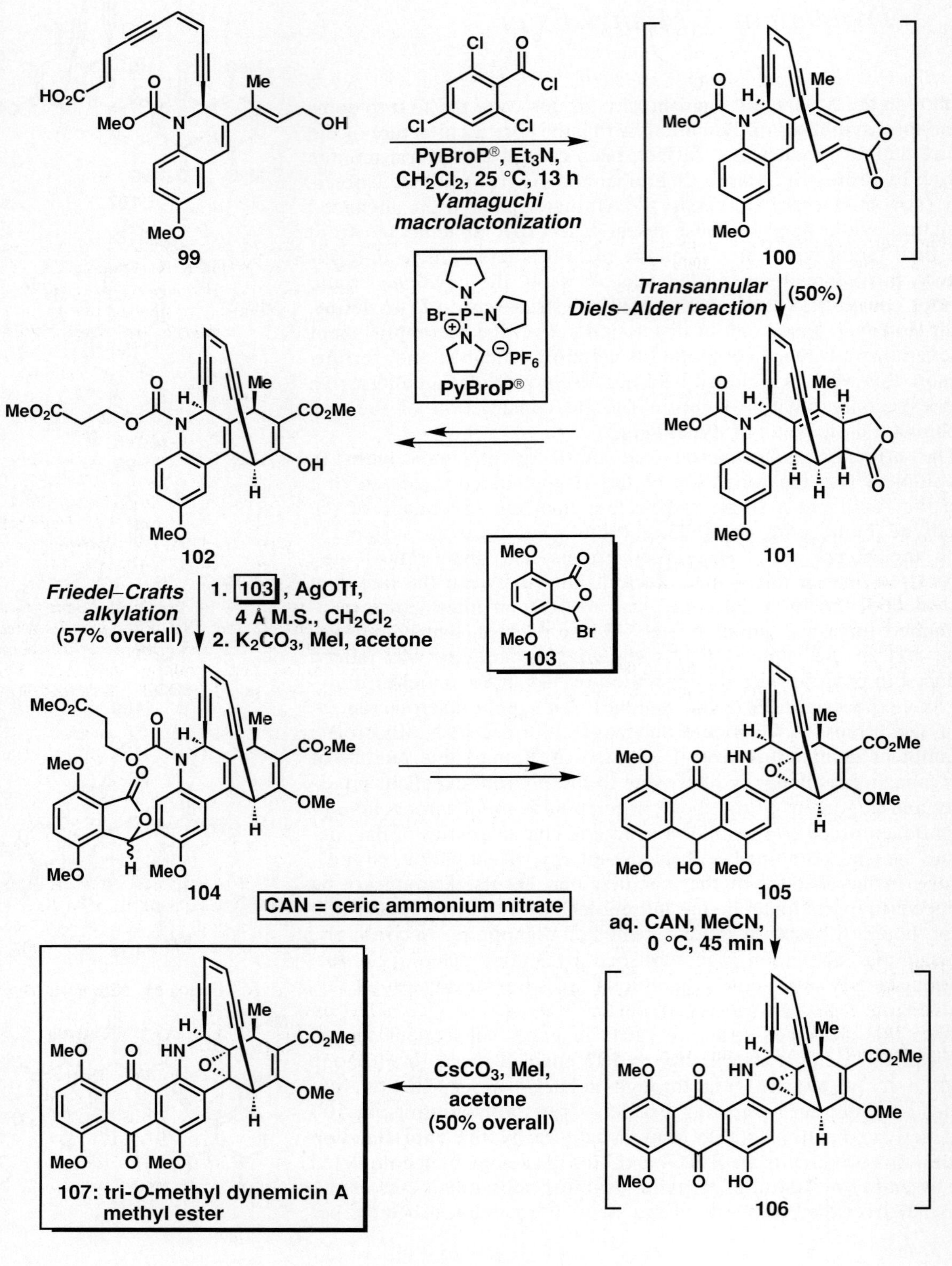

Scheme 16. Schreiber's synthesis of tri-*O*-methyl dynemicin A methyl ester (**107**).

group developed a highly inventive route to attain this advanced structure, affording powerful testimony to the creative sequences that the modern repertoire of synthetic transformations can accomplish when cleverly orchestrated. By the same token, this work also reveals the uncompromising nature of total synthesis in that a simple protecting group can single-handedly block the completion of a given target molecule.

References

1. For some insightful discussions on this concept, see: a) S.L. Schreiber, *Science* **2000**, *287*, 1964; b) S.L. Schreiber, *Bioorg. Med. Chem.* **1998**, *6*, 1127; c) D.T. Hung, T.F. Jamison, S.L. Schreiber, *Chem. Biol.* **1996**, *3*, 623.
2. For a review on the chemistry and biology of the enediynes, see: K.C. Nicolaou, W.-M. Dai, *Angew. Chem.* **1991**, *103*, 1453; *Angew. Chem. Int. Ed. Engl.* **1991**, *30*, 1387.
3. For an account of this reaction, see: a) R.G. Bergman, *Acc. Chem. Res.* **1973**, *6*, 25. For seminal papers regarding its discovery, see: b) R.R. Jones, R.G. Bergman, *J. Am. Chem. Soc.* **1972**, *94*, 660; c) T.P. Lockhart, P.B. Comita, R.G. Bergman, *J. Am. Chem. Soc.* **1981**, *103*, 4082; d) T.P. Lockhart, R.G. Bergman, *J. Am. Chem. Soc.* **1981**, *103*, 4091.
4. A.G. Myers, S.B. Cohen, B.-M. Kwon, *J. Am. Chem. Soc.* **1994**, *116*, 1670.
5. a) M. Konishi, H. Ohkuma, K. Matsumoto, T. Tsuno, H. Kamei, T. Miyaki, T. Oki, H. Kawaguchi, G.D. Van Duyne, J. Clardy, *J. Antibiot.* **1989**, *42*, 1449; b) M. Konishi, H. Ohkuma, T. Tsuno, T. Oki, G.D. Van Duyne, J. Clardy, *J. Am. Chem. Soc.* **1990**, *112*, 3715.
6. a) D.R. Langley, T.W. Doyle, D.L. Beveridge, *J. Am. Chem. Soc.* **1991**, *113*, 4395; b) P.A. Wender, R.C. Kelly, S. Beckham, B.L. Miller, *Proc. Natl. Acad. Sci. U.S.A.* **1991**, *88*, 8835.
7. M.F. Semmelhack, J. Gallagher, D. Cohen, *Tetrahedron Lett.* **1990**, *31*, 1521.
8. For other papers related to the mode of activity of dynemicin A, see the following (and references cited therein): a) Y. Sugiura, T. Shiraki, M. Konishi, T. Oki, *Proc. Natl. Acad. Sci. U.S.A.* **1990**, *87*, 3831; b) J.P. Snyder, G.E. Tipsword, *J. Am. Chem. Soc.* **1990**, *112*, 4040; c) T. Shiraki, Y. Sugiura, *Biochemistry* **1990**, *29*, 9795; d) Y. Sugiura, T. Arakawa, M. Uesugi, T. Shiraki, H. Ohkuma, M. Konishi, *Biochemistry* **1991**, *30*, 2989.
9. For the full account of this total synthesis, see: a) A.G. Myers, N.J. Tom, M.E. Fraley, S.B. Cohen, D.J. Madar, *J. Am. Chem. Soc.* **1997**, *119*, 6072. For the preliminary communication of its completion as well as earlier model studies, see: b) A.G. Myers, M.E. Fraley, N.J. Tom, S.B. Cohen, D.J. Madar, *Chem. Biol.* **1995**, *2*, 33; c) A.G. Myers, M.E. Fraley, N.J. Tom, *J. Am. Chem. Soc.* **1994**, *116*, 11556.
10. For the full account of this total synthesis, see: a) M.D. Shair, T.Y. Yoon, K.K. Mosny, T.C. Chou, S.J. Danishefsky, *J. Am. Chem. Soc.* **1996**, *118*, 9509. For the preliminary communication of its completion as well as earlier model studies, see: b) M.D. Shair, T. Yoon, S.J. Danishefsky, *Angew. Chem.* **1995**, *107*, 1883; *Angew. Chem. Int. Ed. Engl.* **1995**, *34*, 1721; c) M.D. Shair, T. Yoon, T.-C. Chou, S.J. Danishefsky, *Angew. Chem.* **1994**, *106*, 2578; *Angew. Chem. Int. Ed. Engl.* **1994**, *33*, 2477; d) T. Yoon, M.D. Shair, S.J. Danishefsky, *Tetrahedron Lett.* **1994**, *35*, 6259; e) T. Yoon, M.D. Shair, S.J. Danishefsky, G.K. Shulte, *J. Org. Chem.* **1994**, *59*, 3752; f) M.D. Shair, T. Yoon, S.J. Danishefsky, *J. Org. Chem.* **1994**, *59*, 3755. For a personal account of this synthesis, see: g) S.J. Danishefsky, M.D. Shair, *J. Org. Chem.* **1996**, *61*, 16.
11. For an excellent review on the Diels–Alder reactions of isobenzofurans, see: R. Rodrigo, *Tetrahedron* **1988**, *44*, 2093. For the first example of this diene in a natural product total synthesis, see: A.S. Kende, D.P. Curran, Y. Tsay, J.E. Mills, *Tetrahedron Lett.* **1977**, *18*, 3537.
12. a) L. Contreras, C.E. Slemon, D.B. MacLean, *Tetrahedron Lett.* **1978**, *19*, 4237; b) T. Troll, K. Schmid, *Tetrahedron Lett.* **1984**, *25*, 2981; c) M. Iwao, H. Inoue, T. Kuraishi, *Chem. Lett.* **1984**, 1263; d) S.L. Crump, J. Netka, B. Rickborn, *J. Org. Chem.* **1985**, *50*, 2746; e) D. Tobia, B. Rickborn, *J. Org. Chem.* **1987**, *52*, 2611.
13. For a general review on the Diels–Alder reaction in total synthesis, see: K.C. Nicolaou, S.A. Snyder, T. Montagnon, G.E. Vassilikogiannakis, *Angew. Chem.* **2002**, *114,* 1742; *Angew. Chem. Int. Ed.* **2002**, *41*, 1668.
14. R. Yamaguchi, Y. Nakazono, M. Kawanisi, *Tetrahedron Lett.* **1983**, *24*, 1801.
15. Y. Tamura, F. Fukata, M. Sasho, T. Tsugoshi, Y. Kita, *J. Org. Chem.* **1985**, *50*, 2273.
16. For selected examples, see: a) S.A. Hitchcock, S.H. Boyer, M.Y. Chu-Moyer, S.H. Olson, S.J. Danishefsky, *Angew. Chem.* **1994**, *106*, 928; *Angew. Chem. Int. Ed. Engl.* **1994**, *33*, 858; b) S.A. Hitchcock, M.Y. Chu-Moyer, S.H. Boyer, S.H. Olson, S.J. Danishefsky, *J. Am. Chem. Soc.* **1995**, *117*, 5750.

17. K.C. Nicolaou, T.K. Chakraborty, A.D. Piscopio, N. Minowa, P. Bertinato, *J. Am. Chem. Soc.* **1993**, *115*, 4419. For another demonstration in a recent total synthesis, see: D.A. Longbottom, A.J. Morrison, D.J. Dixon, S.V. Ley, *Angew. Chem.* **2002**, *114*, 2910; *Angew. Chem. Int. Ed.* **2002**, *41*, 2786.
18. K. Sonogashira, Y. Tohda, N. Hagihara, *Tetrahedron Lett.* **1975**, *16*, 4467. For a general review on this reaction, see: I.B. Campbell in *Organocopper Reagents* (Ed.: R.J.K. Taylor), IRL Press, Oxford, **1994**, pp. 217–235.
19. J.S. Swenton, C. Shih, C.-P. Chen, C.-T. Chou, *J. Org. Chem.* **1990**, *55*, 2019.
20. R.D. Stephens, C.E. Castro, *J. Org. Chem.* **1963**, *28*, 3313.
21. R. von Schilling, D. Vorlünder, *Ann.* **1899**, *308*, 184.
22. N. Miyaura, T. Yanagi, A. Suzuki, *Synth. Commun.* **1981**, *11*, 513.
23. For an example of the removal of a Boc group merely through heating neat, in the absence of a solvent that can serve as a Brönsted acid, see: V.H. Rawal, R.J. Jones, M.P. Cava, *J. Org. Chem.* **1987**, *52*, 19.
24. S. Cacchi, P.G. Ciattini, E. Morera, G. Ortar, *Tetrahedron Lett.* **1986**, *27*, 5541.
25. M. Yamaguchi, I. Hirao, *Tetrahedron Lett.* **1983**, *24*, 391.
26. a) K.C. Nicolaou, C.-K. Hwang, A.L. Smith, S.V. Wendeborn, *J. Am. Chem. Soc.* **1990**, *112*, 7416; b) K.C. Nicolaou, Y.P. Hong, W.-M. Dai, Z.-J. Zeng, W. Wrasidlo, *J. Chem. Soc., Chem. Comm.* **1992**, 1542; c) K.C. Nicolaou, W.-M. Dai, Y.P. Hong, S.-C. Tsay, K.K. Baldridge, J.S. Siegel, *J. Am. Chem. Soc.* **1993**, *115*, 7944.
27. a) D.H.R. Barton, S.W. McCombie, *J. Chem. Soc., Perkin Trans 1* **1975**, 1574; b) D.H.R. Barton, D. Crich, A. Löbberding, S.Z. Zard, *Tetrahedron* **1986**, *42*, 2329.
28. R.E. Tirpak, R.S. Olsen, M.W. Rathke, *J. Org. Chem.* **1985**, *50*, 4877.
29. R. Barret, M. Daudon, *Tetrahedron Lett.* **1991**, *32*, 2133. One should note that the regioselectivity of ketal formation is different here since compounds with an open amine position, rather than a free phenol, are employed. It is this open site that "latches" onto the hypervalent oxidant.
30. For previous syntheses of such pieces, see: M.P. Sibi, N. Altintas, V. Snieckus, *Tetrahedron* **1984**, *40*, 4593, and references cited therein.
31. a) A.J. Fatiadi, *Synthesis* **1976**, 65; b) A.J. Fatiadi, *Synthesis* **1976**, 133.
32. E.J. Corey, P.L. Fuchs, *Tetrahedron Lett.* **1972**, *13*, 3769.
33. For a review on this reagent, see: J.R. Hanson, *Synthesis* **1974**, 1.
34. M. Guyot, D. Molho, *Tetrahedron Lett.* **1973**, *14*, 3433.
35. Y. Kita, S. Akai, N. Ajimura, M. Yoshigi, T. Tsugoshi, H. Yasuda, Y. Tamura, *J. Org. Chem.* **1986**, *51*, 4150.
36. For other examples, see: a) D.W. Cameron, P.E. Schütz, *J. Chem. Soc. (C)* **1967**, 2121; b) P.L. Julian, W. Cole, G. Diemer, *J. Am. Chem. Soc.* **1945**, *67*, 1721.
37. A.G. Myers, S.B. Cohen, N.J. Tom, D.J. Madar, M.E. Fraley, *J. Am. Chem. Soc.* **1995**, *117*, 7574.
38. a) J. Taunton, J.L. Wood, S.L. Schreiber, *J. Am. Chem. Soc.* **1993**, *115*, 10378; b) J.L. Wood, J.A. Porco, J. Taunton, A.Y. Lee, J. Clardy, S.L. Schreiber, *J. Am. Chem. Soc.* **1992**, *114*, 5898; c) H. Chikashita, J.A. Porco, T.J. Stout, J. Clardy, S.L. Schreiber, *J. Org. Chem.* **1991**, *56*, 1692; d) J.A. Porco, F.J. Schoenen, T.J. Stout, J. Clardy, S.L. Schreiber, *J. Am. Chem. Soc.* **1990**, *112*, 7410.
39. For a recent review of transannular Diels–Alder reactions, see: E. Marsault, A. Toró, P. Nowak, P. Deslongchamps, *Tetrahedron* **2001**, *57*, 4243.

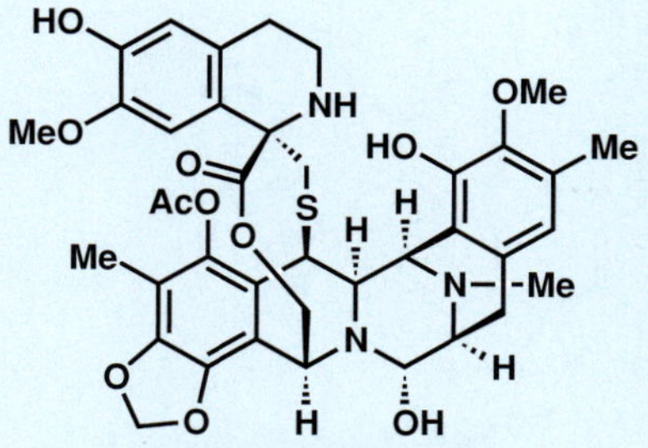

1: ecteinascidin 743

5

E. J. Corey (1996)

Ecteinascidin 743

5.1 Introduction

Key concepts:

- **Pictet–Spengler condensations**
- **Quinone methides**
- **Mannich annulation**
- **Curtius rearrangement**
- **Biomimetic synthesis**

Anyone who has been fortunate enough to go diving in tropical waters has no doubt been entranced by the bright and wondrous assortment of colors found in many of the native sponges and corals. Unfortunately, such beautiful pigmentation also renders these stationary, soft-bodied organisms as highly attractive (and obvious) targets for marine predators. As a result, evolutionary pressure has endowed these species with an elaborate series of chemical defense systems in order to facilitate their survival, most of which rely on several highly toxic compounds meant to kill any creature that ingests them. While such natural selection has served its intended purpose quite effectively up to this point, a fortunate by-product for humanity is the ability of these secondary metabolites to often interact selectively with elements of the cellular machinery in previously unknown ways, thereby pointing to new chemotherapeutic approaches.

Such was the case with a collection of natural products obtained in 1969 from the crude aqueous ethanol extracts of the tunicate *Ecteinascidia turbinata*.[1] Not only did this mixture of compounds display impressive *in vitro* cytotoxicity against a variety of cancer cell lines in the picomolar to low nanomolar range, but they were the most potent marine-based antitumor agents screened between 1972 and 1980 by the National Cancer Institute in the United States.[2] Unfortunately, for nearly two decades following these studies, all attempts to isolate the specific molecular entities responsible for such impressive biological activity proved fruitless as the active components decomposed irrespective of how they were handled. Following the development of new chromatographic technology, however,

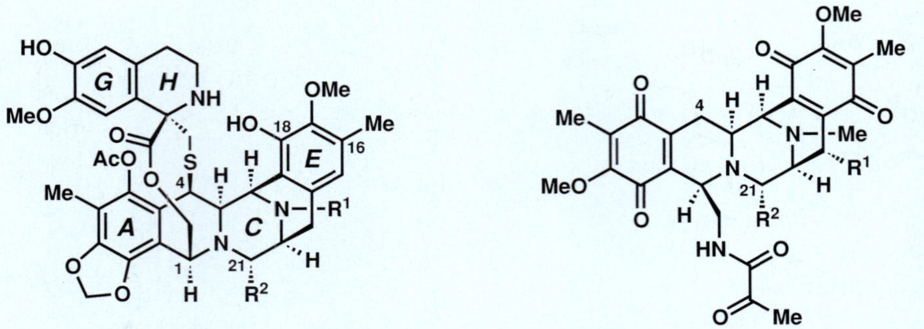

1: ecteinascidin 743, R^1 = Me, R^2 = OH
2: ecteinascidin 729, R^1 = H, R^2 = OH
3: ecteinascidin 745, R^1 = Me, R^2 = H
4: ecteinascidin 770, R^1 = Me, R^2 = CN

5: saframycin A, R^1 = H, R^2 = CN
6: saframycin B, R^1 = H, R^2 = H
7: saframycin C, R^1 = OMe, R^2 = H
8: saframycin S, R^1 = H, R^2 = OH

Scheme 1. Structures of selected members of the ecteinascidin (**1**–**4**) and saframycin (**5**–**8**) families of natural products.

Professor Kenneth Rinehart and his co-workers from the University of Illinois at Urbana-Champaign were able to obtain minute (sub-milligram) quantities of the active constituents in 1986, and, after they were subjected to exhaustive NMR spectroscopic and X-ray crystallographic studies, their complete structures (such as **1**–**4**, Scheme 1) were finally unveiled to the world in 1990.[3] Of these, ecteinascidin 743 (**1**) was the most potent as well as the most abundant.

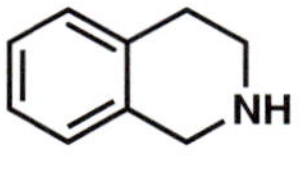

Although the amazing structural constitution of the ecteinascidins was unprecedented at the time of their characterization, as is similarly true today, these natural products can, in fact, be classified within a large family of secondary metabolites that are unified by the presence of several tetrahydroisoquinoline-type subunits (see column figure).[4] In particular, ecteinascidin 743 (**1**) and its relatives (**2**–**4**) bear significant structural homology to the well-known saframycin family of antitumor antibiotics (**5**–**8**, Scheme 1), differing primarily in the oxidation state of their terminal rings in the central pentacyclic array as well as in the degree of functionalization at C-4, which in the case of the ecteinascidins constitutes the attachment point for a ten-membered macrolactone bearing a sulfur atom and an additional tetrahydroisoquinoline system.[5] Indeed, while these two major disparities render the ecteinascidins far more imposing as synthetic targets, the core motifs that are shared by these families of compounds lead to the same mode of antitumor activity based on DNA alkylation within the minor groove.

tetrahydroisoquinoline

As several studies have indicated, it is the presence of either a nitrile or a hydroxy group at C-21 in close proximity to a nitrogen atom that is critical for the biological activity of each of these natural products.[6,7] If these particular functional groups are activated

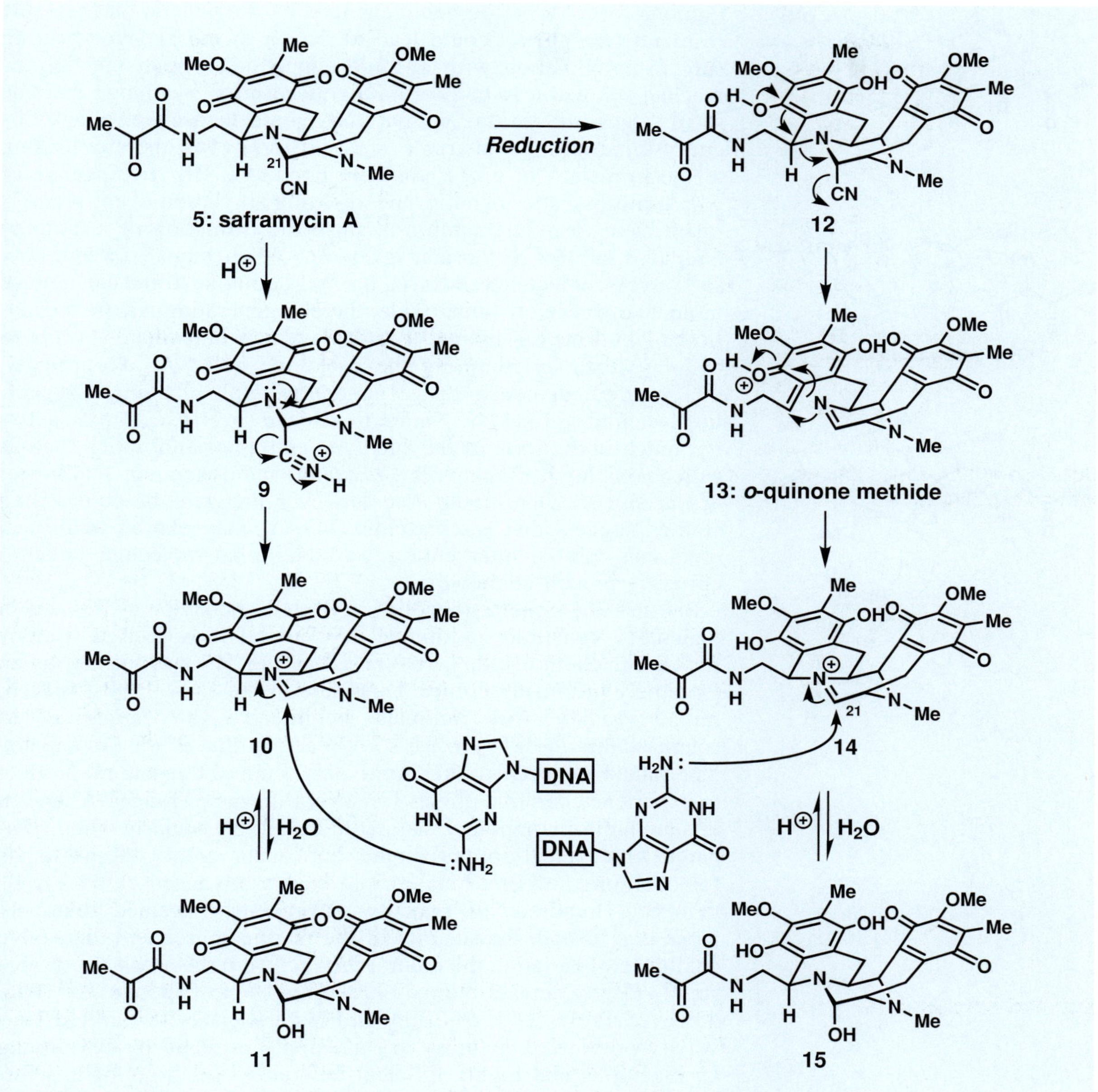

Scheme 2. Postulated mechanisms of cytotoxic action initiated by saframycin A (**5**).

through an event such as protonation, as shown in Scheme 2 using the nitrile within saframycin A (**5**) as a representative model, the neighboring nitrogen atom in **9** is suitably disposed to expel the newly generated leaving group to afford iminium **10**. This reactive species could then be trapped reversibly by water (to form **11**) or, if formed in the nucleus, by the free amine of a guanine residue in DNA.[6] Alternatively, if saframycin A (**5**) is reduced to its hydro-

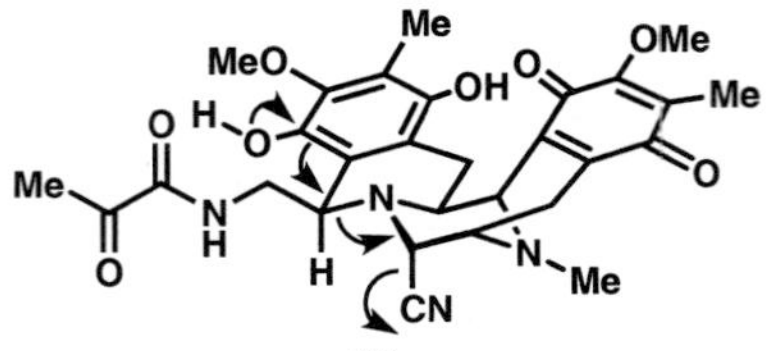

12

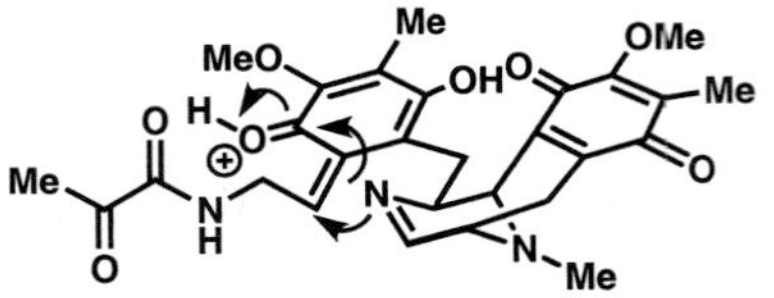

13: *o*-quinone methide

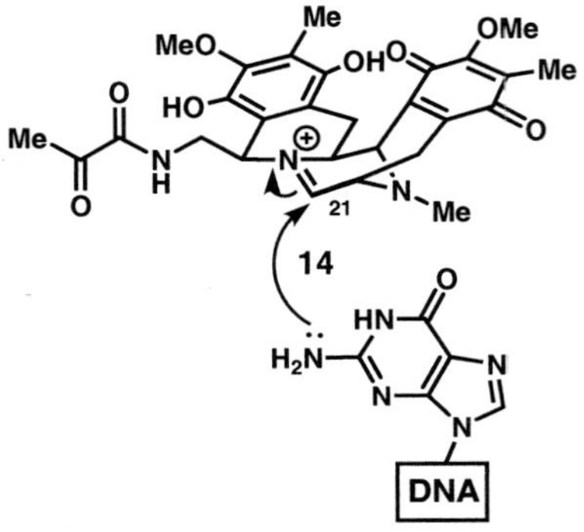

14

quinone form **12** before being engaged by the genetic material, the resultant free phenol could lead to fission of the nitrogen-bearing ring in combination with cyanide expulsion through the defined mechanistic pathway to afford a highly reactive *o*-quinone methide (**13**). A guanine residue within DNA could then either attack this entity directly or its alternate iminium form (**14**), thereby leading to the formation of a new covalent bond at C-21.[7] Irrespective of which mechanistic scenario prevails, this alkylation event alone is insufficient to induce apoptosis since the formation of a diamino aminal is subject to thermal reversion. When bound, though, any saframycin or ecteinascidin in its hydroquinone form can reduce molecular oxygen to superoxide, thereby generating Nature's deadliest of bullets, the hydroxyl radical, which immediately leads to cellular death by rupturing the backbone of DNA. Accordingly, this paradigm demonstrates yet another example of a unique triggering system developed by Nature to achieve specific cytotoxic activity, much in the style of the enediyne antitumor antibiotics such as calicheamicin γ_1^I (Chapter 30, *Classics I*) and dynemicin A (Chapter 4, *Classics II*). One should note, however, that gene-based profiling studies suggest that ecteinascidin 743 (**1**) may exhibit additional modes of cellular intervention; any additional molecular target(s) currently remain unknown.[8]

Because the general model for cytotoxic behavior as defined in Scheme 2 constitutes a unique and potentially effective strategy for cancer chemotherapy, ecteinascidin 743 (**1**) was viewed as an extremely promising clinical candidate in the early 1990s. Indeed, initial studies with animals heightened this appraisal as ecteinascidin 743 (**1**) proved so efficacious that it was anticipated that human subjects would require only 5 mg of this natural product to experience curative effects for several forms of cancer. Although this amount of compound may seem trivial, particularly when compared with typical doses for other anticancer agents, the prospects for obtaining sufficient material to enable advanced clinical trials involving hundreds of patients unfortunately seemed relatively bleak due to both the scarcity of the marine source and the severe challenge of isolating the natural product from the available sponge supply (1 g of ecteinascidin 743 requires 1 ton of sponge). In 1996, however, Professor E. J. Corey and his associates at Harvard University provided a solution to this supply problem by developing an ingenious and highly efficient laboratory synthesis of this formidable synthetic target[9] which has subsequently enabled its large-scale preparation.[10] In this chapter, we will examine this instructive total synthesis of ecteinascidin 743 (**1**) in its optimized form,[11] an achievement whose success ultimately derived from the recognition of several subtle, but crucial, synthetic clues provided by Nature.

5.2 Retrosynthetic Analysis and Strategy

Although the formidable architecture of ecteinascidin 743 (**1**) encompasses a host of complicated problems that are certain to challenge the prowess of any synthetic practitioner, Nature's likely synthesis of this natural product suggests several potential retrosynthetic avenues to render this target more manageable.[12] As shown in Scheme 3, the biosynthesis of ecteinascidin 743 (**1**) can be presumed to commence through the dehydrative dimerization of two tyrosine (**16**) units to afford the symmetric bisamide **17**. Subsequent reduction of this intermediate to an unstable bismethanolamine (**18**) would then enable the construction of the "right-hand" tetrahydroisoquinoline subunit in **19** through a Mannich annulation. In this event, one of the newly generated hydroxy groups in **18** is activated as a leaving group either through protic or enzymatic assistance, and then expelled by an adjacent nitrogen atom to afford an incipient iminium ion which is then engaged by the electron-rich aromatic core. Apart from forging this critical domain, however, this step also serves to desymmetrize the growing molecule as only one aromatic ring within **18** participates in the reaction. Next, a facile Pictet–Spengler cyclization between **19** and an appropriate aldehyde (a reaction that we will discuss in more detail shortly) would complete the second tetrahydroisoquinoline subunit within the core of the ecteinascidins, leaving only incorporation of the ten-membered lactone to establish the entire gross molecular skeleton of this family of natural products. In principle, this final objective could be achieved if **20** were oxidized to its reactive quinone methide congener **21**, the same type of intermediate that was invoked earlier in Scheme 2 as an electrophilic species that could be engaged by DNA. In this case, however, the quinone methide is intercepted by a nucleophilic sulfur atom appended to a tetrahydroisoquinoline system, leading to the incorporation of the third and final such unit as expressed in **22**. Subsequent assembly of the ten-membered lactone ring using functionality derived from the R group at C-1 and adjustment of the oxidation and substitution patterns on each of the three aromatic rings would then complete the biosynthesis of the ecteinascidins.[9]

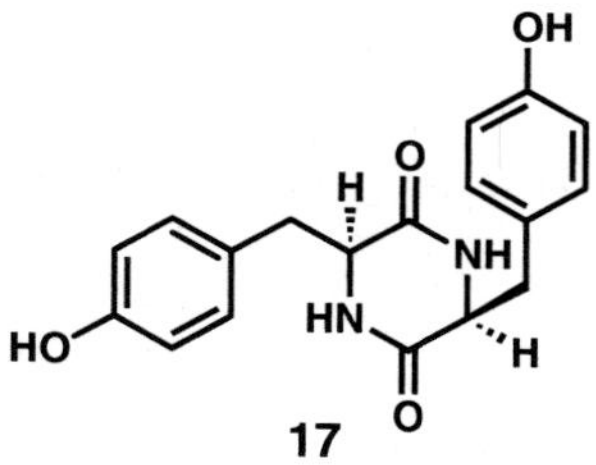

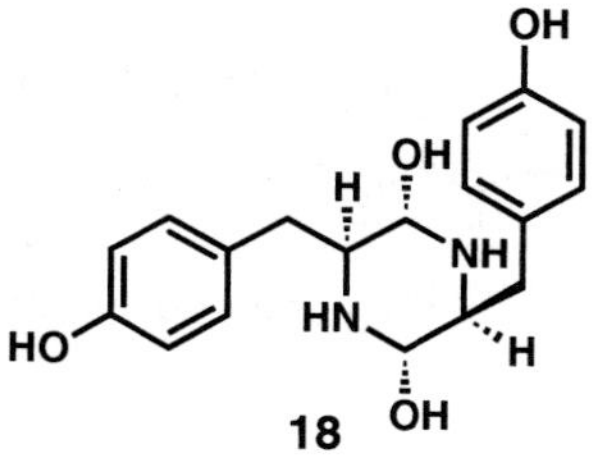

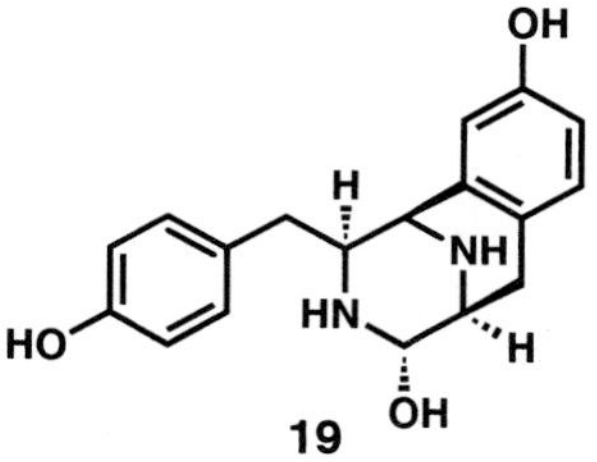

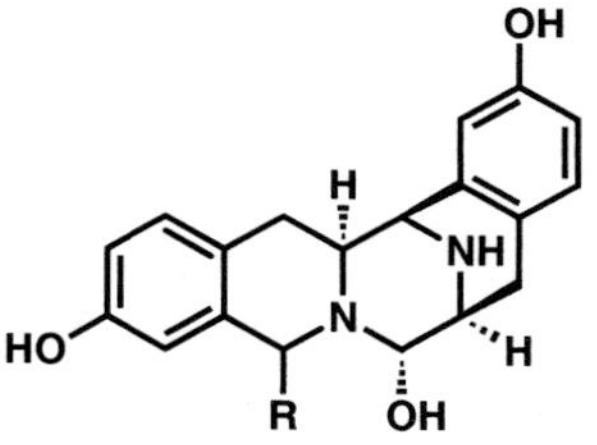

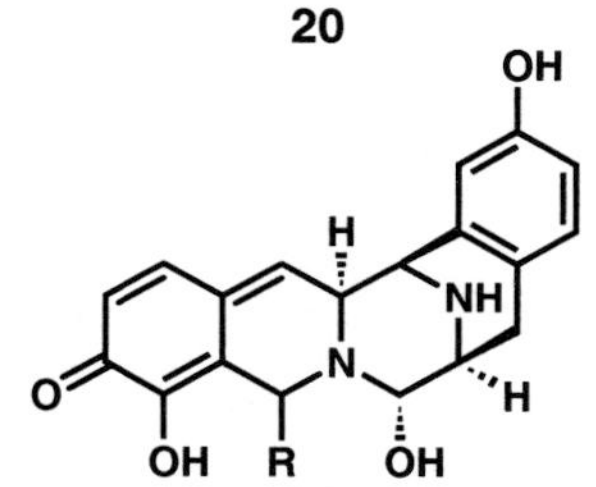

While executing a chemical synthesis directly patterned on this attractive biogenetic scheme would be highly challenging (as it would be hard to garner control of selective conversions such as **17** into **19** which are easily executed with enzymes), several of the key ring forming steps establish a useful framework for how one might consider generating these motifs in the laboratory. Indeed, it was with these general operations in mind that the Corey group developed their synthetic blueprint towards ecteinascidin 743 (**1**). The initial disconnections are shown in Scheme 4.

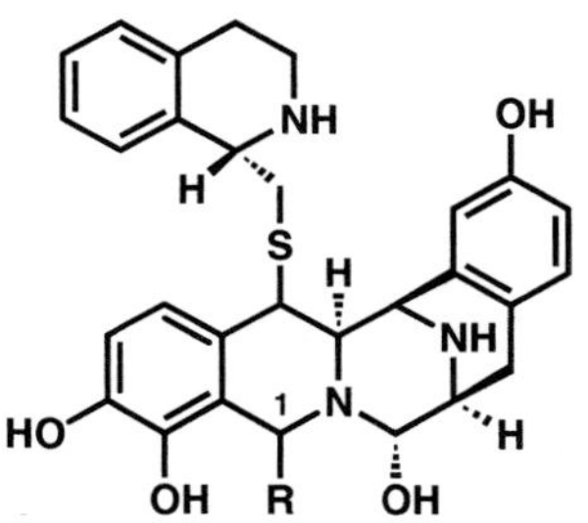

Anticipating that the central pentacyclic core of ecteinascidin 743 (**1**) would need to be constructed prior to the formation of either the ten-membered lactone (ring F) or its appended tetrahydroisoquino-

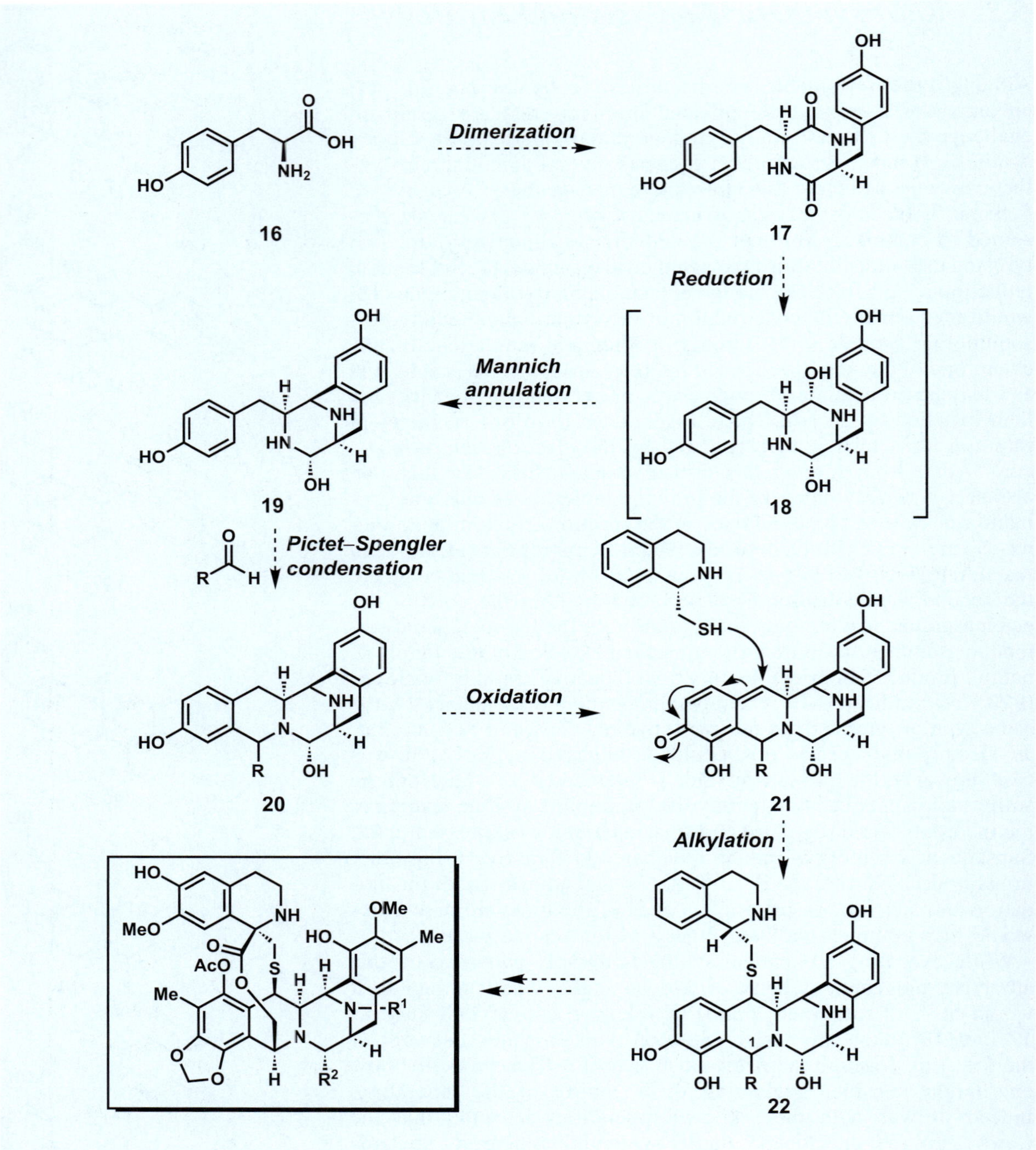

Scheme 3. Postulated biosynthesis of the ecteinascidins.

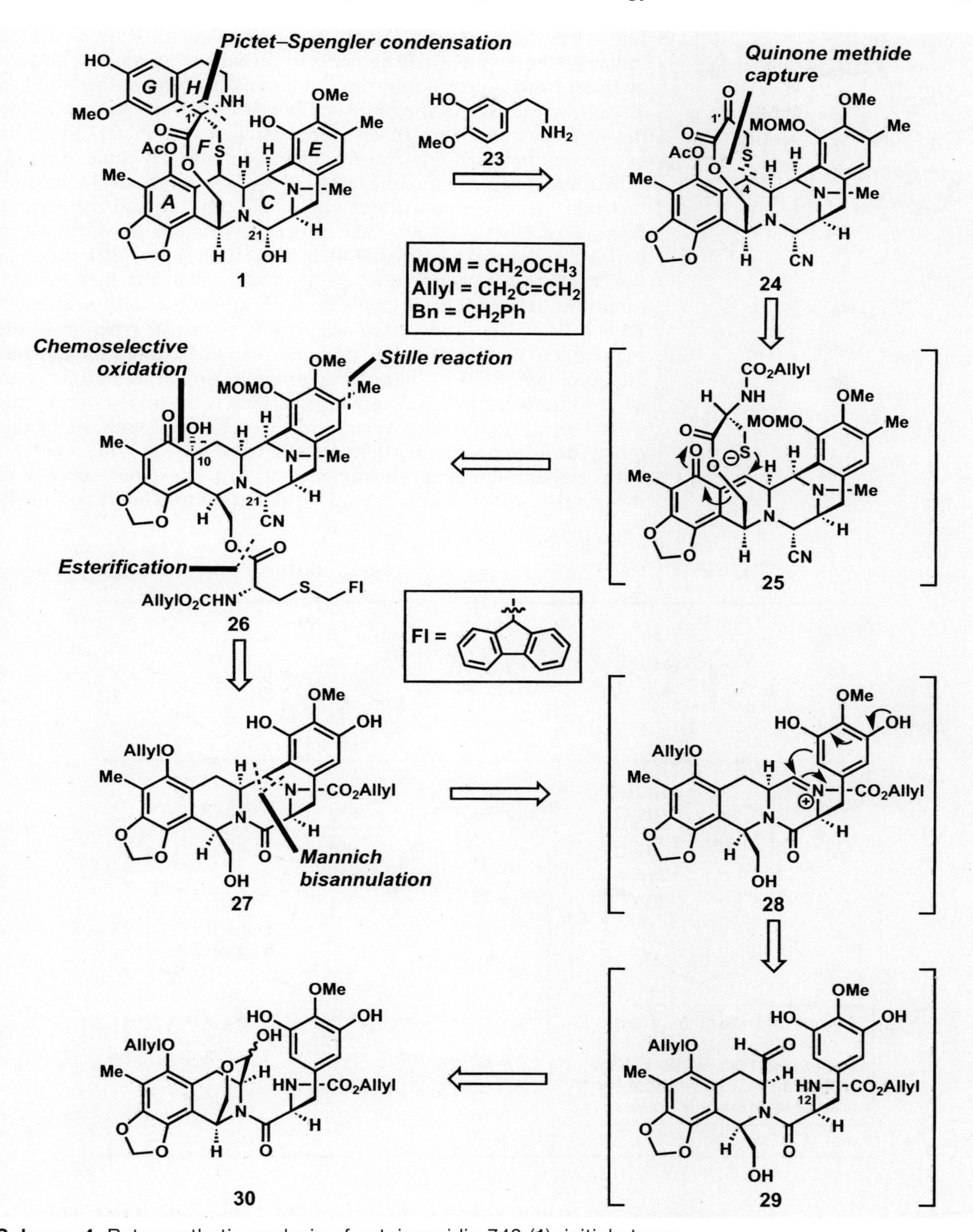

Scheme 4. Retrosynthetic analysis of ecteinascidin 743 (**1**): initial stages.

line system (rings G and H), all the opening simplifications sought to excise these structural segments to reveal such a polycyclic subgoal structure. Accordingly, the GH tetrahydroisoquinoline motif of **1** was cleaved first at the indicated bonds to reveal a ketone within the ten-membered ring of **24**, forecasting that a Pictet–Spengler cyclization between this intermediate and amine **23** could successfully forge this domain in the late-stages of the synthesis. In general, this transformation constitutes one of the most typical methods to fashion such ring systems and proceeds through the mechanism defined in Scheme 5: after the formation of an imine (**32**), protonation leads to an electrophilic species (**33**), which can then be captured through a Mannich cyclization to afford the desired product **35** via **34**.[13] The only drawback to this powerful reaction is the requirement that the aromatic ring must carry activating substituents at appropiate positions in order to provide suitable electronic character for annulation to occur as aromaticity is destroyed during this event. Since the tyrosine system required for the synthesis of this tetrahydroisoquinoline system in ecteinascidin 743 (**1**) contains such electron-donating substituents at two sites, the success of this ring-forming process would virtually seem assured and might

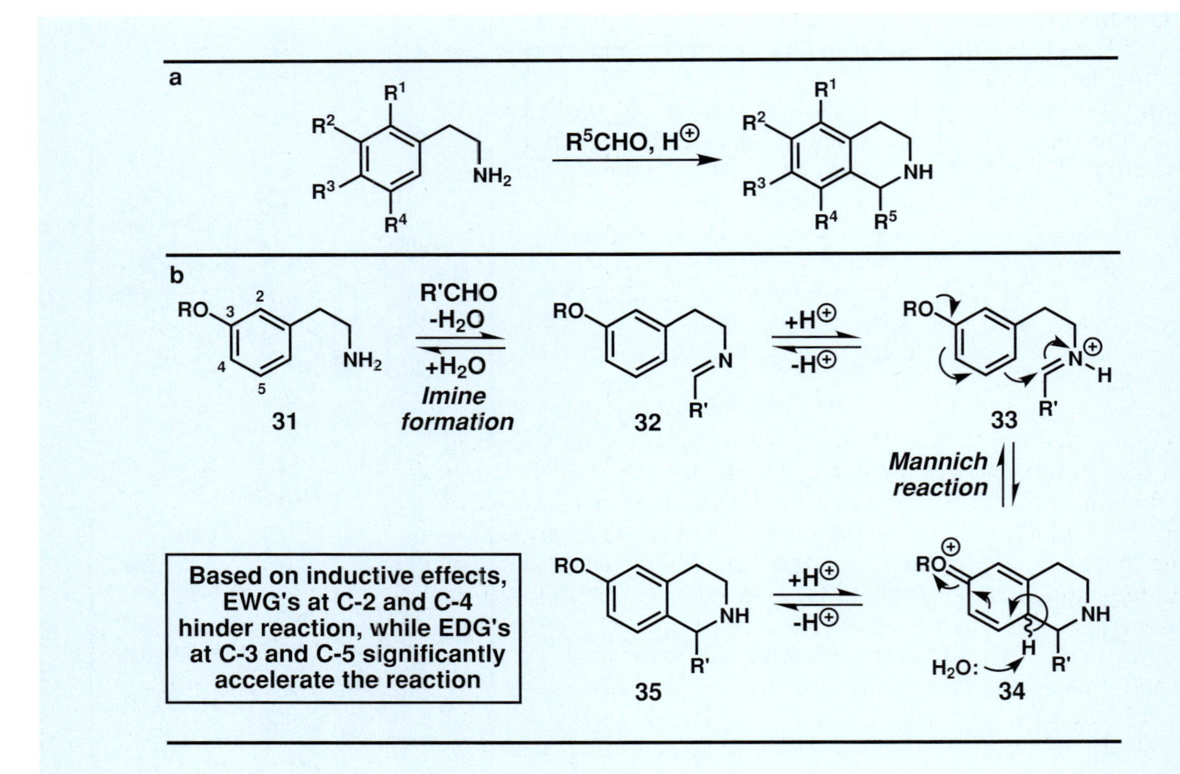

Scheme 5. The Pictet–Spengler reaction (a) and proposed mechanism (b). EDG = electron-donating group, EWG = electron-withdrawing group.

even proceed under exceedingly mild conditions during the actual synthesis.

The major issue clouding this initial disconnection, however, was not the formation of the tetrahydroisoquinoline system itself, but whether or not the spiro linkage at the C-1′ position would be generated in a stereocontrolled manner. Based on the mechanism of the Pictet–Spengler reaction defined in Scheme 5, though, there was reason to believe that such stereoselectivity could be achieved. While the cationic cyclohexadiene system in **34** is likely to constitute a fleeting intermediate since there is a strong driving force to restore its aromaticity to form either **33** or **35**, it is conceivable that this species would be dramatically stabilized if it were stacked above the electron-rich A-ring in **24** (Scheme 4). As a result, such an interaction would direct the orientation of the isoquinoline system in the direction required for the asymmetric generation of the C-1′ quaternary center. Even if all of the material did not cyclize immediately along these lines, since all the steps in the Pictet–Spengler reaction are reversible, as long as the overall event is under thermodynamic control then the preferred spiro linkage should result exclusively because a π-stacked product is more stable than one that lacks this secondary interaction. A second contributing feature for such selectivity derives from steric interactions, in that the alternate spiro linkage would afford a product in which the G- and E-rings and their substituents would be far more crowded than in the desired adduct.

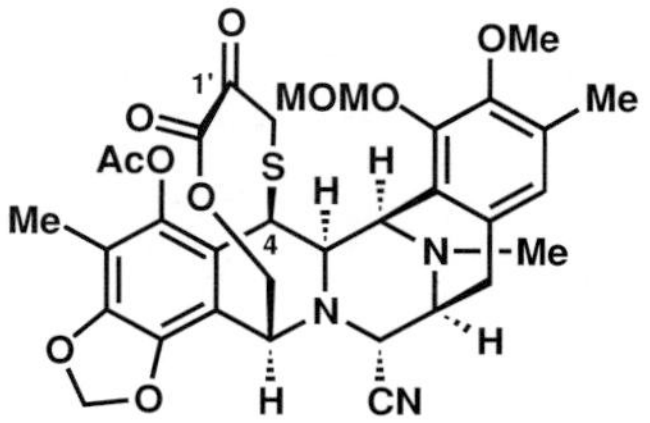

Assuming success in this late-stage operation in terms of stereoselectivity, the only other disconnection separating **1** from **24** was modification of the hydroxy group at C-21 to a nitrile function. Along the lines of Scheme 2, it appeared reasonable to the Corey group that suitable activation of this group in **24**, followed by imine formation as would occur through the participation of the nitrogen atom of the adjacent ring, could then lead to the free alcohol of ecteinascidin 743 (**1**) through attack of nucleophilic water from the more accessible, convex face. In this regard, the nitrile group in **24** could be considered simply as a masked form of this motif to be carried through several steps of the synthesis.

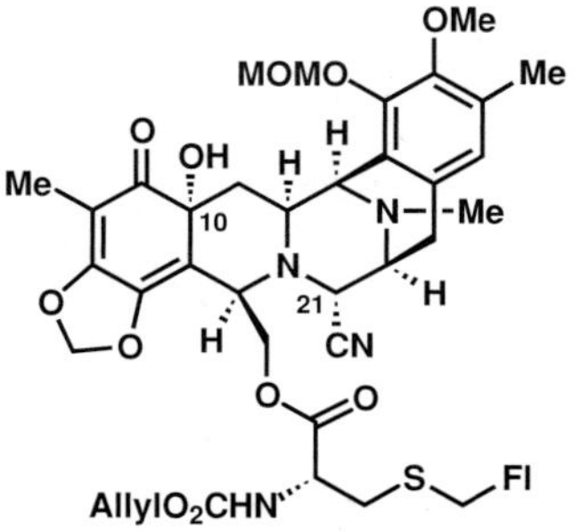

Having excised the G- and H-rings, the opportunity now seemed ripe to disconnect the far more formidable ten-membered F-ring along the lines of the biosynthetic hypothesis defined above in Scheme 3 in which a sulfur nucleophile engaged an intermediate quinone methide. While easily stated, successful application of this strategy to a laboratory synthesis would undoubtedly require the development of an inventive and chemoselective protocol in order to form such a reactive intermediate. In this vein, the Corey group anticipated that if a compound such as **26** could be synthesized, then the C-10 hydroxy group within it could be eliminated under suitable conditions to afford *o*-quinone methide **25**, a species which could then be engaged by a sulfur atom tethered through C-1 to form **24** in a stereospecific manner. Since the nucleophilic thiolate would likely have to be protected throughout the course of

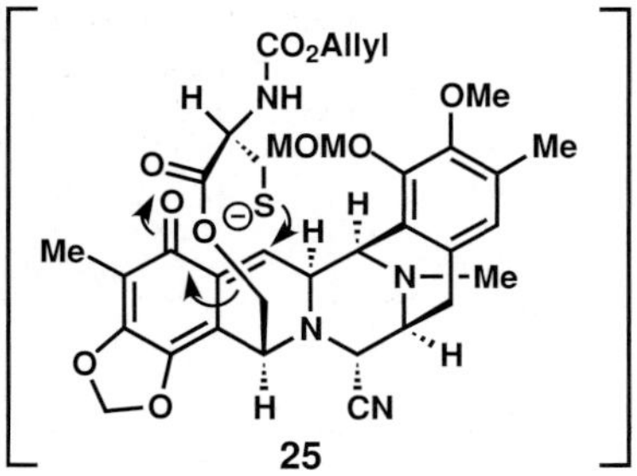

24

26

25

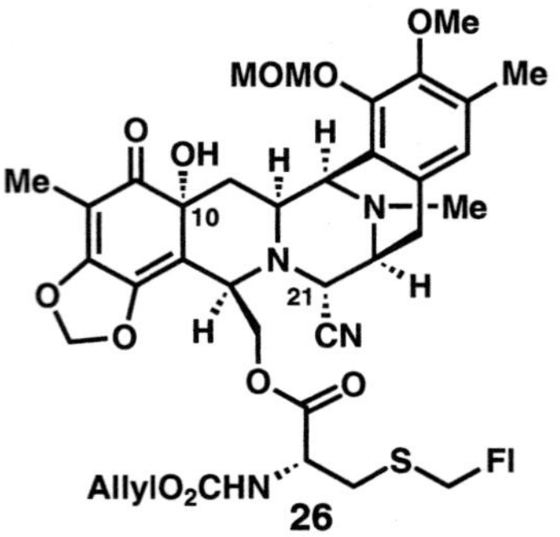

the sequence leading to **26**, however, this nucleophile would need to be generated in the same step as the *o*-quinone methide for this approach to prove successful. Accordingly, as basic conditions could be envisioned to lead to the elimination of the C-10 hydroxy group, the 9-fluorenyl protecting group was selected at this stage for the thiol group, since it too is base-sensitive due to double benzylic activation (as shown in the cleavage mechanism in the neighboring column).[14] Overall, this sequence certainly constitutes a challenging set of operations that would likely require significant empirical investigation to effect. If successful, the cascade would fashion this critical motif in a highly efficient manner. Before continuing with this analysis, one should also note that although the ketone in **24** has been converted into the stable *N*-allyloxycarbonyl in **25** and **26**, these groups are synthetically equivalent since deprotection of the Alloc group, followed by an oxidative transamination reaction with imine formation and hydrolysis, would afford a ketone.

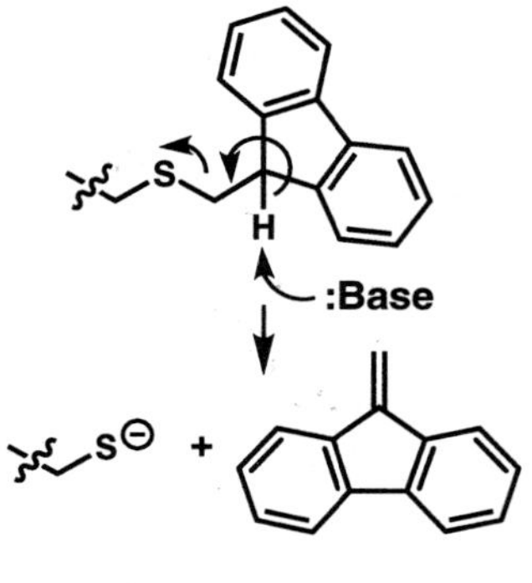

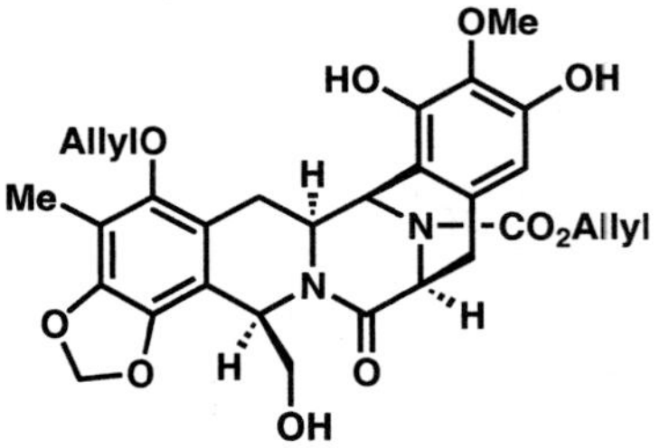

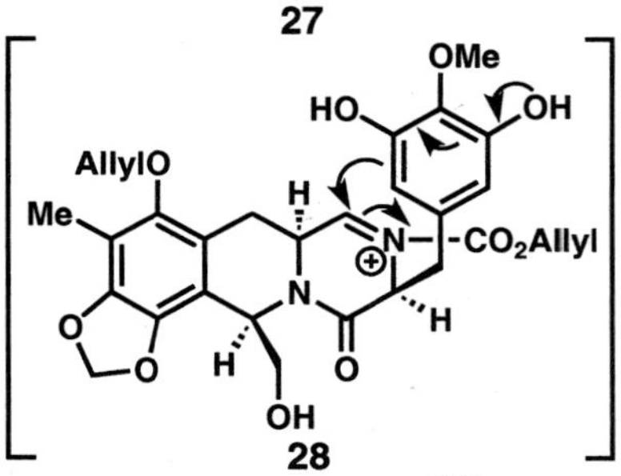

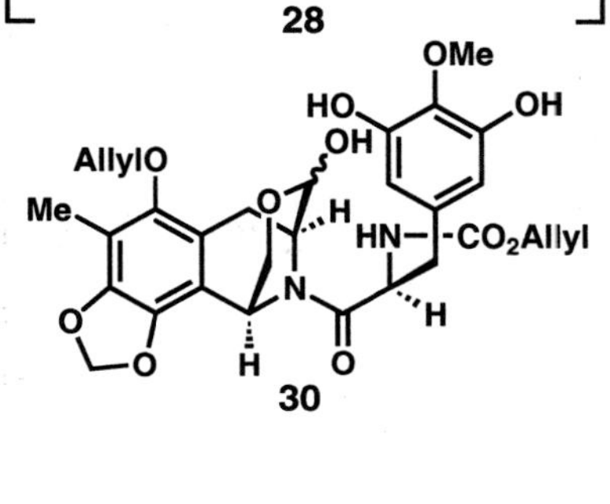

While this initial spate of disconnections has removed a challenging subset of the formidable structural elements of the ecteinascidin architecture, the remaining motifs within **26**, including five rings and six stereocenters, are still quite daunting. However, following several modifications (excision of the ester side chain, removal of the C-10 hydroxy group, adjustment of the C-21 nitrile to a precursor ketone, and conversion of the E-ring methyl group into a hydroxy group), a more manageable pentacyclic array (**27**) resulted which seemed primed for a series of highly productive disconnections based on the biosynthetic hypothesis discussed above. In this regard, it would seem reasonable to presume that the D-ring in **27** could be disassembled directly at this juncture to intermediate **30** by counting on a Mannich-type bisannulation event to forge this motif during the synthesis.[15] Although this transformation is reminiscent of the conversion of **17** into **19** mentioned earlier (cf. Scheme 3), the true brilliance of the design in this context rested in the recognition that the unstable aminal intermediate needed to ultimately generate a reactive iminium electrophile (**28**) for eventual cyclization could be masked as a lactol (**30**). Thus, upon acidic hydrolysis of this motif at the appropriate juncture in the synthesis, aldehyde **29** should be revealed which could form a hemiaminal with the adjacent N-12 amine and thereby initiate the cascade sequence envisioned for the synthesis of **27**.

Having reached this new intermediate, lactol **30** (see Scheme 6) could, of course, be derived from its corresponding lactone (**36**) through a controlled reduction, providing a new subgoal structure whose amide linkage could then be retrosynthetically dissected into two major fragments: amine **37** and carboxylic acid **38**. Exploitation of another Pictet–Spengler disconnection, this time in the intramolecular sense, would then afford **39** from tetrahydroisoquinoline **37** (in which the requisite aldehyde has been masked as a dimethyl acetal in **39**). In the forward direction, this ring-forming process should lead to the stereoselective formation of the C-1 stereogenic center, assuming that **39** was homochiral.

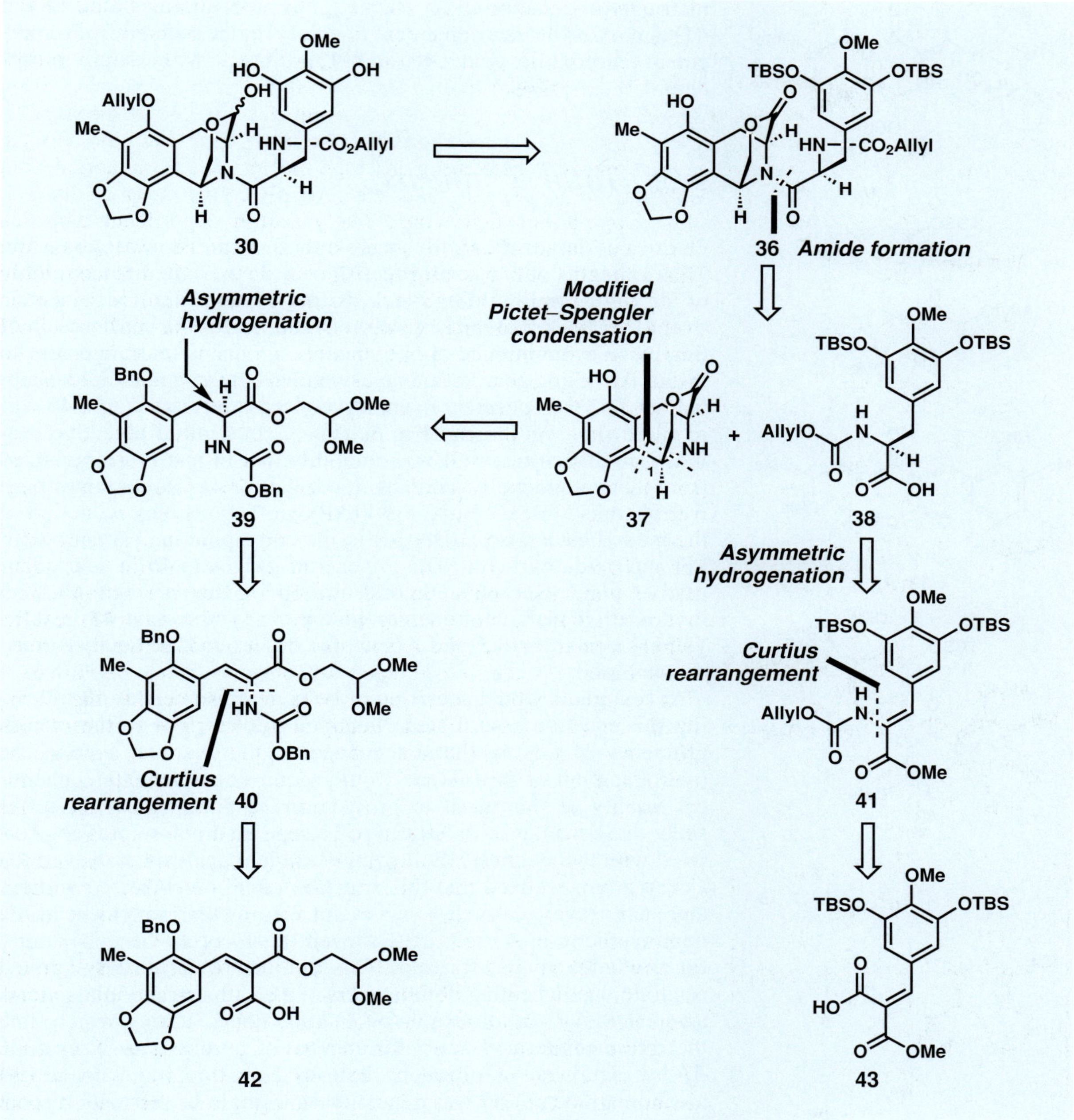

Scheme 6. Retrosynthetic analysis of ecteinascidin 743 (**1**): final stages.

Thus, based on these final simplifications, the synthesis of ecteinascidin 743 (**1**) has been reduced to the creation of two building blocks of relatively similar complexity (**38** and **39**), each bearing only one stereogenic center and a protected amine. In principle, the latter elements could result in both fragments from the asymmetric hydrogenation of an alkene (present in intermediates **40** and **41**) and a Curtius rearrangement of an acyl azide derived from a precursor carboxylic acid (**42** and **43**), events with ample precedent.[16,17]

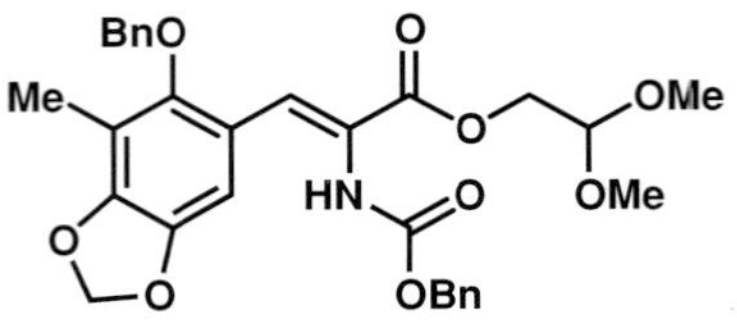

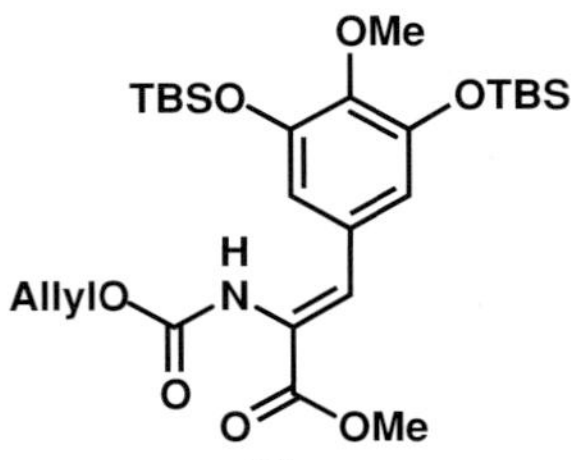

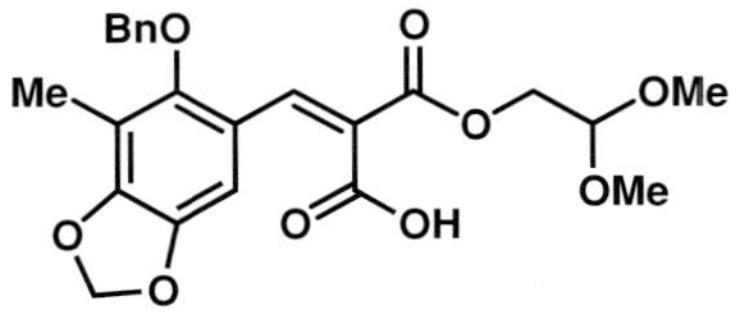

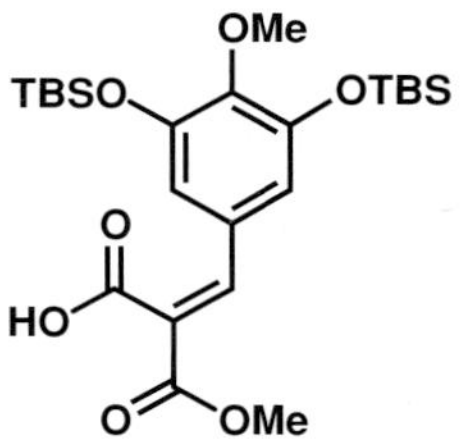

5.3 Total Synthesis

Our discussion of the Corey group's total synthesis of ecteinascidin 743 (**1**) begins with efforts directed towards the expedient assembly of the right-hand building block, E-ring **38**. In order to explore the overall feasibility of the key steps envisioned for the construction of this piece as mentioned above, namely a Curtius rearrangement to install the amide portion and an asymmetric hydrogenation to establish its lone stereocenter, a synthesis of α,β-unsaturated acid **43** was required first. As indicated in Scheme 7, this initial objective provided no difficulties as it was smoothly met in just four operations from the commercially available methyl ester **44**. Silylation of both free phenols (TBSCl, Et_3N, 4-DMAP) was followed by reduction of the methyl ester with DIBAL-H to the corresponding primary alcohol and oxidation with PDC to generate aldehyde **45** in near quantitative yield. Subsequent condensation of this product **45** with monomethyl malonate then completed the synthesis of **43** in 92 % yield as a mixture of *E* and *Z* isomers (1:1) around the newly formed double bond.

At first glance, the generation of both stereoisomers of **43** following this reaction would seem inherently deleterious to the overall efficiency of this synthetic sequence, as the eventual asymmetric hydrogenation of this alkene would require complete stereochemical fidelity of this motif to provide any enantiomeric excess. As such, these isomers would need to be separated before pressing forward with the synthesis. Following sufficient analysis, however, the Corey group realized that this situation could be corrected without any such travail through a successful Curtius rearrangement in the next synthetic operation, as that event could offer the opportunity for productive isomerization. This conjecture fortunately proved accurate. Upon heating both isomers of **43** with diphenylphosphoryl azide (DPPA)[18] in toluene at 70 °C for 2 hours (during which time the intermediate acyl azide **46** underwent pyrolysis to isocyanide **47** by expulsion of nitrogen), followed by trapping with benzyl alcohol, amide (*Z*)-**50** was generated as a single isomer with respect to the geometry around the central C–C double bond. The reason for this fortunate result, however, is not obvious from the defined mechanistic picture in Scheme 7 and is worthy of some additional comment.

As has been demonstrated through countless examples of this rearrangement process in chemical synthesis,[17] the migration reaction that converts an intermediate such as **46** into **47** always proceeds with retention of stereochemical configuration. Thus, follow-

BnO Me O OMe HN O OMe OBn **40**

OMe TBSO OTBS H AllylO N O O OMe **41**

BnO Me O OMe O OH OMe **42**

OMe TBSO OTBS O HO O OMe **43**

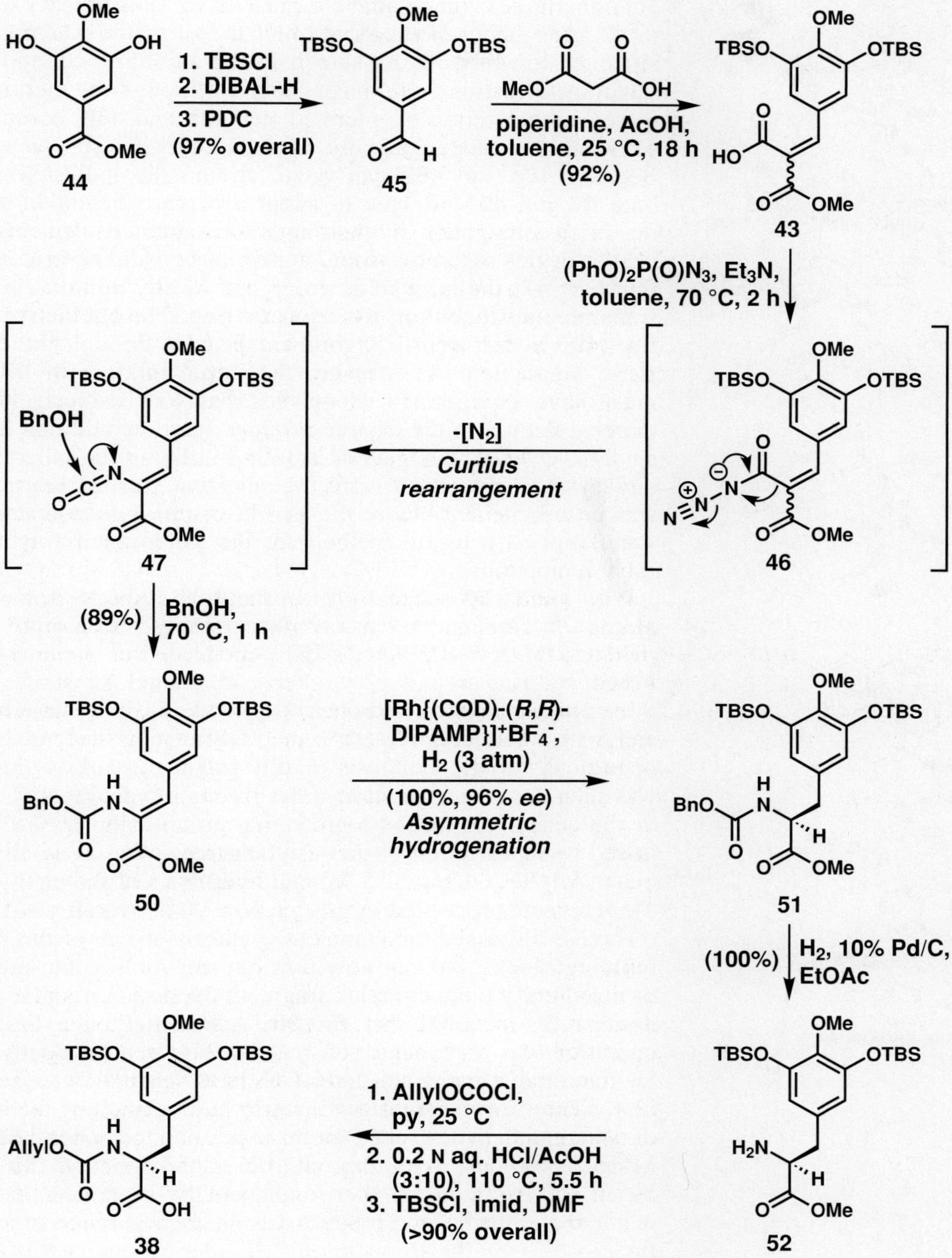

Scheme 7. Synthesis of building block **38**.

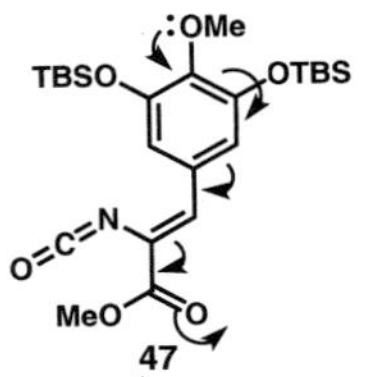

ing the formation of both isocyanide **47** and its *E* isomer (**49**, see column figure), these adducts must have isomerized exclusively to **47** after their generation in order for all of the material to have siphoned towards **50**. As shown in the adjoining column figures, this transformation could have occurred through the thermally promoted isomerization of intermediates **48a** and **48b**, assuming that **47** is the thermodynamically favored entity by several kcal/mol. Because **48a** and **48b** can rotate around the indicated axis and both **47** and **49** will seek to adopt a planar orientation such that maximal conjugation of their many unsaturated elements can be achieved, this assertion would appear to be valid as in a *Z* orientation (i. e. **47**) the isocyanide group can readily point its bulk away from any substituent on the aromatic ring. The alternative possibility (**49**) would incur a significant penalty through the indicated steric interaction. As a result, the initial mixture of **47** and **49** must have equilibrated along this pathway exclusively to **47**, thereby leading to the desired product upon the addition of benzyl alcohol. Overall, this general solution utilizing the inherent topology of the substrate to effect a selective Curtius rearrangement was unprecedented before the advent of this synthesis and clearly stands out as a useful method for the stereospecific synthesis of such compounds.

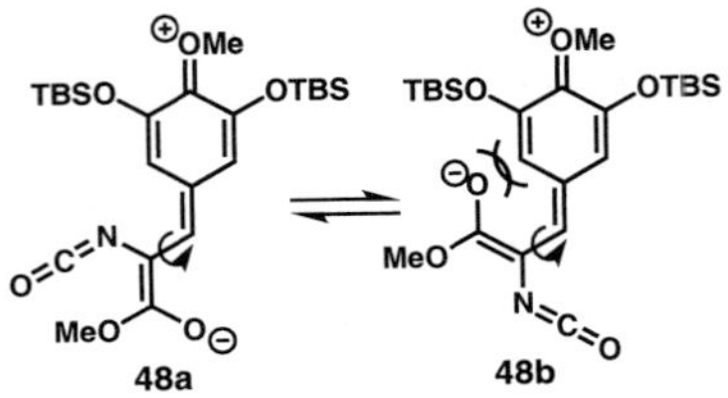

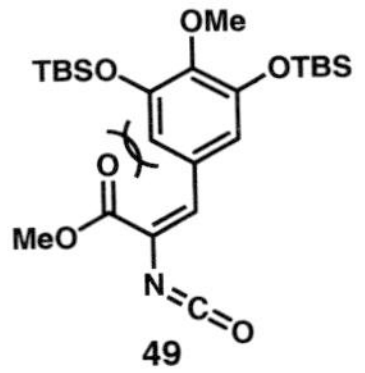

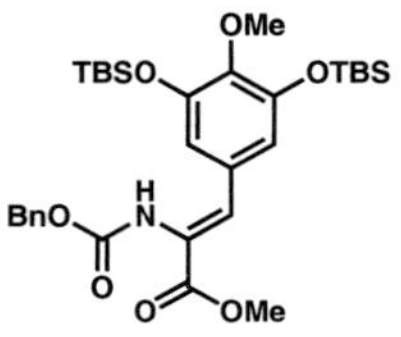

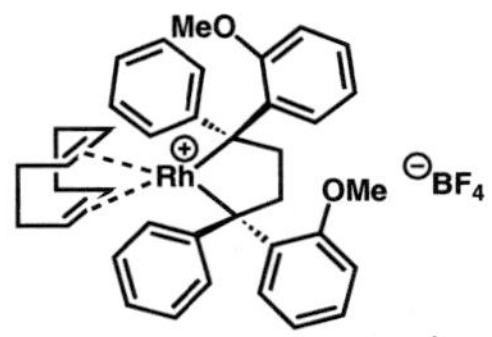

With amide **50** successfully in hand as a single isomer, asymmetric hydrogenation using the famous Monsanto catalyst [Rh{(COD)-(*R*,*R*)-DIPAMP}]$^+$BF$_4^-$ proceeded as previously established by Kagan and co-workers on related substrates[16] under three atmospheres of hydrogen to provide **51** in quantitative yield and with an impressive 96% *ee*. Having now executed the key operations for the synthesis of this fragment, carboxylic acid **38** was then completed in short order through hydrogenative removal of the benzyloxy carbonyl protecting group using 10% Pd/C, followed by reprotection of the resultant free amine as an allyl carbamate (AllylOCOCl, py, 25 °C) and hydrolysis of the methyl ester.[19] These events proceeded in greater than 90% overall yield.

Having discussed the complete synthesis of one of the two main building blocks, we can now turn our attention to the preparation of the slightly more complex fragment **39** based on a plan that predominantly mirrored that for **38**. Accordingly, the first critical operation was a condensation reaction between an aldehyde intermediate and a malonate derivative (see Scheme 9) to generate a new alkene for eventual asymmetric hydrogenation; the synthesis of the aryl aldehyde needed for this operation, compound **58**, is delineated in Scheme 8 starting with the methoxymethyl (MOM)-protected **53**. Although all that fundamentally separated this starting material and the desired product was the incorporation of additional functionality on the aromatic ring, in order to achieve this objective the Corey group needed to apply synthetic technology capable of engendering the requisite site selectivity. In this regard, the inclusion of the MOM protecting group within **53** was far from random in that its presence could be utilized productively to direct the

Rh[(COD)-(R,R)-DIPAMP]$^+$BF$_4^-$

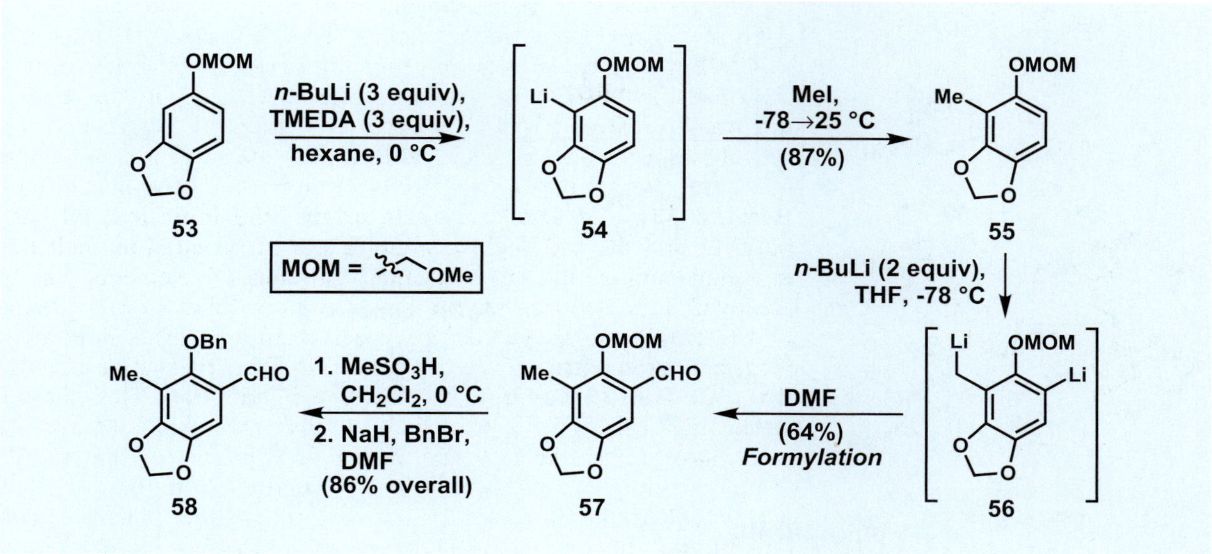

Scheme 8. Synthesis of aryl building block **58**.

specific location of lithiation, which would enable the controlled introduction of these motifs upon the subsequent addition of the appropriate electrophiles. This strategy indeed proved productive, as upon initial treatment of **53** with 3 equivalents of *n*-BuLi in the presence of tetramethylethylenediamine in hexane at 0 °C over 4 hours, lithiation occurred exclusively at the most hindered site on the aromatic ring (to give **54**) since this species could benefit from maximal coordination with the available oxygen atoms. With the addition of MeI and subsequent warming to ambient temperature, **55** was then generated in 87 % yield. A second iteration of this sequence, this time with selective lithiation at the unsubstituted position adjacent to the MOM-protected phenol (**56**), completed the assembly of aldehyde **57** following the reaction of this intermediate with DMF, a well-known formyl cation equivalent. Having accomplished its role as a lithiation director, the MOM group was no longer required for the remainder of the projected synthesis of ecteinascidin 743 (**1**), and, as such, was excised at this juncture with methanesulfonic acid in CH_2Cl_2 at 0 °C and replaced with a benzyl group under standard conditions to afford **58** in 86 % overall yield.

Following the synthesis of allyl 2,2-dimethoxyethyl malonate (**60**) from **59** (Scheme 9), aldehyde **58** was condensed with this intermediate in the presence of AcOH and pyrrolidine in benzene to provide α,β-unsaturated intermediate **61** in near quantitative yield as a mixture of *E* and *Z* isomers. As before, the generation of a mixture of isomers at this point in the synthesis was ultimately of no consequence, as directly after selective cleavage of the allyl protecting group with $Pd(PPh_3)_4$ in buffered media ($Et_3N{\cdot}HCOOH$), Curtius rearrangement of the resultant carboxylic acid (**42**) afforded

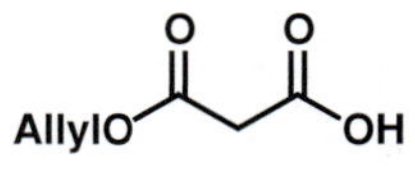

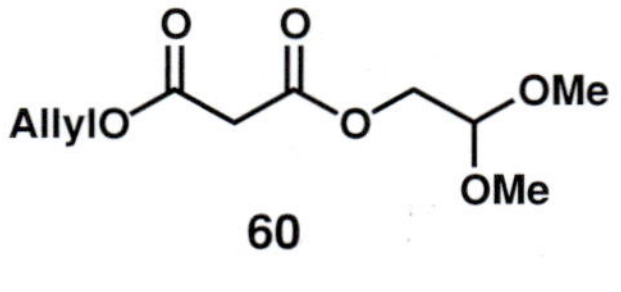

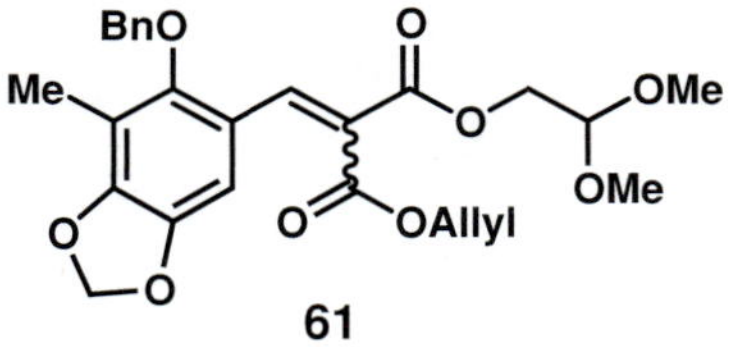

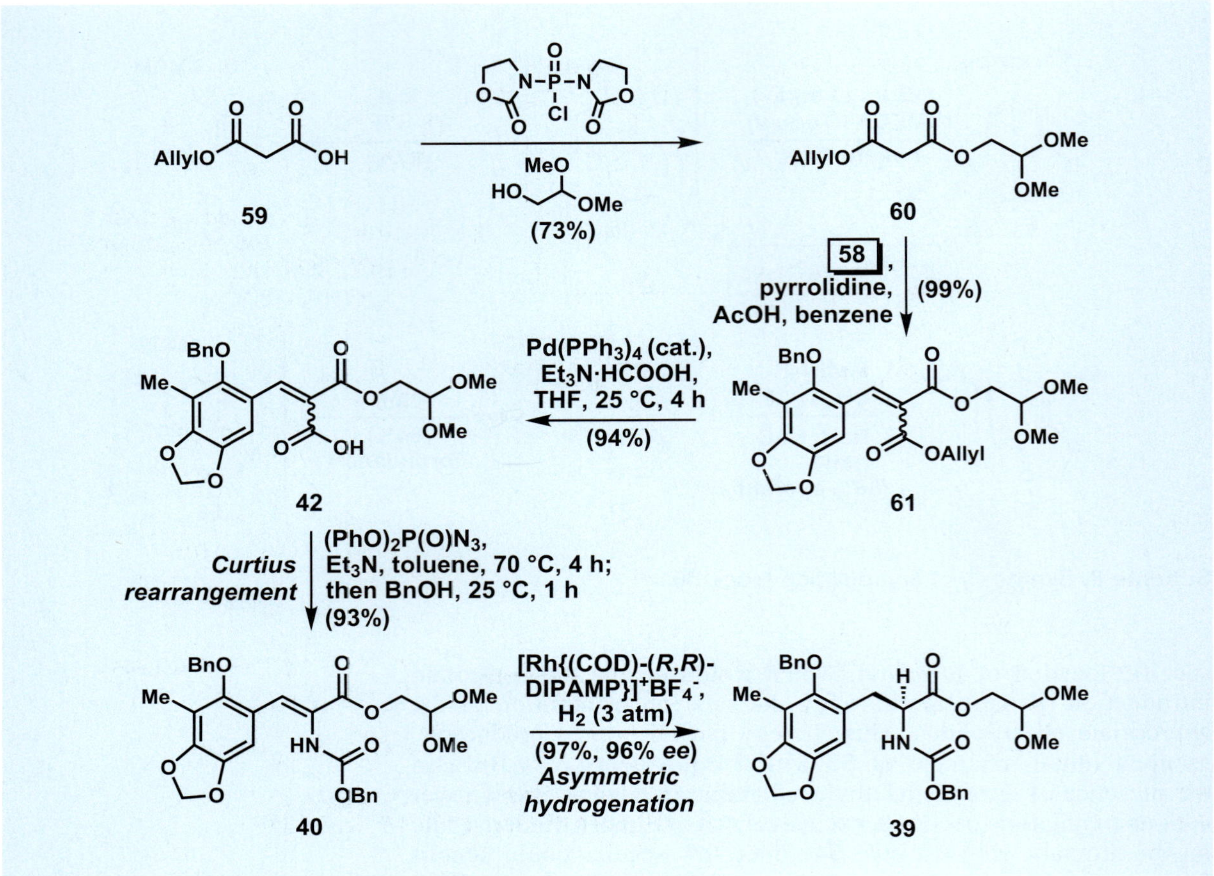

Scheme 9. Synthesis of key intermediate 39.

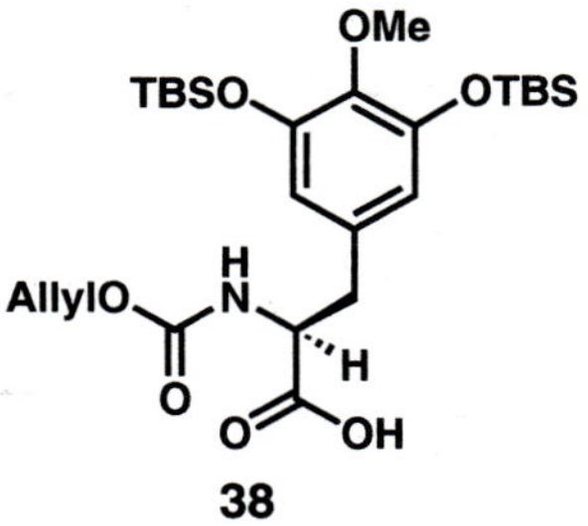

Cbz-protected amine 40 in 87 % overall yield from 61, and as a single stereoisomer, through the same isomerization pathway presented in the context of Scheme 7. Asymmetric hydrogenation of the lone C–C double bond in this intermediate using the identical asymmetric catalyst system encountered earlier then completed the assembly of 39 in 97 % yield and with 96 % *ee*.

With an eye towards merging this half of the molecule with the already synthesized "right-hand" portion (38), compound 39 first had to be converted into the tricyclic lactone 37, a task for which an intramolecular Pictet–Spengler reaction was deemed to be ideal. As shown in Scheme 10, this objective was smoothly met upon initial $BF_3 \cdot OEt_2$-mediated cleavage of the dimethyl acetal, followed by cyclization of the resultant aldehyde (62) to the tetrahydroisoquinoline ring system of 63 in 73 % overall yield over the course of 18 hours in CH_2Cl_2 through the action of $BF_3 \cdot OEt_2$ in the presence of 4-Å molecular sieves. As expected, 63 was generated as a single diastereomer. Subsequent cleavage of the Cbz group guarding the nitrogen atom under standard hydrogenation

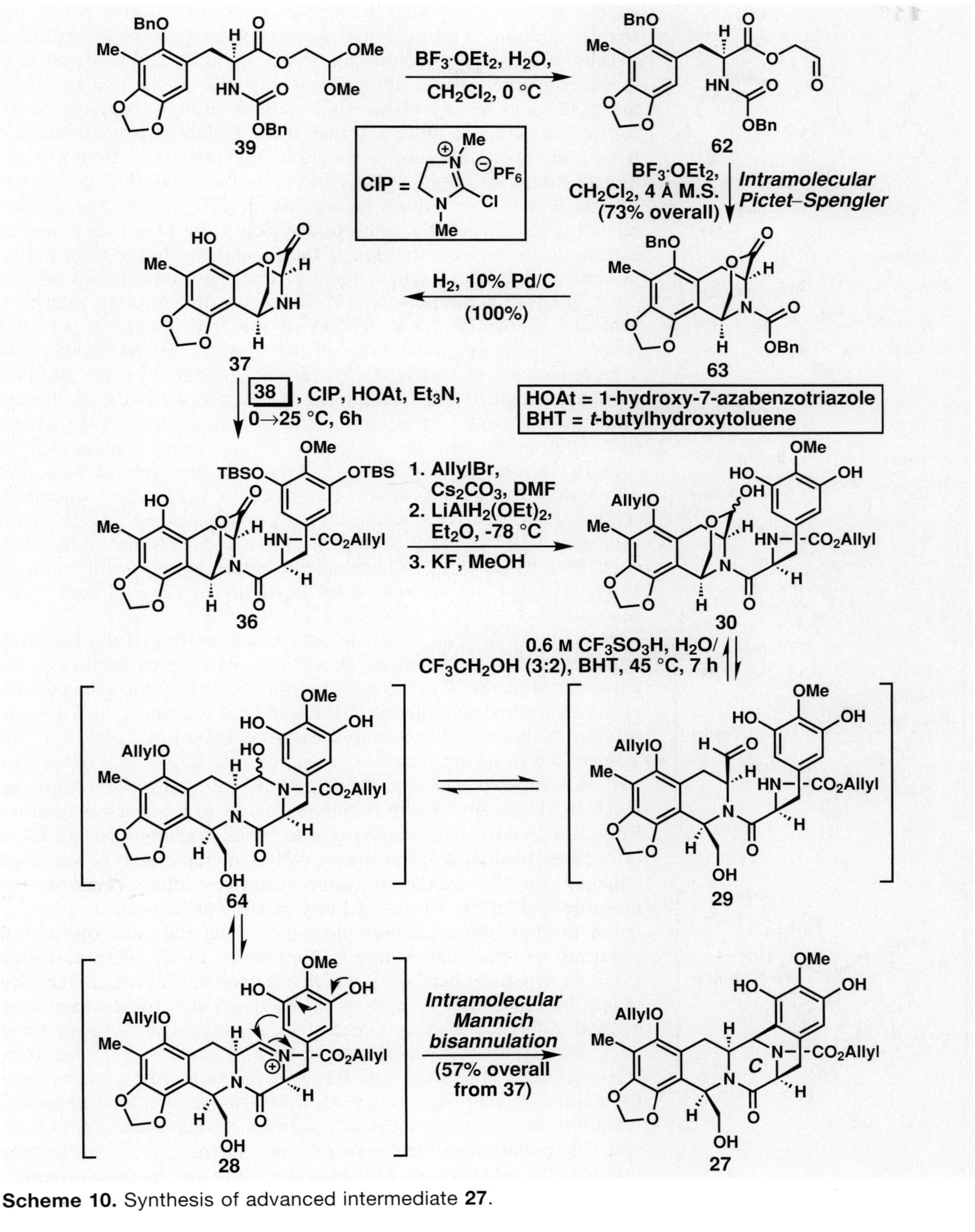

39 → 62: $BF_3 \cdot OEt_2$, H_2O, CH_2Cl_2, 0 °C

CIP = [2-chloro-1,3-dimethylimidazolinium] $PF_6^{\ominus}$

62 → 63: $BF_3 \cdot OEt_2$, CH_2Cl_2, 4 Å M.S. (73% overall) *Intramolecular Pictet–Spengler*

63 → 37: H_2, 10% Pd/C (100%)

37 → 36: 38, CIP, HOAt, Et_3N, 0→25 °C, 6h

HOAt = 1-hydroxy-7-azabenzotriazole
BHT = *t*-butylhydroxytoluene

36 → 30: 1. AllylBr, Cs_2CO_3, DMF 2. $LiAlH_2(OEt)_2$, Et_2O, -78 °C 3. KF, MeOH

30 ⇌ 29: 0.6 M CF_3SO_3H, H_2O/CF_3CH_2OH (3:2), BHT, 45 °C, 7 h

29 ⇌ 64 ⇌ 28

28 → 27: *Intramolecular Mannich bisannulation* (57% overall from 37)

Scheme 10. Synthesis of advanced intermediate **27**.

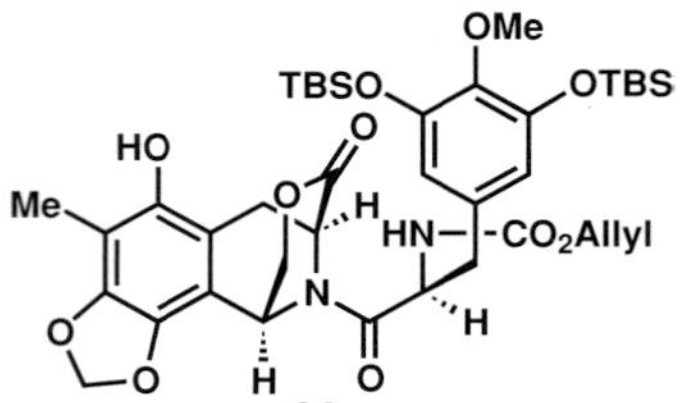

conditions then completed the preparative steps required before the critical reactions at the heart of the overall strategy could be explored.

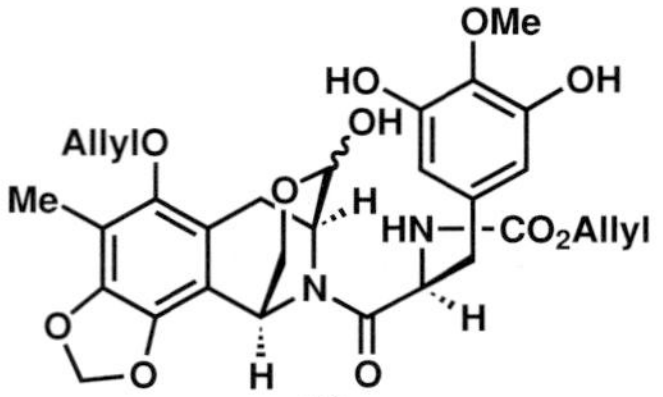

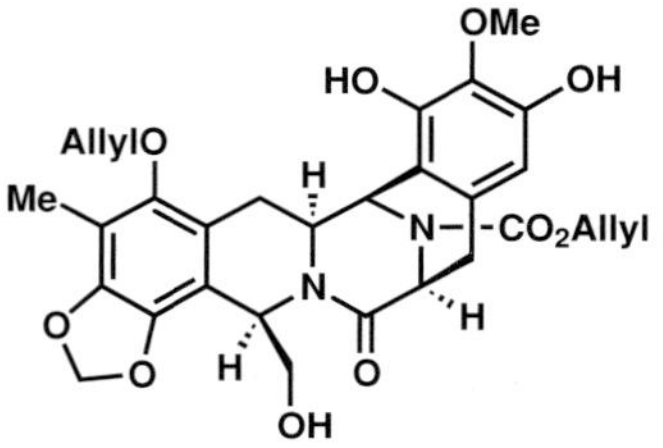

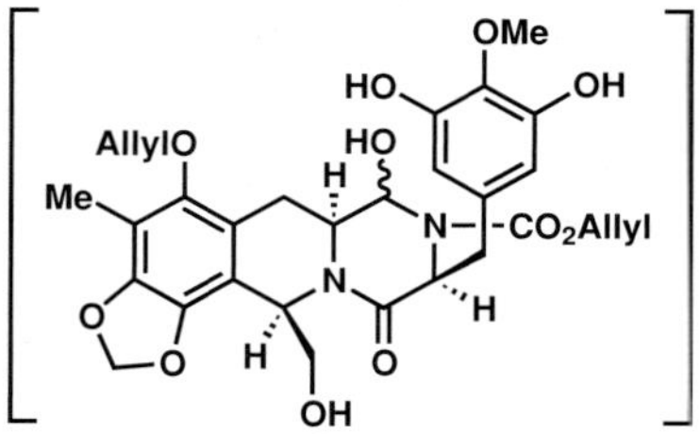

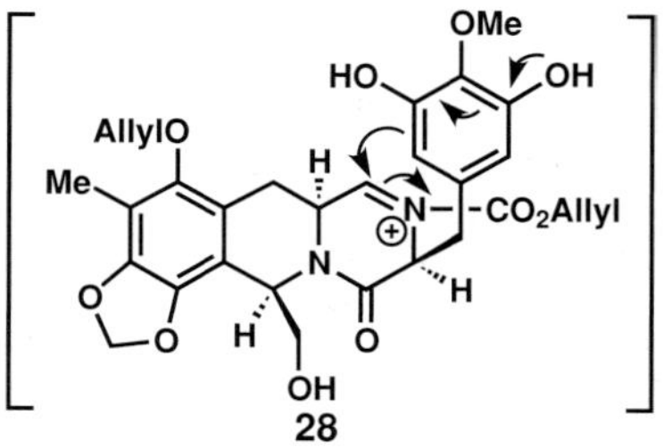

Thus, at this juncture, the previously parallel nature of the synthesis converged into a single linear sequence as the two key fragments (**37** and **38**) were smoothly merged through a peptide bond-forming event by adding a solution of the acylating reagent obtained from **38**, 1-hydroxy-7-azabenzotriazole (HOAt), 2-chloro-1,3-dimethylimidazolidinium hexafluorophosphate (CIP), and Et_3N dropwise at 0 °C to amino lactone **37** in THF.[20] With the desired adduct (**36**) in hand, the opportunity to test the Mannich sequence envisaged for the construction of the second tetrahydroisoquinoline system was nearly reached, merely requiring some initial adjustment of the functionality within **36** to form **30** through relatively standard chemistry: a) protection of the free phenol as an allyl ether; b) selective reduction of the lactone to the lactol with 1.1 equivalents of $LiAlH_2(OEt)_2$ (freshly prepared by the addition of 1 equivalent of ethyl acetate to a solution of $LiAlH_4$ in diethyl ether at 0 °C and stirring for 2 hours);[21] and c) cleavage of the two silyl ethers on the aromatic E-ring using excess KF in MeOH. Before moving forward, however, one should note that although the reducing agent employed in the second operation might seem relatively esoteric, it proved critical for material throughput (92 % for this step) as more conventional reductants, such as DIBAL-H in toluene and triisopropylbutyl aluminum in Et_2O, afforded the desired lactol in yields of only 41 and 73 %, respectively.

With all the desired motifs now in place, probes of the Mannich annulation step commenced in earnest. Most pleasingly, careful exposure of lactol **30** to 0.6 M trifluoromethanesulfonic acid and catalytic *t*-butylhydroxytoluene (BHT, a radical scavenger) in a solvent mixture comprised of water and trifluoroethanol (3:2) at 45 °C for 7 hours led to its anticipated conversion into **27**. As discussed earlier, this most productive event can be rationalized through the mechanistic picture shown in Scheme 10 in which a series of equilibria initiated by acid catalysis led to initial opening of lactol **30** to afford intermediate **29**; the strategically tethered aldehyde was then engaged by the secondary alloc-protected amine to form the unstable aminal **64**, whose hydroxy group was subsequently eliminated through proton capture leading to the iminium species **28** required for eventual Mannich annulation to **27**. The last five steps shown in Scheme 10, which effected the union of the two major building blocks and established all the rings within the ecteinascidin 743 core, proceeded in an impressive overall yield of 57 %, a result that constitutes an average yield of 92 % per step.

Having now appended both tetrahydroisoquinoline systems onto the central C-ring piperazine, the only major synthetic objective remaining to complete the total synthesis of ecteinascidin 743 (**1**) was the formation of the ten-membered macrocyclic F-ring lactone through the envisioned biomimetic *o*-quinone methide capture. Before this operation could be explored, however, several functional group manipulations of **27** were required, most of which are delineated in Scheme 11. First, having served its purpose as a means to link the major building blocks together in the initial creation of **36** (cf. Scheme 10), the C-21 carbonyl functionality was no longer needed and, thus, was converted into a nitrile (i.e. **65**) in 87 % yield through initial formation of an aminal, as effected with $LiAlH_2(OEt)_2$, followed by acid-catalyzed elimination and cyanide capture. As before, this particular reducing reagent proved uniquely

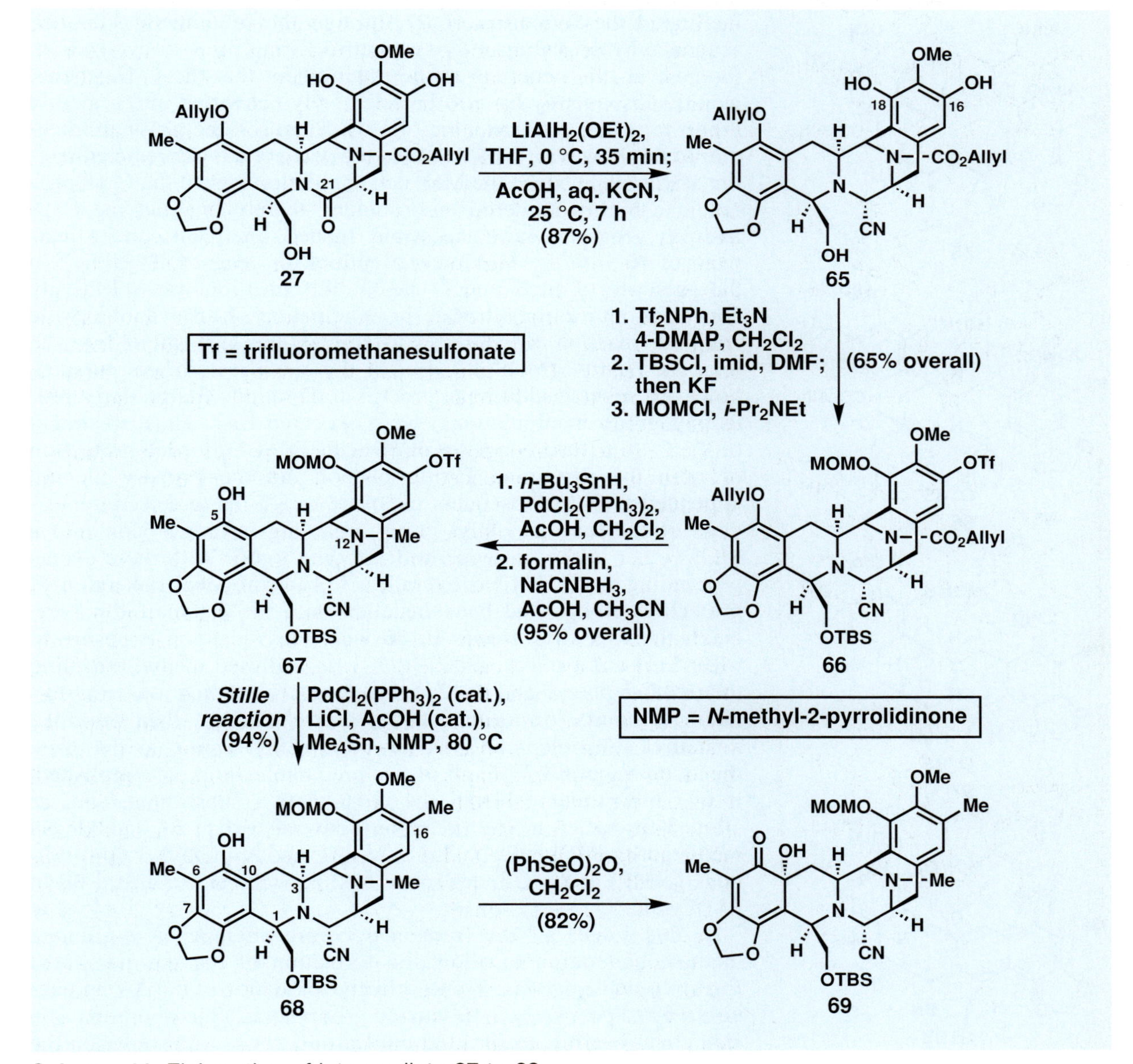

Scheme 11. Elaboration of intermediate **27** to **69**.

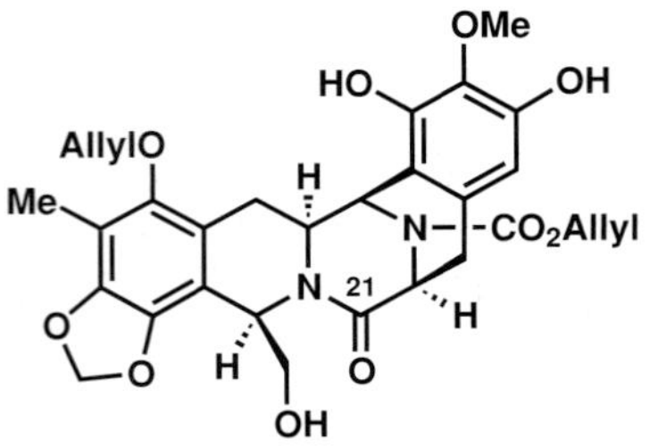

suitable for the controlled, partial reduction of the synthetic intermediate. As such, despite the fact that $LiAlH_2(OEt)_2$ has been employed quite rarely in complex molecule synthesis, its high level of success in two operations within the ecteinascidin context seems to indicate that it is a reducing agent worth examining when more conventional hydride sources fail.

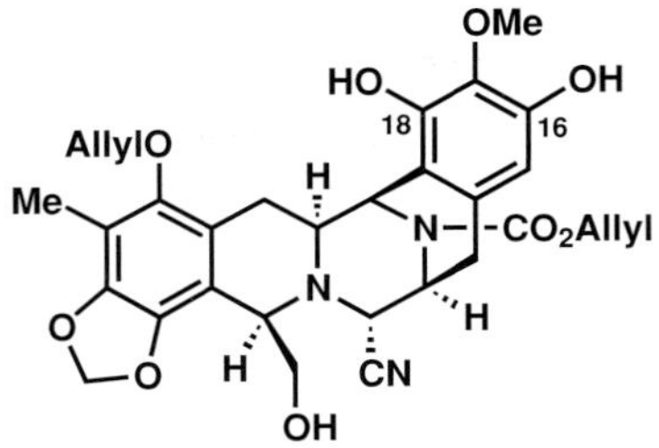

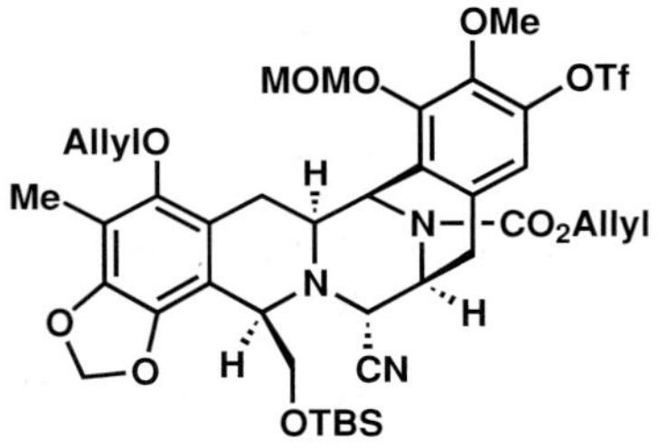

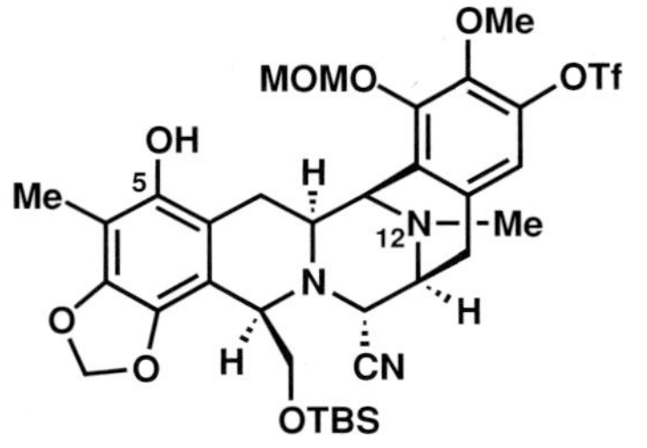

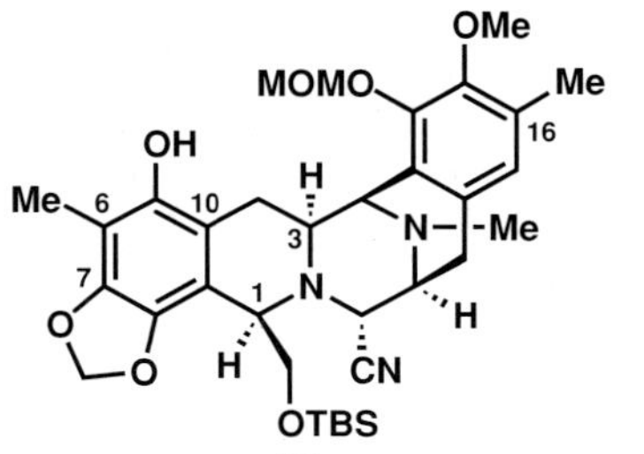

With the C-21 hydroxy group of ecteinascidin 743 installed in a masked form, the functionality in the aromatic E-ring now had to be adjusted to fit that of the final target. For example, although the incorporation of phenols at both the C-16 and C-18 positions facilitated the formation of **27** through the biomimetic Mannich sequence by providing a C_2-symmetric E-ring prior to cyclization, the cost of this benefit was the requirement that the C-16 phenol would subsequently have to be selectively converted into a methyl group to match ecteinascidin 743 (**1**). Based on the gross architecture of **65**, however, such an objective was not considered challenging at this stage, since the Mannich annulation placed the C-18 phenol in a highly hindered environment, thereby making the C-16 hydroxy group far more accessible. Indeed, upon subsequent treatment of **65** with the McMurry–Hendrickson reagent (Tf_2NPh)[22] in the presence of Et_3N and 4-DMAP, this position was selectively converted into an aryl triflate in anticipation of an eventual Stille coupling reaction with Me_4Sn to convert this new motif into the requisite methyl group of **1**. Before this modification was pursued, however, several additional structural alterations (particularly protecting group manipulations) were executed first. Thus, treatment of the C-16 triflate congener of **65** with TBSCl afforded protection of both the remaining E-ring phenol and the primary alcohol appended at C-1. Subsequent exposure to KF led to selective cleavage of the lone aryl silyl group, enabling its conversion into a MOM ether in the final operation leading to **66**. With these events proceeding in 65 % yield overall, a fully orthogonal collection of protecting groups had been installed onto the ecteinascidin core. Next, the two allyl groups in **66** were cleaved upon exposure to n-Bu_3SnH and catalytic $PdCl_2(PPh_3)_2$ in acidified media, affording the free C-5 phenol and N-12 amine. Reductive amination using formalin (a source of formaldehyde) and $NaCNBH_3$ then provided a methyl substituent on this newly unveiled amine to complete the assembly of **67**. With all of this preparative work accomplished, it was now time to install the C-16 methyl substituent, and, as alluded to earlier, a Stille reaction between triflate **67** and Me_4Sn mediated by $PdCl_2(PPh_3)_2$, LiCl, AcOH, and N-methyl-2-pyrrolidinone (NMP) at 80 °C indeed rose to this challenge to afford **68** in 94 % yield.

At this stage, all that remained before attempts at o-quinone methide generation to allow the formation of the ten-membered F-ring could begin was: 1) site-selective oxidation of the A-ring phenol in **68** to provide a C-10 leaving group, and 2) incorporation of a side chain bearing a protected thiol group at C-1. As shown in the final operation presented in Scheme 11, the first of these objectives was smoothly met upon exposure of **68** to benzeneselenenic anhydride in CH_2Cl_2, an event which afforded **69** in 82 % yield as a single stereoisomer with reagent approach controlled by the C-1 and C-3 stereocenters. The exquisite site selectivity of this reaction can be rationalized by assuming that, although oxidation could potentially have occurred at either C-10 or C-6, the electronic properties of the ring (particularly the overriding power of the C-7 phenol as a *para* director) led to the observed product; had there been no such C-7 substituent, oxidation at C-6 would have been equally likely and a mixture of this product and **69** would probably have resulted.[23]

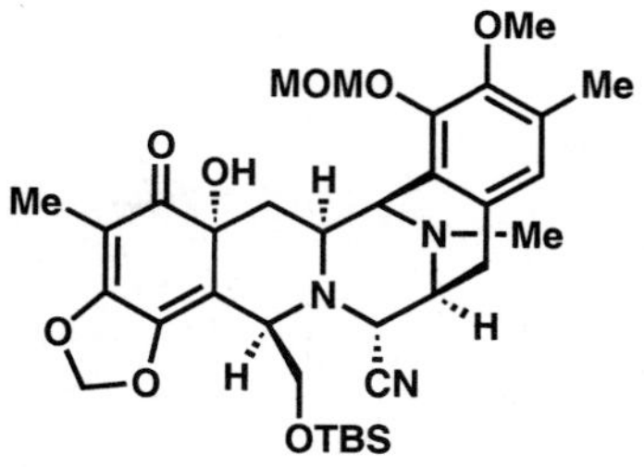

With this critical preparative operation out of the way, efforts were now directed towards appending the thiolate side chain. As shown in Scheme 12, only two steps were required to achieve this objective, namely TBAF-mediated deprotection of the silyl protecting group in **69** followed by a simple esterification between the resultant alcohol and cysteine derivative **70** in the presence of EDC and HOBt. These events afforded **26** in 83 % yield, providing an intermediate with the full complement of functionality needed to test *o*-quinone methide formation and subsequent thiolate capture. How, though, could the C-10 hydroxy group be induced to eliminate under mild conditions on such a complex intermediate? Although far from a simple problem, the Corey group anticipated that by forming a Swern-type intermediate, such as *o*-dimethylsulfonium **71**, the subsequent addition of a suitable base could then afford *o*-quinone methide **72** through the indicated mechanism.

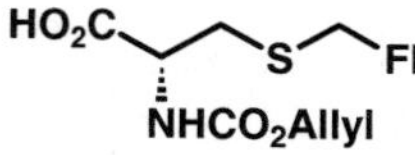

At first, unfortunately, this strategy was met with frustration as attempts to form **71** by the typical Swern combination of oxalyl chloride and DMSO led to several by-products resulting from participation of the displaced chloride ion from the initial reaction of these two reagents. When the source of the Swern reagent was switched to Tf_2O and DMSO, however, the addition of the species thus formed at −40 °C, followed by subsequent stirring for 45 minutes, resulted in no such decomposition, presumably because of the presence of a softer, non-nucleophilic counterion. Assuming then that **71** had indeed been generated, Hünig's base was then added at that temperature, followed by warming to 0 °C to effect the necessary elimination to *o*-quinone methide **72**. Following a quench of any residual Swern reagent with *t*-BuOH, *t*-butyltetramethyl guanidine (**73**, Barton's base)[24] was added to unmask the thiolate by cleaving the 9-fluorenyl group through the process delineated earlier to afford **25**, which then cyclized *in situ* to the ten-membered ring of ecteinascidin 743 (**1**). A trap of the resultant phenolate as its corresponding acetate then completed this amazing one-pot cascade sequence to **74**. Overall, this series of transformations proceeded in 79 % yield from **26**, thus constituting an average yield of 94 % for each step.

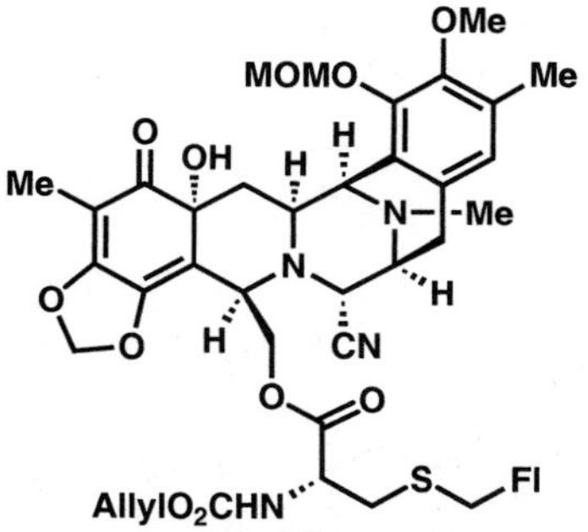

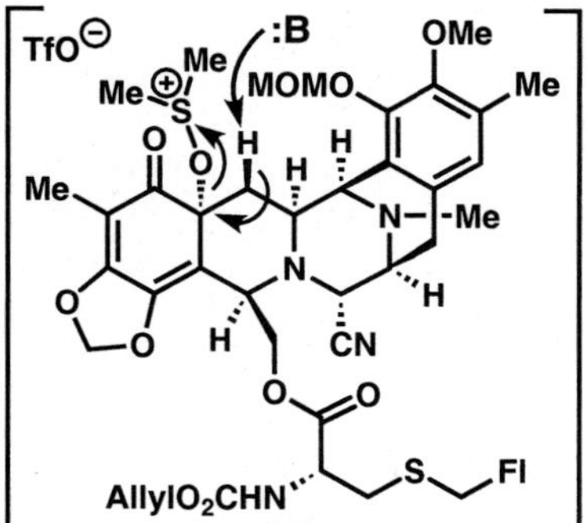

With most elements of ecteinascidin 743 (**1**) now in place, only a few steps remained before the total synthesis would hopefully be successfully concluded. Thus, following deallylation of **74**

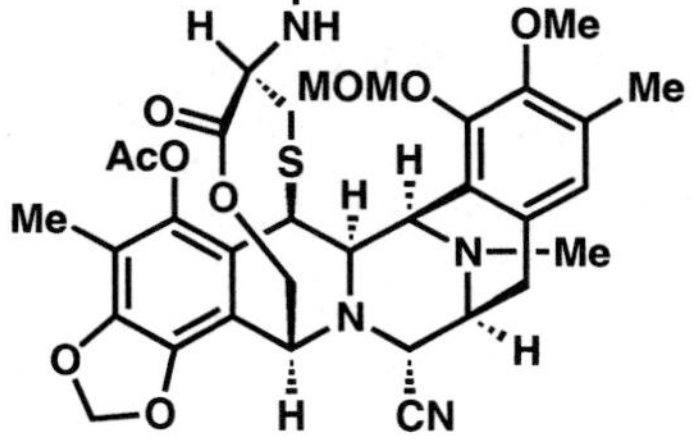

69

70

26

71

74

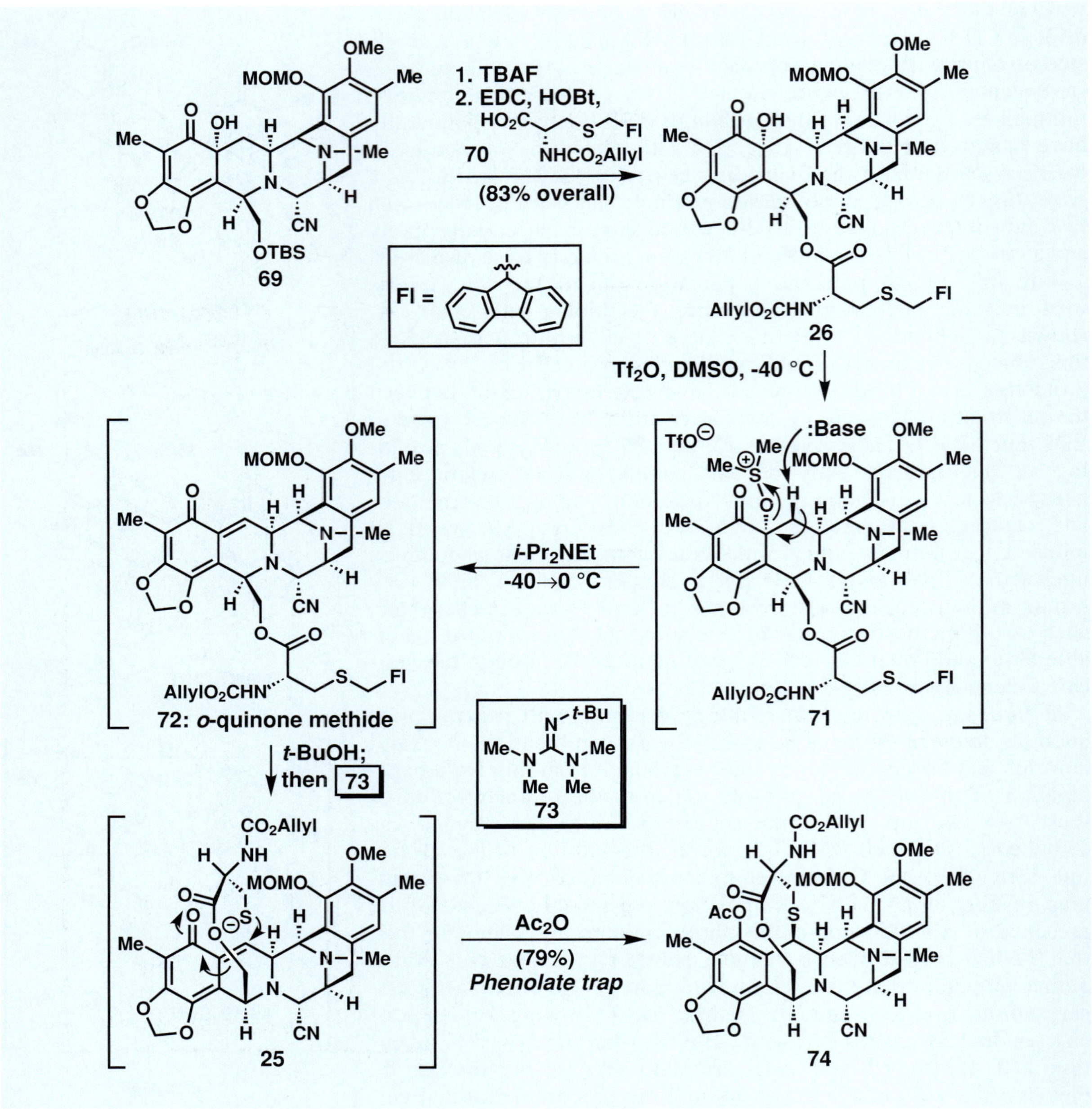

Scheme 12. Key quinone methide capture cascade sequence to advanced intermediate **74**.

(Scheme 13) under similar conditions to those discussed earlier (cf. Scheme 11) to liberate α-amino lactone **75**, subsequent transamination using the methiodide of pyridine-4-carboxaldehyde, DBU, and DMF in CH_2Cl_2 served to unveil α-keto lactone **24** in 70 % yield, presumably through the defined mechanism. With this critical motif in place, all that now remained was to incorporate

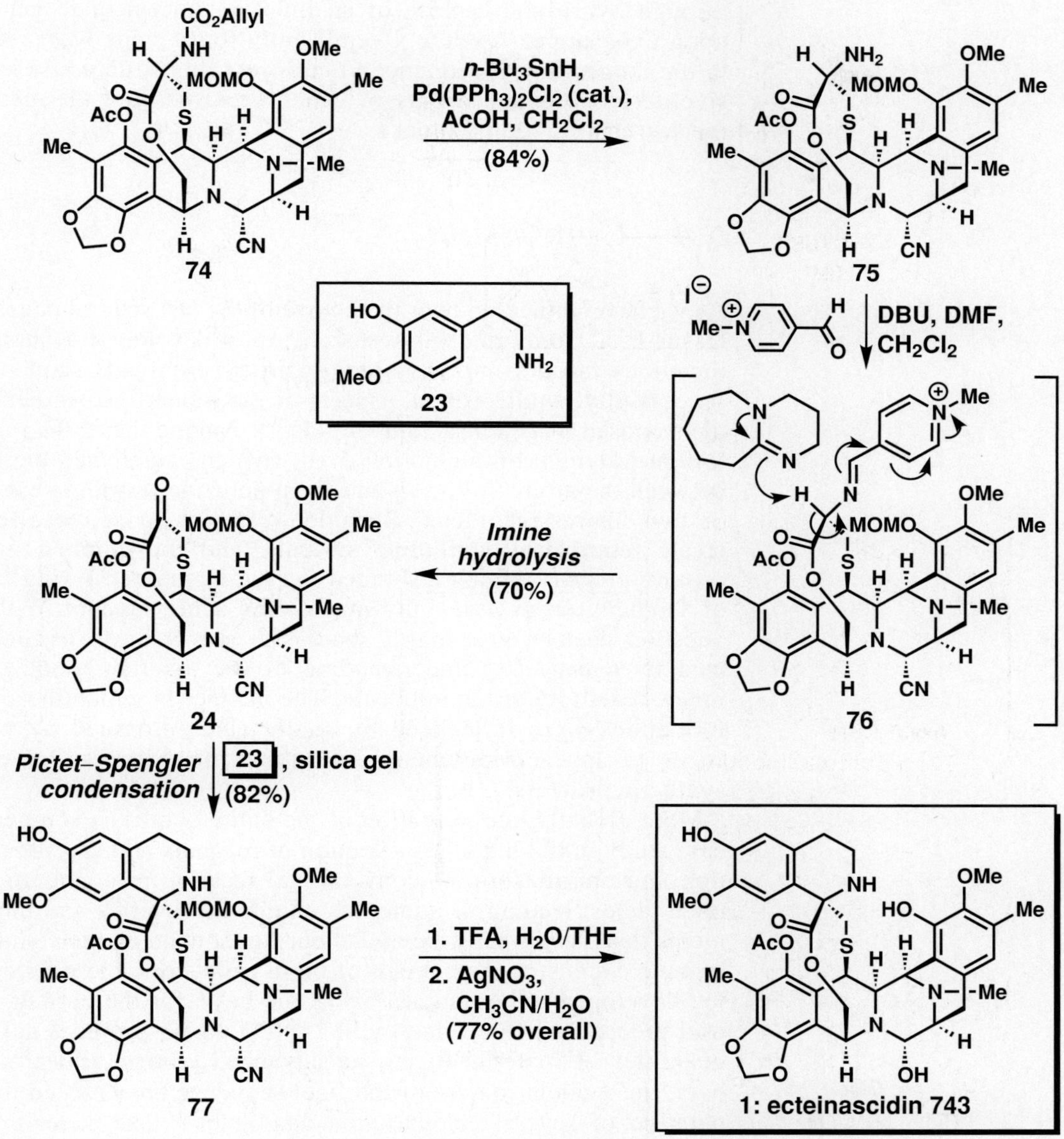

Scheme 13. Final stages and completion of ecteinascidin 743 (**1**).

the final tetrahydroisoquinoline system through a diastereoselective Pictet–Spengler condensation. Most gratifyingly, this event indeed proved stereoselective upon reaction of **24** with amine **23** in the presence of silica gel, affording **77** in 82 % yield. The target molecule **1** was then completed in 77 % yield through TFA-assisted cleavage of the MOM ether in a solvent mixture of H_2O and THF and subsequent exchange of the nitrile group at C-21 for an alcohol through hydrolytic capture of an iminium intermediate generated upon exposure to $AgNO_3$. Overall, only thirty steps were required in the longest linear sequence to construct this structurally impressive natural product, twenty of which occurred after the merger of the two main building blocks.

5.4 *Conclusion*

As we have witnessed over the course of the last several pages, Professor E. J. Corey and his research group developed a number of ingenious reaction protocols and instructive cascade sequences to successfully handle critical aspects of the numerous synthetic problems posed by ecteinascidin 743 (**1**).[9,11] Among these, the intramolecular Mannich bisannulation event (which established the bridge between the aromatic E-ring and the piperazine ring), the execution of two impressive Pictet–Spengler reactions to stereoselectively create tetrahydroisoquinoline systems, and the quinone methide capture process (which fashioned the ten-membered F-ring lactone of **1**) stand out as especially noteworthy achievements. With little question, each of these highly successful conversions was characterized by a masterful understanding of the inherent topology and innate reactivity of the molecule. The manner in which these transformations were influenced by subtle clues provided by Nature, so as to appear biomimetic, constitutes the forefront of current synthetic prowess.

More globally, the execution of the entire synthesis is a beautiful case study in the judicious selection of reagents and careful adaptation and optimization of conventional reaction protocols to maximize yields, leading to numerous unique discoveries and improvements that have greatly expanded our fundamental understanding of organic chemistry. As a result of these endeavors, the efficiency of the developed route now serves as the basis for the current industrial preparation of ecteinascidin 743 (**1**) which provides sufficient quantities of material to pursue advanced clinical trials.[10] Moreover, the modular nature of the final sequence has enabled the preparation of several ecteinascidin analogues whose screening has established a fuller understanding of structure–activity relationships for this important class of antitumor agents and has led to the identification of a simplified analog, phthalascidin 650 (**78**, column figure), with equal potency and greater inherent stability than the parent natural product.[8,25] In due course, more sophisticated testing will reveal the true potential of this new pharmaceutical candidate in the treatment of cancer. At present, one can be certain that the wealth of discoveries that has resulted from this research program is profound. Without a doubt, the state of chemical synthesis would be advanced considerably if all such endeavors were equally successful.

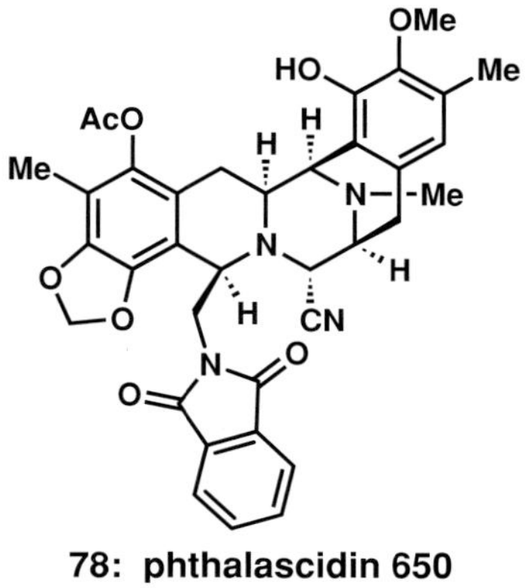

78: phthalascidin 650

5.5 *Fukuyama's Total Synthesis of Ecteinascidin 743 (1)*

In 2002, another impressive total synthesis of ecteinascidin 743 (**1**) was completed by Professor Tohru Fukuyama and several of his coworkers at the University of Tokyo.[26] A collection of the most instructive transformations from this insightful approach, based in part on previous efforts targeting members of the saframycins (cf. Scheme 1),[5] are presented in Schemes 14 and 15.

At the core of this strategy was the development of a set of transformations to forge the tetrahydroisoquinoline ring systems within the central pentacyclic array of ecteinascidin 743 (**1**) using a nonbiomimetic approach (i. e. annulations other than those of the Mannich or Pictet–Spengler type). This objective was ultimately achieved through a successful Heck reaction which converted iodide **84** (Scheme 15) into the bridged piperazine ring system in **85** in 83 % yield, and subsequent spontaneous phenol–aldehyde

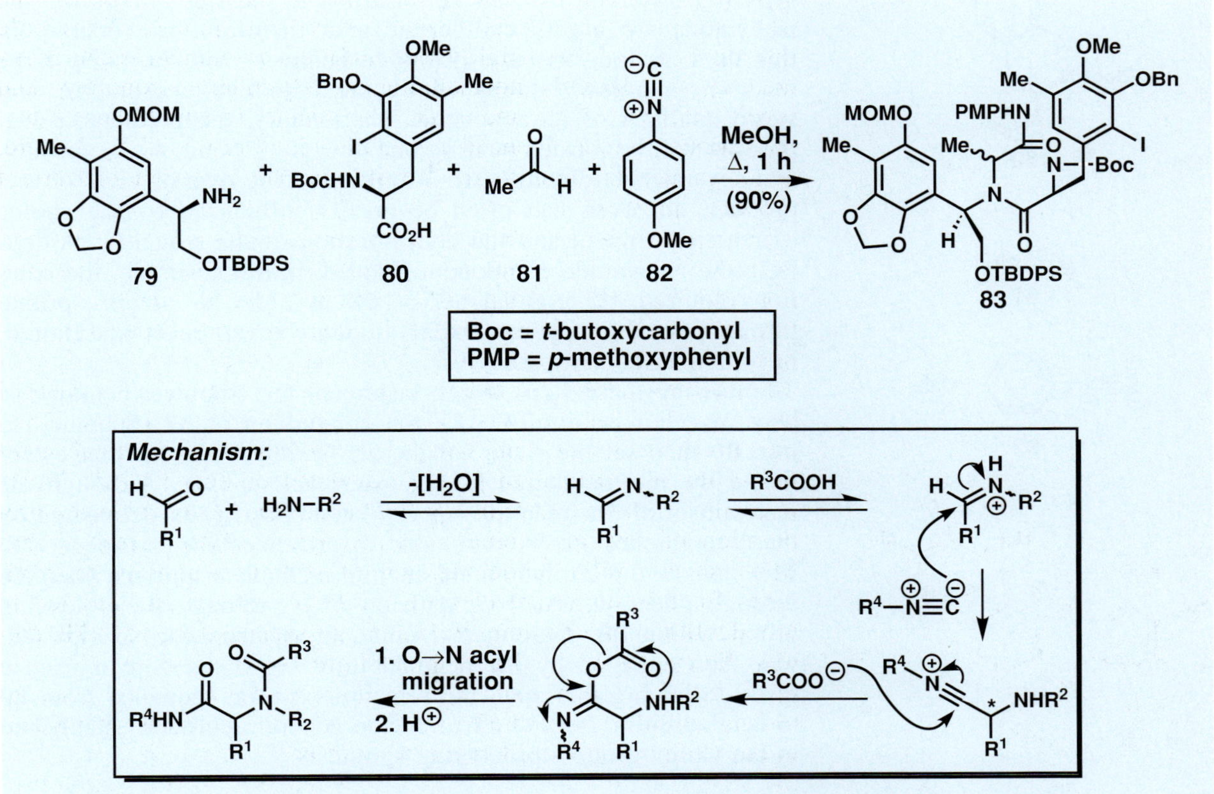

Scheme 14. Fukuyama's approach to ecteinascidin 743 (**1**): an Ugi multicomponent condensation to fashion advanced intermediate **83**.

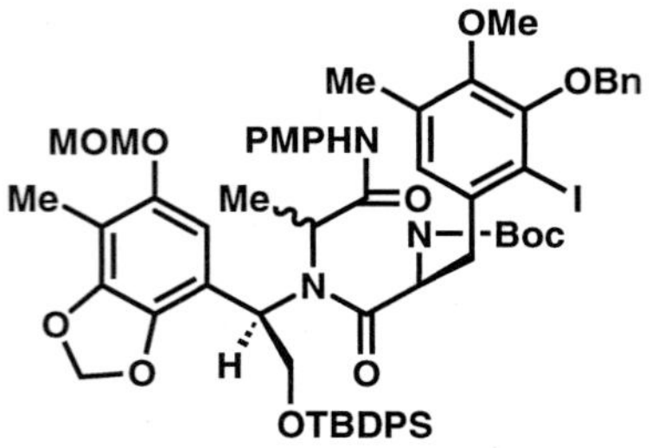

cyclization of **86** to **87** induced by hydrogenolytic cleavage of a benzyl protecting group. Although these reactions are notable for providing an alternate solution for the synthesis of these motifs, the events leading up to this sequence are also significant in that the Fukuyama group was inspired to execute one of the most impressive examples to date of a multicomponent reaction in the context of a total synthesis endeavor with their efficient one-pot construction of **83** (Scheme 14). This event, in which an amine (**79**), a carboxylic acid (**80**), an aldehyde (**81**), and an isocyanate (**82**) were coupled together in 90% yield, is commonly known as the Ugi reaction and has been employed to prepare α-amino acid derivatives since its discovery in 1959.[27]

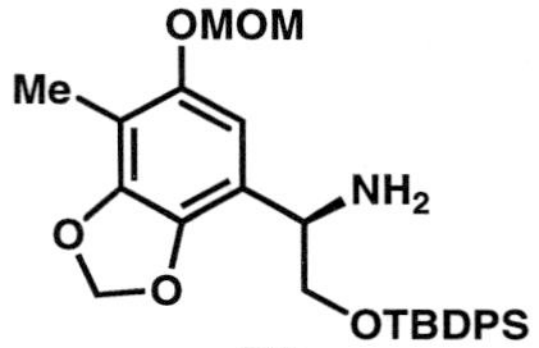

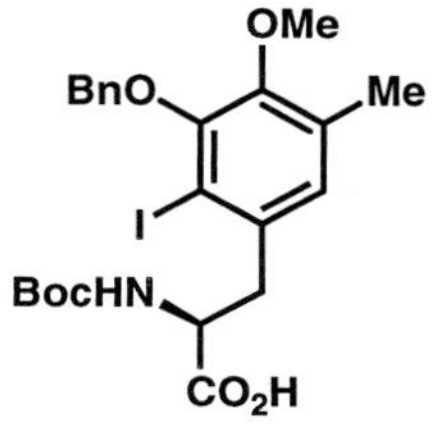

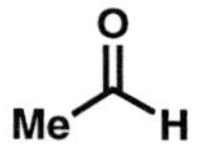

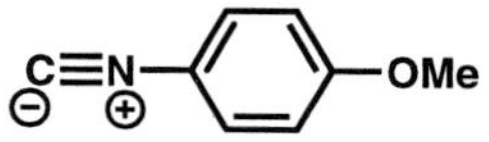

As shown in the mechanistic scheme in the box in Scheme 14, this highly efficient and atom-economic process proceeds through a complex series of consecutive first- and second-order reactions in which the reactive centers of the carboxylic acid and the imine intermediates switch the site of their reactivity several times. Interestingly, although all the steps involving the amine, aldehyde, and carboxylic acid components are reversible (thereby suggesting that numerous reaction pathways are available), few side products are typically observed from these reactions as the steps involving the isocyanide are highly exothermic and, therefore, irreversible. In this vein, as long as the isocyanide behaves within a fairly large reactivity window established by the inductive, mesomeric, and steric qualities of its appended functionality, the predictable Ugi product will be formed bearing one new C–C bond, a stereocenter, and several heteroatom–carbon linkages. The overall yield of this product, however, can often be greatly influenced by the choice of reaction solvent and the concentration of the reagents, particularly the isocyanide component. Indeed, in this example, the reaction required 1.5 equivalents of **82** in order to obtain optimal throughput, although equimolar amounts of all reactants should, in principle, be sufficient.

Following these steps which fashioned the complete pentacyclic core of ecteinascidin 743 (**1**), the conversion of **87** (Scheme 15) into **88** then set the stage for the acid-catalyzed event that established the ten-membered F-ring macrolactone (**89**). Although the mechanism of this reaction has not been specifically delineated by these researchers, it is reasonable to presume that treatment with TFA led to the formation of an intermediate *o*-quinone methide through phenolic-assisted expulsion of a protonated C-4 leaving group within **88**, thereby providing an electrophilic species that was then engaged by the pendant thiol. With the core motifs in place following this step, the remainder of the sequence from **89** to ecteinascidin 743 (**1**) mirrored the general approach established in the Corey group's pioneering synthesis.[9]

OMe Me OBn MOMO PMPHN Me Me O N I -Boc N H O O O OTBDPS
83

OMOM Me NH2 O O OTBDPS
79

OMe BnO Me I BocHN CO2H
80

O Me H
81

C≡N OMe
82

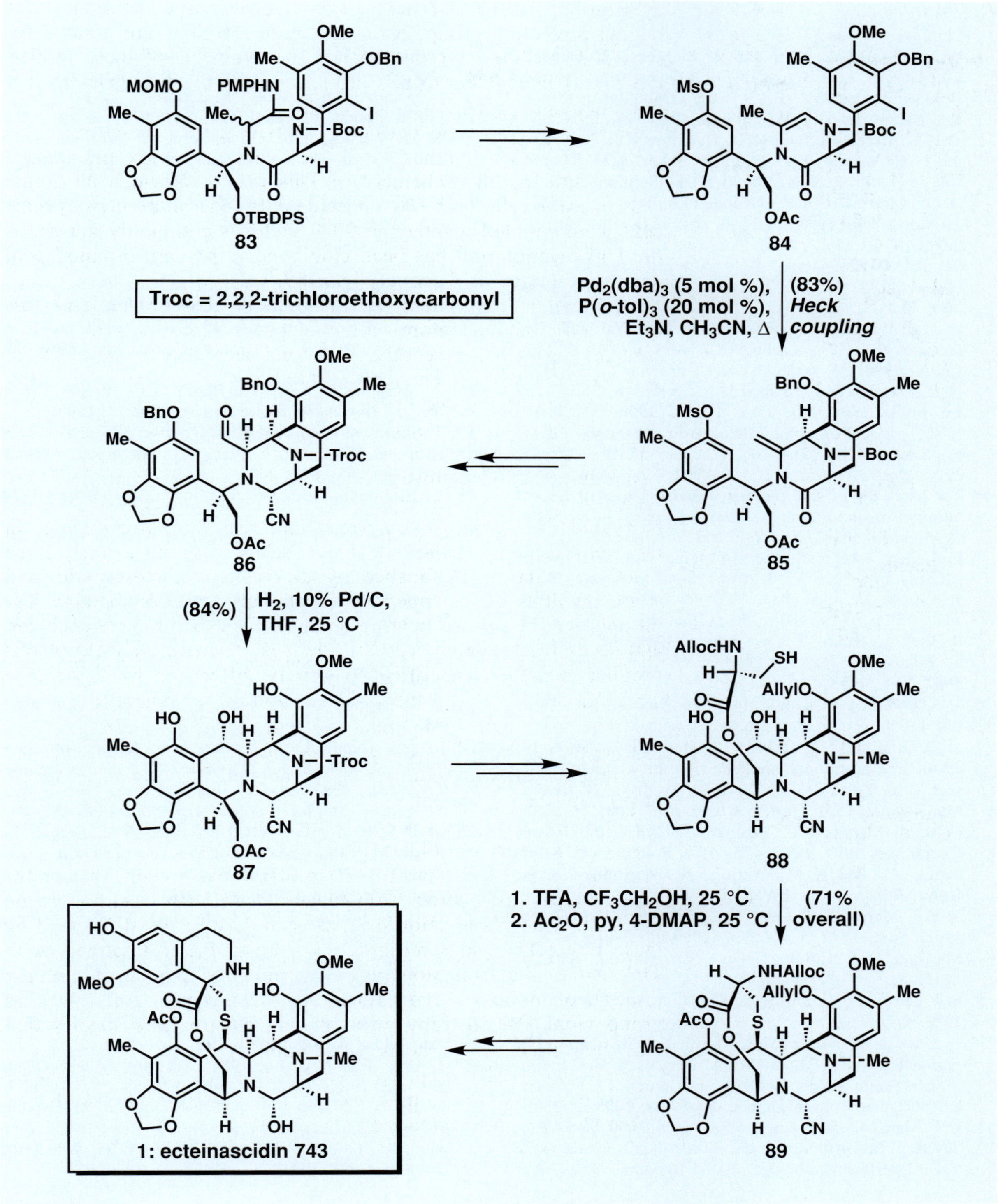

MOMO PMPHN OMe Me OBn Me Me N Boc N H H OTBDPS I

83

MsO OMe Me OBn Me Me N Boc N H H OAc I

84

Troc = 2,2,2-trichloroethoxycarbonyl

$Pd_2(dba)_3$ (5 mol %), $P(o\text{-}tol)_3$ (20 mol %), Et_3N, CH_3CN, Δ (83%) *Heck coupling*

85

86

(84%) H_2, 10% Pd/C, THF, 25 °C

87

88

1. TFA, CF_3CH_2OH, 25 °C
2. Ac_2O, py, 4-DMAP, 25 °C (71% overall)

89

1: ecteinascidin 743

Scheme 15. Key features of Fukuyama's total synthesis of ecteinascidin 743 (**1**).

References

1. M.M. Sigel, L.L. Wellham, W. Lichter, L.E. Dudeck, J.L. Gargus, L.H. Lucas in *Food-Drugs from the Sea, Proceedings, 1969* (Ed.: H.W. Youngken), Marine Technology Society, Washington, **1970**, pp. 281–295.
2. a) K.L. Rinehart, T.G. Holt, N.L. Fregeau, P.A. Keifer, G.R. Wilson, T.J. Perun, R. Sakai, A.G. Thompson, J.G. Stroh, L.S. Shield, D.S. Seigler, L.H. Li, D.G. Martin, C.J.P. Grimmelikhuijzen, G. Gäde, *J. Nat. Prod.* **1990**, *53*, 771; b) K.L. Rinehart, R. Sakai, T.G. Holt, N.L. Fregeau, T.J. Perun, D.S. Seigler, G.R. Wilson, L.S. Shield, *Pure & Appl. Chem.* **1990**, *62*, 1277; c) F. Flam, *Science* **1994**, *266*, 1324.
3. a) A.E. Wright, D.A. Forleo, G.P. Gunawardana, S.P. Gunasekera, F.E. Koehn, O.J. McConnell, *J. Org. Chem.* **1990**, *55*, 4508; b) K.L. Rinehart, T.G. Holt, N.L. Fregeau, J.G. Stroh, P.A. Keifer, F. Sun, L.H. Li, D.G. Martin, *J. Org. Chem.* **1990**, *55*, 4512. X-ray crystallographic data were obtained later: R. Sakai, K.L. Rinehart, Y. Guan, A.H.-J. Wang, *Proc. Natl. Acad. Sci. U.S.A.* **1992**, *89*, 11456.
4. For a review of the chemistry and biology of all the tetrahydroisoquinoline antibiotics, see: J.D. Scott, R.M. Williams, *Chem. Rev.* **2002**, *102*, 1669.
5. For early total syntheses of some members of the saframycins, see: a) T. Fukuyama, L. Yang, K.L. Ajeck, R.A. Sachleben, *J. Am. Chem. Soc.* **1990**, *112*, 3712; b) N. Saito, R. Yamauchi, H. Nishioka, S. Ida, A. Kubo, *J. Org. Chem.* **1989**, *54*, 5391; c) T. Fukuyama, R.A. Sachleben, *J. Am. Chem. Soc.* **1982**, *104*, 4957.
6. J.W. Lown, A.V. Joshua, J.S. Lee, *Biochemistry* **1982**, *21*, 419.
7. a) G.C. Hill, W.A. Remers, *J. Med. Chem.* **1991**, *34*, 1990. For additional studies with the ecteinascidins, see: b) M. Zewail-Foote, V.-S. Li, H. Kohn, D. Bearss, M. Guzman, L.H. Hurley, *Chem. Biol.* **2001**, *8*, 1033; c) B.M. Moore, F.C. Seaman, L.H. Hurley, *J. Am. Chem. Soc.* **1997**, *119*, 5475; d) Y. Pommier, G. Kohlhagen, C. Bailly, M. Waring, A. Mazumder, K.W. Kohn, *Biochemistry* **1996**, *35*, 13303.
8. a) E.J. Martinez, T. Owa, S.L. Schreiber, E.J. Corey, *Proc. Natl. Acad. Sci. U.S.A.* **1999**, *96*, 3496; b) E.J. Martinez, E.J. Corey, T. Owa, *Chem. Biol.* **2001**, *8*, 1151.
9. E.J. Corey, D.Y. Gin, R.S. Kania, *J. Am. Chem. Soc.* **1996**, *118*, 9202.
10. C. Cuevas, M. Pérez, M.J. Martín, J.L. Chicharro, C. Fernández-Rivas, M. Flores, A. Francesch, P. Gallego, M. Zarzuelo, F. de la Calle, J. García, C. Polanco, I. Rodríguez, I. Manzanares, *Org. Lett.* **2000**, *2*, 2545.
11. E.J. Martinez, E.J. Corey, *Org. Lett.* **2000**, *2*, 993.
12. See Ref. 2a for this proposal. Some experimental work toward verifying this and related hypotheses has been performed: a) R.G. Kerr, N.F. Miranda, *J. Nat. Prod.* **1995**, *58*, 1618; b) S. Jeedigunta, J.M. Krenisky, R.G. Kerr, *Tetrahedron* **2000**, *56*, 3303.
13. For a review of the Pictet–Spengler reaction, see: W.M. Whaley, T.R. Govindachari in *Organic Reactions, Vol. 6* (Eds.: R. Adams, et al.), John Wiley & Sons, London, **1951**, pp. 151–190.
14. P.J. Kocienski, *Protecting Groups*, Georg Thieme Verlag, Stuttgart, **1994**, p. 260.
15. For representative reviews of the Mannich reaction in chemical synthesis, see: a) S.K. Bur, S.F. Martin, *Tetrahedron* **2001**, *57*, 3221; b) M. Arend, B. Westermann, N. Risch, *Angew. Chem.* **1998**, *110*, 1096; *Angew. Chem. Int. Ed.* **1998**, *37*, 1044; c) T.A. Blumenkopf, L.E. Overman, *Chem. Rev.* **1986**, *86*, 857.
16. A. Hammadi, J.M. Nuzillard, J.C. Poulin, H.B. Kagan, *Tetrahedron: Asymmetry* **1992**, *3*, 1247.
17. T. Shioiri in *Comprehensive Organic Synthesis*, *Vol. 6* (Eds.: B.M. Trost, I. Fleming), Pergamon, Oxford, **1991**, pp. 806–817.
18. K. Ninomiya, T. Shioiri, S. Yamada, *Tetrahedron* **1974**, *30*, 2151.
19. E.J. Martinez, Ph.D. dissertation, Harvard University, **1999**.
20. For other peptide couplings using CIP, see: a) K. Akaji, N. Kuriyama T. Kimura, Y. Fujiwara, Y. Kiso, *Tetrahedron Lett.* **1992**, *33*, 3177; b) K. Akaji, N. Kuriyama, Y. Kiso, *Tetrahedron Lett.* **1994**, *35*, 3315; c) K. Akaji, N. Kuriyama, Y. Kiso, *J. Org. Chem.* **1996**, *61*, 3350.
21. H.C. Brown, A. Tsukamoto, *J. Am. Chem. Soc.* **1964**, *86*, 1089.
22. a) J.B. Hendrickson, R. Bergeron, *Tetrahedron Lett.* **1973**, *14*, 4607; b) J.E. McMurry, W.J. Scott, *Tetrahedron Lett.* **1983**, *24*, 979. For a review, see: c) W.J. Scott, J.E. McMurry, *Acc. Chem. Res.* **1988**, *21*, 47.
23. a) D.H.R. Barton, A.G. Brewster, S.V. Ley, C.M. Read, M.N. Rosenfeld, *J. Chem. Soc., Perkin Trans. 1* **1981**, 1473; b) D.H.R. Barton, J.-P. Finet, M. Thomas, *Tetrahedron* **1988**, *44*, 6397.
24. D.H.R. Barton, J.D. Elliott, S.D. Géro, *J. Chem. Soc., Perkin Trans 1* **1982**, 2085.
25. E.J. Martinez, E.J. Corey, *Org. Lett.* **1999**, *1*, 75.
26. A. Endo, A. Yanagisawa, M. Abe, S. Tohma, T. Kan, T. Fukuyama, *J. Am. Chem. Soc.* **2002**, *124*, 6552. For earlier studies, see: A. Endo, T. Kann, T. Fukuyama, *Synlett* **1999**, 1103.
27. I. Ugi, *Angew. Chem.* **1982**, *94*, 826; *Angew. Chem. Int. Ed. Engl.* **1982**, *21*, 810. See also A. Tuch, S. Wallé in *Handbook of Combinatorial Chemistry: Drugs, Catalysts, Methods*, *Vol. 2* (Eds.: K.C. Nicolaou, R. Hanko, W. Hartwig), Wiley-VCH, Weinheim, **2002**, pp. 697–700.

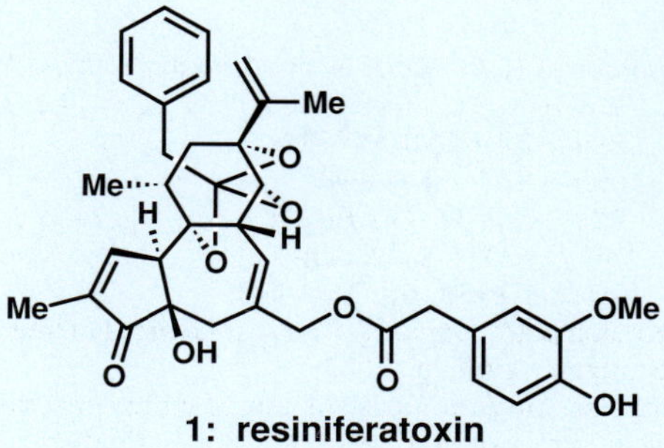

1: resiniferatoxin

6

P. A. Wender (1997)

Resiniferatoxin

6.1 Introduction

Key concepts:

- **1,3-Dipolar cycloadditions**
- **Zirconium-mediated enyne ring closure**
- **Kinetically-controlled enantiomeric enrichment**

If you have ever gone to a restaurant featuring cuisine from Mexico, China, or any other of a host of cultures from the warmer regions of the globe that use chili peppers in their preparations, you have no doubt experienced a hot, numbing sensation in your mouth upon tasting one of the many tempting dishes offered on the menu. The reason for this often painful experience is a compound within these peppers known as capsaicin (**3**, Scheme 1), whose relatively simple molecular structure superficially conceals the complex chain of biochemical events that its presence initiates. Once in the mouth, capsaicin binds to an integral membrane protein receptor found in certain primary afferent sensory neurons, leading to the opening of a cation-selective channel that permits sodium and calcium ions to enter the neuron. This event is the critical requirement for membrane depolarization (i. e. neuronal excitation), dispatching a signal to processing centers in both the spinal cord and the brain which is registered as a perception of pain. Consequently, a series of inflammatory responses is then initiated to modulate this negative sensation.[1]

With prolonged or repeated exposure to capsaicin (**3**), however, one will eventually become insensitive to its effects as excessive cation influx leads to death of the neurons. It is for this reason that people can cultivate a tolerance for spicy food over time, enabling some to have the ability to eat peppers so hot that their mere smell would make the eyes of those less accustomed water profusely. The perhaps more significant feature is this desensitization suggests that capsaicin could serve as an analgesic to mitigate the pain and inflammation associated with several disorders such as

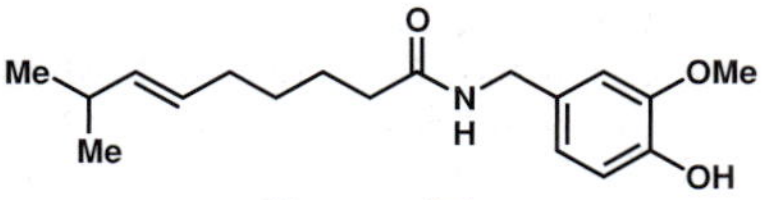

3: capsaicin

1: resiniferatoxin

2: phorbol

3: capsaicin

Scheme 1. The molecular architectures of resiniferatoxin and its structural homologues phorbol (**2**) and capsaicin (**3**).

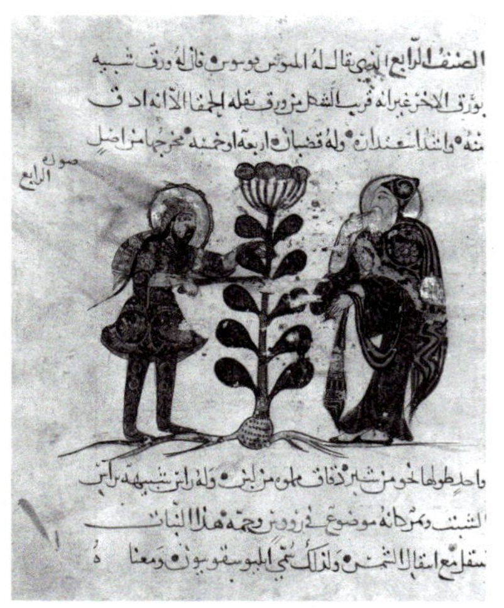

A physician (right) oversees collection of Euphorbium latex by an attendant, as shown in a manuscript dated 1244. Courtesy of the Freer Gallery of Art, Smithsonian Institution.

homovanillyl group

arthritis, diabetic neuropathy, and sympathetic dystrophy.[2] Indeed, over the course of the past several decades capsaicin has been used to desensitize neurons throughout the body. Unfortunately, topical application of this natural product usually leads to severe irritation, making it a difficult medicine for patients to tolerate irrespective of its eventual beneficial effects for pain reduction.

Despite these specific problems with capsaicin, however, the general concept behind using such an analgesic holds great medical promise, and Nature has thankfully provided humanity with several additional compounds possessing similar modes of activity, but with far fewer negative side effects. Foremost among these is a thousand-fold more potent analog of capsaicin (**3**) obtained from the latex of the flowering cactus *Euphorbia resinifera*.[3] Originally named resiniferatoxin (**1**, Scheme 1) upon its full structural elucidation in 1975,[4] this compound has, in fact, been utilized in the medical treatment of pain for close to two thousand years, as documented by sources such as the illuminated Arabic manuscript in the adjoining column. As the molecular connectivities of this natural product reveal, the biological activity of resiniferatoxin is predicated (at least in part) on the presence of the homovanillyl side chain also found in capsaicin (**3**), enabling its recognition by the same neuronal receptors.[5] Intriguingly, although the remainder of the architecture of resiniferatoxin bears striking similarities to a class of compounds typified by phorbol (**2**), the most potent tumor promoters currently known in the literature, resiniferatoxin (**1**) shares none of their co-carcinogenic properties.[6]

In light of the promising medical potential of resiniferatoxin, synthetic chemists have long sought to develop a viable laboratory preparation of this natural product, both to produce the isolate in meaningful quantities and to enable the synthesis of analogs in order to probe the structural features beyond the homovanillyl side chain that confer its unique activity profile. As any student of total synthesis can readily ascertain, however, resiniferatoxin (**1**) provides a particularly challenging synthetic problem through its complicated collection of stereochemical elements, especially the contiguous array of five stereocenters on ring C and the *trans*-fused nature of both the AB- and BC-ring junctions. In addition, resiniferatoxin (**1**) possesses an orthoester motif, a relatively rare functional group in secondary metabolites due to their typical chemical lability (particularly under acidic conditions). Undoubtedly, to conquer such a conglomeration of complexity housed within a compact architecture in a controlled and efficient manner requires the identification of a novel synthetic strategy bolstered by a supply of powerful synthetic transformations. In this chapter, we will analyze just such a sequence as developed by Professor Paul Wender and his co-workers at Stanford University based on an insightful blending of classical and modern synthetic reactions.[7]

6.2 *Retrosynthetic Analysis and Strategy*

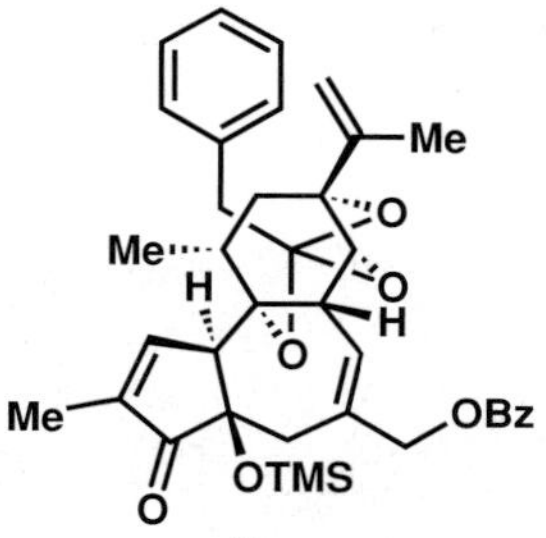

While many complex molecular targets offer a seemingly limitless array of potential disconnections, by contrast several elements within the architecture of resiniferatoxin appear to prescribe a defined sequence of simplifications, at least in the initial stages. First, as alluded to above, because several studies have revealed that the homovanillyl side chain of resiniferatoxin (**1**, Scheme 2) is at least partially responsible for the irritant/analgesic properties of this natural product,[5] it would seem prudent to leave its incorporation until the final stages of the synthesis in order to minimize the degree to which possibly active intermediates would have to be handled. As such, cleavage of this motif through the indicated ester linkage would then provide advanced intermediate **5** (appropriately modified with protecting groups) as a new subgoal structure. A second requisite decision would then be to rupture the potentially sensitive orthoester within **5** to afford **6**, for which one could anticipate an acid-catalyzed union between the C-9 and C-13 alcohol groups and the carbonyl function of the appended phenylacetic acid side chain to create this motif during the synthesis. To obtain this new target (**6**), one could then envision the late-stage introduction of this portion of the molecule through an esterification reaction at the free C-14 hydroxy group of **7**.

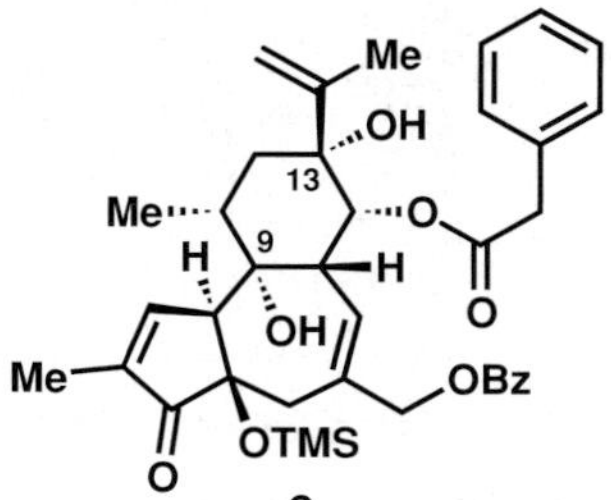

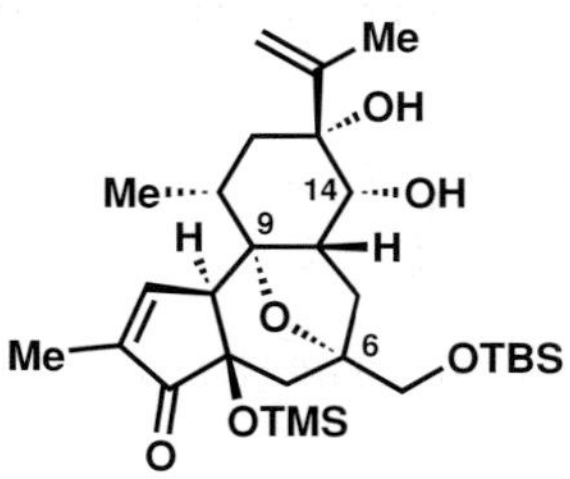

At this juncture, however, any further retrosynthetic simplifications are far from defined by synthetic necessity. As such, the Wender group's most critical insight regarding the resiniferatoxin

Me Me H O O O H OBz OTMS O

5

Me OH 13 Me O 9 H H O OH Me OBz OTMS O

6

Me OH 14 OH Me 9 H H O 6 Me OTBS OTMS O

7

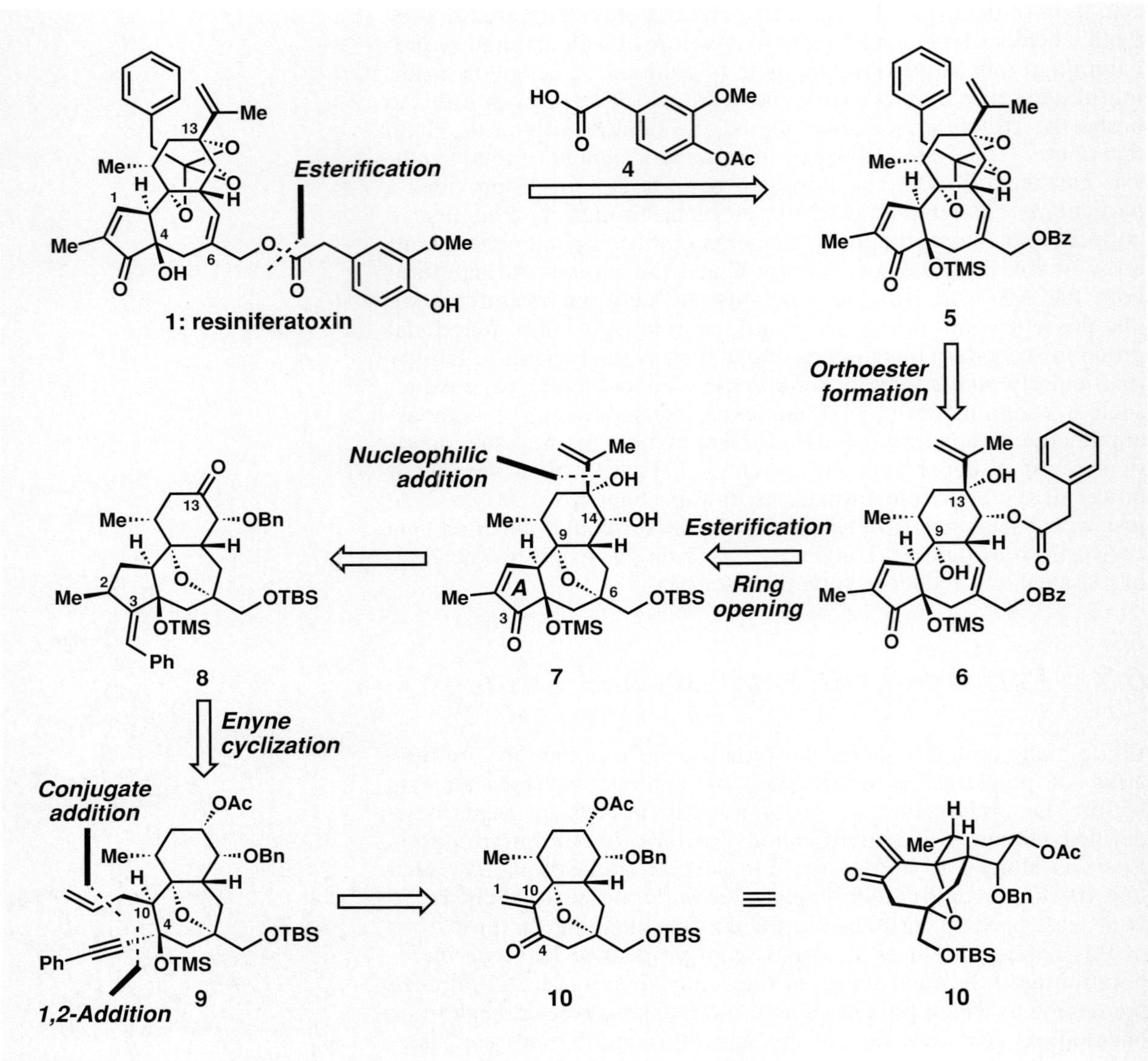

Scheme 2. Wender's retrosynthetic analysis of resiniferatoxin (**1**): initial stages.

architecture was that the C-9 alcohol and the B-ring unsaturation in **6** could arise from a ring-opening elimination of a precursor (**7**) bearing a C6–C9 oxido bridge. Upon initial inspection, the introduction of this motif might seem counterintuitive as an element of retrosynthetic analysis because it creates a more complex synthetic target with two new rings in addition to the central tricycle of resiniferatoxin (**1**). As we shall see, this adjustment was deemed to be absolutely necessary from a strategic standpoint in that it would serve as a means of not only protecting the eventual C-9 hydroxy group throughout the synthesis, but also rigidifying an otherwise highly flexible seven-membered ring to enable the selective introduction of some of its stereochemical information through substrate-controlled operations. Moreover, the presence of this bridge also facilitated the recognition of an insightful disconnection based on a 1,3-dipolar cycloaddition, which we shall encounter shortly.

Before the inherent value of this decision will become readily apparent, however, several additional retrosynthetic adjustments would first have to be implemented. Accordingly, **7** was modified to **8** under the following assumptions: 1) a nucleophile could be induced to stereoselectively engage its C-13 ketone in order to install the isopropenyl side chain, 2) the enone system in the A-ring of **7** could be fashioned from a cyclopentane precursor, and 3) ozonolytic cleavage of the C–C double bond in **8** could lead to the C-3 ketone in **7**. Once again, the latter element of retrosynthetic analysis might seem somewhat mysterious at first glance because it resulted in a more complicated precursor; however, this operation (combined with the alteration of the A-ring alkene) was critical in that it suggested that the A-ring in **8** could be the product of a transition-metal-mediated enyne cyclization of a precursor such as **9**.

As only one of many effective annulation methods currently known in the literature, this general transformation was first discovered by Pauson and co-workers in the early 1970s using cobalt[8] and then extended to group 4 elements such as titanium and zirconium by Nugent and co-workers at DuPont and Negishi and co-workers at Purdue University.[9] As exemplified in Scheme 3 using this latter series of initiators, the reaction proceeds through the initial formation of a metallocene (**12**) upon the addition of a reagent such as $(n\text{-Bu})_2ZrCp_2$ to the alkyne (effected by reductive elimination of the *n*-butyl ligands), followed by intramolecular carbometallation to form a new metallocycle (**13**). A subsequent reaction quench under acidic conditions would then provide **14** as a single olefin isomer, potentially with controlled relative stereochemistry at the starred carbon due to the reversible nature of the conversion of **12** into **13** (if the substrate has a thermodynamic preference).

Significantly, although this particular method of work-up affords the general product needed for the projected synthesis of resini-

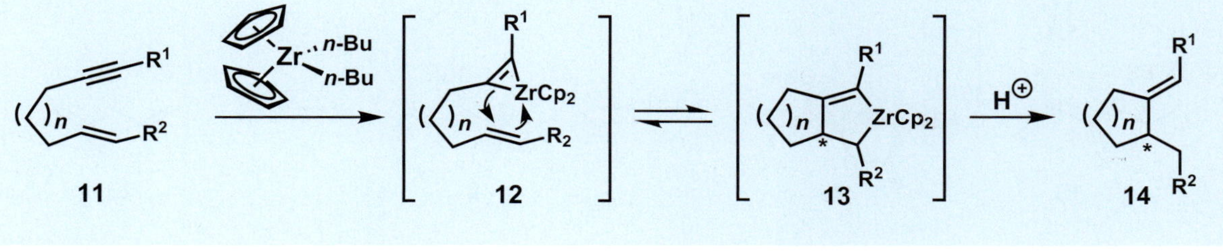

Scheme 3. The mechanism of zirconium-mediated enyne ring closure.

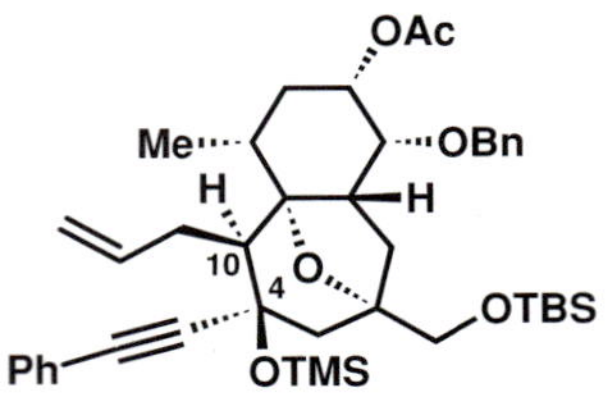

feratoxin (**8**, Scheme 2), the intermediate metallocycle **13** can also be converted into a number of additionally useful products such as cyclopentenones through a Pauson–Khand-type carbonylation using carbon monoxide.[10] The only major cause for concern with this ring-forming reaction is that terminal alkynes cannot be employed because the acidity of their hydrogen atoms will shut down productive reaction pathways; accordingly, the Wender group incorporated a phenyl substituent on the appended alkyne in **9** to overcome this obstacle for the proposed conversion. As a final element of analysis, it should be noted that, although the desired product from this enyne cyclization event (**8**) is drawn as a single C-2 stereoisomer, such stereoselectivity is not necessary since this site will need to eventually become part of a $\Delta^{2,3}$ alkene to match the functional requirements of resiniferatoxin (**1**).

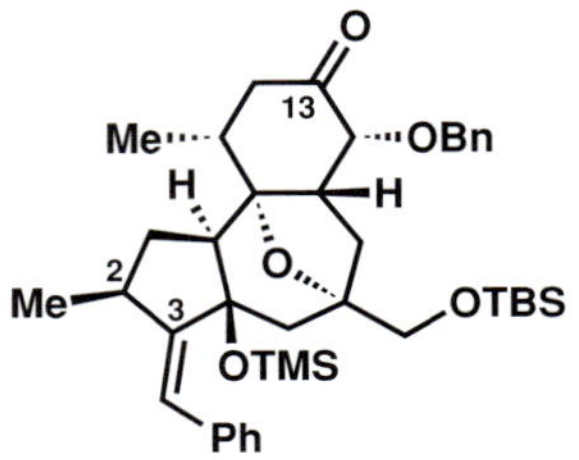

With the A-ring effectively disassembled through this strategy, the *trans* disposition of the vicinal C-10 and C-4 carbon chains on the seven-membered B-ring in **9** now seemed ripe for removal through sequential, stereocontrolled addition reactions to an α,β-unsaturated intermediate such as **10**. In this regard, the previously installed oxido bridge would need to play a crucial role for the overall suc-

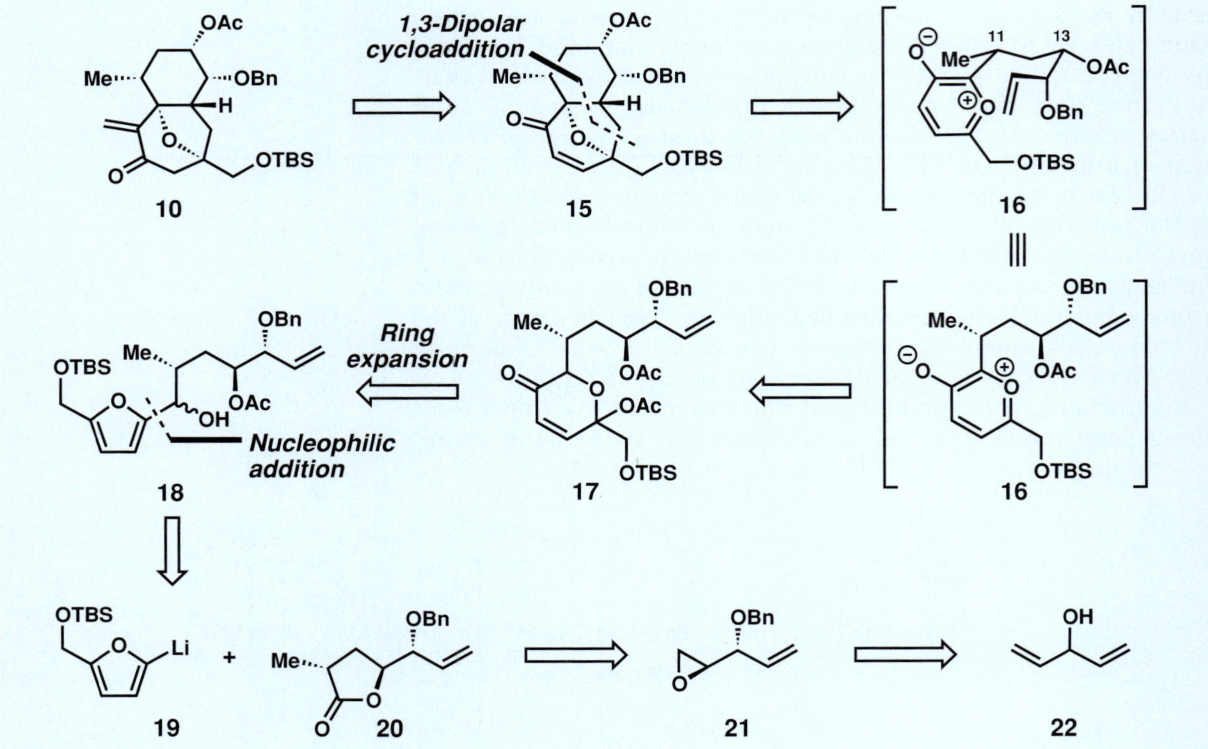

Scheme 4. Wender's retrosynthetic analysis of resiniferatoxin (**1**): final stages.

cess of this strategy during the synthesis by rigidifying the B-ring, as initial Michael addition of a vinyl group to **10** would have to be followed by selective proton capture from the face *syn* to the oxygen bridge. Assuming that these requirements could indeed be met, then 1,2-addition of an acetylenic nucleophile, followed by silyl capture of the resultant alkoxide, would complete the diastereoselective assembly of **9** from **10**.

Following this spate of disconnections, one of the rings from the central tricyclic core of resiniferatoxin (**1**) and two of its perplexing stereocenters have been successfully excised. Several challenging elements still remained for consideration. As shown in Scheme 4, though, if the exocyclic C–C double bond of the α,β-unsaturated system in **10** were modified to an endocyclic variant (**15**), then the majority of the remaining complexity could be dealt with in one fell swoop as this new adduct could conceivably result from an intramolecular 1,3-dipolar cycloaddition of an alkene and a tethered 3-oxidopyrylium motif (**16**). This compound, in turn, could be obtained from pyranone **17**.

As you have surely noted, several examples of 1,3-dipolar cycloaddition reactions have been discussed throughout the course of the *Classics* series due to the amazing library of products that result from their employment.[11] Among the 1,3-dipoles typically employed in these events, of which a representative collection is shown in Scheme 5, oxidopyrylium species (whose resonance form is a carbonyl ylide) constitute a relatively unexplored tool

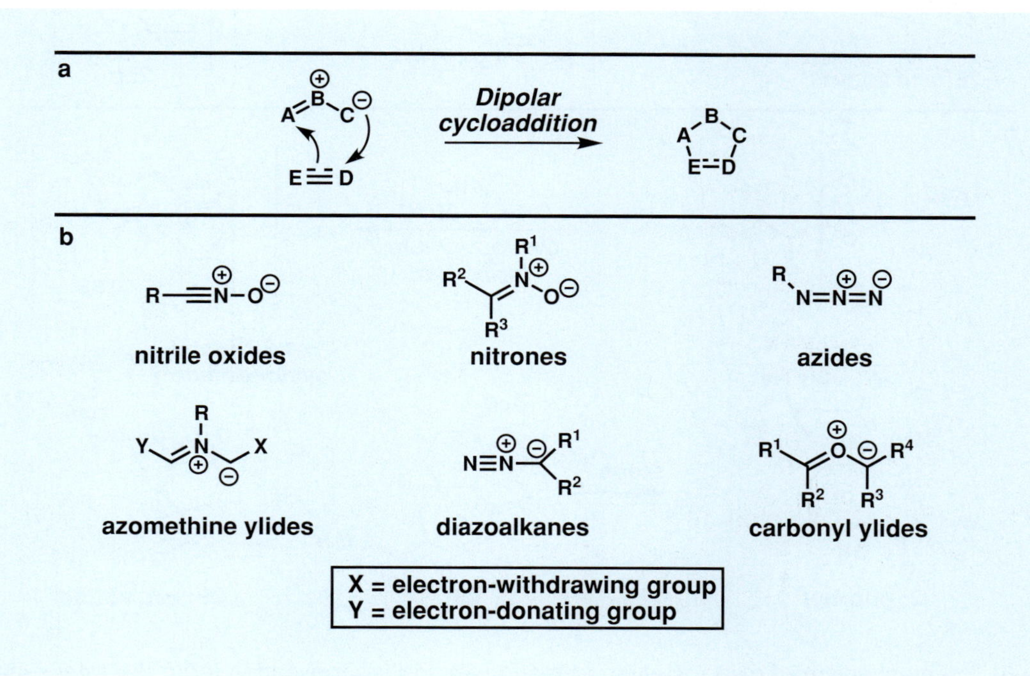

Scheme 5. The dipolar cycloaddition reaction (a) and general classes of 1,3-dipoles (b).

for natural product synthesis,[12] as this dipole has been studied only intermittently over the past four decades in any context.[13]

What is known, however, is these reactive intermediates can be generated quite effectively from starting materials such as **23** upon treatment with base through the mechanism defined in Scheme 6a. Following the addition of an appropriate dipolarophile, the formed 3-oxidopyrylium (**25a** and its resonance form **25b**) can then engage in a productive cycloaddition to forge a bridged bicyclic product **26** with a seven-membered ring; depending on the way in which the participant atoms are counted, this event can be formally considered either as a [3+2] or a higher-order [5+2] cycloaddition reaction. Overall, while this transformation has proven effec-

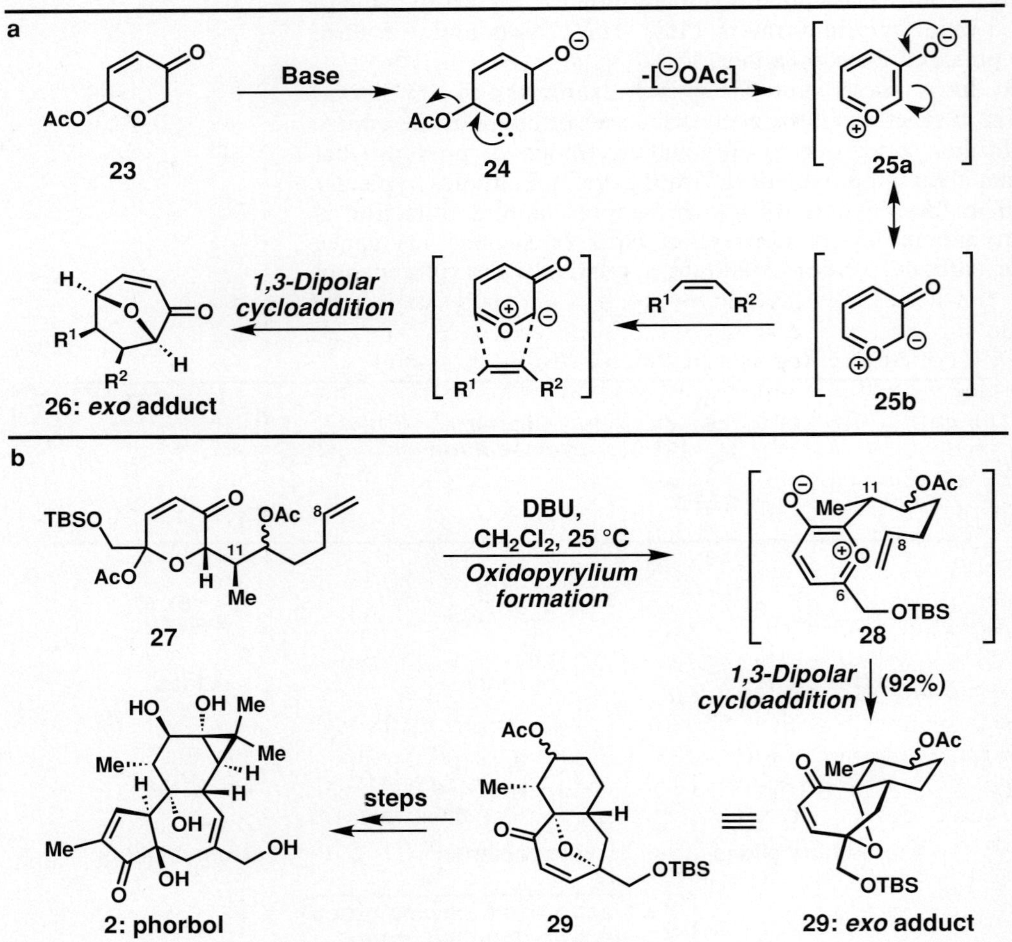

Scheme 6. The mechanism of oxidopyrylium formation (a) and its application (b) to the total synthesis of phorbol (**2**). (Wender, 1989)[14]

tive in intermolecular contexts, intramolecular reactions (as suggested in the proposed conversion of **17** into **15**, cf. Scheme 4) in fact proceed far more smoothly since, apart from the obvious entropic features, there is also less opportunity for non-productive dimerization or intramolecular cyclization reactions of the intermediate carbonyl ylide.[12]

However, the major issue shadowing the application of this reaction in the resiniferatoxin context was not its inherent plausibility, but whether or not the event would prove stereoselective, given that the conversion of **17** into **15** would have to proceed in an *exo* manner and with complete facial selectivity.* The first of these requirements was essentially assured by the intramolecular nature of the reaction, since achieving an *endo* transition state would be nearly impossible with a four-atom tether. In addition, no groups within the reactant were capable of providing the necessary secondary orbital interaction that would favor such a transition state. The second objective would require the adoption of a single transition state for the cycloaddition, an outcome which the Wender group expected would occur based on their 1989 total synthesis of phorbol (**2**, Scheme 6b) where a similar oxidopyrylium-based 1,3-dipolar cycloaddition was applied with great success.[14] Following the synthesis of racemic **27** (with the indicated relative stereochemistry), the addition of DBU at ambient temperature in CH_2Cl_2 afforded a stereoselective synthesis of **29** in 92 % yield. The exquisite nature of this reaction course can be rationalized by assuming that **28** constitutes the favored orientation for this reaction in which the substituents are situated in a pseudochair fashion such that the C-11 methyl group is equatorially disposed, leading to a facially selective addition ordained to proceed in *exo* fashion. Similarly, in the proposed event converting **17** into **15** (see Scheme 4), the C-11 methyl group in conjunction with the C-13 acetate should lead to the predominance of the indicated chair-like reactive conformation **16** which would similarly afford the desired *exo* product. Thus, if successful, this critical event would concurrently form both the B- and C-rings of resiniferatoxin (**1**) and establish the absolute configuration of its C-8 and C-9 stereocenters (assuming a homochiral synthesis of **17**).

From **17**, the remaining retrosynthetic disconnections to readily available starting materials are relatively easy to deduce. First, syntheses of pyranones such as **17** are known to be possible through oxidative ring expansion of furan intermediates such as **18** mediated by a range of electrophilic species like dimethyldioxirane (DMDO), *m*CBPA, and *N*-bromosuccinimide.[15] In turn, the bond between the furan ring and the adjacent hydroxy-substituted carbon atom could then be retrosynthetically cleaved to lithiated species **19**[16] and its

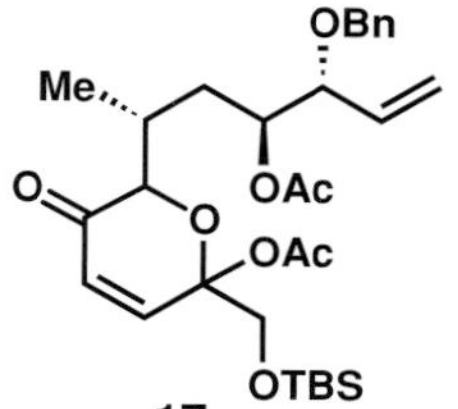

OBn
Me
O
O
OAc
OAc
OTBS
17

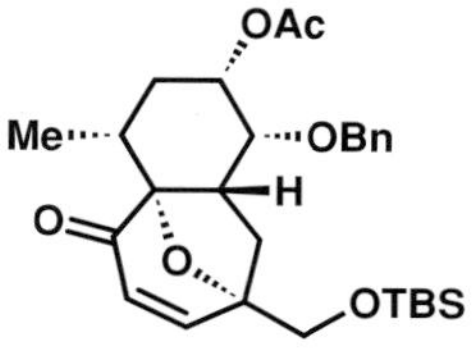

OAc
Me
OBn
H
O
O
OTBS
15

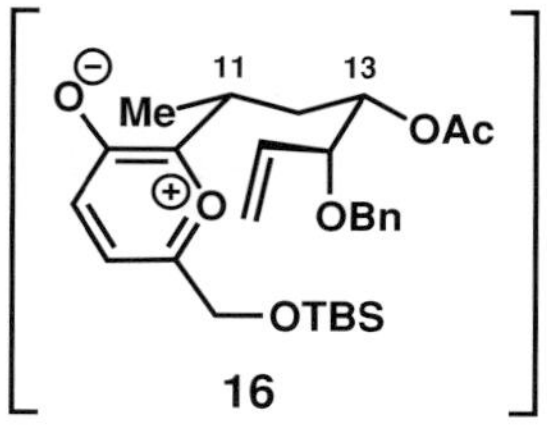

11 13
Me
OAc
O
OBn
OTBS
16

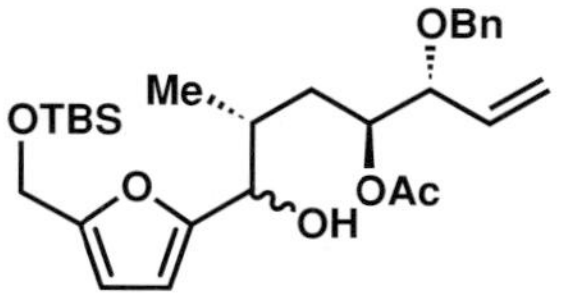

OBn
OTBS
Me
O
OAc
OH
18

* In the parlance of 1,3-dipolar cycloadditions of this type, the *endo* isomer is readily identified as that in which the substituents on the dipolarophile end up *anti* to the oxido bridge. Accordingly, the *exo* isomer is the one in which these substituents are *syn* to the resultant oxygen bridge.

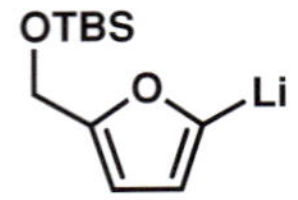

OTBS
Li
19

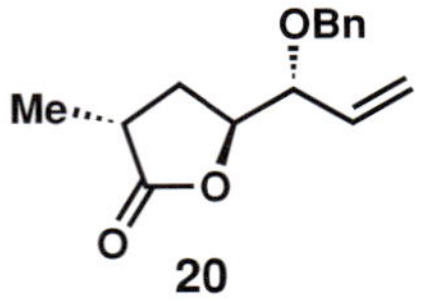

OBn
Me
20

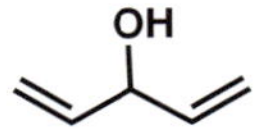

OH
22

electrophilic lactone counterpart **20**.[17] Both these intermediates are known entities available through literature routes, the latter of which could be derived from the simple prochiral starting material **22**.

6.3 Total Synthesis

Our discussion of the Wender group's synthesis of resiniferatoxin (**1**) begins with the efforts targeting the critical lactone precursor **20**, which was known to be accessible from the divinyl alcohol **22** through several different established sequences, as mentioned earlier.[17a,b] Thus, while the majority of the initial steps of this synthetic enterprise are directly patterned from these preparations, the highly instructive nature of some of the transformations merits their full discussion here. First, as shown in Scheme 7, a method was needed to fashion monoepoxide **30** from **22** as this new motif would constitute an electrophile critical for the construction of the remainder of lactone **20**. As indicated, this objective was indeed accomplished in 51 % yield and in 98 % *ee* upon application of a standard Sharpless asymmetric epoxidation (SAE) protocol over the course of 135 hours at −15 °C.[18] The high enantiomeric purity observed in this reaction, however, is far from just the result of a simple asymmetric epoxidation reaction, as the process actually couples stereocontrolled synthesis with an *in situ* kinetic resolution.[19]

As mentioned in Chapter 19 of the first volume of *Classics in Total Synthesis*, an important feature of the SAE reaction is that it imparts a high degree of facial selectivity in the epoxidation event, but yet can also be influenced by any preexisting stereochemical information in the substrate. Both of these characteristics proved critical in the success of the transformation defined in

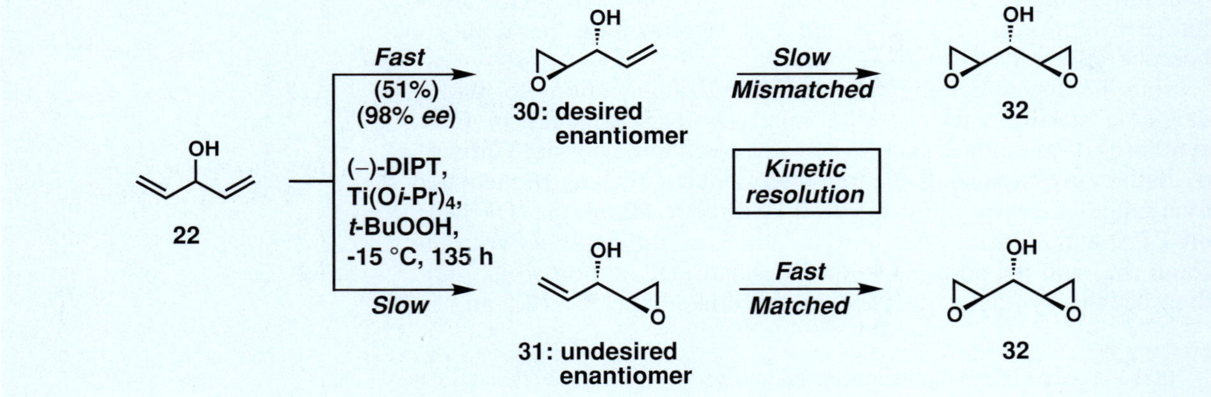

Fast
(51%)
(98% *ee*)
OH
30: desired enantiomer
Slow
Mismatched
32
OH
22
(−)-DIPT, $Ti(Oi\text{-}Pr)_4$, *t*-BuOOH, -15 °C, 135 h
Kinetic resolution
Slow
OH
31: undesired enantiomer
Fast
Matched
32

Scheme 7. Use of the Sharpless asymmetric epoxidation reaction to achieve the synthesis of hydroxy epoxide **30** through kinetically controlled enantiomeric enrichment.

Scheme 7. As one can readily deduce, substrate **22** is itself achiral, yet possesses a prostereogenic center (the central hydroxy-substituted carbon atom) flanked by two enantiotopic vinyl ligands. Based on the inherent preferences of the SAE reaction with the minus form of the diisopropyl tartrate ligand, the desired enantiomer **30** was formed preferentially (as it constitutes the "matched" case), in the opening stages of the reaction along with a small amount of the undesired epoxide **31**. If one were to stop the reaction at this point, however, the separation of these two compounds would be quite cumbersome, requiring chiral HPLC or the attachment of some type of chiral auxiliary onto the free hydroxy group to create a mixture of diastereomers that would be more easily separable. Fortunately, both **30** and **31** bear another vinyl ligand, and thus there is the potential for a second epoxidation reaction to occur. In this case, though, the presence of stereochemical information within **30** and **31** now biases the reaction with this reagent combination such that **31** is the facially matched substrate (and hence a willing participant in another SAE reaction) while **30** is mismatched. As a result, there is a differential reaction rate at which the undesired enantiomer is converted quickly into the bisepoxide **32** while the majority of the desired monoepoxide **30** reacts very slowly, thus remaining behind. As such, this event constitutes a second kinetic resolution that transforms all of the undesired enantiomer into the now easily separable bisepoxide **32**.

With this chiral material (**30**) successfully synthesized, all the necessary stereochemical elements were now in place to enable the incorporation of the remaining stereocenters of resiniferatoxin (**1**) in an entirely guided manner through substrate-controlled events. Thus, proceeding forward with the execution of the synthesis, protection of the free alcohol in **30** (Scheme 8) under standard conditions provided the corresponding benzyl ether (**21**).

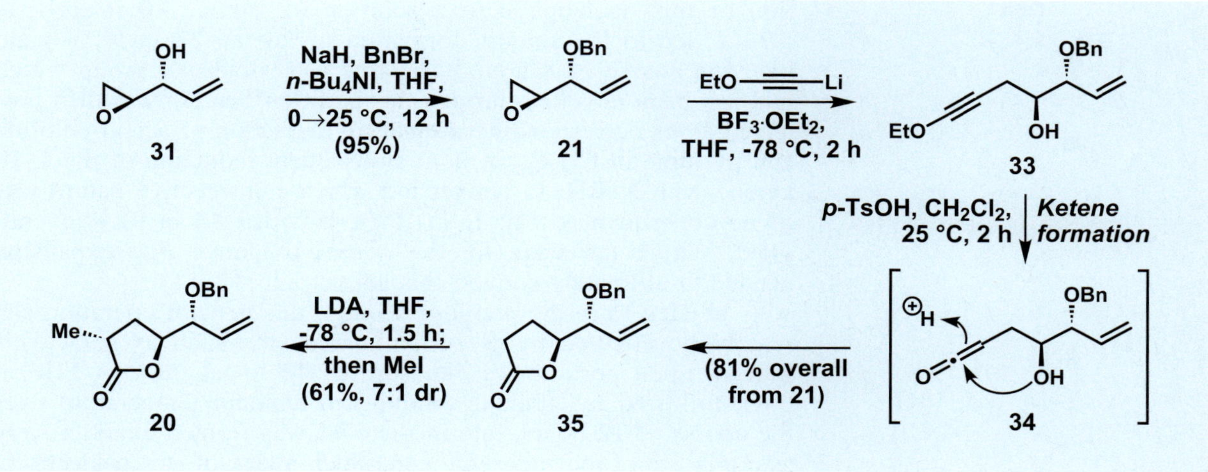

Scheme 8. Synthesis of lactone **20**.

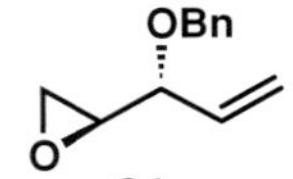

Subsequent addition of an alkynyl borane derived from the initial reaction of lithium ethoxyacetylide and $BF_3{\cdot}OEt_2$ led to site-selective nucleophilic lysis of the epoxide within **21** to afford **33**. Although a lithium acetylide itself should be capable of effecting this same event (typically at an elevated temperature), the conversion of this nucleophile into an alkynyl borane enabled the reaction to proceed under milder conditions with complete regioselectivity, a result that can be rationalized by assuming initial coordination of the reagent to the epoxide (see column figure), thus activating the oxirane for selective intramolecular delivery of the acetylide.[17d,e] With this event smoothly executed, subsequent treatment of **33** with *p*-TsOH in CH_2Cl_2 at ambient temperature then provided lactone **35** in 81 % overall yield for these two steps via ketene **34**. The synthesis of the key building block **20** was then completed, in 61 % yield, through the initial formation of the enolate of **35** with LDA at −78 °C in THF, followed by a facially selective alkylation upon the addition of MeI, presumably guided by the existing stereocenters. Since these stereogenic centers are relatively remote from the reactive site, the overall stereoinduction for this process was only modest (7:1 diastereomeric ratio in favor of the drawn product), but more than sufficient for such a functionalized intermediate.[17a,b]

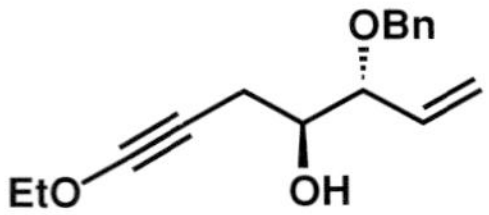

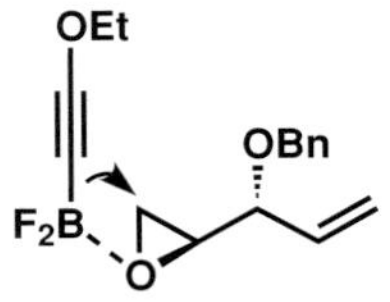

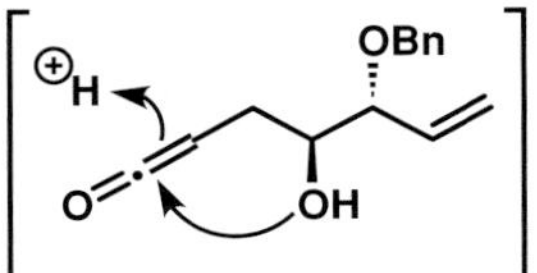

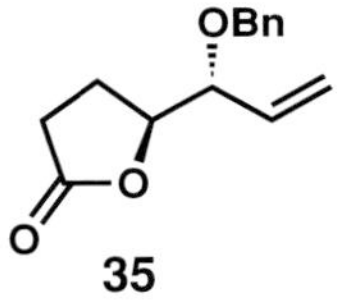

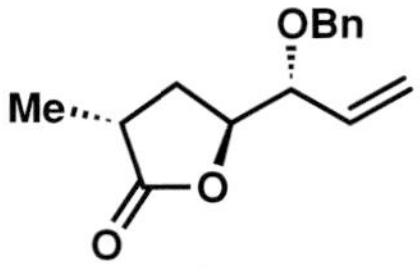

Having executed an efficient and stereocontrolled synthesis of this lactone, the stage was now set to begin exploration of the more challenging steps envisioned for the construction of resiniferatoxin (**1**). The first such operation was a stereoselective intramolecular 1,3-dipolar cycloaddition of a reactive 3-oxidopyrylium and a tethered alkene. In order to explore the viability of this transformation, however, lactone **20** first had to be converted into the key precursor for this event, namely pyranone **17**. As shown in Scheme 9, this requirement did not prove especially challenging to meet. Following the formation of the furanyl lithium species **19**, prepared by treating the known **36**[16] with *n*-BuLi at −78 °C and subsequently stirring the reaction mixture at 0 °C for 30 minutes, the slow addition of this nucleophile to a solution of lactone **20** in THF at −78 °C led to the smooth formation of ketone **37** in 98 % yield over the course of 8 hours. The free C-13 hydroxy group which had just been unveiled through this addition process was then protected as its corresponding acetate by the action of acetyl chloride and pyridine in CH_2Cl_2 at 0 °C. Subsequent reduction of the C-10 ketone with $NaBH_4$ (a conversion whose nonselective nature was of no consequence, vide infra) then provided **18** in 97 % overall yield, setting the stage for the crucial oxidative ring expansion needed to ultimately convert this adduct into **17**.

As alluded to in the previous section, this general operation can readily be effected by electrophilic reagents such as peracids.[15] Indeed, upon portionwise exposure of **18** to *m*CPBA in THF at 0 °C, followed by gradual warming to ambient temperature over the course of 12 hours, intermediate **41** was formed quantitatively as a mixture of four diastereomers. Such a lack of stereoselectivity was not inherently problematic as all the chiral information within

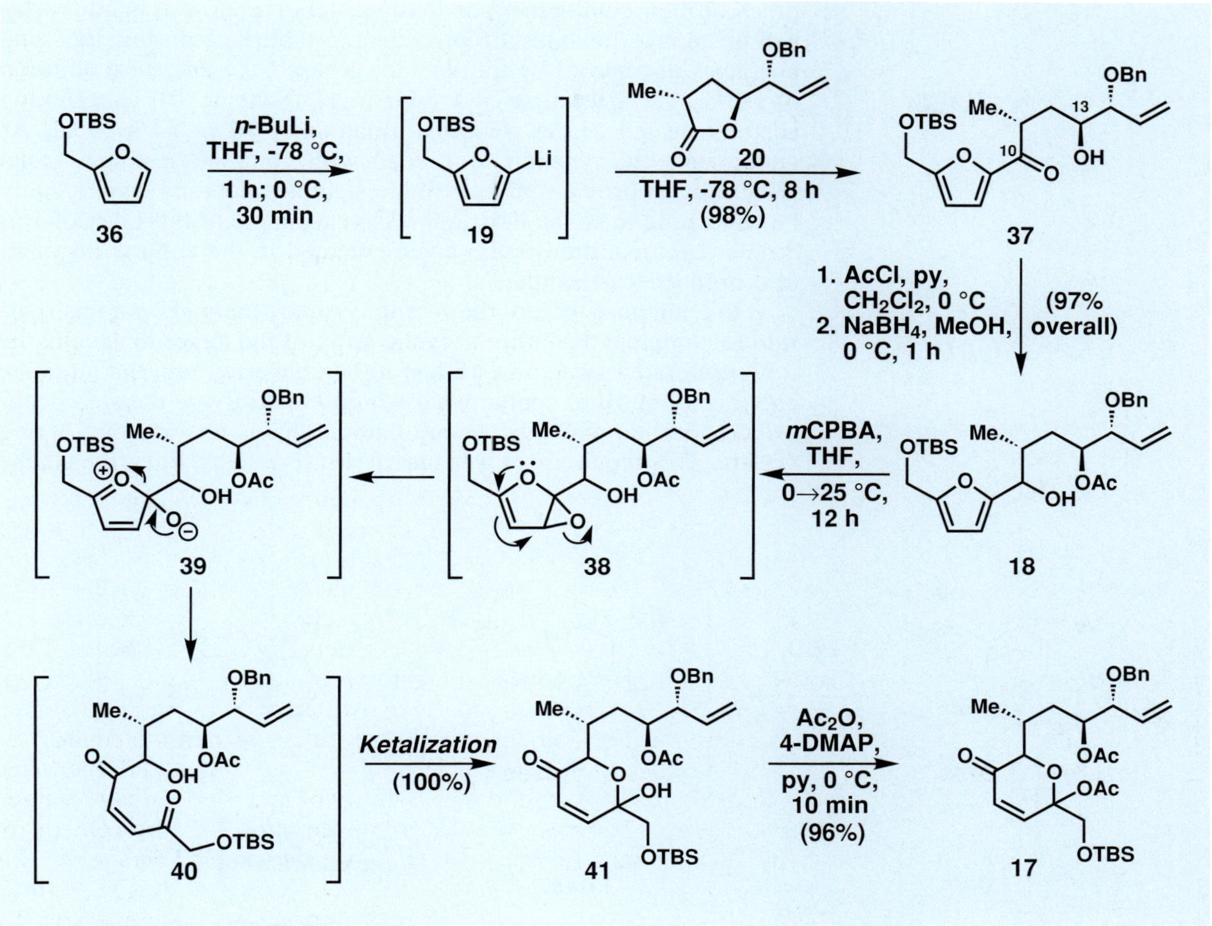

Scheme 9. Elaboration of **20** to pyranone **17**.

the six-membered ring of compound **41** would be destroyed upon successful oxidopyrylium formation. Critical, though, was that the reaction proceeded to provide only one general butenolide structure since this outcome ensured that only one oxidopyrylium would eventually be formed from this intermediate.* With this operation out of the way, a final acetylation of the free hydroxy group to prepare the required leaving group to form an oxidopyrylium species then completed the assembly of **17** in 96 % yield.

Having progressed to this stage, the opportunity to explore the intramolecular 1,3-dipolar cycloaddition strategy previously forecasted to establish both the B- and C-rings of resiniferatoxin and

---

* Actually, although hydroxy-assisted delivery of the peracid to the indicated C–C double bond of **38** ensured the formation of **41**, addition of oxygen to the other olefin would afford a degenerate pathway to the same product.

much of their complement of stereocenters was now at hand. Fortunately, along the lines of precedent established during the total synthesis of phorbol by the Wender group,[14] the dropwise addition of DBU over the course of 1 hour to **17** (Scheme 10) in refluxing acetonitrile led to the smooth formation of **15** in 84% yield. As anticipated, the reaction proceeded with complete *exo* and facial selectivity to provide **15** as a single stereoisomer, and can arguably be contended to be the most impressive example of this class of 1,3-dipolar cycloaddition process yet executed in the context of a natural product total synthesis.

At this juncture, before the A-ring could be merged onto this scaffold to complete the entire tricyclic array of the target molecule, the α,β-unsaturated system of **15** had to be converted into the alternate enone variant **10** to enable what would hopefully be the stereocontrolled addition of the necessary carbon chains to form this A-ring system. This requirement was met in just five steps. First, the alkene

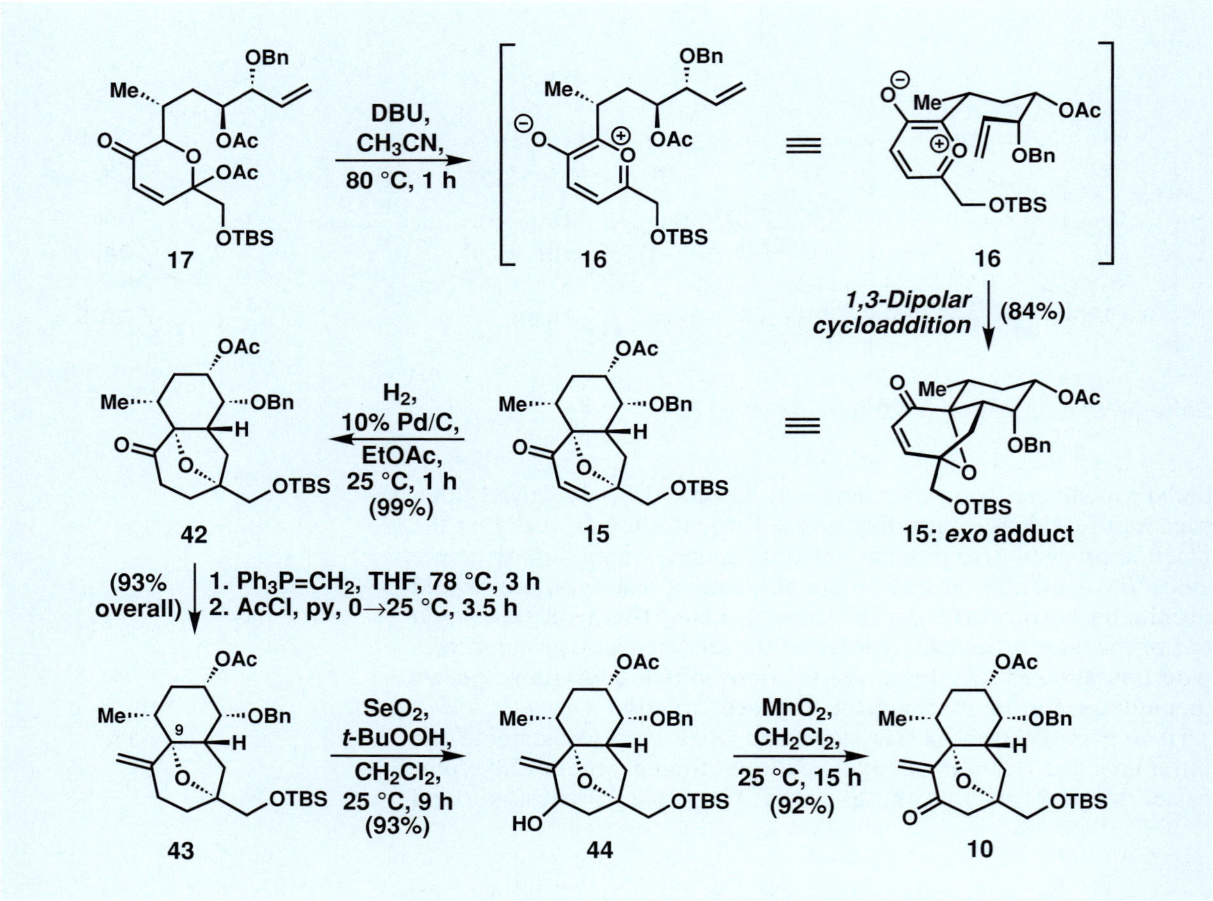

Scheme 10. Elaboration of **17** into bicyclic intermediate **10** through a sequence featuring a 1,3-dipolar cycloaddition.

in **15** was chemoselectively hydrogenated in near quantitative yield (99 %) with catalytic 10 % Pd/C in EtOAc at ambient temperature to afford **42**; intriguingly, no lysis of the benzyl ether was observed under these conditions. Next, the C-10 ketone was readily olefinated with the ylide derived from methylenetriphenylphosphonium bromide (formed with *t*-BuOK), leading to concomitant cleavage of the C-13 acetate group in the same pot. Following reprotection of this position using acetyl chloride to afford **43**, an allylic oxidation was then performed with SeO_2 and *t*-BuOOH in CH_2Cl_2 to provide **44** in 86 % combined yield over these three steps. Subsequent oxidation with MnO_2 then completed the assembly of **10** in 92 % yield. Before pressing forward, a few additional comments regarding the penultimate transformation in Scheme 10 are in order. Although there are several reagent systems capable of effecting allylic oxidations, the combination of a peroxide and catalytic SeO_2 comprises one of the most effective (and therefore most common) mediators for this process, in which SeO_2 serves as the actual oxidant and the peroxide reoxidizes the $Se(OH)_2$ by-product from successful oxidation.[20] Moreover, although the transformation of **43** into **44** does not indicate that site selectivity is possible for this oxidation process because the C-9 center in **43** is fully substituted, selenium-mediated oxidations do follow a well-defined set of reactivity rules when multiple oxidation products are possible. In this regard, numerous precedents[21] have revealed the following: a) oxidation always occurs at the more substituted end of the alkene, b) the reactivity order generally proceeds with trisubstituted olefins as $CH_2 > CH_3 > CH$ while for unsymmetrical 1,1-disubstituted alkenes (e. g. **43**) as $CH > CH_2 > CH_3$, and c) when any double bond is within a ring, oxidation also occurs within the ring.

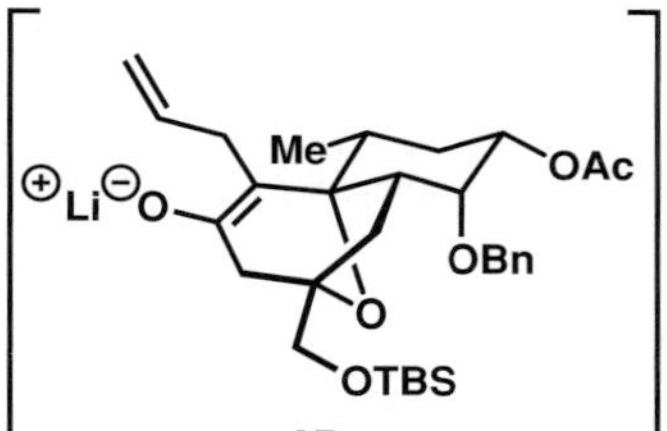

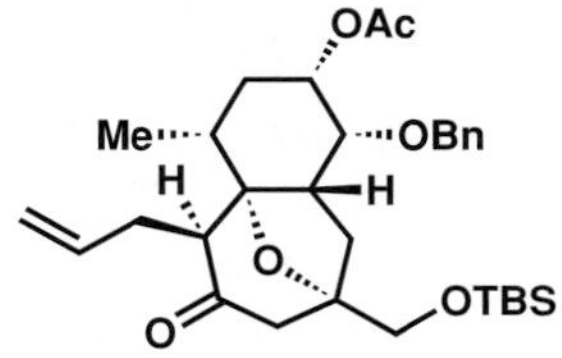

Returning to the synthesis, the formation of enone **10** now provided the opportunity to explore the value of the C6–C9 oxido bridge in restricting the conformational freedom of the seven-membered B-ring, as it was time to add carbon nucleophiles to pave the way for A-ring annulation. Fortunately, as expected based on molecular modeling, the addition of vinyl cuprate (formed by the reaction of CuCN with vinyllithium) in Et_2O at −78 °C to **10** led to the formation of **46** as a single stereoisomer in 96 % yield, with selective proton capture of the incipient intermediate (**45**) occurring on the face *syn* to its oxygen bridge (Scheme 11). With this stereocenter secured, the approach trajectory of the second nucleophilic addition at C-4 was all but assured. Indeed, the addition of lithium phenylacetylide occurred from the desired α-face to provide an intermediate alkoxide, which was quenched with *N*-methyl-*N*-(trimethylsilyl)trifluoroacetamide (MSTFA, a source of TMS groups often used when elimination is possible)[22] to provide **9** in 43 % yield. Since the majority of the remaining material balance from this step was the free C-4 hydroxy congener (**47**, see column figure) of **9**, following chromatographic separation of the two products **47** was then treated with a more standard silylating source (TMSCl) to bolster the overall yield of **9** to 76 % from **46**.

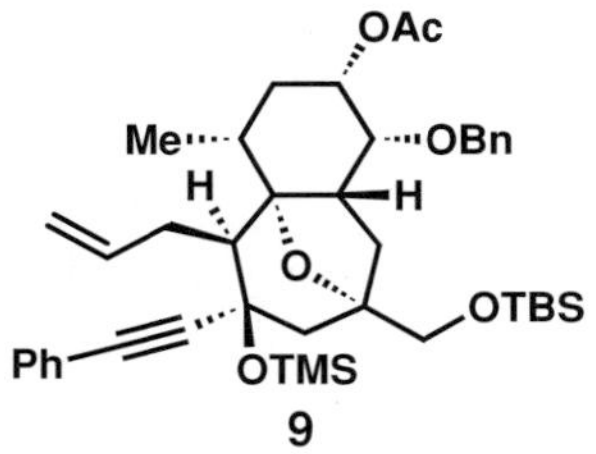

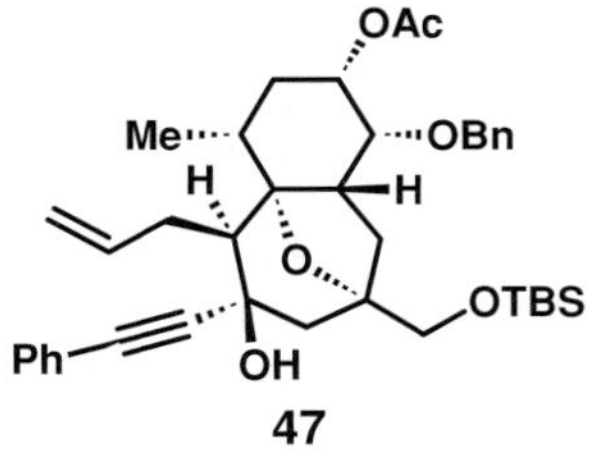

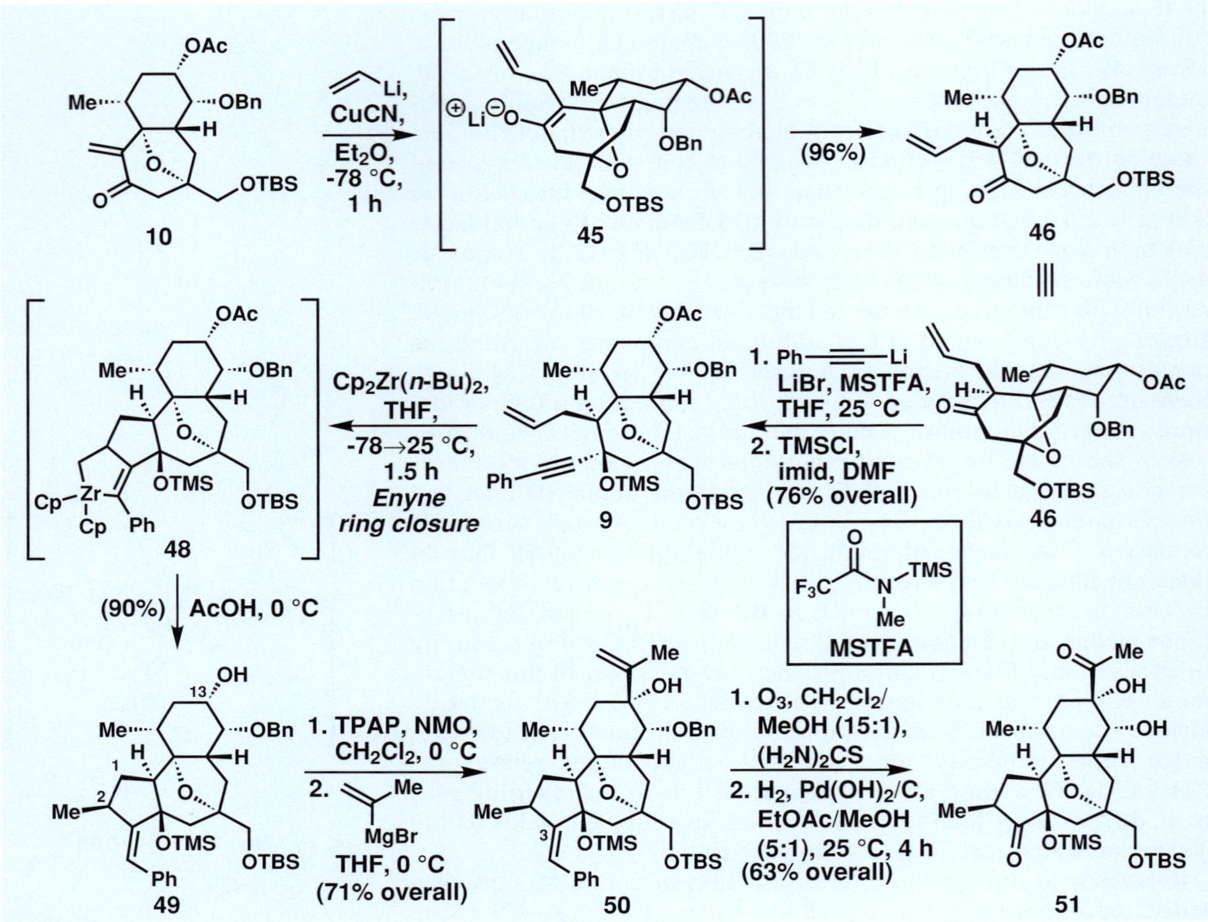

Scheme 11. Conversion of enone **10** into advanced intermediate **51** through zirconium-mediated enyne ring closure.

With an alkyne and alkene successfully appended to the growing resiniferatoxin skeleton, the correct motifs were in place to attempt the construction of the A-ring through a metal-mediated enyne ring closure. Despite the large number of examples of this process in chemical synthesis,[8–10] **9** represented one of the most complicated substrates ever subjected to it. Testament to the powerful nature of the conversion, though, **9** was smoothly transformed into **49** in 90% yield upon treatment with $Cp_2Zr(n\text{-}Bu)_2$ (formed *in situ* by initially reacting Cp_2ZrCl_2 with 2 equivalents of *n*-BuLi)[9d] followed by an AcOH quench of the zirconocene intermediate (**48**). The latter process both installed the desired hydrogen atoms in **49** and led to cleavage of the C-13 acetate group. It is also important to note that although **49** is drawn as a single C-2 stereoisomer in Scheme 11, the exact disposition of the methyl group was not established; given that this position will eventually need to be converted into a C1–C2 olefin, its stereochemical arrangement was, in fact, of no consequence as discussed earlier.

Having now secured all three of the central rings of resiniferatoxin, only a few elements of the target remained to be handled. One of these, the formation of a C-13 quaternary center, was ready to be contended with at this juncture due to the fortuitous loss of the protecting group at that site during the previous operation. Thus, TPAP-mediated oxidation[23] of the hydroxy group at C-13 was followed by the facially selective addition of isopropenylmagnesium bromide to this newly formed ketone in THF at 0 °C, affording **50** in 71 % yield. At this stage, the C-3 alkene that resulted from enyne cyclization was no longer needed. Since a ketone exists in this site in the target molecule, ozonolytic cleavage of this motif was attempted and proceeded smoothly, accompanied by cleavage of the isopropenyl side chain to a methyl ketone. Although this lack of chemoselectivity was unfortunate, its occurrence was unsurprising in that the isopropenyl olefin within **50** is quite accessible. While one could thus envision switching the order of ozonolysis with the addition of the isopropenyl nucleophile, inherently there is no reason to believe that such a nucleophilic addition would prove wholly chemoselective on a bisketone intermediate. Moreover, whereas the side products derived from C-3 addition through such a sequence could never be recycled, the formation of a bisketone in this ozonolytic step from **50** was acceptable in that selective olefination of the carbonyl appended to the C-14 side chain could conceivably recreate the isopropenyl motif of resiniferatoxin (**1**). This operation would have to await several other transformations, though, and the first event executed at this point was simple debenzylation using 20 % $Pd(OH)_2/C$ in a 5:1 mixture of EtOAc and MeOH which completed the assembly of **51** in 63 % overall yield from **50**.

Having completed its role in facilitating the stereoselective assembly of the A-ring, the C6–C9 oxido bridge was now ripe for disassembly, and, therefore, means were next sought to excise it through a ring-opening reaction to afford a free C-9 alcohol. Accordingly, as shown in Scheme 12, following protection of the vicinal alcohols in **51** as a cyclic carbonate by using triphosgene, subsequent selective lysis of the lone TBS group in the presence of a TMS group with 49 % aqueous HF in MeCN at 0 °C afforded **52** in 87 % yield. Although conventional wisdom would seem to dictate that such selectivity should be impossible to achieve, the uniquely hindered nature of the C-4 position is the most plausible explanation for the survival of the TMS group from this necessary, but serendipitous, conversion. With only one silyl group removed, the newly unveiled C-20 alcohol was then converted into an iodide via an intermediate triflate to provide a leaving group for the anticipated elimination reaction. This transformation was effected following precedent[24] established in the realm of carbohydrates using Rieke© zinc, a highly reactive form of zinc powder formed by reducing one of its metal salts in ethereal or hydrocarbon solvents with alkali

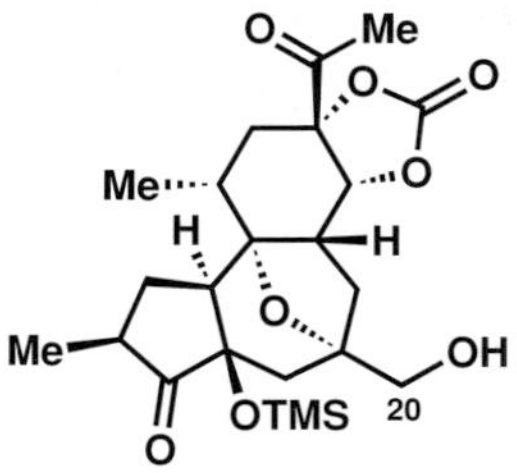

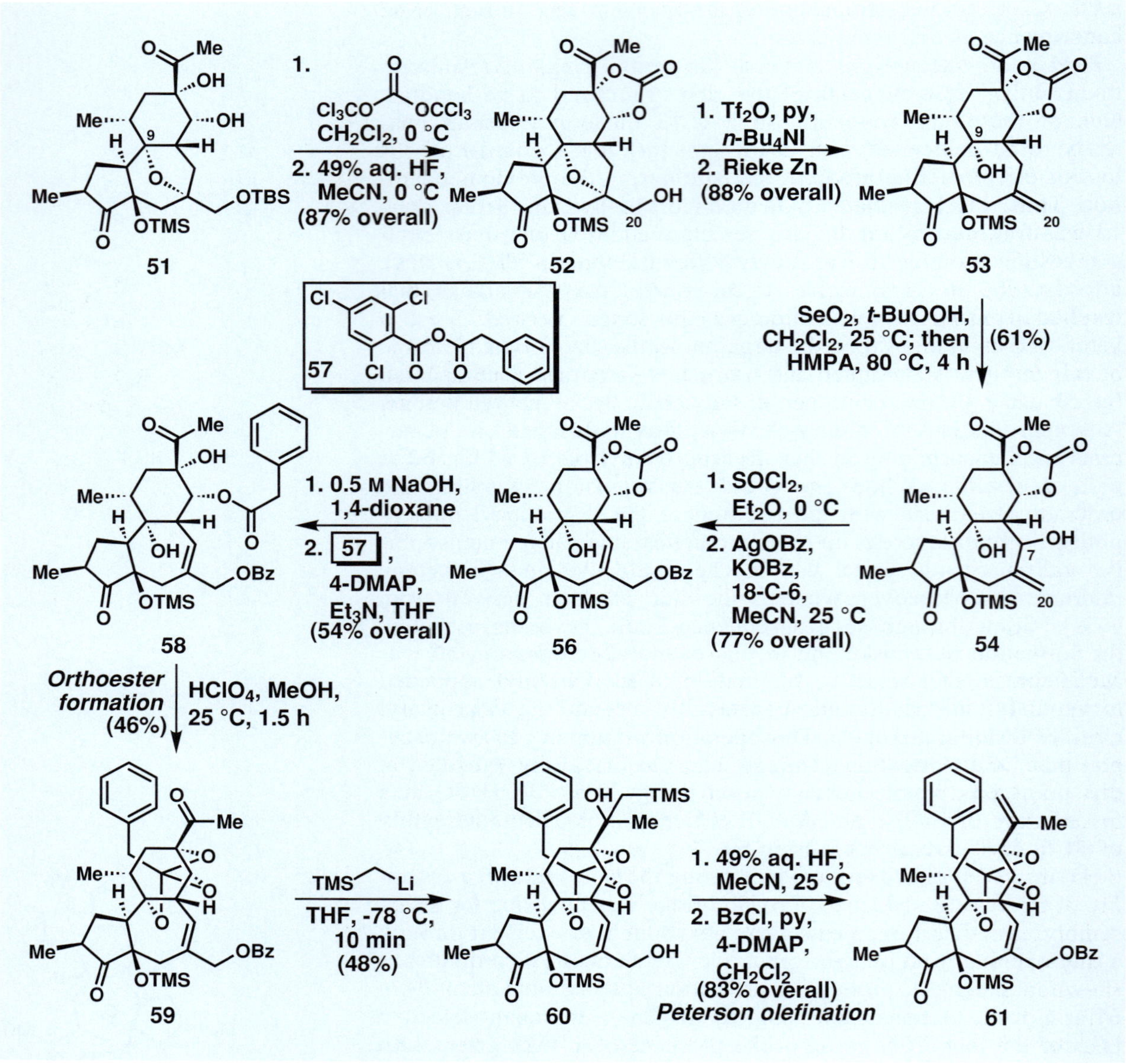

Scheme 12. Synthesis of advanced intermediate **61** from **51**.

metals as reducing agents,[25] in which after successful insertion of this reagent into the C-20 iodide to form an organozinc species, spontaneous elimination led to **53** in 88 % overall yield for these two steps. As an aside, metals prepared by the Rieke process (e. g. zinc and magnesium) are highly effective for the formation of metallated reagents at very low temperatures. Moreover, as can be gleaned from this example, one of their most useful characteristics is their compatibility with a wide variety of functional groups such as chlorides, nitriles, esters, amides, ethers, sulfides, and ketones.[25]

At this juncture, just a few functional group manipulations separated **53** from resiniferatoxin (**1**). As we shall see, though, the complexity of the target molecule rendered these modifications far from trivial. First, having unveiled the C-9 alcohol it was now necessary to modify the B-ring to create endocyclic unsaturation and to reinstall the hydroxy group at C-20 that had been lost in the previous series of events. Accordingly, the repeated application of SeO_2 and *t*-BuOOH to effect an allylic oxidation proved fruitful, since it afforded a new functional handle in **54**. Unfortunately, despite careful tweaking of the reaction conditions to achieve a 61 % yield, the oxidation lacked complete site-selectivity in that the remainder of the material balance was alcohol **55** (see column figure, 38 % yield), a product which could not be recycled. This result is not that surprising given that the selectivity rules defined earlier indicate no predilection for the generation of **54** over **55**, with only the latent differences in steric accessibility enabling the preferential formation of the desired product in this case.

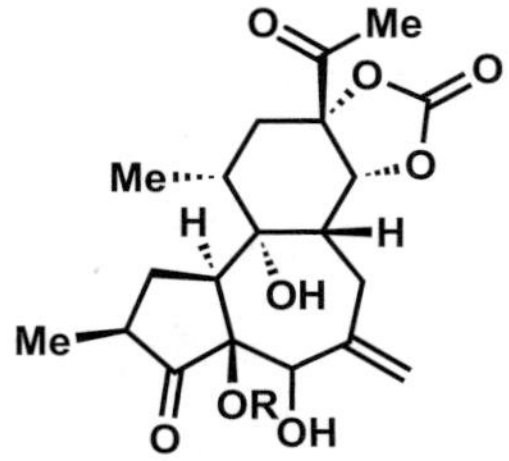

55: R = TMS

Anticipating now that C-20 functionality could be incorporated through an S_N2' reaction, the freshly installed alcohol was converted into a chloride leaving group with $SOCl_2$ in Et_2O at 0 °C. This C-7 motif then obligingly departed in the desired sense in 77 % overall yield upon the addition of KOBz and AgOBz in MeCN in the presence of 18-crown-6. With the B-ring now structurally sound, it was then deemed prudent to fashion the unique orthoester of the target molecule. Following cleavage of the carbonate group in **56**, the more accessible secondary C-14 hydroxy group was engaged in an esterification reaction with **57**, a reactive mixed anhydride of phenylacetic acid formed through its reaction with Yamaguchi's acid chloride (2,4,6-trichlorobenzoyl chloride), to afford ester **58** in 54 % overall yield.[26] Subsequent treatment of **58** with $HClO_4$ in MeOH at 25 °C for 1.5 hours then initiated orthoester assembly to provide **59** in 46 % yield.

Since these events established the last of the ring systems of resiniferatoxin, only two major obstacles remained before the tricyclic core of **1** would be complete: the formation of an enone in ring A and the conversion of the C-14 methyl ketone into an isopropenyl group. The latter task was tackled first, but proved rather challenging to achieve. Attempted olefination of this ketone within **59** under a variety of conventional conditions, such as Wittig olefination or the powerful Tebbe or Nysted variants, failed to deliver the desired product in useful quantities for the synthesis to proceed. Fortunately, application of a Peterson olefination protocol provided a workable entry into these final steps.[27] Thus, the addition of trimethylsilylmethyl lithium to **59** afforded **60** (in which the C-20 benzoate ester was also cleaved). Subsequent addition of 49 % aqueous HF served both to activate the resultant tertiary hydroxy function as a leaving group and to cleave the C–Si bond, resulting in a *syn* elimination reaction leading to olefin formation. Reintroduction of the benzoate ester then completed the assembly of **61** in 83 % overall yield for these two steps. With this task complete, efforts now turned to creating the A-ring enone. Accordingly, as shown in Scheme 13, the C-3 ketone of **61** was converted into its TMS-protected enol tautomer **62** through the action of DABCO and

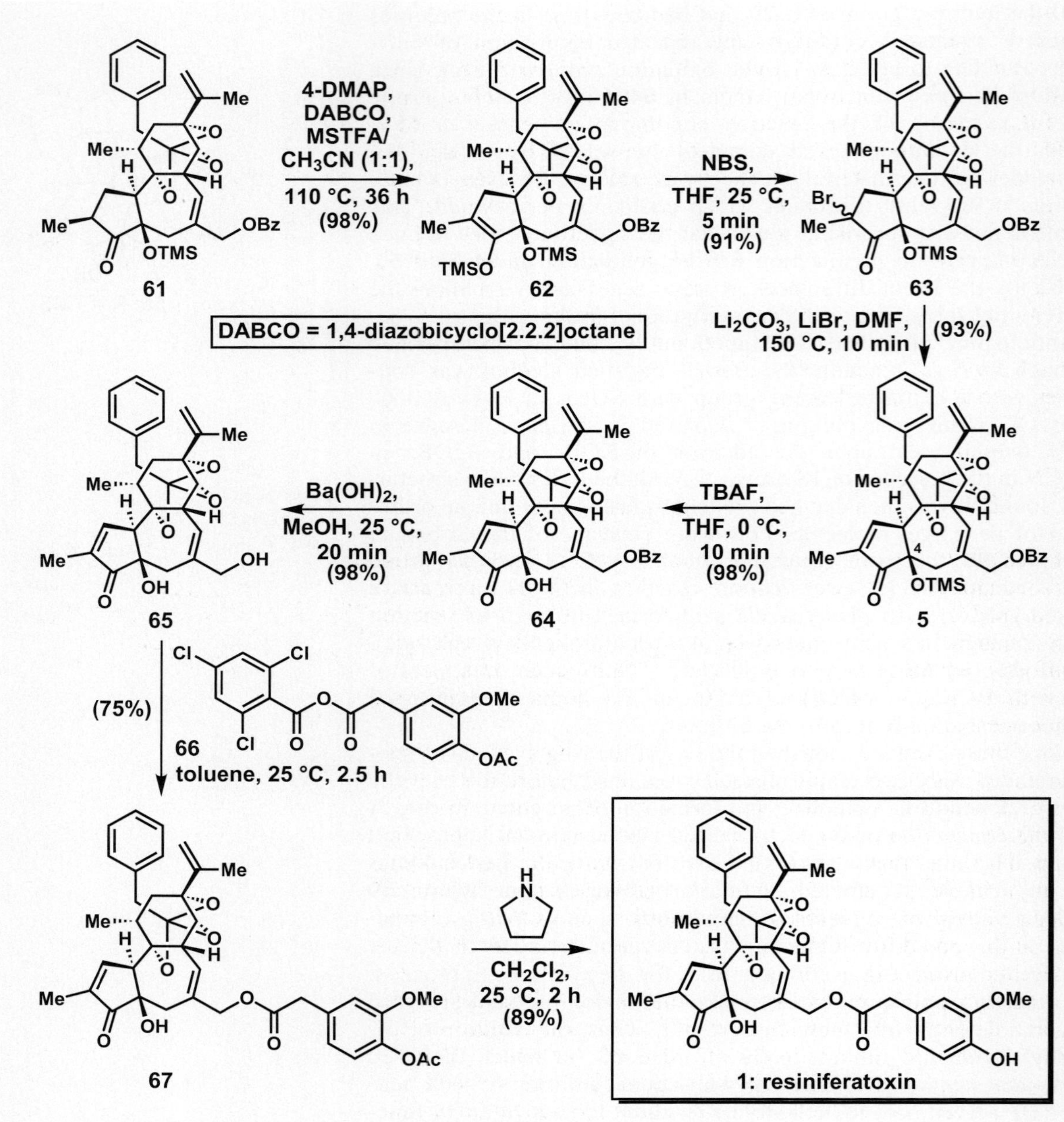

Scheme 13. Final stages and completion of the total synthesis of resiniferatoxin (**1**).

4-DMAP in CH_3CN over the course of 36 hours, once again using MSTFA as a source of the silyl protecting group. This conversion then paved the way for nucleophilic capture of bromine, supplied by *N*-bromosuccinimide (NBS) in THF, to afford **63**. Subsequent elimination of this newly minted motif, induced by Li_2CO_3 and LiBr in DMF at 150 °C, then completed the assembly of α,β-unsaturated ketone **5** in 93 % yield.

Fully synthetic resiniferatoxin (**1**) was nearly at hand, needing only introduction of the homovanillyl side chain before final victory could be declared. Fortunately, this task proved relatively simple to accomplish. Following excision of the previously stubborn TMS group appended to the C-4 alcohol in **5** using TBAF in THF and benzoate ester cleavage as accomplished by $Ba(OH)_2$ in MeOH to generate **65**, the addition of the homovanillyl side chain as its Yamaguchi mixed anhydride (**66**) proceeded in 75 % yield to furnish **67**. With all the requisite structural features now in place, merely an acetate group remained to be excised. Following exposure of **67** to pyrrolidine in CH_2Cl_2 at ambient temperature, this cursory task was smoothly accomplished in 89 % yield. The synthesis of resiniferatoxin (**1**) was finally complete! Overall, the developed route required a total of 44 operations, the majority of which proceeded, impressively, in better than 90 % yield.

6.4 Conclusion

As is true of most contemporary efforts in target-oriented synthesis, the successful route to resiniferatoxin (**1**) described in this chapter was derived from a well-orchestrated amalgamation of classical transformations with state-of-the-art modern synthetic technologies. Within this typical framework, though, perhaps the more significant feature was the deft use of the complexity of the natural product to extend the scope of many of these reactions by demonstrating their virtuosity in highly functionalized settings. In this regard, the intramolecular 1,3-dipolar cycloaddition of a 3-oxidopyrylium ylide with a tethered alkene, which stereoselectively established the BC-ring system, and the subsequent zirconium-mediated enyne cyclization to complete the A-ring are exemplary for their smooth execution in the presence of multiple, highly sensitive functional groups. Both these transformations arguably constitute the most impressive examples of their implementation in the context of natural product synthesis.

In addition to these fundamental contributions, the developed synthesis is also noteworthy for its careful command of stereochemistry throughout its prosecution, as achieved by the combination of asymmetric reactions and exquisite understanding of the inherent topology of the molecule to introduce new stereogenic centers through substrate-controlled events. In particular, the use of an oxido bridge during the majority of the route to ensure conformational control over an otherwise flexible seven-membered ring is a particularly inventive and effective strategy for the creation of new stereochemical elements within the general confines of cyclic stereocontrol. Beyond these instructive features, because resiniferatoxin (**1**) constitutes a promising therapy to modulate pain, the synthetic route developed in the Wender laboratories also paves the way for future studies in chemical biology by enabling the synthesis of structurally modified analogs, an area in which few investigations have been performed to date,[28] but which clearly is an important avenue for further pursuit.

References

1. For lead references, see: a) M. J. Caterina, M. A. Schumacher, M. Tominaga, T. A. Rosen, J. D. Levine, D. Julius, *Nature* **1997**, *389*, 816; b) A. Szallasi, P. M. Blumberg, *Pain* **1996**, *68*, 195.
2. a) C. P. N. Watson, R. J. Evans, V. R. Watt, *Pain* **1989**, *38*, 177; b) D. R. Ross, R. J. Varipapa, *New England J. Med.* **1989**, *321*, 474; c) W. P. Cheshire, C. R. Snyder, *Pain* **1990**, *42*, 307.
3. A. M. Rouhi, *Chem. Eng. News* **1998**, *76* (4), 31.
4. M. Hergenhahn, W. Adolf, E. Hecker, *Tetrahedron Lett.* **1975**, *16*, 1595.
5. a) A. Szallasi, P. M. Blumberg, *Brain Res.* **1990**, *524*, 106; b) A. Szallasi, N. Sharkey, P. M. Blumberg, *Phytotherapy Res.* **1989**, *3*, 253.
6. a) H. zur Hausen, G. W. Bornkamm, R. Schmidt, E. Hecker, *Proc. Natl. Acad. Sci. U.S.A.* **1979**, *76*, 782; b) P. A. Wender, Y. Martin-Cantalejo, A. J. Carpenter, A. Chiu, J. De Brabander, P. G. Harran, J.-M. Jimenez, M. F. T. Koehler, B. Lippa, J. A. Morrison, S. G. Müller, S. N. Müller, C.-M. Park, M. Shiozaki, C. Siedenbiedel, D. J. Skalitzky, M. Tanaka, K. Irie, *Pure & Appl. Chem.* **1998**, *70*, 539.
7. P. A. Wender, C. D. Jesudason, H. Nakahira, N. Tamura, A. L. Tebbe, Y. Ueno, *J. Am. Chem. Soc.* **1997**, *119*, 12976.
8. a) I. U. Khand, G. R. Knox, P. L. Pauson, W. E. Watts, M. I. Foreman, *J. Chem. Soc., Perkin Trans. 1* **1973**, 977; b) M. C. Croudace, N. E. Schore, *J. Org. Chem.* **1981**, *46*, 5357; c) C. Exon, P. Magnus, *J. Am. Chem. Soc.* **1983**, *105*, 2477.
9. a) W. A. Nugent, J. C. Calabrese, *J. Am. Chem. Soc.* **1984**, *106*, 6422; b) T. V. RajanBabu, W. A. Nugent, D. F. Taber, P. J. Fagan, *J. Am. Chem. Soc.* **1988**, *110*, 7128; c) E. Negishi, S. J. Holmes, J. M. Tour, J. A. Miller, F. E. Cederbaum, D. R. Swanson, T. Takahashi, *J. Am. Chem. Soc.* **1989**, *111*, 3336; d) E. Negeshi, F. E. Cederbaum, T. Takahashi, *Tetrahedron Lett.* **1986**, *27*, 2829.
10. a) E. Negishi, S. J. Holmes, J. M. Tour, J. A. Miller, *J. Am. Chem. Soc.* **1985**, *107*, 2568; b) E. Negishi, D. R. Swanson, F. E. Cederbaum, T. Takahashi, *Tetrahedron Lett.* **1987**, *28*, 917.
11. For a recent review, see: S. Karlsson, H.-E. Högberg, *Org. Prep. Proc. Int.* **2001**, *33*, 103. For an excellent treatment of several classes of 1,3-dipolar cycloadditions, see: W. Carruthers, *Cycloaddition Reactions in Organic Synthesis*, Pergamon, Oxford, **1990**, pp. 269–314.
12. a) P. G. Sammes, L. J. Street, *J. Chem. Soc., Chem. Commun.* **1983**, 666; b) P. G. Sammes, L. J. Street, *J. Chem. Soc., Perkin Trans. 1* **1983**, 1261; c) S. M. Bromidge, P. G. Sammes, L. J. Street, *J. Chem. Soc., Perkin Trans. 1* **1985**, 1725.
13. a) T. Do-Minh, A. M. Trozzolo, G. W. Griffin, *J. Am. Chem. Soc.* **1970**, *92*, 1402; b) J. B. Hendrickson, J. S. Farina, *J. Org. Chem.* **1980**, *45*, 3359; c) J. B. Hendrickson, J. S. Farina, *J. Org. Chem.* **1980**, *45*, 3361; d) P. A. Wender, J. L. Mascareñas, *Tetrahedron Lett.* **1992**, *33*, 2115.
14. a) P. A. Wender, H. Y. Lee, R. S. Wilhelm, P. D. Williams, *J. Am. Chem. Soc.* **1989**, *111*, 8954; b) P. A. Wender, H. Kogen, H. Y. Lee, J. D. Munger, R. S. Wilhelm, P. D. Williams, *J. Am. Chem. Soc.* **1989**, *111*, 8957. An asymmetric synthesis of phorbol was also accomplished by the Wender group: c) P. A. Wender, K. D. Rice, M. E. Schnute, *J. Am. Chem. Soc.* **1997**, *119*, 7897.
15. H. Heaney, J. S. Ahn in *Comprehensive Heterocyclic Chemistry II, Vol. 2* (Eds.: A. R. Katritzky, C. W. Rees, E. F. V. Scriven), Pergamon, Oxford, **1996**, pp. 310–312. For one of the earliest examples of this process, see: Y. Lefebvre, *Tetrahedron Lett.* **1972**, *13*, 133.
16. a) F. E. McDonald, C. B. Connolly, M. M. Gleason, T. B. Towne, K. D. Treiber, *J. Org. Chem.* **1993**, *58*, 6952; b) J. A. Marshall, D. J. Nelson, *Tetrahedron Lett.* **1988**, *29*, 741.
17. a) D. Askin, R. P. Volante, R. A. Reamer, K. M. Ryan, I. Shinkai, *Tetrahedron Lett.* **1988**, *29*, 277; b) S. L. Schreiber, D. B. Smith, *J. Org. Chem.* **1989**, *54*, 9; c) S. L. Schreiber, T. Sammakia, D. E. Uehling, *J. Org. Chem.* **1989**, *54*, 15. For the selective addition of alkynyl nucleophiles to epoxides, see: d) M. Yamaguchi, I. Hirao, *Tetrahedron Lett.* **1983**, *24*, 391. For a review of selective epoxide opening, see: e) J. Gorzynski Smith, *Synthesis* **1984**, 629.

18. For a review, see: T. Katsuki, V. S. Martin in *Organic Reactions, Vol. 48* (Eds.: L. A. Paquette et al.), John Wiley & Sons, New York, **1996**, pp. 1–299.
19. a) Y. Gao, R. M. Hanson, J. M. Klunder, S. Y. Ko, H. Masamune, K. B. Sharpless, *J. Am. Chem. Soc.* **1987**, *109*, 5765; b) S. L. Schreiber, T. S. Schreiber, D. B. Smith, *J. Am. Chem. Soc.* **1987**, *109*, 1525.
20. M. A. Umbreit, K. B. Sharpless, *J. Am. Chem. Soc.* **1977**, *99*, 5526.
21. U. T. Bhalerao, H. Rapoport, *J. Am. Chem. Soc.* **1971**, *93*, 4835 and references cited therein.
22. Early examples of the use of this reagent include: a) H. Gleispach, *J. Chromatogr.* **1974**, *91*, 407; b) M. Donike, J. Zimmermann, *J. Chromatogr.* **1980**, *202*, 483.
23. For an excellent review of this oxidant, see: S. V. Ley, J. Norman, W. P. Griffith, S. P. Marsden, *Synthesis* **1994**, 639.
24. B. Bernet, A. Vasella, *Helv. Chim. Acta* **1984**, *67*, 1328.
25. R. D. Rieke, *Aldrichimica Acta* **2000**, *33*, 52 and references cited therein.
26. J. Inanaga, K. Hirata, H. Saeki, T. Katuki, M. Yamaguchi, *Bull. Chem. Soc. Jpn.* **1979**, *52*, 1989.
27. D. J. Peterson, *J. Org. Chem.* **1968**, *33*, 780. For a comprehensive review, see: D. J. Ager in *Organic Reactions, Vol. 38* (Eds.: L. A. Paquette et al.), John Wiley & Sons, New York, **1990**, pp. 1–224.
28. a) C. S. J. Walpole, R. Wrigglesworth, S. Bevan, E. A. Campbell, A. Dray, I. F. James, K. J. Masdin, M. N. Perkins, J. Winter, *J. Med. Chem.* **1993**, *36*, 2381; b) J. Lee, G. Acs, P. M. Blumberg, V. E. Marquez, *Bioorg. Med. Chem. Lett.* **1995**, *5*, 1331; c) C. S. J. Walpole, S. Bevan, G. Bloomfield, R. Breckenridge, I. F. James, T. Ritchie, A. Szallasi, J. Winter, R. Wrigglesworth, *J. Med. Chem.* **1996**, *39*, 2939; d) G. Appendino, G. Cravotto, G. Palmisano, R. Annunziata, A. Szallasi, *J. Med. Chem.* **1996**, *39*, 3123; e) J. Lee, S.-U. Park, J.-Y. Kim, J.-K. Kim, J. Lee, U. Oh, V. E. Marquez, M. Beheshti, Q. J. Wang, S. Modarres, P. M. Blumberg, *Bioorg. Med. Chem. Lett.* **1999**, *9*, 2909.

1: R = H, epothilone A
2: R = Me, epothilone B

7

K. C. Nicolaou (1997)

Epothilones A and B

7.1 Introduction

The past few years have witnessed an exponential level of growth in our comprehension of the complex biochemical machinery that underlies numerous diseases such as cancer and AIDS, due primarily to advances in the areas of molecular and cellular biology. Although such knowledge has afforded numerous new biological targets for potential intervention by exogenous pharmaceutical agents, developing compounds to selectively perturb these biomolecules has proven challenging. This unfortunate state of affairs is certainly not due to a lack of effort or an inability on the part of chemists to construct diverse molecular structures. Rather, it reflects the inherent difficulty in anticipating the key molecular motifs that may prove critical for achieving a specific and/or unique interaction with a particular protein or gene sequence. Fortunately, Nature frequently provides essential clues to overcome such barriers in the form of novel molecular architectures in its natural products. Classic examples of such breakthroughs would, of course, include salicylic acid (later to become aspirin)[1] and penicillin,[2] molecules whose structures and mode of action have clearly shaped subsequent developments in the fields of pain relief and infectious disease.

In the domain of cancer therapy, Nature has been an especially generous benefactor. The first volume of *Classics in Total Synthesis* describes two signature events in this regard: first, the isolation of Taxol™, an agent which induces apoptosis in dividing cells by effecting tubulin polymerization, and second, the discovery of the enediyne class of antitumor antibiotics such as calicheamicin γ_1^I which effect lethal, double-strand cuts on DNA through the forma-

Key concepts:

- **Olefin metathesis: history and utility**
- **Alkyne metathesis**
- **Solid-phase organic synthesis (SPOS)**
- **Combinatorial chemistry**

tion of damaging benzenoid diradicals.[3] In both cases, the natural product stimulated a renaissance in cancer research and chemotherapy by providing strong support for certain mechanisms of intervention, leading not only to the development of clinically used anticancer agents, but also to a plethora of new knowledge in chemistry, biology, and medicine. Unfortunately, neither these natural products nor their analogues constitute the long sought and likely mythical "magic bullet" capable of treating all forms of cancer. As such, additional compound classes possessing unprecedented chemical biology profiles will be required for some time to come in order to continue to advance the state of cancer treatment.

Thankfully, in the early 1980s Nature once again gifted humanity with another lead in the form of epothilones A and B (**1** and **2**, respectively), polyketides first isolated from the culture extract of the cellulose-degraded myxobacterium *Sorangium cellulosum* strain So ce90 found in soil collected from the banks of the Zambesi River in South Africa.[4] The potential role of these agents in cancer chemotherapy, however, initially remained latent as early studies by Höfle, Reichenbach, and their collaborators at the Gesellchaft für Biotechnologische Forschung (GBF) in Germany sought primarily to explore their role as potential pesticides based on their unique and narrow spectrum of antifungal activity.[5] This situation changed when scientists from Merck in the United States independently isolated epothilones A and B and made the remarkable discovery that these compounds kill tumor cells with extremely high efficiency.[6] Indeed, these agents have IC_{50} values in the low nanomolar range that rival, and in some cases exceed, those of Taxol™ in the majority of the cell lines used in a typical first round anticancer activity screen. More significantly, ensuing studies revealed that these natural products kill cells by the same mode of action as Taxol™ and that they exhibit remarkable efficacy against several Taxol™-resistant tumor cell lines, thereby leading to the expectation that they could be developed into highly valuable new weapons for the treatment of cancer. Accordingly, the disclosure of this promising pharmacological behavior in the patent literature, coupled with the release of the complete structural assignment of the epothilones in the July issue of *Angewandte Chemie* in 1996,[5b] prompted numerous researchers from academic institutions and the pharmaceutical industry to embark immediately on research programs directed toward their total syntheses.

Testament to the urgency of finding new anticancer agents and the skill of synthetic chemists, in late 1996 the first total synthesis of epothilone A was accomplished by Danishefsky and co-workers at Columbia University,[7] followed shortly by independent total syntheses from the Nicolaou[8] and Schinzer[9] laboratories. Since then, over twenty additional and often highly insightful and creative routes to these agents have been reported by these and other synthetic chemists from around the world.[10] Taken collectively, these accomplishments embody a particularly accurate barometer of the current state of chemical synthesis, as the epothilones have served as a premier stage upon which to demonstrate the efficacy of new synthetic methodology in such diverse fields as asymmetric catalysis, metal-mediated couplings, macrocycle-forming reactions, enzymatic synthesis, and antibody catalysis. Beyond these fundamental and far-reaching applications, these syntheses have also provided diverse classes of structural analogues in quantities sufficient to enable more thorough investigations of the epothilones' chemical biology and pharmacology, including advanced clinical trials.

Rather than delve comprehensively into this voluminous body of work, a task which has already been accomplished successfully in several review articles,[10] we have opted in this chapter for a more selective and instructive focus based on analyzing only those strategies towards the epothilones that applied the olefin metathesis reaction as a key C–C bond-forming step. Indeed, this group of transformations perhaps constitutes the most important addition to the repertoire of synthetically useful reactions during the last decade, as it has substantially broadened and even altered the way in which we currently think about retrosynthetic analysis and synthetic strategy in general.[11] Because of the emerging importance of the olefin metathesis reaction in chemical synthesis, a topic which is unchartered territory in this series, we felt that a brief history of its evolution and a sampling of some of its applications in the synthesis of complex molecules would be appropriate and instructive.

7.1.1 Olefin Metathesis: A Brief History

Derived from the Greek words *meta* (change) and *thesis* (position), olefin metathesis, like ionic metathesis, is the exchange of the parts of two substances, AB + CD $\rightarrow$ AC + BD, where the reactants are olefins. Scheme 1a provides an illustration of such a reaction between two terminal alkenes and indicates that metal carbenes initiate such exchange reactions. Unlike other transformations which rearrange C–C double bonds, such as the Diels–Alder reaction or the Cope rearrangement, olefin metathesis is generally thought to involve a more complex series of steps than a single pericyclic process. As shown in Scheme 1b, in the opening step a metal-based carbene species (the catalyst) adds to an olefinic substrate through what constitutes a formal [2+2] process to afford a metallocyclobutane intermediate. A retro [2+2] cycloaddition in the opposite direction then effects the loss of an olefin, in this case ethylene, leading to the formation of a new metal carbene species which can then add to another olefin molecule to generate a new metallocycle. This intermediate then breaks down into a new olefinic product, concomitantly regenerating the starting metal carbene catalyst, which, in turn, re-enters the cycle to effect further conversion.

The wealth of synthetic transformations that can be accomplished when this reaction is applied to appropriate substrates is astonish-

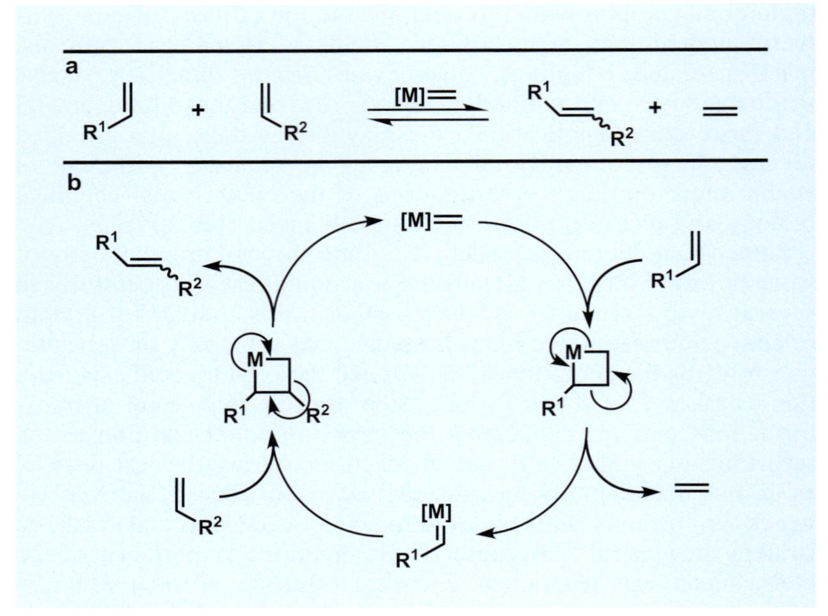

Scheme 1. The olefin metathesis reaction: generalized process (a) and mechanism (b). Note: although not specifically defined on this scheme, it is important to recognize that each of these steps is reversible.

ing. Consider, for example, the possible metathesis reactions available to a diolefinic substrate such as **3** (Scheme 2) in which the terminal double bonds are united by a chain, possibly including a heteroatom. Upon exposure of **3** to an appropriate metal carbene catalyst, one would expect initial insertion to lead to **4** (with regioselectivity governed primarily by the substitution pattern of the olefin, particularly at the terminal position, favoring less-substituted alkenes). Complex **4** is a reactive species that can then engage in several divergent metathesis reactions based on the specific experimental conditions employed and the properties of the original substrate (**3**). If the starting diolefin is present in sufficiently high concentrations, then a likely outcome would be for the metal carbene **4** derived from **3** to react with an additional molecule of **3** in an intermolecular reaction, affording a new product **6** ($n = 2$) and ethylene. Repetition of this sequence with the incorporation of additional monomeric units of **3** until catalysis stops would then be termed acyclic-diene metathesis polymerization (ADMET). Alternatively, with an appropriate chain between the double bonds, **4** can engage in an intramolecular olefin metathesis process to afford a ring-closed adduct **5**. The latter reaction, more commonly referred to as ring-closing metathesis (RCM), can effect the formation of both conventional rings (5-, 6-, and 7-membered) as well as macrocyclic systems, and is a favored pathway over ADMET when the reaction is performed under high dilution conditions. In general,

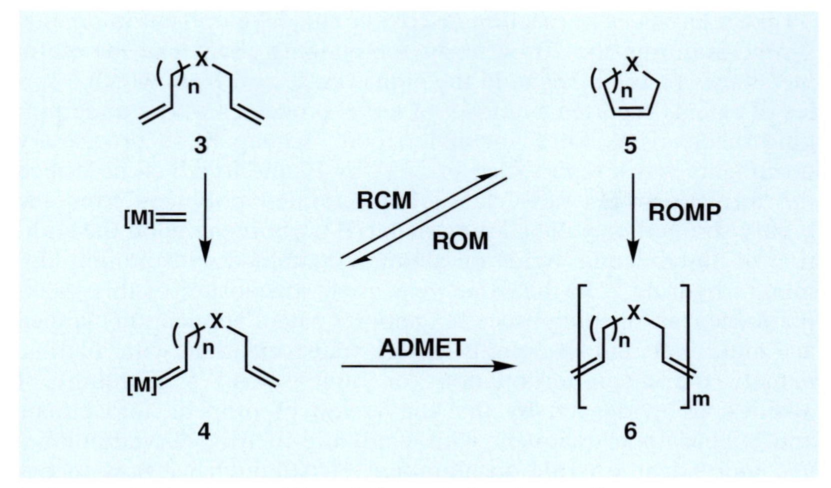

Scheme 2. Versions of the olefin metathesis reaction.
RCM = ring-closing metathesis, ROM = ring-opening metathesis,
ADMET = acyclic-diene metathesis polymerization,
ROMP = ring-opening metathesis polymerization.

however, whether or not a product such as **5** is isolated at the end of the reaction depends greatly on its thermodynamic stability. For instance, if the formation of an endocyclic olefin leads to a highly strained ring system (such as a 4- or 8-membered ring), then reversible ring-opening metathesis (ROM) would lead back to the starting carbene **4**, eventually funneling all of the material into the ADMET pathway to afford **6** rather than **5**. A corollary to this discussion would be to start a metathesis reaction with a strained cyclic ring system such as cyclobutene (**5**, $n = 0$), which could then undergo ring-opening metathesis followed by insertion of the new carbene into another molecule of **5** in a new ring-opening metathesis event, ultimately leading to polymer **6**. This route from cyclic olefins to polymers has been dubbed with the acronym ROMP (ring-opening metathesis polymerization).

Although our discussion thus far has proceeded only in general terms, soon to be clarified with actual substrates, we often become too focused on the practical applications of a reaction at the expense of an appreciation for the extensive effort and insightful moments that honed the transformation into the sharp synthetic tool we currently take for granted. Indeed, despite the fact that the overall picture of the olefin metathesis reaction presented in the preceding paragraphs suggests that the synthetic community has a lucid understanding of the transformation, it took decades of fundamental research before such clarity was attained. So, although we will engage in examples of metathesis shortly, we felt that it would be appropriate to highlight first some of the key events in the storied evolution of the olefin metathesis timeline that have led to our current picture of the reaction.

Like many tales of reaction discovery and development in organic synthesis during the 20th century, the opening chapter of the olefin metathesis reaction began in the industrial sector, from which a series of patents reported a number of novel processes whose underpinning mechanisms were not understood. Among these proprietary documents was a report filed in 1957 by Eleuterio which described the formation of a new class of unsaturated polymers from the highly strained bicyclic starting material norbornene upon the addition of molybdenum oxide on alumina combined with lithium aluminum hydride.[12] In the same year as the disclosure of this unexplainable ring-opening process,* another patent application claimed an additional and seemingly novel transformation with olefins, namely the disproportionation (or double-bond scrambling) of alkenes, as evidenced by the conversion of propene into ethene and butene upon treatment with a mixture of triisobutylaluminum and molybdenum oxide on alumina.[13,14] Although we now recognize that the polymerization of cyclic alkenes to polyalkenemers and the disproportionation of acyclic alkenes are the same reaction, it took nearly a decade before this crucial connection was first made by Calderon and co-workers at the Goodyear Tire and Rubber Company.[15] It was also these authors who christened olefin disproportionation and ring-opening polymerization as "olefin metathesis." Despite having a new name, however, the underlying mechanism of the process remained a mystery. Consequently, the search for new reagents to effect the reaction essentially constituted alchemy with one distinction: successfully identified systems did turn alkenes into gold in the form of commercially valuable polymers possessing unique physical and chemical properties.

In the confusion that characterizes poorly understood science, several "running" mechanistic hypotheses were in existence during this early period of olefin metathesis explorations. At first, these proposals questioned whether olefin metathesis exchanged alkyl or alkylidene (alkyl carbene) groups.[16] Eventually, Mol and Calderon initiated a series of experiments between 1968 and 1972 using isotopically labeled alkenes which demonstrated conclusively that the groups that were interchanged by olefin metathesis were alkylidenes.[15b,17] The mechanism by which this interchange occurred, particularly in terms of the role played by the metal species, remained conjecture. Some early thoughts, shown in Scheme 3, included pathways along which [2+2] cycloadditions gave cyclobutanes coordinated to metals[18] and pathways in which metals and olefins combined to give metallocyclopentane intermediates.[19] While

* Interestingly, this same norbornene polymer had been patented two years earlier by Anderson and Merckling at DuPont using a mixture of titanium compounds as the metathesis initiator (see: U. S. Patent 2,721,189, **1955**; [*Chem. Abst.* **1956**, *50*, 14596]). At the time, however, they did not realize that this polymer was unsaturated, a result that would not be disclosed until much later: W. L. Truett, D. R. Johnson, I. M. Robinson, B. A. Montague, *J. Am. Chem. Soc.* **1960**, *82*, 2337.

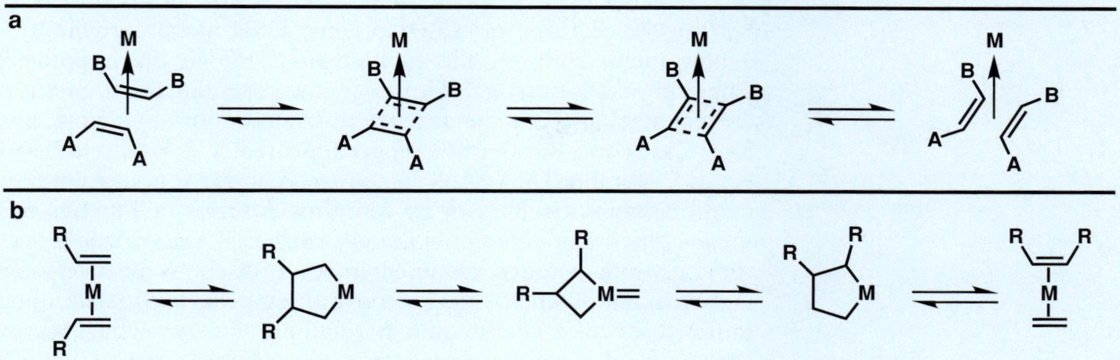

Scheme 3. Early mechanistic conjectures for the olefin metathesis reaction: original pairwise exchange proposal (a) and a metallocyclopentane alternative (b).

these hypotheses accounted for the transformation, they eluded experimental support.

Amazingly, the event most critical to elucidating the mechanism of the olefin metathesis reaction and engendering its development into a useful synthetic transformation had occurred long before olefin metathesis even had a name through the preparation of the first isolable metal carbenes by E. O. Fischer and A. Maasböl in 1964.[20] The connection between these species and metathesis, however, did not come until 1970 when two researchers in France, Y. Chauvin and J. Hérisson, suggested that metal carbenes were the metal-carrying participants in olefin metathesis.[21] The essence of their conception is the picture in Scheme 1b: metathesis is the result of metal-based carbenes adding first to alkenes to form metallocyclobutane-like adducts, followed by the subtraction of different metal carbenes and new olefin products. Although this postulate could explain the products formed by metathesis reactions, the weakness of this proposal was that it could not be reconciled with the product distributions obtained from certain reactions, especially metathesis between cyclic olefins and terminal alkenes, which resulted in non-statistical ratios of the three possible products favoring the cross-adduct (a result that could be explained by the alternatives in Scheme 3). Thus, this mechanistic scheme was treated with a certain (and understandable) level of skepticism by the chemical community. Indeed, during the next four years not even one chemist openly supported it in the literature. Some researchers concluded that this scheme must be incorrect due to the problem stated above,[22a] while others were willing to accept that it was no more likely than alternative alkylidene exchange pathways such as those in Scheme 3.[22b–e]

It would not be long before enlightenment was achieved as definitive evidence for the original Chauvin proposal came in late 1974 and early 1975. First, C. P. Casey and T. J. Burkhardt at the University of Wisconsin, Madison demonstrated that an alkylidene could be transferred from an olefin to an isolated metal carbene to form a new metal carbene, the fundamental step in the hypothesis of Chauvin and Hérisson.[23] However, the essential requirement for a correct mechanism is the absence of contradictory evidence. Shortly after Casey and Burkhardt's report appeared, T. J. Katz and co-workers at Columbia University began to clear the fog surrounding the metal carbene mechanism by reporting a series of kinetics experiments which explained how the postulate of Chauvin could account for the product ratios obtained in reactions between cycloalkenes and terminal alkenes, thus reconciling empirical evidence obtained in the absence of kinetic data that had previously seemed incongruent. Katz also demonstrated that the related reactions of cycloalkenes with internal alkenes (as shown in the experiment in Scheme 4a) initially give products other than those expected according to the proposals in Scheme 3, but in accordance with Chauvin's postulate.[24] Just a few weeks later, R. H. Grubbs (then at Michigan State University, now at the California Institute of Technology) reported similar experiments showing that such anomalous products were among the isotopically labeled ethenes resulting from metathesis reactions to form cyclohexene, as shown in Scheme 4b.[25]

With a singular mechanistic picture for olefin metathesis in hand, the transformation could be approached far more rationally, especially in terms of developing new reagents capable of effecting

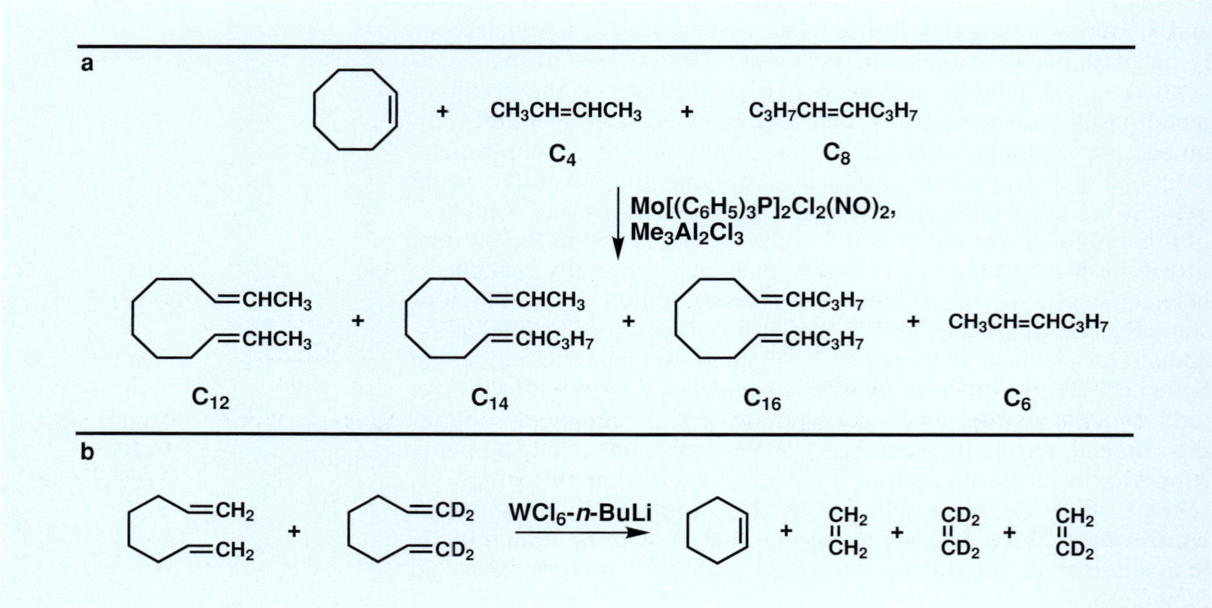

Scheme 4. Validation for the Chauvin mechanism: kinetics studies by Katz (a) and isotopic-labeling studies by Grubbs (b).

the reaction. In this regard, an important implication of the mechanism proposed by Chauvin is isolable metal carbenes alone, without the added Lewis acid co-catalysts as employed before, should induce olefins to undergo metathesis. While a seemingly obvious statement today, in 1975 (when Katz first made this suggestion) there was no proof that metal carbenes alone were capable of initiating metathesis. In fact, several elegant experiments that might have provided such a demonstration did not.[26] As such, the demonstration by Katz and co-workers in 1976 that two single-component, well-defined tungsten carbenes (**7** and **8**, Scheme 5) without added coactivators could initiate olefin metathesis was a pivotal event.[27] Indeed, these catalysts induced both cyclic and acyclic olefins to metathesize with high degrees of regio- and stereospecificity, and, most importantly, afforded far fewer side reactions and more reliable conversions than systems composed of combinations of transition metals and Lewis acids. The significance of this finding is perhaps best summarized in a 1995 review article by Grubbs et al. in which it was noted that, "It is fair to say that the search for and discovery of single-component catalysts has initiated a new period in the evolution of olefin metathesis catalysis."[11e]

Thus, from 1976 onwards, the use of multi-component catalyst systems gradually lost its prominence, and the field of olefin meta-

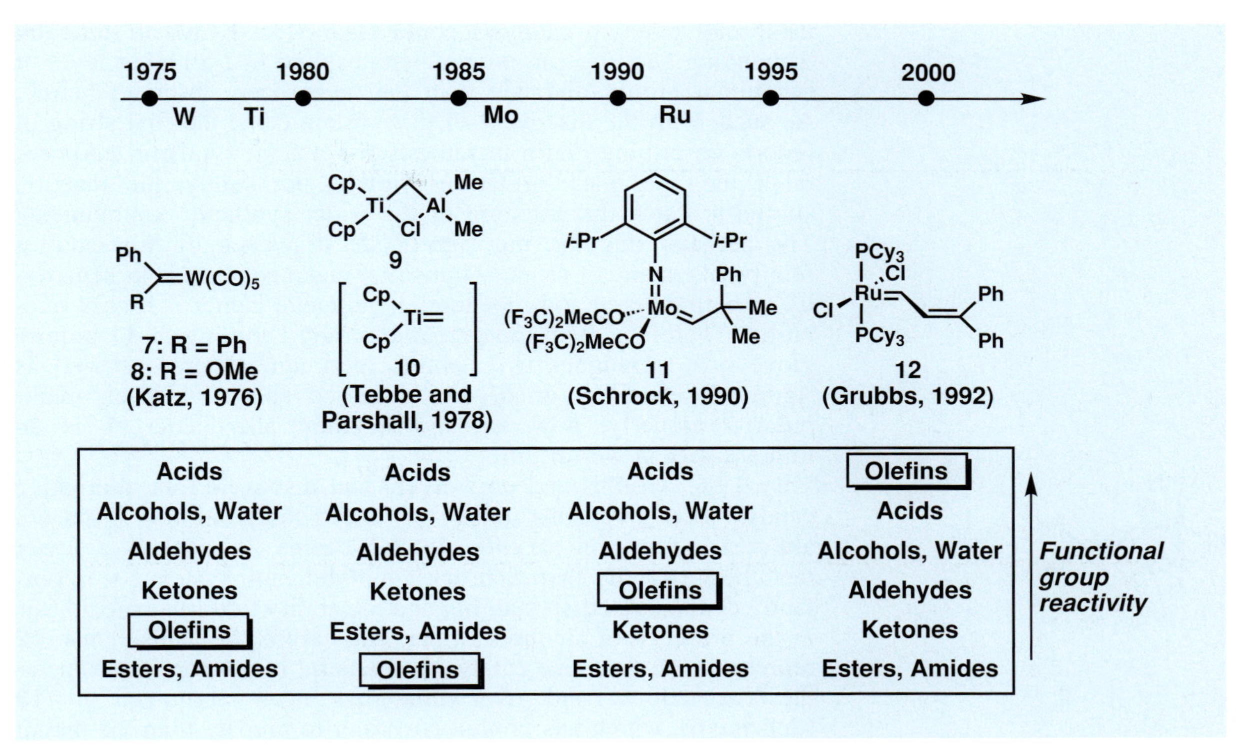

Scheme 5. The evolution of single-component olefin metathesis catalysts and their functional group tolerance.

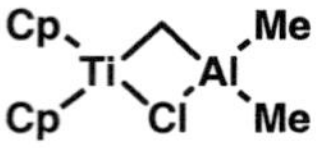

thesis entered into a rational phase of research that focused on elucidating the features necessary to design more active, single-component carbene initiators with greater functional group tolerance. While this objective is simply stated, it took years of research to open avenues useful for the synthesis of complex molecules. An early success, at least in terms of generating a carbene catalyst for metathesis from a different transition metal, came with the disclosure by Tebbe and Parshall of their titanium reagent **9** (Scheme 5), which forms methylidene **10** in solution.[28] Overall, however, this and related agents afforded no clear tactical advantage over the previously disclosed tungsten carbenes, as their functional group tolerance was, in fact, inferior. As indicated in the table in Scheme 5, the Tebbe reagent (**9**) reacts with almost all other functional groups in preference to olefins, accounting for why it is currently widely employed to methylenate carbonyl compounds, but is not a popular catalyst to initiate metathesis of complex molecules.

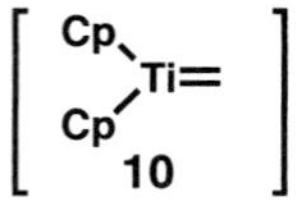

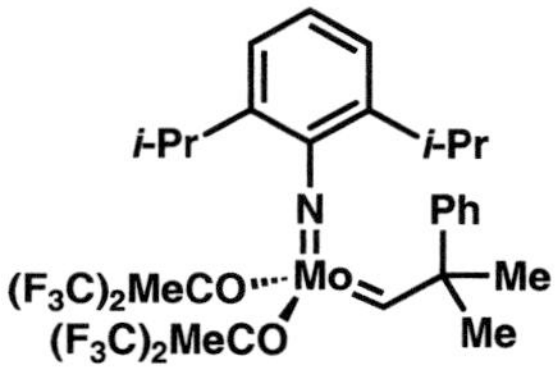

The first critical and groundbreaking advance in catalyst design came nearly a decade after Katz's initial tungsten carbenes, in the form of a family of molybdenum catalysts such as **11** (Scheme 5) which had far higher levels of metathesis activity than encountered previously.[29] First identified by R. R. Schrock at the Massachusetts Institute of Technology, initial experimentation with **11** established that it could induce metathesis of both terminal and internal olefins and effect ROMP of low-strained cyclic monomers. Furthermore, these molybdenum catalysts could also effect RCM of sterically demanding and electron poor substrates, and had a higher level of functional group tolerance than had ever been observed before. As such, with the discovery of this system came the first string of reports describing olefin metathesis for general synthetic purposes other than polymer formation. Slowly, but surely, the reaction started to catch the attention of the wider synthetic community.[30] The perhaps singular, but significant, drawback of this catalyst family, like many of its early transition metal cousins, is its sensitivity to both oxygen and moisture, as its metal center is highly oxophilic. Therefore, both the preparation and handling of **11** require glove box techniques to establish inert atmospheres as well as rigorously purified, dried, and degassed solvents. When manipulated properly, however, molybdenum alkylidene **11** is an impressively powerful tool.

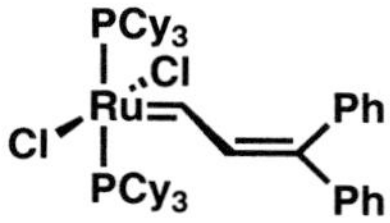

By 1992, Grubbs and co-workers had discovered an alternative catalyst that overcame many of these shortcomings.[31] Indeed, although ruthenium alkylidene **12** (Scheme 5) displays a lower metathesis activity than Schrock's molybdenum systems, it importantly demonstrated air stability and the ability to initiate metathesis in the presence of alcohols, water, and carboxylic acids. Thus, **12** represents the first true catalyst for general bench top olefin metathesis reactions, and over time has been optimized to **13** (Scheme 6), which has proven far easier to prepare than the parent structure **12** and constitutes the current "gold standard" with which all new catalyst systems are compared.[32] Without question, this

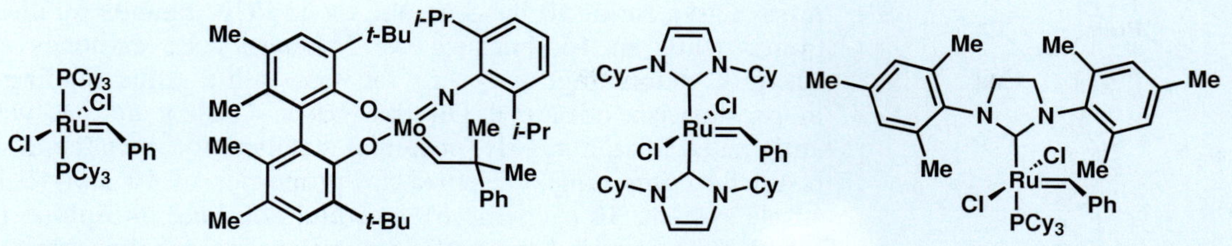

13 (Grubbs, 1995)
14 (Schrock and Hoveyda, 1998)
15 (Herrmann, 1998)
16 (Grubbs, 1999)

Scheme 6. Recent developments in olefin metathesis catalysts.

class of catalysts has opened up new vistas in synthetic applications, most notably in the total synthesis of complex natural products.[11] As an interesting aside, one should note that, in general, the trend for metathesis activity of late-transition-metal catalysts such as **11** can be correlated to the basicity of the attached ligands; in this regard, metathesis activity generally increases upon appending more electron-withdrawing ligands to the metal center. Application of this "rule" to ruthenium systems, however, proved inadequate as the most effective initiators, **12** and **13**, have bulky and strongly basic (σ-donating) phosphine ligands![11a]

The practice of catalyst design (or redesign in some cases) continues to be vibrant, and seldom does a month pass without the disclosure of a new catalyst for metathesis applications. Many of these novel systems are inspired by, or at least derive from, efforts to facilitate the construction of highly functionalized, complex molecules. Given the established ability of metathesis to effect numerous transformations difficult to achieve otherwise, this trend is only likely to continue. Among the alternative catalysts, several of the more important systems complete the lineup in Scheme 6. First in this regard is a variant of **11** developed by Schrock and Hoveyda in which the molybdenum center is coordinated to a chiral derivative of BINOL, creating a homochiral environment that can induce catalytic asymmetric olefin metathesis.[33] As an important proof-of-principle, this catalyst has inspired others to search for alternative systems capable of inducing asymmetry in substrate classes that have proven recalcitrant with **14**.

Equally significant are the last two catalyst systems shown in Scheme 6. As mentioned above, strongly basic ligands promote the activity of ruthenium-based metathesis systems of the type defined by **13**; as such, the discovery by Herrmann and co-workers in 1998 that a modified variant of **13** bearing more strongly σ-donating *N*-heterocyclic carbene ligands (**15**, Scheme 6) could initiate metathesis is quite important.[34] Despite this potentially beneficial feature, however, the overall metathesis activity of **15** is only a slight improvement over that of **13**. This result reflects the fact that, in its active form, **13** has lost one of its PCy_3 ligands by dissociation,[35] while in **15**, because the *N*-heterocyclic carbenes are so strongly σ-donating, they are far less labile (thus leading to a more sluggish initiator). On this basis, Grubbs and co-workers anticipated that a highly powerful catalyst for metathesis could result by combining the beneficial properties of **13** and **15** into a single system, **16** (Scheme 6), in which the lone phosphine ligand could dissociate, but the better σ-donating imidazoline-type ligand would remain attached to the ruthenium core.[36] Indeed, catalyst **16** engenders metathesis with particularly high levels of activity, and even enables trisubstituted olefins to form by RCM in situations that would be impossible to effect with any of the earlier ruthenium-based catalysts.[37] In addition, unlike catalysts **12** and **13** which are somewhat thermally unstable, carbene **16** is much longer-lived at elevated temperatures. In this regard, the bulky mesityl ligands likely shield the metal center from reaction with air, thus decreasing the rate of catalyst decomposition. The true power of catalyst **16** is only now being fully explored, but early evidence suggests that its unique reactivity profile nicely complements that of **11** and **13** and promises new applications in chemical synthesis (vide infra).

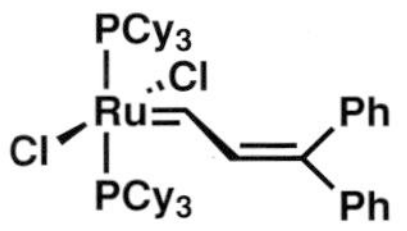

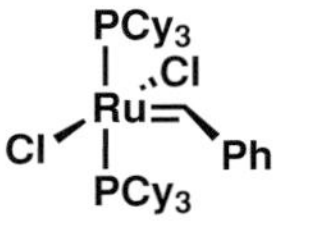

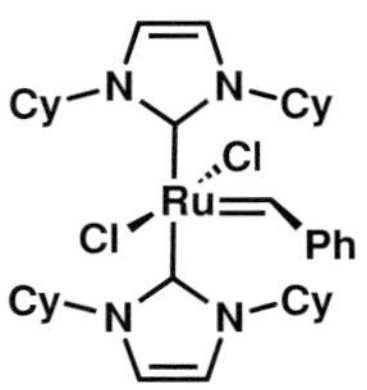

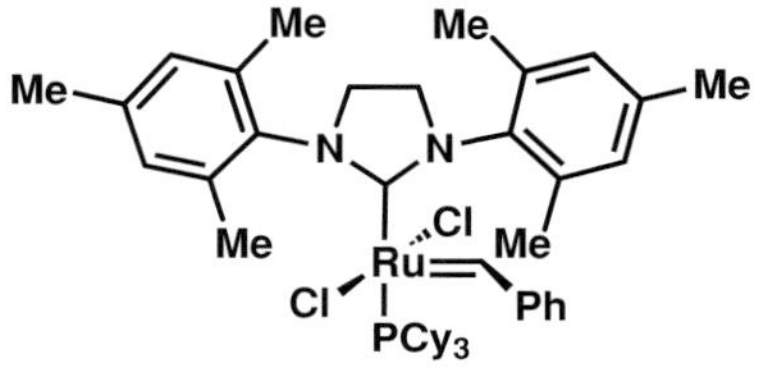

7.1.2 *Olefin Metathesis: Case Studies*

Having traversed some of the key events in the history of olefin metathesis, it is now appropriate to discuss some of the resultant fruits of that early labor in the form of practical applications in organic synthesis. Since the general reaction was born in the industrial sector, we felt it appropriate to commence with some examples of commercial processes. Among several of the profitable industrial procedures that benefit from olefin metathesis, one of the oldest is the Phillips triolefin process (Scheme 7a) which utilizes a molybdenum-based catalyst system to convert propene (**17**) into a mixture of 2-butene (**18**) and ethene (**19**).[38] These products are then used as monomers for polymer synthesis as well as for general use in petroleum-related applications. The reverse reaction can also be employed to prepare propene for alternative uses.

Apart from the synthesis of such simple compounds, far more complex entities in the form of high molecular weight polymers reflect the bulk of current metathesis applications. A representative example along these lines is the synthesis of the ring-opened metathesis polymer **22** of norbornene (**20**) which proceeds as shown in Scheme 7b.[39] Over 45,000 tons of this polymer are produced annually by this, so called, Norsorex process.[11d] Among different potential precursors for ROMP, norbornene is particularly easy to convert into polymers because the metathesis relieves the strain of the ring system. Moreover, "living" norbornene polymer (with the carbene still in place as in **21**) can become a scaffold

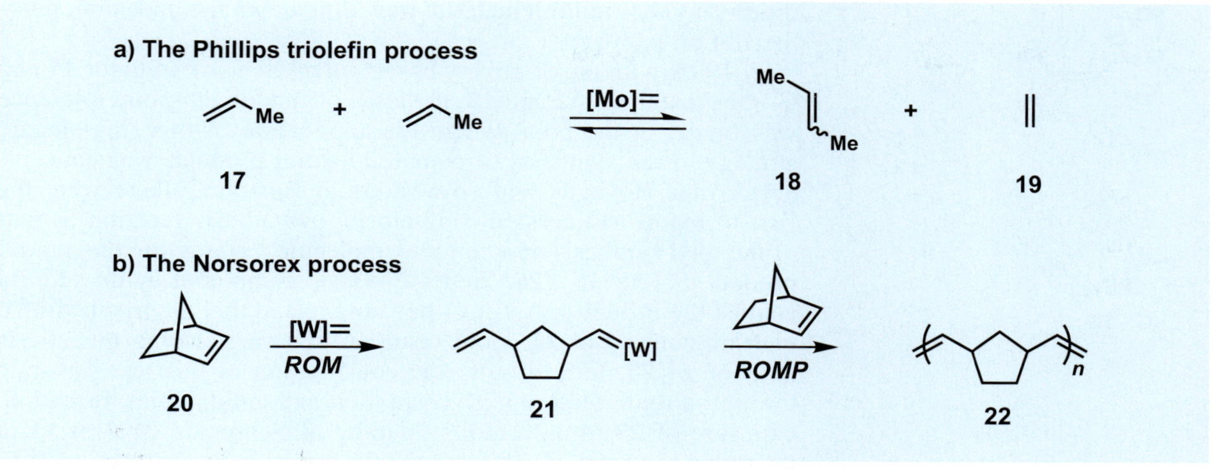

Scheme 7. Industrial olefin metathesis applications: the Phillips triolefin process for the production of butene and ethene (a) and the Norsorex process for the ring-opening metathesis polymerization (ROMP) of norbornene (b).

for block co-polymers if a second cyclic olefin is added after all the initial norbornene has reacted. Besides commercial utility, the ring-opening metathesis polymerization of norbornene is also important as it conventionally serves as the standard reaction for comparing the activities of new catalyst systems. One should also note that these industrial applications are the only arena in which mixed or heterogeneous metathesis catalyst systems are still employed, primarily because they cost little and the substrates to which they are applied lack functionality.

Because our main focus lies in the realm of complex molecules, particularly natural products, for the remainder of this section we will limit our discussion to examples in this field. Based on the demonstrated power of olefin metathesis to effect ring closures, one of the earliest applications sought by synthetic chemists was the construction of macrocyclic systems. In principle, although olefin metathesis is reversible, and, therefore, under thermodynamic control, such ring closures should be possible because if RCM generates two molecules from one (based on the evaporative loss of a molecule of ethene), then a large enough gain in entropy should result (despite the decrease in chain flexibility) to sufficiently drive the reaction forward. The initial proof-of-principle experiments were reported in 1980 in two separate publications that appeared almost simultaneously, first by D. Villemin and then by J. Tsuji and S. Hashiguchi, using simple terminal olefins that had their alkyl chains linked by an ester.[40] Even though the then-available catalyst systems prevented exploration of more functionalized systems, these authors reported the formation of 15-, 16-, 19-, and 21-membered rings (albeit in modest yields). These experiments also established the need to perform RCM under high-dilution conditions in order to limit material travelling down the undesired polymerization pathway.

By 1995, with the discovery of the potent Schrock initiator **11** and Grubbs' catalysts **12** and **13**, the level of functional group tolerance was finally at the point where the application of this ring-closure strategy to the synthesis of complex natural products was conceivable. Amir Hoveyda and co-workers at Boston College were the first to report a successful ring-closing metathesis reaction as part of the total synthesis of a complex molecule.[41] Targeting the natural product Sch38516 (**25**, Scheme 8), a compound with activity against the influenza A virus, they anticipated that the trisubstituted olefinic compound **24** could result from a ring-closing metathesis reaction of **23**. Indeed, after the construction of this test substrate through a route featuring glycosidation and amide bond formation, exposure of **23** to a 20 mol % loading of Schrock's catalyst **11** in warm benzene (60 °C) effected the desired metathesis reaction, giving **24** in 90 % yield.

Several features of this reaction are worth further consideration. First, one should recognize that although the then-available ruthenium alkylidenes such as **12** and **13** have a level of functional group tolerance compatible with **23**, their reactivity as metathesis catalysts has, in general, proven too low for them to effect the formation of macrocyclic trisubstituted olefins. Accordingly, at the

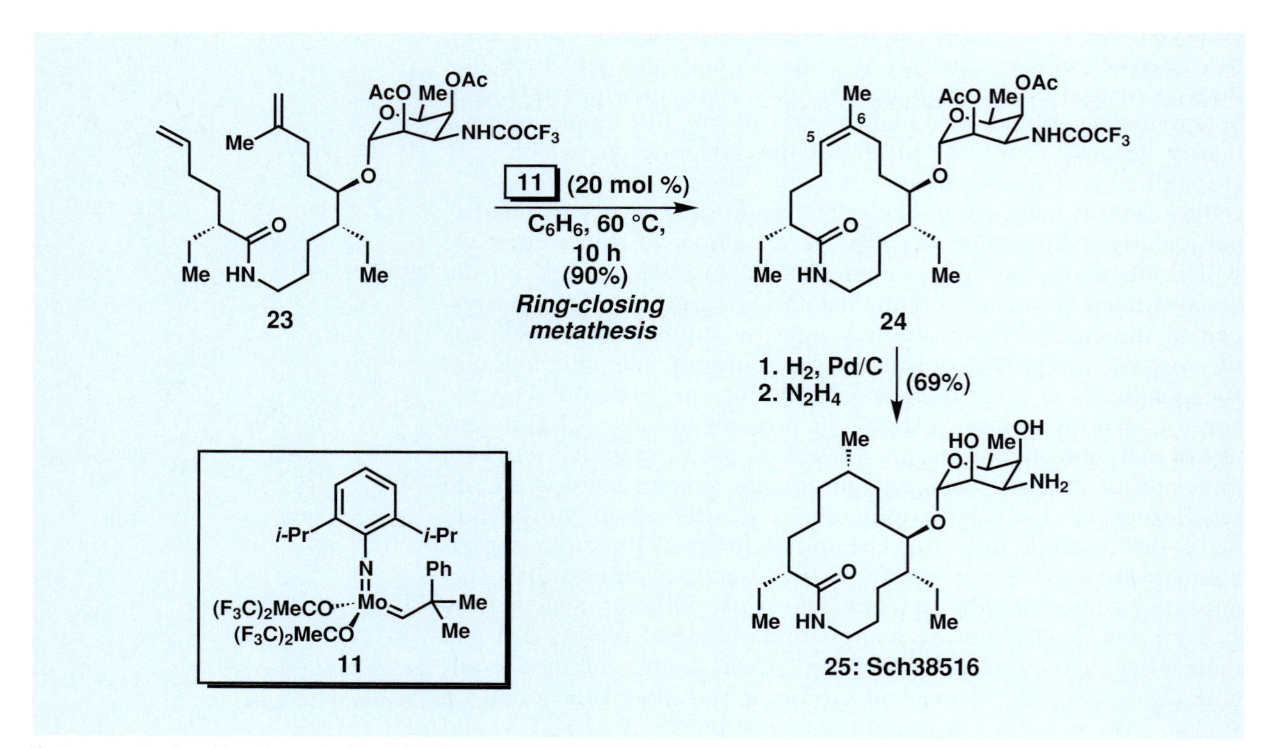

Scheme 8. Application of ring-closing metathesis in the total synthesis of Sch38516 (**25**). (Hoveyda et al., 1996)[41]

time this work was performed, the Schrock catalyst **11** was uniquely suited for the transformation, although more recent additions to the ruthenium family (e. g. catalyst **16**) have since proven competent in this regard. Moreover, attachment of the sugar portion onto **23** prior to effecting RCM was a critical element of the successful strategy, since, even though the same metathesis reaction worked with equal facility in the absence of the carbohydrate moiety, the resultant macrocyclic aglycon had such a low solubility in most organic media that subsequent glycosidation proved impossible to effect.[41a] Finally, the RCM approach was uniquely suited to deliver the trisubstituted olefin of **24** stereoselectively, as more classical synthetic approaches based upon macrolactamization suffered from initial ineffective formation of the C5–C6 bond. Indeed, with this region of the natural product devoid of additional stereocenters, all efforts to form this system through Wittig olefination chemistry proved to be both low-yielding and non-stereoselective. It is, however, equally instructive to point out here that although the metathesis reaction gave the proper geometry of the final olefin in this instance, not all reactions afford equal levels of stereocontrol (vide infra).

An equally didactic example of ring-closing metathesis as a critical design element resides in the approach of T. Müller and A. Fürstner to tricolorin A (**29**) which targeted **28** for practical purposes since this compound had already been advanced to the natural product (Scheme 9).[42] Although one could certainly construct the 19-membered ring in **28** through macrolactonization, a reaction that is successful,[43] Fürstner's strategy sought to employ RCM as a means to enable the formation of larger, more lipophilic backbones to explore the demonstrated antitumor activity of tricolorin A. As anticipated, once **26** was in hand, RCM proceeded smoothly upon application of only 5 mol% of initiator **13** in refluxing CH_2Cl_2, affording the desired system **27** as a mixture of *E* and *Z* isomers. As a further illustration of the importance of catalyst development to applications in synthesis, neither the free OH nor any other member of the functional group array on the disaccharide core interfered with the metathesis reaction; these groups would have hindered the reaction had either the tungsten or molybdenum systems discussed earlier been employed instead.

More globally, when this work is considered in combination with additional RCM studies (many executed by Fürstner and co-workers), a general trend is established which reveals that it is not necessarily the conformational predisposition of the original substrate or the ring size of the product that defines whether or not a given RCM reaction will be successful.[44] Rather, the site of ring closure and its surroundings are pivotal. In this regard, the steric effects of substituents neighboring the participating olefins and the presence of nearby polar functionalities provide the greatest degree of RCM predictability. In terms of this latter feature, it should be considered that once the carbene catalyst has added to one of the reacting olefins, a Lewis basic functional group within the substrate, such as a carbonyl, can form a chelate complex with the metal. If this moiety

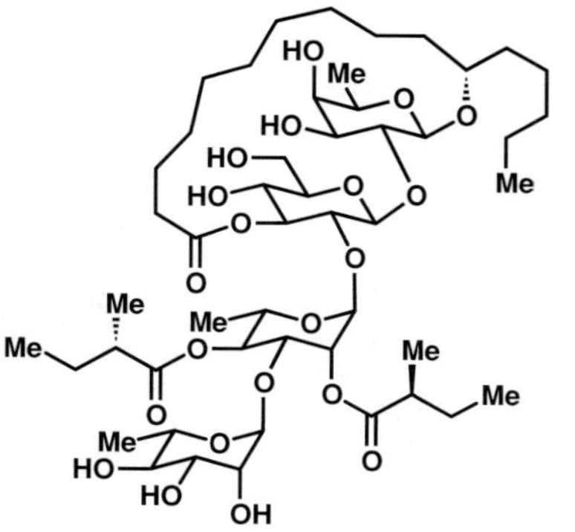

29: tricolorin A

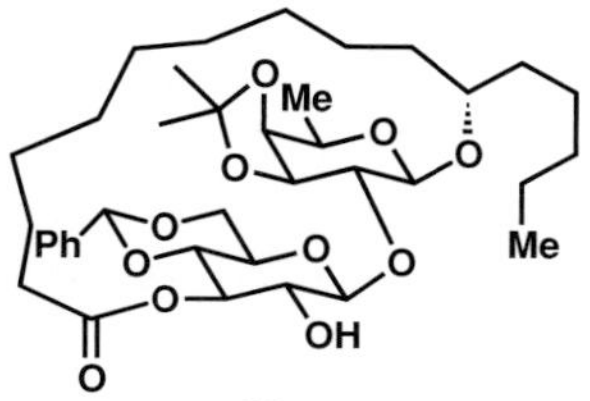

28

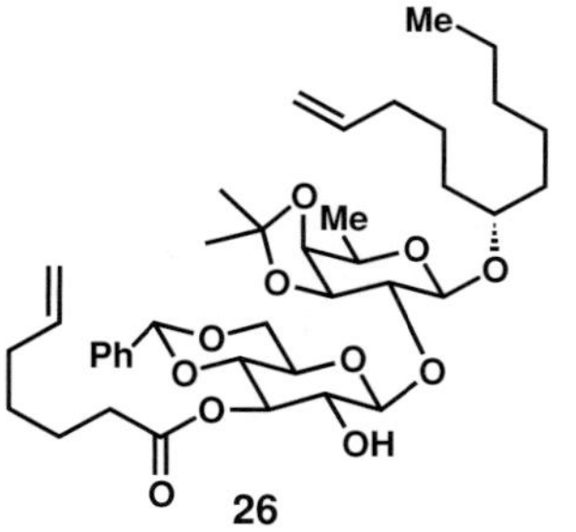

26

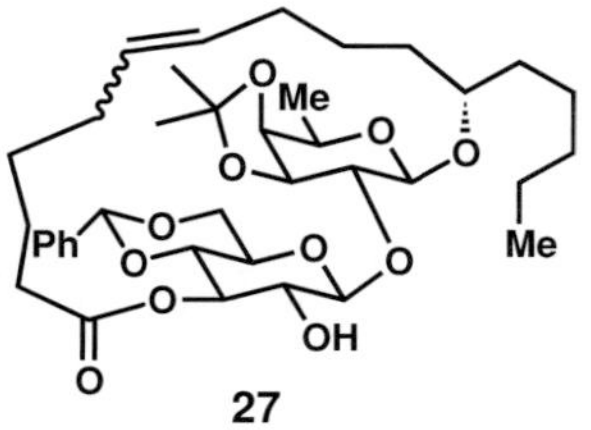

27

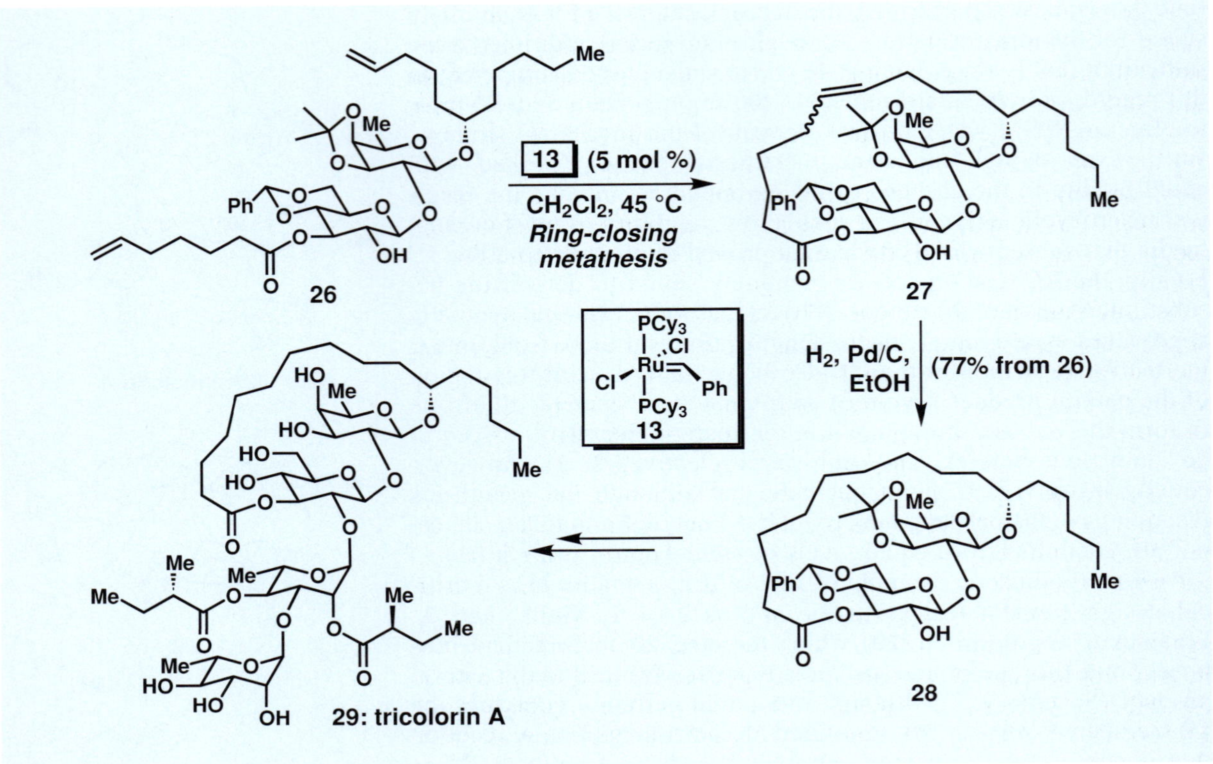

Scheme 9. Ring-closing olefin metathesis in a formal total synthesis of tricolorin A (**29**). (Fürstner et al., 1998)[42]

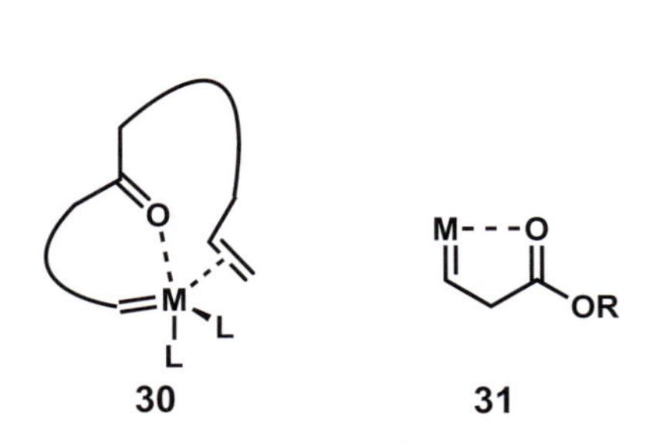

is sufficiently far from the reacting olefin, then this occurrence would likely facilitate RCM by reducing the chain mobility and, therefore, the entropy of the metathesis transition state **30** (column figure). Accordingly, this line of thought offers a reasonable hypothesis for the formation of diverse macrocyclic rings of various sizes with essentially equal facility. Conversely, if the same carbonyl is close to the reacting olefin, as in **31**, then a highly stable complex due to coordination of the ester group to the proximal carbene moiety could result, shutting down metathesis completely. In this sense, the particular positioning of the reacting olefins for RCM in the tricolorin A context was certainly choreographed to take these considerations into account.

As mentioned earlier, when metathesis reactions are performed on diolefinic substrates under more concentrated conditions, the predominant product is not a cyclic species, but instead a dimerized or, as more appropriately defined, a cross-metathesis adduct. For several years now, numerous researchers have been attempting to employ this variant of olefin metathesis to generate acyclic olefins by reacting two different olefinic monomers, providing a transformation that might compete with more conventional approaches to these compounds such as Wittig olefination. Invariably, however, mixtures of products result, primarily due to competitive homodimerization (i. e., self-metathesis) since one of the reacting olefins is likely to have greater metathesis activity than its partner. As an additional concern, the stereochemistry of the resultant heterodimer is often uncontrolled, and usually unpredictable. Without question, this area of research is one of the next frontiers of metathesis chemistry, and some useful strides are being made through the design of new catalysts and a more thorough comprehension of substrate scope. In general, though, the process has yet to be applied in complex situations.[45]

In stark contrast to this scenario, because the dimerization of a single monomer is quite facile, this process has been applied on several occasions for the synthesis of medicinal agents related to natural products. The dimerization of drugs constitutes a relatively new design strategy for enhancing biological activity and is based on the idea that once a dimer is delivered to the cellular target, the concentration of the bioactive molecule will be higher since effectively two molecules are present instead of one.[46] Moreover, this dimerization approach can, in principle, afford a means to combat resistance developed to a particular drug by providing a new ligand predicated on a similar mode of action as the original agent, but which the bacterial machinery has not yet figured out how to circumvent. A signature example based on this concept is Schreiber's synthesis of "FK1012," the dimer of the natural product FK506 (**32**, Scheme 10), an established immunosuppressant.[47]

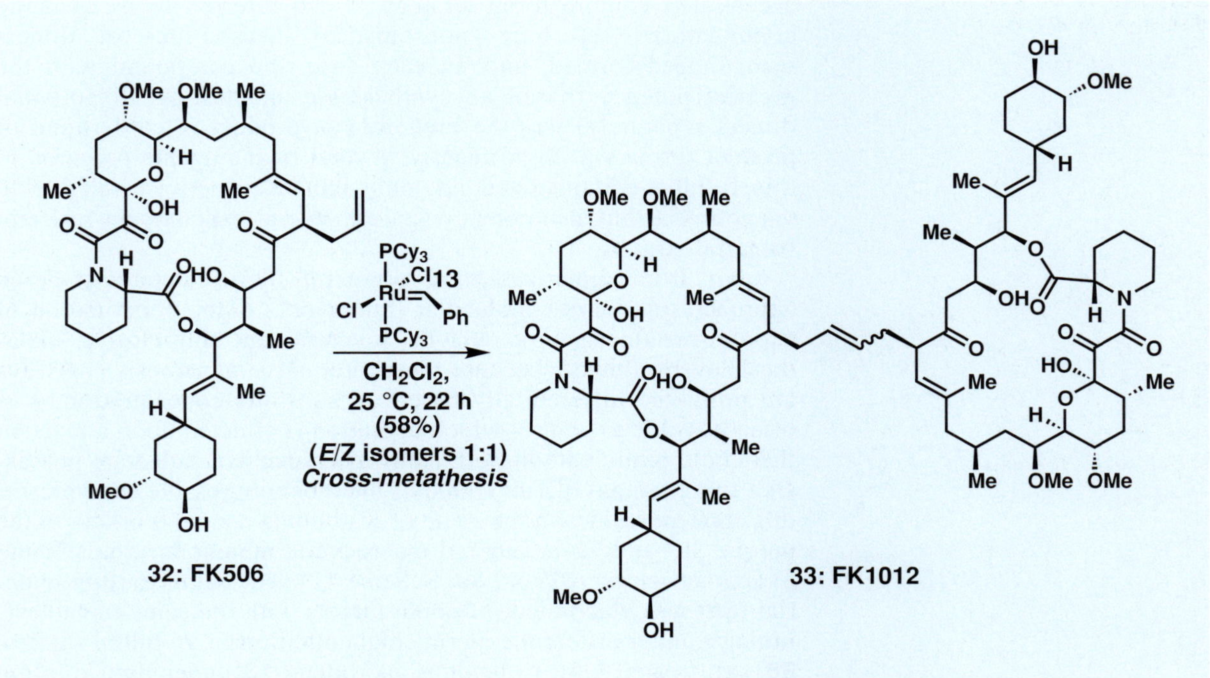

Scheme 10. Dimerization of FK506 through olefin cross-metathesis. (Schreiber and Diver, 1997)[47]

Indeed, upon reaction of **32** with catalyst **13** at ambient temperature, the desired dimer **33** was formed by cross-metathesis in 58 % yield. Although the material return is modest, this result is still remarkable because there are so many seemingly deleterious coordinating functional groups near the terminal olefin, including two free hydroxy groups and a ketone.

A more recent application of a similar dimerization-type approach to combat drug resistance was performed by Nicolaou and co-workers on vancomycin-like monomers **34** (Scheme 11, see Chapter 9 for a more comprehensive discussion of this significant natural product).[48] Among several of the noteworthy features of the developed chemistry to reach **35** was the employment of a phase-transfer reagent ($C_{15}H_{25}NMe_3Br$) to encapsulate the ruthenium catalyst **13** and enable it to carry out its function in aqueous media at 23 °C. Because these reaction parameters essentially constitute ambient conditions, it was decided to extend the initial homodimerization approach to include the selective formation of heterodimers by adding combinations of different substrates of type **34** in the presence of the biological target of vancomycin, a terminal D-Ala-D-Ala peptide subunit. Since it had already been established that two monomers of vancomycin could bind simultaneously (and reversibly) through hydrogen bonds with this target,[49] this design assumed that those monomers within the collection of examined substrates which bound tightest to this peptide chain would be captured by cross-metathesis as a new dimer, thus leading to the formation of highly active antibacterial agents. Upon execution of this target-accelerated combinatorial strategy,[50] also referred to as dynamic combinatorial screening, nonstatistical distributions of dimers were indeed formed, and, in each case, the compound with the greatest potency (based on synthesizing and testing all potential dimers separately) was the predominant product in each round of product formation. Significantly, several of the agents prepared in this fashion demonstrated not only enhanced activity relative to vancomycin, but also potency against several vancomycin-resistant bacterial strains.

Apart from dimerization, a powerful olefin metathesis-based approach to complex molecular frameworks is the combination of a cross-metathesis process with a subsequent ring-closing metathesis event. Indeed, because these alternative metathesis pathways are promoted differentially based on substrate concentration, it is often possible to control which operation is effected upon a reactant that could participate in both pathways (such as a substrate possessing two terminal olefins). One of the most impressive examples of this concept resides in the route of Kakinuma and co-workers to the unique 36- and 72-membered macrocyclic membrane lipids found in archaebacteria (**37** and **39**, Scheme 12). The same starting material (**36**) was elaborated into both targets with the same metathesis initiator under different experimental conditions.[51] In initial studies, **36** was treated with ruthenium alkylidene **13** under high dilution conditions suitable for ring-closing metathesis in refluxing

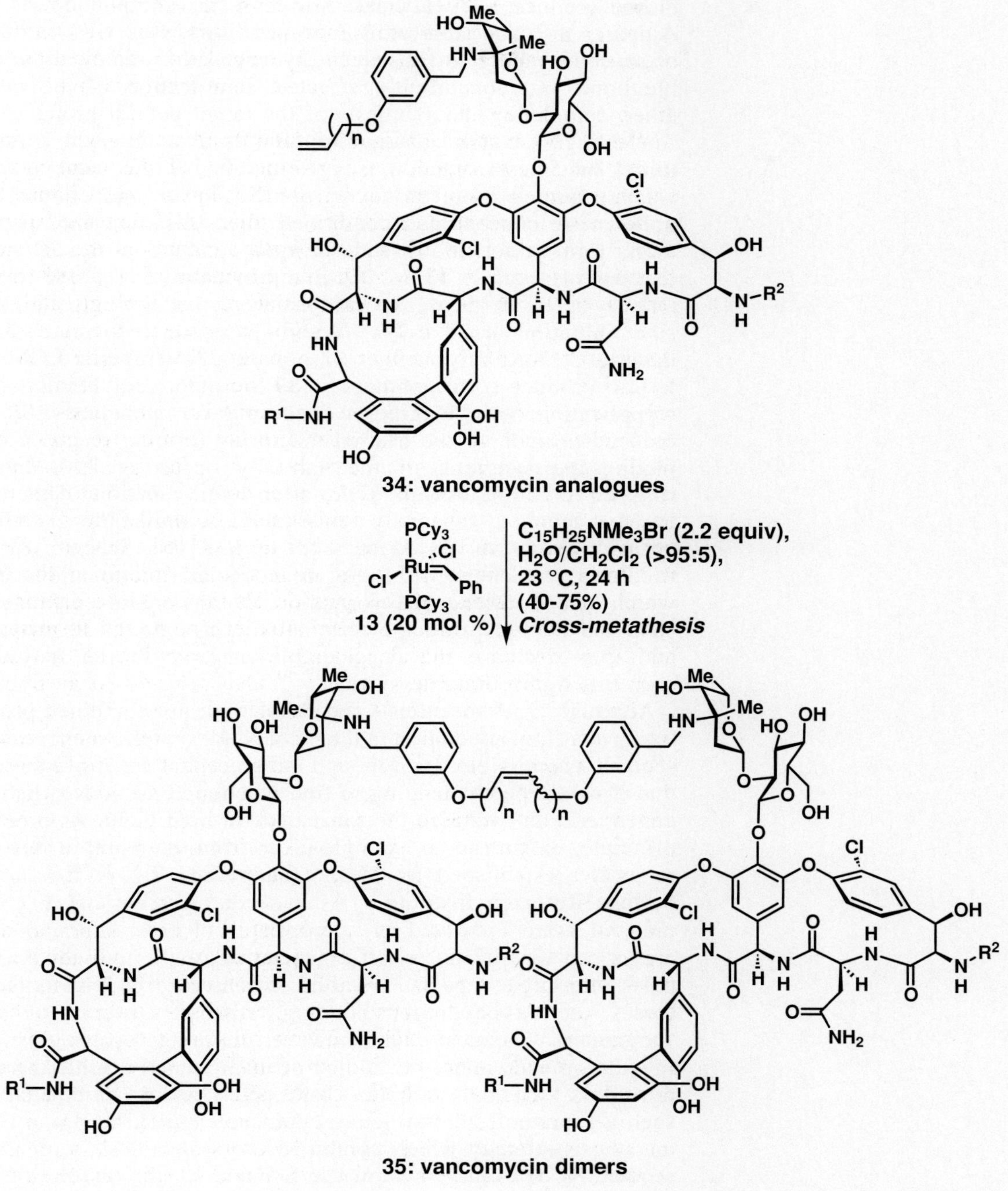

34: vancomycin analogues

13 (20 mol %)

$C_{15}H_{25}NMe_3Br$ (2.2 equiv), H_2O/CH_2Cl_2 (>95:5), 23 °C, 24 h (40-75%) ***Cross-metathesis***

35: vancomycin dimers

Scheme 11. Dimerization of vancomycin derivatives through cross-metathesis. (Nicolaou et al., 2000)[48]

CH_2Cl_2, leading to the smooth formation of the 36-membered ring-closed product in 79% yield. Although this compound was produced as a 7:1 mixture of *E*:*Z* isomers, this issue was ultimately of no consequence as subsequent hydrogenation removed the double bond and concurrently effected deprotection of the benzyl ether, completing the synthesis of the target natural product (**37**). Interestingly, a small amount of **38** was also obtained from the initial metathesis reaction, suggesting that if the same reaction was performed more slowly (i.e. at a lower temperature) and under more concentrated conditions, then this dimerized product might form preferentially. Indeed, upon application of a 20 mol % loading of catalyst **13** to **36** at a concentration of 108 mM in CH_2Cl_2 at 25 °C, these expectations were met through the exclusive formation of **38** in 81% yield (based on recovered starting material). Amazingly, neither **37** nor the 72-membered RCM diolefinic product corresponding to **39** were formed. Finally, upon reapplication of high dilution conditions for metathesis, **38** was induced to undergo the desired macroring-forming reaction, completing the synthesis of the naturally occurring 72-membered ring **39** in 34% overall yield after hydrogenation. This latter RCM reaction is especially remarkable, as unlike the paradigms defined above in the context of Sch38516 (Scheme 8) and tricolorin A (Scheme 9), there are no polar functionalities in **38** which can facilitate the formation of an organized transition state that would bring the terminal olefins into close proximity and thus decrease the considerable entropic barrier associated with this highly flexible system.

Although cross-metathesis and RCM were not combined productively in a single operation in the preceding example, such a cascade sequence is possible, even though the precedent set above suggests that it would be challenging to find reaction conditions suitable to engender both events in the same flask in high yield. As proof-of-principle, we turn to an example taken from the realm of designed molecules established by M.R. Ghadiri and T.D. Clark at The Scripps Research Institute.[52] As a general precept, when certain molecules are dissolved in appropriate solvents, supramolecular arrays can form through self-organization of the individual molecules into larger regular assemblies as enforced by intermolecular forces such as hydrogen bonding.[53] In the absence of some "permanent" chemical event, however, these aggregates exist only in solution and cannot be studied or manipulated outside the reaction flask. In this example, Clark and Ghadiri anticipated that such a supramolecular structure could be captured using an olefin metathesis strategy which combined cross-metathesis with a subsequent RCM event. As shown in Scheme 13, the dissolution of a flat macrocyclic peptide subunit containing two L-homoallylglycine residues in $CDCl_3$ was believed to lead to two interconverting hydrogen-bonded diastereomeric cylindrical ensembles, one of which is **40**, reminiscent of the tertiary peptide structure of antiparallel β-sheets. Proof of this conjecture was obtained upon addition

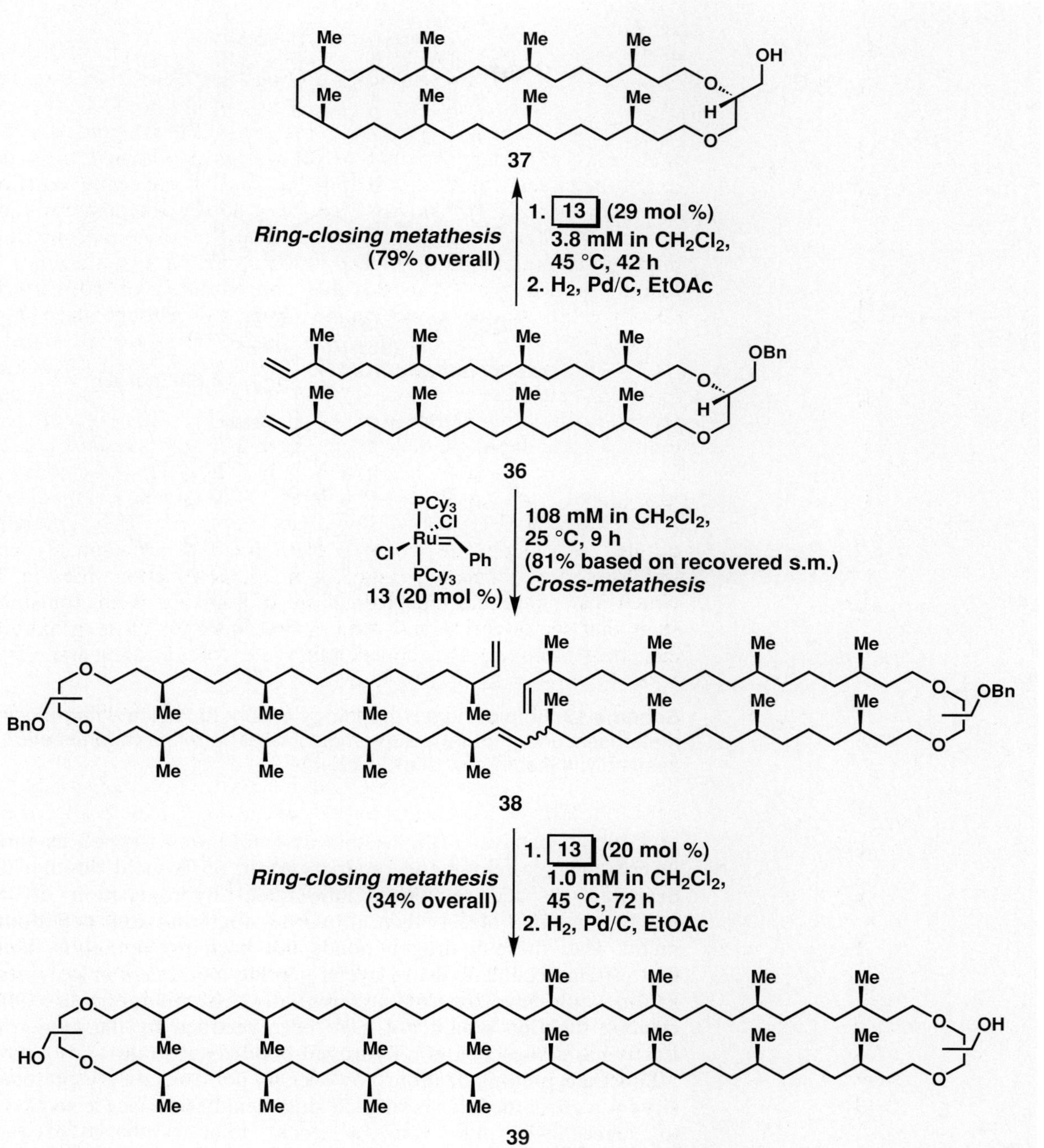

Scheme 12. The olefin metathesis-based synthesis of the 36- and 72-membered macrocyclic membrane lipids found in archaebacteria **37** and **39**. (Kakinuma et al., 1998)[51]

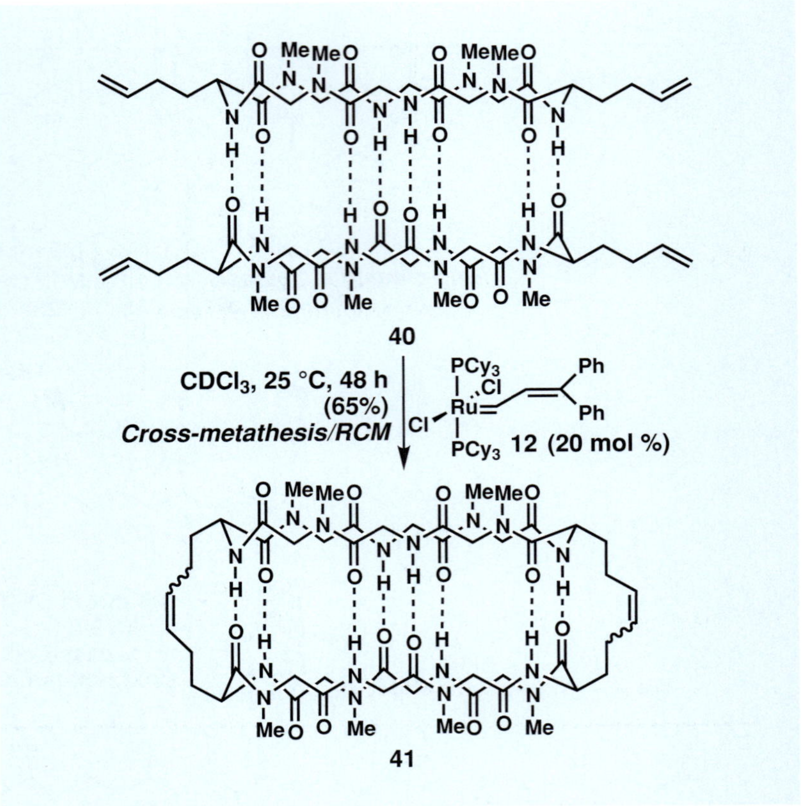

Scheme 13. Employing a H-bonding network to facilitate ring-closing metathesis and capture a supramolecular β-barrel peptidomimetic assembly. (Ghadiri and Clark, 1995)[52]

of Grubbs' catalyst (**12**), as this assembly was trapped as three inseparable diastereomeric products **41** in 65 % yield through the designed cascade sequence. Subsequent hydrogenation of the newly generated olefins then afforded a single β-barrel peptidomimetic. Had these hydrogen bonds not been present, it is likely that intramolecular RCM of cyclic peptide monomers or polymerization would have occurred preferentially (assuming, of course, that catalyst function would not have been retarded by the numerous Lewis basic sites in a non-hydrogen-bonded substrate).

Substrate preorganization, however, is not a necessary prerequisite for a cross-metathesis/ring-closing metathesis cascade sequence to succeed. In this vein, a recent total synthesis of (−)-cylindrocyclophane A (**45**, Scheme 14a) by A. B. Smith and his group at the University of Pennsylvania is particularly instructive.[54] Having already synthesized this target through conventional, stepwise approaches, Smith and co-workers turned to the far more challenging, but inherently more tantalizing, prospect of effecting the synthesis of the [7,7]-paracyclophane structure comprising the

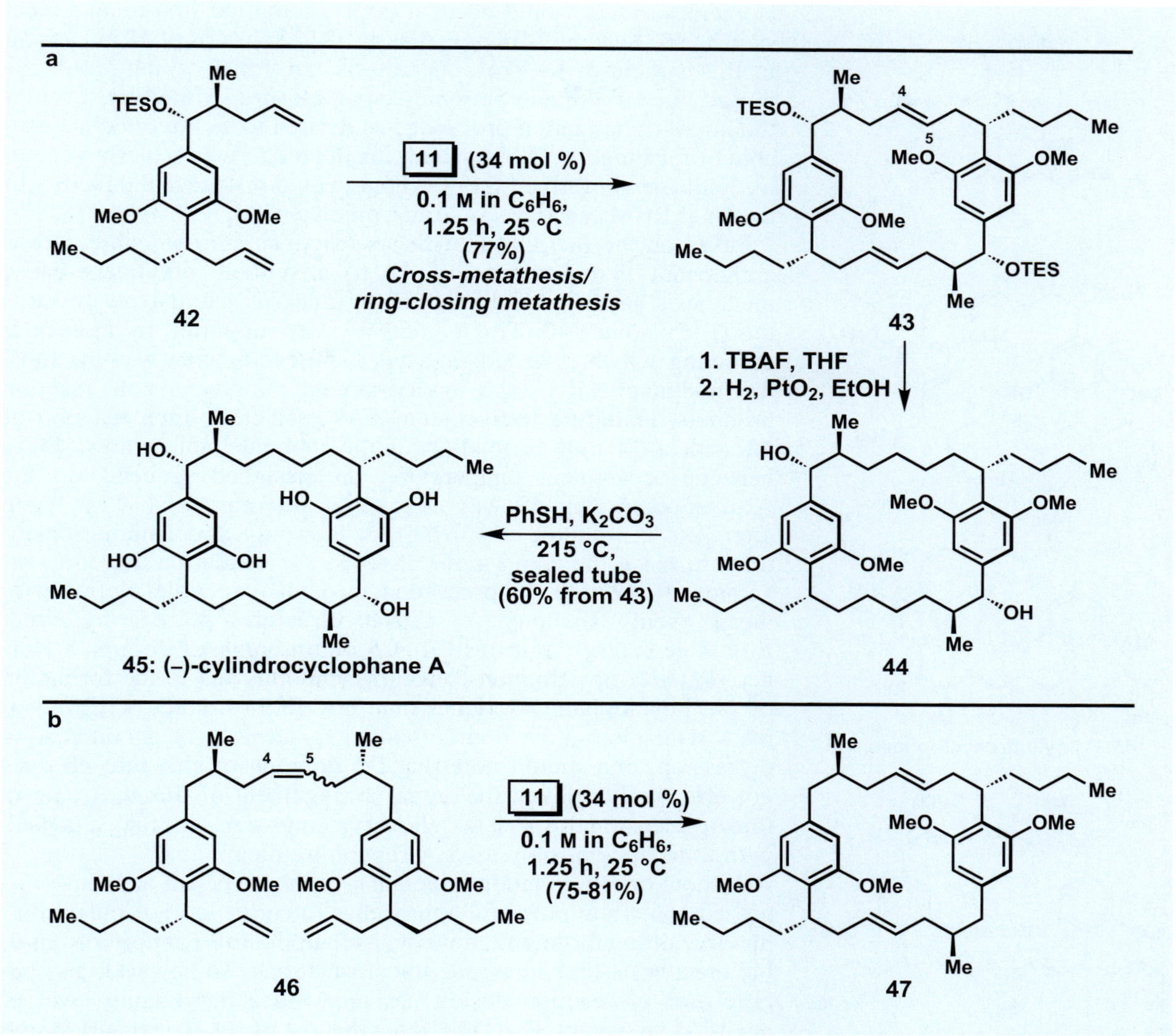

Scheme 14. Exploitation of the reversible nature of olefin metathesis to synthesize (–)-cylindrocyclophane A (**45**) based on thermodynamic product stability (a) and experimental validation of the reversible nature of the employed olefin metathesis reaction (b). (Smith et al., 2001)[54]

core of the natural product through a one-pot dimerization of two appropriately functionalized resorcinol monomers. More specifically, it was anticipated that an initial olefin cross-metathesis reaction between two molecules of **42**, followed by a productive ring-closing metathesis event, could afford the complete [7,7]-cyclophane skeleton **43** in a single operation. While this strategy is certainly appealing on paper, several features of the design appear potentially troublesome. Specifically, instead of a [7,7]-cyclophane, one might ask whether the alternate possibility of forming the [8,6]-

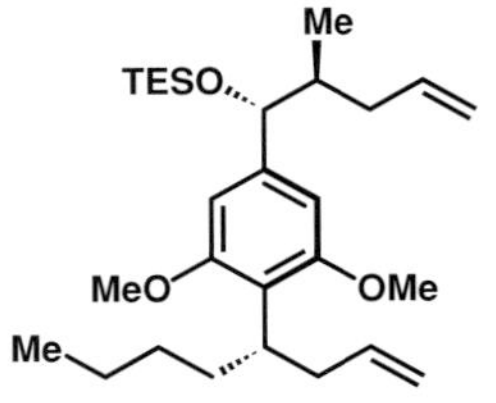

isomeric product would occur if dimerization led first to an adduct such as **46** (Scheme 14b; note that the OTES groups of **42** are absent in this structure) by cross-metathesis, an entity which would be poised for subsequent intramolecular closure. Moreover, even if the initial dimerization proceeded as desired to an intermediate capable of forming the [7,7]-system, the degree to which olefin geometry in the newly formed C4–C5 bond would influence the facility of the final RCM reaction could not predicted.

Based on the fact that metathesis reactions are reversible, it was anticipated that exposure of **42** to a suitable alkylidene-based metathesis catalyst would funnel the material exclusively towards the (*E*,*E*)-isomer **43**. This conjecture was supported by molecular modeling which revealed that this [7,7]-cyclophane was the most thermodynamically stable entity among all conceivable reaction products, including [8,6]-systems. As predicted, upon reaction of **42** with a 34 mol % loading of the molybdenum catalyst **11** in benzene at ambient temperature, the designed sequence to the 22-membered ring **43** was effected in an impressive 77 % yield after just 75 minutes of stirring, leaving only three routine operations to reach the target molecule (**45**). As validation that the conversion of **42** into **43** proceeded through reversible olefin-metathesis events, treatment of related structures **46** bearing either *E* or *Z* geometry (with this C4–C5 olefin prepared through a Horner–Wadsworth–Emmons reaction) smoothly led to the formation of [7,7]-cyclophane **47** rather than any [8,6]-product or [7,7]-isomers with alternative double-bond geometries. As an interesting digression, one should note that the drawing of structure **45** does not effectively convey the actual arrangement of this molecule in three-dimensional space as the aryl rings are, in fact, stacked, with a distance of only 7.65 Å separating them.

Without question, tandem sequences of the type we have just witnessed constitute powerful approaches to construct molecular complexity, although they do not encapsulate the full panoply of possible metathesis-based cascade transformations. To be sure, this specific area of reaction design has received a burgeoning level of attention in recent years, a trend that is likely to expand in the future, particularly in terms of combining metathesis with other reactions in the current synthetic repertoire. An early example of such a sequence can be found in a strategy established by the Nicolaou group which accomplished the formation of complex polyether frameworks representative of those possessed by diverse natural products such as brevetoxin B (see Chapter 37 in *Classics I*).[55] Thus, upon treatment of olefinic ester **48** (Scheme 15) with stoichiometric amounts of the Tebbe reagent (**9**), the titanium carbene **10** derived from this titanocene effected initial methylenation of the ester group to generate enol ether **49**, as would be expected based on the established preference of this reagent to engage carbonyl functionalities before olefins. With excess Tebbe reagent in solution, however, subsequent olefin metathesis between the newly generated olefin and the neighboring alkene ensued, affording poly-

42

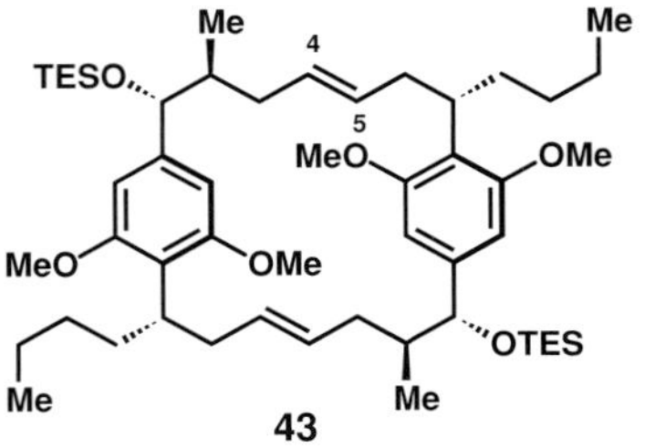

43

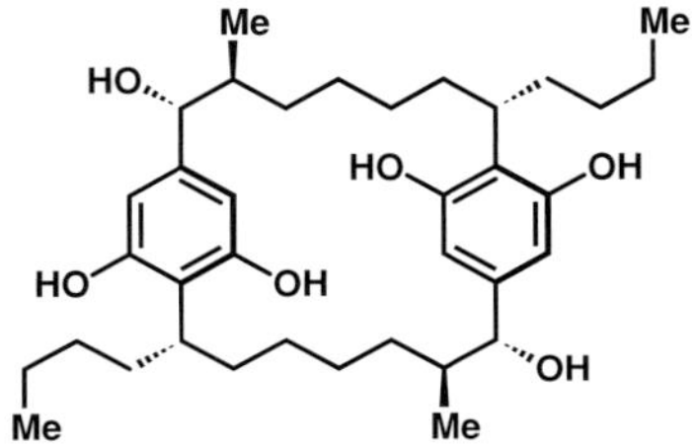

45: (–)-cylindrocyclophane A

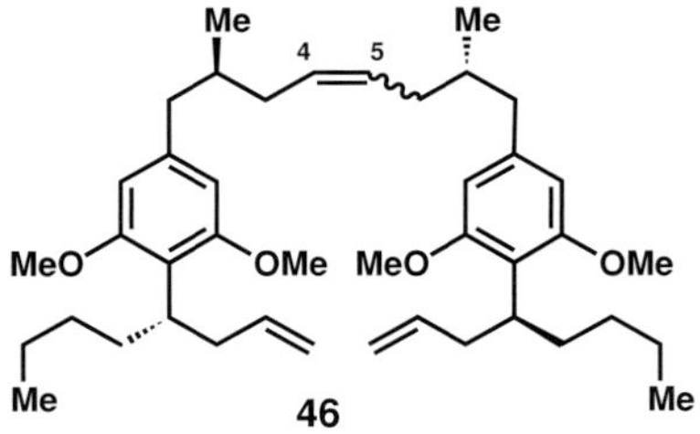

46

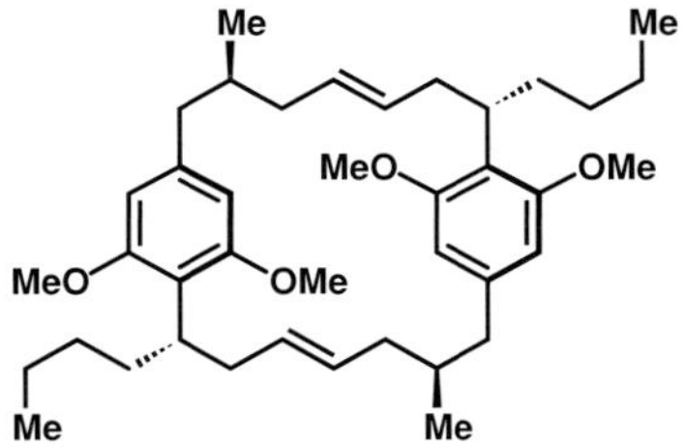

47

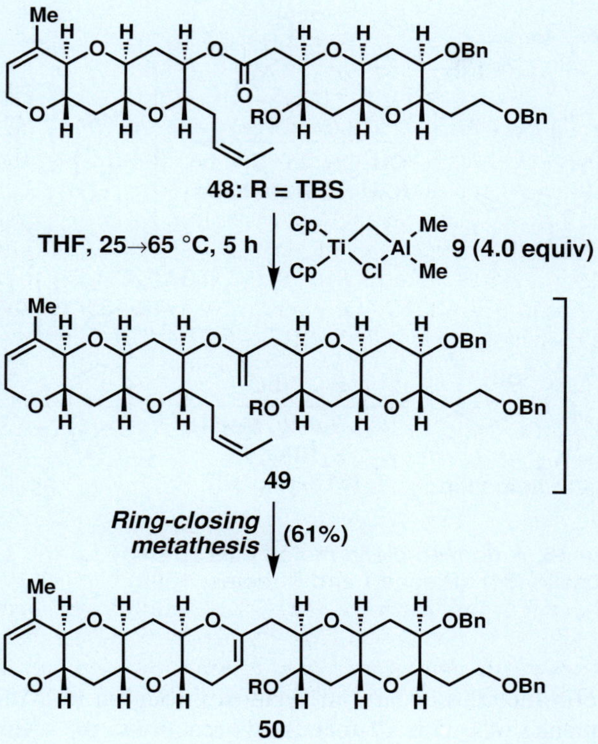

48: R = TBS

THF, 25→65 °C, 5 h; Cp2Ti(μ-Cl)(μ-CH2)AlMe2 9 (4.0 equiv)

49

Ring-closing metathesis (61%)

50

Scheme 15. Synthesis of complex polyether frameworks by tandem methylenation/olefin metathesis. (Nicolaou et al., 1996)[55]

ether **50** in 61 % yield. Since the initial disclosure of this transformation, the developed technology has been applied to several of the ring systems embedded within the structure of the complex marine natural product maitotoxin.[56] Beyond direct practical application, although the overall reaction is not catalytic with respect to the metal complex, it does illustrate the value of olefin metathesis in the synthesis of polyether-containing natural products, a strategy which has been utilized with great success on several occasions. Among these, the recent total synthesis of ciguatoxin CTX 3C by the Hirama group in which metathesis constituted the key transformation to form cyclic ethers of various sizes is noteworthy.[57]

We are nearly at the end of our initial foray into metathesis chemistry and we felt that it would be appropriate to close with two additional cascade sequences, starting with a recent total synthesis of (−)-halosalin (**54**, Scheme 16) by S. Blechert and R. Stragies.[58] In the designed strategy to this target molecule, after a stereoselective synthesis of homochiral **51**, it was envisioned that this compound could be induced to form **52** directly with one of the avail-

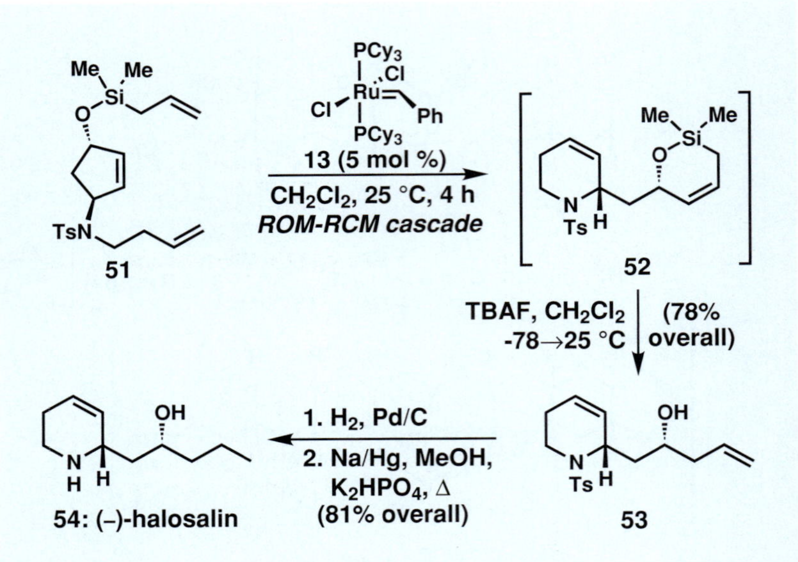

Scheme 16. A domino olefin metathesis strategy for the synthesis of (–)-halosalin (**54**). (Blechert and Stragies, 1999)[58]

able olefin metathesis catalyst systems through a ROM/RCM domino sequence. As with all metathesis reactions, the ultimate distribution of products is predicated on the overall differences in enthalpy, and so it was anticipated that the depicted reaction should succeed based on the relative thermodynamic stability of unsaturated six-membered rings compared to their five-membered counterparts. Most gratifyingly, upon application of only 5 mol % of ruthenium catalyst **13**, the expected product **52** was obtained smoothly. After subsequent treatment with TBAF *in situ* to cleave the temporary silicon tether, **53** was ultimately isolated from the reaction mixture in 78 % yield. One should realize that whether ring-opening metathesis or carbene addition to a terminal olefin occurred first, these alternate scenarios in fact are degenerate pathways to the same product. The true strength of the approach in this context rests in the transfer of easily installed stereochemical information on a wholly carbon-based precursor to a heterocyclic product in which such chiral information would otherwise be challenging to encode with equal facility.

As a final entry in this section, we felt it appropriate to turn to an example that perhaps reflects a likely focus of metathesis explorations in the future. While all of our preceding examples have illustrated metathesis reactions exclusively of olefinic substrates, other unsaturated precursors can engage in productive metathesis events when suitably activated by appropriate metal-carbene complexes. One of these processes, the enyne metathesis reaction,[59] is shown in Scheme 17. After insertion of the ruthenium alkylidene

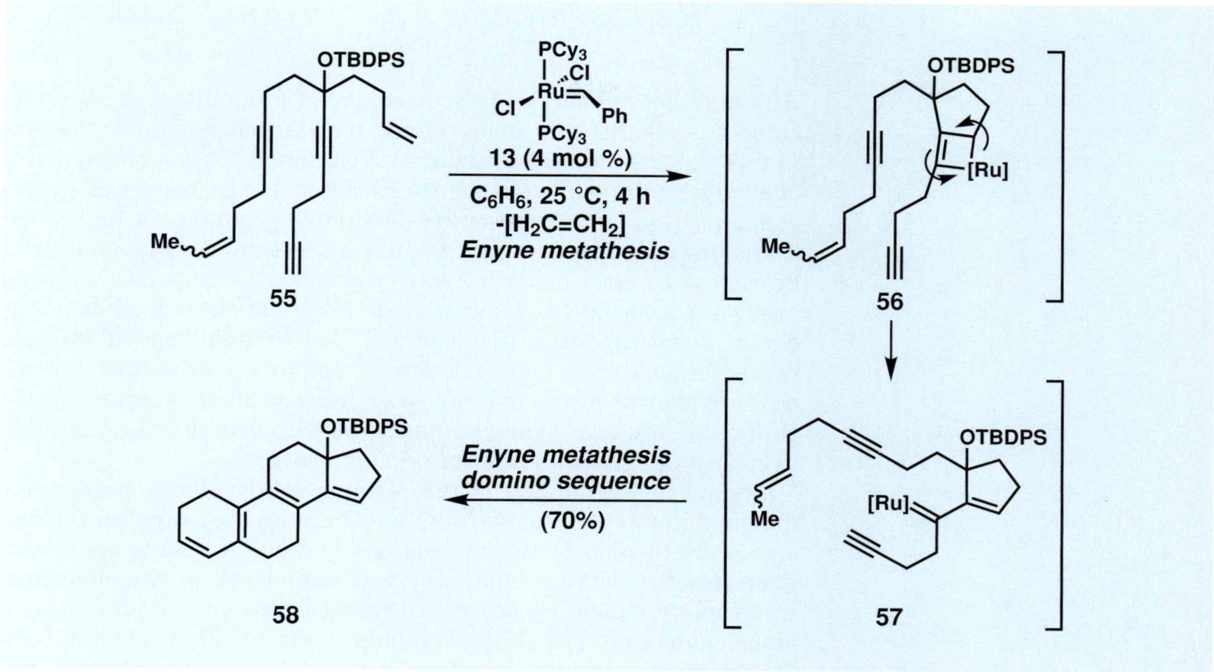

Scheme 17. Use of a domino enyne ring-closing metathesis sequence for the construction of steroid-type polycycles. (Grubbs et al., 1998)[60]

into the terminal (most reactive) double bond of **55**, enyne metathesis/cyclization led, via **56**, to intermediate **57** bearing a new carbene center. The latter species was now poised to initiate a series of reactions with the remaining three unsaturated sites, ultimately affording a complete steroid framework (**58**) in 70 % overall yield.[60] As such, each alkyne served as a metathesis relay point for subsequent cyclization. The efficiency of this particular enyne domino sequence, which was reported by the Grubbs group, bodes well for its future in complex molecule construction (examples of which we will examine more closely in Chapter 16). Historically, we should note that the first enyne metathesis reaction was carried out by Katz and Sivavec in 1985.[61] The related alkyne metathesis process (which Katz also predicted could occur)[24a] will be encountered in due course (vide infra).

Having discussed both the historical and practical aspects of the olefin metathesis reaction, we are now ready to embark on the journey at the heart of this chapter, namely the total synthesis of epothilone A (**1**).

7.2 Retrosynthetic Analysis and Strategy

Although epothilone A (**1**, Scheme 18) is a relatively small molecule in comparison to many of the targets that comprise chapters in this text, its molecular framework certainly offers a considerable challenge to the synthetic chemist based on its dense array of stereocenters and potentially sensitive functional groupings. Characteristic features along these lines include a 16-membered macrolactone bearing a *cis* epoxide, two hydroxy groups, two secondary methyl groups, a *gem*-dimethyl-substituted carbon atom, as well as a side chain consisting of a trisubstituted double bond and a thiazole ring. Surely, these elements afford ample opportunities for the development of new synthetic strategies and methodologies in the context of any total synthesis endeavor with benefits likely spilling over into the general realm of organic synthesis.

Beyond this beneficial feature, from a retrosynthetic perspective the epothilones are wonderful targets because they offer an impressive array of possible disconnections that could lead to successful strategies for sculpting the structural framework of this molecule. In terms of means by which to generate the macrocyclic system alone, numerous possibilities should come to mind, not the least of which is a macrolactonization or an aldol-based macroring-forming reaction. While both of these approaches have indeed proven quite effective in the hands of several investigators to reach epothilone A, our focus in this chapter will be on metathesis-based macrocyclization routes.[10] As such, we shall examine in detail the first total synthesis of epothilone A (**1**) that used this strategy, as brought to fruition by Nicolaou and co-workers in 1997,[8] and, in later sections, touch upon other equally instructive ring-closing metathesis approaches to this family of molecules.

As with all well-conceived synthetic plans, the original Nicolaou retrosynthetic blueprint for epothilone A (**1**), shown in Scheme 18, sought a design that was flexible, as convergent as possible, and which aimed to test and develop new synthetic strategies and methods. However, beyond these standard goals of modern synthetic engineering, because the epothilones are such important targets from the chemical biology and drug discovery perspectives, an additional design element included the ability to deliver as many structural variations as possible to explore structure–activity relationships (SAR). Accordingly, unlike most synthetic plans in which the controlled installation of stereochemical elements is deemed of critical importance, obtaining mixtures of isomers along the route was viewed as an advantage since it would provide additional structural analogues which might show unique activity profiles relative to the parent natural product, and thereby shed light on the molecular basis of the biological activity of the epothilones.

Guided by this set of construction principles, the first retrosynthetic simplification consisted of excising the C12–C13 epoxide

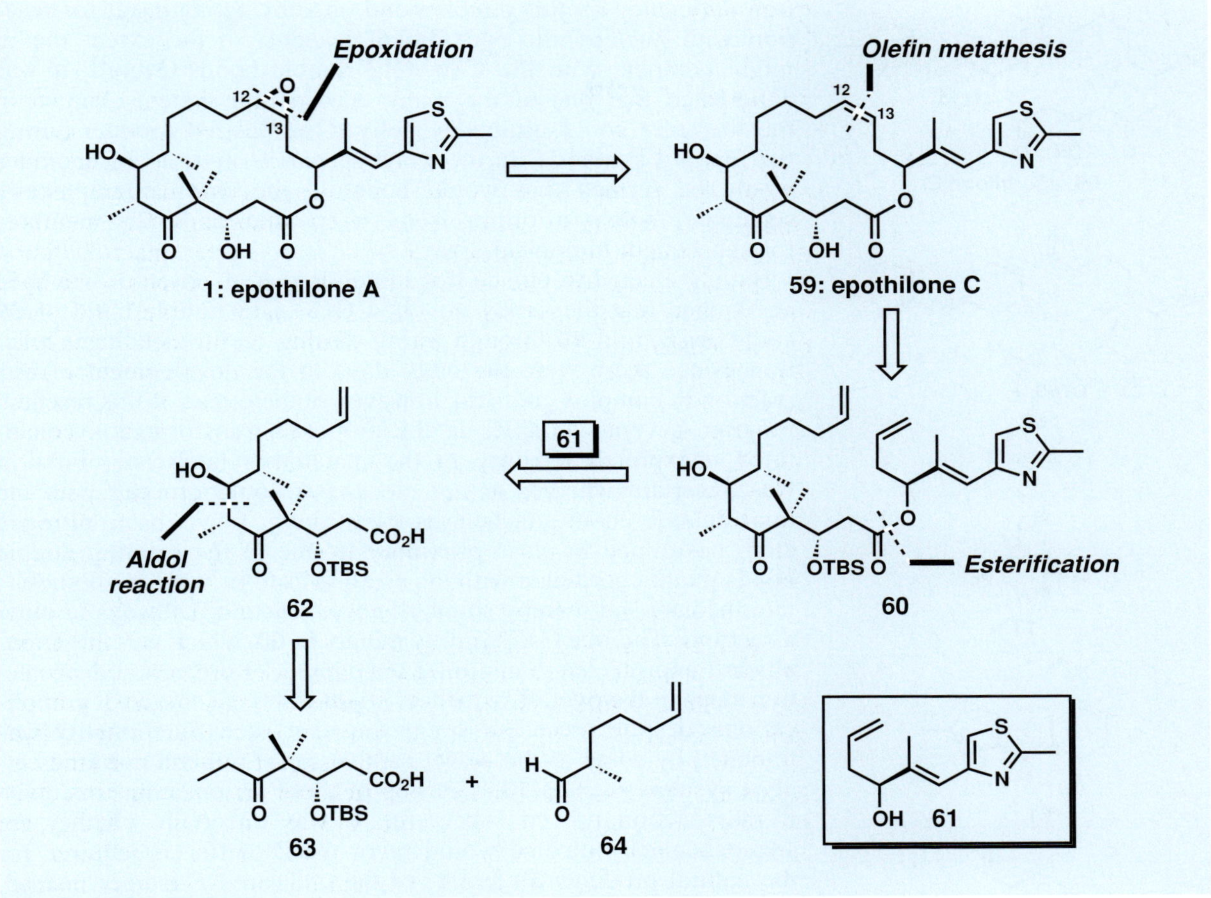

Scheme 18. Retrosynthetic analysis of epothilone A (**1**).

oxygen atom, leading to olefin **59** as an advanced subtarget. Because **59** is also a member of the epothilone family (epothilone C), a successful synthetic route to this compound would thus afford access to two natural products, assuming that the epoxidation reaction required to complete epothilone A (**1**) could be achieved. While it was always envisioned that such an epoxidation could be effected, the more pressing concern of this strategy was whether this conversion would proceed with the requisite level of stereocontrol. For certain, the inherent molecular architecture of **59** would play a leading role in this task, although its overall effect could not be anticipated on the basis of first principles. Apart from this issue, as an additional element of challenge, this endeavor would also require the chemoselective engagement of the C12–C13 double bond in the presence of two oxidizable heteroatoms in the heterocyclic ring and the trisubstituted double bond of the thiazole side chain. Even though the latter double bond is certainly more hindered, the neighboring thiazole ring confers a reasonable degree of elec-

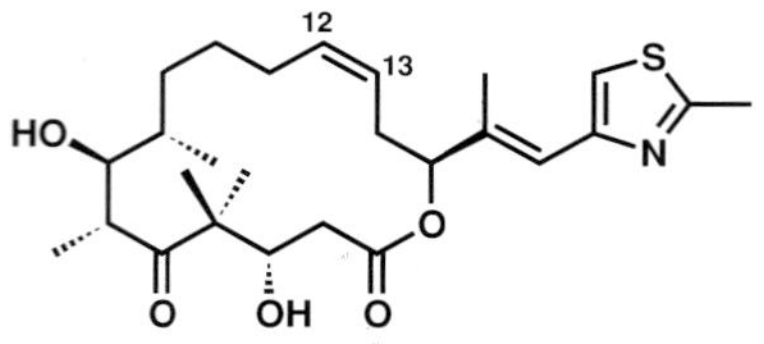

59: epothilone C

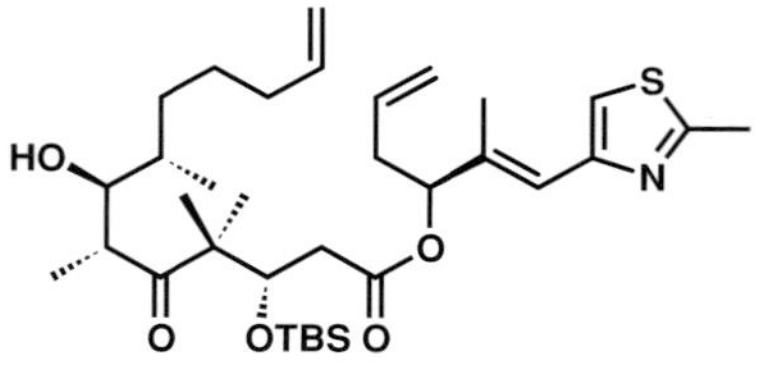

60

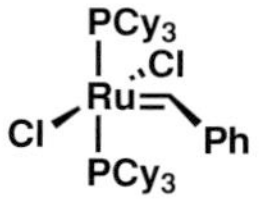

13

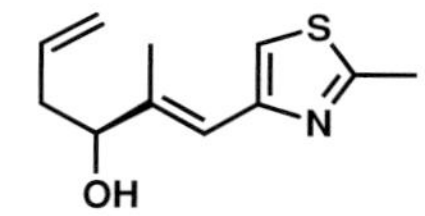

61

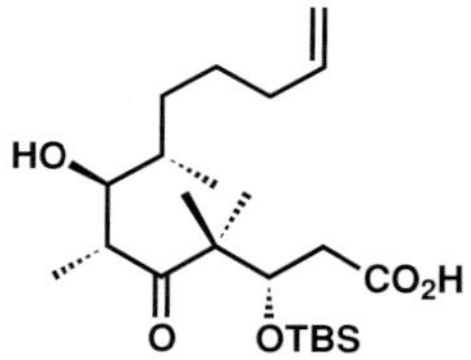

62

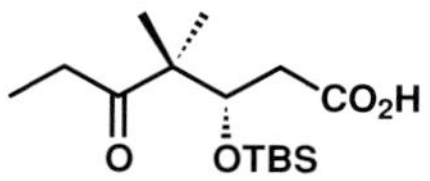

63

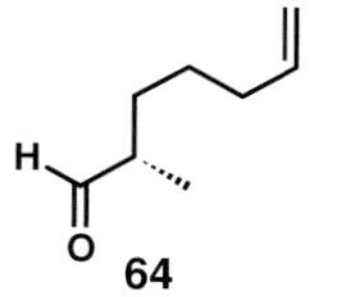

64

tron deficiency on this double bond, thereby activating it for reaction with nucleophilic epoxidation reagents to the extent that it might compete with the C12–C13 double bond. Overall, it was anticipated that one of the many epoxidation systems known in the literature could ultimately deliver the desired product during the forward synthesis. In the worst scenario, mixtures of epoxides would be formed that would hopefully be chromatographically separable, thereby affording isomeric epoxides as library members for subsequent biological assays.

Having elected to pursue this initial course of action, it was then envisioned that the newly unveiled C12–C13 double bond of **59** could arise from **60** through a ring-closing olefin metathesis reaction. Since these were the early days in the development of this reaction in complex contexts, however, at the outset of this research program several variables in the proposed transformation constituted unexplored territory in the metathesis landscape. First, it was uncertain whether or not metathesis could proceed with the thiazole side chain attached, as the strongly Lewis basic nitrogen atom positioned in close proximity to one of the reacting double bonds might coordinate with the catalyst itself or with an alkylidene intermediate and thereby shunt effective reaction pathways in other directions. The free C-7 hydroxy moiety in **60**, which was intentionally left unprotected to minimize the number of protection/deprotection steps in the overall sequence, might also be a source of complications, despite precedent suggesting that such functionality was tolerated by some of the newer (at that time) ruthenium-based catalyst systems such as **13**.[62] On top of these serious concerns, even if the reaction proved successful, it was uncertain whether the stereochemical outcome would favor the *Z* olefin, as required for the natural products **59** and **1**, or the undesired *E* isomer instead. When these troubling issues are considered collectively, the designed olefin metathesis strategy certainly seemed to promise an exciting undertaking that would hopefully extend (or, if unsuccessful, define) some of the frontiers of olefin metathesis in complex molecular contexts.

At this juncture in the analysis, although the overall molecular complexity of **60** versus the original target molecule has not been greatly simplified in terms of number of atoms or stereocenters, the remaining design elements were relatively easy to deduce. Thus, striving for convergence, the retrosynthetic arrow now touched the ester linkage in **60**, leading to carboxylic acid **62** and the much smaller thiazole side chain **61**. Next, it was reasoned that the C-7 hydroxy group and its neighboring C-6 methyl group arranged in a *syn* fashion in **62** could result from a diastereoselective aldol-type union of ketoacid **63** and aldehyde **64**. One should note, however, that this key coupling was anticipated to proceed with relating not absolute stereocontrol. Since chiral aldehydes afford poor aldol stereoselection (cf. Chapter 3). Again, such a result was viewed favorably based on arguments of library construction for chemical biology studies. These final retrosynthetic simplifications have afforded three fragments of roughly equal size and stereochemical complexity. As such, this plan certainly reduces the synthetic challenges associated with **1** to far smaller and more manageable fragments, and concurrently affords a simple means to generate analogues, especially of the C-15-linked side chain, which is designed to be incorporated near the end of the synthetic sequence.

7.3 Total Synthesis

7.3.1 Solution-Phase Synthesis of Epothilone A (1)

Based on the retrosynthetic analysis presented above, the first critical task on the way to testing the olefin metathesis-based strategy for the total synthesis of epothilone A (**1**) was the construction of the three key building blocks **61**, **63**, and **64**. Focusing our attention first on the preparation of aldehyde **64**, the Nicolaou group developed several approaches to this building block based on asymmetric synthesis with chiral auxiliaries; one of these solutions is detailed in Scheme 19.

In general, although face-selective alkylations of chiral enolates constitute one of the most important methods for asymmetric bond formation, particularly for the generation of acyclic stereocenters α to carbonyl groups, relatively few of the available protocols engender high-yielding and stereoselective reactions with non-activated alkyl halides. A stark exception to this trend resides in the family of chiral sultam derivatives developed by Oppolzer and co-workers, e. g. **65**, which have found widespread utility in organic synthesis.[63] The power of this methodology is appropriately demonstrated here where exposure of **65** to NaHMDS in THF at −78 °C smoothly effected the formation of the chiral enolate (*Z*)-**67** whose structure is shown in the column figure. Upon addition of 5-iodo-1-pentene in the presence of HMPA, the enolate then selectively reacted with this unactivated primary alkyl iodide exclusively from the bottom face, with the alternative possibility effectively prevented by the *gem*-dimethyl substituents situated

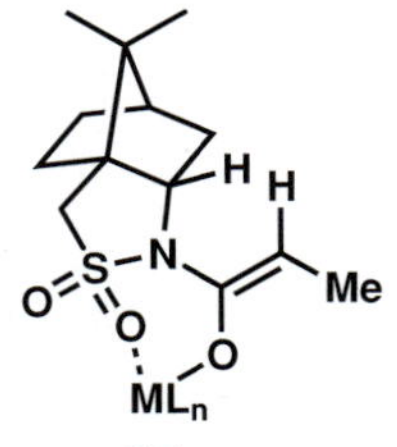

H H Me O=S N O O ML_n

67

65 — NaHMDS, THF, -78 °C; n-C_5H_9I, HMPA, -78→25 °C, 5 h — 66 — 1. $LiAlH_4$ 2. TPAP, NMO (57% from 65) — 64

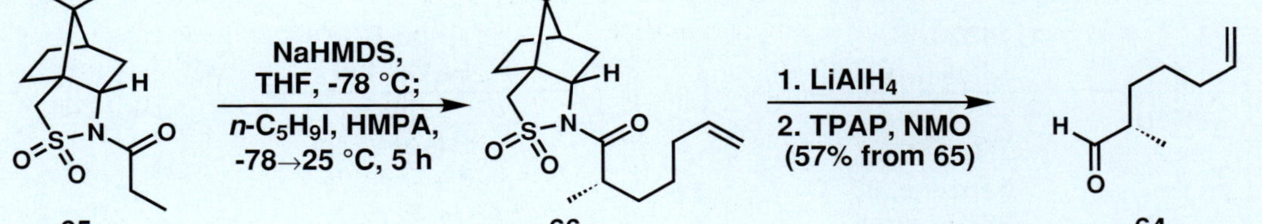

Scheme 19. Synthesis of key building block **64**.

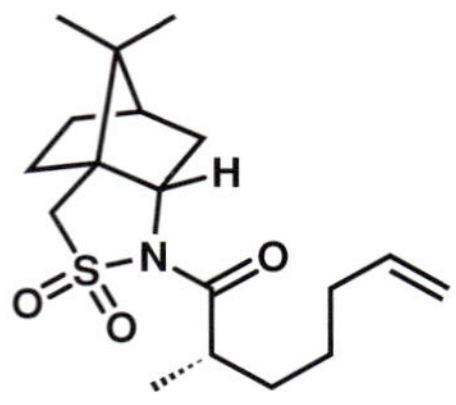

66

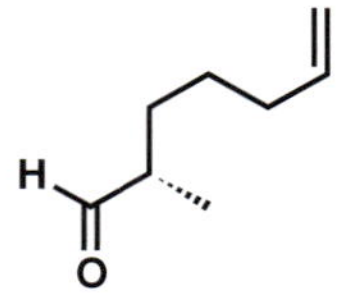

64

at the apex of the [2.2.1]-bicyclic ring system. Accordingly, this auxiliary afforded **66** as a single stereoisomer in high yield. With success achieved in this initial step, the electron-deficient chiral auxiliary was then smoothly excised with $LiAlH_4$ in THF at −78 °C, providing a primary alcohol which was oxidized with TPAP/NMO to complete the synthesis of the desired aldehyde **64** in 57 % overall yield for the three steps.

With the synthesis of one fragment accomplished, routes to the other two pieces, **61** and **63**, were sought next. As shown in Scheme 20, each of these syntheses was predicated on employing an asymmetric allylboration[64] reaction as the key operation to fashion their stereogenic centers. Thus, dealing first with the optically active ketoacid **63**, the synthesis of this building block began with a starting material derived from the chiral pool following an approach which anticipated the requisite carboxylic acid arising from the oxidation of an aldehyde precursor obtained through ozonolytic cleavage of a terminal double bond. To execute this strategy successfully, the asymmetric addition of an allyl chain onto this starting material was required first. This goal was admirably met upon the addition of H. C. Brown's chiral (+)-Ipc_2B(allyl) reagent[65] to **68** in Et_2O at −100 °C, affording the desired allylated product with greater than 98 % *ee* (based on Mosher's ester analysis method). With the required stereogenic center secured, the newly formed hydroxy group was then protected as its corresponding TBS ether with TBSOTf to provide **69** in 73 % overall yield after these two operations. Arrival at carboxylic acid **63** was finally accomplished, in 84 % yield, through the conventional manipulations of ozonolytic cleavage followed by oxidation of the resultant aldehyde with $NaClO_2$.

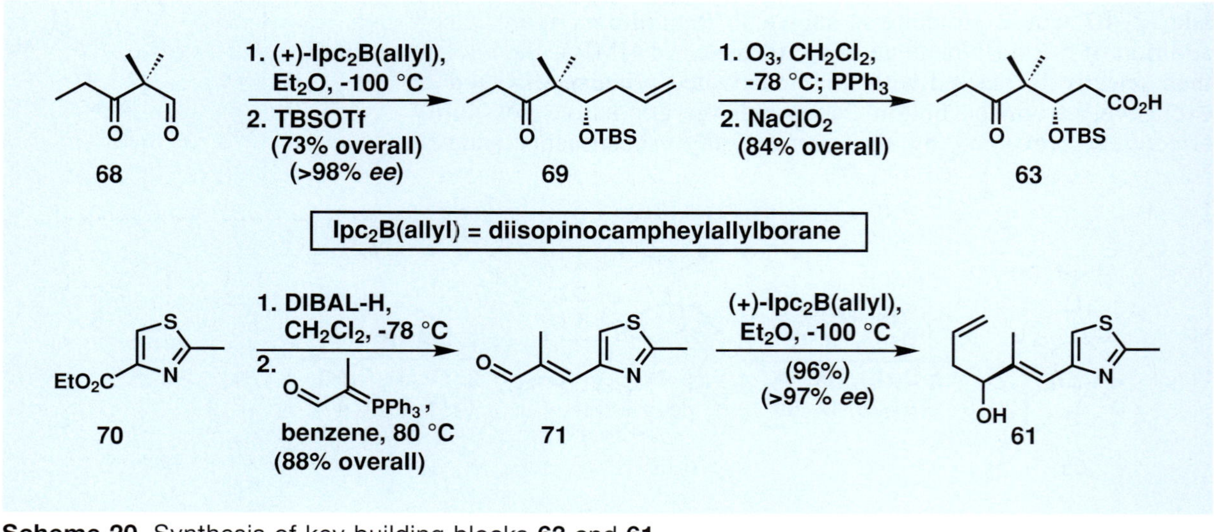

68 → 1. (+)-Ipc_2B(allyl), Et_2O, -100 °C; 2. TBSOTf (73% overall) (>98% *ee*) → 69 → 1. O_3, CH_2Cl_2, -78 °C; PPh_3; 2. $NaClO_2$ (84% overall) → 63

Ipc_2B(allyl) = diisopinocampheylallylborane

70 → 1. DIBAL-H, CH_2Cl_2, -78 °C; 2. PPh_3, benzene, 80 °C (88% overall) → 71 → (+)-Ipc_2B(allyl), Et_2O, -100 °C (96%) (>97% *ee*) → 61

Scheme 20. Synthesis of key building blocks **63** and **61**.

The remaining fragment, thiazole **61**, was readily synthesized through the three-step protocol defined in the lower half of Scheme 20 starting with the known thiazole ester derivative **70**.[66] In the opening operation, conversion of the ester into the corresponding aldehyde was achieved through a controlled DIBAL-H reduction at −78 °C employing CH_2Cl_2 as solvent, and was then followed by a stereoselective Wittig olefination with the stabilized ylide $Ph_3P{=}C(Me)CHO$ in benzene at 80 °C, ultimately providing the (*E*)-α,β-unsaturated aldehyde **71** in 88 % overall yield. One should note that although the Wittig reagent itself possesses a reactive carbonyl function, the steric bulk of the triphenylphosphine portion of the ylide prevents self-condensation, thereby enabling this reagent to be utilized successfully. At this stage, the separation

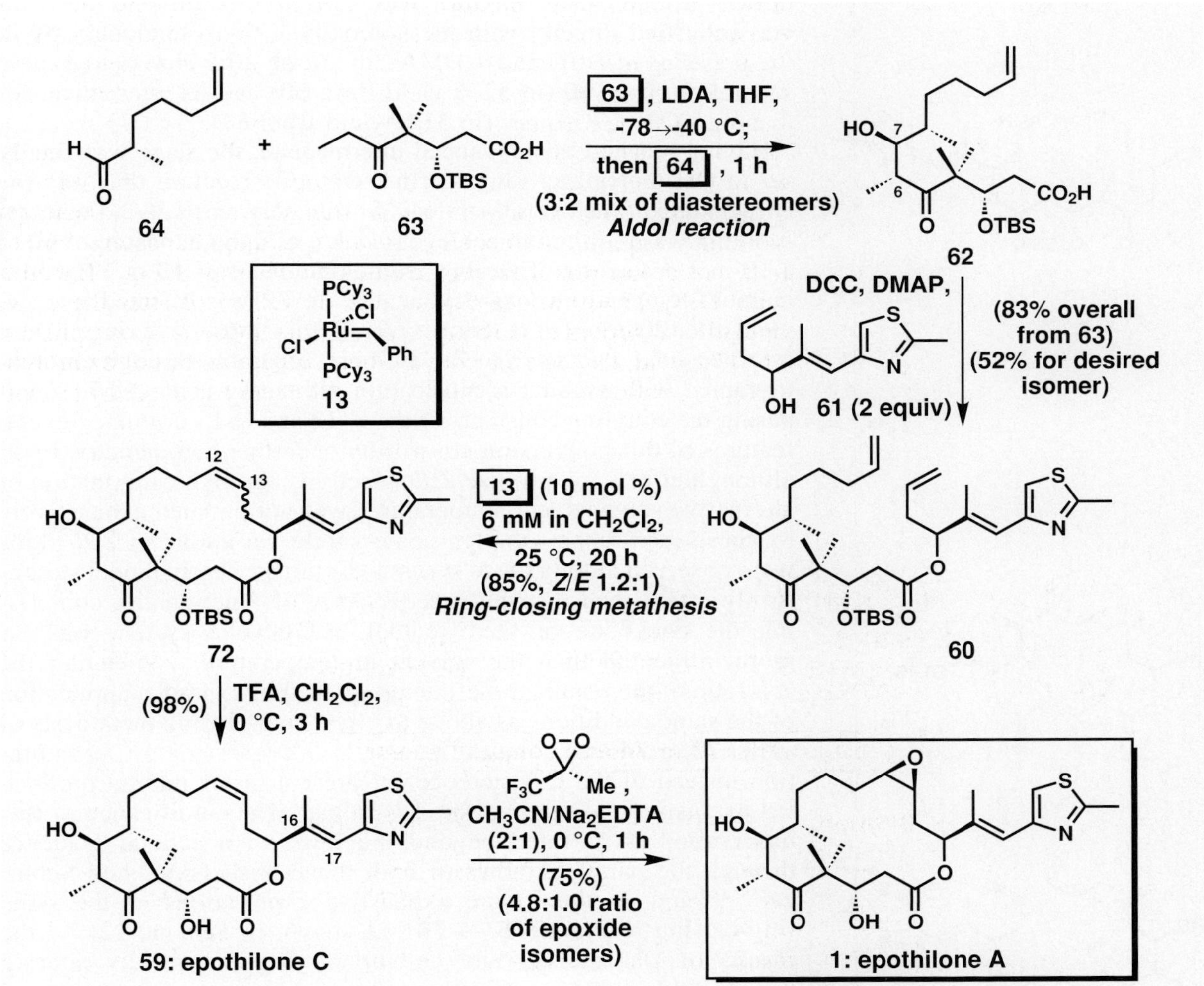

Scheme 21. Completion of the total synthesis of epothilone A (**1**).

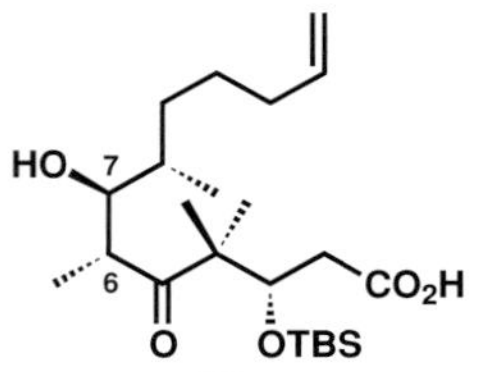

between **71** and **61** should be relatively obvious, merely requiring a stereoselective addition of (+)-Ipc_2B(allyl). Indeed, upon addition of this borane to **71** in a mixture of Et_2O and pentane at −100 °C, **61** was obtained in 96 % yield and with greater than 97 % *ee*.

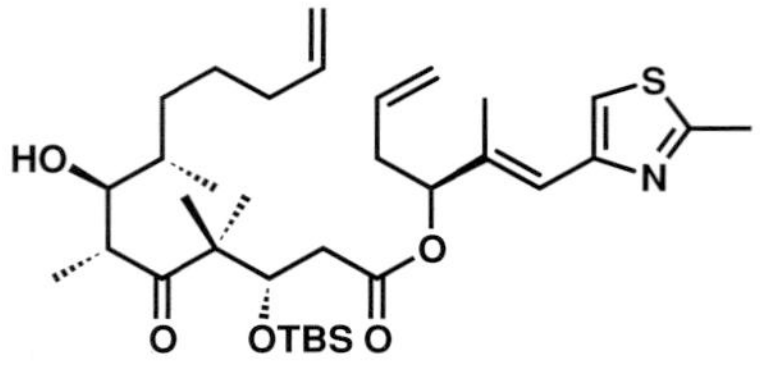

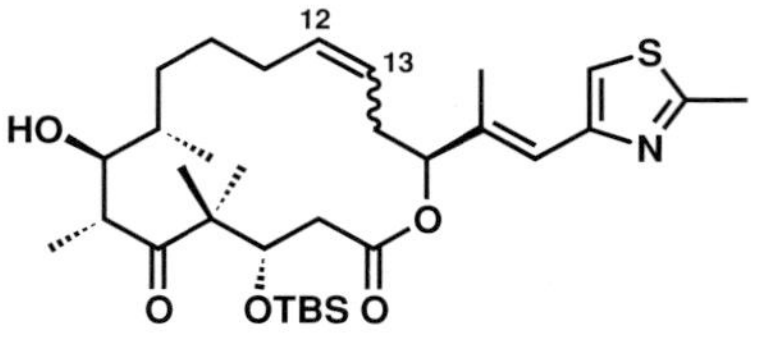

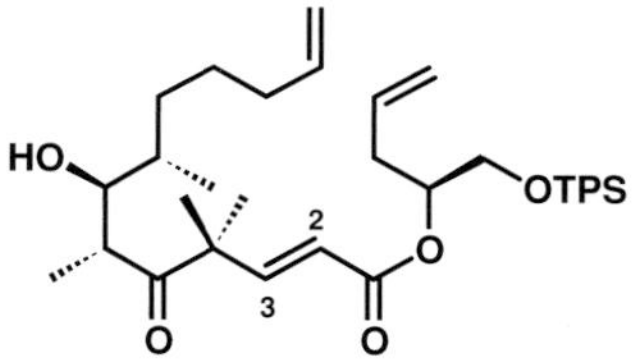

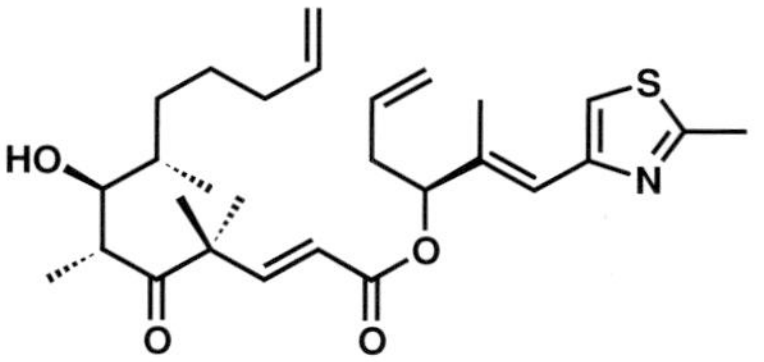

With this preliminary phase of synthetic work accomplished, the more intriguing elements of the anticipated epothilone A synthesis could now begin in earnest. As shown in Scheme 21, the initial merger of fragments **61**, **63**, and **64** into advanced intermediate **60** proceeded quite smoothly. Treatment of **63** with 2.3 equivalents of LDA in THF at −78 °C with warming to −40 °C over one hour (to effect both deprotonation of the free acid and formation of the enolate) was followed by the addition of 1.6 equivalents of aldehyde **64**, leading to a facile aldol reaction which provided **62** and its C6−C7 *syn* diastereomer in a 3:2 ratio that favored the desired drawn product. This mixture was carried forward and the acid was esterified directly with the homoallylic thiazole alcohol **61** in the presence of DCC and 4-DMAP to afford, after chromatographic separation, pure **60** (in 52 % yield from **63**) and its alternative *syn* disposed C6−C7 isomer (in 31 % yield from **63**).

Having reached this advanced intermediate, the stage was finally set to test the ring-closing olefin metathesis reaction that was the cornerstone of the overall strategy. Fortunately, virtually no reaction scouting was required to achieve success, as upon exposure of **60** to a 10 mol % loading of Grubbs' ruthenium catalyst **13** in CH_2Cl_2 at ambient temperature macrocyclization to **72** was effected in 85 % yield after 20 hours of reaction. Overall, this impressive ring closure afforded a 1.2:1 mixture of *Z*:*E* isomers, separable through chromatography, with product identification ultimately achieved by comparing the coupling constants of the C-12 and C-13 protons. Several features of this conversion are worthy of further commentary. First, although efforts to optimize *Z*:*E* selectivity through examination of alternative solvents and temperatures was not pursued exhaustively by these researchers (though some subtle variations in *Z*:*E* ratios were observed among those systems examined), early model studies firmly established that the specific array of functionality comprising the backbone of the eventual macrocyclic system was far more influential than the solvent or temperature in dictating the *Z*:*E* ratio of the resultant olefinic products. For example, application of the same conditions as above to effect ring-closing metathesis of either **73** or **74** (see column figures), both possessing $\Delta^{2,3}$ unsaturation instead of the C-3 stereocenter present in the natural product, led exclusively to the undesired *E* isomer. The significance of this observation is elevated beyond the level of anecdotal evidence through the parallel studies of both the Danishefsky[67] and Schinzer[68] groups in their more exhaustive explorations of the same ring-closing reaction (**75** → **76**), as shown in Scheme 22. As the results of Danishefsky and co-workers suggest, merely altering the stereochemistry at C-3 from the α-disposition of the natural product to the opposite β-orientation had a profound, but negative impact on the *Z*:*E* ratio. Similarly, modifying the C-5 ketone to

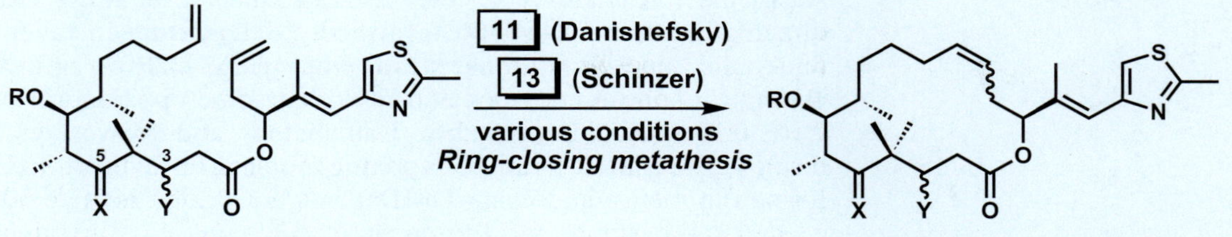

Investigator	R	X	Y	Solvent	T (°C)	t (h)	*Z*/*E* Ratio	Yield (%)
Danishefsky	TBS	α-OTPS, β-H	α-OH	C_6H_6	55	2	1:3	86
	TBS	α-OTPS, β-H	α-OTES	C_6H_6	55	2	1:5	80
	TBS	α-OTPS, β-H	β-OH	C_6H_6	55	2	1:9	81
	TBS	O	α-OTBS	C_6H_6	55	2	5:3	86
	H	O	α-OH	C_6H_6	55	2	1:2	65
	TBS	O	β-OTBS	C_6H_6	55	2	1:2	88
Schinzer	TBS	O	α-OTBS	CH_2Cl_2	25	16	1.7:1	94

Scheme 22. Olefin metathesis studies exploring the stereoselectivity in the formation of the epothilone scaffold.

any reduced congener dramatically eroded *Z* selectivity. The example from the Schinzer laboratories, when compared with the same substrate used by Danishefsky and co-workers, provides additional support that changing reaction parameters such as solvent, temperature, or even metathesis catalyst leads only to a slightly altered ratio of olefinic products. Taken cohesively, these studies indicate that, while the controlled formation of the *E* olefin of the epothilone A macrocycle is a realistic goal, obtaining the *Z* olefin exclusively would appear to be impossible upon adherence to this strategy armed only with current metathesis technology.

As mentioned earlier, however, mixtures of olefin isomers were acceptable for biological testing purposes, and since the *Z* and *E* isomers of **72** (Scheme 21) were separable by chromatography, pure (*Z*)-**72** was advanced through the final stages needed to reach epothilone A (**1**). Thus, proceeding with the synthesis, the lone protecting group in **72** was cleaved by treatment with TFA in CH_2Cl_2 at 0 °C giving **59** in near-quantitative yield, leaving epoxidation as the only step remaining to complete the molecule. As anticipated, optimization was required before this reaction provided the desired stereoisomer with a reasonable degree of selectivity. Under the best conditions, treatment of **59** with 1,1,1-trifluorodimethyldioxirane (generated *in situ* by reacting 1,1,1-trifluoroacetone with OXONE®) afforded a 4.8:1.0 mixture of **1** and its C12–C13 stereoisomeric epoxide in 75 % overall yield after just

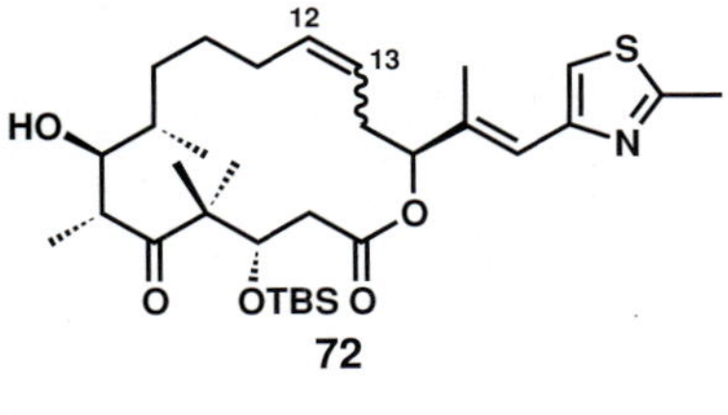

72

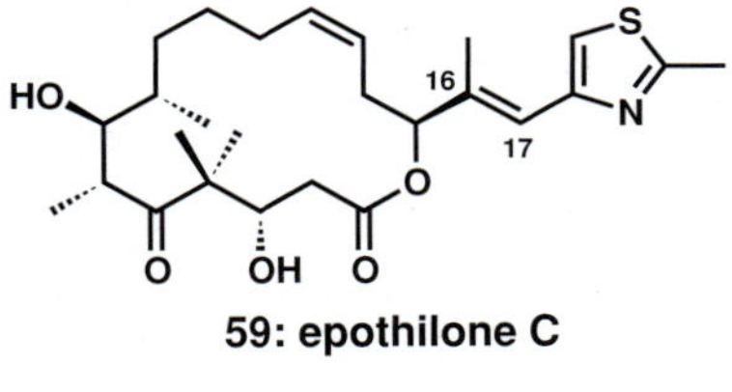

59: epothilone C

one hour of stirring at 0 °C in a 2:1 mixture of CH_3CN and Na_2EDTA. Alternatively, the use of the structurally related dimethyldioxirane (DMDO) furnished a 3:1 mixture in favor of **1** under the same conditions, while employing *m*CPBA afforded a 1:1 ratio of these epoxides and provided some epoxidation of the C16–C17 olefin. Remarkably, Danishefsky and co-workers have disclosed that a 16:1 ratio of epoxide isomers in favor of **1** can be formed in 49 % yield using DMDO in CH_2Cl_2 at −35 °C.[67]

7.3.2 Solid-Phase Synthesis of Epothilone A (1)

Although the preceding discussion has carried us through the original Nicolaou route to epothilone A (**1**), the story of the developed chemistry did not stop once the spectroscopic properties of the synthetic and natural material were shown to be identical. Indeed, subsequent work sought to apply this chemistry to generate large libraries of epothilone analogues for biological screening. One way this objective could potentially be implemented would be to substitute the three main building blocks in the original synthesis with ones bearing slightly altered functionality such as those defined in Scheme 23. If these fragments were joined in a combinatorial fashion, it would be possible to effect the construction of 45 unique compounds (3 × 3 × 5) in principle, assuming that these

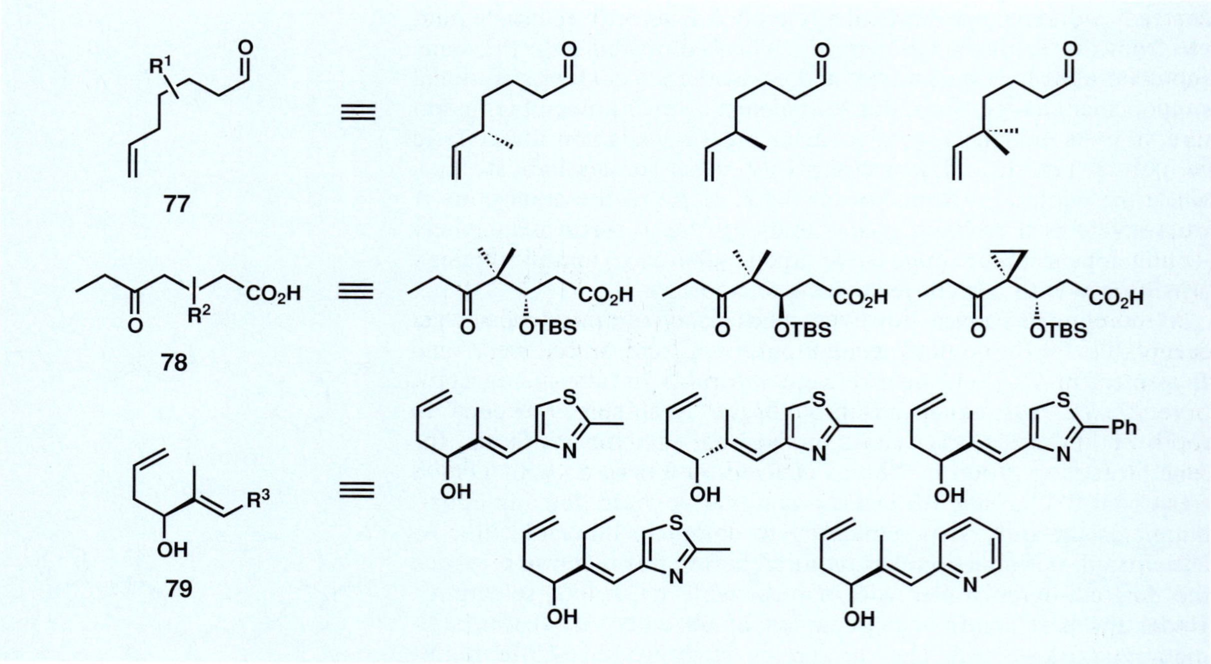

Scheme 23. Representative examples of building blocks (**77**, **78**, and **79**), which could be employed to construct a library of epothilone A (**1**) analogues.

syntheses succeeded with equal proficiency. Moreover, if the aldol and metathesis steps proceeded with similar levels of stereocontrol, then four isomeric forms of each specific combination of three building blocks would be obtained, thus providing, in theory, a total of 180 compounds for testing.

While one could certainly attempt to pursue such a library through parallel solution-phase synthesis, it would likely take several months to complete the project due primarily to the numerous and time-consuming purification steps that would undoubtedly be required. As such, the Nicolaou group sought an alternative method to achieve this set of objectives in a more timely and less exhausting manner by implementing a general strategy based on solid-phase chemistry,[69] a technique pioneered by R. Bruce Merrifield in the 1960s which garnered him a Nobel prize in 1984.[70]

In his groundbreaking studies, Merrifield sought to fashion long chains of polypeptides in a universal manner from the readily available collections of amino acid building blocks. His key insight rested in the idea that by ligating one of the starting building blocks to a solid resin platform such as polystyrene (which provides a large anchor that can be seen with the naked eye), reactions could be driven to completion using excess amounts of reagents, and then the purification of products could be performed simply by washing away excess reagents with appropriate solvents. As a result, laborious extraction and chromatographic separations were avoided, leading to enhanced yields in each step because material was not lost through these manipulations. Most impressively, the strategy worked well with the existing chemical methods that had been developed for conventional solution phase synthesis of peptides, and once a desired polypeptide chain had been synthesized, the product could then be cleaved from the bead-like polystyrene resin through a simple chemical "deprotection." Based on the success and speed of this approach, the high-throughput combinatorial generation of libraries of peptides, and, shortly thereafter, oligonucleotides, became a common enterprise and has since had a profound impact on chemistry, biology, and biotechnology.[71]

In more recent times, the solid-phase chemistry technique has been extended to the more general and vast field of chemical synthesis, and has gained considerable utility in library construction beyond conventional parallel synthesis protocols through an approach known as "split-and-pool," or "split synthesis," first introduced by Furka et al. in 1988.[72] The split-and-pool concept, which is shown in cartoon format in Scheme 24, is particularly powerful despite the relatively simple nature of the design. For the sake of argument, imagine that you wanted to synthesize a library of polypeptides composed of three unique amino acid building blocks. Starting with a suitable amount of an appropriately functionalized resin, you first divide the solid support into three separate, but equal, pools (the split step) and then chemically append a different amino acid onto the resin in each pool (the introduction of diversity

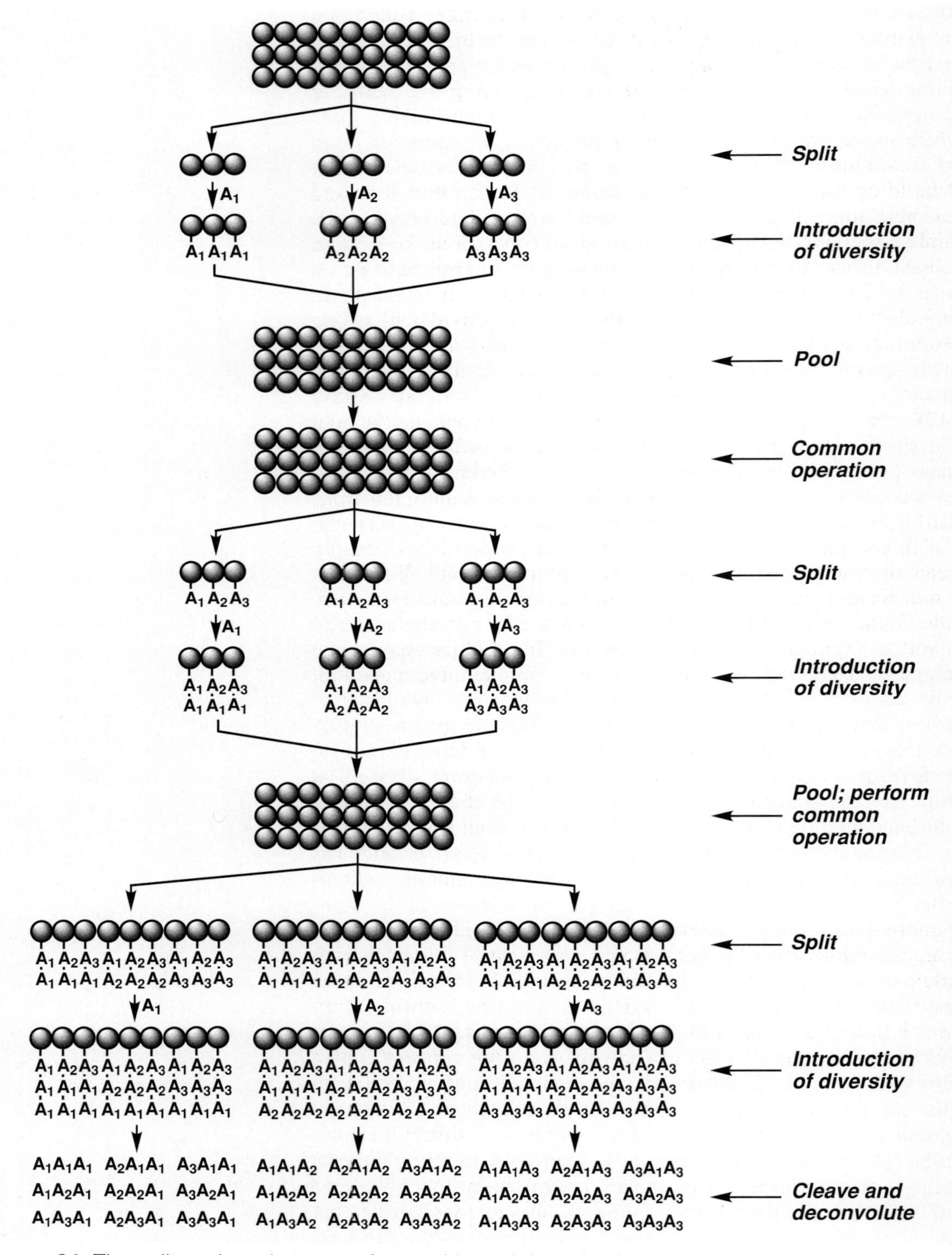

Scheme 24. The split-and-pool strategy for combinatorial synthesis.

step). Seeking now to add a second amino acid, you then recombine your resins into a single pool so that you can perform a common operation such as a deprotection step to reveal a free carboxylic acid. With the attached substrate now ready for coupling with the amino group of the second amino acid, you again divide your resin into three separate pools which results in an approximately equal distribution of resin carrying each of the individual amino acid building blocks used in the first step. Addition of a different amino acid to each pool then produces the complete array of all possible dipeptides. Repeating this set of operations one more time would then afford the entire collection of 27 tripeptide products ($3 \times 3 \times 3$), entities which can be freed from the resin in a final cleavage step. Impressively, this method provides a powerful solution to combinatorial library generation, with two major, but manageable, caveats. First, because each diversity step divides the resin into increasingly smaller portions, the amount of resin initially needed (where substrate loading is typically on the order of 0.5 to 1.0 mmol g^{-1} resin) to yield a sufficient amount of each product for screening (~1 to 5 mg) could become quite large and unwieldy to handle in early steps. Furthermore, in order to identify each of the final products cleaved from the resin at the end of this sequence, some sort of "tag" is needed, a label that can be read through a process known as "deconvolution". While we will not discuss such deconvolution strategies in detail here (they can be found in appropriate primary and secondary sources),[69,73] suffice it to say that numerous solutions to this problem have been developed based on both chemical and nonchemical encoding techniques.

The solid-phase synthesis of epothilone A (**1**) and analogues thereof by the Nicolaou group serves as a useful vehicle to illustrate the efficiency of this technique.[74] To apply the power of Furka's method to generate libraries of compounds related to this natural product, however, a solid-phase route would have to be established first since chemistry in solution cannot always be directly transferred to the solid-phase. Towards this end, the Nicolaou group initially examined the general approach applied above in the solution-phase total synthesis, altered only with the initial incorporation of a modified building block to enable attachment of the epothilone template to the solid support. As shown in Scheme 25, the synthesis began with the preparation of an appropriate resin, starting from chloromethylpolystyrene (**80**). Thus, hoping to eventually fashion an ylide-based reagent that could add to **82** to form the olefin **83** which would ultimately constitute one of the alkenes in the metathesis reaction, **80** was first converted into **81** by conventional operations. Most gratifyingly, upon treatment of this new resin with NaHMDS, the first building block (**82**) of the synthesis was loaded onto the resin through a Wittig olefination, affording **83** in $>70\,\%$ yield (as determined by cleaving the product off some of the resin and measuring mass recovery). After the next two operations in which **83** was smoothly converted into **84**,

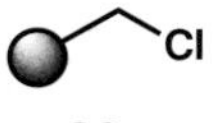

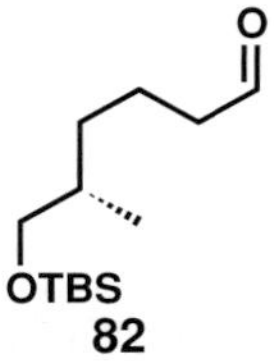

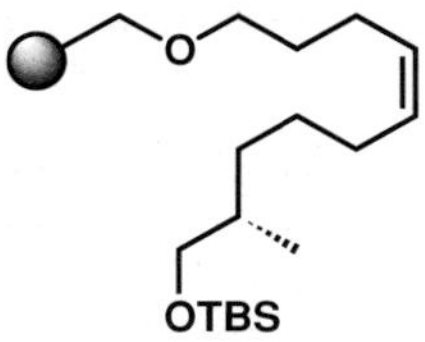

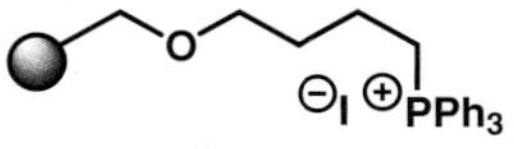

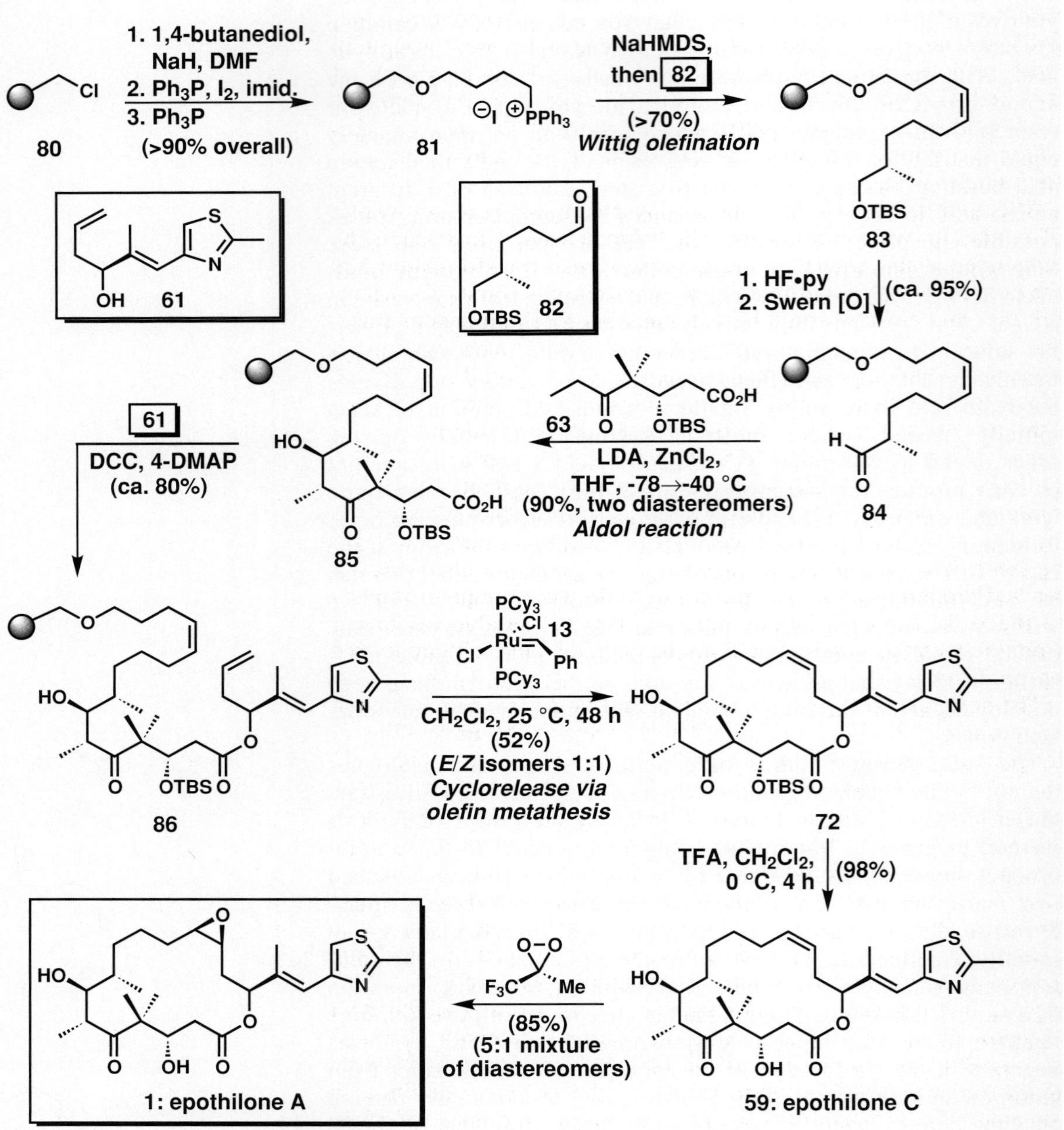

Scheme 25. Solid-phase total synthesis of epothilone A (**1**).

namely HF•py-induced cleavage of the silyl ether and subsequent Swern oxidation of the resultant alcohol to the corresponding aldehyde, the next piece (**63**) was incorporated, as before, through an aldol reaction promoted by LDA and $ZnCl_2$. With a mixture of C-6 and C-7 stereoisomeric products obtained in a 1:1 ratio, the final building block **61** was then attached through the same esterification conditions, providing the key substrate **86** ready for the olefin metathesis step.

Although success in this ring-closing metathesis reaction would, of course, provide the macrocyclic system needed to fashion the remainder of the natural product, the designed transformation had the potential of proving notable for other reasons. First, if successful, this ring closure would be accompanied by simultaneous traceless release of the desired product from the resin, meaning that no remnants of the original tether which united the epothilone scaffold to the polystyrene support would remain. This outcome would be in contrast to that of most conventional approaches in which some signature of the original tether usually endures, whether as a hydroxy group or as another functional handle.[75] More importantly, however, cleavage in this manner imparts a safety feature to this cyclorelease strategy: only material that is capable of undergoing metathesis will be freed from the resin. As a result, any precursor that had not reacted properly during a step leading to **86** would remain attached to the solid support, thereby ensuring that the products obtained from the metathesis step would not be contaminated with undesired by-products.

The designed strategy worked remarkably well as, upon treatment of precursor **86** with the same ruthenium catalyst **13** under slightly longer reaction times than in solution, cyclorelease from the resin was indeed effected, leading to the free epothilone A macrocycles as the expected mixture of four products (including **72**) in 52 % combined yield. These isomers could then be separated by high pressure liquid chromatography (HPLC), leaving only application of the same solution-phase steps of deprotection and epoxidation to provide the fully elaborated products. As such, a viable solid-phase synthesis of epothilone A had been developed to which the power of split-and-pool synthesis could now be applied. In this regard, one can consider steps such as the Wittig olefination with compounds of type **82** as an introduction of a diversity step, and subsequent operations such as the HF•py deprotection and Swern oxidation as pooled steps in terms of the picture described in Scheme 24. As matters transpired, upon incorporation of three sets of building blocks (of the types defined in Scheme 23), the strategy proved highly successful, affording a diverse set of structural congeners of epothilone A.[76]

Employing the above combinatorial strategy as well as conventional solution-based approaches, researchers around the world have synthesized hundreds of epothilone analogues. Their screening has established a strong foundation for an SAR profile. As summarized in Scheme 26, these studies reveal the inherent molecular

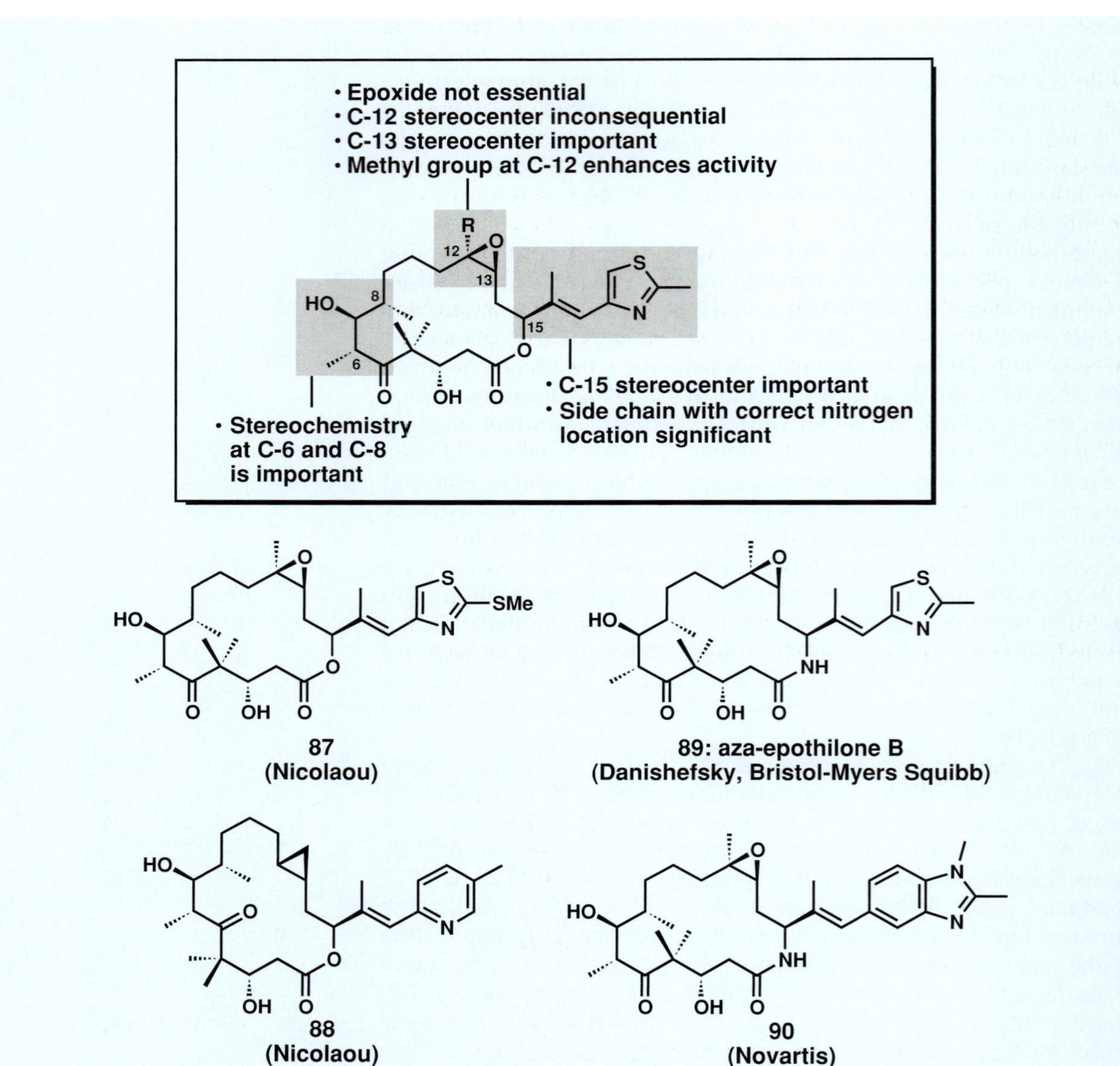

Scheme 26. Structure–activity relationships (SAR) for the epothilones and examples of novel analogues (**87**–**90**).

architecture of the epothilones as more or less essential for biological activity, although there are a few isolated examples of alternative stereochemistries being tolerated, such as compound **88**. Some of the more interesting and active families of analogues include those in which the thiazole side chain has been modified, as in **87**, and the aza-analogue of epothilone B (**89**, currently in clinical trials).[10a]

7.3.3 *Alternative Metathesis Approaches to the Epothilones*

Although the syntheses described above proved valuable in providing numerous analogues for chemical biology studies on the epothilone class, they also left room for improvement. Specifically, the relatively poor *Z*:*E* ratios obtained in the ring-closing metathesis reactions in both the solution- and solid-phase approaches to epothilone A (**1**) by Nicolaou and co-workers, a result corroborated by the additional systems and reaction conditions probed by the groups of Danishefsky and Schinzer (cf. Scheme 22), certainly define a practical limitation of the olefin metathesis reaction step in delivering the natural geometry at the C12–C13 junction. The same general problem was encountered in Danishefsky's route to epothilone B (**2**, Scheme 27) where an equimolar mixture of *E* and *Z* isomers still resulted despite the presence of an extra methyl group present at C-12 in **91** that could have potentially conferred greater selectivity in the olefin metathesis ring closure leading to **92**.[77] Similar findings were later disclosed by Grieco and May in a related approach to epothilone B using the same highly active catalyst (**11**) employed by the Danishefsky group.[78]

When such problems are encountered in synthesis, a golden opportunity exists to develop insightful and thought-provoking solutions. The recent successful metathesis-based approach by

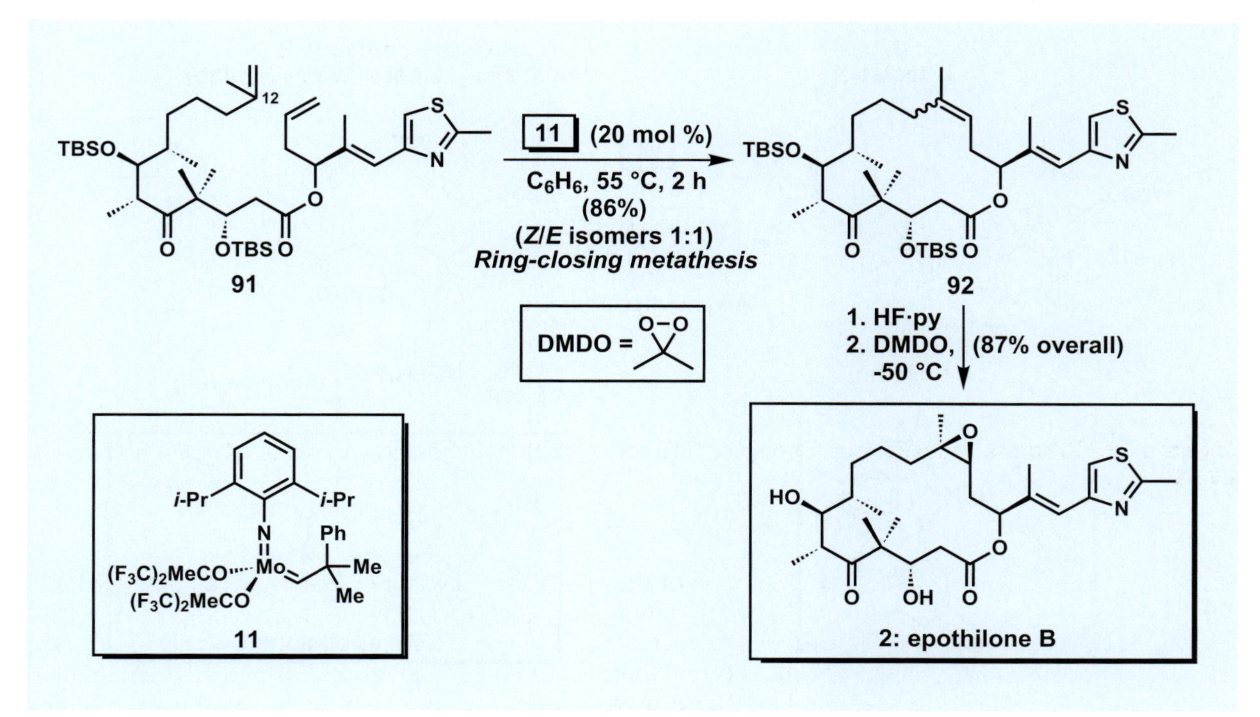

Scheme 27. Danishefsky's total synthesis of epothilone B (**2**) using molybdenum olefin-metathesis catalyst **11**.[67]

Fürstner and co-workers to epothilone C (**59**, Scheme 28), and therefore formal total synthesis of epothilone A (**1**), is one such example.[79] Over the course of the past five years, alkyne metathesis has become one of the most important extensions of the original olefin metathesis reaction for complex molecule construction.[80] First conceived of as a synthetic possibility by Katz in 1975[24a] and experimentally verified by Schrock through the development of an appropriate catalyst system in 1981,[81] the alkyne metathesis reaction in the epothilone context was intended by Fürstner to effect the formation of the C12–C13 alkyne **95** from precursor **93**. Assuming that such a rigid alkyne-containing macrocycle could be fashioned, its selective hydrogenation under Lindlar conditions (palladium on $BaSO_4$ poisoned with quinoline) would then be expected to afford the coveted *Z* isomer exclusively. Most pleasingly, this strategy proved rewarding as application of Fürstner's catalyst **94** (10 mol %)[82] to **93** in a solvent mixture of toluene and CH_2Cl_2 at 80 °C smoothly gave the desired alkyne metathesis product **95** in 80 % yield. With this adduct in hand, the anticipated hydrogenolysis and fluoride-induced deprotections of the TBS ethers proceeded without incident, ultimately leading to epothilone C (**59**) in 79 % yield as a single stereoisomer.

In addition to this elegant reformulation of the metathesis strategy toward the epothilones, an equally intriguing and challenging

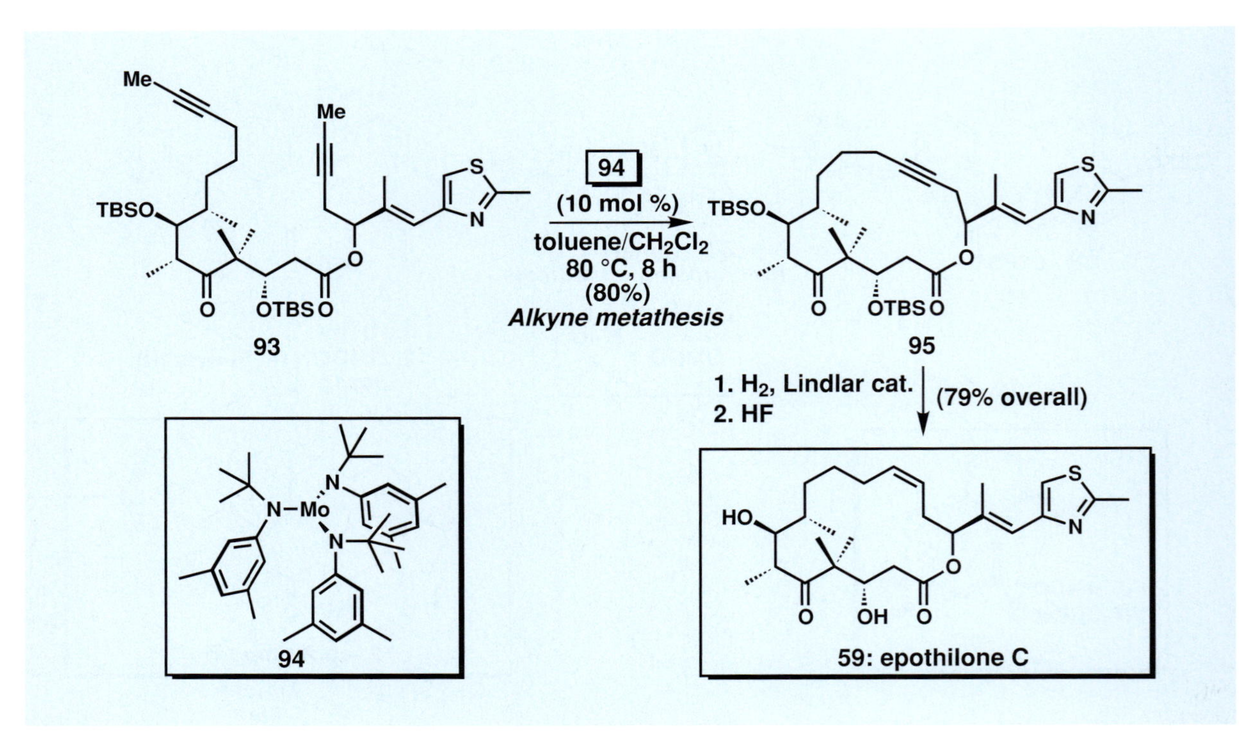

Scheme 28. Fürstner's total synthesis of epothilone C (**59**) using ring-closing alkyne metathesis with molybdenum-based catalyst **94**.[79]

metathesis-based proposition was advanced by Danishefsky and coworkers during their early efforts towards epothilone B (**2**) based on the premise of effecting a macrocyclic ring closure not at the relatively obvious C12–C13 site, but instead at a remote, unfunctionalized domain of the molecule using a substrate such as **96**. As delineated in Scheme 29, while such a strategy would most likely provide a mixture of C9–C10 *Z*:*E* isomers in the resultant product (**97**), this result could then be obviated by subsequent hydrogenation. Unfortunately, application of a variety of catalysts selected from the then-available repertoire, including **11** and **13**, to epoxide-bearing intermediate **96** failed to engender any productive metathesis, leading largely to polymerized and/or decomposed material. Several attempts with related structural congeners of this diolefinic starting material similarly failed. What is important to realize, however, is the overall approach was not suspect; rather, the right catalyst had not been developed at the time the original explorations were performed. The veracity of this statement was beautifully illustrated in 2002 when S. Sinha and J. Sun at The Scripps Research Institute revisited the same problem, this time employing the "second-generation" ruthenium alkylidene **16**.[83] Most gratifyingly, upon exposing **96** to 30 mol % of this catalyst in refluxing CH_2Cl_2, the previously recalcitrant ring-closing metathesis was smoothly brought about in an impressive 89 % yield after 48 hours. Subsequent hydrogenation of the resulting disubstituted olefin proceeded smoothly without

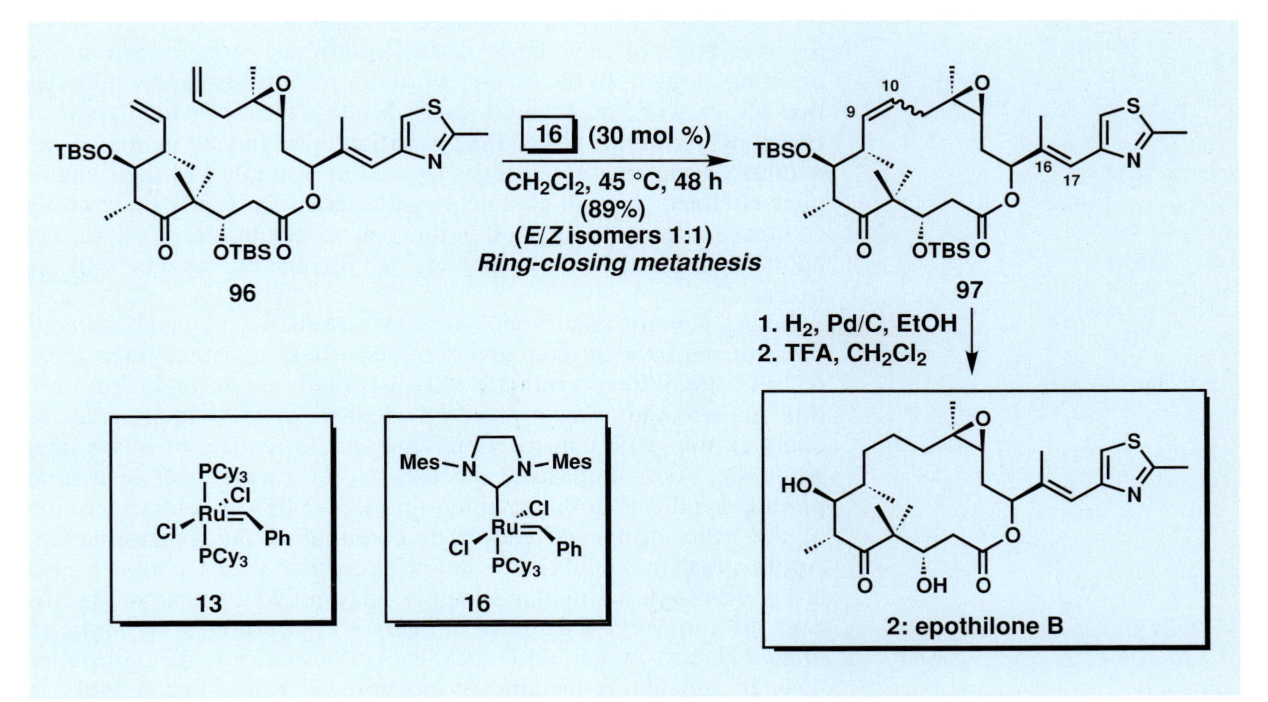

Scheme 29. Sinha's successful execution of the Danishefsky approach to synthesize epothilone B (**2**) by effecting macrocyclization at C9–C10 through ring-closing metathesis.[83]

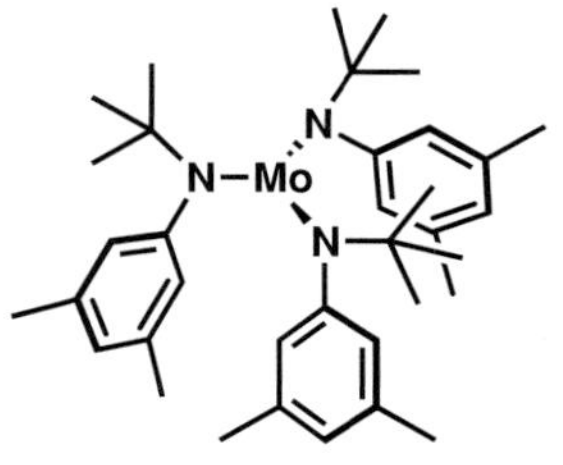

competitive reduction of the C16–C17 alkene, and upon TFA-induced cleavage of the TBS ethers, the desired natural product (**2**) was obtained.

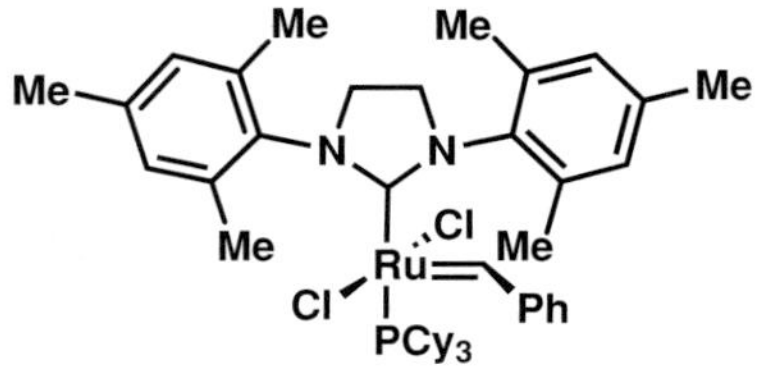

While both this synthesis and the previous example by Fürstner are certainly important for the epothilone field by providing stereospecific constructions of **1** and **2**, the perhaps more important feature is their beautiful illustration of the synergistic relationship between catalyst design and subsequent application in complex molecule synthesis. Key to the success of each approach was the development of a catalyst system that could reduce these creative metathesis designs to practice. As the scope of **94** and **16** as metathesis initiators is explored further, buttressed by the addition of more powerful catalysts than these systems, the range of metathesis-based applications in natural product synthesis should become even more spectacular.

7.4 Conclusion

Even though the numerous case studies of the olefin metathesis reaction defined in the beginning of this chapter are certainly illustrative of the value of this transformation to construct complex molecular frameworks, the collection of epothilone syntheses described in the latter half perhaps offers a more comprehensive display of the virtuosity of metathesis-based processes. In many ways, the relevance of metathesis to the epothilone story constitutes a defining moment in the history of olefin metathesis as the culmination of years of fundamental research into a clear practical application with profound biomedical ramifications. Indeed, some of the naturally occurring epothilones as well as a number of their analogues obtained through chemical synthesis are currently undergoing advanced clinical trials, and, in due course, the true promise for the epothilone family of compounds as therapeutic agents will be revealed.

Taking a more panoramic view of metathesis, as catalysts continue to improve in their levels of selectivity and reactivity, there is little doubt that synthetic chemists will increasingly embrace this transformation as a standard element in their approaches to complex molecule construction. One should realize, however, that while we have emphasized the use of olefin metathesis in natural product synthesis in this chapter, this aspect reflects only a fraction of the wide utility and enormous potential of the transformation. Applications in the fabrication of designed molecules, a topic hinted at here through a singular example, may prove even more significant in future investigations. In many ways, although we strive to mimic Nature in her efficiency to synthesize organic molecules, through the olefin metathesis reaction we now have a tool for which Nature appears to have no direct rival.

References

1. a) K. D. Rainsford, *Aspirin and the Salicylates*, Thetford Press, Thetford, **1984**, p. 335; b) G. Weissmann, *Sci. Am.* **1991**, *264*, 84.
2. a) J. C. Sheehan, *The Enchanted Ring: The Untold Story of Penicillin*, MIT Press, Cambridge, **1982**, p. 224; b) S. Selwyn, R. W. Lacey, M. Bakhtiar, *The Beta-Lactam Antibiotics: Penicillins and Cephalosporins in Perspective*, Hodder and Stoughton, London, **1980**, p. 364.
3. For some reviews of these two agents, see: a) K. C. Nicolaou, R. K. Guy, P. Potier, *Sci. Am.* **1996**, *274*, 94; b) K. C. Nicolaou, W. M. Dai, R. K. Guy, *Angew. Chem.* **1994**, *106*, 38; *Angew. Chem. Int. Ed. Engl.* **1994**, *33*, 15; c) K. C. Nicolaou, A. L. Smith, E. W. Yue, *Proc. Natl. Acad. Sci. U. S. A.* **1993**, *90*, 5881.
4. G. Höfle, N. Bedorf, K. Gerth, H. Reichenbach, DE-B 4,211,055, **1993** [*Chem. Abstr.* **1993**, *119*, 180598].
5. a) K. Gerth, N. Bedorf, G. Höfle, H. Irschik, H. Reichenbach, *J. Antibiot.* **1996**, *49*, 560; b) G. Höfle, N. Bedorf, H. Steinmetz, D. Schomburg, K. Gerth, H. Reichenbach, *Angew. Chem.* **1996**, *108*, 1671; *Angew. Chem. Int. Ed. Engl.* **1996**, *35*, 1567.
6. D. M. Bollag, P. A. McQueney, J. Zhu, O. Hensens, L. Koupal, J. Liesch, M. Goetz, E. Lazarides, C. M. Woods, *Cancer Res.* **1995**, *55*, 2325.
7. A. Balog, D. Meng, T. Kamenecka, P. Bertinato, D.-S. Su, E. J. Sorensen, S. J. Danishefsky, *Angew. Chem.* **1996**, *108*, 2976; *Angew. Chem. Int. Ed. Engl.* **1996**, *35*, 2801. For previous model studies, see: a) D. Meng, E. J. Sorensen, P. Bertinato, S. J. Danishefsky, *J. Org. Chem.* **1996**, *61*, 7998; b) P. Bertinato, E. J. Sorensen, D. Meng, S. J. Danishefsky, *J. Org. Chem.* **1996**, *61*, 8000.
8. Z. Yang, Y. He, D. Vourloumis, H. Vallberg, K. C. Nicolaou, *Angew. Chem.* **1997**, *109*, 170; *Angew. Chem. Int. Ed. Engl.* **1997**, *36*, 166. For earlier model studies, see: K. C. Nicolaou, Y. He, D. Vourloumis, H. Vallberg, Z. Yang, *Angew. Chem.* **1996**, *108*, 2554; *Angew. Chem. Int. Ed. Engl.* **1996**, *35*, 2399. For the full account of this work, see: K. C. Nicolaou, Y. He, D. Vourloumis, H. Vallberg, F. Roschangar, F. Sarabia, S. Ninkovic, Z. Yang, J. I. Trujillo, *J. Am. Chem. Soc.* **1997**, *119*, 7960.
9. D. Schinzer, A. Limberg, A. Bauer, O. M. Böhm, M. Cordes, *Angew. Chem.* **1997**, *109*, 543; *Angew. Chem. Int. Ed. Engl.* **1997**, *36*, 523.
10. a) K. C. Nicolaou, A. Ritzén, K. Namoto, *Chem. Commun.* **2001**, 1523; b) K.-H. Altmann, M. Wartmann, T. O'Reilly, *Biochim. Biophys. Acta* **2000**, *1470*, M79; c) J. Mulzer, *Monatsh. Chem.* **2000**, *131*, 205; d) K. C. Nicolaou, D. Hepworth, N. P. King, M. R. V. Finlay, *Pure & Appl. Chem.* **1999**, *71*, 989; e) C. R. Harris, S. J. Danishefsky, *J. Org. Chem.* **1999**, *64*, 8434; f) K. C. Nicolaou, F. Roschanger, D. Vourloumis, *Angew. Chem.* **1998**, *110*, 2120; *Angew. Chem. Int. Ed.* **1998**, *37*, 2014.
11. a) T. M. Trnka, R. H. Grubbs, *Acc. Chem. Res.* **2001**, *34*, 18; b) A. Fürstner, *Angew. Chem.* **2000**, *112*, 3140; *Angew. Chem. Int. Ed.* **2000**, *39*, 3012; c) R. H. Grubbs, S. Chang, *Tetrahedron* **1998**, *54*, 4413; d) M. Schuster, S. Blechert, *Angew. Chem.* **1997**, *109*, 2124; *Angew. Chem. Int. Ed. Engl.* **1997**, *36*, 2036; e) R. H. Grubbs, S. J. Miller, G. C. Fu, *Acc. Chem. Res.* **1995**, *28*, 446. For applications of olefin metathesis to polymer synthesis, see: R. H. Grubbs, W. Tumas, *Science* **1989**, *243*, 907.
12. H. S. Eleuterio, U. S. Patent 3,074,918, **1963** [*Chem. Abstr.* **1961**, *55*, 84720].
13. E. F. Peters, B. L. Evering, U. S. Patent 2,963,447, **1960** [*Chem. Abstr.* **1961**, *55*, 29435].
14. For some other early examples of metathesis in both the patent and the scientific literature, see: a) R. L. Banks, G. C. Bailey, *Ind. Eng. Chem. Prod. Res. Develop.* **1964**, *3*, 170; b) G. Natta, G. Dall'Asta, G. Mazzanti, *Angew. Chem.* **1964**, *76*, 765; *Angew. Chem. Int. Ed. Engl.* **1964**, *3*, 723; c) F. W. Michelotti, W. P. Keaveney, *J. Polym. Sci. Part A* **1965**, *3*, 895; d) G. Natta, G. Dall'Asta, L. Porri, *Makromol. Chem.* **1965**, *81*, 253; e) E. A. Zuech, W. B. Hughes, D. H. Kubicek, E. T. Kittleman, *J. Am. Chem. Soc.* **1970**, *92*, 528.
15. a) N. Calderon, H. Y. Chen, K. W. Scott, *Tetrahedron Lett.* **1967**, 3327; b) N. Calderon, E. A. Ofstead, J. P. Ward, W. A. Judy, K. W. Scott, *J. Am. Chem. Soc.* **1968**, *90*, 4133; c) N. Calderon, *Acc. Chem. Res.* **1972**, *5*, 127.
16. For some discussion of this topic, see: R. H. Grubbs, *Prog. Inorg. Chem.* **1978**, *24*, 1.
17. a) A. Clark, C. Cook, *J. Catal.* **1969**, *5*, 420; b) J. C. Mol, J. A. Moulijn, C. Boelhouwer, *J. Chem. Soc. Chem. Commun.* **1968**, 633; c) J. C. Mol, G. T. Visser, C. Boelhouwer, *J. Catal.* **1970**, *17*, 114; d) G. Dall'Asta, G. Motroni, *Eur. Polym. J.* **1971**, *7*, 707; e) G. Dall'Asta, G. Motroni, L. Motta, *J. Polym. Sci. Part A-1* **1972**, *10*, 1601.
18. a) C. P. C. Bradshaw, E. J. Howman, L. Turner, *J. Catal.* **1967**, *7*, 269; b) C. T. Adams, S. G. Brandenberger, *J. Catal.* **1969**, *13*, 360.
19. a) R. H. Grubbs, T. K. Brunck, *J. Am. Chem. Soc.* **1972**, *94*, 2538; b) C. G. Biefield, H. A. Eick, R. H. Grubbs, *Inorg. Chem.* **1973**, *12*, 2166.
20. E. O. Fischer, A. Maasböl, *Angew. Chem.* **1964**, *76*, 645; *Angew. Chem. Int. Ed. Engl.* **1964**, *3*, 580.
21. J.-L. Hérisson, Y. Chauvin, *Makromol. Chem.* **1970**, *141*, 161.
22. a) J. C. Mol, J. A. Moulijn, *Advan. Catal.* **1975**, *24*, 131; b) W. B. Hughes, *Organometal. Chem. Synth.* **1972**, *1*, 341; c) J. M. Basset, G. Coudurier, R. Mutin, H. Praliaud, Y. Trambouze, *J. Catal.* **1974**, *34*, 196; d) J. Levisalles, H. Rudler, D. Villemin, *J. Organomet. Chem.* **1975**, *87*, C7; e) N. Calderon in *Chemistry of Double-Bonded Functional Groups*,

Part 2 (Ed.: S. Patai), Wiley, New York, **1977**, pp. 913–964.

23. C. P. Casey, T. J. Burkhardt, *J. Am. Chem. Soc.* **1974**, *96*, 7808.
24. a) T. J. Katz, J. McGinnis, *J. Am. Chem. Soc.* **1975**, *97*, 1592; b) T. J. Katz, R. Rothchild, *J. Am. Chem. Soc.* **1976**, *98*, 2519; c) T. J. Katz, J. McGinnis, *J. Am. Chem. Soc.* **1977**, *99*, 1903; d) T. J. Katz, *Advan. Organomet. Chem.* **1977**, *16*, 283.
25. a) R. H. Grubbs, P. L. Burk, D. D. Carr, *J. Am. Chem. Soc.* **1975**, *97*, 3265; b) R. H. Grubbs, D. D. Carr, C. Hoppin, P. L. Burk, *J. Am. Chem. Soc.* **1976**, *98*, 3478.
26. See reference 23 and C. P. Casey, H. E. Tuinstra, M. C. Saeman, *J. Am. Chem. Soc.* **1976**, *98*, 608. In both of these studies, metal carbenes were combined with an excess of olefins, but no disproportionations were reported.
27. a) J. McGinnis, T. J. Katz, S. Hurwitz, *J. Am. Chem. Soc.* **1976**, *98*, 605; b) T. J. Katz, J. McGinnis, C. Altus, *J. Am. Chem. Soc.* **1976**, *98*, 606; c) T. J. Katz, S. J. Lee, N. Acton, *Tetrahedron Lett.* **1976**, *17*, 4247; d) T. J. Katz, N. Acton, *Tetrahedron Lett.* **1976**, *17*, 4251; e) S. J. Lee, J. McGinnis, T. J. Katz, *J. Am. Chem. Soc.* **1976**, *98*, 7818.
28. a) F. N. Tebbe, G. W. Parshall, D. W. Ovenall, *J. Am. Chem. Soc.* **1979**, *101*, 5074; b) F. N. Tebbe, G. W. Parshall, G. S. Reddy, *J. Am. Chem. Soc.* **1978**, *100*, 3611. For a review of this reagent, see: R. H. Grubbs, S. H. Pine in *Comprehensive Organic Synthesis*, *Vol. 5* (Eds.: B. M. Trost, I. Fleming), Pergamon, Oxford, **1991**, pp. 1115–1127.
29. a) R. R. Schrock, J. S. Murdzek, G. C. Bazan, J. Robbins, M. DiMare, M. O'Regan, *J. Am. Chem. Soc.* **1990**, *112*, 3875; b) G. C. Bazan, E. Khosravi, R. R. Schrock, W. J. Feast, V. C. Gibson, M. B. O'Regan, J. K. Thomas, W. M. Davis, *J. Am. Chem. Soc.* **1990**, *112*, 8378; c) G. C. Bazan, J. H. Oskam, H. N. Cho, L. Y. Park, R. R. Schrock, *J. Am. Chem. Soc.* **1991**, *113*, 6899. This catalyst family was first introduced in the following seminal paper: C. J. Schaverien, J. C. Dewan, R. R. Schrock, *J. Am. Chem. Soc.* **1986**, *108*, 2771. For a related catalyst design applied to tungsten, see: F. Quignard, M. Leconte, J.-M. Basset, *Chem. Commun.* **1985**, 1816.
30. For a useful starting point, see: R. R. Schrock, *Tetrahedron* **1999**, *55*, 8141.
31. S. T. Nguyen, L. K. Johnson, R. H. Grubbs, J. W. Ziller, *J. Am. Chem. Soc.* **1992**, *114*, 3974.
32. a) P. Schwab, M. B. France, J. W. Ziller, R. H. Grubbs, *Angew. Chem.* **1995**, *107*, 2179; *Angew. Chem. Int. Ed. Engl.* **1995**, *34*, 2039; b) P. Schwab, R. H. Grubbs, J. W. Ziller, *J. Am. Chem. Soc.* **1996**, *118*, 100.
33. J. B. Alexander, D. S. La, D. R. Cefalo, A. H. Hoveyda, R. R. Schrock, *J. Am. Chem. Soc.* **1998**, *120*, 4041. For a recent review of this area of research, see: A. H. Hoveyda, R. R. Schrock, *Chem. Eur. J.* **2001**, *7*, 945.
34. T. Weskamp, W. C. Schattenmann, M. Spiegler, W. A. Herrmann, *Angew. Chem.* **1998**, *110*, 2631; *Angew. Chem. Int. Ed.* **1998**, 37, 2490.
35. a) E. L. Dias, S. T. Nguyen, R. H. Grubbs, *J. Am. Chem. Soc.* **1997**, *119*, 3887; b) J. A. Tallarico, P. J. Bonitatebus, M. L. Snapper, *J. Am. Chem. Soc.* **1997**, *119*, 7157.
36. a) M. Scholl, S. Ding, C. W. Lee, R. H. Grubbs, *Org. Lett.* **1999**, *1*, 953; b) M. Scholl, T. M. Trnka, J. P. Morgan, R. H. Grubbs, *Tetrahedron Lett.* **1999**, *40*, 2247. For related studies with similar ligands concurrent to this work, see: c) J. Huang, E. D. Stevens, S. P. Nolan, J. L. Petersen, *J. Am. Chem. Soc.* **1999**, *121*, 2674 and references therein; d) L. Ackermann, A. Fürstner, T. Weskamp, F. J. Kohl, W. A. Herrmann, *Tetrahedron Lett.* **1999**, *40*, 4787 and references therein. One should note, however, that preeminence in terms of discovery for this final group of catalysts constitutes a fierce argument; the issue is raised and discussed in the following review article: W. A. Hermann, *Angew. Chem.* **2002**, *114*, 1342; *Angew. Chem. Int. Ed.* **2002**, *41*, 1290.
37. For some early studies with this catalyst, see: a) A. K. Chatterjee, R. H. Grubbs, *Org. Lett.* **1999**, *1*, 1751; b) A. K. Chatterjee, J. P. Morgan, M. Scholl, R. H. Grubbs, *J. Am. Chem. Soc.* **2000**, *122*, 3783.
38. Phillips Petroleum Company, *Hydrocarbon Process* **1967**, *46*, 232.
39. R. F. Ohm, *Chemtech* **1980**, *10*, 183.
40. a) D. Villemin, *Tetrahedron Lett.* **1980**, *21*, 1715; b) J. Tsuji, S. Hashiguchi, *Tetrahedron Lett.* **1980**, *21*, 2955.
41. a) Z. Xu, C. W. Johannes, A. F. Houri, D. S. La, D. A. Cogan, G. E. Hofilena, A. H. Hoveyda, *J. Am. Chem. Soc.* **1997**, *119*, 10302; b) Z. Xu, C. W. Johannes, S. S. Salman, A. H. Hoveyda, *J. Am. Chem. Soc.* **1996**, *118*, 10926; c) A. F. Houri, Z. Xu, D. A. Cogan, A. H. Hoveyda, *J. Am. Chem. Soc.* **1995**, *117*, 2943.
42. A. Fürstner, T. Müller, *J. Org. Chem.* **1998**, *63*, 424.
43. a) D. P. Larson, C. H. Heathcock, *J. Org. Chem.* **1996**, *61*, 5208; b) D. P. Larson, C. H. Heathcock, *J. Org. Chem.* **1997**, *62*, 8406; For the first total synthesis of tricolorin A, see: c) S.-F. Lu, Q. O'yang, Z.-W. Guo, B. Yu, Y.-Z. Hui, *Angew. Chem.* **1997**, *109*, 2442; *Angew. Chem. Int. Ed. Engl.* **1997**, *36*, 2344; d) S.-F. Lu, Q. O'yang, Z.-W. Guo, B. Yu, Y.-Z. Hui, *J. Org. Chem.* **1997**, *62*, 8400.
44. For representative studies, see: a) A. Fürstner, K. Langemann, *J. Org. Chem.* **1996**, *61*, 3942; b) A. Fürstner, *Synlett* **1999**, 1523. Early expressions of this idea can be found in reference 11e and citations therein.
45. See reference 11 and S. E. Gibson, S. P. Keen, *Top. Organo. Chem.* **1998**, *1*, 155.
46. For representative references, see: a) J. P. Mackay, U. Gerhard, D. A. Beauregard, R. A. Maplestone, D. H. Williams, *J. Am. Chem. Soc.* **1994**, *116*, 4573; b) P. J. Loll, A. E. Bevivino, B. D. Korty, P. H. Axel-

sen, *J. Am. Chem. Soc.* **1997**, *119*, 1516; c) T. Staroske, D.P. O'Brien, T.J.D. Jørgensen, P. Roepstorff, D.H. Williams, A.J.R. Heck, *Chem. Eur. J.* **2000**, *6*, 504; d) S.J. Sucheck, A.L. Wong, K.M. Koeller, D.D. Boehr, K. Draker, P. Sears, G.D. Wright, C.-H. Wong, *J. Am. Chem. Soc.* **2000**, *122*, 5230; e) D. Bradley, *Drug Discovery Today* **2000**, *5*, 44; f) M. Mammen, S.-K. Choi, G.M. Whitesides, *Angew. Chem.* **1998**, *110*, 2908; *Angew. Chem. Int. Ed.* **1998**, 37, 2754.
47. S.T. Diver, S.L. Schreiber, *J. Am. Chem. Soc.* **1997**, *119*, 5106.
48. K.C. Nicolaou, R. Hughes, S.Y. Cho, N. Winssinger, H. Labischinski, R. Endermann, *Chem. Eur. J.* **2001**, *7*, 3824.
49. a) D.H. Williams, A.J. Maguire, W. Tsuzuki, M.S. Westwell, *Science* **1998**, *280*, 711; b) U. Gerhard, J.P. Mackay, R.A. Maplestone, D.H. Williams, *J. Am. Chem. Soc.* **1993**, *115*, 232; c) J.P. Mackay, U. Gerhard, D.A. Beauregard, M.S. Westwell, M.S. Searle, D.H. Williams, *J. Am. Chem. Soc.* **1994**, *116*, 4581.
50. For an entry to this field, see the following articles and references cited therein: a) J.-M. Lehn, A.V. Eliseev, *Science* **2001**, *291*, 2331; b) G.R.L. Cousins, R.L.E. Furlan, Y.-F. Ng, J.E. Redman, J.K.M. Sanders, *Angew. Chem.* **2001**, *113*, 437; *Angew. Chem. Int. Ed.* **2001**, *40*, 423; c) T. Giger, M. Wigger, S. Audétat, S.A. Benner, *Synlett* **1998**, 688; d) H. Hioki, W.C. Still, *J. Org. Chem.* **1998**, *63*, 904.
51. K. Arakawa, T. Eguchi, K. Kakinuma, *J. Org. Chem.* **1998**, *63*, 4741.
52. T.D. Clark, M.R. Ghadiri, *J. Am. Chem. Soc.* **1995**, *117*, 12364.
53. For examples from some specific laboratories, see: a) G.M. Whitesides, E.E. Simanek, J.P. Mathias, C.T. Seto, D.N. Chin, M. Mammen, D.M. Gordon, *Acc. Chem. Res.* **1995**, *28*, 37; b) S. Anderson, H.L. Anderson, J.K.M. Sanders, *Acc. Chem. Res.* **1993**, *26*, 469 and references cited therein.
54. a) A.B. Smith, C.M. Adams, S.A. Kozmin, D.V. Paone, *J. Am. Chem. Soc.* **2001**, *123*, 5925; b) A.B. Smith, C.M. Adams, S.A. Kozmin, *J. Am. Chem. Soc.* **2001**, *123*, 990.
55. K.C. Nicolaou, M.H.D. Postema, C.F. Claiborne, *J. Am. Chem. Soc.* **1996**, *118*, 1565.
56. K.C. Nicolaou, M.H.D. Postema, E.W. Yue, A. Nadin, *J. Am. Chem. Soc.* **1996**, *118*, 10335.
57. M. Hirama, T. Oishi, H. Uehara, M. Inoue, M. Maruyama, H. Oguri, M. Satake, *Science* **2001**, *294*, 1904.
58. R. Stragies, S. Blechert, *Tetrahedron* **1999**, *55*, 8179.
59. For some discussion, see: *Topics in Organometallic Chemistry, Vol. 1* (Ed.: A. Fürstner), Springer-Verlag, Berlin, **1998**, p. 231. For a representative application, see: M. Mori, T. Kitamura, N. Sakakibara, Y. Sato, *Org. Lett.* **2000**, *2*, 543.
60. W.J. Zuercher, M. Scholl, R.H. Grubbs, *J. Org. Chem.* **1998**, *63*, 4291. For earlier examples of the same related concept, see: S.-H. Kim, W.J. Zuercher, N.B. Bowden, R.H. Grubbs, *J. Org. Chem.* **1996**, *61*, 1073.
61. T.J. Katz, T.M. Sivavec, *J. Am. Chem. Soc.* **1985**, *107*, 737.
62. G.C. Fu, S.T. Nguyen, R.H. Grubbs, *J. Am. Chem. Soc.* **1993**, *115*, 9856.
63. W. Oppolzer, R. Moretti, S. Thomi, *Tetrahedron Lett.* **1989**, *30*, 5603.
64. For a general review of this area of research, see: a) H.C. Brown, P.K. Jadhav, B. Singram, *Mod. Synth. Methods* **1986**, *4*, 307; b) P.V. Ramachandran, *Aldrichimica Acta* **2002**, *35*, 23.
65. U.S. Racherla, H.C. Brown, *J. Org. Chem.* **1991**, *56*, 401.
66. M.W. Bredenkamp, C.W. Holzapfel, W.J. van Zyl, *Synth. Commun.* **1990**, *20*, 2235.
67. D. Meng, P. Bertinato, A. Balog, D-S. Su, T. Kamenecka, E.J. Sorensen, S.J. Danishefsky, *J. Am. Chem. Soc.* **1997**, *119*, 10073.
68. D. Schinzer, A. Bauer, O.M. Böhm, A. Limberg, M. Cordes, *Chem. Eur. J.* **1999**, *5*, 2483.
69. For a general entry to this field, see: *Handbook of Combinatorial Chemistry: Drugs, Catalysts, Methods, Vol. 1 and 2* (Eds.: K.C. Nicolaou, R. Hanko, W. Hartwig), Wiley-VCH, Weinheim, **2002**, p. 1114.
70. R.B. Merrifield, *J. Am. Chem. Soc.* **1963**, *85*, 2149.
71. R.B. Merrifield, *Angew. Chem.* **1985**, *97*, 801; *Angew. Chem. Int. Ed. Engl.* **1985**, *24*, 799.
72. Á. Furka, F. Sebestyén, M. Asgedmo, G. Dibó, *Int. J. Peptide Prot. Res.* **1991**, *37*, 487.
73. For representative primary literature, see: a) K.S. Lam, S.E. Salmon, E.M. Hersh, V.J. Hruby, W.M. Kazmierski, R.J. Knapp, *Nature* **1991**, *354*, 82; b) R.A. Houghten, C. Pinilla, S.E. Blondelle, J.R. Appel, C.T. Dooley, J.H. Cuervo, *Nature* **1991**, *354*, 84; c) K.C. Nicolaou, X.-Y. Xiao, Z. Parandoosh, A. Senyei, M.P. Nova, *Angew. Chem.* **1995**, *107*, 2476; *Angew. Chem. Int. Ed. Engl.* **1995**, *34*, 2289; d) A. Studer, S. Hadida, R. Ferritto, S.-Y. Kim, P. Jeger, P. Wipf, D.P. Curran, *Science* **1997**, *275*, 823. For some secondary sources, see: e) T. Krämer, V.V. Antonenko, R. Mortezaei, N.V. Kulikov in *Handbook of Combinatorial Chemistry: Drugs, Catalysts, Methods, Vol. 1* (Eds.: K.C. Nicolaou, R. Hanko, W. Hartwig), Wiley-VCH, Weinheim, **2002**, pp. 170–189; f) D.J. Gravert, K.D. Janda, *Chem. Rev.* **1997**, *97*, 489.
74. K.C. Nicolaou, N. Winssinger, J. Pastor, S. Ninkovic, F. Sarabia, D. Vourloumis, Z. Yang, T. Li, P. Giannakakou, E. Hamel, *Nature* **1997**, *387*, 268.
75. S. Bräse, S. Dahmen in *Handbook of Combinatorial Chemistry: Drugs, Catalysts, Methods, Vol. 1* (Eds.: K.C. Nicolaou, R. Hanko, W. Hartwig), Wiley-VCH, Weinheim, **2002**, pp. 59–169.
76. K.C. Nicolaou, D. Vourloumis, T. Li, J. Pastor, N. Winssinger, Y. He, S. Ninkovic, F. Sarabia, H. Vallberg, F. Roschangar, N.P. King, M.R.V. Finlay, P. Giannakakou, P. Verdier-Pinard, E. Hamel, *Angew.*

Chem. **1997**, *109*, 2181; *Angew. Chem. Int. Ed. Engl.* ***1997***, *36*, 2097. For additional library construction by parallel solution-phase synthesis, see: K. C. Nicolaou, H. Vallberg, N. P. King, F. Roschangar, Y. He, D. Vourloumis, C. G. Nicolaou, *Chem. Eur. J.* **1997**, *3*, 1957.
77. D. Meng, D.-S. Su, A. Balog, P. Bertinato, E. J. Sorensen, S. J. Danishefsky, Y.-H. Zheng, T.-C. Chou, L. He, S. B. Horwitz, *J. Am. Chem. Soc.* **1997**, *119*, 2733.
78. S. A. May, P. A. Grieco, *Chem. Commun.* **1998**, 1597.
79. A. Fürstner, C. Mathes, K. Grela, *Chem. Commun.* **2001**, 1057.
80. U. H. F. Bunz, L. Kloppenburg, *Angew. Chem.* **1999**, 111, 503; *Angew. Chem. Int. Ed.* **1999**, *38*, 478.
81. J. H. Wengrovius, J. Sancho, R. R. Schrock, *J. Am. Chem. Soc.* **1981**, *103*, 3932.
82. A. Fürstner, C. Mathes, C. W. Lehmann, *J. Am. Chem. Soc.* **1999**, *121*, 9453.
83. J. Sun, S. C. Sinha, *Angew. Chem.* **2002**, *114*, 1439; *Angew. Chem. Int. Ed.* **2002**, *41*, 1381.

1: manzamine A

8

J. D. Winkler (1998)
S. F. Martin (1999)

Manzamine A

8.1 Introduction

While our abilities as synthetic chemists have sharpened considerably over the past few decades to the point where the total synthesis of most natural products can be achieved with existing methodology, every so often a construct is isolated from Nature's library of chemical diversity whose architecture defies conventional synthetic techniques. Whether it is their possession of a novel molecular motif or a particularly challenging array of stereochemical elements, these natural products serve as an engine that drives the general field of organic synthesis forward by forcing the practitioner to devise new synthetic methods and insightful strategies to effect their construction.

Key concepts:

- **Photoaddition/ fragmentation/ Mannich cyclizations**
- **Ring-closing metathesis**
- **Tandem Stille/Diels–Alder reactions**

In 1986, the first member of a family of such structures, manzamine A (**1**), was isolated from marine sponges of the *Haliclona* and *Pellina* genera off the coast of Okinawa and fully characterized due to its attractive cytotoxicity profile in several preliminary assays.[1] While such biological activity suggests that investing time, money, and effort in an endeavor towards the total synthesis of **1** would likely provide eventual dividends in the fields of cancer research and chemical biology, the unique molecular architecture of this target would seem to guarantee an immediate and perhaps greater return in terms of fundamental method development in chemical synthesis. Indeed, with an unprecedented pentacyclic core comprised of 13- and 8-membered rings attached to a central pyrrolo[2,3-*i*]isoquinoline system and an array of five stereocenters, this molecule would appear to possess a level of complexity capable of thwarting any orthodox plan of attack to effect its assembly. To make matters worse, although several related structural congeners of **1**, such as manzamine B (**2**, Scheme 1), were subsequently reported in the literature,[2] for several years following their isolation there seemed to be no clear biogenetic route that could account for the construction of these compounds in Nature. As such, synthetic practitioners had few real clues to guide their design of potential schemes to **1**.

In 1992, however, this bleak situation began to brighten when Sir Jack Baldwin and his co-worker Roger Whitehead at the University of Oxford advanced an intriguing biosynthetic hypothesis which could account, on paper, for the formation of the polycyclic framework of virtually all the manzamine alkaloids.[3] In their elegant and insightful conception, which is shown in Scheme 1 for manzamine B (**2**), initial retrosynthetic excision of the β-carboline

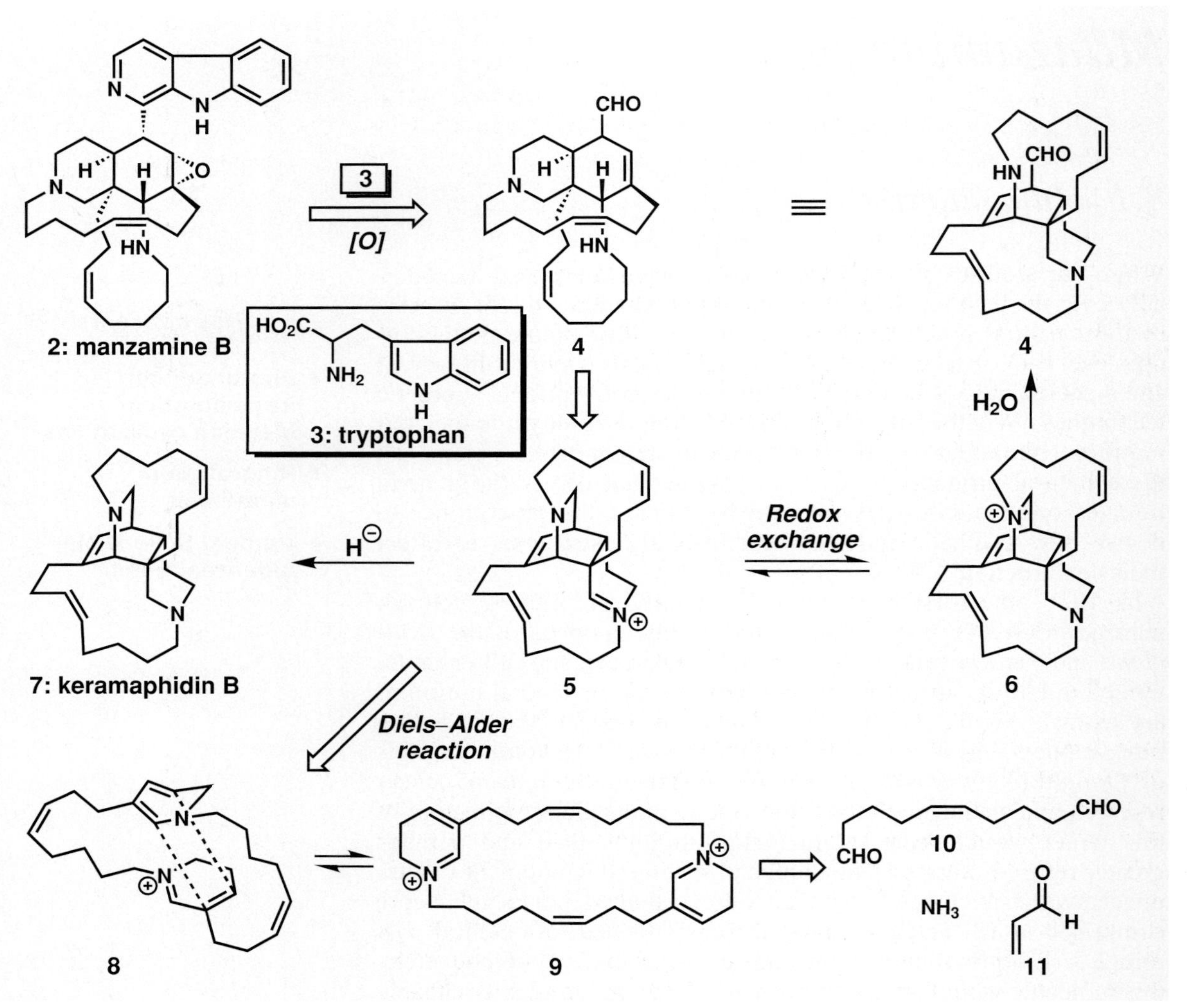

Scheme 1. The Baldwin–Whitehead hypothesis for the biogenesis of manzamine B (**2**) and related alkaloids.

portion as the amino acid tryptophan (**3**) would lead back to aldehyde **4**, an intermediate that could then be simplified to **5** through the loss of a molecule of water and a redox exchange event. While these initial simplifications certainly seem reasonable on the basis of first principles, their validity is significantly enhanced by the fact that the reduced form of **5** and **6** has the same structure as the marine-derived natural product keramaphidin B (**7**). With the manzamine problem now redefined as the synthesis of **5**, careful evaluation of this intermediate suggested that it could arise from an *endo* selective intramolecular Diels–Alder reaction of bishydropyridine salt **8**. This compound, in turn, could be derived, via its tautomeric form **9**, from the union of two molecules each of the symmetrical dialdehyde **10**, acrolein (**11**), and ammonia.[4] As such, this analysis has uncovered a hidden symmetry in the architecture of the manzamine alkaloids that had previously escaped detection.

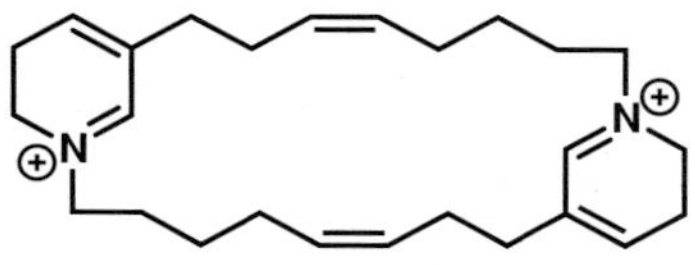

While certainly appealing, the critical element required to establish the validity of any biosynthetic hypothesis is substantive proof. In 1999, the Baldwin group obtained the first empirical evidence for their proposal by achieving a total synthesis of keramaphidin B (**7**) based on the intramolecular Diels–Alder reaction at the heart of their biogenetic scheme.[5] As shown in Scheme 2, after preparing key intermediate **9** in just eight synthetic operations from phosphorane **12**, dissolution of this compound in buffered aqueous methanol for one hour followed by treatment with $NaBH_4$ at −78 °C provided a small, but detectable, amount of keramaphidin B (**7**), presumably through the intermediacy of **8** and **5**. Although the yield for the final steps is quite low, it is reasonable to assume that in Nature there is enzymatic assistance, either in the form of a "Diels–Alderase" or a species that merely encapsulates **8**, which could greatly improve the outcome by limiting the conformational mobility of this reactive intermediate and/or directing its motifs into proper alignment for pericyclic reaction.[6] Such participation would appear to be crucial for the legitimacy of this biogenetic proposal, as in its absence the major reaction product observed by Baldwin and his co-workers was **15**, the result of preferential reduction of both iminium ions in **9** instead of productive Diels–Alder cycloaddition.

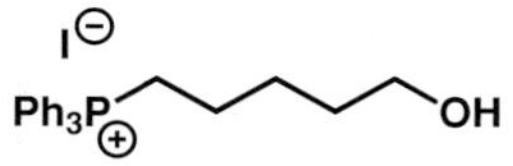

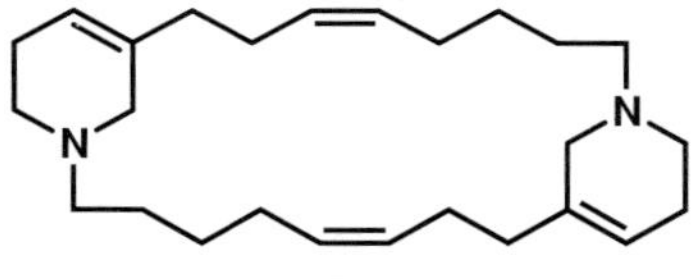

9

$Ph_3P^{\oplus}$ $I^{\ominus}$ OH

12

15

Although Baldwin and Whitehead's perceptive biosynthetic hypothesis serves as a useful framework from which to develop a coherent synthetic approach towards manzamine A (**1**), other elegant solutions to the problems posed by this complex molecular architecture based on stepwise chemical synthesis are also conceivable. In this chapter, we shall focus in detail on two such constructions of manzamine A from the groups of Jeffrey Winkler at the University of Pennsylvania[7] and Stephen Martin at the University of Texas, Austin.[8] While both serve to illustrate the current state-of-the-art in methods for complex molecule construction through routes that effectively combine modern and classical reactions, perhaps the more significant feature of each approach resides in the

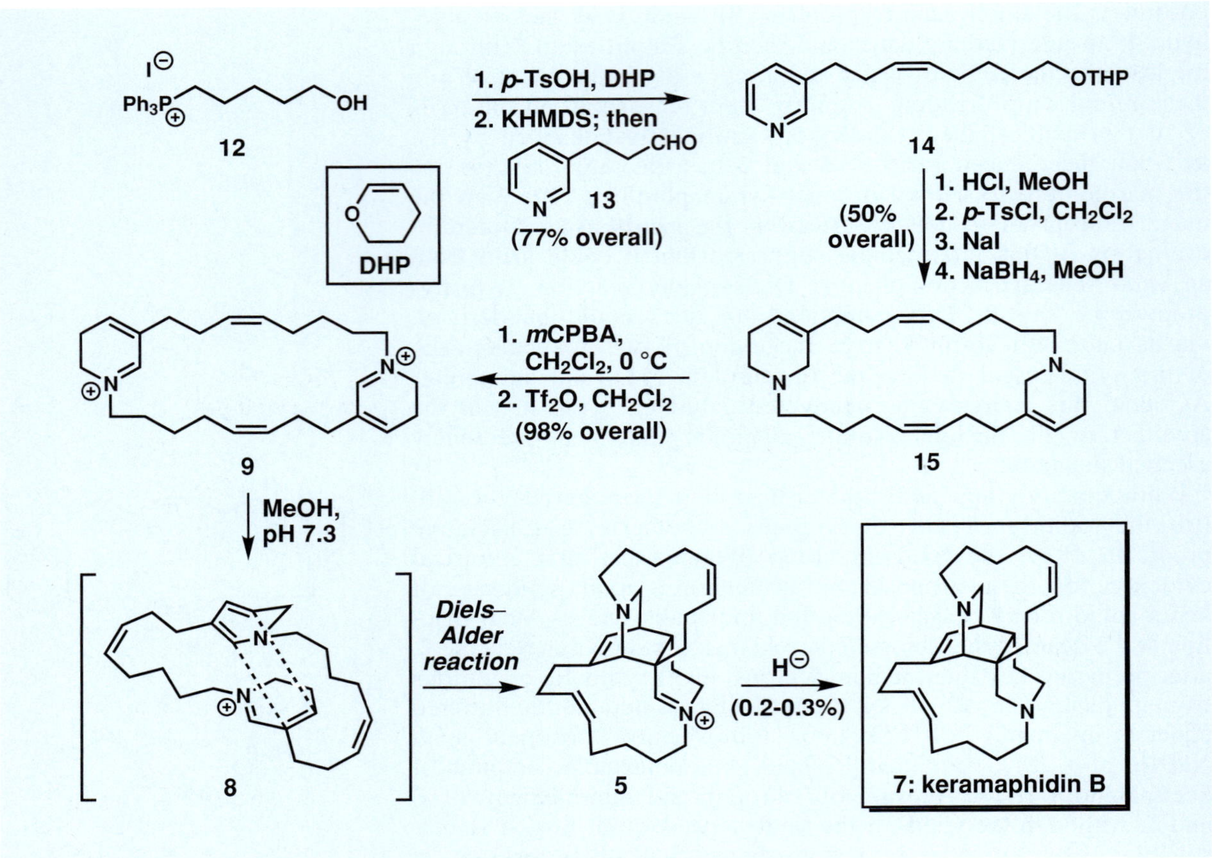

Scheme 2. Biomimetic total synthesis of keramaphidin B (**7**). (Baldwin et al., 1999)[5]

successful development of a novel and instructive cascade sequence to handle the complex stereochemical requirements of the target's central core.

8.2 Retrosynthetic Analysis and Strategy

8.2.1 Winkler's Synthetic Approach to Manzamine A (1)

Apart from the biogenetic scheme described above, the only other useful piece of information that could direct the development of synthetic routes to manzamine A was provided by the Kobayashi group who demonstrated that the related natural product ircinal A (**17**, Scheme 3) could be converted into manzamine A (**1**) through initial condensation with tryptamine (**16**) and subsequent acid-induced Pictet–Spengler cyclization (see Chapter 5 for a discussion of this reaction), followed by DDQ-mediated oxidation to complete the fully aromatic β-carboline system.[9] Although this discovery

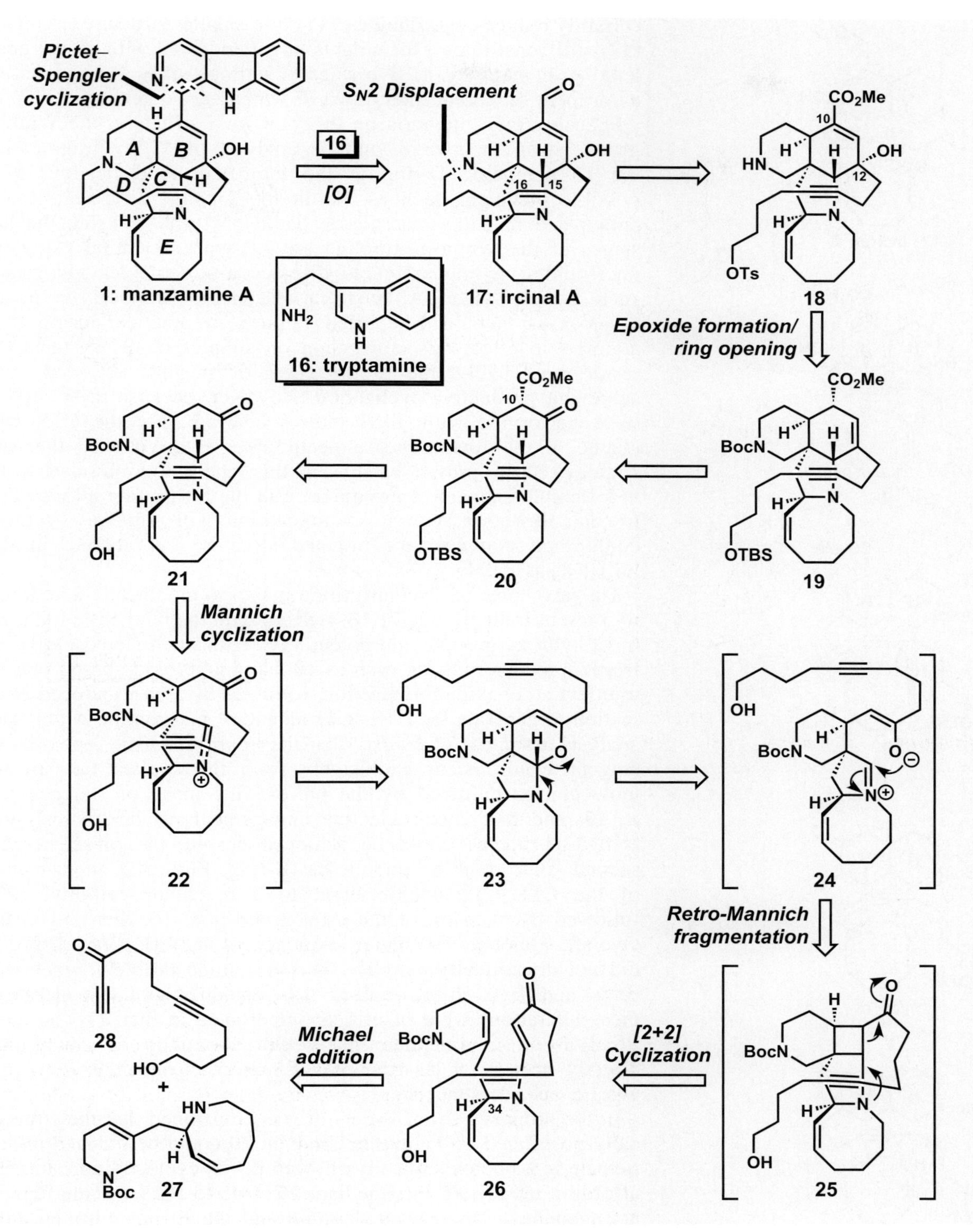

Scheme 3. Winkler's retrosynthetic analysis of manzamine A (**1**).

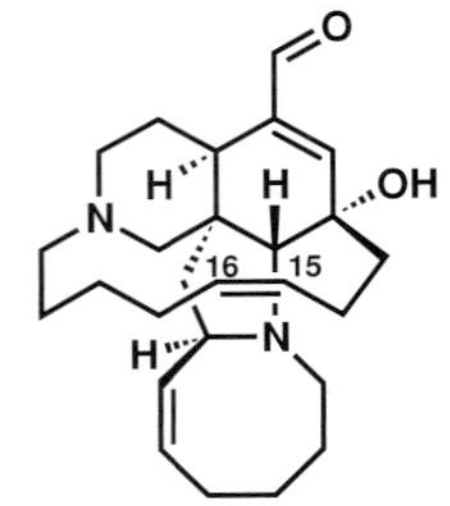

17: ircinal A

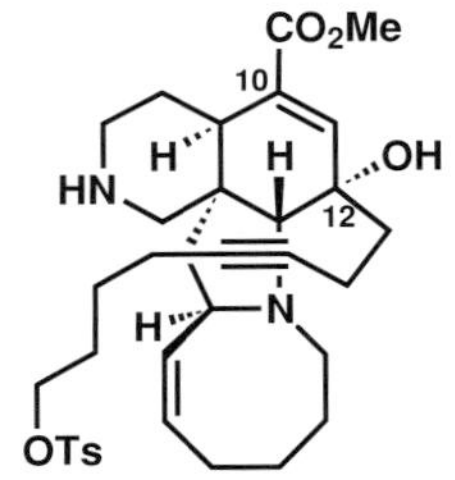

18

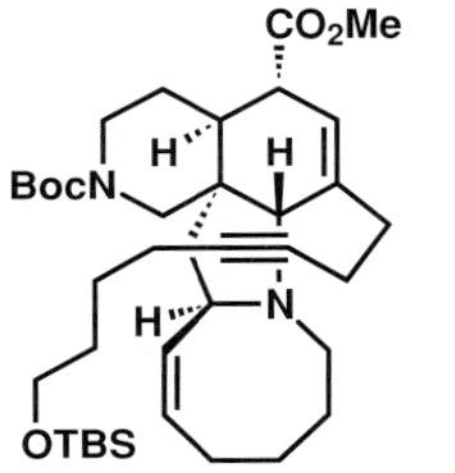

19

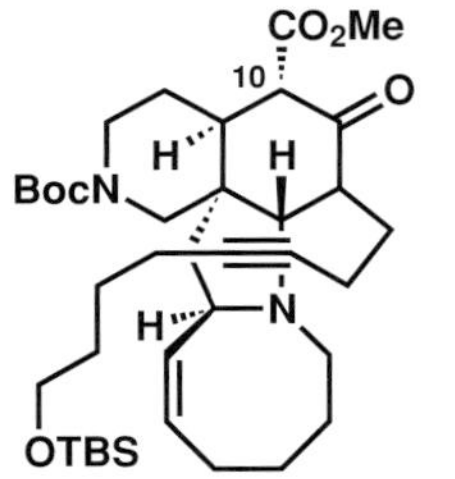

20

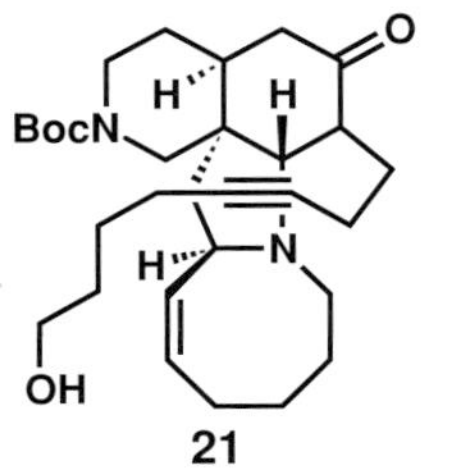

21

certainly reduces manzamine A (**1**) to a smaller structure, ircinal A (**17**) still constitutes a formidable target endowed with all the challenging elements of **1**: five rings of various sizes and a daunting assortment of stereocenters, two of which are quaternary.

Focusing their attention on this new goal structure, the Winkler group began their retrosynthetic analysis of **17** by fragmenting the 13-membered D-ring at the indicated bond, leading to a possible precursor such as **18**. In the forward direction, it was anticipated that this macrocycle could be formed during the late stages of the synthesis through an S_N2-type alkylation using the nucleophilic A-ring amine to displace a suitable leaving group, such as a tosylate. Although a reasonable proposal on paper, other researchers had reported failures in practice during their attempts to effect this conversion on simpler model systems,[10] a result most likely due to the energetic penalty accessed upon achieving productive cyclization. However, because these studies were performed with alkyl tethers bearing only the C15–C16 alkene of the final target molecule, it was envisioned that this strategy could prove feasible if this chain incorporated a far less flexible element of unsaturation in the form of an alkyne. Following alkylation, partial hydrogenation with Lindlar's catalyst could then complete the targeted structure (**17**) based on this initial disassembly.

The next spate of disconnections sought to modify the functionality present in the B-ring of **18**, starting with the recognition that the C-12 hydroxy group could potentially be installed stereoselectively from a precursor alkene such as **19** using a peroxide-based reagent to effect an epoxidation reaction, followed by a base-induced elimination to generate the C10–C11 alkene of manzamine A (**1**). This analysis assumed, of course, that the approach of the epoxidation reagent would occur exclusively from the convex face of the molecule, as enforced by the pre-existing array of stereocenters in **19**, and that chemoselective engagement of the trisubstituted C10–C11 alkene could be achieved despite the presence of a second C–C double bond in the E-ring. From **19**, modification of the C11–C12 double bond to a precursor carbonyl (**20**), followed by excision of the ester group at C-10, then led to **21**. Overall, although this latter sequence of modifications from **18** did not dramatically simplify the manzamine problem, since four rings and five chiral centers still remained for consideration, these alterations were of critical importance in that **21** was identified as an intermediate that could potentially be constructed directly from a far less complex compound, triene **26**, through a unique cascade sequence.

In the proposed set of events, it was envisioned that the strategically positioned C–C double bonds in **26** could be induced to participate in a photochemically allowed [2+2] cycloaddition, thereby affording tetracyclic intermediate **25**. Due to the particular spatial arrangement of the carbonyl group and the E-ring nitrogen atom in this strained product, this new compound was not expected to prove isolable, but instead to spontaneously undergo a retro-Mannich fragmentation that would afford ketoiminium intermediate **24**, an entity that would then likely cyclize *in situ* to aminal **23**. Finally, it was anticipated that suitable adjustment of the reaction media (whether in terms of solvent, temperature, or pH) could then effect the conversion of **23** into **21**, via **22**, through a Mannich cyclization. If this appealing and highly elegant strategy could be put into practice, then this cascade sequence would achieve the concomitant formation of two dissimilar rings and cast the complete ABCE tetracyclic core of manzamine A. Of equal significance, this approach would result in the orchestrated installation of four new stereocenters with absolute control as conducted by the stereogenic center in the starting material **26**.

Achieving success in this daring endeavor, however, was far from guaranteed. Although the Winkler group, among others, has employed similar photoaddition/fragmentation/Mannich cyclization sequences to access diverse polycyclic architectures, the present context would pose the most rigorous test of the overall reaction process to date.[11] Perhaps most troublesome was the fact that no precedent existed to suggest that tertiary vinylogous amides such as **26** could be employed in the initial [2+2] photoaddition necessary to set the cascade into motion. Indeed, all previous examples of this chemistry had relied exclusively on starting materials that contained secondary vinylogous amides, compounds well-precedented to participate in intramolecular [2+2] cyclizations because the hydrogen atom of the amide can provide internal hydrogen bonding to stabilize the excited state of the acyclic chromophore. Since a tertiary vinylogous amide lacks this critical hydrogen, the outcome of the overall sequence was potentially in jeopardy.

Nevertheless, from a retrosynthetic perspective it seemed relatively easy to test this approach since it was envisioned that **26** could arise from a Michael-type addition of the basic nitrogen atom in **27** to the α,β-unsaturated system of alkyne **28**. Based on this final disassembly, the manzamine problem has been dramatically simplified to the construction of two relatively simple building blocks with roughly equal numbers of carbon atoms. The only real issue left for the Winkler group to address was the stereoselective incorporation of the stereogenic center in **27**, as this homochiral element was needed to dictate the installation of all remaining stereocenters of the targeted structure. Fortunately, with numerous asymmetric reactions, as well as plentiful sources of enantiopure starting materials from the chiral pool, this requirement was not anticipated to be problematic.

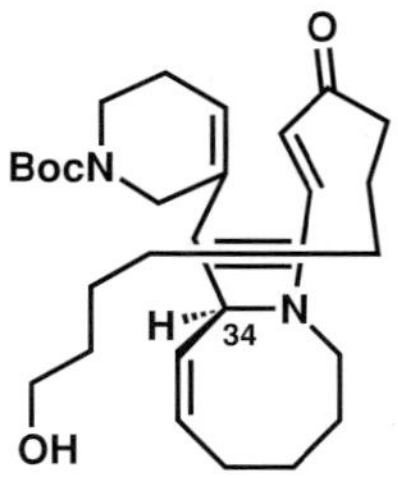

BocN O H 34 N OH **26**

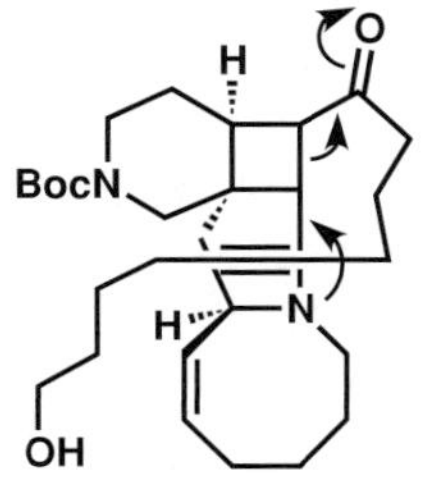

H O BocN H N OH **25**

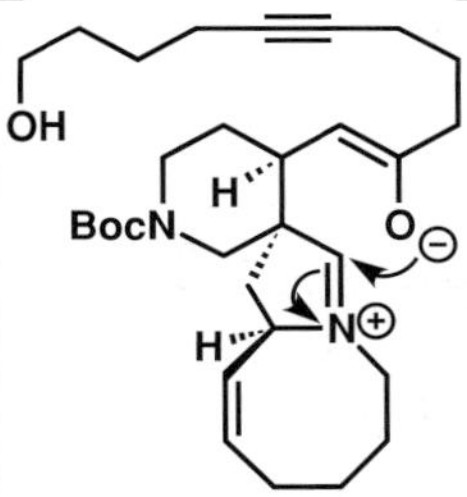

OH H BocN O⊖ H N⊕ **24**

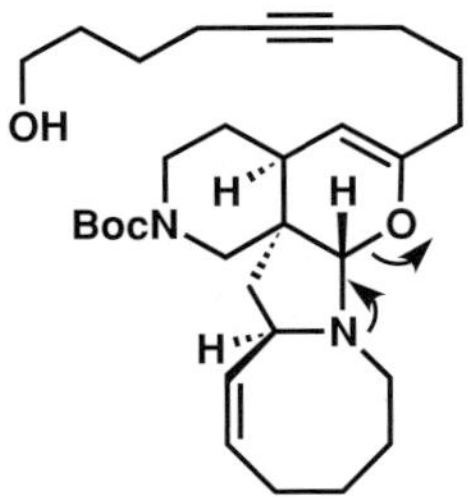

OH H H BocN O H N **23**

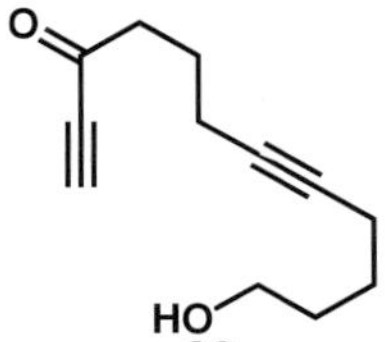

O HO **28**

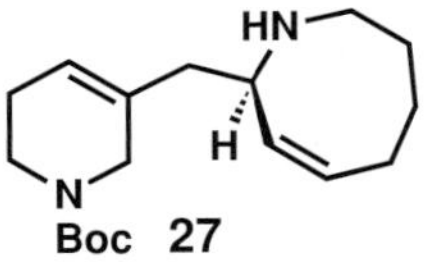

HN N H Boc **27**

8.2.2 Martin's Synthetic Approach to Manzamine A (1)

Concurrent with the development of this set of disconnections for the synthesis of manzamine A (**1**), the Martin group established a similarly creative, but unique, blueprint for the construction of this challenging molecular architecture (which is delineated in Scheme 4). Taking advantage of the known conversion of **17** into **1** established by Kobayashi and co-workers, the initial retrosynthetic disconnections of this strategy sought to disassemble the larger 13- and 8-membered D- and E-rings from the pyrrolo-isoquinoline core of ircinal A (**17**). Instead of anticipating an alkylation strategy to fashion these domains during the synthesis, these researchers viewed the lone alkene bond in each ring as a signal to adopt two different ring-closing metathesis (RCM) reactions of suitable diolefinic precursors as a means to construct them. Thus, setting this insight into motion, the E-ring olefin was retrosynthetically cleaved first, leading to **29** in which one of the terminal olefin chains could conceivably be appended just prior to metathesis through an amide coupling reaction, with the other alkene already present as an inherent part of the core structure. Following some additional, but subtle, modifications of protecting groups, a second RCM-based disconnection of the D-ring in **29** then led to triene **30**.

As the examples of ring-closing metathesis encountered during our discussion of the epothilones in Chapter 7 demonstrate, this C–C bond-forming transformation is unquestionably a powerful method to fashion both medium-sized as well as macrocyclic rings on a diverse collection of substrates, including natural products.[12] While these precedents were certainly encouraging for the likely realization of the projected ring closures in the present context, the Martin group could draw an additional measure of comfort from the successful metathesis-based formation of both the 8- and 13-membered rings of manzamine during model studies (carried out by their group and by Pandit and co-workers).[8c,13] If experience is any guide, however, one must always remember a fundamental tenet of total synthesis endeavors: "model is model, and real is real," meaning that since model systems never fully mirror the actual test substrate, surprises are a frequent occurrence. For example, in the projected metathesis reaction from **30** to forge the D-ring in **29**, an additional C–C double bond is present which could react with either of the peripheral alkenes to provide two additional ring-closed products. Although metathesis in these cases would result in 11- and 9-membered rings, less-favored products since they possess far lower flexibility and greater transannular steric repulsions than the desired product, the possibility of their formation cannot be entirely excluded. Apart from these alternative intramolecular metathesis pathways, a perhaps more pressing issue is the opportunity that the extra alkene in **30** could provide for nonproductive polymerization, even under high dilution conditions, should the formation of the 13-membered ring prove to be relatively sluggish.

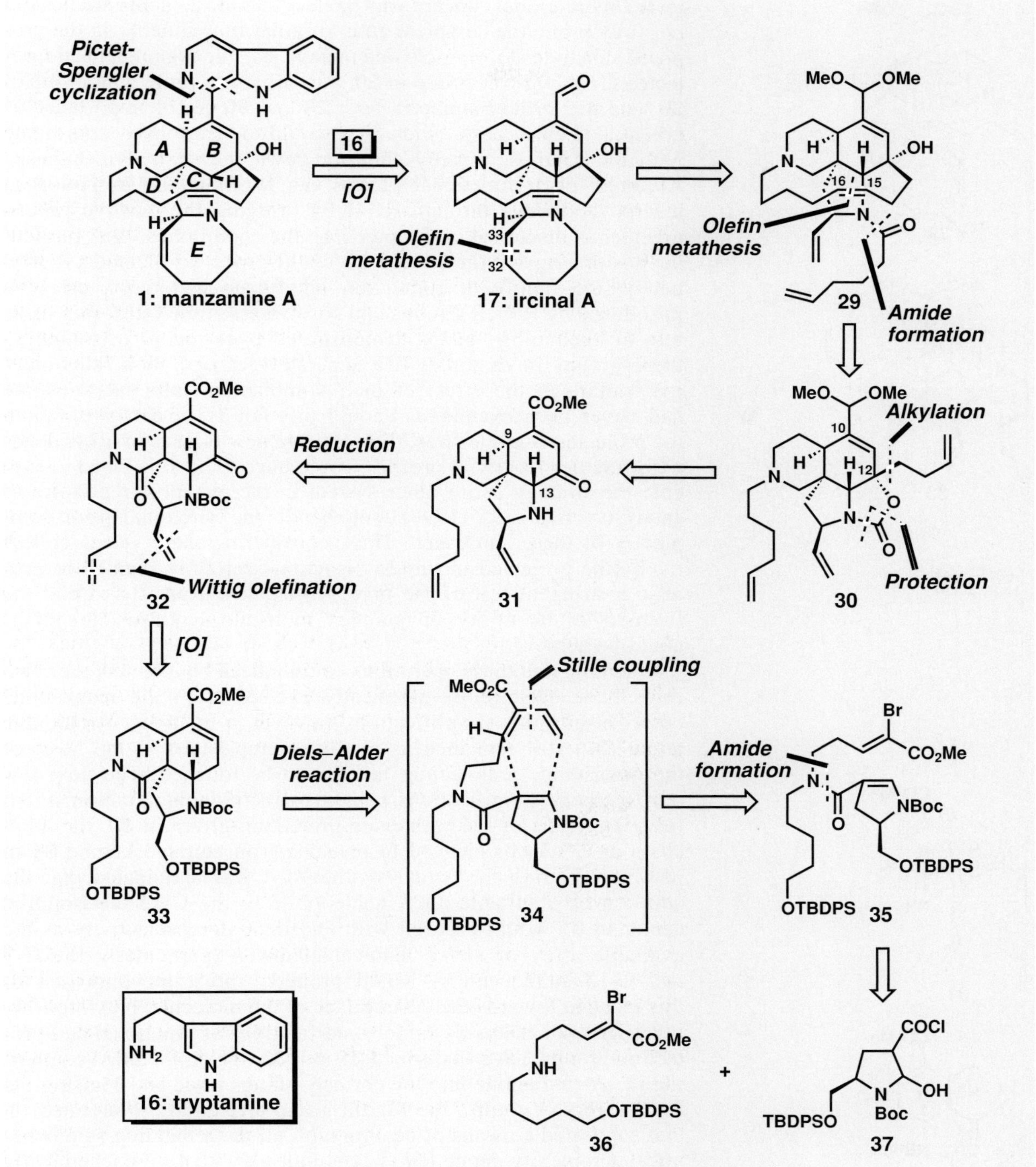

Scheme 4. Martin's retrosynthetic analysis of manzamine A (**1**).

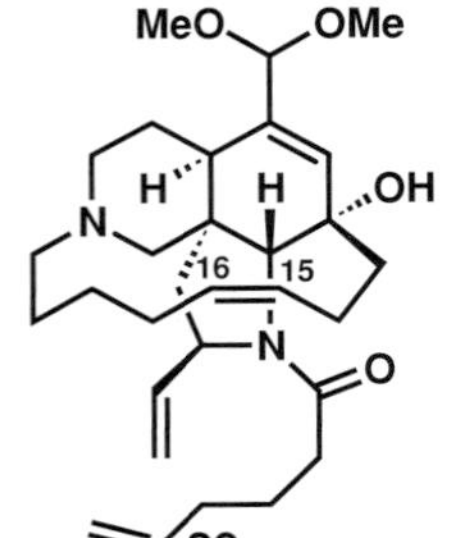

Of a more global concern was the fact that these model studies did not fully probe the complete array of polar functionality in the proposed substrates for ring-closing metathesis. For example, the unprotected C-12 hydroxy group in **29**, the carbamate protecting group in **30**, and the tertiary amine in both **29** and **30** are of importance as potential Lewis basic sites that could deleteriously coordinate with the catalyst, thereby shutting down the desired metathesis. Although polar functional groups can facilitate the formation of macrocyclic rings through RCM by bringing the reactive groups together as discussed in Chapter 7 on the epothilones, their particular positioning within **29** and **30** could prove problematic. Fortunately, the family of ruthenium alkylidene metathesis catalysts introduced by Robert Grubbs and co-workers at the California Institute of Technology had a demonstrated track-record of tolerating these groups in examples that separately carried such functionality,[14] although the effect of their combination into one substrate had never been examined. Should this family of catalysts fail in the projected ring-closures, then entirely new classes of alkylidenes with superior reactivity profiles would have to be developed to execute the strategy as no other system in the available repertoire of catalysts (circa 1995–1998) could handle the functional group complexity of these substrates. Thus, considering these issues collectively, the projected metathesis reactions certainly seemed to promise a stringent test of the process that would greatly extend the frontiers of its utility in complex molecule synthesis if success could be achieved.

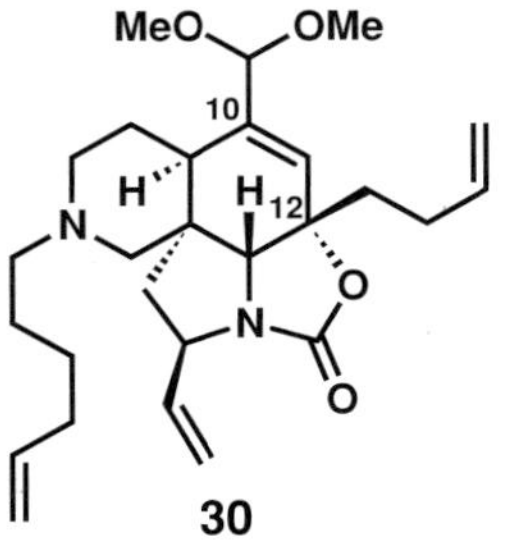

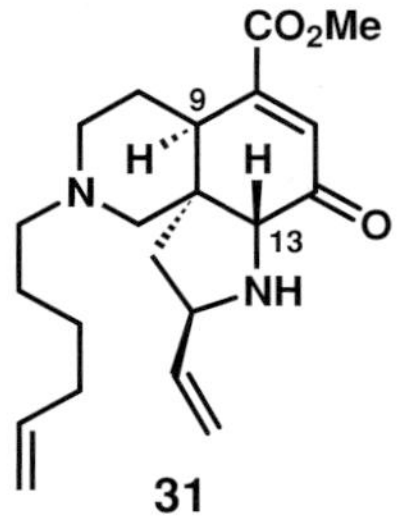

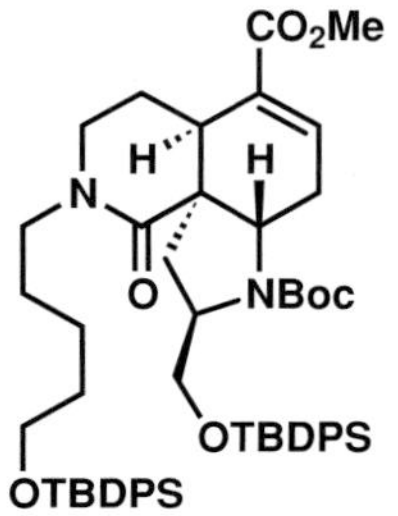

Assuming that these operations could indeed be effected, then two rings have been retrosynthetically excised from the manzamine core. Despite this simplification, however, a relatively formidable target (**30**) still remained as all five stereocenters at the heart of the manzamine architecture had yet to be touched. The next few retrosynthetic modifications sought to rectify this situation. First, following lysis of the carbamate protecting group in **30**, the alkyl chain at C-12 was cleaved to reveal α,β-unsaturated ketone **31** in its wake. During the actual synthesis, it was anticipated that the addition of a suitable alkyl nucleophile to the C-12 electrophilic center in **31** would proceed with excellent stereoselectivity as the available array of stereochemical elements (particularly the C-9 and C-13 stereocenters) would properly direct the approach of this reactant towards the desired face of the molecule. Next, following a series of relatively minor modifications of functional and protecting groups that transformed **31** into **33** (which included converting the A-ring amine into the corresponding amide and excising the C-12 carbonyl group), the Martin group arrived at an intermediate that suggested a means of dealing with all the remaining stereochemical complexity in one fell swoop. Indeed, with a stereochemically rich six-membered ring bearing a single element of unsaturation, an intramolecular Diels–Alder reaction would appear to be an obvious retrosynthetic transform to apply to **33**. Accordingly, in the synthetic direction a successful [4+2] cyclization of **34** in an *endo* fashion (with respect to the five-membered ring) would lead to installation of the requisite B-ring C–C double bond destined to become the C10–C11 alkene function of manzamine A (**1**). In addition, this reaction would concurrently establish three new stereocenters, assuming that the lone, but somewhat distal, chiral center in **34** could sufficiently bias the facial presentation of the dienophile to the diene in the transition state for the intended cycloaddition.

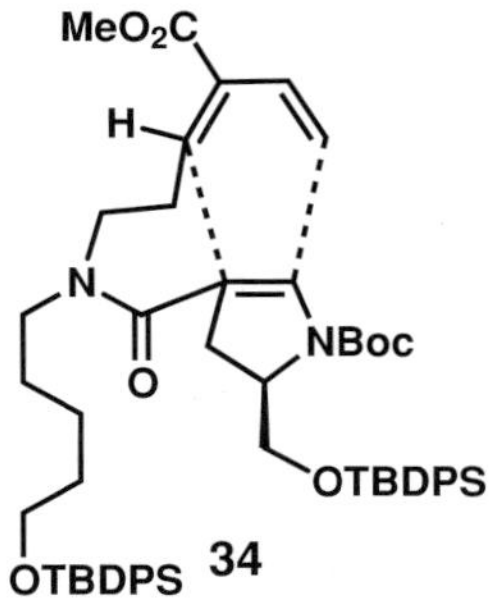

While the power of the Diels–Alder reaction has been applied in countless ways to enable the synthesis of complex molecules,[6a] the present context would in fact constitute relatively unchartered territory for the transformation. Not only was precedent for amide-linked intramolecular Diels–Alder reactions sparse,[15] but the designed sequence additionally required the incorporation of an *E* olefin in the diene, a motif that is often difficult to construct. In addition, with the use of a dienophile that is formally a vinylogous imide and the presence of an ester group on the diene system (required to eventually create the β-carboline system of **1**), the stage would appear to be set for an inverse-electron-demand Diels–Alder cycloaddition (see Chapter 2 for a discussion of this reaction),[16] a relatively challenging reaction to achieve in an all-carbon context such as **34**. Although these elements afforded a high level of challenge, the Martin group further increased the difficulty by anticipating that the diene system in **34** could be fashioned in the same operation as the Diels–Alder reaction through a Pd-mediated coupling, such as the Stille reaction of vinyl halide **35** with vinyl tri-*n*-butyltin, since both of these transformations could be promoted thermally.[17] This unprecedented domino sequence is certainly enticing on paper; if it could prove successful in the laboratory, then it would enable early access to the complete ABC portion of the manzamine core.

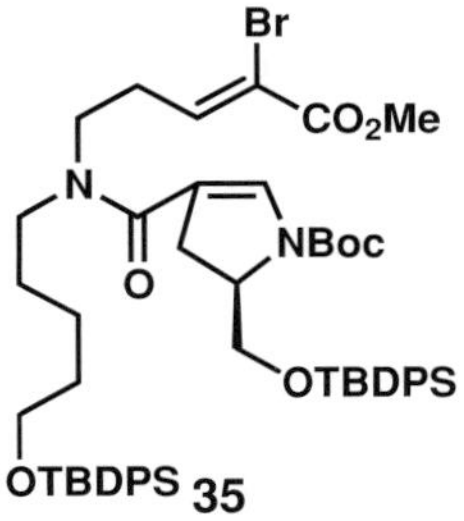

With the essential elements of the complex manzamine architecture now handled through this designed cascade, a final disconnection of the amide linkage in **35** revealed amine **36** and acid chloride derivative **37** as the subtargets needed to commence a serious drive towards **1**. Thus, as in the Winkler analysis, the synthetic sequence would rely on the creation of a single stereocenter in these building blocks with high fidelity, that is, the defined stereogenic center in **37**, to ensure a stereoselective synthesis overall. Interestingly, the center chosen for this purpose in both analyses is the same carbon atom in the final product (C-34). As we shall see, however, the methods employed in the two approaches to incorporate this stereocenter proved to be far different.

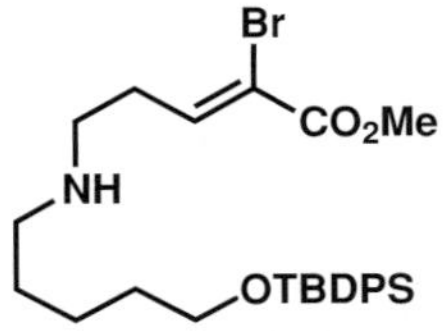

MeO_2C H N O NBoc OTBDPS OTBDPS **34**

Br CO_2Me N O NBoc OTBDPS OTBDPS **35**

Br CO_2Me NH OTBDPS **36**

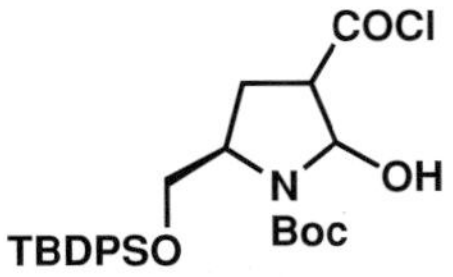

COCl OH N Boc TBDPSO **37**

8.3 Total Synthesis

8.3.1 Winkler's Total Synthesis of Manzamine A (1)

Synthetic studies towards manzamine A (**1**) commenced with exploration of routes to the essential building block **27** containing the final $\Delta^{32,33}$ E-ring alkene bond, starting with commercially available pyridine-3-methanol (**38**).[7b] As shown in Scheme 5, preliminary efforts sought to convert this material into a suitably functionalized six-membered ring (**40**), which could ultimately be employed as the electrophile in an alkylation reaction with a chiral enolate to generate the lone stereocenter in **27** critical to the execution of the entire strategy. As such, following initial quaternization of the pyridine nitrogen atom with benzyl bromide to form an *N*-benzylated salt, treatment with $NaBH_4$ effected facile double reduction to the corresponding tetrahydropyridine. This intermediate was then converted into **39** in 50 to 60 % yield by directly exchanging the benzyl protecting group with a methyl carbamate. In this event, the slow introduction of methyl chloroformate to a solution of the benzylamine in benzene and $NaHCO_3$ at room temperature led to the initial addition of the carbamate to the amine, effecting quaternization of the nitrogen atom attended by the concomitant release of chloride anion; upon heating the solution at reflux, the liberated chloride ion then attacked the benzylic position (see column figure), leading to the release of benzyl chloride and the formation of the desired carbamate (**39**) over the course of 16 hours. Significantly, although these conditions could conceivably have led to side products since the hydroxy group in **38** was left unprotected, no such difficulties were encountered. Having formed the key elements of this initial fragment, final modifications to reach **40** proved relatively easy to achieve. First, the methyl carbamate was exchanged for a *t*-butyl carbamate (Boc) protecting group in a two-step operation involving KOH-mediated cleavage of the methyl carbamate accompanied by subsequent treatment with $(Boc)_2O$. These events were then followed by conversion of the alcohol into the corresponding bromide under standard conditions (Br_2, Ph_3P, imidazole), leading to the formation of **40** in 83 % overall yield from **39**.

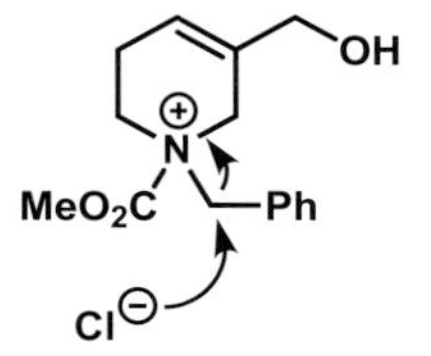

With this piece synthesized, the opportunity to install the stereogenic center in **27** selectively through a diastereoselective alkylation reaction was now at hand. For this purpose, the Winkler group decided to enlist a widely applicable and useful method developed by Andrew Myers and co-workers at Harvard University for the synthesis of α-amino acids in which pseudoephedrine glycinamide (**41**), an inexpensive and commercially available material, serves as a chiral auxiliary capable of controlling the selectivity of nucleophile addition.[18] Indeed, this synthetic technology proved highly efficacious as treatment of **41** with two equivalents of LDA in the presence of LiCl in THF at −78 °C, followed by the slow addition

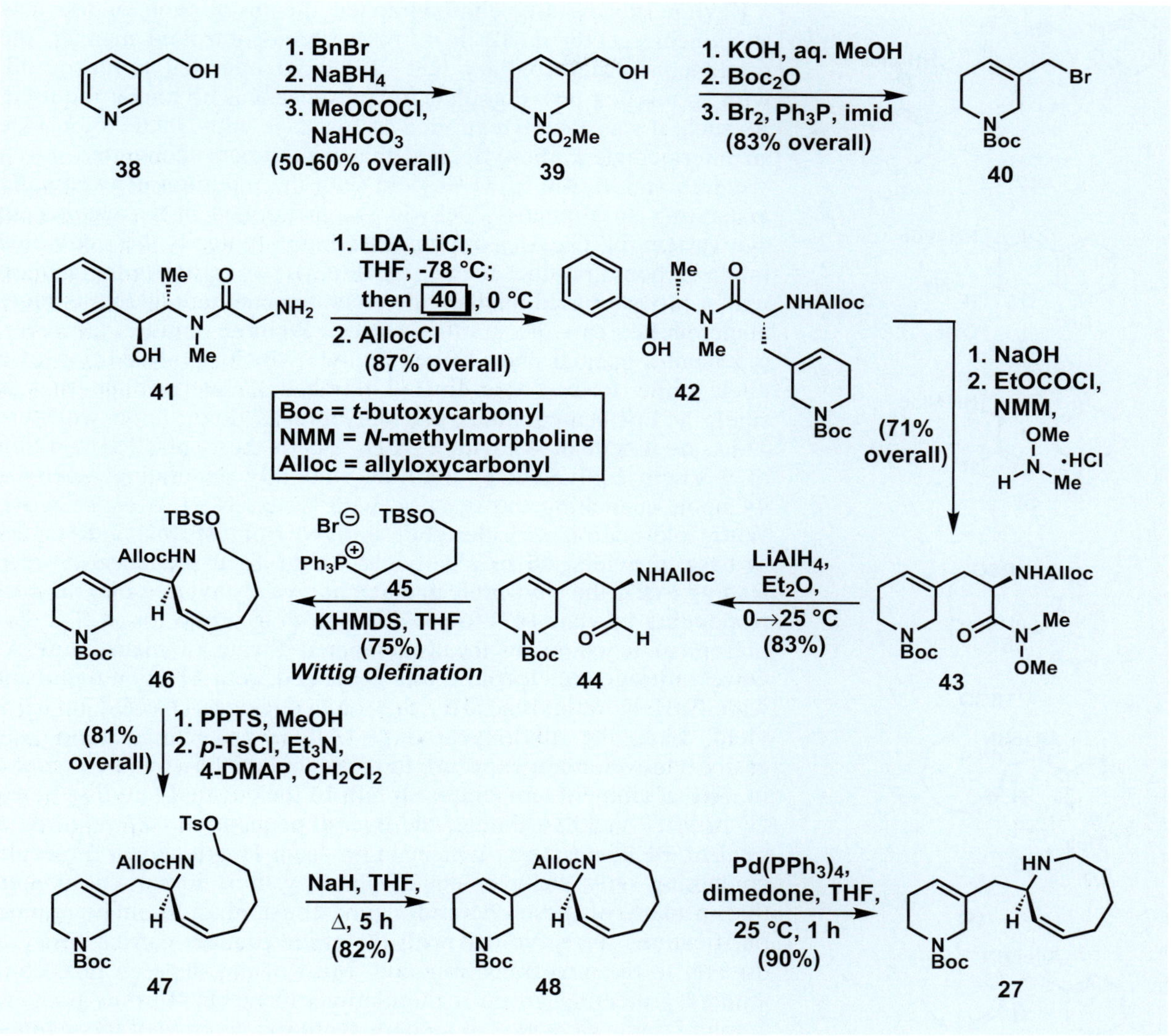

Scheme 5. Winkler's synthesis of key intermediate **27**.

of **40** at 0 °C, led to the desired alkylation product **42** in high diastereomeric purity (>99 % after recrystallization). Importantly, one should note that in this event the addition of base at −78 °C first deprotonated both the alcohol and the free amine in **41**; warming the solution to 0 °C and stirring for a short period of time prior to the addition of **40** then effected equilibration of the N,O-dianion to the corresponding *Z* enolate through C to N proton transfer, leading to the observed C-alkylated product once **40** was introduced to the reaction vessel. Following isolation, in the next synthetic operation the free amine was directly protected under standard conditions with allyl chloroformate to afford **42** in 87 % overall yield.

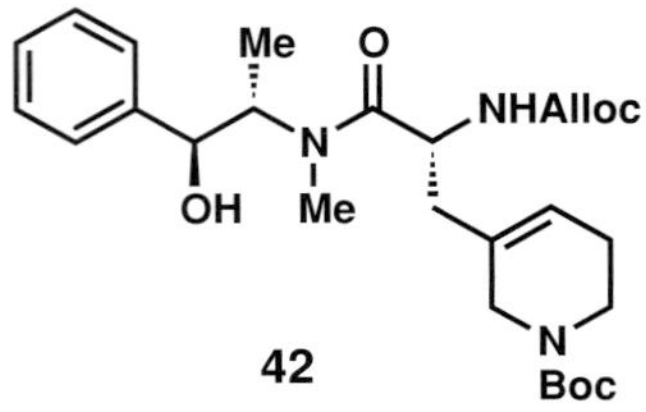

Having successfully choreographed the installation of the new stereogenic center in **42** in a highly stereocontrolled manner, the pseudoephedrine auxiliary had served its intended purpose and, with its mission now complete, its presence was no longer required. As such, it was cleanly extruded at this stage with NaOH to provide an intermediate carboxylic acid that was directly converted into a Weinreb amide (**43**) in 71 % yield with the intention of eventually generating an aliphatic aldehyde. As mentioned in the chapter on cytovaricin in *Classics I*, amides cannot typically be converted into a carbonyl product such as an aldehyde or ketone upon reaction with a lithium-based nucleophile since the resultant adduct is more nucleophilic than the starting amide. Weinreb amides, however, represent a special class of compounds which upon addition of a nucleophile form a coordinated tetrahedral intermediate that is stable at low temperatures and only breaks down upon work-up. Thus, treatment of **43** with $LiAlH_4$ led to the controlled addition of a single equivalent of hydride, smoothly generating aldehyde **44** upon quenching the reaction with water. Next, a *cis*-selective Wittig olefination with the ylide derived from **45** (using KHMDS as base) provided **46** in 75 % yield. In the final sequence of steps leading to **27**, the silyl-protected alcohol was converted into the corresponding tosylate (**47**) to provide a leaving group for an S_N2 displacement to complete the 8-membered E-ring of manzamine A. Once a nitrogen nucleophile was generated from **47** upon treatment with NaH in refluxing THF, this objective was effected in 82 % yield. Last, the allyloxycarbonyl (Alloc) protecting group was easily removed upon exposure to catalytic $Pd(PPh_3)_4$ and dimedone in THF at ambient temperature to afford the coveted building block **27** in 90 % yield. Although the overall sequence to **27** required a total of 14 linear steps, which might seem lengthy for a molecule containing only one element of chirality, it is important to note that in many of these operations time-consuming chromatographic purifications were avoided with the crude product carried forward directly to the next transformation. Most of the steps, in fact, constituted protecting group manipulations to reach **40**, thus perhaps emphasizing a weakness in modern synthesis in terms of the manner in which the whims of protecting groups can often dictate the length of synthetic routes. In general, however, the operations proceeded in quite high yield, and, as a result, multigram quantities of **27** could readily be processed to facilitate the synthetic explorations necessary to complete the total synthesis of manzamine A.

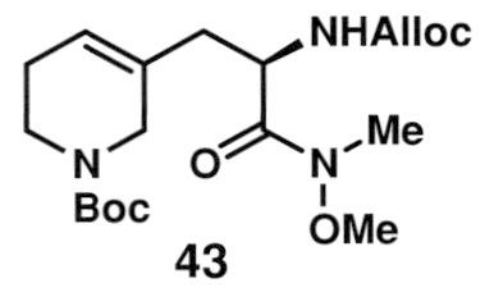

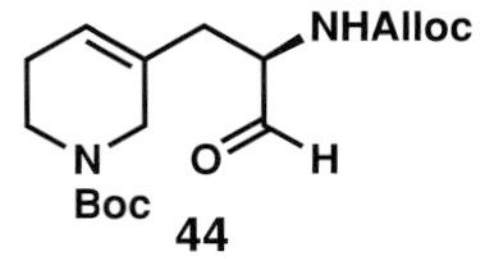

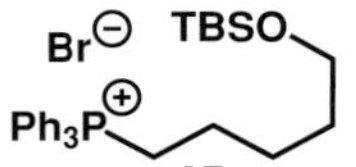

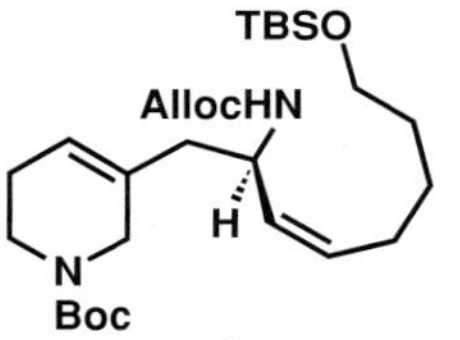

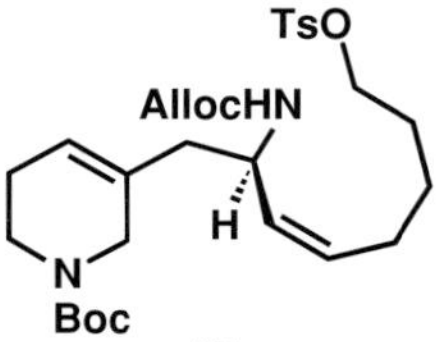

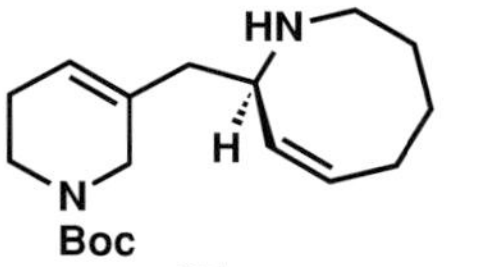

The preparation of the other essential building block needed to carry out the critical aspects of the synthetic plan, acetylenic ketone **28**, was achieved in two steps in 74 % overall yield from the known methyl 10-hydroxy-5-decynoate (**49**)[19] through initial conversion of the ester into the corresponding Weinreb amide (**50**), followed by the controlled addition of a single equivalent of ethynylmagnesium bromide as shown in Scheme 6.

With both **27** and **28** in hand, these fragments were smoothly joined in a Michael addition simply by stirring at ambient tempera-

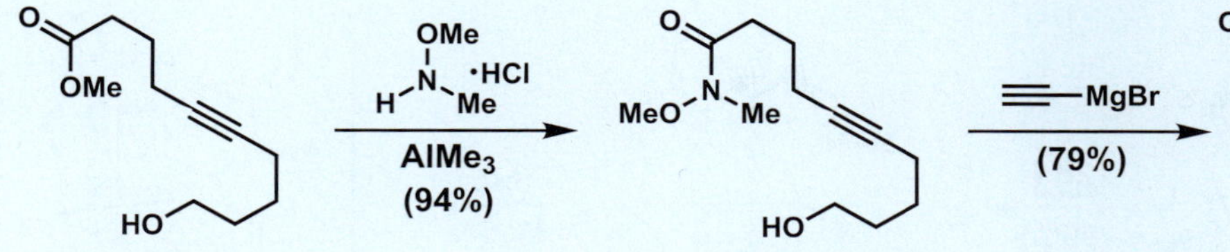

Scheme 6. Winkler's synthesis of building block **28**.

ture in CH_2Cl_2 to afford key intermediate **26** (Scheme 7) in near quantitative yield (99 %), thereby setting the stage for the key cascade sequence of reactions (photoaddition/fragmentation/Mannich closure) which lay at the heart of this synthetic enterprise. Despite the previously mentioned lack of precedent for tertiary vinylogous amide participation in [2+2] cycloadditions, exposure of an acetonitrile solution of **26** to ultraviolet light for 6 hours smoothly led to the formation of aminal **23**, presumably through the sequence discussed earlier with initial photoaddition followed by the retro-Mannich fragmentation of **25** and intramolecular nucleophilic attack of the enolate at the iminium site of intermediate **24**. Upon treatment of **23** with AcOH in pyridine at reflux over 4 hours, conversion into **22** followed by Mannich ring closure completed the synthesis of tetracycle **21** in 20 % overall yield from **26**, accounting for an average yield of 60 % per step in this series of events. As expected, **21** was produced as a single stereoisomer, presumably due to the overwhelming influence of the lone C-34 stereocenter in **26** in directing the formation of the resultant stereogenic centers during this elegant cascade sequence. One should note that although the stereochemistry at C-12 is not defined in Scheme 7, a single stereochemical arrangement resulted at this site, but was not assigned at this stage of the synthesis since it would need to be destroyed during subsequent operations to match the functional requirements of manzamine A (**1**). Moreover, one should recognize that although the fully substituted amide proved amenable to the cascade sequence, the success of the strategy was highly context dependent. For example, efforts to effect the same set of reactions with a substrate bearing an alkyl tether destined to become the D-ring (**51**, see column figure) failed despite numerous attempts.[7b]

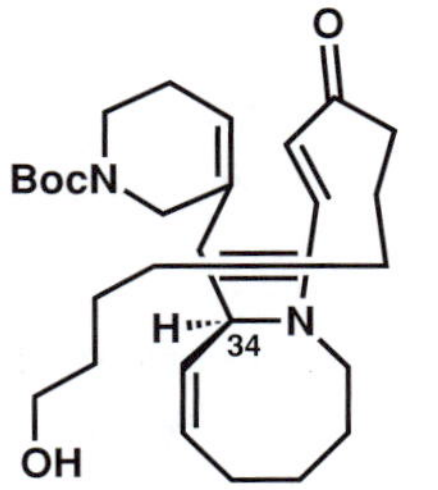

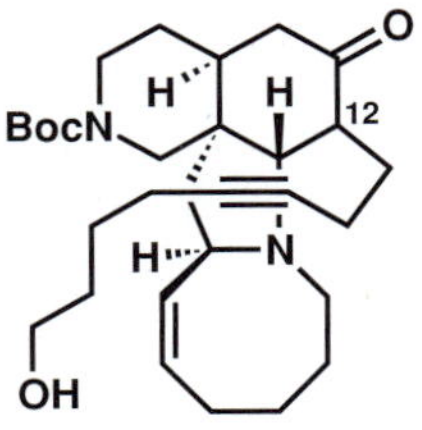

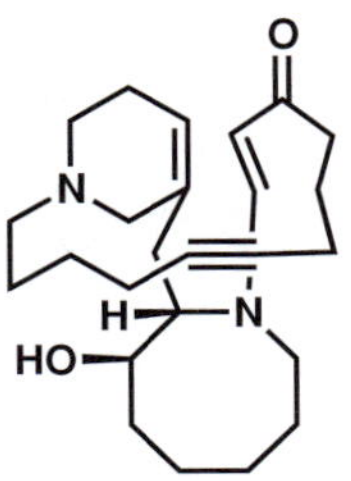

Having achieved the critical designed sequence of reactions, all that essentially remained for the Winkler group to complete their total synthesis was to elaborate the A- and B-rings in **21** to include the final array of functionality present in manzamine A (**1**). Towards this end, the free hydroxy group of the C-12-tethered alkyne chain was protected as a TBS ether to afford **52** in 87 % yield. A methyl ester was then installed on the B-ring at C-10 using LiHMDS in the presence of HMPA at −78 °C to generate the kinetic enolate from

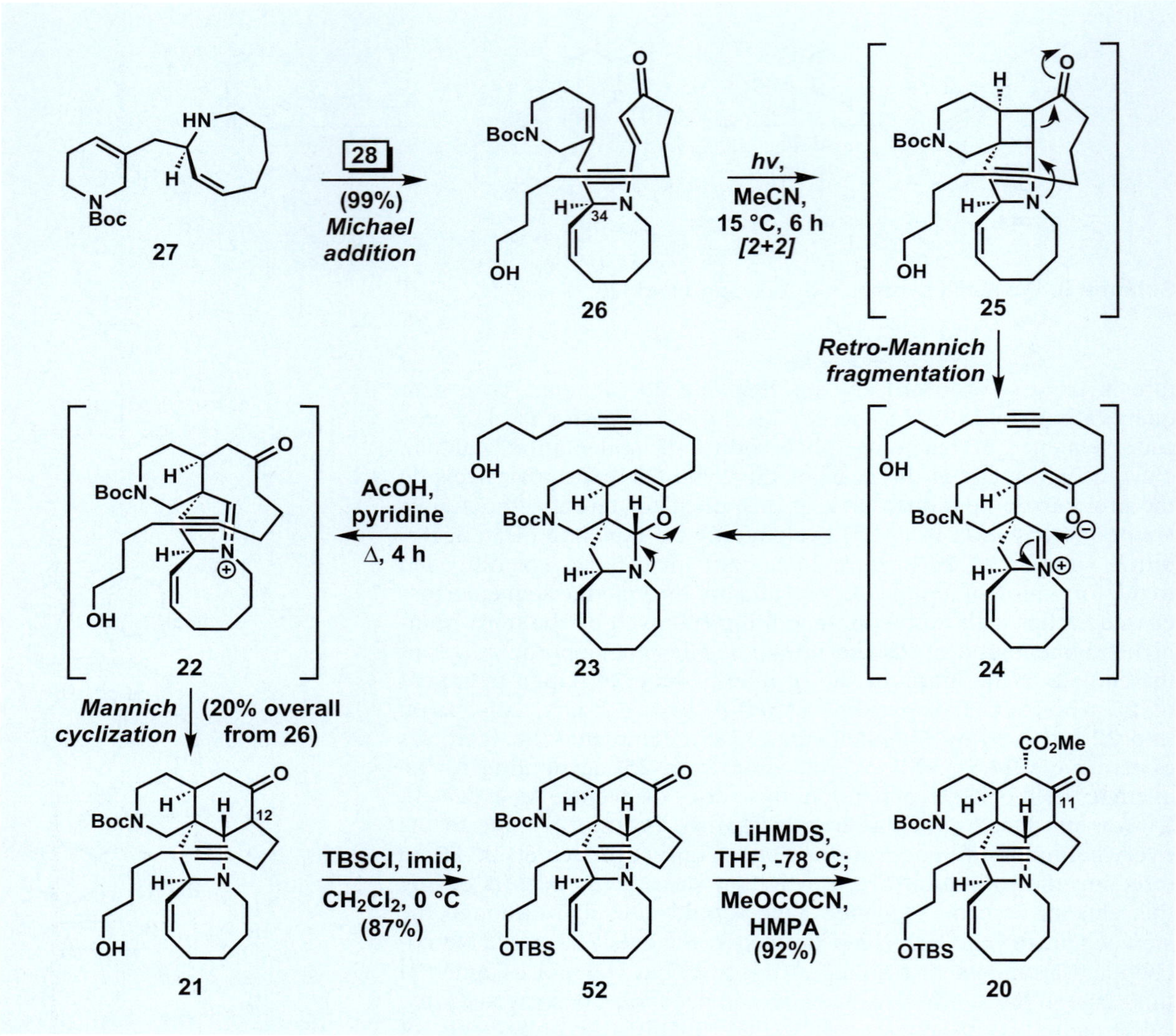

Scheme 7. Winkler's synthesis of advanced intermediate **20**.

52, which was immediately quenched with methyl cyanoformate (Mander's reagent). As one would expect, approach of this electrophile occurred from the more accessible convex face of the AB ring system, providing **20** with the absolute stereochemistry shown. Since this center would be destroyed later in the synthetic sequence, however, the selectivity of this operation was actually of no consequence.

The next set of operations sought to establish the final format of the C-12 stereocenter bearing a hydroxy group, since it was the only element of chirality not controlled by the cascade sequence. Thus, the C-11 ketone in **20** (Scheme 8) was reduced using

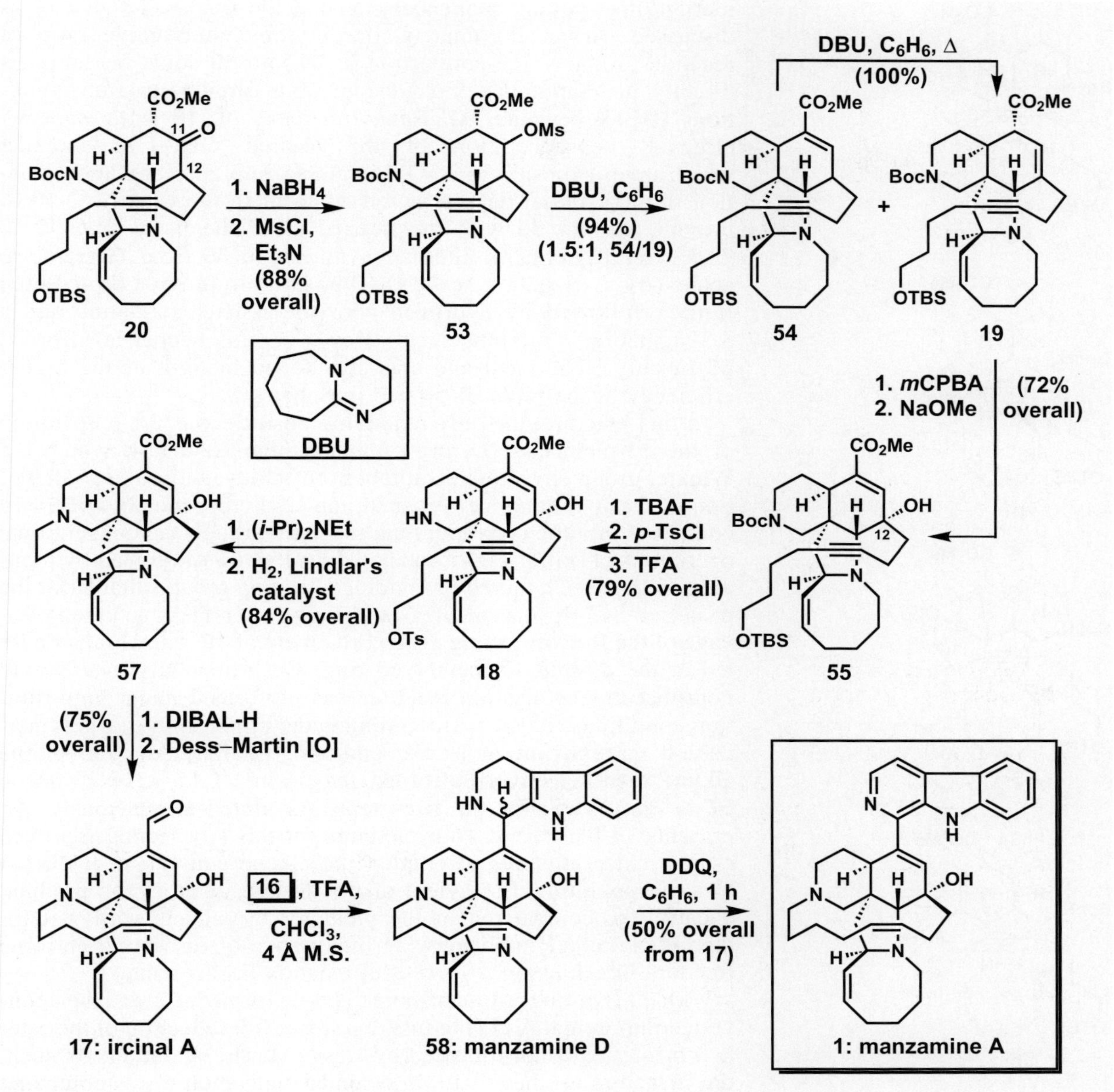

Scheme 8. Completion of Winkler's synthesis of manzamine A (**1**).

$NaBH_4$, with conversion of the resultant alcohol into the corresponding mesylate (**53**) proceeding in 88 % yield. This newly installed leaving group was then immediately eliminated upon treatment with DBU at room temperature in benzene, affording a 1.5:1 mixture of the α,β- and β,γ-unsaturated esters **54** and **19**, with the minor isomer being the desired compound for the envisioned epoxidation/ring-opening sequence to install the final C-12 alcohol as discussed earlier. Fortunately, despite this unfavorable ratio of reaction products, the conversion of **54** into **19** could be achieved directly in quantitative yield using more forcing reaction conditions (DBU, benzene, Δ). Next, treatment of **19** with *m*CPBA effected the generation of the desired C11–C12 epoxide, which upon exposure to NaOMe entered into an elimination reaction through initial deprotonation adjacent to the ester at C-10 to provide alcohol **55** with the desired stereochemistry at C-12 in 72 % overall yield. An alternate synthesis of **55** from **19** was also achieved by selenation at the C-10 position to provide **56** (column figure), followed by hydrogen peroxide-assisted oxidation and a 3,3-sigmatropic rearrangement. However, this sequence afforded **55** in only 37 % yield and was abandoned in light of the higher efficiency of the route illustrated in Scheme 8.

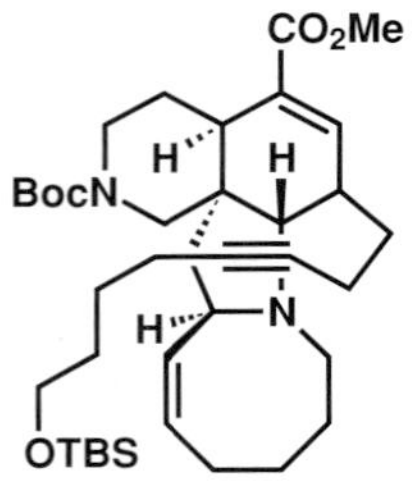

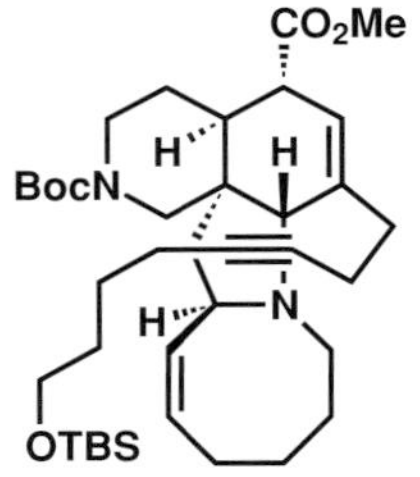

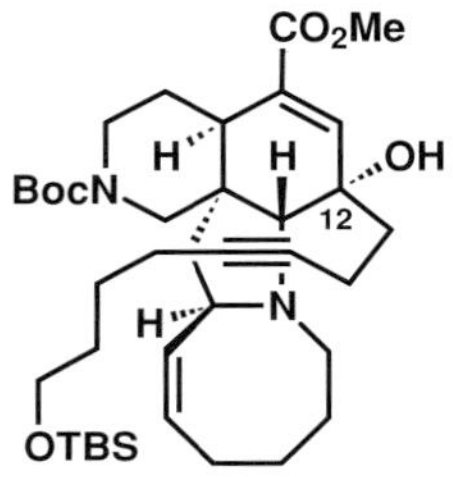

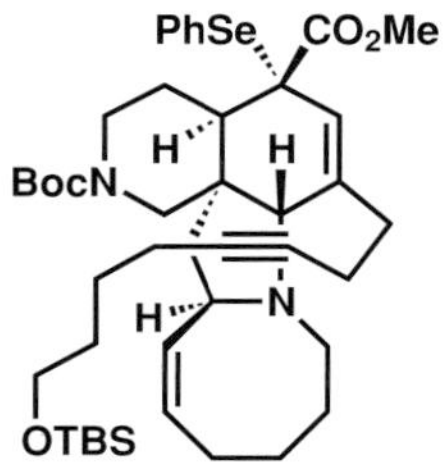

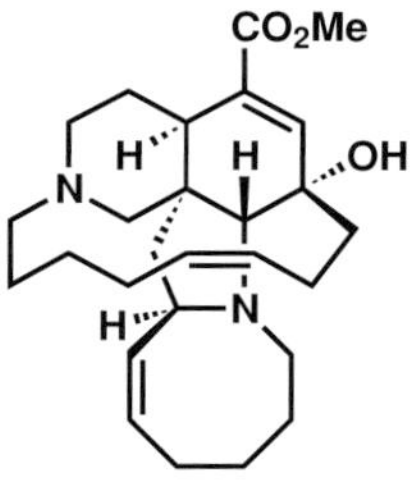

At this juncture, the only remaining obstacle was the preparation of the 13-membered D-ring of manzamine A, a task which the Winkler group envisioned could best be achieved through *N*-alkylation of the A-ring amine. As such, the silyl ether moiety in **55** was converted into the corresponding tosylate through the conventional operations of initial TBAF-mediated desilylation followed by treatment with *p*-TsCl, thereby providing a leaving group suitable for the projected S_N2 displacement. Gratifyingly, after TFA-mediated cleavage of the Boc protecting group, treatment of **18** with Hünig's base led to the desired 13-membered ring. Experimentally, one should note that this cyclization reaction was performed under high dilution conditions (0.001 M) to minimize the opportunity for polymerization through intermolecular addition. Lindlar reduction of the alkyne then selectively afforded the desired C15–C16 *Z* alkene **57** in 84 % yield for the two steps. As hinted at previously, the presence of the C15–C16 triple bond for this key sequence proved critical since attempts to employ the alkene equivalent of **18** led to a disappointing 12 % yield of the desired macrocyclic product. Clearly, the comparison of the yield for the alkene with that for the alkyne tether implicates the importance of restricting entropic freedom in achieving a successful reaction in this context.

With all five rings of manzamine A now formed, **57** was then converted into ircinal A (**17**) in just two steps: full reduction of the ester to a primary alcohol followed by Dess–Martin oxidation. As such, the first total synthesis of this secondary metabolite was complete. Finally, following the procedure of Kobayashi and co-workers,[9] **17** was then transformed into manzamine A (**1**) in 50 % overall yield upon Pictet–Spengler cyclization with tryptamine (**16**) in the presence of TFA to afford the natural product manzamine D (**58**), followed by DDQ-mediated oxidation. This elegant sequence provided synthetic manzamine A (**1**) in a total of 31 linear steps, with just 17 of these operations occurring after the union of building blocks **27** and **28**.

8.3.2 *Martin's Total Synthesis of Manzamine A (1)*

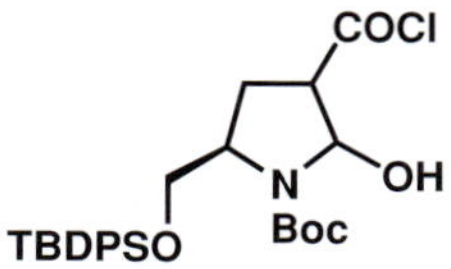

As a prerequisite to testing the critical elements of their designed strategy towards manzamine A (**1**), the Martin group initially needed to develop synthetic routes for their projected building blocks **36** and **37**. Fortunately, both proved relatively easy to acquire. Focusing our attention first on amine **36**, the synthesis of this compound began as shown in Scheme 9 by elaborating commercially available 5-amino-1-pentanol (**59**) in anticipation of an eventual Wittig reaction to fashion the trisubstituted alkene stereoselectively.[8] Thus, following initial protection of both the amine and alcohol functionalities in **59** using standard protocols, treatment of **60** with acrolein in the presence of acid led to conjugate addition, generating Wittig olefination precursor **61** in 69 % overall yield from **59**. Subsequent exposure of this aldehyde to stabilized ylide **62** in CH_2Cl_2 then led smoothly to trisubstituted alkene **63**, predominantly as the anticipated *Z* isomer (84 %) with only a small amount of the chromatographically separable *E* isomer obtained (7 %). The overall selectivity of this conversion (12:1) should serve as a reminder that for almost half a century the Wittig reaction has proven to be one of the most powerful and reliable methods for the regio- and stereocontrolled synthesis of C–C double bonds; with several modifications to the original reaction conditions, particularly as developed by Schlösser and Still,[20] both *cis*- and *trans*-disubstituted or trisubstituted olefins can be synthesized in high stereochemical fidelity essentially at will.

Having now secured the most challenging structural element of this fragment, the Boc protecting group in **63** was then removed upon treatment with TMSOTf in the presence of 2,6-lutidine, with subsequent exposure to *p*-TsOH in MeOH completing the synthesis

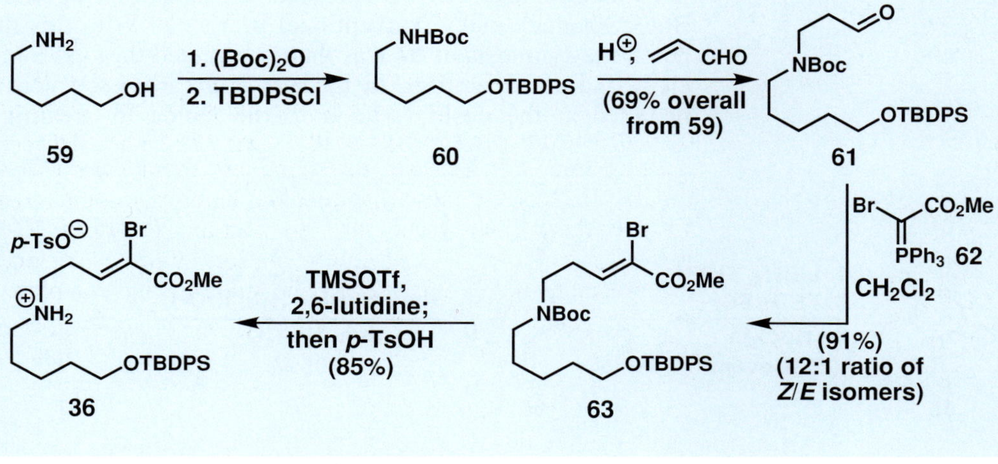

COCl
TBDPSO N OH
Boc
37

59: NH2, OH
1. $(Boc)_2O$
2. TBDPSCl
60: NHBoc, OTBDPS
$H^{\oplus}$, CHO
(69% overall from 59)
61: O, NBoc, OTBDPS
Br, CO_2Me, PPh_3 **62**
CH_2Cl_2
(91%)
(12:1 ratio of *Z*/*E* isomers)
63: Br, CO_2Me, NBoc, OTBDPS
TMSOTf, 2,6-lutidine; then *p*-TsOH (85%)
36: *p*-$TsO^{\ominus}$, Br, CO_2Me, $^{\oplus}NH_2$, OTBDPS

Scheme 9. Martin's synthesis of key intermediate **36**.

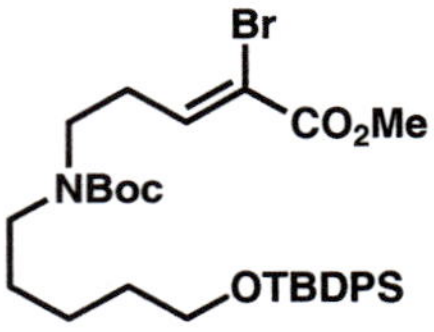

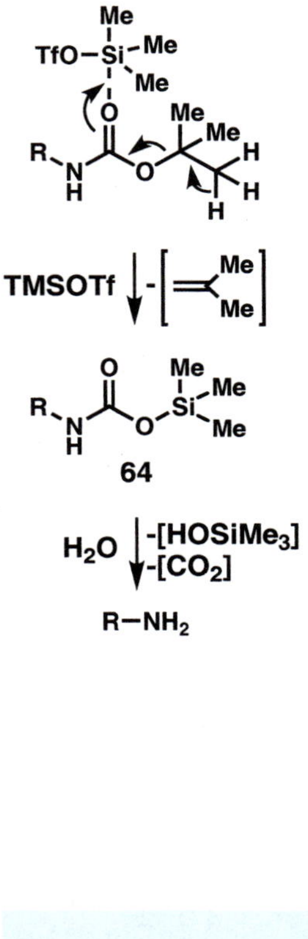

of **36** by affording this key building block as a crystalline tosylate salt in 85 % yield.[21] Before moving forward, this final deprotection step is worthy of some additional comment. As you probably already know, the particular value of the Boc protecting group relative to all other carbamate alternatives resides in its stability to a wide range of basic reaction conditions (since the bulk of the *t*-butyl group prevents nucleophilic lysis) and its relative lability when exposed to TFA or acids of similar strengths. However, for substrates that are acid sensitive such as silyl ether **63**, which would suffer Si–O bond cleavage, more selective means of Boc deprotection are required for the power of this protecting group to be fully realized. In this regard, the discovery that Boc groups can be removed merely through exposure to a silyl triflate constitutes a particularly effective solution to this problem, as the reaction conditions avoid the use of a protic acid. As shown in the figures in the neighboring column, the power of this deprotection method lies in the fact that the silyl triflate serves to convert the Boc group into an *N*-silyloxycarbonyl compound, a species that can be viewed as a masked form of the extremely unstable *N*-carboxylate ion. Indeed, when TMSOTf is employed in the reaction (as used in the conversion of **63** into **36**), the resultant silyloxycarbonyl system **64** (column figure) proves to be so labile that it cleaves *in situ* due to adventitious water or upon standard aqueous work-up, leading directly to the desired free amine with the concomitant loss of carbon dioxide and trimethylsilanol. We shall encounter this methodology again with another acid-sensitive substrate during our discussions of vancomycin (Chapter 9).

Returning now to the synthesis, carboxylic acid derivative **37** was accessed in just four operations as shown in Scheme 10, starting with commercially available (*R*)-5-(methoxycarbonyl)-2-pyrrolidinone (**65**). Initial manipulation of this substrate into **66** was achieved in 71 % overall yield upon complete reduction of the methyl ester with $LiBH_4$, followed by protection of the resultant primary alcohol as a TBDPS ether and the lactam nitrogen atom as its Boc carbamate using conventional protocols. With this material in hand, the synthesis of **37** was then completed through a highly efficient one-pot operation. In the event, a carboxylic acid was first installed at the desired site α to the amide by treating **66** with

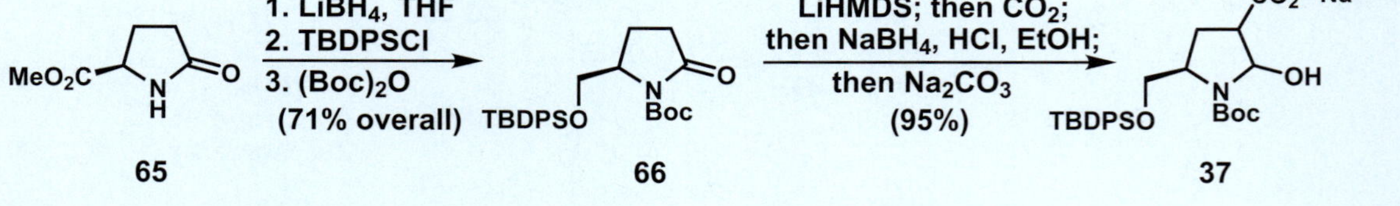

Scheme 10. Martin's synthesis of key intermediate **37**.

LiHMDS at −78 °C in THF and quenching the resultant anion with CO_2. Upon completion of this preliminary conversion, acidification of the reaction media followed by the addition of $NaBH_4$ at 0 °C effected the reduction of the amide carbonyl to complete the functional group array of the desired building block (**37**). Finally, treatment with Na_2CO_3 afforded the final product in a crystalline form suitable for long-term storage. One should note that this final event of the sequence was critical, as the free acid variant of **37** suffered from facile decarboxylation upon standing.

With the synthesis of both **36** and **37** achieved, exploration of the most challenging transformations in the projected synthesis of manzamine A could now begin in earnest. As shown in Scheme 11, initial efforts in this direction proceeded admirably as these building blocks were readily joined to form amide **35** upon reaction of **37** first with oxalyl chloride, to generate the corresponding acid chloride, followed by the slow addition of **36** as its free base. Having forged this important union, the stage was finally set to examine the viability of the critical domino Stille/Diels–Alder sequence that was anticipated to deliver the ABC core in a stereoselective manner. The cascade sequence proceeded exactly as planned, as upon reaction of **35** with vinyl tri-*n*-butyltin in the presence of catalytic $Pd(PPh_3)_4$ in refluxing toluene over the course of 30 hours, the desired cycloadduct **33** was produced in 68 % yield as a single stereoisomer! The level of success in this reaction (both in terms of yield and stereocontrol) is remarkable not only because of the numerous challenges which potentially could have clouded its progress as discussed earlier, but also because it was the first notable example in which a Stille coupling reaction and a [4+2] cycloaddition had been marshaled into a cascade sequence for a total synthesis endeavor.

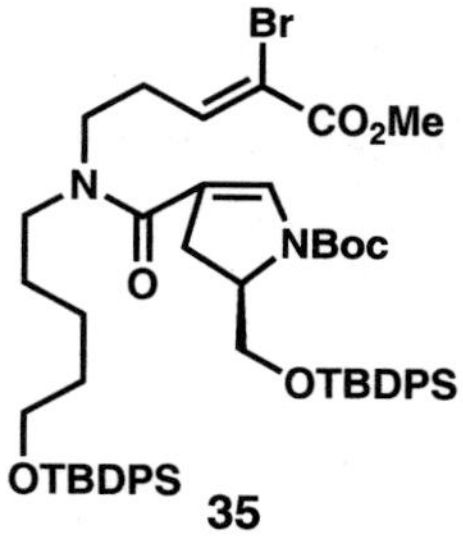

With many of the key structural elements of manzamine A now in place, all that remained for the Martin group was to effect some slight adjustments of the resident set of functional groups in **33** and then decorate their modified core with suitable olefin chains to explore the sequential metathesis reactions envisioned for the construction of the two remaining rings of the final target. Towards this end, in anticipation of adding a tethered alkene at C-12 through nucleophilic addition, the allylic methylene group in **33** first had to be oxidized to the corresponding α,β-unsaturated system (**67**). Although there are numerous reagent combinations which can effect this general type of conversion, the only tenable synthetic solution in this context proved to be a modified variant of the standard CrO_3 oxidation developed by Corey and Fleet,[22] and later modified by Salmond,[23] in which **33** was stirred with an excess of both CrO_3 (20 equiv) and 3,5-dimethylpyrazole (30 equiv) in CH_2Cl_2 at room temperature for 2 days. These conditions provided **67** in 63 % overall yield (80 % based on recovered starting material). Several features of this transformation are worthy of comment. First, although several different roles for the 3,5-dimethylpyrazole in this reaction have been invoked, all concur that this reagent helps

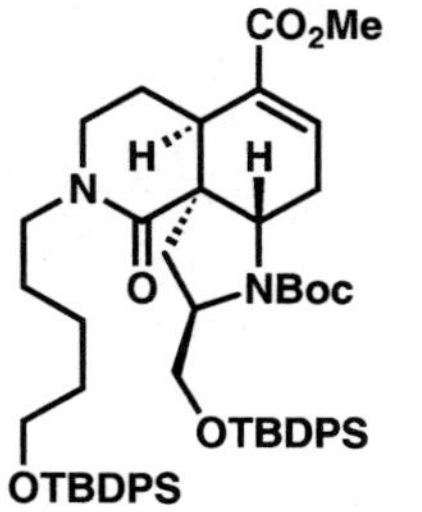

Br CO2Me N O NBoc OTBDPS OTBDPS **35**

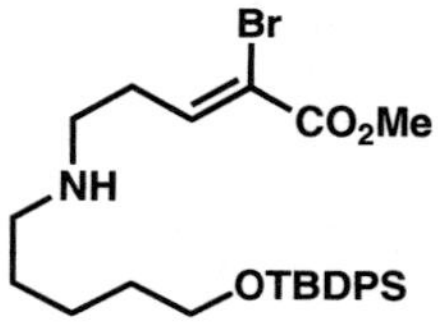

Br CO2Me NH OTBDPS **36**

CO2Me H H N O NBoc OTBDPS OTBDPS **33**

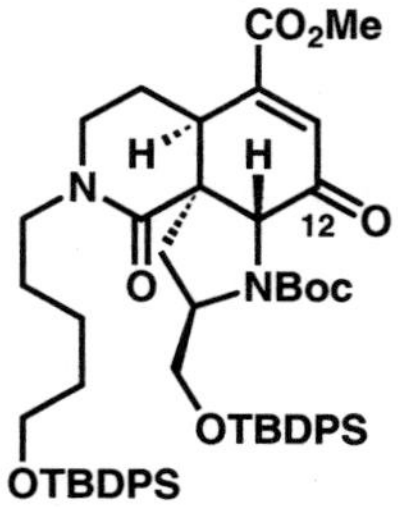

CO2Me H H N 12 O O NBoc OTBDPS OTBDPS **67**

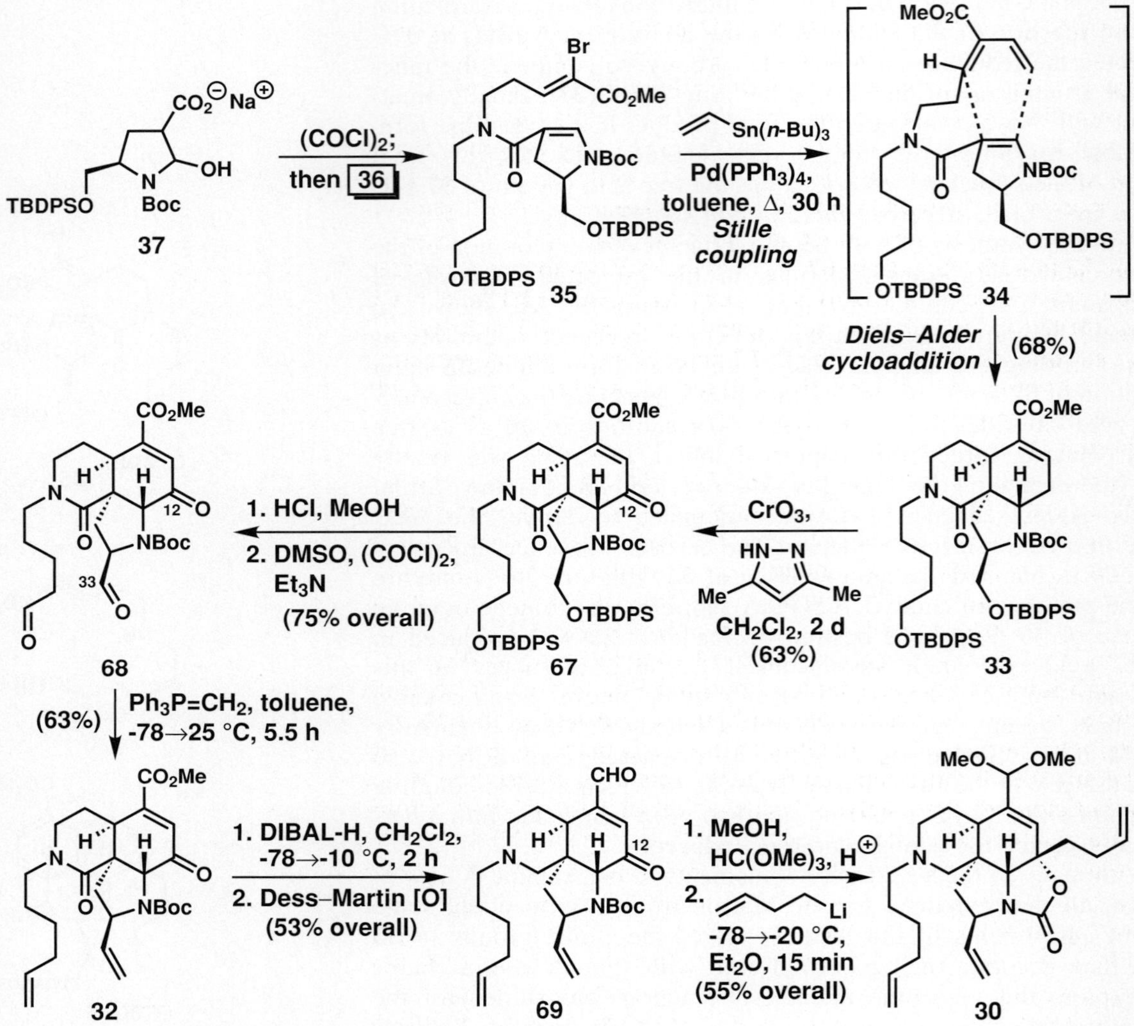

Scheme 11. Martin's synthesis of advanced intermediate **30**.

to solubilize the CrO_3 and that a 1:1 complex between CrO_3 and the pyrazole is the active oxidant in the process. Although this statement would suggest that only a stoichiometric amount of 3,5-dimethylpyrazole would be required to execute the reaction smoothly, in this case additional pyrazole was added to minimize the formation of imide by-products arising from concomitant oxidation of the methylene units adjacent to the A-ring amide; in fact, CrO_3/3,5-dimethylpyrazole is one of the few reagent combinations known that can effect this relatively challenging transformation.[24] Finally, while the mechanistic details of the oxidation remain to be fully clarified, current evidence suggests that the reaction proceeds through an ene-type addition to the olefin.

Having effected the conversion of **33** into **67**, it was now time to install the olefin side chains to test the metathesis reactions projected for the assembly of the remaining ring systems of manzamine A. As such, the silyl ethers were simultaneously cleaved upon exposure to HCl, with subsequent Swern oxidation [DMSO, $(COCl)_2$, Et_3N] affording dialdehyde **68**. This intermediate was then readily olefinated in tandem fashion using methylene triphenylphosphorane, generated under lithium-free conditions, to produce diene **32** in 47 % yield from **67**. Care in these conditions was critical, as alternative bases and solvents (such as *n*-BuLi/THF and NaH/DMSO) led to incomplete methylenation of the C-33 aldehyde as well as some olefination of the C-12 ketone. Next, global reduction of the carbonyl groups, including full reduction of the A-ring amide to the corresponding amine, was effected with excess DIBAL-H. Subsequent chemoselective re-oxidation of the two allylic alcohols upon treatment with Dess–Martin periodinane then completed the synthesis of diene **69** in 53 % overall yield. Finally, following selective protection of the newly-formed aldehyde as its corresponding dimethyl acetal through a standard technique, the remaining tethered olefin was installed in 55 % yield to provide **30** through the stereoselective 1,2-addition of 4-butenyllithium to the α,β-unsaturated ketone. In this transformation, the C-12 alkoxide generated through the addition process engaged the proximal Boc protecting group to form a new cyclic carbamate with the expulsion of *t*-butoxide. Although we previously mentioned the reluctance of Boc groups to be attacked by nucleophilic reagents, this final cyclization is not contradictory to that analysis because that discussion only applies to intermolecular situations. Indeed, as this example illustrates, the hard nucleophile used for the alkylation did not touch the suitably electrophilic Boc group (an intermolecular reaction); only when a new nucleophile was generated within **69** was the Boc group engaged.[25]

Beyond this unique feature of chemical reactivity, a few additional words about 4-butenyllithium are in order before continuing our discussion of the synthesis. As we have seen on numerous occasions, alkenyl- and aryllithium species are easily generated through lithium–halogen exchange using a base such as *n*-BuLi, typically in just a few minutes at low temperature. The same principle does not hold true for the preparation of primary alkyllithium species such as 4-butenyllithium. Prior to 1990, a typical preparation of these reagents relied upon reacting lithium metal directly with a suitable alkyl halide, conversions which typically took several days to complete at ambient temperature and that often resulted in the formation of significant amounts of undesired side products through processes such as β-elimination or S_N2-based dimerizations between successfully lithiated compound and starting material. As such, applications for these species in synthesis were, in general, limited due to the challenges imposed in executing their preparation. In 1990, however, these problems were circumvented when Negishi and co-workers discovered that primary alkyl iodides (but not bromides or chlorides) could be converted cleanly into the corresponding alkyllithium species without any side products upon treatment with 2 equivalents of *t*-BuLi in Et_2O at −78 °C for 30 minutes, followed by slow warming to ambient temperature over an additional hour.[26] The amount of *t*-BuLi is critical to ensure complete lithiation as the *t*-BuI that is produced from successful exchange with the first equivalent of base will then undergo β-elimination through a reaction with the remaining *t*-BuLi; in the absence of the second equivalent, the less basic alkyllithium species can perform the same task. One should also realize that alkyl bromides react through a different mechanistic pathway with *t*-BuLi (single-electron transfer) and are not universally successful, whereas chlorides are entirely inert to the reaction conditions. As a final note, the choice of Et_2O as solvent is critical for the success of the technique, not only for its general compatibility with *t*-BuLi, but also because of how it affects the aggregation state of the generated alkyllithium nucleophiles.

Returning now to the synthesis, with triene **30** synthesized, the necessary groundwork had been laid to examine the first ring-closing metathesis reaction. Gratifyingly, following exposure of **30** (Scheme 12) to a 13 mol % loading of the ruthenium alkylidene initiator **70**[27] in CH_2Cl_2 under high dilution conditions (0.005 M) to minimize polymerization, RCM to the anticipated 13-membered E-ring was smoothly achieved after 3 hours at reflux, ultimately affording an 8:1 mixture of *Z*/*E* isomers which were separable by column chromatography. The desired *Z* olefin **71** was isolated in 67 % yield, and, importantly, no alternative products corresponding to the formation of 9- or 11-membered rings were identified. Interestingly, previous model studies of this catalyst system (**70**) on substrates related to this transformation indicated that both of the tertiary amines required protonation prior to RCM in order to prevent them from acting as Lewis bases capable of bonding to the metal center of the catalyst and thereby shutting down the productive reaction pathway, but this precaution proved unnecessary for successful conversion in this case. Additionally, while the model studies by Pandit and co-workers to form the D-ring by metathesis similarly demonstrated *Z* selectivity,[13] one should realize that these fortunate results are relatively unique to the manzamine context as most RCM-based macrocyclizations provide products with predominantly *E* geometry.[28]

With success in this endeavor, only one more ring remained before the entire manzamine architecture would be in hand. Thus, pressing forward to test the final ring-closing metathesis step which would hopefully solve this problem, base-induced cleavage of the cyclic carbamate was followed by *N*-acylation with 5-hexenoic acid chloride to afford **29** in 75 % overall yield. While the smooth nature of these conversions was satisfying, a perhaps more exciting, but unanticipated, finding was that **29** was crystal-

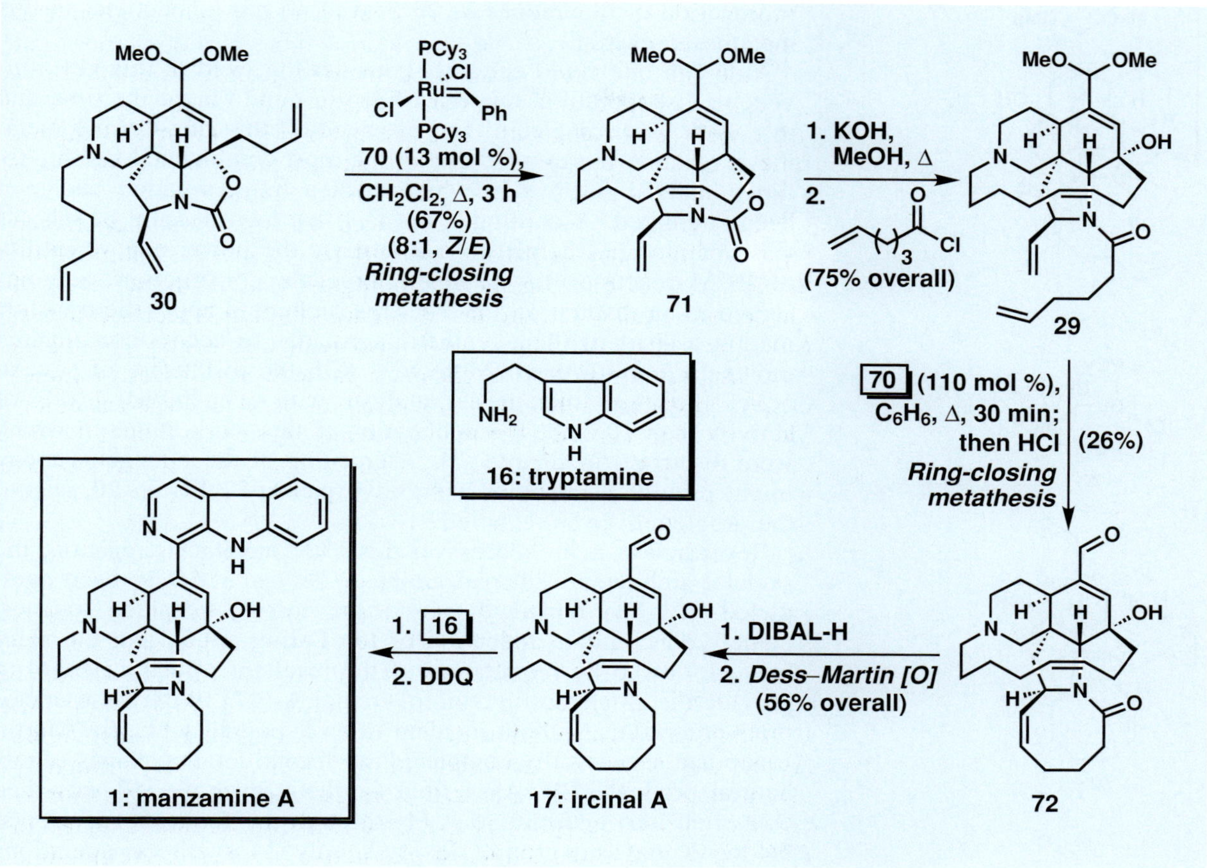

Scheme 12. Completion of Martin's synthesis of manzamine A (**1**).

line. As a result, the Martin group could use X-ray crystallographic analysis to confirm beyond all doubt that the stereochemical outcome of the crucial Stille/Diels–Alder sequence as well as the first olefin metathesis reaction had occurred exactly as NMR spectroscopic data had suggested.

Up to this point, the overall sequence had proceeded with few major synthetic hurdles which proved unyielding to sufficient experimentation. Unfortunately, this pattern was about to be broken. Despite the ease with which the 8-membered azocene E-ring was fashioned by metathesis in model studies related to the manzamine problem, the conversion of **29** into **72** occurred under optimized conditions (110 mol% of **70**, 0.004 M in C_6H_6, Δ, 30 minutes) in a modest yield of only 26%, with the final product isolated after acid-catalyzed removal of the dimethyl acetal protecting group. This result is likely the product of alternative metathesis pathways available due to the proximal olefin in the 13-membered D-ring and not some other inherent polar functional group in the test substrate. Indeed, protection of the free hydroxy group as well as protonation of the amines in **29** proved no more beneficial in this metathesis reaction.

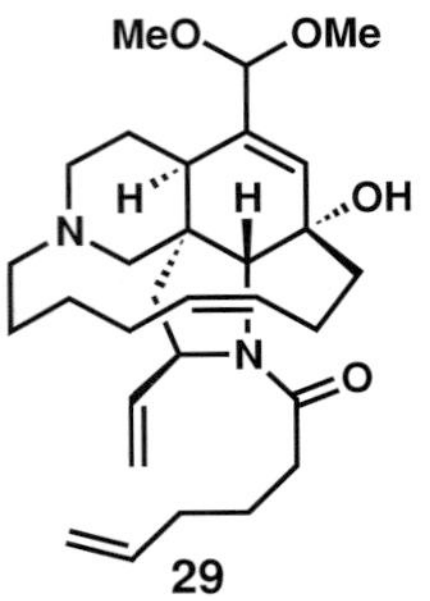

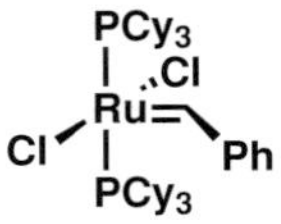

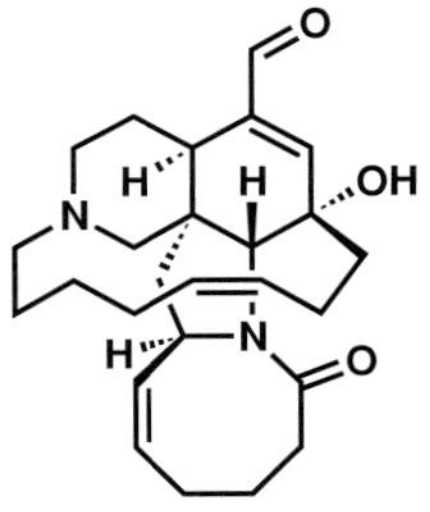

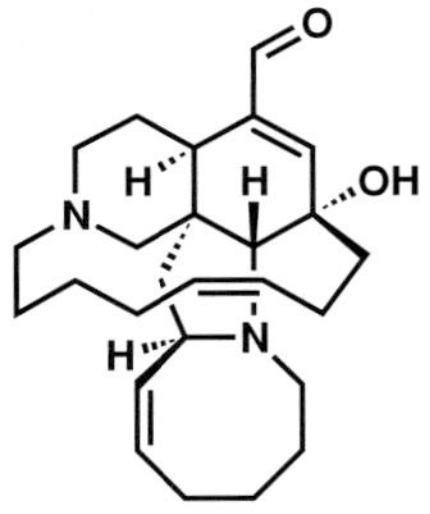

Although one could certainly consider the yield in this key conversion disappointing, one must keep in mind that at the time that this work was completed (and even today) this ring-closing metathesis reaction represented one of the most complicated and highly functionalized settings in which such a transformation had ever been attempted. Accordingly, the fact that any measure of success was obtained has helped to demonstrate the power and versatility of RCM reactions for complex molecule construction, and has served as an inspiration for other researchers to apply the transformation as part of their synthetic strategies to access complicated molecular architectures. Moreover, with the disclosure of several novel classes of metathesis catalysts with even higher levels of activity than **70** since the publication of this work (some of which were discussed in Chapter 7), a screening of these reagents today might provide a far more impressive yield of **72** from **29**, should the problem ever be revisited.

Nevertheless, with success in this final metathesis reaction, the skeletal architecture of manzamine A (**1**) had virtually been completed. All that remained were some cursory finishing touches, namely concomitant reduction of the D-ring amide and the aldehyde, followed by reoxidation of the resultant primary alcohol to provide the aldehyde function of ircinal A (**17**). With these operations proceeding without incident in 56 % overall yield, the Martin group had successfully completed the second total synthesis of this natural product (**17**). As a final synthetic exercise, **17** was then converted into manzamine A (**1**) through the sequence developed by the Kobayashi group.[9] In total, only 24 synthetic operations from commercially available materials were required to complete manzamine A (**1**) by this approach, with the longest linear sequence consisting of just 21 steps.

29

70

72

17: ircinal A

8.4 Conclusion

During a 1965 lecture describing a successful total synthesis of the natural product colchicine, R. B. Woodward eloquently formulated the fundamental appeal of endeavors in total synthesis when he noted that

> *... although the specific objective in synthetic work is defined with unique precision, the manner of reaching it most emphatically is not. It would be possible to synthesize a molecule ... in countless different ways, no one of which would resemble any other except in its outcome. Much of the charm and fascination of this kind of work lies in the free reign which the imagination may be permitted in planning the adventure, as well as in executing it.*[29]

While several chapters in this text offer powerful testimony to support these assertions, the total syntheses of manzamine A developed by the Winkler and Martin research groups perhaps best encapsulate the meaning of these remarks. Indeed, although certain concepts were mirrored in both of these syntheses, such as the development of a novel cascade sequence to synthesize the pyrrolo-isoquinoline core based on only one stereocenter to guide the absolute stereochemistry of the final product, no single aspect of the execution proved similar. Such creativity is witness both to the ingenuity and ability of the synthetic practitioners who performed this research, as well as to the wealth of transformations in our current synthetic arsenal which can be combined in many novel ways. Without question, each of these concise and elegant syntheses sets an impressively high standard for future synthetic efforts targeting members of this growing class of important natural products, though it would seem almost certain that the potential for fundamental discoveries from synthetic endeavors toward the manzamine alkaloids is far from exhausted.[30]

References

1. a) R. Sakai, T. Higa, C. W. Jefford, G. Bernardinelli, *J. Am. Chem. Soc.* **1986**, *108*, 6404; b) H. Nakamura, S. Deng, J. Kobayashi, Y. Ohizumi, Y. Tomotake, T. Matsuzaki, Y. Hirata, *Tetrahedron Lett.* **1987**, *28*, 621.
2. a) M. Tsuda, J. Kobayashi, *Heterocycles* **1997**, *46*, 765; b) T. Ichiba, R. Sakai, S. Kohmoto, G. Saucy, T. Higa, *Tetrahedron Lett.* **1988**, *29*, 3083.
3. J. E. Baldwin, R. C. Whitehead, *Tetrahedron Lett.* **1992**, *33*, 2059.
4. A modified form of the Baldwin–Whitehead hypothesis has also been proposed to explain the origins of a number of structurally related natural products, though the experimental evidence supporting this alternative has not been advanced to an intermediate as complex as keramaphidin B: a) K. Jakubowicz, K. B. Abdeljelil, M. Herdemann, M.-T. Martin, A. Gateau-Olesker, A. Al Mourabit, C. Marazano, B. C. Das, *J. Org. Chem.* **1999**, *64*, 7381; b) M. Herdemann, A. Al-Mourabit, M.-T. Martin, C. Marazano, *J. Org. Chem.* **2002**, *67*, 1890.
5. a) J. E. Baldwin, T. D. W. Claridge, A. J. Culshaw, F. A. Heupel, V. Lee, D. R. Spring, R. C. Whitehead, R. J. Boughtflower, I. M. Mutton, R. J. Upton, *Angew. Chem.* **1998**, *110*, 2806; *Angew. Chem. Int. Ed.* **1998**, *37*, 2661; b) J. E. Baldwin, T. D. W. Claridge, A. J. Culshaw, F. A. Heupel, V. Lee, D. R. Spring, R. C. Whitehead, *Chem. Eur. J.* **1999**, *5*, 3154.
6. The existence of "Diels–Alderases" has been debated for some time in relation to several natural products, and, although often assumed, evidence remains inconclusive. For commentary on this issue, see: a) K. C. Nicolaou, S. A. Snyder, T. Montagnon, G. Vassilikogiannakis, *Angew. Chem.* **2002**, *114*, 1742; *Angew. Chem. Int. Ed.* **2002**, *41*, 1668; b) G. Pohnert, *ChemBioChem* **2001**, *2*, 873; c) S. Laschat, *Angew. Chem.* **1996**, *108*, 313; *Angew. Chem. Int. Ed. Engl.* **1996**, *35*, 289. For perhaps the best (and singular) example of a true Diels–Alderase to date, see: K. Auclair, A. Sutherland, J. Kennedy, D. J. Witter, J. P. Van den Heever, C. R. Hutchinson, J. C. Vederas, *J. Am. Chem. Soc.* **2000**, *122*, 11519.
7. a) J. D. Winkler, J. M. Axten, *J. Am. Chem. Soc.* **1998**, *120*, 6425. For earlier model studies by the Winkler group, see: b) J. D. Winkler, J. Axten, A. H. Hammach, Y.-S. Kwak, U. Lengweiler, M. J. Lucero, K. N. Houk, *Tetrahedron* **1998**, *54*, 7045; c) J. D. Winkler, J. Stelmach, M. G. Siegel, N. Haddad, J. M. Axten, W. P. Dailey, *Isr. J. Chem.* **1997**, *37*, 47; d) J. D. Winkler, M. G. Siegel, J. E. Stelmach, *Tetrahedron Lett.* **1993**, *34*, 6509.
8. a) S. F. Martin, J. M. Humphrey, A. Ali, M. C. Hillier, *J. Am. Chem. Soc.* **1999**, *121*, 866; b) J. M. Humphrey, Y. Liao, A. Ali, T. Rein, Y.-L. Wong, H.-J. Chen, A. K. Courtney, S. F. Martin, *J. Am. Chem. Soc.* **2002**, *124*, 8584. For earlier model studies by the Martin group, see: c) S. F. Martin, H.-J. Chen, A. K. Courtney, Y. Liao, M. Pätzel, M. N. Ramser, A. S. Wagman, *Tetrahedron* **1996**, *52*, 7251; d) S. F. Martin, Y. Liao, Y. Wong, T. Rein, *Tetrahedron Lett.* **1994**, *35*, 691; e) S. F. Martin, T. Rein, Y. Liao, *Tetrahedron Lett.* **1991**, *32*, 6481.
9. K. Kondo, H. Shigemori, Y. Kikuchi, M. Ishibashi, T. Sasaki, J. Kobayashi, *J. Org. Chem.* **1992**, *57*, 2480.

10. D. de Oliveira Imbroisi, N. S. Simpkins, *J. Chem. Soc., Perkin Trans 1* **1991**, 1815.
11. For an excellent review on the use of this type of cascade sequence towards the total synthesis of natural products, see: J. D. Winkler, C. Mazur Bowen, F. Liotta, *Chem. Rev.* **1995**, *95*, 2003.
12. For recent reviews on olefin metathesis, see: a) T. M. Trnka, R. H. Grubbs, *Acc. Chem. Res.* **2001**, *34*, 18; b) A. Fürstner, *Angew. Chem.* **2000**, *112*, 3140; *Angew. Chem. Int. Ed.* **2000**, *39*, 3012; c) R. H. Grubbs, S. Chang, *Tetrahedron* **1998**, *54*, 4413; d) M. Schuster, S. Blechert, *Angew. Chem.* **1997**, *109*, 2124; *Angew. Chem. Int. Ed. Engl.* **1997**, *36*, 2036; e) R. H. Grubbs, S. J. Miller, G. C. Fu, *Acc. Chem. Res.* **1995**, *28*, 446.
13. B. C. Borer, S. Deerenberg, H. Bieräugel, U. K. Pandit, *Tetrahedron Lett.* **1994**, *35*, 3191.
14. G. C. Fu, S. T. Nguyen, R. H. Grubbs, *J. Am. Chem. Soc.* **1993**, *115*, 9856.
15. For examples of such Diels–Alder reactions, see: a) G. A. Kraus, J. Raggon, P. J. Thomas, D. Bougie, *Tetrahedron Lett.* **1988**, *29*, 5605; b) M. E. Kuehne, W. G. Bornmann, W. G. Earley, I. Marko, *J. Org. Chem.* **1986**, *51*, 2913; c) D. J. Morgans, G. Stork, *Tetrahedron Lett.* **1979**, *20*, 1959.
16. For selected reviews of inverse-electron-demand Diels–Alder reactions, see: a) S. Jayakumar, M. P. S. Ishar, M. P. Mahajan, *Tetrahedron* **2002**, *58*, 379; b) M. Behforouz, M. Ahmadian, *Tetrahedron* **2000**, *56*, 5259; c) D. L. Boger, *Chemtracts: Org. Chem.* **1996**, 149; d) D. L. Boger, *J. Heterocycl. Chem.* **1996**, *33*, 1519; e) D. L. Boger in *Comprehensive Organic Synthesis, Vol. 5* (Ed.: B. Trost), Pergamon, Oxford, **1991**, pp. 451–512; f) D. L. Boger, *Chem. Rev.* **1986**, *86*, 781; g) S. M. Weinreb, *Acc. Chem. Res.* **1985**, *18*, 16; h) D. L. Boger, *Tetrahedron* **1983**, *39*, 2869.
17. For reviews of cascade-based sequences, including those incorporating the Diels–Alder reaction, see: a) R. A. Bunce, *Tetrahedron* **1995**, *51*, 13103; b) K. C. Nicolaou, T. Montagnon, S. A. Snyder, *Chem. Commun.* **2003**, 551.
18. A. G. Myers, J. L. Gleason, T. Yoon, *J. Am. Chem. Soc.* **1995**, *117*, 8488.
19. W. Nowak, H. Gerlach, *Liebigs Ann. Chem.* **1993**, 153.
20. a) M. Schlösser, K. F. Christmann, *Angew. Chem.* **1966**, *78*, 115; *Angew. Chem. Int. Ed. Engl.* **1966**, *5*, 126; b) M. Schlösser, G. Müller, K. F. Christmann, *Angew. Chem.* **1966**, *78*, 677; *Angew. Chem. Int. Ed. Engl.* **1966**, *5*, 667; c) C. Sreekumar, K. P. Darst, W. C. Still, *J. Org. Chem.* **1980**, *45*, 4260.
21. M. Sakaitani, Y. Ohfune, *J. Org. Chem.* **1990**, *55*, 870.
22. E. J. Corey, G. W. J. Fleet, *Tetrahedron Lett.* **1973**, *14*, 4499.
23. W. G. Salmond, M. A. Barta, J. L. Havens, *J Org. Chem.* **1978**, *43*, 2057.
24. G. Blay, L. Cardona, B. García, C. L. García, J. R. Pedro, *Tetrahedron Lett.* **1997**, *38*, 8257.
25. For an excellent discussion of this topic, see: C. Agami, F. Couty, *Tetrahedron* **2002**, *58*, 2701.
26. E. Negishi, D. R. Swanson, C. J. Rousset, *J. Org. Chem.* **1990**, *55*, 5406. For the application of *t*-BuLi in the synthesis of aryl- and alkenyllithiums, see: a) E. J. Corey, D. J. Beames, *J. Am. Chem. Soc.* **1972**, *94*, 7210; b) D. Seebach, H. Neumann, *Chem. Ber.* **1974**, *107*, 847.
27. a) P. Schwab, M. B. France, J. W. Ziller, R. H. Grubbs, *Angew. Chem.* **1995**, *107*, 2179; *Angew. Chem. Int. Ed. Engl.* **1995**, *34*, 2039; b) P. Schwab, R. H. Grubbs, J. W. Ziller, *J. Am. Chem. Soc.* **1996**, *118*, 100.
28. For examples and discussion of this concept, see: A. Fürstner, T. Dierkes, O. R. Thiel, G. Blanda, *Chem. Eur. J.* **2001**, *7*, 5286 and references cited therein.
29. R. B. Woodward, *Harvey Lecture Ser.* **1963–1964**, *59*, 31.
30. For excellent reviews summarizing the synthetic efforts of other groups toward the manzamines up to the time of the publication of these total syntheses, see: a) N. Matzanke, R. J. Gregg, S. M. Weinreb, *Org. Prep. Proc. Int.* **1998**, *30*, 1; b) E. Magnier, Y. Langlois, *Tetrahedron* **1998**, *54*, 6201; c) M. Tsuda, J. Kobayashi, *Heterocycles* **1997**, *46*, 765. Additional manzamine alkaloids continue to be identified, providing new targets for synthetic explorations: K. A. El Sayed, M. Kelly, U. A. K. Kara, K. K. H. Ang, I. Katsuyama, D. C. Dunbar, A. A. Khan, M. T. Hamann, *J. Am. Chem. Soc.* **2001**, *123*, 1804.

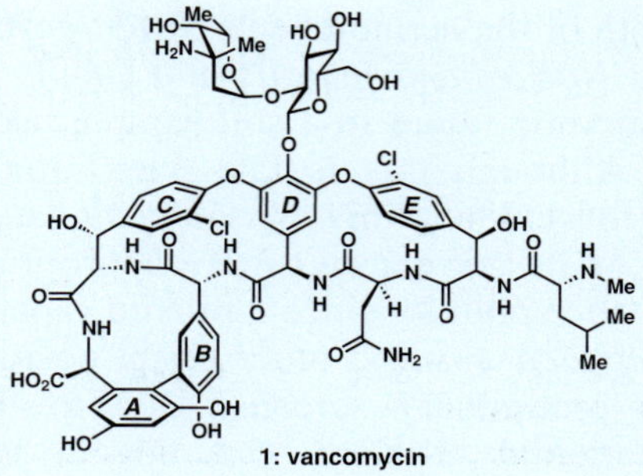

1: vancomycin

9

K. C. Nicolaou (1998)
D. A. Evans (1998)

Vancomycin

9.1 Introduction

Sir Alexander Fleming's discovery of penicillin in 1928 initiated a revolution in the practice of medical care by affording the first weapon that could effectively combat several deadly strains of bacteria, but mankind's war with bacteria is far from over. In fact, its ending may never be reached due to the laws of evolutionary pressure.[1] Because bacteria are armed with rapid growth cycles, and thus the means to mutate in a relatively short period of time, those colonies that survived the lethal force of penicillin have since developed into stronger and more potent bacterial strains capable of wreaking even greater havoc on human health than their predecessors. The only real means available for mankind to combat these threats is the deployment of new antibacterial agents possessing greater activity and novel modes of action; fortunately, the past several decades have witnessed the development of several such classes of antibiotics. However, none of these therapeutics has proven capable of killing all bacteria and each of these additional weapons will eventually become obsolete as those microbes that survive will develop the ability to better evade and resist their potency in subsequent rounds of attack. The magnitude and severity of this problem is perhaps underscored best by the fact that for some of the most lethal strains of bacteria currently facing humanity, we have merely a single line of chemical defense, which, if broken, could prove catastrophic.

The main subject of this chapter is one of these remaining stalwart defenses, an antibiotic known as vancomycin (**1**), which constitutes the only agent presently capable of combating deadly methicillin-resistant strains of *Staphylococcus aureus*.[2] This structurally com-

Key concepts:

- **Bacterial resistance**
- **Atropisomerism**
- **Sharpless asymmetric aminohydroxylation (AA)**
- **Cyclic bisaryl ether construction**
- **Amino acid synthesis**
- **Peptide synthesis**

plex natural product was first isolated in 1956 by scientists at Eli Lilly from the fermentation broth of the actinomycete *Streptomyces orientalis* (later reclassified as *Nocardia orientalis* and finally as *Amycolatopsis orientalis*), a microbe found in a soil sample taken from the jungles of Borneo. Although this isolate was initially assigned the lackluster name of O5865, this trivial designation was quickly dropped in favor of its current and more appropriate title as a derivative of the verb "to vanquish" since it proved capable of killing every strain of *Staphylococcus* and other Gram-positive bacteria thrown at it during preliminary screens. Due to this impressive spectrum of bactericidal efficacy, vancomycin was approved for clinical use just two years after its discovery. In the ensuing decades, it has become one of the most valuable antibiotics in our current arsenal and has unquestionably saved countless lives.[2]

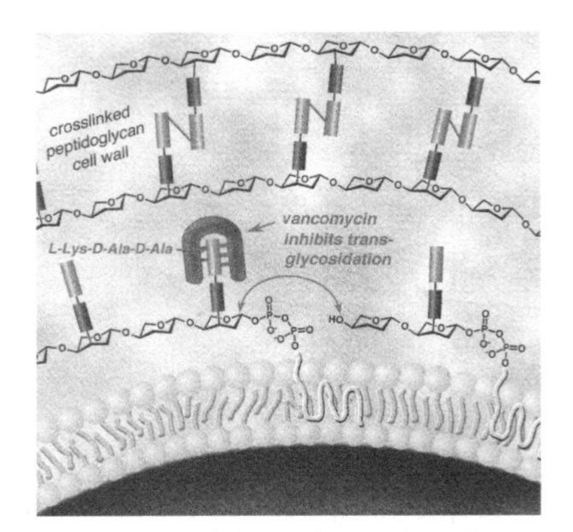

Figure 1. Mechanism of action for vancomycin (**1**) and other glycopeptide antibiotics.

Similar to many other antibiotics, the potency of vancomycin derives from its ability to interfere with an essential biochemical process needed for bacteria to survive, namely cell wall biosynthesis. As you might already know, these cell walls are composed of a combination of sugars and proteins that are carefully interwoven to create a framework capable of maintaining the structural integrity of the bacterium. However, since the cell wall exists in a constant state of flux between its formation and disassembly, this structural domain constitutes an excellent point for antibiotic intervention in that these agents can shift the balance towards deconstruction, and, thus, bacterial death through membrane lysis. As shown in Figure 1, vancomycin achieves this result by binding to the protein portion of the growing peptidoglycan, an event that prevents transglycosidases from polymerizing the carbohydrate units needed to form the backbone of the cell wall. The reason for the particular effectiveness of vancomycin is the strength of its molecular recognition for this peptide chain, which, as we know from the pioneering studies of Dudley Williams and his collaborators, is the product of a series of five well-defined hydrogen bonds with the terminal L-Lys-D-Ala-D-Ala sequence of these building blocks (see Scheme 1).[3]

Unfortunately, such specificity also implies that bacteria need only make a few minor structural modifications to thwart this antibiotic. This conjecture is proving dangerously true with several recent reports, many from Christopher Walsh's group at the Harvard University Medical School, describing the existence of *Staphylococcus aureus* strains unimpeded by vancomycin, whose sole difference from susceptible bacteria is the exchange of their terminal D-alanine residues for L-lactates.[4] Because of this one simple change, vancomycin can now form only four hydrogen bonds with this new target sequence because a hydrogen bond donor (an NH) has been replaced with an acceptor (an O). While this result reflects the loss of just 20 % of vancomycin's noncovalent interactions, it leads to a weakening of its binding strength (and thus its potency) by a factor of 1000.

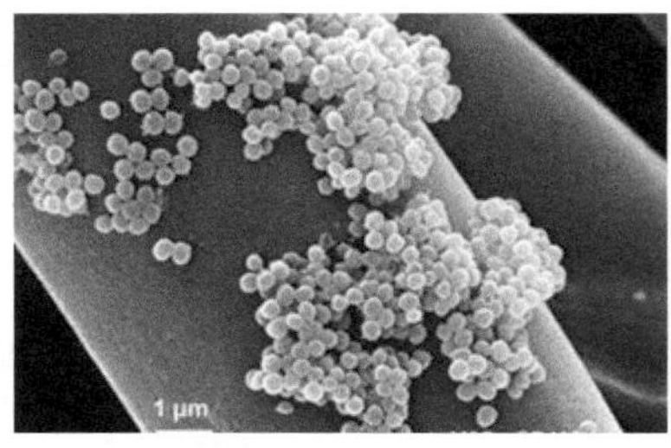

Staphylococcus aureus
Courtesy of Laetitia Benard (GRAM EA 2656 and UMR 6634)

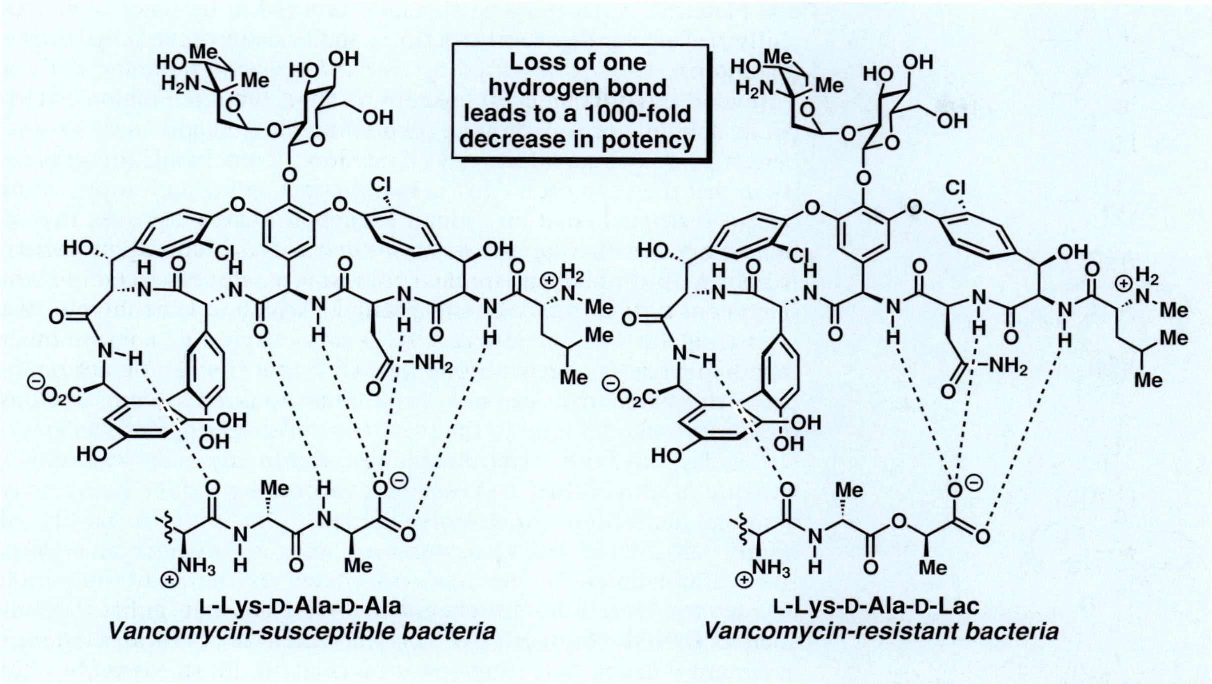

Loss of one hydrogen bond leads to a 1000-fold decrease in potency

L-Lys-D-Ala-D-Ala
Vancomycin-susceptible bacteria

L-Lys-D-Ala-D-Lac
Vancomycin-resistant bacteria

Scheme 1. The hydrogen-bond network between vancomycin and its biological target L-Lys-D-Ala-D-Ala and the development of resistance to vancomycin.

In response to this ingenious mode of bacterial defiance, chemists are continuing their search for new families of antibiotics, but they are also seeking to modify vancomycin (**1**) at the molecular level with the hope of restoring its efficacy against these new resistant strains.[5] To achieve this latter objective, access to a synthetic route for this natural product would unquestionably be beneficial in the efforts to introduce such structural adjustments. However, just as bacteria are proving resilient to the mode of action of vancomycin, its highly complex structure has similarly frustrated prodigious efforts to accomplish its total synthesis ever since its complete molecular constitution was disclosed in the early 1980s.[6] At the heart of this challenge (see Scheme 2) is an unparalleled conglomeration of sensitive functional groups and overall stereochemical complexity that includes a β-linked glycoside, an amide bond in a non-natural *s-cis* configuration, and 18 stereogenic centers (9 of which reside on the aglycon portion). While the stereoselective formation of such features is a standard problem in any target-oriented endeavor, this general issue takes on a slightly different tinge with vancomycin as three of its seven amino acid subunits are arylglycines, building blocks whose lone stereogenic centers are readily epimerized upon exposure to base. Thus, once any of these amino acids has been installed into a synthetic intermediate en route to vancomycin (**1**), the remaining (and likely lengthy) sequence to

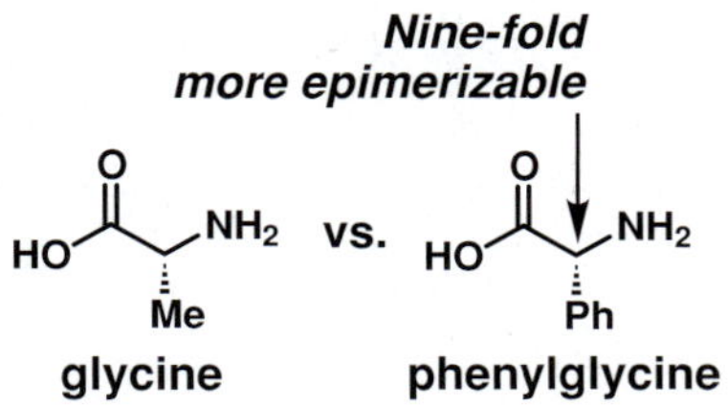

Nine-fold more epimerizable

glycine vs. phenylglycine

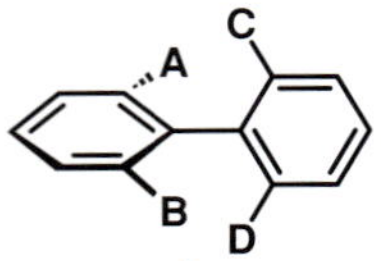

3

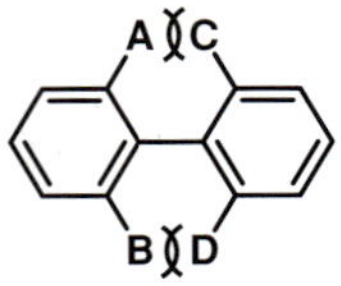

4

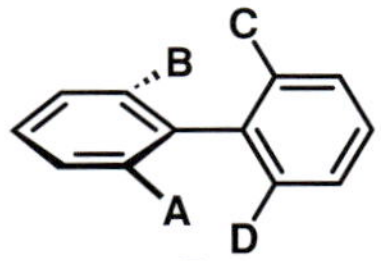

5

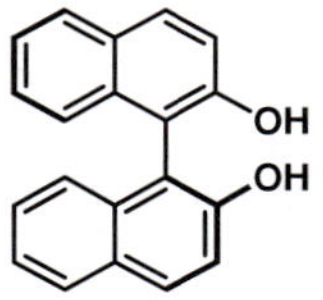

6: (S)-BINOL

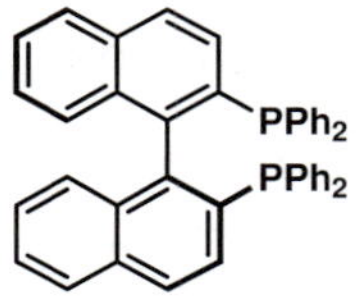

7: (S)-BINAP

complete the target must be specially tailored in terms of both protecting groups and general reaction conditions to prevent their racemization.

Beyond this fundamental concern, each of the three major macrocyclic subunits of vancomycin (two 16-membered and one 12-membered) possesses an additional element of stereochemical complexity in that they can each exist in two distinct and enantiotopic forms due to restricted rotation around their biaryl axes (as indicated in Scheme 2 with circular arrows on these linkages). This phenomenon is known as atropisomerism, and is most typically observed in compounds of type **3** (see column figures) in which at least three of the sites assigned with the letters A to D are occupied by a group other than hydrogen (as occurs along the AB biaryl linkage of vancomycin). Atropisomerism can also exist in constrained systems such as the bisaryl ethers found in the two 16-membered rings of vancomycin, as long as both enantiomers are isolable and each possesses a half-life of longer than 1000 seconds (according to Oki's arbitrarily assigned definition for atropisomerism).[7] In fact, the ability of highly substituted biaryl systems to exist as distinct and long-lived atropisomers is the basis for many of the most important asymmetric reactions in chemical synthesis that enlist ligands such as BINOL (**6**) or BINAP (**7**), since their axial chirality confers asymmetry to the transition states involved in these events.[8]

The only real technique available to convert one atropisomeric form into another (i. e. **3** into **5**) is the application of sufficient thermal activation to overcome the rotational energy barrier between the two, which, depending on the system, can require 30–40 kcal/mol. Importantly, an atropisomer exposed to these conditions will exist as an equimolar mixture of both possible atropisomeric

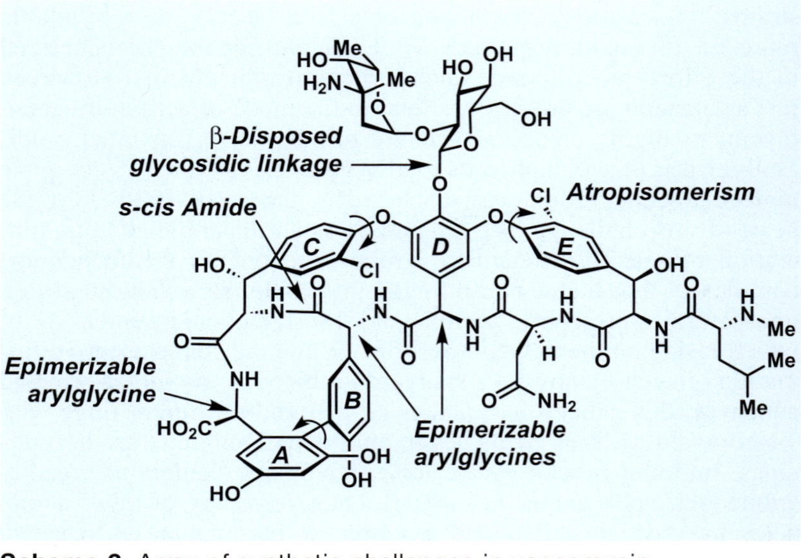

Scheme 2. Array of synthetic challenges in vancomycin.

forms unless there is an inherent reason for a thermodynamic preference. Thus, just like stereocenters, unless an atropisomer is generated enantiospecifically from the outset, chances are generally low that a mixture can be converted into a single atropisomeric form. Unfortunately, although we currently possess a collection of powerful transformations such as Suzuki or Stille couplings to fashion the systems that often exist as atropisomers, we have not yet developed sufficient methodology capable of selectively creating a single atropisomer. As a representative example of this conundrum, consider the total synthesis of the anti-HIV agents michellamines A and B (**11** and **12**) by Bringmann and co-workers in which a tandem Suzuki reaction between bisaryl triflate **8** and 2 equivalents of boronic acid **9** was employed to fashion the complete skeletal framework of the target molecules (Scheme 3).[9] This step, while highly convergent, lacked complete atropselectivity in that following two deprotection operations, **11** and **12** were obtained in a ratio of 1:2.5. Since both final atropisomeric products were natural products, the general absence of atropselectivity in this synthesis was acceptable, and even beneficial, from a tactical standpoint. For vancomycin, however, for which only a single atropiso-

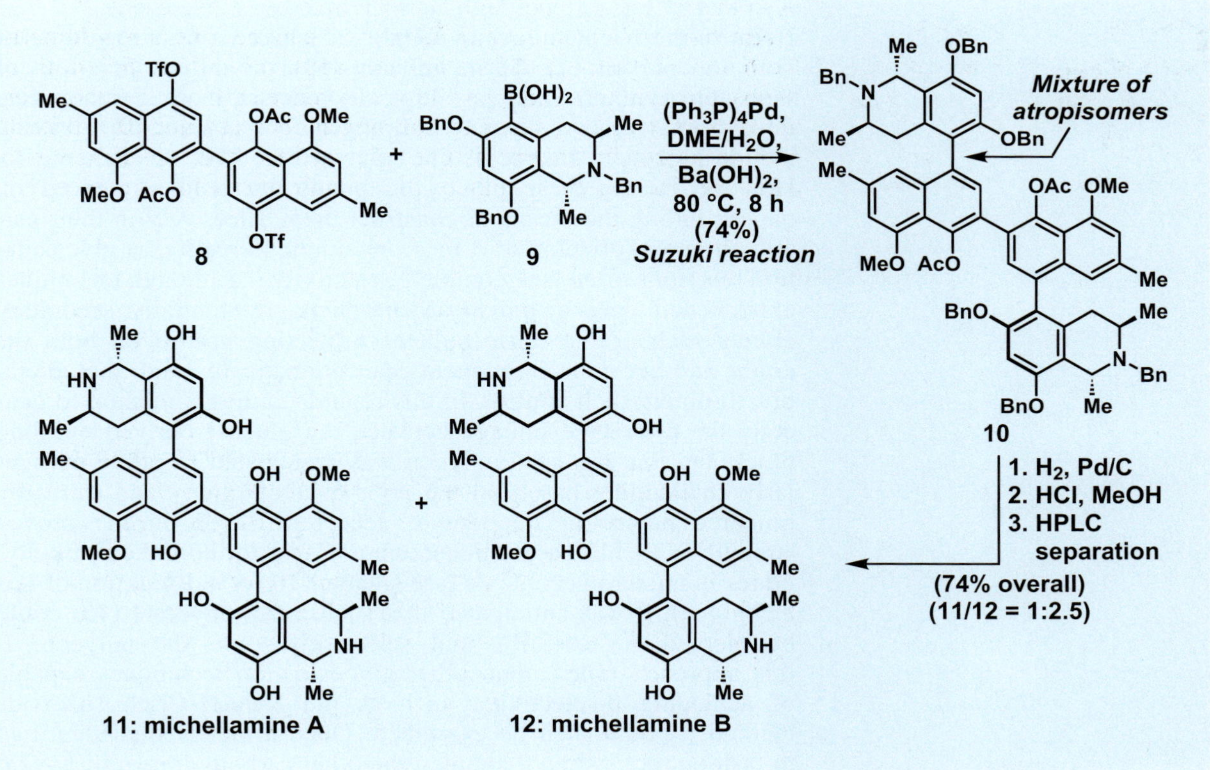

Scheme 3. The challenges imposed by atropisomers in the total synthesis of natural products, as evidenced by Bringmann's route to the michellamines (**11** and **12**).

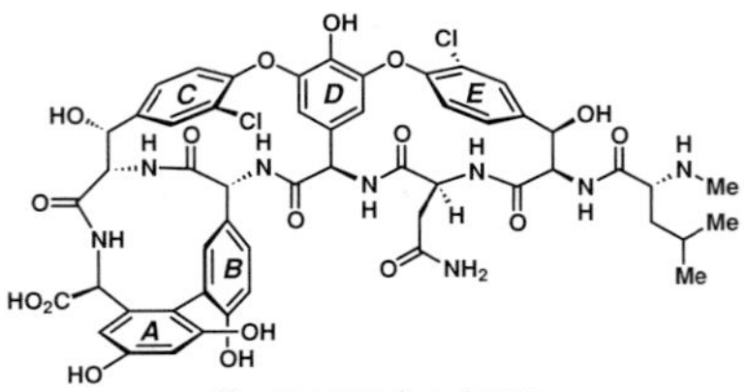

meric form out of eight total possibilities reflects the conformational disposition of the natural product, such a lack of control in generating any of the three atropisomeric axes of vancomycin would be inherently deleterious to the elegance and efficiency of the overall synthetic sequence.

Fortunately, this issue, along with all the other challenges discussed above, was successfully addressed in 1998 and 1999 with the first chemical synthesis of vancomycin (**1**) by the Nicolaou group at The Scripps Research Institute,[10,11] and its aglycon (**13**, see column figure) by David Evans and his co-workers at Harvard University.[12,13] For the remainder of this chapter, we will analyze these unique and insightful solutions in detail, each of which required the investment of several years in basic research to develop the special synthetic strategies and tactics that were ultimately called upon to tackle the most nefarious architectural elements of vancomycin.

13: vancomycin aglycon

9.2 *Retrosynthetic Analysis and Strategy*

9.2.1 *Nicolaou's Synthetic Approach to Vancomycin (1) and the Vancomycin Aglycon (13)*

Because the Nicolaou group sought to achieve a total synthesis of both vancomycin (**1**) and its aglycon (**13**), the initial operations of their retrosynthetic analysis logically excised both carbohydrate units from **1** as two separate building blocks (**14** and **15**), leaving behind a protected form of the aglycon core (**16**, see Scheme 4). However, while these glycosidic bonds are readily cleaved on paper, finding the means to construct them selectively in their natural anomeric orientation is far from simple in reality, as glycosidation reactions often lack complete selectivity.[14] Although techniques exist to deliver α- or β-disposed products preferentially, secondary effects such as the steric bulk of protecting groups on both the donor and acceptor components can dramatically, and unpredictably, diminish such control. In this regard, finding a method to generate the β-directed linkage between the glucose-derived building block and the protected aglycon was anticipated to prove particularly challenging based on the poor nucleophilicity and sterically hindered nature of the phenol acceptor. Based on the proven strength of trichloroacetimidate donors[15] to afford β-linked carbohydrates in numerous contexts (see Chapter 10 for a discussion of this technique), it was anticipated that such a carbohydrate (**15**) could be merged successfully, and selectively, onto the aglycon. If this approach failed, then other glycosidation techniques capable of achieving β-selectivity, such as the Kahne–Crich sulfoxide method,[16] could then be examined. Once a means was identified to fashion the desired β-anomer, the other carbohydrate unit, L-vancosamine, could be merged to complete the vancomycin skeleton through one of the many α-directing anomeric substituents such

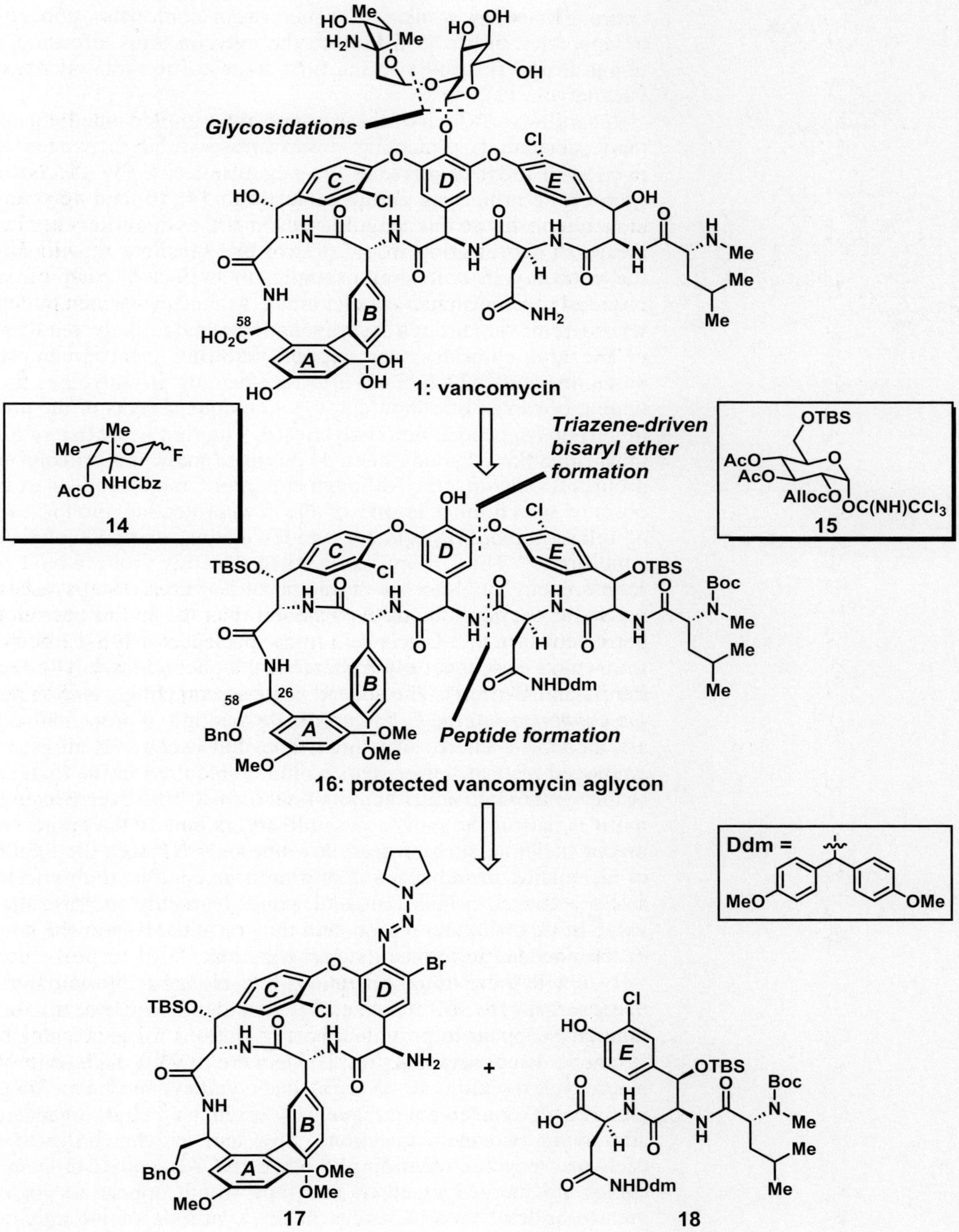

Scheme 4. Nicolaou's retrosynthetic analysis of vancomycin (**1**): initial stages.

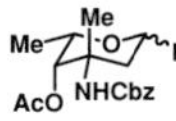

as the fluoride in **14**.[17] Alternatively, and more convergently, the entire glycosidic domain of vancomycin could also conceivably be appended in block fashion to the aglycon, thus affording additional design flexibility in the final stages of the total synthesis of vancomycin (**1**).

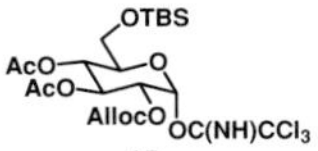

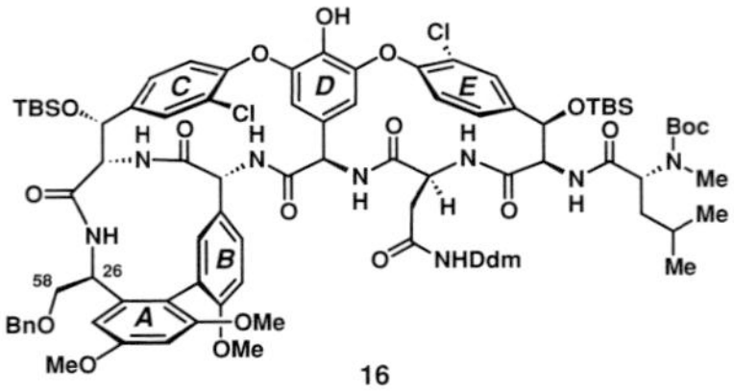

Regardless of which of these options was implemented, the global (and substantial) protecting group array would then need to be excised in order to arrive at the target molecule (**1**). Accordingly, each of the protecting groups attached to **14**, **15**, and **16** was judiciously chosen at this stage based on the assumption that certain means of deprotection would destroy the sensitive functionality in the vancomycin core. For example, the selected groups avoided obviously problematic choices such as numerous benzyl ethers, whose removal through hydrogenation would likely lead to loss of the aryl chlorides, and groups requiring cleavage by strong bases, reagents which could also racemize the arylglycines. In particular, however, one should note two unique aspects of the protecting groups appended onto aglycon **16**. The first is the use of a Ddm (4,4′-dimethoxydiphenylmethyl) group to mask the primary amide group of vancomycin. Although this motif might appear to be an obscure selection, it is one of the few protecting groups capable of reliably guarding primary amides against a wealth of reaction conditions.[18] Thus, it was envisioned that this group would prove robust enough to survive the numerous synthetic steps needed to reach **16**, but could be cleaved smoothly at the end of the synthesis upon exposure to a Lewis acid in the presence of EtSH (conditions that would also hopefully rupture all the phenolic methyl ethers on the A- and B-rings). The second critical protecting group feature is the conversion of the C-58 carboxylic acid into a protected secondary alcohol, a choice one might question because if an ester was employed instead, fewer steps would be required in the final stages of the synthesis to adjust it to its final format. However, because this motif is part of the most accessible arylglycine in the entire vancomycin skeleton, such protection would only enhance the likelihood of racemizing the adjacent C-26 proton. In contrast to this scenario, a less activated primary alcohol would markedly increase the pK_a value of this benzylic proton, and thus raise the bar for the strength of base needed to induce its epimerization.

Even with these initial operations implemented, a formidable synthetic target (**16**) still remained for consideration whose size would inherently appear to provide a host of options for reasonable retrosynthetic disconnections. Yet, if one were to view each of the three macrocycles within **16** as a distinct synthetic problem, then the plethora of conceivable sequences could be better organized by addressing two main questions: how and in what order to form each macrocycle. Accordingly, since the AB and C-O-D macrocycles are merged together, and thus would appear to possess a greater bulk of the challenges in the synthesis of the aglycon, it seemed prudent to disassemble the D-O-E macrocycle first. Still, the question remained as to how this general approach could be implemented best. Based on a series of model studies performed by the Nicolaou group in combination with progress toward vancomycin (**1**) and other glycopeptide antibiotics reported by several other researchers,[2,11] the optimal answer to this query appeared to be initial peptide formation between the carboxylic acid in a tripeptide such as **18** and the free amine of an advanced AB/C-O-D system (**17**), followed by macrocyclization through bisaryl ether formation to complete the 16-membered D-O-E ring. While the first of these reactions was almost all but assured for the forward synthesis based on the powerful arsenal of peptide-forming reagents,[19] the second requirement left several different possibilities open, each with unique advantages and disadvantages.

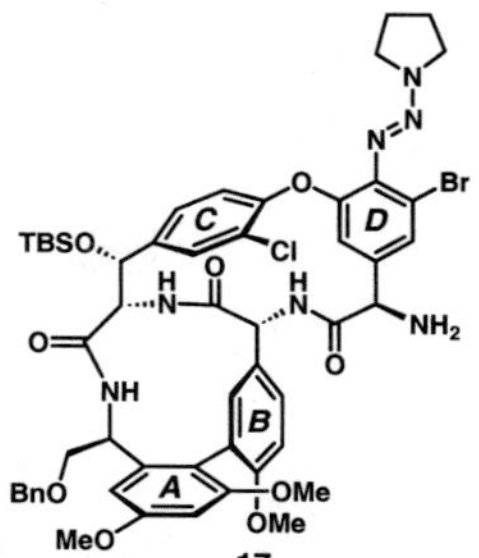

As shown in Scheme 5a, one of the most powerful methods to form bisaryl ethers is nitro-group activated nucleophilic aromatic substitution ($\mathrm{S_NAr}$). Indeed, as indicated by the model studies executed by Beugelmans, Zhu, and co-workers (**19**→**21**[20] and **23**→**24**[21]), this reaction can proceed smoothly and in high yield to afford bisaryl ether systems pertinent to vancomycin (**1**). Equally significant, although a base is required to effect this macrocyclization, the conditions are mild enough such that no epimerization of the C-36 arylglycine center was observed in either case. Since a nitro group is employed to drive the reaction, as a consequence this motif must then be exchanged following macrocyclization through a Sandmeyer reaction to incorporate the final chlorine substituents of the C- and/or E-rings. While this requirement is not problematic, the larger issue brought into stark relief by these examples is the potential for this method of ring closure to provide atropselectivity. Unfortunately, the cyclization of **19** resulted in an equimolar mixture of both atropisomeric products, while the impressive tandem macrocyclization of **23** placed the nitro groups in their non-natural configuration in both 16-membered ring systems. While each of these results was discouraging, as the first was nonselective and the second could not possibly lead to vancomycin (**1**), the latter did indicate that the formation of a single atropisomer through bisaryl ether closure was possible.

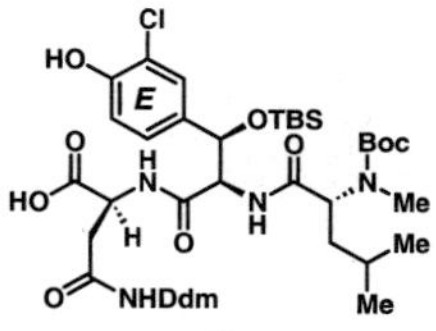

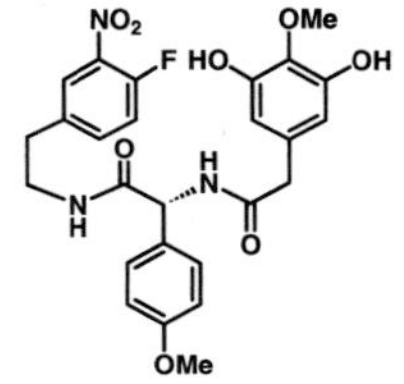

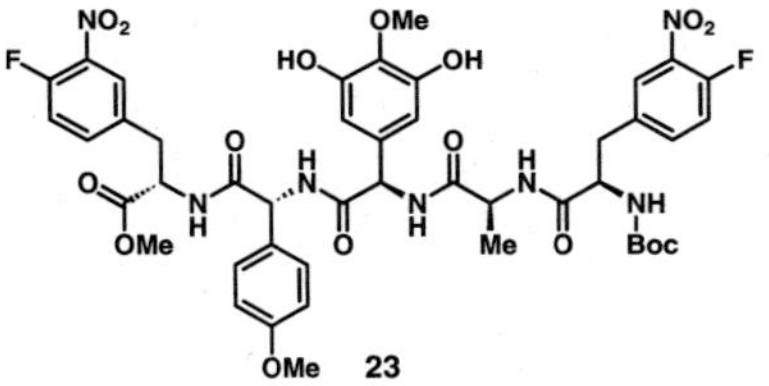

Apart from this powerful approach, a more "biomimetically" inspired bisaryl ether macrocyclization would rely on an oxidative coupling reaction between a halogenated phenol donor and an aryl halide acceptor mediated by a reagent such as thallium(III) nitrate.[22] Although this activator provides quite mild reaction conditions, the major handicap incurred by adopting this method is the requirement that the donor must possess a 2,6-dihalogenation pattern for proper activation. As such, one of these halogens must be removed selectively following cyclization to afford ring systems relevant to vancomycin. As Yamamura has illustrated with a creative solution, however, this potential weakness can be overcome by using a mixed halogenated substrate (**25**) for the macrocyclization step, followed by hydrogenative removal of the more labile bromine atom to generate a monochlorinated product (atropisomers **28** and **29**).[23] As an alternative cyclization technique, one could also

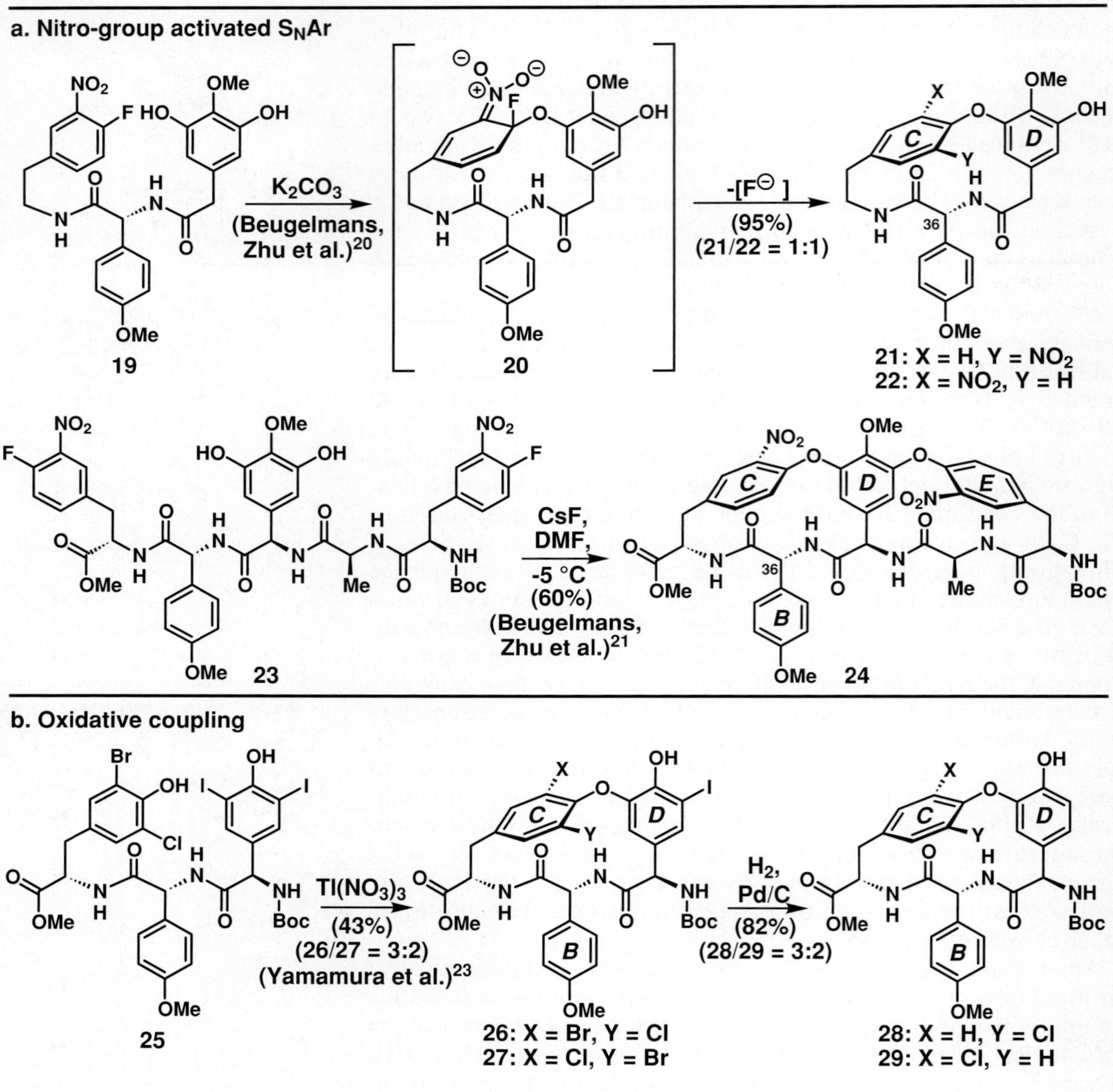

a. Nitro-group activated S_NAr

b. Oxidative coupling

Scheme 5. Methods of synthesizing cyclic bisaryl ethers.

employ a copper-catalyzed Ullmann-type coupling reaction to form the bisaryl ethers of vancomycin. Dale Boger and his group at The Scripps Research Institute first demonstrated the feasibility of this general approach through the conversion of model system **30** into **31** in 59 % yield.[24] However, successful macrocyclization in this case was attended, to a minor extent, by glycine epimerization (~5 %), a problematic result because this amino acid is nine times

c. Classical Ullmann reaction applied to bisaryl ether synthesis

30 → 31: NaH (3 equiv), $CuBr{\cdot}Me_2S$, py, Δ (59%) (5% epimerization of alanine residue) (Boger et al.)[24]

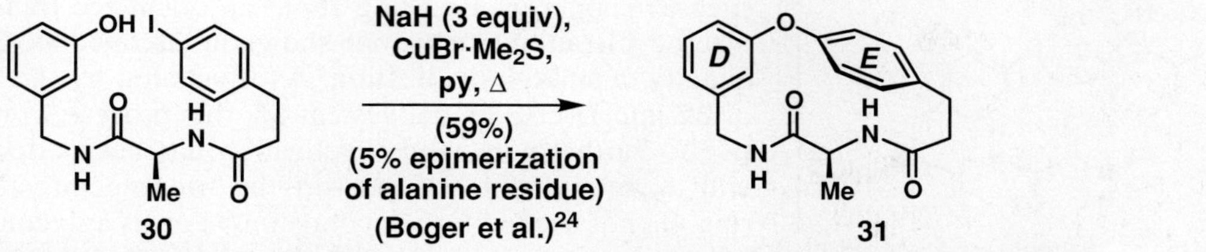

d. Triazene-driven bisaryl ether synthesis

32 → [33]: K_2CO_3, $CuBr{\cdot}Me_2S$, py (Nicolaou et al.)[25]

[33] → [34] → 35 (54%)

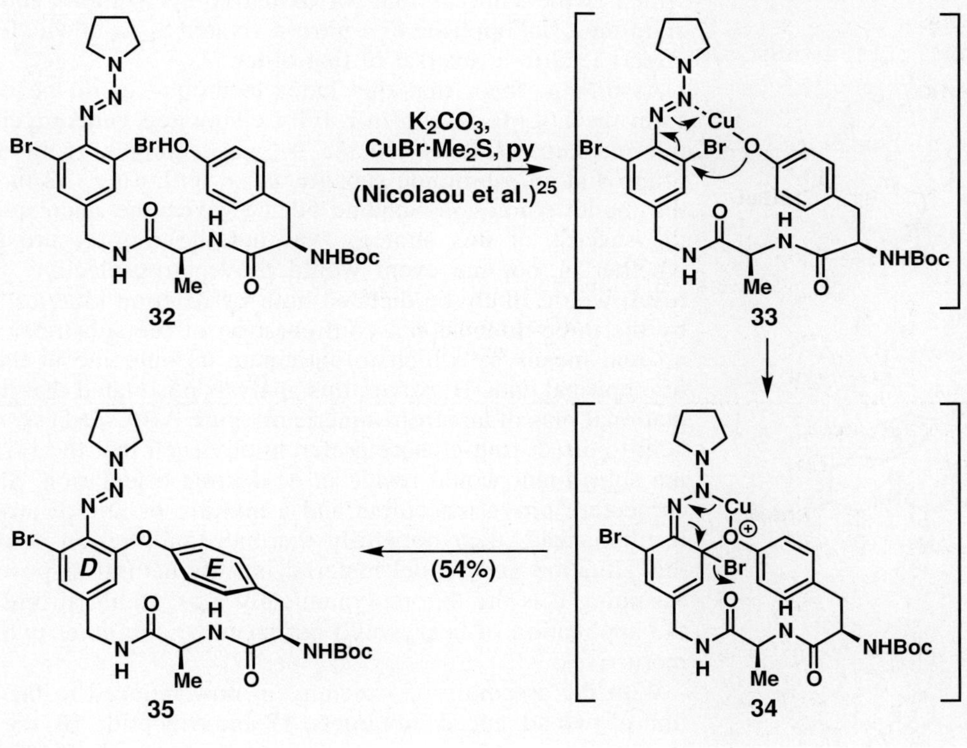

Scheme 5. Continued.

less activated than an arylglycine for racemization. As such, conditions milder than those employed in this example would most likely be needed to implement this strategy effectively in vancomycin-relevant systems in order to achieve acceptable stereocontrol.

While each of these three methods for bisaryl ether synthesis (among several others) could potentially be applied in the macrocyclization step needed to construct the D-O-E system, the Nicolaou group decided instead to develop their own means to form such systems using a triazene moiety on the aromatic acceptor to drive the ring closure (see Scheme 5d).[25] This approach was predicated on the idea that since this motif would likely constitute an electron sink, the addition of an appropriate metal cation to the reaction mixture, such as copper, could bring about an organized transition state for macrocyclization like the one shown in intermediate **33**, and, consequently, a smooth cyclization. As illustrated by the transformation of **32** into D-O-E model system **35**, this process works quite effectively. Furthermore, once cyclization has been effected, treatment with aqueous acid can convert the triazene directly into the D-ring phenol needed for vancomycin (**1**) or its aglycon (**13**). Incidentally, it is worth noting that the reactivity order for halide displacement on the acceptor ring with this methodology is I>Br≫Cl>F, which is the same as that for oxidative cyclizations and Ullmann couplings, but opposite to *o*-nitro-activated S_NAr in which inductive effects lead to a reversal of that order.

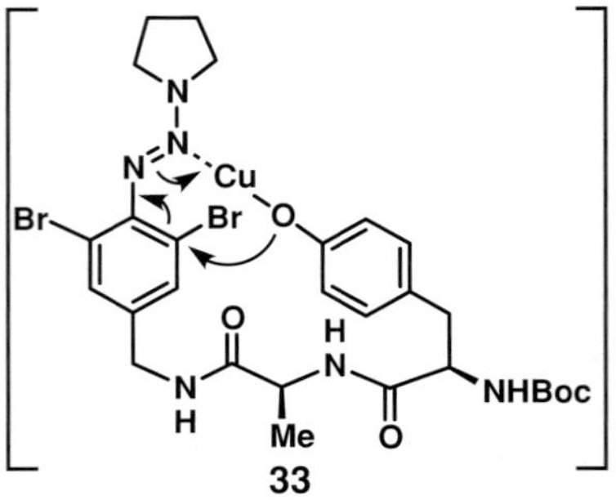

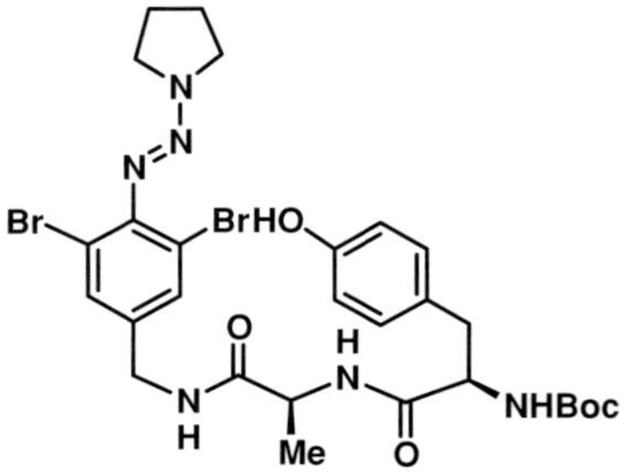

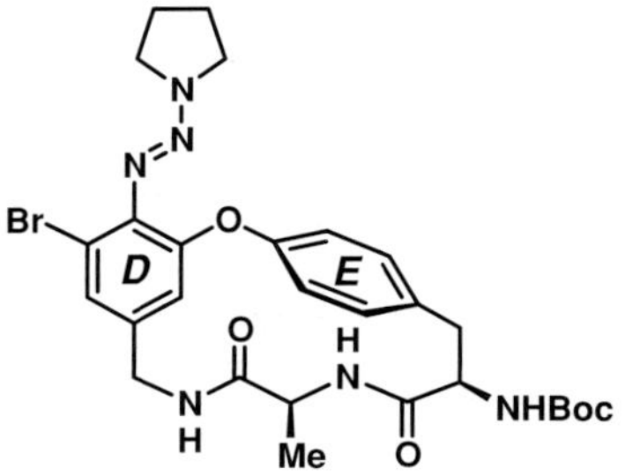

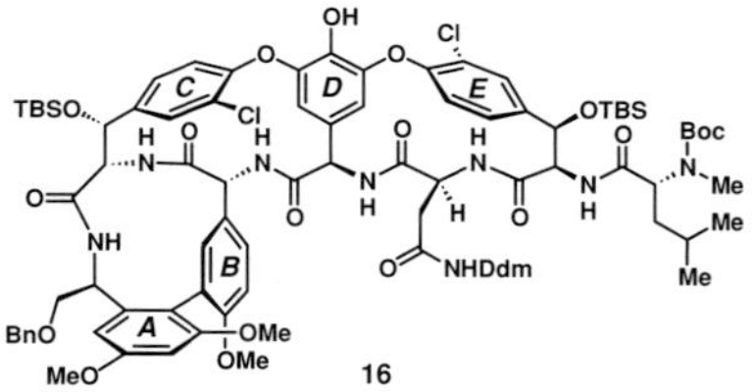

Assuming, then, that this latter technique could be enlisted to form the D-O-E system in a fully elaborated vancomycin system, intermediate **17** (see Scheme 4) was equipped with a triazene group and an *o*-bromide substituent on its D-ring. As indicated by the model studies in Scheme 5d, however, the main question for the success of this strategy was not the ring-closure itself, but whether or not the event would prove atropselective. Since this result would likely be dictated both by reaction kinetics as well as by the three-dimensional configuration of the substrate, there was no real means by which to anticipate its outcome in the absence of empirical data. However, this analysis postulated that the conformational bias of an atropisomerically pure AB/C-O-D system might help to direct ring-closure preferentially such that the E-ring chlorine substituent would reside in its desired orientation. Should this conjecture prove inaccurate and a mixture of D-O-E atropisomers result instead, then hopefully thermal equilibration could correct this situation and funnel material into its natural disposition (**16**), assuming it is the thermodynamically most stable atropisomer and the application of heat would not racemize the other two atropisomeric axes.

With the assembly of vancomycin now reduced to the construction of two advanced subtargets, **17** and tripeptide **18**, the Nicolaou group first sought to tackle retrosynthetically the larger and far more challenging of these two goal structures, AB/C-O-D intermediate **17** (see Scheme 6). Once again viewing each macrocycle within this target structure as a separate synthetic problem, in principle, either of these large rings could be fashioned first, with the second then built upon it with the functionalities of its properly decorated substructure. Apart from simplicity, such a strategy would also serve a tactical purpose in that the rigidity of the first macrocyclic ring would likely facilitate the formation of the second by decreasing the entropic freedom of the reactive motifs needed for macrocyclization, bringing them into a favorable conformation for a productive union. To best take advantage of this idea, it was decided first to open the AB macrocycle, since a 12-membered ring is far more

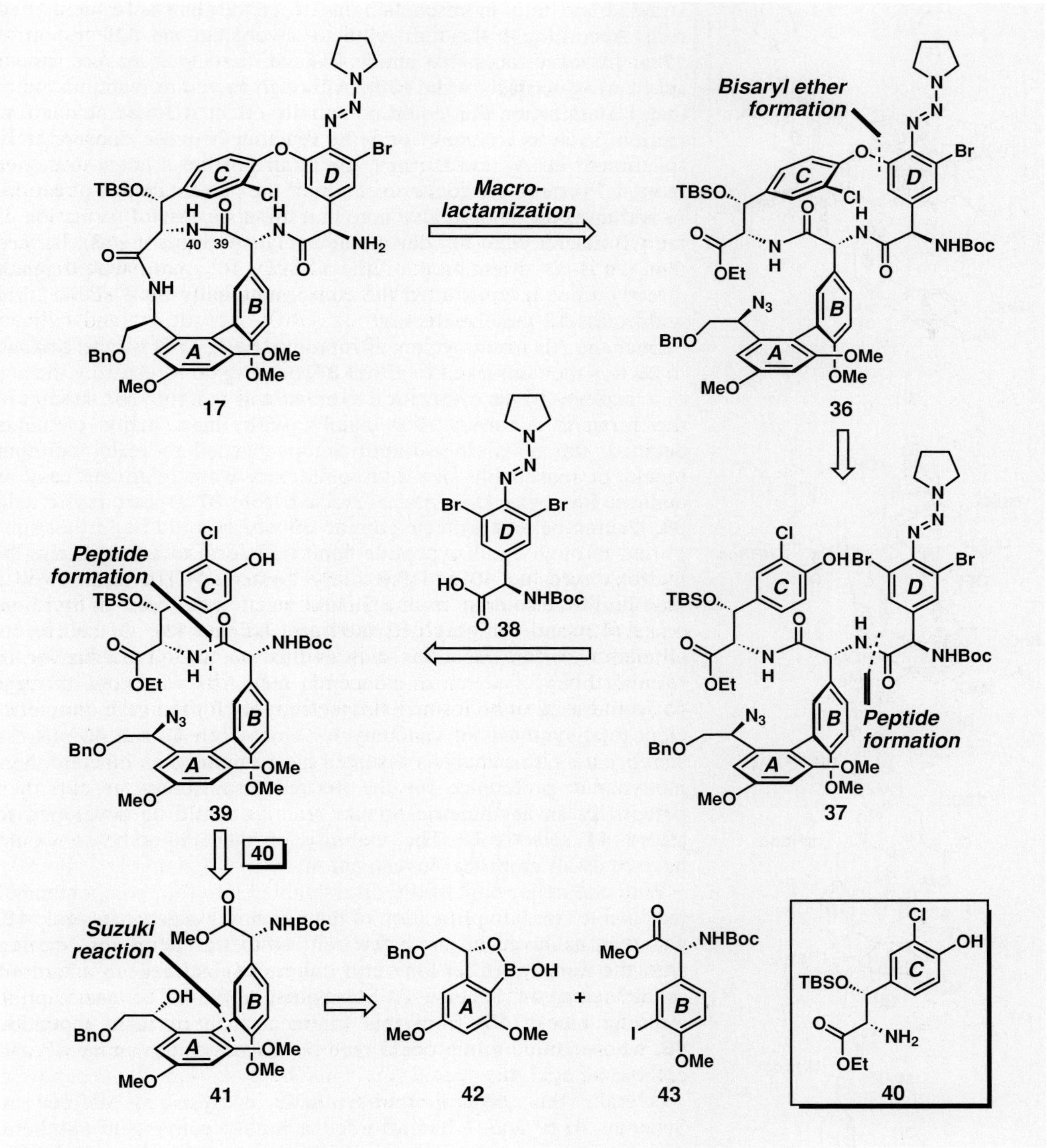

Scheme 6. Nicolaou's retrosynthetic analysis of AB/C-O-D fragment **17**.

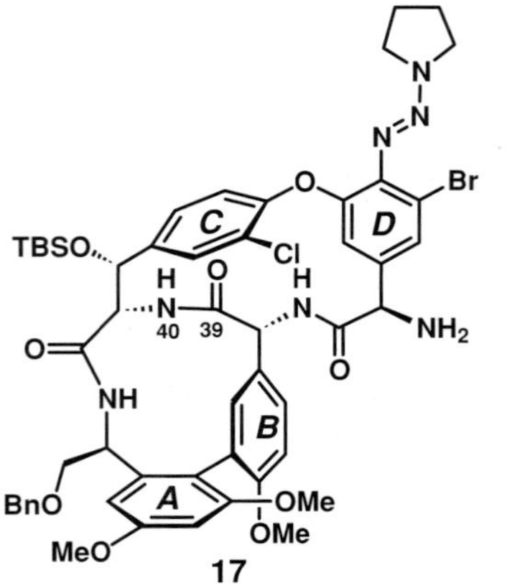

strained and thus more challenging to create than a 16-membered ring. Accordingly, the retrosynthetic sword cut the AB system in **17** at its most accessible amide linkage, leaving a masked amino acid surrogate in its wake (**36**). Although there are reactions other than lactamization that could potentially effect AB macrocycle formation, such as a Suzuki or Stille reaction between appropriately functionalized A- and B-rings, the failure of these latter strategies in model systems led to the selection of the present course of action. In addition, one should also note that upon successful formation of the AB macrocycle (**17**) during the actual synthesis it was assumed that the *s-cis* orientation of the C-39/N-40 amide would result directly since it constituted the conformationally most stable form within this 12-membered ring.

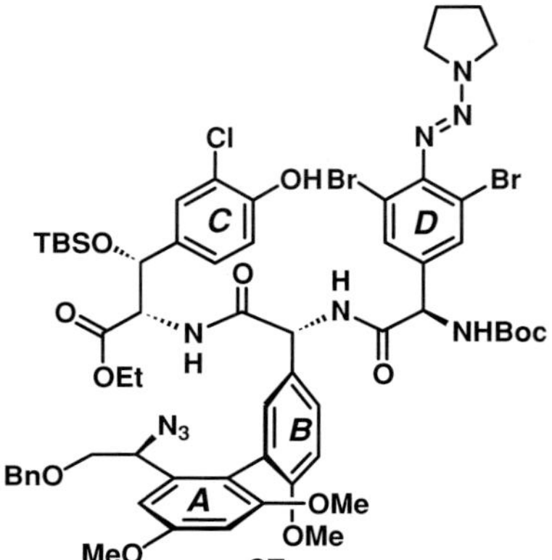

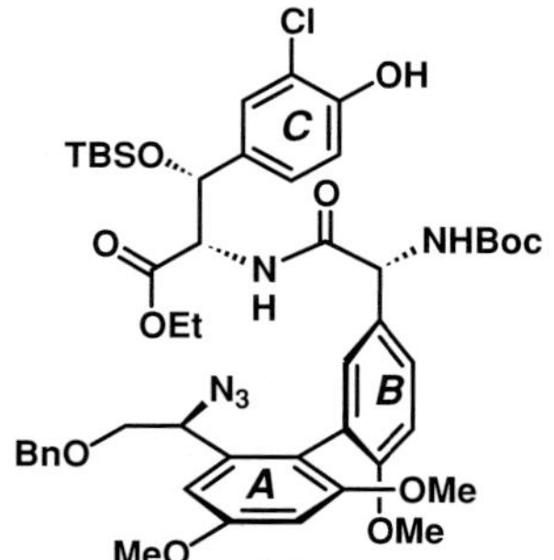

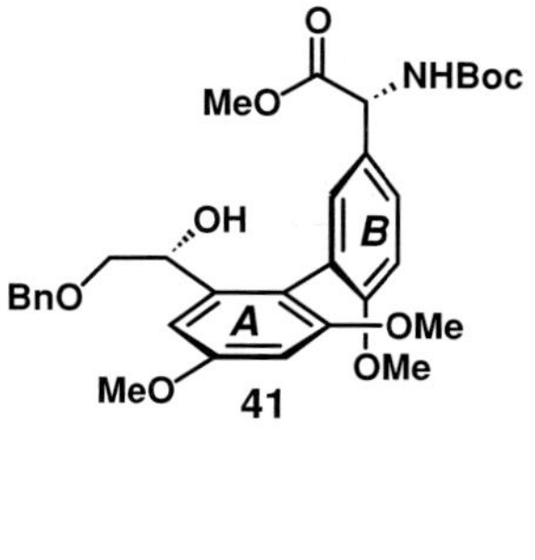

Once the AB macrocycle was ruptured, the C-O-D bisaryl linkage in **36** was then unlocked to afford **37**, once again forecasting the use of a triazene-driven cyclization to create this macrocyclic subunit in the forward synthesis. Fortunately, with these major elements decided, the remaining simplifications needed to reach building blocks of reasonable size and complexity were relatively easy to deduce. First, the D-ring was excised from **37** as carboxylic acid **38**, leaving behind protected amine **39** which could be further simplified through another peptide-band transform to afford C-ring β-hydroxy tyrosine **40** and AB biaryl system **41**. In turn, **41** was then projected to arise from a Suzuki reaction between an aryl boronate (**42**) and a highly activated aryl halide (**43**). Of course, as alluded to earlier, reactions such as this one are quite effective in forming biaryl systems in a racemic form. As we know, though, **41** would need to be formed atropselectively for the sake of an efficient total synthesis of vancomycin. Since both **42** and **43** possess stereocenters, this analysis assumed there might be an inherent thermodynamic preference for the desired atropisomer; if not, then potentially an asymmetric Suzuki reaction could be developed to afford **41** selectively. The viability of these hypotheses would have to await empirical investigation.

With one major target fully disassembled into four components of reasonable size, simplification of the remaining piece, tripeptide **18**, was then achieved in just a few retrosynthetic operations, starting with the removal of its lone aryl chlorine substituent to afford **44** as delineated in Scheme 7. Subsequent removal of one peptide building block (**45**) from this fragment then revealed dipeptide **46**, whose amide group could then be dissected into amine **47** and carboxylic acid **48**.

Overall, this general retrosynthetic analysis as defined in Schemes 4, 6, and 7 has afforded a highly convergent synthetic plan for vancomycin (**1**) requiring the stereospecific creation of two functionalized carbohydrates and seven amino acid building blocks of varying complexity. With little question, many of their stereocenters could derive from commercially available members of the chiral pool. For some of the more complex of these fragments, the Nicolaou group envisioned that a series of asymmetric

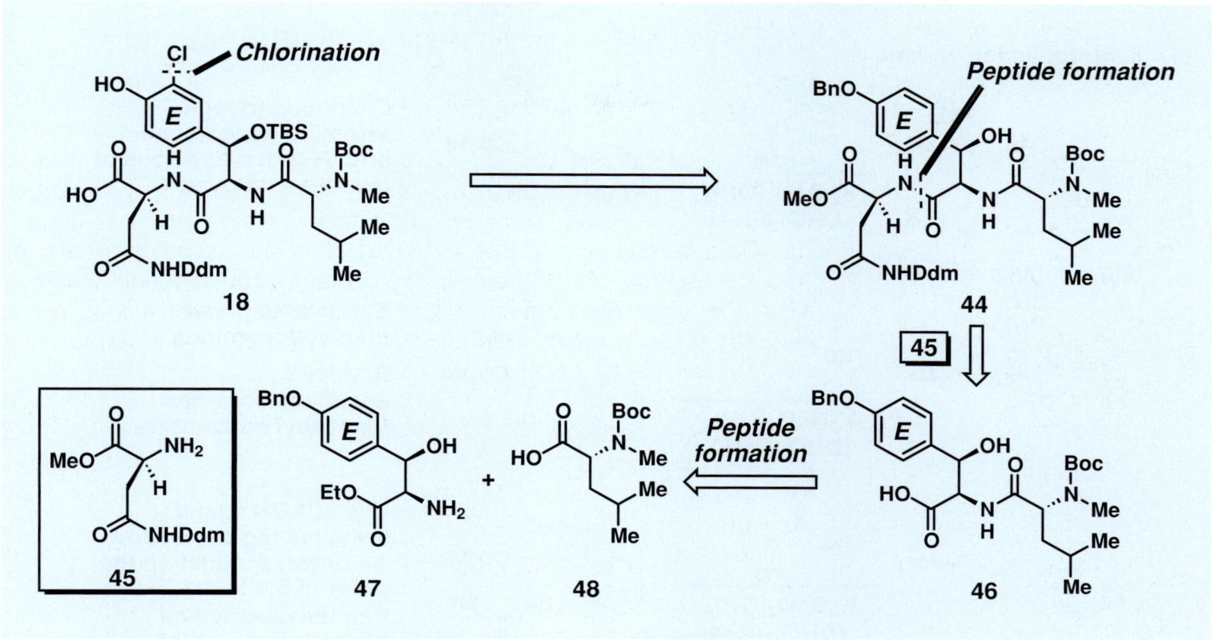

Scheme 7. Nicolaou's retrosynthetic analysis of tripeptide **18**.

reactions, especially those developed by K. Barry Sharpless, could prove quite valuable in fashioning their stereochemical arrangements from achiral starting materials. Of particular concern was the task of efficiently fashioning the adjacent amino alcohol motifs within the two β-hydroxy tyrosine pieces (building blocks **40** and **47**), since traditional routes to such compounds are lengthy due to their reliance on multiple protecting group manipulations. Fortunately, in 1996, Sharpless unveiled a new asymmetric reaction, an osmium-catalyzed aminohydroxylation (AA), which looked as if it could prove especially valuable in creating these building blocks more expeditiously.[26,27]

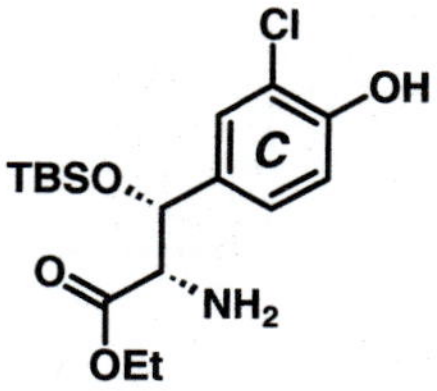

As shown in Scheme 8, the Sharpless AA reaction enables the stereospecific delivery of a protected nitrogen (either as a sulfonamide or as a carbamate) and an alcohol to the same side of a double bond in a single step, governed by the same facial selectivity rules as the Sharpless asymmetric dihydroxylation (AD) reaction discussed extensively in Chapter 35 of *Classics I*.[28] Although experience has revealed that this reaction is not as general as the venerable Sharpless AD, it works particularly well for those substrates that would be needed in the vancomycin context as indicated by the generalized entries in Scheme 8. While typical AA conditions in the presence of the standard PHAL ligand lead to products in which the protected nitrogen atom is α to the phenyl ring, the alternate regioselectivity (which is required for vancomycin) can be obtained merely by exchanging this motif for a quinone ligand, such as

40

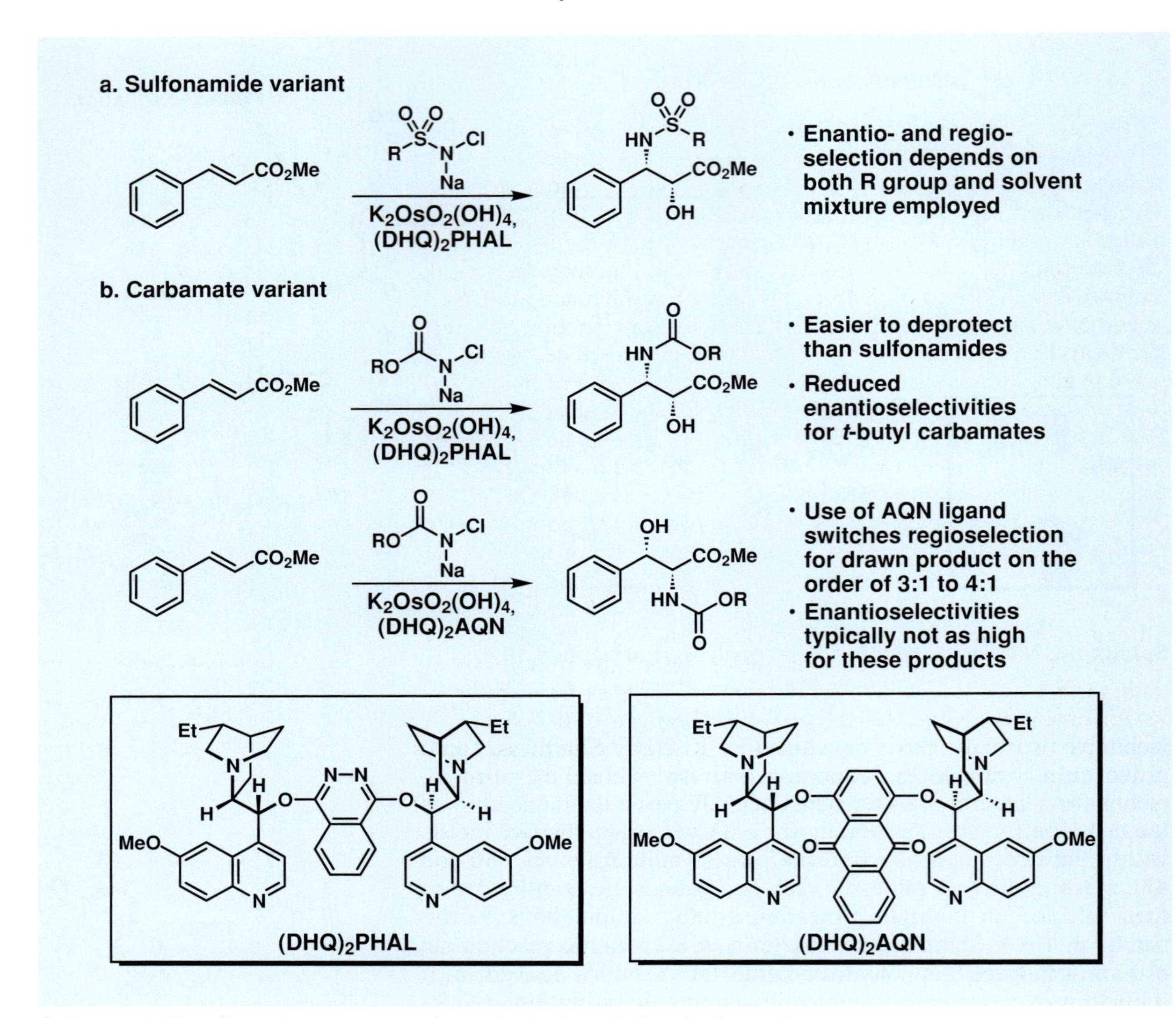

Scheme 8. The Sharpless asymmetric aminohydroxylation (AA) reaction.

AQN.[29] Although this variant of the catalytic AA reaction usually affords 1,2-amino alcohol products with decreased enantioselectivity relative to those reactions using PHAL, it is currently one of the most concise solutions for the stereospecific synthesis of β-hydroxy tyrosines.

9.2.2 *Evans' Synthetic Approach to the Vancomycin Aglycon (13)*

Although complementary in some minor respects to that of the Nicolaou analysis, the synthetic approach developed by Evans' group to vancomycin aglycon (**13**) sought quite unique solutions to its numerous synthetic challenges, particularly in regard to the order and method of forming each macrocycle. In the initial stages, these researchers similarly excised the D-O-E macrocycle from the central framework, anticipating the use of a nitro-activated S_NAr macrocyclization rather than a triazene-driven reaction to create its bisaryl ether, thus leading to advanced intermediates **49** and **50** as new goal structures (see Scheme 9). The selection of this method of cyclization was predicated upon both the successes delineated in Scheme 5 as well as their own productive experiences with this reaction in their 1997 total synthesis of orienticin C aglycon (**55**, vancomycin aglycon minus the aryl chlorine atoms),[30] the first member of this class of glycopeptide antibiotics to be synthesized in the laboratory. As mentioned earlier, by adopting this strategy, a Sandmeyer reaction would then be required to convert the nitro group (in what would hopefully be the correct D-O-E atropisomer) into an aryl chloride. Apart from these design elements, as we already know from the discussion above, an equally important consideration at this stage was the selection of an appropriate protecting group array to respond tactically to critical aspects of the chemically sensitive functional groups of vancomycin. While general aspects of the groups appended in this manner onto both **49** and **50** mirror the rationales mentioned in the Nicolaou analysis, one critical feature of this approach, the conversion of the C-58 carboxylic acid into a methyl amide, is worthy of additional commentary. As has been demonstrated in numerous contexts, such protection effectively deactivates the carboxylic acid, thereby assuring that the neighboring proton at C-26 would not be overly acidic and thus prone to epimerization during the synthetic sequence.[31] Moreover, the robustness of this amide to multifarious reaction conditions could almost guarantee its survival in the steps leading up to **13**; the price for this behavior was the relatively vigorous conditions that would be required to dismantle it on a highly functionalized substrate at the end of the sequence. Based on several studies carried out in their laboratories, the Evans group hoped that this potential problem could be circumvented by first appending an electron-withdrawing group onto its free secondary position (such as a nitrosamide, see the adjoining column), since the resultant product would be more susceptible to nucleophilic lysis.[32,33] To implement this strategy successfully, however, the C-58 methyl amide would have to be engaged preferentially by a nitrosating reagent in the presence of a host of primary and secondary amides.

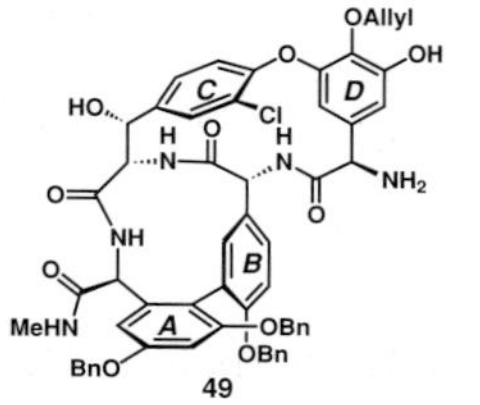

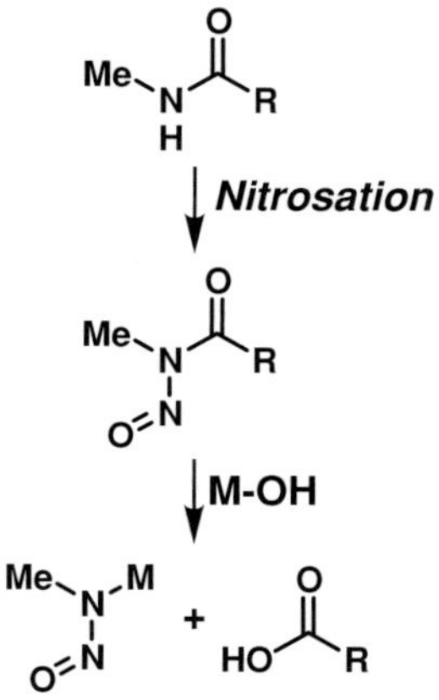

Having dissected vancomycin aglycon (**13**) into two major pieces, the simpler of these, tripeptide **50** (Scheme 9), was then disconnected into three peptide building blocks (**52**, **53**, and **54**) through

49

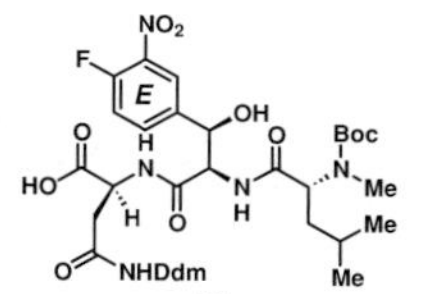

50

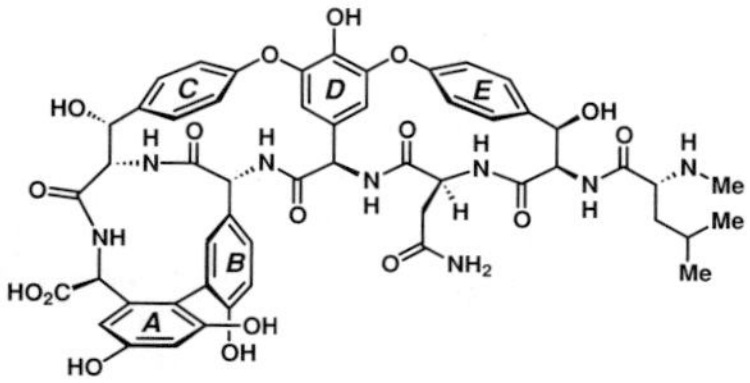

55: orienticin C aglycon

Nitrosation

M-OH

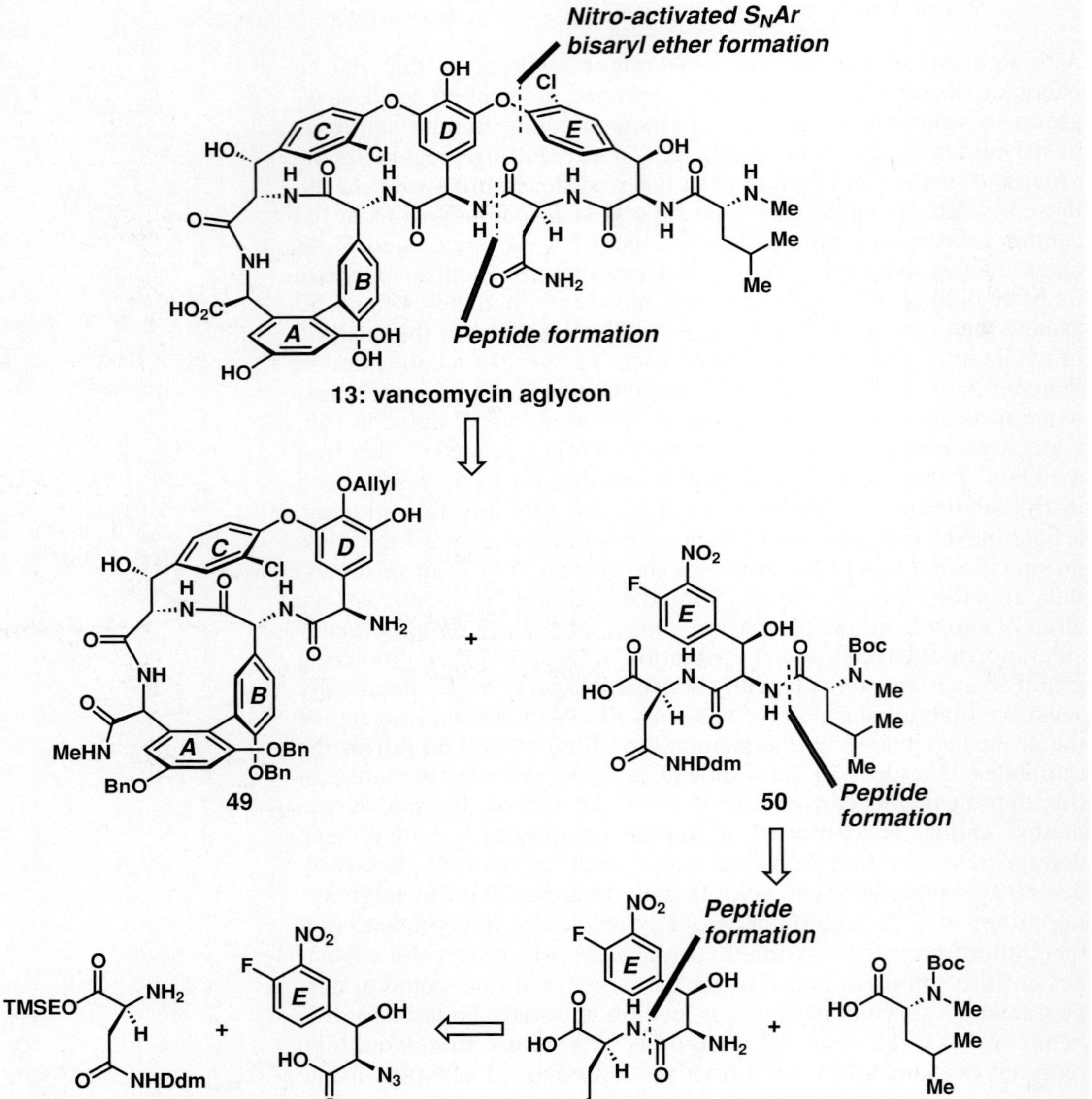

Scheme 9. Evans' retrosynthetic analysis of vancomycin (**1**): initial stages.

two different amide bond formation transforms, leaving the more challenging AB/C-O-D system (**49**) for further consideration. As shown in Scheme 10, this fragment was evaluated along quite unique lines relative to that presented earlier in Scheme 6. Rather than open the AB macrocycle first, the Evans group decided to

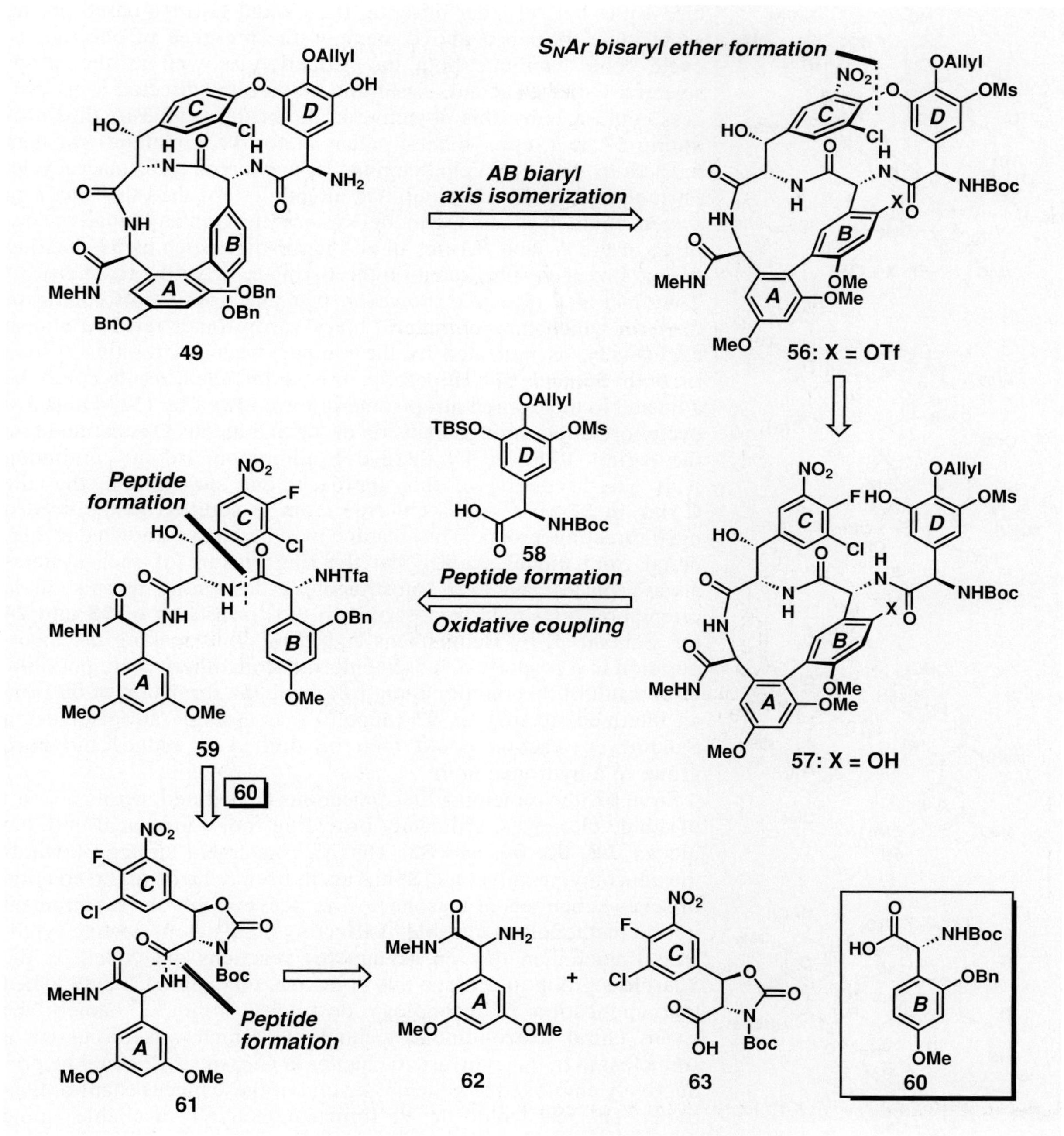

Scheme 10. Evans' retrosynthetic analysis of AB/C-O-D fragment **49**.

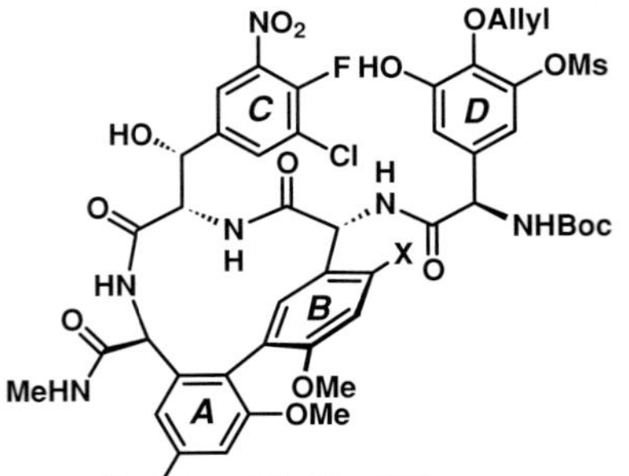

57: X = OH

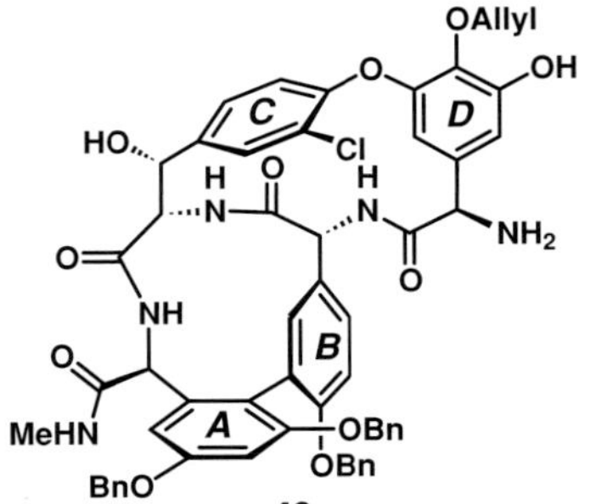

49

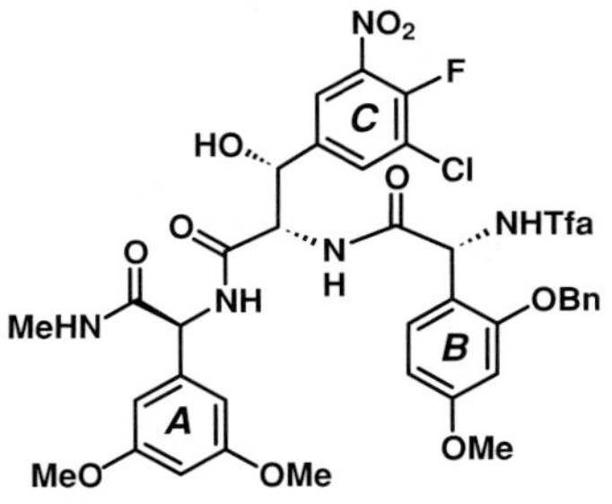

59

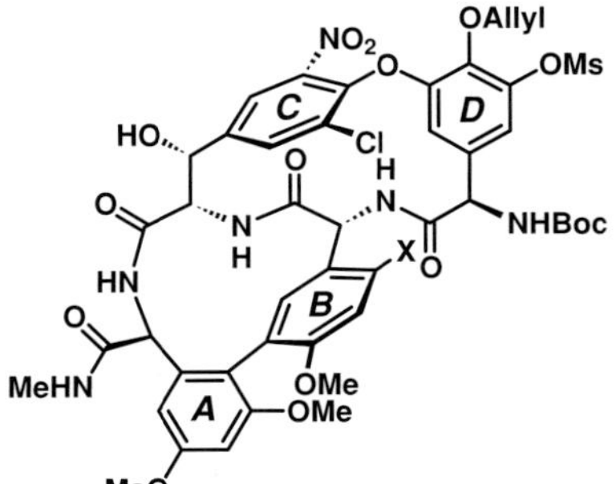

56: X = OTf

cleave the bisaryl ether bridging the C- and D-rings based on the same logic described above; namely, the presence of one macrocycle could facilitate both the formation as well as the atropselectivity of the second. Assuming that a nitro-directed S_NAr process could achieve this objective in the actual synthesis, thus providing **57** as a retrosynthetic precursor for **49**, a method was then needed to form the challenging 12-membered AB macrocycle. Fortunately, the Evans group had already established that this ring system could be created in model systems through an oxidative coupling of the A- and B-rings in an intermediate such as **59** (bearing an extra oxygen substituent on the B-ring to drive the reaction).[13cd] They had also observed, however, that this reaction afforded products in which the non-natural biaryl atropisomer resulted almost exclusively, as indicated by the unique structures for this system in both **56** and **57**. Hopefully, this unintended result could be adjusted to the desired atropisomeric form (**49**) after C-O-D macrocycle formation through excision of the extraneous O substituent on the B-ring, followed by thermal equilibration. Before continuing with our discussion of this approach, one should note that the C-ring in **57** possesses a chlorine atom in addition to the needed nitro-directing group. This feature was based on knowledge garnered from model studies that the ring closure of such systems always placed the nitro substituent in the wrong atropisomeric orientation, as similarly observed in the conversion of **23** into **24** (cf. Scheme 5) by Beugelmans and Zhu,[21] thus making the incorporation of a properly disposed chlorine atom otherwise impossible. As a result of this modification, following the formation of **56** from an intermediate such as **57** (hopefully as a single atropisomer), a Sandmeyer reaction could then be drafted to reduce the nitro group to a hydrogen atom.

From **57**, the remaining disconnections constituted merely a series of amide cleavages, ultimately providing four amino acid building blocks: **58**, **60**, **62**, and **63**. Thus, a convergent strategy towards the vancomycin aglycon (**13**) has again been reduced to the creation of seven subtargets of reasonable size, leaving only the requirement to find methodology capable of effecting their stereoselective synthesis. Rather than rely on asymmetric reactions developed by the Sharpless group to achieve this objective, this approach anticipated that application of technology developed in-house, namely the Evans chiral oxazolidinone technology (which was discussed at some length in the cytovaricin chapter in *Classics I*),[34] could be productively employed to generate several of these pieces enantioselectively. The rest would derive from commercially available amino acids.

Before we begin our discussion of how these two distinct strategies were reduced to practice, it is important to reiterate that both required considerable model system exploration and general strategy feasibility studies before they reached their full maturation as expressed here.[11,13] While space constraints severely limit the degree to which we can touch upon these instructive investigations, we strongly recommend a careful reading of the primary literature[10-13] as well as several review articles[2,5] to gain a fuller appreciation of the truly evolutionary nature of these research programs.

9.3 *Total Synthesis*

9.3.1 *Nicolaou's Total Synthesis of the Vancomycin Aglycon (13)*

Similar to most endeavors in natural product total synthesis, the Nicolaou group began their studies towards the vancomycin aglycon (**13**) with building block construction to set the foundation that would hopefully enable the facile construction of the molecular architecture of the target through the designed sequence.[10] While the synthetic routes to such pieces are typically presented in a perfunctory manner in communications and even in full papers, such a delivery fails to reflect the significant investment in research time that is often required to access these preliminary targets. Indeed, each of the four aryl building blocks (**38**, **40**, **42**, and **43**) envisioned for the construction of the AB/C-O-D subunit (**17**) presented its own unique set of challenges that prevented any target from proving wholly pedestrian to synthesize, and, in some cases, required several months of investigation. The final, optimized sequences to these pieces are delineated in Schemes 11–14, and, for lack of a better selection criterion, we will discuss these fragments in alphabetical order according to their ring designations within vancomycin (**1**).

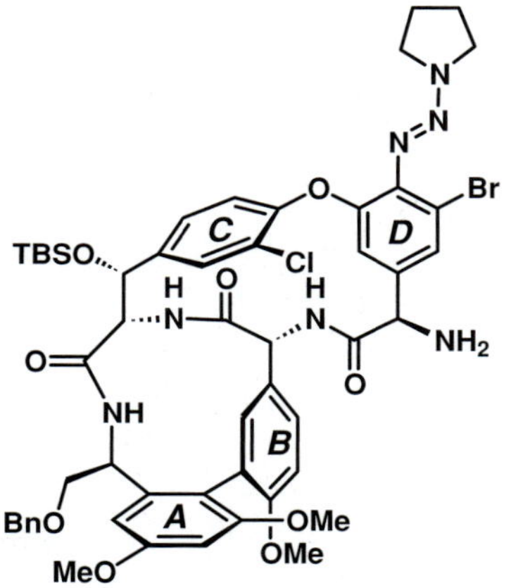

Accordingly, first for our consideration is the A-ring boronic acid **42** (Scheme 11), a fragment whose synthesis logically began from commercially available 3,5-dimethoxybenzaldehyde (**64**) since this material already possessed the desired phenolic protecting groups for the envisioned sequence. The price for this convenience was finding the means by which to install a lone stereogenic center on this substrate in order to elaborate it to **42**. Fortunately, this requirement could be met through initial Wittig homologation using the ylide derived from methylenetriphenylphosphonium bromide to generate a styrene intermediate,

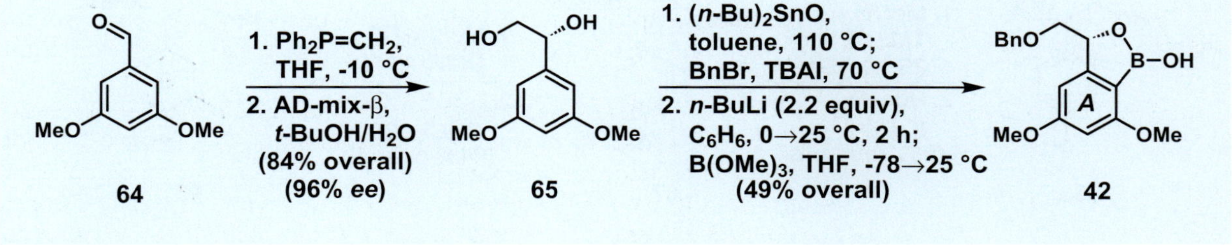

Scheme 11. Nicolaou's synthesis of A-ring amino acid building block **42**.

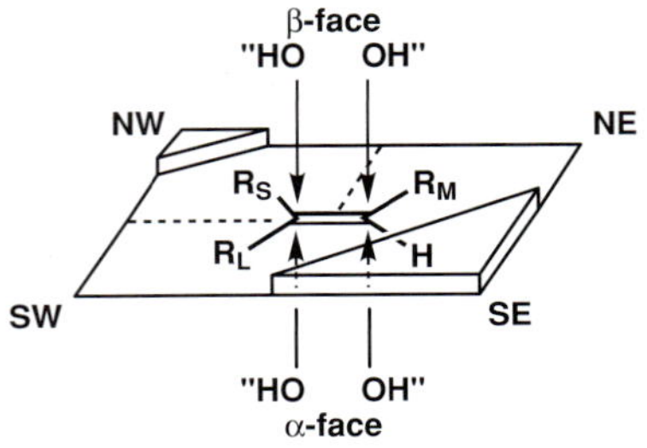

β-face
"HO OH"
NW NE
R_S R_M
R_L H
SW SE
"HO OH"
α-face

Figure 2. Mnemonic device for the Sharpless AD of olefins.

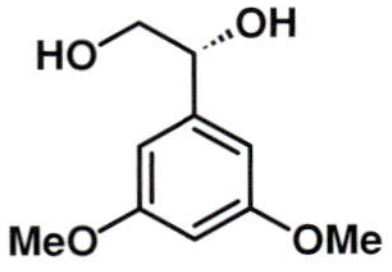

HO OH
MeO OMe
65

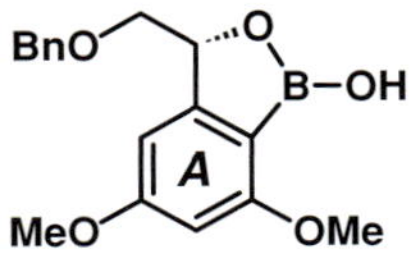

BnO O B–OH A
MeO OMe
42

followed by a Sharpless AD reaction[28] with the commercial AD-mix-β, which gave rise to diol **65** in 84 % overall yield and 96 % *ee*. The exquisite stereocontrol effected by this latter reaction reflects the well-documented tenet that styrenes are the best substrates for the Sharpless AD protocol in terms of both catalytic activity and enantioselectivity, with a rationale for the observed facial selectivity provided in Figure 2 (once again, see Chapter 35 of *Classics 1* for further discussion). At this stage, only two operations separated this intermediate (**65**) from the final building block (**42**). First, the more accessible terminal alcohol was selectively engaged as a benzyl ether by using tin acetal methodology (see Chapter 10 for a discussion of this technique), leaving behind an unprotected benzylic alcohol that could be used productively in the next step to direct the site of lithiation to complete the desired boronic acid.[35] Indeed, treatment of this intermediate with 2.2 equivalents of *n*-BuLi, followed by a quench of the resultant dilithiated species with freshly distilled $B(OMe)_3$ at −78 °C and an eventual acidic workup provided the A-ring building block (**42**) in 49 % overall yield.[36]

The synthesis of the B-ring arylglycine (**43**) was somewhat more straightforward, with the lone stereocenter of this piece obtained from the chiral pool in the form of (D)-4-hydroxyphenylglycine (**66**) as shown in Scheme 12. As such, only a few functional group manipulations were required to properly adjust this material to its requisite format, starting with esterification of the carboxylic acid with MeOH, promoted by the Lewis acidic additive TMSCl. Subsequent protection of the free amine in the resultant product as a *t*-butyl carbamate (Boc) under standard conditions then accomplished the synthesis of **67** in 93 % yield, leaving only phenolic protection and selective iodination to complete **43**. These operations were smoothly executed in 84 % overall yield through initial methylation of the phenolic residue in **67** using MeI and K_2CO_3 in DMF, followed by CF_3COOAg-accelerated iodination in $CHCl_3$. Although the silver salt employed in this final step is relatively expensive, its use was critical in order to ensure complete iodination; "cheaper" protocols (such as NaI, NaOH, and NaOCl in MeOH) on phenol **67**

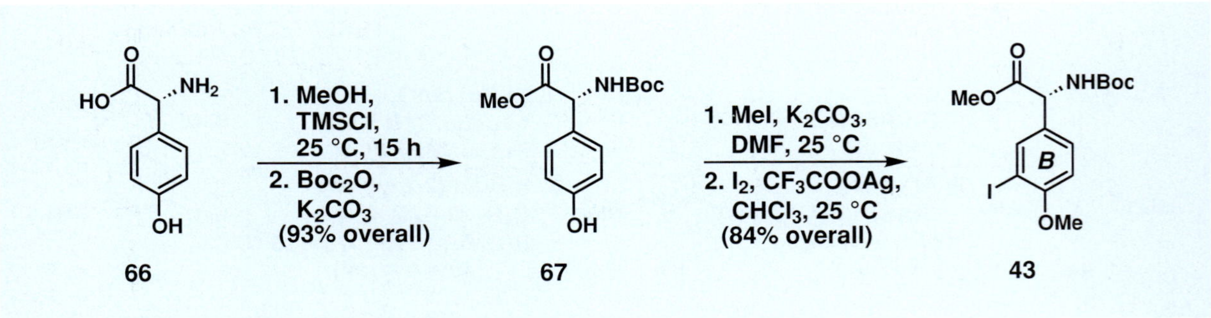

HO O NH2 OH
66
1. MeOH, TMSCl, 25 °C, 15 h
2. Boc_2O, K_2CO_3 (93% overall)
MeO O NHBoc OH
67
1. MeI, K_2CO_3, DMF, 25 °C
2. I_2, CF_3COOAg, $CHCl_3$, 25 °C (84% overall)
MeO O NHBoc B I OMe
43

Scheme 12. Nicolaou's synthesis of B-ring amino acid building block **43**.

failed to reach completion, leaving behind unhalogenated starting material that could not be readily separated from its iodinated counterpart.

With the syntheses of these two fragments discussed, we now turn our attention to the highly challenging β-hydroxy tyrosine amino acid building block **40** (Scheme 13), a compound whose 1,2-amino alcohol motif possesses a *syn*-(*S*,*R*) configuration. As mentioned earlier, it was anticipated that both these stereogenic centers might arise in a single operation from a cinnamate derivative through application of a Sharpless AA reaction. To test this conjecture, *p*-hydroxybenzaldehyde (**68**) was first benzylated and then homologated through a Horner–Wadsworth–Emmons (HWE) reaction to provide the needed olefinic test substrate (**69**). Most gratifyingly, this intermediate smoothly entered into the desired AA pathway to afford Cbz-protected amino alcohol **70** as the desired regioisomer in 45 % yield and 87 % *ee*. Although the overall yield for this event might seem modest at first glance, one should consider that an alternative sequence using an equally enantioselective transformation on substrate **69**, namely Sharpless AD, required three additional synthetic operations before **70** was obtained.[37] As such, the concise nature of the AA approach (both in terms of steps as well as time invested) was certainly worth its material return. From **70**, completion of the C-ring fragment (**40**) required silylation of the benzylic alcohol, hydrogenative removal of both the phenolic benzyl and Cbz groups, and, finally, phenol-directed monochlorination of the aromatic ring as effected with SO_2Cl_2 in Et_2O/CH_2Cl_2 (1:10) at 0 °C. Overall, these three steps proceeded without incident in a combined yield of 76 %.

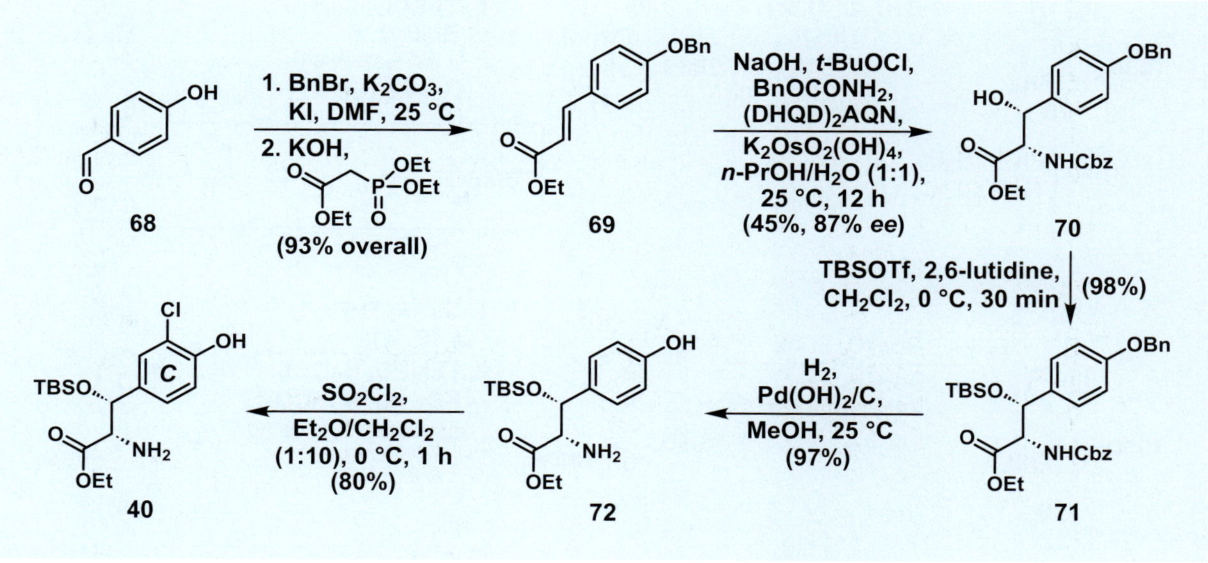

Scheme 13. Nicolaou's synthesis of C-ring amino acid building block **40**.

Finally, we turn to the remaining fragment required to implement the critical elements of the strategy envisioned for the construction of the AB/C-O-D macrocyclic system, the D-ring aryltriazene **38**. As delineated in Scheme 14, this piece ultimately required the greatest number of linear steps to acquire among these building blocks; despite its length, each operation of the route was high-yielding, with the Sharpless AD reaction once again productively encoding chiral information. The sequence began with esterification of *p*-aminobenzoic acid **73** by way of an intermediate acid chloride

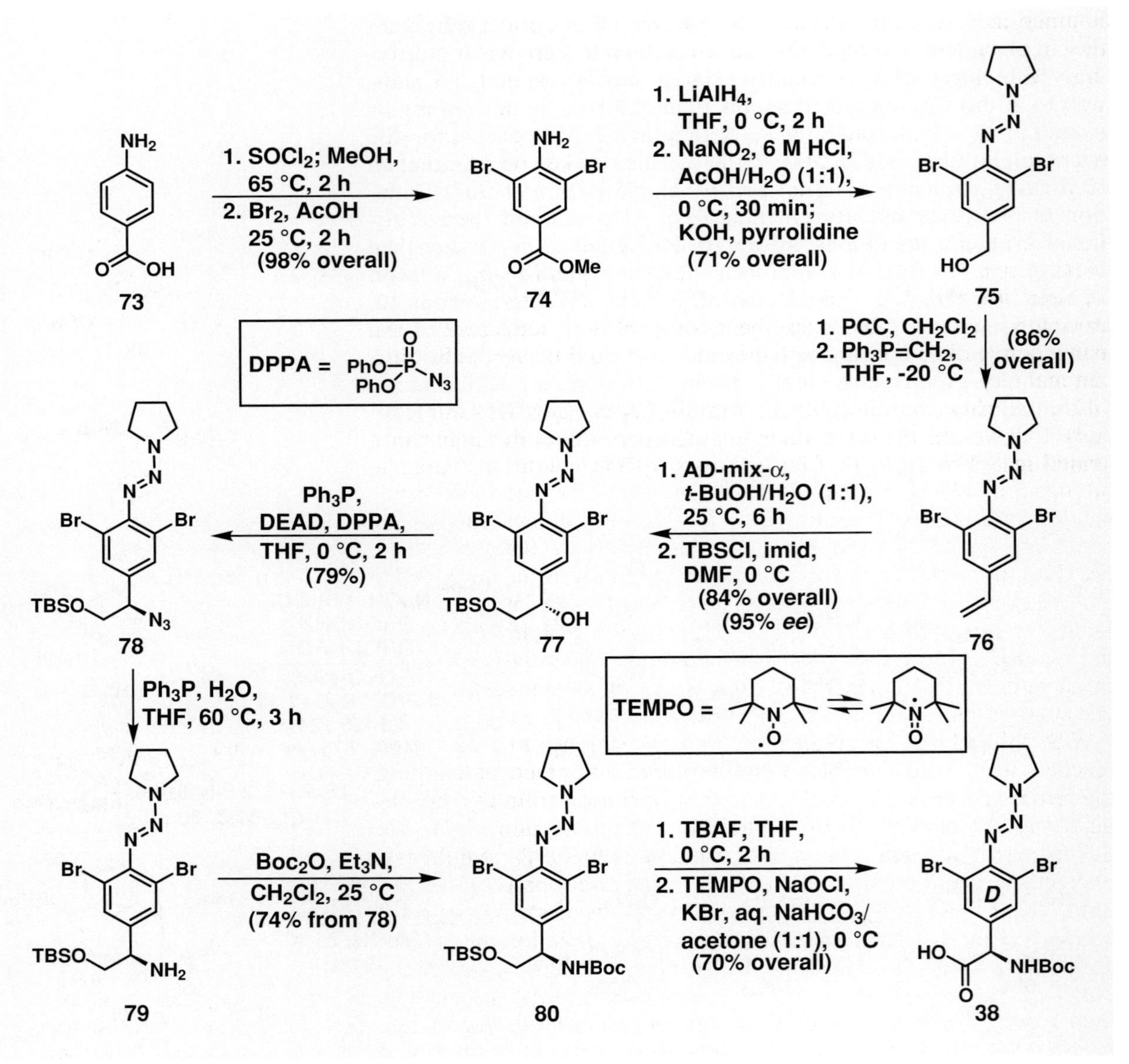

Scheme 14. Nicolaou's synthesis of D-ring amino acid building block **38**.

generated through initial reaction with $SOCl_2$. A subsequent tandem bromination then afforded 2,6-dibromide **74** in near quantitative yield for these two steps. At this stage of the projected route to **38**, it was deemed prudent to convert the aromatic amine into the triazene motif, whose ultimate role would be to engender the formation of both bisaryl ether macrocycles within the vancomycin aglycon (**13**). Thus, once the ester in **74** was fully reduced to a primary alcohol with $LiAlH_4$, the free amine of the resultant product was then diazotized using $NaNO_2$ and HCl (6 M) in a mixture of AcOH/H_2O (1:1) at 0 °C. When the resulting diazonium salt was treated with nucleophilic pyrrolidine in basified media, triazene **75** was obtained in 71 % overall yield from **74**. With this critical functionality in place, the next objective was to install the stereocenter of the targeted building block through a Sharpless AD reaction. Accordingly, the benzylic alcohol in **75** was oxidized to an aldehyde with PCC and then converted into a styrene (**76**) upon successful Wittig olefination. As expected, this substrate (**76**) was smoothly dihydroxylated with AD-mix-α in a 1:1 mixture of *t*-BuOH and H_2O at ambient temperature over the course of 6 hours, providing the desired enantiomer in 95 % yield and 95 % *ee*. From this intermediate, merely a few functional group modifications were required to complete the target (**38**); since these operations proceeded uneventfully over the course of six additional steps using conventional protocols, they do not require any additional commentary.

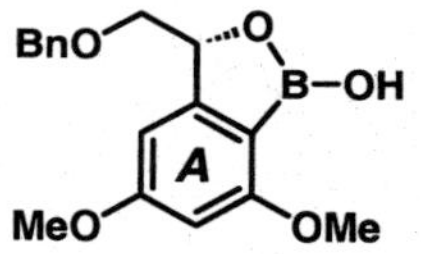

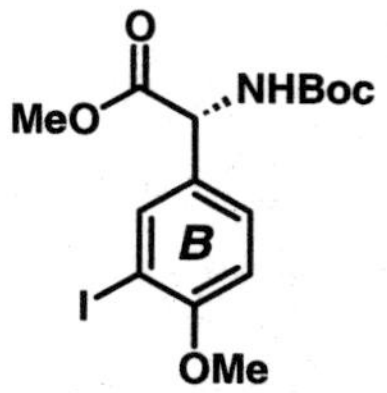

Once all these building blocks were in hand, effort was immediately focused on effecting their union according to the plan delineated in Scheme 6. The first task was to find conditions capable of forming the AB biaryl axis through a Suzuki reaction. Fortunately, little reaction scouting was required to achieve success as treatment of **42** and **43** (see Scheme 15) with $Pd(PPh_3)_4$ and Na_2CO_3 at 90 °C in a solvent combination of toluene, MeOH, and H_2O (10:1:0.5) provided a mixture of both atropisomeric Suzuki products (**81** and **41**) in 84 % combined yield. As hoped, the desired atropisomer (**41**) was formed preferentially in a 2:1 ratio, and could be separated from **81** through standard column chromatography.

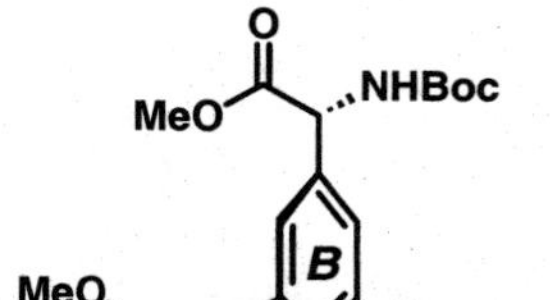

With this adduct secured, a preliminary staging area had been reached from which to begin the explorations aimed at merging the remaining building blocks. The first operation from this beachhead was the conversion of the benzylic alcohol group in **41** (see Scheme 16) into an azide upon reaction with DPPA (diphenylphosphoryl azide) using a standard Mitsunobu protocol.[38] Not only did this step provide the needed inversion of stereochemistry at this site, but it also incorporated a masked handle for an eventual amine that could be carried for several operations before being unveiled. Subsequent lysis of the methyl ester in this intermediate then afforded carboxylic acid **82**, a compound poised for a coupling reaction with the C-ring amine building block (**40**) to further extend the ever-growing vancomycin scaffold. As indicated in Scheme 16,

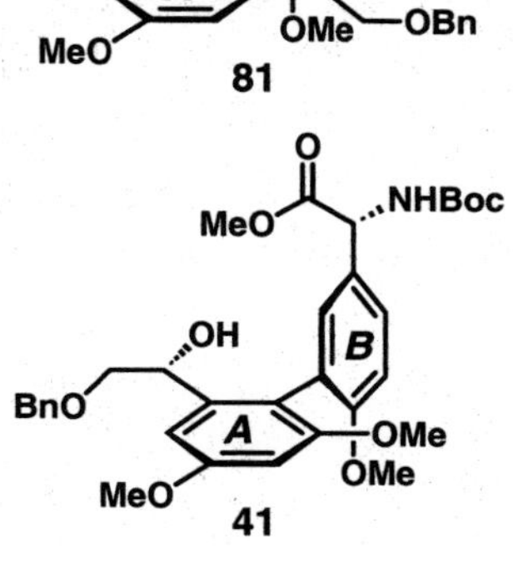

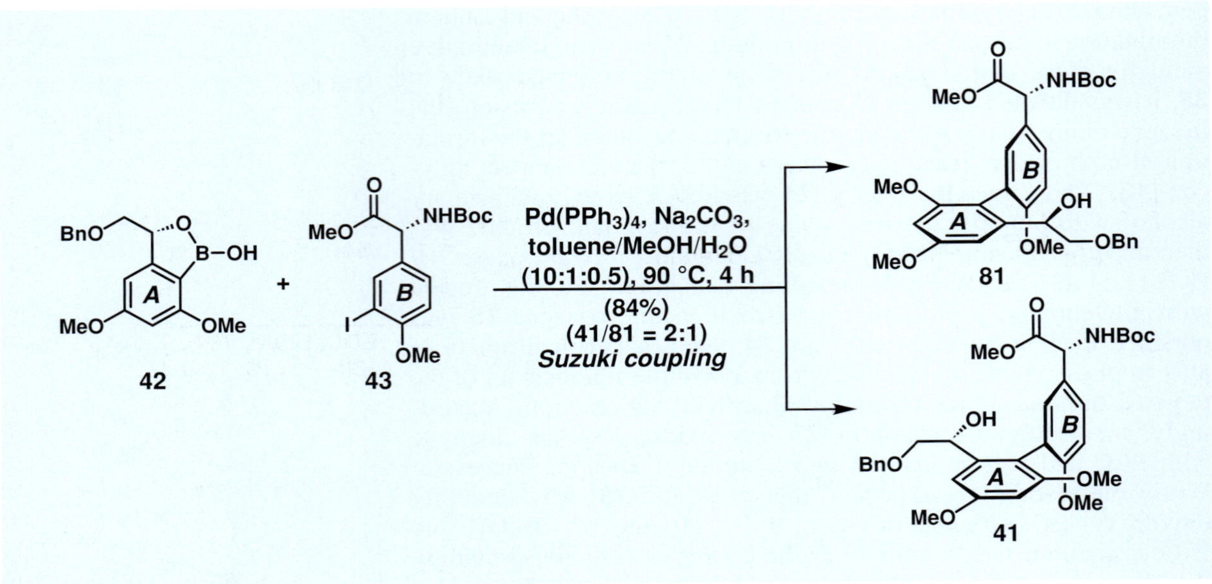

Scheme 15. Nicolaou's construction of the AB-biaryl system (**41**) through a Suzuki reaction.

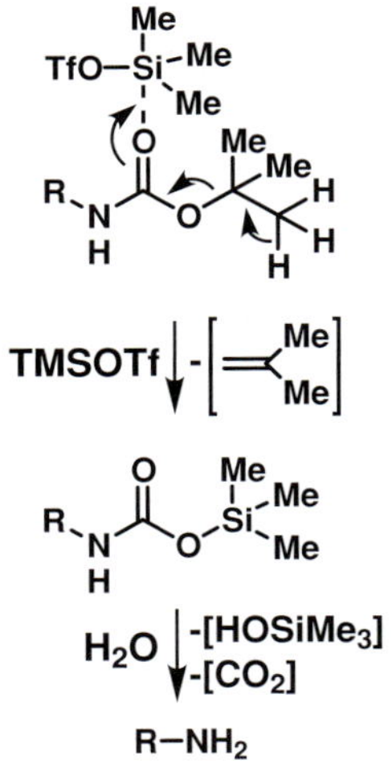

this event proceeded smoothly in 87 % yield over the course of 12 hours at low temperature (−10 °C) in THF using the common amide-bond generating reagents EDC (for peptide coupling) and HOAt (an additive to prevent racemization of epimerizable centers). In order to append the final building block onto **83**, a technique was now required to excise its lone Boc protecting group to provide the requisite free amine. Although simply stated, standard protic acid conditions to effect this cleavage (i. e. TFA) could not be employed in this instance since the presence of other acid-sensitive groups all but assured that the deprotection would not be wholly chemoselective. Fortunately, an alternative and milder protocol using the Lewis acidic TMSOTf and 2,6-lutidine rose to the challenge, affording the desired product (**39**) in 90 % yield. Since this methodology is discussed at some length in Chapter 8, we will not comment any further on its value at this point, except to note that a mechanism to account for its effectiveness is provided in the adjacent column.[39] With a free amine now revealed in **39**, the final building block (**38**) was then smoothly merged to provide **37** in 90 % yield, again using EDC and HOAt to govern the union.

Having stitched together the four aryl building blocks, all that essentially remained from **37** to complete the AB/C-O-D fragment were two macrocyclization events. The first of these closures, the formation of the C-O-D bisaryl ether system, was mildly effected through treatment of this compound with $CuBr{\cdot}Me_2S$ and K_2CO_3 in a mixture of pyridine and MeCN at 82 °C. Unfortunately, this macrocyclization failed to afford any atropselectivity in that both **36** and **84** were obtained in a 1:1 ratio in 67 % combined yield.

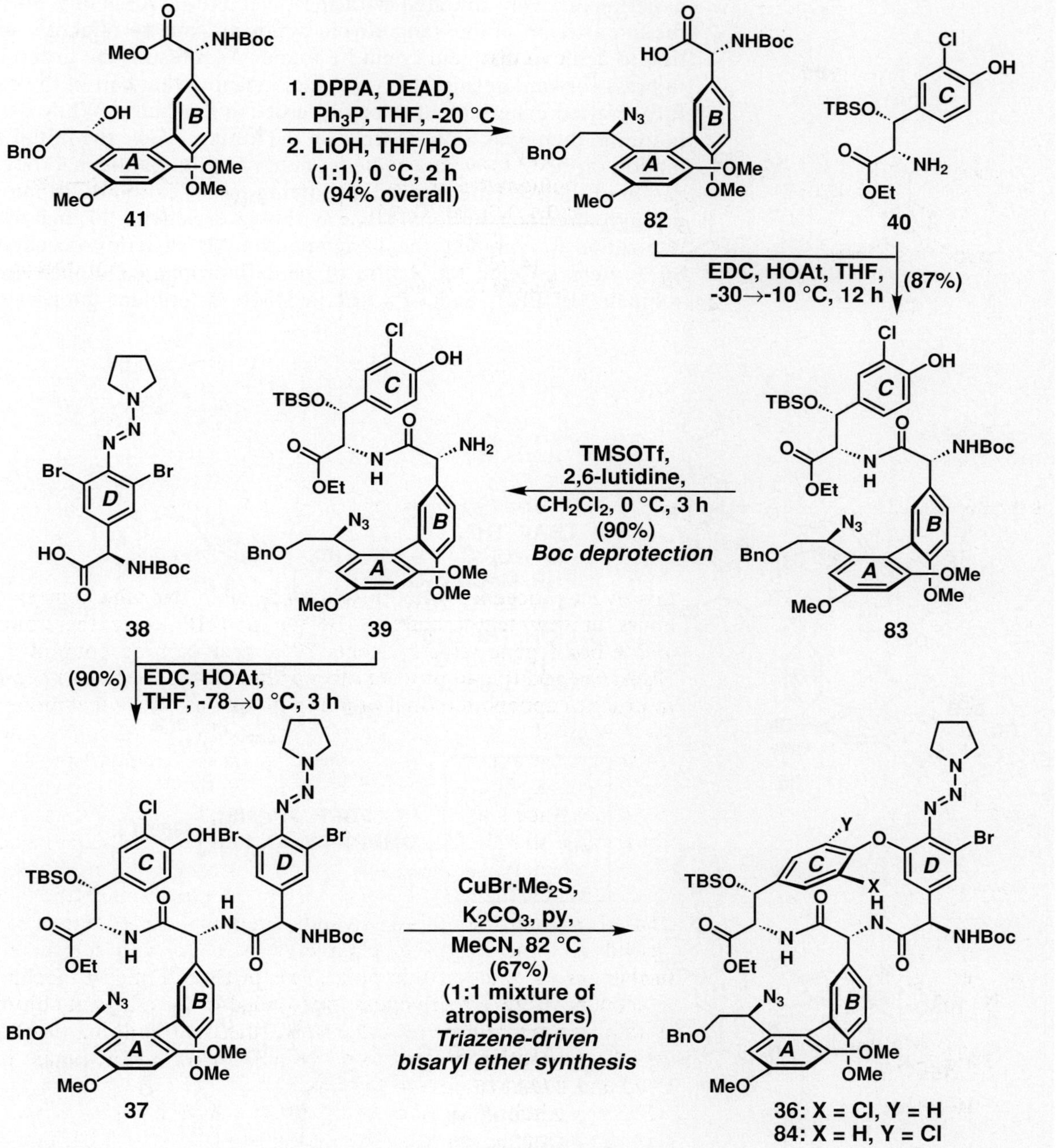

Scheme 16. Nicolaou's formation of the C-O-D ring system (**36**) of vancomycin.

Following chromatographic separation of these products, several experiments were initiated with the objective of funneling **84** into the mainstream of the vancomycin synthesis, but no effective solution to achieve this goal could be found. As a result, the group had to press forward despite this outcome, meaning that half of the carefully crafted material had to be discarded at this stage. While disappointing, there was a ray of light on the horizon since the initial formation of the C-O-D ring system greatly facilitated the macrocyclization of the AB ring. Indeed, following modification of **36** into **85** through three conventional steps as shown in Scheme 17, macrolactamization to complete the 12-membered AB system proceeded in 86 % yield through the action of pentafluorophenyl diphenylphosphinate (FDPP)[40] and i-Pr_2NEt in DMF at ambient temperature.

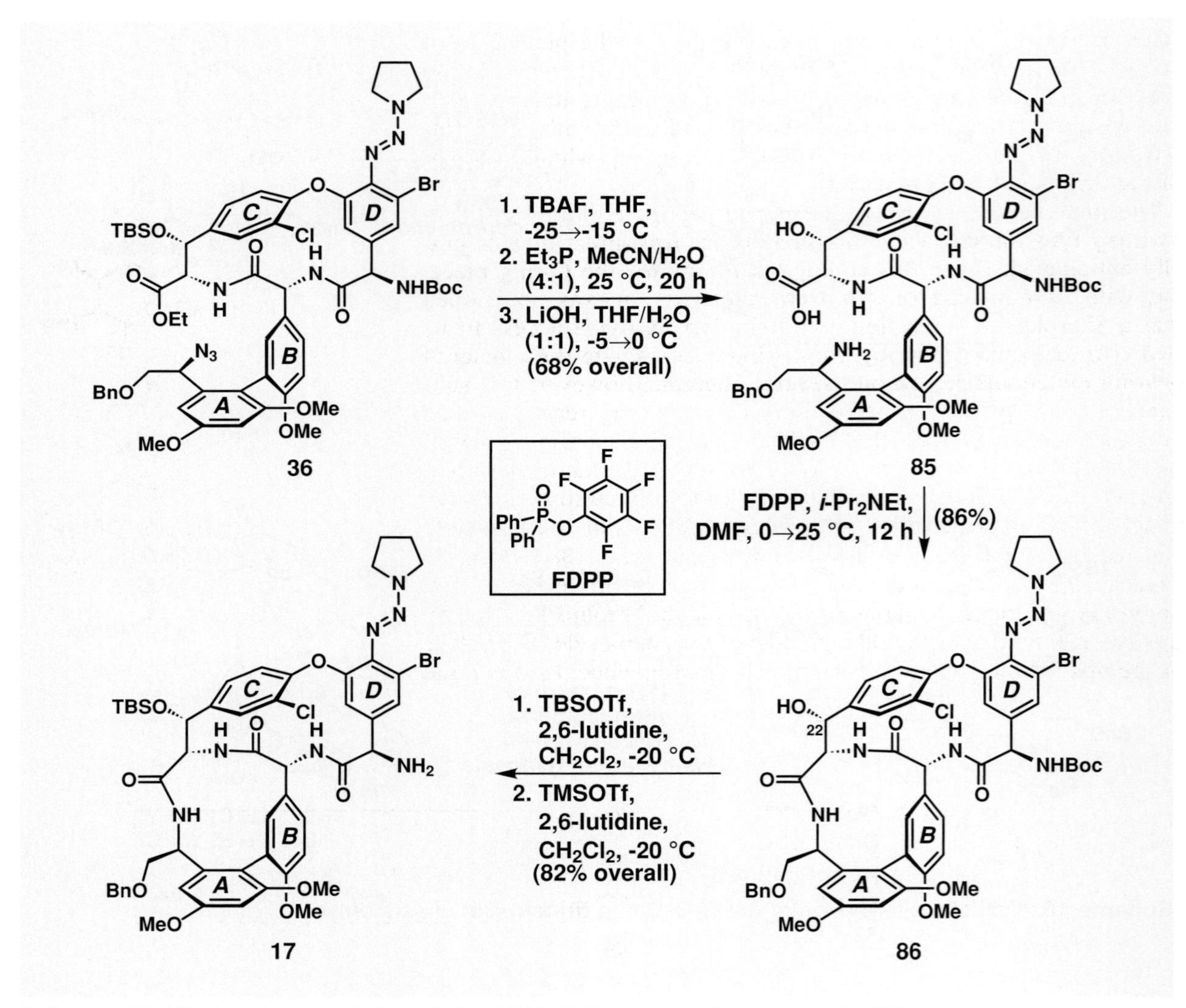

Scheme 17. Nicolaou's completion of the AB/C-O-D system of vancomycin (**17**).

Compared to the yields observed in several other macrolactamizations leading to 12-membered rings, including vancomycin model systems, this result is particularly admirable.[41] With the entire AB/C-O-D skeleton in place following the formation of **86**, only two operations were required to complete **17**, namely silyl reprotection of the C-22 alcohol (which was deprotected in the first step in Scheme 17 to enable subsequent methyl ester hydrolysis as the latter motif was otherwise too sterically shielded) and TMSOTf-induced Boc cleavage. These transformations proceeded in 82% overall yield.

Although achieving the synthesis of this advanced intermediate was certainly pleasing, before final victory could be declared it still remained to establish whether the remaining D-O-E macrocycle could be glued onto this framework. Efforts to achieve this objective began immediately, starting with the synthesis of the amino acids needed to forge tripeptide **18** according to the plan laid out earlier. The first of these, amine **45** (Scheme 18), was readily acquired from the known asparagine-derived starting material **87**[42] through simple alkylation under basic conditions with MeI, followed by hydrogenative removal of the Cbz group which was part of the original starting material.

The route needed to access the second peptide building block, β-hydroxy tyrosine **47**, was unfortunately far less direct than originally anticipated. Since this compound resembled the C-ring piece, but with *anti* instead of *syn* stereochemistry, it was envisioned that a Sharpless AA reaction could similarly provide access to its two critical stereocenters by employing a cinnamate-type material bearing *cis* geometry. Despite several attempts, however, this substrate did not prove amenable to this asymmetric reaction, most probably due to the fact that the stereochemical configuration of its double bond led to greatly reduced reactivity, and, therefore, poor enantioselectivity. Consequently, a more circuitous, but ultimately effective, sequence to **47** was developed using the Sharpless AD reaction on a *trans*-oriented olefinic substrate (**88**), with the needed amine arising through a nucleophilic substitution of a leaving group generated from one of the two alcohol groups inserted by this key reaction. One should take note of the chemoselective nature of the nosylation needed to accomplish this sequence, as it is quite

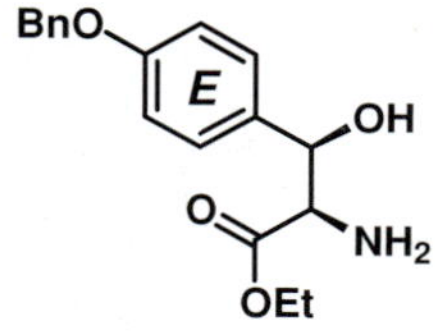

BnO E OH NH_2 O OEt
47

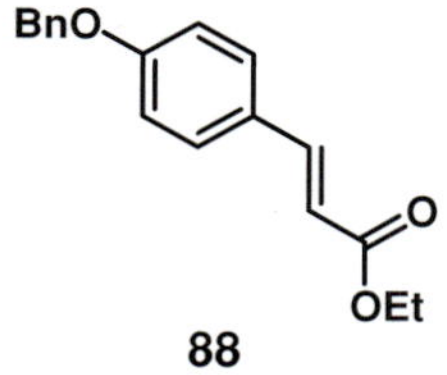

BnO O OEt
88

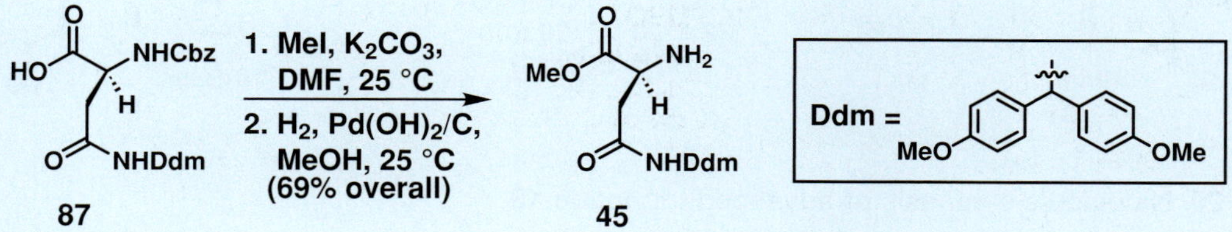

HO O NHCbz H O NHDdm **87** — 1. MeI, K_2CO_3, DMF, 25 °C; 2. H_2, $Pd(OH)_2/C$, MeOH, 25 °C (69% overall) → MeO O NH_2 H O NHDdm **45**

Ddm = MeO ... OMe

Scheme 18. Nicolaou's synthesis of amino acid building block **45**.

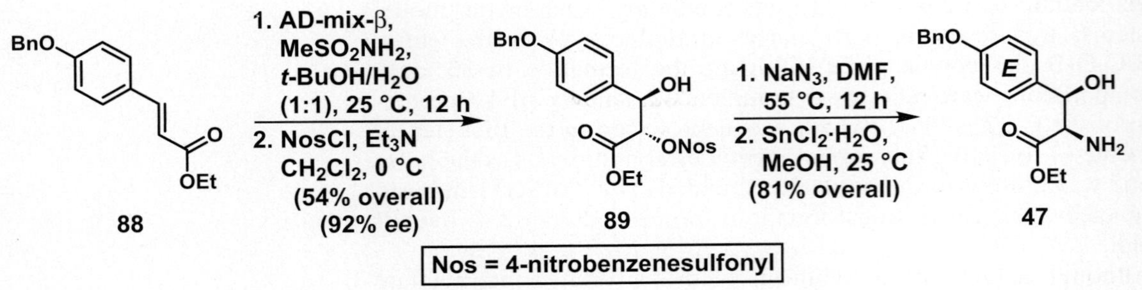

Nos = 4-nitrobenzenesulfonyl

Scheme 19. Nicolaou's synthesis of E-ring amino acid building block **47**.

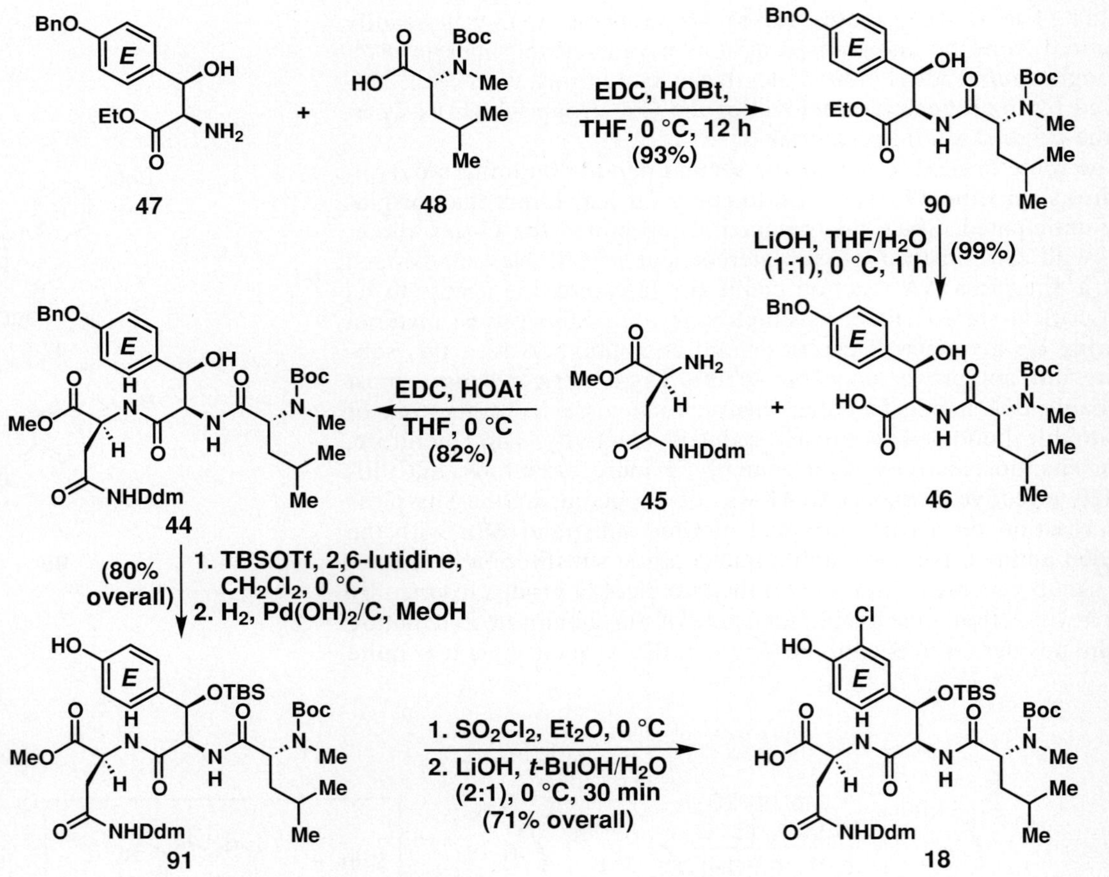

Scheme 20. Nicolaou's synthesis of advanced tripeptide **18**.

striking. Moreover, it is important to realize that this leaving group was selected over its more common relatives (such as a tosylate or a mesylate) as its electron-withdrawing nitro group provided markedly easier displacement.

With these fragments in hand, efforts could now commence on their merger into tripeptide **18**, since the final amino acid (**48**) is a commercially available material. As shown in Scheme 20, carboxylic acid **48** was first combined with E-ring building block **47** through a peptide-coupling reaction in THF at 0 °C using the by now familiar reagent EDC in the presence of HOBt. With this reaction providing **90** in 93 % yield, subsequent hydrolysis of the ethyl ester then set the stage for the final addition of the asparagine-derived building block (**45**). Having thus synthesized tripeptide **44**, only a few cursory functional manipulations remained to complete the targeted array of the functionality of this fragment as expressed in **18**. First, the hydroxy group was protected as a silyl ether, and then the phenolic benzyl ether was cleaved to afford **91** thereby enabling a subsequent monochlorination reaction with SO_2Cl_2. With these events smoothly executed, a final hydrolysis of the terminal methyl ester then provided carboxylic acid **18** in 57 % overall yield from **44**, an intermediate ready to be combined with the AB/C-O-D framework (**17**).

As originally forecasted, the amalgamation of these two advanced intermediates into **92** (Scheme 21) proceeded without incident in 86 % yield through the action of EDC and HOBt over the course of 12 hours, leaving only the construction of one bisaryl ether to complete the last of the three macrocyclic subunits of vancomycin. Despite the high hopes for atropselectivity in this reaction, upon exposure of **92** to the standard triazene cyclization conditions of $CuBr{\cdot}Me_2S$ and K_2CO_3 in pyridine and MeCN, the D-O-E system was formed with the rather disappointing result of a 3:1 ratio of isomers in favor of the non-natural product (**94**). Fortunately, however, this twist of fate did not mortally wound the strategy, as the outcome could be ameliorated through thermal equilibration using conditions developed by the Boger group in model systems to combat the same problem.[43] Thus, following the chromatographic separation of **93** and **94**, the latter was heated at 140 °C for 4 hours in 1,2-dichlorobenzene to afford a mixture of **94** and **93** in an approximate ratio of 3:2 and an overall combined yield of 80–84 %. Repetition of this cycle served to greatly enrich the supplies of the correct atropisomer (**93**), thereby fueling the studies toward the final drive for the vancomycin aglycon (**13**).

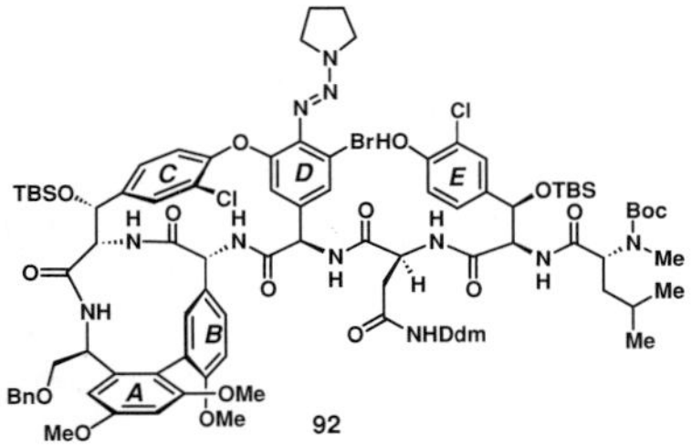

The first critical objective in this campaign was finding a method to convert the aryl triazene into a phenol, since its mission in directing the formation of both bisaryl ether systems was now complete. Although model studies indicated its relatively facile removal,[25] this motif unfortunately proved far more resilient within the context of **93** as direct techniques to effect its transformation into a phenol under various acidic conditions universally failed, leading only to reduced product (i. e. a hydrogen atom instead of the triazene

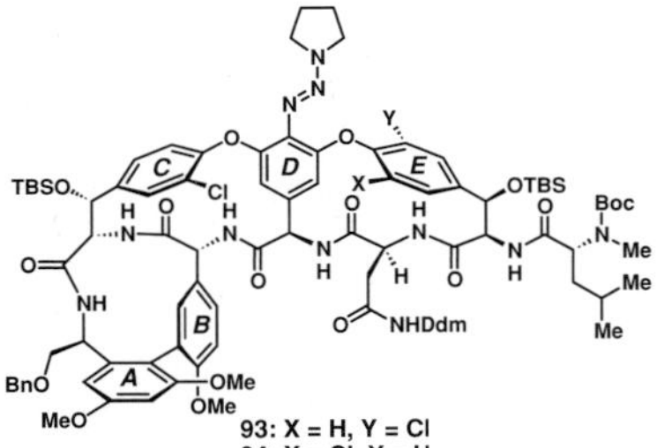

92

93: X = H, Y = Cl
94: X = Cl, Y = H

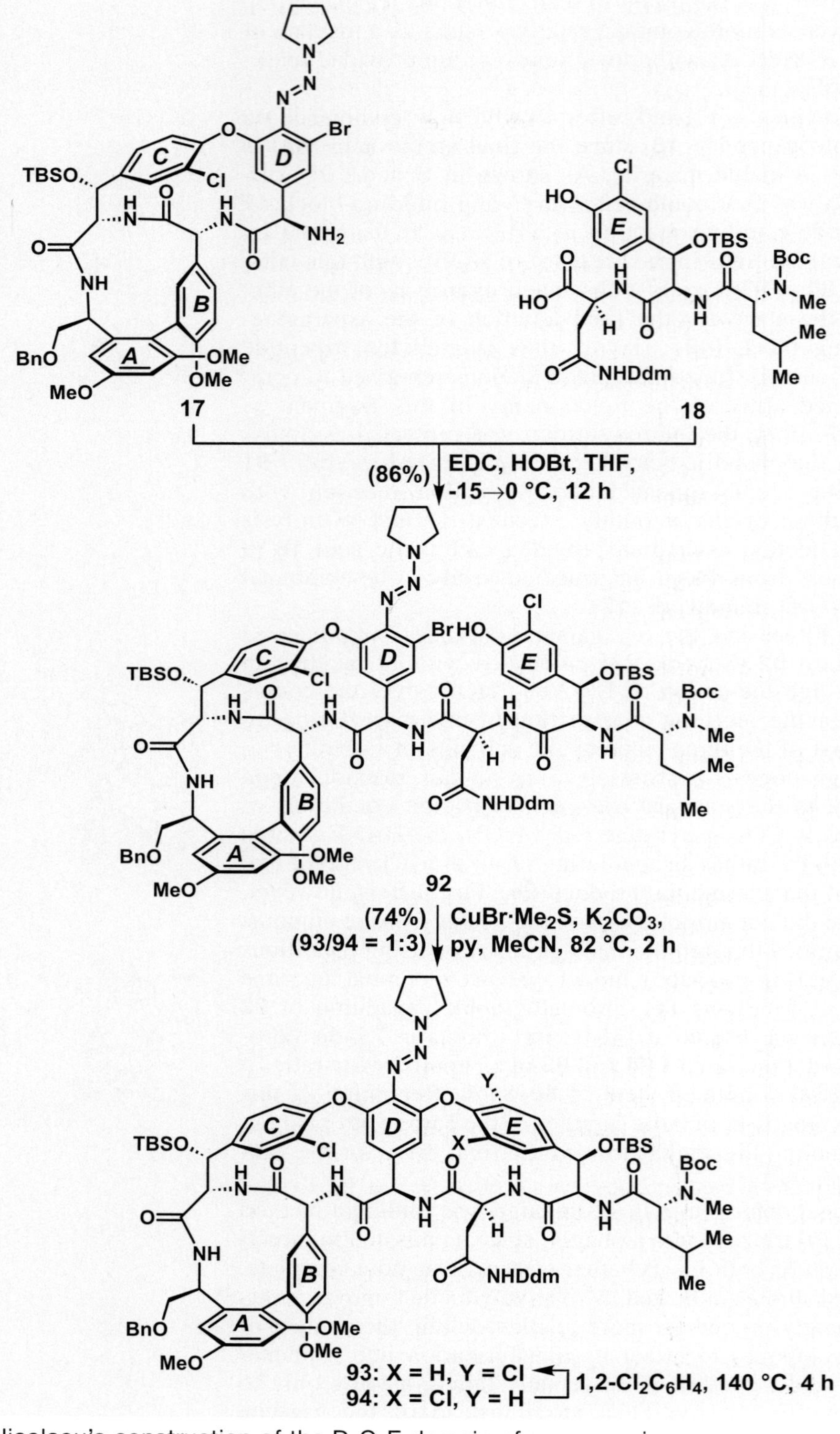

Scheme 21. Nicolaou's construction of the D-O-E domain of vancomycin.

group). Such stubbornness, of course, required the development of a more indirect route to achieve the same objective, and, after a significant amount of exploration, it was determined that the best course of action was the initial exchange of the triazene for iodine, followed by its transformation into a phenol.[44] The first step of this sequence was accomplished using TMSI (as formed with NaI and TMSCl) in the presence of additional elemental iodine in MeCN at ambient temperature over the course of 15 minutes, providing a 1:1 mixture of **95** (Scheme 22) and its reduced congener in approximately 70 % yield. Significantly, in the absence of additional iodine, almost no halogenated material was produced, as TMSI itself proved incapable of intercepting the likely intermediate benzenoid radical generated during the course of this reaction. With an iodine in place (**95**), its conversion into a phenol was then accomplished through a one-pot, three-step process involving initial iodine-magnesium exchange, followed by boronation and oxidation. Overall, these operations effected the transformation of **93** into **16** in 34 % overall yield, with the remaining material balance constituting the reduced form of **16**.

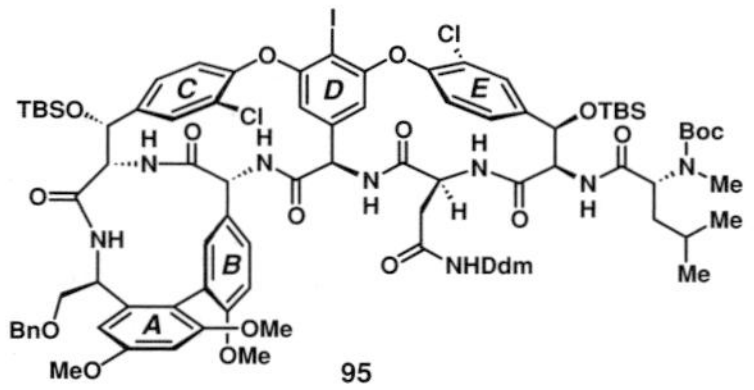

With the triazene finally removed, the remaining steps from **16** to the vancomycin aglycon (**13**) involved a series of protecting and functional group manipulations. Scheme 23 presents the optimized sequence that ultimately delivered the targeted product. First, careful hydrogenation, so as not to reduce the aryl chlorides, effected the removal of the benzyl protecting group from the C-58 alcohol function. Before this position was oxidized to a carboxylic acid, however, the phenolic group required protection, a goal that was accomplished through methylation (in the hopes that this group could be removed in the same operation that would winnow the three A- and B-ring methyl ethers later in the sequence). With this transformation achieved, the requisite carboxylic acid was then formed through a standard two-step oxidation protocol using Dess–Martin periodinane and $KMnO_4$, and subsequently captured as a methyl ester to facilitate its isolation. Overall, these five steps leading to **96** proceeded in a cumulative yield of 74 %.

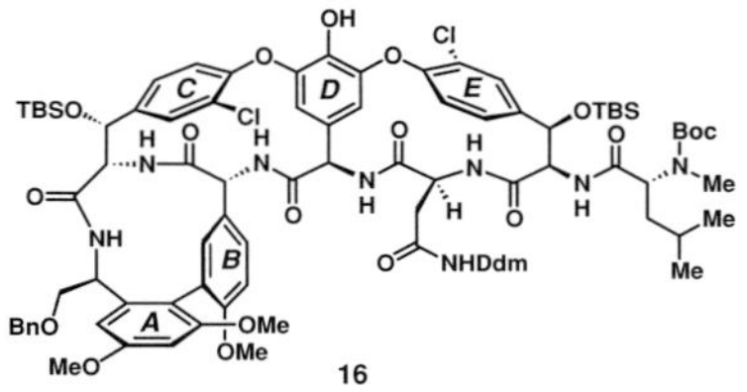

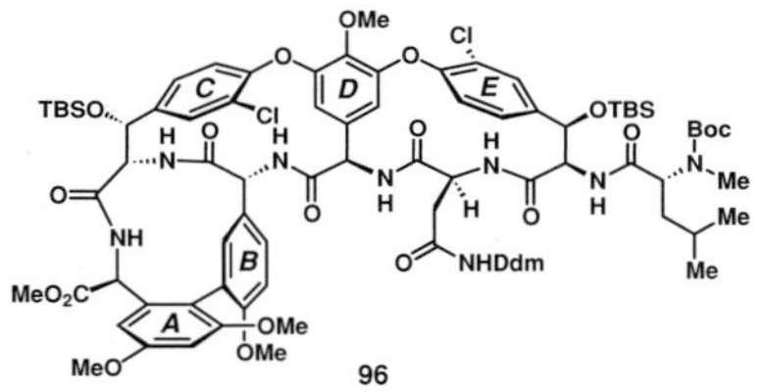

At this stage, only global deprotection was required to finish the synthesis of the vancomycin aglycon (**13**). Despite the seemingly divergent conglomeration of protecting groups that remained (one Boc carbamate, two silyl ethers, a Ddm group, and five methyl groups), it was anticipated they could all succumb to cleavage in a single operation through the synergistic action of $AlBr_3$ and EtSH. This conjecture proved accurate as exposure of **96** to this combination led to the isolation of the vancomycin aglycon (**13**) in yields that varied from 30–50 %. Alternatively, a stepwise procedure involving the initial excision of the silyl ethers using pyridine-buffered HF•py (to reduce the acidity of this reagent), followed by treatment with $AlBr_3$ and EtSH, enhanced the overall throughput to 62 %. The synthesis of **13** was finally complete.

95

16

96

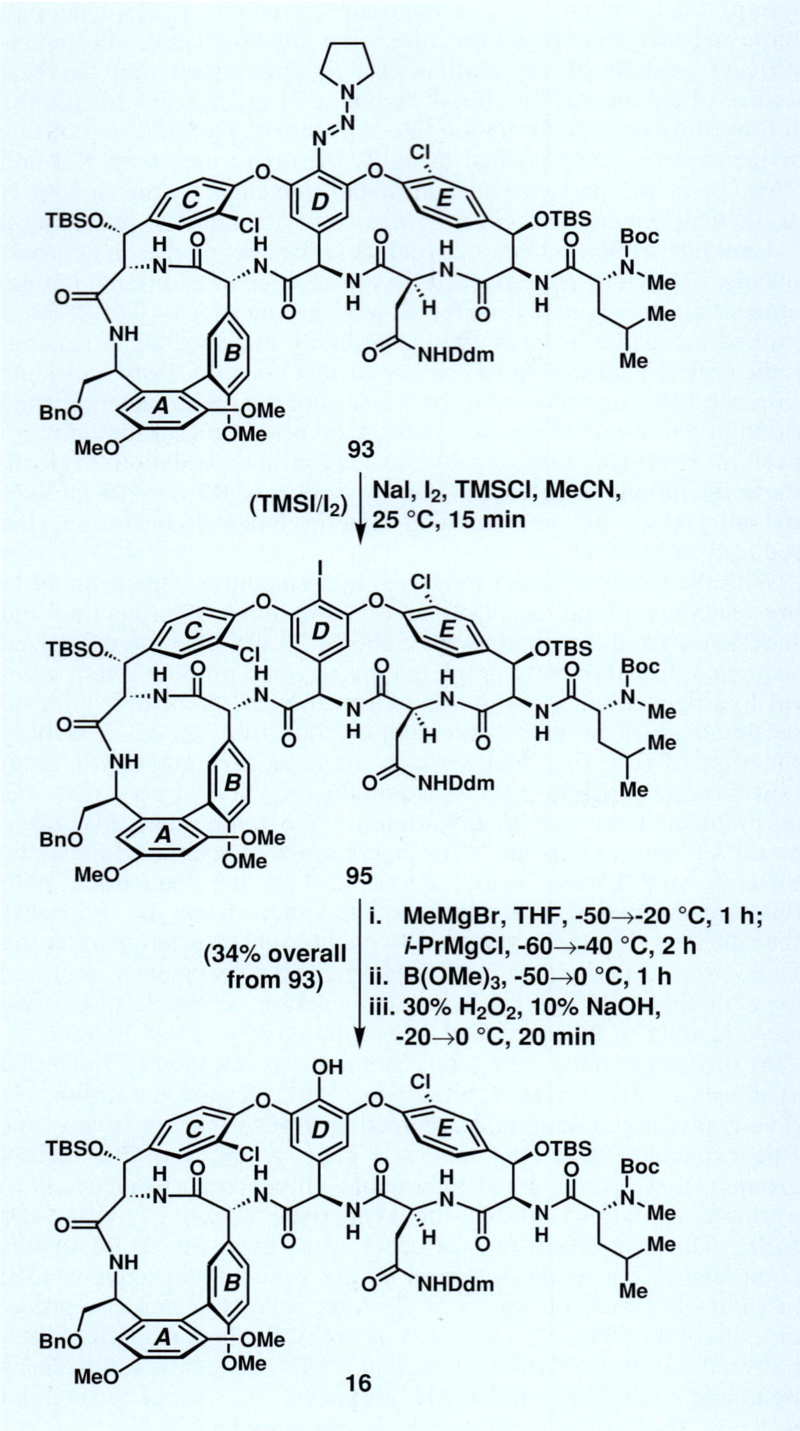

93

(TMSI/I_2) NaI, I_2, TMSCl, MeCN, 25 °C, 15 min

95

(34% overall from 93)

i. MeMgBr, THF, -50→-20 °C, 1 h; *i*-PrMgCl, -60→-40 °C, 2 h

ii. $B(OMe)_3$, -50→0 °C, 1 h

iii. 30% H_2O_2, 10% NaOH, -20→0 °C, 20 min

16

Scheme 22. Removal of the triazene moiety from **93**.

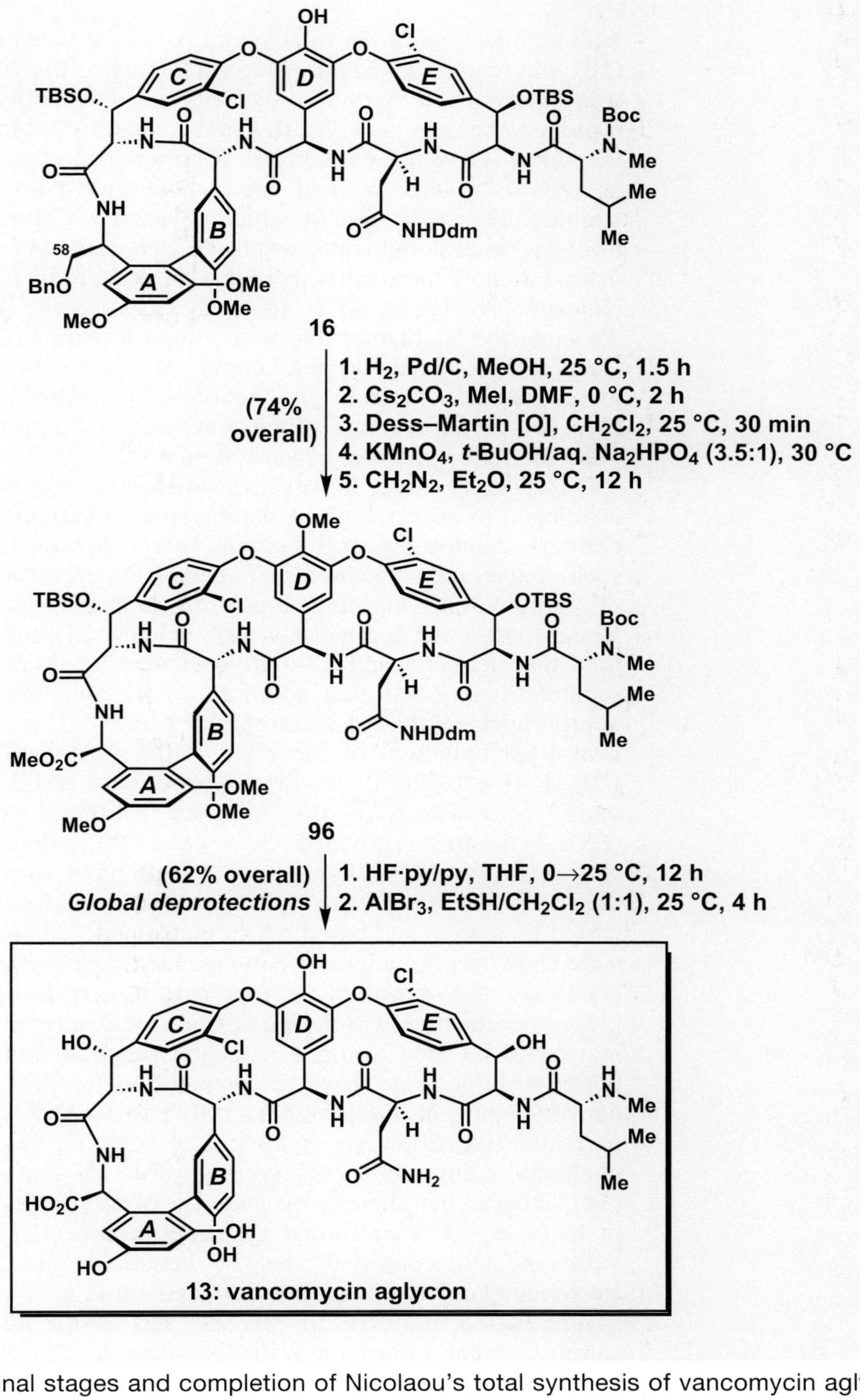

16

(74% overall)
1. H_2, Pd/C, MeOH, 25 °C, 1.5 h
2. Cs_2CO_3, MeI, DMF, 0 °C, 2 h
3. Dess–Martin [O], CH_2Cl_2, 25 °C, 30 min
4. $KMnO_4$, *t*-BuOH/aq. Na_2HPO_4 (3.5:1), 30 °C
5. CH_2N_2, Et_2O, 25 °C, 12 h

96

(62% overall)
Global deprotections
1. HF·py/py, THF, 0→25 °C, 12 h
2. $AlBr_3$, EtSH/CH_2Cl_2 (1:1), 25 °C, 4 h

13: vancomycin aglycon

Scheme 23. Final stages and completion of Nicolaou's total synthesis of vancomycin aglycon (**13**).

9.3.2 Nicolaou's Total Synthesis of Vancomycin (1)

Although achieving the total synthesis of the vancomycin aglycon (**13**) was certainly a major accomplishment, the Nicolaou group was still enticed by the possibility of developing a laboratory route to vancomycin (**1**) itself. From a chemical biology standpoint, this objective would be of equal, if not greater, importance than the aglycon as modification of the carbohydrate units could provide simple access to a host of structural analogs. Moreover, material to create such compounds would be easy to access through direct degradation of the readily available natural product (**1**) because its exposure to TFA at 50 °C smoothly afforded the aglycon (**13**).[45] As such, the Nicolaou route to **1** sought to start from synthetic **13** (vide infra), whose supplies could, of course, be enriched from the degradation reaction just mentioned. Attachment of the D-glucose and L-vancosamine subunits would then be required to complete the total synthesis of vancomycin (**1**).

Before proceeding, however, sequences would first need to be developed to access both of the envisioned carbohydrate coupling partners, namely glycosyl fluoride **14** and trichloroacetimidate **15**; such routes are delineated in Schemes 24 and 25. The synthesis of carbohydrate building block **14** began from a homochiral straight-chain starting material (**97**, Scheme 24) under the assumption that its C-3 and C-4 stereocenters could arise from two separate stereocontrolled additions of suitable nucleophiles onto appropriately activated electrophilic centers. Thus, following the controlled reduction of the ester within **97** to an aldehyde with DIBAL-H at −78 °C in CH_2Cl_2, the *anti* addition of an acyl anion equivalent[46] in the form of 2-lithio-2-ethylvinyl ether (EVE-Li)[47] then provided enol ether **98**, which upon acidic workup provided methyl ketone **99** in 65 % overall yield. The drawn diastereomer was favored to the extent of 85 % *de*, as would be anticipated based on the presumed Felkin–Anh transition state shown in the adjacent column. Although the resultant stereochemistry at C-4 was in fact opposite to that desired in the final piece, the outcome was designed since the center in its present format could be used to direct a second substrate-controlled nucleophilic addition step. Thus, in preparation for this event, **99** was converted into an oxime ether (**100**) based on the belief that this particular electrophilic group could both be intercepted by a nucleophile and could concurrently afford the amine of the target (**14**). Indeed, the subsequent addition of allylmagnesium bromide in Et_2O at −35 °C afforded key intermediate **101** in 95 % yield with complete control of the C-3 center. At this stage, the C-4 stereocenter in **101** was then inverted through a standard oxidation/reduction protocol to provide **102**. With all the requisite stereochemical information finally secured, the remaining steps to **14** were mere functional group manipulations as delineation in Scheme 24.

TIPSO Me H H OEt

OH Me TIPSO OEt

98

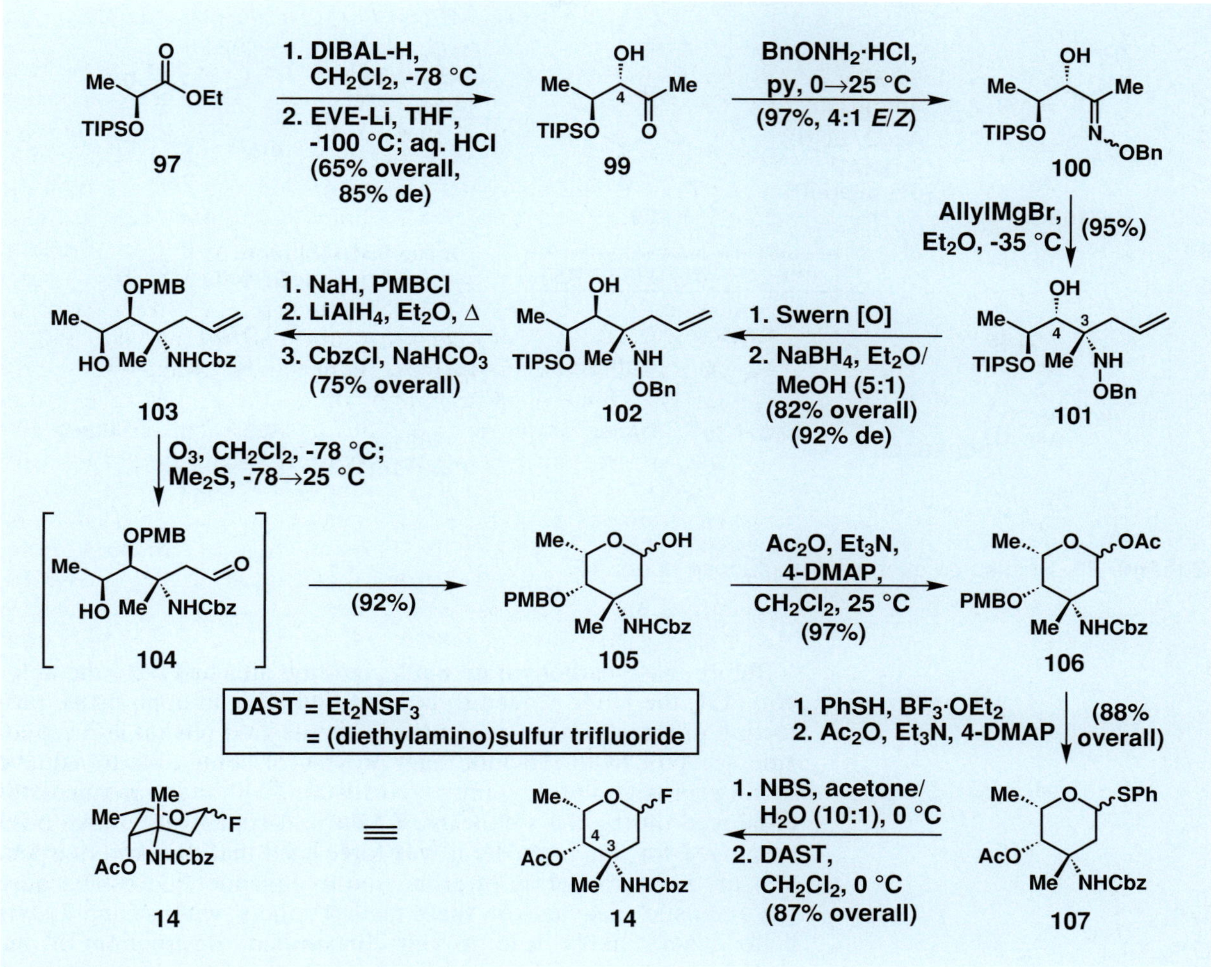

Scheme 24. Nicolaou's synthesis of vancosamine donor **14**.

The other carbohydrate building block (**15**) proved far easier to acquire, since it is a glucose derivative and thus numerous commercially-available starting materials could be enlisted to provide the majority of its stereocenters. In this case, the Nicolaou group selected D-glucal (**108**), whose hydroxy groups were initially differentiated through sequential protecting steps to provide **109**. Next, a facially selective dihydroxylation employing OsO_4 in a solvent mixture of acetone/H_2O (9:1) led to **110** in 84 % yield. Selective protection of the hydroxy group at C-2 as an allyloxy carbonate was then achieved through tin acetal methodology, and, finally, the anomeric position was converted into the desired trichloroacetimidate donor (**15**) using trichloroacetonitrile and DBU in CH_2Cl_2 at $-10\,°C$. With this final step executed under thermodynamic control, selectivity for the desired α-disposed product was excellent (~14:1).

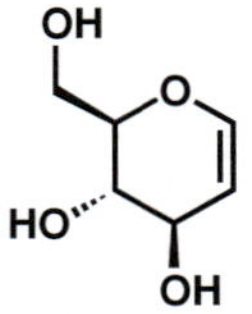

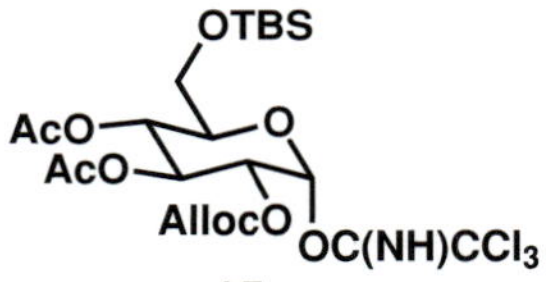

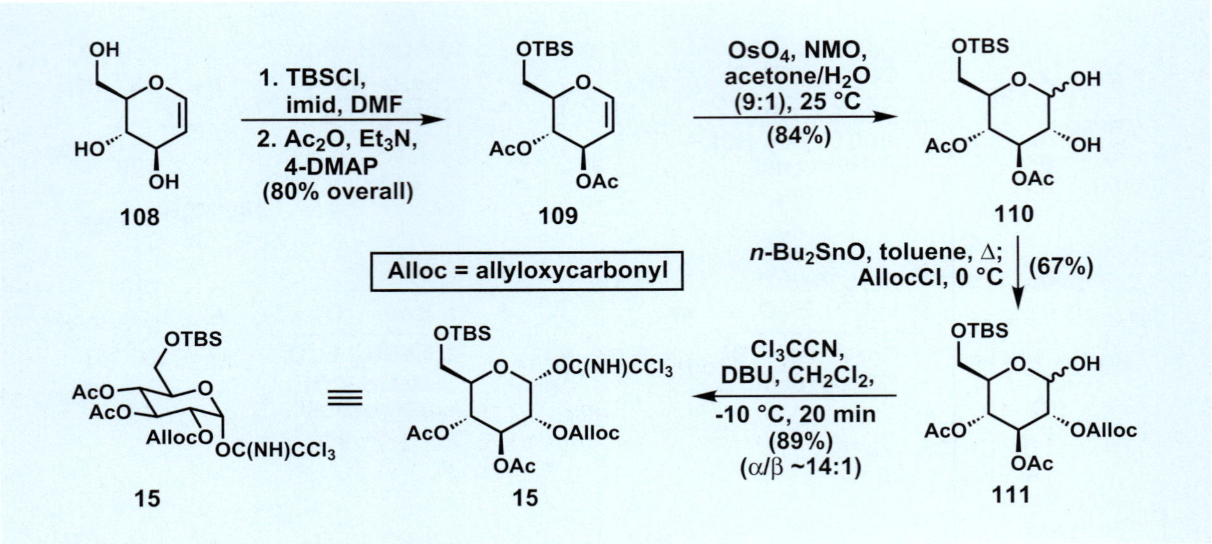

Scheme 25. Nicolaou's synthesis of glucose donor **15**.

Before these carbohydrate units could be attached onto the aglycon (**13**), the latter needed to be modified with an appropriate protecting group regime to present only the desired phenol as a glycoside acceptor, with the additional proviso of being able to achieve facile deprotection following glycosidation. Although intermediates employed during the synthesis of **13** could conceivably have been employed for this purpose, it was forecasted that they did not possess an acceptable array of protection to engender success because the requisite cleavage of their methyl ethers with strong Lewis acids would likely lead to the concomitant destruction of the newly formed glycosidic bonds. As such, a new protecting group ensemble was appended onto the fully deprotected aglycon as shown in Scheme 26. Thus, TBS ethers were attached to every alcohol and phenol within **13** using TBSOTf in the presence of 2,6-lutidine, and then the accessible C-58 carboxylic acid was converted into a methyl ester with diazomethane to provide **112** in 66% yield for these two steps. Of course, by executing this latter protection, strongly basic conditions could not be employed in any subsequent deprotection step as the C-26 center might racemize. Protection of the secondary amine in **112** as a benzyl carbamate, followed by selective excision of the D-ring phenolic TBS group using $KF{\cdot}Al_2O_3$ in MeCN, then afforded a substrate (**113**) ready to be merged with carbohydrate units **14** and **15**.

Pressing forward, the proposed glycosidation of **113** (Scheme 27) with trichloroacetimidate donor **15** was effected precisely as hoped through activation with catalytic $BF_3{\cdot}OEt_2$ in CH_2Cl_2 at $-78\,°C$ over the course of 6 hours, providing the β-disposed product in 82% yield along with a small amount of the readily separable

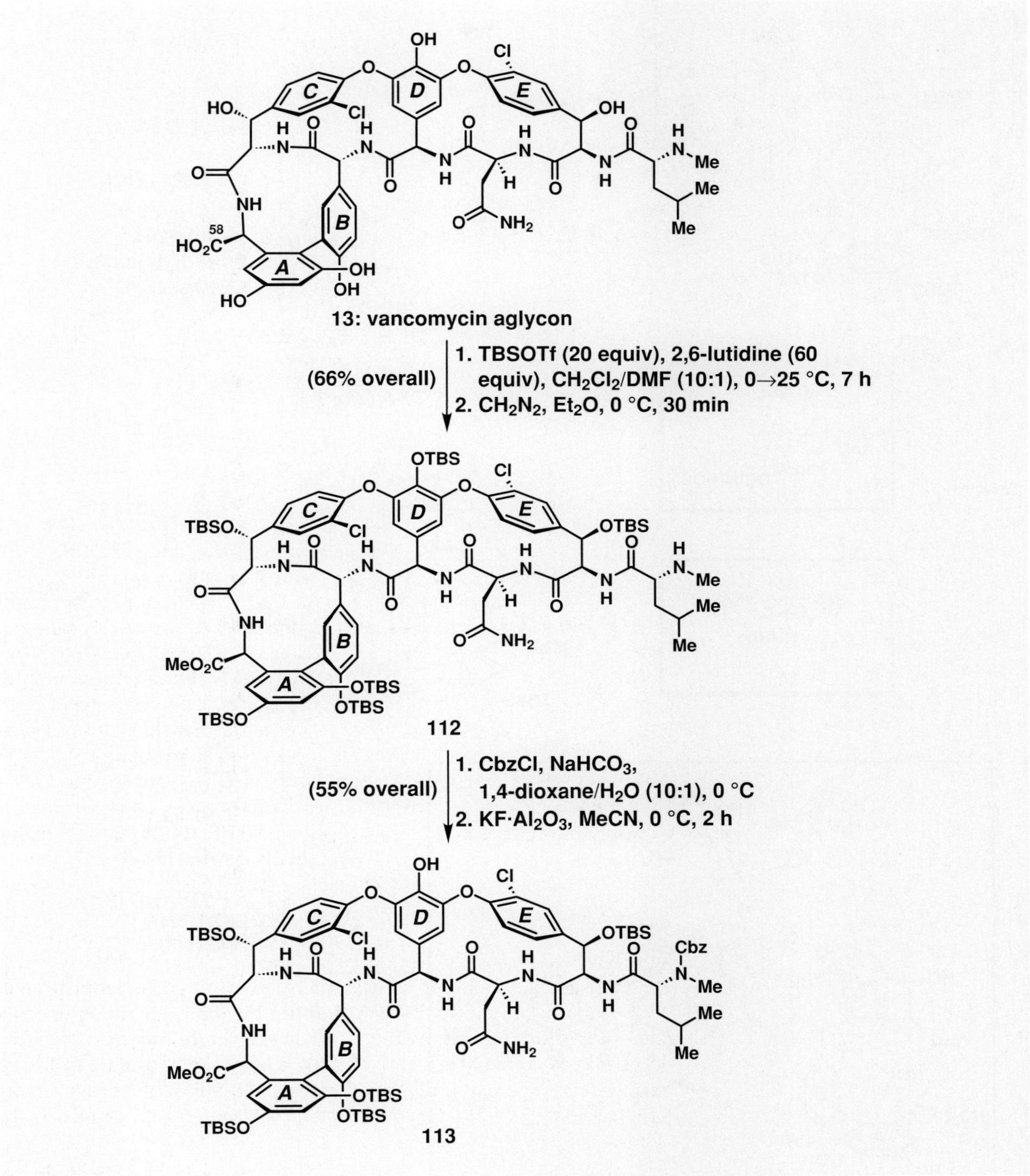

13: vancomycin aglycon

(66% overall)
1. TBSOTf (20 equiv), 2,6-lutidine (60 equiv), CH_2Cl_2/DMF (10:1), 0→25 °C, 7 h
2. CH_2N_2, Et_2O, 0 °C, 30 min

112

(55% overall)
1. CbzCl, $NaHCO_3$, 1,4-dioxane/H_2O (10:1), 0 °C
2. KF·Al_2O_3, MeCN, 0 °C, 2 h

113

Scheme 26. Nicolaou's protection of vancomycin aglycon (**13**) in advance of glycosylation.

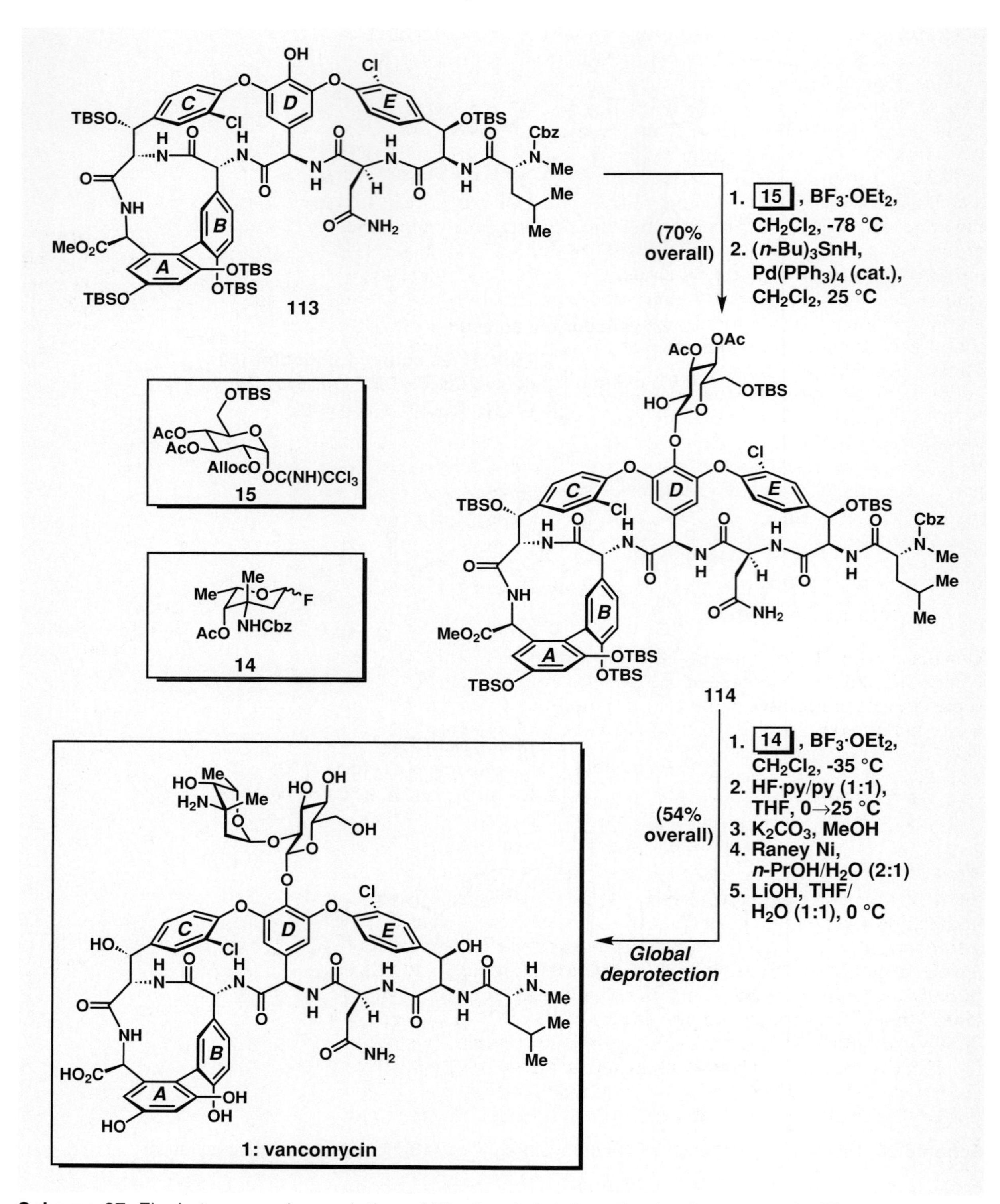

Scheme 27. Final stages and completion of Nicolaou's total synthesis of vancomycin (**1**).

α-anomeric isomer. Seeking to append the second carbohydrate, the C-2 alcohol of the glucose motif was then liberated through palladium-mediated cleavage of its allyl carbonate, providing **114**. Reaction of this compound with glycosyl fluoride **14** in CH_2Cl_2, once again using $BF_3{\cdot}OEt_2$ as an activator, most gratifyingly provided the complete vancomycin architecture in 84 % yield with 8:1 α/β selectivity for its new glycosidic linkage. Intriguingly, the initial attachment of the glucose building block onto **114** was critical for the selectivity of this operation. In the absence of a bulky anomeric group on this sugar, its coupling with **14** led to a significantly reduced preference for the α-anomer. Accordingly, block-type approaches were not pursued since the two carbohydrates could not first be joined together with acceptable levels of anomeric selectivity. Only deprotections remained at this stage, and in four additional operations all the resident groups appended to the vancomycin skeleton were successfully cleaved, affording synthetic **1** in 54 % overall yield from **114**. Of note in these final steps is the use of Raney Ni to effect the cleavage of the two Cbz groups, as more typical deprotection through hydrogenation was attended by partial cleavage of the two aryl chlorine substituents of vancomycin.[48]

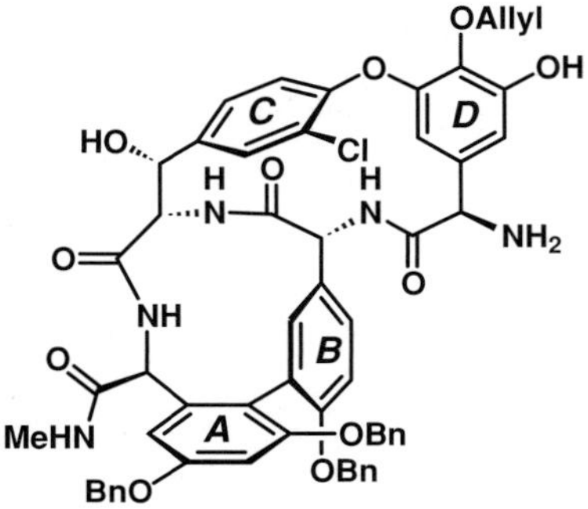

9.3.3 *Evans' Total Synthesis of the Vancomycin Aglycon (13)*

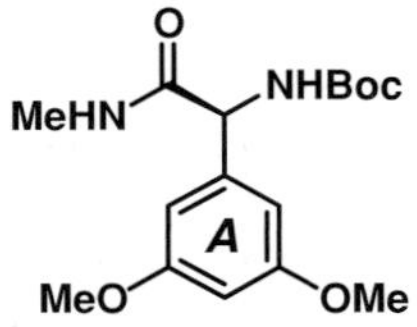

Our discussion of the Evans approach to vancomycin aglycon (**13**) begins, as did the Nicolaou analysis, with a presentation of the routes developed to access the four building blocks envisioned for the construction of the critical AB/C-O-D fragment (**49**, cf. Scheme 9). As mentioned earlier, chiral oxazolidinone technology was the essential tool that enabled the controlled installation of most of the stereocenters within these targets.

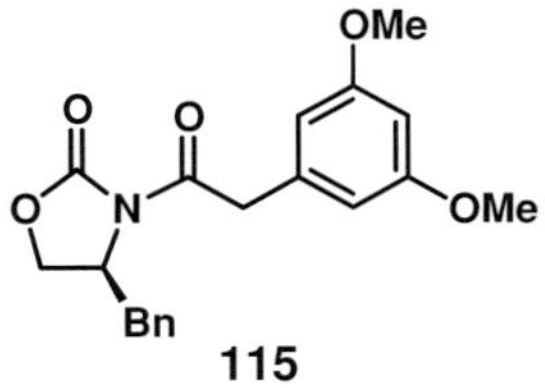

First up for consideration is how such auxiliaries can engender access to arylglycine residues such as the A-ring building block **62**. As shown in Scheme 28, it was expected that following treatment of a (3,5-dimethoxyphenyl)acetic acid derivative bearing a homochiral oxazolidinone (**115**) with a base such as KHMDS to afford a rigidly framed *Z* enolate (**116**), the addition of an appropriate electrophilic source of nitrogen (such as trisyl azide) could then provide the amine portion of the critical stereocenter of the arylglycine.[49] Indeed, this conjecture proved accurate as **117** was formed in 78 % yield with 80 % *de*. With the stereogenic center of this building block secured, the final steps in the sequence to **62** involved reduction of the newly installed azide function with *in situ* protection of the resultant amine with Boc, cleavage of the chiral auxiliary with LiOH, and capture of the resultant carboxylic acid as a methyl amide.

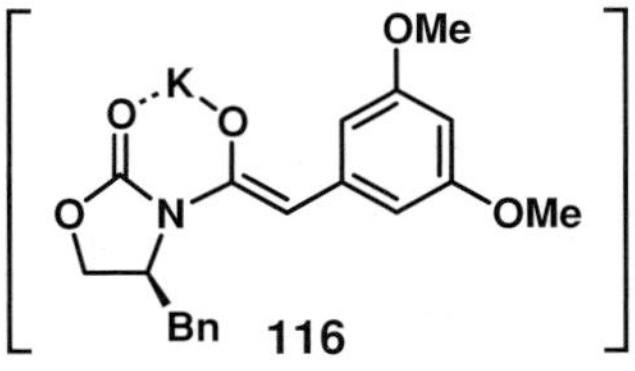

The synthesis of the B-ring arylglycine piece proceeded along similar lines as the A-ring fragment, but since this piece possessed

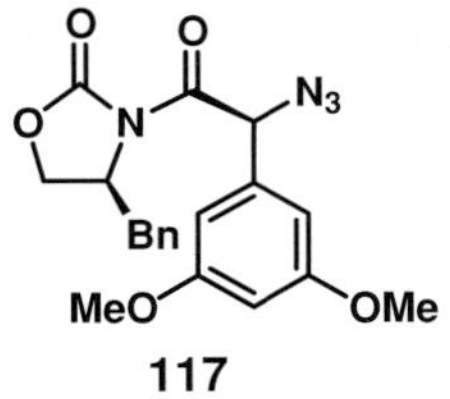

OAllyl OH O D C Cl HO H O H N N NH2 O O NH B O MeHN A OBn OBn BnO

49

O MeHN NHBoc A MeO OMe

62

OMe O O O N OMe Bn

115

OMe K O O O N OMe Bn **116**

O O N N3 O Bn MeO OMe

117

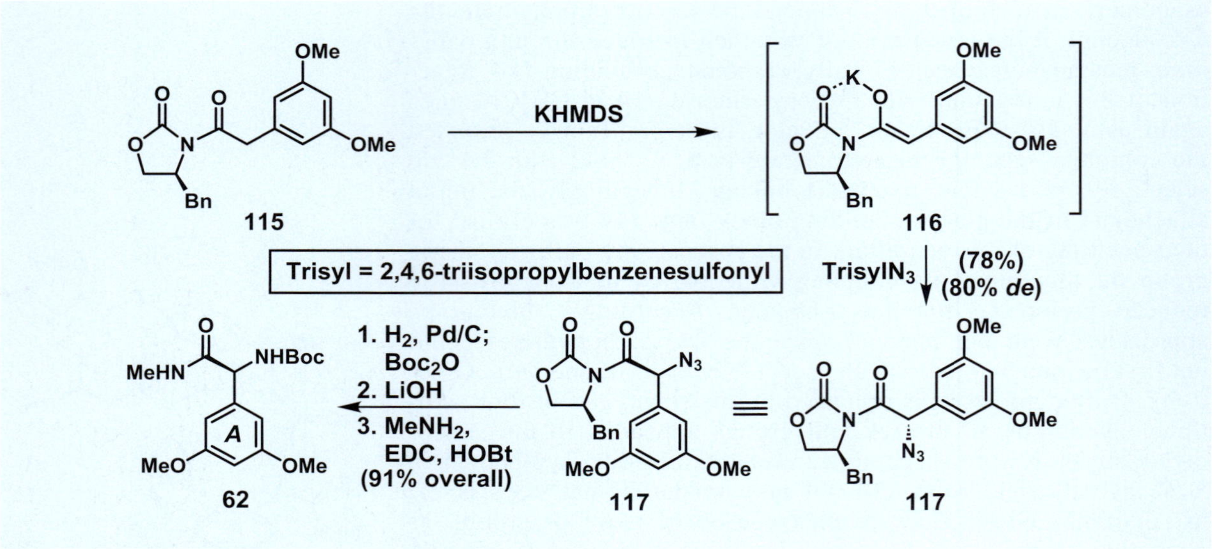

Scheme 28. Evans' synthesis of A-ring building block **62**.

R instead of *S* stereochemistry, it required a slightly different approach from the same type of starting phenylacetic acid derivative (**118**, Scheme 29). Thus, following the formation of the *Z* enolate (**119**), instead of trapping this intermediate with an electrophilic nitrogen source that would ultimately afford the incorrect amine stereochemistry, it was treated with NBS to supply a bromide leaving group that could enable the subsequent introduction of nitrogen through an S_N2 reaction. With this operation affording **120** as the desired diastereomer (7.3:1 diastereoselectivity), in the next step nucleophilic azide, as supplied by tetramethylguanidinium azide (TMGA),[50] then furnished **121** with the desired *R* stereochemical format. Having now properly assembled the lone stereocenter, only reduction of the azide followed by *in situ* protection with Boc and subsequent LiOH-mediated cleavage of the chiral auxiliary were required to complete the assembly of **60**. Overall, these four steps proceeded in a combined 54 % overall yield from **118**. Taken together, the efficient and highly enantioselective construction of both **60** and **62** amply illustrates the power of this particular auxiliary approach to afford both stereoisomers of an α-amino acid. Equally important, the method includes conditions capable of effecting the smooth cleavage of the chiral directing group (for use in additional reactions) without any racemization of the substrate.

Apart from yielding arylglycines, chiral oxazolidinones also provide a powerful tool with which to stereoselectively fashion protected β-hydroxy tyrosines such as the C-ring building block **63**.[51] As indicated in Scheme 30, the critical element of this variant of substrate-controlled enantioselective synthesis is the inventive use

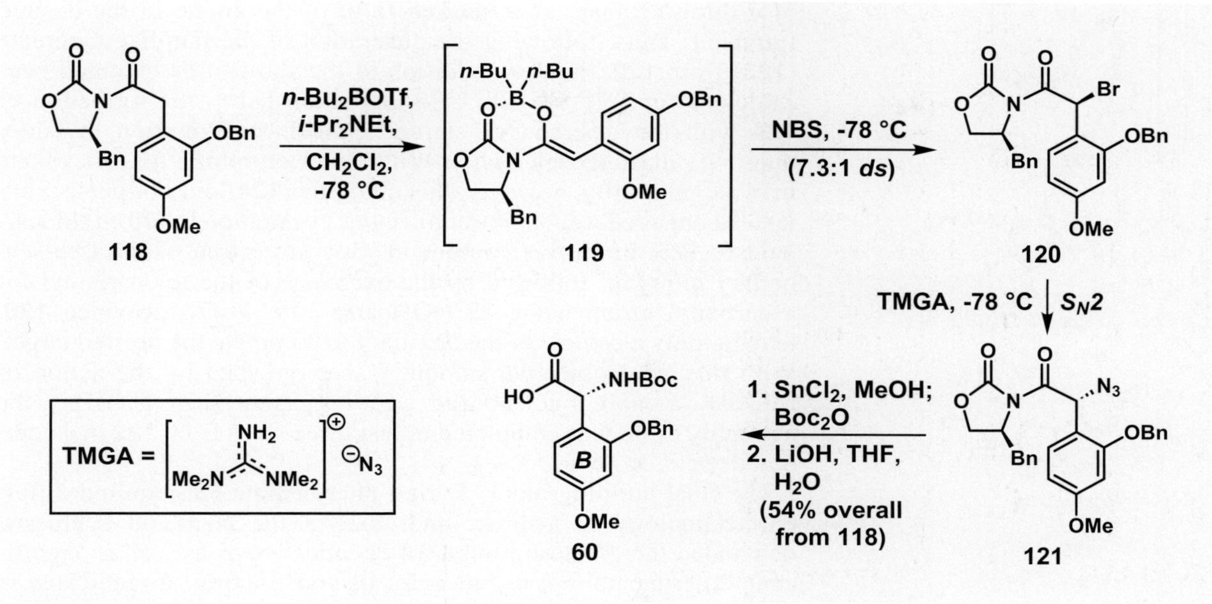

Scheme 29. Evans' synthesis of B-ring building block **60**.

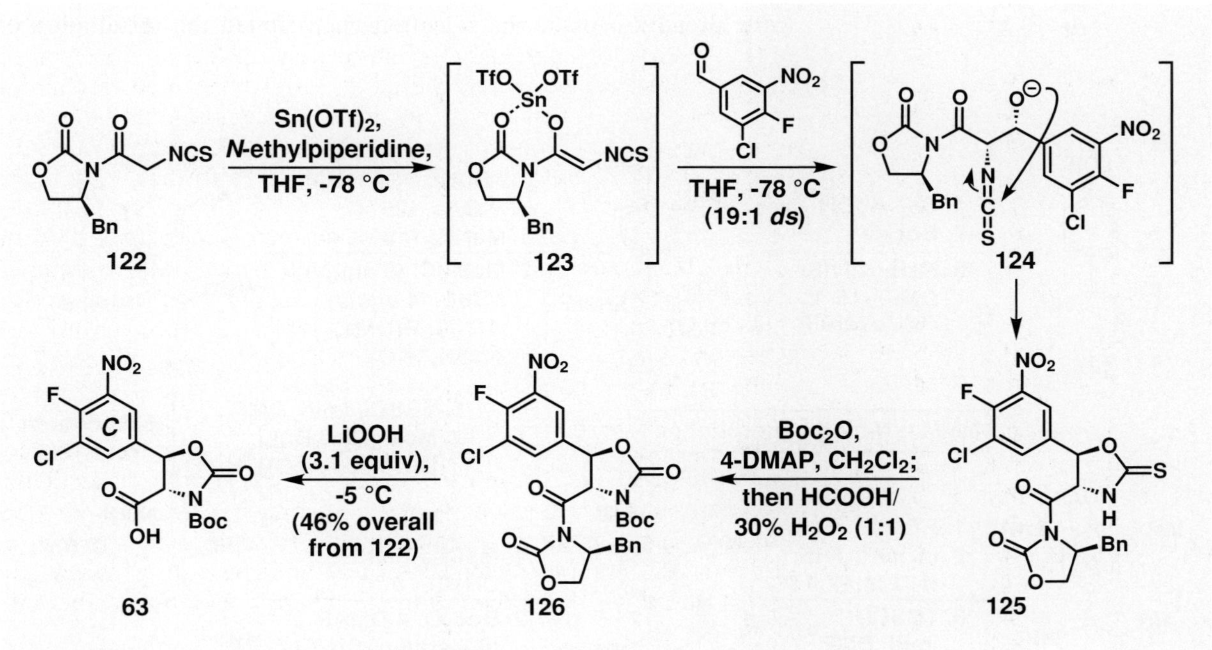

Scheme 30. Evans' synthesis of C-ring building block **63**.

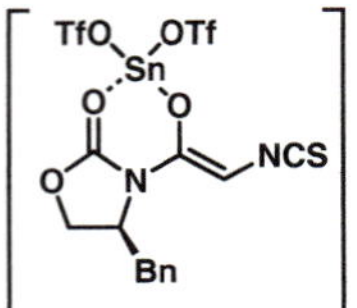

of a thioisocyanate as a masked form of the amine in the desired fragment. Thus, following the generation of the familiar *Z* enolate (**123**) from **122**, the slow addition of the shown trisubstituted benzaldehyde in THF at −78 °C to this anion led to the formation of **124** with the desired *syn* stereochemical configuration in better than 19:1 diastereoselectivity. With the electrophilic thioisocyanate in close proximity, however, this alkoxide (**124**) did not persist, but instead engaged it in an intramolecular cyclization to afford thiocarbamate **125**. From this compound, Boc protection of the free secondary nitrogen, followed by the exchange of the thiocarbonyl for a carbonyl group using HCOOH and 30% H_2O_2, provided **126**, leaving only cleavage of the auxiliary to complete the desired target. With this final objective smoothly accomplished by the action of LiOOH (a more nucleophilic cleaving agent than LiOH),[52] the assembly of **63** was completed in just three steps from **122** in a combined yield of 46%.

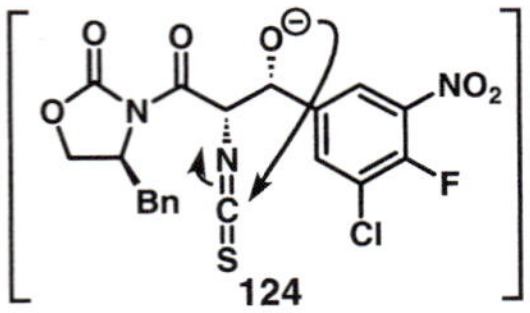

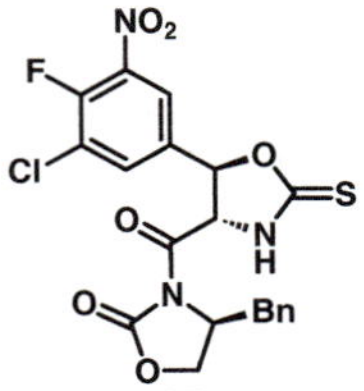

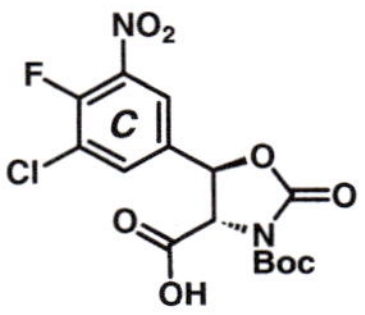

The final building block, D-ring intermediate **58**, required different technology to acquire, and, like in the Nicolaou synthesis, demanded the greatest number of chemical steps as well as significant experimental optimization for its construction. As indicated in Scheme 31, synthetic endeavors towards this fragment began with a commercial source of its stereogenic center in the form of 4-hydroxyphenylglycine (**127**). While this starting material was certainly an attractive one, since most of the atoms of the targeted building block were already in place, its selection necessitated the installation of

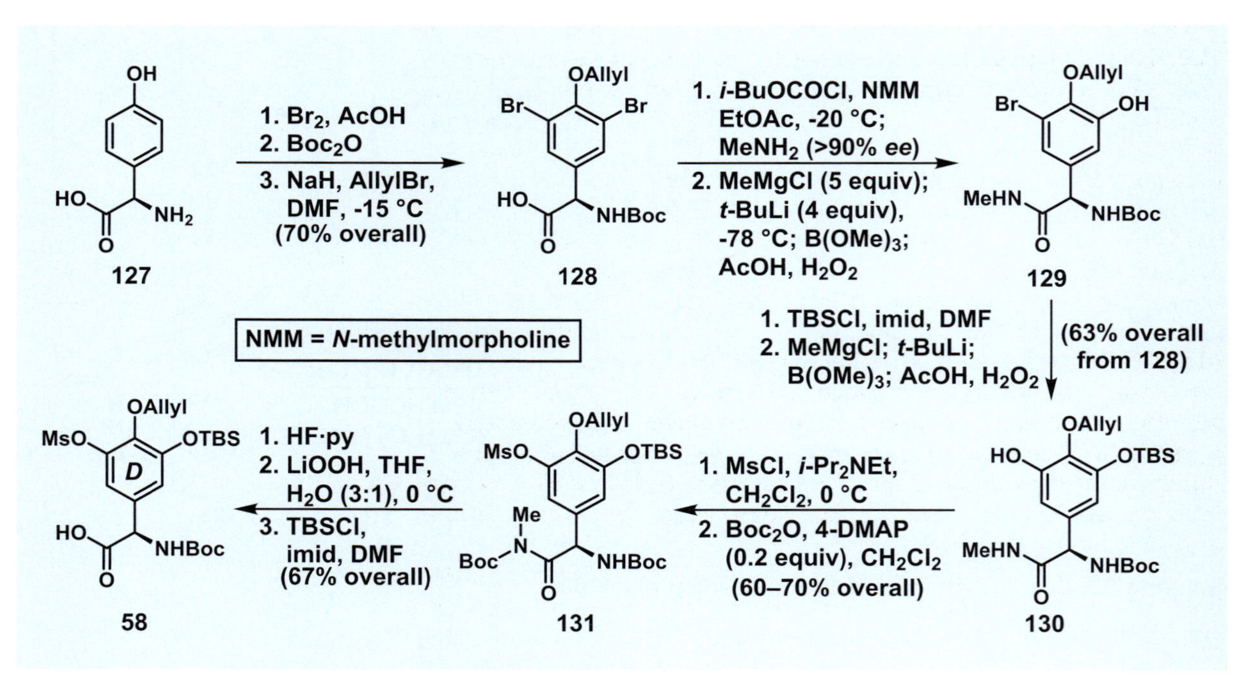

Scheme 31. Evans' synthesis of D-ring building block **58**.

two additional phenolic groups to serve as donors in the proposed S_NAr reactions that would hopefully lead to both bisaryl ether macrocycles in the vancomycin aglycon (**13**). The challenge in this endeavor resided not so much in the introduction of the phenol functions, but in achieving protecting group manipulations along the sequence. In the first step, tandem dibromination of **127** with Br_2 followed by Boc protection of the free amine and engagement of the lone phenol as an allyl ether afforded **128** in 70 % overall yield. The smooth prosecution of these steps left only carboxylic acid protection before the halides could be converted into phenolic OH groups. To meet this objective, the Evans group elected to incorporate a methyl amide as the protective device. Although we have already seen one means to effect the installation of such a motif in the steps converting **117** into **62** (cf. Scheme 28), application of these conditions here unfortunately led to significant racemization of the stereogenic center of the arylglycine. Although this substrate is not so different from **117**, in this case the electronic properties of the aromatic ring in combination with a prolonged reaction time to form the activated ester most likely induced methylamine to act as a base rather than as a nucleophile. As such, an alternative solution was required, and after significant exploration it was determined that the use of *i*-BuOCOCl and the weakly basic *N*-methylmorpholine (NMM) to quickly form an isobutyl mixed anhydride in EtOAc at −20 °C,[53] followed by the subsequent addition of $MeNH_2$, could accomplish the desired protection with less than 5 % racemization.

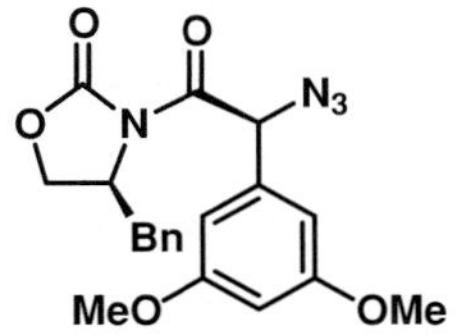

With this transformation achieved, effort could now be directed towards effecting mono-oxygenation of this product to phenol **129**. This objective was met, but, once again, a fair degree of reaction scouting was required to achieve maximum throughput. The optimal set of conditions was found to be initial Grignard formation with an excess of MeMgCl (5 equiv) at −78 °C in THF, followed by transmetallation to a lithiated nucleophile upon the addition of 4 equivalents of *t*-BuLi and then a standard $B(OMe)_3$ quench and oxidative workup. One should note that the initial use of excess MeMgCl was required to induce complete monodeprotonation of the methyl amide protected starting material, a condition that in turn necessitated the addition of excess *t*-BuLi, as the residual MeMgCl reacted with this lithium source to form a new metal species that was inactive for metal exchange. As a general tenet, although direct lithium insertion into dihalogenated substrates is possible, it is generally far easier to achieve cleaner, and selective, lithiation through initial magnesium insertion followed by transmetallation, as executed here. Indeed, attempts to effect direct lithiation failed to deliver **129**.

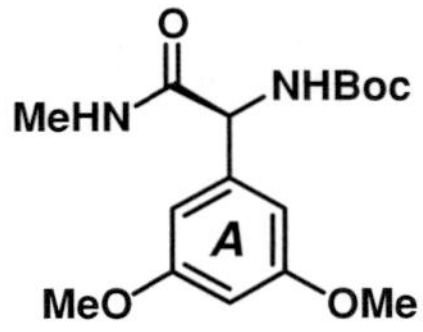

Returning now to the synthesis of fragment **58**, following silylation of the newly installed phenolic OH group in **129** under standard conditions (TBSCl, imid, DMF), a second iteration of the same phenol-forming reaction then provided **130** (Scheme 31). At this stage, only protection of the new OH group and cleavage of the methyl

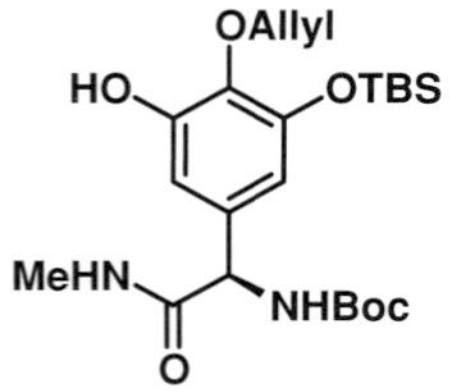

OAllyl HO OTBS MeHN NHBoc O **130**

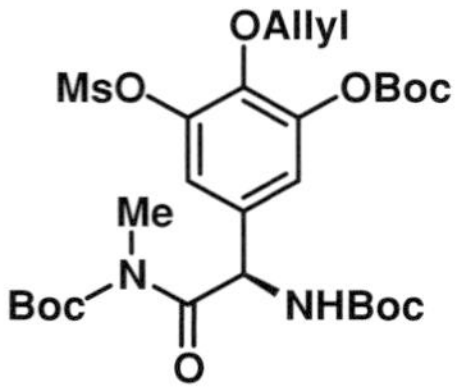

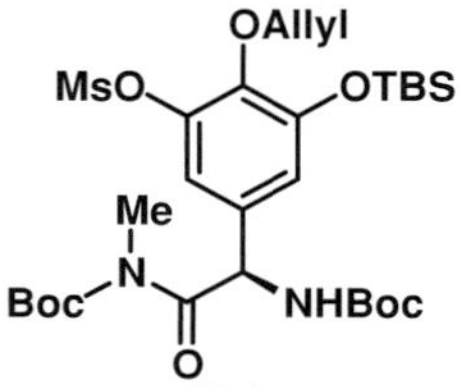

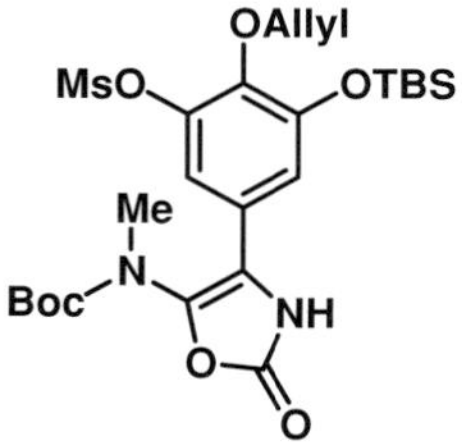

OAllyl MsO OBoc Me N Boc NHBoc O **132**

OAllyl MsO OTBS Me N Boc NHBoc O **131**

OAllyl MsO OTBS Me N Boc NH O O **133**

OAllyl MsO OTBS D HO NHBoc O **58**

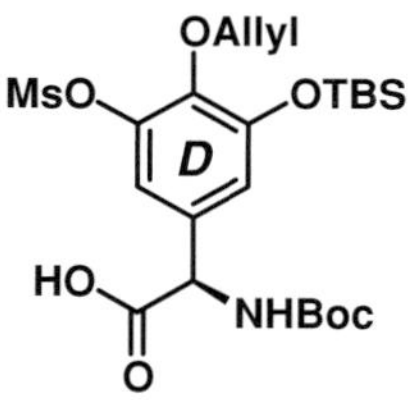

amide protecting group were required to complete the targeted **58**. Despite the simply stated nature of these conversions, in practice they proved far less facile to execute than originally anticipated. For example, although initial mesylation of **130** proved smooth with MsCl and Hünig's base in CH_2Cl_2, upon subsequent attempted conversion of the *N*-methyl amide into a Boc-protected imide (to facilitate its nucleophilic lysis along the lines mentioned earlier with nitrosoamides) a fair amount of silyl deprotection was observed, leading to significant amounts of by-product **132** (see column figure). When the stoichiometric equivalents of 4-DMAP were decreased to catalytic amounts, however, this unanticipated reaction could be suppressed, leading to **131** in yields that ranged from 60 to 70 % for these two steps. Since 4-DMAP alone is not sufficient to induce desilylation, it is reasonable to presume that the array of electron-withdrawing groups on the aromatic ring within this substrate, particularly the mesylate group, rendered the TBS group more susceptible to nucleophilic attack due to the incipient stabilization of the phenolate anion generated upon its cleavage. With **131** in hand, direct lysis of the methyl amide with LiOOH was then attempted, but a considerable amount of oxazolinone **133** (see column figure) was observed instead, presumably due to the impact of the relatively electron-poor aromatic ring in enabling facile enolization. It was, therefore, anticipated that the generation of a free OH group could decrease this effect through superior electron donation and thus shut down this alternative, undesired pathway. As indicated by the successful conversion of **131** into **58** over three steps in 67 % overall, this insightful conjecture proved accurate. Thus, the D-ring building block was finally in hand, despite several substrate-induced challenges encountered en route to its assembly.[34]

With the construction of the A-, B-, C-, and D-ring amino acids complete, the fundamental elements of the forward strategy could now be explored. As shown in Scheme 32, their preliminary union into advanced intermediate **59** proceeded quite smoothly. First, following lysis of the lone Boc group in **62** with TFA in the presence of Me_2S in CH_2Cl_2, the resultant amine was readily combined with C-ring carboxylic acid **63** with EDC and HOBt to afford **61** in 72 % yield. In preparation for the eventual incorporation of the B-ring (**60**), the cyclic carbamate was ruptured with Li_2CO_3, and then the residual Boc group was cleaved as before (TFA, Me_2S) to reveal a reactive amine handle. The combination of EDC and HOBt then ensured the facile addition of the B-ring (**60**) onto the growing vancomycin skeleton to provide **134** in 91 % overall yield from **61**. A final exchange of the Boc carbamate in **134** for a trifluoroacetate alternative led to **59**, setting the stage for the first critical operation of the envisioned total synthesis of the vancomycin aglycon (**13**), namely oxidative cyclization to generate the AB macrocycle.

As delineated in Scheme 33, the coveted ring closure of **59** proceeded as designed through a previously developed recipe which

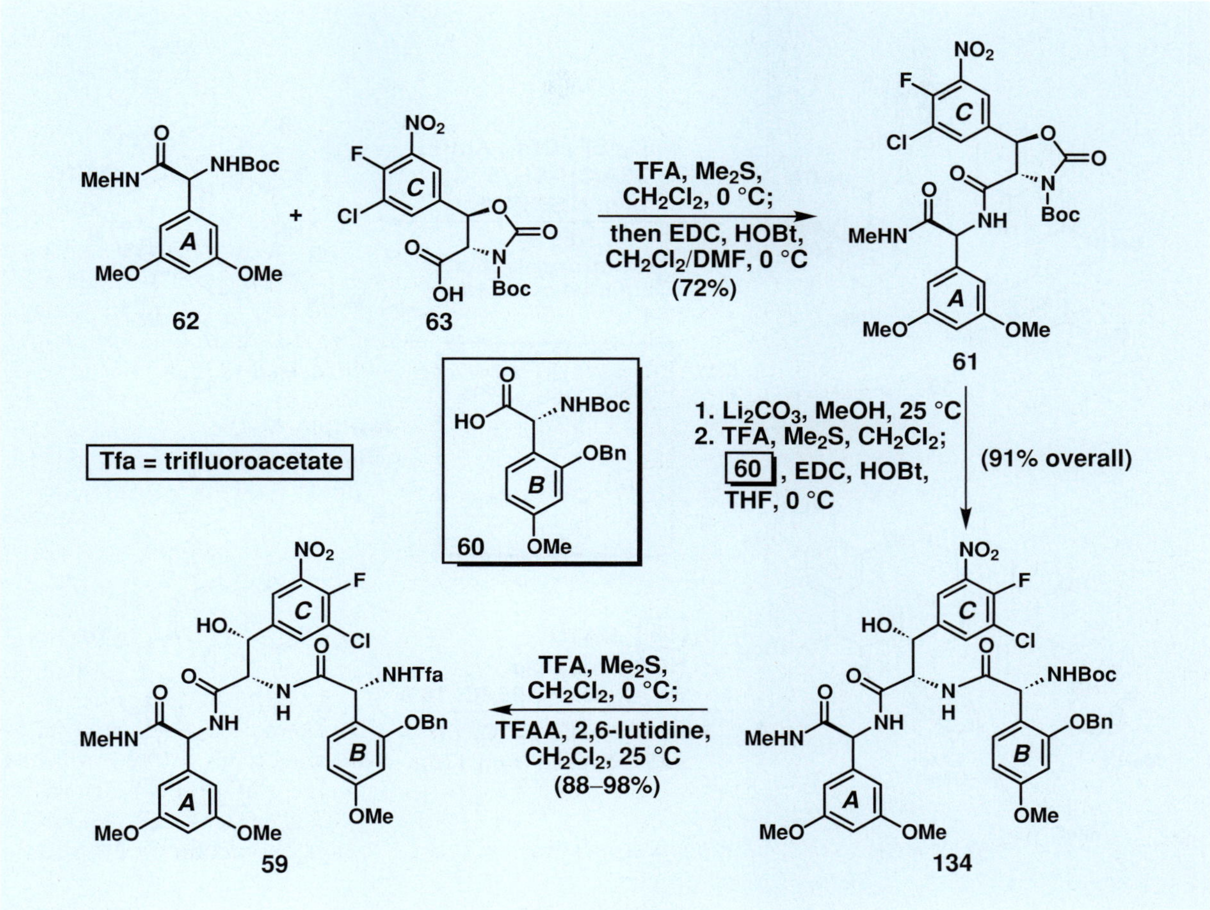

Scheme 32. Evans' synthesis of advanced intermediate **59**.

called for a combination of VOF_3, $BF_3 \cdot OEt_2$, and $AgBF_4$ in a solvent mixture of TFA and CH_2Cl_2 at 0 °C.[30] When the reaction was quenched with $NaBH(OAc)_3$, **135** was pleasingly obtained in 65 % yield with 19:1 atropselectivity in favor of the drawn product in which the C-34 benzyl ether had also been cleaved.* As noted above when discussing the planning stages of the synthesis, the

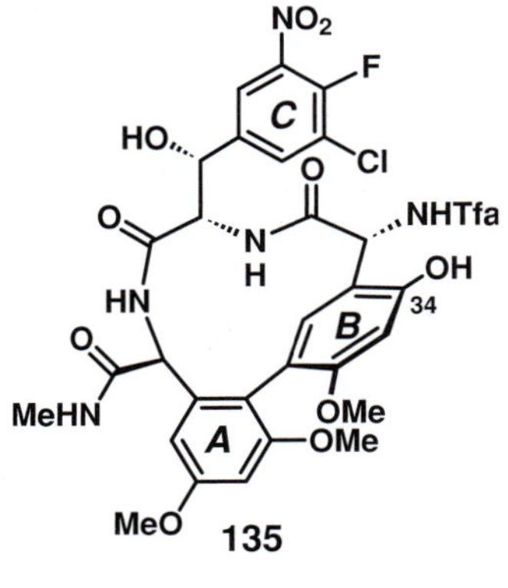

* This oxidative closure likely proceeds through a radical cation mechanism. Moreover, we should note that although the reagent combination employed might seem highly contrived, each reactant played a unique and necessary role. For example, the Ag^{I} salt served to sequester traces of chloride ion arising from commercial TFA, while the $BF_3 \cdot OEt_2$ served to prevent competitive attack of oxygen nucleophiles onto the B-ring radical cation intermediate prior to ring-closure. The final reaction with $NaBH(OAc)_3$ enabled the effective quench of the cation formed from the product (**135**) which could also be attacked by nucleophiles to afford unwanted side products.

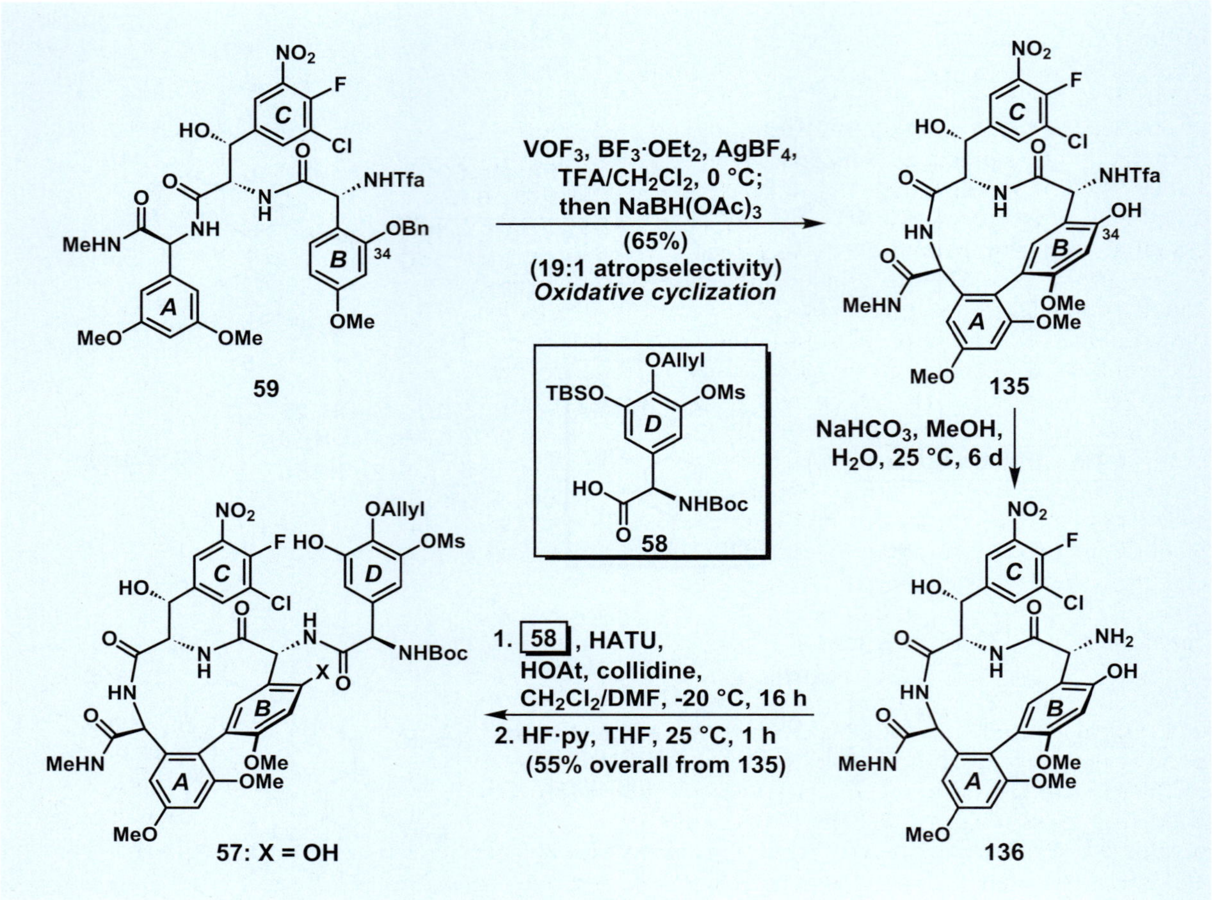

VOF_3, $BF_3 \cdot OEt_2$, $AgBF_4$, TFA/CH_2Cl_2, 0 °C; then $NaBH(OAc)_3$ (65%) (19:1 atropselectivity) *Oxidative cyclization*

59 → 135

$NaHCO_3$, MeOH, H_2O, 25 °C, 6 d

135 → 136

1. 58, HATU, HOAt, collidine, CH_2Cl_2/DMF, -20 °C, 16 h
2. HF·py, THF, 25 °C, 1 h (55% overall from 135)

136 → 57: X = OH

Scheme 33. Evans' synthesis of advanced intermediate **57**.

near exclusive formation of the undesired atropisomer from this sequence was anticipated from studies towards other glycopeptide antibiotics, but since it did not reflect the structure of vancomycin, hopefully it could be corrected through a thermal equilibration. Before pursuing this crucial objective, however, the construction of the C-O-D macrocycle was sought first. Consequently, the trifluoroacetamide group within **135** was cleaved with $NaHCO_3$ over the course of 6 days in a mixture of MeOH and H_2O to provide an intermediate (**136**) which was then merged with the D-ring building block **58** in the presence of an alternative peptide-forming reagent (HATU) in combination with HOAt and collidine. Deprotection of the silyl ether on the newly installed D-ring with HF•py in THF at ambient temperature finally completed the assembly of **57** in 55 % overall yield from **135**. As such, attempts at effecting bisaryl ether formation through a nitro-driven S_NAr reaction could now begin.

As expected, treatment of **57** (Scheme 34) with Na_2CO_3 in DMSO at ambient temperature, followed by a quench with the McMurry–Hendrickson reagent (Tf_2NPh)[55] to form a C-34 triflate, led to the desired macrocyclic product (**56**) with an impressive degree of atropselectivity (5:1). This fortuitous outcome presumably derived both from the favorable conformational biases of the substrate (which included the incorrect AB atropisomeric axis) and the reaction kinetics as dictated by the method of ring closure. With this essential step now behind them, the remaining objectives for the Evans group to reach **49** included alteration of a few protecting and functional groups, but, more importantly, the conversion of the AB macrocycle into its desired atropisomeric form. Thus, after removal of the nitro group on the C-ring through initial reduction (Zn, AcOH, EtOH, 40 °C) and a subsequent diazotization/reduction, the C-34 aryl triflate along with the phenolic allyl group were removed following brief exposure to a palladium catalyst in the presence of formic acid (which serves as a reducing agent, not as a proton donor), providing **137** in 77 % overall yield from **56**. At this stage, it was anticipated that after the methyl ethers on rings A and B were cleaved the rotational barrier for isomerization of the AB axis would be sufficiently reduced, hopefully enabling access to the correct AB/C-O-D ring system upon thermal activation. Most pleasingly, following adjustment of the protecting group array on **137**, sequential treatment with $AlBr_3$ and EtSH at 0 °C afforded triphenol **138**, which upon gentle heating in MeOH at 55 °C underwent isomerization with almost complete selectivity for **139**! With the desired AB/C-O-D atropisomer (**139**) finally in hand, its A- and B-ring OH groups were benzylated, the D-ring pivaloate ester was exchanged for an allyl ether, and the mesylate and trifluoroacetate groups were removed to provide **49**. As such, the latter compound was now suitably outfitted to pursue its union with the remaining tripeptide portion of the vancomycin aglycon (**13**).

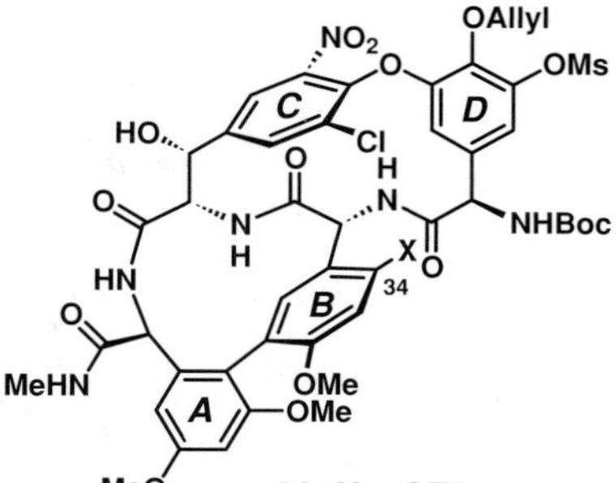

56: X = OTf

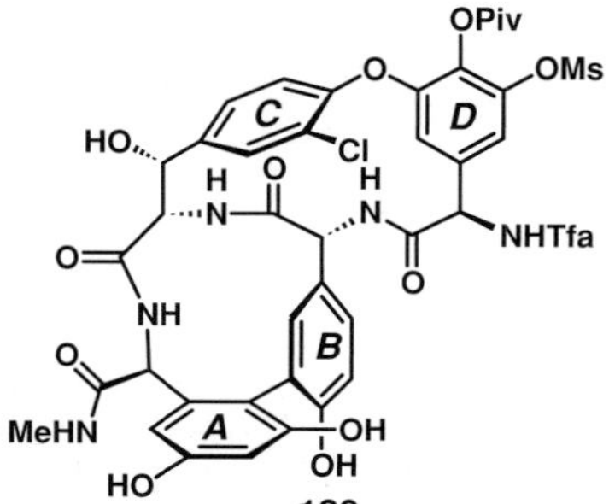

139

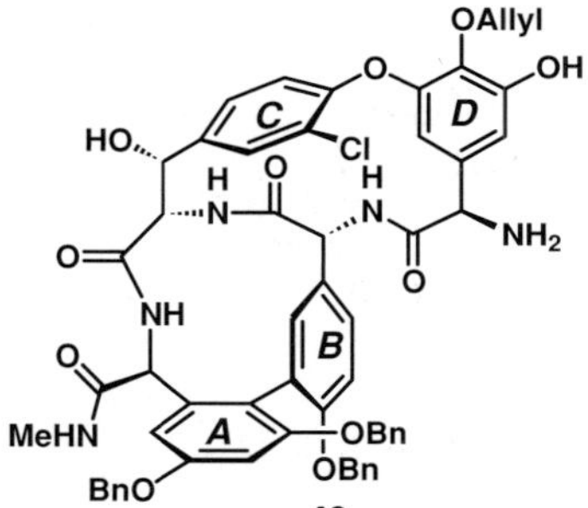

49

Before we proceed to the discussion of these final operations, however, we still need to describe the synthesis of this requisite tripeptide segment (**50**). The initial fragment needed for this assembly, the asparagine-derived amino acid **53** (see Scheme 35), proved quite easy to acquire from the previously mentioned intermediate **81** through two relatively standard operations which proceeded in 69 % overall yield. While this chemistry does not merit extensive discussion, the selection of a (2-trimethylsilyl)ethyl ester protecting group is worthy of some commentary due to the fact that it is not commonly encountered in natural product synthesis. This rarity greatly obscures its value since this moiety, like other silyl groups, is readily cleaved under acidic conditions or with fluoride ion, leading to a cascade that affords the free carboxylic acid attended by the loss of ethylene gas as shown in the neighboring column.[56] As such, they can serve as ideal companions to silyl ethers in late-stage deprotection regimes attempting to effect global cleavage in a minimal number of steps or to provide an orthogonal form of deprotec-

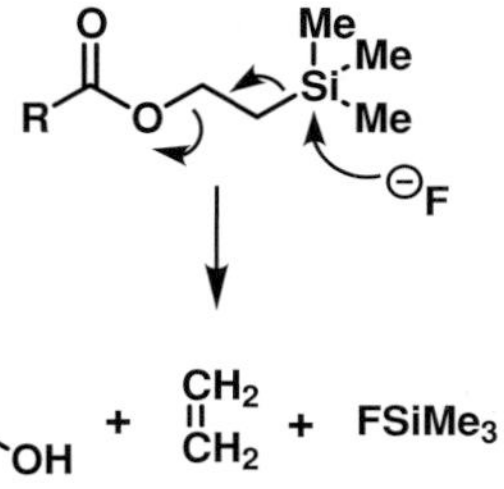

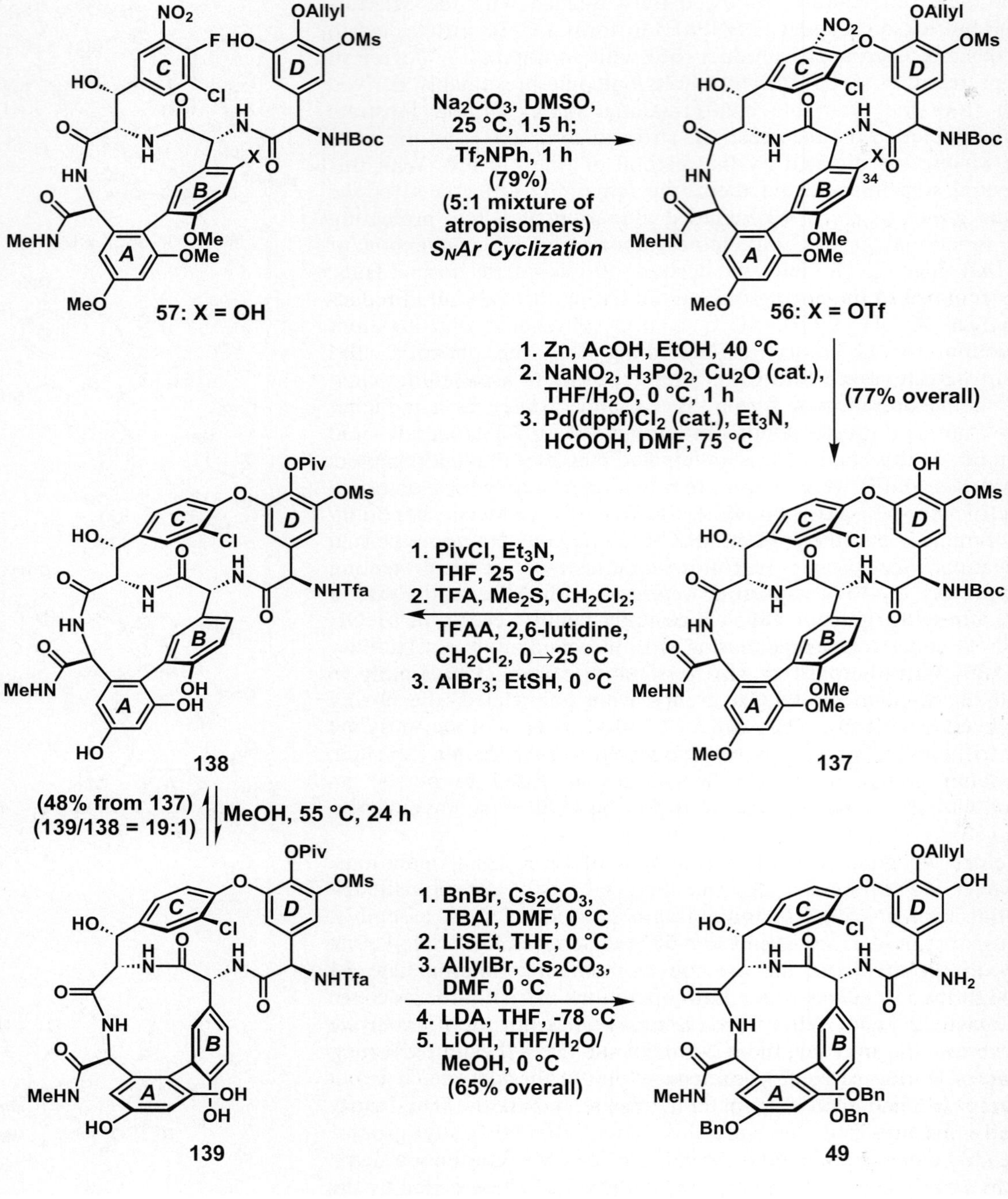

Scheme 34. Evans' construction of the AB/C-O-D domain of vancomycin (49).

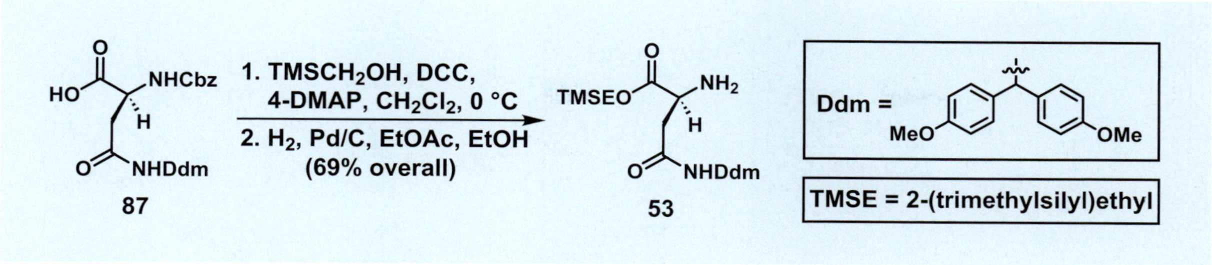

Scheme 35. Evans' synthesis of building block **53**.

tion within the collection of more standard carboxylic acid protecting groups.

Returning to our main task, the E-ring β-hydroxy tyrosine (**54**) was accessed, unsurprisingly, using chiral oxazolidinone technology. However, as this piece possesses alternate stereochemistry from that of its C-ring counterpart, a twist on the approach defined for that fragment (i. e. **63**) was required to achieve its construction. Accordingly, rather than start with a thiocyanate derivative which would provide an amine directly with 1,2-*syn* stereochemistry, the synthesis commenced instead with an alkyl bromide (**140**, Scheme 36) anticipating that this group could eventually be displaced with azide to afford the desired 1,2-*anti* arrangement of stereochemistry in the targeted building block (**54**). Indeed, the synthesis of 1,2-*syn* product **142** via intermediate *Z* enolate **141**, the subsequent S_N2 displacement of the bromine substituent with azide as supplied by TMGA, and the hydrolysis of the auxiliary group (LiOH, H_2O_2, THF, H_2O, 0 °C) led to the formation of **54** with the desired stereochemistry. With this compound in hand, the remaining operations to tripeptide **50** are familiar and therefore do not really require further commentary. The final steps towards vancomycin aglycon (**13**) could now begin in earnest.

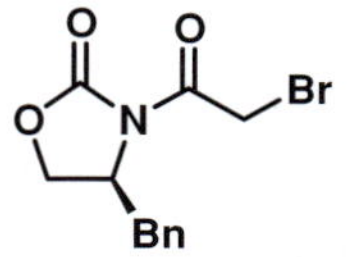

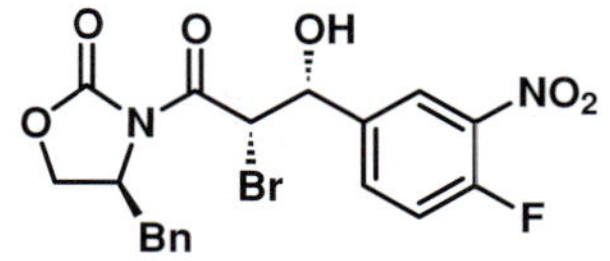

As with the Nicolaou synthesis of **13**, these concluding operations began with the union of the two advanced intermediates through peptide coupling, and, as shown in Scheme 37, **49** and **50** were neatly combined into **144** under the influence of EDC and HOBt in 86 % yield. The success of this operation then paved the way for the final S_NAr macrocyclization event. Unlike the triazene-driven macrocyclization discussed earlier, which led to a disappointing distribution of products, the use of CsF and DMSO at ambient temperature in this case afforded a satisfying 5:1 mixture of atropisomers (**145** and **146**) in favor of the desired isomer in a total combined yield of 95 %.

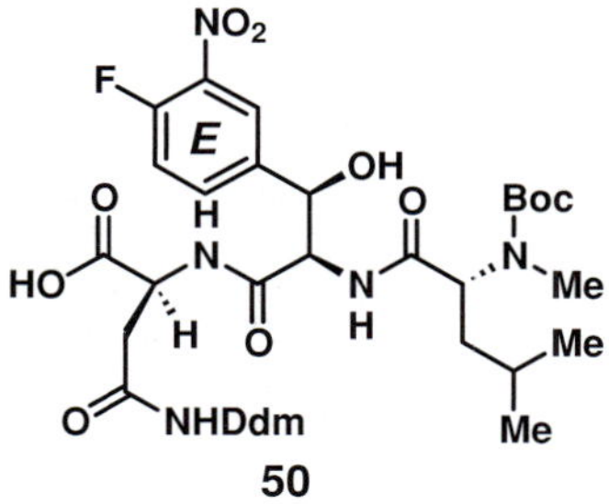

Once ample supplies of the needed atropisomer were in hand, the Evans group then moved forward to accomplish the final goal, which meant that they had to deprotect the array of groups on **145** (Scheme 38) and convert the nitro group of the E-ring into a chlorine residue through a Sandmeyer reaction. The latter objective was accomplished using a slightly modified diazotization/reduction

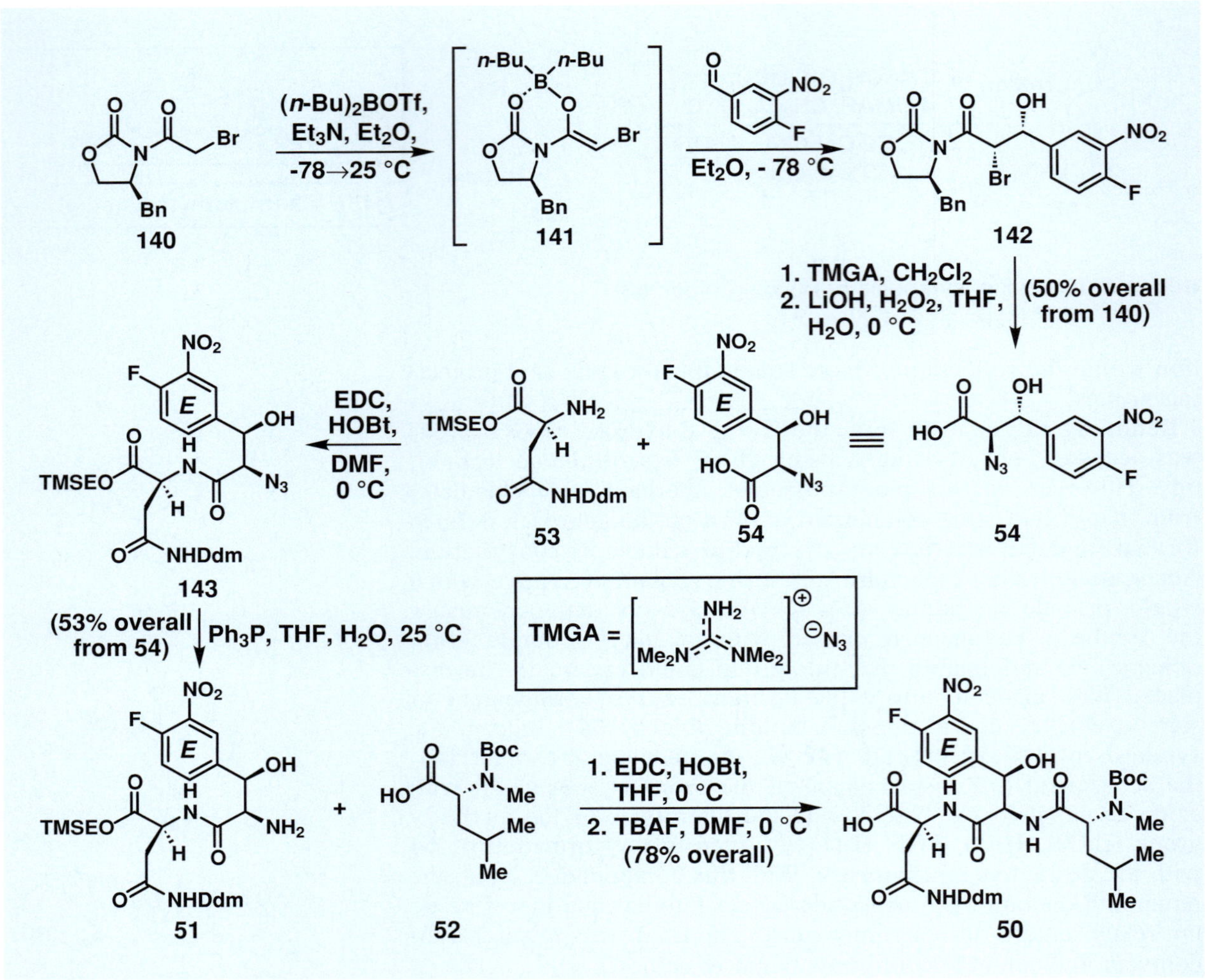

Scheme 36. Evans' synthesis of advanced tripeptide 50.

step in 81 % yield over two operations. Seeking now to begin the removal of the protecting groups assembled on the aglycon core, the methyl amide was then excised in 68 % yield through initial chemoselective nitrosation in the presence of seven other amide linkages followed by basic hydrolysis as effected with LiOOH. The relative facility of this cleavage protocol in the presence of a wealth of chemically sensitive functionality certainly confers new value to methyl amide protection of carboxylic acids within the context of complex molecule construction. Vancomycin aglycon (**13**) was then completed in three additional steps from **147**, starting with removal of the phenolic allyl protecting group with catalytic $Pd(PPh_3)_4$ and hydrogenative severing of the three benzyl ethers through transfer hydrogenation conditions employing 1,4-cyclohex-

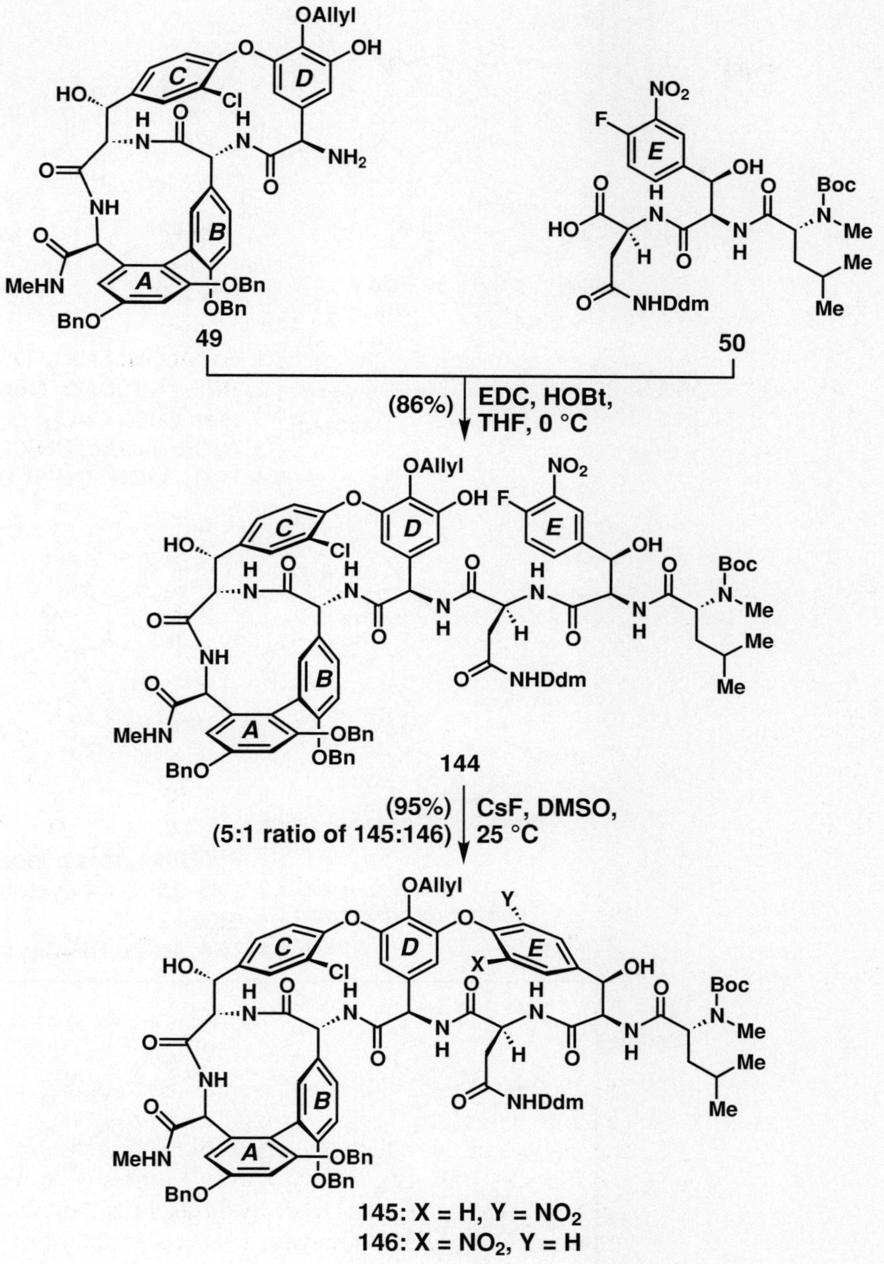

Scheme 37. Evans' construction of the D-O-E domain of vancomycin.

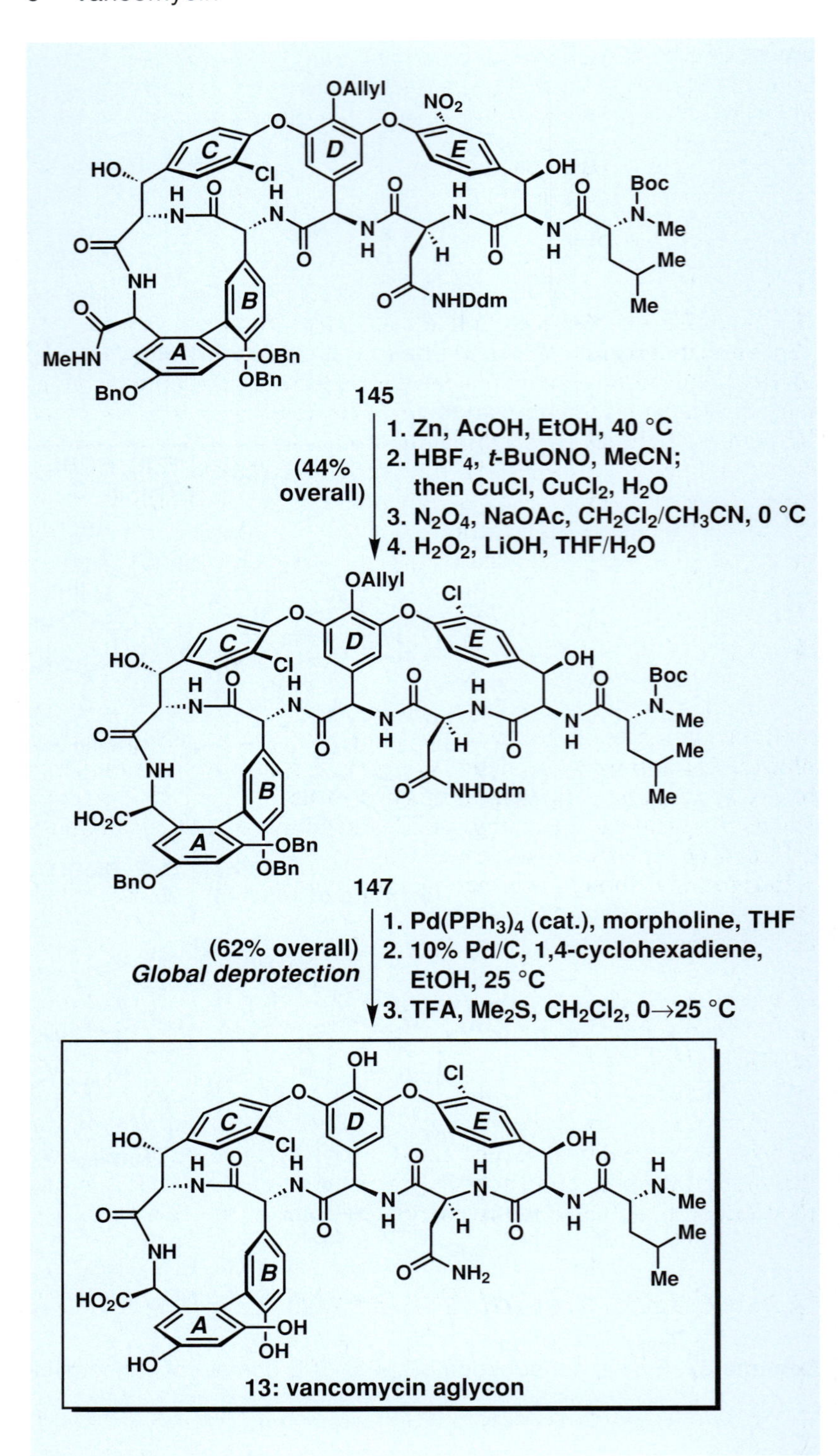

Scheme 38. Final stages and completion of Evans' total synthesis of the vancomycin aglycon (**13**).

adiene as a hydrogen source to prevent loss of the aryl chlorine substituents. Finally, TFA/Me_2S-mediated scission of the acid-labile Ddm and Boc protecting groups completed the synthesis of **13** in 62 % overall yield for these final three steps.

9.4 Conclusion

In many respects, the accomplishment of two total syntheses of the vancomycin aglycon (**13**) in 1998, as well as vancomycin (**1**) itself in 1999, constitutes a new plateau for the power of synthetic chemistry as this target long thwarted efforts to effect its assembly in the laboratory. Critical to reaching this peak was overcoming many steep inclines and seemingly impassable barriers in the form of novel molecular motifs and stereochemical features through both the development and application of several unique synthetic strategies and tactics. In this regard, the creation of a new method by which to form macrocyclic bisaryl ethers, driven by aryl triazenes, in combination with efficient amino acid syntheses derived from the power of the Sharpless AD and AA reactions, stand as the critical tools that enabled the Nicolaou group to make their ascent. For the Evans team, effectively using their chiral oxazolidinone methods to enantioselectively fashion four amino acid building blocks, developing an oxidative coupling protocol to form hindered biaryl systems, and finding the means to accomplish the selective and facile cleavage of methyl amides on polyfunctional substrates afforded an equally effective pathway.

Intriguingly, though, neither journey was completely smooth in that each had to confront at least one unintended, and potentially disastrous, roadblock that resulted from their selected synthetic sequence. For the Nicolaou approach, this challenge was the unanticipated recalcitrance of the D-ring triazene to be converted into a phenol, a problem that ultimately required the development of a novel reaction protocol to effect this substitution. In the Evans route, this roadblock was the initial near exclusive formation of the incorrect AB atropisomer, which, although problematic on the surface, was creatively turned into a blessing by using it to direct the highly atropselective formation of the C-O-D macrocycle and then effecting its selective isomerization later in the sequence.

9.5 Boger's Total Synthesis of the Vancomycin Aglycon (13)

In 1999, a third total synthesis of the vancomycin aglycon (**13**) was achieved by Professor Dale Boger and his group at The Scripps Research Institute based on a unique design for determining the order of macrocycle construction relative to those already presented

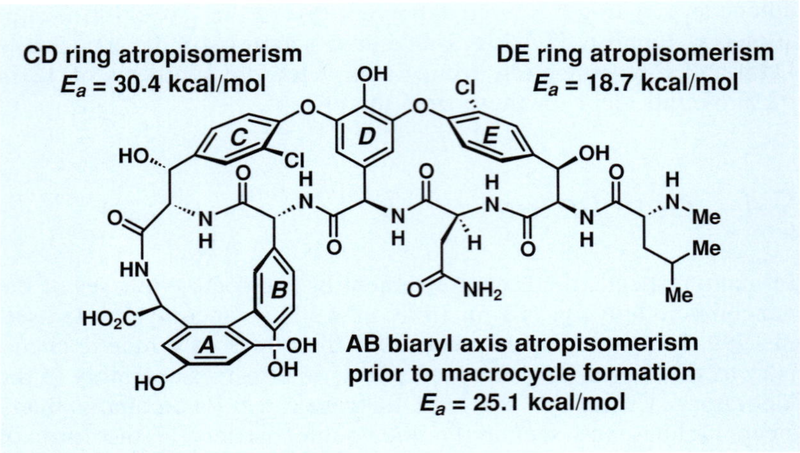

Scheme 39. Boger's study on the experimental energies required to effect equilibration of the atropisomers that can be formed around the individual vancomycin ring systems.

in this chapter.[57,58] Rather than leave the atropselectivity of each macrocyclization to chance, a result dictated by both the method employed and the unique array of functionality present in the substrate, the Boger group decided to order their ring-closures rationally such that each could be thermally equilibrated once formed in order to enhance the amount of the desired atropisomer before moving forward. In order to achieve this objective, however, they needed to assess the thermal barriers required for isomerizing each atropisomeric axis so that they would not destroy already established, atropisomerically pure macrocycles. Following several months of model studies to acquire such information, the answer eventually emerged through the results shown in Scheme 39. As indicated, among all three macrocyclic systems the C-O-D bisaryl ether required the most energy to isomerize (30.4 kcal/mol), while the D-O-E system required the least activation (18.7 kcal/mol). In between these disparate values was an E_a barrier of 25.1 kcal/mol for an AB biaryl system in the absence of its complete macrocycle. Accordingly, these results indicated that if a C-O-D system were formed first, followed by the AB biaryl axis and, finally, the D-O-E bisaryl ether ring, then each system could be thermally funneled into the desired atropisomeric form following its initial synthesis without impacting any previously equilibrated systems. Intriguingly, this order of operations does not mirror that of either the Nicolaou or Evans approaches.

The successful prosecution of this strategy is shown in Schemes 40 and 41, starting with the synthesis of the C-O-D macrocycle. Thus, once **148** was in hand, its treatment with a mixture of K_2CO_3 and $CaCO_3$ effected a nitro-driven S_NAr reaction, which provided both possible atropisomeric products (**149** and **150**) in a 1:1 ratio.

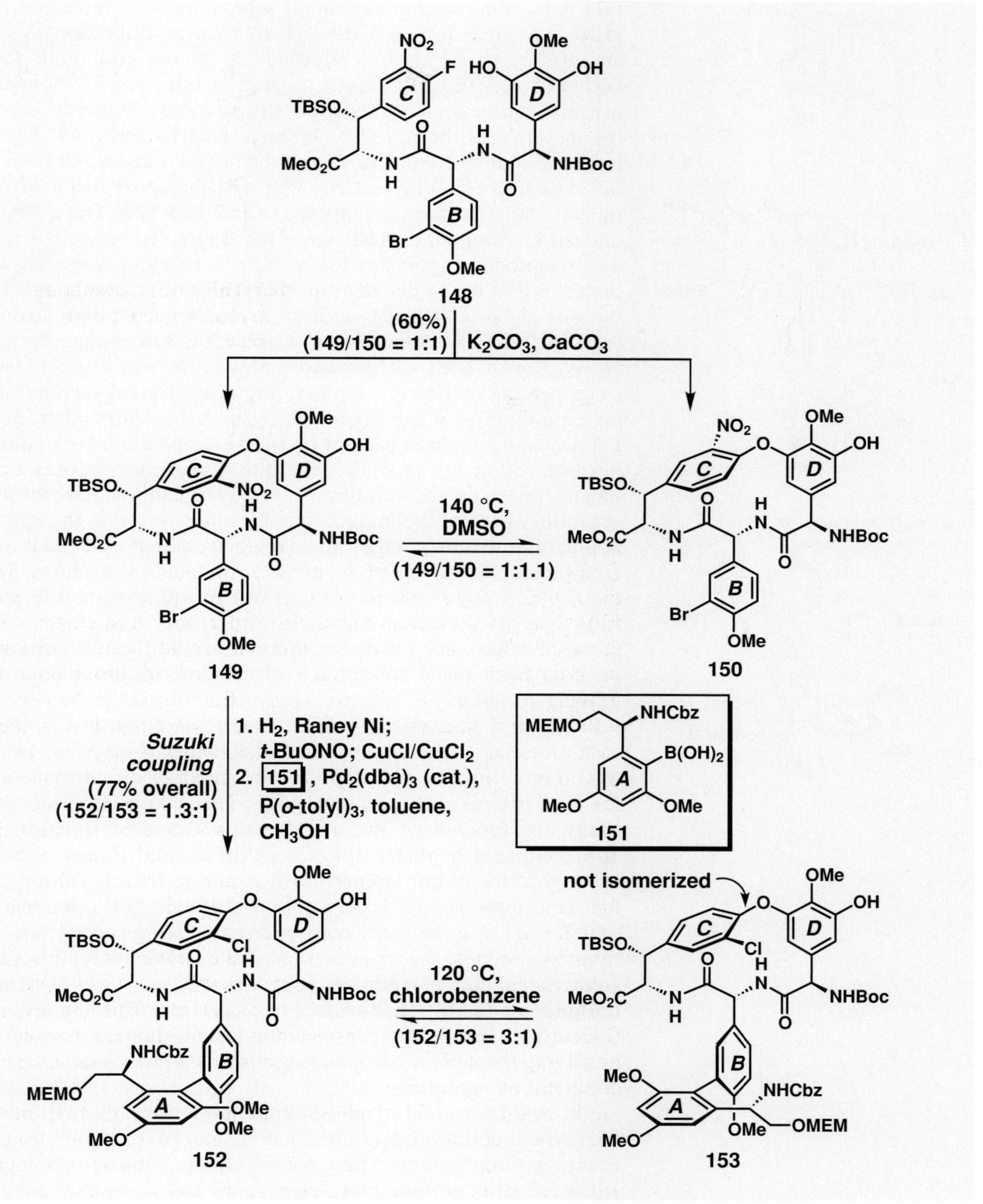

Scheme 40. Boger's sequence to access advanced intermediate **152** in atropisomerically pure form through a series of thermal equilibrations.

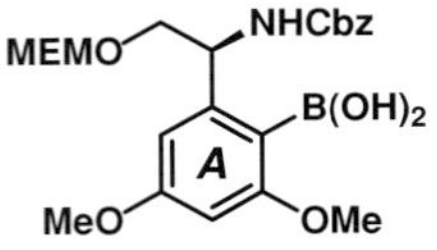

Following their chromatographic separation, the undesired isomer (**150**) was then recycled through thermal equilibration in DMSO at 140 °C, an event that afforded an almost equimolar ratio of both **149** and **150**. Continued separation followed by thermal atropisomerization of the unwanted atropisomer eventually converted the majority of the initially obtained material into **149**. Next, following Sandmeyer substitution of the nitro group in **149** for a chlorine, a Suzuki coupling reaction with **151** then provided a 1.3:1 mixture of the AB biaryl atropisomers **152** and **153**. Once again, the undesired compound (**153**) could be thermally converted into the desired product upon heating at 120 °C in chlorobenzene, in this instance in a much better ratio (**152/153** = 3:1), without affecting the already established C-O-D bisaryl ether axis. Two features of the reactions discussed above deserve further commentary. First, the choice of both solvent and temperature was critical. In some cases heating at the same temperature in a different solvent afforded no equilibration whatsoever, while in successfully identified solvents heating at either a lower or higher temperature led to increased decomposition. As such, these results suggest that there is a careful balance between the conditions employed and material throughput using this general technique, thereby indicating that the successful application of this strategy in any other context will likely require significant amounts of protocol scouting and optimization. Second, the Boger group could have conceivably equilibrated **149** and **150** following the conversion of their nitro groups into chlorine substituents if it was not for their observation that the nitro compounds were far more easily subjected to atropisomerization (i. e. possessed a lower E_a value).

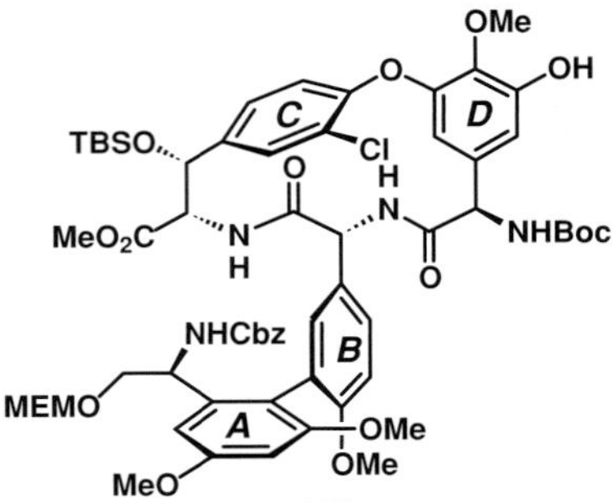

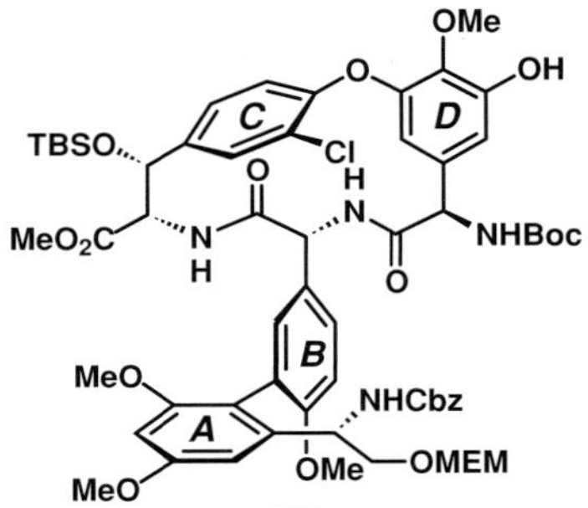

In the final bisaryl ether ring closure (see Scheme 41), the same conditions as those employed in the Evans route were used, but this time provided a delightful 8:1 mixture of atropisomers in favor of the desired product (**155**). This exquisite result is most likely the product of the unique set of resident functionality in **154** compared to that employed in the related Evans cyclization. Moreover, the minor isomer (**156**) could be recycled through thermal equilibration to a 1:1 mixture of **155** and **156** upon heating at 140 °C in 1,2-dichlorobenzene, thereby leaving only a few operations to complete the aglycon (**13**). Taken cohesively, this alternate solution to the vancomycin aglycon (**13**) was quite effectively designed, with only the first macrocyclization affording no inherent selectivity. Through successive rounds of thermal isomerization, however, more than sufficient material supplies were garnered to complete the campaign.

Although this total synthesis concludes our presentation of this intriguing natural product, the final chapter of its exploration is certainly far from written. Many new synthetic tools and strategies are still needed to provide more efficient solutions to the vancomycin architecture, and a wealth of fundamental knowledge in the area of chemical biology remains to be discovered. Hopefully, studies aimed in this latter direction will not only reveal new insight into

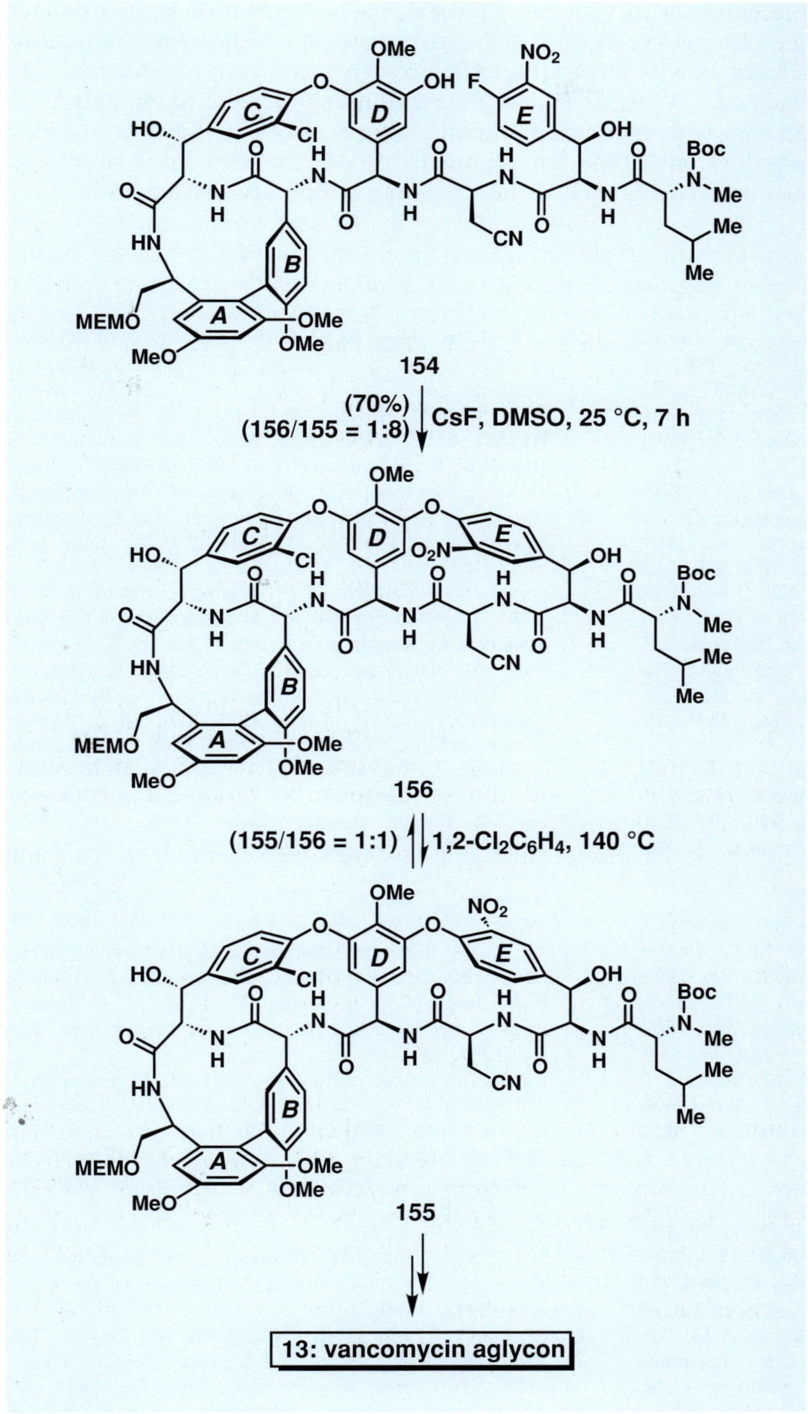

154

(70%)
(156/155 = 1:8)
CsF, DMSO, 25 °C, 7 h

156

(155/156 = 1:1)
1,2-$Cl_2C_6H_4$, 140 °C

155

13: vancomycin aglycon

Scheme 41. Final ring closure and equilibration process in Boger's total synthesis of the vancomycin aglycon (**13**).

the behavior of vancomycin itself, but will generate a blueprint for the rational creation of vancomycin analogs whose structural modifications will restore the efficacy of vancomycin against resistant bacteria.[59] With these microbes continuing to adapt to our antibiotic arsenal with increasing resourcefulness, access to such knowledge provides one of the few glimmers of hope that new lines of defense can be erected, even if they are only temporary barricades.

References

1. a) J. C. Sheehan, *The Enchanted Ring: The Untold Story of Penicillin*, MIT Press, Cambridge, **1982**, p. 224; b) S. Selwyn, R. W. Lacey, M. Bakhtiar, *The Beta-Lactam Antibiotics: Penicillins and Cephalosporins in Perspective*, Hodder and Stoughton, London, **1980**, p. 364.
2. For a general review on vancomycin, see: K. C. Nicolaou, C. N. C. Boddy, S. Bräse, N. Winssinger, *Angew. Chem.* **1999**, *111*, 2230; *Angew. Chem. Int. Ed.* **1999**, *38*, 2096 and references cited therein. Of additional interest, we suggest the following: K. C. Nicolaou, C. N. C. Boddy, *Sci. Am.* **2001**, *284*, 46.
3. D. H. Williams, M. P. Williamson, D. W. Butcher, S. J. Hammond, *J. Am. Chem. Soc.* **1983**, *105*, 1332. For a recent review, see: D. H. Williams, B. Bardsley, *Angew. Chem.* **1999**, *111*, 1264; *Angew. Chem. Int. Ed.* **1999**, *38*, 1172.
4. For recent reviews of this research, see: a) B. K. Hubbard, C. T. Walsh, *Angew. Chem.* **2003**, *115*, 752; *Angew. Chem. Int. Ed.* **2003**, *42*, 730; b) C. T. Walsh, S. L. Fisher, I.-S. Park, M. Prahalad, Z. Wu, *Chem. Biol.* **1996**, *3*, 21; c) G. D. Wright, C. T. Walsh, *Acc. Chem. Res.* **1992**, *25*, 468.
5. For reviews of synthetic approaches to vancomycin see Ref. 2 and: a) A. V. Rama Rao, M. K. Gurjar, K. L. Reddy, A. S. Rao, *Chem. Rev.* **1995**, *95*, 2135; b) J. Zhu, *Synlett* **1997**, 133.
6. a) C. M. Harris, T. M. Harris, *J. Am. Chem. Soc.* **1982**, *104*, 4293; b) G. M. Sheldrick, P. G. Jones, O. Kennard, D. H. Williams, G. A. Smith, *Nature* **1978**, *271*, 223; c) D. H. Williams, J. R. Kalman, *J. Am. Chem. Soc.* **1977**, *99*, 2768; d) F. J. Marshall, *J. Med. Chem.* **1965**, *8*, 18.
7. M. Oki, *Top. Stereochem.* **1983**, *14*, 1.
8. M. McCarthy, P. J. Guiry, *Tetrahedron* **2001**, *57*, 3809.
9. a) G. Bringmann, R. Götz, P. A. Keller, R. Walter, M. R. Boyd, F. Lang, A. Garcia, J. J. Walsh, I. Tellitu, K. V. Bhaskar, T. R. Kelly, *J. Org. Chem.* **1998**, *63*, 1090. For the initial isolation of this family of natural products, see: b) M. R. Boyd, Y. F. Hallock, J. H. Cardellina, K. P. Manfredi, J. W. Blunt, J. B. McMahon, R. W. Buckheit, G. Bringmann, M. Schäffer, G. M. Cragg, D. W. Thomas, J. G. Jato, *J. Med. Chem.* **1994**, *37*, 1740. For another total synthesis of these natural products, see: c) T. R. Hoye, M. Chen, L. Mi, O. P. Priest, *Tetrahedron Lett.* **1994**, *35*, 8747.
10. a) K. C. Nicolaou, H. Li, C. N. C. Boddy, J. M. Ramanjulu, T.-Y. Yue, S. Natarajan, X.-J. Chu, S. Bräse, F. Rübsam, *Chem. Eur. J.* **1999**, *5*, 2584; b) K. C. Nicolaou, C. N. C. Boddy, H. Li, A. E. Koumbis, R. Hughes, S. Natarajan, N. F. Jain, J. M. Ramanjulu, S. Bräse, M. E. Solomon, *Chem. Eur. J.* **1999**, *5*, 2602; c) K. C. Nicolaou, A. E. Koumbis, M. Takayanagi, S. Natarajan, N. F. Jain, T. Bando, H. Li, R. Hughes, *Chem. Eur. J.* **1999**, *5*, 2622; d) K. C. Nicolaou, H. J. Mitchell, N. F Jain, T. Bando, R. Hughes, N. Winssinger, S. Natarajan, A. E. Koumbis, *Chem. Eur. J.* **1999**, *5*, 2648. For the initial disclosure of the total synthesis of vancomycin aglycon, see: e) K. C. Nicolaou, S. Natarajan, H. Li, N. F. Jain, R. Hughes, M. E. Solomon, J. M. Ramanjulu, C. N. C. Boddy, M. Takayanagi, *Angew. Chem.* **1998**, *110*, 2872; *Angew. Chem. Int. Ed.* **1998**, *37*, 2708; f) K. C. Nicolaou, N. F. Jain, S. Natarajan, R. Hughes, M. E. Solomon, H. Li, J. M. Ramanjulu, M. Takayanagi, A. E. Koumbis, T. Bando, *Angew. Chem.* **1998**, *110*, 2879; *Angew. Chem. Int. Ed.* **1998**, *37*, 2714; g) K. C. Nicolaou, M. Takayanagi, N. F. Jain, S. Natarajan, A. E. Koumbis, T. Bando, J. M. Ramanjulu, *Angew. Chem.* **1998**, *110*, 2881; *Angew. Chem. Int. Ed.* **1998**, *37*, 2717. For the initial disclosure of the total synthesis of vancomycin, see: h) K. C. Nicolaou, H. J. Mitchell, N. F. Jain, N. Winssinger, R. Hughes, T. Bando, *Angew. Chem.* **1999**, *111*, 253; *Angew. Chem. Int. Ed.* **1999**, *38*, 240.
11. For earlier model studies, see: a) K. C. Nicolaou, H. J. Mitchell, F. L. van Delft, F. Rübsam, R. M. Rodríguez, *Angew. Chem.* **1998**, *110*, 1972; *Angew. Chem. Int. Ed.* **1998**, *37*, 1871; b) K. C. Nicolaou, X.-J. Chu, J. M. Ramanjulu, S. Natarajan, S. Bräse, F. Rübsam, C. N. C. Boddy, *Angew. Chem.* **1997**, *109*, 1551; *Angew. Chem. Int. Ed. Engl.* **1997**, *36*, 1539; c) K. C. Nicolaou, J. M. Ramanjulu, S. Natarajan, S. Bräse, H. Li, C. N. C. Boddy, F. Rübsam, *Chem. Commun.* **1997**, 1899.
12. a) D. A. Evans, M. R. Wood, B. W. Trotter, T. I. Richardson, J. C. Barrow, J. L. Katz, *Angew. Chem.* **1998**, *110*, 2864; *Angew. Chem. Int. Ed.* **1998**, *37*, 2700; b) D. A. Evans, C. J. Dinsmore, P. S. Watson, M. R. Wood, T. I. Richardson, B. W. Trotter, J. L. Katz, *Angew. Chem.* **1998**, *110*, 2868; *Angew. Chem. Int. Ed.* **1998**, *37*, 2704.

13. For earlier model studies, see: a) D. A. Evans, C. J. Dinsmore, A. M. Ratz, *Tetrahedron Lett.* **1997**, *38*, 3189; b) D. A. Evans, P. S. Watson, *Tetrahedron Lett.* **1996**, *37*, 3251; c) D. A. Evans, C. J. Dinsmore, D. A. Evrard, K. M. DeVries, *J. Am. Chem. Soc.* **1993**, *115*, 6426; d) D. A. Evans, C. J. Dinsmore, *Tetrahedron Lett.* **1993**, *34*, 6029.
14. a) B. G. Davis, *J. Chem. Soc., Perkin Trans. 1* **2000**, 2137; b) G.-J. Boons, *Contemp. Org. Synth.* **1996**, *3*, 173; c) G.-J. Boons, *Tetrahedron* **1996**, *52*, 1095. For a more specialized review of glycosidations in the context of total synthesis, see: K. C. Nicolaou, H. J. Mitchell, *Angew. Chem.* **2001**, *113*, 1624; *Angew. Chem. Int. Ed.* **2001**, *40*, 1576.
15. R. R. Schmidt in *Comprehensive Organic Synthesis, Vol. 6* (Eds.: B. M. Trost and I. Fleming), Pergamon, Oxford, **1994**, pp. 33–64.
16. a) D. Kahne, S. Walker, Y. Cheng, D. Van Engen, *J. Am. Chem. Soc.* **1989**, *111*, 6881; b) D. Crich, S. Sun, *J. Am. Chem. Soc.* **1998**, *120*, 435.
17. K. Toshima, K. Tatsuta, *Chem. Rev.* **1993**, *93*, 1503.
18. For some discussion, see: J. L. Radkiewicz, H. Zipse, S. Clarke, K. N. Houk, *J. Am. Chem. Soc.* **1996**, *118*, 9148 and references therein. The Ddm group is, in fact, relatively underappreciated. For example, it fails to appear in the treatise on protecting groups used by all students of organic chemistry: T. W. Greene, P. G. M. Wutz, *Protective Groups in Organic Synthesis*, John Wiley & Sons, New York, **1999**, p. 779.
19. For a general review on the formation of peptide bonds, see: G. Benz in *Comprehensive Organic Synthesis, Vol. 6* (Eds.: B. M. Trost, I. Fleming), Pergamon, Oxford, **1991**, pp. 381–417.
20. R. Beugelmans, M. Bois-Choussy, C. Vergne, J.-P. Bouillon, J. Zhu, *Chem. Commun.* **1996**, 1029.
21. R. Beugelmans, J. Zhu, N. Husson, M. Bois-Choussy, G. P. Singh, *J. Chem. Soc., Chem. Commun.* **1994**, 439.
22. a) D. A. Evans, J. A. Ellman, K. M. DeVries, *J. Am. Chem. Soc.* **1989**, *111*, 8912; b) Y. Suzuki, S. Nishiyama, S. Yamamura, *Tetrahedron Lett.* **1989**, *30*, 6043.
23. H. Konishi, T. Okuno, S. Nishiyama, S. Yamamura, K. Koyasu, Y. Terada, *Tetrahedron Lett.* **1996**, *37*, 8791.
24. D. L. Boger, Y. Nomoto, B. R. Teegarden, *J. Org. Chem.* **1993**, *58*, 1425 and references cited therein.
25. a) K. C. Nicolaou, C. N. C. Boddy, S. Natarajan, T.-Y. Yue, H. Li, S. Bräse, J. M. Ramanjulu, *J. Am. Chem. Soc.* **1997**, *119*, 3421. For the later use of this technology to achieve atropselective ring closure, see: b) K. C. Nicolaou, C. N. C. Boddy, *J. Am. Chem. Soc.* **2002**, *124*, 10451.
26. a) G. Li, H.-T. Chang, K. B. Sharpless, *Angew. Chem.* **1996**, *108*, 449; *Angew. Chem. Int. Ed. Engl.* **1996**, *35*, 451; b) G. Li, K. B. Sharpless, *Acta Chem. Scand.* **1996**, *50*, 649.
27. For selected examples of recent papers in the AA field, see: a) Z. P. Demko, M. Bartsch, K. B. Sharpless, *Org. Lett.* **2000**, *2*, 2221; b) L. K. Goossen, H. Liu, K. R. Dress, K. B. Sharpless, *Angew. Chem.* **1999**, *111*, 1149; *Angew. Chem. Int. Ed.* **1999**, *38*, 1080; c) K. L. Reddy, K. B. Sharpless, *J. Am. Chem. Soc.* **1998**, *120*, 1207; d) K. L. Reddy, K. R. Dress, K. B. Sharpless, *Tetrahedron Lett.* **1998**, *39*, 3667; e) M. Bruncko, G. Schlingloff, K. B. Sharpless, *Angew. Chem.* **1997**, *109*, 1580; *Angew. Chem. Int. Ed. Engl.* **1997**, *36*, 1483; f) G. Li, H. H. Angert, K. B. Sharpless, *Angew. Chem.* **1996**, *108*, 2995; *Angew. Chem. Int. Ed. Engl.* **1996**, *35*, 2813.
28. H. C. Kolb, M. VanNieuwenhze, K. B. Sharpless, *Chem. Rev.* **1994**, *94*, 2483.
29. H. Becker, K. B. Sharpless, *Angew. Chem.* **1996**, *108*, 447; *Angew. Chem. Int. Ed. Engl.* **1996**, *35*, 448.
30. a) D. A. Evans, C. J. Dinsmore, A. M. Ratz, D. A. Evrard, J. C. Barrow, *J. Am. Chem. Soc.* **1997**, *119*, 3417; b) D. A. Evans, J. C. Barrow, P. S. Watson, A. M. Ratz, C. J. Dinsmore, D. A. Evrard, K. M. DeVries, J. A. Ellman, S. D. Rychnovsky, J. Lacour, *J. Am. Chem. Soc.* **1997**, *119*, 3419.
31. For challenges in employing secondary amides as protecting groups, see: H. Kung, H. Waldmann in *Comprehensive Organic Synthesis, Vol. 6* (Eds.: B. M. Trost, I. Fleming), Pergamon, Oxford, **1991**, pp. 631–646.
32. D. A. Evans, P. H. Carter, C. J. Dinsmore, J. C. Barrow, J. L. Katz, D. W. Kung, *Tetrahedron Lett.* **1997**, *38*, 4535.
33. For earlier expressions of this same concept, see: a) J. Garcia, J. González, R. Segura, F. Urpí, J. Vilarrasa, *J. Org. Chem.* **1984**, *49*, 3322; b) R. Berenguer, J. Garcia, J. Vilarrasa, *Synthesis* **1989**, 305.
34. For an early discussion of Evans' oxazolidinone chiral auxiliaries, see: D. A. Evans, J. V. Nelson, T. R. Taber, *Top. Stereochem.* **1982**, *13*, 1.
35. For a review on directed *o*-lithiation, see: V. Snieckus, *Chem. Rev.* **1990**, *90*, 879.
36. For other methods to asymmetrically synthesize arylglycines, see: R. M. Williams, J. A. Hendrix, *Chem. Rev.* **1992**, *92*, 889.
37. A. V. Rama Rao, T. K. Chakraborty, K. L. Reddy, A. S. Rao, *Tetrahedron Lett.* **1994**, *35*, 5043.
38. For a review of the Mitsunobu reaction, see: O. Mitsunobu, *Synthesis* **1981**, 1.
39. M. Sakaitani, Y. Ohfune, *J. Org. Chem.* **1990**, *55*, 870.
40. S. Chen, J. Xu, *Tetrahedron Lett.* **1991**, *32*, 6711.
41. For example, macrocyclizations of the D-O-E macrocycle (14-membered ring) which should, in principle, be easier, afforded low to modest yields of products: a) M. J. Crimmin, A. G. Brown, *Tetrahedron Lett.* **1990**, *31*, 2017 and ensuing paper; b) A. J. Pearson, H. Shin, *J. Org. Chem.* **1994**, 59, 2314. In model systems (Ref. 10a and 11c), the Nicolaou group achieved far inferior yields during the same type of macrocyclization. Attempts to form the AB system by itself through macrocyclization failed: A. G. Brown, M. J. Crimmin, P. D. Edwards, *J. Chem. Soc., Perkin Trans. 1*, **1992**, 123.

42. W. Koenig, R. Geiger, *Chem. Ber.* **1970**, *103*, 2041.
43. D.L. Boger, O. Loiseleur, S.L. Castle, R.T. Beresis, J.H. Wu, *Bioorg. Med. Chem. Lett.* **1997**, *7*, 3199.
44. H. Ku, J.R. Barrio, *J. Org. Chem.* **1981**, *46*, 5239.
45. R. Nagarajan, A.A. Schabel, *J. Chem. Soc., Chem. Commun.* **1988**, 1306.
46. M. Hirama, I. Nishizaki, T. Shigemoto, S. Itô, *J. Chem. Soc., Chem. Commun.* **1986**, 393.
47. a) J.E. Baldwin, G.A. Höfle, O.W. Lever, *J. Am. Chem. Soc.* **1974**, *96*, 7125; b) R.K. Boeckman, K.J. Bruza, *J. Org. Chem.* **1979**, *44*, 4781.
48. For another synthesis of vancomycin by attaching sugars to aglycon derived from the natural product, see: C. Thompson, M. Ge, D. Kahne, *J. Am. Chem. Soc.* **1999**, *121*, 1237.
49. a) D.A. Evans, T.C. Britton, J.A. Ellman, R.L. Dorow, *J. Am. Chem. Soc.* **1990**, *112*, 4011; b) D.A. Evans, D.A. Evrard, S.D. Rychnovsky, T. Früh, W.G. Whittingham, K.M. DeVries, *Tetrahedron Lett.* **1992**, *33*, 1189.
50. A.J. Papa, *J. Org. Chem.* **1966**, *31*, 1426.
51. D.A. Evans, A.E. Weber, *J. Am. Chem. Soc.* **1987**, *109*, 7151.
52. D.A. Evans, T.C. Britton, J.A. Ellman, *Tetrahedron Lett.* **1987**, *28*, 6141.
53. G.W. Anderson, J.E. Zimmerman, F.M. Callahan, *J. Am. Chem. Soc.* **1967**, *89*, 5012.
54. For the description of elements of this synthesis, see: a) J.A. Ellman, Ph.D. Thesis, Harvard University, 1989; b) C.J. Dinsmore, Ph.D. Thesis, Harvard University, 1999.
55. a) J.B. Hendrickson, R. Bergeron, *Tetrahedron Lett.* **1973**, *14*, 4607; b) J.E. McMurry, W.J. Scott, *Tetrahedron Lett.* **1983**, *24*, 979. For a review, see: W.J. Scott, J.E. McMurry, *Acc. Chem. Res.* **1988**, *21*, 47.
56. For discussion on this protecting group, see: P.J. Kocienski, *Protecting Groups*, Georg Thieme Verlag, Stuttgart, **1994**, pp. 132–134.
57. a) D.L. Boger, S. Miyazaki, S.H. Kim, J.H. Wu, S.L. Castle, O. Loiseleur, Q. Jin, *J. Am. Chem. Soc.* **1999**, *121*, 10004; b) D.L. Boger, S. Miyazaki, S.H. Kim, J.H. Wu, O. Loiseleur, S.L. Castle, *J. Am. Chem. Soc.* **1999**, *121*, 3226. For a review of this synthesis and other work from the Boger group, see: D.L. Boger, *Med. Res. Rev.* **2001**, *21*, 356.
58. For earlier studies, see Ref. 43 and: a) D.L. Boger, S.L. Castle, S. Miyazaki, J.H. Wu, R.T. Beresis, O. Loiseleur, *J. Org. Chem.* **1999**, *64*, 70; b) D.L. Boger, S. Miyazaki, O. Loiseleur, R.T. Beresis, S.L. Castle, J.H. Wu, Q. Jin, *J. Am. Chem. Soc.* **1998**, *120*, 8920; c) D.L. Boger, R.M. Borzilleri, S. Nukui, *Bioorg. Med. Chem. Lett.* **1995**, *5*, 3091.
59. For some examples, see: a) K.C. Nicolaou, S.Y. Cho, R. Hughes, N. Winssinger, C. Smethurst, H. Labischinski, R. Endermann, *Chem. Eur. J.* **2001**, *7*, 3798; b) K.C. Nicolaou, R. Hughes, S.Y. Cho, N. Winssinger, H. Labischinski, R. Endermann, *Chem. Eur. J.* **2001**, *7*, 3824.

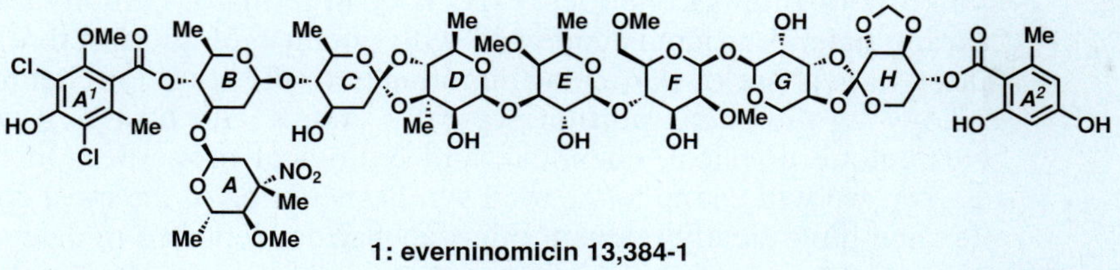

1: everninomicin 13,384-1

10

K. C. Nicolaou (1999)

Everninomicin 13,384-1

10.1 Introduction

Oligosaccharides are the most structurally and functionally diverse biopolymers, with well-established roles in molecular processes critical to eukaryotic biology and disease, including cell–cell recognition, cellular transport, and adhesion.[1] Fundamentally, such variation defines polysaccharides as ideal carriers of biological information, because unlike peptides or nucleic acids in which such content is determined primarily by the number and sequence of monomeric units, the storage capacity of carbohydrates is amplified significantly through their inherent structural features. Indeed, with up to five different sites upon which to append glycosyl units and the opportunity to assume either α- or β-anomeric stereochemistry for each glycosidic linkage, three sugar units can be joined in 120 isomeric forms, whereas three amino acids or base pairs provide a maximum of six different linear combinations. With the addition of even more fragments, this divergence reaches exponential proportions. Unfortunately, with increased complexity also comes a concurrent escalation in synthetic difficulty. This statement is manifested in Nature by the fact that multiple sets of diverse enzymes are required to achieve the synthesis of a unique polysaccharide, while far simpler and cohesive enzymatic systems can generate numerous and distinct families of polypeptides or nucleic acids. Similarly, while laboratory methods to prepare proteins and artificial DNA are well-established and are even automated, the controlled synthesis of polysaccharides still constitutes an enterprise laden with difficulty and intensive effort, burdened by reliance on protecting group chemistry.

Key concepts:

- **Glycosidation techniques**
- **Orthoester formation**
- **DAST-mediated migrations**
- **Design of protecting group ensembles**

Over the course of the past quarter century, however, synthetic chemists have made considerable strides in the stereoselective synthesis of oligosaccharides, especially in terms of carbohydrate-bearing natural products endowed with potent biological activities. These targets are of particular importance to the study of chemical biology as the sugar portion typically endows its host molecule with unique physical, chemical, and biological properties. In this chapter, we will examine the total synthesis of one of the most complex and biologically relevant oligosaccharides reported to date, the antibiotic everninomicin 13,384-1 (**1**), as achieved by the Nicolaou group in 1999.[2] Our goal in this analysis is not only to convey the current state-of-the-art in carbohydrate synthesis, but also to highlight the methodological advances that can result when the practitioner of chemical synthesis is confronted with unique and challenging molecular architectures, such as **1**, that defy conventional synthetic technologies.

Before embarking on this synthetic odyssey, we feel that it would be instructive to delve first into some background material by highlighting a few of the most commonly used *O*-glycosidation methods, paying particular attention to those approaches which proved invaluable to the everninomicin 13,384-1 (**1**) project. While this excursion might seem ancillary to the final goal, as some examples of these techniques have been encountered within several chapters of both volumes of *Classics*, we believe that it is justified in order to enable a fuller appreciation of this important area of research as well as of the work at hand. Our treatment, however, is far from encyclopedic. As a result, we highly encourage you to turn to several well-organized and comprehensive review articles on this subject for further enrichment.[3] Additionally, we apologize to those investigators whose instructive works are not specifically highlighted here as a result of limited space.

10.1.1 Glycosidation Methods in Carbohydrate Synthesis

The issue of controlling anomeric stereochemistry in glycosidation reactions constitutes a subject that has occupied the hearts and minds of synthetic chemists for well over one hundred years, but, despite extensive efforts, a universal method to handle this problem still remains elusive. The reason for this failure is not due to a lack of ingenuity, but instead reflects the fact that glycosidations do not proceed through a consistent and unified mechanistic pathway. This statement is perhaps best illustrated through a discussion of the classical Koenigs–Knorr glycosidation technique, first described in 1901, which is still representative of the majority of approaches to achieve glycosyl bond formation today.[4]

As shown in Scheme 1, this reaction commences with activation of the anomeric halide, typically through the addition of a heavy-metal salt (i. e. an appropriate Lewis acid), to generate a leaving

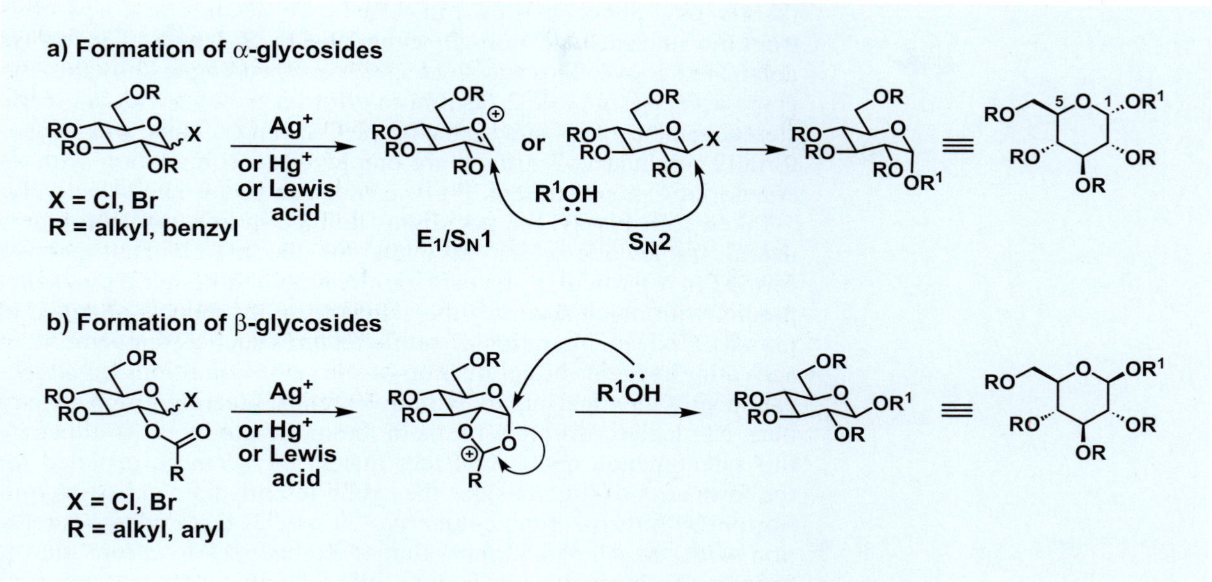

Scheme 1. General methods for the stereoselective synthesis of α- and β-glycosides.

group. What occurs next, however, is highly dependent on the nature of the substituent at C-2. When this position is occupied by an innocuous moiety such as an alkyl ether, or is deoxygenated, the alcohol of interest (R^1OH) will add in an axial fashion to afford the α-disposed glycosyl product as shown in Scheme 1a (note that the α-anomer is readily identified as the product with anomeric stereochemistry opposite to that of the C-5 center, while the β-anomer has the alternate *syn* stereochemical relationship).* As suggested by the drawn sequence, this latter event proceeds either through an E_1/S_N1 pathway, invoking an oxonium intermediate formed via anchimeric assistance, or through S_N2 displacement of the thermodynamically less stable β-conformer generated through *in situ* anomerization of the starting material. Which mechanism is correct is of little consequence since the results are degenerate by virtue of the anomeric effect. Most likely, though, both scenarios contribute as most reactions in organic synthesis occur somewhere along the spectrum of reactivity, rarely demonstrating unique character (because a substitution reaction almost never exhibits pure S_N2 behavior in a strict sense).

In contrast to this sequence of events, when the C-2 position is occupied by an ester, a phenylthio, or a phenylseleno group, a preliminary intramolecular cyclization occurs prior to the addition of

---

* For a comprehensive explanation of carbohydrate nomenclature, see: A. D. McNaught, *Carbohydr. Res.* **1997**, *297*, 1. Therein, designations of α- and β-anomeric stereochemistry are explained for substrates that lack a C-5 stereocenter.

the glycosyl donor as shown in Scheme 1b, leading to a new electrophilic intermediate through either the E_1/S_N1 or S_N2 pathways delineated above. This transient species is then engaged by the glycosyl acceptor in an S_N2 fashion to afford a product with anomeric stereochemistry opposite to that of the C-2 center. Thus, when equatorially oriented C-2 groups are employed in conjunction with an equatorial C-5 substituent, β-glycosides can be formed selectively.

Taken collectively, the paradigms defined in Scheme 1 provide a useful framework which accounts for the predominant anomer formed in a particular glycosidation reaction, although they cannot predict with much accuracy the actual ratio of products obtained in practice. Indeed, alteration of subtle features such as temperature or activating reagent, or application of the same conditions on stereochemically identical sugars with divergent protecting group ensembles, can lead to different ratios of anomeric products. To illustrate this phenomenon better, consider that in the scenario depicted for the formation of β-glycosides the cyclic intermediate exists in equilibrium with its oxonium counterpart, a species that undergoes reaction with the alcohol donor almost exclusively to afford the α-anomer. As a result, any feature such as solvent, temperature, or protecting group array that would favor the existence of an oxonium species would therefore erode β-selectivity in this reaction. Although these issues can prove troublesome, over the course of the past few decades several glycosidation methods similar in mechanism to the Koenigs–Knorr process have been developed, most utilizing modified anomeric substituents, that provide generally reliable methods to achieve the predominant formation of a particular anomer. For the remainder of this section, we will discuss some of the most powerful of these techniques in the context of complex natural product total synthesis.

Although glycosyl bromides and chlorides are viable anomeric substituents as defined in Scheme 1, they are, in general, only of historical value as the original participants in the Koenigs–Knorr reaction due to their limited thermal stability, their facile hydrolysis, and the rather harsh conditions required for their synthesis. In contrast, glycosyl fluorides display excellent stability profiles relative to their halogen cousins based on the much higher strength of the C–F bond. However, with increased bond strength an enhanced level of activation is necessary to engender participation in glycosidation reactions. Indeed, although glycosyl fluorides were first reported in 1923, they remained almost a forgotten entity in oligosaccharide synthesis because activators developed over the ensuing fifty years failed to facilitate their participation as glycosyl donors.[5] This bleak picture changed in 1981 with the discovery by Mukaiyama and co-workers that a mixture of $AgClO_4$ and $SnCl_2$ could activate these donors and smoothly effect glycosylations with suitable acceptors.[6] Since then, numerous related activating systems[7] have been developed, paving the way for subsequent applications in complex contexts, often with remarkable levels of success.

The total synthesis of α-cyclodextrin (**6**, Scheme 2) by Ogawa and Takahashi benefited significantly from this methodology.[8] Striving for a convergent synthesis of this cyclic hexameric oligosaccharide, glycosyl fluoride **3** was coupled with disaccharide acceptor **2** in an early key step under the activating influence of AgOTf and $SnCl_2$ to afford tetramer **4** in 80% yield as a 1.8:1 mixture of anomers favoring the desired α-glycoside. Although higher α selectivity was anticipated from this event, the steric bulk neighboring the free hydroxy group in **2** is the likely culprit for the relatively modest anomeric selectivity observed in this coupling. After several

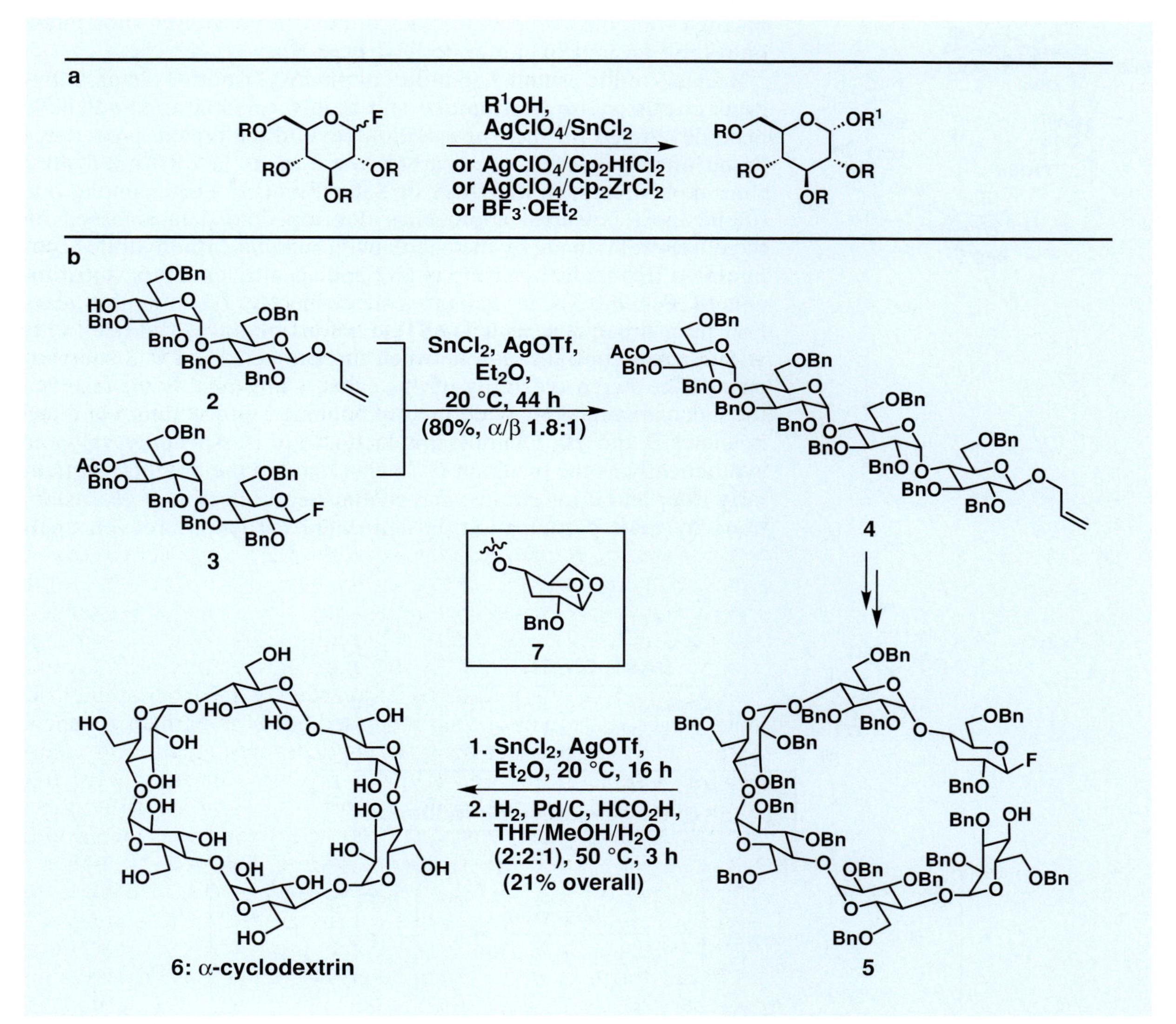

Scheme 2. The glycosyl fluoride method for glycoside formation (a) and its application (b) to the synthesis of α-cyclodextrin (**6**). (Ogawa, 1987)[8]

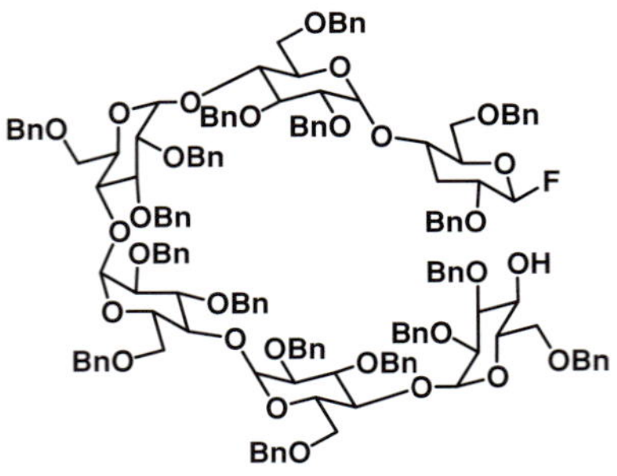

5

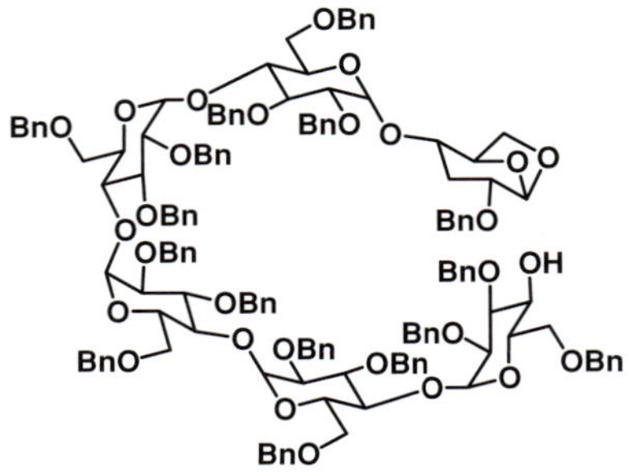

7

additional transformations had transpired to reach **5**, including the merger of another disaccharide unit by means of glycosyl fluoride technology, the stage was then set for a macrocyclization event based on this glycosidation method. Although success in this task would require overcoming a substantial entropic barrier in such a large and flexible system, upon exposure of **5** to the same set of activating conditions as before with prolonged stirring at ambient temperature, the desired cyclic hexamer was indeed obtained in 21% yield exclusively as the requisite α-anomer. Although the yield for this challenging macroring-forming glycosidation is not overwhelming, it elicits admiration to have occurred at all, as the major side product formed during this reaction, saccharide **7**, resulted from intramolecular attack of the C-6 benzyloxy substituent onto the transient oxonium derived from **5**.

Because of the established utility of glycosyl fluorides in carbohydrate chemistry, several unique and highly versatile methods have been developed for their preparation beyond the typical procedures based on the reaction of a carbohydrate in its lactol form with a fluoride source such as DAST or SelectFluor™.[9] For example, one of the most powerful approaches developed to date is based on the observation made by the Nicolaou group that carbohydrates containing a free hydroxy group at C-2 and specific anomeric substituents (**8**, Scheme 3) can undergo a stereospecific 1,2-migration reaction upon treatment with DAST to afford glycosyl fluorides (**11**) with a *trans* configuration between the anomeric and C-2 substituents.[10] The stereoselectivity of this event is presumably the result of the mechanistic pathway defined in Scheme 3 proceeding via intermediates **9** and **10**. Overall, this technology is especially valuable synthetically as the resultant C-2 substituent in the product, particularly ester and thiol groups, can enable stereoselective β-glycosidations by their participation as delineated earlier. Moreover, apart

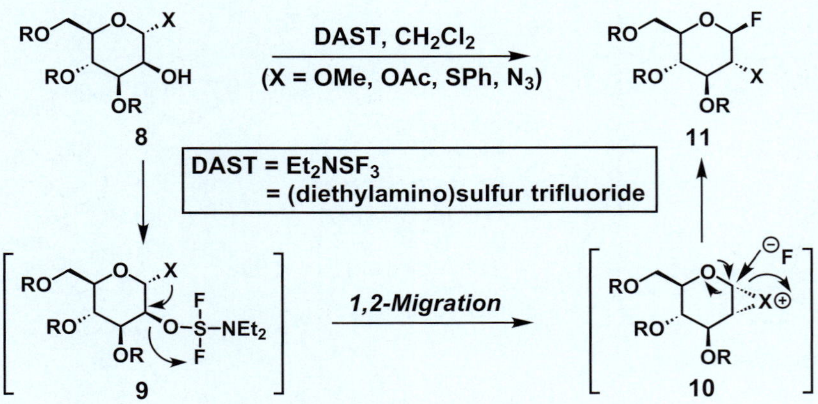

DAST, CH_2Cl_2 (X = OMe, OAc, SPh, N_3)

DAST = Et_2NSF_3 = (diethylamino)sulfur trifluoride

1,2-Migration

Scheme 3. DAST-promoted 1,2-migrations in carbohydrates.

from serving as a directing group, the thiophenyl moiety can also be reductively cleaved after glycosidation to afford a 2-deoxyglycoside.

Glycosyl fluorides have also been prepared from glycosyl sulfides, another class of potent glycosyl donors, upon treatment with DAST and an activating reagent such as *N*-bromosuccinimide. The utility of this particular process, a reaction also pioneered by the Nicolaou group, is illustrated in their partial synthesis of avermectin $B_{1\alpha}$ (**18**, Scheme 4)[11] in which glycosyl acceptor **12** was converted into a highly competent glycosyl donor through initial silyl protection followed by conversion of the anomeric phenylsulfide into the corresponding α-oriented glycosyl fluoride (**13**). Upon treatment of a mixture of **12** and **13** with $SnCl_2$ and $AgClO_4$, **13** was activated selectively, thereby becoming the glycosyl donor. This donor was then merged smoothly with acceptor **12** to form **14** as the α-anomer exclusively in 65 % yield. With this first round of coupling achieved, the thiophenyl group of **14** was then transformed into its more potent glycosyl fluoride counterpart **15** with DAST and NBS, generating a new donor that was then activated under

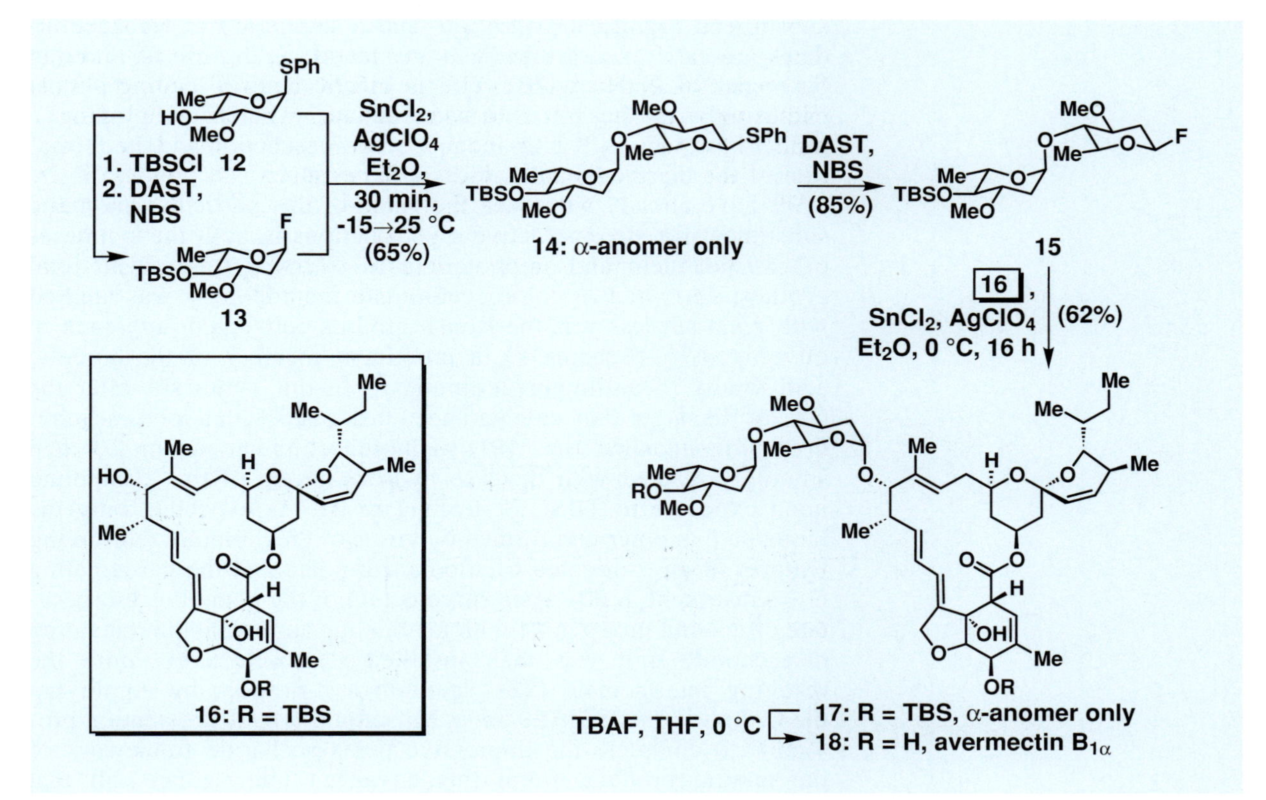

Scheme 4. The "two-stage activation" method for oligosaccharide synthesis: the use of thioglycosides as convenient sources of fluorinated glycosyl donors and application of the technology to the partial synthesis of avermectin $B_{1\alpha}$ (**18**). (Nicolaou, 1984)[11]

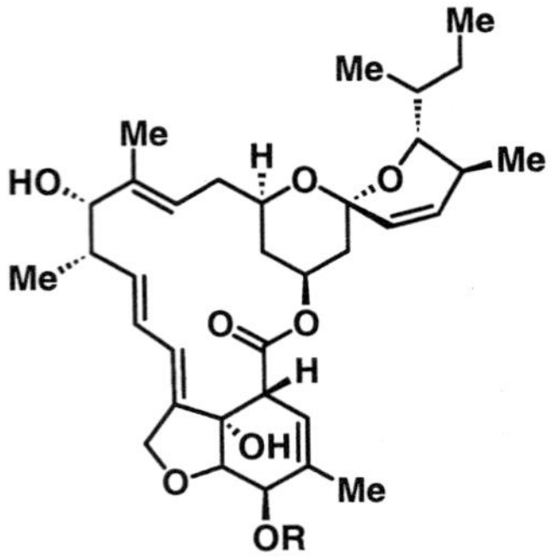

16: R = TBS

the same $AgClO_4/SnCl_2$ conditions and joined with the avermectin aglycon **16** to afford avermectin $B_{1\alpha}$ (**18**) after deprotection. Overall, this general strategy is defined as the "two-stage activation method" for oligosaccharide synthesis as the anomeric position is activated twice (first the thiophenyl with NBS to form a fluoride and then with a metal complex or a Lewis acid to effect glycosidation) during each round of coupling. While powerful in its own right, this method has found widespread application in iterative, block-type approaches to polysaccharide construction.

Besides glycosyl fluorides, another highly versatile and widely utilized family of glycosyl donors is the trichloroacetimidates, originally introduced by R. R. Schmidt and co-workers in 1980.[3d,12] Readily prepared from a precursor lactol by treatment with base and trichloroacetonitrile, these entities smoothly participate in glycosidation reactions upon activation with $BF_3{\cdot}OEt_2$ or a silyl triflate.[13] Unlike the divergent E_1/S_N1 and S_N2 mechanisms defined in Scheme 1, however, trichloroacetimidates typically follow S_N2-based pathways in their reactions with glycosyl acceptors. As such, they are highly valuable synthetic intermediates for β-glycosidations, especially for mannose-type sugars in which C-2 participation cannot result in β-selectivity due to the axial orientation of this substituent. Significantly, both α- and β-anomeric trichloroacetimidates are easily accessible from any lactol, as the use of a strong base such as NaH or DBU effects kinetic control leading almost exclusively to the β-trichloroacetimidate, whereas employing a milder base such as K_2CO_3 induces a slower reaction and the formation of the thermodynamic, α-disposed product.

We have already witnessed the value of this particular anomeric substituent for stereoselective glycosidations in both the syntheses of calicheamicin and amphotericin in *Classics I*. A recent total synthesis in which trichloroacetimidate methodology was applied with great success was the Roush group's convergent approach to olivomycin A (Scheme 5), a prominent member of the aureolic acid family of antitumor antibiotics.[14] In this synthesis, after the core of the target molecule had been fashioned with a lone carbohydrate unit attached (i.e. **19**), trichloroacetimidate donor **20** then smoothly reacted with the free hydroxy group of that compound upon exposure to TBSOTf, leading to **21**. As expected, only the requisite β-anomer was formed by virtue of the mutually reinforcing features of an α-oriented trichloroacetimidate and the participating C-2 substituent in **20**. After deprotection of the phenolic chloroacetate protecting group in **21** with methanolic ammonia, the remaining disaccharide unit was then installed stereoselectively onto the resulting intermediate (**22**), again in a β-fashion, by employing the C-2 phenylselenide **23** and a Mitsunobu-type glycosidation protocol[15] to complete the impressive pentasaccharide framework of the natural product. From this advanced intermediate, all that remained to complete the total synthesis of olivomycin A (**24**) was global deprotection and excision of the superfluous C-2 directing groups.

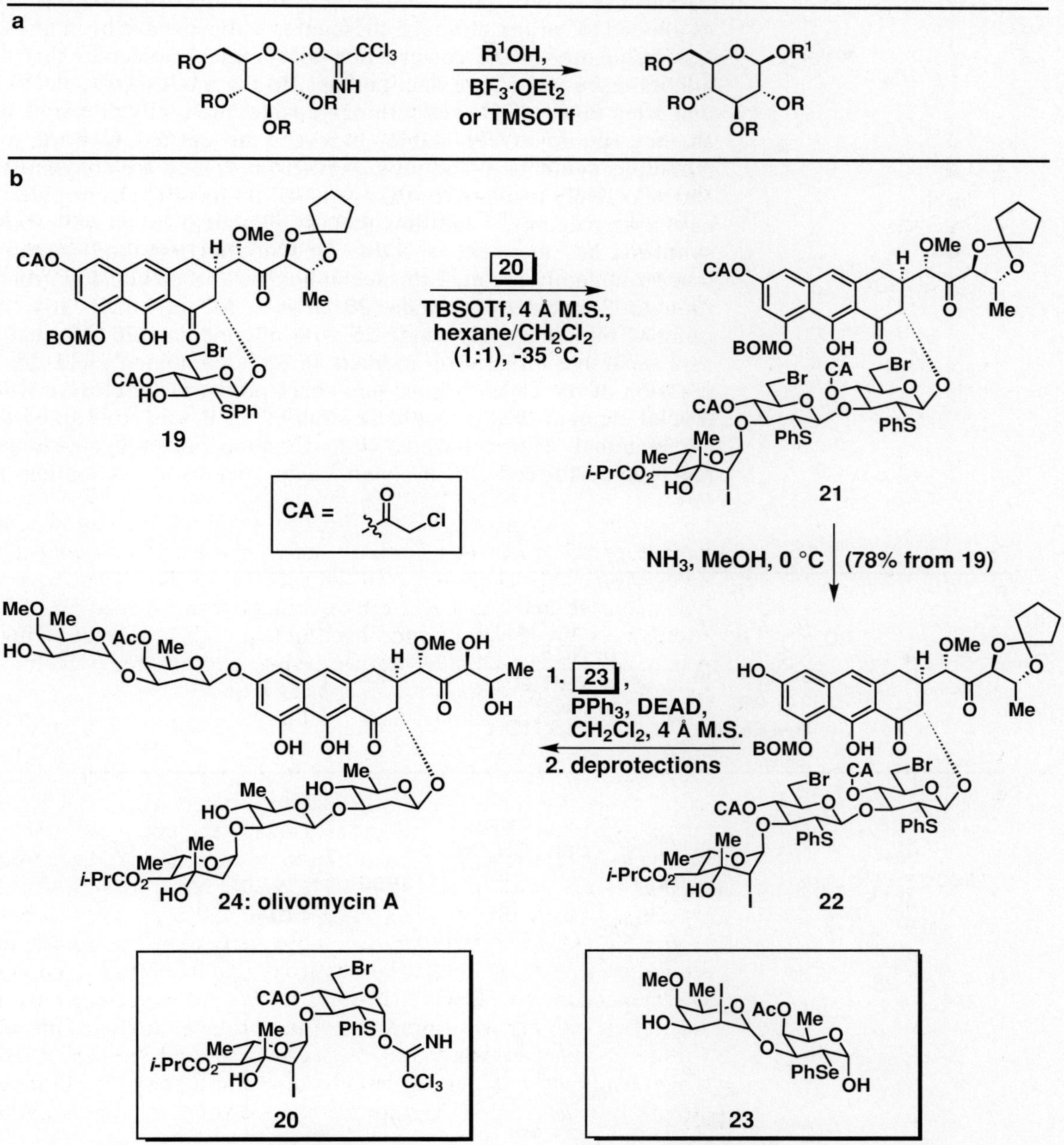

Scheme 5. The trichloroacetimidate method for glycoside formation (a) and its application (b) to the total synthesis of olivomycin A. (Roush, 1999)[14]

Apart from glycosyl fluorides and trichloroacetimidates, thioglycosides can also be effectively employed in glycosidation reactions as alluded to earlier, although these intermediates have been utilized less frequently in the construction of complex molecules than the donors discussed above due, perhaps, to the modest α/β selectivity that often results.[16] However thioglycosides are easily prepared, and as they are relatively stable, they can be carried forward over numerous synthetic operations. Furthermore, they are conveniently and selectively activated with several sulfur-specific electrophilic or oxidizing reagents,[17] and thus their employment in natural product synthesis has met with success. One of the first total syntheses that prominently featured this technology was the Nicolaou group's route to *O*-mycinosyltylonide (**28**, Scheme 6)[18] in which early coupling of thiophenyl glycoside **25** with intermediate **26** afforded **27** as a separable mixture of anomers in 85 % combined yield. Incorporation of the carbohydrate side chain prior to ring closure was a crucial element of this synthetic strategy, as it was envisioned that an eventual Horner–Wadsworth–Emmons macrocycle-forming reaction performed on an open chain precursor resembling the

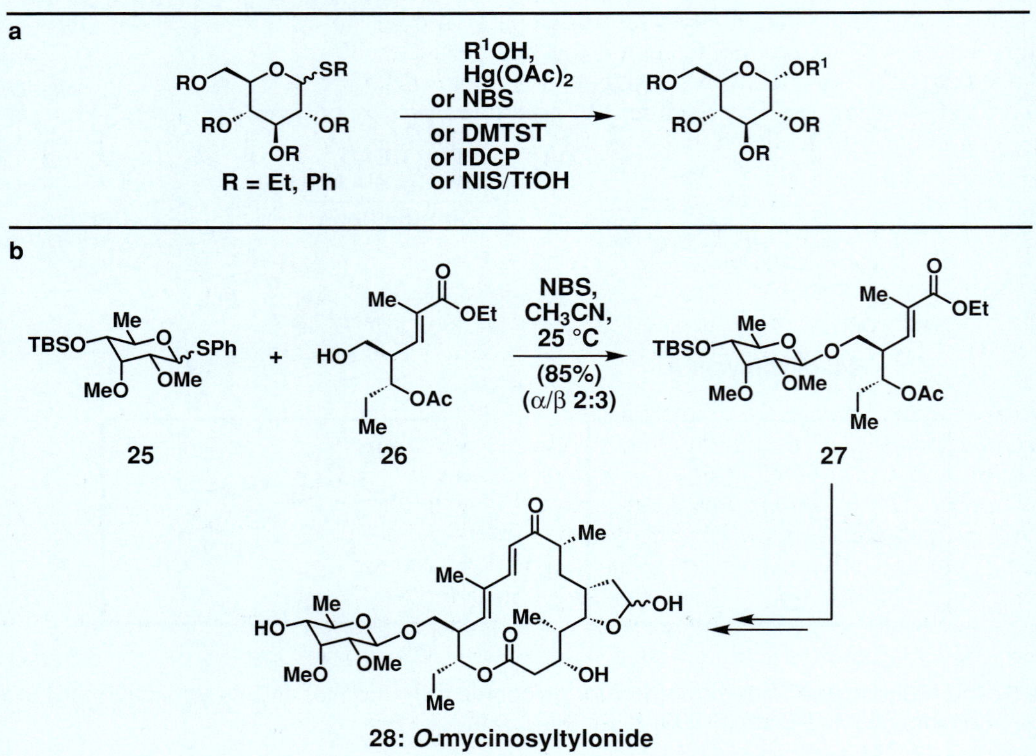

a

R = Et, Ph

R^1OH, $Hg(OAc)_2$ or NBS or DMTST or IDCP or NIS/TfOH

b

25 + **26** → **27**

NBS, CH_3CN, 25 °C (85%) (α/β 2:3)

28: *O*-mycinosyltylonide

Scheme 6. The use of thioglycosides as glycosyl donors (a) and application (b) of the method in the total synthesis of *O*-mycinosyltylonide (**28**). (Nicolaou, 1982)[18]

final natural product as much as possible in its substituents would be entropically favored for closure. Indeed, with the sugar attached, the planned macrocyclization to **28** proceeded smoothly in high yield and stereoselectivity.

Independent of their direct participation as donors, glycosyl sulfides can also be converted into their more reactive sulfoxide counterparts, entities which have proven to be highly valuable glycosyl donors for the synthesis of β-disposed oligosaccharides, particularly of the mannose type. The concept that anomeric-type sulfoxides could be displaced upon Lewis acid activation by nucleophiles was first demonstrated in 1986 by the Nicolaou group in their synthesis of didehydrooxocane ring systems crucial for success in the brevetoxin B project (see *Classics I*, Chapter 37).[19] In 1989, D. Kahne and his associates at Princeton University established the sulfoxide-based glycosidation reaction, a process that was later extended considerably by D. Crich and his group at the University of Chicago.[20] As shown in Scheme 7a, upon reaction of the glycosyl sulfoxide with triflic anhydride at low temperature, the sulfoxide moiety is activated, becoming a leaving group that is then eliminated through anchimeric assistance with the concurrent generation of triflic acid. In the absence of the glycosyl acceptor, this acid adds to the oxonium species to form an α-disposed glycosyl triflate, an intermediate verified by low-temperature NMR spectroscopic studies. Upon addition of the alcohol acceptor, the triflate is then displaced in an S_N2 fashion to afford the desired β-glycoside. One should note that the 4,6-*O*-benzylidene protecting group in Scheme 7a is not a random choice, but is in fact a requisite element of functionality to achieve high β-selectivity. This group both locks the conformation of the glycosyl donor and helps to ensure that it reacts through the triflate form, which would otherwise be in equilibrium with the alternative oxonium structure that would afford an α-disposed glycoside if it constituted the reactive species. In this vein, the benzylidene moiety acts as an electron-withdrawing group, thereby destabilizing the formation of a charged oxonium intermediate and shifting the equilibrium towards the side of the glycosyl triflate and the observed β-selectivity.

A recent synthesis of the β-mannan hexasaccharide unit **35** (Scheme 7b) found in the cell wall of bacterial antigens by Crich and co-workers is an excellent demonstration of this technology.[21] Using both benzylidene and 4,6-PMP protecting groups (which are purportedly more electron withdrawing than simple 4,6-*O*-benzylidenes), highly selective 1→4 and 1→3 β-linkages were formed during critical operations in the drive to reach **34**. Subsequent coupling with additional building blocks in a convergent manner based on sulfoxide chemistry led to the generation of **35** in relatively short order, with excellent β-selectivity observed in each glycosidation step.

In rounding out this foray into glycosidation chemistry, it would be most appropriate to close with the glycal method, a procedure that has found numerous applications in total synthesis endeavors. In this general strategy, pioneered by Lemieux and co-workers

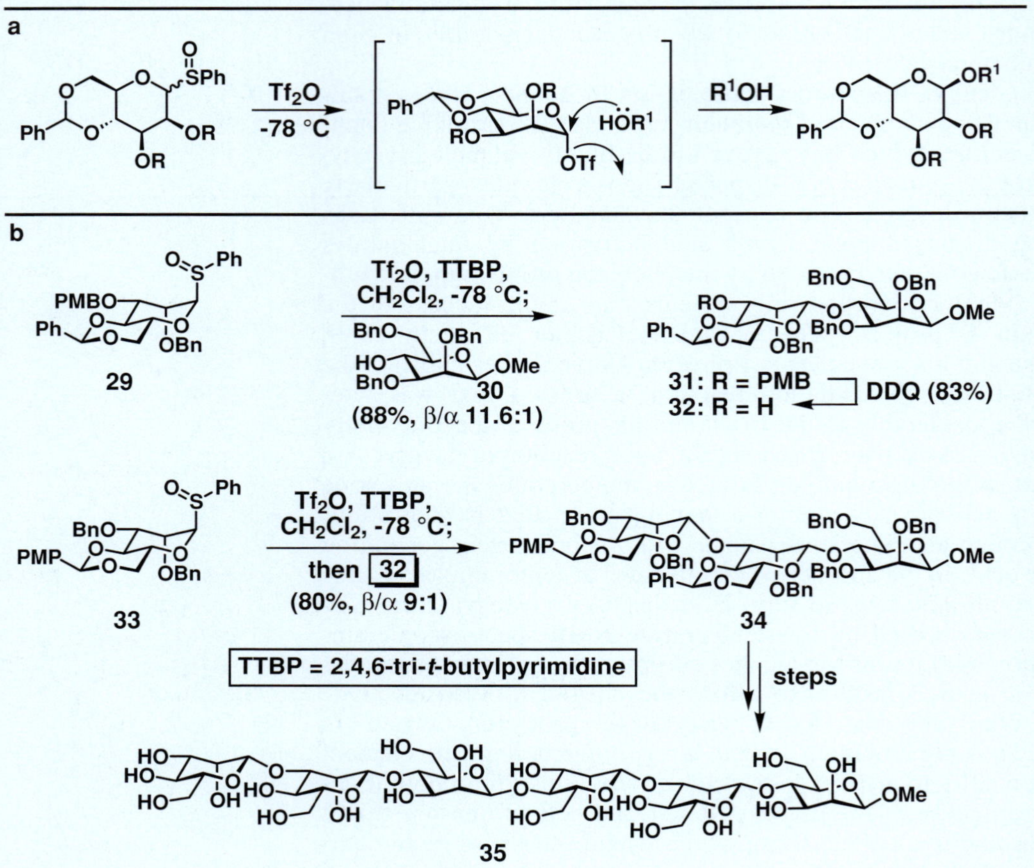

Scheme 7. The sulfoxide method for glycoside bond formation (a) and its application (b) to the synthesis of a hexasaccharide β-mannan found in bacterial antigens (**35**). (Crich, 2001)[21]

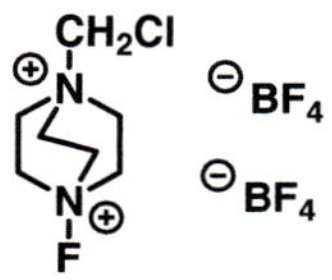

in 1964 and subsequently extended by the work of the Tatsuta and Thiem groups,[22] reaction of a glycal intermediate with a reagent such as SelectFluor™, *N*-iodosuccinimide, or dimethyldioxirane in the presence of a Lewis acid leads to an intermediate electrophilic species that can be engaged by a glycosyl acceptor molecule to provide a product that is *trans* functionalized in the anomeric and C-2 positions. The synthesis of avermectin $A_{1\alpha}$ (**41**, Scheme 8) by Danishefsky and co-workers is particularly illustrative of this glycosidation method.[23] Upon activation of glycal **37** with NIS, the addition of carbohydrate unit **36** cleanly afforded disaccharide **38** exclusively as the requisite α-anomer. After executing four additional steps in which the anomeric methoxy group was eliminated to form a new glycal intermediate (**39**), the stage was set for a second coupling, this time with avermectin aglycon **16**

SelectFluor™

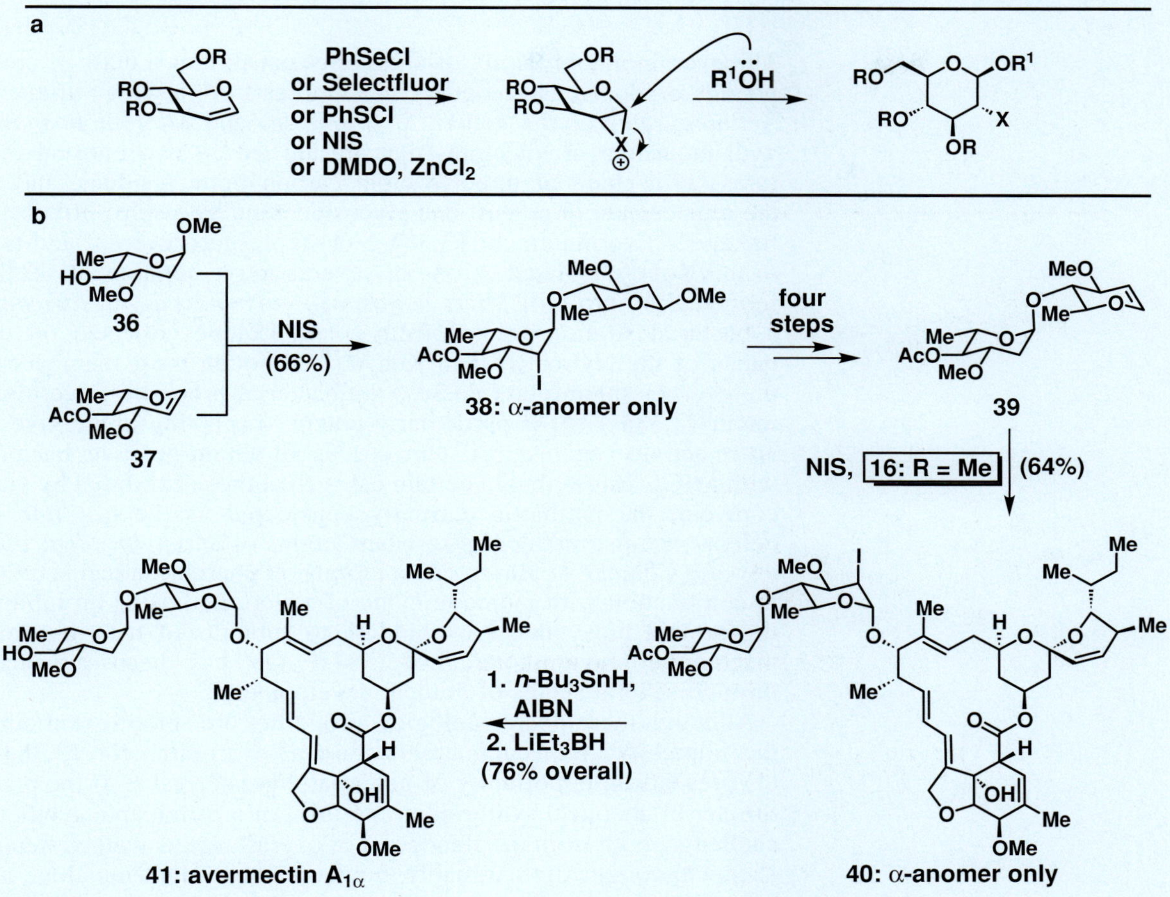

Scheme 8. The glycal method for oligosaccharide synthesis (a) and its application (b) to the synthesis of avermectin $A_{1\alpha}$ (**41**). (Danishefsky, 1987)[23]

(R = Me, cf. Scheme 4) upon activation with NIS. As expected, the desired product **40** was generated in 64 % yield, exclusively as the α-anomer. With these glycosidations complete, the natural product (**41**) was but two steps away through dehalogenation induced by n-Bu_3SnH and AIBN, followed by acetolysis upon exposure to $LiEt_3BH$.

While this concludes our condensed survey of glycosidation techniques, we hope that your comprehension of the remainder of this chapter, where most of the approaches defined above will be revisited, will now be sufficiently facilitated. Thus, without further delay, we turn to the main theme of this chapter's story, the total synthesis of everninomicin 13,384-1 (**1**).[2]

10.2 Retrosynthetic Analysis and Strategy

The everninomicin family of antibiotics constitutes a class of compounds produced as secondary metabolites from a wide variety of *Actinomycetes* of the genera *Streptomyces* and *Micromonospora*, with molecular architectures that are unified by two unusual features: 1) a chain of three to eight carbohydrate residues, and 2) the replacement of at least one glycoside bond by a spiro-orthoester linkage.[24] Everninomicin 13,384-1 (**1**) represents a recent addition to this family, isolated as one of several active components in the fermentation broth of *Micromonospora carbonacea var. africana*, a bacterial strain extracted from a soil sample collected on the banks of the Nyiro River in Kenya.[25] Although most members of the everninomicin class possess antibacterial properties, everninomicin 13,384-1 (**1**) is particularly potent, displaying impressive *in vitro* activity against resistant strains of Gram-positive bacteria with MIC_{90} values that in certain cases rival those exhibited by vancomycin, the antibiotic currently considered as the last line of defense against methicillin-resistant strains of *Staphylococcus aureus* (see Chapter 9). Based on its promising pharmaceutical activity, in conjunction with a unique mode of action predicated on inhibiting protein biosynthesis by binding to subunits of the ribosomal machinery,[26] everninomicin 13,384-1 (**1**) has been advanced through several stages of clinical development.[27]

Although these pharmacological properties are certainly enticing, the impressive molecular architecture of everninomicin 13,384-1 (**1**) presents an opportunity of potentially greater value to the practitioner of chemical synthesis in the form of a formidable synthetic challenge, both from the standpoint of overall size as well as stereochemical complexity. Arguably, however, these elements alone are not uniquely striking, as several total syntheses of natural products consisting of more carbohydrate units and stereogenic centers than **1** have been achieved over the course of the past decade.[28] Rather, the true call to arms issued by this structure rests in its unique series of interconnectivities, which includes two highly sensitive orthoesters strategically positioned in the heart of the molecule, as well as a 1,1′-disaccharide bridge composed of two β-mannoside bonds joining rings F and G. Clearly, any carefully considered plan of attack to reach the everninomicin 13,384-1 (**1**) structure not only would have to address the stereoselective generation of these uncommon linkages, an issue for which general synthetic technology was lacking at the outset of these studies, but also would require precise timing in their incorporation such that they could survive subsequent elaboration to the target molecule once in place.

Aside from these potentially thorny problems, from a retrosynthetic standpoint everninomicin 13,384-1 (**1**) is perhaps an ideal candidate for a convergent synthesis, as the modular nature of the target leads to the appealing prospect of synthesizing each component ring system and then assembling larger segments from these individual building blocks. While such an analysis clearly reduces many of the synthetic challenges posed by **1** to much smaller and more manageable tasks, each ring synthesis would still likely constitute an enterprise fraught with unique difficulties, as the constituents of **1** include unconventional fragments such as a nitrosugar (evernitrose, ring A), two highly substituted aromatic rings, and two 2,6-dideoxyglycosides (rings B and C). The major drawback incurred by adopting such a retrosynthetic strategy, particularly on a target the size of everninomicin 13,384-1 (**1**), is the seemingly limitless collection of conceivable synthetic blueprints. Fortunately, the number of potentially viable routes to this target is modulated significantly by several pieces of information garnered during the isolation, characterization, and degradation studies of **1**.[25]

First, since it had been established that the CD orthoester moiety in **1** is highly sensitive to acidic conditions,[24] any synthetic plan should strive to reserve its incorporation until the tail end of the synthesis. Consequently, this linkage was the first strategic site to be disassembled in the retrosynthetic analysis, thereby leading to fragments **42** and **43** (see Scheme 9). In the forward direction it was anticipated that the fusion of these two pieces could be achieved stereoselectively by employing a method of orthoester formation first pioneered by P. Sinaÿ and co-workers,[29] appropriately modified to suit the problem at hand, as shown in its generalized form in Scheme 10. In this protocol, after a glycosidation event in which the 2-phenylseleno group in the glycosyl donor (**51**) guides the resultant formation of anomeric stereochemistry as expressed in **53**, exposure to an oxidant such as $NaIO_4$ then effects conversion to the corresponding selenoxide (**54**). Rather than isolate this intermediate, the seleno functionality is then thermally eliminated in a *syn* fashion, affording a glycal intermediate (**55**). With a neighboring hydroxy group in close proximity, this newly installed olefin does not persist, but instead serves as a willing participant in a 5-*exo*-trig cyclization to afford the desired orthoester **56**. Overall, while there are several different methods to install such 2-phenylseleno groups onto appropriately functionalized precursors, the present analysis was predicated upon the employment of a DAST-induced 1,2-migration reaction (see Scheme 3) on a carbohydrate of type **49** to construct the glycosyl donor **51** stereoselectively, as required for the planned orthoester formation.[10] At the outset of these studies, however, phenylselenides represented untested entities for such reactions. As such, the ability to fashion the functionality present in ring C of **42** (Scheme 9) through such synthetic technology constituted a crucial lynchpin of the everninomicin synthetic plan.

At this stage, close scrutiny of the larger of these two new fragments suggested that the EF glycosidic linkage in **43** would likely be the most straightforward to construct under any circumstances, and as a result, this bond was retrosynthetically cleaved to reveal DE disaccharide **44** and FGHA^2 aggregate **45** as new subgoal struc-

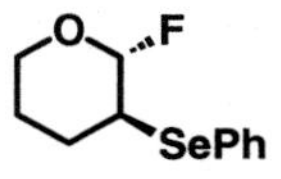

O F SePh
51

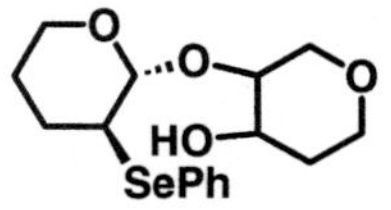

O O O HO SePh
53

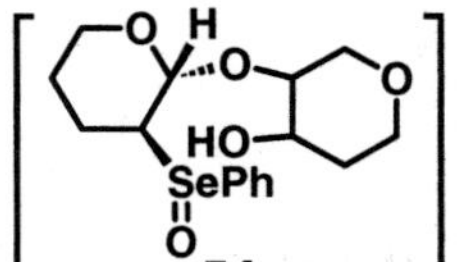

O H O O HO SePh O
54

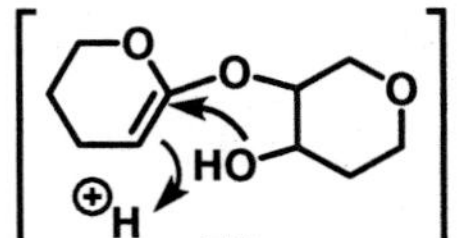

O O O HO ⊕H
55

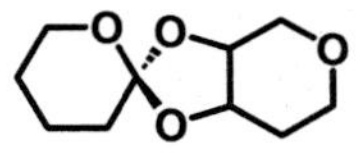

O O O O
56

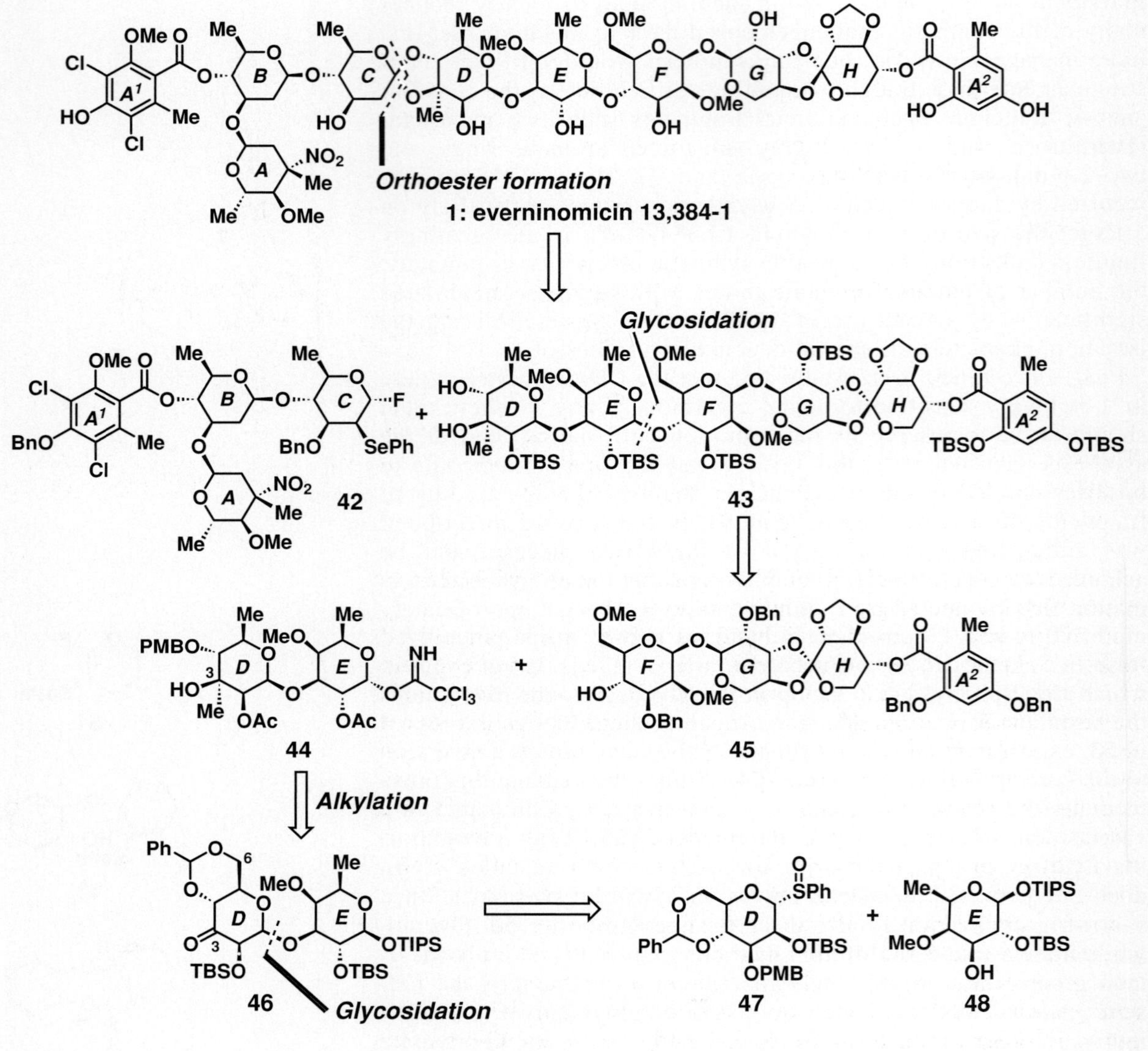

Scheme 9. Retrosynthetic analysis of everninomicin 13,384-1 (1).

tures. Synthetically, it was anticipated that the use of a trichloroacetimidate donor bearing a participating 2-acetoxy group (as drawn in **44**) would permit the control of the requisite β-anomeric stereochemistry upon reaction with glycosyl acceptor **45** in the presence of a suitable activator. Moreover, apart from providing a highly convergent synthetic blueprint, this synthetic design also provided an added benefit in the form of sequence flexibility in the final stages of the synthesis. For example, a suitably functionalized DE segment could also first be coupled to $A^1B(A)C$ assembly **42**, and then

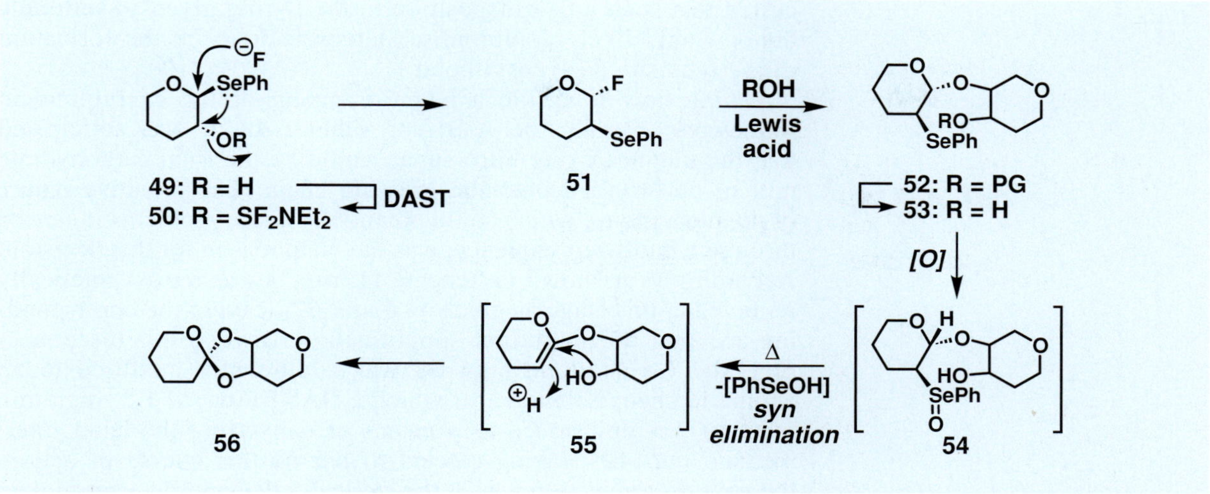

Scheme 10. Orthoester formation via 1,2-phenylseleno migration followed by glycosylation and ring closure after *syn* elimination. PG = protecting group. Note: substituents on carbohydrate rings have been deleted for clarity.

joined with the remaining FGHA[2] portion should problems be encountered in the original ordering of events.

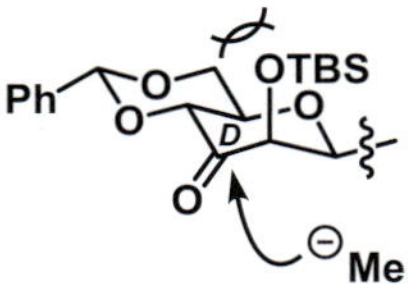

Having decided upon these initial, but highly crucial simplifications, what remained to be defined was the most effective set of disconnections to enable the facile synthesis of each of the three major fragments (**42**, **44**, and **45**) from smaller, individual ring systems. Starting with the easiest segment to consider along these lines, the DE disaccharide **44** (Scheme 9), it was envisioned that the critical β-stereochemistry of its lone glycosyl bond could be fashioned through application of the Kahne–Crich glycosyl sulfoxide method[20] by employing a donor such as **47**, which carries the requisite C-4/C-6 benzylidene protecting group, in conjunction with an appropriately functionalized acceptor, E-ring **48**. Since the D-ring fragment is endowed with mannose-type stereochemistry, for which the Kahne–Crich methodology is ideal, this glycosidation reaction was anticipated to proceed smoothly with excellent β-selectivity. In order to reach this stage from **44**, the unique C-3 tertiary alcohol center in ring D of this compound would have to be disassembled first, leading to a precursor D-ring ketone such as **46**. Although this analysis requires the stereoselective addition of an appropriate methyl nucleophile onto this intermediate, molecular models of the most stable conformer of **46** suggested that a bulky C-2 D-ring protecting group (such as TBS) could reasonably be expected to shield the top face of this substrate, thereby guiding nucleophilic approach from the opposite side of the molecule as illustrated in the column figure. As an aside, while incorporating this quaternary center prior to joining the D- and E-ring monosaccharides might provide a more expedient synthesis by reducing the overall number of linear steps necessary to reach **44**, it was forecasted that such a C-3 disposition in the D-ring glycosyl sulfoxide donor would likely compromise stereoselectivity in the formation of the requisite β-glycosyl bond.

Aiming now to tackle a larger fragment of the everninomicin framework, namely the $A^1B(A)C$ subunit **42**, it was anticipated that the unique A-ring nitro sugar would be the final carbohydrate unit to be installed onto the growing chain. The sensitive nature of the nitro group, which could potentially cause problems if carried through a multistep sequence, was the main reason for this decision. Accordingly, as shown in Scheme 11, ring A was retrosynthetically removed at this stage as glycosyl donor **57**, leaving the corresponding acceptor **58** for further simplification. Beyond this disconnection, the C-ring domain in **58** was additionally modified to an anomeric phenylselenide, for which a DAST-induced 1,2-migration reaction was envisioned as a means of converting the latter intermediate into **42**. Having elected to pursue this course of action, the next disconnection was at the sterically demanding ester moiety linking the A^1 and B rings. While a simple operation on paper, the actual construction of this bond was expected to be anything but trivial, as unanticipated difficulties had been encountered by other researchers in their attempts to forge this A^1B union in model systems using conventional forms of the A^1-ring, such as carboxylic acids or acid chlorides, under a plethora of reaction conditions.[30] Based on this background information, the present analysis relied on the assumption that a suitably activated acyl fluoride (**59**) might overcome these issues, enabling a high-yielding union with the BC alcohol **60** at the right moment in the synthesis.

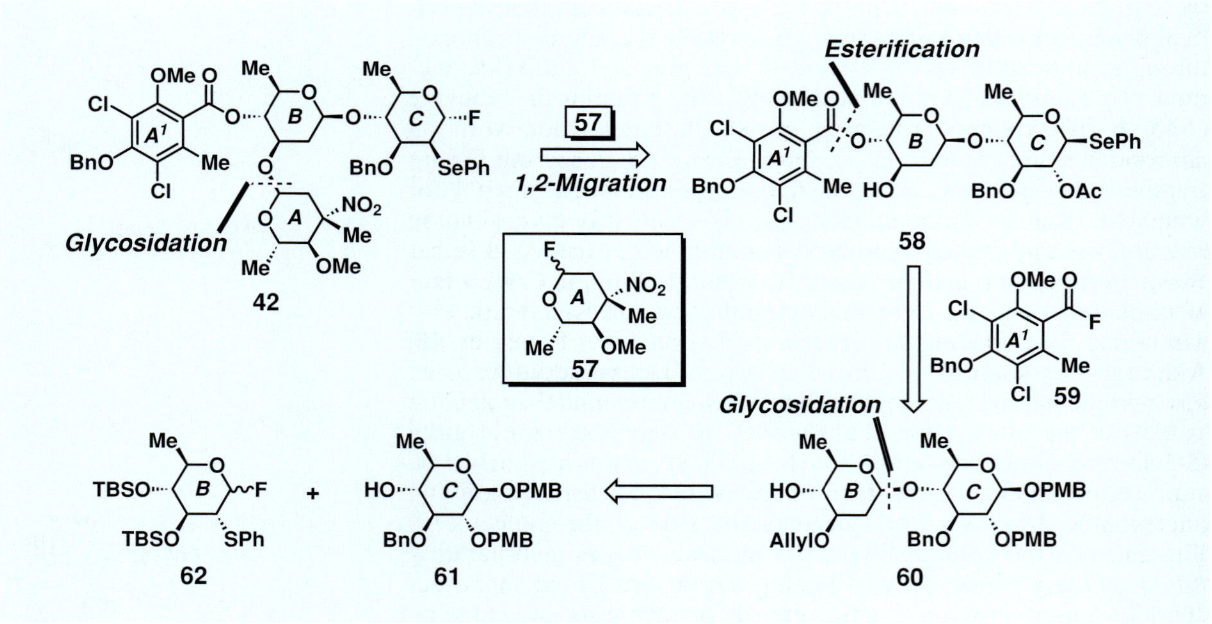

Scheme 11. Retrosynthetic analysis of the everninomicin $A^1B(A)C$ fragment **42**.

Assuming that this esterification reaction would proceed as expected, the $A^1B(A)C$ ensemble **42** has been simplified to the point at which only one disconnection of note remained, namely breaking of the BC fragment (**60**) into its constituent B- and C-ring carbohydrate units. Because the everninomicin structure dictates that these saccharides be linked in a β-fashion with the additional proviso of C-2 deoxygenation in ring B, it was anticipated that both of these issues could be addressed concurrently by incorporating a 2-thiophenyl moiety in **62**. Indeed, this auxiliary should serve as a directing group during the glycosidation reaction with **61**, and, after its mission in this operation was accomplished, it could then be excised with Raney Ni to afford the desired 2-deoxyglycoside system. As a final comment for this subunit, the neighboring arrangement of the fluoro and thiophenyl groups in **62** should signal the potential origin of this intermediate from a 2-hydroxyphenylthioglycoside through a DAST-orchestrated 1,2-migration reaction.

At this point, only the $FGHA^2$ segment (**45**, see Scheme 12) requires further discussion, and we have deliberately left its analysis for last since this fragment includes several of the most challenging features of the everninomicin skeleton, namely the GH orthoester and the exclusively β-linked 1,1′ FG disaccharide domain. Mindful of the difficulties that might be encountered in constructing these particular connections, it was decided to deal with them as early in the synthesis as possible, hoping to avoid potential pitfalls once advanced synthetic intermediates had been reached. Thus, based on this design criterion, it was envisioned that the A^2 aryl system (**63**) could be appended onto the appropriately functionalized FGH assembly **64** at some point in the late stages of the drive to complete **45**. Similar to the fully substituted A^1-ring in the form of **42** (Scheme 11), the Nicolaou group anticipated that an acid fluoride would be required to achieve ester formation in this highly hindered context. Pursuing this retrosynthetic course of action further, the GH orthoester was disconnected next, again forecasting application of a sequence involving an initial stereoselective 1,2-migration to forge an appropriate glycosyl fluoride endowed with a C-2 selenophenyl group **66**, followed by the orthoester protocol of Sinaÿ and co-workers.[29] While this general scheme was expected to succeed, several questions clouded a general feeling of security. First, it was unclear which alcohol group of the G-ring would be the best candidate at which to append the H-ring glycoside for orthoester formation, as both the C-3 and the C-4 hydroxy groups in **67** represented reasonable participants. Additionally, although molecular modeling studies of the likely transition states in the conversion of **65** into **64** indicated that the stereochemistry of the product would be favored in the reaction, such analysis did not preclude the possibility of surprises resulting from the unique framework of the FG system. Thus, in order to best deal with these concerns, the synthesis of the ring G portion of **67** was designed in a flexible manner such that either a free C-3 or C-4 hydroxy group could be generated as a glycosyl acceptor for reaction with **66**, affording either **65** or its

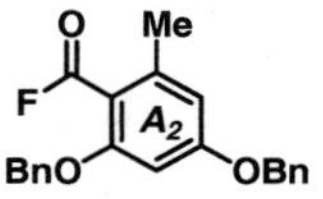

O Me F A2 BnO OBn
63

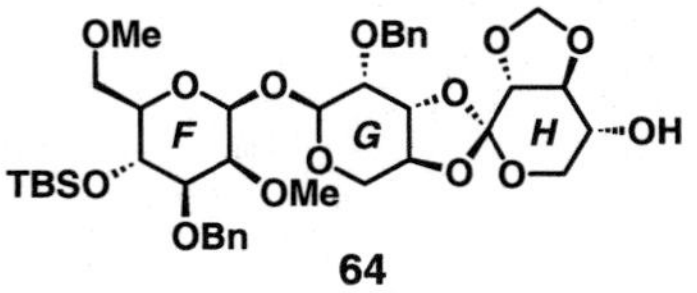

OMe OBn O O F G H OH TBSO OMe OBn
64

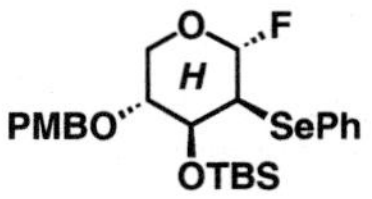

O F H PMBO SePh OTBS
66

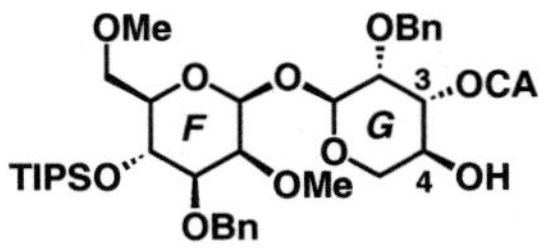

OMe OBn 3 OCA F G TIPSO OMe 4 OH OBn
67

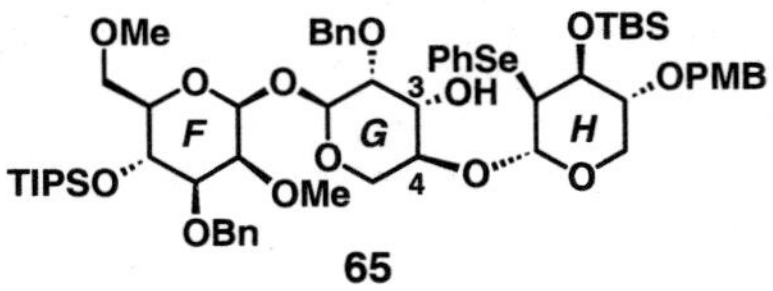

OMe BnO OTBS PhSe 3 OH OPMB F G H TIPSO OMe 4 OBn
65

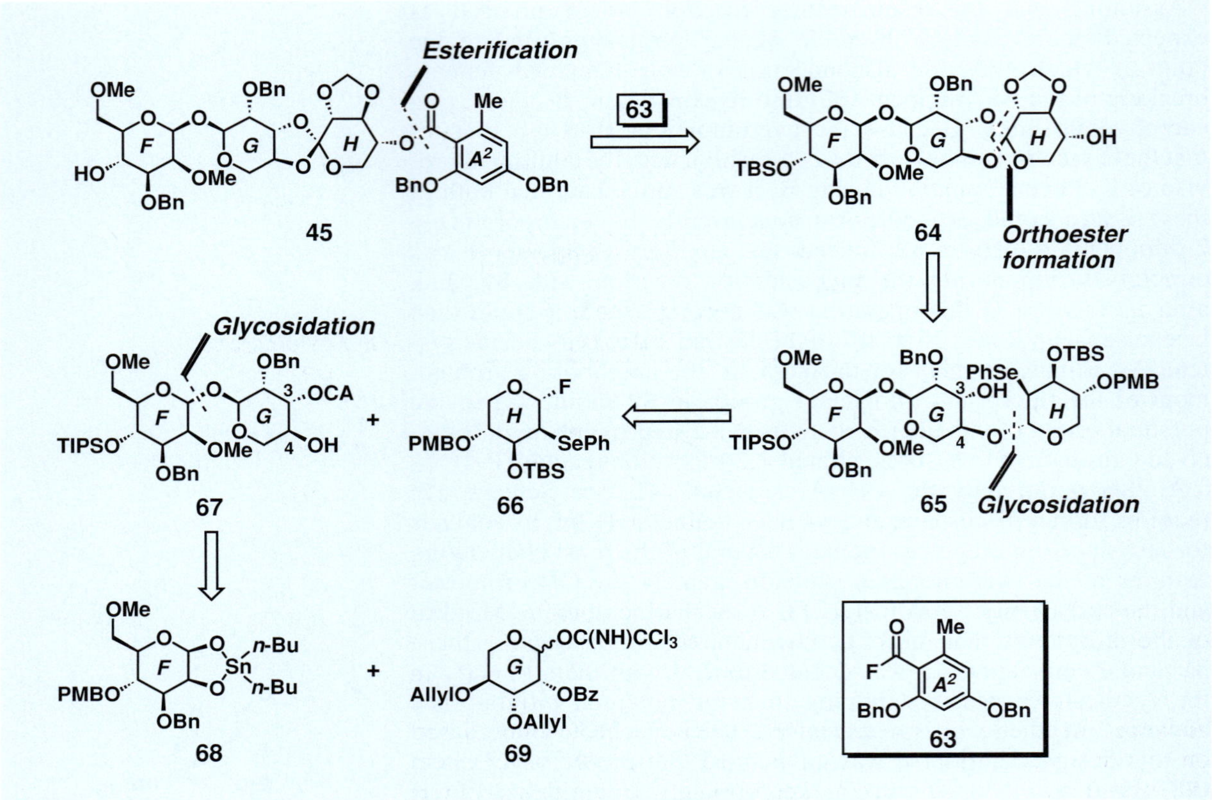

Scheme 12. Retrosynthetic analysis of the everninomicin FGHA^2 fragment 45.

C-3 linked congener in the hope that one of these precursors would provide **64** with the stereochemical arrangements of everninomicin.

Following this spate of disconnections, all that remained to consider at this juncture was the nontrivial issue of forming the 1,1′-FG disaccharide, linked exclusively in a β-fashion. Of course, to achieve such a union would require careful supervision of stereochemistry at both anomeric centers while the new bridge is formed. However, while such control is well-precedented for glycosyl donors, a commensurate level of predictability with glycosyl acceptors does not currently exist. The inherent challenges presented by this portion are reflected in some initial model studies directed toward the synthesis of systems of this type. As shown in Scheme 13, conventional approaches using a trichloroacetimidate donor (**71**) in conjunction with an acceptor of type **70** (R = Bn) afforded exclusively the undesired α-glycosidic bond on the glycosyl acceptor (i. e. **72**), as would be expected by the overwhelming influence of the anomeric effect. In fact, any method in which the anomeric stereochemistry of the glycosyl acceptor remained unfixed led exclusively to the α-anomer. With the problem redefined in this manner, however, came an idea that could potentially solve this problem.

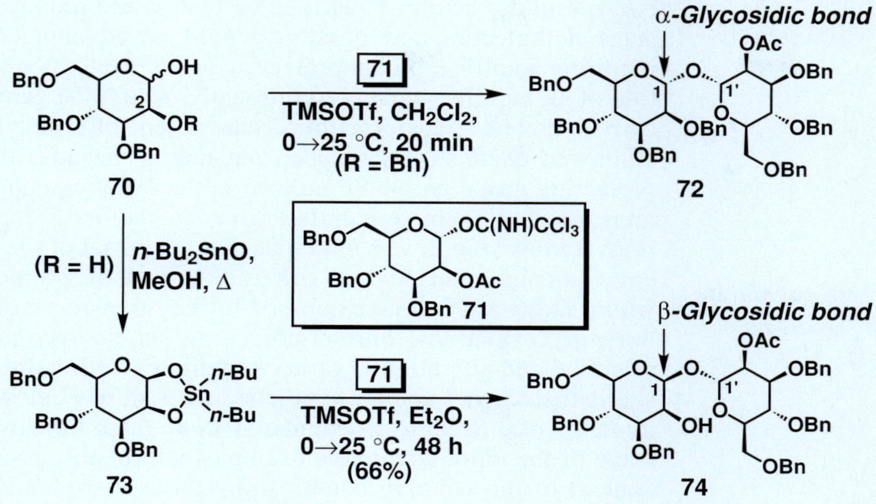

Scheme 13. Testing methodology for the stereoselective construction of 1,1′-disaccharide systems: synthesis of model 1,1′-disaccharides **72** and **74**.

Viewing this reaction as essentially a selective protection of the anomeric position in the acceptor glycoside, the Nicolaou group anticipated that success could be realized by locking this position as a tin acetal,[31] a functionality with well-documented displacement by alcohols with retention of initial stereochemistry. As proof of principle for this idea, **70** (R = H) was converted into a five-membered cyclic stannane (**73**) possessing a β-disposed anomeric center by treatment with n-Bu_2SnO, and upon coupling with **71** in the presence of TMSOTf, the desired 1β,1α′-linkage was indeed formed in an acceptable 66 % yield.[32] Thus, based on this success, the 1,1′ β-linked system of everninomicin would appear attainable from F-ring tin acetal **68** (see Scheme 12) and the G-ring trichloroacetimidate **69** bearing a participating C-2 benzoate ester group.

As a final comment on the designed synthetic strategy towards the total synthesis of everninomicin 13,384-1 (**1**), one should note that we have deliberately side-stepped the complicated issue of protecting group selection on each of the projected building blocks and their subsequent synthetic intermediates. Such ignorance, however, should not lull one into falsely believing that this facet of the design stage was in any way glossed over, as it represented one of the most fundamental challenges of this molecule. Several issues along these lines are worthy of mention. First, while it goes without saying that all synthetic endeavors require a careful balance between protecting group stability over the course of several synthetic operations and relative ease of removal at the desired juncture in the synthesis, this general concern is quite pressing with everninomicin as the final natural product had demonstrated sensitivity to media that is even mildly acidic. Thus, a route that would require as few late-stage deprotections as possible would be advantageous, thereby requiring a unified, global protecting group strategy in the construction of all building blocks. Additionally, within the general realm of carbohydrates, transformations can often fail not because the employed chemistry is suspect, but merely because the particular protecting group ensemble utilized in the event can engender problems based on their steric bulk or overall stability to reaction conditions. In this vein, it was anticipated at the outset of the everninomicin campaign that several different arrays of protecting groups would likely have to be examined for key transformations, particularly glycosidations. Furthermore, one should recognize that the selection and alteration of protecting groups within the above retrosynthetic schemes and the synthetic routes shown below are the culmination of difficulties encountered along these lines, and, although some of the more instructive examples will be discussed within the context of the winning combination, the reader should turn to the full account[2a–c] of this work to appreciate more fully the true levels of havoc that protecting group chemistry wreaked upon the campaign to synthesize **1**. Without question, this feature serves as another reminder of certain weaknesses in our present day ability to fully predict synthetic transformations.

10.3 Total Synthesis

10.3.1 Synthesis of the $A^1B(A)C$ Fragment

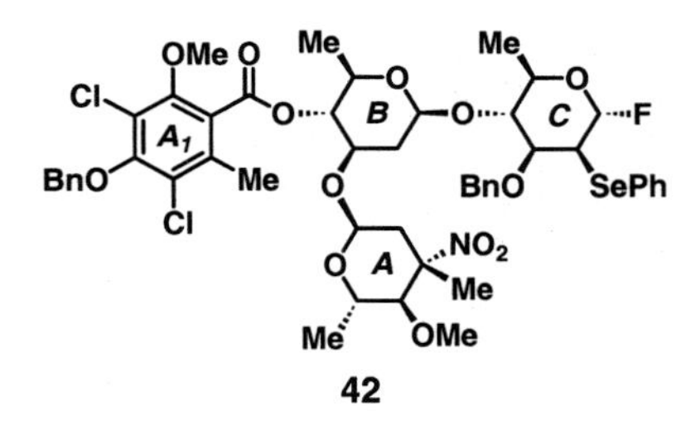

In the synthetic direction, our discussion opens with the synthesis of the major left-hand portion of everninomicin 13,384-1 (**1**), the $A^1B(A)C$ fragment **42**, starting with the preparation of the unique C-3 branched A-ring dideoxy-L-sugar **57** (evernitrose, Scheme 14). Overall, the synthetic approach towards this building block was predicated on the assumption that this carbohydrate could be accessed from a homochiral straight chain precursor **80** in which the C-3 and C-4 stereocenters might arise from two separate stereocontrolled additions of suitable nucleophiles onto appropriately activated electrophilic centers. Moreover, it was envisioned that the nitro group in **57** could result from ozonolysis of an oxyamine intermediate, a well-precedented conversion in the literature.[33]

Thus, the synthesis of this key building block **57** commenced with the controlled reduction of the TIPS-protected ethyl L-lactate **75** to the corresponding aldehyde using DIBAL-H at −78 °C in Et_2O, followed by the stereoselective *anti* addition of an acyl anion equivalent[34] in the form of EVE-Li,[35] providing **76** in 63 % yield. The drawn diastereomer was favored to the extent of 85 % *de*, as would be anticipated based on the presumed Felkin–Ahn transition state shown in the column figure. At this juncture, in anticipation of a second nucleophilic addition, this time onto an oxime ether,

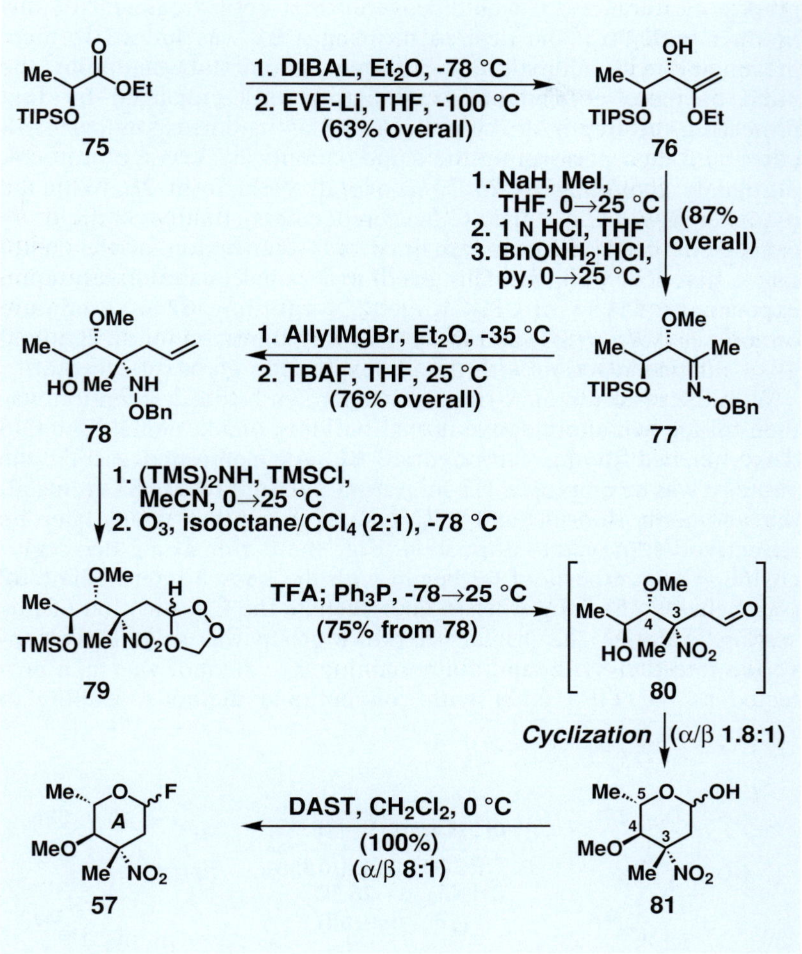

Scheme 14. Synthesis of the A-ring nitrosugar building block **57**.

76 was elaborated into **77** through the simple operations of methyl ether formation, acid-induced hydrolysis of the alkyl enol ether, and oxime capture of the resultant ketone by condensation with *O*-benzylhydroxyamine hydrochloride in pyridine, ultimately providing a 4:1 mixture of *E* and *Z* isomers. Next, the addition of allylmagnesium bromide to **77** occurred smoothly and in a stereocontrolled fashion in Et_2O at −35 °C, affording key intermediate **78** in 76 % yield after TBAF-mediated cleavage of the bulky TIPS group. Having reached this stage, it was expected that upon exposure of **78** to ozone, concomitant formation of both the aldehyde and nitro groups could be achieved, leading to intermediate **80** which would spontaneously cyclize to carbohydrate **81**. While oxidation of the oxyamine to a nitro group indeed occurred under these conditions, surprisingly, a stable intermediate ozonide was isolated instead of the aldehyde after work-up with Me_2S and silica-gel chromatography. Fortunately, it was discovered that upon treatment of this product with Ph_3P the desired nitrosugar **81** was indeed formed, presumably via aldehyde **80**. From an operational standpoint, the yield of these conversions was significantly enhanced by first protecting the free hydroxy group in **78** as its corresponding TMS ether, and then performing the same ozonolysis/cleavage protocol, ultimately providing **81** in 75% overall yield from **78**. With the requisite C-3, C-4, and C-5 stereocenters finally secured as expressed in **81**, all that remained was conversion of the lactol into a glycosyl fluoride. This event proceeded quantitatively upon exposure to DAST in CH_2Cl_2 at 0 °C, affording **57** as a mixture of anomers ($\alpha/\beta = 8:1$) and thereby setting the stage for an eventual glycosidation reaction with the corresponding glycosyl acceptor.

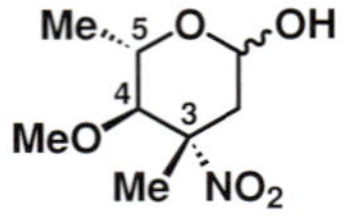

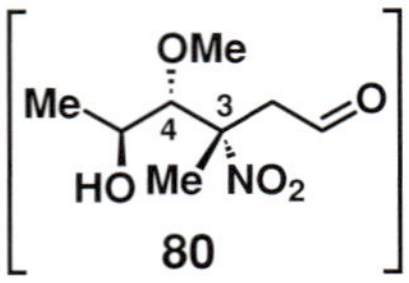

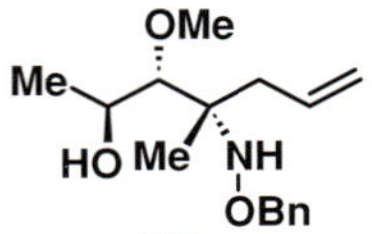

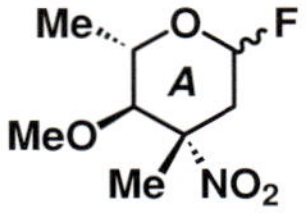

With the uncommon A-ring nitrosugar synthesized, attention was then turned to a more conventional building block, namely the C-6 deoxygenated B-ring carbohydrate **62**. As mentioned earlier, the strategy was to employ a 1,2-migration initiated by DAST to install the anomeric fluoride and a C-2 substituent that could later be reductively removed to ultimately forge the B-ring as a 2-deoxyglycoside. The synthesis of **62** began with the known intermediate **82** (see Scheme 15).[36] In order to deoxygenate the C-6 position of this starting material, the primary hydroxy group was selectively converted into a tosylate, and the remaining free alcohol was then protected as the TIPS ether with conventional methods, leading to

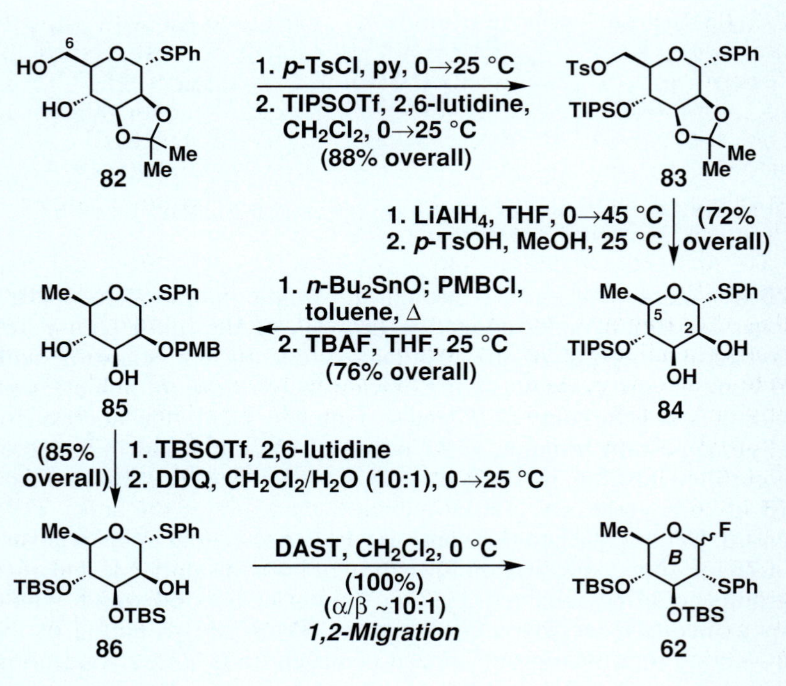

Scheme 15. Synthesis of the B-ring carbohydrate building block **62**.

intermediate **83**. With these objectives smoothly achieved in 88 % overall yield, the tosylate was then reductively removed upon treatment with $LiAlH_4$ in warm THF, affording the C-5 methyl group of the B-ring as it appears in the targeted natural product.

In anticipation of the key 1,2-migration which would hopefully complete this building block, several functional and protecting group manipulations were now required. As such, the acetonide was smoothly solvolyzed upon treatment with *p*-TsOH in MeOH, conveniently affording **84** in 72 % overall yield from **83**. Selective protection of the newly unveiled C-2 hydroxy group as a PMB ether was then accomplished in a two-step, one-pot protocol involving initial tin acetal formation upon treatment with *n*-Bu_2SnO in refluxing toluene,[31] followed by the addition of PMBCl. While such tin acetal methodology is typically employed to differentiate axially versus equatorially disposed hydroxy groups, resulting in the selective protection of the equatorial group, in this case the axially-oriented C-2 group was engaged in preference to the equatorial C-3 alcohol. As such, the bulky TIPS group in **84** must have effectively shielded the C-3 position, sufficiently biasing the encounter with PMBCl to afford solely the desired C-2 protected product. Although the role played by the large silicon group in this event was well-appreciated, its size was expected to engender problems in subsequent transformations and, therefore, it was expelled at this stage with the aid of TBAF to provide **85** in 76 % overall yield. With diol **85** now in hand, both of the free hydroxy groups were protected by exposure to TBSOTf and 2,6-lutidine, and the C-2 PMB group was then removed upon treatment with DDQ to provide **86** in 85 % overall yield. Having finally established the desired array of functionality, it was time to attempt the requisite 1,2-migration leading to **62**. Most gratifyingly, the anticipated conversion occurred quantitatively in the presence of DAST in CH_2Cl_2 at 0 °C, affording the B-ring building block **62** with the predicted inversion of stereochemistry at C-2 as a 10:1 mixture of anomeric fluorides favoring the α-configuration.

As the stereochemical elements of the desired C-ring building block (**61**) are similar to those of D-glucose, it was hypothesized that this fragment could be constructed in short order from the readily available glycal derivative **87**,[37] as shown in Scheme 16. While the stereochemical relationships between **87** and **61** are obvious, a method to chemoselectively differentiate the two free hydroxy groups in **87** represented a critical qualification for using this enticing starting material, as a free alcohol group at C-4 would eventually be required for glycosidation chemistry. Fortunately, several techniques already existed to achieve such an operation, foremost of which is tin acetal methodology similar to that used for the construction of the B-ring subunit.[31] Thus, after formation of the requisite stannane upon treatment with *n*-Bu_2SnO in refluxing toluene, selective protection of the hydroxy group at C-3 was achieved upon subsequent addition of benzyl bromide and *n*-Bu_4NI, again the result of greater steric accessibility of this

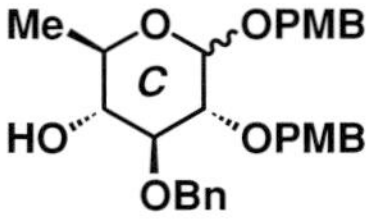

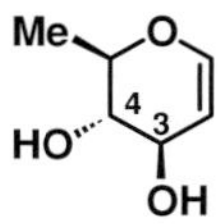

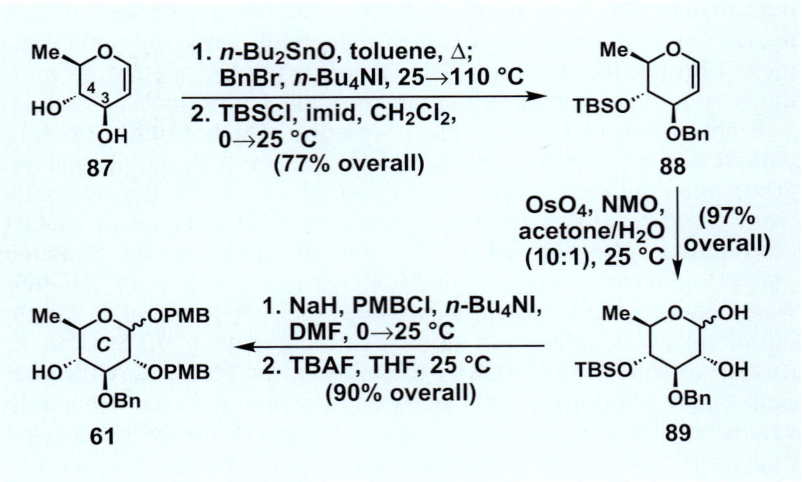

Scheme 16. Synthesis of the C-ring carbohydrate building block **61**.

particular position in the glycal system. With this initial protection accomplished, the remaining hydroxy group was then converted into its corresponding TBS ether with TBSCl and imidazole, providing **88** in 77 % overall yield for these two steps. At this juncture, it was time to incorporate the remaining hydroxy groups of the targeted building block onto **88**, a goal that was easily met through an osmium-mediated dihydroxylation of the more accessible face of glycal derivative **88**; this event afforded the desired C-2 stereochemistry in **89**, but as a 1:1 mixture of anomers. Based on the projected synthetic sequence, though, this uncontrolled C-1 stereochemistry was deemed inconsequential as it would eventually be destroyed. As such, only two operations separated **89** from **61**, namely protection of both of the newly installed hydroxy groups as their corresponding PMB ethers followed by TBAF-mediated lysis of the C-4 silyl ether. These events proceeded smoothly in a combined yield of 90 %.

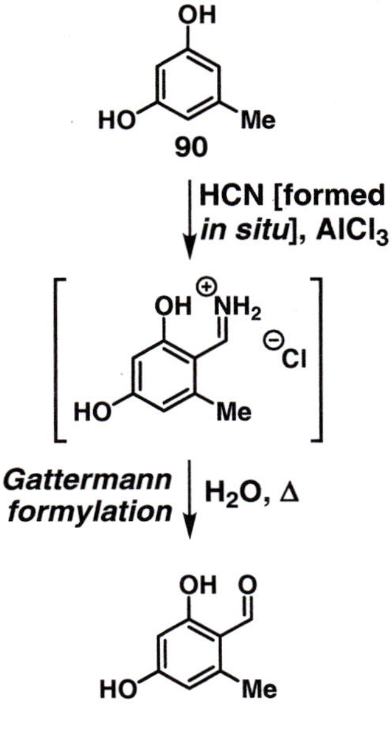

With all three carbohydrates in hand, only the synthesis of the A^1-ring acyl fluoride building block **59** remained before operations could begin to join these units into the larger $A^1B(A)C$ fragment destined for elaboration into everninomicin 13,384-1 (**1**). The preparation of such highly substituted aromatic systems, however, is often far from trivial, typically requiring lengthy protection/deprotection sequences to enable differentiation of the open sites on the aryl core. In the present circumstance, it was envisioned that commercially available orcinol (**90**, Scheme 17) could potentially constitute a useful starting point for **59**, based on the premise that selective formylation of only one of the three free positions of the aromatic ring could be easily effected. Fortunately, application of Gattermann-type formylation conditions of $Zn(CN)_2$, $AlCl_3$, and HCl to **90**,[38] followed by treatment with H_2O at 100 °C, provided an intermediate aldehyde (see adjoining column figure for a putative

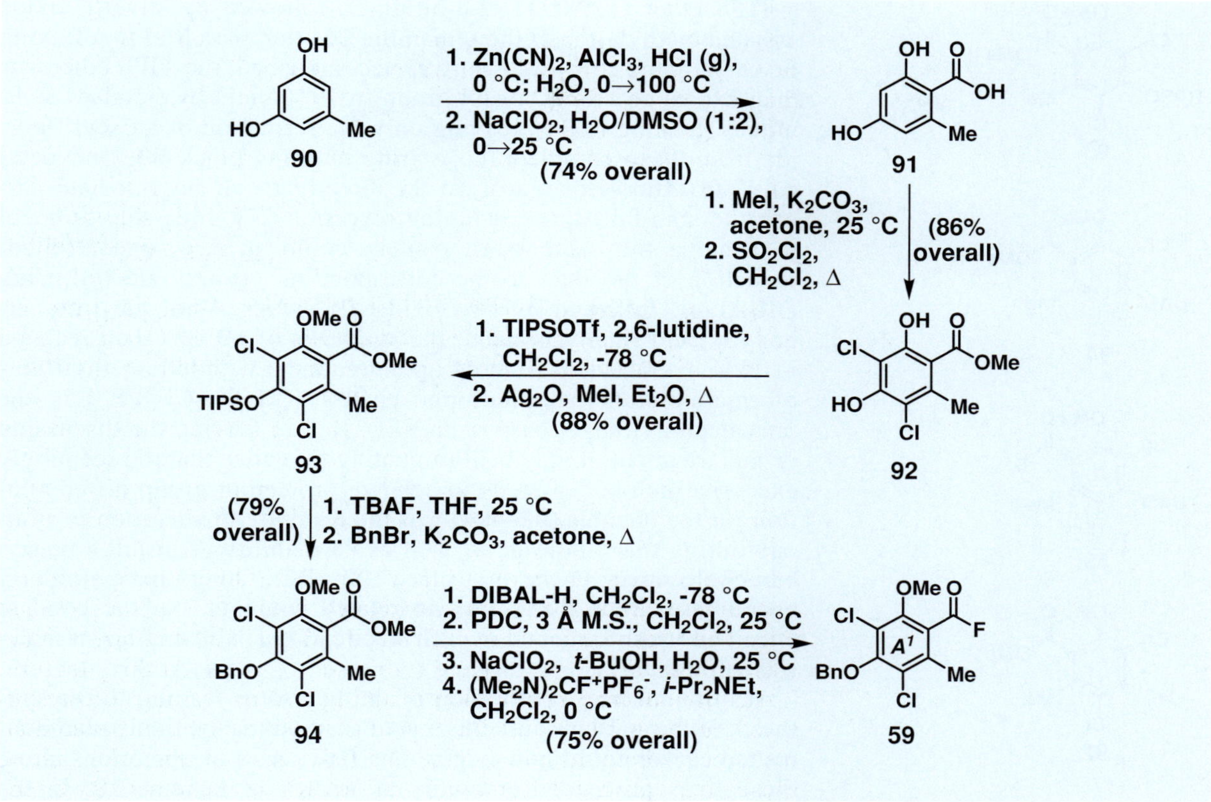

Scheme 17. Synthesis of the A[1]-ring dichloroisoeverninic acyl fluoride building block **59**.

mechanism) which was immediately converted into the corresponding desired carboxylic acid **91** in 74% overall yield by oxidation with $NaClO_2$. With this initial substitution accomplished, it was anticipated that the two aryl chlorine residues in **59** could then be installed onto the remaining open sites of **91**. Indeed, after conversion of the carboxylic acid into the corresponding methyl ester, a phenol-assisted bischlorination was achieved smoothly with sulfuryl chloride in refluxing CH_2Cl_2, providing **92** in 86% yield. As an aside, while other reagents such as chlorine gas could achieve the same chlorination with equal facility, the present set of conditions was selected to enable facile and operationally simple throughput of sufficient quantities of material for late-stage explorations.

Having successfully loaded all six aromatic substituents onto **90**, all that remained to access **59** were some subtle modifications; despite the straightforward nature of this statement, however, a rather lengthy sequence was required to complete these tasks. The first prerequisite step, selective and sequential protection of the phenolic groups in **92**, was achieved without incident through initial protection of the more activated, non-hydrogen bonded phenol as a TIPS ether (TIPSOTf, 2,6-lutidine), followed by silver(I) oxide assisted methylation of the remaining position with MeI in refluxing Et_2O. With the construction of **93** accomplished, the TIPS ether was then exchanged for a benzyl group in 79 % yield by standard techniques (leading to **94**), leaving only the formation of an acyl fluoride from **94** to complete the A^1-ring building block **59**. Unexpectedly, all efforts to hydrolyze **94** directly to an intermediate carboxylic acid failed, necessitating reversion to a three-step protocol to achieve the same goal, namely complete (i. e. uncontrolled) reduction of the ester to the corresponding primary alcohol using DIBAL-H, followed by two oxidation events. With the free carboxylic acid finally in hand, the synthesis of **59** was then realized in 75 % overall yield from **94** upon treatment with tetramethylfluoroformamidinium hexafluorophosphate[39] $[(Me_2N)_2CF^+PF_6^-]$ in the presence of Hünig's base (i-Pr_2NEt). Before leaving the discussion of this fragment, it may be illuminating to realize that the seemingly excessive number of steps to achieve protecting group differentiation for the phenols (**92**→**94**) was the result of an alteration in strategy during the synthesis, as well as an inability to install a benzyl ether selectively in the first place (**92**→**93**). Others have similarly encountered such difficulties in related contexts, but the good to excellent yields obtained in each step readily enabled efficient material throughput.

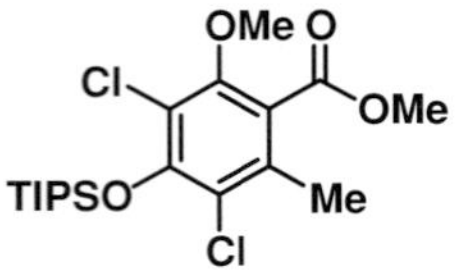

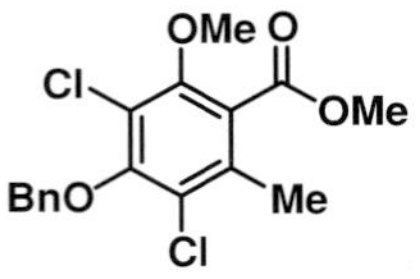

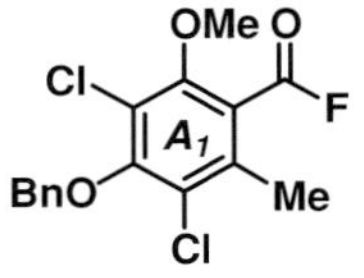

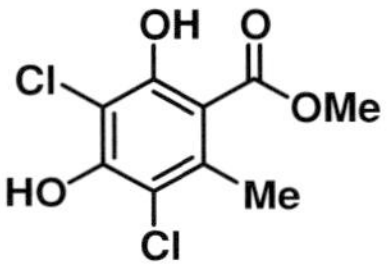

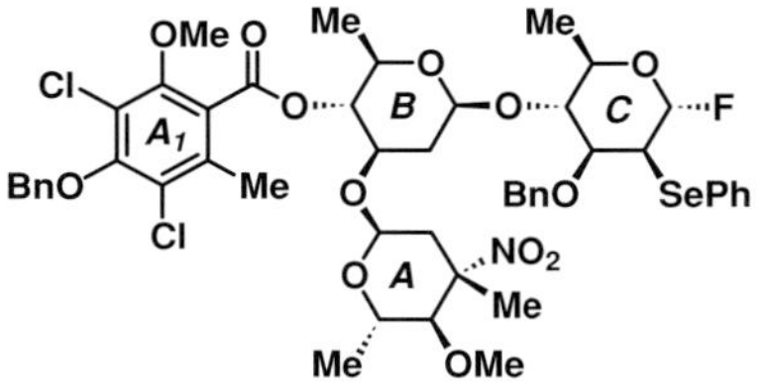

At this juncture, exploration of the key steps leading to the synthesis of the $A^1B(A)C$ assembly **42** based on the synthetic plan delineated earlier could now begin. The first series of operations along these lines proceeded smoothly as shown in Scheme 18. In the opening event, $SnCl_2$-mediated coupling of the B-ring glycosyl donor **62** with C-ring acceptor **61** proceeded under mild conditions in a 1:1:1 mixture of $CH_2Cl_2/Et_2O/Me_2S$ at −10 °C to afford a disaccharide linked exclusively in a β-fashion in 71 % yield.[10] As an aside, one should note that while the particular mixture of solvents used in this transformation might appear relatively esoteric, this combination proved critical for preventing unwanted PMB cleavage and facilitating the solubilization of both reactants. Having admirably performed its β-directing role in the glycosidation reaction, the B-ring phenylsulfide was no longer required, and, as such, it was reductively cleaved upon exposure to Raney Ni to afford 2-deoxyglycoside **95**. At this stage, the protecting group array on the B-ring of **95** had to be adjusted to enable the eventual acceptance of the A^1-ring aromatic system. Towards this end, both the TBS ethers were cleaved from **95** with TBAF, and selective protection of the alcohol group at C-3 as the corresponding allyl ether was achieved through application of tin acetal methodology. The remaining free hydroxy group was then activated for coupling with **59** through treatment with n-BuLi in THF at −78 °C; upon the addition of acyl fluoride **59** to the resultant anion at 0 °C, esterification proceeded smoothly in 99 % yield. In a final modification to prepare the newly generated A^1BC system for glycosidation with the nitrosugar A-ring unit, the allyl group was selectively removed

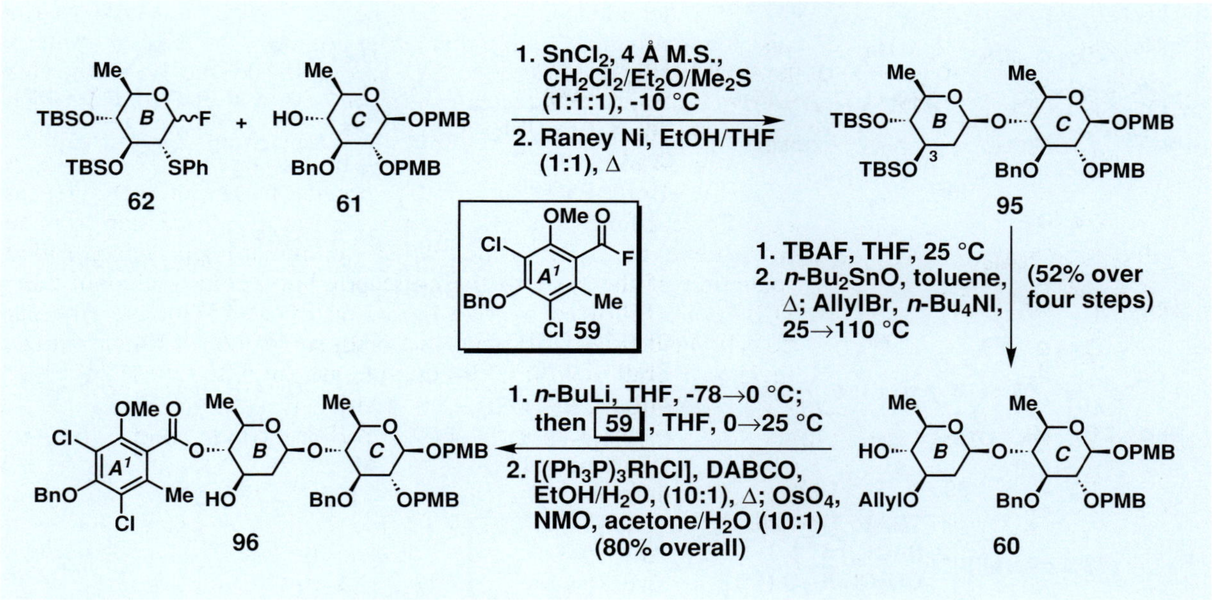

Scheme 18. Construction of the A[1]BC fragment **96**.

using a two-step procedure involving preliminary activation with a combination of Wilkinson's catalyst $[(Ph_3P)_3RhCl]$ and DABCO followed by oxidative cleavage with OsO_4/NMO, affording **96** in 80 % overall yield from **60**.

Having progressed this far, the synthesis of the $A^1B(A)C$ fragment **42** was well within reach, with only incorporation of the A-ring nitrosugar and modification of the C-2 and C-3 positions in ring C of **96** remaining as critical steps. However, the ordering of these events constituted two divergent, but seemingly equally viable, pathways (at least on paper). Initial efforts focused on the use of a free hydroxy group in **96** as a glycosyl acceptor upon activation of the A-ring glycosyl fluoride **57**. While this coupling proceeded smoothly and the requisite α-anomer was formed, all efforts to subsequently install the required phenylseleno functionality onto the C-ring of any intermediate after this glycosidation failed. In light of these unsuccessful attempts, the only conceivable alternative at this advanced stage (apart from formulating a new synthetic plan of attack) was switching the ordering of these events. Fortunately, this approach nicely overcame this roadblock. Thus, protection of the free B-ring hydroxy group of **96** (Scheme 19) as a bulky TIPS ether was followed by removal of both PMB groups with PhSH and $BF_3{\cdot}OEt_2$ in CH_2Cl_2 at −35 °C to afford **97** in 77 % overall yield. As a prelude to incorporating a phenylseleno group stereoselectively at the anomeric position of ring C in advance of reaction with A-ring sugar **57**, protection of the hydroxy group at C-2 in ring C of **97** was required. This task was achieved by tandem protection of both hydroxy groups as their acetates with Ac_2O followed

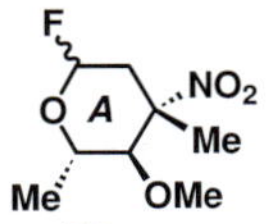

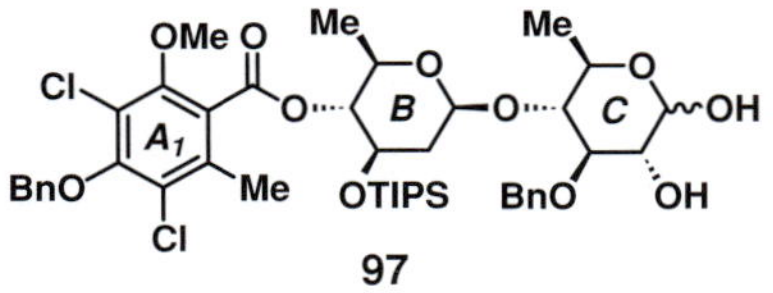

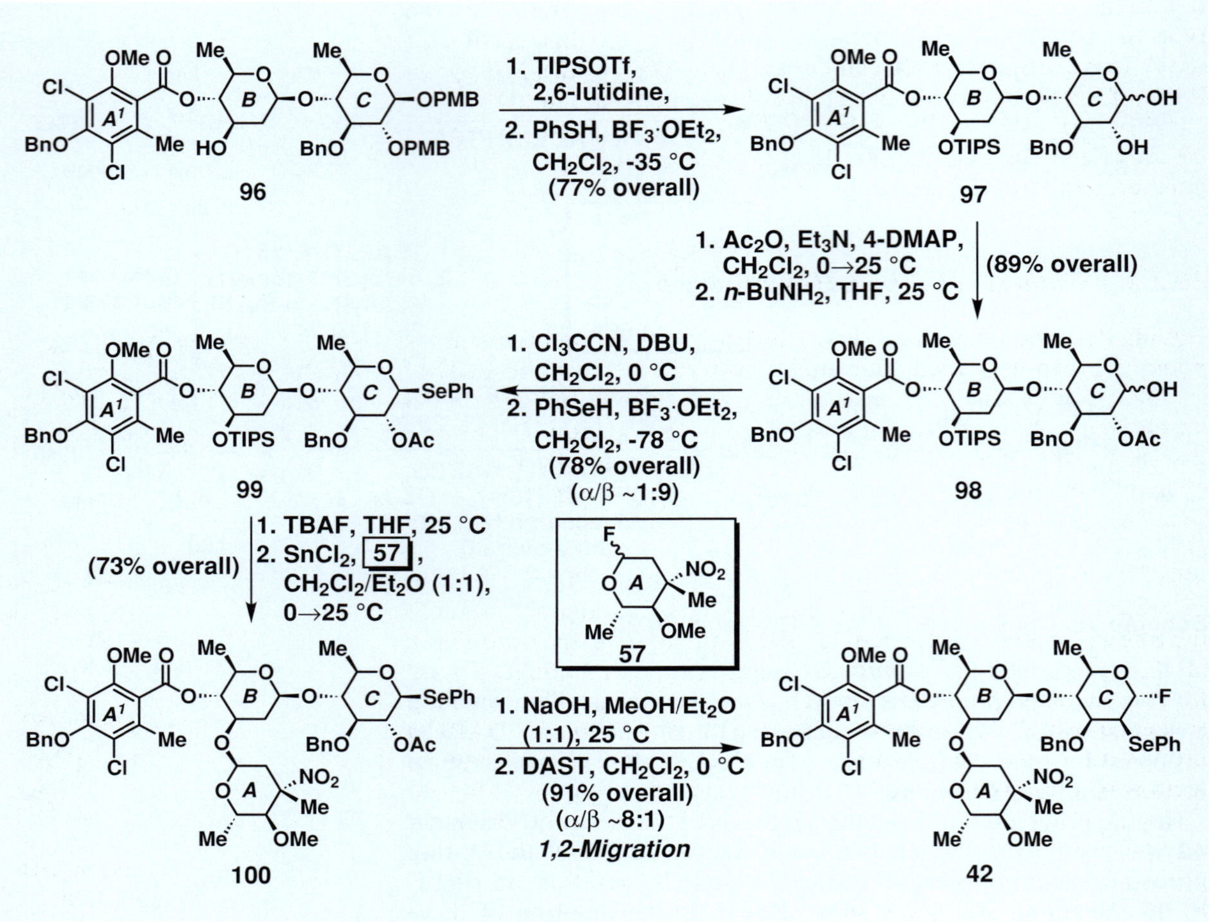

Scheme 19. Construction of A^{1}B(A)C fragment **42** of everninomicin.

by chemoselective lysis of the more reactive anomeric acetate through treatment with n-$BuNH_2$ in THF at ambient temperature. With **98** in hand, conversion of the lactol into its trichloroacetimidate congener with trichloroacetonitrile and DBU was followed by the addition of PhSeH[40] under the activating influence of $BF_3 \cdot OEt_2$ in CH_2Cl_2 at −78 °C. These operations afforded the desired β-phenylseleno glycoside **99** in 78 % overall yield from **98** as a 9:1 mixture of β- to α-anomers.

With this functionality installed, the opportunity to incorporate the A-ring nitrosugar was nearly at hand, necessitating only the removal of the TIPS group from **99** to provide a suitable glycosyl acceptor for eventual coupling. Once this goal was achieved by exposure of **99** to TBAF, the successful formation of oligosaccharide **100** was smoothly realized upon glycosidation with evernitrose glycosyl fluoride **57** activated by $SnCl_2$, leading to an 8:1 mixture favoring the desired α-anomer at the newly formed glycoside bond. At this stage, only a DAST-mediated 1,2-migration of the type discussed earlier (Scheme 3) stood between this version of the $A^1B(A)C$ segment (**100**) and its ultimate format (**42**). Most gratifyingly, after NaOH-mediated acetolysis, exposure to DAST in CH_2Cl_2 at 0 °C effected the desired formation of glycosyl fluoride **42**, with the forecasted inversion of stereochemistry at both the anomeric and the C-2 positions in ring C.

10.3.2 Synthesis of the FGHA² Fragment

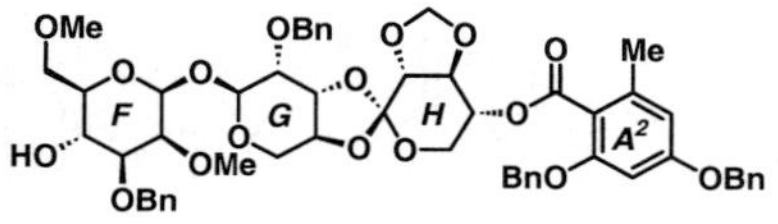

Having described the synthesis of the left-hand portion of everninomicin, we can now focus our attention on the pentacyclic $FGHA^2$ fragment **45** bearing the β-linked 1,1′ glycosidic linkage and a stereochemically challenging orthoester. In anticipation of the critical steps that would hopefully lead to the construction of these elements, our analysis commences with the syntheses of the four requisite building blocks for this assembly as delineated in Schemes 20–23.

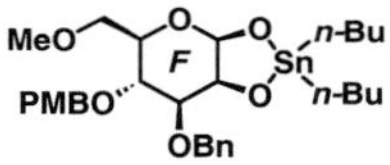

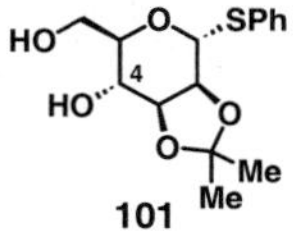

Efforts towards the F-ring glycoside **68** (Scheme 20) began with the selection of a starting material from the chiral pool, namely the known mannose intermediate **101**, which possessed essentially all the stereocenters of the projected target, but still required several functional and protecting group manipulations. The opening sequence sought to achieve selective incorporation of a PMB group on the secondary hydroxy group at C-4. As such, initial protection of the more accessible primary alcohol group at C-6 with TBSOTf was followed by engagement of the remaining hydroxy group (C-4) as the planned PMB ether under conditions that generated the more active iodide alkylating agent *in situ* (by adding NaH, PMBCl, and n-Bu_4NI in this order) to facilitate reaction at this relatively hindered position. Finally, upon silyl deprotection with TBAF in THF at ambient temperature, alcohol **102** was isolated in a combined yield of 88 % over these three operations. At this juncture, the alcohol group at C-6 was then engaged as a methyl ether through a standard protocol, and the acetonide was cleaved under acidic conditions to afford **103** in 81 % overall yield. Although one might envision that this rather lengthy protection/deprotection sequence to form **103** could have been avoided by methylation of the C-6 hydroxy group in **101**, achieving such chemoselective protection of the primary position in practice proved challenging.

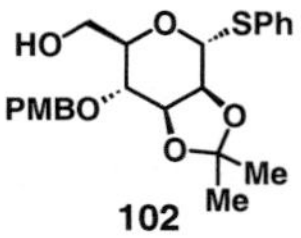

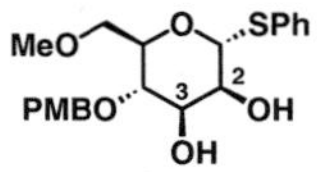

From **103**, the key F-ring building block was only three steps away. First, treatment with n-Bu_2SnO in refluxing toluene afforded the 2,3-tin acetal which reacted exclusively at the C-3 position upon addition of benzyl bromide and catalytic n-Bu_4NI, thereby leading to the corresponding phenylthioglycoside with only the C-2 hydroxy group free. Of course, the observed chemoselectively was in accordance with the well-established precedent that equatorially disposed alcohol groups are engaged in preference to those in an axial orientation in such reactions. Next, oxidative release of the thioglycoside was

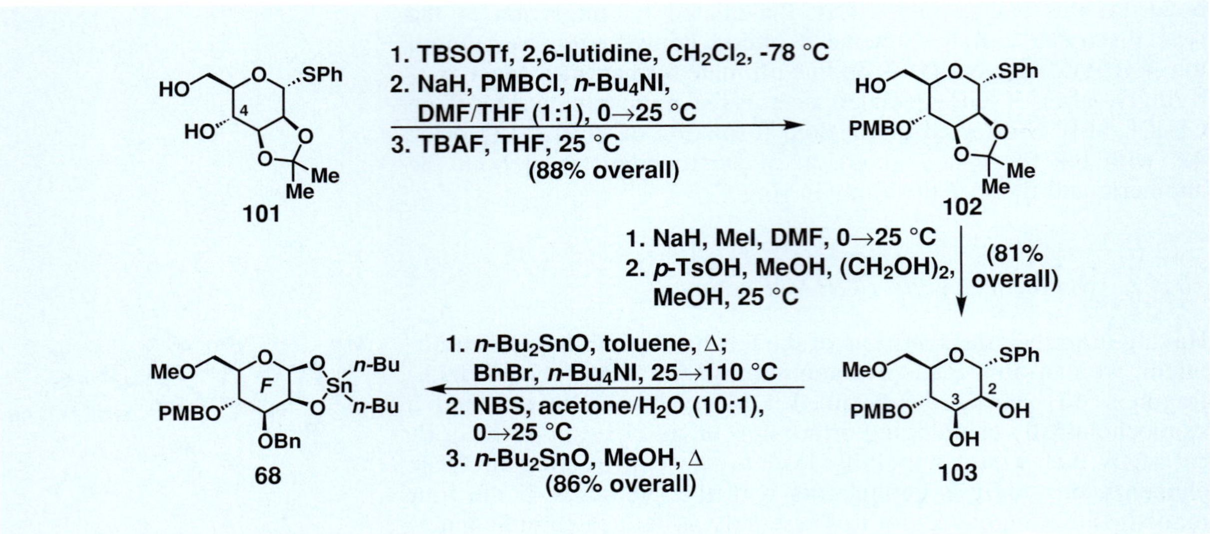

Scheme 20. Synthesis of the F-ring tin acetal building block **68**.

achieved upon activation with NBS in aqueous acetone, and, with a 1,2-diol now revealed, stannane **68** was readily obtained upon a final reaction with n-Bu_2SnO in refluxing methanol.

The synthesis of the coupling partner for the F-ring, the trichloroacetimide G-ring fragment **69**, was perhaps the most challenging of the four building blocks required to build the $FGHA^2$ assembly. Starting with the C_2-symmetric diisopropyl L-tartrate (**104**, Scheme 21), perallylation was followed by $LiAlH_4$-mediated reduction of both terminal esters to the corresponding alcohols. Upon treatment of this diol intermediate with one equivalent of NaH in the presence of TBDPSCl, desymmetrization was achieved through monosilylation, affording **105** in 81 % yield overall for these three steps. If one envisioned that the two protected hydroxy groups in this product would eventually constitute the C-3 and C-4 stereocenters in **69**, then a one-carbon extension onto **105** would be required to fashion an intermediate such as **112**, which could cyclize to the desired carbohydrate ring system. In this regard, application of the TMS thiazole homologation method of Dondoni and co-workers, wherein an aromatic thiazole ring serves as the synthetic equivalent of a formyl group, proved quite fruitful.[41] Thus, to append this ring system onto **105**, Swern oxidation of the latter to the corresponding aldehyde was followed by the addition of 2-TMS-thiazole, effecting *in situ* generation of a thiazole nucleophile, which resulted in the formation of an equimolar mixture of adducts **106** and **107**. Although some degree of selectivity was expected from this addition process, its stereo-randomness was of no real consequence as **106** and **107** were separable by column chromatography, and the undesired epimer **107** could be converted into the desired compound **106** in 67 % yield through a conventional two-stage

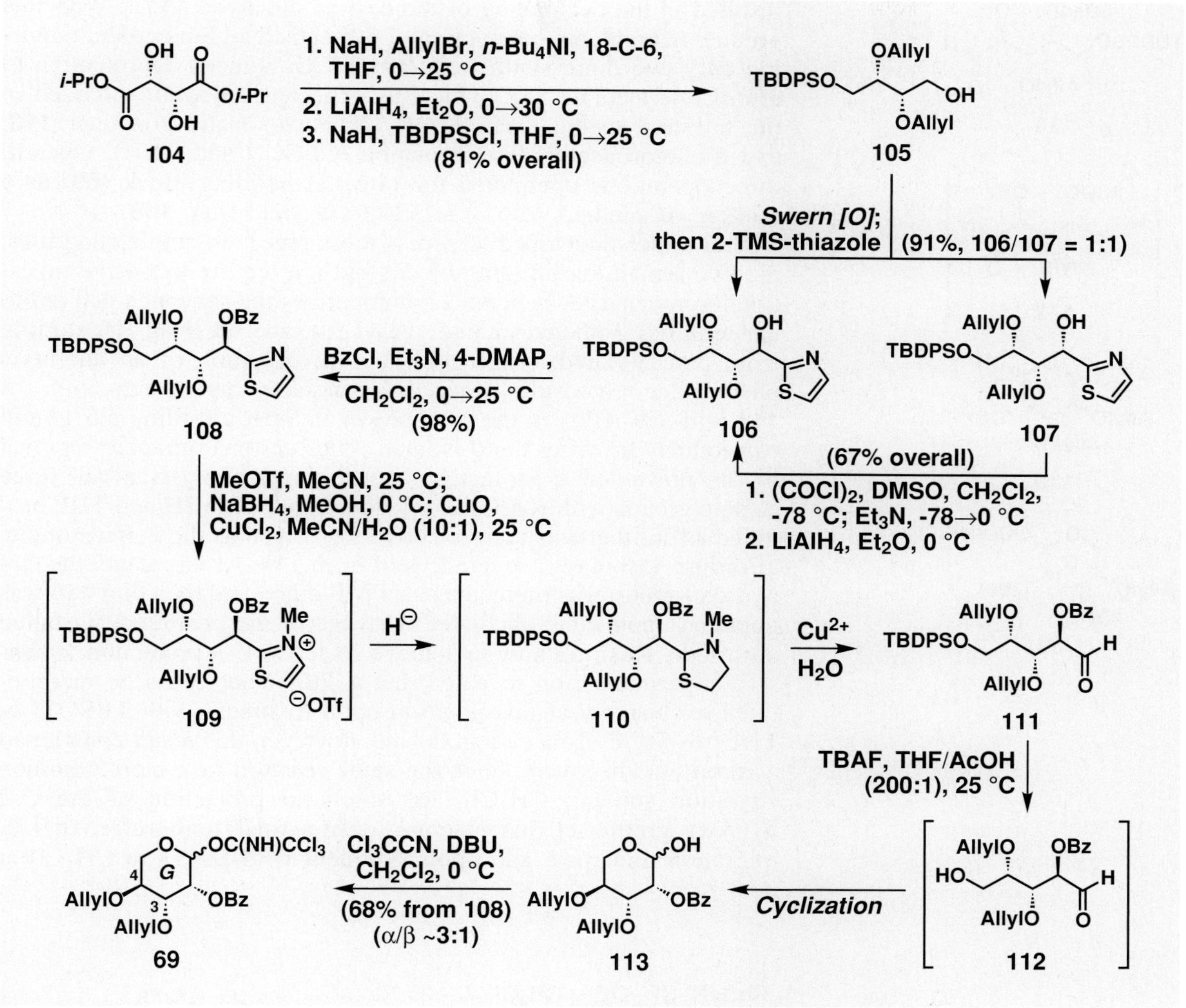

Scheme 21. Synthesis of the G-ring carbohydrate building block **69**.

recycling protocol consisting of Swern oxidation and reduction with $LiAlH_4$.

Having finally secured the stereochemistry of C-2 in **106**, the newly generated hydroxy group was converted into a benzoate, setting the stage for the critical series of events that would unravel the thiazole to the desired aldehyde. In this three-step, one-pot reaction, treatment of **108** with MeOTf in acetonitrile at ambient temperature led to methylation of the aromatic nitrogen atom to afford an activated intermediate (**109**) which was smoothly reduced to the fully hydrogenated system **110** upon subsequent addition of $NaBH_4$. Finally, treatment of the latter intermediate with Cu(II), delivered in the form of CuO and $CuCl_2$, effected oxidative cleavage and resulted in the decloaking of the desired aldehyde **111**.[41] With this product in hand, the completion of **69** seemed all but certain, requiring only two more synthetic steps: TBAF-induced deprotection to afford intermediate **112**, which, as expected, smoothly cyclized in the buffered media (THF/AcOH, 200:1) to lactol congener **113**, and tricholoroacetimidate formation (Cl_3CCN and DBU). Overall, these operations completed the targeted building block (**69**) as a mixture of anomers ($\alpha/\beta = 3:1$) in 68 % yield from **108**.

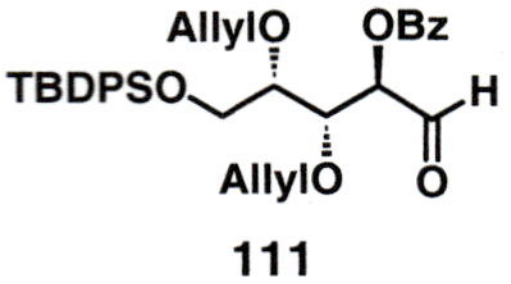

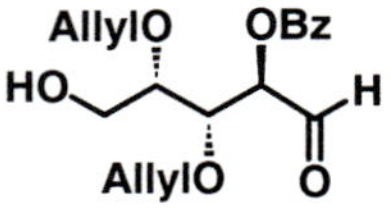

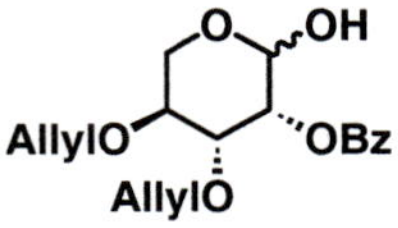

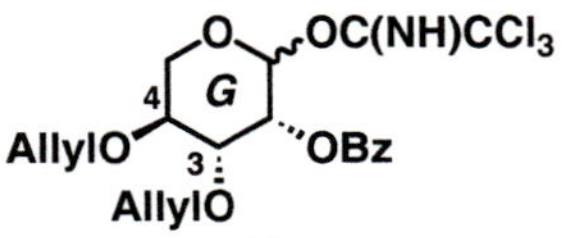

With routes developed to access these pieces in ample quantities, the two remaining building blocks did not require extensive investigation to acquire. Scheme 22 summarizes the sequence that led to the required 2-phenylselenoglycosyl fluoride **66** (ring H), starting from peracetylated xylose **114**. Initial formation of an anomeric phenylselenoglycoside proceeded smoothly upon activation of **114** with $BF_3{\cdot}OEt_2$ in the presence of PhSeH, affording a 5:1 ratio of products favoring the β-isomer. After chromatographic removal of the unwanted α-anomer, exhaustive methanolysis of all three acetate groups with K_2CO_3 in a 1:1 mixture of MeOH and THF provided a triol that was then selectively engaged as the 2,3-acetonide, affording **115** in 69 % overall yield from **114**. At this stage, the C-4 hydroxy group was protected as a PMB ether, and then diol **116** was obtained upon PPTS-mediated cleavage of the previously installed acetonide. Desiring now to achieve selective C-3 protection, extensive experimentation revealed that a TBS ether could be incorporated exclusively at this position upon treatment with TBSOTf in THF at −78 °C. To a certain degree, however, this result constituted a fortuitous discovery since the same reaction in a more common silylation solvent, CH_2Cl_2, led solely to protection of the C-2 hydroxy group. At this juncture, only a final stereoselective 1,2-migration remained, and upon treatment with DAST in CH_2Cl_2 at

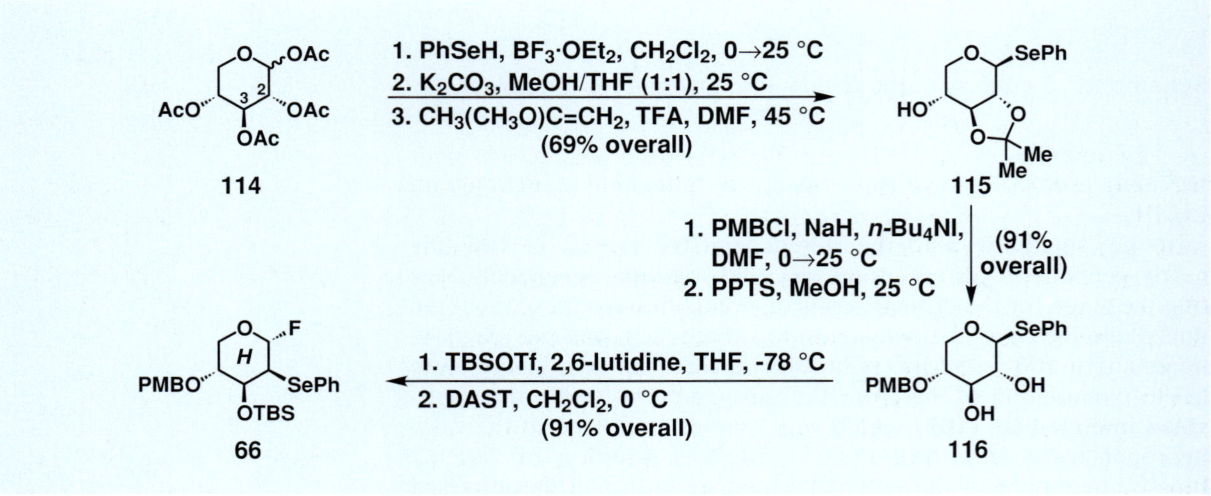

Scheme 22. Synthesis of the H-ring carbohydrate building block **66**.

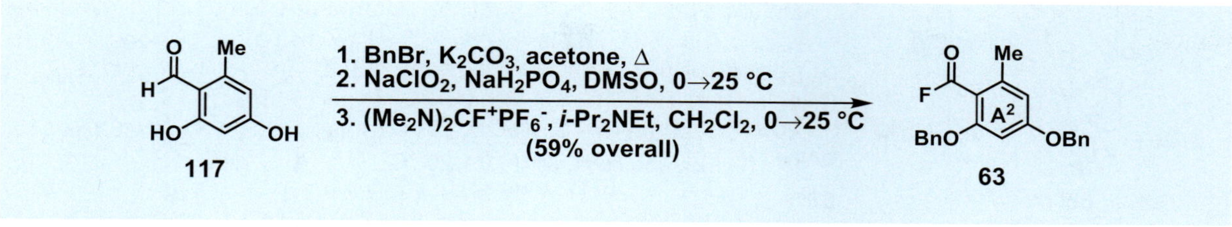

Scheme 23. Synthesis of the A^2-ring acyl fluoride building block **63**.

0 °C this step was instigated to afford 2-phenylselenoglycoside **66** in 91 % yield from **116**.

The construction of the final piece, acyl fluoride **63**, proceeded quite readily along the lines of the A^1-ring fragment construction already discussed. As shown in Scheme 23, after benzylation of salicylaldehyde derivative **117** (obtained from Gattermann formylation of **90** as shown in Scheme 17), oxidation to the corresponding carboxylic acid with $NaClO_2$ was followed by acyl fluoride formation with $(Me_2N)_2CF^+PF_6^-$ in the presence of Hünig's base, affording the chromatographically stable **63** in 59 % overall yield.

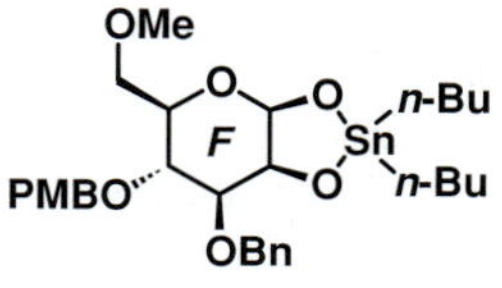

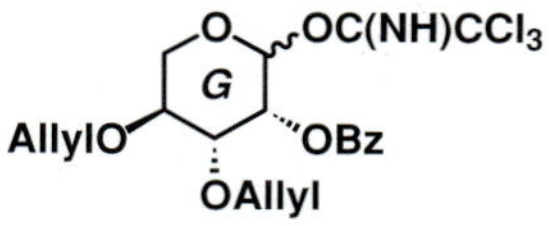

With the four building blocks in hand, efforts were immediately expended to effect their union, starting with explorations to form the 1,1′ FG glycosidic linkage. As delineated in Scheme 24, the desired coupling between tin acetal **68** and trichloroacetimidate **69** proceeded precisely as anticipated from the earlier model studies, and, after acidic work-up and careful methylation of the resultant alcohol (MeI, NaH, DMF), **118** was isolated in 64 % yield exclusively as the requisite 1β,1′α-linked disaccharide. With this key step out of the way, all efforts concentrated on advancing this intermediate (**118**) to the next critical operation, the formation of the GH orthoester. Thus, a sequence of protecting group modifications were required as defined by the series of five operations that effected the conversion of **118** into **121**, operations which merit no additional commentary due to their conventional nature and smooth execution (see Scheme 24 for details).

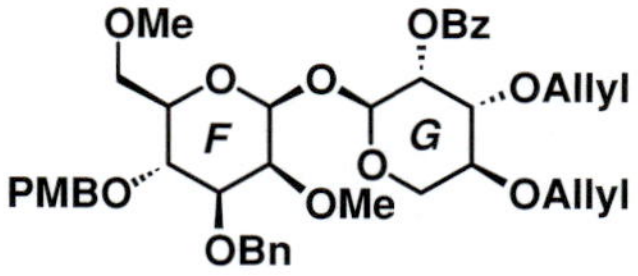

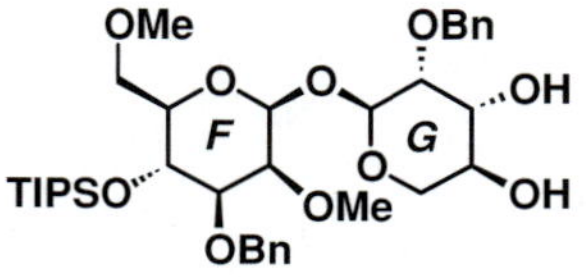

Numerous methods were then tested to realize selective protection of only one of the hydroxy groups of diol **121** in anticipation of eventual coupling with the H-ring fragment **66**. Invariably, a mixture was obtained with the best-yielding reaction affording a regioisomeric mixture of chloroacetates (CA) **122** and **67** in a 1:1 ratio. As mentioned earlier, however, since it was unknown whether a C-3 or C-4 hydroxy group on ring G (in an intermediate of type **65**) would provide the correct stereochemistry during orthoester formation, this nonselective protecting step was considered productive because it provided access to both G-ring glycosyl acceptors. So, seeking to test a free C-3 hydroxy group in the last step of orthoester construction (i. e. first linking the G- and H-ring saccharides through the C-4 position of ring G), **67** (R = CA) was coupled to **66** upon activation with $SnCl_2$ in Et_2O to afford the desired trisaccharide as a single anomer. Subsequent treatment with K_2CO_3 in

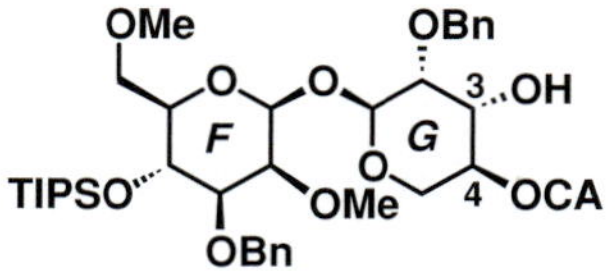

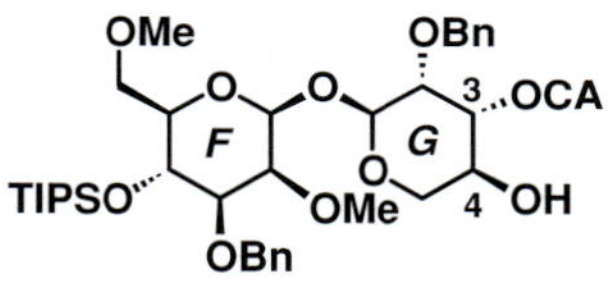

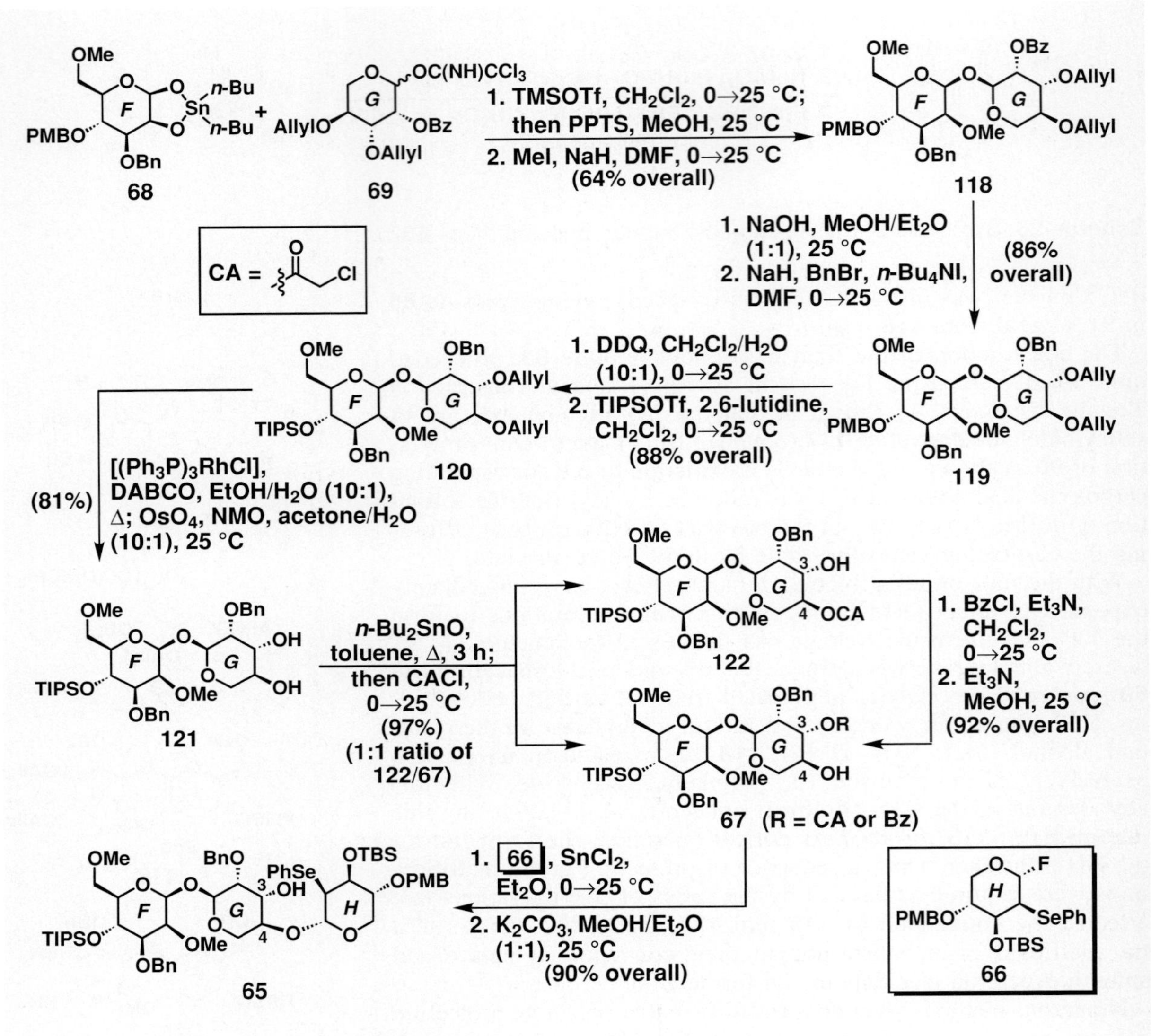

OMe
F
PMBO
OBn
n-Bu
Sn
n-Bu
68
+
OC(NH)CCl_3
G
AllylO
OBz
OAllyl
69
1. TMSOTf, CH_2Cl_2, 0→25 °C; then PPTS, MeOH, 25 °C
2. MeI, NaH, DMF, 0→25 °C
(64% overall)
OMe
OBz
OAllyl
F
G
PMBO
OMe
OAllyl
OBn
118
1. NaOH, MeOH/Et_2O (1:1), 25 °C
2. NaH, BnBr, n-Bu_4NI, DMF, 0→25 °C
(86% overall)
CA = (chloroacetyl, C(=O)CH_2Cl)
OMe
OBn
OAlly
F
G
PMBO
OMe
OAlly
OBn
119
1. DDQ, CH_2Cl_2/H_2O (10:1), 0→25 °C
2. TIPSOTf, 2,6-lutidine, CH_2Cl_2, 0→25 °C
(88% overall)
OMe
OBn
OAllyl
F
G
TIPSO
OMe
OAllyl
OBn
120
[$(Ph_3P)_3$RhCl], DABCO, EtOH/H_2O (10:1), Δ; OsO_4, NMO, acetone/H_2O (10:1), 25 °C
(81%)
OMe
OBn
OH
F
G
TIPSO
OMe
OH
OBn
121
n-Bu_2SnO, toluene, Δ, 3 h; then CACl, 0→25 °C
(97%)
(1:1 ratio of 122/67)
OMe
OBn
3 OH
F
G
TIPSO
OMe
4 OCA
OBn
122
1. BzCl, Et_3N, CH_2Cl_2, 0→25 °C
2. Et_3N, MeOH, 25 °C
(92% overall)
OMe
OBn
3 OR
F
G
TIPSO
OMe
4 OH
OBn
67 (R = CA or Bz)
1. 66, $SnCl_2$, Et_2O, 0→25 °C
2. K_2CO_3, MeOH/Et_2O (1:1), 25 °C
(90% overall)
OMe
BnO
PhSe
3 OH
OTBS
OPMB
F
G
H
TIPSO
OMe
4
OBn
65
F
H
PMBO
SePh
OTBS
66

Scheme 24. Assembly of the everninomicin FGH fragment **65**.

an equal volume mixture of MeOH and Et_2O at ambient temperature then led to cleavage of the chloroacetate group, affording 2-phenylselenoglycoside **65** in 90 % yield over the two steps. Before moving on, one should note that the overall material throughput to **65** was significantly enhanced by the discovery that the C-4 protected **122** could be funneled through the same sequence once it had first been converted into the benzoylated variant of **67** (as shown in Scheme 24).

Having toiled extensively to reach this advanced synthetic intermediate, the opportunity to harvest some fruits from that labor in

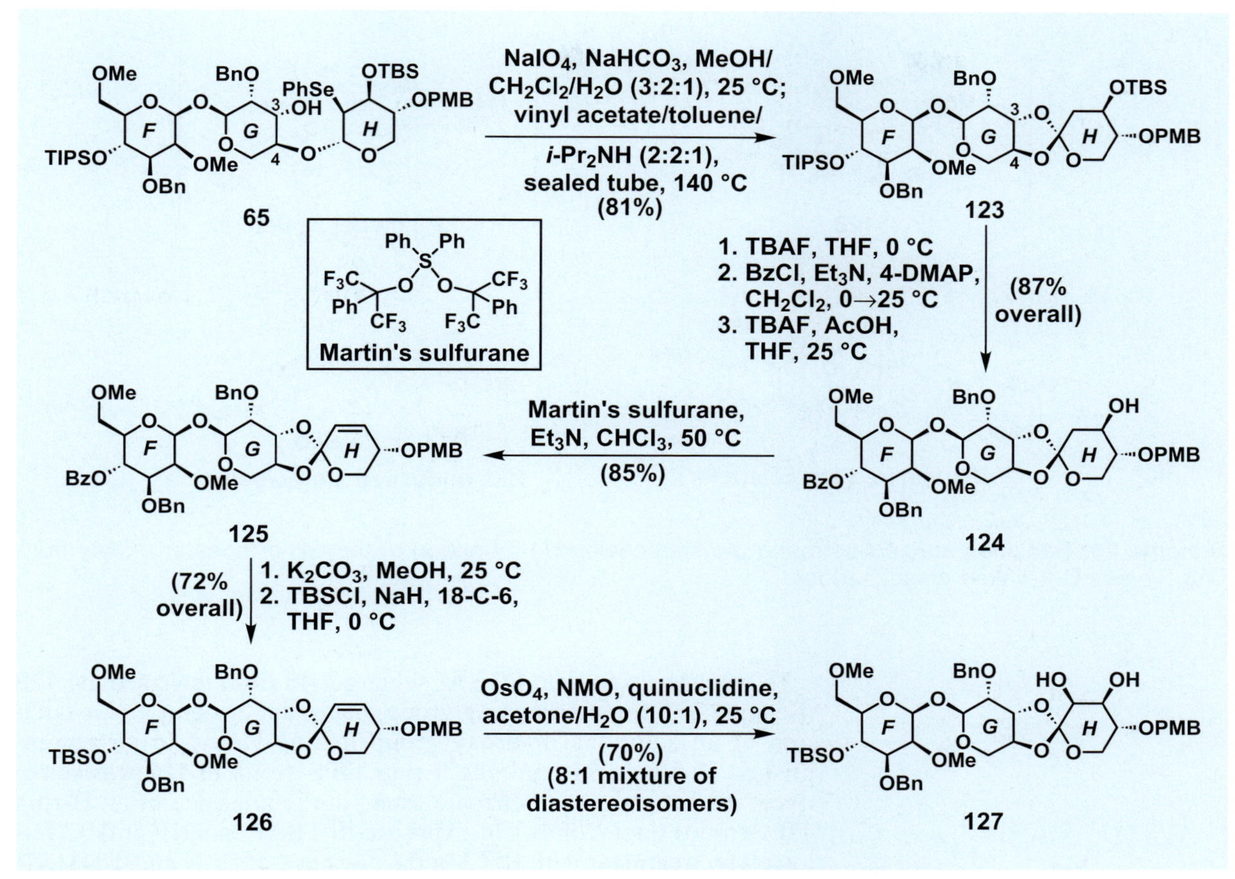

Scheme 25. Elaboration of trisaccharide **65** to the FGH orthoester fragment **127**.

the form of successful orthoester formation was at hand. Gratifyingly, application of the conditions developed by Sinaÿ and coworkers[29] to **65** smoothly afforded **123** in 81 % yield (Scheme 25), an adduct possessing the orthoester stereochemistry desired for the everninomicin framework. Importantly, one should note that glycosidic tethering through the C-4 hydroxy group of ring G was critical to achieving this result, as model studies indicated that the alternative scenario (i. e. using a C-3 linked trisaccharide) afforded predominantly the orthoester with the opposite stereochemistry.

These interesting observations are summarized in the transition state models **128** and **129** shown in Scheme 26, thereby accounting for the formation of these two epimeric orthoesters (**123** and **130**). It is also instructive to recognize that the mixture of solvents used in this reaction, particularly the addition of vinyl acetate and *i*-Pr_2NH, was predicated on the known ability of these species to trap the PhSeOH formed upon selenoxide elimination, thus preventing side reactions that might otherwise have occurred in its presence.

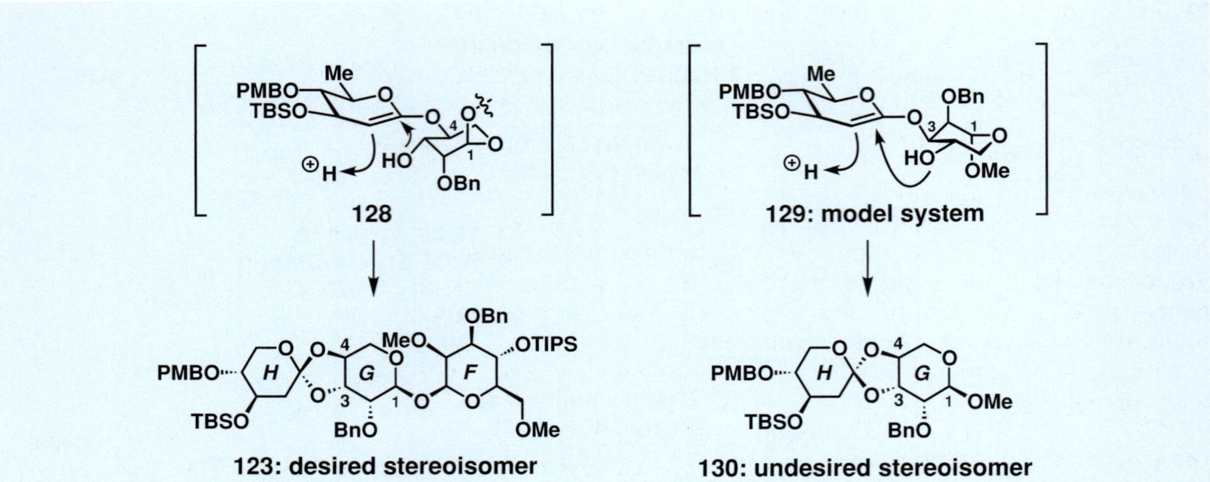

Scheme 26. Transition states illustrating the stereoselective formation of the GH orthoester moiety from C-3- versus C-4-linked disaccharides.

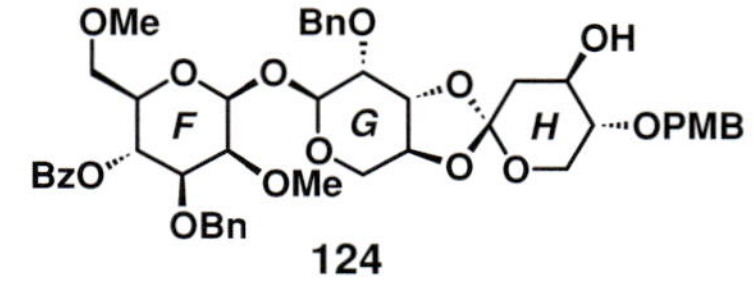

To reach the complete FGHA[2] subtarget **45** from intermediate **123** (Scheme 25), some protecting group manipulations and the installation of an additional hydroxy group in ring H had to be accomplished. Towards this end, the F-ring TIPS group in **123** was selectively exchanged for a benzoate ester (in the presence of an H-ring TBS group) upon controlled exposure to TBAF in THF at 0 °C followed by treatment with BzCl in the presence of Et_3N and 4-DMAP. With these opening steps successfully accomplished, reaction with buffered TBAF (to prevent hydrolysis of the benzoate ester) at ambient temperature then effected lysis of the previously untouched TBS group, leading to hydroxybenzoate **124** in 87 % overall yield for the three steps. With the aim of employing a dihydroxylation reaction to generate a *cis*-diol system in ring H, the free C-4 hydroxy group of this system was then smoothly eliminated with Martin's sulfurane[42] to give glycal precursor **125** in 85 % yield. At this stage, stability-related issues of subsequent intermediates once again dictated an exchange of protecting groups, this time in ring F. Thus, the benzoate group of **125** was cleaved upon exposure to K_2CO_3 in MeOH and replaced with a TBS group by treating the resultant alcohol with NaH and TBSCl in the presence of 18-crown-6, furnishing olefinic substrate **126**. With an appropriate collection of protecting groups finally in place, the desired *cis*-1,2-diol system was now installed by treating **126** to OsO_4/NMO in the presence of activating quinuclidine[43] to afford diol **127** in 70 % yield as an 8:1 diastereomeric mixture favoring the drawn product.

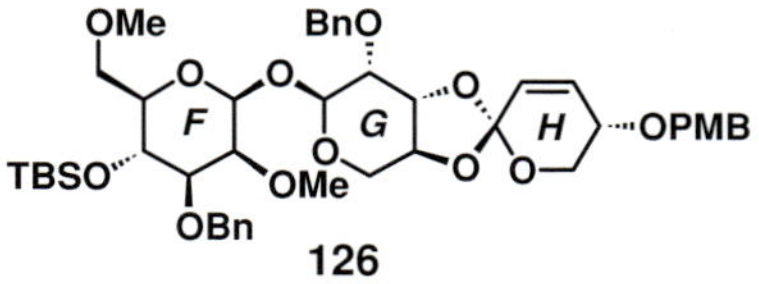

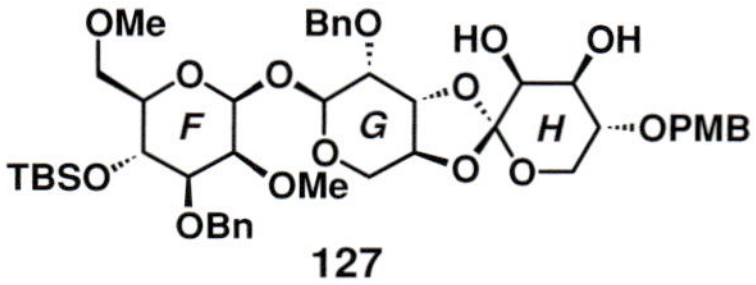

Unfortunately, although these steps proceeded without incident, their overall yield proved capricious to scale-up, due in large part to the extreme sensitivity of the H-ring allylic orthoester during the conversion of **125** into **126**. As such, an alternate and improved synthesis of **127** from **125** was sought and found as shown in Scheme 27. In this route, dihydroxylation of the C–C double bond in ring H occurred first under conditions similar to those described above, and was followed by benzoate cleavage and temporary protection of the newly minted diol as a carbonate through the action of triphosgene and pyridine. After silylation of the remaining hydroxy group to provide **132**, removal of the carbonate group completed the required functional group manipulations to generate **127** in significantly higher overall yield than the original route, despite being two steps longer.

With sufficient amounts of **127** now available, the sequence to **45** was completed in a few additional steps as shown in Scheme 28. First, in order to invert the C-2 stereocenter in ring H to its destined spatial disposition in everninomicin 13,384-1 (**1**), selective protection of the C-3 alcohol as a benzoate ester through the employment of tin acetal methodology[44] was followed by the standard operations of Dess–Martin oxidation[45] and subsequent reduction with Li(t-BuO)$_3$AlH, affording **133** with the desired stereochemistry in 78 % overall yield. Although the latter two transformations are relatively routine in traditional settings, a nearly exhaustive set of oxidizing conditions (such as Swern oxidation, TPAP/NMO, n-Bu_2SnO/Br_2) and reducing agents ($NaBH_4$, $Na(OAc)_3BH$, $LiAlH_4$, K-Selectride, and L-Selectride) were explored essentially in complete combinatorial fashion before this winning set of reagents was identified. Next, having completed its supporting role in these operations, the benzoate ester was then hydrolyzed from **133** and

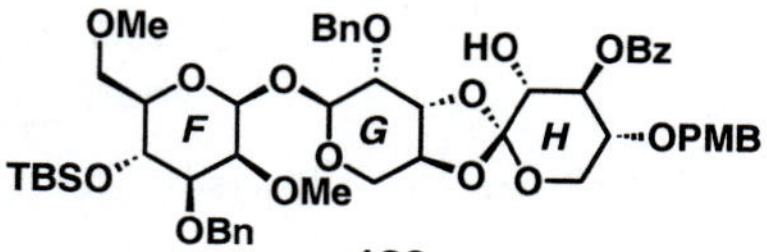

OMe BnO HO OBz F G H OPMB TBSO OMe OBn **133**

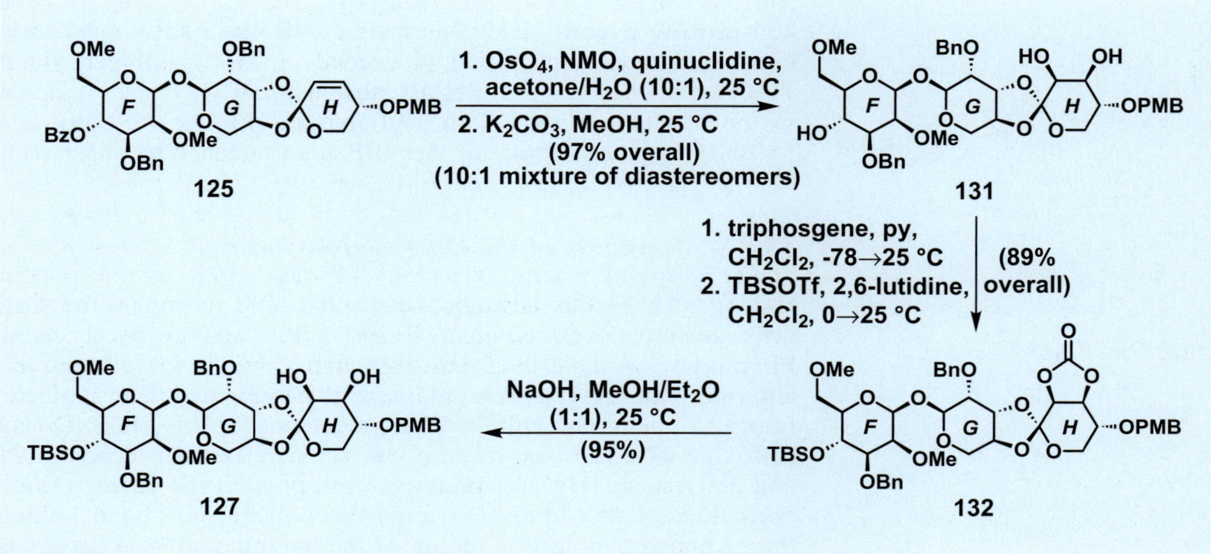

OMe OBn F G H OPMB BzO OMe OBn **125**

1. OsO_4, NMO, quinuclidine, acetone/H_2O (10:1), 25 °C
2. K_2CO_3, MeOH, 25 °C
(97% overall)
(10:1 mixture of diastereomers)

OMe BnO HO OH F G H OPMB HO OMe OBn **131**

1. triphosgene, py, CH_2Cl_2, -78→25 °C
2. TBSOTf, 2,6-lutidine, CH_2Cl_2, 0→25 °C
(89% overall)

OMe BnO F G H OPMB TBSO OMe OBn **132**

NaOH, MeOH/Et_2O (1:1), 25 °C
(95%)

OMe BnO HO OH F G H OPMB TBSO OMe OBn **127**

Scheme 27. Alternative synthesis of the FGH fragment **127**.

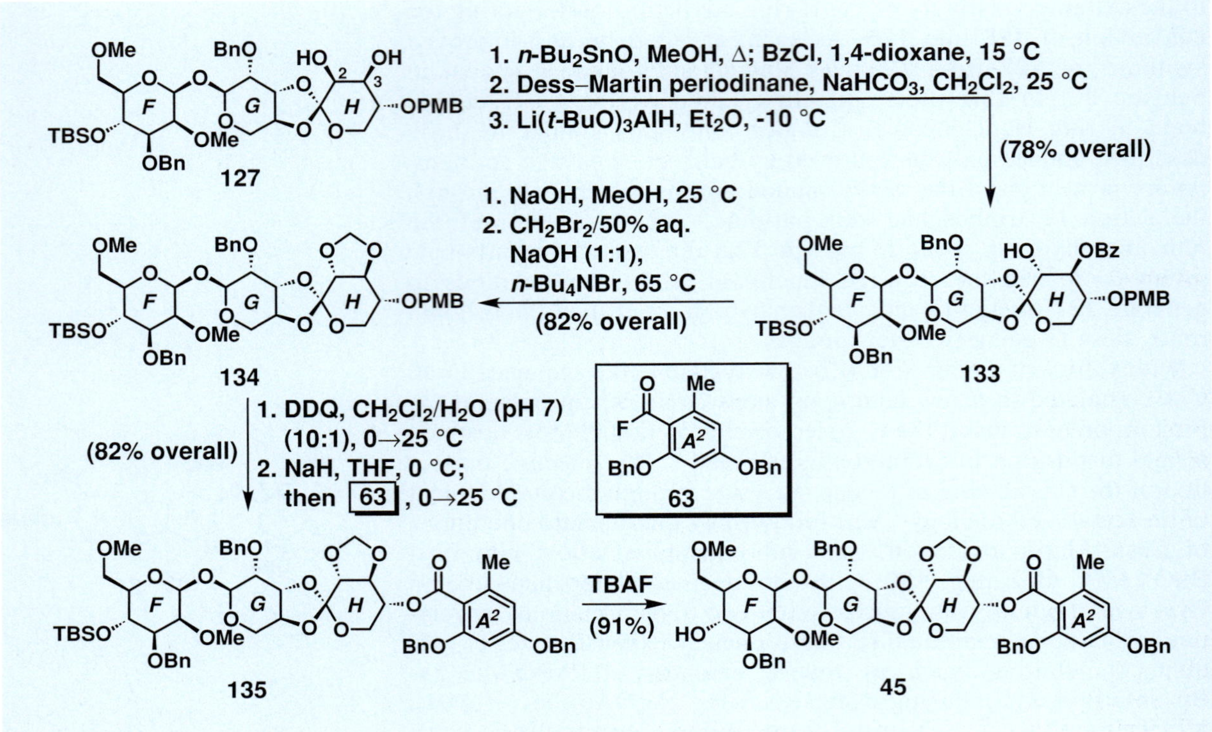

Scheme 28. Completion of the synthesis of the FGHA^2 fragment **45**.

the two adjacent hydroxy groups were bridged by a methylene spacer upon the slow addition of this compound to a mixture of aqueous NaOH, CH_2Br_2, and n-Bu_4NBr at 65 °C,[46] leading to acetal **134** in 82 % overall yield. Intermediate **45** was finally reached in three additional steps which proceeded in 75 % combined yield: 1) selective cleavage of the PMB ether with DDQ, 2) esterification of the resultant alcohol with acyl fluoride **63** (the A^2-ring), and 3) TBAF-induced rupture of the TBS ether attached to ring F.

10.3.3 Synthesis of the DE Fragment

Having reached this advanced stage, the time to pursue the final drive towards everninomicin 13,384-1 (**1**) was nearly at hand. First, however, the critical DE dissacharide bridge to unite the left and right-hand fragments would have to be prepared. The synthetic routes to obtain the carbohydrate precursors for this piece, D-ring sulfoxide **47** and E-ring alcohol **48**, are delineated in Schemes 29 and 30, respectively. In anticipation of applying the Kahne–Crich glycosidation procedure[20] to forge the β-mannoside bond linking the carbohydrates in this region of the everninomicin skeleton, as mentioned earlier, efforts towards **47** commenced from the known

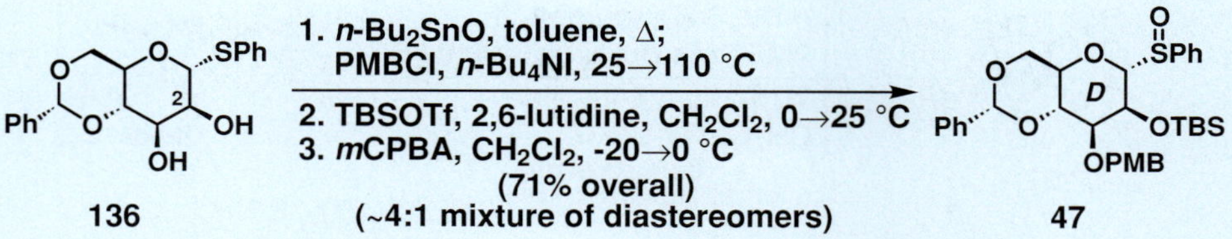

136

1. n-Bu_2SnO, toluene, Δ; PMBCl, n-Bu_4NI, 25→110 °C
2. TBSOTf, 2,6-lutidine, CH_2Cl_2, 0→25 °C
3. mCPBA, CH_2Cl_2, -20→0 °C
(71% overall)
(~4:1 mixture of diastereomers)

47

Scheme 29. Synthesis of the D-ring carbohydrate building block **47**.

mannose derivative **136** (Scheme 29)[47] which already possessed the C-4/C-6 benzylidene protecting group required for execution of the projected glycosidation step. From the latter compound, sulfoxide **47** was readily accessed through selective protection of the C-3 hydroxy group as a PMB ether using an intermediate tin acetal (with selectivity based on equatorial versus axial differentiation), followed by silylation of the remaining alcohol (TBSOTf, 2,6-lutidene) and *m*CPBA-mediated oxidation of the anomeric phenylsulfide to the corresponding sulfoxide. Overall, these three operations proceeded quite smoothly in a combined yield of 71 % from **136**, affording **47** as a 4:1 mixture of chromatographically separable diastereomers.

Efforts towards the E-ring glycosyl acceptor (**48**, Scheme 30) began with a similarly functionalized starting material as that employed for the synthesis of **47**, namely galactose-derived diol **137**.[48] As with many of the other carbohydrate starting materials obtained from the chiral pool for the synthesis of everninomicin, this building block provided all the requisite stereochemical elements desired for the final intermediate. However, several functional and protecting group exchanges were necessary. In the initial operations towards this piece, using chemistry which by now should be quite familiar, the C-3 hydroxy group of **137** was selectively engaged as a PMB ether following the formation of a tin acetal species. The remaining hydroxy moiety was then converted into its TBS-protected congener (**138**). To deoxygenate the C-6 position, the 4,6-*O*-benzylidene moiety was cleaved under conditions employing $Zn(OTf)_2$ and EtSH, an event that was immediately followed by selective formation of a tosylate by reacting the primary alcohol with *p*-TsCl, leading to intermediate **139** in 75 % yield. Upon exposure of **139** to $LiAlH_4$, this newly installed leaving group obligingly departed to afford the deoxygenated product. Upon methylation of the free secondary hydroxy group at C-4, **140** was isolated in 85 % overall yield. Having progressed this far, the chores that remained included alteration of the anomeric position and deprotection of the PMB group at C-3 to provide **48**. These operations proceeded smoothly over the course of three steps: NBS-induced hydrolysis of the phenylthioglycoside, TIPS protection of the resultant lactol, and DDQ-initiated cleavage of

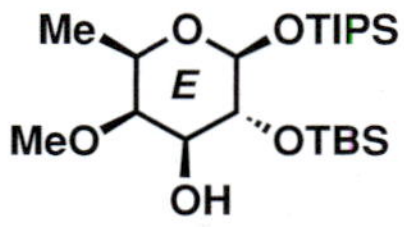

48

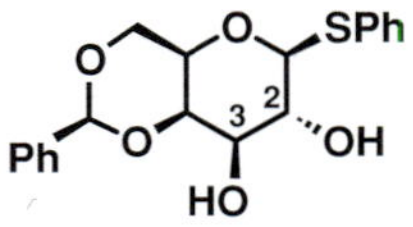

137

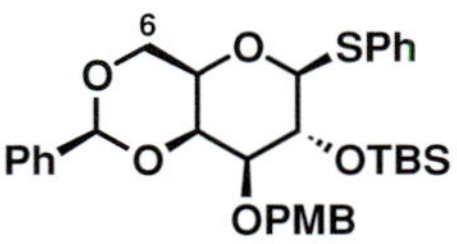

138

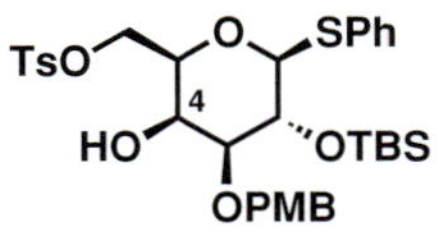

139

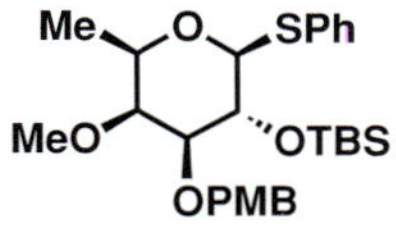

140

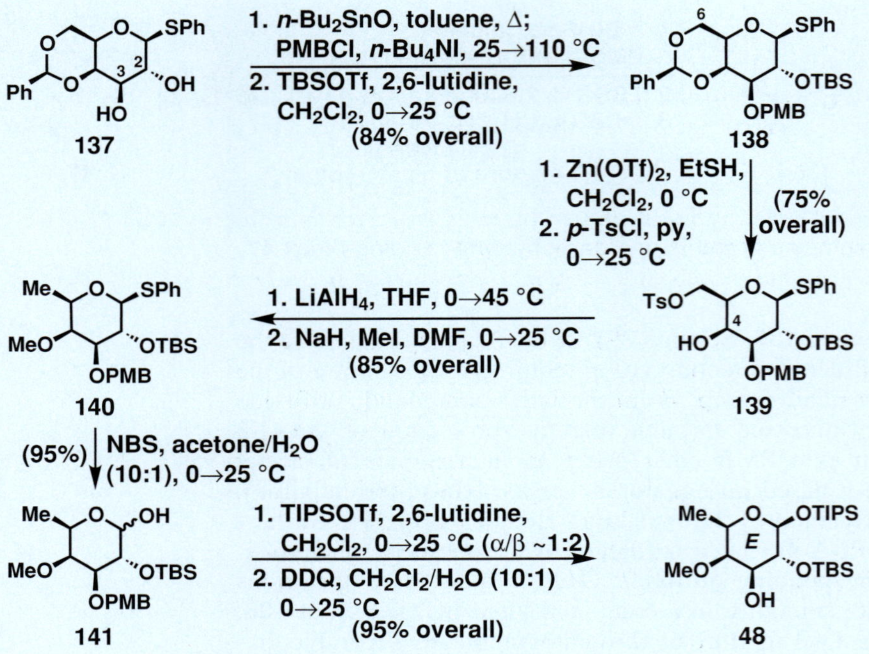

Scheme 30. Synthesis of the E-ring carbohydrate building block **48**.

the C-3 PMB group. Although **48** was ultimately isolated as a 1:2 mixture of α- and β-anomers, separation was not required as both of these diastereomers could be taken through the remainder of the sequence, as will be explained shortly.

The time for joining these freshly molded fragments and elaborating the product to the DE disaccharide unit **44** was now at hand. These efforts would require the stereoselective installation of a methyl group at the C-3 position of ring D. Before this challenge could be tackled, the fragments first needed to be coupled stereoselectively in a β-fashion. Most gratifyingly, exposure of **47** to the conditions of Crich and co-workers (Tf_2O and DTBMP in CH_2Cl_2 at $-78\,°C$), followed by the addition of **48**, smoothly provided a major product bearing exclusively the β-mannoside linkage from which the lone PMB group was subsequently cleaved by the action of DDQ to furnish **142** in 67 % overall yield. With a free hydroxy group now unveiled, it was time to attempt the construction of the quaternary center at C-3 bearing the required methyl and alcohol groups. Thus, to generate an electrophile for nucleophilic interception, the hydroxy group at C-3 was oxidized with TPAP in the presence of NMO. Upon treatment with MeLi in Et_2O at $-78\,°C$, the desired quaternary carbon center in **143** was forged as a single stereoisomer. As discussed earlier in the planning phases of the synthesis, the equatorially oriented β-mannoside bond and the bulky, axially disposed TBS group at the C-2 position in ring D likely provided a decisively biased encounter for the incoming nucleophile during this reaction, with its approach occurring as drawn in the column figure. It is worth noting that in contrast to this analysis, had the glycosidic linkage been of the α-configuration, it is probable that no stereofacial control would have been observed from this reaction due to opposing 1,2- and 1,3-interactions.

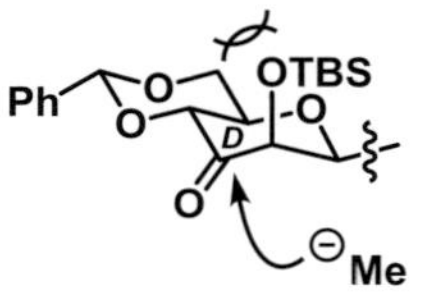

With **143** in hand, the next set of operations sought to deoxygenate the C-6 position of ring D. In line with methods utilized for the same purpose in the synthesis of the E-ring precursor, its benzylidene group was initially cleaved through hydrogenation in EtOAc

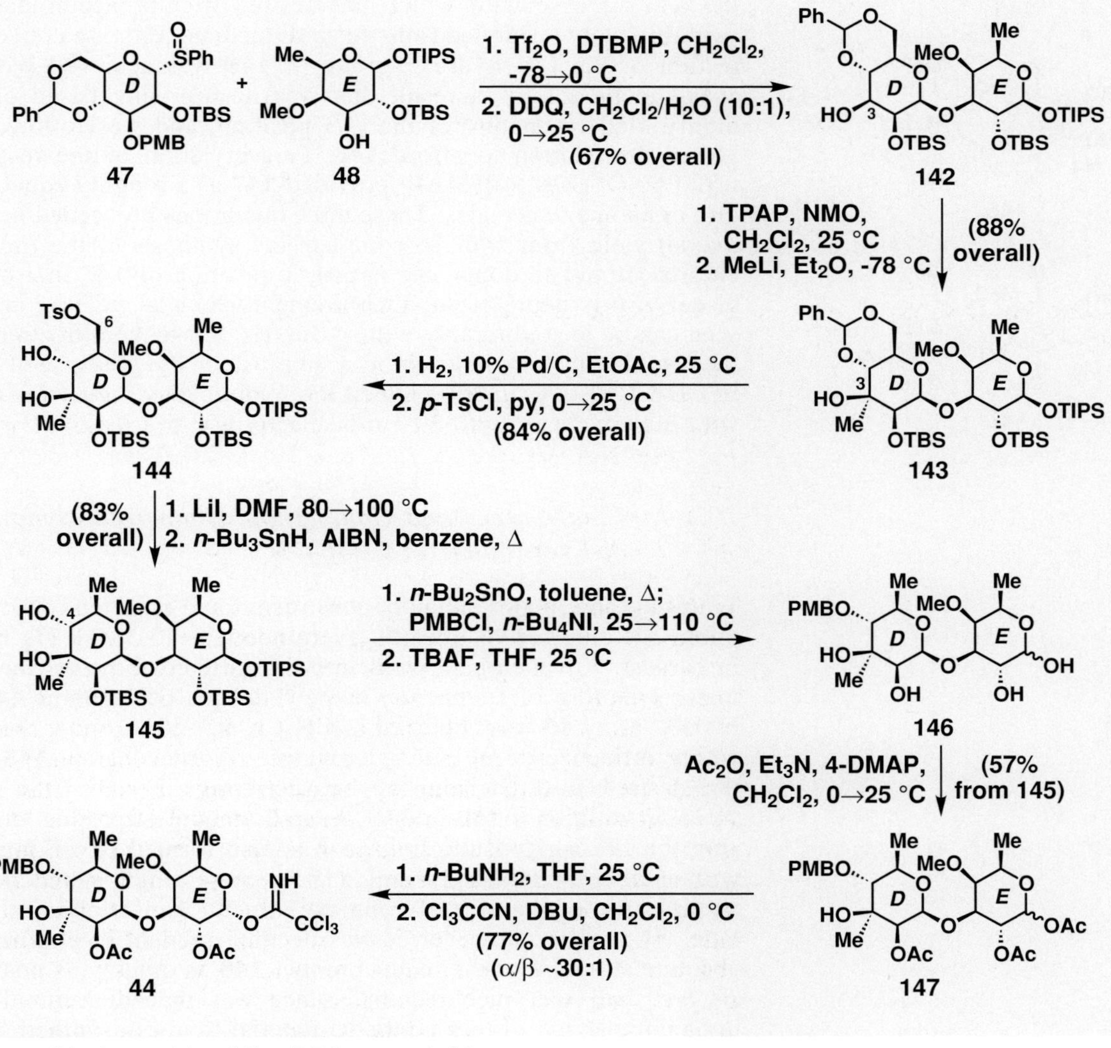

Scheme 31. Assembly of the DE fragment **44**.

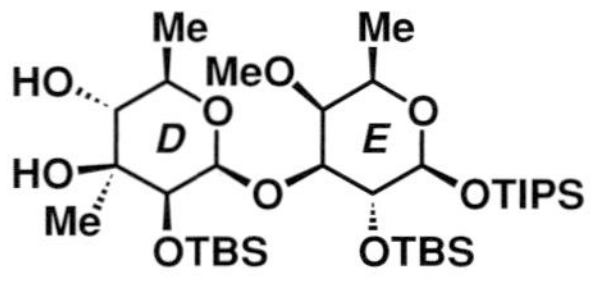

at 25 °C and the resulting primary alcohol was converted into a tosylate. At this stage, however, all attempts to use conventional reducing agents led to decomposition and TBS migration, rather than deoxygenation. To circumvent this minor stumbling block, the tosylate was converted into the corresponding iodide by a Finkelstein reaction (LiI, DMF, 100 °C), and subsequent reduction with n-Bu_3SnH and the radical initiator AIBN afforded **145** in 69 % overall yield from **143**. Having finally succeeded in effecting deoxygenation at C-6, numerous methods were now attempted to protect the free tertiary alcohol, but this task proved unachievable with both standard and unconventional reagent combinations. In the face of these obstacles, it was postulated that this position could remain unprotected throughout the remainder of the sequence. Fortunately, this synthetic gamble turned out to be rewarding, as evidenced by the relative ease with which the desired trichloroacetimidate **44** was ultimately fashioned from this intermediate. First, selective protection of the C-4 hydroxy group of **145** was achieved with tin acetal methodology, a result that was unsurprising based on the highly hindered nature of the C-3 position, and was followed by global desilylation to afford **146**. Peracetylation of the resulting triol (Ac_2O, Et_3N, 4-DMAP) provided **147** as a roughly equal mixture of anomeric acetates. These three operations proceeded in 57 % overall yield from **145**. To complete the synthesis of the final trichloroacetimidate donor, the anomeric position of **147** first had to be selectively deprotected, a requirement which as we have already seen can be met admirably with n-$BuNH_2$. Subsequent treatment of the resulting lactol with trichloroacetonitrile in the presence of DBU in CH_2Cl_2 at 0 °C finally yielded the long-coveted α-anomer of **44** with high exclusivity (>30:1 α/β selectivity).

Me Me HO MeO O D E O HO O OTIPS Me OTBS OTBS **145**

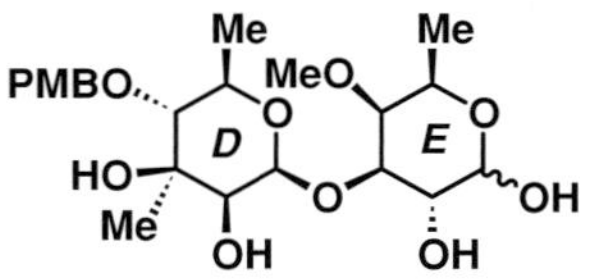

Me Me PMBO MeO O D E O HO O OH Me OH OH **146**

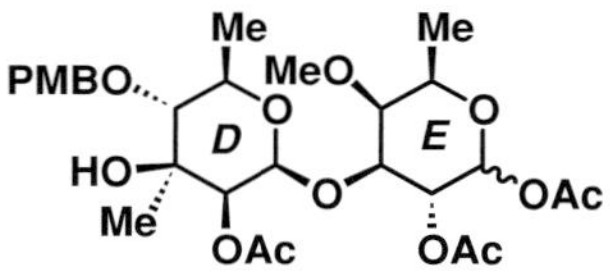

Me Me PMBO MeO O D E O HO O OAc Me OAc OAc **147**

10.3.4 Final Stages and Completion of the Total Synthesis of Everninomicin 13,384-1

With the three major subunits constructed and now ready for coupling, the final drive towards everninomicin 13,384-1 (**1**) began in earnest. Advancing forward into unchartered territory, the first major coupling of fragments, namely that of DE subunit **44** and $FGHA^2$ array **45** was achieved in CH_2Cl_2 at −20 °C under the activating influence of $BF_3 \cdot OEt_2$, providing oligosaccharide **148** with the desired β-stereochemistry between rings E and F in 55 % yield. In addition to this adduct, a small amount of product possessing the EF α-glycosidic linkage was also formed (5 %), together with another unidentified isomer (18 %) possessing β-stereochemistry. Because of the lack of complete structural information of this latter by-product, however, it was deemed prudent to confirm the absolute structure of the major product **148** as quickly as possible. As we shall see, such reconnaissance was indeed accomplished upon comparison of this synthetic material to a semisynthetic sample derived from the natural product, but the fragment would have

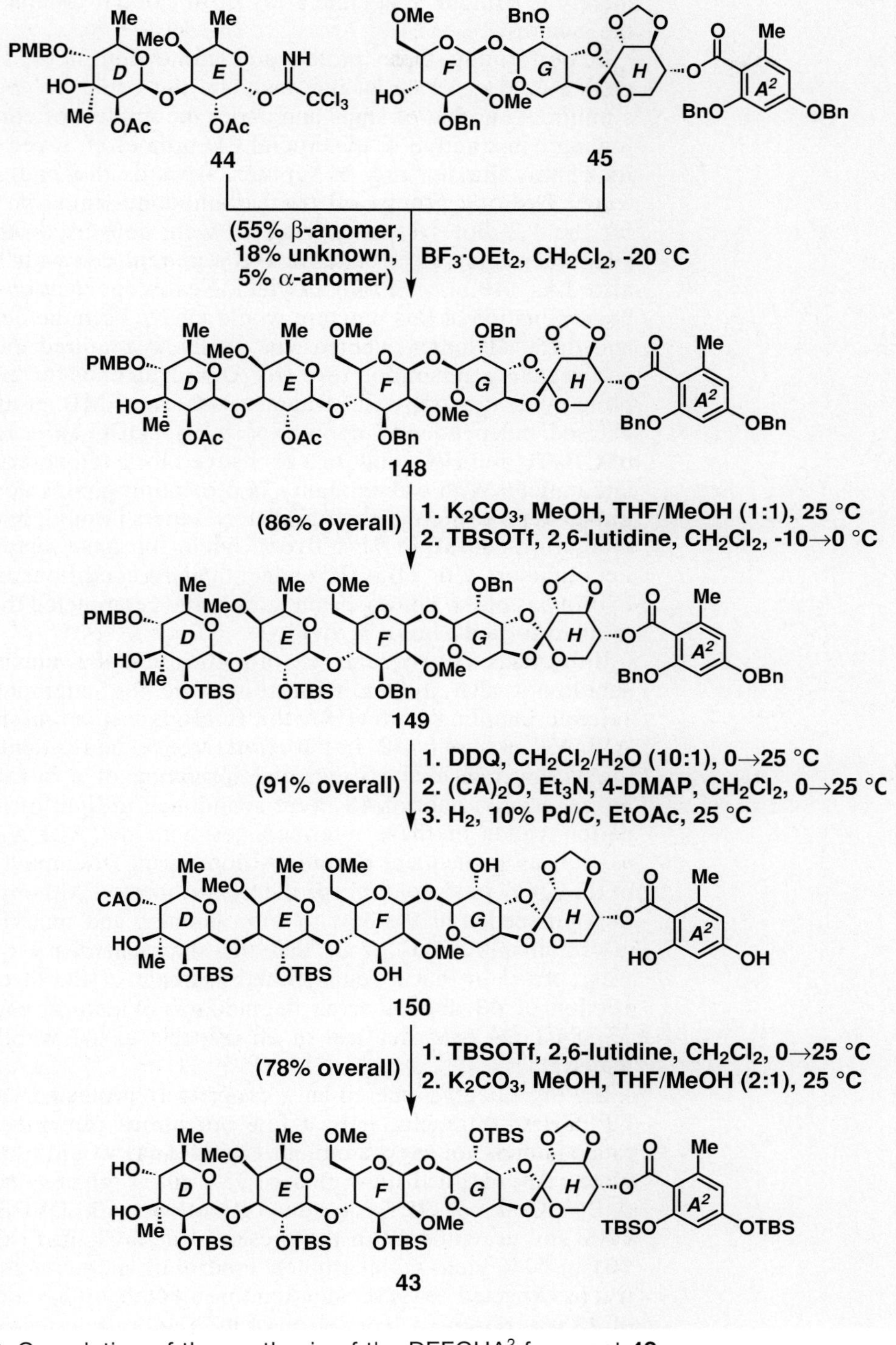
44
45
(55% β-anomer,
18% unknown,
5% α-anomer)
BF3·OEt2, CH2Cl2, -20 °C
148
(86% overall)
1. K2CO3, MeOH, THF/MeOH (1:1), 25 °C
2. TBSOTf, 2,6-lutidine, CH2Cl2, -10→0 °C
149
(91% overall)
1. DDQ, CH2Cl2/H2O (10:1), 0→25 °C
2. (CA)2O, Et3N, 4-DMAP, CH2Cl2, 0→25 °C
3. H2, 10% Pd/C, EtOAc, 25 °C
150
(78% overall)
1. TBSOTf, 2,6-lutidine, CH2Cl2, 0→25 °C
2. K2CO3, MeOH, THF/MeOH (2:1), 25 °C
43

Scheme 32. Completion of the synthesis of the DEFGHA^2 fragment **43**.

to be advanced first to **43** through a series of operations designed merely to provide a suitable array of functional groups to complete the synthesis.

In performing these protection/deprotection steps, the ultimate goal was to generate an intermediate that could be deprotected in a minimal number of steps and under the mildest of conditions following construction of the crucial CD orthoester, given the proclivity of this functionality to rupture. Towards this end, all the protected hydroxy groups on the carbohydrate rings of **148**, except for the 1,2-diol system on ring D, were converted into the same silyl ether. First, the acetates in **148** were replaced with TBS groups, affording **149** in 86 % overall yield. Next, cognizant of the fact that hydrogenation at this juncture would cleave both the benzyl groups and the PMB ether, whereas the synthesis required the controlled and selective generation of a free D-ring alcohol for eventual coupling with the $A^1B(A)C$ fragment **42**, the PMB group was first excised independently upon exposure to DDQ in a 10:1 mixture of CH_2Cl_2 and H_2O, and then exchanged for a temporary chloroacetate moiety. With orthogonality in protecting groups now achieved, global deprotection of the four benzyl ethers through hydrogenation then afforded **150** in 91 % overall yield for these three operations. Persilylation with TBSOTf under standard conditions, followed by hydrolysis of the chloroacetate group, then completed the manipulations to form **43** in 78 % yield.

If the particular sequence to form **43** from **148** seems lengthy, one should note that, in fact, it constituted the third attempt to prepare a suitable coupling partner for the final glycosidation step with the $A^1B(A)C$ assembly **42**. Initial efforts to use the resident functionality within **148**, that is, eventual formation of a hexabenzyl or a hexaacetate variant of **43**, were abandoned in light of low glycosidation yields of these intermediates with the $A^1B(A)C$ fragment as well as subsequent decomposition during attempted application of the Sinaÿ protocol to forge the CD orthoester. Although these studies proceeded at the cost of time, money, and material, they did afford unambiguous proof that the stereochemistry drawn in **43** was correct in that a semisynthetic sample of the hexabenzylated version of **43** derived from degradation of natural everninomicin 13,384-1 (**1**) was identical in all respects to the wholly synthetic material.

Having finally garnered an appropriately protected variant of the $DEFGHA^2$ fragment, only a few operations remained before the total synthesis of everninomicin 13,384-1 (**1**) would finally be complete. As outlined in Scheme 33, initial glycosylation of the $A^1B(A)C$ donor **42** proceeded smoothly with $DEFGHA^2$ hexa-TBS diol acceptor **43** in the presence of $SnCl_2$ in Et_2O, affording **151** in 70 % yield with complete control of anomeric stereochemistry; as expected, no glycosidation through the tertiary alcohol center of **43** was observed. Formation of the final orthoester site was then achieved with a level of facility commensurate to that accomplished with the GH system by oxidation of the selenide to the corresponding selenoxide using $NaIO_4$, followed by heating in a 2:2:1 mixture of vinyl acetate/toluene/*i*-Pr_2NH in a sealed tube at 140 °C for 12 h. The stereoselective nature of this reaction, which afforded the fully protected everninomicin skeleton **152** in 65 % yield, was expected based on the assumed conformation for orthoester formation (**153**, column figure).

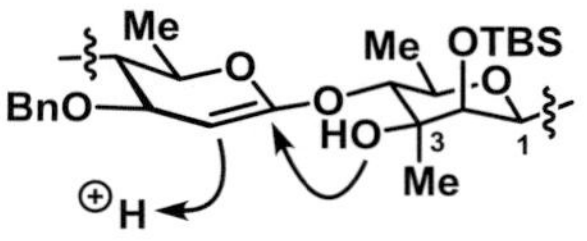

Me Me OTBS BnO O O O HO 3 1 Me ⊕H **153**

(70%) $SnCl_2$, Et_2O, 0→25 °C

42 **43** **151**

(65% overall) 1. $NaIO_4$, $MeOH/CH_2Cl_2/H_2O$ (3:2:1), 25 °C 2. vinyl acetate/toluene/*i*-Pr_2NH (2:2:1), sealed tube, 140 °C, 12 h

152

(75% overall) 1. H_2, 10% Pd/C, $NaHCO_3$, *t*-BuOMe, 25 °C 2. TBAF, THF, 25 °C

1: everninomicin 13,384-1

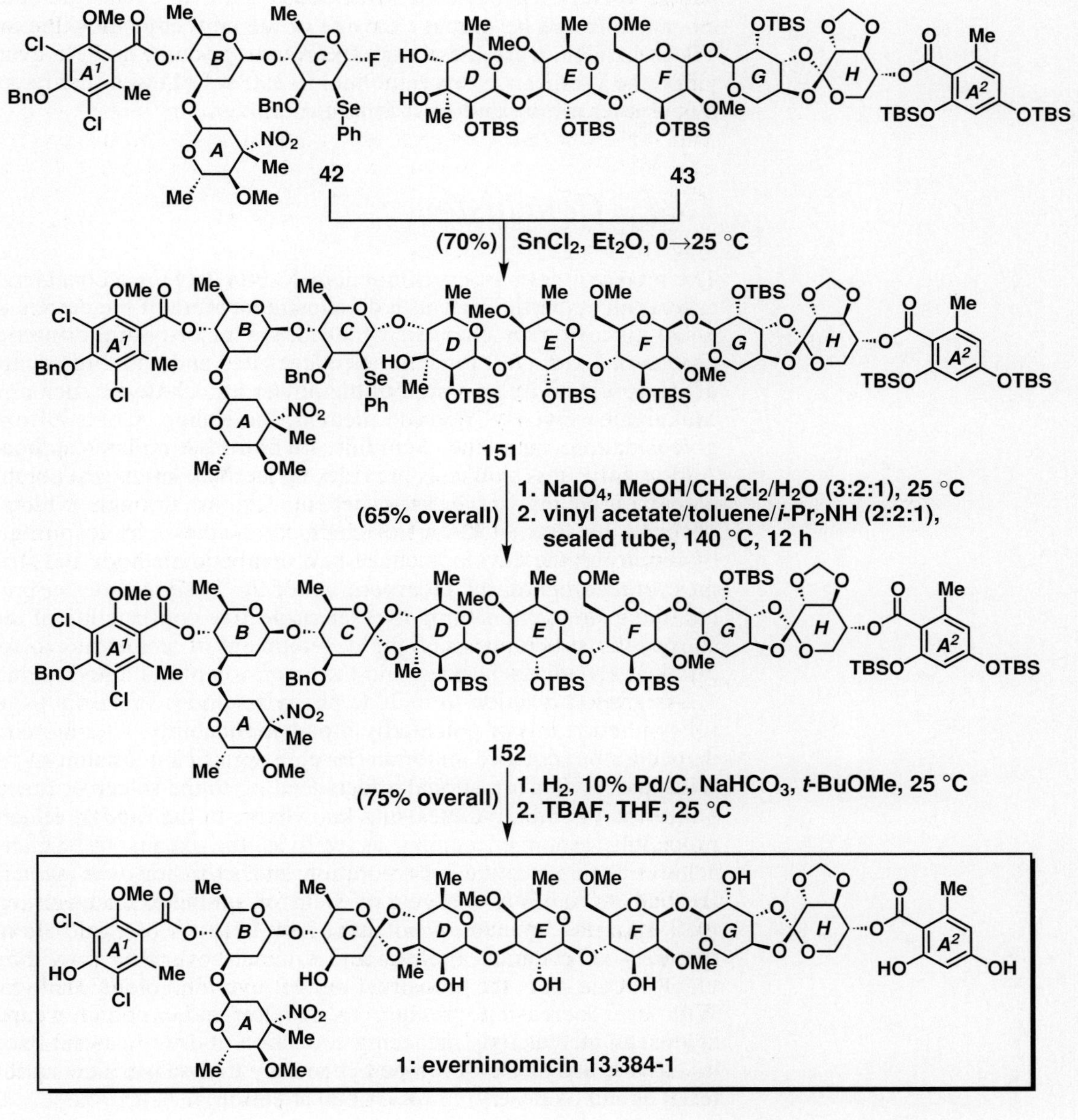

Scheme 33. Final stages and completion of the total synthesis of everninomicin 13,384-1 (**1**).

Although only deprotection steps separated **152** from everninomicin 13,384-1 (**1**), extensive experimentation was required with a gamut of solvents, catalysts, buffers, and other typical experimental parameters in order to define a suitable set of reaction conditions that could effect these operations without damaging the sensitive CD orthoester site and/or the fickle chloro and nitro groups. Eventually it was determined that the optimal conditions to complete the target molecule (**1**), and avoid these problems, were initial hydrogenation in $NaHCO_3$-buffered *t*-BuOMe at 25 °C to excise the benzyl ethers, followed by global cleavage of the silyl ethers using excess TBAF in THF. With these steps executed smoothly in 75 % overall yield, the synthesis of everninomicin 13,384-1 (**1**) was finally completed after a long and hard-fought campaign.

10.4 Conclusion

The total synthesis of everninomicin 13,384-1 (**1**) is a considerable achievement, particularly as a diagnostic measure of the current status of glycosidation chemistry. Indeed, the majority of the anomeric centers in **1** were fashioned in excellent yield and with high degrees of stereoselectivity by virtue of the power of techniques such as the Mukaiyama glycosyl fluoride method, the Kahne–Crich sulfoxide glycosidation, and the Schmidt trichloroacetimidate approach. Additionally, this synthesis provides further validation that complex natural products serve as excellent forums through which to advance the state of the art of chemical synthesis by inspiring, or by requiring, the development of new synthetic methods and strategies. In this regard, the incorporation of the Sinaÿ orthoester protocol into a strategy enabling the stereoselective construction of these spirocyclic structural units, the development of techniques to form β-linked 1,1′-disaccharides, and the extension of the DAST-induced 1,2-migration reaction to include phenylselenides stand out as useful synthetic tools of potentially broad applicability. At a more fundamental, but equally important level, a significant amount of light was shed on conformational effects leading to the selective functionalization of carbohydrates, and knowledge in the field of effective protecting group ensembles as well as the means by which to achieve their selective incorporation and extrusion was garnered. Beyond its value to the general field of synthetic chemistry, this work has also opened up potential new avenues of exploration in the areas of chemical biology and drug discovery by providing a viable route for the construction of everninomicin analogues. With ever-increasing pressure to develop more potent weapons against an increasingly menacing collection of drug-resistant bacterial strains, use of the developed chemistry to discover new antibacterial agents is deserving of serious attention.[49]

References

1. For some lead references in this area, see: a) *Carbohydrates in Chemistry and Biology, Part I: Chemistry of Saccharides, Vol. 1–2* (Eds.: B. Ernst, G.W. Hart, P. Sinaÿ), Wiley-VCH, Weinheim, **2000**; b) *Carbohydrates in Chemistry and Biology, Part II: Biology of Saccharides, Vol. 3–4* (Eds.: B. Ernst, G.W. Hart, P. Sinaÿ), Wiley-VCH, Weinheim, **2000**; c) C.R. Bertozzi, L.L. Kiessling, *Science* **2001**, *291*, 2357; d) K.J. Yarema, C.R. Bertozzi, *Curr. Opin. Chem. Biol.* **1998**, *2*, 49; e) C.-H. Wong, *Pure & Appl. Chem.* **1997**, *69*, 419.
2. a) K.C. Nicolaou, R.M. Rodríguez, H.J. Mitchell, H. Suzuki, K.C. Fylaktakidou, O. Baudoin, F.L. van Delft, *Chem. Eur. J.* **2000**, *6*, 3095; b) K.C. Nicolaou, H.J. Mitchell, R.M. Rodríguez, K.C. Fylaktakidou, H. Suzuki, *Chem. Eur. J.* **2000**, *6*, 3116; c) K.C. Nicolaou, H.J. Mitchell, R.M. Rodríguez, K.C. Fylaktakidou, H. Suzuki, S.R. Conley, *Chem. Eur. J.* **2000**, *6*, 3149; d) K.C. Nicolaou, H.J. Mitchell, H. Suzuki, R.M. Rodríguez, O. Baudoin, K.C. Fylaktakidou, *Angew. Chem.* **1999**, *111*, 3523; *Angew. Chem. Int. Ed.* **1999**, *38*, 3334; e) K.C. Nicolaou, R.M. Rodríguez, K.C. Fylaktakidou, H. Suzuki, H.J. Mitchell, *Angew. Chem.* **1999**, *111*, 3529; *Angew. Chem. Int. Ed.* **1999**, *38*, 3340; f) K.C. Nicolaou, H.J. Mitchell, R.M. Rodríguez, K.C. Fylaktakidou, H. Suzuki, *Angew. Chem.* **1999**, *111*, 3535; *Angew. Chem. Int. Ed.* **1999**, *38*, 3345. For earlier model studies, see: K.C. Nicolaou, R.M. Rodríguez, H.J. Mitchell, F.L. van Delft, *Angew. Chem.* **1998**, *110*, 1975; *Angew. Chem. Int. Ed.* **1998**, *37*, 1874.
3. a) B.G. Davis, *J. Chem. Soc., Perkin Trans. 1* **2000**, 2137; b) G.-J. Boons, *Contemp. Org. Synth.* **1996**, *3*, 173; c) G.-J. Boons, *Tetrahedron* **1996**, *52*, 1095; d) R.R. Schmidt in *Comprehensive Organic Synthesis, Vol. 6* (Eds.: B.M. Trost, I. Fleming), Pergamon, Oxford, **1994**, pp. 33–64; e) K. Toshima, K. Tatsuta, *Chem. Rev.* **1993**, *93*, 1503. For a more specialized review of glycosidations in the context of total synthesis, see: K. C. Nicolaou, H. J. Mitchell, *Angew. Chem.* **2001**, *113*, 1624; *Angew. Chem. Int. Ed.* **2001**, *40*, 1576.
4. W. Koenigs, E. Knorr, *Ber. Dtsch. Chem. Ges.* **1901**, *34*, 957.
5. D.H. Brauns, *J. Am. Chem. Soc.* **1923**, *45*, 833.
6. T. Mukaiyama, Y. Murai, S. Shoda, *Chem. Lett.* **1981**, 431.
7. a) Wm. Rosenbrook, Jr., D.A. Riley, P.A. Lartey, *Tetrahedron Lett.* **1985**, *26*, 3; b) G.A. Olah, J.T. Welch, Y.D. Vankar, M. Nojima, I. Kerekes, J.A. Olah, *J. Org. Chem.* **1979**, *44*, 3872; c) T. Mukaiyama, Y. Hashimoto, S. Shoda, *Chem. Lett.* **1983**, 935; d) Y. Araki, K. Watanabe, F.-H. Kuan, K. Itoh, N. Kobayashi, Y. Ishido, *Carbohydr. Res.* **1984**, *127*, C5; e) H. Kunz, W. Sager, *Helv. Chim. Acta* **1985**, *68*, 283; f) S. Hashimoto, M. Hayashi, R. Noyori, *Tetrahedron Lett.* **1984**, *25*, 1379.
8. Y. Takahashi, T. Ogawa, *Carbohydr. Res.* **1987**, *164*, 277.
9. a) G.H. Posner, S.R. Haines, *Tetrahedron Lett.* **1985**, *26*, 5; b) M. Burkart, Z. Zhang, S.-C. Hung, C.-H. Wong, *J. Am. Chem. Soc.* **1997**, *119*, 11743.
10. K.C. Nicolaou, T. Ladduwahetty, J.L. Randall, A. Chucholowski, *J. Am. Chem. Soc.* **1986**, *108*, 2466.
11. K.C. Nicolaou, R.E. Dolle, D.P. Papahatjis, J.L. Randall, *J. Am. Chem. Soc.* **1984**, *106*, 4189.
12. R.R. Schmidt, J. Michel, *Angew. Chem.* **1980**, *92*, 763; *Angew. Chem. Int. Ed. Engl.* **1980**, *19*, 731. Interestingly, the parent glycosyl acetimidates originally introduced by Sinaÿ do not work as well as the trichloroacetimidate congeners: a) J.-R. Pougny, P. Sinaÿ, *Tetrahedron Lett.* **1976**, *17*, 4073; b) J.-R. Pougny, J.-C. Jacquinet, M. Nassr, D. Duchet, M.-L. Milat, P. Sinaÿ, *J. Am. Chem. Soc.* **1977**, *99*, 6762; c) P. Sinaÿ, *Pure & Appl. Chem.* **1978**, *50*, 1437.
13. R.R. Schmidt, *Angew. Chem.* **1986**, *98*, 213; *Angew. Chem. Int. Ed. Engl.* **1986**, *25*, 212.
14. W.R. Roush, R.A. Hartz, D.J. Gustin, *J. Am. Chem. Soc.* **1999**, *121*, 1990.
15. W.R. Roush, X.-F. Lin, *J. Am. Chem. Soc.* **1995**, *117*, 2236.
16. P.J. Garegg, *Adv. Carb. Chem. Biochem.* **1997**, *52*, 179.
17. R.J. Ferrier, R.W. Hay, N. Vethaviyasar, *Carbohydr. Res.* **1973**, *27*, 55.
18. K.C. Nicolaou, M.R. Pavia, S.P. Seitz, *J. Am. Chem. Soc.* **1982**, *104*, 2027; K.C. Nicolaou, S.P. Seitz, M.R. Pavia, *J. Am. Chem. Soc.* **1982**, *104*, 2030.
19. a) K.C. Nicolaou, M.E. Duggan, C.-K. Hwang, *J. Am. Chem. Soc.* **1986**, *108*, 2468; b) K.C. Nicolaou, C.V.C. Prasad, C.-K. Hwang, M.E. Duggan, C.A. Veale, *J. Am. Chem. Soc.* **1989**, *111*, 5321.
20. a) D. Kahne, S. Walker, Y. Cheng, D. Van Engen, *J. Am. Chem. Soc.* **1989**, *111*, 6881; b) D. Crich, S. Sun, *J. Am. Chem. Soc.* **1998**, *120*, 435.
21. D. Crich, H. Li, Q. Yao, D.J. Wink, R.D. Sommer, A.L. Rheingold, *J. Am. Chem. Soc.* **2001**, *123*, 5826.
22. a) R.U. Lemieux, S. Levine, *Can. J. Chem.* **1964**, *42*, 1473; b) R.U. Lemieux, B. Fraser-Reid, *Can. J. Chem.* **1965**, *43*, 1460; c) K. Tatsuta, K. Fujimoto, M. Kinoshita, S. Umezawa, *Carbohydr. Res.* **1977**, *54*, 85; d) J. Thiem, H. Karl, J. Schwentner, *Synthesis* **1978**, 696; e) J. Thiem, H. Karl, *Tetrahedron Lett.* **1978**, 4999.
23. a) S.J. Danishefsky, D.M. Armistead, F.E. Wincott, H.G. Selnick, R. Hungate, *J. Am. Chem. Soc.* **1987**, *109*, 8117; b) S.J. Danishefsky, H.G. Selnick, D.M. Armistead, F.E. Wincott, *J. Am. Chem. Soc.* **1987**, *109*, 8119; c) S.J. Danishefsky, D.M. Armistead, F.E. Wincott, H.G. Selnick, R. Hungate, *J. Am. Chem. Soc.* **1989**, *111*, 2967.

24. For reviews on the orthosomicin class, see: a) P. Juetten, C. Zagar, H. D. Scharf, *Recent Prog. Chem. Synth. Antibiot. Relat. Microb. Prod.* **1993**, 475 and references cited therein; b) D. E. Wright, *Tetrahedron* **1979**, *35*, 1207.
25. a) A. K. Ganguly, B. Pramanik, T. C. Chan, O. Sarre, Y.-T. Liu, J. Morton, V. F. Girijavallabhan, *Heterocycles* **1989**, *28*, 83; b) J. A. Maertens, *Curr. Opin. Anti-Infect. Invest. Drugs* **1999**, *1*, 49; c) J. A. Maertens, *IDrugs* **1999**, *2*, 446.
26. a) L. Belova, T. Tenson, L. Xiong, P. M. McNicholas, A. S. Mankin, *Proc. Natl. Acad. Sci. U.S.A.* **2001**, *98*, 3726; b) F. M. Aarestrup, L. B. Jensen, *Antimicrob. Agents Chemother.* **2000**, *44*, 3425; c) H. Wolf, *FEBS Lett.* **1973**, *36*, 181.
27. For initial chemical modifications of everninomicin 13,384-1 and the resultant biological activity of these analogs, see: a) A. K. Ganguly, V. M. Girijavallabhan, G. H. Miller, O. Z. Sarre, *J. Antibiot.* **1982**, *35*, 561; b) A. K. Ganguly, O. Sarre, S. Szmulewicz, U. S. Patent **1975**, 3,920,629 [*Chem. Abstr.* **1976**, *84*, 90526]; c) A. K. Ganguly, V. M. Girijavallabhan, O. Sarre, H. Reimann, U. S. Patent **1978**, 4,129,720 [*Chem. Abstr.* **1979**, *90*, 168928]; d) A. K. Ganguly, O. Sarre, U. S. Patent **1975**, 3,915,956 [*Chem. Abstr.* **1976**, *84*, 122237]; e) A. K. Ganguly, J. L. McCormick, T.-M. Chan, K. Saksena, P. R. Das, *Tetrahedron Lett.* **1997**, *38*, 7989; f) A. K. Ganguly, J. L. McCormick, T.-M. Chan, A. K. Saksena, P. R. Das, U. S. Patent **1998**, 5,763,600 [*Chem. Abstr.* **1998**, *129*, 397317]; g) A. K. Ganguly, J. L. McCormick, A. K. Saksena, P. R. Das, T.-M. Chan, *Bioorg. Med. Chem. Lett.* **1999**, *9*, 1209.
28. For some representative examples, see: a) K. C. Nicolaou, N. Watanabe J. Li, J. Pastor, N. Winssinger, *Angew. Chem.* **1998**, *110*, 1636; *Angew. Chem. Int. Ed.* **1998**, *37*, 1559; b) K. C. Nicolaou, T. J. Caulfield, H. Kataoka, N. A. Stylianides, *J. Am. Chem. Soc.* **1990**, *112*, 3693.
29. M. Trumtel, P. Tavecchia, A. Veyrières, P. Sinaÿ, *Carbohydr. Res.* **1990**, *202*, 257.
30. a) P. Jütten, H.-D. Scharf, G. Raabe, *J. Org. Chem.* **1991**, *56*, 7144; b) J. Dornhagen, H.-D. Scharf, *Tetrahedron* **1985**, *41*, 173; c) P. Jütten, J. Dornhagen, H.-D. Scharf, *Tetrahedron* **1987**, *43*, 4133.
31. a) A. David in *Preparative Carbohydrate Chemistry* (Ed.: S. Hanessian), Marcel Dekker, Inc., New York, **1997**, pp. 69–83; b) T. B. Grindley, *Adv. Carb. Chem. Biochem.* **1998**, *53*, 17.
32. a) K. C. Nicolaou, F. L. van Delft, S. R. Conley, H. J. Mitchell, Z. Jin, R. M. Rodríguez, *J. Am. Chem. Soc.* **1997**, *119*, 9057; b) K. C. Nicolaou, K. C. Fylaktakidou, H. J. Mitchell, F. L. van Delft, R. M. Rodríguez, S. R. Conley, Z. Jin, *Chem. Eur. J.* **2000**, *6*, 3166.
33. P. S. Bailey, J. E. Keller, *J. Org. Chem.* **1968**, *33*, 2680.
34. M. Hirama, I. Nishizaki, T. Shigemoto, S. Itô, *J. Chem. Soc., Chem. Commun.* **1986**, 393.
35. a) J. E. Baldwin, G. A. Höfle, O. W. Lever, Jr., *J. Am. Chem. Soc.* **1974**, *96*, 7125; b) R. K. Boeckman, Jr., K. J. Bruza, *J. Org. Chem.* **1979**, *44*, 4781.
36. A. Ya Chernyak, K. V. Antonov, N. K. Kochetkov, *Biorg. Khim.* **1989**, *15*, 1113.
37. C. Czernecki, K. Vijayakumaran, G. Ville, *J. Org. Chem.* **1986**, *51*, 5472.
38. G. Solladié, A. Rubio, M. C. Carreño, J. L. García Ruano, *Tetrahedron: Asymmetry* **1990**, *1*, 187. For a review of the Gattermann formylation reaction, see: W. E. Truce in *Organic Reactions, Vol. 9* (Eds.: R. Adams, et al.), John Wiley & Sons, London, **1957**, pp. 37–72.
39. L. A. Carpino, A. El-Faham, *J. Am. Chem. Soc.* **1995**, *117*, 5401.
40. S. Mehta, B. M. Pinto, *J. Org. Chem.* **1993**, *58*, 3269.
41. a) A. Dondoni, G. Fantin, M. Fogagnolo, A. Medici, P. Pedrini, *J. Org. Chem.* **1988**, *53*, 1748; b) A. Dondoni, A. Marra, D. Perrone, *J. Org. Chem.* **1993**, *58*, 275.
42. J. C. Martin, R. J. Arhart, *J. Am. Chem. Soc.* **1971**, *93*, 4327.
43. a) F. He, Y. Bo, J. D. Altom, E. J. Corey, *J. Am. Chem. Soc.* **1999**, *121*, 6771; b) E. J. Corey, S. Sarshar, M. D. Azimioara, R. Newbold, M. C. Noe, *J. Am. Chem. Soc.* **1996**, *118*, 7851.
44. X. Wu, F. Kong, *Carbohydr. Res.* **1987**, *162*, 166.
45. a) D. B. Dess, J. C. Martin, *J. Org. Chem.* **1983**, *48*, 4155; b) D. B. Dess, J. C. Martin, *J. Am. Chem. Soc.* **1991**, *113*, 7277; c) S. D. Meyer, S. L. Schreiber, *J. Org. Chem.* **1994**, *59*, 7549.
46. K. S. Kim, W. A. Szarek, *Synthesis* **1978**, 48.
47. T. Oshitari, M. Shibasaki, T. Yoshizawa, M. Tomita, K. Takao, S. Kobayashi, *Tetrahedron* **1997**, *53*, 10993.
48. K. C. Nicolaou, C. W. Hummel, Y. Iwabuchi, *J. Am. Chem. Soc.* **1992**, *114*, 3126.
49. For an instructive review of this growing area of research, see: T. K. Ritter, C.-H. Wong, *Angew. Chem.* **2001**, *113*, 3616; *Angew. Chem. Int. Ed.* **2001**, *40*, 3508.

1: trichodimerol

2: bisorbicillinol

11

K. C. Nicolaou (1999)

Bisorbicillinoids

11.1 Introduction

Although our ability to synthesize molecules has progressed dramatically over the course of the past several decades, organic synthesis is still in its infancy compared to Nature's deft preparation of a dazzling variety of complex molecules. Taking the same number and types of constituent atoms, Nature combines them with seemingly limitless variation, in the process creating molecular architectures that not even the most imaginative of chemists could be expected to conjure on their own. Within this collection of diverse and interesting structures, however, is a special group of natural products whose domains suggest a likely biogenetic origin as dimerization of suitably activated monomers; an assortment of such compounds (**3**–**8**) is shown in Scheme 1.[1] Targets such as these have always been eyed by synthetic chemists with a unique level of admiration because of their inherent aesthetic appeal, since they often possess an axis or point of symmetry. Equally significant, they typically offer an unparalleled degree of challenge to the practitioner in that effecting the critical dimerization step has historically constituted one of the most difficult transformations to achieve in the laboratory. In this chapter, we will analyze in detail a set of impressive dimerization reactions that led to several members of the bisorbicillinoid family of natural products, such as the cage-like C_2-symmetric compound trichodimerol (**1**) and the unique dodecaketide bisorbicillinol (**2**), as developed by the Nicolaou and Corey research groups in 1999.[2,3] Not only was keen theoretical and experimental insight required to achieve success, but, even more impressively, these seemingly disparate natural products were synthesized from the same monomeric

Key concepts:

- **Dimerization reactions**
- **Biomimetic synthesis**
- **Cascade reactions**

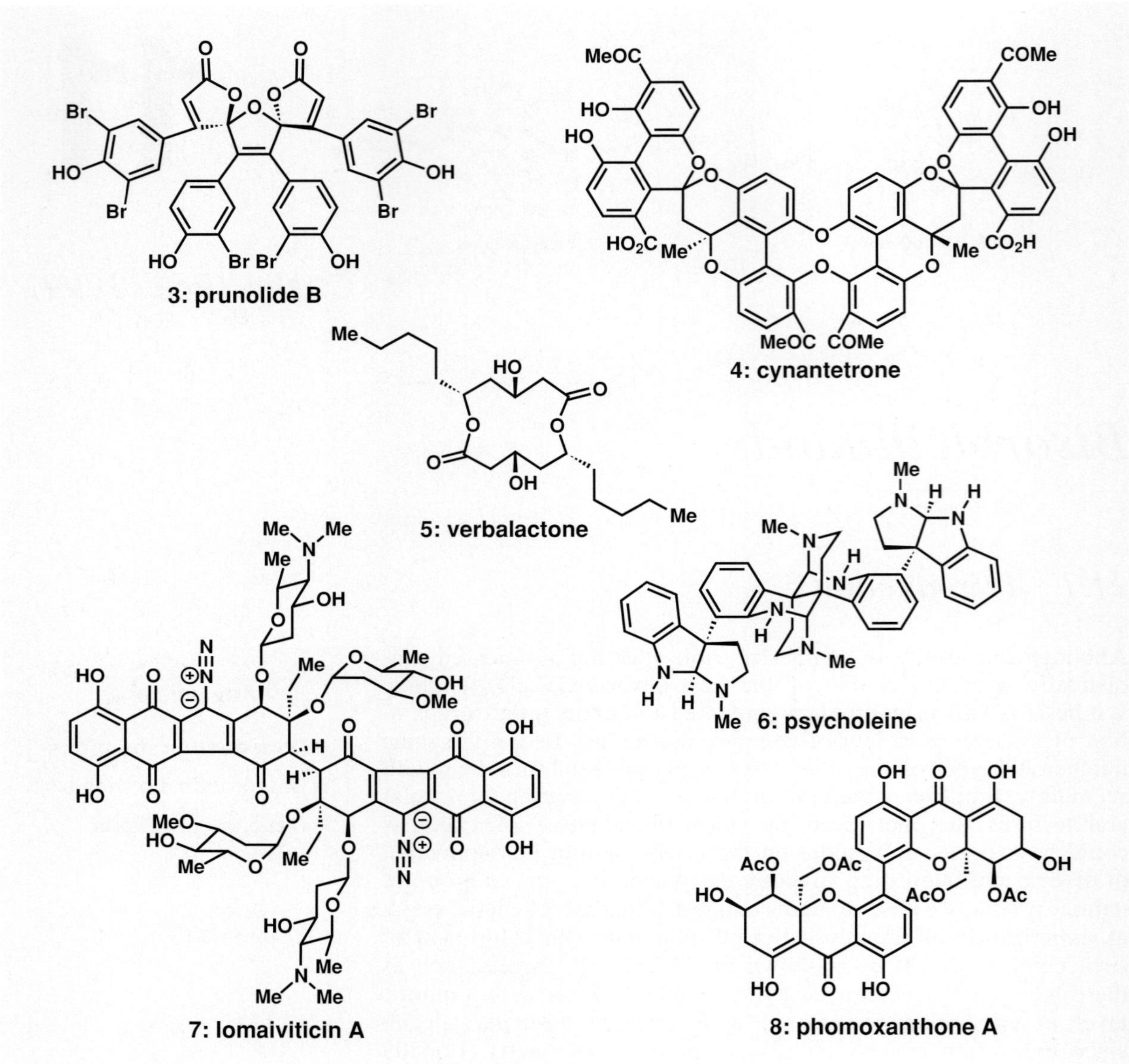

3: prunolide B
4: cynantetrone
5: verbalactone
6: psycholeine
7: lomaiviticin A
8: phomoxanthone A

Scheme 1. Structures of some novel natural products whose architectures can be formally considered as dimeric.

starting material through subtle modification of reaction conditions!

As a prelude to this discussion, we thought that it would be instructive to set the stage briefly with two examples of recent dimerization-based total syntheses in order to establish some of the inherent challenges in these approaches to construct molecular complexity. We start with the natural product hybocarpone (**17**, Scheme 2), a C_2-symmetric compound whose biosynthetic origin

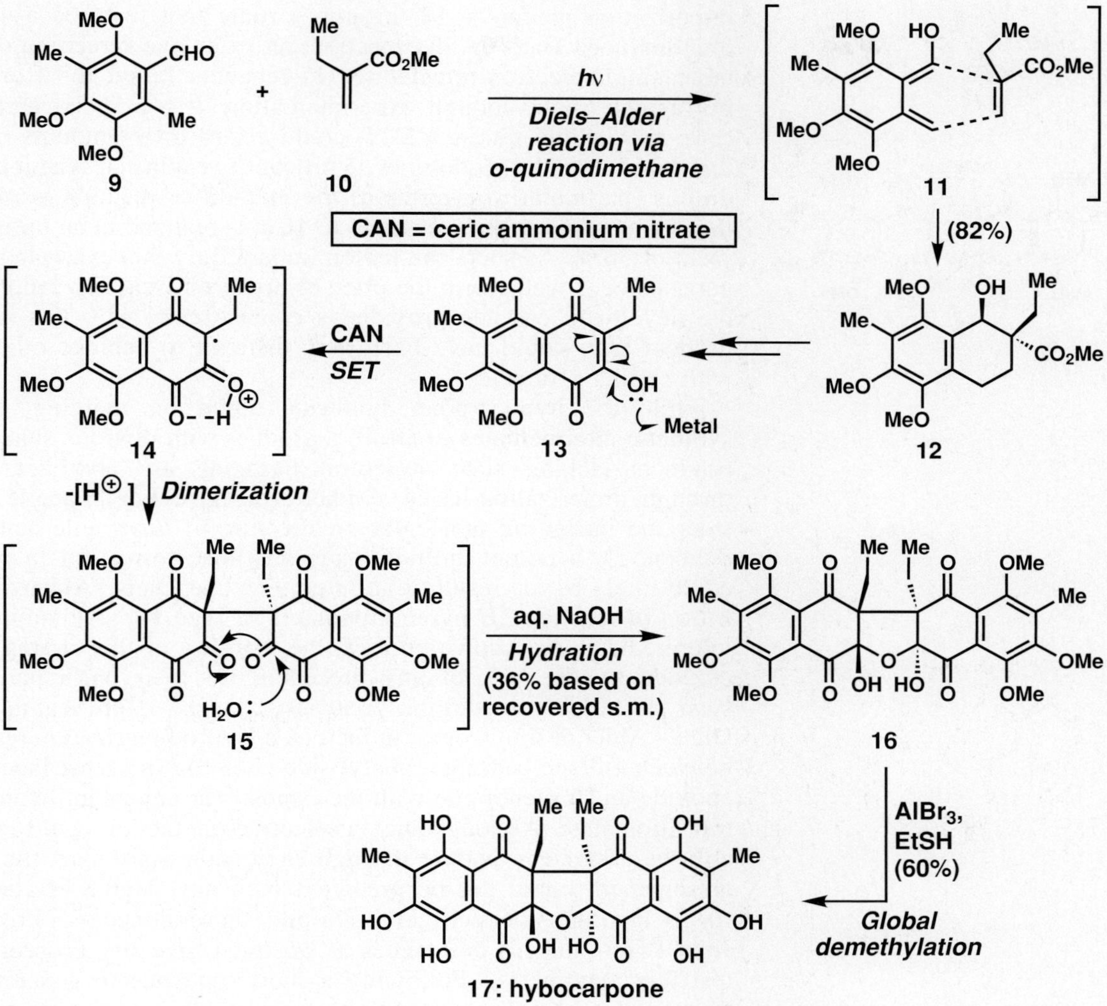

Scheme 2. Dimerization-based total synthesis of hybocarpone (**17**). (Nicolaou, 2001)[4]

could conceivably be the radical-based dimerization of a highly reactive intermediate such as **14** (obtained from a precursor naphtharazin such as **13**) to afford **15**, followed by a hydration event.[4] Although this bold process forming a hindered carbon–carbon bond and installing four stereogenic centers is easily presented on paper, putting strategies like this one into practice is often incredibly difficult. Not only must one find/develop a suitable method to generate reliably a fleeting monomeric intermediate such as **14**, but that species must additionally be formed under appropriate reaction conditions (including solvent polarity, overall concentration, and temperature) to induce it to enter into the desired dimerization pathway. These issues engendered great frustration for the Nicolaou group in this synthesis. Following the preparation of dimerization precursor **13** through a route that featured a unique photoinduced Diels–Alder reaction, an extensive screening of suitable single-electron-transfer (SET) reagents failed to afford **16**. Fortunately, with enough experimentation, it was discovered that ceric ammonium nitrate (CAN) could afford trace amounts of this dimerized material; following significant tweaking of reaction conditions (particularly in terms of the method of reaction work-up), the desired dimerization cascade to **16** was realized in an optimized yield of 36 %. As such, the lesson imparted by this example is that persistence is well worth the price of numerous reaction failures, as the developed cascade provided a concise entry into this natural product that would have been very difficult to achieve otherwise with similar efficiency.

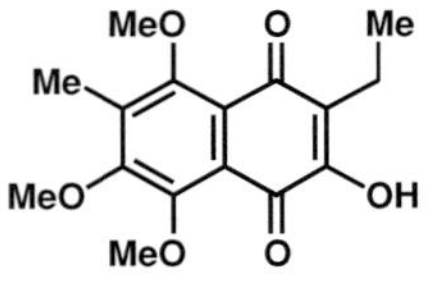

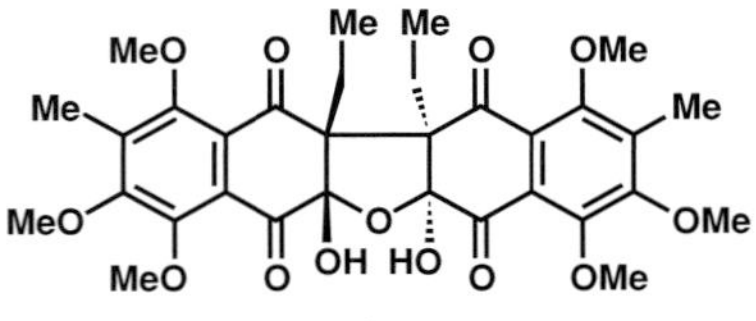

While the hybocarpone synthesis relied on merging achiral synthetic intermediates to afford a product with obvious symmetry, products lacking such structural harmony are also accessible through dimerization-based synthetic designs. For example, consider the intriguing molecular architecture of torreyanic acid (**21**, Scheme 3), a potent antitumor agent whose formation in Nature could likely be the result of an intermolecular Diels–Alder dimerization of reactive 2*H*-pyran monomers of type **19**.[5] Although such intermediates are known entities, the viability of this strategy was clouded by the lack of any precedent for their participation in such [4+2] cycloaddition reactions. More significantly, the Diels–Alder design would, in fact, be a heterodimerization reaction between a diene bearing a pentyl side chain on the same face as its epoxide and a dienophile with the opposite arrangement in an *endo* transition state. Although such a selective interaction could reasonably be assured in Nature through enzymatic assistance, the same arrangement could not be predicted with any degree of certainty for a laboratory experiment. Despite these concerns, Professor John Porco and his colleagues at Boston University proceeded to test this hypothesis.[6] Following a short sequence to generate **18**, these researchers discovered that, upon exposure of this compound to Dess–Martin periodinane and subsequent standing on silica gel for 1 hour to induce a domino oxidation/6π-electron electrocyclization event leading to reactive intermediate **19**, racemic **20** was indeed isolated from the reaction mixture in 39 % yield along with another dimer (not shown) corresponding to an *exo*-based Diels–Alder reaction (41 % yield). Although the desired isomer (**20**) was not formed exclusively, the elegance of the approach and the extension of this [4+2] cycloaddition into previously unknown contexts defines new frontiers for such dimerization reactions and enhances the viability of the original biosynthetic hypothesis for this unique epoxyquinoid natural product.

With a better understanding of the issues associated with dimerization-based syntheses, we shall now proceed with a discussion of the molecules comprising the main theme of this chapter.

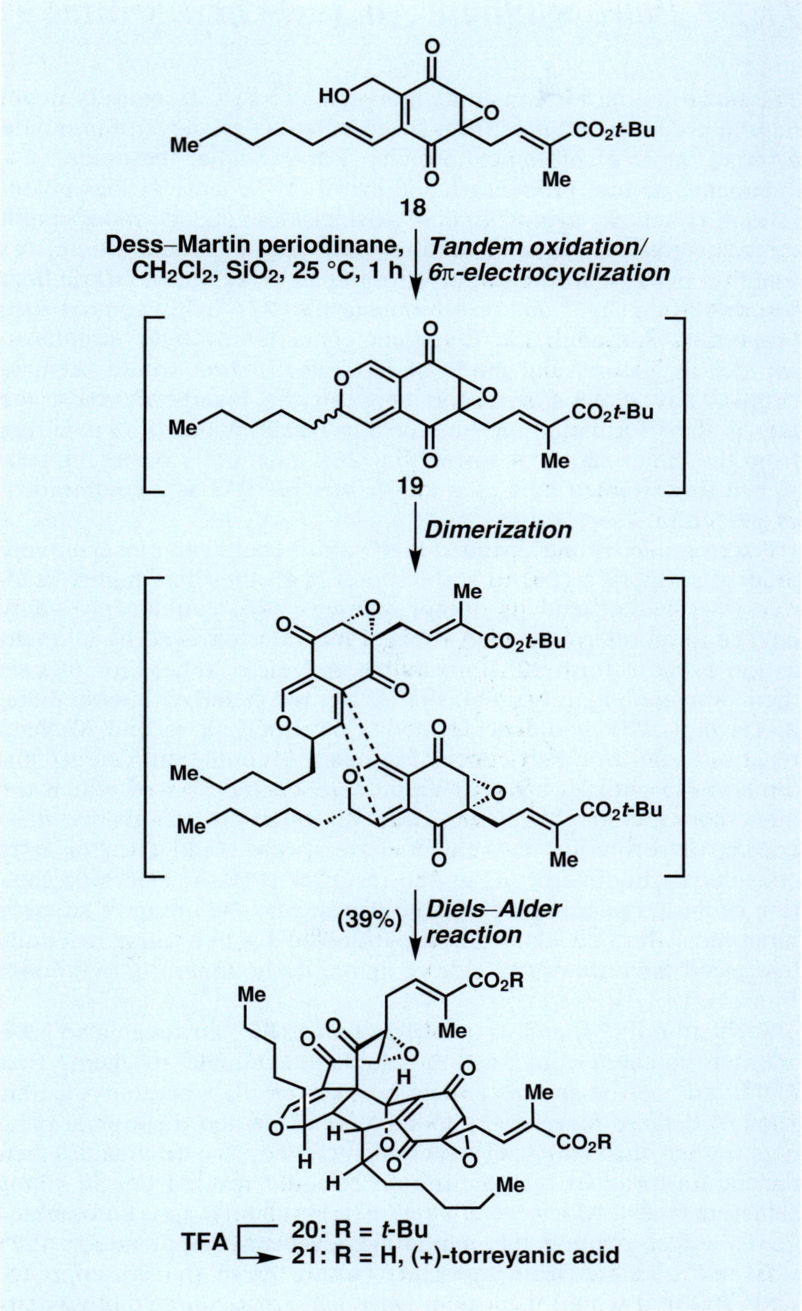

Scheme 3. Biomimetically inspired total synthesis of (±)-torreyanic acid (**21**). (Porco et al., 2000)[6]

11.2 Retrosynthetic Analysis and Strategy

The bisorbicillinoids constitute a diverse class of structurally novel natural products obtained from several distinct species of fungi with a broad range of biological activity. For example, the unique C_2-symmetric natural product trichodimerol (**1**, Scheme 4)[7] has potent inhibitory activity against lipopolysaccharide-induced production of tumor necrosis factor α, bisvertinol (**23**)[8] and its structural relatives function in the inhibition of β-(1,6)-glucan biosynthesis, while both bisorbicillinol (**2**)[7c,9] and bisorbibutenolide (**27**)[9] exhibit antioxidant properties. Although the divergent character of their structures, sources in Nature, and mode of biological action would seem to conceal any direct connection between the bisorbicillinoids, the fact is their formation can universally be rationalized as resulting from the dimerization of sorbicillin (**26**), a naturally occurring substance first isolated by Cram and Tishler in 1948 as a contaminant of penicillin.[10]

For example, if one assumed that Nature could enantioselectively oxidize sorbicillin (**26**) to sorbicillinol (**24**), then this highly reactive intermediate and its quinol tautomer (**25**) could conceivably engage in an intermolecular Michael reaction followed by a ketalization event to form **22**. If a suitable enzyme or other biomolecule then delivered hydrogen to site A of this putative intermediate, bisvertinol (**23**) would be formed. Conversely, a second Michael reaction/ketalization sequence along path B could complete trichodimerol (**1**) with its array of eight stereocenters, six of which are quaternary. As an interesting aside, although not strictly a dimerization of sorbicillin-derived monomers, one could also formally consider trichodimerol (**1**) as the result of a [4+4] electrocyclization of an intermediate of type **29** (formed by the union of suitable monomers through ketalization), followed by two aldol reactions to append the unsaturated side chains of trichodimerol (see column figures).[2c]

Sorbicillinol (**24**) and its quinol tautomer (**25**), however, also have another potential dimerization pathway available to them, first identified by Abe and his collaborators from the University of Shizuoka.[9] If these pieces were to serve as diene and dienophile partners rather than Michael reaction/ketalization participants, then the natural product bisorbicillinol (**2**) could result from an *endo*-selective Diels–Alder event which would establish four new stereogenic centers. Since the natural product bisorbibutenolide (**27**) was also obtained from the same culture broth (*trichoderma* sp. USF-2690), it would then seem reasonable to presume that this latter secondary metabolite could be derived from **2** through a ring-contraction reaction involving one of the tertiary alcohol groups and the proximal carbonyl group as defined by the mechanistic arrows in Scheme 4. Overall, this rearrangement should be an entropically favored process since it would be attended by a significant relief in the ring strain and overall steric crowding of **2**.

Aldol reaction
1: trichodimerol
28 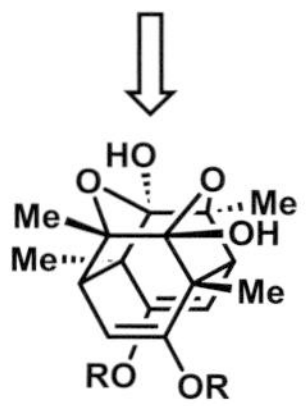 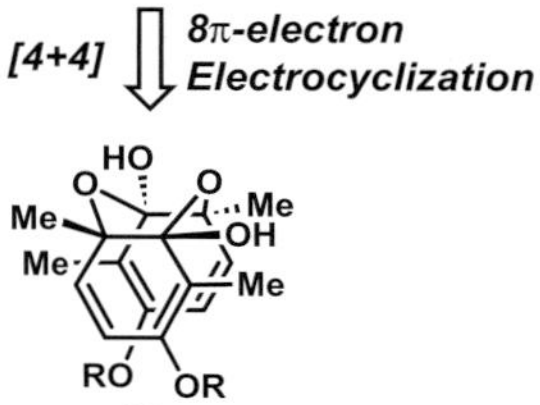
[4+4] 8π-electron Electrocyclization
29

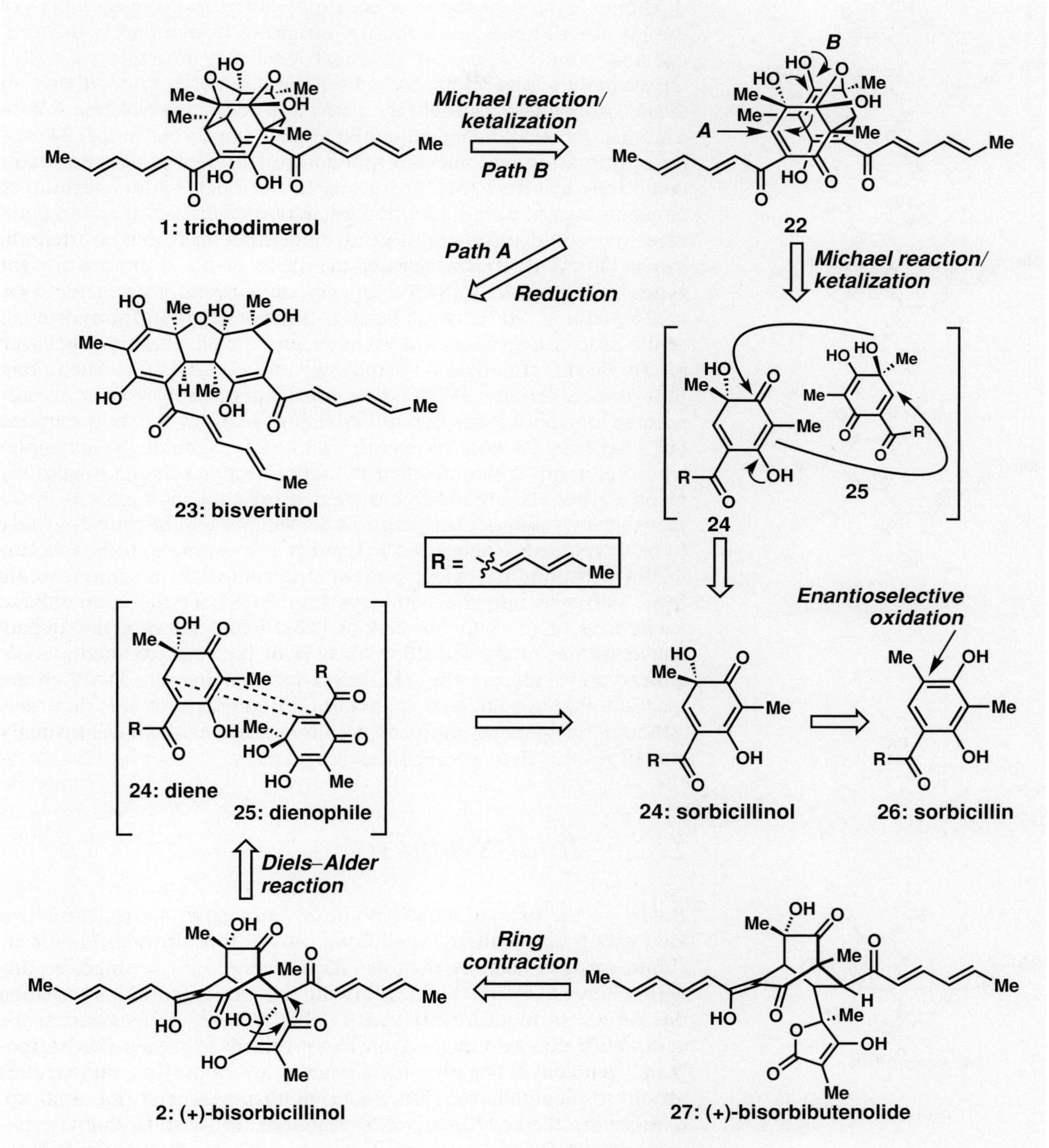

1: trichodimerol

Michael reaction/ ketalization

Path B

22

Path A

Reduction

Michael reaction/ ketalization

23: bisvertinol

24

25

R =

Enantioselective oxidation

24: diene

25: dienophile

24: sorbicillinol

26: sorbicillin

Diels–Alder reaction

Ring contraction

2: (+)-bisorbicillinol

27: (+)-bisorbibutenolide

Scheme 4. Proposed biogenesis of members of the bisorbicillinoids from sorbicillin (**26**).

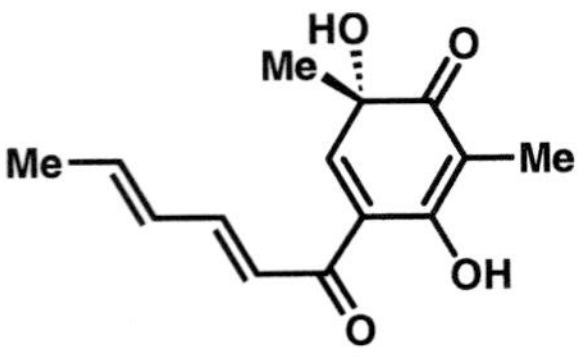

Although the biogenetic connections between these members of the bisorbicillinoids are certainly intriguing from a purely theoretical standpoint,[11] at a more practical level they provide reasonable retrosynthetic blueprints for an efficient laboratory synthesis of these compounds. As such, the pathways defined in Scheme 4 indicate that one would need a method to generate sorbicillinol (**24**) and its equilibrating tautomer **25** in a controlled manner under suitable conditions to effect their merger either in Diels–Alder fashion to generate bisorbicillinol (**2**), or a ketalization/Michael reaction pathway that could ultimately afford bisvertinol (**23**) and/or trichodimerol (**1**). With routes already available in the literature for the synthesis of sorbicillin (**26**),[12] its conversion into a protected form of **24** (such as **30**, column figure), followed by careful hydrolysis of the protecting group, would thus seem to be a workable approach to initiate the dimerization sequences that could lead to such complex dodecaketides. While the same experimental issues already pointed out above for the dimerization steps in the hybocarpone (**17**, Scheme 2) and torreyanic acid (**21**, Scheme 3) examples would certainly be confronted by adopting this scheme toward the bisorbicillinoids, an additional element of challenge existed in the fact that two unique dimerization conditions would probably have to be developed to access both **1** and **2**. For example, basic reaction media with minimal water present (to promote ketalization) might favor entrance into the pathways implicated for the formation of trichodimerol (**1**) and bisvertinol (**23**), while altering the overall concentration of **24** and **25** or the pH of the reaction media could conceivably enhance the likelihood of inducing the Diels–Alder reaction that would lead to bisorbicillinol (**2**). As we shall see, although this general approach would indeed bear fruit, many unexpected results also accompanied the journey.

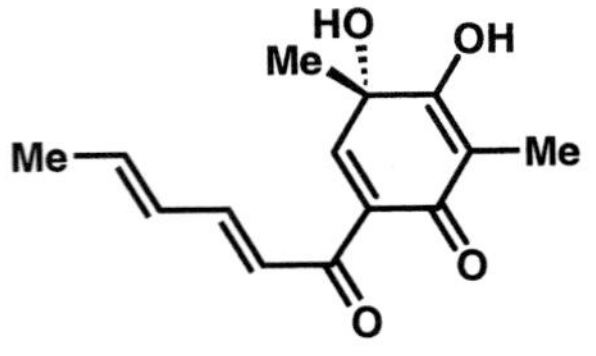

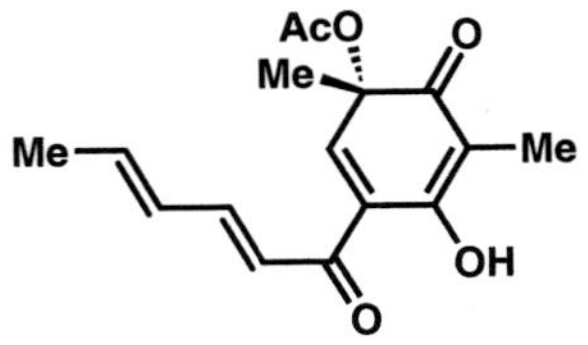

11.3 Total Synthesis

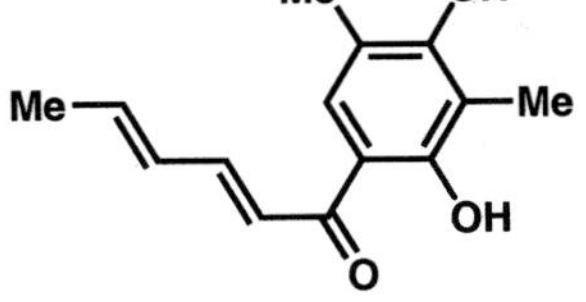

Based on the biosynthetic hypotheses delineated above, the Nicolaou group began their expeditions toward the bisorbicillinoids by synthesizing α-acetoxy dienone **30** (Scheme 5),[2] an objective that was achieved in 40 % yield following numerous failures by treating the known sorbicillin (**26**)[12] with dry $Pb(OAc)_4$ in degassed acetic acid. With this key material in hand, efforts to mimic the biosynthetic pathways began with attempts to hydrolyze the acetate group to generate the reactive quinol monomers (**24** and **25**) required for the key dimerization reactions. Most gratifyingly, exposure of a 0.05 M solution of racemic **30** in THF/H_2O (9:1) to 10 equivalents of solid KOH at 0 °C for 2 hours, followed by an aqueous HCl quench, led to (±)-bisorbicillinol (**2**) in 40 % yield. Although the diastereocontrol observed in the formation of **2** would seem to require an *endo* selective Diels–Alder cycloaddition, attempts were made to verify this reaction course by performing the transformation again in deuterated solvent to enable its analysis by ^{1}H NMR spectroscopy. As expected, this study revealed that exposure of **30** to base caused deacetylation, generating a quinolate system which rapidly equilibrated to a roughly equal mixture (3:2) of two distinct species (presumably the anionic forms of **24** and **25**). Upon acidification of the reaction media, the fleeting quinols were generated which then united immediately in a Diels–Alder reaction to form **2** (as only signals corresponding to the cycloaddition product were observed).

Although this experiment indicated the likely mechanistic pathway for the formation of this polycyclic natural product, the more

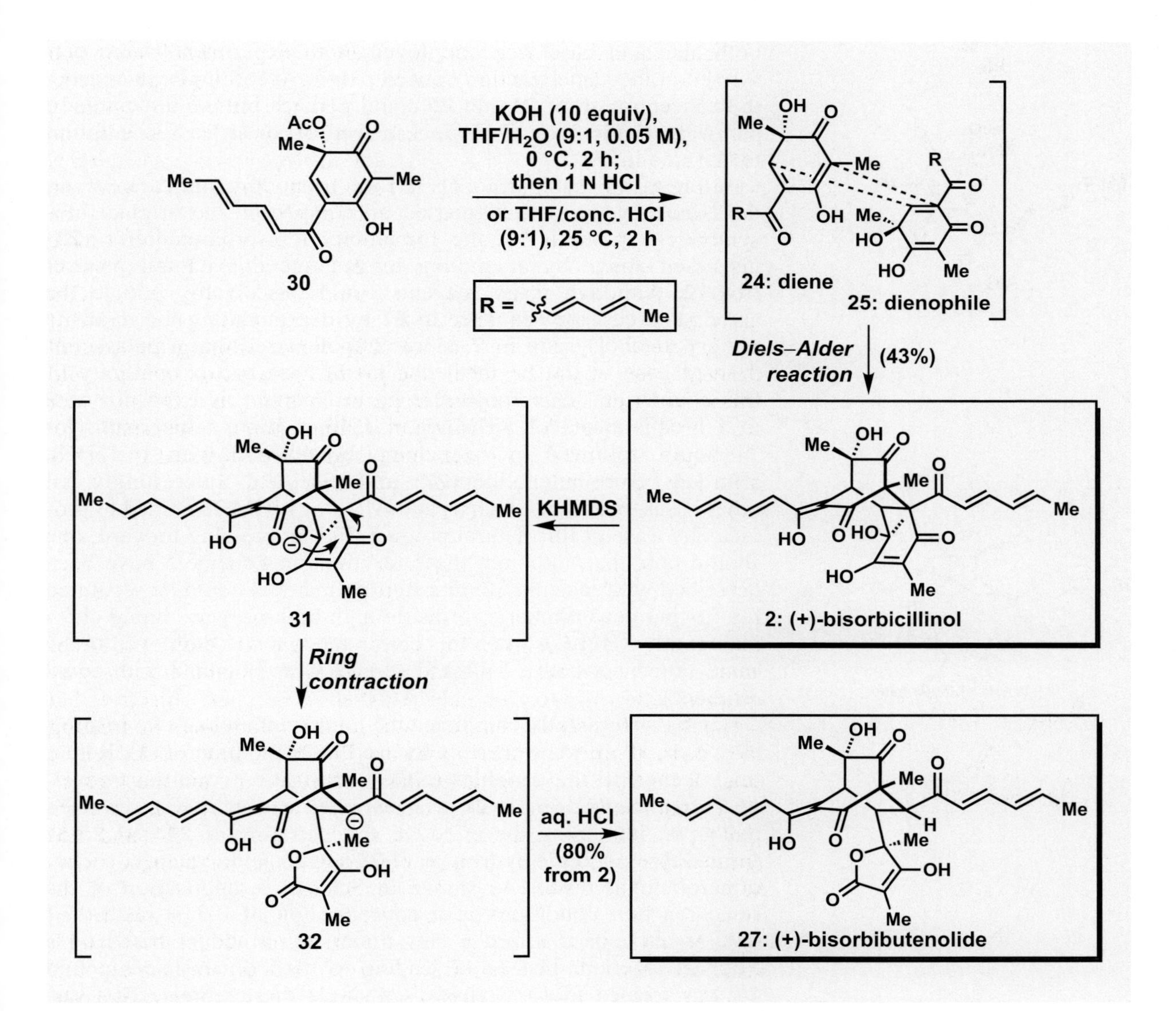

Scheme 5. Biomimetic total synthesis of (+)-bisorbicillinol (**2**) and (+)-bisorbibutenolide (**27**).

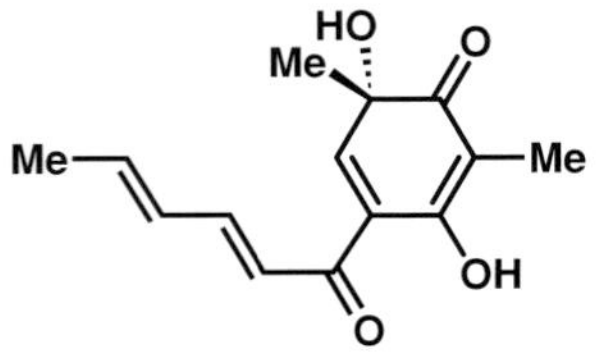

significant finding was the appearance of **2** only after the reaction mixture had been acidified. As such, this observation would seem to suggest that the free quinols **24** and **25**, and not their quinolate counterparts, are the active participants entering into any dimerization reaction. Thus, bisorbicillinol (**2**) should also be formed simply through acidic hydrolysis of the acetate group in **30**, as the reactive quinols **24** and **25** would be generated directly. In line with this expectation, treatment of **30** with concentrated HCl in THF at ambient temperature indeed afforded **2** in 43 % yield. Surprisingly, however, the formation of trichodimerol (**1**) was not observed from either of these initial sets of conditions, even though both quinols **24** and **25** would undoubtedly be required to generate this natural product. While these protocols were relatively disparate in that both acid and base were employed, both experiments were performed at the same reaction concentration. Accordingly, increasing the concentration of **24** and **25** could perhaps initiate dimerization pathways to this natural product instead of, or at least in addition to, **2** (vide infra).

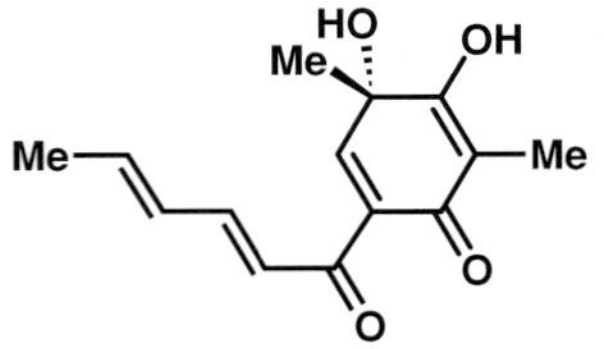

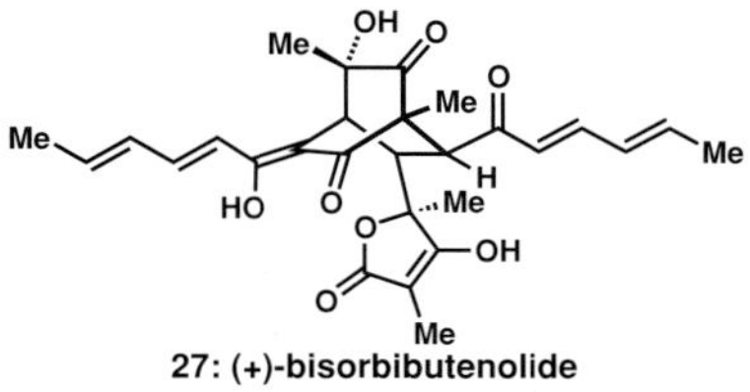

Before testing such protocols to form trichodimerol (**1**), however, the Nicolaou group first focused on completing the original biosynthetic hypothesis for the formation of bisorbibutenolide (**27**) discussed above by attempting to generate this natural product from **2**. Although many reagents could conceivably initiate the anticipated cascade sequence to **27** by deprotonating the requisite tertiary alcohol group in **2**, it was hypothesized that a potassium-derived base would be ideal due to the size and oxophilicity of this counterion. This conjecture proved correct as exposure of **2** to 1.1 equivalents of KHMDS in THF at room temperature for 24 hours, followed by quenching the reaction with 1 N HCl, afforded bisorbibutenolide (**27**) in 80 % yield. Interestingly, all other bases examined (such as NaHMDS or LiHMDS) failed to provide any trace of this natural product. Before pressing forward, one should note that, although these reactions in Scheme 5 have been described with racemic **30**, this starting material could be separated into its pure enantiomeric forms through high-pressure liquid chromatography (HPLC); when this chiral material was employed in the same sequence, both (+)-**2** and (+)-**27** were obtained with equal efficiency.

Having successfully completed the total syntheses of two bisorbicillinoids, all efforts were now focused on trichodimerol (**1**). Rather than change all the variables in the dimerization conditions developed for bisorbicillinol (**2**), as mentioned earlier it was reasoned that upon increasing the effective concentration of **24** and **25** as formed through basic hydrolysis of **30** and an acidic quench, trichodimerol might result. As shown in Scheme 6, application of the same reaction conditions at a concentration of 0.3 M instead of 0.05 M did indeed afford a new dimerization adduct based on a Michael reaction. Instead of leading to trichodimerol, compound **34** was formed in 65 % yield as a single diastereomer. This surprising result can be rationalized with the indicated in which

HO Me O Me Me OH O
24

HO Me OH Me Me O O
25

AcO Me O Me Me OH O
30

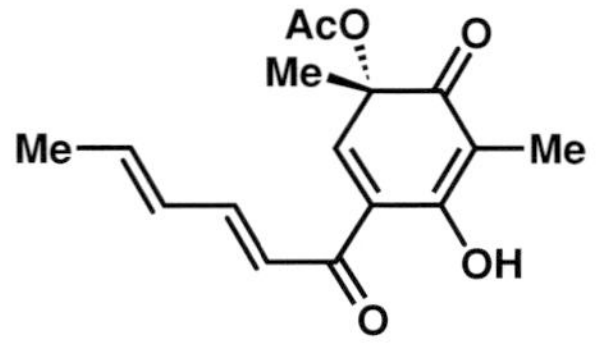

27: (+)-bisorbibutenolide

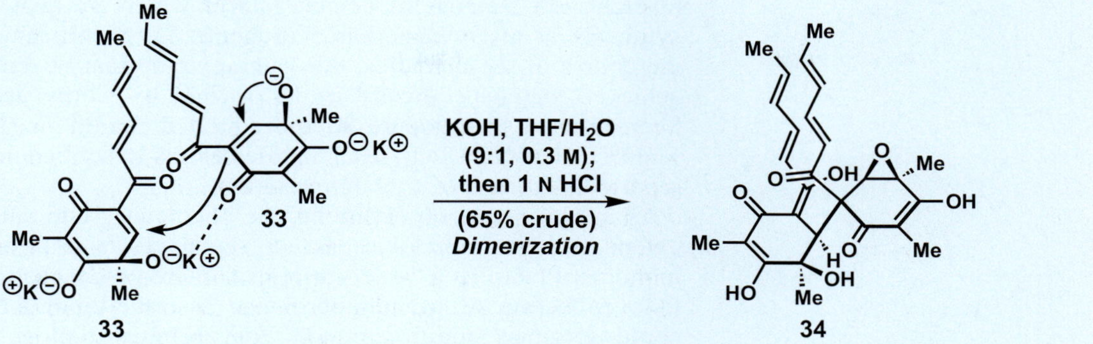

Scheme 6. The effect of concentration on the dimerization of sorbicillinol monomers.

chelation of both monomeric dianions of **33** mechanism by potassium, to minimize steric and anionic interactions, served to establish which face of **33** would be attacked by the incoming quinolate. Although the resultant product (**34**) certainly constitutes an interesting entity for which no similar secondary metabolite has yet been identified from a bacterial culture, at first its formation did not seem to point directly to a means for the formation of trichodimerol (**1**). With careful consideration of the collected data, however, came the solution that ultimately led to the total synthesis of this natural product.

Since only bisorbicillinol (**2**) resulted after quinols **24** and **25** had been formed at a modest concentration while **34** was generated almost exclusively when their quinolate counterparts were unveiled at a much higher concentration, these findings seemed to suggest that concentration changes would not likely lead to the synthesis of trichodimerol (**1**). What feature, then, could be modified to achieve success? Since the assumed biomimetic formation of **1** requires a reversible set of Michael reactions and ketalizations, the presence of excess water (as occurred in the formation of **2** by quenching with dilute aqueous HCl) could, perhaps, have prevented the synthesis of **1** by hindering the critical ketalization steps needed for its creation. Thus, altering the manner in which the monomeric quinols were formed through a different reaction quench with minimal (i. e. no more than stoichiometric) water present might provide a reasonable course of action to overcome this difficulty. Most gratifyingly, this hypothesis proved correct as treatment of **30** with finely powdered $CsOH{\cdot}H_2O$ in MeOH for 7 hours, followed by slow neutralization with $NaH_2PO_4{\cdot}H_2O$ over 12 hours, provided **1** in a yield of 16 % after isolation; most of the remaining material balance was accounted for through the concurrent formation of sorbicillin (**26**, 12 % yield) as well as the Diels–Alder dimer, bisorbicillinol (**2**, 22 % yield). While the numerical yield for trichodimerol (**1**) might seem low at first glance, based on the number of events

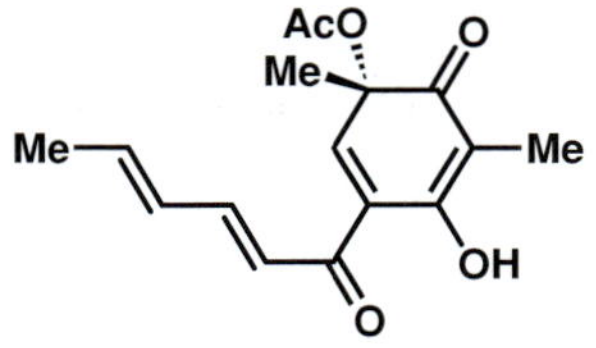

30

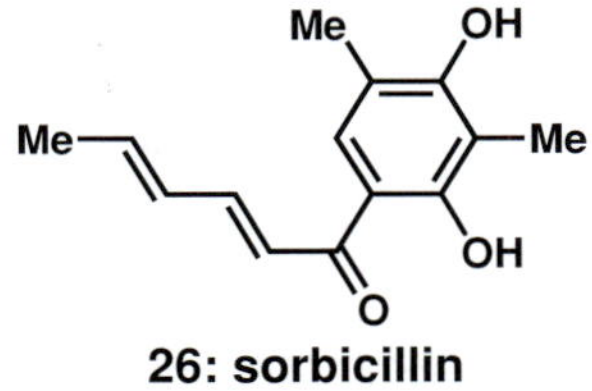

26: sorbicillin

that were required to effect its formation, including the generation of eight new stereogenic centers (six of which are quaternary), its synthesis at all is remarkable (Scheme 7). Significantly, during the course of these studies, this same extraordinary event was also achieved stereospecifically in 10% yield by Corey and Barnes-Seeman using enantiopure **30** under related conditions (NaOMe in MeOH; then NaH_2PO_4, HCl in MeOH) as described in the first reported synthesis of trichodimerol (**1**).[3]

As a final set of observations, the Nicolaou group subsequently extended these impressive cascade sequences to different starting monomers bearing a wide variety of altered side chains, leading to a collection of trichodimerol and bisorbicillinol analogs that could serve as additional tools for chemical biology studies.[2a] Beyond this important development, following the reported syntheses of these members of the bisorbicillinoids, in 2000 Abe

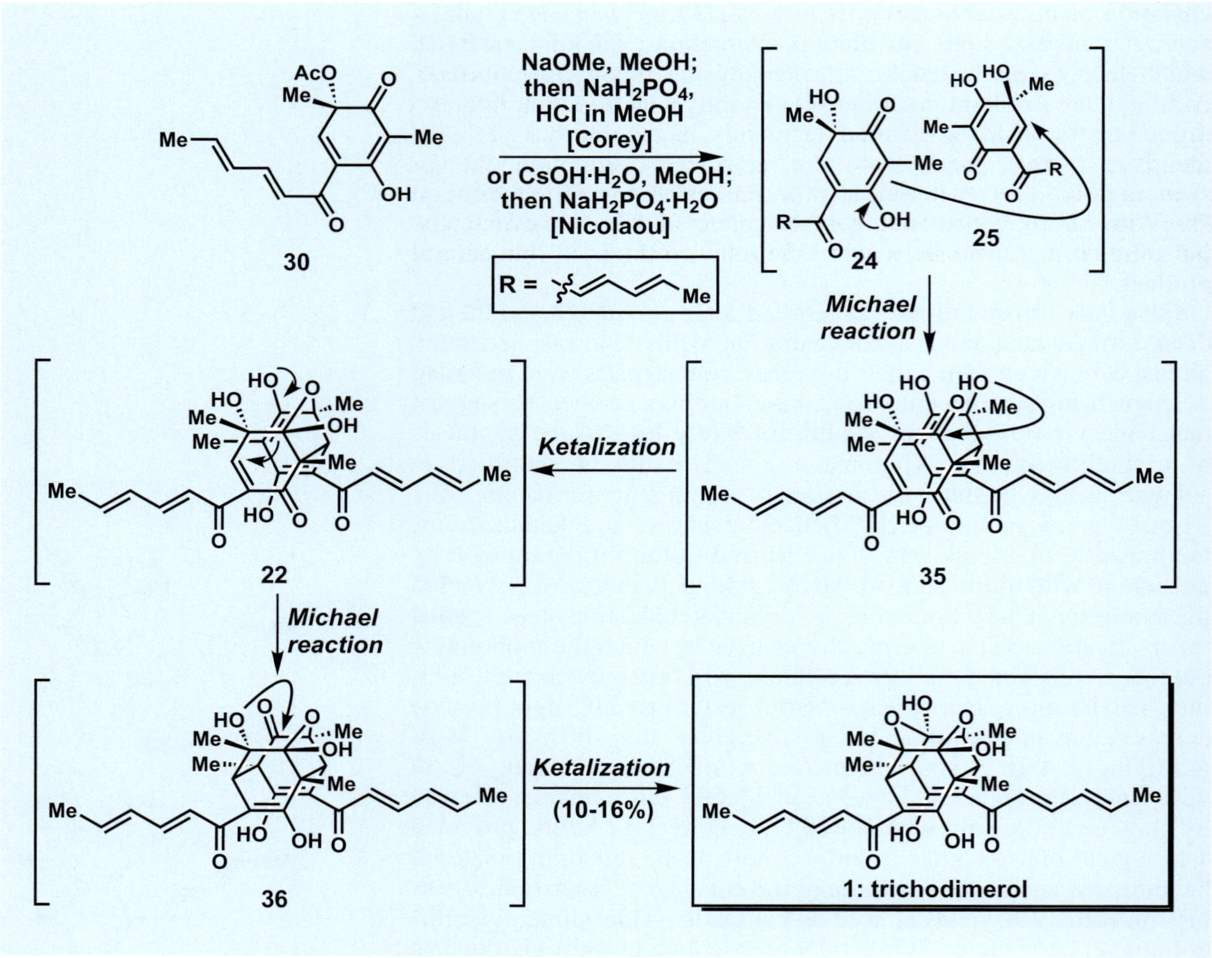

Scheme 7. Biomimetic total synthesis of trichodimerol (**1**).

and co-workers were able to isolate the postulated reactive intermediate sorbicillinol (**24**) invoked in all of the preceding schemes from the same fungus as trichodimerol, thereby affording further proof for the series of biosynthetic relationships defined in Scheme 4.[13]

11.4 Conclusion

As this chapter has amply illustrated, dimerization-based approaches to fashion molecular complexity are an aesthetically pleasing and powerful technique to create heavily functionalized molecular scaffolds in a minimum of synthetic steps. While such efficiency admirably mirrors the ease with which Nature constructs such natural products, the brevity of the developed sequence often belies the difficulty which is confronted in finding laboratory conditions to effect the key dimerization reaction. Indeed, it is the combination of keen analytical skill in identifying unanticipated by-products from numerous failed reactions and exquisite experimental technique in properly modulating reaction conditions that ultimately affords a protocol to induce monomeric material to travel down the desired dimerization pathway. While tremendous travail is often required to reach this point, the effort is worthwhile as what results is a beautiful synthesis and a level of efficiency that frequently cannot be matched by other, stepwise approaches.

References

1. Prunolide B: A. R. Carroll, P. C. Healy, R. J. Quinn, C. J. Tranter, *J. Org. Chem.* **1999**, 64, 2680; cynantetrone: P.-L. Huang, S.-J. Won, S.-H. Day, C.-N. Lin, *Helv. Chim. Acta* **1999**, *82*, 1716; verbalactone: P. Magiatis, D. Spanakis, S. Mitaku, E. Tsitsa, A. Mentis, C. Harvala, *J. Nat. Prod.* **2001**, *64*, 1093; psycholeine: A. C. Lebsack, J. T. Link, L. E. Overman, B. A. Stearns, *J. Am. Chem. Soc.* **2002**, *124*, 9008; lomaiviticin A: H. He, W.-D. Ding, V. S. Bernan, A. D. Richardson, C. M. Ireland, M. Greenstein, G. A. Ellestad, G. T. Carter, *J. Am. Chem. Soc.* **2001**, *123*, 5362; phomoxanthone A: M. Isaka, A. Jaturapat, K. Rukseree, K. Danwisetkanjana, M. Tanticharoen, Y. Thebtaranonth, *J. Nat. Prod.* **2001**, *64*, 1015.
2. a) K. C. Nicolaou, G. Vassilikogiannakis, K. B. Simonsen, P. S. Baran, Y.-L. Zhong, V. P. Vidali, E. N. Pitsinos, E. A. Couladouros, *J. Am. Chem. Soc.* **2000**, *122*, 3071; b) K. C. Nicolaou, K. B. Simonsen, G. Vassilikogiannakis, P. S. Baran, V. P. Vidali, E. N. Pitsinos, E. A. Couladouros, *Angew. Chem.* **1999**, *111*, 3762; *Angew. Chem. Int. Ed.* **1999**, *38*, 3555. For earlier studies, see: c) K. C. Nicolaou, R. Jautelat, G. Vassilikogiannakis, P. S. Baran, K. B. Simonsen, *Chem. Eur. J.* **1999**, *5*, 3651.
3. D. Barnes-Seeman, E. J. Corey, *Org. Lett.* **1999**, *1*, 1503.
4. K. C. Nicolaou, D. Gray, *Angew. Chem.* **2001**, *113*, 783; *Angew. Chem. Int. Ed.* **2001**, *40*, 761.
5. J. C. Lee, G. A. Strobel, E. Lobkovsky, J. Clardy, *J. Org. Chem.* **1996**, *61*, 3232.
6. C. Li, E. Lobkovsky, J. A. Porco, *J. Am. Chem. Soc.* **2000**, *122*, 10484. In later studies targeting a related natural product, the homodimerization of 2*H*-pyran monomers in a Diels–Alder event was also demonstrated to be possible: M. Shoji, J. Yamaguchi, H. Kakeya, H. Osada, Y. Hayashi, *Angew. Chem.* **2002**, *114*, 3324; *Angew. Chem. Int. Ed.* **2002**, *41*, 3192.
7. a) R. Andrade, W. A. Ayer, P. P. Mebe, *Can. J. Chem.* **1992**, *70*, 2526; b) Q. Gao, J. E. Leet, S. T. Thomas, J. A. Matson, D. P. Bancroft, *J. Nat. Prod.* **1995**, *58*, 1817; c) R. Andrade, W. A. Ayer, L. S. Trifonov, *Can. J. Chem.* **1996**, *74*, 371; d) G. A. Warr, J. A. Veitch, A. W. Walsh, G. A. Hesler, D. M. Pirnik, J. E. Leet, P.-F. M. Lin, I. A. Medina, K. D. McBrien, S. Forenza, J. M. Clark, K. S. Lam, *J. Antibiot.* **1996**, 234; e) N. Abe, T. Murata, A. Hirota, *Biosci. Biotechnol. Biochem.* **1998**, *62*, 2120.
8. a) L. S. Trifonov, H. Hilpert, P. Floersheim, A. S. Dreiding, D. M. Rast, R. Skrivanova, L. Hoesch, *Tetrahedron* **1986**, *42*, 3157; b) M. Kontani, Y. Sakagami, S. Marumo, *Tetrahedron Lett.* **1994**, *35*, 2577.
9. N. Abe, T. Murata, A. Hirota, *Biosci. Biotechnol. Biochem.* **1998**, *62*, 661.
10. a) D. J. Cram, M. Tishler, *J. Am. Chem. Soc.* **1948**, *70*, 4238; b) D. J. Cram, *J. Am. Chem. Soc.* **1948**, *70*, 4240.
11. We should note that many additional biosynthetic pathways for members of the bisorbicillinoids have been advanced, many of which can be found in the references cited above. Although we have not defined these alternative routes in this chapter, they are equally interesting and worthy of consideration. For the earliest of such proposals, see: L. S. Trifonov, A. S. Dreiding, L. Hoesch, D. M. Rast, *Helv. Chim. Acta* **1981**, *64*, 1843.
12. a) J. F. W. McOmie, M. S. Tute, *J. Chem. Soc.* **1958**, 3226; b) F. Bigi, G. Casiraghi, G. Casnati, S. Marchesi, G. Sartori, C. Vignali, *Tetrahedron* **1984**, *40*, 4081.
13. N. Abe, O. Sugimoto, K. Tanji, A. Hirota, *J. Am. Chem. Soc.* **2000**, *122*, 12606.

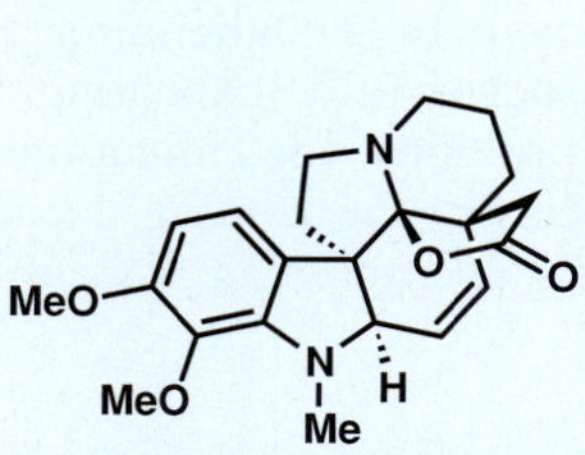

MeO MeO N Me H N O O

1: aspidophytine

12

E. J. Corey (1999)

Aspidophytine

12.1 Introduction

Key concepts:

- **Cascade reactions**
- **Vilsmeier-Haack formylation**
- **CBS reduction**

As several of the total syntheses described both in this book as well as in the preceding volume of *Classics in Total Synthesis* have demonstrated, cascade transformations offer a magical level of power that is arguably unsurpassed by any other weapon in our current synthetic arsenal. Indeed, cascade reactions have been utilized in the context of some of the most significant achievements in total synthesis, including landmark accomplishments such as Sir Robert Robinson's elegant biomimetic formation of tropinone[1] through a series of Mannich reactions in 1917, and W. S. Johnson's synthesis of progesterone in 1971 which defined polyolefinic cation–π cyclizations as an invaluable synthetic tool.[2] Despite this long and proud lineage, however, the full potential of cascade sequences has barely been tapped.

With ever-increasing pressure to fashion diverse molecular architectures rapidly through efficient and atom-economical[3] processes with high degrees of selectivity, cascade reactions are destined to become an integral design aspiration of most synthetic endeavors. In order to push the state-of-the-art of these sequences, contemporary as well as future generations of synthetic practitioners will require increasingly precise mechanistic and kinetic understanding of organic transformations combined with a large dose of intellectual flexibility and creativity. In this chapter, we will explore an example of one such advance in the form of a highly concise and elegant total synthesis of aspidophytine (**1**), achieved by E. J. Corey and members of his group at Harvard University in 1999, in which a highly instructive cascade sequence efficiently forged the entire carbogenic skeleton of the target molecule.[4] Without question, this particular series of chemical events reflects keen mechanistic analysis of a challenging problem and acute experimental skill in successfully executing the designed plan. More significantly, it provides an indication of the anticipated power of artificial synthesis in the coming decades as it further strives to replicate Nature's adeptness in fashioning its own molecules.

12.2 Retrosynthetic Analysis and Strategy

For hundreds of years, various civilizations throughout Mexico and Central America have used the ground leaves of the plant *Haplophyton cimicidum* as an insecticidal powder, primarily to control the local cockroach population.[5] Only in more recent times have the active constituents of this powerful formulation, more colorfully known to the natives as "la hierba de la cucaracha," been determined. These compounds include several biologically active indole alkaloids such as haplophytine (**2**, Scheme 1), a molecule that was isolated and fully characterized from the concoction after years of intense effort by the research groups of H. R. Snyder, M. P. Cava, P. Yates, and D. E. Zacharias.[6] Although the architectural skeleton of this particular natural product (**2**) is quite daunting, an important initial design clue for synthetic avenues towards its polycyclic framework was provided by the discovery that haplophytine could be smoothly converted into aspidophytine (**1**) in good yield upon treatment with a mineral acid at elevated temperatures.[6g,i] As such, although no product corresponding to the other domain of **2** has yet been isolated, aspidophytine (**1**) represents an obvious initial target to develop some of the strategies needed to accomplish the total synthesis of haplophytine (**2**), and constitutes a conceivable biogenetic precursor for this natural product.

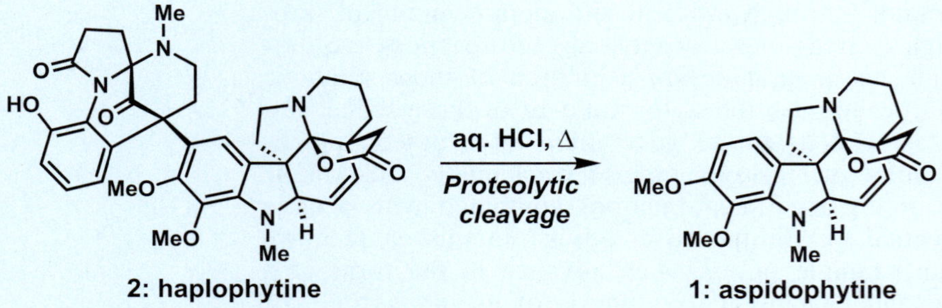

Scheme 1. The structures of haplophytine (**2**) and aspidophytine (**1**), biologically active indole alkaloids obtained from the dried leaves of the plant *Haplophyton cimicidum*.

While the synthetic problems associated with structure **2** are reduced considerably through such an analysis, the remaining challenges posed by aspidophytine's (**1**) compact skeleton of six rings and four adjacent stereogenic centers (three of which are quaternary carbon atoms) are hardly trivial, making it an attractive target for total synthesis in its own right. Indeed, for over 25 years aspidophytine (**1**) lingered as an unsolved synthetic puzzle in the chemical literature, with no viable route available for its construction. The efforts of E. J. Corey and his group rectified this situation, and simultaneously provided a spectacular paradigm of cascade reactions particularly relevant to the total synthesis of indole alkaloid natural products.[4]

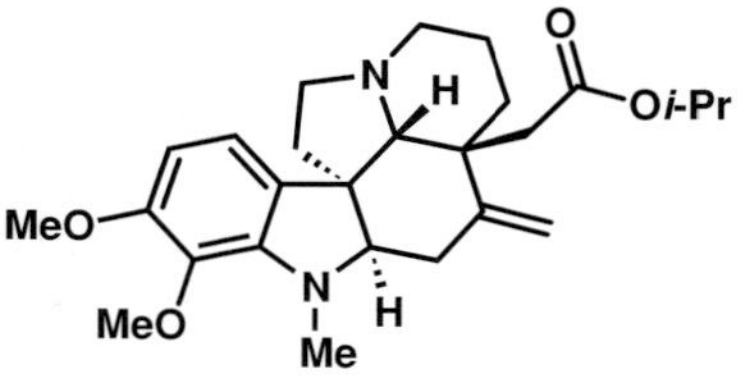

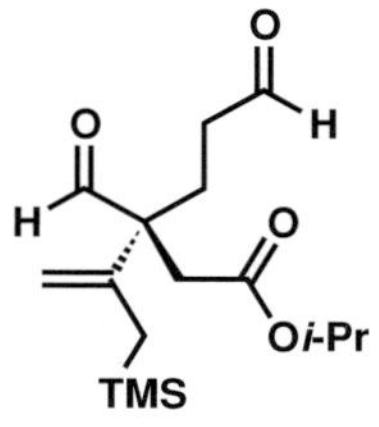

In assessing plausible retrosynthetic transforms to apply to aspidophytine (**1**), the Corey group initially turned to some additional important clues provided by the early isolation and characterization studies. Because the carbinolamine lactone moiety in **1** was readily cleaved upon either hydrogenolysis or exposure to sodium borohydride,[6f,g,i] it was anticipated that the seemingly labile nature of this functionality would best be controlled by reserving its incorporation until the late stages of the synthesis. Thus, starting from a pentacyclic compound such as **3** (Scheme 2), deprotection of the isopropyl ester group and subsequent lactonization onto a reactive iminium species formed by a suitable reagent could serve as a conceivable set of conditions to complete the hexacyclic framework of the target molecule in the forward direction. The other transform connecting aspidophytine (**1**) and precursor **3** is modification of the internal alkene to an exocyclic double bond. Although the synthetic operations necessary to excise the extra carbon atom present within **3** and simultaneously introduce the required endocyclic unsaturation in **1** would likely prove straightforward, as several different routes could be anticipated to succeed in this task, the incorporation of this seemingly extraneous functional handle in **3** represents a critical feature of the synthetic strategy. Indeed, with this exocyclic olefin in place, it was envisioned that **3** could arise from building blocks **8** and **9** in a single operation!

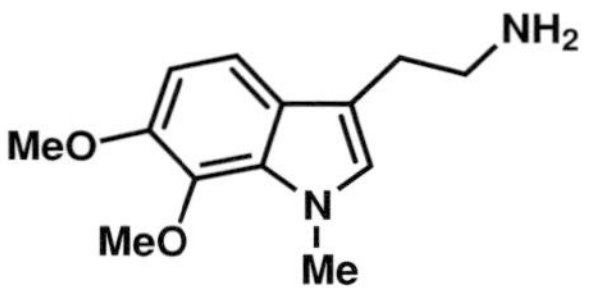

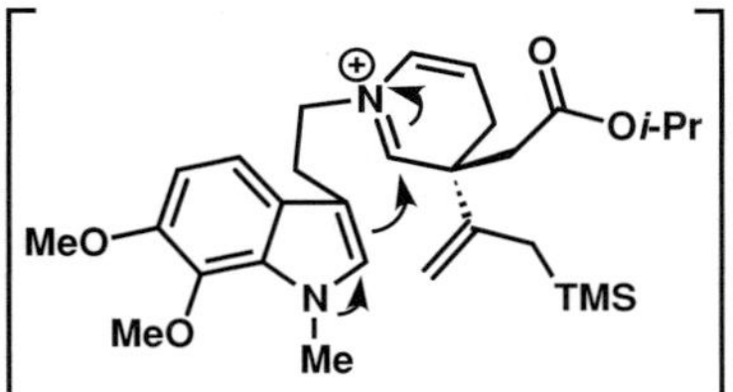

In the proposed sequence of events, mixing dialdehyde **8** with tryptamine derivative **9** in the synthetic direction was anticipated to afford **7** through two consecutive condensation reactions, certainly a reasonable supposition on the basis of first principles. With a reactive iminium species formed by this initial set of operations, a facile cyclization was then expected to ensue based on the participation of the lone pair of electrons on the indole nitrogen atom. This event would forge the new five-membered ring in **6** with concomitant installation of two stereogenic centers whose absolute stereochemistry would, presumably, be governed by the pre-existing homochiral stereocenter present in **8**. Having successfully formed this intermediate (**6**), it was then surmised that the strategically positioned TMS group could be cleaved in media of suitable acidity, a process that would initiate a second cyclization leading to **5** bearing five of the six rings present in the final target

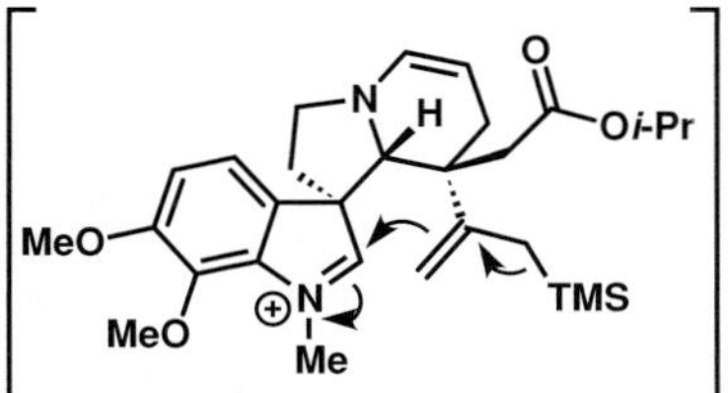

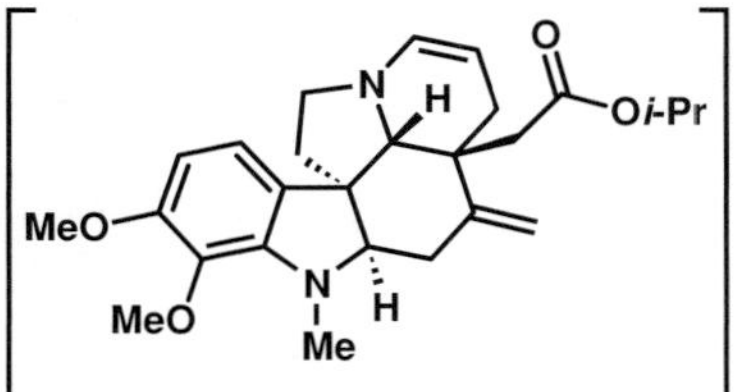

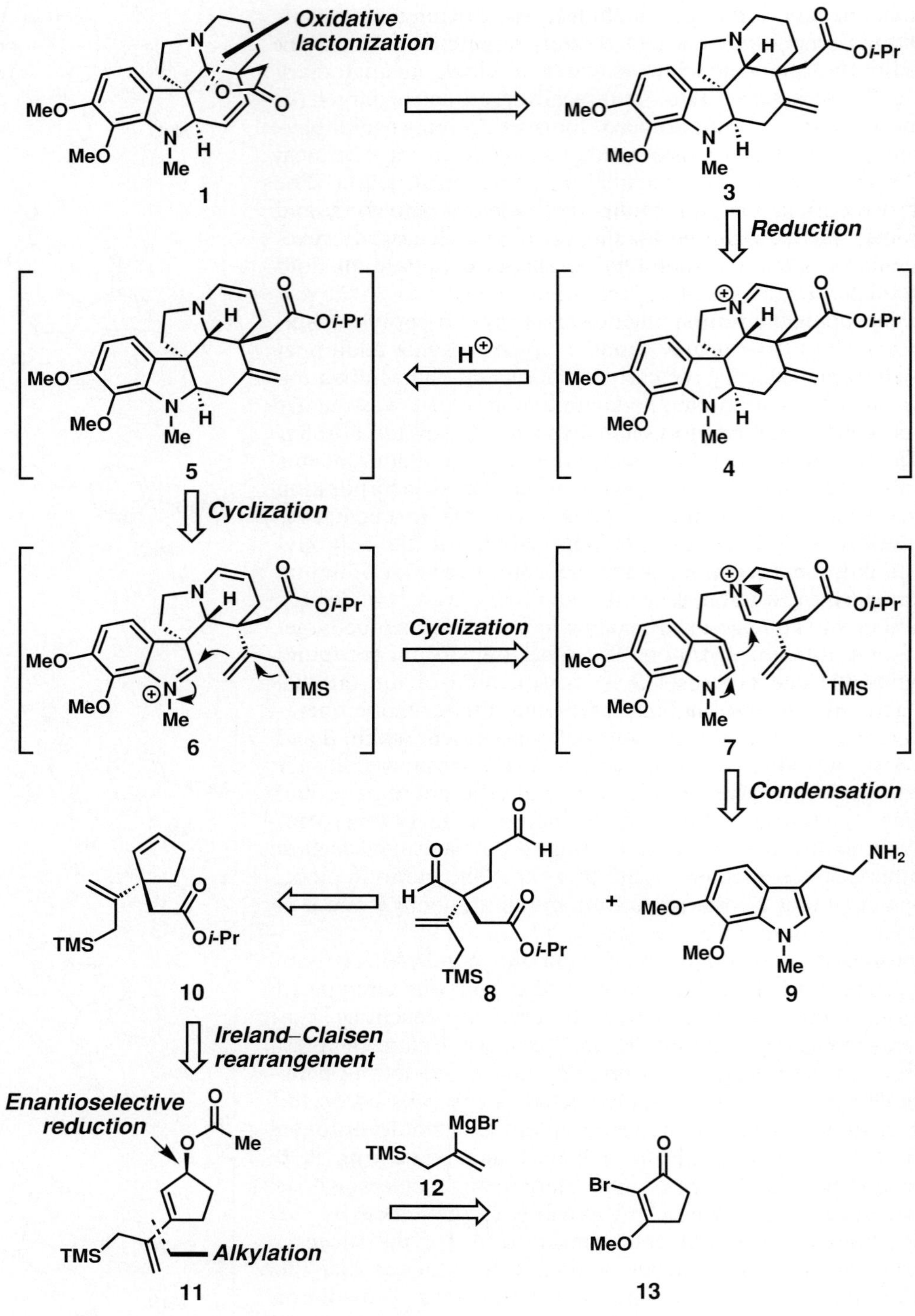

Scheme 2. Corey's retrosynthetic analysis of aspidophytine (**1**).

molecule. With the pH value of the solution still below 7, this intermediate was not expected to persist, but, instead, to accept a proton, leading to iminium species **4** whose conversion into **3** could then be envisioned upon treatment with a suitable reducing agent such as $NaCNBH_3$. On paper, this highly appealing (if not spectacular) cascade sequence leading to the establishment of three new rings and three stereogenic centers seems both reasonable and mechanistically sound. The moment of truth for all synthetic chemistry, however, lies in its execution. Careful control of experimental conditions would undoubtedly be required for success in this scenario, particularly in terms of balancing such obvious features as reaction temperature, stoichiometry, and pH. More subtle facets, such as the precise timing of reagent addition once certain key intermediates were formed (assuming these fleeting species could even be detected), would likely prove equally crucial for optimizing the formation of **3**, and preventing the reaction from proceeding along undesired pathways. Granted, never before had such a sequence been demonstrated. Instead of leading to hesitation, such knowledge should serve aş the impetus to forge ahead and attempt execution of the idea in the laboratory, for only by confronting the inherent challenges posed by such daring synthetic strategies, rather than opting for safer and more conventional stepwise approaches, can the field of total synthesis be advanced decisively.

If success in this tandem reaction pathway could be achieved, then the synthetic problem posed by aspidophytine (**1**) would be dramatically reduced to the synthesis of intermediates **8** and **9**, entities possessing roughly equal numbers of carbon atoms. With several routes to tryptamine derivatives such as **9** already known in the literature,[7] its synthesis was not expected to provide any particular challenges, and as such will not be analyzed here in greater detail. Developing a sequence to synthesize the other key building block (**8**) was not nearly as obvious or pedestrian. First, it was anticipated that a precursor cyclopentene ring such as **10** could serve as a suitable masking device for the two aldehyde groups in **8**, functionalities which could be revealed in the forward direction at the appropriate juncture simply through a dihydroxylation/oxidative cleavage protocol. Key to the success of this plan would be the chemoselective reaction of only the endocyclic olefin in **10** with OsO_4, a result which seemed reasonable based on well-established precedent that strained, but less sterically encumbered, double bonds are dihydroxylated faster than those that are unactivated and hindered.[8] The next synthetic insight was that the lone stereogenic center in **10** could arise stereoselectively from an Ireland–Claisen rearrangement[9] of homochiral **11**, an intermediate in which the allylic alcohol might be generated asymmetrically from a precursor ketone such as **13**. With numerous chiral reducing agents known in the literature, many of which were pioneered by the Corey group,[10] success in this latter task was forecasted with confidence. Finally, removal of the TMS-based side chain from **11** as the Grignard reagent **12** completed retrosynthetic simplification to **13**. Overall, the majority of the steps leading to **10** had been verified in alternate contexts, and, as a result, were anticipated to succeed admirably when called upon during the synthesis.[11]

12.3 Total Synthesis

The total synthesis of aspidophytine (**1**) commenced with the preparation of the key tryptamine building block **9**, as shown in Scheme 3. Although we have already confronted a typical means of indole preparation in Woodward's total synthesis of strychnine,[12] namely the Fischer indole synthesis, in this context an alternative approach based on reductive cyclization was employed instead to fashion this heterocycle. Starting from known intermediate **14**,[13]

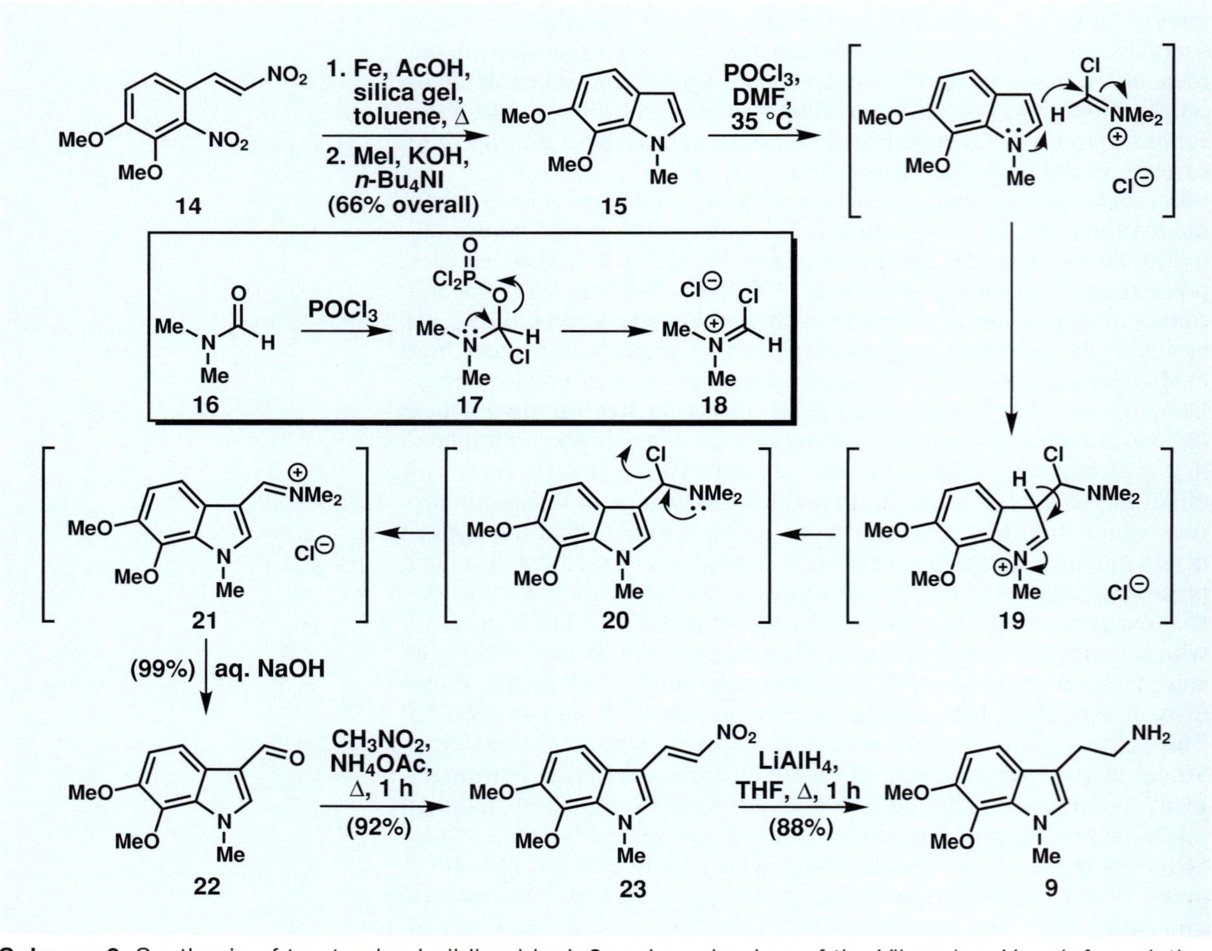

Scheme 3. Synthesis of tryptamine building block **9** and mechanism of the Vilsmeier–Haack formylation reaction.

treatment with iron powder and acetic acid in refluxing toluene smoothly afforded the desired 6,7-dimethoxyindole through a sequence likely to involve initial reduction of both nitro groups to their amino counterparts, followed by 5-*exo*-trig ring closure of the aromatic amine onto the adjacent vinyl function with eventual expulsion of NH_4OAc to effect aromatization.[14] Key to obtaining consistent and reasonable yields of this indole product was the addition of silica gel to the reaction mixture,[15] a modification which is rationalized as preventing intermolecular side reactions between polar intermediates and residual starting material by localizing charged species on the silica-gel surface, thus forcing the reaction to proceed through the desired intramolecular pathway. As an aside, other methods to effect the same reductive cyclization include the classical techniques of hydrogenation over Pd/C[16] and exposure to zinc granules in refluxing aqueous AcOH.[17] Overall, selecting between these alternative methods to achieve optimal yields is generally predicated upon the particular substitution pattern in the targeted indole. With this step smoothly accomplished, the so-formed 6,7-dimethoxyindole was then readily *N*-methylated under phase-transfer conditions (MeI, KOH, and *n*-Bu_4NI), providing **15** in 66 % overall yield for the two steps.

Striving now for the homologation of **15** to tryptamine **9**, an initial extension of a carbonyl group to form **22** was smoothly achieved in 99 % yield using the classical Vilsmeier–Haack formylation protocol.[18] Although we have seen this reaction before, working through the mechanism of this useful transformation at this juncture might be instructive. Addition of $POCl_3$ to DMF (**16**) leads to the formation of a chloroiminium ion (**18**) by the indicated pathway (see inset box, Scheme 3). With this highly reactive electrophile generated in the presence of **15**, nucleophilic attack by the indole portion onto **18** ensues, providing **19** which then converts into **21**, via **20**, through rearomatization. Finally, upon exposure to aqueous NaOH, the latter species is readily hydrolyzed to the desired 3-formylindole **22**. One should note that this method generally constitutes the most commonly employed aromatic formylation technique, particularly for electron-rich substrates. Moreover, numerous variants of this reaction have been developed over the years to generate species **18** by alternative means, typically by substituting $POCl_3$ with phosgene or triflic anhydride.[19] Having achieved the synthesis of **22**, only two operations remained to complete **9**, namely initial condensation with nitromethane to form **23** followed by complete reduction of the newly installed vinyl nitro group with $LiAlH_4$ in refluxing THF, events which proceeded in an overall yield of 81 %.

With one key building block in hand, we now turn to the more challenging dialdehyde precursor **8** whose construction is outlined in Scheme 4. Starting from the known cyclopentenone **13**,[11] alkylation with Grignard reagent **12** (generated *in situ* from the corresponding vinyl halide precursor)[20] in the presence of $CeCl_3$ smoothly provided **25** in 82 % yield through the intermediacy of

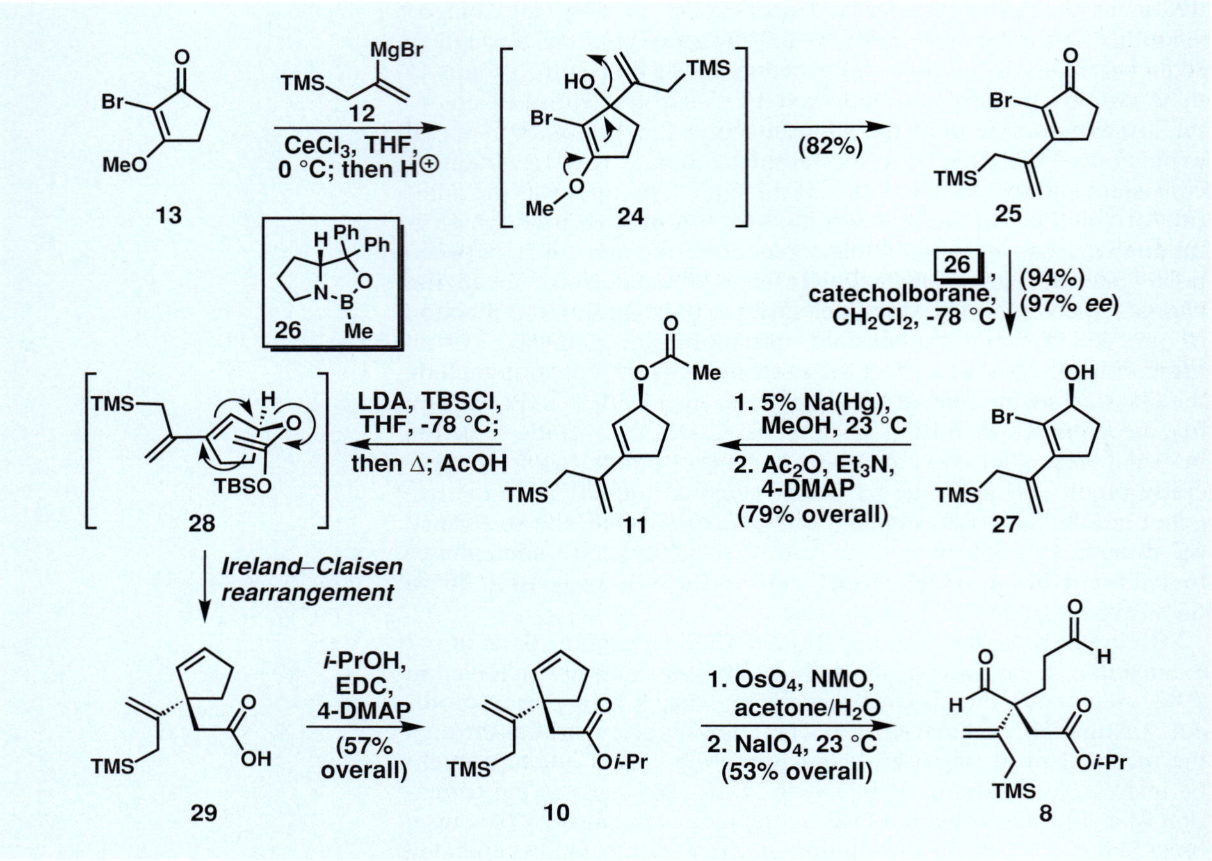

Scheme 4. Synthesis of key intermediate **8** based on an Ireland–Claisen rearrangement.

tertiary alcohol **24**. Although this reaction appears to be rather straightforward, several features of the conversion merit further discussion, especially with regard to the numerous, but non-obvious roles played by $CeCl_3$ in the process. Without question, the addition of $CeCl_3$ to **13** with stirring at ambient temperature for several hours prior to the addition of **12** was crucial to achieving selective 1,2-addition by virtue of activating carbonyl coordination which is not possible in the absence of the lanthanide.[11] Furthermore, the presence of $CeCl_3$ was equally critical for obtaining high yields of the alkylation product by reducing the propensity for undesired or abnormal side reactions that are often observed in Grignard additions, such as competitive enolization, which could certainly occur with **13**. This beneficial feature was first observed by Imamoto, who rationalized the result on the grounds that the strong basicity of the Grignard reagent was weakened in the presence of $CeCl_3$ (potentially by transmetallation occurring in advance of nucleophilic addition), thereby enabling a more controlled reaction with the carbonyl compound to occur.[21] Moreover, after the formation of **24**, the residual cerium salts probably facilitated the subsequent elimination–hydrolysis sequence to **25** by serving as a weak Lewis acid.[22]

Having achieved this initial alkylation, the next planned synthetic operation was asymmetric reduction of the ketone in **25**. Among the numerous tools capable of accomplishing such reactions, chiral oxazaborolidines are among the most practical and versatile catalysts currently available.[10] First introduced in a seminal paper by Corey, Bakshi, and Shibata (CBS) in 1987,[23] this reducing system proved readily amenable to the present context. Thus, reaction of **25** in CH_2Cl_2 at −78 °C with a slight excess of catecholborane in the presence of 10% of chiral catalyst **26** (derived from (*R*)-diphenylprolinol and methylboronic acid)[24] provided the desired product **27** in both excellent yield and high enantiomeric excess. Because of the importance of this reaction in organic synthesis, it would be prudent to delve into it in greater detail before pressing onward.

Scheme 5 outlines a reasonable mechanistic proposal accounting for the catalytic enantioselective reduction of ketones such as **25** by oxazaborolidines of type **26**. Upon addition of a borane reagent,

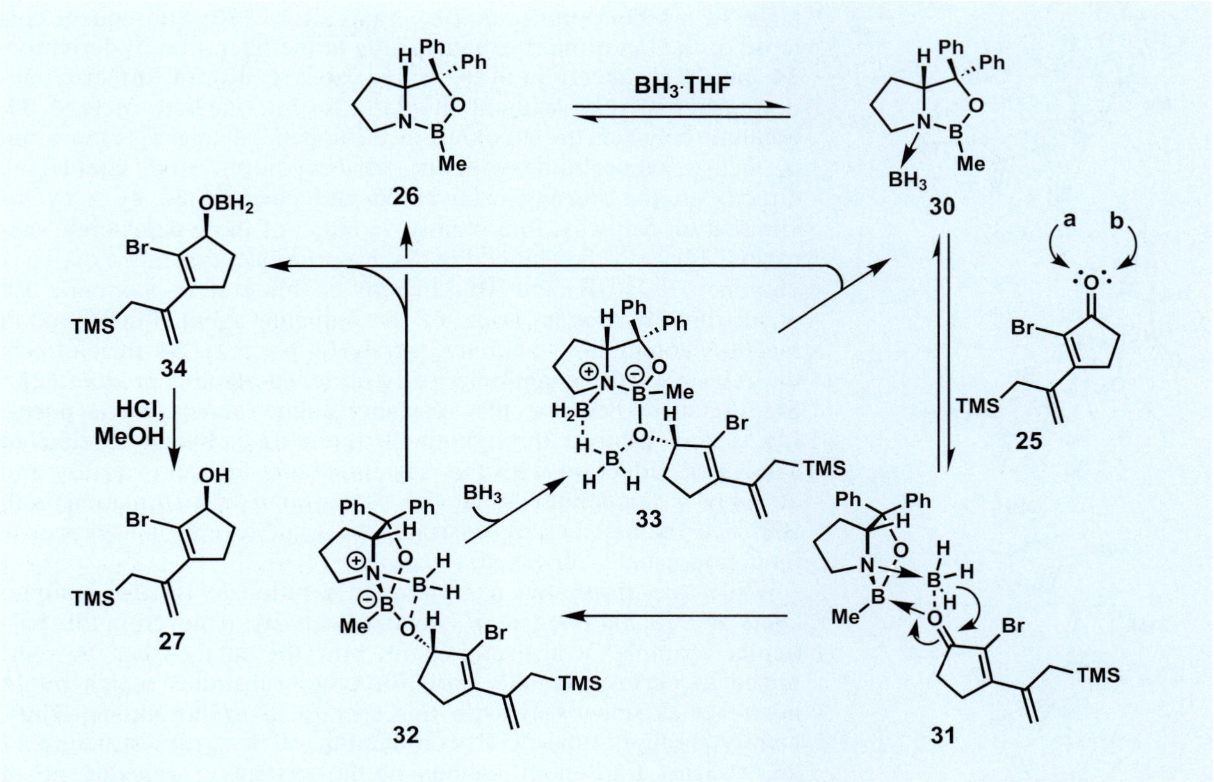

Scheme 5. Mechanism of the asymmetric reduction of carbonyl compounds (e. g. **25**) with borane and chiral oxazaborolidine catalysts (e. g. **26**).

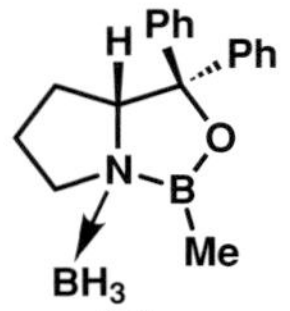

illustrated in this case with BH_3·THF instead of catecholborane for overall ease of presentation, rapid and reversible complexation of the borane to the Lewis basic nitrogen atom occurs from the uniquely accessible β-face, leading to the *cis*-fused complex **30** (an entity whose existence and stereochemical identity has been verified in related contexts by X-ray crystallographic analysis).[25] Importantly, this initial coordination redistributes electron density such that the borane reagent is activated as a hydride donor and the Lewis acidity of the endocyclic boron atom is enhanced, thereby facilitating subsequent complexation to the ketone substrate (**25**). Based on the inherent homochirality of the oxazaborolidine catalyst in conjunction with the differential accessibility of the lone pairs in **25** (a versus b) as enforced by the bromine substituent adjacent to the carbonyl function, the union of **30** and **25** proceeds with the near-exclusive formation of **31**, an intermediate with minimized steric interactions.

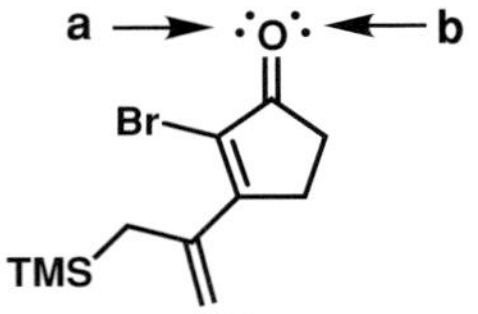

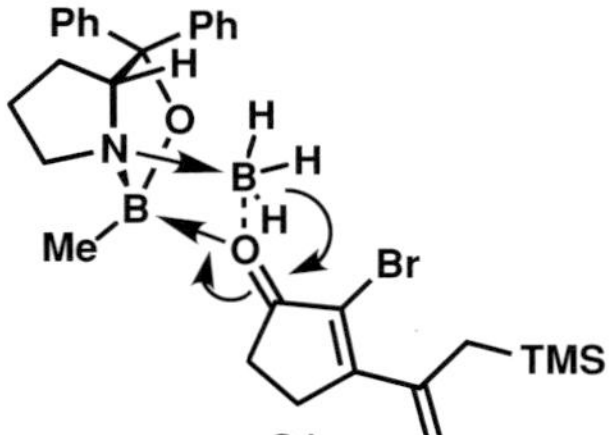

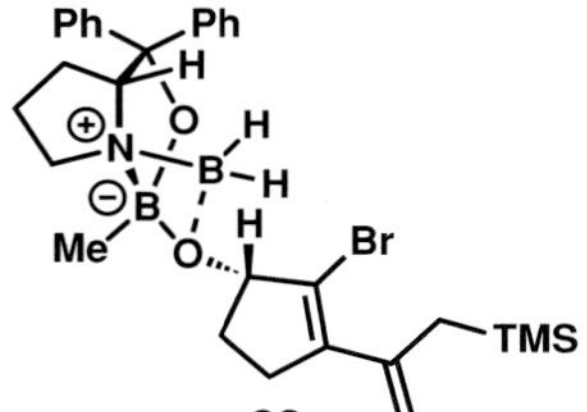

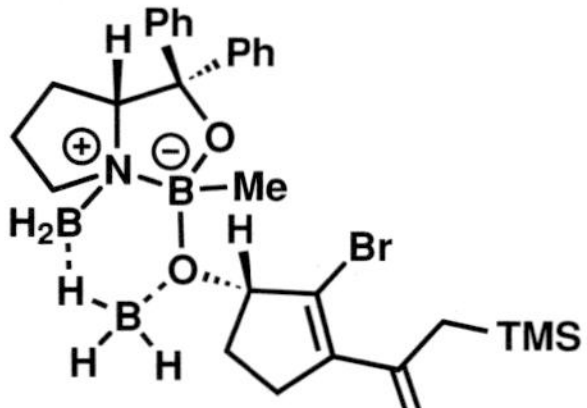

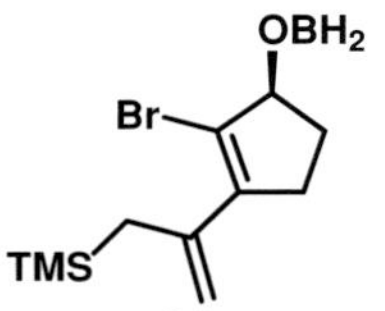

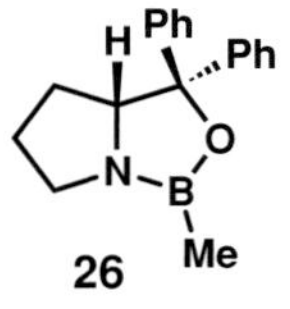

While groups other than bromine could have performed this role equally well, the choice of this atom in this case was predicated upon the anticipation that it could be easily excised later in the aspidophytine synthesis once its mission had been fulfilled.[26] With **31** formed, a facially selective hydride transfer through a six-membered transition state then provides complex **32**, which upon acceptance of a second molecule of borane affords **33**. Subsequent collapse of the latter intermediate results in the liberation of derivative **34** and the regeneration of the active catalyst (**30**) for further reduction. Alternatively, without invoking an intermediate of type **33**, bonding between the alkoxide in **32** appended to the boron atom of the oxazaborolidine with the adjacent boron atom could lead directly to the starting catalyst **26** and borinate **34** by a cycloelimination pathway. Importantly, in either of these degenerate scenarios, **34** can be envisioned to disproportionate, forming a dialkoxyborane [$(RO)_2BH$] and BH_3, thereby enabling atom-economic use of all three hydrogen atoms of the reducing agent. Finally, upon reaction completion, acid hydrolysis of borinate **34** then effects the release of the enantiomerically enriched alcohol product (**27**). Significantly, because the oxazaborolidine serves as a potent Lewis acid catalyst throughout the event (i. e. ligand-accelerated catalysis), reduction with this system occurs far more readily and at a lower temperature than the background transformation with BH_3·THF alone, thereby ensuring the high levels of asymmetric induction usually observed.

While this discussion has provided a rationale for the catalytic cycle and the high degree of enantioselectivity achieved in this particular example, it also anticipates how the catalyst can be optimized to perform equally well for other substrates which might not react as selectively with this specific oxazaborolidine. Thus, merely through superficial examination of the gross structure of **26**, at least four modifications of the system to generate novel catalysts should be apparent: 1) altering the appendages on the oxazaborolidine ring, 2) changing the *gem*-diphenyl substituents,

3) exchanging the group on the endocyclic boron atom, and 4) using different hydride sources. Indeed, all these features have been systematically examined by the Corey group and based on the collected results, after sufficient optimization some variant of this catalyst system can be obtained to reduce essentially any ketone stereoselectively.[10]

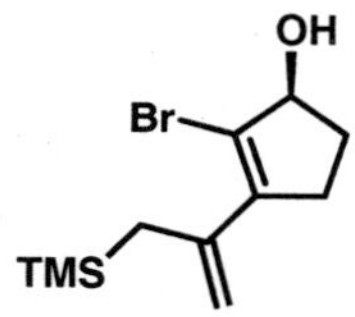

Having completed this lengthy, but hopefully educational, digression, we are now ready to return to the synthesis of aspidophytine. With the asymmetric preparation of **27** achieved, the disposable bromine atom had served its intended purpose and, since its presence was no longer required, it was readily extruded upon exposure to sodium/mercury amalgam (see Scheme 4). Subsequent acetylation, in anticipation of the proposed Ireland–Claisen sequence, then provided acetate **11** in 79 % overall yield for the two operations. In the next step, formation of silyl enol ether **28** proceeded smoothly at −78 °C in the presence of LDA and TBSCl in THF. Upon subsequent heating at reflux, followed by quenching with AcOH, carboxylic acid **29** was obtained as a crude oil, the product of a [3,3]-sigmatropic rearrangement.[12] Immediate esterification with 2-propanol as promoted by EDC and 4-DMAP then generated **10** in 57 % yield for these two steps. As would be expected, because the Ireland–Claisen reaction is a pericyclic process, the stereochemical information initially present in **11** was communicated in this sequence with complete fidelity, providing **10** as a single stereoisomer. Having reached this advanced stage, only dihydroxylation of the endocyclic double bond and subsequent $NaIO_4$-mediated cleavage of the resulting diol separated compound **10** from the coveted intermediate **8**. Thankfully, these operations proceeded smoothly in a combined yield of 53 %.

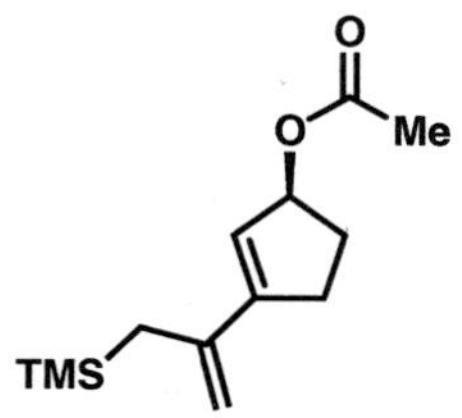

With both dialdehyde **8** and tryptamine **9** synthesized, the stage was now set for the critical operation of uniting them as a prelude to the final steps leading to aspidophytine (**1**). After an arduous exploration of reaction conditions, the projected domino sequence was indeed realized in an impressive overall yield of 66 %. In the event, as illustrated in Scheme 6, a solution of tryptamine **9** in anhydrous MeCN was added to an equimolar amount of dialdehyde **8** in MeCN. After five minutes of stirring at ambient temperature, this solution was then cooled to 0 °C and added dropwise through a cannula to a vigorously agitated solution of trifluoroacetic anhydride (2.0 equiv) in MeCN at 0 °C. Once two additional hours had passed, during which time conversion into **4** likely occurred by the pathway discussed earlier (Scheme 2), addition of excess $NaCNBH_3$ (5.0 equiv) at 0 °C completed the sequence, furnishing pentacyclic intermediate **3**.

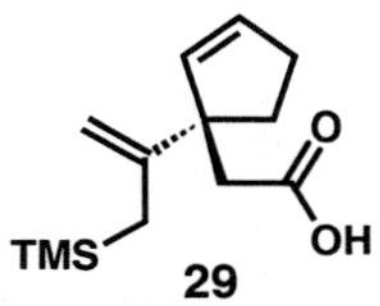

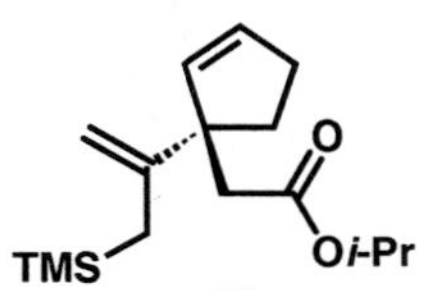

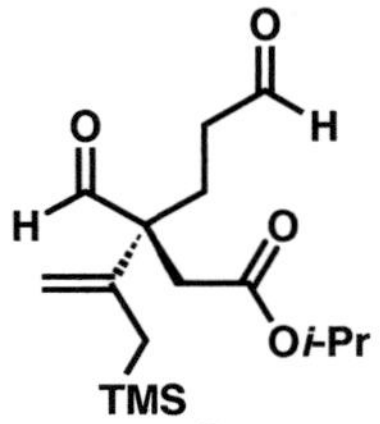

At this juncture, with the entire carbocyclic core of aspidophytine established, an advanced staging area had been reached from which the total synthesis could hopefully be completed after a few finishing touches. Specifically, what remained was the fusion of the γ-lactone and the excision of the unwanted carbon atom with the introduction of a double bond inside the adjacent cyclohexane ring.

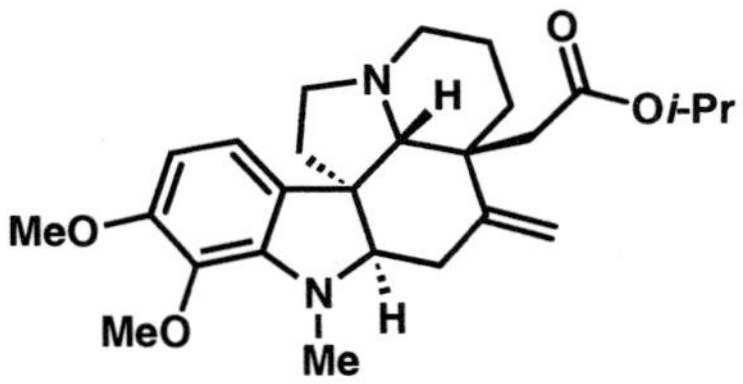

OH Br TMS **27**

O Me O TMS **11**

TMS H O TBSO **28**

O OH TMS **29**

O Oi-Pr TMS **10**

O H O H O Oi-Pr TMS **8**

O N H Oi-Pr MeO MeO N Me H **3**

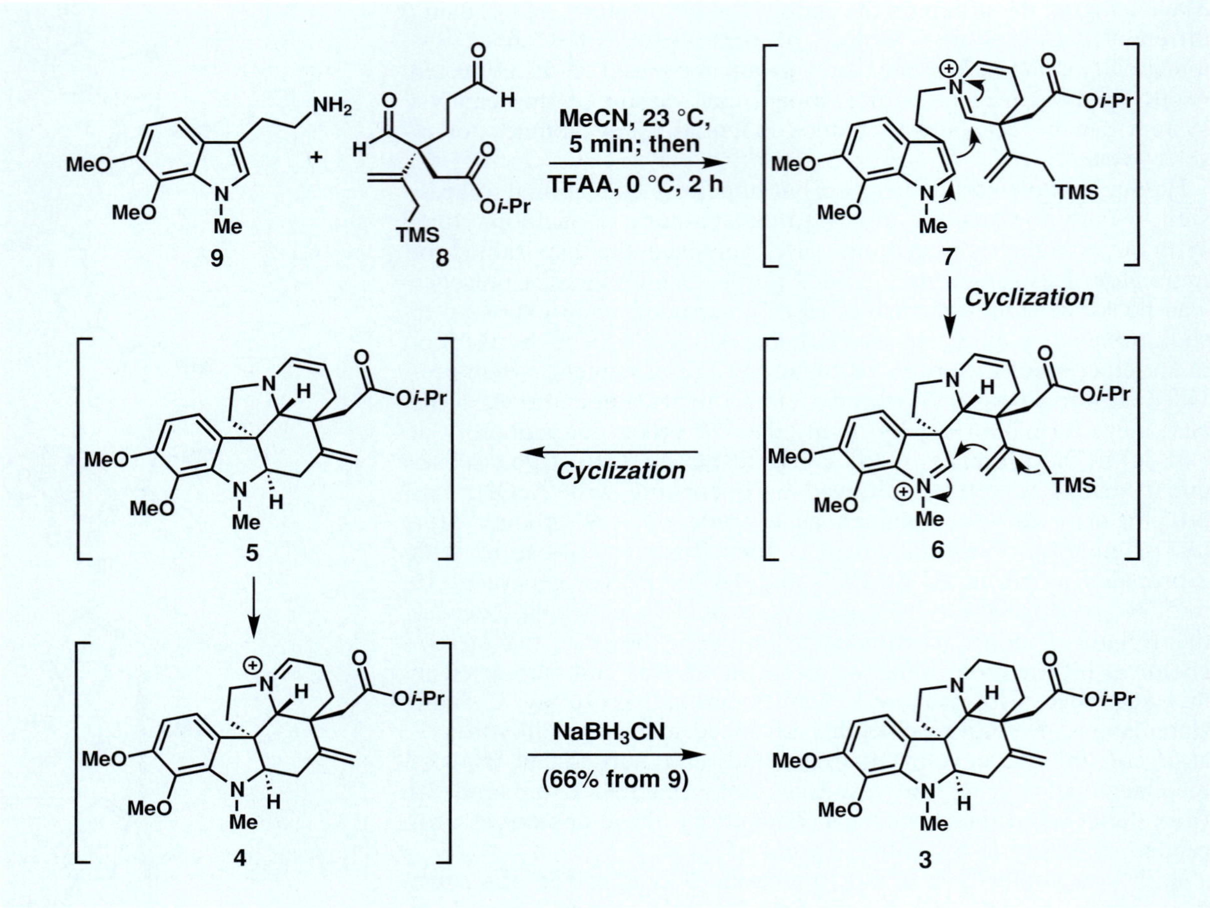

Scheme 6. The cascade sequence leading to the pentacyclic core (**3**) of aspidophytine from building blocks **8** and **9**.

Initial efforts targeted lactone formation. As originally envisioned, after basic cleavage of the isopropyl protecting group, exposure of the resultant free carboxylic acid to potassium ferricyanide effected oxidative lactonization to **36** in 79 % overall yield from **3** (Scheme 7). Presumably, treating the intermediate salt with this mild oxidant effected formation of zwitterionic intermediate **35** by initially converting the tertiary amine into an iminium ion, with subsequent carboxylate attack at the newly-generated electrophile leading to the installation of the final ring of the hexacyclic natural product target (**1**).

With **36** in hand, the only remaining task was the removal of one carbon atom and the creation of an element of unsaturation. Although originally anticipated as a set of relatively facile conversions, the complex aspidophytine framework greatly exacerbated the difficulties of the proposed transformations beyond initial expectations. The first defined goal, oxidative cleavage of the exo-

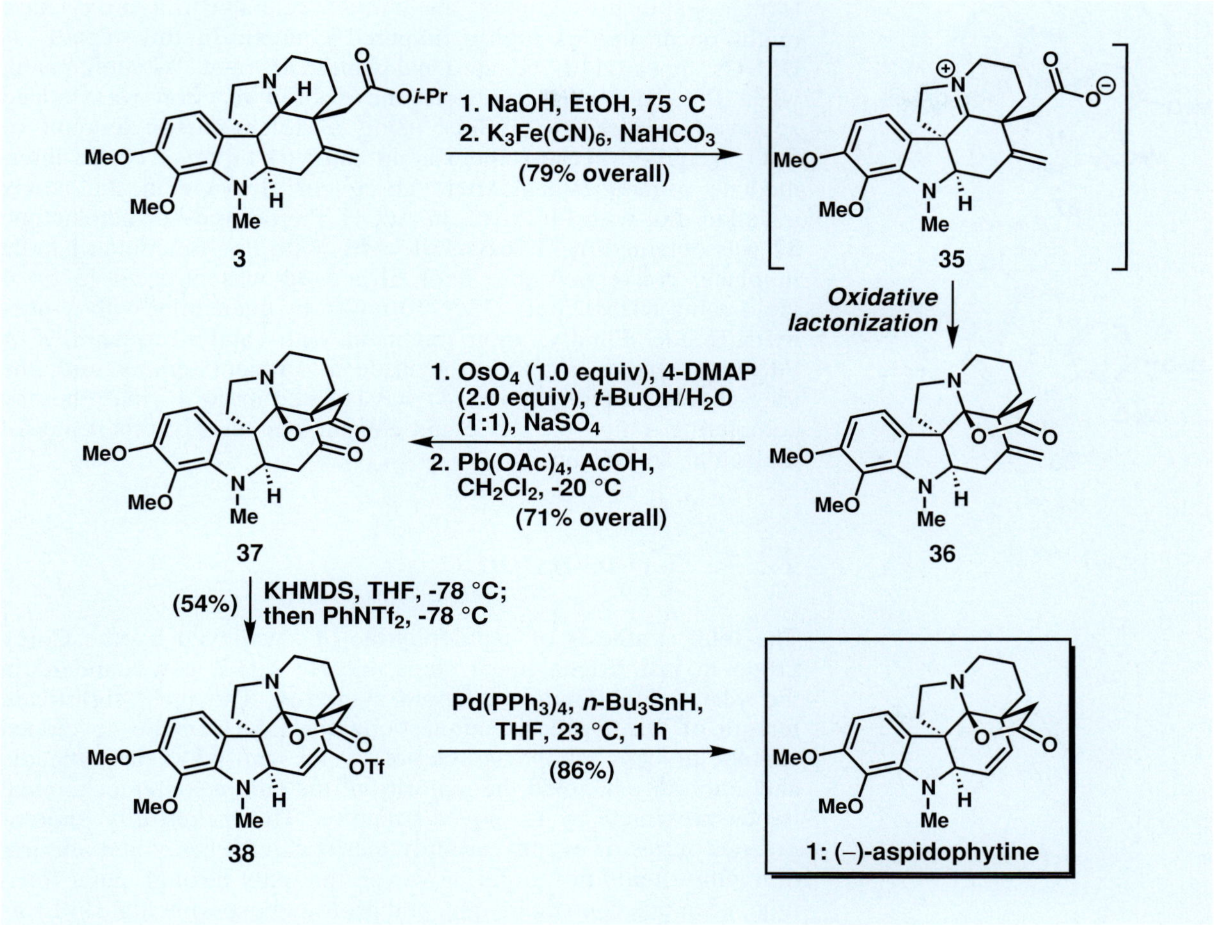

Scheme 7. Final stages and completion of the total synthesis of (–)-aspidophytine (**1**).

cyclic double bond in **36** to the corresponding ketone (**37**), proved impossible to effect with traditional OsO_4-based systems, a phenomenon most likely due to effective shielding of both faces of the terminal double bond by the resident polycyclic architecture. Even application of standard techniques to enhance the reactivity of osmium for dihydroxylation and to facilitate catalytic turnover, (for example, the addition of a tertiary nitrogen-containing activator such as pyridine or quinuclidine) similarly failed.[27] Faced with this unanticipated challenge, a new dihydroxylation system was required to advance the synthesis. Fortunately, based on extensive mechanistic studies previously undertaken by the Corey group exploring the activating influence of various amine-based ligands on osmium,[28] a potential solution was already in hand. Namely, it was hypothesized that by employing a ligand endowed with the properties of stronger basicity than pyridine and greater steric modesty than quinuclidene in precisely a 2:1 ratio with OsO_4, to form a stable bis-complex, an amine-accelerated dihydroxylation might occur in this highly hindered context. In this regard, 4-DMAP appeared to be an ideal candidate, and, as anticipated, when this base was employed the desired reaction was indeed achieved in just ten minutes using a stoichiometric amount of OsO_4 (required because there is no catalytic turnover of the intermediate osmate ester). After subsequent cleavage of the newly installed diol with $Pb(OAc)_4$ in AcOH, the desired nor-ketolactone **37** was obtained in 71 % overall yield. With this functional handle in place, conversion into enol triflate **38** was effected in 54 % yield using KHMDS at −78 °C followed by quenching with *N*-phenyltriflimide. Finally, upon treatment with catalytic quantities of $Pd(PPh_3)_4$ and tri-*n*-butyltin hydride at ambient temperature, the long-coveted aspidophytine (**1**) was isolated in 86 % yield, thereby completing a highly concise and elegant synthesis of this beautiful molecular architecture.

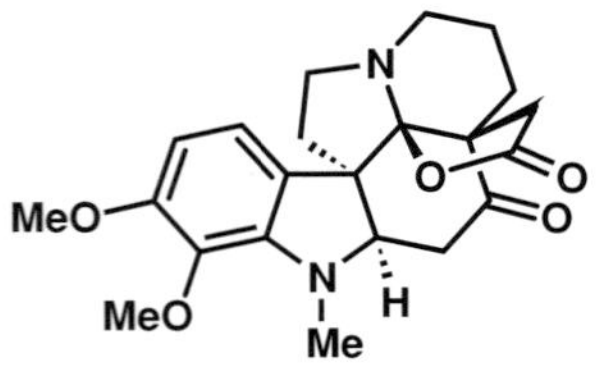

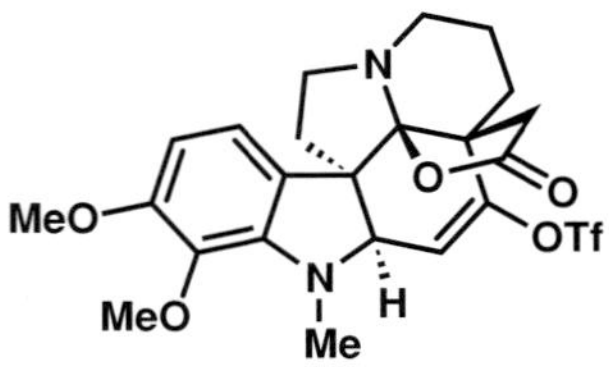

12.4 Conclusion

The total synthesis of aspidophytine (**1**), achieved by the Corey group in just fifteen linear steps, clearly sets a new standard in the vast field of indole alkaloid synthesis. The most significant feature of this synthesis unquestionably resides in the developed cascade process which cast the polycyclic scaffold of the molecule and admirably handled the majority of the complex stereochemical issues presented by its novel structure. This beautifully choreographed series of events certainly merits careful study and admiration. One should not forget, however, that only through inner fortitude, keen mechanistic insight, and profound experimental skill was such an elegant cascade sequence achieved. As such, the success of the work described in this chapter should serve as a source of inspiration to those individuals faced with similar challenges to persevere in their efforts to develop even more amazing reaction cascades.

References

1. R. Robinson, *J. Chem. Soc.* **1917**, 762.
2. a) W. S. Johnson, M. B. Gravestock, R. J. Parry, R. F. Myers, T. A. Bryson, D. H. Miles, *J. Am. Chem. Soc.* **1971**, *93*, 4330; b) W. S. Johnson, M. B. Gravestock, B. E. McCarry, *J. Am. Chem. Soc.* **1971**, *93*, 4332; c) M. B. Gravestock, W. S. Johnson, B. E. McCarry, R. J. Parry, B. E. Ratcliffe, *J. Am. Chem. Soc.* **1978**, *100*, 4274.
3. B. M. Trost, *Science* **1991**, *254*, 1471.
4. F. He, Y. Bo, J. D. Altom, E. J. Corey, *J. Am. Chem. Soc.* **1999**, *121*, 6771.
5. D. G. Crosby in *Naturally Occurring Insecticides* (Eds.: M. Jacobsen, D. G. Crosby), Marcel Dekker, New York, **1991**, p. 213.
6. a) E. F. Rogers, H. R. Snyder, R. F. Fischer, *J. Am. Chem. Soc.* **1952**, *74*, 1987; b) H. R. Snyder, R. F. Fischer, J. F. Walker, H. E. Els, G. A. Nussberger, *J. Am. Chem. Soc.* **1954**, *76*, 2819; c) H. R. Snyder, R. F. Fischer, J. F. Walker, H. E. Els, G. A. Nussberger, *J. Am. Chem. Soc.* **1954**, *76*, 4601; d) H. R. Snyder, H. F. Strohmayer, R. A. Mooney, *J. Am. Chem. Soc.* **1958**, *80*, 3708; e) M. P. Cava, S. K. Talapatra, K. Nomura, J. A. Weisbach, B. Douglas, E. C. Shoop, *Chem. Ind. (London)* **1963**, 1242; f) M. P. Cava, S. K. Talapatra, P. Yates, M. Rosenberger, A. G. Szabo, B. Douglas, R. F. Raffauf, E. C. Shoop, J. A. Weisbach, *Chem. Ind. (London)* **1963**, 1875; g) I. D. Rae, M. Rosenberger, A. G. Szabo, C. R. Willis, P. Yates, D. E. Zacharias, G. A. Jeffrey, B. Douglas, J. L. Kirkpatrick, J. A. Weisbach, *J. Am. Chem. Soc.* **1967**, *89*, 3061; h) D. E. Zacharias, *Acta Crystallogr., Sect. B* **1970**, *26*, 1455; i) P. Yates, F. N. MacLachlan, I. D. Rae, M. Rosenberger, A. G. Szabo, C. R. Willis, M. P. Cava, M. Behforouz, M. V. Lakshmikantham, W. Zeiger, *J. Am. Chem. Soc.* **1973**, *95*, 7842. Interest in these compounds still persists today in terms of isolation efforts and biosynthetic pathway studies; for a recent example, see: M. A. Mroue, K. L. Euler, M. A. Ghuman, M. Alam, *J. Nat. Prod.* **1996**, *59*, 890.
7. R. J. Sundberg in *Comprehensive Heterocyclic Chemistry II, Vol. 2* (Eds.: A. R. Katritzky, C. W. Rees, E. F. V. Scriven), Pergamon, London, **1996**, pp. 167–170.
8. M. Schröder, *Chem. Rev.* **1980**, *80*, 187.
9. R. E. Ireland, R. H. Mueller, A. K. Willard, *J. Am. Chem. Soc.* **1976**, *98*, 2868.
10. E. J. Corey, C. J. Helal, *Angew. Chem.* **1998**, *110*, 2092; *Angew. Chem. Int. Ed.* **1998**, *37*, 1986.
11. C. P. Jasperse, D. P. Curran, *J. Am. Chem. Soc.* **1990**, *112*, 5601.
12. a) R. B. Woodward, M. P. Cava, W. D. Ollis, A. Hunger, H. U. Daeniker, K. Schenker, *J. Am. Chem. Soc.* **1954**, *76*, 4749; b) R. B. Woodward, M. P. Cava, W. D. Ollis, A. Hunger, H. U. Daeniker, K. Schenker, *Tetrahedron* **1963**, *19*, 247; c) K. C. Nicolaou, E. J. Sorensen, *Classics in Total Synthesis*, VCH, Weinheim, **1996**, pp. 21–40.
13. F. Benington, R. D. Morin, L. C. Clark, Jr., *J. Org. Chem.* **1959**, *24*, 917.
14. R. J. Sundberg in *Comprehensive Heterocyclic Chemistry II, Vol. 2* (Eds.: A. R. Katritzky, C. W. Rees, E. F. V. Scriven), Pergamon, London, **1996**, pp. 121–131.
15. A. K. Sinhababu, R. T. Borchardt, *J. Org. Chem.* **1983**, *48*, 3347.
16. C. F. Huebner, H. A. Troxell, D. C. Schroeder, *J. Am. Chem. Soc.* **1953**, *75*, 5887.
17. P. J. Harrington, L. S. Hegedus, *J. Org. Chem.* **1984**, *49*, 2657. For a recent example of such a reductive cyclization with subsequent formation of a tryptamine derivative, see: K. C. Nicolaou, S. A. Snyder, K. B. Simonsen, A. E. Koumbis, *Angew. Chem.* **2000**, *112*, 3615; *Angew. Chem. Int. Ed.* **2000**, *39*, 3473.
18. G. F. Smith, *J. Chem. Soc.* **1954**, 3842.
19. O. Meth-Cohn, S. P. Stanforth in *Comprehensive Organic Synthesis, Vol. 2* (Eds.: B. M. Trost, I. Fleming), Pergamon, Oxford, **1991**, pp. 777–794.
20. a) B. M. Trost, T. A. Grese, D. M. T. Chan, *J. Am. Chem. Soc.* **1991**, *113*, 7350; b) H. Nishiyama, H. Yokoyama, S. Narimatsu, K. Itoh, *Tetrahedron Lett.* **1982**, *23*, 1267.
21. T. Imamoto, N. Takiyama, K. Nakamura, T. Hatajima, Y. Kamiya, *J. Am. Chem. Soc.* **1989**, *111*, 4392.
22. M. T. Crimmins, D. Dedopoulou, *Synth. Commun.* **1992**, *22*, 1953.
23. E. J. Corey, R. K. Bakshi, S. Shibata, *J. Am. Chem. Soc.* **1987**, *109*, 5551.
24. E. J. Corey, R. K. Bakshi, S. Shibata, C.-P. Chen, V. K. Singh, *J. Am. Chem. Soc.* **1987**, *109*, 7925.
25. E. J. Corey, M. Azimioara, S. Sarshar, *Tetrahedron Lett.* **1992**, *33*, 3429.
26. For examples of similarly incorporated steric bulk in such cyclopentenone systems effecting control in the CBS reduction, including bromine as a substituent, see: a) E. J. Corey, A. V. Gavai, *Tetrahedron Lett.* **1988**, *29*, 3201; b) E. J. Corey, R. K. Bakshi, *Tetrahedron Lett.* **1990**, *31*, 611; c) E. J. Corey, K. S. Rao, *Tetrahedron Lett.* **1991**, *32*, 4623; d) E. J. Corey, H. Kigoshi, *Tetrahedron Lett.* **1991**, *32*, 5025.
27. This approach was originally pioneered during World War II: R. Criegee, B. Marchand, H. Wannowius, *Justus Liebigs Ann. Chem.* **1942**, *550*, 99. For an excellent review on asymmetric dihydroxylation, including other tricks for activation, see: H. C. Kolb, M. S. VanNieuwenhze, K. B. Sharpless, *Chem. Rev.* **1994**, *94*, 2483.
28. E. J. Corey, S. Sarshar, M. D. Azimioara, R. C. Newbold, M. C. Noe, *J. Am. Chem. Soc.* **1996**, *118*, 7851.

1: CP-225,917 (phomoidride A)

2: CP-263,114 (phomoidride B)

13

K. C. Nicolaou (1999)

CP-Molecules

13.1 Introduction

Key concepts:

- **Arndt–Eistert homologation**
- **Intramolecular Diels–Alder reactions**
- **Evolution of synthetic strategies**
- **Cascade reactions**
- **Hypervalent iodine reagents**

When one encounters a total synthesis in the primary literature, the accomplishment is often presented as a "fait accomplit," such that the work is described as a series of conversions smoothly advancing starting materials and synthetic intermediates into the target molecule. While this delivery style communicates a clear and concise view of the successful solution, it unfortunately also conveys a somewhat stilted view of the actual progression of research programs in complex molecule synthesis because no synthetic endeavor ever proceeds with such clarity from an initial paper-based design. Indeed, the final strategy always results from a constantly evolving synthetic approach whose rate of mutation is dictated by the frequency with which substrates exhibiting unanticipated or undesired chemical reactivity are encountered.[1] In this chapter, we hope to impart at least a fraction of this inherent and dynamic quality of complex molecule construction by analyzing the process through which a synthetic strategy towards two formidable targets, the CP-molecules (**1** and **2**), reached maturation in the laboratories of the Nicolaou group at The Scripps Research Institute.[2]

Both of the CP-molecules, CP-225,917 (**1**) and CP-263,114 (**2**), were originally obtained in the early 1990s by Dr. Takushi Kaneko and his associates at Pfizer Central Research in Groton, Connecticut, from the fermentation broth of an unidentified fungus that grew on a Juniper tree in Texas. While such a beginning is quite typical for most secondary metabolites, these compounds would soon be elevated into a position of prominence once initial screening revealed that they could inhibit two critical enzymes in medically relevant pathways.[3] The first of these cellular targets, squalene synthase, is an essential participant in the biosynthesis of cholesterol, while the second, protein farnesyl transferase, mediates the addition of a farnesyl group to several proteogenic products of the *Ras* oncogene family such as the tumor promoter p21.

With little question, such an intriguing combination of enzyme inhibitory properties renders these compounds as highly opportunistic targets for total synthesis. However, a perhaps more compelling reason to undertake such a campaign resides in the manner in which their constituent atoms have been woven together into a uniquely compact and intricate polycyclic architecture. For example, while both **1** and **2** contain a total of 31 carbon atoms, Nature requires less than half of these (15) along with just three oxygen atoms to create the multifarious five-, six-, seven-, and nine-membered rings that comprise the heart of their structures. While both the sheer density and the manner in which this cyclic complexity has been fused provides a formidable synthetic challenge, the fact that these ring systems also contain an array of relatively rare and sensitive functionalities, such as a maleic anhydride, either a γ-hydroxylactone (as in **1**) or an internal ketal (as in **2**), and an anti-Bredt bridgehead olefin, elevates the task of their creation to high levels of difficulty. The last of these features, in particular, could likely instigate several synthetic dilemmas as its presence imposes a high degree of strain on the entire molecule due to the fact that its substituents cannot adopt a coplanar alignment. Although Bredt's original 1927 "rule" on the forbidden nature of such structures indicates that they should be impossible to synthesize (a conjecture which is predictive in most instances), a few exceptions have been noted in systems in which the size of the largest ring containing the double bond is at least eight-membered, and, more typically, nine-membered, as in the CP-molecules.[4] Only in these examples is such strain not inherently deleterious to the integrity of the bicycle.

Of course, the amalgamation of such ring systems and stereochemical complexity into a single and relatively small architectural unit begs the question as to how such structures are ever formed in Nature. Although many researchers have advanced different hypotheses to answer this query,[5] one of the best empirically supported pictures derives from the work of Professor Gary Sulikowski and his collaborators at Texas A&M University.[6] In their biogenetic proposal, which is shown in Scheme 1, the key structural elements of the CP-molecules result from the union of two monomeric synthons of type **3**, starting with their merger into enol **6** through a Michael addition effected by enzymatic activation of one substrate (**5**) and decarboxylation of the other (**4**). This adduct (**6**) can then rearrange into the near-complete CP architecture (**9**) by one of two mechanistically degenerate pathways. The first option, path a, projects a second Michael addition onto the neighboring maleic anhydride which initiates a subsequent [3,3] sigmatropic rearrangement of the Cope type, while an equally compelling alternative, path b, reflects an

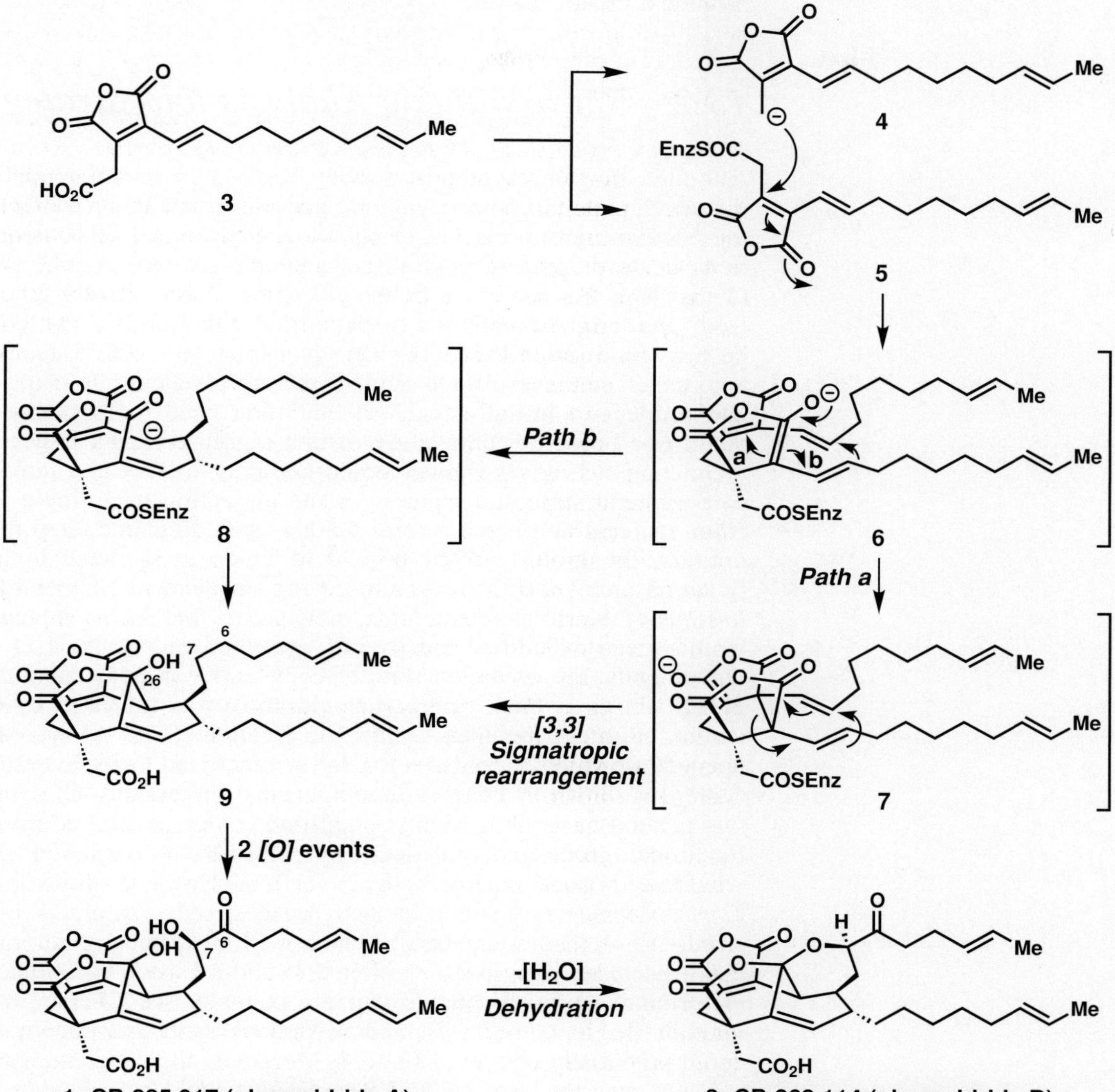

Scheme 1. Decarboxylative dimerization pathway for the biosynthesis of the CP-molecules (**1** and **2**). (Sulikowski, 2002)[6]

initial anionic rearrangement to **8** followed by the addition of the resultant carbon nucleophile to the proximal C-26 carbonyl, thereby completing the bicyclo[4.3.1] skeleton. From this advanced structure (**9**), only two oxidation events are required to create the proper functionalization at C-6 and C-7 corresponding to CP-225,917 (**1**); a final dehydration merging the γ-hydroxylactone and the adjacent alcohol group at C-7 into a six-membered pyran ring would then provide CP-263,114 (**2**).

13.2 Retrosynthetic Analysis and Strategy

Although biogenetic proposals often serve as a useful guide for the development of a retrosynthetic plan towards a target molecule, the Nicolaou group elected to pursue a different set of considerations in developing their strategy to synthesize the CP-molecules (**1** and **2**).[7,8] As shown in Scheme 2, since it was already known from the original isolation papers[3] that CP-225,917 (**1**) could be converted into CP-263,114 (**2**) upon treatment with a catalytic amount of methanesulfonic acid, it therefore seemed both logical and prudent to initially focus all attention on **1**.* How, though, could one begin the unraveling of this highly compact molecular architecture? Logical choices could certainly include the removal of peripheral structural features or the dissection of a single ring from its central bicyclic core. Rather than implementing these options, the group elected instead to guide their initial simplifications along the lines of introducing an element of symmetry to some of the subgoal structures, anticipating that such a maneuver would greatly facilitate the overall retrosynthetic analysis of the target. Thus, the oxidation state at C-27 in **1** was adjusted to that of **10**, whose terminal carboxylic acid chain was truncated by one carbon atom and reduced to **11**. The lactol motif of **11** was then opened, affording access to a locally symmetrized C-14 quaternary center (as shown in **12** with dimethyl acetal protection). Of course, this general analysis is highly simplified in that several additional functional group manipulations and changes in oxidation state would need to accompany these conversions. However, these objectives do appear reasonable in their broadest sense as the classical Arndt–Eistert homologation[9] (which will be discussed in more detail later) has long been employed to add a carbon unit adjacent to terminal carboxylic acids (for an example, see Chapter 8 on vitamin B_{12} in *Classics I*), and several oxidants are known that could potentially convert **10** into **1**. Moreover, the differential convex and concave faces of the CP-molecules could be anticipated to enable selective manipulation of the substituents at C-14 in **12** to

---

* One should note that the absolute configuration of the CP-molecules is, in fact, opposite to that depicted throughout these pages; at the start of this program, however, this was not known. The drawn structures correspond to the antipode created during this synthetic program, which ultimately enabled an assignment to be made. Later in the chapter, when total syntheses from other groups are discussed, the structures will similarly reflect the antipode prepared in each laboratory. In addition, although the Nicolaou group developed both racemic and asymmetric syntheses of the CP-molecules, only the latter will be presented here.

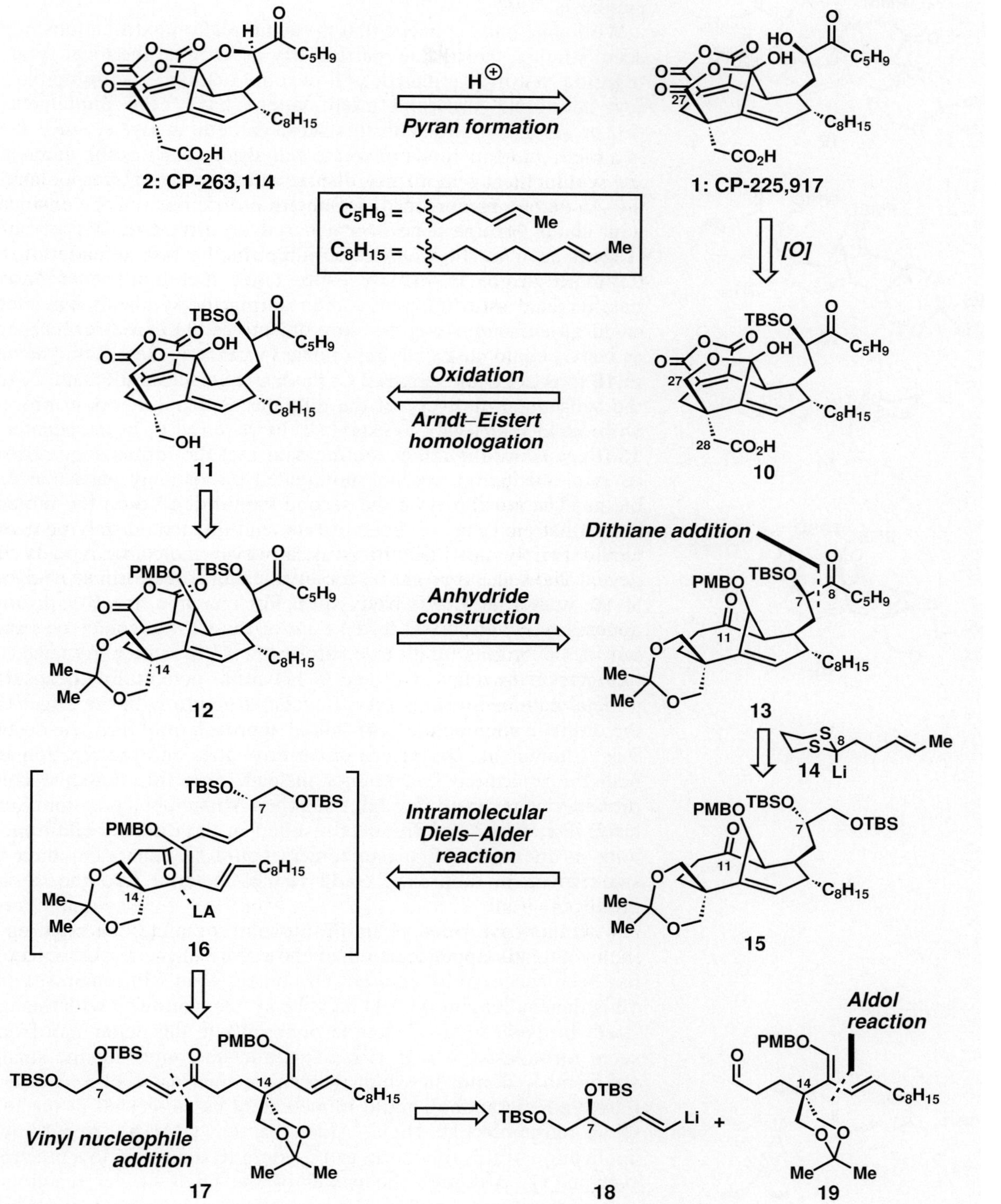

O, H, O, C_5H_9, O, O, O, O, O, C_8H_{15}, CO_2H

2: CP-263,114

H⊕

Pyran formation

O, HO, O, O, OH, C_5H_9, O, O, 27, C_8H_{15}, CO_2H

1: CP-225,917

C_5H_9 = Me

C_8H_{15} = Me

[O]

O, TBSO, O, O, OH, C_5H_9, O, O, C_8H_{15}, OH

11

Oxidation

Arndt–Eistert homologation

O, TBSO, O, O, OH, C_5H_9, O, O, 27, C_8H_{15}, 28, CO_2H

10

O, PMBO, TBSO, O, O, C_5H_9, O, O, C_8H_{15}, Me, Me, 14, O

12

Anhydride construction

Dithiane addition

O, PMBO, TBSO, 7, 8, C_5H_9, O, 11, O, C_8H_{15}, Me, Me, O

13

S, S, 8, Me, 14, Li

TBSO, 7, OTBS, PMBO, O, O, C_8H_{15}, O, Me, Me, 14, LA, O

16

Intramolecular Diels–Alder reaction

PMBO, TBSO, 7, OTBS, O, 11, O, C_8H_{15}, Me, Me, O

15

OTBS, TBSO, 7, O, PMBO, 14, C_8H_{15}, O, O, Me, Me

Vinyl nucleophile addition

17

OTBS, TBSO, 7, Li

18

\+

Aldol reaction

O, PMBO, 14, C_8H_{15}, O, O, Me, Me

19

Scheme 2. Nicolaou's retrosynthetic analysis of the CP-molecules (**1** and **2**).

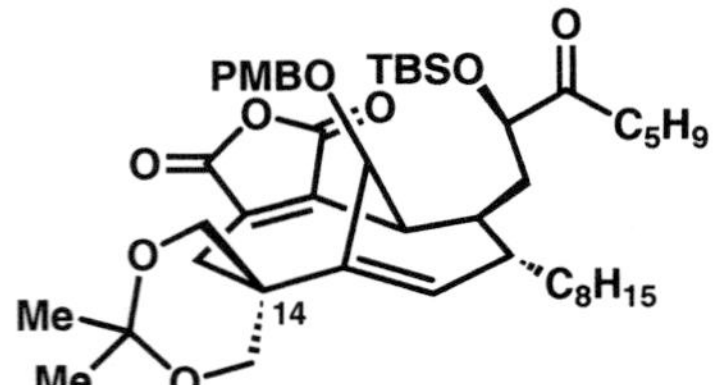

match the chiral format of **1** and **2** at the appropriate juncture in the synthesis.

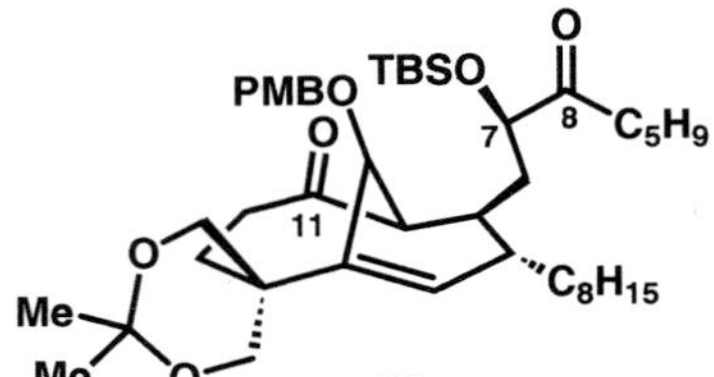

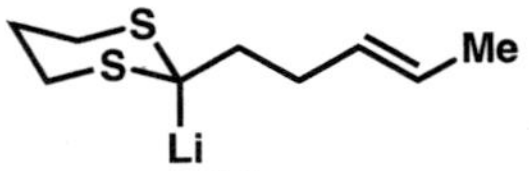

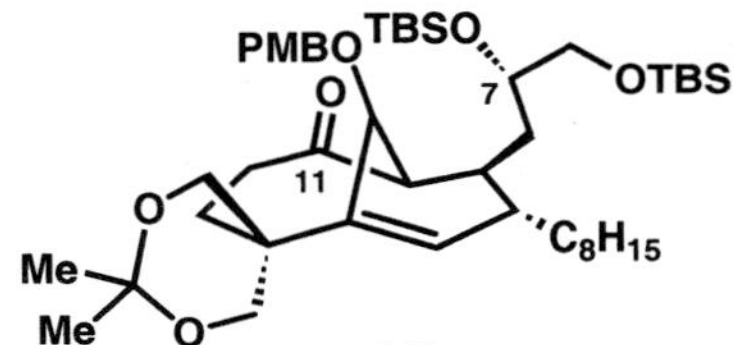

While one must concede that these particular modifications might seem strange, their value, particularly in terms of the local symmetrization that they effected, will become readily discernible once a few additional structural modifications have been implemented. So, pressing forward with this retrosynthetic analysis, since there is a plethora of methods to create anhydrides such as the one residing within **12**, this motif was disassembled from the bicyclic core to a C-11 ketone precursor (**13**). Subsequent excision of the exocyclic acyl chain through a transform based on dithiane addition[10] of a reagent such as **14** would then simplify the task of creating the CP-molecules to **15**. Although this latter operation seems innocuous, its successful implementation during the synthesis was predicated on accomplishing two key objectives: 1) that a reagent such as $NaIO_4$ could cleave an unprotected variant of the 1,2-diol system in **15** to provide the required C-7 aldehyde electrophile, and, 2) that the subsequent delivery of the dithiane nucleophile could proceed with facial selectively to establish the required C-7 stereocenter in **13**. Because of the rather routine nature of the former requirement, its implementation was not anticipated to pose any particular problems. The feasibility of the second would depend on the inherent conformational bias of the substrate, and thus could only be evaluated experimentally. Significantly, however, molecular models suggested that success might be achieved using some form of chelation to the numerous Lewis basic sites. For example, the C-7 position appears relatively distal to the core of the CP-molecules on paper, but in reality it is relatively close to the C-11 center. As such, the two carbonyl groups (C-7 and C-11) could potentially encapsulate a metal cation, forming a rigid cyclic structure with the effect that the dithiane nucleophile (**14**) should approach only from the desired face. Should this conjecture prove erroneous and the reaction furnish the undesired C-7 epimer instead, then this outcome could probably be reversed through a standard Mitsunobu inversion. Similarly, if a mixture of products resulted in the dithiane addition, as long as the two C-7 diastereomers could be separated, then the same inversion approach could funnel material into the desired product.

With this first round of simplifications complete, the majority of the cyclic bulk appended to the central core of the target molecules has been successfully excised. Of course, what still remains within **15** is the challenging [4.3.1] bicyclic system endowed with the anti-Bredt bridgehead C–C double bond. While this latter motif does seem formidable, if it is viewed within the context of the smaller six-membered ring in which it resides, then a possible course of retrosynthetic action should immediately be suggested in the form of an intramolecular Diels–Alder reaction (IMDA). Accordingly, application of this transform to **15** would lead directly to a precursor such as **17**. Although the power of the Diels–Alder reaction to create complicated polycyclic architectures is rarely paralleled,[11]

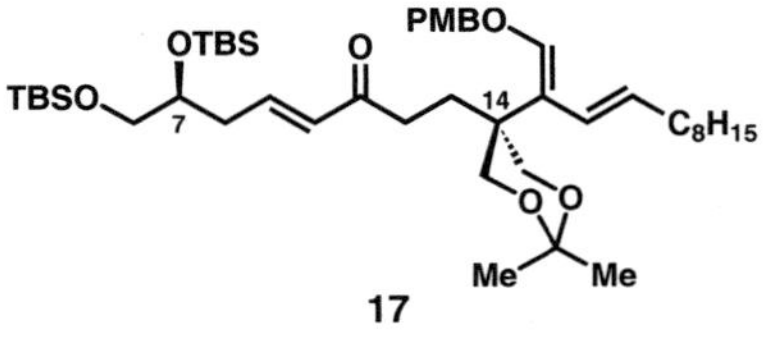

such Type II [4+2] cycloadditions[12] in which the dienophile is tethered to the C-2 position of the diene system (as shown in the neighboring column figures) are one of the few methods capable of reliably forming anti-Bredt olefins. As such, the Nicolaou group could derive some measure of comfort in the ability of this reaction to stitch together the basic core of the CP-molecules upon which the remainder of the strategy could be applied. As an additional benefit of adopting this particular method to generate the bicyclic system, the cycloaddition could potentially proceed with asymmetric induction since the lone stereocenter planted within **17** could lead to the exclusive adoption of a transition state such as **16**. Alternatively, if the resident functionality itself could not lead to such facial selectivity, then perhaps the numerous Lewis basic sites within **17** could be called upon to coordinate an appropriate external ligand to achieve the same objective. Of course, while all endeavors in modern target-oriented synthesis should strive to create a single stereoisomer of a natural product, this standard goal assumed a much higher level of importance for the CP-molecules since their absolute configuration was unknown at the start of this program. Accordingly, access to a pure sample of either antipode through chemical synthesis would permit a definitive structural assignment to be made.

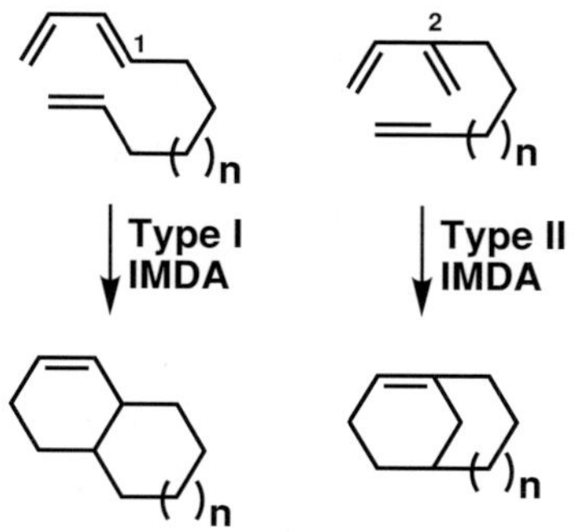

Having now reduced the challenging architecture of the CP-molecules to structure **17**, the benefit of previously symmetrizing the C-14 center should finally be clear as it opened several avenues to create this new subtarget using simple materials. For example, the addition of a vinyl nucleophile such as **18** to aldehyde **19**, followed by an oxidation step to complete the requisite α,β-unsaturated system, could constitute one approach to deliver this key Diels–Alder precursor (**17**). Had the C-14 carbon in **17** still been homochiral, such facility would not exist since its specific positioning makes an efficient asymmetric synthetic route, or the identification of a suitable starting material that could be elaborated to such an adduct, quite difficult to envision.

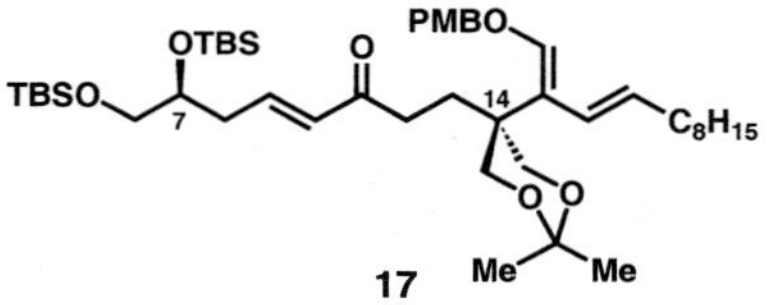

Overall, this plan to access the CP-molecules (**1** and **2**) reflects careful analysis of the numerous problems posed by its unique molecular connectivities, and, on the basis of existing literature precedent, would appear to be a sound approach worthy of empirical pursuit. However, the pathway chartered by this analysis did not ultimately prove capable of delivering either **1** or **2**, though it did provide a general road-map. Its failure was not due to some major flaw in its appreciation for the known fundamentals of organic chemistry, but instead was the product of a series of unlikely pitfalls encountered during "conventional" operations in which steric shielding and the instability of some of the already installed functionality wreaked havoc. Rather than provide more details now and spoil key elements of the story ahead, we shall, instead, commence our discussions with the implementation of the strategy laid out in Scheme 2, providing revised analyses at those junctures where obstacles dictated a change of course.

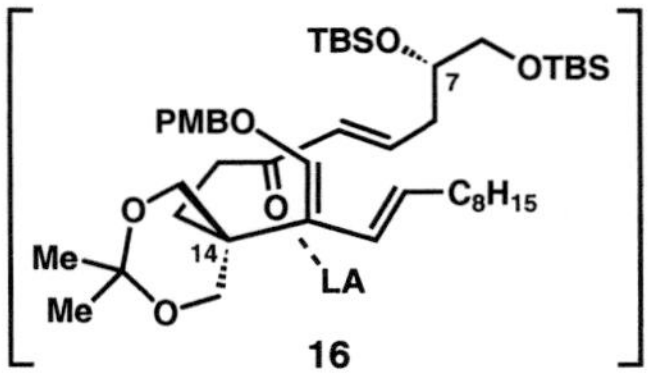

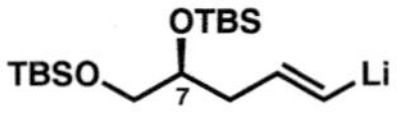

1 2 n n Type I IMDA Type II IMDA n n

OTBS O PMBO TBSO 7 14 C_8H_{15} O O Me Me **17**

TBSO 7 OTBS PMBO O C_8H_{15} O Me Me 14 O LA **16**

OTBS TBSO 7 Li **18**

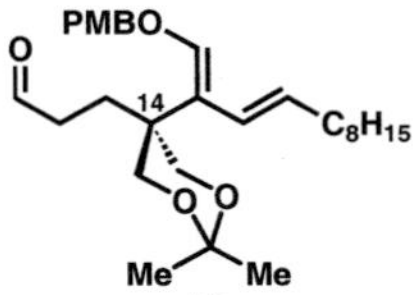

O PMBO 14 C_8H_{15} O O Me Me **19**

13.3 Total Synthesis

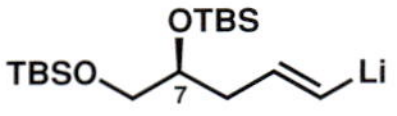

Seeking to test the viability of the Diels–Alder reaction that would hopefully forge the majority of the core carbon framework of the CP architecture, the Nicolaou group began their efforts with attempts to create both **18** and **19**, the precursors which, as mentioned above, were defined as the critical building blocks for the key Diels–Alder precursor (**17**). Accordingly, in order to generate the lengthy C_8H_{15} carbon chain of **19**, efforts began by constructing aldehyde **22** (Scheme 3), anticipating that this piece could be merged with the remainder of **19** through an aldol condensation. The synthesis of this piece began with cyclooctene (**20**), a compound whose double bond could be ozonolyzed and concurrently desymmetrized to aldehyde acetal **21** following methodology originally established by S. L. Schreiber.[13] Having accomplished this highly efficient, one-pot transformation, a subsequent Wittig olefination under Schlösser-modified conditions[14] to generate an *E*-disposed olefin, followed by acid-catalyzed cleavage of the now extraneous dimethyl acetal, then completed **22**. As an aside, although the same fragment could also be accessed from a commercially available straight chain diol, the number of protecting group manipulations required to achieve the needed differentiation of its terminal groups made this route far longer than that depicted in Scheme 3.

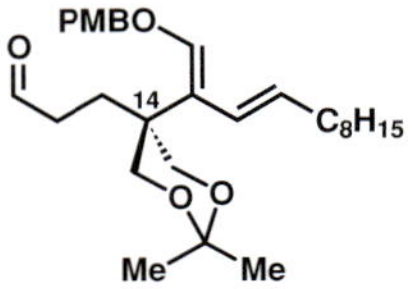

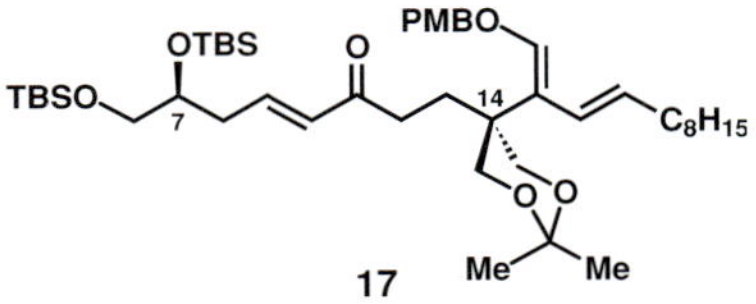

With this substrate in hand, a suitable coupling partner for an eventual aldol reaction had to be found in order to attempt the formation of **19**. For this purpose, it was forecasted that a compound such as imine **26** (Scheme 4) would be an effective choice, as it could likely enable the facile arrival at **19** following its merger with **22**.[15] This conjecture indeed proved accurate. The synthesis of **26** began from the readily available dimethyl malonate (**23**) using iterative alkylations to append two functionalized side chains to the central methylene group of this compound. With these operations leading to the assembly of **24** in 93 % overall yield, the next steps sought to convert the two methyl esters into an acetonide since the carbon atom to which these groups are appended is destined to become the C-14 quaternary center of the CP-molecules (the position which the analysis above had so carefully sought to symmetrize). As indicated, after an initial reduction of

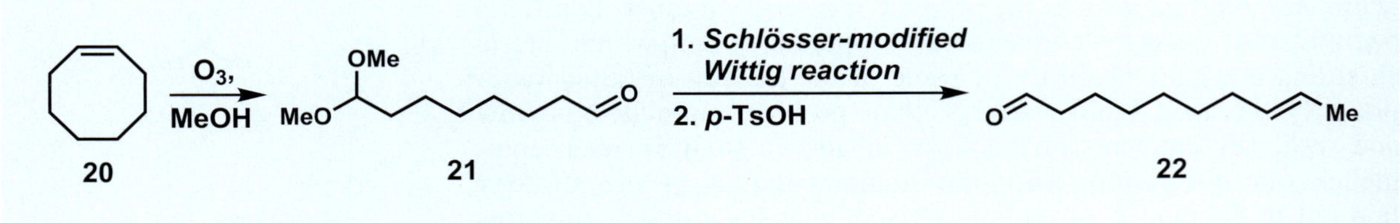

Scheme 3. Synthesis of aldehyde chain **22**.

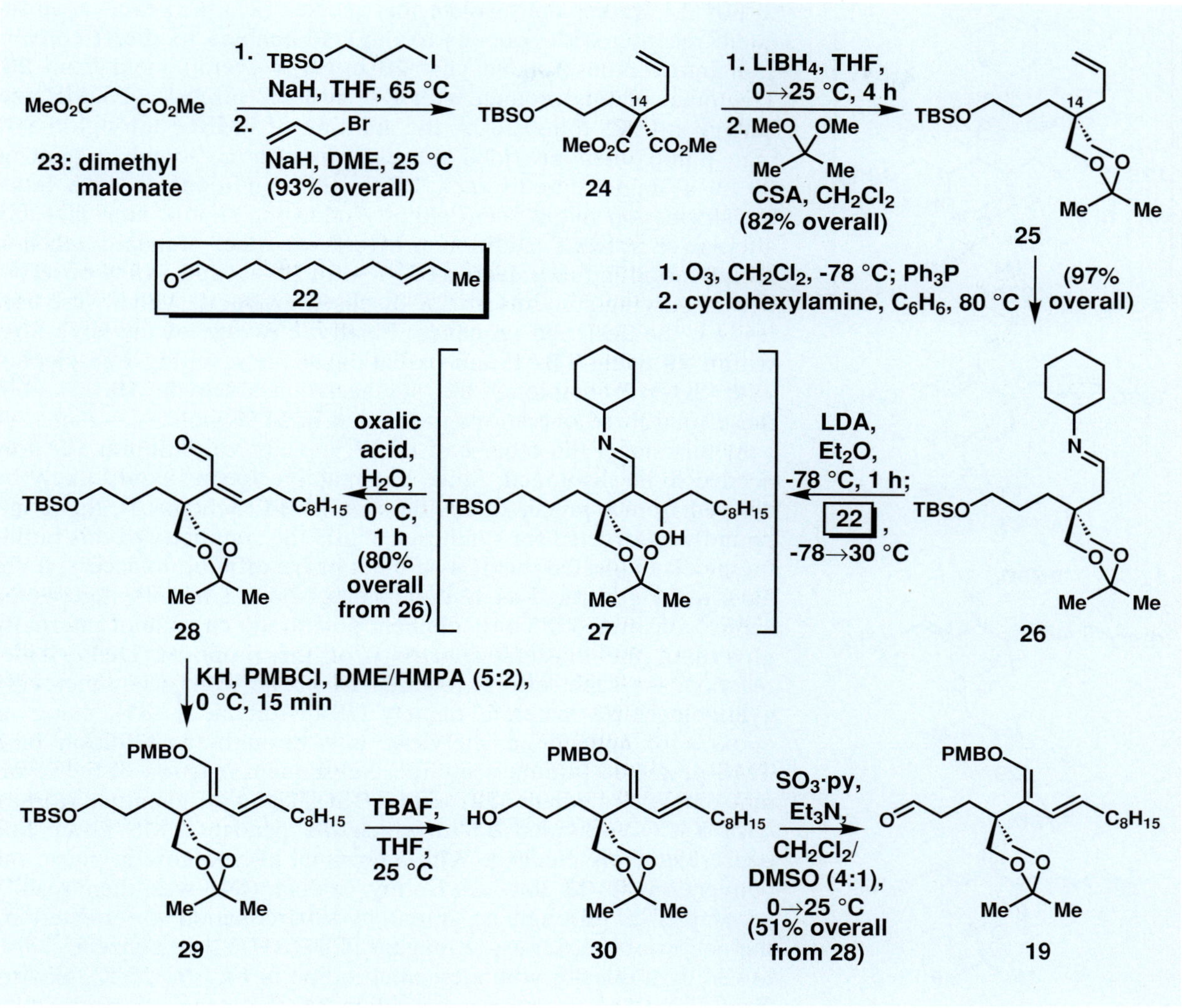

Scheme 4. Synthesis of key building block **19**.

24 with $LiBH_4$ in THF, acetal capture of the resultant diol was smoothly accomplished with 2,2-dimethoxypropane under the activating influence of catalytic CSA, furnishing **25** in 82 % yield for these two operations. The synthesis of imine **26** was then accomplished by initial ozonolytic cleavage of the terminal alkene within **25**, followed by reaction of the resultant aldehyde with cyclohexylamine in refluxing benzene using a Dean–Stark trap to drive the reaction to completion by the azeotropic removal of water.

Seeking now to complete the gross architecture of **19**, compound **26** was first deprotonated at the position adjacent to its imine motif using LDA in Et_2O at −78 °C, and then treated with aldehyde **22** to form the desired aldol product (**27**) upon warming the reaction to 30 °C.* Rather than isolate this species (**27**), however, a subsequent reaction with aqueous oxalic acid enabled its direct conversion into α,β-unsaturated enal **28** in 80 % overall yield from **26**. Treatment of this product with KH in a 5:2 mixture of DME and HMPA at 0 °C, followed by the addition of PMBCl through a syringe pump, then provided the electron-rich *E*,*Z*-diene system of **29** as a single stereoisomer. While the conditions for this latter transformation might seem slightly odd, one should note that any alternative solvent, addition protocol, or other standard reaction modification led to a decreased yield of **29** as well as the observation of varying amounts of *E*,*E*-disposed products, which were useless for the designed sequence. Finally, cleavage of the silyl ether within **29** using TBAF and oxidation of the resulting free alcohol with SO_3·py completed the synthesis of aldehyde **19**. Overall, these final three operations proceeded in 51 % yield.

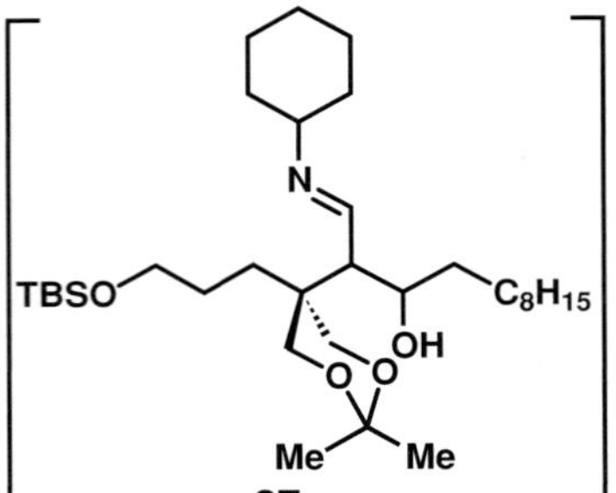

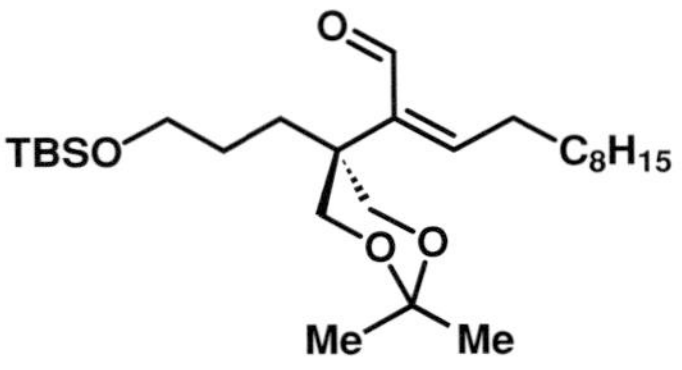

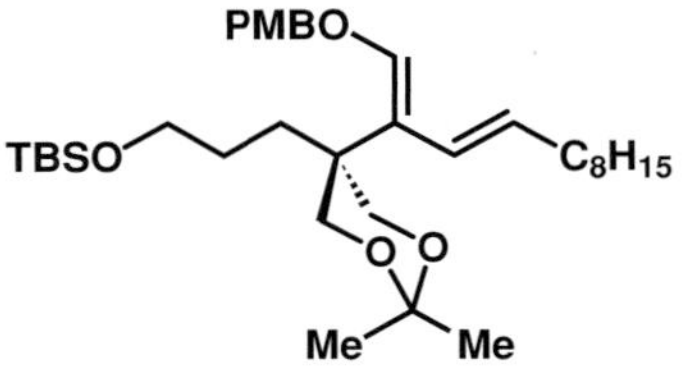

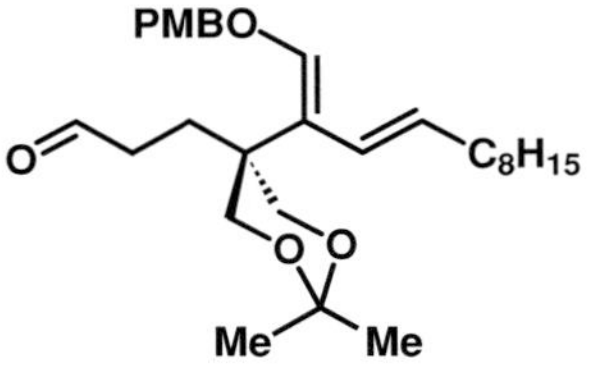

At this stage, the other half of **17**, namely vinyllithium **18**, now needed to be fashioned. Since this reactive species could likely be derived from a precursor iodide such as **34** (Scheme 5), this compound was targeted for synthesis. While the simplicity of this building block suggested that it would not prove difficult to access, it did have to be generated as a single stereoisomer since its stereogenic center constituted the only element potentially capable of internally governing the diastereoselectivity of the proposed Diels–Alder reaction. As such, efforts towards **34** began from a commercially available chiral material, namely (*R*)-(+)-glycidol (**31**), using its epoxide to append an acetylene unit through the addition of a TMS-protected lithium acetylide. Subsequent reaction of the resultant diol intermediate (**32**) with TBSOTf and an eventual workup with K_2CO_3 provided **33** in which the pendant TMS group had been selectively cleaved. With a terminal alkyne now in place, the conversion of **33** into an *E*-vinyl iodide (**34**) was then readily accomplished through an initial hydrozirconation as effected by the action of the Schwartz reagent [$Cp_2Zr(H)Cl$] in benzene,[16] followed by a quench with elemental iodine in CCl_4 at 25 °C. Having finally realized a synthesis of iodide **34**, the stage was now set to merge the two main fragments. Pleasingly, following the activation of **34** to nucleophile **18** with *n*-BuLi, the slow addition of this species to a solution of aldehyde **19** in THF at −78 °C achieved the goal, furnishing the anticipated product (**35**) in 90 % yield. A final oxidation using Dess–Martin periodinane (DMP) then completed the construction of the desired Diels–Alder precursor (**17**).

With this compound in hand, attempts could now begin to examine the viability of the Diels–Alder approach to cast the bicyclic [4.3.1] core of the CP-molecules. Fortunately, preliminary probes of the reaction using Me_2AlCl as a catalyst in CH_2Cl_2 at −10 °C

* As this example demonstrates, imines are particularly effective substrates for α-alkylation through enolate chemistry. In contrast, an aldehyde would not be feasible in this scenario.

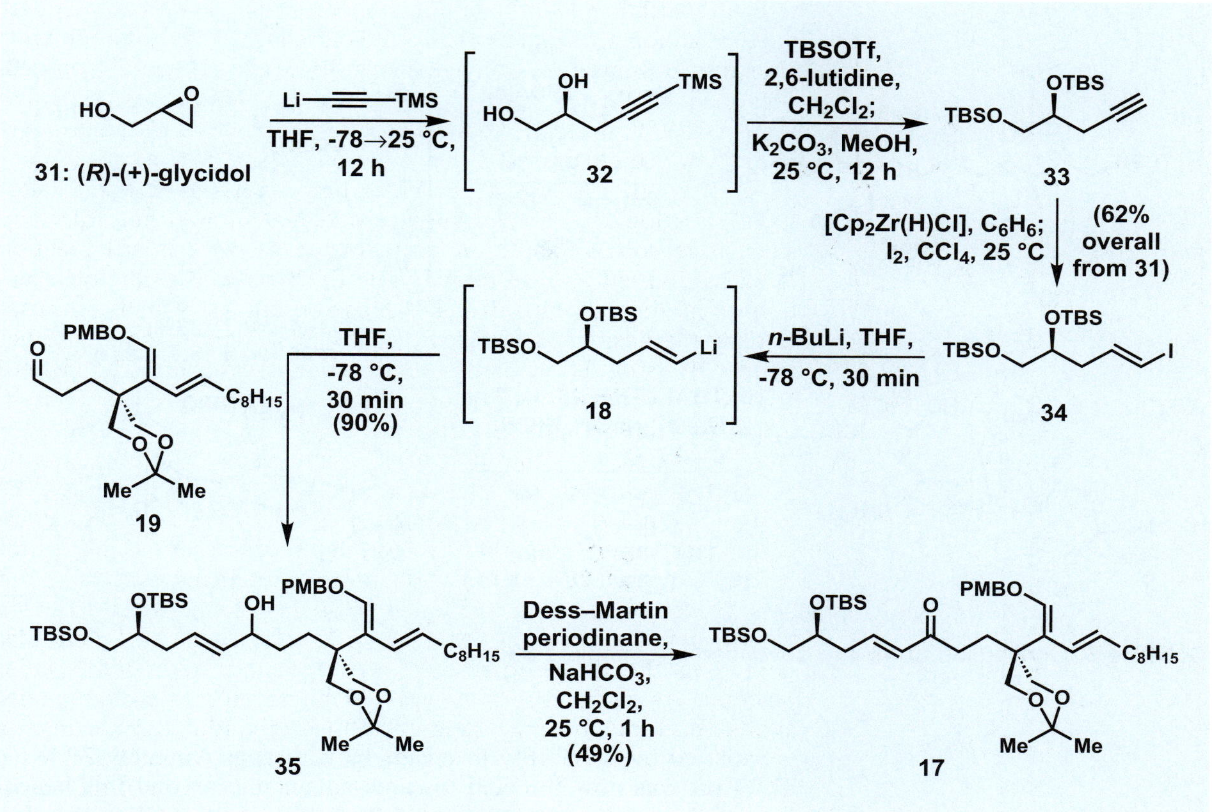

Scheme 5. Synthesis of Diels–Alder precursor **17**.

revealed that the desired transformation of **17** into **15** (see Scheme 6) was indeed quite facile, providing this compound and its diastereomer (**37**, see column figure) in a 2:1 ratio and in 90 % yield. Although the predominance of the desired stereoisomer (**15**) from this initial experiment was certainly pleasing, considerable effort had to be expended in order to enhance its diastereoselectivity. After a nearly exhaustive screening of reaction solvents, temperatures, and aluminum-based Lewis acids (including several chiral Lewis acids), it was discovered that optimal selectivity for **15** could be achieved with a 20 mol % loading of catalyst **36** in toluene at −80 °C, ultimately providing a reproducible 5.7:1 ratio of **15** and **37** in 88 % combined yield. Interestingly, in the absence of the relatively remote C-7 stereocenter in **17** (i. e. an achiral substrate), no chiral Lewis acid could be found that could engender useful levels of diastereoselectivity. Moreover, one should note that although deprotection of the silyl protected alcohols within **17** should, in principle, provide a substrate with better ability to chelate Lewis acids (and thus one that might lead to enhanced diastereoselectivity), such a compound consistently decomposed when examined in the context of this key C–C bond-forming process.

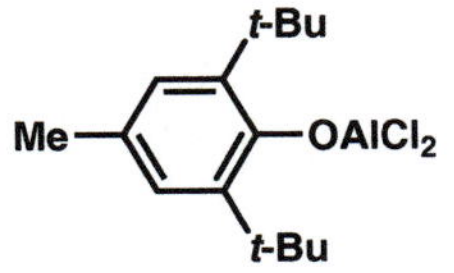

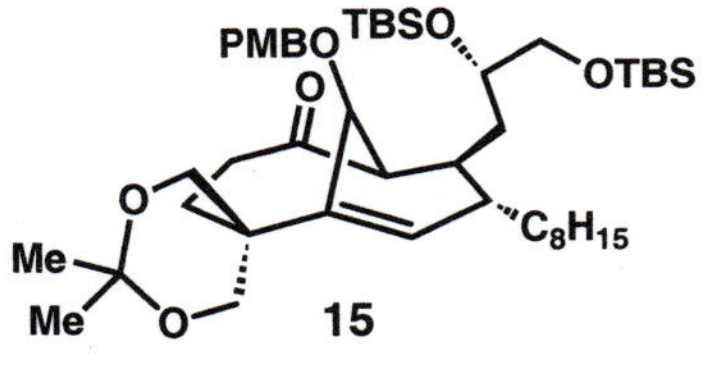

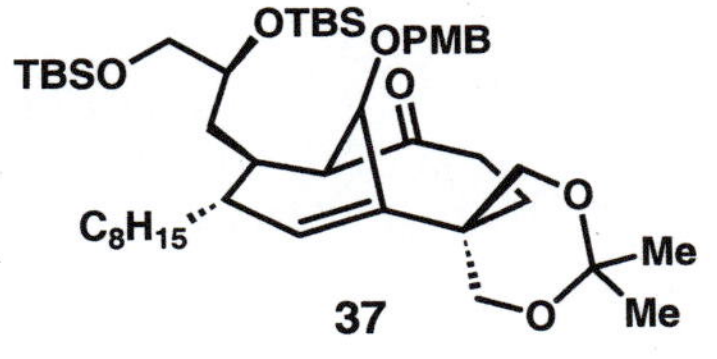

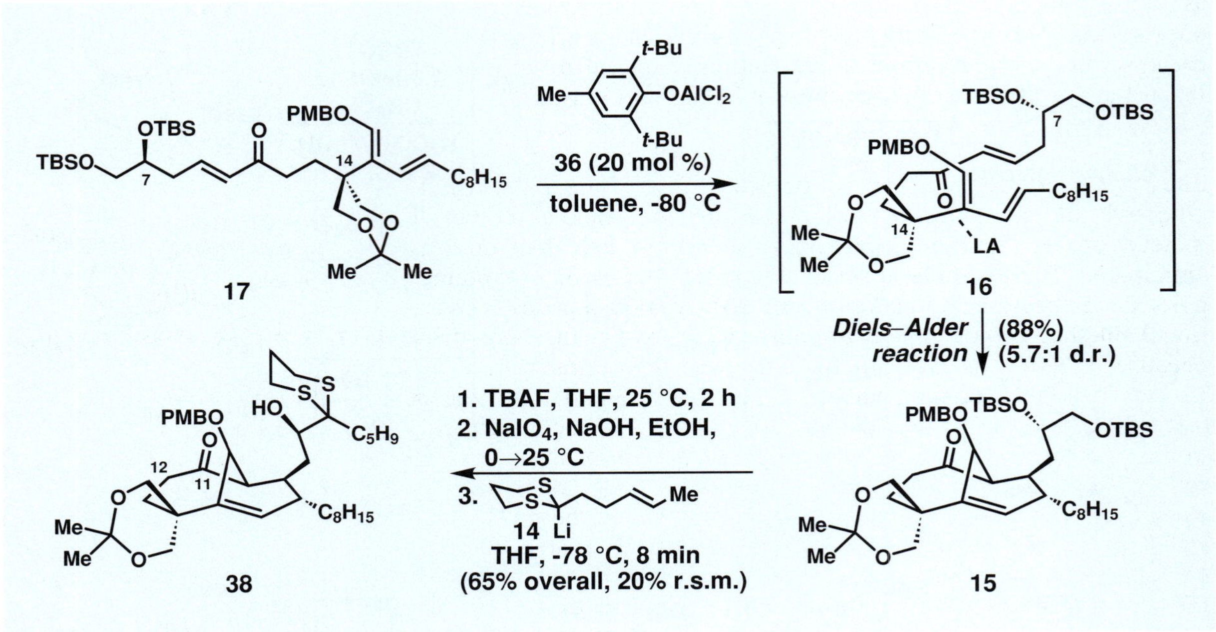

Scheme 6. Conversion of 17 into advanced intermediate 38.

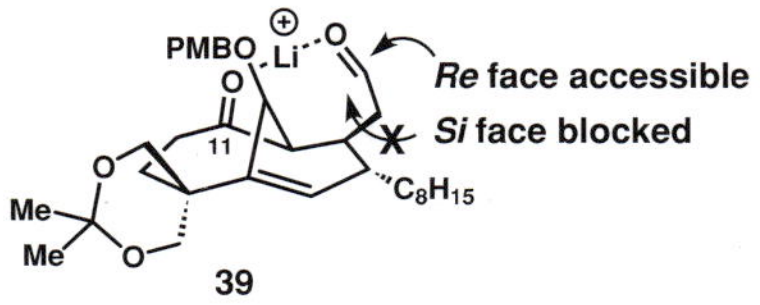

Satisfied by the ability to create 15 with approximately 70% *ds*, all effort was now directed towards advancing beyond this beachhead and tackling the remaining challenges of the CP-molecules, starting with the installation of the upper C_5H_9 side chain and its adjacent C-8 ketone. As mentioned earlier, this task was to rely on the addition of a dithiane nucleophile. Thus, to prepare the electrophilic C-7 aldehyde that would be needed for this purpose, the two TBS ethers within 15 were removed through the action of TBAF, and the resultant diol was then successfully cleaved with $NaIO_4$ under basic conditions to provide the requisite substrate. Upon the addition of dithiane 14 to this aldehyde in THF at −78 °C and a subsequent reaction quench after just 8 minutes of stirring, the expected coupling product (38) with the desired stereochemistry at C-7 was formed as the major product with 11:1 selectivity. Although this result was fortunate, as it obviated the need for additional steps seeking to effect the epimerization of this center, the reason for its accomplishment is far from random. Indeed, if lithium chelation occurred between the C-11 ketone and the C-7 aldehyde to frame a rigid eight-membered ring (as shown in reactive conformation 39 in the neighboring column figure), then approach of 14 from the desired *Re* face would be favored because the alternative *Si* face would be effectively blocked. As evidence for the validity of this mechanistic conjecture, a 7:1 ratio of products in favor of the undesired, *S*-configured, C-7 alcohol resulted when the same dithiane nucleophile (14) was added to a substrate such as **40** in which such chelation cannot exist. Thus, strikingly, it appears that only the ability of a lithium cation to bridge the two carbonyl units was required to resoundingly overturn the inherent facial bias of the substrate for this nucleophilic addition.

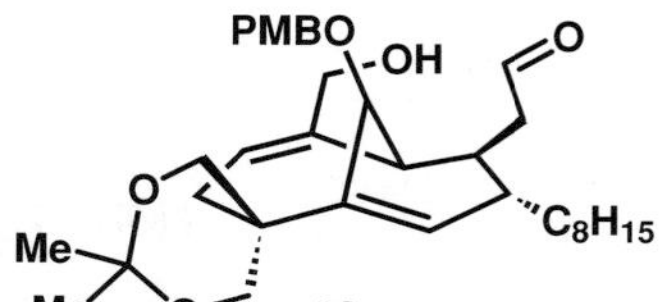

While progressing the synthesis to this stage was certainly satisfying in terms of its relative brevity (only 15 steps in the longest linear sequence) as well as its stereoselectivity, the next critical objective of the planned route, creating the maleic anhydride moiety, proved far more challenging to accomplish than initially anticipated. For example, a series of model studies on compounds possessing the same C-11 ketone as **38** revealed that no conventional method could open a pathway to reach the anhydride because, as soon as any functionality was incorporated at C-12, its steric bulk combined with that at C-11 blocked any reagent from accessing these sites further. Although the reader is directed to the primary literature for a complete discussion of the numerous approaches that were attempted along these lines,[7] one series of experiments is worthy of more careful evaluation here. While it failed to deliver the desired anhydride, its unanticipated outcome served as the seed for an idea that provided a ray of hope that the developed synthetic route would not have to be entirely abandoned.

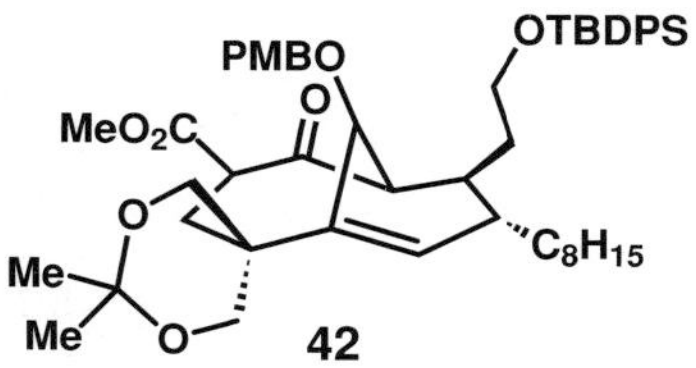

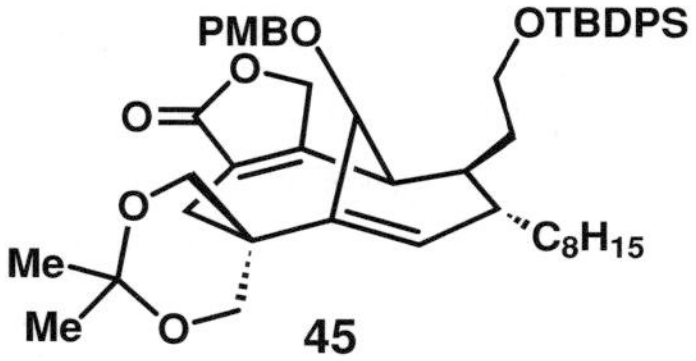

As shown in Scheme 7a, following the formation of **42** through a quench of the lithium enolate of **41** with methylcyanoformate (Mander's reagent), a subsequent attempt to effect an epoxidation using Corey's relatively small sulfur-based ylide did not lead to the anticipated compound (**43**), but instead to several side products. Among these, however, was lactone **45**, the first structure obtained up to that point which remotely resembled the anhydride, but whose formation at first seemed quite mysterious. Thoughtful analysis eventually pointed to the likelihood that **45** resulted from an initial E_2-type β-elimination of the hydrogen atom adjacent to the ester functionality in **43** (induced by the excess base used to form the sulfur ylide originally), followed by nucleophilic attack of the resultant alkoxide (**44**) onto the proximal ester. As one might expect, as soon as this picture became clear, immediate attempts were made to optimize the reaction, expecting that **45** could be directly oxidized to the complete maleic anhydride (**46**). Unfortunately, no conditions could be found that were capable of reliably delivering more than the 10% yield of **45** that was obtained in the first experiment.

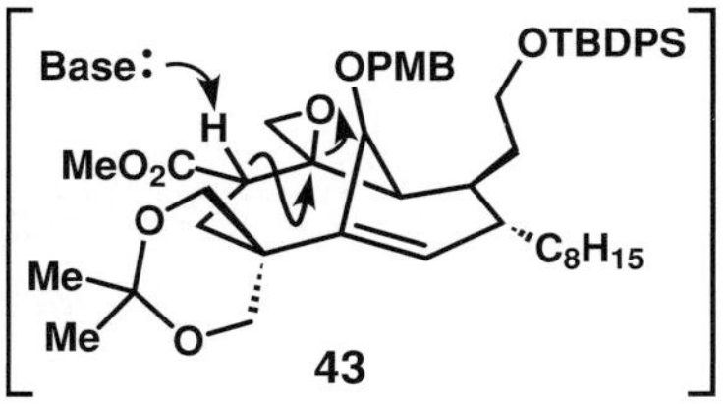

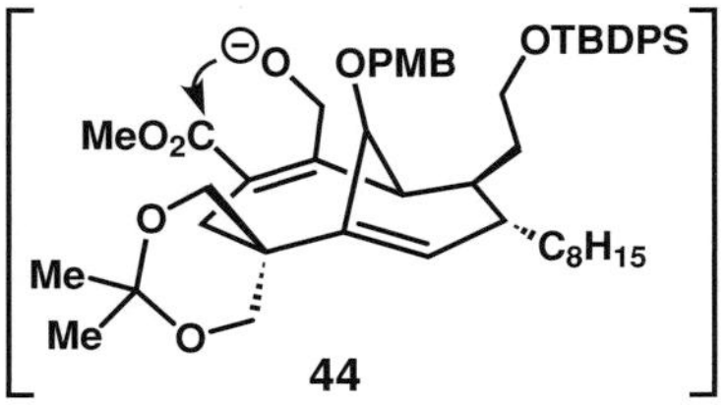

Since the low-yielding nature of this reaction seemed to have its deliterious origins in the epoxide-forming step, and not the later operations of the sequence, the Nicolaou group reasoned that perhaps a better approach would be to create the epoxide ring from a diol such as **48** (see Scheme 7b). Moreover, if the ester within **42** was exchanged for a cyanide, perhaps the entire anhydride (**47**) could be formed in a single pot, since the same β-elimination of **49** to **50** and subsequent 5-*exo*-dig cyclization onto this alternate C-12 substituent would afford a carbocyclic iminobutenolide (**51**).

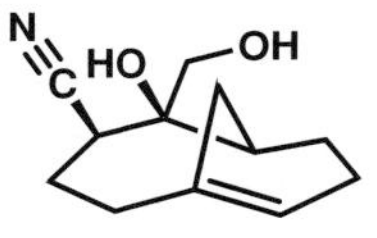

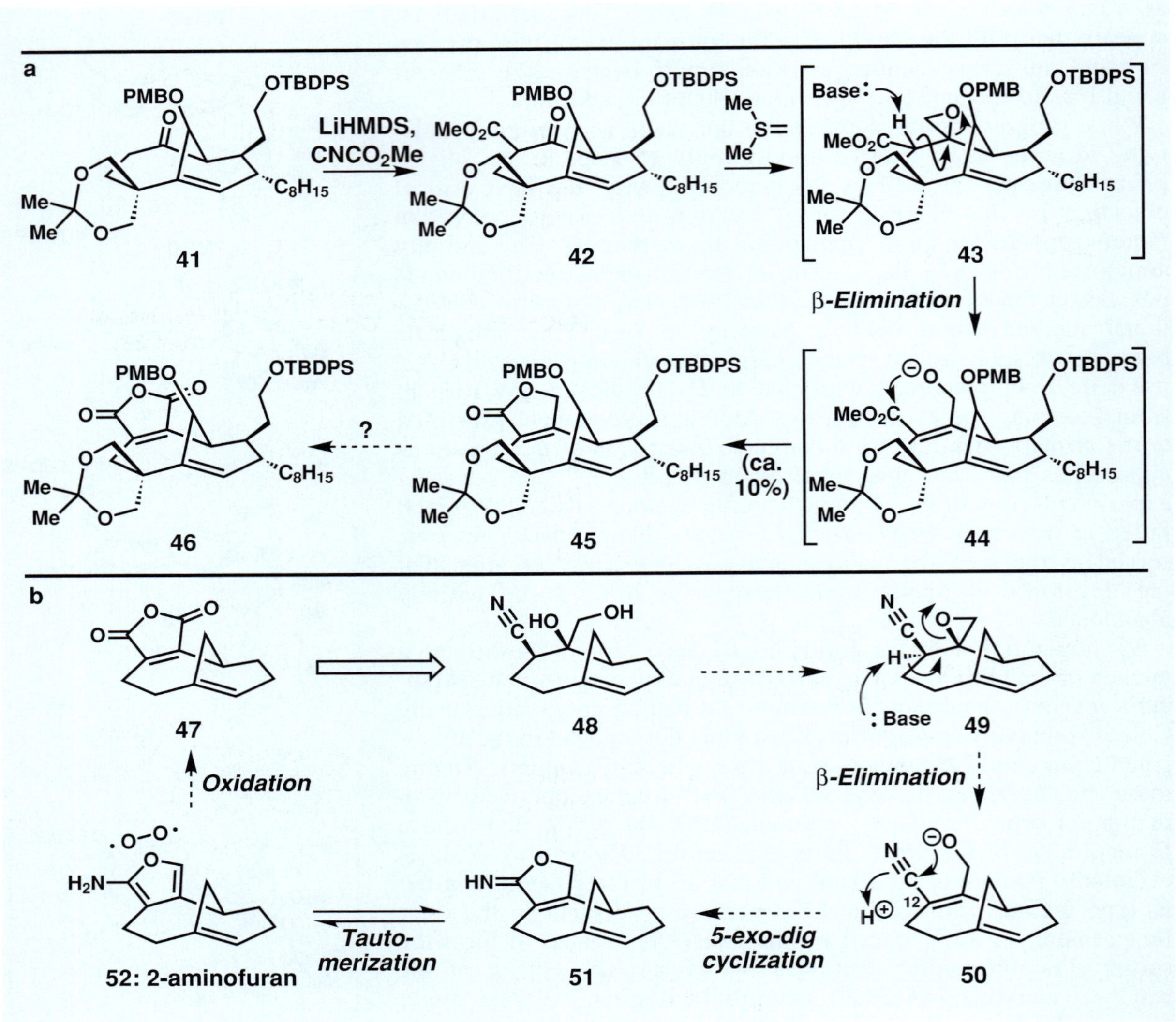

Scheme 7. Attempts to create the maleic anhydride (a) led to the design of a unique approach (b) to complete the task.

While seeking to access this compound might seem far from obvious, based on Dewar's theoretical work[17] it was anticipated that **51** would readily tautomerize to a 2-aminofuran (**52**), a likely fleeting, but electron-rich heterocycle[18] that could potentially undergo auto-oxidation in air to reveal the coveted maleic anhydride (**47**) directly. Although a risky strategy with only sparse literature support, the successful formation of **45** from **42** indicated that the initial stages of the proposed cascade were reasonable; implementation of its remaining steps would, unquestionably, extend the frontiers of organic synthesis by providing new insight

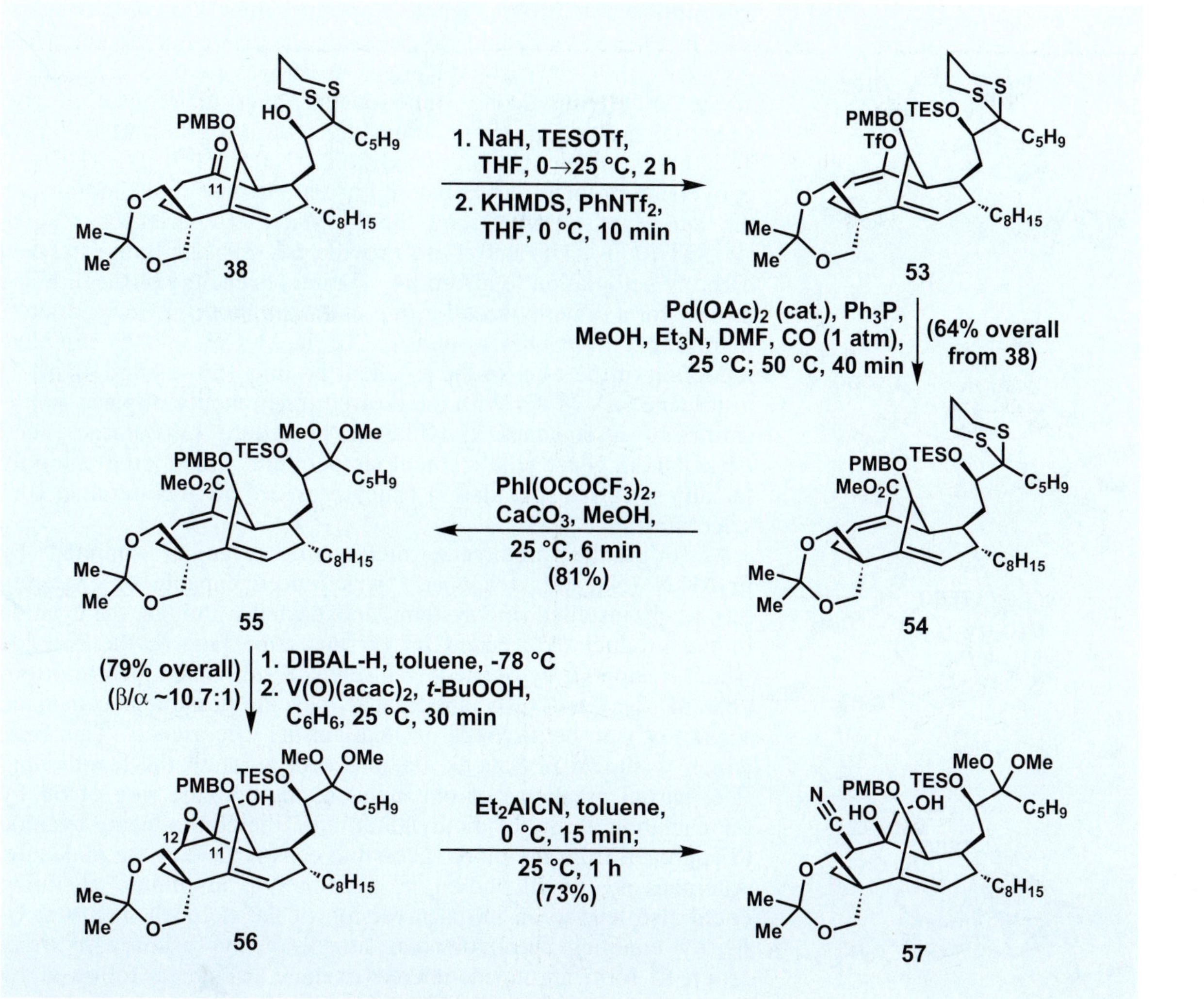

Scheme 8. Preparation of advanced intermediate **57** from **38**.

into chemical reactivity. Accordingly, all effort now sought to convert **38** into an intermediate of type **48** in order to probe the validity of this new approach.*

As indicated in Scheme 8, preparing the test substrate for this novel reaction cascade, diol **57**, did not prove overly challenging, although one major surprise was encountered in the course of its construction. The first steps of the sequence sought to install a C11–C12 epoxide stereoselectively (as expressed in **56**), since

* Significantly, although the cyanide in **48** is drawn with specific stereochemistry, either stereoisomer would be acceptable in that the drawn structure could lead to **50** through a concerted E_2-type β-elimination pathway, while its undrawn epimer could afford access to the same compound through dehydration of the neighboring tertiary alcohol.

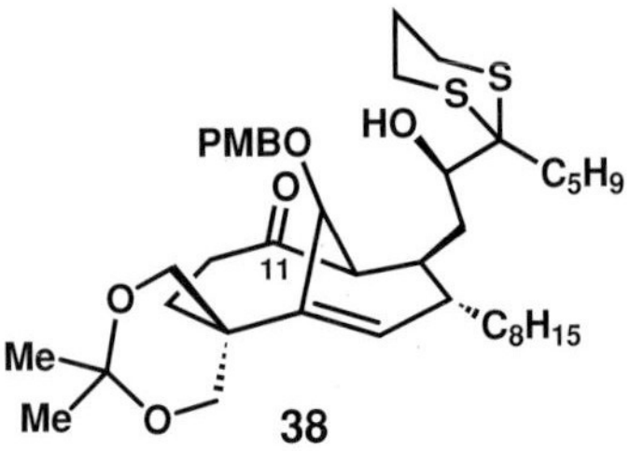

this motif would provide an electrophilic center that could conceivably be opened by cyanide to complete the desired compound (**57**). As such, the C-11 carbonyl group in **38** was used as a handle to create an allylic alcohol functionality over the course of five steps: 1) protection of the exocyclic hydroxy group at C-7 as a TES ether under standard conditions (NaH, TESOTf, THF), 2) conversion of the C-11 ketone into its enol triflate congener through the action of KHMDS and the McMurry–Hendrickson reagent ($PhNTf_2$)[19] in THF at 0 °C to provide **53**, 3) palladium-catalyzed carboxymethylation to afford **54**, 4) direct exchange of the dithiane moiety for a dimethylacetal group using the Stork–Zhao protocol[20] involving $PhI(OCOCF_3)_2$ and $CaCO_3$ in MeOH, and 5) complete reduction of the ester in the resultant product (**55**) using DIBAL-H in toluene at −78 °C. With the desired functionality in place, application of a standard epoxidation procedure [$V(O)(acac)_2$ and *t*-BuOOH in benzene at ambient temperature][21] then led to a highly facially selective epoxidation which provided **56** in better than 10:1 selectivity.*

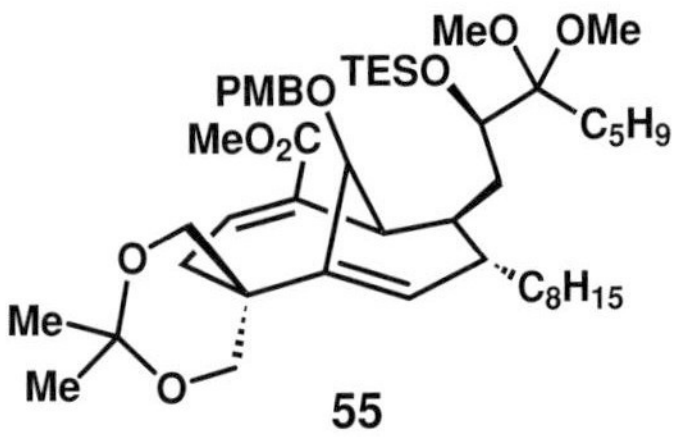

As originally predicted, nucleophilic cyanide, supplied by Et_2AlCN (Nagata's reagent),[22] was indeed capable of engaging this newly installed ring system. Unexpectedly, though, the cyanide in the product (**57**) ended up on the same face as the epoxide which it supposedly opened! While this result is certainly unorthodox and was not fully anticipated, several possible mechanistic scenarios can be invoked to explain its occurrence. The first, which is shown in Scheme 9a, proceeds through the lengthening (i. e. partial breaking) of one bond in the epoxide ring of **56** by the chelating ability of diethylaluminum, thereby inducing cyanide to approach from the more accessible convex face of the molecule. Alternatively, as illustrated in Scheme 9b, aluminum chelation could also lead to an initial inversion of the stereochemistry at C-12 by enabling intramolecular attack of the primary hydroxy group to form a four-membered oxetane, which, if followed by S_N2 delivery of the cyanide ion, would account for a double-inversion pathway to **57**. Finally, if the cyanide did indeed effect epoxide lysis through a standard S_N2 reaction, then the resultant stereocenter must have epimerized *in situ* under the force of the reaction conditions to give rise to **57** (Scheme 9c).

Irrespective of which of these pathways (or others not delineated) was responsible for the stereospecific construction of **57**, with this substrate prepared, the opportunity to test the designed cascade sequence that would hopefully lead to the long-coveted maleic anhydride was now at hand. Amazingly, following several weeks of optimization, this process was reduced to practice, ultimately

PMBO HO O 11 S S C_5H_9 C_8H_{15} O Me Me O **38**

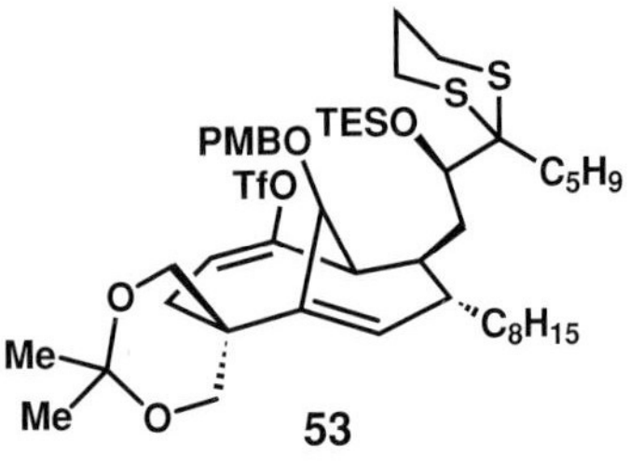

PMBO TESO TfO S S C_5H_9 C_8H_{15} O Me Me O **53**

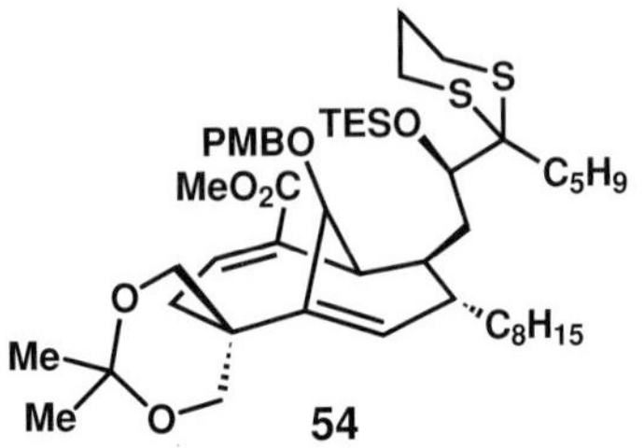

PMBO TESO MeO_2C S S C_5H_9 C_8H_{15} O Me Me O **54**

PMBO TESO MeO_2C MeO OMe C_5H_9 C_8H_{15} O Me Me O **55**

* As an aside, in the absence of the upper C_5H_9 side chain, the selectivity for this process was only 3.5:1, providing further evidence of the closely allied nature of the substituents in the CP-molecules and the effect that their presence or absence can have on certain conversions. Several more examples of this trend will be encountered before the synthesis of the CP-molecules is complete.

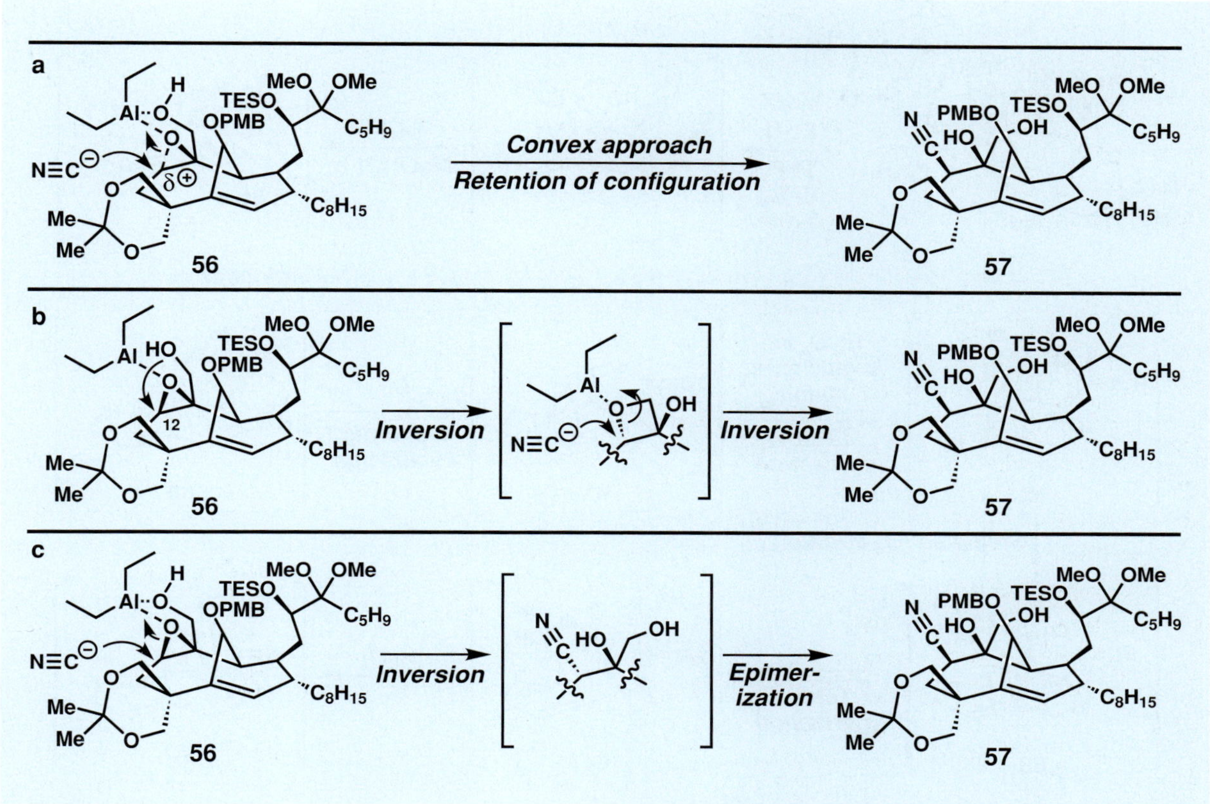

Scheme 9. Degenerate mechanistic rationales for the stereoselective addition of cyanide to epoxide 56.

affording the anhydride ring in 56 % yield. In the opening stages of this instructive domino sequence, the more accessible primary hydroxy group within **57** (Scheme 10) was engaged as a mesylate under standard conditions (MsCl, Et_3N, THF, 0 °C) to provide a leaving group for epoxide formation. Upon treatment of this intermediate (**58**) with K_2CO_3 in MeOH at ambient temperature to generate the epoxide system (**59**), the basic reaction conditions then set into motion a series of events which presumably led to iminobutenolide **61** via intermediate **60** through the same β-elimination/5-*exo*-dig cyclization pathway discussed earlier. Next, in order to enhance the energetically favored tautomerization of this intermediate (**61**) to its 2-aminofuran counterpart (**62**), the reaction medium was then slightly acidified with oxalic acid and exposed to air so that once **62** was generated it could harvest the triplet oxygen that would enable its direct conversion into hydroperoxide **63**. Once formed, subsequent tautomerization of this species to **64**, followed by the loss of water and hydrolytically assisted expulsion of ammonia, then completed this cascade leading to **66**. As proof for this rather complex mechanistic picture, an isolated epoxide bearing alternate C-11 stereochemistry (**11-*epi*-59**, see column figure) failed

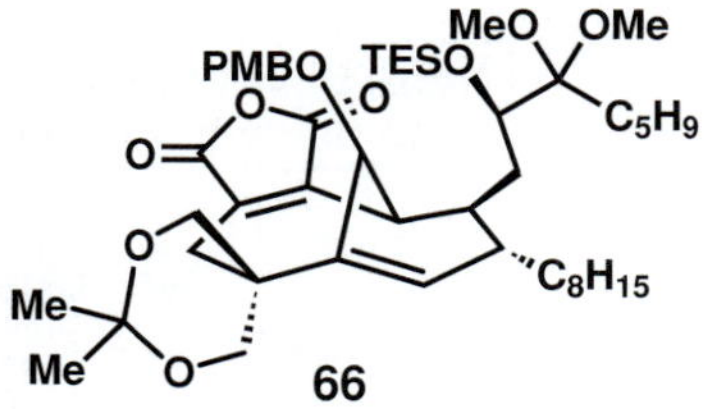

66

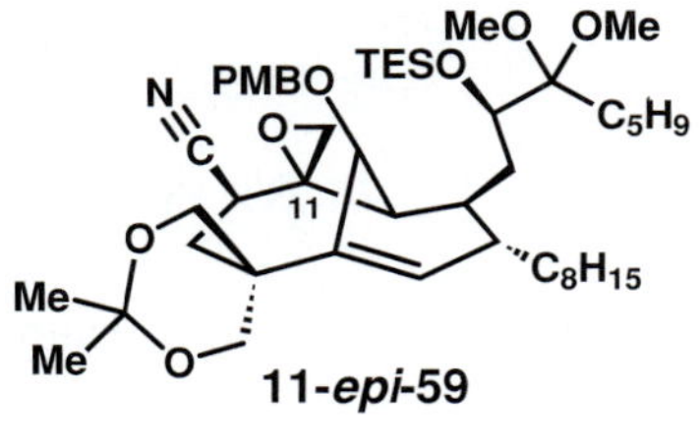

11-*epi*-59

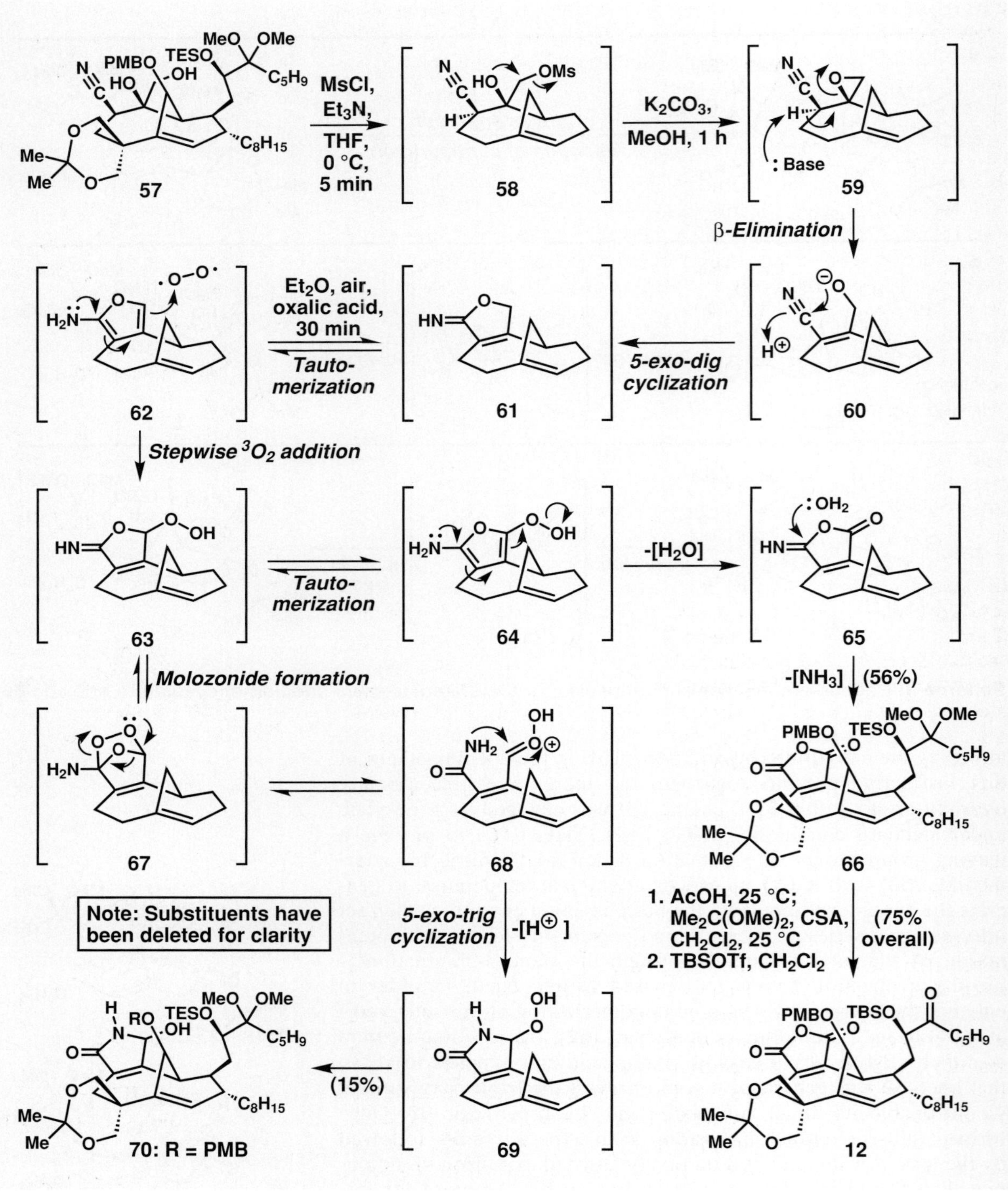

MeO OMe
PMBO TESO
N HO OH
C C_5H_9
O C_8H_{15}
Me
Me O
57

MsCl,
Et_3N,
THF,
0 °C,
5 min

N
C HO OMs
H
58

K_2CO_3,
MeOH, 1 h

N
C O
H
: Base
59

β-*Elimination*

N O
C
H⊕
60

5-exo-dig
cyclization

HN O
61

Et_2O, air,
oxalic acid,
30 min

Tauto-
merization

H_2N O ·O-O·
62

Stepwise $^3O_2$ *addition*

HN O O OH
63

Tauto-
merization

H_2N O O OH
64

-[H_2O]

: OH_2
HN O O
65

-[NH_3] (56%)

MeO OMe
PMBO TESO
O O
O C_5H_9
C_8H_{15}
Me
Me O
66

Molozonide formation

H_2N O O O
67

OH
NH_2 O⊕
O
68

Note: Substituents have
been deleted for clarity

5-exo-trig
cyclization
-[H⊕]

1. AcOH, 25 °C;
$Me_2C(OMe)_2$, CSA,
CH_2Cl_2, 25 °C
2. TBSOTf, CH_2Cl_2
(75%
overall)

H OH
N O
O
69

(15%)

MeO OMe
H RO TESO
N OH
O C_5H_9
C_8H_{15}
Me
Me O
70: R = PMB

O
PMBO TBSO
O O
O C_5H_9
C_8H_{15}
Me
Me O
12

Scheme 10. The cascade sequence developed for the construction of the maleic anhydride moiety of the CP-molecules.

to enter into the same reaction sequence when exposed to commensurate reaction conditions, thereby providing solid evidence that the conversion of **59** into **61** proceeded through E_2 collapse rather than through an E_1 pathway. Moreover, although iminobutenolide **61** could not be isolated owing to its reactivity, it was characterized in solution by NMR spectroscopic analysis. Intriguingly, besides the desired compound (**66**), one other side product was also obtained in 15 % yield, namely hydroxyamide **70**. Presumably, one can envision an additional pathway available for **63** that would lead to molozonide **67** rather than to **64**. If this event occurred, a cascade reminiscent of the standard ozonolysis reaction[23] could then convert this adduct into amide **68**, a species poised for a subsequent intramolecular 5-*exo*-trig cyclization with its internal electron sink to form **69**. A final disproportionation step, a well-documented mechanistic process,[24] would then account for the formation of **70**.

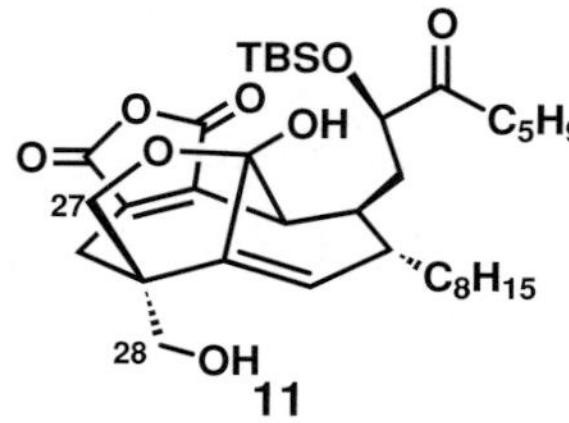

Having finally identified a solution for the construction of the maleic anhydride (**66**), effort could now be directed towards the final ring system not yet engraved onto the bicyclic core: the γ-hydroxylactone. If this motif could be forged, then only a single carbon atom would have to be added in order to complete the structure of CP-225,917 (**1**). Thus, following the elaboration of **66** to **12** (Scheme 10), a crystalline intermediate whose X-ray crystallographic analysis verified the stereochemical course of all the steps up to this point, the next set of operations sought to convert **12** into **11** as shown in Scheme 11. As indicated, this goal was reached through DDQ-mediated cleavage of the PMB ether, oxidation of the resultant alcohol with PDC to form **71**, and, finally, acid-catalyzed cleavage of the dimethyl acetal. Although the overall yield of 43 % for these rather conventional operations might seem unduly low, it primarily reflects the difficulty in removing the PMB group cleanly from **12**, thereby illustrating how simple protecting group manipulations can often impact the efficiency of synthetic sequences.

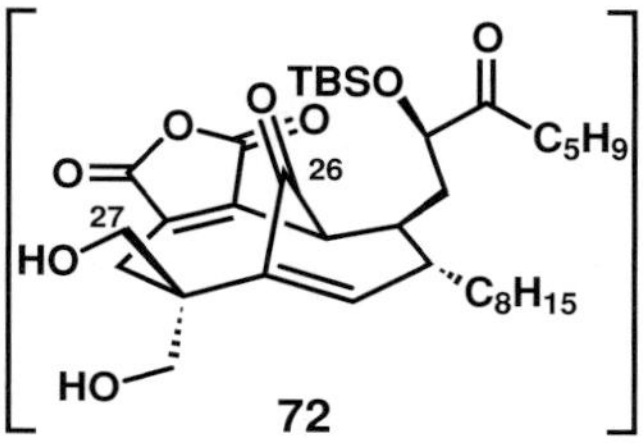

The more important feature to note, however, is the fact that **11**, while stable in its closed lactol form, can also exist fleetingly in its ring-opened tautomeric form (i. e. **72**). Indeed, such ring-chain tautomerism[25] was deemed to be the critical feature that could enable a successful oxidation of the C-27 methylene within **11** to match that of the CP-molecules in the next step. For example, if the C-27 position of an open congener of **11** (such as **74**) could be captured through oxidation to the corresponding aldehyde (**75**), then the addition of a molecule of water to either of the newly unveiled carbonyls in this product should induce a cascade that would lead to the γ-hydroxylactol system of **77**. At first, though, no conventional oxidation protocol could be found to successfully reduce this idea to practice. For example, after converting **11** into **73**, attempted oxidation with DMP at ambient temperature led only to recovered starting material. Reasoning that at ambient temperature the substrate preferred to reside ex-

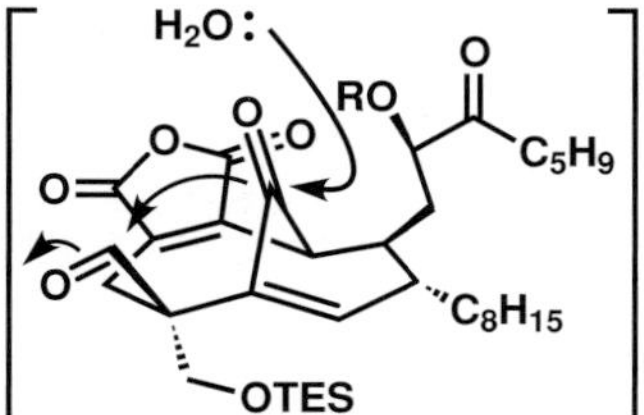

TBSO O OH C_5H_9 27 C_8H_{15} 28 OH
11

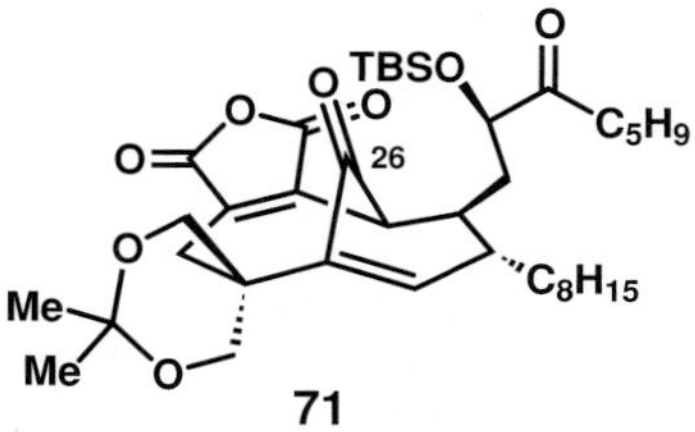

O TBSO O 26 C_5H_9 C_8H_{15} Me Me O
71

O TBSO O 26 27 HO C_5H_9 C_8H_{15} HO
72

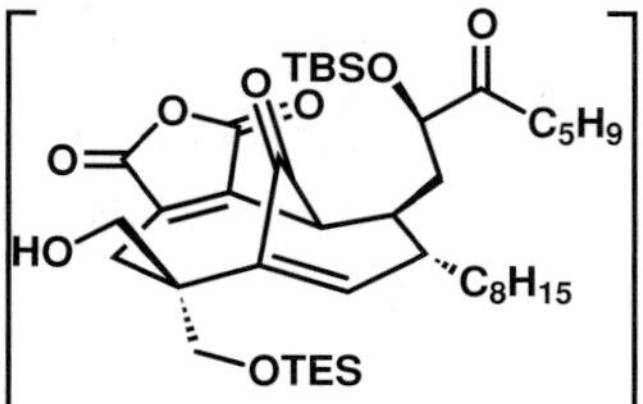

O TBSO O HO C_5H_9 C_8H_{15} OTES
74

H_2O: RO O C_5H_9 C_8H_{15} OTES
75: R = TBS

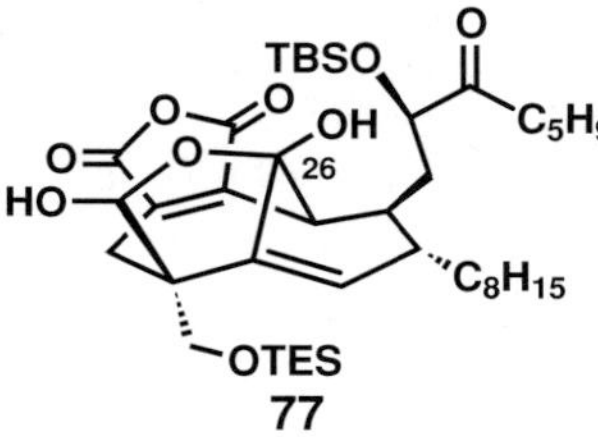

TBSO O OH 26 HO C_5H_9 C_8H_{15} OTES
77

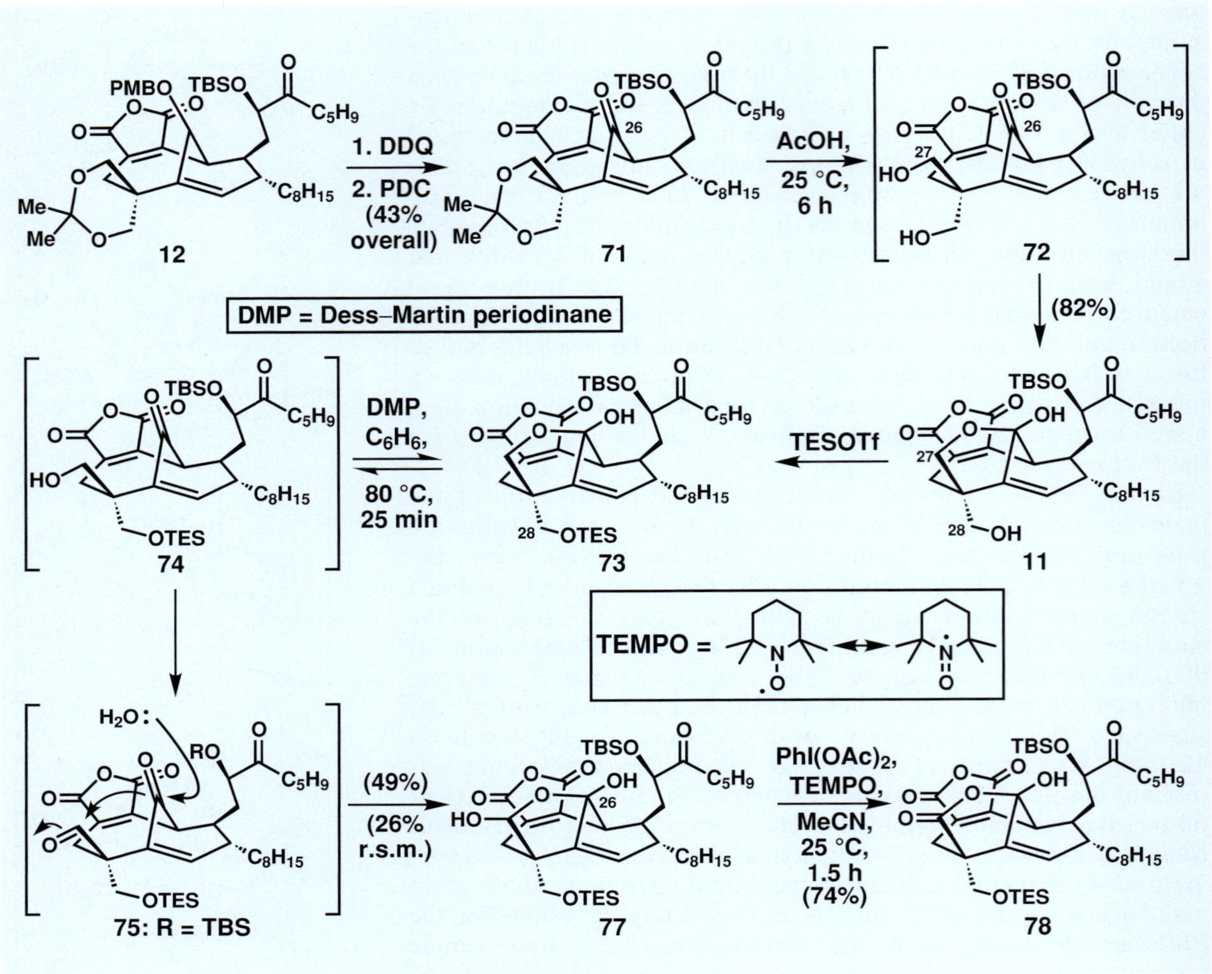

Scheme 11. Construction of key intermediate **78** from **12**.

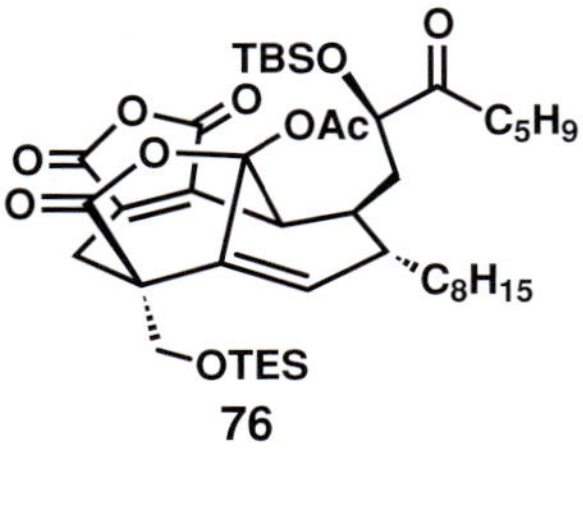

clusively in its closed, unreactive lactol form, the decision was made to warm the reaction mixture, hoping that the open tautomer would then persist long enough to be intercepted by DMP. Interestingly, although this is a simple and standard reaction modification, it constituted new ground for DMP as no literature precedent existed to indicate that this oxidant could be successfully employed above 25 °C, a result most likely due to fears that this oxidant would not be stable under such conditions.[26] Nevertheless, treatment of **73** with DMP in benzene at 80 °C did lead to the smooth, and safe, formation of the desired product in 49 % yield. The remaining material balance was recovered starting material (26 %) and a small amount (9 %) of side product **76** (see column figure) which resulted from the attack of **75** by acetic acid instead of by water. Even though compound **76** was useless for the continued progression of the synthesis because its acetate group could not be removed, the fact that its C-27 position was fully oxidized was quite intriguing considering that none of the material with the desired C-26 hydroxy group (i. e. **77**) was isolated as its γ-hydroxylactone congener (i. e. **78**). Thus, this result suggested that when the C-26 did not bear a free hydroxy group (in other words, the OH function was protected), oxidation at C-27 of the γ-hydroxylactol to the lactone form was relatively easy to achieve in light of the fact that DMP at 80 °C did not effect this same reaction on **77**. Although certainly an interesting insight into a unique aspect of chemical reactivity within the context of the CP-molecules, this finding was also quite troubling for the viability of the projected route because it indicated that finding any method capable of oxidizing **77** to **78** would likely be far from trivial. This prediction rang true through a wide-ranging panel of oxidants, until it was fortunately discovered that TEMPO in the presence of $PhI(OAc)_2$ in MeCN could rise to the occasion and complete the assembly of γ-hydroxylactone **78** in 74 % yield over the course of 1.5 hours.[27]

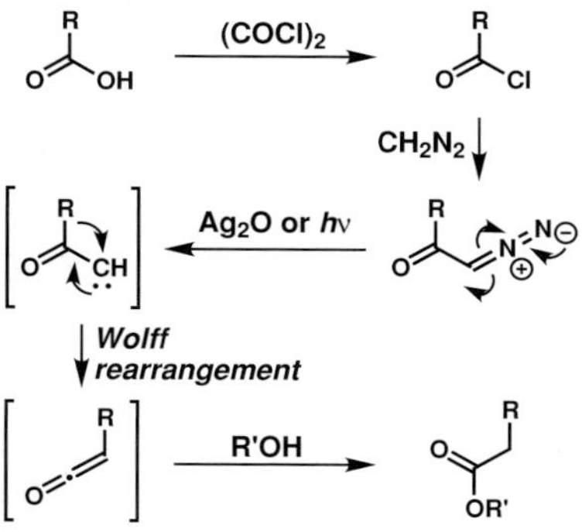

With this ring system finally installed, all that formally remained to complete the target structures was the extension of the sidearm at C-14 by one carbon atom. As already mentioned, the process forecasted to accomplish this critical step was an Arndt–Eistert homologation, a mechanistically intricate operation whose key steps are delineated in Figure 1. As shown, this classic transformation commences with the initial conversion of a free terminal carboxylic acid into an acid chloride, a step that suitably activates the substrate for subsequent diazoketone formation using CH_2N_2. Once this motif is in place, the homologation is then completed through a Wolff rearrangement promoted either by a silver-based reagent or by light; one should take careful note of the homology between these final mechanistic steps and those of the Curtius rearrangement (which was discussed in Chapter 5). Thus, in order to instigate the key steps of this general reaction course, the TES-protected alcohol within **78** required transformation into a free carboxylic acid, a provision that would normally be quite simple to accomplish. Stunningly, however, while the conversion of **78** into aldehyde **80** (Scheme 12) proceeded smoothly, attempts to oxidize this intermediate further to its corresponding carboxylic acid (**81**) led to complete decomposition, presumably through the indicated radical-based decarboxylation pathway. As a result, this stumbling block all but assured that a new strategy would need to be developed in order to circumvent this problem and complete the synthesis of the CP-molecules.

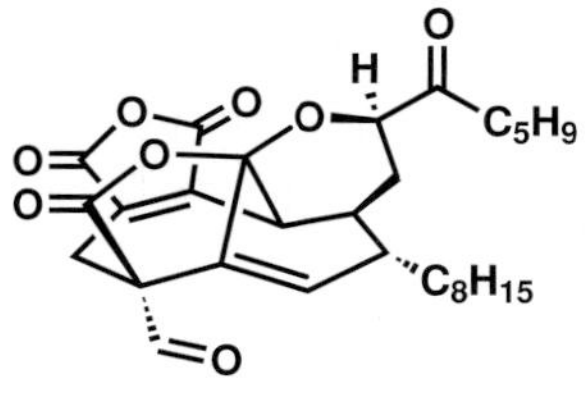

The central issue in contemplating such a redesign is how far to retreat within the developed sequence, in that one would desire to capitalize on the already developed chemistry to the greatest degree possible. Accordingly, the simplest, and perhaps most tactically logical approach to explore next would be merely to switch the order of the two main events needed to elaborate an advanced compound such as **12** to the final targets (**1** and **2**). In other

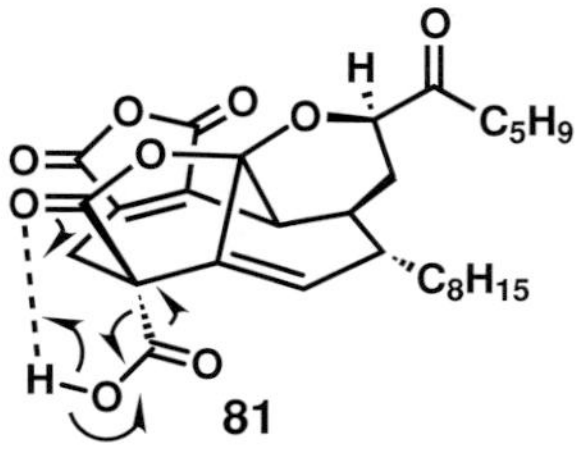

(COCl)$_2$ CH$_2$N$_2$ Ag$_2$O or hν Wolff rearrangement R'OH

Figure 1. The Arndt–Eistert homologation sequence.

80

81

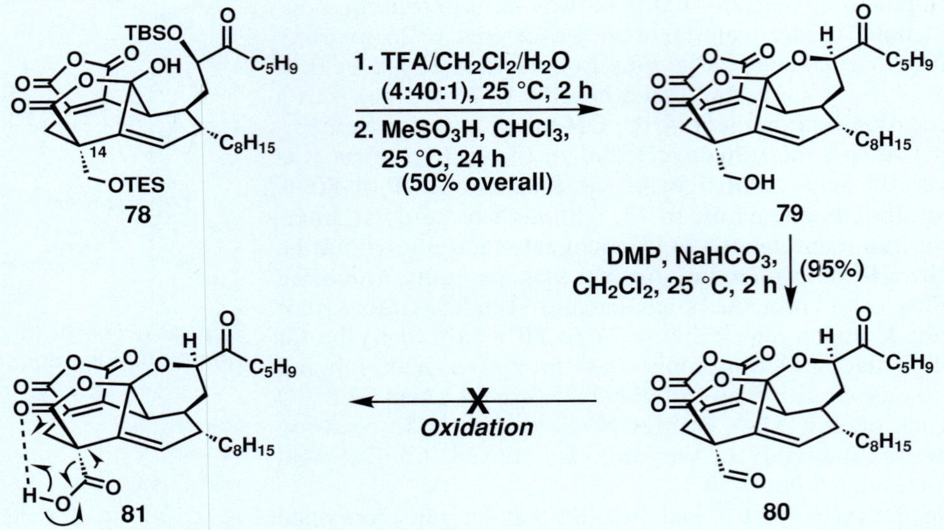

Scheme 12. Attempts to synthesize carboxylic acid **81** led to a roadblock.

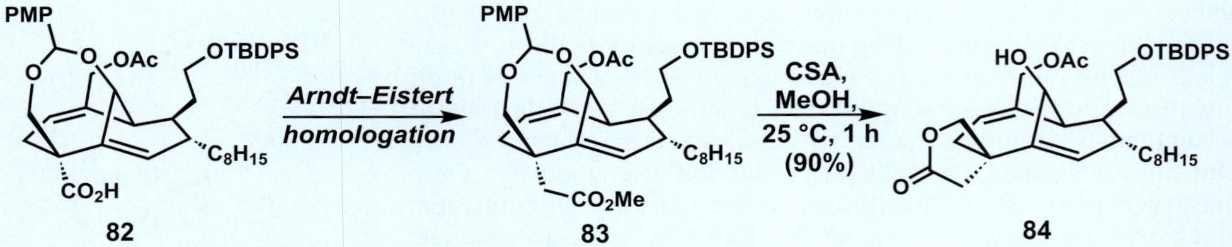

Scheme 13. Attempts to proceed through an alternate strategy led to a new synthetic challenge.

words, execute an Arndt–Eistert homologation first, and then create the γ-hydroxylactone ring system of the CP-architecture through a sequence that would mirror that of Scheme 11. While simply stated, preliminary explorations of this alternate strategy on model systems quickly revealed one major flaw in that following Arndt–Eistert homologation of **82** to **83** (Scheme 13), subsequent unraveling of its benzylidene acetal (to begin constructing the γ-hydroxylactone) led exclusively to compounds such as **84** in which the primary alcohol had engaged the proximal ester at C-14 to generate a lactone. While this motif looks innocent since it should be subject to ring opening with nucleophiles, extensive investigations towards this end revealed no species capable of effecting its scission. As such, the securely "locked" nature of this structure was a potentially fatal blow for this revised strategy.

With challenges, however, also come the inspiration for innovative solutions. Here, the Nicolaou group anticipated that perhaps one viable alternative which could breathe new life into this approach might constitute exchanging the ester within **83** for an amide, since this alternate functionality would likely be resistant to lactone formation under acidic conditions. This idea was tested immediately, starting with the conversion of **12** into **87** through the same general sequence discussed earlier, with the critical step being the formation of benzylidene acetal **86** from **85** using DDQ (a reaction discussed at length in Chapter 3). Unfortunately, upon attempted Arndt–Eistert homologation of the carboxylic acid group of this adduct (**87**), the required acid chloride could not be formed, presumably due to steric shielding from the substituents on the CP-core and/or the buried nature of this motif on the con-

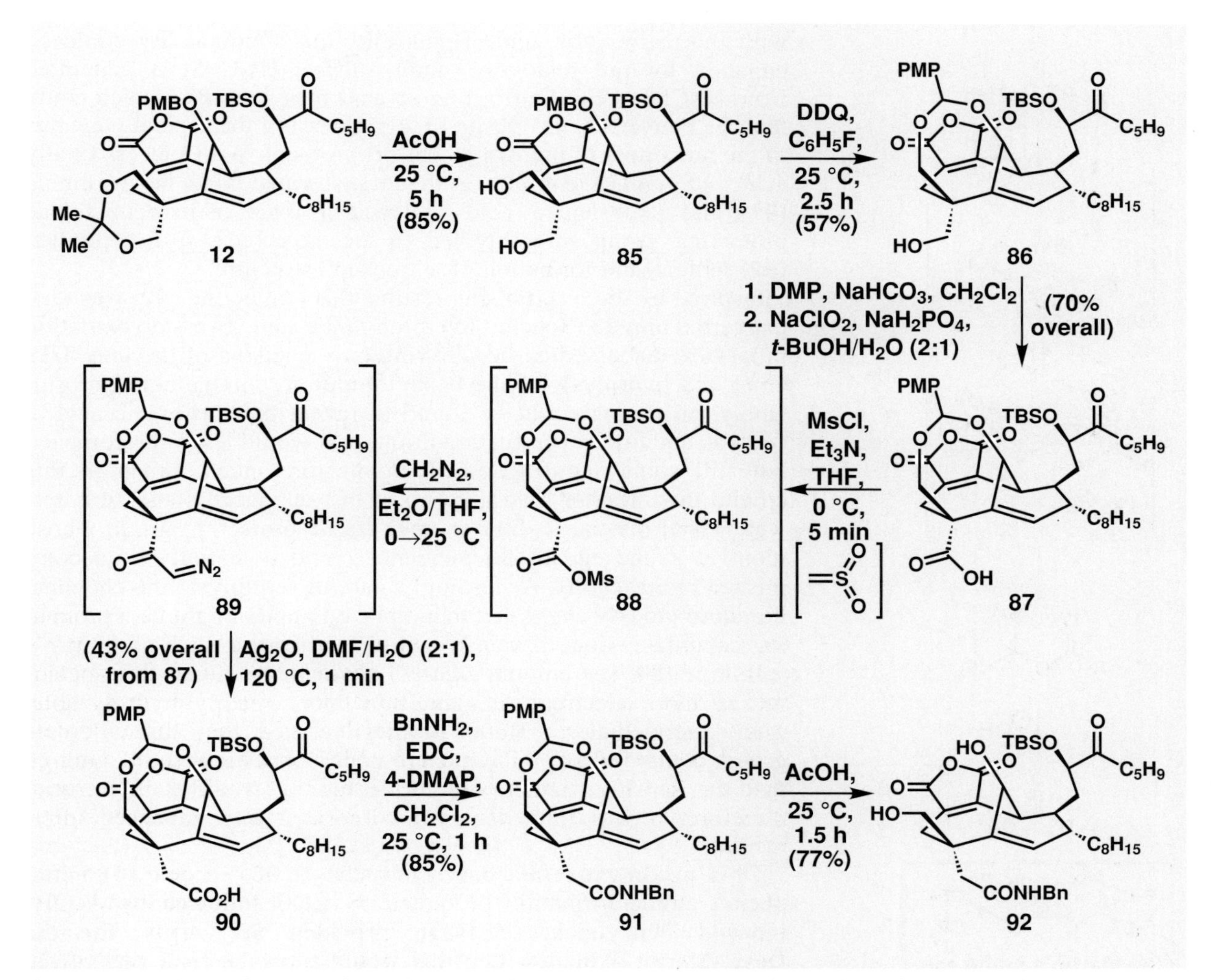

Scheme 14. Use of a benzyl amide protecting group prevents the formation of a "locked" structure.

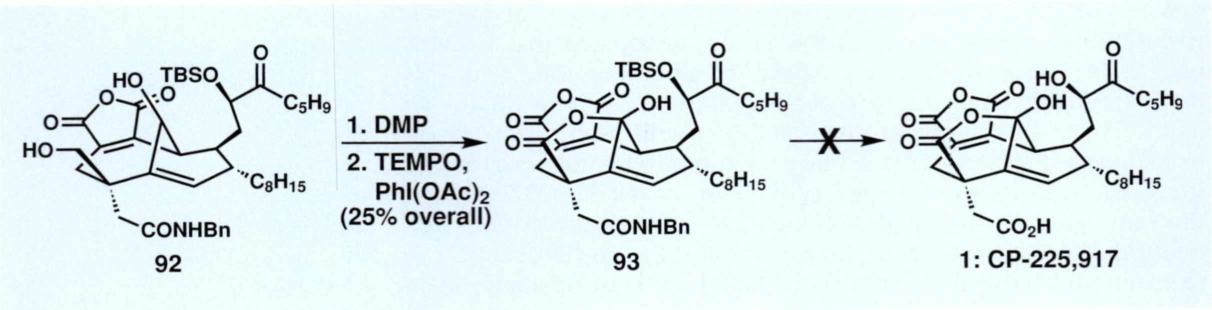

Scheme 15. Failed hydrolysis of benzyl amide **93**.

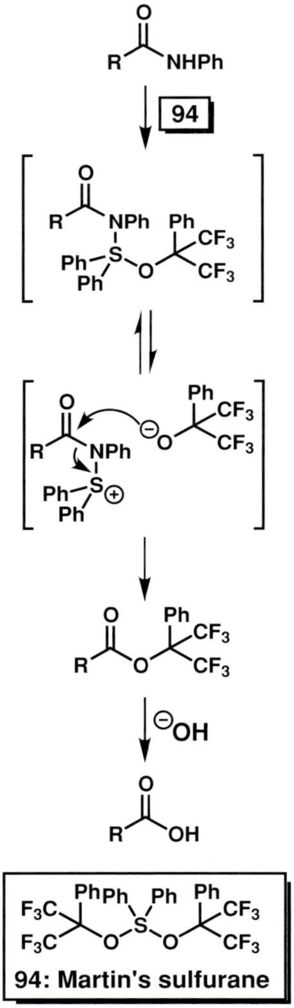

94: Martin's sulfurane

cave side of the molecule which prevented approach by the reagent. Numerous other standard acyl activators similarly failed to react with this carboxylic acid. Thankfully, this functionality could be engaged by the relatively small sulfene [$H_2C{=}SO_2$] generated from MsCl and Et_3N, affording an acyl mesylate (**88**) which could then be converted into **90** in 43 % yield using the typical elements of the remainder of the Arndt–Eistert protocol. With success finally achieved in this event, **90** was then transformed into a benzyl amide (**91**), and a subsequent acid-catalyzed cleavage of its benzylidene protecting group smoothly led to the expected 1,4-diol product (**92**) without the formation of a "locked" structure.

Inspired by these promising results, this compound (**92**) was then converted into **93** (Scheme 15) through the same two-step oxidation procedure discussed earlier, leaving only excision of the lone TBS ether and hydrolysis of the benzyl amide to complete **1**. Unfortunately, no means could be found to reveal the carboxylic acid at C-28 as the most powerful conditions that would leave the compact, but still somewhat fragile, CP-architecture intact, failed in this crucial task. Rather than give up at this advanced stage, attempts were made instead to find another amide protecting group whose removal at the end of the sequence could potentially be accomplished more readily. Accordingly, careful scrutiny of the chemical literature quickly suggested that a phenyl amide might be a promising candidate, since it was known that a reagent such as Martin's sulfurane (**94**, see column figures) could first convert this species into a more electrophilic, and thus more readily hydrolyzable, ester intermediate.[28] Model studies verified that this alternate group could indeed be removed under conditions mild enough that the sensitive maleic anhydride moiety would survive, and, therefore, it was immediately incorporated into advanced intermediates.

Thus, needing to retreat only a few steps to **90** (Scheme 16), initial phenyl amide formation promoted by EDC followed by AcOH-induced benzylidene cleavage provided **95**, ready for the Dess–Martin oxidation step that would form the final ring of the CP-molecules. Surprisingly, though, after the central C-26 alcohol

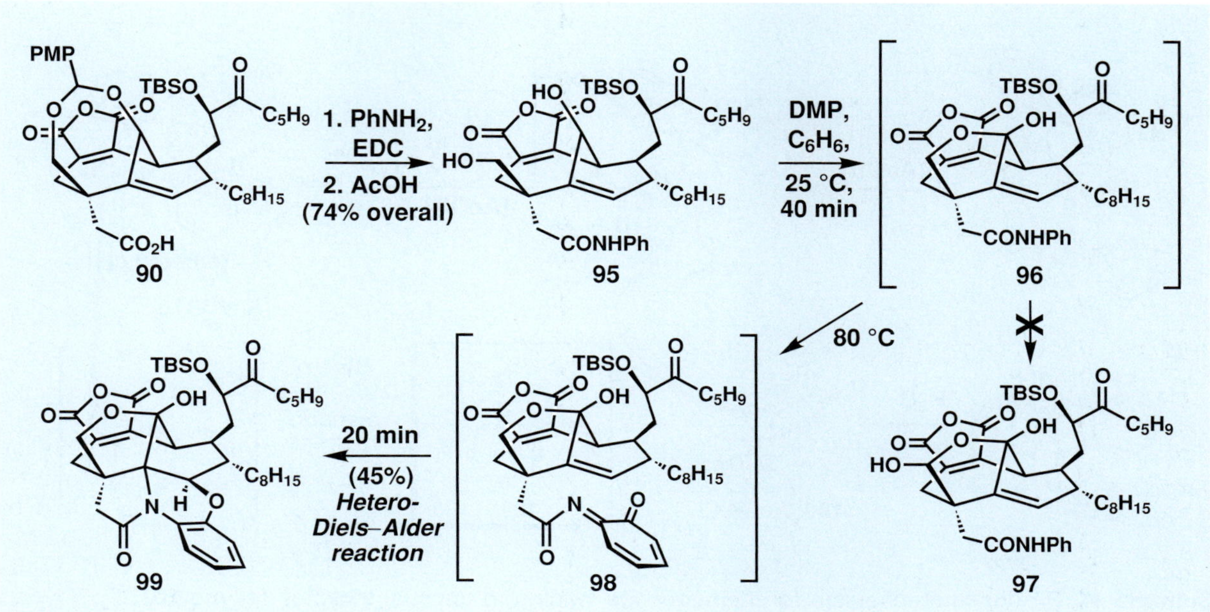

Scheme 16. Unexpected results upon attempted oxidation of **95**.

was oxidized with DMP at ambient temperature to afford lactol **96**, subsequent treatment with additional DMP at 80 °C did not lead to the desired product (**97**); instead, the major isolated product was the unique heterocycle **99**. This unexpected outcome can be rationalized through the mechanistic picture shown in Scheme 17. As drawn, following substrate activation by DMP, subsequent reaction of the resulting intermediate with Ac-IBX (**100**, a compound formed by the hydration of DMP) generated an intermediate *o*-azaquinone; this motif then served as a highly competent diene in an intramolecular hetero-Diels–Alder reaction with the adjacent olefinic dienophile. Although this result was yet another dead-end for the campaign to synthesize the CP-molecules, the unique mechanism responsible for its execution constituted a new reaction pathway for hypervalent iodine reagents,[26d] thereby providing insight for the rational design of a series of synthetic methodologies with broad utility.[29]

Irrespective of the importance of this finding in leading to new fields of investigation, this discovery did not bring the total synthesis of the CP-molecules any closer to fruition. A suitable protecting group for the carboxylic acid still remained to be identified. After pursuing a few additional ideas that similarly led to stumbling blocks, one final solution was inspired by Sir Derek Barton's imaginative protection of carboxylic acids as masked heterocycles during his synthetic studies towards penicillin.[30] As an illustration of this powerful concept, if a carboxylic acid is converted into an indoline amide (see column figures), the resultant species will possess

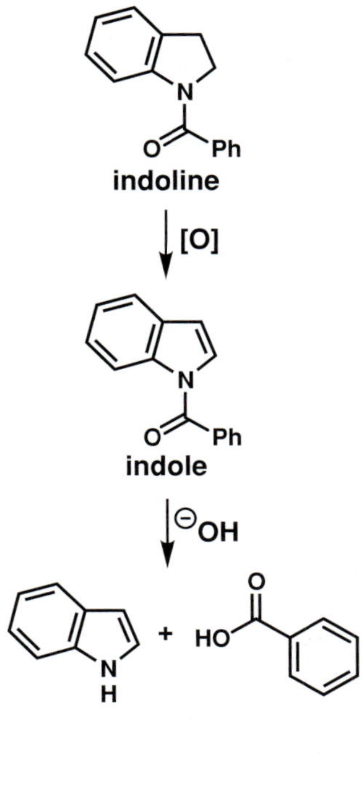

indoline

[O]

indole

$^{\ominus}$OH

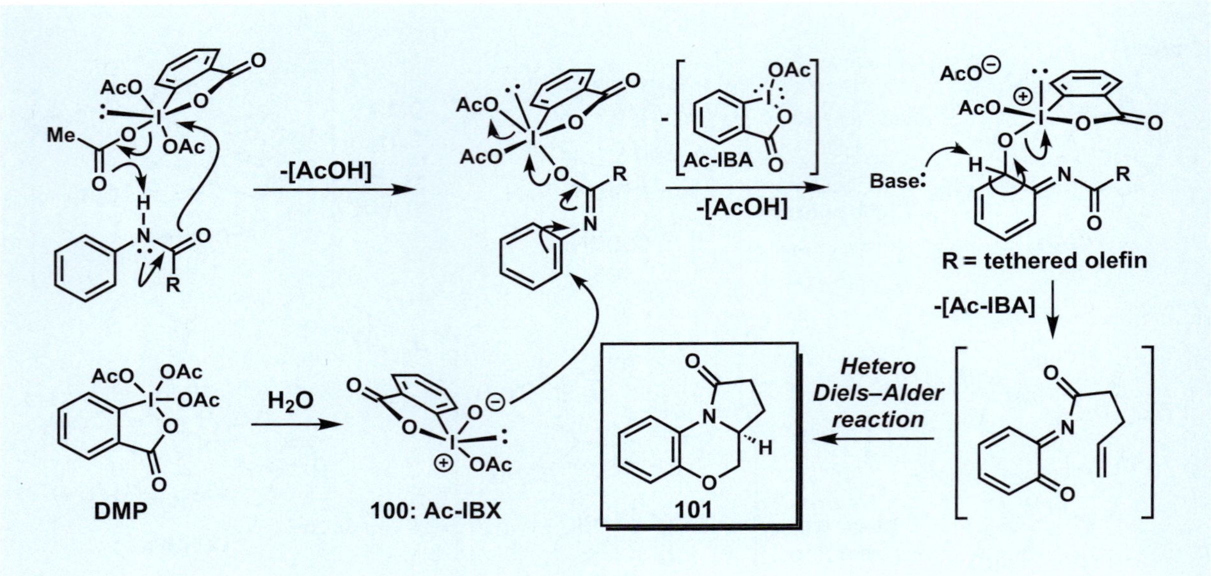

Scheme 17. Proposed mechanism for the sequence leading to unusual tricycles of type **101**.

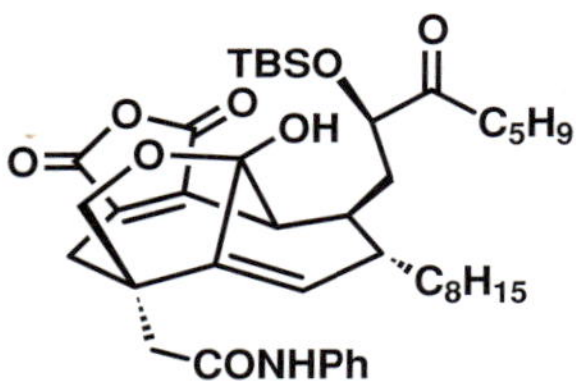

96

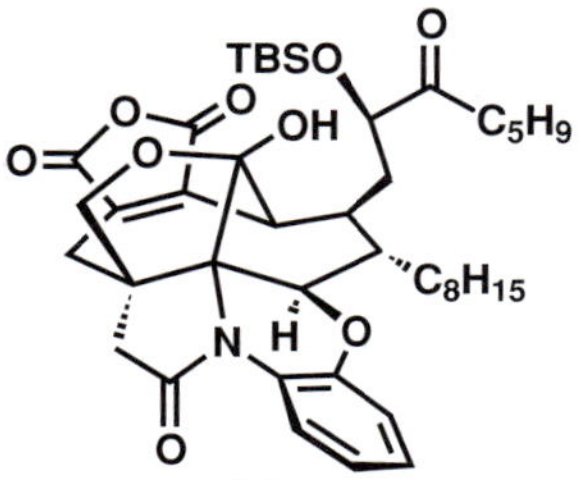

99

such poor electrophilicity that it can resist hydrolysis under virtually any set of basic conditions. However, if this protecting group is oxidized to its heterocyclic indole counterpart, then this resilience will be lost. For this reason, an indoline certainly appeared to be an attractive protecting group for advancing the synthesis forward since it would likely be inert for the entire synthetic sequence, but could be easily activated for what would hopefully be a facile cleavage at the end of the sequence. Moreover, because the nitrogen atom of an indoline amide does not possess a hydrogen atom, the critical participant in the oxidative conversion of **96** into polycycle **99**, this side reaction could also be averted.

Despite the promising nature of this idea, this strategy was thwarted soon after its inception because **102** could not be advanced to **103** (Scheme 18). As such, a maze had seemingly been entered from which there appeared no obvious escape since any minor change in one part of the molecule instigated a problem with some aspect of the developed sequence. In truth, the answer needed to break free from this quagmire could be found within the architecture of the CP-molecules themselves, as revealed by several clues gathered during the synthesis. More specifically, whenever compounds with the closed pyran architecture corresponding to **2** had been formed (such as **79** in Scheme 12), these intermediates always proved far more stable and easier to purify than their open lactol congeners. Accordingly, instead of targeting **1**, perhaps the same general approach using Barton's masked-heterocycle protection could be applied towards structures that would lead to **2** instead. Not only would this strategy obviate a series of protecting group

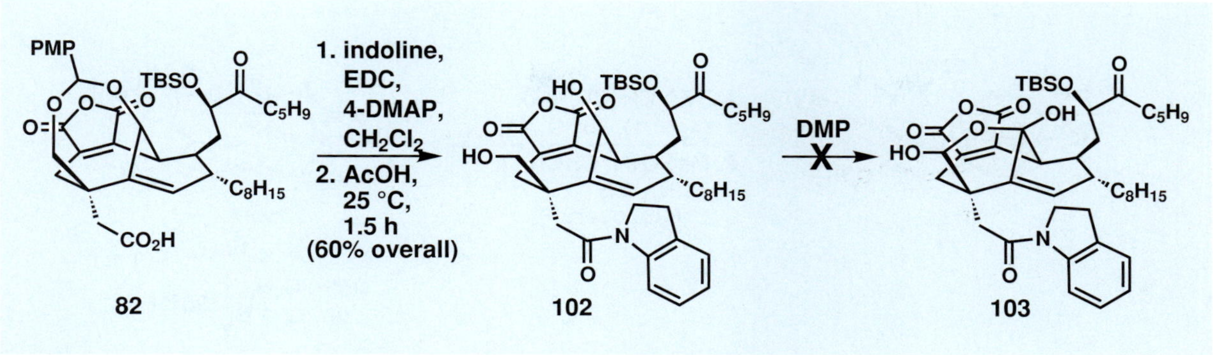

Scheme 18. Another approach thwarted by a late-stage roadblock.

manipulations incurred by attempting to construct **1**, but it would also likely improve material throughput due to the seemingly more favorable physical properties of structures resembling **2**. Of course, since no conditions had been identified that could convert **2** into **1**, the selection of this strategy meant that only one of the CP-molecules could be prepared. Based on the number of failures that had already been confronted, this outcome would still be welcome.

As indicated in Scheme 19, this new strategy was put into play by initially converting **77** into **104** through the conditions seen earlier in Scheme 12, thereby setting the stage for a subsequent Dess–Martin oxidation, which smoothly provided aldehyde **105**. Although this operation seems simple enough, it is important to note that the use of benzene as solvent was critical for its chemoselectivity. For example, in CH_2Cl_2 a fair amount of material (about 25 %) was obtained with C-27 fully oxidized (a transformation readily accomplished in this case since the pyran ring was present, vide supra). Of course, this alternate product was useless because earlier approaches had shown that attempting to oxidize such compounds to their C-14 carboxylic acid counterparts would only lead to decomposition (cf. Scheme 12). However, these investigations are quite important since they indicate that the reactivity of DMP can be tuned merely based on solvent choice, a fact worth remembering should your attempts to effect an oxidation with this reagent go astray.

With an effective synthesis of **105** accomplished, its free lactol was then protected as a TBS ether to prevent the formation of any C-14 "locked" structures (cf. Scheme 13). Oxidation of the resulting aldehyde to a carboxylic acid then proceeded smoothly with $NaClO_2$ in 66 % overall yield. Subsequent application of the previously developed Arndt–Eistert homologation protocol based on acyl mesylates,[31] installation of the indoline amide (EDC, 4-DMAP, CH_2Cl_2), and cleavage of the TBS protecting group (TFA/CH_2Cl_2/H_2O, 4:40:1) then led to the assembly of **109** in 29 % yield overall from **106**. Finally, oxidation at C-27 with DMP at

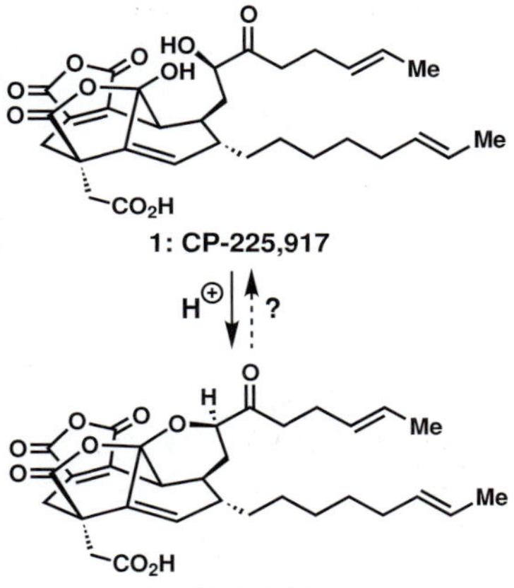

1: CP-225,917

2: CP-263,114

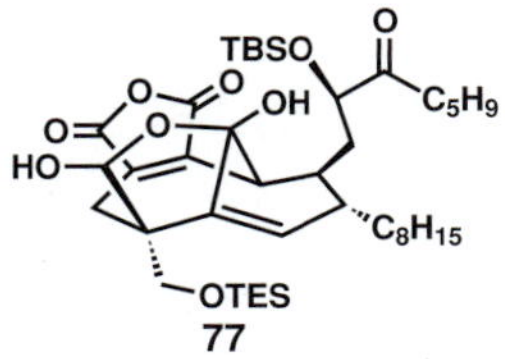

77

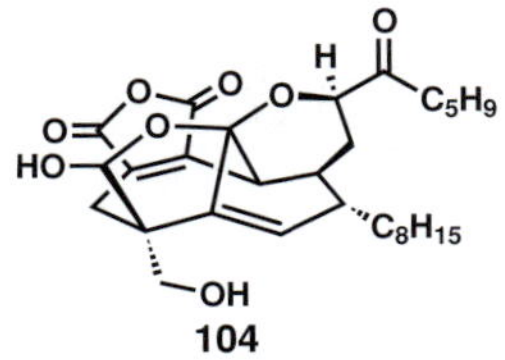

104

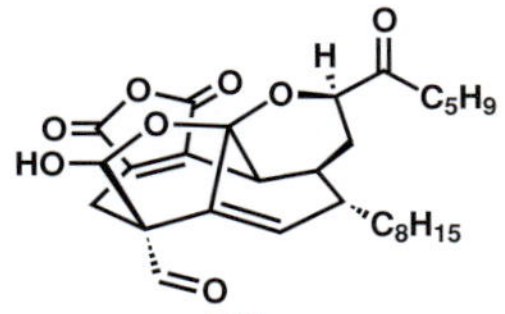

105

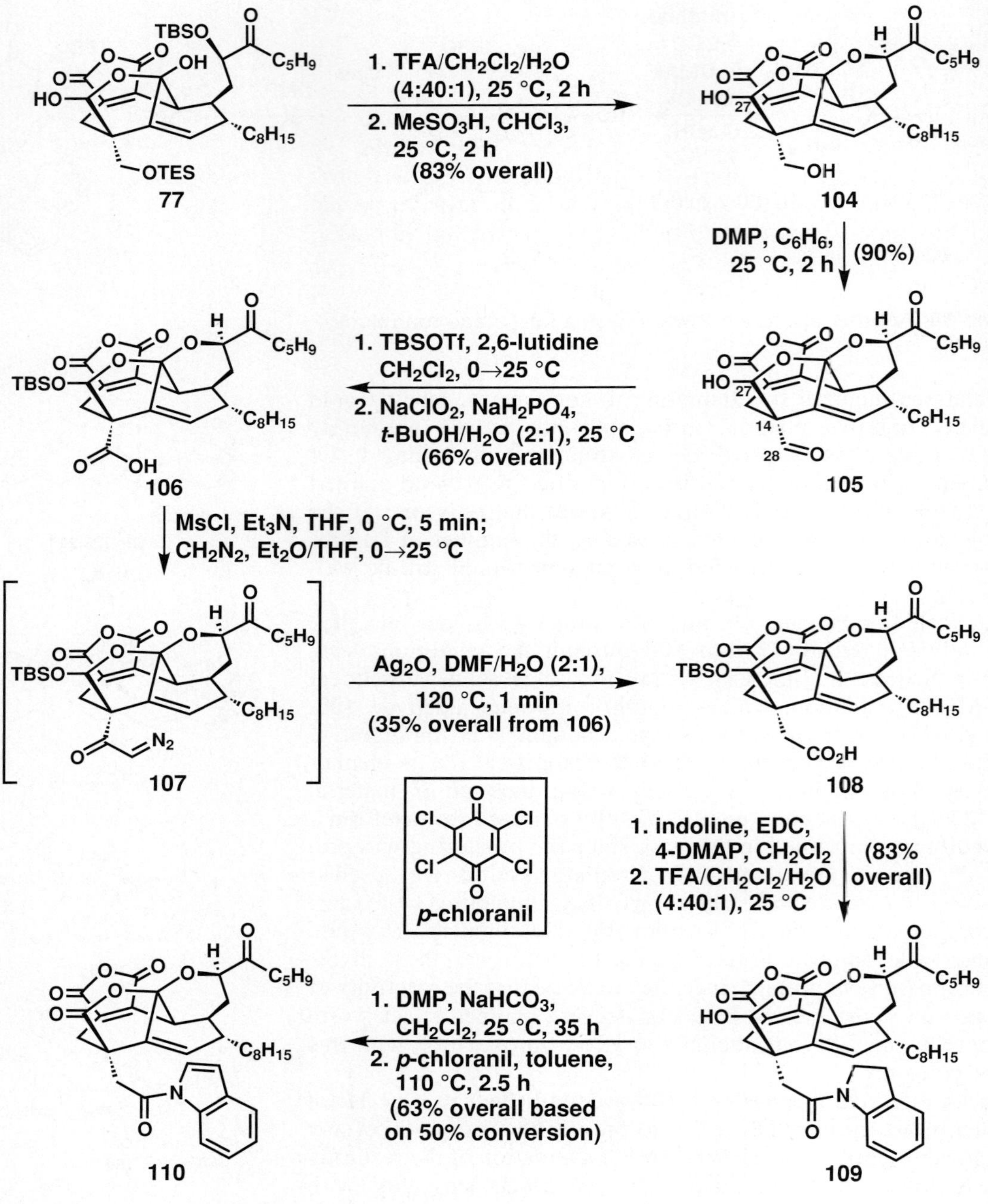

77 → 104: 1. $TFA/CH_2Cl_2/H_2O$ (4:40:1), 25 °C, 2 h; 2. $MeSO_3H$, $CHCl_3$, 25 °C, 2 h (83% overall)

104 → 105: DMP, C_6H_6, 25 °C, 2 h (90%)

105 → 106: 1. TBSOTf, 2,6-lutidine CH_2Cl_2, 0→25 °C; 2. $NaClO_2$, NaH_2PO_4, *t*-BuOH/H_2O (2:1), 25 °C (66% overall)

106 → 107: MsCl, Et_3N, THF, 0 °C, 5 min; CH_2N_2, Et_2O/THF, 0→25 °C

107 → 108: Ag_2O, DMF/H_2O (2:1), 120 °C, 1 min (35% overall from 106)

108 → 109: 1. indoline, EDC, 4-DMAP, CH_2Cl_2; 2. $TFA/CH_2Cl_2/H_2O$ (4:40:1), 25 °C (83% overall)

109 → 110: 1. DMP, $NaHCO_3$, CH_2Cl_2, 25 °C, 35 h; 2. *p*-chloranil, toluene, 110 °C, 2.5 h (63% overall based on 50% conversion)

p-chloranil

Scheme 19. Successful synthesis of indole amide **110** with a complete CP architecture.

ambient temperature, followed by a second oxidation step using *p*-chloranil[32] to convert the indoline amide into its aromatic indole congener, afforded **110** in 63 % yield (based on 50 % conversion) for these two steps. As such, only hydrolysis of the indole group remained to complete the total synthesis of the CP-molecules.

Astonishingly, attempted hydrolysis of this group from **110** using LiOH in a mixture of THF and H_2O at ambient temperature, followed by a quench with NaH_2PO_4, did not lead to **2**, but instead provided CP-225,917 (**1**) directly in 72 % yield! This amazing result can potentially be rationalized through the degenerate mechanistic pathways shown in Scheme 20 (**111**→**116** and **112**→**116**) in which the following events occurred: the maleic anhydride was temporarily masked as its open dianion, the C-29 carboxylic acid was deprotonated, the γ-hydroxylactone was opened by base, and then

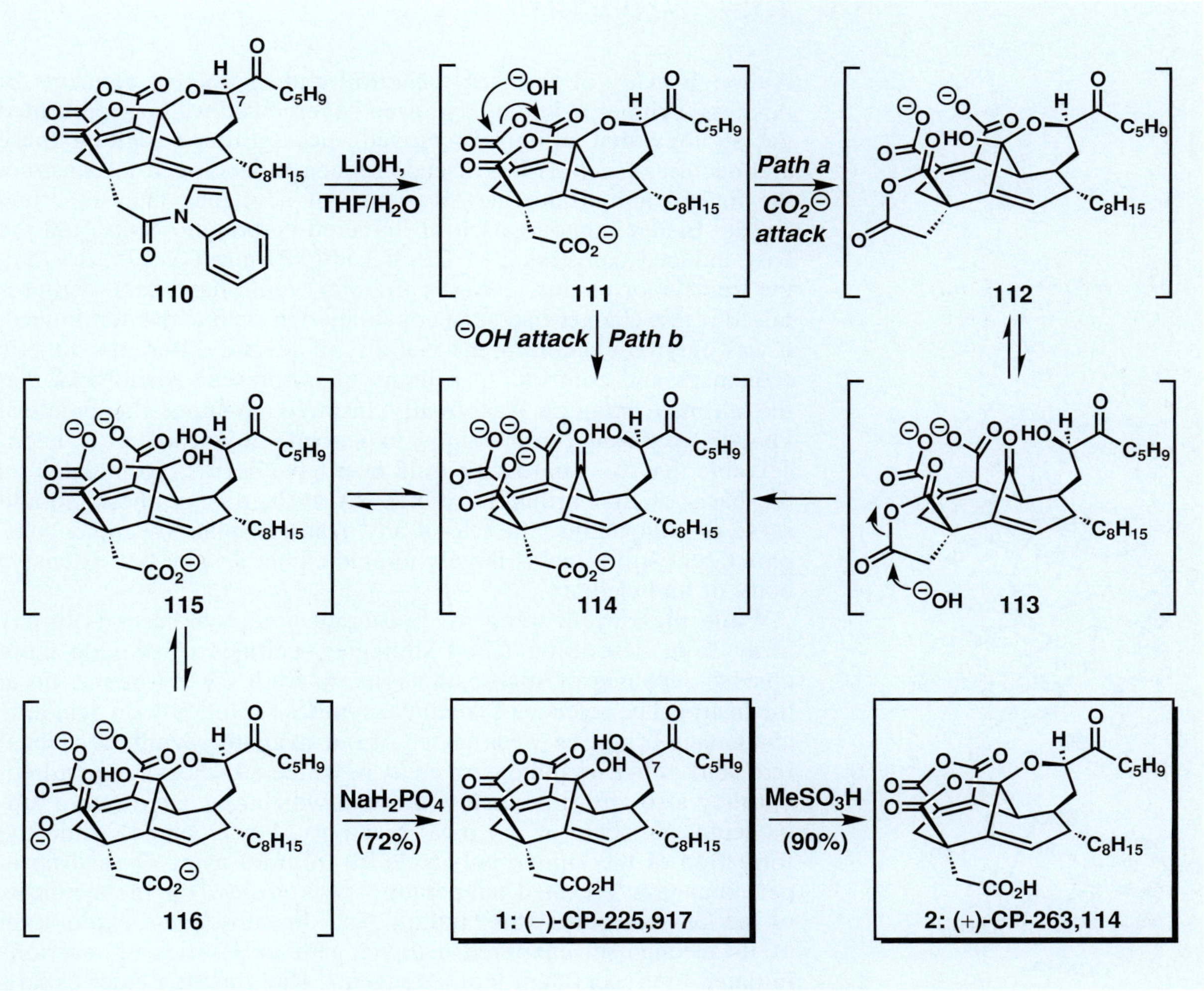

Scheme 20. Final stages and completion of Nicolaou's total synthesis of the CP-molecules (**1** and **2**).

both the anhydride and C-29 carboxylic acid were reconstituted through an acidic reaction quench. Importantly, virtually no epimerization of the potentially labile C-7 center occurred during this reaction process. At this stage, having found a means to generate **1**, a final conversion using the known Pfizer conditions[3] then completed the synthesis of the second CP-molecule, CP-263,114 (**2**), in 90 % yield from **1**. As a concluding exercise, comparison of the optical rotations of these synthetic materials established that they were opposite in sign to that of the natural products, a result deriving from the arbitrary selection of (*R*)-(+)-glycidol (**31**, Scheme 5) at the start of the total synthesis. Thus, the absolute configuration of the CP-molecules is, in fact, the opposite to that depicted in the structures drawn above.

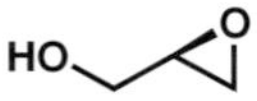

13.4 Conclusion

Although many of the most beautiful and instructive elements of this total synthesis would have been covered had we only presented the strategy that ultimately proved successful in reaching these architectures [such as the cascade sequence developed to construct the maleic anhydride, the novel use of acyl mesylates to effect Arndt–Eistert homologation of hindered carboxylic acids, and the base-induced conversion of CP-263,114 (**2**) into CP-225,917 (**1**)], the true flavor of this research program would have been compromised if this chapter had been constructed in such a manner. Indeed, it was only by examining the viability of several different synthetic sequences and confronting a litany of unforeseen roadblocks that the ultimate strategy was finally inspired. Without the practical knowledge gleaned from combat experience, if you will, it is inconceivable that this approach would ever have been so formulated on the basis of first principles alone. As such, this synthesis should serve to indicate that the fate of any synthetic plan towards a complex target still remains largely unpredictable despite our extensive body of knowledge.

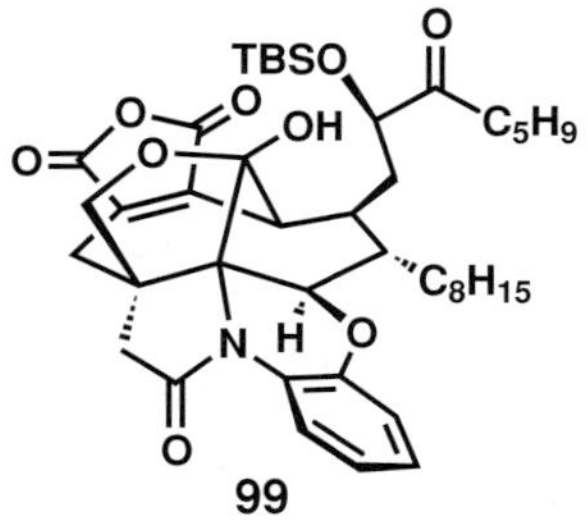

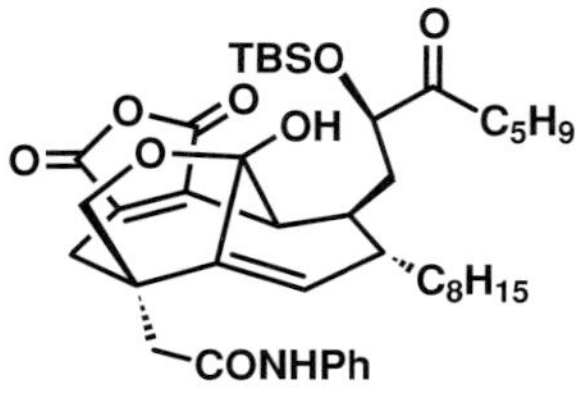

While the current trend for presenting total syntheses is to shy away from describing failed strategies, perhaps to provide more concise accounts of such campaigns, in truth these troubles lie at the heart of the science of chemical synthesis. Not only do synthetic challenges bring the weaknesses of our available synthetic arsenal into stark relief by indicating areas in which advances are required, but they also afford a window through which one can observe fundamental elements of chemical reactivity. For example, while the formation of the unique polycycle **99** from **96** using Dess–Martin periodinane at elevated temperatures was useless for the synthesis of the CP-molecules, this "failure" and the subsequent exploration of its mechanism chartered a direct path to a series of reactions initiated by hypervalent iodine reagents. Had this particular experiment never been performed and the rationale for its outcome not explored, this field of study might have remained latent for many years to come, if ever uncovered at all.

One can only hope that the instructive nature of the challenges confronted in this synthesis will encourage chemists in the future to be more willing to share such unique moments of their synthetic endeavors with the rest of the community. Not only would such an approach effectively convey the vitality of this field of research to the next generation of synthetic practitioners, but it would also ensure the more rapid extension of basic knowledge in chemical synthesis in directions that, for the present, can barely be imagined.

13.5 Fukuyama's Total Synthesis of the CP-Molecules

One year after the Nicolaou group completed their total synthesis of the CP-molecules, the second laboratory route to these formidable targets was concluded in May 2000 by Professor Tohru Fukuyama and his group at the University of Tokyo.[33] The key elements of their effective synthetic strategy, which targeted CP-263,114 (**2**), are shown in Scheme 21. As in the Nicolaou approach, these researchers sought to employ a substrate-controlled intramolecular Diels–Alder reaction to create the bicyclic [4.3.1] core and its anti-Bredt olefin. However, rather than rely on a stereocenter relatively remote to the cycloaddition participants, they projected that a stereogenic center far closer to the dienophile in the form of a C-12 center bearing an Evans chiral auxiliary (**117**) could accomplish this objective. This approach proved fruitful in that upon activation of this substrate with $ZnCl_2{\cdot}OEt_2$ in a solvent mixture comprised of CH_2Cl_2 and a small amount of pyridine, the *endo* Diels–Alder product (**119**) was smoothly generated with almost complete facial selectivity for the drawn diastereoisomer (~20:1). Importantly, one should realize that although typically it is the stereogenic center within such an auxiliary that governs stereoselectivity, in this case this feature was not active; it was the size of the entire motif that enabled the C-12 center to govern the diastereoselectivity of this event effectively.[34] As an aside, in the absence of pyridine, an acid-catalyzed olefin isomerization occurred instead, which prevented access to the desired Diels–Alder pathway.

With this essential operation implemented, the crude Diels–Alder product (**119**) was then subjected to reaction with the lithium anion derived from allyl mercaptoacetate, an event that provided **120** in 53 % overall yield from **117**. Seeking next to create the maleic anhydride moiety, treatment of this product (**120**) with DBU in THF at ambient temperature then resulted in an intramolecular aldol reaction that furnished **121** in 93 % yield as a single stereoisomer. Following decarboxylation at C-31 and dehydration of the alcohol group at C-11 to generate a thiobutenolide, a few

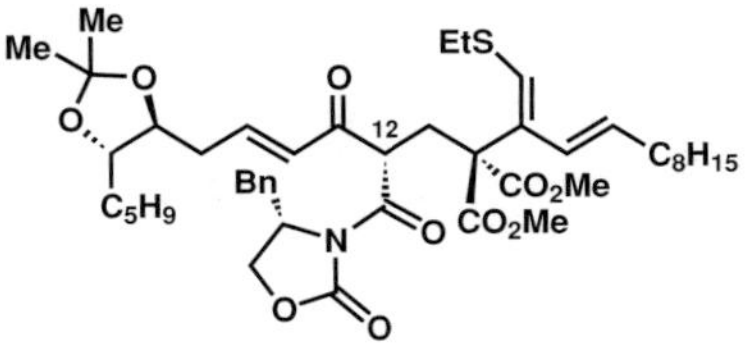

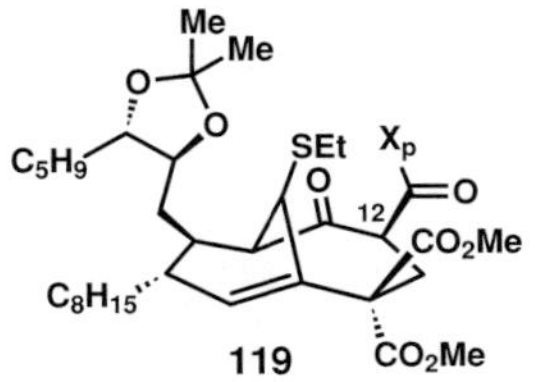

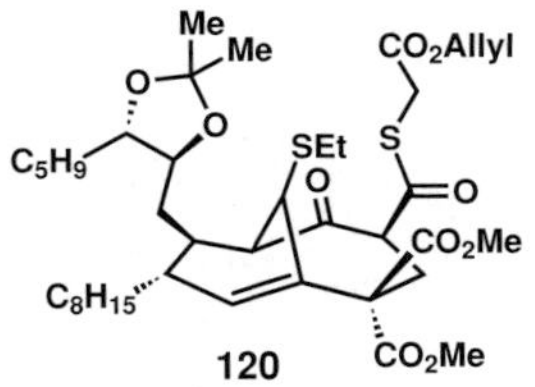

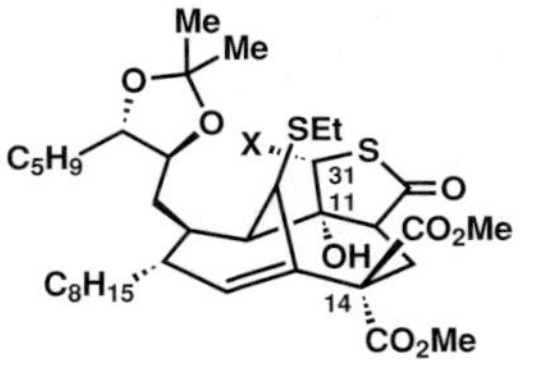

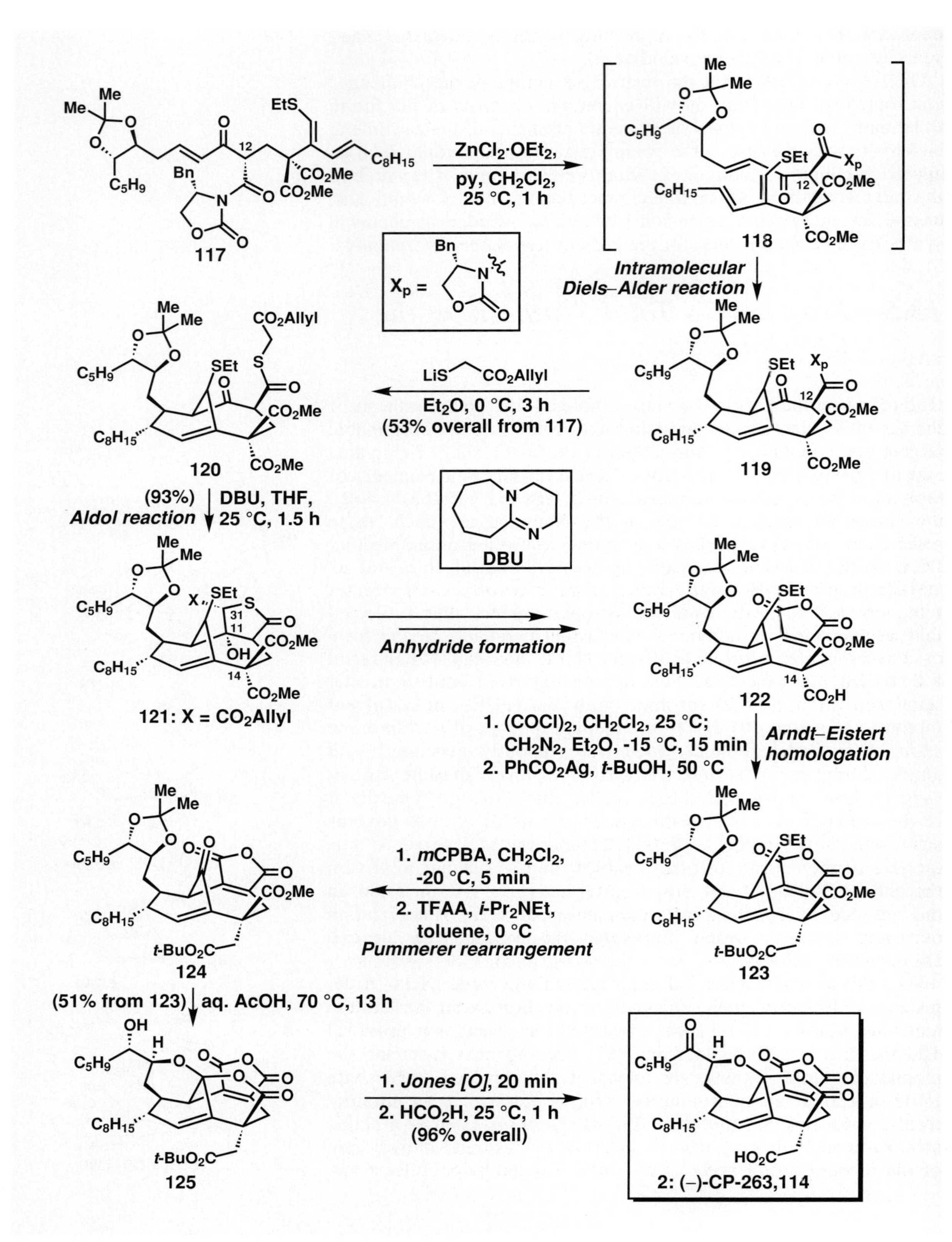

Scheme 21. Fukuyama's total synthesis of (–)-CP-263,114 (**2**).

additional operations then led to anhydride **122** in which the more accessible of the two methyl esters at C-14 had also been hydrolyzed. As such, with this ring appended onto the bicyclic core, only two key objectives remained to complete the total synthesis of **2**, namely homologation of the carboxylic acid side chain at C-14 by one carbon atom and the construction of two more rings. Fortunately, the first of these requirements proved relatively easy to accomplish using the standard conditions of the Arndt–Eistert reaction, in which **122** was converted into **123** by way of an acid chloride intermediate generated with oxalyl chloride. Accordingly, although the unique spatial environment of this same carboxylic acid on several intermediates of the Nicolaou synthesis could not be induced to react with this protocol, in this case some aspect of the particular arrangement of functional groups in **122** enabled the conventional technique to succeed.

Having accomplished this goal, the apical thioether at C-26 was then carefully oxidized to a sulfoxide (without overoxidation to the sulfone), enabling a subsequent TFAA-promoted Pummerer rearrangement[35] to generate ketone **124**. With this motif unveiled, subsequent cleavage of the dimethyl acetal with AcOH was attended by productive cyclization and dehydration to afford the entire gross architecture of the target molecule expressed within **125** in 51 % yield from **123**. Finally, Jones oxidation of the now free secondary alcohol followed by formic acid mediated lysis of the *t*-butyl ester completed the synthesis of **2** in 96 % overall yield. As a final note, although these researchers did not synthesize the other CP-molecule, CP-225,917 (**1**), their total synthesis of **2** is also a formal synthesis of **1** in that the Nicolaou route established a means to effect this conversion (cf. Scheme 20).[7]

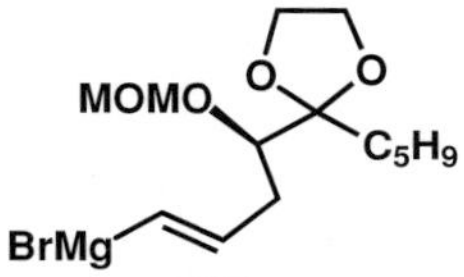

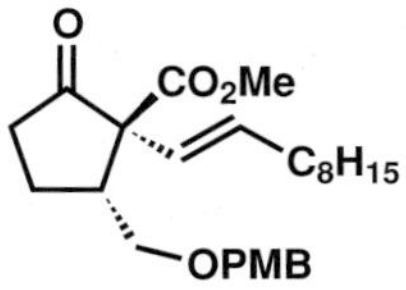

13.6 *Shair's Total Synthesis of the CP-Molecules*

Just a few weeks after the Fukuyama group reported their total synthesis of **2**, Professor Matthew Shair and his group at Harvard University completed the third laboratory route to these fascinating compounds.[36] In most respects, this approach constitutes a significant departure from that of either the Nicolaou or Fukuyama syntheses. In the early stages of their campaign, these researchers created the central bicyclic architecture not through an intramolecular Diels–Alder reaction, but instead through a clever cascade sequence initiated by the addition of vinyl Grignard reagent **127** (Scheme 22) to the ketone in homochiral **126**. With this event providing **128**, the stage was set for a subsequent anion-accelerated oxy-Cope rearrangement via a chair-like transition state to generate an eight-membered ring in the form of **129**.[37] However, this compound did not prove to be isolable, as it was suitably disposed to

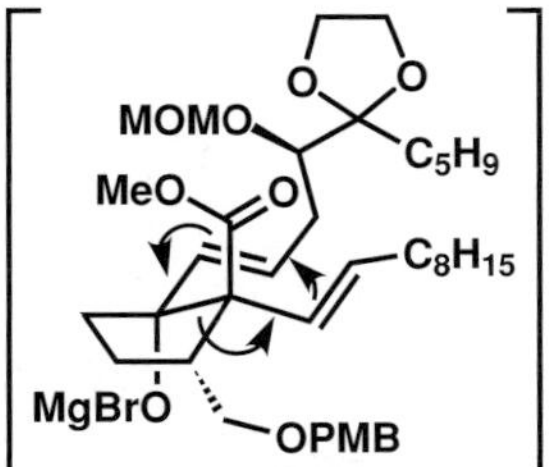

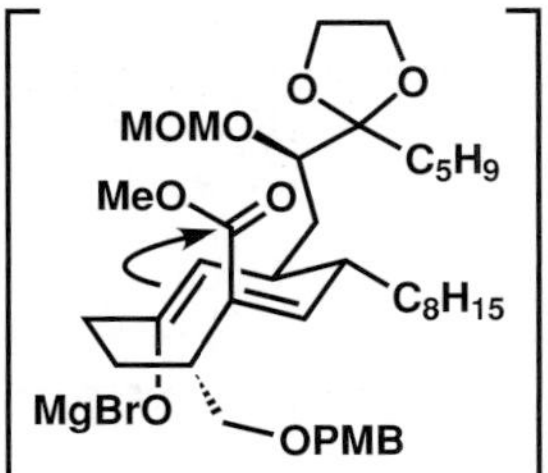

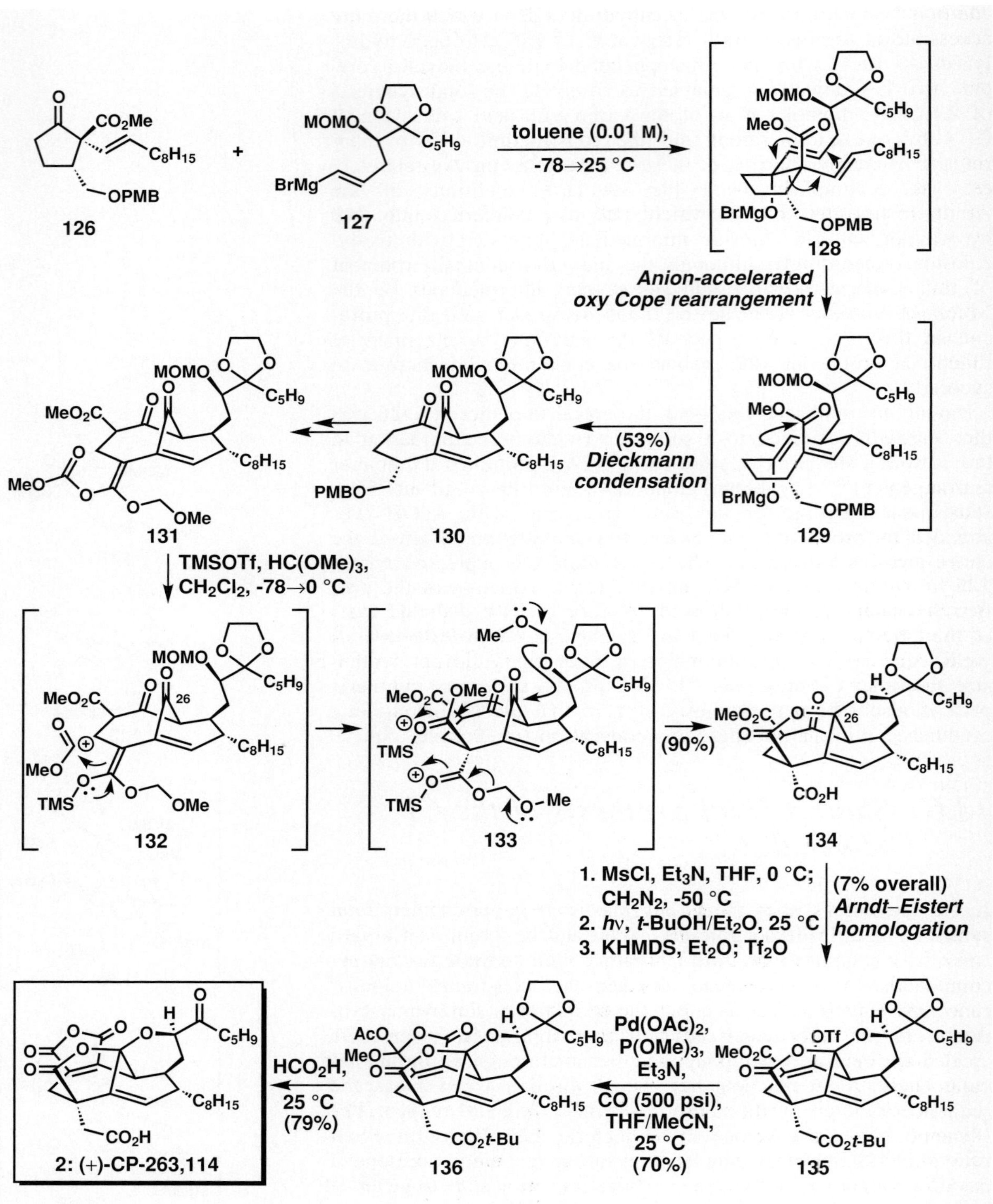

Scheme 22. Shair's total synthesis of (+)-CP-263,114 (**2**).

participate in a terminating Dieckmann condensation which gave rise to the complete [4.3.1] bicycle of the CP-molecules (**130**). Overall, this truly amazing chain of events proceeded in 53 % yield, a result that accounted for an average conversion of 81 % per step. Not only did the realization of this sequence enable material to be readily advanced to this staging area for the final drive to complete the molecule, but it also afforded empirical evidence for some aspects of the biogenetic hypothesis shown in Scheme 1, as its steps are reminiscent of the operations postulated for the conversion of **6** into **9**.

This impressive "biomimetic" cascade would not be the only domino sequence that would denote this total synthesis. Indeed, after **130** had been elaborated into **131**, treatment of the latter compound with trimethylorthoformate and a Lewis acid in the form of TMSOTf led to the one-pot assembly of **134** in 90 % yield, in which a quaternary center and two new rings had been created concurrently! A plausible mechanism to account for this highly efficient sequence might constitute initial ionization of the enol carbonate with TMSOTf to provide silylketene acetal **132**, a species poised for reaction with the liberated acyl cation that would then lead to the assembly of **133**. Since excess TMSOTf was in solution, both MOM groups within this new intermediate were cleaved with the effect that the resultant hydroxy group at C-7 could initiate the shown cyclization cascade that ultimately afforded the lactone and pyran systems of **134**. Moreover, by having effected the deprotection of the carboxylic acid at C-14 during the course of this productive sequence, the required extension by one carbon atom to match the side chain of the target molecules could be attempted in the next operation through an Arndt–Eistert reaction. As with the Nicolaou synthesis, this particular position resisted activation by conventional protocols, necessitating application of the acyl mesylate technique developed by the Nicolaou group in order to achieve diazoketone formation. Unlike the Nicolaou or Fukuyama variants of this reaction, however, photochemical conditions, rather than a silver species, were employed to complete the Wolff rearrangement that concluded the intended homologation.

With this task accomplished, only the construction of the maleic anhydride portion remained, an objective that both the Fukuyama and Nicolaou approaches had tackled early on following the assembly of the bicyclic core of the CP-molecules. In this case, however, leaving this problem pending until the end proved to be a wise design element in that, after the formation of an enol triflate (**135**), subsequent palladium-mediated carboxymethylation under a CO atmosphere at rather high pressure (500 psi) provided **136** in 70 % yield. The degree of success for this reaction is noteworthy because it is an impressive example of the transformation on a highly complex intermediate, and attempts by the Nicolaou group to execute this same transformation on simpler intermediates failed (an approach that we did not discuss directly). Finally, treatment with formic acid then cleaved the protecting groups to afford synthetic **2** in 79 % yield.

13.7 Danishefsky's Total Synthesis of the CP-Molecules

The fourth, and final, total synthesis of the CP-molecules accomplished to date was reported at the end of 2000 by Professor Samuel Danishefsky and several of his collaborators at Columbia University.[38] Testament to the power of modern synthetic organic chemistry and the ingenuity of its practitioners, this route (see Scheme 23) provided yet another novel means by which to fashion the architecture of the CP-molecules. Although the first objective in this enterprise was to create the bicyclic core, as all the other approaches had done, in this case that goal was accomplished by an initial aldol condensation between lithium enolate **138** (generated with LDA in THF at −78 °C) and furanoaldehyde **137**, followed by a Heck cyclization of the resultant TBS-trapped intermediate (**139**) to generate **140**. Although this product lacked the anti-Bredt olefin, the regiospecific installation of its lone element of endocyclic unsaturation enabled facile addition of both side chains at C-3 and C-4 (**141**→**143**), with eventual formation of the anti-Bredt system (**145**) through an E_2 elimination of the mesylate within **144** as promoted by DBU in toluene at 80 °C. Apart from the sheer creativity of these operations, one should also take special note that the maleic anhydride moiety was masked as a silylated furan within these intermediates. Only this approach incorporated this ring system in the operations which led to the construction of the bicyclic core, while all the other strategies appended such a motif onto an already established [4.3.1] system.

Having progressed to this juncture, the remaining steps sought to generate the final two ring systems of **2** and fully elaborate both the side chains at C-3 and C-4 to their proper disposition as in the target molecules. Thus, following the formation of **147** in which the phenyl sulfoxide served as a masked leaving group for an eventual cyclization, an OsO_4-mediated dihydroxylation of the terminal olefin provided a diol product whose alcohol group at C-7 immediately cyclized onto the proximal apical ketone to form hemiacetal **148**. While this reaction was effective in terms of yield (70 %), the facial approach of the osmium reagent occurred only from the undesired α-face to afford C-7 stereochemistry epimeric to that of the targeted natural products. Since numerous attempts to convert this material into that with the desired stereochemical configuration failed (such as equilibration and epimerization), the synthesis was prosecuted to its logical conclusion to afford **152** (i. e. 7-*epi*-CP-263,114). Following further investigation, however, a sequence was eventually identified that could transform this product into a 1:1 mixture of CP-225,917 (**1**) and its C-7 epimer as shown in Scheme 24. Although the route needed to ultimately implement this objective might seem rather long, its length reflects the extreme difficulty in rupturing the pyran system without disturbing several

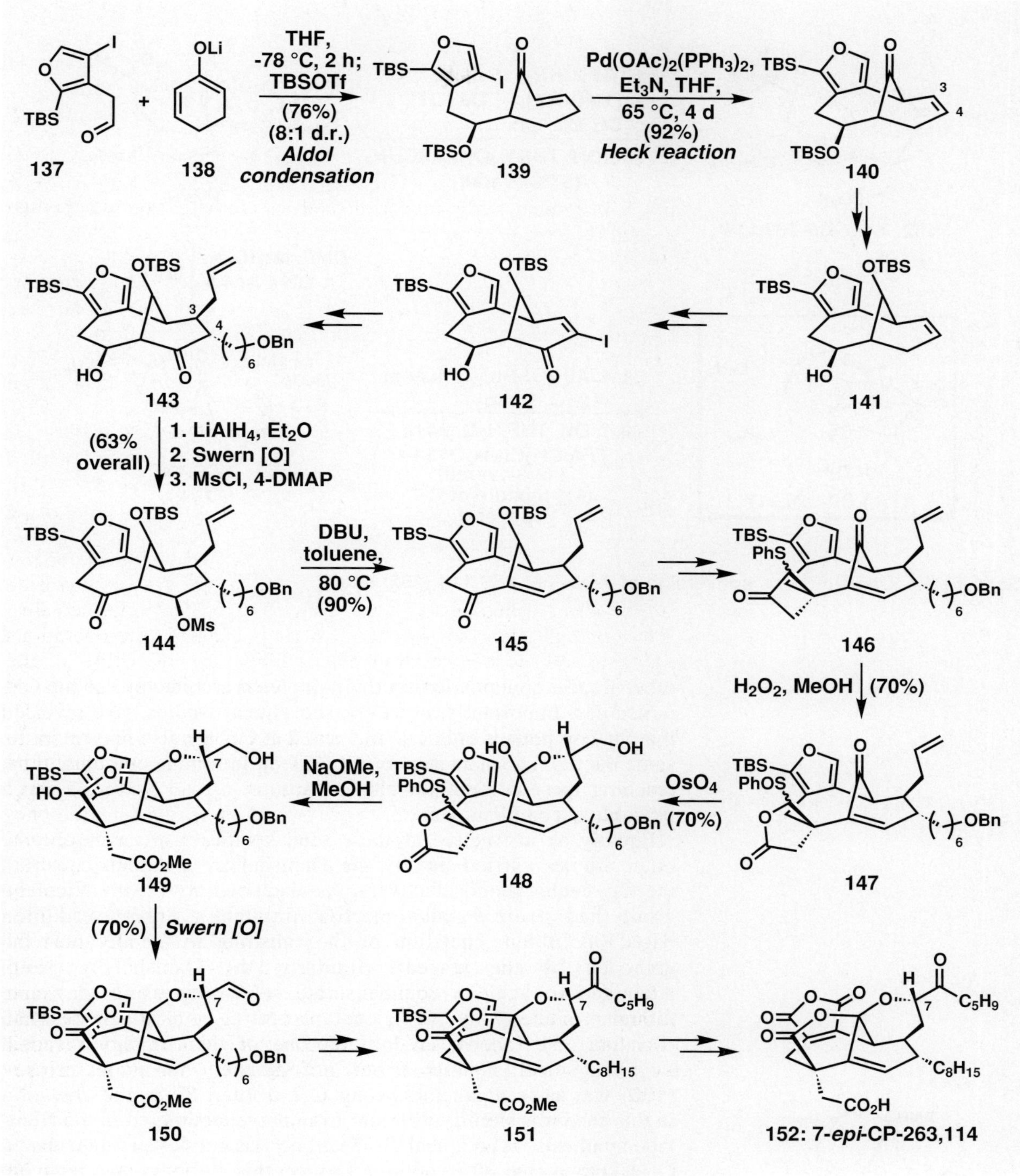

Scheme 23. Danishefsky's total synthesis of 7-*epi*-CP-263,114 (**152**).

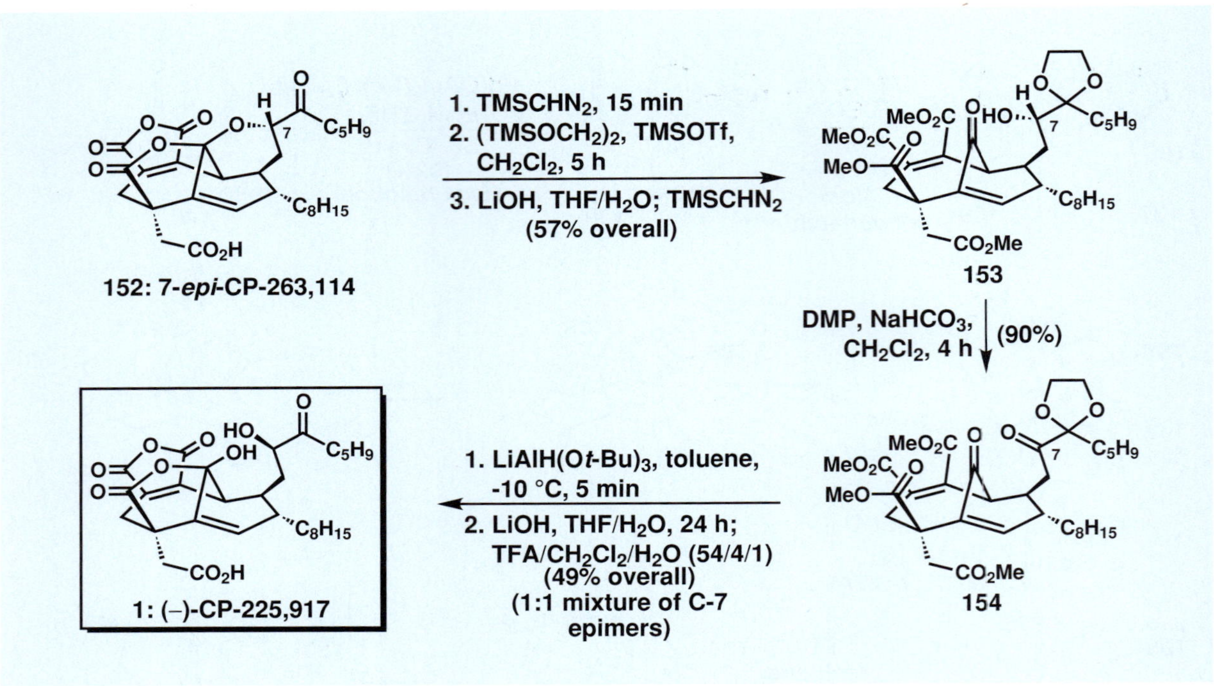

Scheme 24. Danishefsky's conversion of 7-*epi*-CP-263,114 (**152**) into **1**.

other fragile domains within the completed architecture of the CP-molecules. Importantly, however, subsequent studies have revealed that the compounds epimeric to **1** and **2** at C-7 are also present in the same bacterial fermentation broth,[39] albeit in minute amounts; thus, reaching these compounds also constitutes a total synthesis and a laudable accomplishment.

Equally as instructive as these total syntheses are a number of other studies carried out by the Danishefsky group to construct the C-7 center stereoselectively. As discussed above, the Nicolaou group had realized this objective through a dithiane addition based on lithium chelation of the substrate to facially bias the approach by the reagent. Similarly, the Danishefsky group attempted to achieve commensurate selectivity using the same dithiane nucleophile (**14**), but operating on aldehyde **155** (Scheme 25). Indeed, an initial probe of this strategy revealed exquisite selectivity (10:1), but, unfortunately, the major product (**156**) was once again the wrong C-7 epimer. Since the chelation in this case was slightly different from that encountered in the Nicolaou synthesis, as an apical C-27 ketone was employed rather than a C-11 ketone (see **39** in column figure), this carbonyl was reduced, and the two phenyl sulfide diastereomers of the resulting hydroxy compound were separated to provide two new homochiral test substrates (**157** and **159**). Application of the same reaction conditions

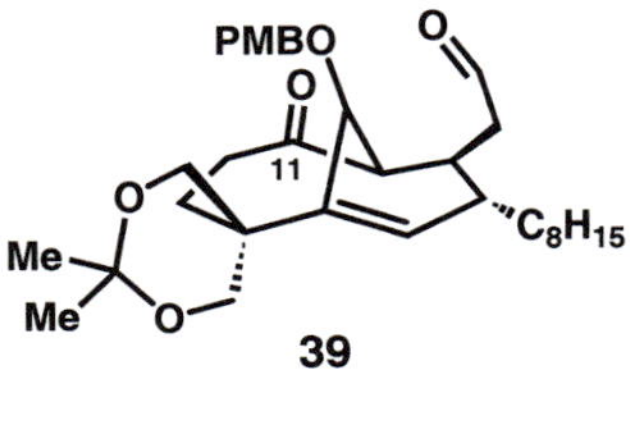

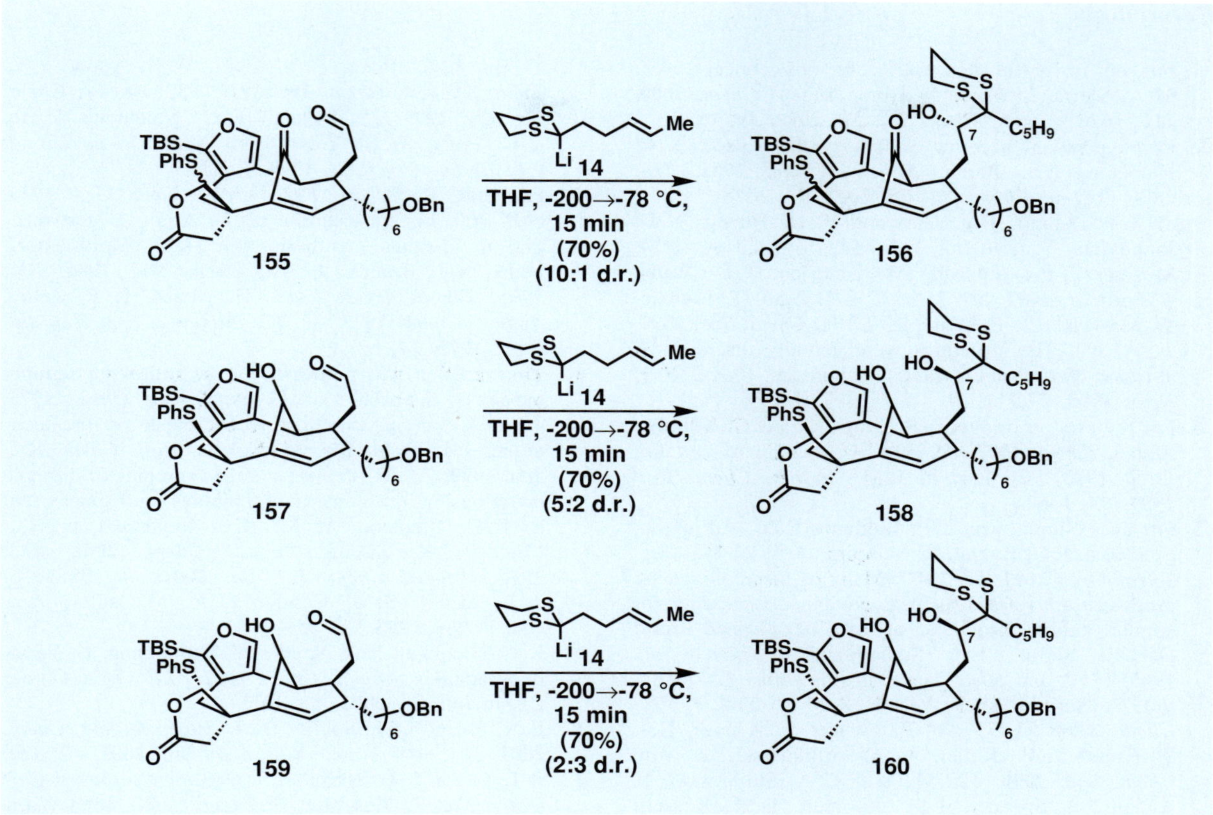

Scheme 25. Danishefsky's studies on C-7 stereocontrol.

to these intermediates now revealed a truly amazing finding in that they led to markedly different degrees of selectivity for the C-7 stereocenter (5:2 versus 2:3). Although the products of this sequence were not ultimately converted into the targeted natural products, these results provide a striking example of how stereochemical elements quite remote from the reactive center can dramatically influence the course of a reaction. This observation is well worth remembering since synthetic chemists often tend to emphasize the importance of substituents proximal to the reacting center, ignoring those more remotely disposed.[40]

References

1. For an insightful description of this concept, see: M. A. Sierra, M. C. de la Torre, *Angew. Chem.* **2000**, *112*, 1628; *Angew. Chem. Int. Ed.* **2000**, *39*, 1538.
2. For a personal account of this synthesis, see: K. C. Nicolaou, P. S. Baran, *Angew. Chem.* **2002**, *114*, 2800; *Angew. Chem. Int. Ed.* **2002**, *41*, 2678.
3. a) T. T. Dabrah, H. J. Harwood, L. H. Huang, N. D. Jankovich, T. Kaneko, J.-C. Li, S. Lindsey, P. M. Moshier, T. A. Subashi, M. Therrien, P. C. Watts, *J. Antibiot.* **1997**, *50*, 1; b) T. T. Dabrah, T. Kaneko, W. Massefski, E. B. Whipple, *J. Am. Chem. Soc.* **1997**, *119*, 1594. The structures were actually disclosed in a report two years earlier: S. Stinson, *Chem. Eng. News* **1995**, *73*(21), 29.
4. For reviews on bridgehead olefins, see: a) G. Köbrich, *Angew. Chem.* **1973**, *85*, 494; *Angew. Chem. Int. Ed. Engl.* **1973**, *12*, 464; b) P. M. Warner, *Chem. Rev.* **1989**, *89*, 1067.
5. For an example, see: J. E. Baldwin, R. M. Adlington, F. Roussi, P. G. Bulger, R. Marquez, A. V. W. Mayweg, *Tetrahedron* **2001**, *57*, 7409. Many of these biogenetic proposals are related to ideas advanced far earlier for similar natural products, as can be gleaned from: D. H. R. Barton, J. K. Sutherland, *J. Chem. Soc.* **1965**, 1769, and other papers in this series.
6. a) G. A. Sulikowski, F. Agnelli, R. M. Corbett, *J. Org. Chem.* **2000**, *65*, 337; b) P. Spencer, F. Agnelli, H. J. Williams, N. P. Keller, G. A. Sulikowski, *J. Am. Chem. Soc.* **2000**, *122*, 420; c) G. A. Sulikowski, F. Agnelli, P. Spencer, J. M. Koomen, D. H. Russell, *Org. Lett.* **2002**, *4*, 1447; d) G. A. Sulikowski, W. Liu, F. Agnelli, R. M. Corbett, Z. Luo, S. J. Hershberger, *Org. Lett.* **2002**, *4*, 1451.
7. For the full account of this work, see: a) K. C. Nicolaou, J. Jung, W. H. Yoon, K. C. Fong, H.-S. Choi, Y. He, Y.-L. Zhong, P. S. Baran, *J. Am. Chem. Soc.* **2002**, *124*, 2183; b) K. C. Nicolaou, P. S. Baran, Y.-L. Zhong, K. C. Fong, H.-S. Choi, *J. Am. Chem. Soc.* **2002**, *124*, 2190; c) K. C. Nicolaou, Y.-L. Zhong, P. S. Baran, J. Jung, H.-S. Choi, W. H. Yoon, *J. Am. Chem. Soc.* **2002**, *124*, 2202. For the initial communications, see: d) K. C. Nicolaou, P. S. Baran, Y.-L. Zhong, H.-S. Choi, W. H. Yoon, Y. He, K. C. Fong, *Angew. Chem.* **1999**, *111*, 1774; *Angew. Chem. Int. Ed.* **1999**, *38*, 1669; e) K. C. Nicolaou, P. S. Baran, Y.-L. Zhong, K. C. Fong, Y. He, W. H. Yoon, H.-S. Choi, *Angew. Chem.* **1999**, *111*, 1781; *Angew. Chem. Int. Ed.* **1999**, *38*, 1676.
8. For earlier model studies, see: a) K. C. Nicolaou, M. W. Härter, L. Boulton, B. Jandeleit, *Angew. Chem.* **1997**, *109*, 1243; *Angew. Chem. Int. Ed. Engl.* **1997**, *36*, 1194; b) K. C. Nicolaou, M. H. D. Postema, N. D. Miller, G. Yang, *Angew. Chem.* **1997**, *109*, 2922; *Angew. Chem. Int. Ed. Engl.* **1997**, *36*, 2821; c) K. C. Nicolaou, P. S. Baran, R. Jautelat, Y. He, K. C. Fong, H.-S. Choi, W. H. Yoon, Y.-L. Zhong, *Angew. Chem.* **1999**, *111*, 532; *Angew. Chem. Int. Ed.* **1999**, *38*, 549; d) K. C. Nicolaou, Y. He, K. C. Fong, W. H. Yoon, H.-S. Choi, Y.-L. Zhong, P. S. Baran, *Org. Lett.* **1999**, *1*, 63.
9. F. Arndt, B. Eistert, *Ber. Dtsch. Chem. Ges.* **1935**, *68B*, 200. For an example of the Arndt–Eistert reaction in chemical synthesis, see: A. B. Smith, B. H. Toder, S. J. Branca, *J. Am. Chem. Soc.* **1984**, *106*, 3995. For a review, see: H. Meier, K.-P. Zeller, *Angew. Chem.* **1975**, *87*, 52; *Angew. Chem. Int. Ed. Engl.* **1975**, *14*, 32.
10. This reaction was pioneered in the following seminal paper: D. Seebach, E. J. Corey, *J. Org. Chem.* **1975**, *40*, 231. For an excellent recent paper on the technique, see: a) M. Ide, M. Nakata, *Bull. Chem. Soc. Jpn.* **1999**, *72*, 2491. For selected examples of dithiane couplings in the context of complex molecules, see: b) K. C. Nicolaou, Y. Li, K. C. Fylaktakidou, H. J. Mitchell, K. Sugita, *Angew. Chem.* **2001**, *113*, 3972; *Angew. Chem. Int. Ed.* **2001**, *40*, 3854; c) A. B. Smith, S. M. Condon, J. A. McCauley, *Acc. Chem. Res.* **1998**, *31*, 35.
11. K. C. Nicolaou, S. A. Snyder, T. Montagnon, G. Vassilikogiannakis, *Angew. Chem.* **2002**, *114*, 1742; *Angew. Chem. Int. Ed.* **2002**, *41*, 1668.
12. B. R. Bear, S. M. Sparks, K. J. Shea, *Angew. Chem.* **2001**, *113*, 864; *Angew. Chem. Int. Ed.* **2001**, *40*, 820.
13. R. E. Claus, S. L. Schreiber in *Organic Syntheses, Collective Vol. 7*, (Eds.: J. P. Freeman, et al.), John Wiley & Sons, New York, **1990**, pp. 168–171. For a review of this general concept, see: C. S. Poss, S. L. Schreiber, *Acc. Chem. Res.* **1994**, *27*, 9.
14. M. Schlösser, K. F. Christmann, *Liebigs Ann.* **1967**, *708*, 1.
15. G. Wittig, H. Reiff, *Angew. Chem.* **1968**, *80*, 8; *Angew. Chem. Int. Ed. Engl.* **1968**, *7*, 7.
16. J. Schwartz, J. A. Labinger, *Angew. Chem.* **1976**, *88*, 402; *Angew. Chem. Int. Ed. Engl.* **1976**, *15*, 333.
17. N. Bodor, M. J. S. Dewar, A. J. Harget, *J. Am. Chem. Soc.* **1970**, *92*, 2929.
18. For studies on these species, see: a) D. J. Lythgoe, I. McClenaghan, C. A. Ramsden, *J. Heterocycl. Chem.* **1993**, *30*, 113; b) K. Ito, K. Yakushijin, *Heterocycles* **1978**, *9*, 1603; c) K. Yakushijin, M. Kozuka, H. Furukawa, *Chem. Pharm. Bull.* **1980**, *28*, 2178. For the use of these compounds in synthesis, see: d) A. Padwa, M. Dimitroff, B. Liu, *Org. Lett.* **2000**, *2*, 3233.
19. a) J. B. Hendrickson, R. Bergeron, *Tetrahedron Lett.* **1973**, *14*, 4607; b) J. E. McMurry, W. J. Scott, *Tetrahedron Lett.* **1983**, *24*, 979. For a review, see: W. J. Scott, J. E. McMurry, *Acc. Chem. Res.* **1988**, *21*, 47.
20. G. Stork, K. Zhao, *Tetrahedron Lett.* **1989**, *30*, 287.
21. K. B. Sharpless, R. C. Michaelson, *J. Am. Chem. Soc.* **1973**, *95*, 6136.

22. a) W. Nagata, M. Yoshioka, T. Okumura, *Tetrahedron Lett.* **1966**, *7*, 847; b) G. Liu, T.C. Smith, H. Pfander, *Tetrahedron Lett.* **1995**, *36*, 4979.
23. C. Geletneky, S. Berger, *Eur. J. Org. Chem.* **1998**, 1625.
24. L.L. Smith, M.J. Kulig, *J. Am. Chem. Soc.* **1976**, *98*, 1027.
25. a) J.E. Whiting, J.T. Edward, *Can. J. Chem.* **1971**, *49*, 3799; b) C.D. Hurd, W.H. Saunders, *J. Am. Chem. Soc.* **1952**, *74*, 5324.
26. For early literature pertaining to this compound, see: a) D.B. Dess, J.C. Martin, *J. Org. Chem.* **1983**, *48*, 4155; b) D.B. Dess, J.C. Martin, *J. Am. Chem. Soc.* **1991**, *113*, 7277; c) S.D. Meyer, S.L. Schreiber, *J. Org. Chem.* **1994**, *59*, 7549. For an advanced treatise on hypervalent iodine reagents, see: d) A. Varvoglis, *Hypervalent Iodine in Organic Synthesis*, Academic Press, San Diego, **1996**, p. 256.
27. a) P.L. Anelli, C. Biffi, F. Montanari, S. Quici, *J. Org. Chem.* **1987**, *52*, 2559; b) A. De Mico, R. Margarita, L. Parlanti, A. Vescovi, G. Piancatelli, *J. Org. Chem.* **1997**, *62*, 6974.
28. a) J.A. Franz, J.C. Martin, *J. Am. Chem. Soc.* **1973**, *95*, 2017; b) J.C. Martin, J.A. Franz, R.J. Arhart, *J. Am. Chem. Soc.* **1974**, *96*, 4604.
29. For an entry into this area of research, see the following and references therein: a) K.C. Nicolaou, P.S. Baran, Y.-L. Zhong, K. Sugita, *J. Am. Chem. Soc.* **2002**, *124*, 2212; b) K.C. Nicolaou, K. Sugita, P.S. Baran, Y.-L. Zhong, *J. Am. Chem. Soc.* **2002**, *124*, 2221; c) K.C. Nicolaou, P.S. Baran, Y.-L. Zhong, S. Barluenga, K.W. Hunt, R. Kranich, J.A. Vega, *J. Am. Chem. Soc.* **2002**, *124*, 2233; d) K.C. Nicolaou, T. Montagnon, P.S. Baran, Y.-L. Zhong, *J. Am. Chem. Soc.* **2002**, *124*, 2245.
30. M.J.V. de Oliveira Baptista, A.G.M. Barrett, D.H.R. Barton, M. Girijavallabhan, R.C. Jennings, J. Kelly, V.J. Papadimitriou, J.V. Turner, N.A. Usher, *J. Chem. Soc., Perkin Trans. 1* **1977**, 1477.
31. For the extension of this technique to numerous substrates, see: K.C. Nicolaou, P.S. Baran, Y.-L. Zhong, H.-S. Choi, K.C. Fong, Y. He, W.H. Yoon, *Org. Lett.* **1999**, *1*, 883.
32. For example using this reagent to create *N*-based heterocycles, see: a) L. Vo-Quang, Y. Vo-Quang, *J. Heterocycl. Chem.* **1982**, *19*, 145; b) B.E. Landberg, J.W. Lown, *J. Chem. Soc., Perkin Trans. 1* **1975**, 1326; c) R. Huisgen, M. Seidel, G. Wallbillich, H. Knupfer, *Tetrahedron* **1962**, *17*, 3.
33. N. Waizumi, T. Itoh, T. Fukuyama, *J. Am. Chem. Soc.* **2000**, *122*, 7825.
34. For another example of such stereoselectivity, see: D.A. Evans, D.H.B. Ripin, J.S. Johnson, E.A. Shaughnessy, *Angew. Chem.* **1997**, *109*, 2208; *Angew. Chem. Int. Ed. Engl.* **1997**, *36*, 2119.
35. For a review on the Pummerer rearrangement, see: O. De Lucchi, U. Miotti, G. Modena in *Organic Reactions, Vol. 40* (Eds.: L.A. Paquette, et al.), John Wiley & Sons, New York, **1991**, pp. 157–406.
36. C. Chen, M.E. Layton, S.M. Sheehan, M.D. Shair, *J. Am. Chem. Soc.* **2000**, *122*, 7424. For earlier model studies, see: C. Chen, M.E. Layton, M.D. Shair, *J. Am. Chem. Soc.* **1998**, *120*, 10784.
37. For other approaches to the CP-molecules using an oxy-Cope rearrangement, see: a) M.M. Bio, J.L. Leighton, *J. Am. Chem. Soc.* **1999**, *121*, 890; b) P.W.M. Sgarbi, D.L.J. Clive, *Chem. Commun.* **1997**, 2157.
38. a) Q. Tan, S.J. Danishefsky, *Angew. Chem.* **2000**, *112*, 4683; *Angew. Chem. Int. Ed.* **2000**, *39*, 4509. For previous model studies, see: b) O. Kwon, D.-S. Su, D. Meng, W. Deng, D.C. D'Amico, S.J. Danishefsky, *Angew. Chem.* **1998**, *110*, 1978; *Angew. Chem. Int. Ed.* **1998**, *37*, 1877; c) O. Kwon, D.-S. Su, D. Meng, W. Deng, D.C. D'Amico, S.J. Danishefsky, *Angew. Chem.* **1998**, *110*, 1981; *Angew. Chem. Int. Ed.* **1998**, *37*, 1880; d) D. Meng, S.J. Danishefsky, *Angew. Chem.* **1999**, *111*, 1582; *Angew. Chem. Int. Ed.* **1999**, *38*, 1485; e) D. Meng, Q. Tan, S.J. Danishefsky, *Angew. Chem.* **1999**, *111*, 3393; *Angew. Chem. Int. Ed.* **1999**, *38*, 3197.
39. P. Spencer, F. Agnelli, G.A. Sulikowski, *Org. Lett.* **2001**, *3*, 1443.
40. For an elegant example of this concept, see: P. Linnane, N. Magnus, P. Magnus, *Nature* **1997**, *385*, 799.

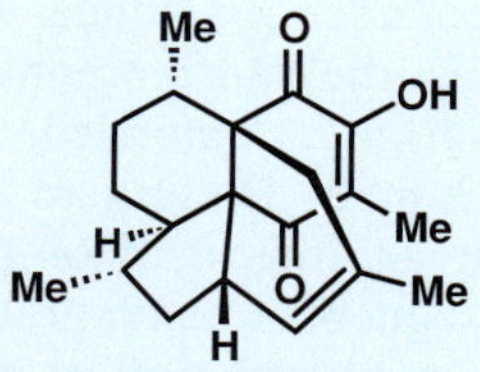

1: colombiasin A

14

K. C. Nicolaou (2001)

Colombiasin A

14.1 Introduction

Key concepts:

- Biomimetic synthesis
- Tsuji-Trost reaction
- Carroll rearrangement
- Quinone-based Diels–Alder reactions
- Diene protecting groups

The *Pseudopterogorgia* class of corals, which are found throughout the Caribbean, has long been a source of intricate and complex bioactive secondary metabolites with diverse applications, including use as antiinflammatory agents and analgesics.[1] In 2000, a new natural product from this family of marine organisms was introduced to the scientific community in the form of colombiasin A (**1**), a novel diterpenoid which demonstrated inhibitory activity against *Mycobacterium tuberculosis* H37Rv.[2] The structure of colombiasin A is made up of 20 carbon atoms arranged in a compact tetracyclic framework, with a periphery adorned by four methyl groups, two carbonyl groups, two carbon–carbon double bonds, and a hydroxy group. This unprecedented molecular architecture certainly offers a considerable challenge for synthetic practitioners. Moreover, its six stereogenic centers, two of which are adjacent quaternary carbon atoms strategically positioned in the nucleus of the molecule, define imposing elements that could potentially resist any carefully considered synthetic plan of attack.

Fortunately, due to the numerous natural products isolated from *Pseudopterogorgia* corals, cohesive biosynthetic pathways have been proposed to account for the formation of the various terpenoid structures observed, providing invaluable clues for the design of potential synthetic routes to **1**. For example, in 1988, A. D. Rodríguez and his co-workers from the University of Puerto Rico disclosed the structure of elisabethin A (**3**, Scheme 1),[3] a compound that they believed might be derived from another isolate, bicyclic quinone **2**,[4] through a C1–C9 cyclization event. Despite the fact that these metabolites were obtained from morphologically similar but unique species of *Pseudopterogorgia* corals, the structural homology shared by **2** and **3** certainly offers credence to the plausibility of such a biosynthetic connection. Since colombiasin A (**1**) was obtained from precisely the same organism as **3**, *Pseudopterogorgia elisabethae*, Rodríguez extended his analysis to the idea that oxidation of C-12 in **3**, leading to the hydroxy group drawn in **4**, would provide an intermediate that could be converted into **1**. In the proposed sequence, activation of the newly installed oxygen atom either by phosphorylation or protonation, followed by base-catalyzed removal of the proton at C-2 and intramolecular alkylation of the resultant enolate, would generate colombiasin A (**1**).[2] Although this hypothesis is reasonable, evidence is circumstantial, as **4** has not been observed as a distinct entity and biosynthetic feeding experiments have not, as yet, been performed to verify that **3** can be converted into **1**.

An equally enticing interpretation for the formation of colombiasin A (**1**), first advanced by the Nicolaou group in 2001, relies instead upon dehydration of the hydroxy group at C-9 in **2** with concurrent isomerization leading to the diene system **5**. Due to the proximal nature of the putative diene and dienophile units in **5**, a productive Diels–Alder cycloaddition would then lead directly to **1** with concomitant formation of the adjacent and daunting quaternary centers. In this chapter, we shall focus our attention

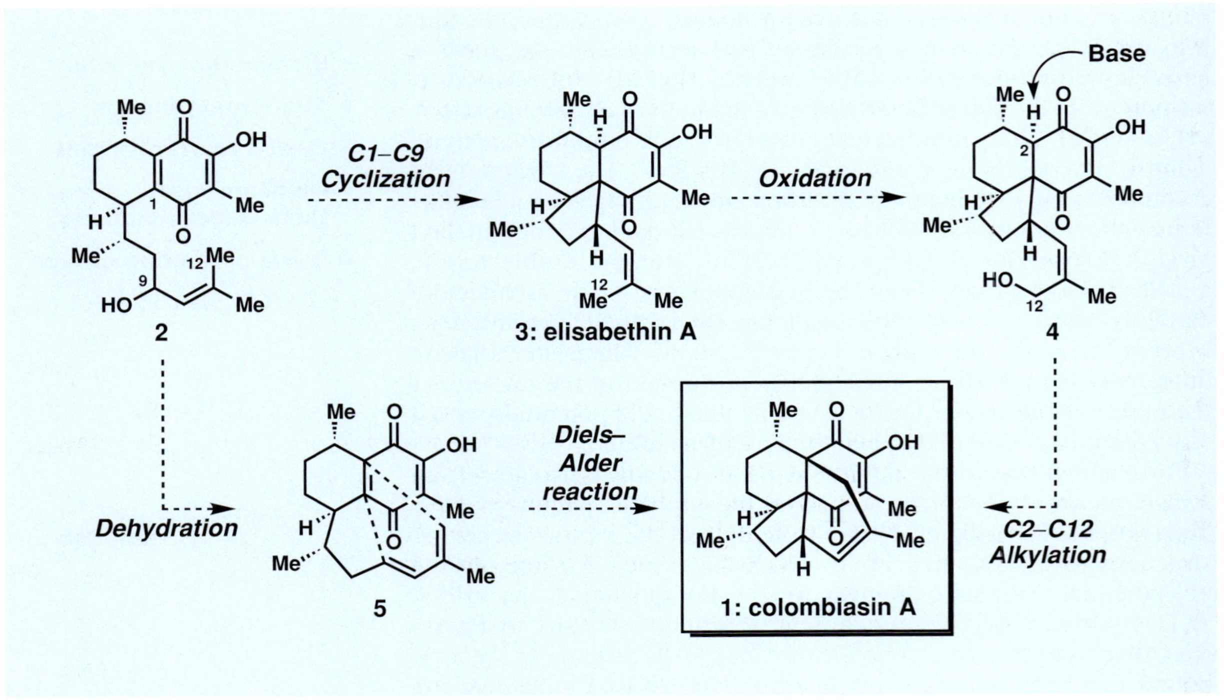

Scheme 1. Plausible biosynthetic pathways leading to the generation of colombiasin A (**1**) from other natural products isolated from various members of the *Pseudopterogorgia* genus.

on the successful realization of this latter biosynthetic hypothesis through an efficient and asymmetric total synthesis of colombiasin A (**1**) as developed in the Nicolaou laboratories.[5]

14.2 Retrosynthetic Analysis and Strategy

With the goal of performing a late-stage Diels–Alder cycloaddition as a means to forge the C- and D-rings of colombiasin A (**1**), a strategy based on testing the biogenetic route delineated above, the Nicolaou group first decided to define **6** (Scheme 2) as the key target of their synthetic endeavors to **1**. This molecule is essentially homologous with **5** (Scheme 1), altered only through the addition of a methyl ether protecting group on the C-16 alcohol and the incorporation of a hydroxy group at C-5. Although these structural modifications represent potentially superfluous chemical moieties for the projected Diels–Alder reaction, their inclusion in this analysis was deemed strategically justified for the successful execution of transformations leading to **6**. Thus, with a secure route to **6**, completion of the synthesis of colombiasin A was anticipated through a thermally induced Diels–Alder cycloaddition, with the proviso of using Lewis acid activation if the reaction proved lethargic. However, based on the relative proximity of the diene and dienophile, elements that certainly seemed capable of achieving competent orbital alignment based on molecular models, a facile cycloaddition was expected. As such, a successful reaction at this stage would provide direct access to the entire skeletal framework of colombiasin A, while also accommodating the stereoselective generation of both quaternary centers (assuming an *endo* specific Diels–Alder cycloaddition). Subsequent C-5 deoxygenation using a procedure such as the radical-based Barton–McCombie reaction,[6] followed by demethylation of the C-16 methyl ether, would then complete the synthesis of **1** from **6** based on this initial disassembly.

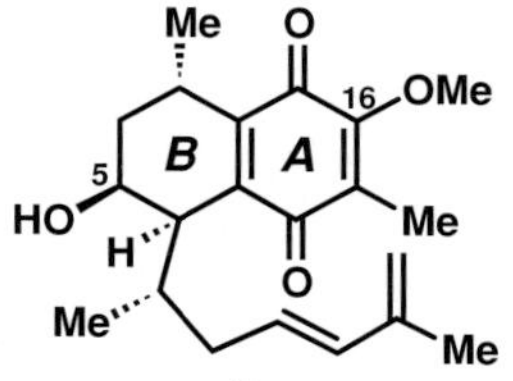

Having reduced the synthetic problem to the creation of quinone **6**, further analysis suggested the diene portion could be installed stereoselectively by Wittig olefination of the precursor aldehyde functionality in **8** using the ylide derived from phosphonium salt **7**. Additionally, based on the large body of literature that amply illustrates the difficulty in handling isolated quinone adducts (due to their noted instability and proclivity to undergo side reactions),[7] it was decided to mask this moiety as the aromatic A-ring shown in **8** rather than carry this reactive functionality through the subsequent steps of the synthesis. A demethylation–oxidation sequence induced by AgO in the presence of a mineral acid such as HNO_3 was envisioned as a suitable set of conditions capable of revealing the quinone at the appropriate juncture in the forward direction of the synthesis.[8] From **8**, simplification to **9** provided a compound whose C-6 appendage seemed ripe for disconnection. Although one might envision that the installation of this carbon

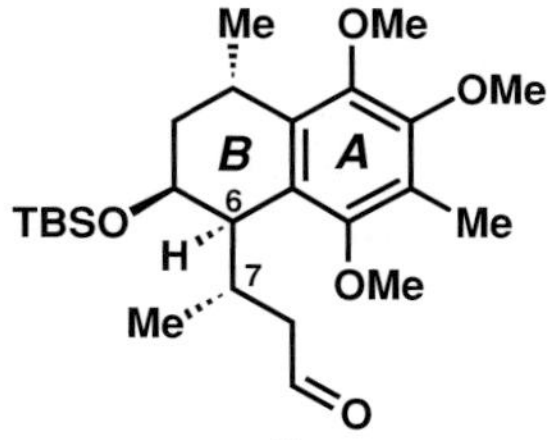

Me O 16 OMe B A 5 HO H Me Me Me
6

Me OMe OMe B A 6 TBSO H 7 Me OMe Me O
8

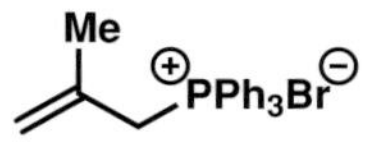

Me ⊕PPh3Br⊖
7

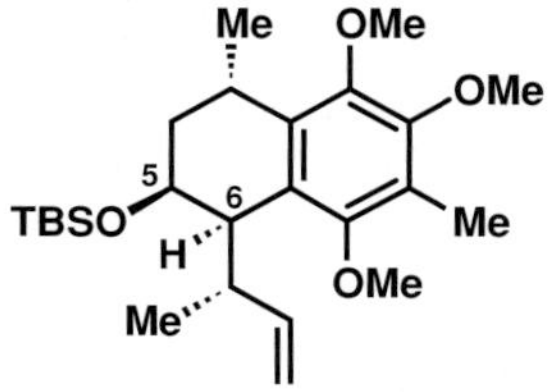

Me OMe OMe 5 6 TBSO H Me OMe Me
9

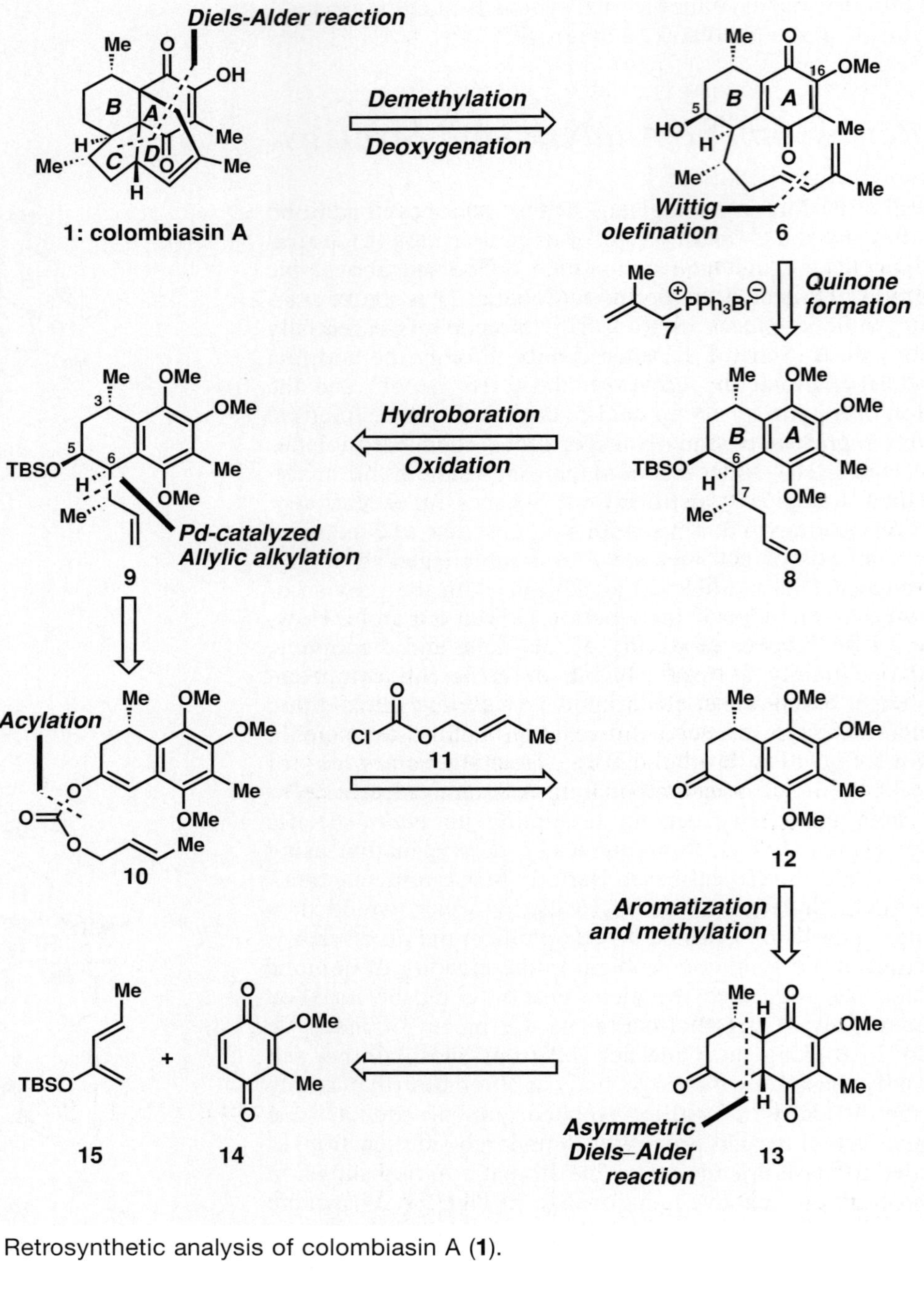

Scheme 2. Retrosynthetic analysis of colombiasin A (**1**).

chain could be achieved through generation of the thermodynamic enolate from a compound such as **12**, followed by the addition of a suitable alkyl halide, numerous studies towards this end indicated that such an event would be difficult.[9] Even if these experiments had been successful, the alkylation would most likely not have proceeded with the requisite stereoselective generation of both the C-6 and C-7 centers.

An alternative, more classical approach for carbon–carbon-bond formation in this context could potentially rely on employment of the Carroll rearrangement.[10] As shown in Scheme 3, this transformation can be considered simply as a modified Claisen sigmatropic rearrangement, whereby heating a compound such as **16** at temperatures typically in the range of 180 to 210 °C is ultimately expected to lead to the formation of **19** through the indicated mechanism. Starting in the early 1970s, however, Professors J. Tsuji and B. M. Trost independently disclosed a series of seminal papers which greatly extended the ability of chemists to forge such carbon–carbon bonds on activated methylene scaffolds based on the observation that vinyl β-ketoesters (such as **16**) react with catalytic amounts of Pd(0) to form π-allyl complexes *in situ* such as **20**. These species can then react with carbon-based soft nucleophiles, leading to products such as **19**.[11]

Several features of this general transformation, more commonly known as the Tsuji–Trost reaction, are of note. First, from an historical perspective, this reaction reflects the first demonstration of a metallated species acting as an electrophile, countering decades of research which indicated that such entities only behaved as nucleophiles. On a more practical level, the Tsuji–Trost reaction proceeds under neutral and mild conditions (typically at temperatures

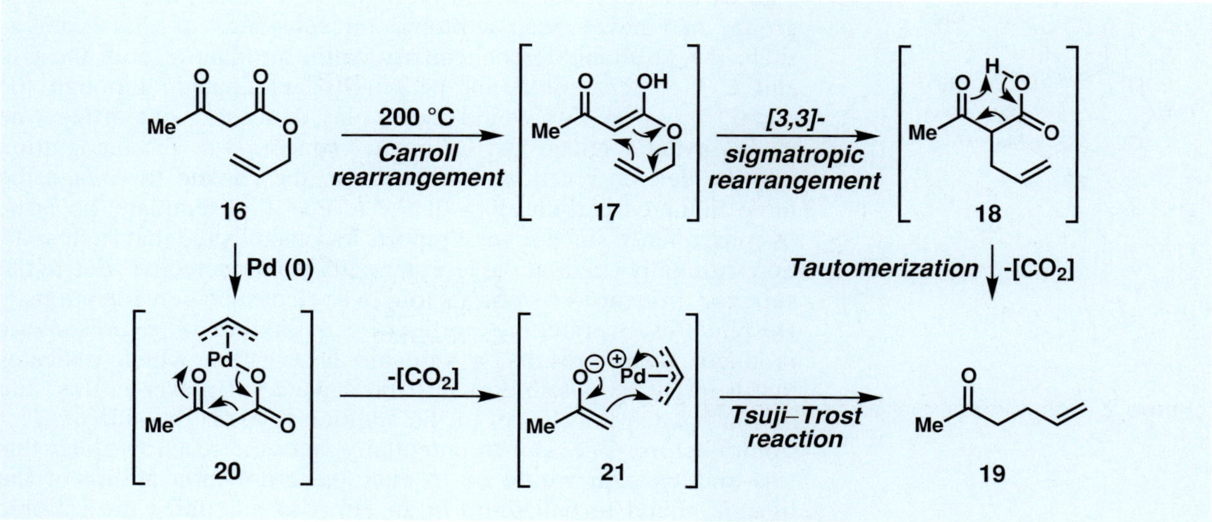

Scheme 3. Carbon–carbon bond formation through the Carroll rearrangement and the Tsuji–Trost reaction.

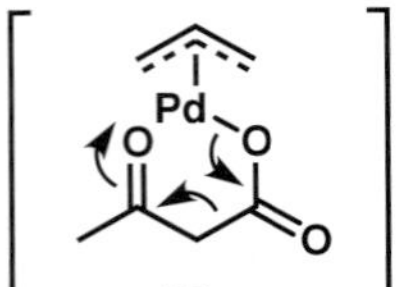

20

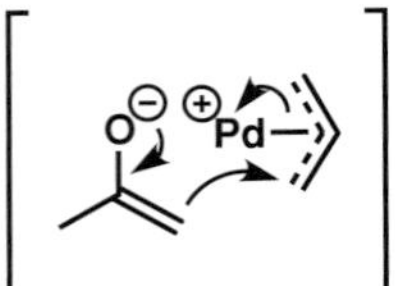

21

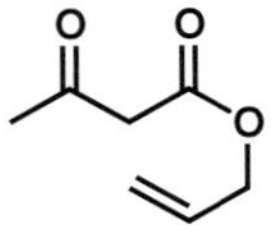

16

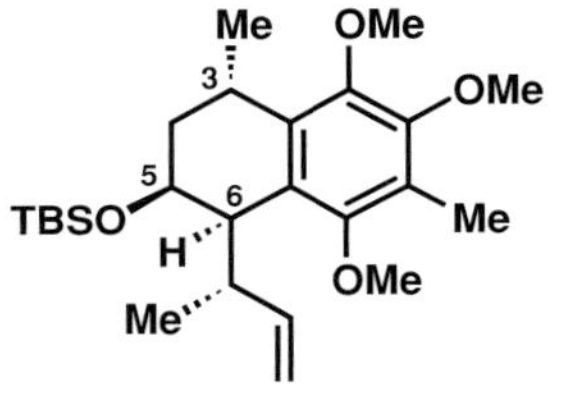

9

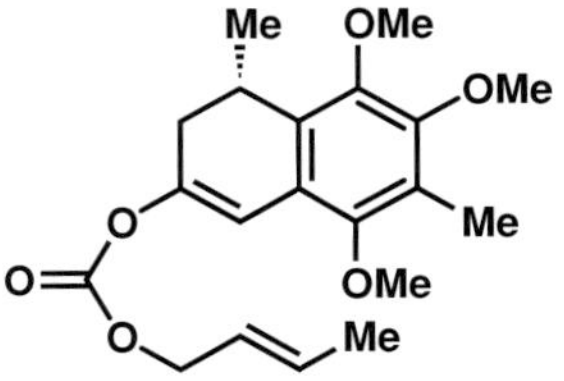

10

between 0 and 25 °C), thereby decreasing the potential for side product formation through processes such as olefin isomerization, ester hydrolysis, or retro-Dieckmann or retro-Michael reactions (as pertinent to substrates possessing β-ketoesters) typically observed in Carroll-type rearrangements. In addition, because palladium attachment to generic allyl groups is a reversible process, tethering of the allyl portion through an ester linkage greatly enhances the overall rate of this reaction, as loss of carbon dioxide (illustrated in the conversion of **20** into **21**) prevents reversion of the π-allyl complex back to starting material. Last, although the Carroll rearrangement and the Tsuji-Trost reaction of **16** result in the same product, the distinct mechanisms for these conversions have significant implications for the contexts in which they would likely prove useful. For example, in substrates in which enol tautomerization is not possible (as required for the Carroll rearrangement), such as α,α-disubstituted malonate derivatives, the palladium-mediated process is uniquely suited to achieve the desired transformation leading to a new quaternary carbon center.[12]

Based on the demonstrated power of palladium-catalyzed allylic alkylation in numerous contexts and on the considerations discussed above, this method was projected as the means by which to install the C-6 side chain in **9** (Scheme 2) from a precursor crotyl enol carbonate **10**. In general, enol carbonates are a proven substrate class for the Tsuji–Trost reaction, and their overall utility derives from their ability to enable regioselective enolate formation upon exposure to palladium complexes.[11g] Despite the large body of work dedicated to the study of this allylation methodology, however, several features of the intended reaction for colombiasin A lacked literature precedent, raising numerous concerns which shadowed the viability of the proposed strategy. For example, the use of a crotyl enol carbonate, which would be required to install the C-7 methyl group, had never been explored on substrates of this type; as such, the resultant stereochemistry of the product at both the C-6 and C-7 centers could not be readily anticipated, although the methyl group at C-3 would likely play at least a supporting role in this event's selectivity. Of greater concern was the recognition that the desired reaction would require the enolate to engage the more hindered carbon atom of the η^3-Pd–allyl template, an issue of consequence since several reports had established that such additions normally occur at the less sterically encumbered site due to the superior influence of steric factors over electronic considerations.[13] Despite this overarching preference, a mixture of regioisomeric products always results, a valuable observation which indicates that it might be possible to shift the balance of power in this "tug of war" match for control of the addition towards the side of electronic factors. One way to potentially coax the reaction along this less-trodden path would be to alter the σ-donation ability of the ligands bound to palladium in an effort to maximize the cationic character of the palladium–π-allyl complex, thereby creating an electronic influence that could overwhelm steric factors. Although at the outset of this synthesis no systematic studies along these lines had yet been performed, there was evidence that such a strategy could indeed level the playing field between these competing factors.[14]

Clearly, the ability to garner control of this particular reaction process successfully represented a potential Achilles' heel in the proposed plan to synthesize colombiasin A. From a retrosynthetic perspective, however, it seemed relatively simple to test this allylation strategy, since it was envisioned that **10** could be generated in short order through *O*-acylation of the enolate derived from **12** using (*E*)-crotyl chloroformate (**11**, Scheme 2). Unmasking the quinone unit in **12** then led to the realization that the bicyclic template **13** could arise from a Diels–Alder union of Danishefsky-type diene **15** and quinone **14**. The regioselectivity necessary to arrive at the desired product (**13**) was anticipated based on the different levels of steric accessibility and electron deficiency of the two carbon–carbon double bonds in **14**, in conjunction with the fact that one of the two carbonyl groups of the quinone is part of a vinylogous ester domain. As such, the proposed sequence would utilize exactly the same carbon atoms for the dienophile portion in both Diels–Alder events. Apart from this interesting feature, the particular advantage of incorporating a Diels–Alder reaction early in the synthetic blueprint was the opportunity that it provided for an enantioselective total synthesis, one that is dependent on asymmetric induction in the presence of an enantiomerically pure Lewis acid catalyst. Moreover, because the absolute configuration of the natural product was not established in the original isolation report,[2] the ability to install the C-3 methyl group stereoselectively would enable access to both antipodes, resolving this issue conclusively if a crystalline intermediate could be obtained to verify the stereochemistry.

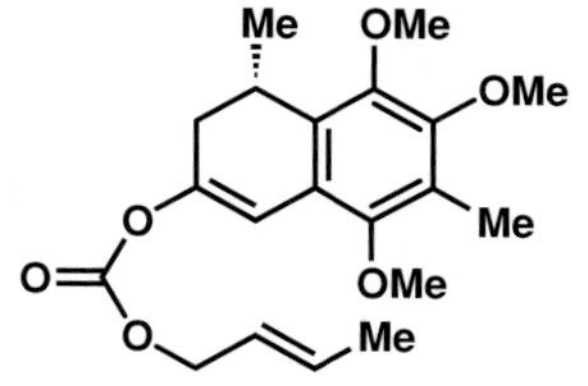

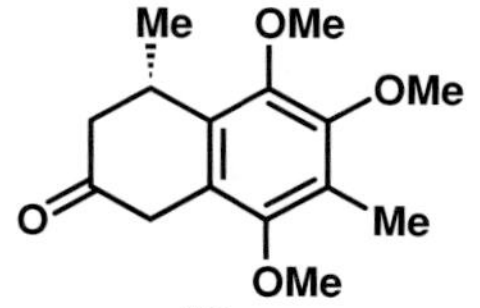

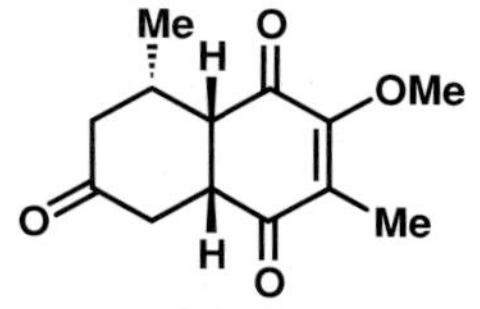

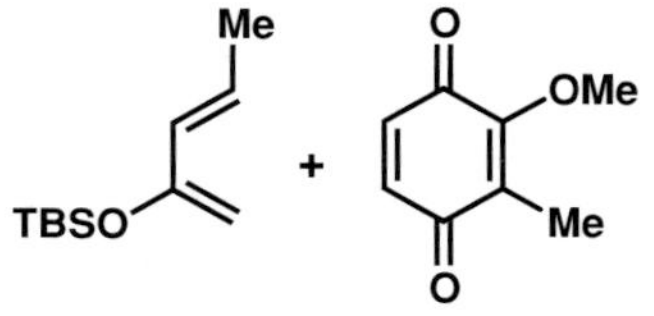

As most students of organic chemistry should recognize, the use of a quinone-based Diels–Alder reaction to form an initial array of rings and stereocenters, elements that pave the way for subsequent stereoselective elaboration to the final target molecule, represents a popular strategy that has been used for over 50 years in the context of some of the most notable achievements in total synthesis, including cortisone and cholesterol (Woodward and co-workers, 1952),[15] reserpine (Woodward and co-workers, 1958),[16] tetrodotoxin (Kishi and co-workers, 1972),[17] gibberellic acid (Corey and co-workers, 1978),[18] and myrocin C (Danishefsky and co-workers, 1994).[19] However, none of these elegant examples proceeded in an asymmetric fashion, and at the outset of synthetic studies for colombiasin A, only one catalyst system had been proven to be capable of achieving asymmetry in Diels–Alder reactions involving quinones.[20] As such, were that catalyst to fail in the selective union of **14** and **15**, then entirely new systems would have to be developed to implement the strategy successfully.

14.3 Total Synthesis

Synthetic endeavors towards colombiasin A (**1**) began with efforts directed at effecting the first projected asymmetric Diels–Alder reaction, as shown in Scheme 4. After *ortho* methylation of 1,2,4-trimethoxybenzene (**22**),[21] formation of quinone **14** was achieved from **23** by using an oxidative demethylation procedure established by Rapoport and Snyder based on treatment with AgO in the presence of HNO_3.[8] This set of conditions is often the method of choice for quinone generation in contexts far more complex than the one at hand, particularly since undesired oxidative side products are rarely observed. Significantly, AgO does not engage isolated methoxyl groups that lack an *ortho*- or *para*-substituted methoxy neighbor, and, as this example demonstrates, selectivity can be achieved when several different products could arise based on the multiple methoxy groups arranged in an *ortho*/*para* format.

With the dienophile in hand, Danishefsky-type diene **15** (Scheme 4) was prepared simply by trapping the enolate of *trans*-3-penten-2-one (generated through exposure to Et_3N in CH_2Cl_2 at −78 °C) with TBSOTf. Having synthesized both **14** and **15**, the stage was set to attempt the Diels–Alder cycloaddition. Initial efforts to forge this merger began with the use of Mikami's catalyst, a titanium complex of presumed structure **25** generated through the reaction of (*S*)-BINOL (**24**) with $(i\text{-PrO})_2TiCl_2$. As alluded to earlier, this system represents the only Lewis acid reported up to 2001 capable of inducing useful levels of asymmetry in Diels–Alder reactions with quinones.[20] Gratifyingly, exposure of **14** and **15** to a 30 mol % loading of catalyst **25** in toluene at −60 °C, followed by slow warming to −10 °C over 7 hours, led to the formation of **26** in 85 % yield along with a small amount of the undesired regioisomer **27**; of even greater significance was the fact that the desired [4+2] cycloadduct **26** was formed with high enantiomeric excess (>94 % *ee*).

Several features of this reaction merit further discussion, especially in regard to the observed regio- and stereoselectivity of the process. When the Diels–Alder union was performed at ambient temperature in the absence of **25**, only the desired regioisomer **26** was obtained, a result that can be readily rationalized from a frontier molecular orbital perspective through analysis of optimal alignment of orbital coefficients between the dienophile and the diene. With the addition of a Lewis acid such as **25**, however, the polarity of the orbitals can be altered significantly, sometimes even resulting in a complete reversal in regioselectivity![22] In the context of this reaction, the observed product mixture can be explained by invoking two different possible transition states involving catalyst **25** which depend on the carbonyl group of the quinone that it coordinates. Normally, one would expect titanium to coordinate preferentially to the vinylogous ester carbonyl group since it has

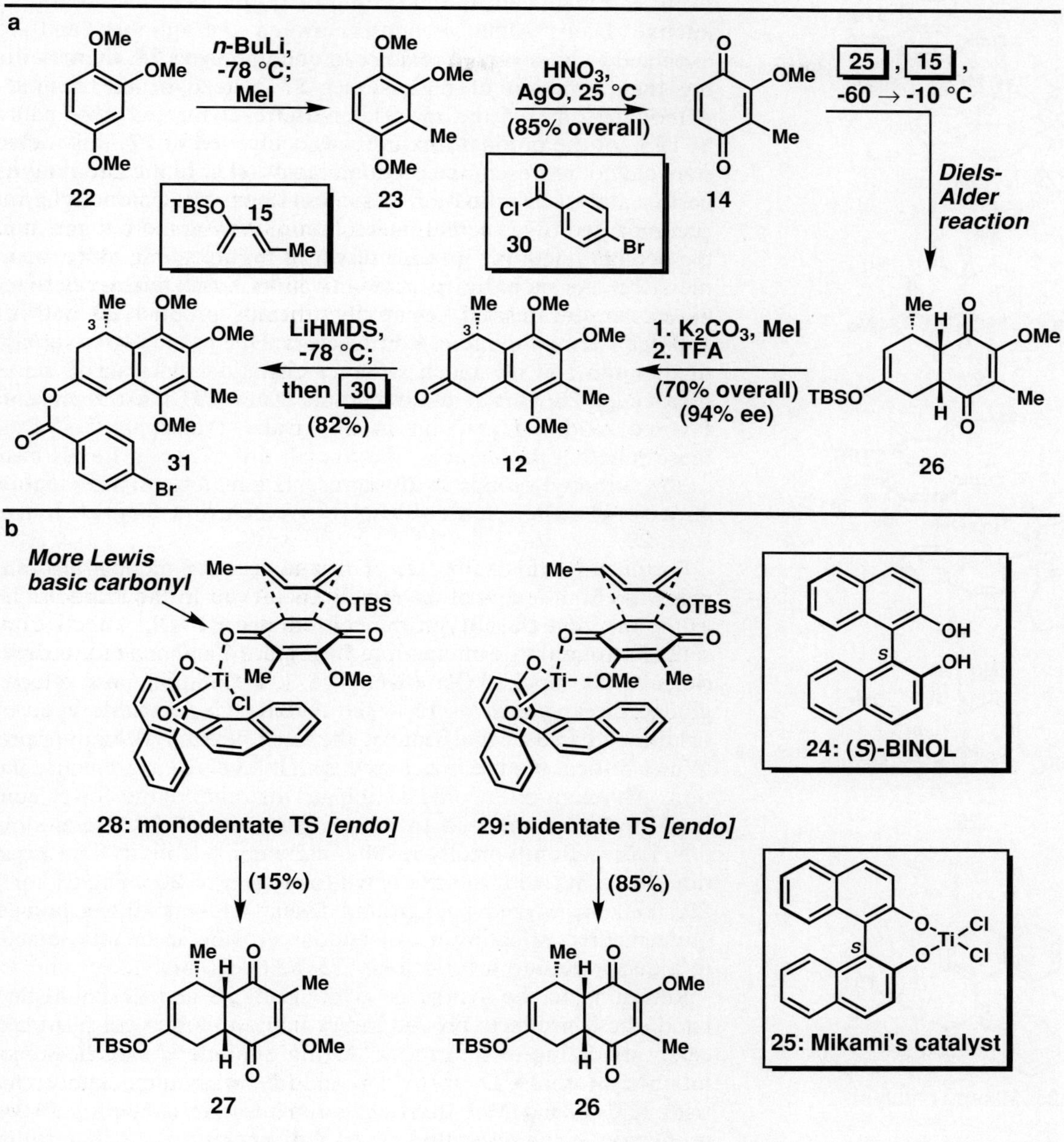

Scheme 4. Synthesis of key intermediate **12** (a) and proposed *endo* transition states leading to products **26** and **27** in the Diels–Alder cycloaddition (b).

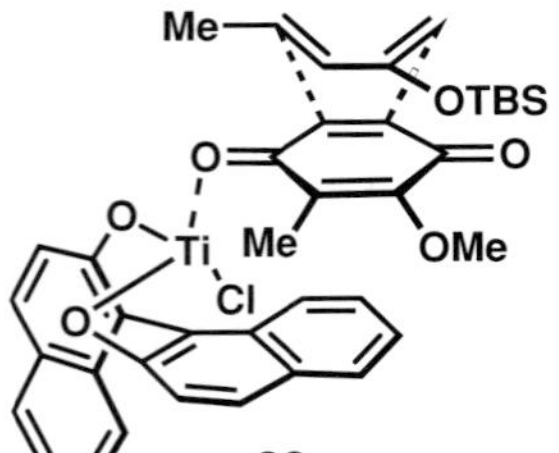

greater Lewis basicity, as shown in transition state **28** (Scheme 4b). In this arrangement, the matching of orbital coefficients for a productive Diels–Alder reaction between the quinone and **15** is expected to be reversed relative to uncomplexed **14**, thereby dictating the formation of regioisomer **27** instead of the desired **26**. Although some of the material must react through this pathway, as 15 % of the product mixture was composed of **27**, this transition state cannot represent the predominant model. In the alternative scenario, catalyst coordination to the less Lewis basic carbonyl group is accompanied by a second interaction involving the oxygen atom of the pendant methoxy group, creating a highly stable bidentate complex (**29**). As such, Lewis acid activation in this manner is expected to bolster the desired selectivity towards product **26** with diene approach along the indicated trajectory. Due to the high oxophilicity of titanium and the much stronger chelation that can be achieved relative to **28**, this latter transition state (**29**) must represent the favored mode of reaction in this Diels–Alder process. Despite these inherent preferences, the overall difference in Lewis basicity of the carbonyl groups in **28** represents a feature of sufficient influence so as to interfere with exclusive reaction through transition state **29**.

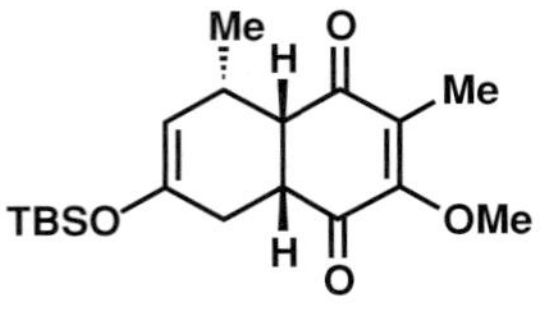

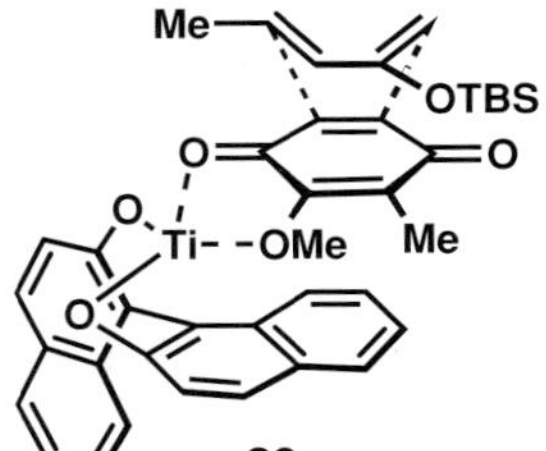

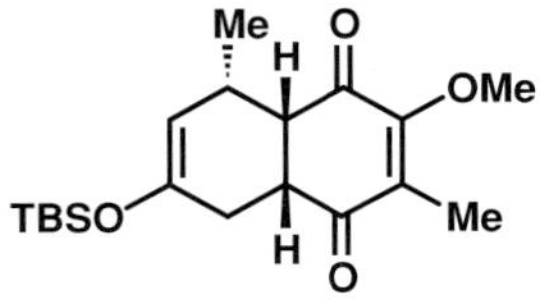

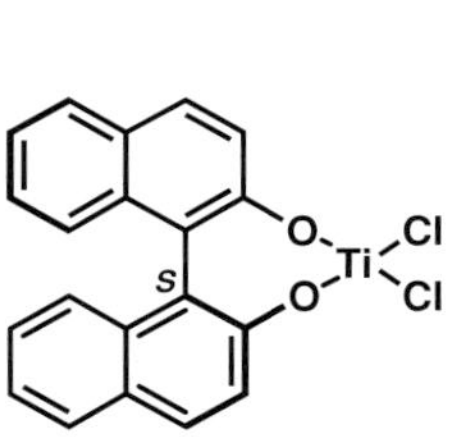

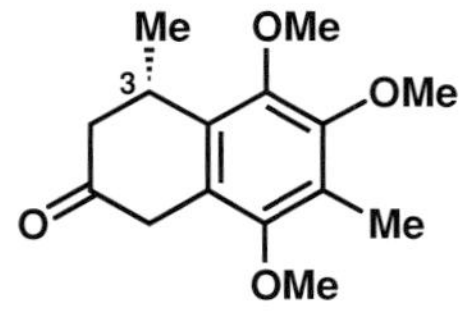

Fortunately, this same set of transition state models also anticipates the high levels of asymmetry achieved in the transformation, as the inherent chirality of the coordinated BINOL, effectively presented through $\pi-\pi$ interactions between **14** and one of the naphthol rings of **25**, blocked the lower face of the quinone and selectively guided the approach of **15** to the reactive C–C double bond of **14** (as drawn in **29**). Significantly, the level of catalyst loading proved to be a critical feature in achieving high levels of asymmetric induction. Although one would anticipate that employing fewer equivalents of **25** would lead to greater amounts of **26** (since slowing down a reaction typically results in greater selectivity), an expectation borne out with the exclusive formation of **26** using 5 mol % of **25**, the uncatalyzed background Diels–Alder addition proceeded quite readily, effecting a tremendous erosion in enantioselectivity to a disappointing level of only 15 %.

Returning to the synthesis, while both **26** and **27** could be isolated, these products proved rather labile to silica gel purification, easily oxidizing to reform the A-ring quinone. As such, the crude mixture of these Diels–Alder adducts was immediately treated with K_2CO_3 and MeI to effect quinol formation with subsequent protection as the bismethyl ether. This operation was then followed by acidic hydrolysis of the silyl enol ether, leading to a mixture of **12** and the corresponding isomer resulting from **27** (not drawn) in 70 % overall yield for these three operations. At this stage, the two products could be separated chromatographically, providing enantiomerically pure **12** which was carried forward. Although the chirality of **12** could be predicted from the assumed transition states in Scheme 4, it was deemed prudent to confirm this conjecture by other means before pressing on with the synthesis. Fortunately, acylation of the lithium enolate of **12** with 4-bromobenzoyl chloride (**30**) afforded a solid product **31** from which crystals suitable for X-ray crystallographic analysis were obtained. This analysis conclusively established the anticipated *S* configuration of the C-3 stereocenter for this series of compounds.

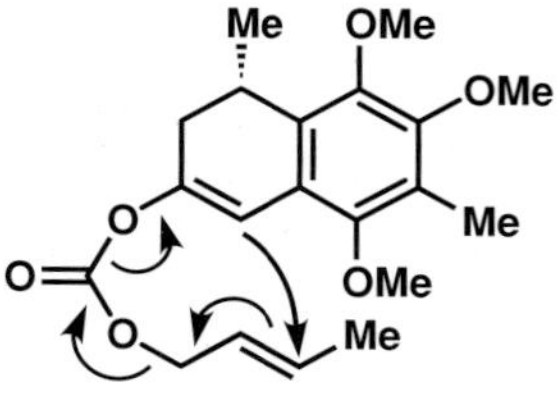

With this first key goal achieved, the next hurdle in the synthesis of colombiasin A was application of the Tsuji–Trost reaction to affix the alkyl side chain at C-6. The initial requirement to explore this possibility, formation of enol carbonate **10**, was smoothly carried out in 94 % yield by *O*-acylation of the enolate derived from **12** using (*E*)-crotyl chloroformate (**11**), as shown in Scheme 5a. Next, as a preliminary foray into allylation, exposure of a solution of **10** in THF to catalytic amounts of $Pd(PPh_3)_4$ at ambient temperature led to the rapid formation of two new products, with consumption of the starting material after just 15 minutes. Unfortunately, only after fully elaborating these products to the complete colombiasin skeletal framework and comparing their spectra to that of the natural product was it revealed that the newly formed adducts were **32** and **33**, generated in a 1:2.4 ratio favoring the desired olefin **33**, but with the incorrect stereochemistry at C-7. To ensure that the initial geometry of the crotyl group of **10** was not responsible for the observed stereochemical outcome at C-7 in **33**, the same reaction was performed with the *Z* isomer of **10**. Once again, the exact same ratio of isomers favoring **33** was obtained without correction of the stereochemistry at C-7.

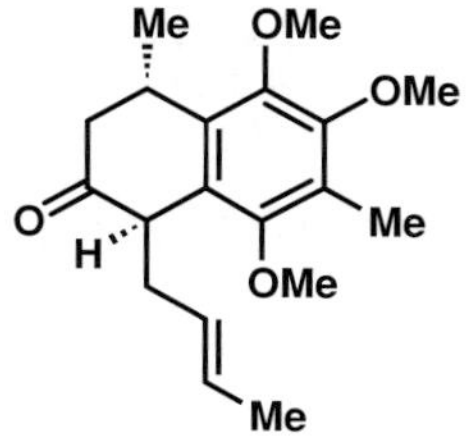

These two experiments can be mechanistically rationalized by assuming that after oxidative addition of Pd(0) to the crotyl group of enol carbonate **10**, with ensuing loss of carbon dioxide, a *syn*-$[Pd(\eta^3\text{-crotyl})]L_2$ complex **E** (Scheme 5b) was generated along with an enolate. Because both **10** and its *Z* isomer provided the same ratio of **32** and **33**, the *anti*-configured complex **A**, formed from the *Z* isomer of **10**, must have rapidly equilibrated to the sterically more favored *syn*-η^3-crotyl palladium complex **E** through π-σ-π isomerization,[23] proceeding through the series of intermediates shown in Scheme 5b.* Addition of the carbon-based nucleophile then occurred either at site a or b in **E**. Although it was originally anticipated that nucleophilic attack would occur preferentially at the less-sterically hindered, unsubstituted allyl terminus (carbon a), use of Ph_3P as a ligand for palladium must have enabled sufficient build-up of positive charge on the metal center and the crotyl ligand in **E** so as to direct attack predominantly at carbon b where such cationic character was better stabilized. Indeed, this latter supposition was confirmed when application of stronger σ-donating ligands relative to Ph_3P, such as the more basic 1,2-bis(diphenylphosphino)ethane and $P(Oi\text{-}Pr)_3$, led to a remarkable erosion in selectivity for **33**.[24] Thus, taken as a whole, the highly different product ratios

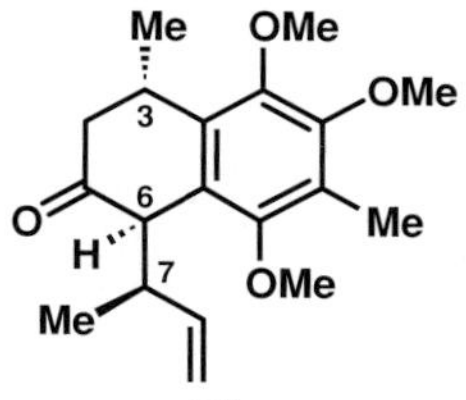

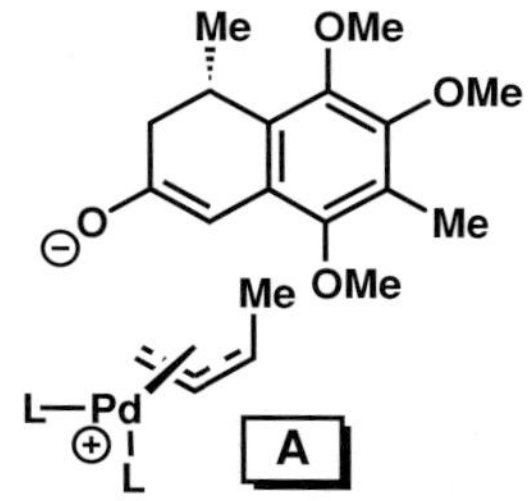

Me OMe OMe Me OMe Me O O O **10**

Me OMe OMe Me OMe O H Me **32**

Me OMe OMe Me OMe 3 6 7 O H Me **33**

Me OMe OMe Me OMe O Me OMe L–Pd L **A**

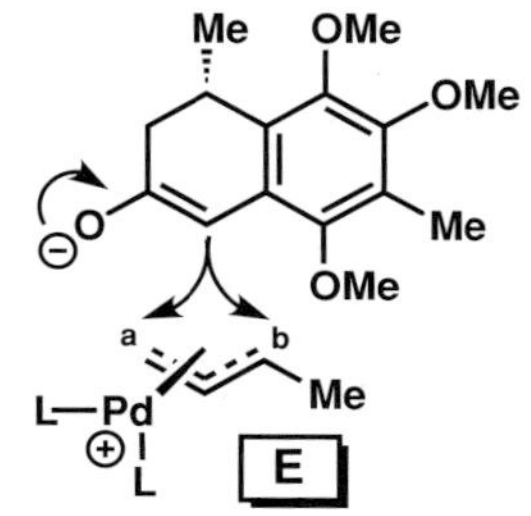

Me OMe OMe Me OMe O a b Me L–Pd L **E**

* Here, the designation of *syn* or *anti* refers to the spatial location of the methyl group relative to the designated H atom in the boxed structure in Scheme 5b.

obtained with these three different ligands (see Table in Scheme 5a) indicates that the regioselectivity of nucleophile addition in the Tsuji–Trost reaction can be tuned beyond mere steric control of the process, with the possibility of achieving a major reaction product derived from the addition to the more substituted carbon atom, a result that certainly went against prevailing wisdom! We

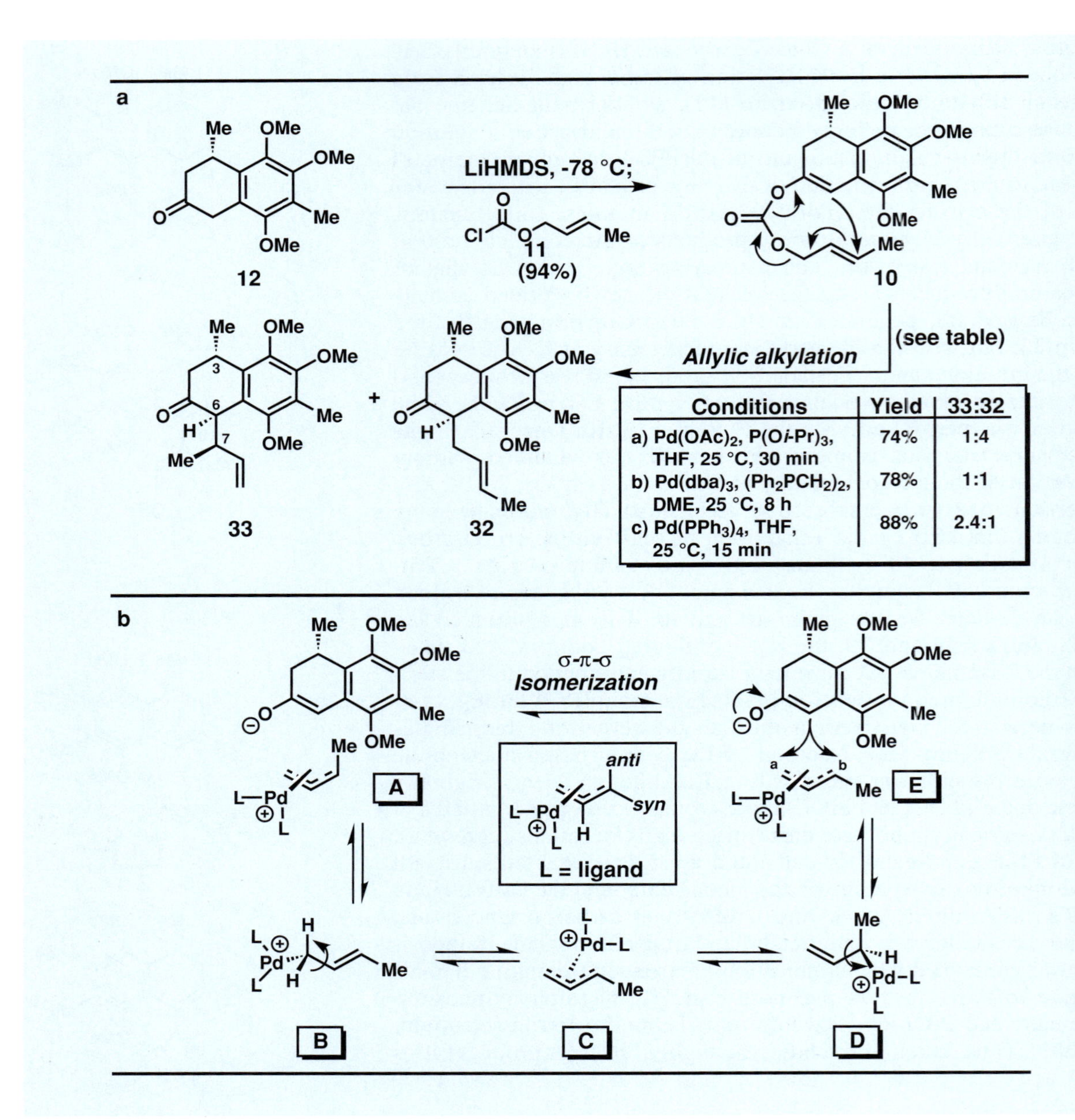

Conditions	Yield	33:32
a) $Pd(OAc)_2$, $P(O\textit{i}\text{-}Pr)_3$, THF, 25 °C, 30 min	74%	1:4
b) $Pd(dba)_3$, $(Ph_2PCH_2)_2$, DME, 25 °C, 8 h	78%	1:1
c) $Pd(PPh_3)_4$, THF, 25 °C, 15 min	88%	2.4:1

Scheme 5. Formation of intermediates **32** and **33** (a) and mechanistic rationale for the outcome of the Pd-catalyzed crotylation of enol carbonate **10** (b).

should note, however, that the physical steric bulk of the ligands cannot be wholly excluded as a contributing and modulating factor for the observed ratios of **32** and **33**, although the electronic properties of the bound ligands would appear to be the predominant forces in dictating the product distribution based on the tabulated results in Scheme 5.

While these discussions have accounted for the observed product distribution of **32** and **33**, we have not yet engaged the issue of their absolute stereochemistry. Since only a single diastereomer of **33** was formed at both C-6 and C-7, despite the assumed equimolar presence of both enantiomeric forms of the *syn* complex **E**, the enolate must have discriminated between the two enantiotopic faces of the palladium–crotyl complex, presumably due to the influence of the lone stereogenic center C-3. Clearly, this inherent bias would likely plague the formation of the desired C-7 epimer in any strategy based on the incorporation of a side chain at C-6. As a final comment, although this reaction can be formally considered as a sigmatropic rearrangement along the lines of a Claisen reaction through the arrows drawn for the conversion of **10** into **33** in Scheme 5a, the preceding discussion certainly indicates that such a simple view is inaccurate and misleading since that model cannot successfully predict the observed products or their stereochemistry.

Given the overall success of the Tsuji–Trost conversion, despite the unavoidable incorporation of incorrect C-7 stereochemistry, the Nicolaou group pressed forward, setting correction of the C-7 stereocenter as the next objective. After significant route scouting, it was determined that the desired C-7 epimer could best be obtained through the sequence delineated in Scheme 6. First, reduction of the C-5 carbonyl group in **33** was smoothly achieved through exposure to $NaBH_4$, with the observed stereoselectivity presumably controlled by the bulkiness of the alkyl chain at C-6. Subsequent silylation then completed the synthesis of **34** in 91 % overall yield for these two steps. Of course, since there is no oxygen atom at this position in the final target molecule, the facial discrimination in the first operation was of no consequence, since the alcohol group will be extruded later. At this stage, because it was envisioned that epimerization of the C-7 center could likely be achieved through basic equilibration, the pK_a of the C-7 proton had to be lowered so that it could be abstracted selectively by conventional bases. Towards this end, the terminal C–C double bond in **34** was converted into the corresponding aldehyde **35** by initial osmium-mediated dihydroxylation, followed by oxidative cleavage with $NaIO_4$ supported on silica gel.[25] As an aside, this modification of conventional glycol cleavage with $NaIO_4$ represents a very useful procedure, as $NaIO_4$ alone is soluble only in water/alcohol mixtures or in THF. By appending $NaIO_4$ onto silica gel (which is presumably achieved through interactions between the oxidant and the hydroxy groups on the surface),[26] the reagent readily performs oxidations in nonpolar solvents such as CH_2Cl_2 or benzene, and the pure aldehyde product can be obtained simply by filtration and

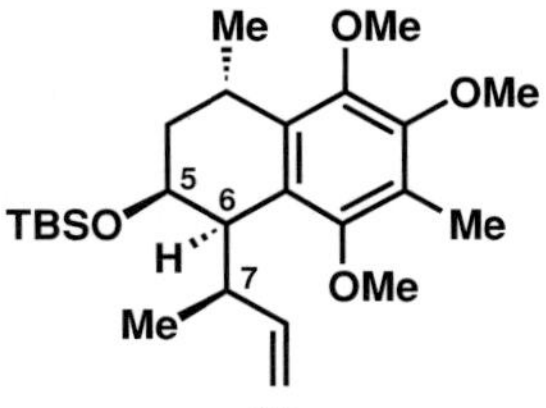

Me OMe OMe TBSO 5 6 H 7 Me Me OMe

34

Me OMe OMe TBSO H 7 Me Me OMe O

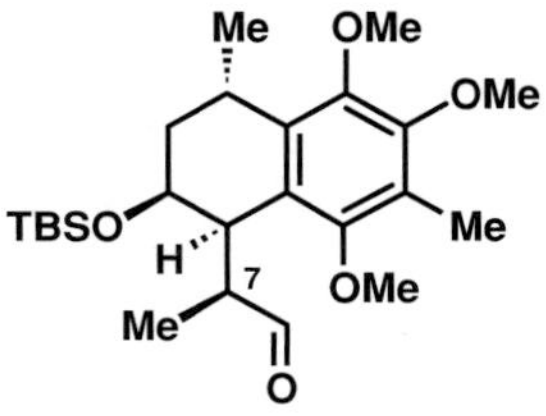

35

removal of the solvent. These features are particularly valuable for cases involving the formation of labile or water-soluble aldehydes.

Nevertheless, with the acidity of the C-7 proton now enhanced, exposure of **35** to NaOMe in MeOH effected thermodynamic equilibration to a 2:1 mixture of **35** and **36**, favoring the original, undesired C-7 epimer **35** (Scheme 6). After chromatographic separation of the two isomers, the recovered starting material (**35**) was cycled again through the same process in near quantitative yield; ultimately, conversion into **36** was slightly more than 50 %. With the correct stereochemistry at C-7 finally secured, **36** was then advanced to aldehyde **8** in anticipation of eventual Wittig homologation through three simple operations which proceeded through intermediate **9** in a combined yield of 77 %: olefination with the ylide derived from methyl triphenylphosphonium bromide, hydroboration/oxidation, and oxidation with PCC.

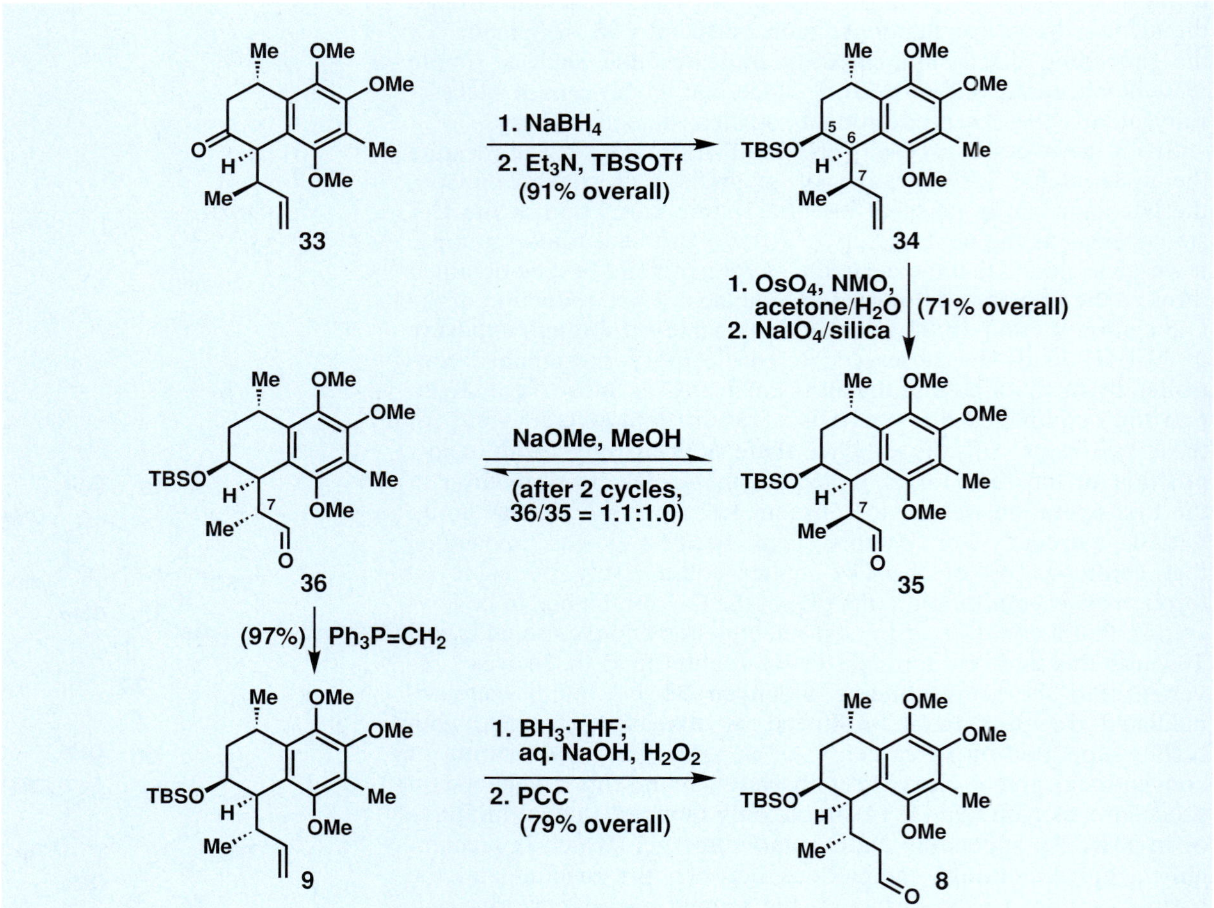

Scheme 6. Successful epimerization at C-7 and elaboration to key intermediate **8**.

Having labored extensively to establish three of the stereocenters of colombiasin A, the opportunity to harvest the fruit of that labor through the final proposed Diels–Alder reaction was nearly at hand. Accordingly, the requisite diene system was smoothly fashioned through the addition of a solution of aldehyde **8** in THF to the semistabilized ylide generated from **7** upon exposure to *n*-BuLi, providing the desired aromatic diene **37** as a 3:1 mixture of *E*/*Z* isomers (Scheme 7). With this product in hand, efforts were immediately expended to reveal the camouflaged A-ring quinone by using the oxidative demethylation procedure of Rapoport and Snyder (AgO and HNO_3).[8] Unfortunately, only traces of the desired

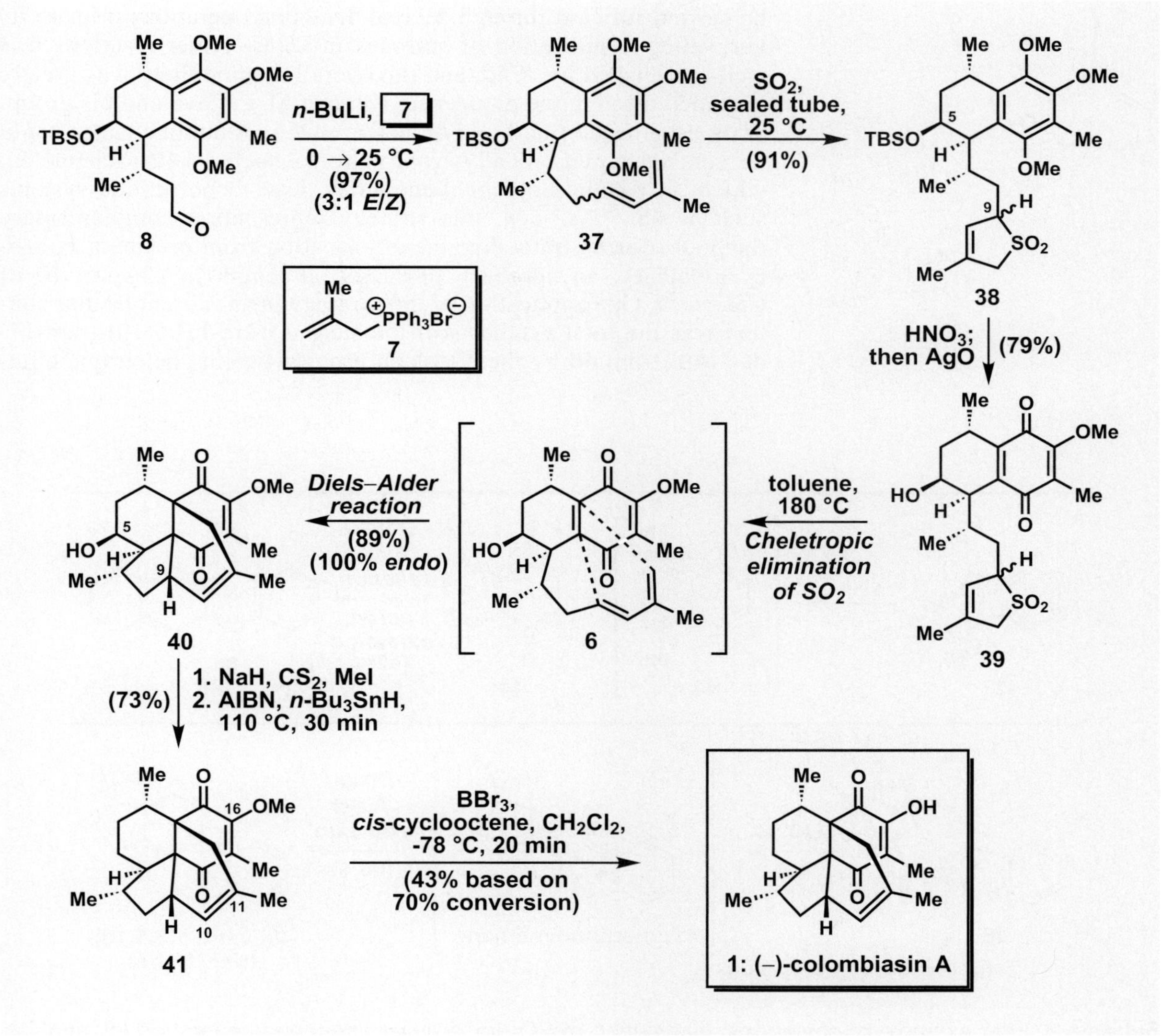

Scheme 7. Final stages and completion of the total synthesis of (–)-colombiasin A (**1**).

diene-quinone were obtained, likely due to side reactions caused by the diene system during the oxidation step, as many unidentified by-products lacking this portion of the molecule were observed.

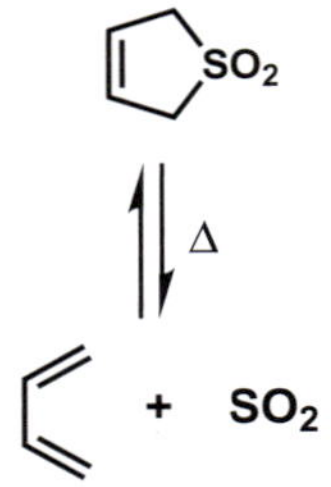

To circumvent this stumbling block, application of a classical masking trick in Diels–Alder chemistry proved highly rewarding. Since the pioneering work of Staudinger in the early part of the last century,[27] it has been well-documented that SO_2 can add cleanly to a diene system, providing an adduct that can be reverted back into the original diene through a thermally induced retro-Diels–Alder process which leads to the expulsion, or, as more commonly defined, cheletropic elimination, of SO_2. More importantly, the initially formed sulfone is inert to numerous reaction conditions, making it an excellent masking device for a diene system that can be carried forward through several synthetic operations prior to its unraveling. Application of sulfones in Diels–Alder reactions was well-established by 1950, and the overall methodology was greatly extended through the pioneering work of M. P. Cava and his group. They established that highly reactive *ortho*-quinodimethanes could be generated pyrolytically from sulfones such as **42** (Scheme 8), which, in the presence of dienophiles, lead to polycyclic systems such as **45**.[28] As such, this strategy represents a complementary method to *ortho*-quinodimethane formation from precursor benzocyclobutenes, an approach discussed at length in Chapter 10 of *Classics I*. One application of this strategy in an intramolecular context was the total synthesis of the steroid estra-1,3,5(10)-trien-17-one (**48**) from **46** by the Nicolaou group wherein cheletropic elim-

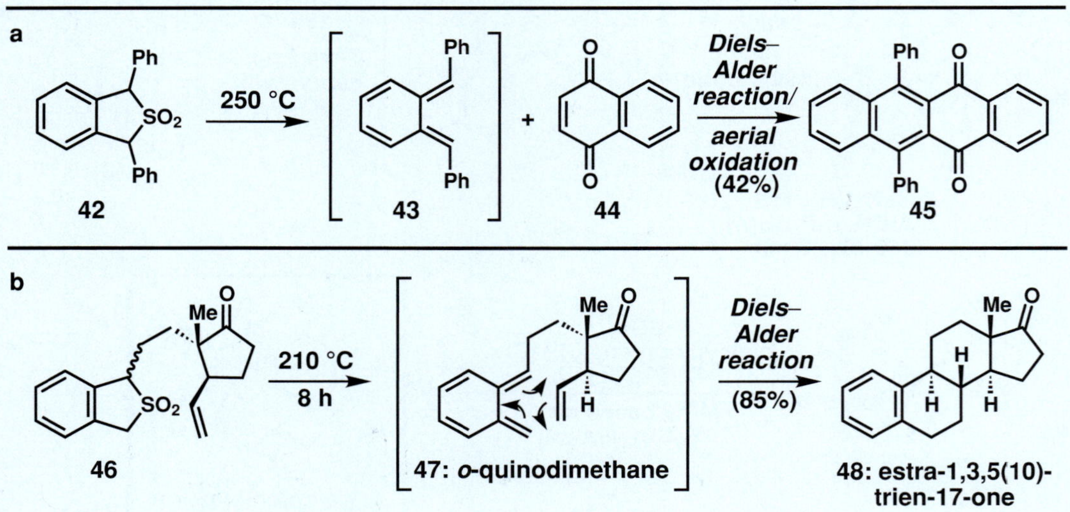

Scheme 8. First example of cheletropic elimination of SO_2 for *o*-quinodimethane generation (a), and application of the strategy as the key step in Nicolaou's total synthesis of estra-1,3,5(10)-trien-17-one (b).

ination of SO_2 was readily effected in high yield under conditions that required neither a catalyst nor high dilution.[29]

Following this historical precedent, a sulfone was readily appended onto the diene system of **37**, affording **38** in 91 % yield (Scheme 7).[30] One should note that in this transformation both isomers obtained from the prior Wittig olefination were employed, resulting in the mixture of stereoisomers at C-9 in **38**. Based on the assumption that eventual thermal extrusion of SO_2 from either diastereomer would provide the requisite *trans*-substituted olefin in the diene desired for Diels–Alder reaction, this lack of selectivity was not deemed problematic.[31] However, the pressing issue at this juncture remained the generation of the quinone. Fortunately, with the diene now protected, oxidative demethylation smoothly furnished **39** in 79 % yield with concomitant TBS deprotection.

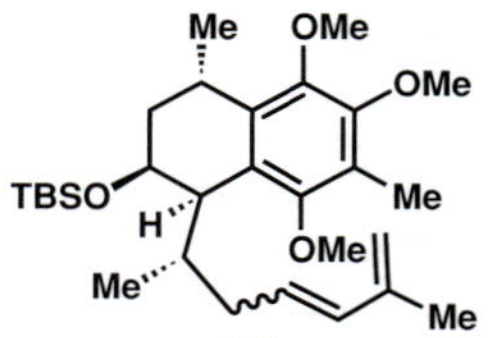

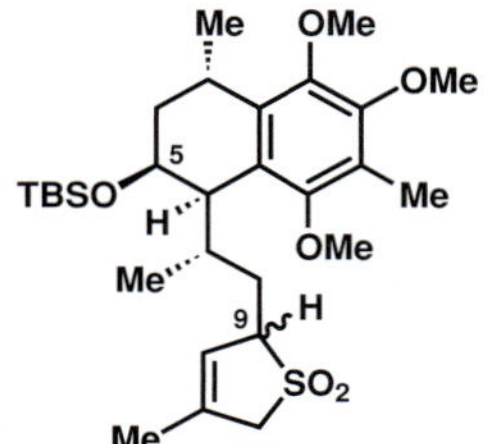

Having successfully unveiled the dienophile, it was now time to unmask the diene. Heating a solution of **39** in toluene in a sealed tube protected from light at 180 °C effected cheletropic elimination of SO_2, providing putative intermediate **6** which readily engaged in the anticipated Diels–Alder union leading to the desired polycyclic product **40** in 89 % yield, solely as the *endo* isomer. During the experiment, light was excluded from the reaction medium because, in earlier studies, the reaction of the C-7 epimer **49** under the same conditions in the presence of normal room light led exclusively to **50** (Scheme 9), presumably the product of a stepwise [2+2] cycloaddition involving 1,4-diradical intermediates with the stereochemistry of the product dictated by the unique colombiasin framework.[32] The pivotal role played by light in this process was confirmed when the complete conversion of **49** into **50** could be accomplished after only 15 minutes of exposure to visible light at ambient temperature.

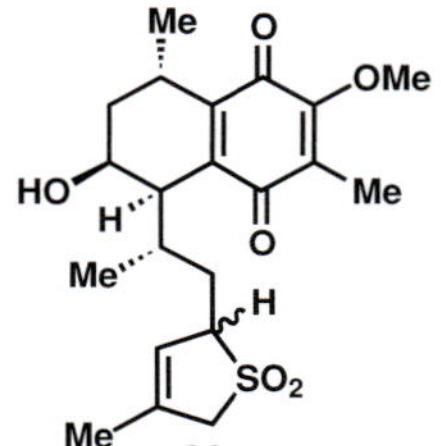

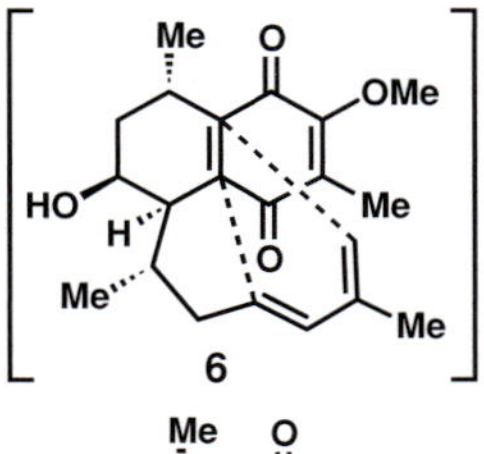

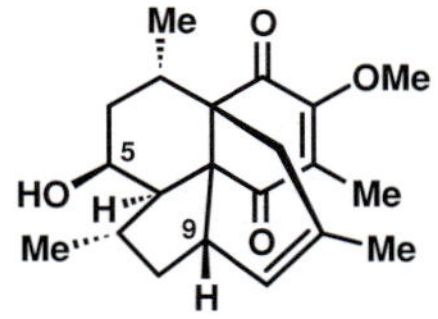

With the formation of the entire colombiasin skeleton accomplished through the successful generation of **40** (Scheme 7), including the two quaternary carbon atoms and the additional three stereocenters present in the target molecule, only a few minor details remained before arriving at colombiasin A (**1**). First, the extraneous hydroxyl group at C-5 was reductively removed through the use of the Barton–McCombie protocol[6] in which the initial for-

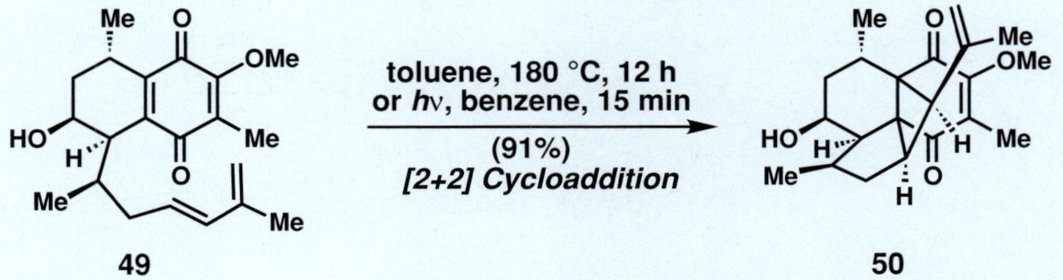

Scheme 9. Unique polycyclic product **50** formed through a photoinduced [2+2] cycloaddition.

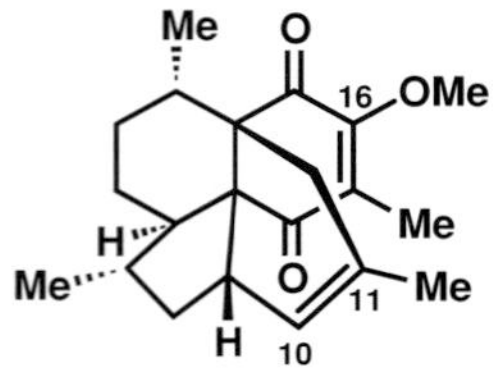

mation of a xanthate ester (NaH; CS_2; MeI) was followed by treatment with n-Bu_3SnH in the presence of AIBN as a radical initiator to furnish **41** in 73 % overall yield. Finally, the synthesis of colombiasin A (**1**) was completed upon cleavage of the methyl ether at C-16 by controlled exposure to BBr_3 and *cis*-cyclooctene in CH_2Cl_2 at −78 °C. In the absence of a "sacrificial" or competitive olefin such as *cis*-cyclooctene, this demethylation was also attended by acid-induced migration of the C−C double bond in the D-ring to form the apparently more stable $\Delta^{11,12}$ isomer. Before leaving this synthesis, however, one should note that the signs of the optical rotations of natural colombiasin A (**1**) and the final synthetic material matched, thereby establishing the absolute configuration of the natural product as drawn throughout the schemes in this chapter.

14.4 Conclusion

The described synthesis of colombiasin A (**1**), using the Diels−Alder reaction as the primary tool for the construction of its molecular complexity, not only provided validation for a putative biosynthetic route to this fascinating natural product, but also advanced our knowledge regarding the generality and scope of this reaction process to fashion complicated molecules asymmetrically.[33] In addition, the extension of the highly useful palladium-mediated Tsuji−Trost allylation to a new class of substrates, as well as the demonstration that the regioselectivity of enolate addition to the palladium−crotyl complex could be altered simply through ligand substitution, was a noteworthy achievement of this endeavor.[34]

References

1. a) S. A. Look, W. Fenical, R. S. Jacobs, J. Clardy, *Proc. Natl. Acad. Sci. U.S.A.* **1986**, *83*, 6238; b) A. D. Rodríguez, C. Ramírez, I. I. Rodríguez, C. L. Barnes, *J. Org. Chem.* **2000**, *65*, 1390.
2. A. D. Rodríguez, C. Ramírez, *Org. Lett.* **2000**, *2*, 507.
3. A. D. Rodríguez, E. González, S. D. Huang, *J. Org. Chem.* **1998**, *63*, 7083.
4. C. A. Harvis, M. T. Burch, W. Fenical, *Tetrahedron Lett.* **1988**, *29*, 4361.
5. a) K. C. Nicolaou, G. Vassilikogiannakis, W. Mägerlein, R. Kranich, *Angew. Chem.* **2001**, *113*, 2543; *Angew. Chem. Int. Ed.* **2001**, *40*, 2482; b) K. C. Nicolaou, G. Vassilikogiannakis, W. Mägerlein, R. Kranich, *Chem. Eur. J.* **2001**, *7*, 5359.
6. a) D. H. R. Barton, S. W. McCombie, *J. Chem. Soc., Perkin Trans 1* **1975**, 1574; b) D. H. R. Barton, D. Crich, A. Löbberding, S. Z. Zard, *Tetrahedron* **1986**, *42*, 2329.
7. For instructive examples of this topic, see: a) M. Breuning, E. J. Corey, *Org. Lett.* **2001**, *3*, 1559; b) A. Terada, Y. Tanoue, A. Hatada, H. Sakamoto, *Bull. Chem. Soc. Jpn.* **1987**, *60*, 205. For a review of methods in quinone synthesis, including the difficulty in their handling, see: P. J. Dudfield in *Comprehensive Organic Synthesis, Vol. 7* (Ed.: B. M. Trost, I. Fleming), Pergamon, Oxford, **1991**, pp. 345–356.
8. a) C. D. Snyder, H. Rapoport, *J. Am. Chem. Soc.* **1972**, *94*, 227; b) J. R. Luly, H. Rapoport, *J. Org. Chem.* **1981**, *46*, 2745.
9. For similar difficulties in a related context, see: L. A. Paquette, F. Gallou, Z. Zhao, D. G. Young, J. Liu, J. Yang, D. Friedrich, *J. Am. Chem. Soc.* **2000**, *122*, 9610.
10. a) M. F. Carroll, *J. Chem. Soc.* **1940**, 704; b) M. F. Carroll, *J. Chem. Soc.* **1940**, 1266; c) M. F. Carroll, *J. Chem. Soc.* **1941**, 507; d) W. Kimel, A. C. Cope, *J. Am. Chem. Soc.* **1943**, *65*, 1992.
11. For selected reviews, see: a) B. M. Trost, *Acc. Chem. Res.* **1980**, *13*, 385; b) J. Tsuji, I. Minami, *Acc. Chem. Res.* **1987**, *20*, 140; c) J. Tsuji, *Organic Synthesis with Palladium Compounds*, Springer-Verlag, West Berlin, **1980**, pp. 37–51; d) B. M. Trost, T. R. Verhoeven in *Comprehensive Organometallic Chemistry II, Vol. 8* (Eds.: E. W. Abel, F. G. A. Stone, G. Wilkinson), Pergamon, Oxford, **1982**, pp. 799–938. For selected original papers with leading references, see: e) B. M. Trost, T. R. Verhoeven, *J. Am. Chem. Soc.* **1976**, *98*, 630; f) J. Tsuji, I. Minami, I. Shimizu, *Tetrahedron Lett.* **1983**, *24*, 1793; g) J. Tsuji, Y. Ohashi, I Minami, *Tetrahedron Lett.* **1987**, *28*, 2397.
12. For a more comprehensive discussion of this process, see: J. Tsuji, *Palladium Reagents and Catalysts*, John Wiley & Sons, Chichester, **1995**, p. 560.
13. B. M. Trost, M.-H. Hung, *J. Am. Chem. Soc.* **1984**, *106*, 6837.
14. B. Åkermark, S. Hansson, B. Krakenberger, A. Vitagliano, K. Zetterberg, *Organometallics* **1984**, *3*, 679.
15. R. B. Woodward, F. Sondheimer, D. Taub, K. Heusler, W. M. McLamore, *J. Am. Chem. Soc.* **1952**, *74*, 4223.
16. R. B. Woodward, R. E. Bader, H. Bickel, A. J. Frey, R. W. Kierstead, *Tetrahedron* **1958**, *2*, 1.
17. a) Y. Kishi, M. Aratani, T. Fukuyama, F. Nakatsubo, T. Goto, S. Inoue, H. Tanino, S. Sugiura, H. Kakoi, *J. Am. Chem. Soc.* **1972**, *94*, 9217; b) Y. Kishi, T. Fukuyama, M. Aratani, F. Nakatsubo, T. Goto, S. Inoue, H. Tanino, S. Sugiura, H. Kakoi, *J. Am. Chem. Soc.* **1972**, *94*, 9219.
18. a) E. J. Corey, R. L. Danheiser, S. Chandrasekaran, P. Siret, G. E. Keck, J.-L. Gras, *J. Am. Chem. Soc.* **1978**, *100*, 8031; b) E. J. Corey, R. L. Danheiser, S. Chandrasekaran, G. E. Keck, B. Gopalan, S. D. Larsen, P. Siret, J.-L. Gras, *J. Am. Chem. Soc.* **1978**, *100*, 8034.
19. M. Y. Chu-Moyer, S. J. Danishefsky, G. K. Schulte, *J. Am. Chem. Soc.* **1994**, *116*, 11213.
20. a) K. Mikami, M. Terada, Y. Motoyama, T. Nakai, *Tetrahedron: Asymmetry* **1991**, *2*, 643; b) K. Mikami, Y. Motoyama, M. Terada, *J. Am. Chem. Soc.* **1994**, *116*, 2812; c) J. D. White, Y. Choi, *Org. Lett.* **2000**, *2*, 2373. For earlier preparations of related chiral catalysts, see: d) K. Narasaka, M. Inoue, N. Okada, *Chem. Lett.* **1986**, 1109; e) K. Narasaka, M. Inoue, T. Yamada, *Chem. Lett.* **1986**, 1967.
21. M. C. Carreño, J. L. García Ruano, M. A. Toledo, A. Urbano, *Tetrahedron: Asymmetry* **1997**, *8*, 913.
22. For instructive examples of this concept, see reference 17 and a) W. Kreiser, W. Haumesser, A. F. Thomas, *Helv. Chim. Acta* **1974**, *57*, 164; b) T. A. Engler, R. Iyengar, *J. Org. Chem.* **1998**, *63*, 1929.
23. For a review, see: K. Vrieze in *Dynamic Nuclear Magnetic Resonance Spectroscopy* (Eds.: M. Jackman, F. A. Cotton), Academic Press, New York, **1975**, pp. 441–487.
24. P. D. Dias, M. E. Minas de Piedade, J. A. Martinho Simões, *Coord. Chem. Rev.* **1994**, *135/136*, 737.
25. Y.-L. Zhong, T. K. M. Shing, *J. Org. Chem.* **1997**, *62*, 2622.
26. D. N. Gupta, P. Hodge, J. E. Davies, *J. Chem. Soc., Perkin Trans. 1* **1981**, 2970.
27. a) H. Staudinger, German Patent 506,839 [*Chem. Abstr.* **1913**, *25*, 522]; b) H. Staudinger, B. Ritzenthaler, *Chem. Ber.* **1935**, *68B*, 455. For a review on the process in Diels–Alder reactions applied to natural product total synthesis, see: A. Ichihara, *Synthesis* **1987**, 207.
28. a) M. P. Cava, A. A. Deana, *J. Am. Chem. Soc.* **1959**, *81*, 4266; b) M. P. Cava, M. J. Mitchell, A. A. Deana, *J. Org. Chem.* **1960**, *25*, 1481.
29. a) K. C. Nicolaou, W. E. Barnette, *J. Chem. Soc. Chem. Commun.* **1979**, 1119; b) K. C. Nicolaou, W. E. Barnette, P. Ma, *J. Org. Chem.* **1980**, *45*, 1463.

30. S. Yamada, T. Suzuki, H. Takayama, K. Miyamoto, I. Matsunaga, Y. Nawata, *J. Org. Chem.* **1983**, *48*, 3483.
31. For instructive mechanistic commentary, see: a) W. L. Mock, *J. Am. Chem. Soc.* **1975**, *97*, 3666; b) W. L. Mock, *J. Am. Chem. Soc.* **1975**, *97*, 3673.
32. D. I. Schuster, G. Lem, N. A. Kaprinidis, *Chem. Rev.* **1993**, *93*, 3.
33. For general reviews on the Diels–Alder reaction in total synthesis, including asymmetric synthesis, see: a) K. C. Nicolaou, S. A. Snyder, T. Montagnon, G. Vassilikogiannakis, *Angew. Chem.* **2002**, *114*, 1742; *Angew. Chem. Int. Ed.* **2002**, *41*, 1668; b) E. J. Corey, *Angew. Chem.* **2002**, *114*, 1724; *Angew. Chem. Int. Ed.* **2002**, *41*, 1650.
34. Since this chapter was written, an additional total synthesis of colombiasin A was achieved: A. I. Kim, S. D. Rychnovsky, *Angew. Chem.* **2003**, *115*, 1305; *Angew. Chem. Int. Ed.* **2003**, *42*, 1267. A total synthesis of elisabethin A has also been accomplished: T. J. Heckrodt, J. Mulzer, *J. Am. Chem. Soc.* **2003**, *125*, 4680.

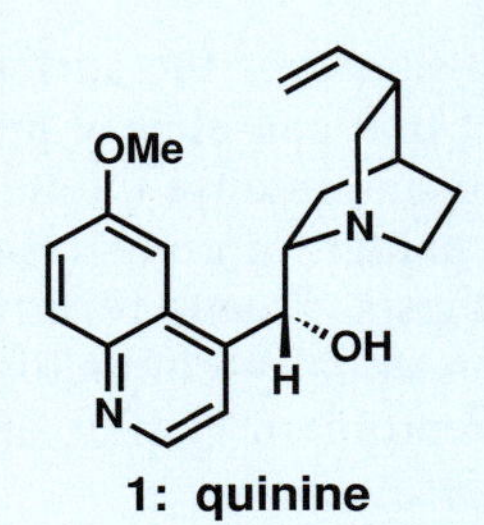

15

G. Stork (2001)

Quinine

15.1 Introduction

Key concepts:

- **Synthesis evolution**
- **Strategy design**

Western culture first became aware of quinine in the 1630s when Spanish missionaries stationed in parts of the Incan Empire (present day Peru) learned from the local inhabitants that extracts from the bark of a tree they called "quina" could treat certain fevers and ailments which we would now characterize as malaria. When these remedies proved invaluable a few years later in saving the life of the Countess of Chinchon, the consort of the Spanish Viceroy of Peru (for whom the chinchona alkaloids are likely named), word spread to European courts of this new medicine, and quina bark was soon exported across the Atlantic Ocean where those few who could afford this scarce and expensive curative benefited from its powers. Although malaria was a significant problem throughout Europe, acceptance of quinine was not universal as it was "marketed" with the name "Jesuit's powder" after the Roman Catholic missionaries who had first discovered it. As a result, nobles in Protestant Europe, particularly England, considered it in bad form to use any medicine so clearly associated with the papacy, and resorted instead to more classical therapies of the period such as bloodletting or drinking mercury. While such reluctance is believed to have led indirectly to the death of several important historical figures such as Oliver Cromwell, once European nations expanded their empires into additional regions where malaria was prevalent, like the Indian subcontinent, quinine became a widely embraced medicine which to this day is still utilized in many parts of the world. Its true impact in this regard is challenging to estimate, but it is not unreasonable to assume that quinine has been responsible for saving hundreds of thousands of lives. Without doubt, this agent has profoundly shaped the development of modern civilization over the course of the past four centuries.[1,2,4]

Although this short anecdotal sketch barely scratches the surface of the interesting history of quinine, a subject that can clearly provide numerous opportunities for book contracts and intense intellectual discourse, our primary objective in this chapter, of course, lies exclusively in the realm of quinine's total synthesis. True to form, in this venue quinine serves as the source of an equally compelling story in that it has constituted a synthesis problem unlike any other encountered during the past two centuries.

On the surface, quinine (**1**) appears no different than the other molecular targets discussed in this book–smaller, perhaps, than many of these natural products, but still laden with stereochemical complexity in the form of four stereogenic centers. Indeed, the majority of the objectives behind undertaking its total synthesis mirror those discussed throughout the *Classics* series. For example, several researchers have sought to use quinine (**1**) as a platform to develop unique synthetic strategies. Others have viewed the quinoline and quinuclidene systems as excellent proving grounds for new methodology in heterocyclic chemistry, while some have desired to make sufficient quantities of quinine (**1**) for pharmaceutical applications. What ultimately renders quinine so synthetically unique is over 150 years of prodigious effort was required before a stereoselective total synthesis of this compound was finally achieved in 2001 by Gilbert Stork and his colleagues at Columbia University.[3,4] Given our current synthetic prowess in which syntheses of natural products with far greater structural complexity than quinine (**1**) are routinely achieved just a few years following their structural characterization, such a statement would seem incredibly hard to believe. The source of quinine's synthetic challenge over the ages, however, was not the lack of suitable technology to fashion its key structural elements, but rather the consistent application of a singular general synthetic strategy which made an asymmetric synthesis virtually impossible to achieve. Before we get to the conclusion of this interesting story and analyze the Stork group's unique approach to this agent, we feel that setting the stage with a brief history of the synthetic studies targeting this extraordinary natural product is necessary to fully appreciate the logic behind their final design.

15.1.1 Initial Synthetic Forays

Throughout the 19th century (as well as most of history, for that matter), the supply of quinine was always less than its demand in the Western world because of the dual challenges of procuring large quantities from typically far away natural sources and maintaining a consistent raw material pipeline, as wars were frequent in the territories from which quinine was obtained.[2a] This supply issue proved to be especially fatal during the American Civil War

(1861–1865) when the lack of sufficient quantities of quinine was believed to have led to the death of more soldiers stationed in the Southern states from malaria than from wounds sustained in battle.[1] It was during this period, though, that chemists first began to contemplate a solution to this general availability issue in the form of an effective laboratory route to quinine. While the value of synthetic chemistry was only barely perceived by the greater scientific community, achievements such as the synthesis of urea (Wöhler, 1828)[5] and acetic acid (Kolbe, 1845)[6] may have provided some measure of confidence to early chemists that other natural products such as quinine could be synthesized, though tremendous travail would likely be required. Indeed, at the time the only available aids were a select number of raw chemicals, a small collection of reaction conditions such as "heating" and distillation, an authentic sample of the natural product (which was first isolated in 1820 by Pelletier and Caventou), and the molecular formula of quinine (determined by Strecker in 1854).[7]

The first pioneer in the quest for synthetic quinine was August Wilhelm Hofmann, the German director of the new Royal College of Chemistry in London, who published the idea in 1849 that quinine could be synthesized using materials derived from readily available coal tar.[1] In 1856, he was able to convince his 18-year old pupil, William Henry Perkin, to undertake this project armed with the notion that the reaction of two molecules of *N*-allytoluidine (**2**, Scheme 1) with oxygen through ". . . the discovery of an appropriate metamorphic process" could lead to quinine (**1**) following the loss of water.[8] Although this equation was certainly mathematically sound, the actual molecular constitution of both quinine and the starting material shown in Scheme 1 reveal the obvious futility of Perkin's endeavors for a total synthesis. Significantly, however, this equation fails to indicate the valuable discovery that he would make in the process.

Perkin began his investigations in his home laboratory over the Easter vacation by first treating **2** with potassium dichromate, a reaction that ultimately generated an intractable reddish-brown pre-

Sir William Henry Perkin
Courtesy of the William Haynes Portrait Collection, Chemist's Club Archives, Chemical Heritage Foundation Image Archives

2: *N*-allyltoluidine ($C_{10}H_{13}N$) + **2: *N*-allyltoluidine** ($C_{10}H_{13}N$) + 3O --?, -[H_2O]--> **1: quinine** ($C_{20}H_{24}N_2O_2$)

Scheme 1. Initial forays into the laboratory synthesis of quinine: the "mathematical approach" by Sir William Henry Perkin (1856).

cipitate from which he could find no quinine. Undeterred by this initial failure, Perkin decided that "screening" more conditions was necessary and next tested the same oxidation reaction using aniline as his starting material. Instead of a crude brown product, this time he synthesized a black solid. While most of us today would be apt to throw such a "tar" into the trash, Perkin instead tried to extract this residue with several agents, and, when he used alcohol, he observed the formation of a brilliant purple solution that could readily dye fabrics. Upon concentration of this alcoholic media, he obtained a beautiful crystalline product. Although not quinine, this compound, which he called "mauveine",* launched the beginning of the chemical industry throughout the world and led to great commercial profit for Perkin.[1]

"mauveine" (R = H or Me)

Hofmann and Perkin, though, were not alone in performing scientific research on quinine during this period. Across the English Channel, the famous French scientist Louis Pasteur was attempting to employ this natural product as a suitable base to resolve racemic acids, a goal that he ultimately realized in 1853 (without quinine) in his successful resolution of tartaric acid into its two enantiomeric forms. During the course of these investigations, however, Pasteur observed that the reaction of quinine (**1**, Scheme 2) with aqueous sulfuric acid led to the formation of a new product which we now know as quinotoxine (**3**).[10,11] While Pasteur himself never deduced the molecular basis for this event, as the structure of quinine was unknown (we can now rationalize the reaction through the mechanism shown involving intermediates **4** and **5**), this degradation reaction served as the catalyst that eventually enabled chemists to slowly unravel the structure of quinine. By 1908, the German chemist Paul Rabe reported the correct molecular connectivities of this natural product,[12] thereby enabling the chemical community to pursue more rationally designed synthetic studies towards quinine than those initiated by Perkin. Rabe would play a leading role in this regard, and spent most of the next quarter of a century in pursuit of the total synthesis of quinine.

Louis Pasteur

As the absolute configuration of quinine was still unknown and stereochemical analysis did not exist, Rabe's ultimate objective was unfortunately far in advance of the realm of possibility. Even if he had successfully achieved the synthesis of the molecular structure of quinine, he would have conceivably produced sixteen different products, of which quinine would have required selective isolation from the other fifteen stereoisomers. Despite these overarching challenges, Rabe was able to effect a partial synthesis of quinine (**1**) from Pasteur's degradation product quinotoxine (**3**, cf. Scheme 2).[13] This remarkable achievement is documented in Scheme 3 wherein, following bromination of the free amine in the piperidine ring of **3**

Paul Rabe

* Following extensive studies, it turns out that Perkin's starting aniline was, in fact, contaminated with *o*- and *p*-toluidine impurities which proved critical in the formation of mauveine; as shown in the column figure, mauveine actually constitutes a mixture of two compounds.[9]

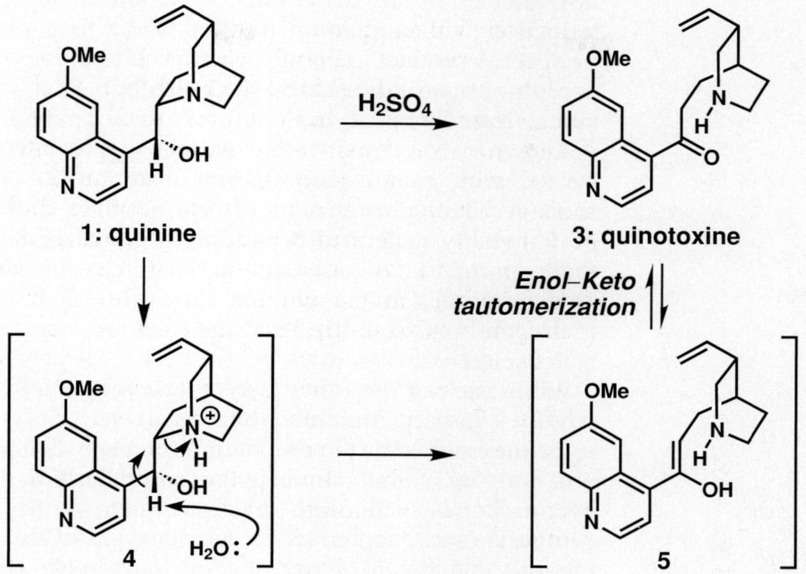

Scheme 2. Pasteur's acid-catalyzed rearrangement of quinine (**1**) to quinotoxine (**3**) as first characterized by Rabe.

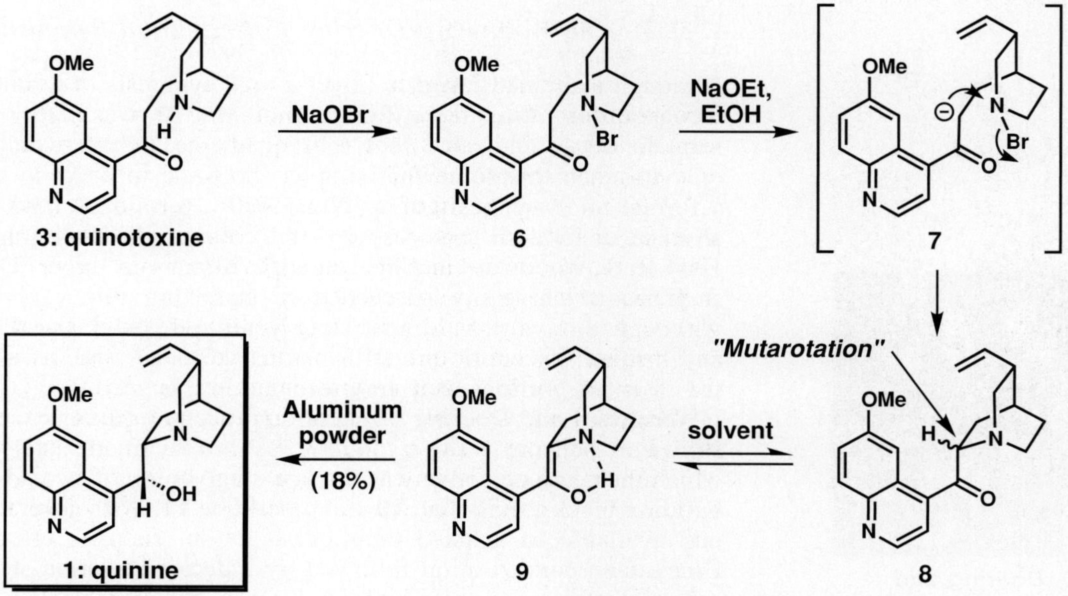

Scheme 3. Rabe's reconstitution of quinine (**1**) from quinotoxine (**3**). 1918

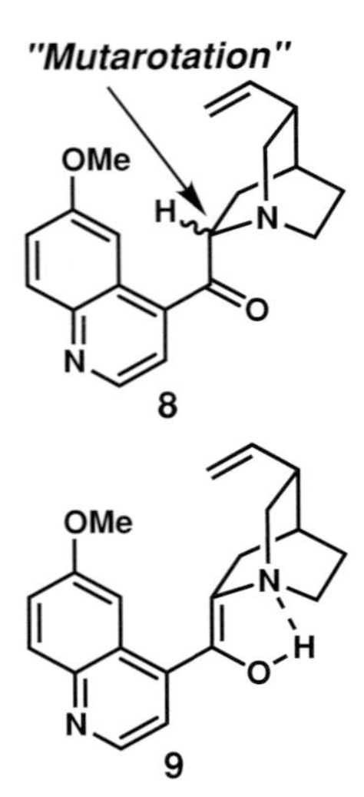

with NaOBr, exposure of the resultant product (**6**) to NaOEt in ethanol effected basic dehydrohalogenation to afford **8**. Upon final reduction with aluminum powder, Rabe then obtained a mixture of several products, among which was quinine (**1**). Though Rabe could not have rationalized the collection of products he obtained since stereochemical and conformational analysis were essentially unknown at the time, today we can appreciate that although he started with enantiopure starting material (**3**), at the end of his sequence, he had produced all four possible C8–C9 stereoisomers of **1**. Freshly generated **8** underwent epimerization at C-9 through facile *in situ* keto-enol tautomerization (what has been termed a "mutarotation" in the quinine literature) as its enol form (**9**) is hydrogen-bond-stabilized and the final reduction reaction was non-stereoselective.

While each of these initial synthetic forays fell short of affording wholly synthetic quinine, the endeavors were far from fruitless since they served as the foundation for a substantial body of our current body of synthetic knowledge, particularly in the field of heterocyclic chemistry. Indeed, reactions such as the Skraup quinoline synthesis (see Chapter 18 for a discussion of this general transformation) and much of our general knowledge about pyridine and quinoline ring systems had their inception from this initial work related to quinine.[2a] Equally significant, these early studies towards quinine also established the groundwork necessary to enable the next generation of synthetic practitioners to achieve the first total synthesis of **1**.

15.1.2 The Woodward/Doering Formal Total Synthesis

Although Rabe had failed to effect a total synthesis of quinine, his reconstitution of this target from quinotoxine (**3**) was highly significant in that it indicated that subsequent researchers would need only to reach this somewhat simpler molecule in order to declare a formal total synthesis of **1**. While still a formidable task in the absence of modern spectroscopy and column chromatography, in 1944 R. B. Woodward and his student William von Eggers Doering at Harvard University succeeded in achieving this very goal.[14] Although this work is almost sixty years old, its classical design and brilliant execution are still instructive today, and, as a result, the essential portions of it are presented in this section.[15]

W. v. E. Doering and R. B. Woodward
Courtesy of Time Life Syndication

Woodward and Doering's general approach to quinotoxine (**3**) is shown in Scheme 4 (as a modern day retrosynthetic analysis) in which their key objective was the late-stage union of two advanced building blocks, quinoline **10** and piperidine **11**. With several methods available to achieve their combination, such as effecting a Dieckmann condensation followed by a decarboxylation step, this proposal was quite reasonable and would then enable use of the Rabe route to complete quinine (**1**).[13] The true challenge for this analysis, however, was achieving the selective synthesis of building

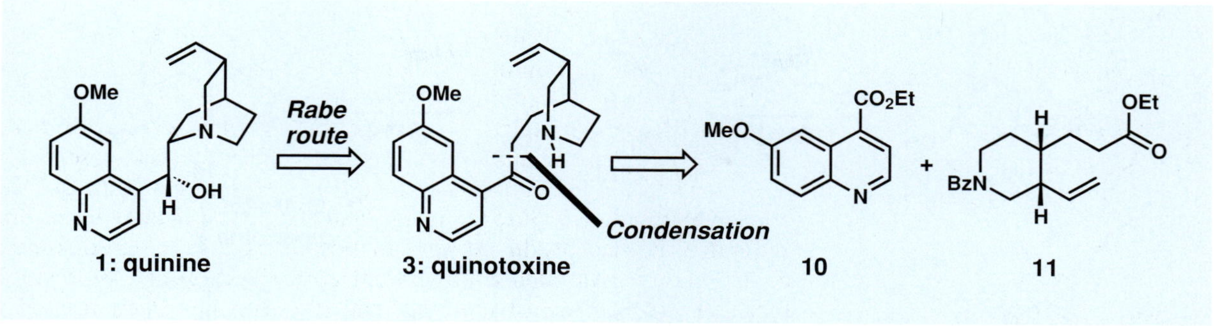

Scheme 4. General retrosynthetic plan utilized by Woodward and Doering in their formal total synthesis of quinine (**1**).

block **11** (the ethyl ester of a quinine degradation product known as homomeroquinene) in which the two side chains would need to be incorporated in a *cis* arrangement.

At first glance, one might imagine that this goal could be accomplished by the separate introduction of appropriate alkyl tethers on a simple piperidine ring. While feasible, the potential length of such a sequence and the fact that it would require carrying reactive functionality throughout the synthesis would seem to obviate its practicality. Faced with this realization, Woodward and Doering pioneered a strategy that would become one of the most important approaches for the synthesis of complex molecules: the formation, modification, and eventual cleavage of carbon frameworks in cyclic settings to generate acyclic stereochemical elements.[16] We have already witnessed the ingenuity and utility of this concept in Woodward's synthesis of both strychnine and reserpine in the first volume of *Classics*. For quinine, this strategy was expressed in the idea that a 2-methylcyclohexanone ring system appended in a *cis* fashion to a piperidine (such as **19**, Scheme 5) could serve as a masked form of both the ester and vinyl side chains of **11** to be unveiled through an appropriate oxidative ring cleavage process. As a result, the critical requirement for the success of this approach would be finding conditions to effect such an opening that would also be mild enough to prevent any equilibration of the piperidine substituents, as the thermodynamically favored 1,2-*trans* system would result and thereby destroy the carefully crafted *cis* arrangement of these portions at the ring fusion in **19**.

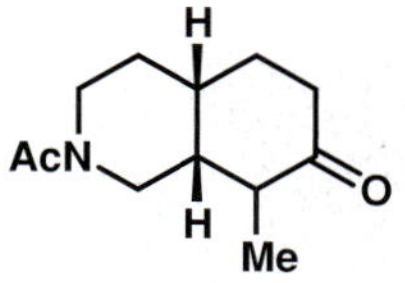

In order to examine the viability of this approach, efforts were first directed towards the stereospecific synthesis of **19** following a route which anticipated that hydrogenation of the aromatic system in an intermediate such as **18** could deliver hydrogen in a facially selective manner to afford the requisite stereocenters of the target (in a relative sense). As shown in Scheme 5, while **19** was produced through such a reaction, the final hydrogenation of **18** did not proceed stereoselectively since double-bond isomerization of partially reduced intermediates probably occurred at the elevated tempera-

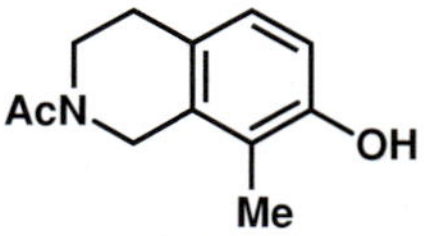

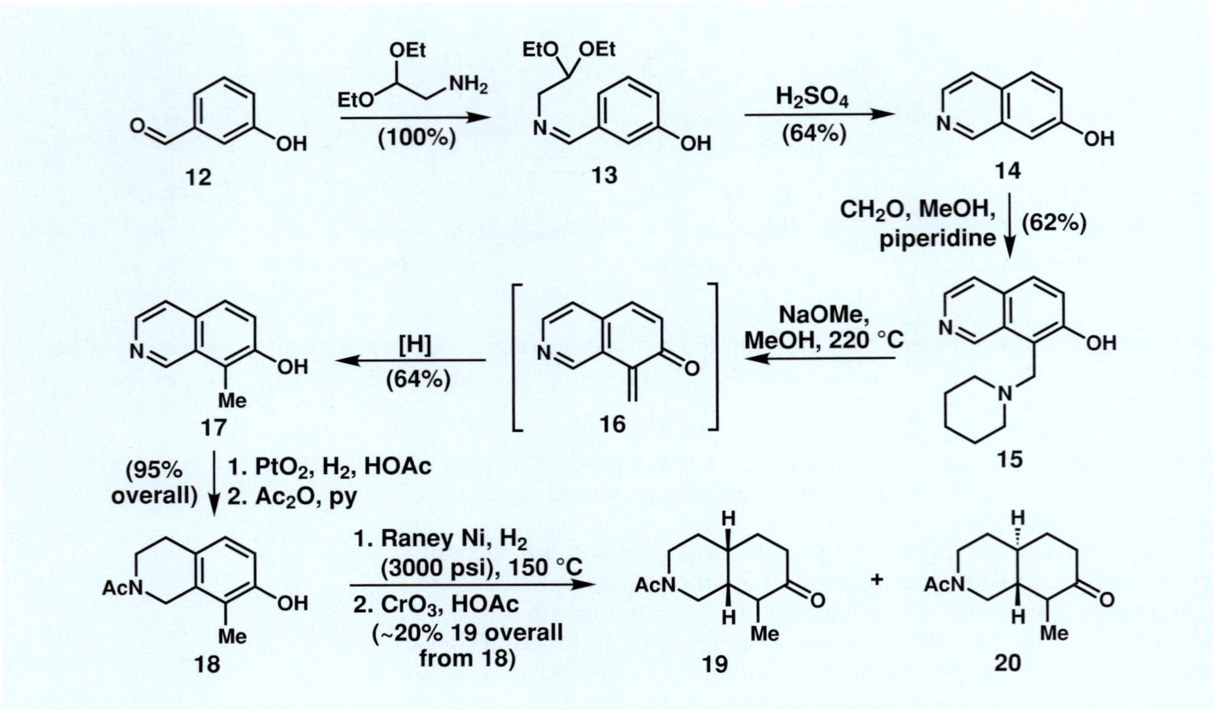

Scheme 5. Woodward and Doering's synthesis of advanced intermediate **19**.

tures employed (150 °C), also giving rise to **20**. Despite the nonselective nature of this reaction, however, **19** could be obtained in pure form in 20 % yield (as a racemic mixture of *cis*-fused products) following a tedious separation protocol. Before pressing on with the synthesis, though, it is instructive to note that with the formation of key intermediate **15**, Woodward and Doering had already incorporated all the carbon atoms that would become the eventual building block in just three steps. While a seemingly insignificant observation, the rapid construction of the carbon framework in his target molecules would be similarly echoed in many of Woodward's later syntheses and constitutes a distinctive hallmark of his synthetic strategies toward complex molecular architectures.

With **19** synthesized, the next step was ring opening, and, as shown in Scheme 6, this conversion was brilliantly achieved upon application of a nitrite ester cleavage protocol (a reaction that had previously been used by Rabe during some of his additional degradation work).[15] As illustrated, exposure of **19** to ethyl nitrite in the presence of NaOEt in EtOH converted the initial ketone into an ester with attendant oxidation of the adjacent carbon atom to an oxime, leading to **25** in 68 % yield through a remarkable process that can be rationalized by invoking intermediates **21** through **24**. Although one might expect such strongly basic conditions to epimerize the carefully installed stereocenters in **19**, the formation of

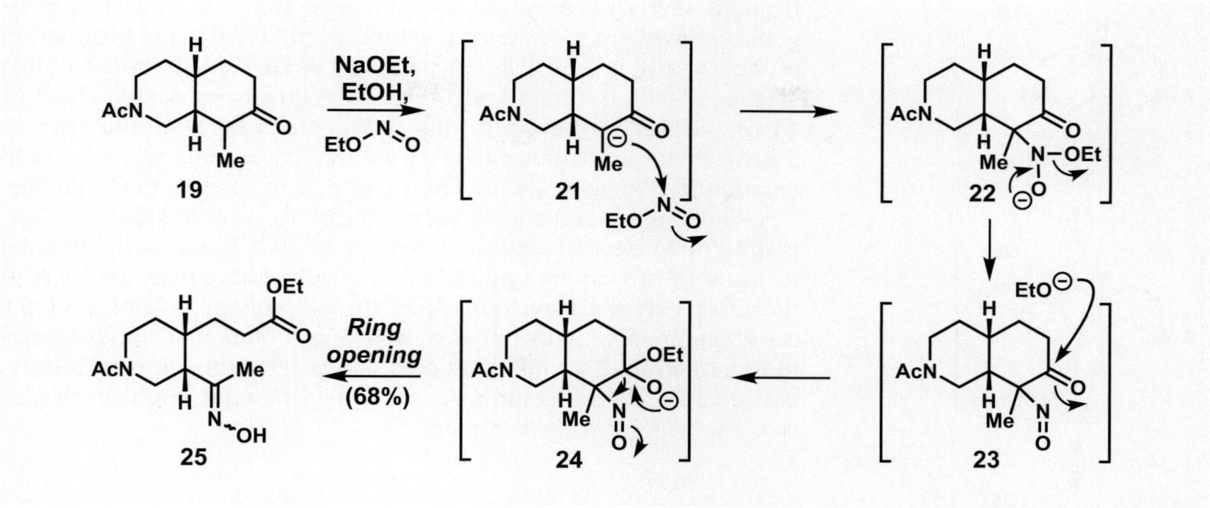

Scheme 6. Woodward and Doering's opening of cyclic template **19** to provide advanced intermediate **25**.

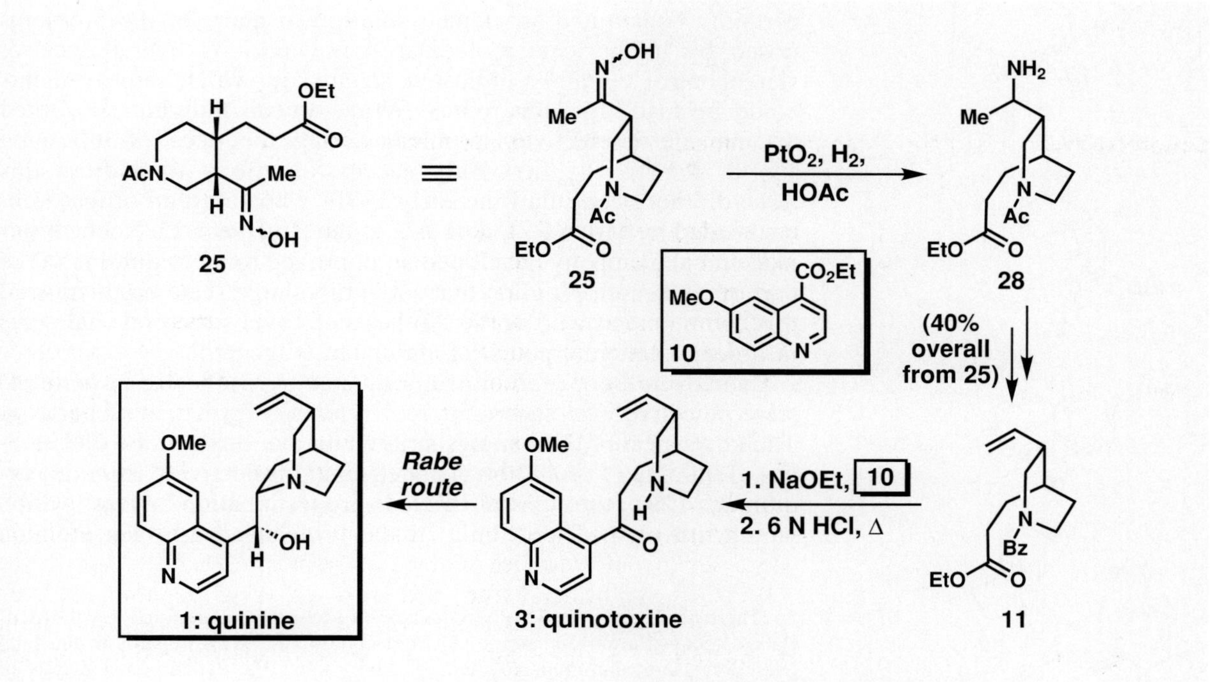

Scheme 7. Completion of Woodward and Doering's formal total synthesis of quinine (**1**).

an oxime adjacent to the piperidine ring was an ingenious design element because the most acidic proton in the resultant product (**25**) resided on the oxygen atom of the oxime. Thus, as shown in the column figures on the next page, under basic conditions, **26** (whose resonance, or canonical, form is **27**) was formed exclusively from **25**, thereby preventing proton release from the adjacent stereogenic carbon center of the ring, which would have led to destruction of the initially installed stereochemistry in **19**. Equally brilliant, this cleavage reaction directly afforded the requisite ester side chain of **11** (cf. Scheme 4) and left only reduction of the oximino ester to a primary amine followed by an *in situ* base-induced elimination reaction (E_2) to generate the vinyl group of this key target molecule.

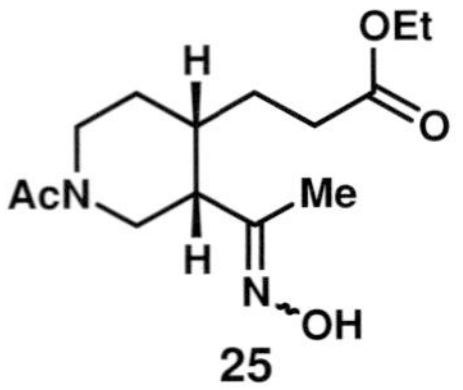

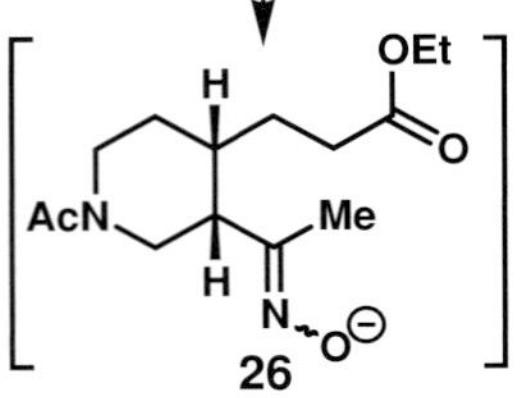

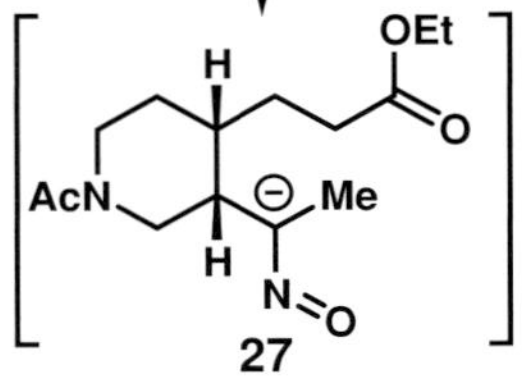

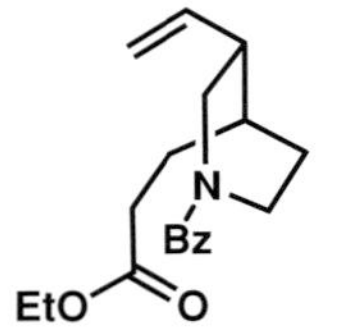

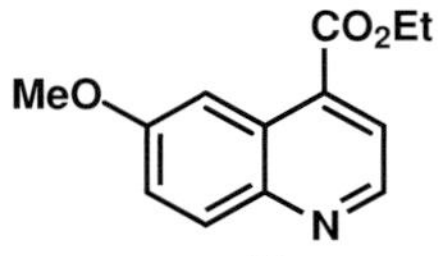

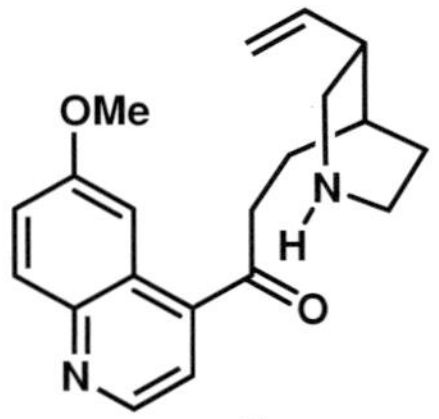

As indicated in Scheme 7, these objectives were met with the formation of **11** over several steps, via **28**, in 40 % yield. Following the reaction of this compound (**11**) in a Dieckmann condensation with the readily synthesized quinoline **10**, subsequent treatment of the resultant product with hot 6 N HCl then completed the synthesis of quinotoxine (**3**), which was obtained as its natural epimer following selective crystallization. As such, the first total synthesis of quinine (**1**) was formally complete.*

AcN H H Me N OH OEt O **25**

AcN H H Me N O⊖ OEt O **26**

AcN H H ⊖ Me N O OEt O **27**

N Bz EtO O **11**

MeO N CO$_2$Et **10**

OMe N N H O **3**

15.1.3 *The Hoffmann-La Roche Total Synthesis*

Although the Woodward/Doering formal synthesis of quinine (**1**) certainly constituted an elegant solution to many of the problems posed by its intricate molecular framework, its overall lack of stereocontrol certainly indicated an area in which improvements could be made by later routes. While several individuals worked on this stereoselectivity problem during the decades following World War II, the first truly successful efforts to address this issue did not occur until the early 1970s when a group of researchers headed by Milan R. Uskoković at the Hoffmann-La Roche Pharmaceutical Company developed an improved route to quinine (**1**) as part of a research program that sought the large scale production of this compound as well as the synthesis of novel structural analogues in order to generate potential antimalarial agents.[18]

Their overall conception of a sequence to synthesize quinine (**1**) stereoselectively is shown in retrosynthetic format in Scheme 8. The key aspect of their analysis rested on the idea that the C-9 alcohol of quinine (**1**) could be installed enantioselectively from deoxyquinine (**29**) through a SET/disproportionation event which would utilize the steric bulk of the bridgehead nitrogen atom in

* There has been some debate in the current literature concerning the validity of Rabe's reconstitution of quinine (**1**) from quinotoxine (**3**) in regards to the final reduction using aluminum powder, especially since the experimental details of the conversion were not explicitly defined in the original 1918 report and were only later discussed in a 1932 paper by the same author.[13] This issue is clearly of consequence because if this reaction did not proceed as written, then the Woodward/Doering route would not constitute a formal synthesis of quinine (**1**), but merely a synthesis of quinotoxine (**3**) since the Harvard researchers did not repeat Rabe's chemistry. While we do not wish to engage directly in revisionist commentary about whether or not this conversion is valid, we do think it important to note that Woodward and Doering were not alone in basing their synthetic work on the assumption that the Rabe route indeed led to the generation of quinine.[17]

the quinuclidine ring to govern the approach of molecular oxygen to the opposite face following formation of a C-9 anion from **29**. The success of this reaction would, of course, be predicated on the degree of facial bias that a lone pair of electrons could provide in a variant of asymmetric induction that is more classically effected with a stereocenter adjacent to a carbonyl group, with the additional proviso of selective abstraction of a proton from the C-9 position in **29**. While the former concern could only be probed empirically, the latter issue was assured based on the well-documented tenet that the acidity of hydrogen atoms on methyl or methylene groups in the 2- or 4-position of quinoline or pyridine ring systems is relatively high because of the resultant stabilization achieved by delocalization of the anionic center through the aromatic system onto the electronegative nitrogen atom (as shown in the column figures).[15] Indeed, it was this very acidity that enabled Pasteur's conversion of quinine (**1**) into quinotoxine (**3**) to proceed with dilute sulfuric acid, in which case H_2O served as the base that abstracted a proton from this same site (cf. Scheme 2).

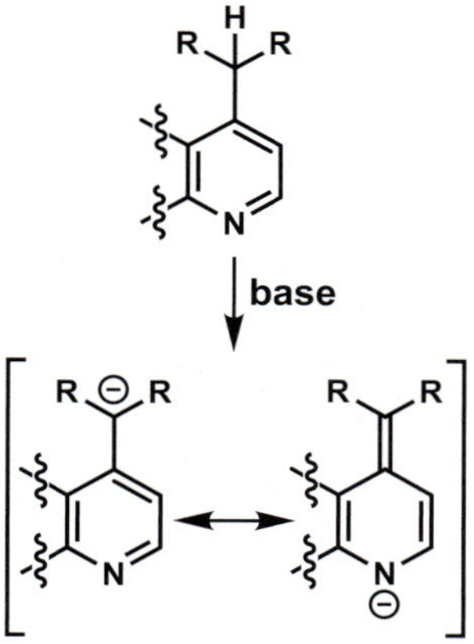

Assuming that this initial disassembly could handle the selective formation of the C-9 stereocenter, the Hoffmann-La Roche group then anticipated that they could install the C-8 stereocenter in **29** in a controlled manner upon application of a cyclization reaction between the nitrogen atom in **30** and a neighboring group that could be suitably displaced. Although this disconnection involved

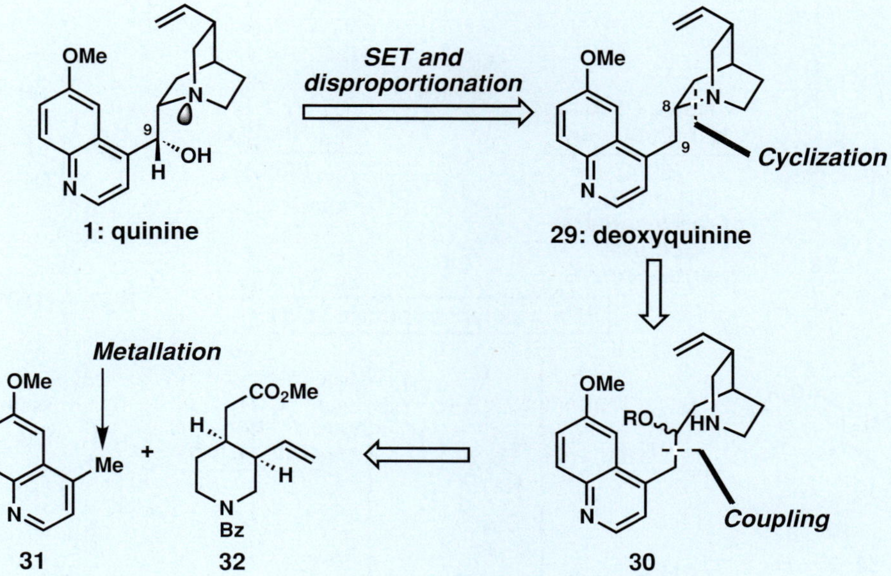

Scheme 8. General retrosynthetic blueprint utilized by researchers at Hoffmann-La Roche to synthesize quinine (**1**). SET = single-electron-transfer.

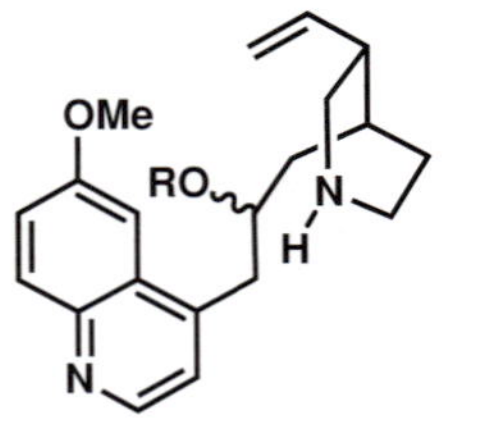

30

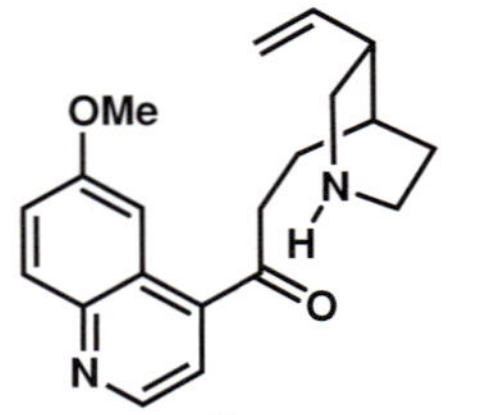

3

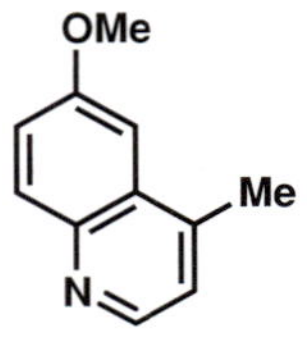

31

the same bond forged by the Rabe reconstitution of quinine (**1**) from quinotoxine (**3**), this approach anticipated that the stereochemistry at C-8 would not be similarly racemized once formed by avoiding the presence of a C-9 carbonyl. In turn, following a series of minor functional group changes, **30** might then arise from a reaction between a metallated nucleophile generated from the known quinoline **31** and a protected form of meroquinene (**32**), a degradation product of the related alkaloid natural product cinchonine. As we shall see, this general analysis proved effective, but still failed to deliver one of the quinine's stereocenters selectively.

The stereoselective route developed to access the key 1,2-*cis* functionalized building block **32** is shown in Scheme 9, starting with an enantiomerically pure bicyclic adduct **33** bearing a single stereogenic center. As in the Woodward–Doering route to quinine (**1**), the use of such a bicyclic template to commence synthetic studies was predicated on ultimately effecting a ring-opening reaction to reveal an appropriately functionalized monocyclic product. Instead of a nitrite ester cleavage reaction, however, the anticipated reaction to achieve this goal here was the pyrolytic fragmentation of *N*-nitrosolactam **36**. Thus, **33** was first converted into amide **34** in 63 % yield through a Schmidt rearrangement as effected with NaN_3 in the presence of hot polyphosphoric acid; only minor amounts of the regioisomeric product obtained by insertion into the α,β-unsaturated system were observed. Subsequent stereoselective hydrogenation of the C–C double bond, as guided by the single stereogenic center in **34**, followed by reaction with N_2O_4, completed

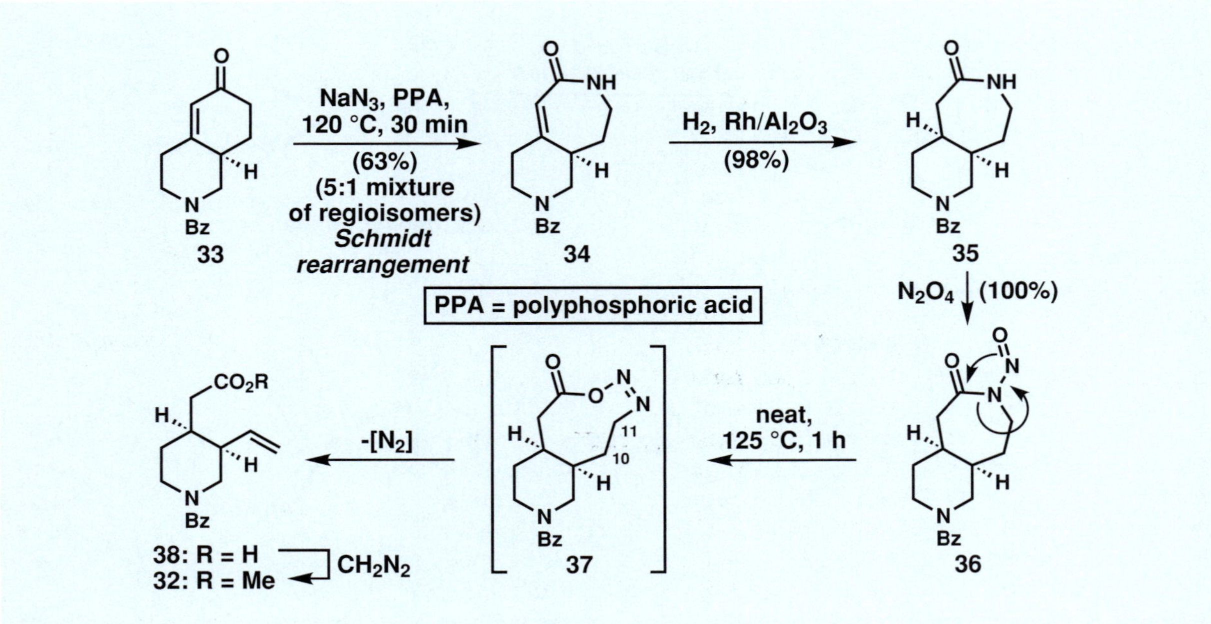

33 → 34: NaN_3, PPA, 120 °C, 30 min (63%) (5:1 mixture of regioisomers) *Schmidt rearrangement*

34 → 35: H_2, Rh/Al_2O_3 (98%)

35 → 36: N_2O_4 (100%)

36 → 37: neat, 125 °C, 1 h

37 → 38: -[N_2]

38: R = H; 32: R = Me (CH_2N_2)

PPA = polyphosphoric acid

Scheme 9. The Hoffmann-La Roche synthesis of quinine: synthesis of intermediate **32**.

the assembly of **36** in near quantitative yield. This compound was then directly heated to 125 °C in the absence of solvent and "... after about 10 s, a relatively violent reaction occurred which was accompanied by a dense cloud of white smoke and change in color from the characteristic yellow-green of the starting material to a dark brown."[18f] After an additional hour of heating at this temperature, followed by treatment with CH_2N_2, the desired ring-opened ester **32** was obtained in 48% yield along with a small amount of lactone **39** (column figure). While these two products are relatively disparate, they can both be accounted for through a common reaction intermediate (**37**) by invoking two different modes for the expulsion of nitrogen based on the participation of the oxygen atoms. If the lone pair of electrons of one of these atoms abstracts a hydrogen atom from C-10, then the intermediate acid **38** would be obtained through an elimination reaction, while nucleophilic attack of the masked acid on the adjacent C-11 would directly afford lactone **39**.[19]

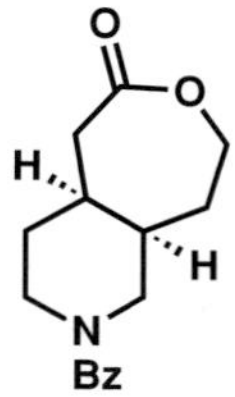

With success in the ring fragmentation reaction and sufficient amounts of **32** available to press forward, the opportunity to explore the stereoselectivity of the oxygenation reaction at the core of the entire strategy was nearing. To reach this critical stage, the known quinoline **31** was selectively lithiated by the action of LDA in THF at −78 °C at the anticipated position *para* to the nitrogen atom of the ring, and following the addition of **32** to this anion, intermediate **40** was obtained smoothly in 78% yield (Scheme 10). In advance of the cyclization reaction necessary to complete the remaining quinuclidine part, treatment with DIBAL-H in toluene at −78 °C effected cleavage of the benzoate ester as well as reduction of the C-8 ketone, and was followed by chemoselective acetylation of the alcohol using glacial acetic acid as activated by $BF_3 \cdot OEt_2$ to afford **41**. With this intermediate synthesized, cyclization to deoxyquinine (**29**) was then achieved in 79% yield upon heating a solution of **41** at reflux with acetic acid buffered by NaOAc in benzene. Unfortunately, this reaction did not prove stereospecific because the conditions used to effect the conversion in fact afforded a reversible pathway between **29** and **42**, leading to scrambling of the C-8 stereochemistry. Indeed, although **29** and its C-8 epimer could be chromatographically separated, exposure of either pure product to the same reaction conditions resulted in a 1:1 mixture of both C-8 products of deoxyquinine. Despite this regrettable outcome, the designed oxygenation process fortunately resulted in the stereoselective formation of the C-9 alcohol of quinine (**1**) based on the initial stereochemistry of the C-8 center in **29**, with the approach of molecular oxygen presumably governed by the presence of the lone pair of electrons on the nitrogen atom. As a result, only two of four possible C-8/C-9 stereoisomers of **1** resulted from this sequence, with (−)-quinine (**1**) obtained in 32% yield for this final operation [72% overall yield for (−)-**1** and its C-8 epimer].

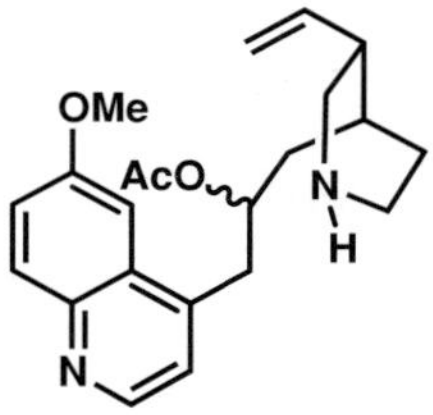

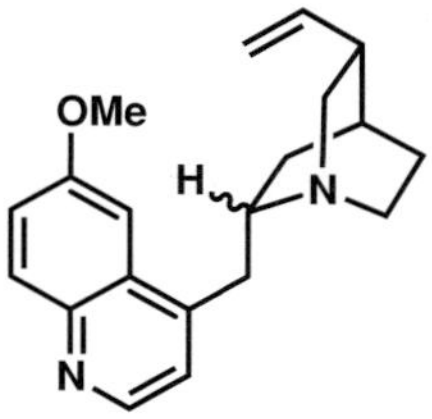

39

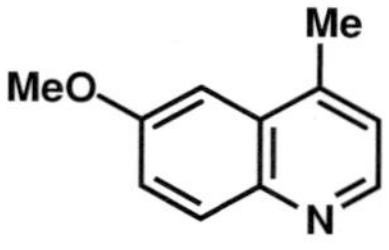

31

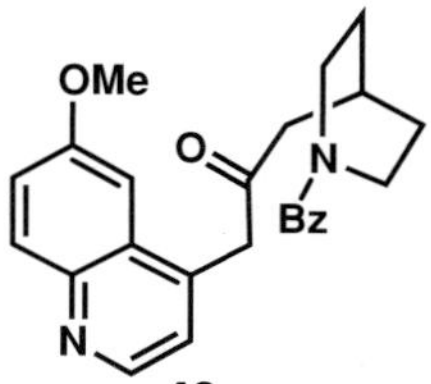

40

41

29: deoxyquinine

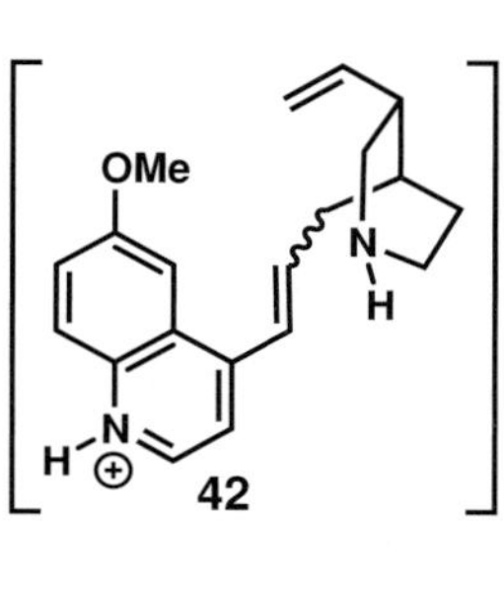

42

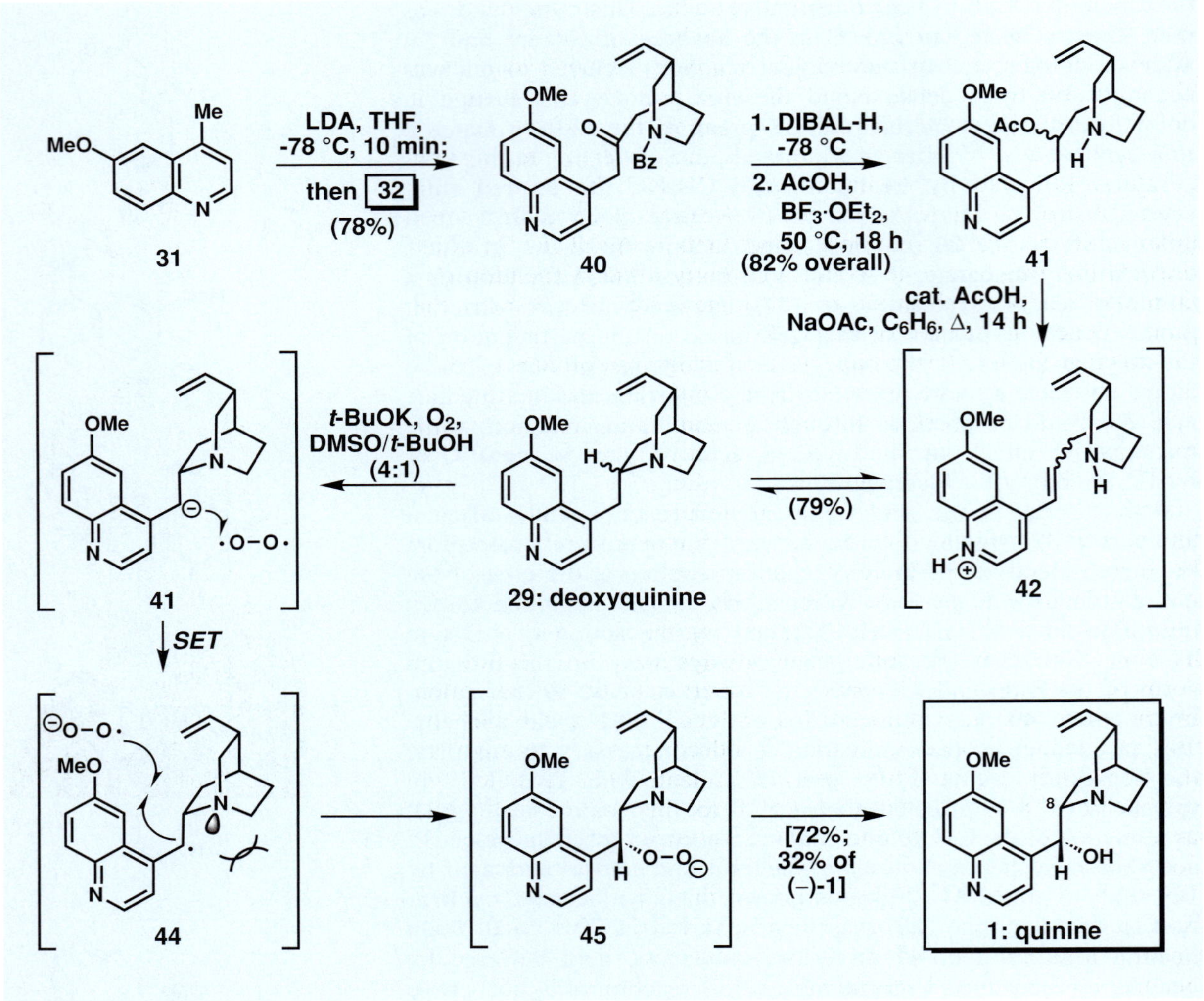

Scheme 10. Completion of the Hoffmann-La Roche synthesis of quinine (1).

15.2 Retrosynthetic Analysis and Strategy

Although the Hoffmann-La Roche synthesis failed to control the stereochemistry upon the formation of the C-8 center, the stereoselectivity of the final oxygenation reaction indicated that if one could generate deoxyquinine (**29**) with its correct absolute stereochemistry, then one would be able to synthesize (−)-quinine (**1**) stereoselectively. Of course, the Hoffmann-La Roche design sought to achieve this very goal, but, unfortunately, failed due to the nonselective nature of the reaction utilized to fashion the quinuclidene ring, a result also observed during the earlier Rabe approach based on a cyclization at the same site. The lack of stereoselectivity observed in these closures (deriving from what might be termed the "Rabe disconnection" of quinine) is relatively obvious to deduce with conformational analysis, as the two pseudo-chair forms of the synthetic precursor shown in the neighboring column are of similar energy with no obvious features that differentiate them in a thermodynamically controlled reaction. As such, both are equally conceivable transition states for the cyclization reaction, and equimolar mixtures of the possible C-8 products result. Thus, in order to avoid this outcome, one would seemingly need to abandon this particular site of cyclization and seek new ways to fashion the quinuclidene ring of deoxyquinine (**29**).

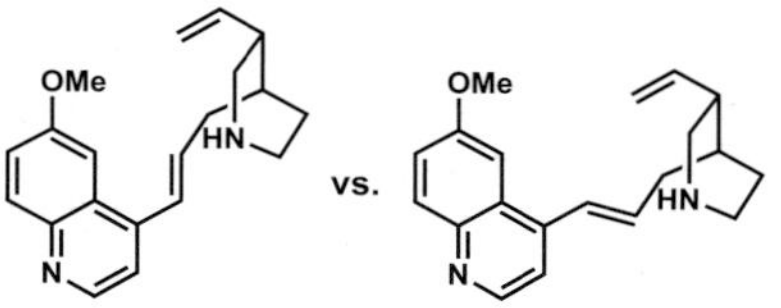

In this regard, the Stork group's critical insight for quinine (**1**) was that an alternative ring closure predicated on an S_N2 alkylation of a piperidine precursor such as **46** bearing an appropriate tether with a leaving group should lead to the stereoselective generation of **29** (as shown in Scheme 11).[3] Although we often seek retrosynthetic "simplifications" in our analyses of molecules, this disconnection does not really afford an intermediate with reduced complexity, as the new goal structure would require the stereoselective formation of a trisubstituted piperidine ring (a seemingly more challenging molecular entity than the disubstituted ring obtained by the Rabe disconnection). However, one should always heed to the following words of wisdom first set to paper by Robert Ireland: "All too often the most convenient way to draw a molecule on paper belies the most efficient synthetic approach."[15] Indeed, if **46** is redrawn in its corresponding chair form, one can immediately appreciate the logic behind this new design in that all three piperidine substituents require a conformation in which they are equatorially oriented to minimize 1,3-diaxial strain as drawn. As such, it would be reasonable to presume that the C-8 stereocenter in **46**, and thus ultimately the C-8 center in (−)-quinine (**1**), might arise from the facially selective delivery of a hydride equivalent onto a suitable imine precursor (**47**) in an axial fashion.

G. Stork

Assuming this sequence to be feasible, the formation of **47** would then seem conceivable through a simple imine condensation reaction between a free amine and a ketone, as might occur following Staudinger reduction of the azide in **48**. From this new goal structure, reduction of the C-9 carbonyl to the free alcohol in **49** suggested a synthetic pathway similar to that of the other approaches already discussed in which a nucleophile generated from **31** could engage an appropriate building block, such as **50**, to fashion this advanced intermediate. Thus, based on this line of reasoning, only the formation of the two stereocenters of **50** remained to be negotiated during the synthesis to achieve the stereoselective formation of (−)-quinine (**1**).

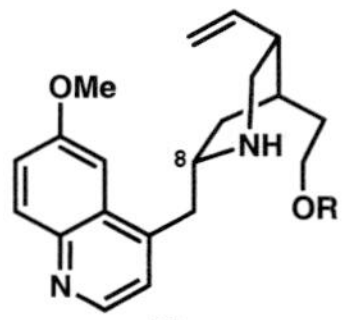

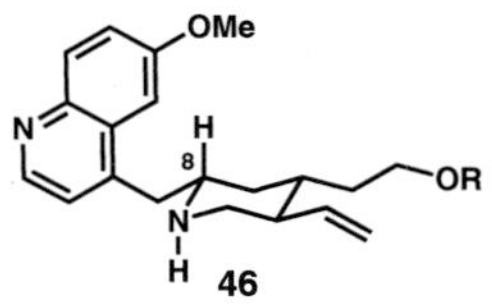

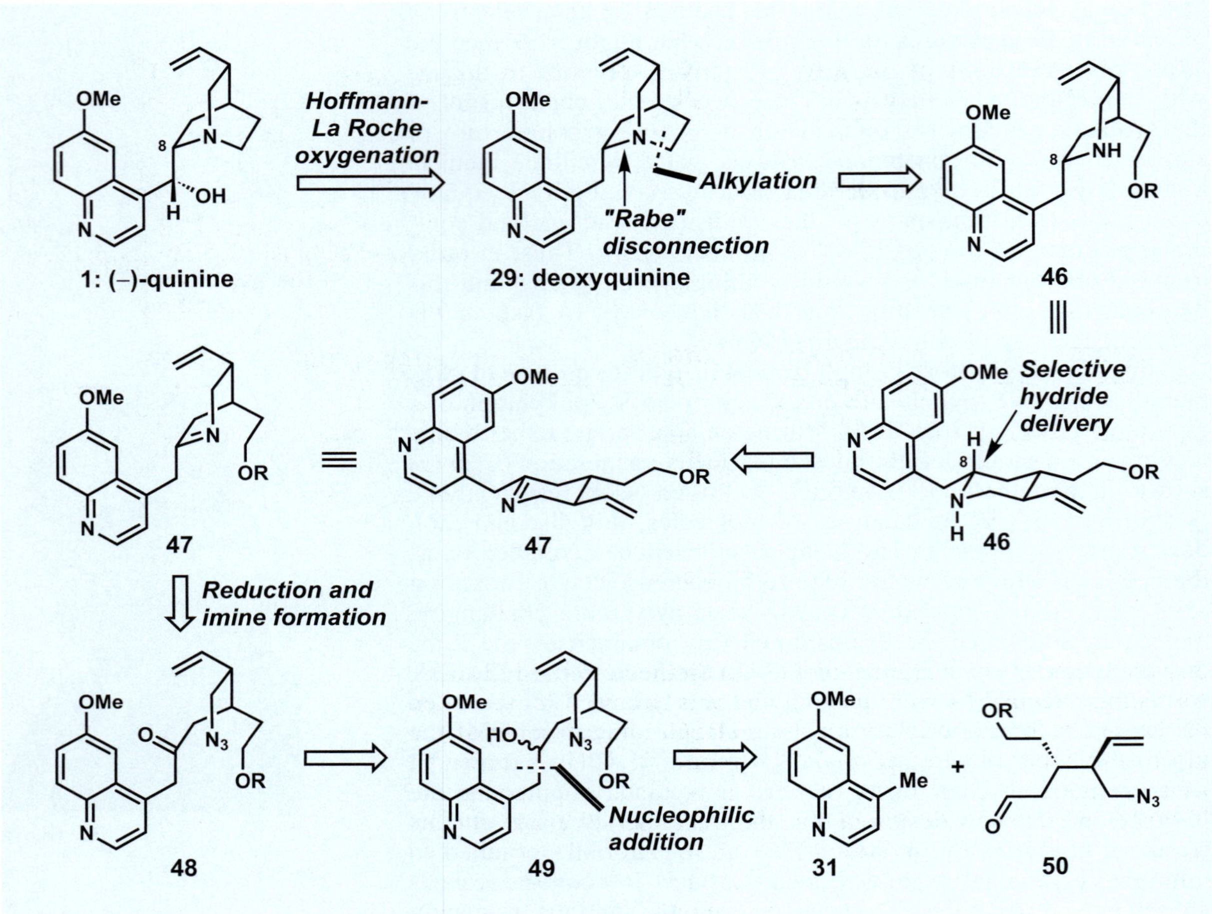

Scheme 11. Stork's retrosynthetic analysis of (–)-quinine (**1**).

15.3 Total Synthesis

The Stork group's efforts toward (–)-quinine (**1**) began with the synthesis of azido aldehyde **50** utilizing a route which anticipated that the stereogenic center in the known (*S*)-vinylbutyrolactone (**51**, Scheme 12)[20] could direct an asymmetric alkylation reaction that would lead to the stereocontrolled installation of the second stereocenter of the targeted building block through the formation of an intermediate such as **55**. Such a strategy, of course, is based on the general tenet of cyclic stereocontrol to generate new stereogenic centers. Unfortunately, initial experiments that sought to alkylate **51** directly with TBDPS-protected iodoethanol to form **55** were thwarted, thereby necessitating reversion to a more circuitous, but ultimately effective, pathway to this intermediate based on acyclic stereoinduction. Accordingly, to fashion a precursor for this alterna-

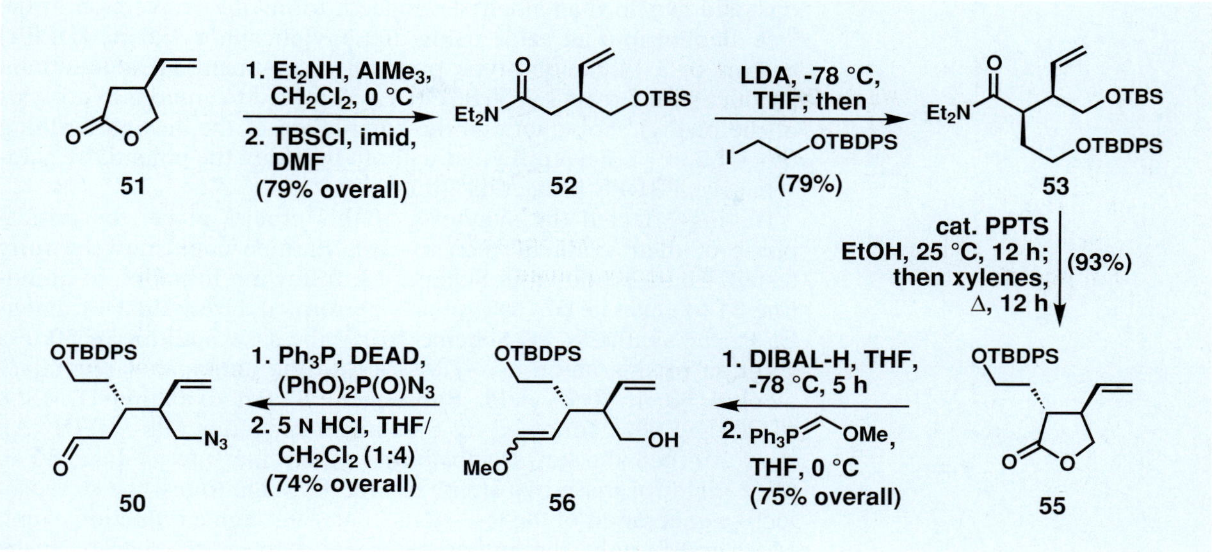

Scheme 12. Stork's synthesis of building block **50**.

tive approach, initial treatment of **51** with diethylamine led to amide formation attended by ring opening and was followed by silylation of the newly formed primary alcohol (TBSCl, imidazole, DMF) to afford the linear product **52** in 79 % overall yield. Upon exposure of this intermediate to LDA in THF at −78 °C, the addition of TBDPS-protected iodoethanol to the resultant lithium enolate afforded the desired alkylation product (**53**) in 79 % yield and with greater than 20:1 diastereoselectivity; such exquisite control can be rationalized by assuming formation of a *Z* enolate (as always occurs with acyclic amides) and electrophile approach from the face occupied by the smaller vinyl group in the drawn orientation where allylic strain is minimized (**54**, see column figure). Subsequent treatment of **53** with catalytic PPTS in EtOH at ambient temperature then led to selective lysis of the TBS group, and following heating in refluxing xylenes over the course of 12 hours, the originally desired **55** was finally obtained.

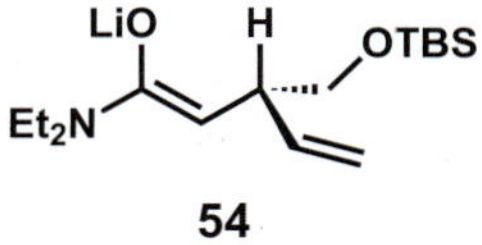

With a solution obtained for the incorporation of both stereocenters of this key building block, the synthesis of **50** was nearly at hand, requiring merely a few functional group manipulations and the addition of a single carbon atom. The initial operations from **55** sought to solve the latter of these problems through initial controlled reduction to a lactol intermediate, followed by Wittig homologation of the open chain equilibrium form of this lactol using methoxymethylene triphenylphosphorane, steps which generated **56** in 75 % yield. As most students of organic chemistry should realize, this particular ylide is especially effective for achieving the one-carbon homologation of aldehydes, as the methyl ether in this context is readily hydrolyzed with water upon treatment with aqueous acid to afford an aldehyde. Indeed, following conversion of the free alcohol into an azide using diphenylphosphoryl azide (DPPA) as part of a Mitsunobu-type protocol,[21] treatment of the resultant product with 5 N HCl in THF/CH_2Cl_2 (1:4) led to smooth hydrolysis of the methyl enol ether and the completion of the desired building block **50** in 74 % overall yield without touching the potentially acid-sensitive TBDPS protecting group.

Having effected the synthesis of this crucial piece, the critical phase of their synthetic plan towards quinine could now be fully tested. Thus, as shown in Scheme 13, following lithiation of quinoline **31** to generate **57** (as similarly performed during the Hoffmann-La Roche synthesis, cf. Scheme 10),[18g] the slow addition of **50** to a solution of this anion at -78 °C led to the anticipated secondary alcohol **58** in 70 % yield. Subsequent Swern oxidation [DMSO, $(COCl)_2$, Et_3N], followed by Staudinger reduction (Ph_3P, THF, Δ), then afforded the key trisubstituted piperidine intermediate **60** in 69 % yield for these two steps, setting the stage to test the stereoselective generation of the C-8 stereocenter through a reduction event. Most gratifyingly, the anticipated axial delivery of hydride to the imine group of **60** indeed occurred upon reaction with $NaBH_4$ in MeOH/THF, presumably the result of the adoption of a single conformational form that placed all three substituents of the piperidine ring in an equatorial position as discussed earlier, leading to **61** in 91 % yield. With the step that stood as the lynchpin of the entire strategy smoothly executed, the stereocontrolled synthesis of quinine now seemed almost assured with only the formation of the

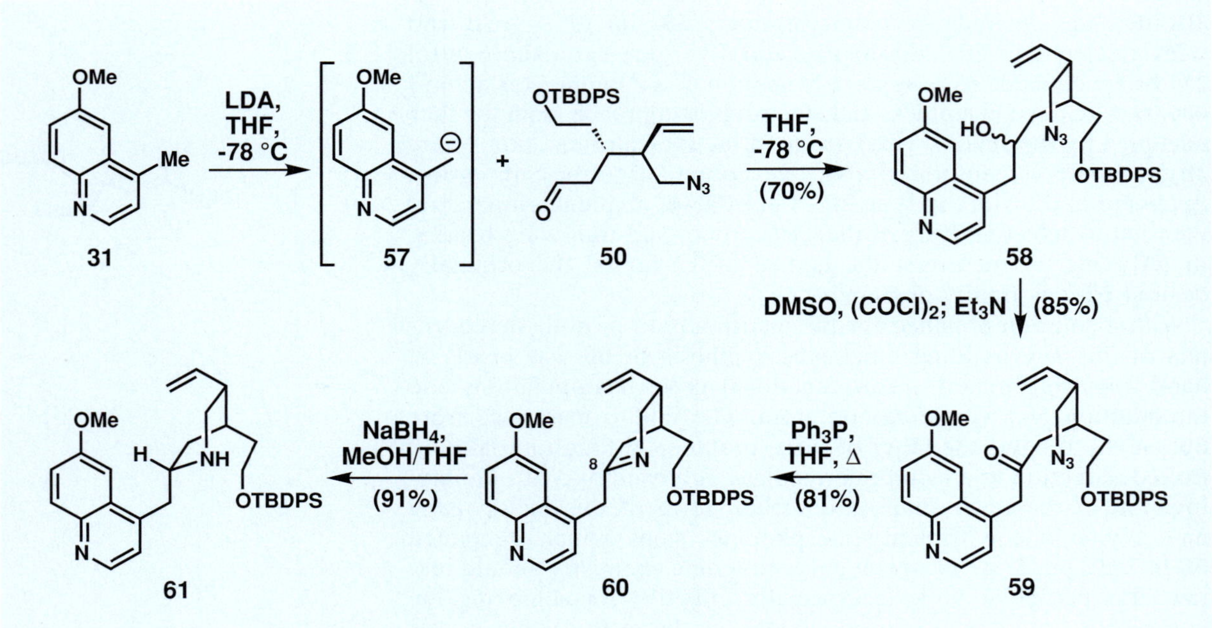

Scheme 13. Stork's synthesis of advanced intermediate **61**.

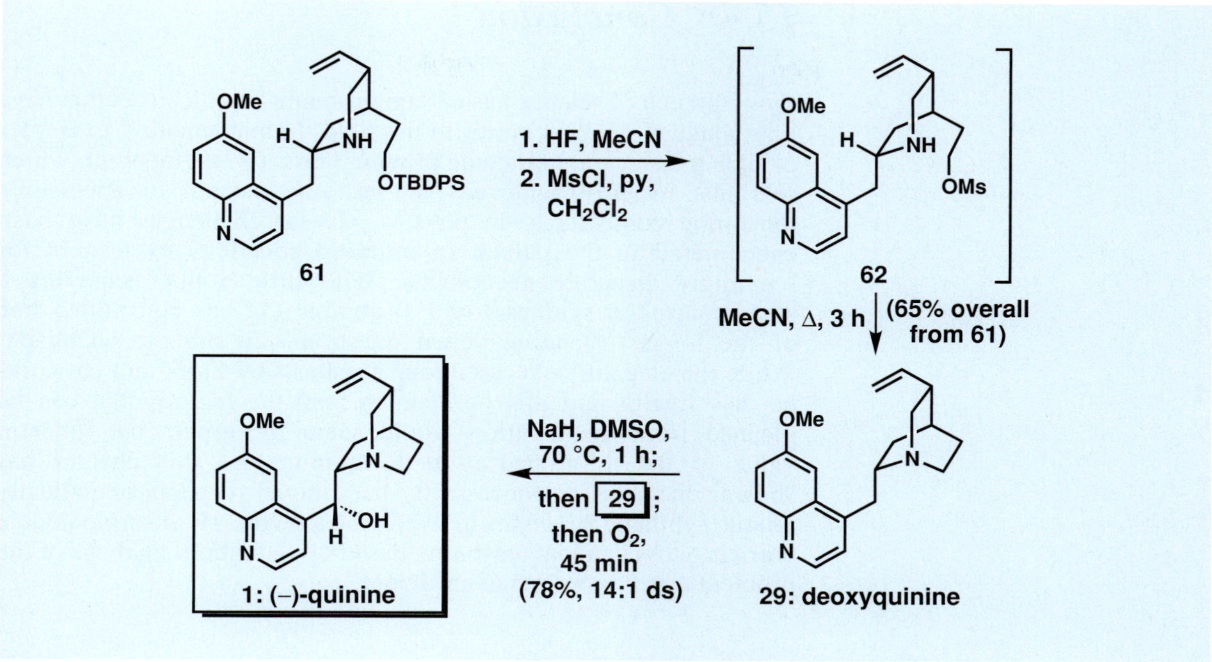

Scheme 14. Final stages and completion of Stork's total synthesis of (–)-quinine (**1**).

final quinuclidene ring and application of the Hoffmann-La Roche oxygenation protocol remaining to complete (−)-quinine (**1**).

Fortunately, these final steps proved relatively easy to execute. As shown in Scheme 14, following conversion of the protected alcohol **61** into mesylate **62** through conventional protocols, subsequent heating in MeCN at reflux for 3 hours effected the requisite *N*-alkylation needed to complete the quinuclidene ring. As such, this transformation led to the first enantioselective synthesis of deoxyquinine (**29**) in 65 % overall yield from **61**. Significantly, no temporary protection of the piperidine amine was required for the success of this sequence. Finally, following a modified format of the Hoffmann-La Roche oxygenation process which employed the "dmsyl" anion (generated by heating DMSO with NaH) instead of *t*-BuOK as base, (−)-quinine (**1**) was obtained in 78 % yield as a 14:1 mixture of diastereomers. Overall, this elegant sequence, which accomplished the first stereoselective synthesis of (−)-quinine (**1**), required only 13 linear steps and proceeded in an impressive combined yield of 15 %.[4]

15.4 Conclusion

Every branch of science has certain problems or objectives that have long plagued the best minds in the field. In mathematics, examples of such puzzles might include proving Fermat's last theorem (which was just recently achieved) or verifying several of Riemann's remaining conjectures. In physics, similar challenges have been encountered in the pursuit of unifying equations to account for seemingly disparate phenomena. With little doubt, achieving a stereocontrolled synthesis of (–)-quinine (**1**) has constituted one of the longest standing "open questions" in organic chemistry. While the elegantly executed total synthesis by Stork and co-workers has finally laid this problem to rest, the lessons that can be gleaned from the solution would seem to surpass the inherent value of the developed route to quinine. As this chapter has shown, one does not necessarily need novel reagents or radically unique synthetic strategies to overcome a seemingly insurmountable barrier; pure chemical reasoning based on thoughtful analysis of the problem can be more than sufficient.

References

1. Much of the interesting history of quinine and malaria is recounted in S. Garfield, *Mauve*, Faber and Faber, London, **2000**, p. 224. We highly recommend this text for all those interested in this molecule and the birth of modern industrial chemistry.
2. For general articles on the early work towards quinine, see: a) R. B. Turner, R. B. Woodward in *The Alkaloids, Vol. 3* (Ed.: R. H. F. Manske), Academic Press, New York, **1953**, pp. 1–63; b) M. R. Uskoković, G. Grethe in *The Alkaloids, Vol. 14* (Ed.: R. H. F. Manske), Academic Press, New York, **1973**, pp. 181–223.
3. G. Stork, D. Niu, R. A. Fujimoto, E. R. Koft, J. M. Balkovec, J. R. Tata, G. R. Dake, *J. Am. Chem. Soc.* **2001**, *123*, 3239.
4. For some brief discussions of the Stork synthesis of quinine and the history of this molecule, see: a) G. Appendino, F. Zanardi, G. Casiraghi, *Chemtracts: Org. Chem.* **2002**, *15*, 175; b) S. M. Weinreb, *Nature* **2001**, *411*, 429; c) P. Ball, *Chem. Br.* **2001**, *37*, 26.
5. F. Wöhler, *Ann. Phys. Chem.* **1828**, *12*, 253.
6. H. Kolbe, *Ann. Chem. Pharm.* **1845**, *54*, 145.
7. A. Strecker, *Annalen* **1854**, *91*, 155.
8. W. H. Perkin, *J. Chem. Soc.* **1896**, *69*, 596.
9. O. Meth-Cohn, M. Smith, *J. Chem. Soc., Perkin Trans. 1* **1994**, 5.
10. a) L. Pasteur, *Compt. rend.* **1853**, *37*, 110; b) L. Pasteur, *Annalen* **1853**, *88*, 209.
11. The structure of quinotoxine was first documented by P. Rabe, *Annalen* **1909**, *365*, 366.
12. a) P. Rabe, *Chem. Ber.* **1908**, *41*, 62; b) P. Rabe, E. Ackerman, W. Schneider, *Chem. Ber.* **1907**, *40*, 3655.
13. a) P. Rabe, K. Kindler, *Chem. Ber.* **1918**, *51*, 466; b) P. Rabe, *Annalen* **1932**, *492*, 242.
14. a) R. B. Woodward, W. E. Doering, *J. Am. Chem. Soc.* **1944**, *66*, 849; b) R. B. Woodward, W. E. Doering, *J. Am. Chem. Soc.* **1945**, *67*, 860.
15. For an insightful and educational account of this synthesis, see: R. E. Ireland, *Organic Synthesis*, Prentice-Hall, Englewood Cliffs, **1969**, pp. 123–139.
16. For a short discussion of this topic and some of R. B. Woodward's other contributions, see: G. Stork, *Nature* **1980**, *284*, 383. An additional source of excellent information on Woodward, including aspects of the quinine total synthesis, can be found in the text: *Robert Burns Woodward: Architect and Artist in the World of Molecules* (Eds.: O. T. Benfey, P. J. T. Morris), Chemical Heritage Foundation, Philadelphia, **2001**, p. 470.
17. M. Prostenik, V. Prelog, *Helv. Chim. Acta* **1943**, *26*, 1965.
18. a) M. Uskoković, J. Gutzwiller, T. Henderson, *J. Am. Chem. Soc.* **1970**, *92*, 203; b) J. Gutzwiller, M. Uskoković, *J. Am. Chem. Soc.* **1970**, *92*, 204; c) M. Uskoković, C. Reese, H. L. Lee, G. Grethe, J. Gutzwiller, *J. Am. Chem. Soc.* **1971**, *93*, 5902; d) G. Grethe, H. L. Lee, T. Mitt, M. R. Uskoković, *J. Am. Chem. Soc.* **1971**, *93*, 5904; e) G. Grethe, H. L. Lee, T. Mitt, M. R. Uskoković, *Helv. Chim. Acta* **1973**, *56*, 1485; f) M. R. Uskoković, T. Henderson, C. Reese, H. L. Lee, G. Grethe, J. Gutzwiller, *J. Am. Chem. Soc.* **1978**, *100*, 571; g) J. Gutzwiller, M. R. Uskoković, *J. Am. Chem. Soc.* **1978**, *100*, 576; h) G. Grethe, H. L. Lee, T. Mitt, M. R. Uskoković, *J. Am. Chem. Soc.* **1978**, *100*, 581; i) G. Grethe, H. L. Lee, T. Mitt, M. R. Uskoković, *J. Am. Chem. Soc.* **1978**, *100*, 589.
19. E. H. White, *J. Am. Chem. Soc.* **1955**, *77*, 6014.
20. a) K. Kondo, E. Mori, *Chem. Lett.* **1974**, 741; b) F. Ishibashi, E. Taniguchi, *Bull. Chem. Soc. Jpn.* **1988**, *61*, 4361.
21. B. Lal, B. N. Pramanik, M. S. Manhas, A. K. Bose, *Tetrahedron Lett.* **1977**, *18*, 1977.

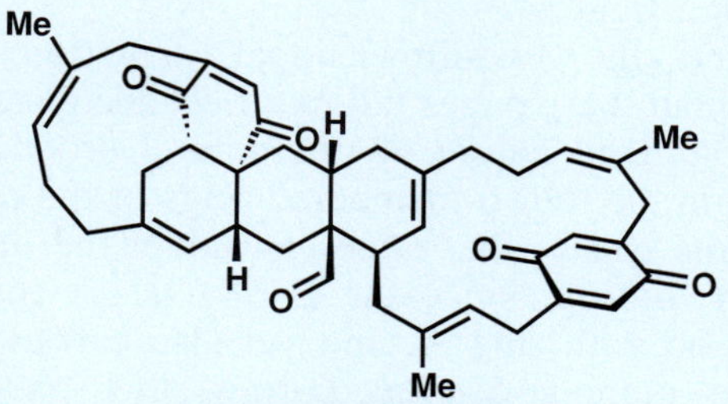

16

1: longithorone A

M. D. Shair (2002)

Longithorone A

16.1 Introduction

Among the various classes of natural products, perhaps none is as extensive as the collection of structures that derive from the union and rearrangement of terpene building blocks. Indeed, whether in the form of a relatively small compound such as pinene or a far more complex and biochemically relevant steroid hormone, these secondary metabolites are ubiquitous throughout Nature.[1] Despite their widespread presence, however, their structures are far from ordinary, as many possess molecular architectures so exotic that even the most inventive of chemists would have a hard time imagining them prior to their full structural elucidation.

One such natural product, longithorone A (**1**), was first reported to the chemical community in 1994 by Professor F. J. Schmitz and several of his co-workers at the University of Oklahoma.[2] Although this isolate from the tunicate *Aplydium longithorax* possesses only weak cytotoxicity (ED_{50} = 10 µg/mL) compared to some of the other sponge-derived agents discussed in this book, such as ecteinascidin 743 (Chapter 5) or diazonamide A (Chapter 20), its unique conglomeration of rings and stereochemical complexity more than makes up for its lack of biological activity, and thus makes it a highly desirable synthetic target. At its most basic level, the carbocyclic skeleton of longithorone A comprises an extraordinary array of 6-, 10-, and 16-membered rings adorned by six stereocenters, two of which are quaternary. While each of these cyclic systems seem relatively disparate when inspected individually, global scrutiny in fact reveals a certain degree of structural harmony in that longithorone A (**1**) can be viewed as the amalgamation of two smaller macrocyclic subunits, each com-

Key concepts:

- **Biomimetic synthesis**
- **Atropisomerism**
- **Enyne metathesis**
- **Organozinc reagents**
- **Transannular Diels–Alder reactions**

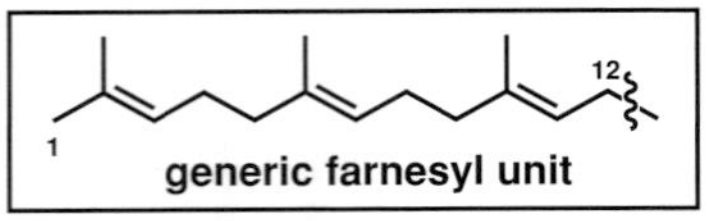

generic farnesyl unit

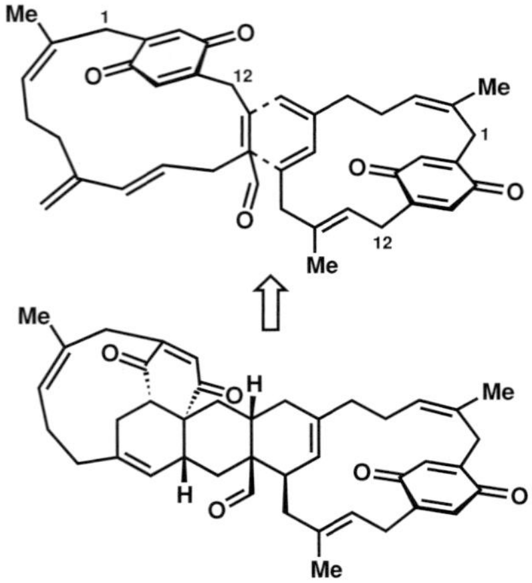

1: (–)-longithorone A

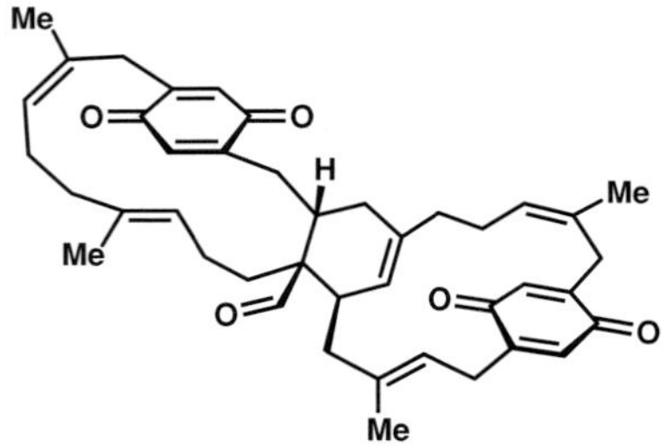

6: longithorone I

prised of a farnesyl unit attached to the 2- and 5-positions of a *para*-quinone moiety (see column figures).

What this analysis and the two-dimensional depiction of longithorone A (**1**) throughout these pages fail to reflect adequately, however, is the severely strained nature of this natural product's structure, particularly within the four 6-membered rings of the central core whose attachments ensure that none can adopt the most stable chair conformation. Indeed, since the B-ring is *cis* fused with the C-ring, *trans* fused with ring A, and yet also carries an attachment point for the quinone-derived D-ring, this pattern forces both the A- and B-rings to reside in distorted boat conformations while the C- and D-rings exist as mutated half-chairs. Apart from providing an unprecedented spatial arrangement, such constraint also gives rise to an element of chirality known as atropisomerism, as the quinone ring within the 16-membered [12]-paracyclophane (ring E) has no rotational freedom (see Chapter 9 for a discussion of this topic).

While these connectivities are difficult to fathom, the means by which Nature might create such a molecular framework is far less mysterious. With three 6-membered ring systems each bearing a lone element of unsaturation, the Diels–Alder reaction would appear to be an obvious candidate for installing most of the polycyclic complexity of longithorone A.[3] Thus, as originally proposed by Schmitz,[2] the unique fusion of the A- and B-rings, as well as the 10-membered macrocycle of longithorone A (**1**, Scheme 1), could potentially arise concurrently in a successful transannular Diels–Alder reaction between the diene in **2** and the proximal C–C double bond of a quinone in the larger [12]-cyclophane system.[4] In turn, the 6-membered C-ring could be envisioned to arise through an *endo* and facially selective [4+2]-cycloaddition between dienophile **3** and diene **4**. This biogenetic hypothesis leading to two structurally related [12]-cyclophanes certainly seems reasonable on paper, and further support for its validity comes from the existence of two additional terpenoid natural products from the same biological source. The first, longithorone B (**5**),[5] resembles the proposed diene **4**, but its quinone is of the opposite atropisomeric configuration, while the second, longithorone I (**6**), displays a structure similar to the presumed initial Diels–Alder product, but, in this case, cannot participate in the second [4+2]-cycloaddition reaction due to its lack of a diene system.[2b]

If this evidence were not incriminating enough, additional substantiation for its likelihood was provided in 2002 by Professor Matthew Shair and two of his graduate students at Harvard University through the first chemical synthesis of longithorone A (**1**) predicated on these same Diels–Alder reactions.[6] This accomplishment is the subject of this chapter, an achievement whose success ultimately relied not only on the power of [4+2]-cycloadditions, but also on an inventive strategy to atropselectively construct both [12]-paracyclophanes at the heart of Schmitz' biosynthetic hypothesis.

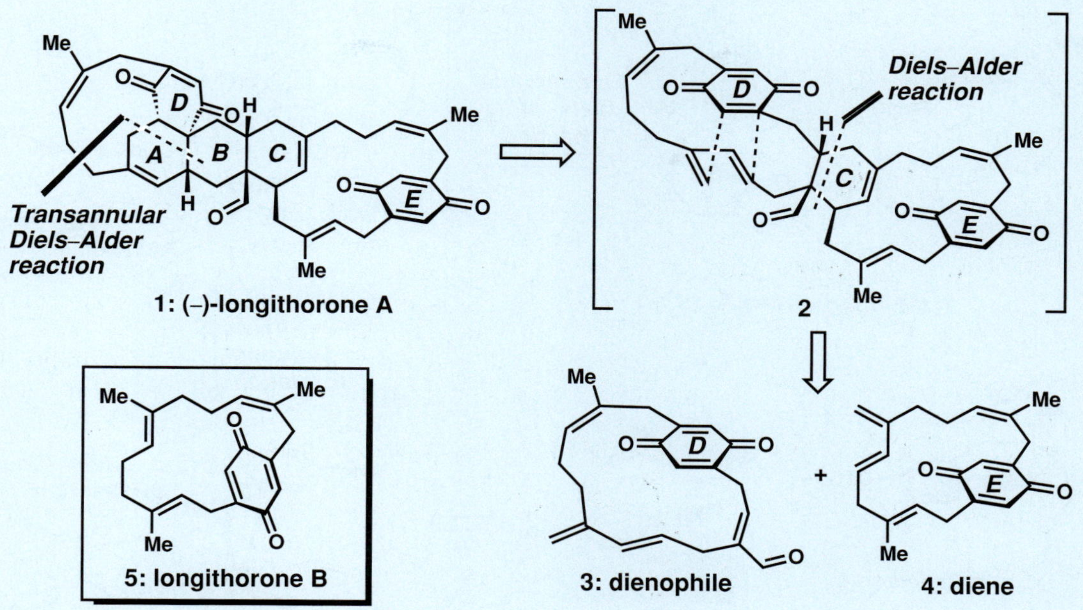

Scheme 1. Schmitz' biogenetic hypothesis for the formation of (–)-longithorone A (**1**).

16.2 Retrosynthetic Analysis and Strategy

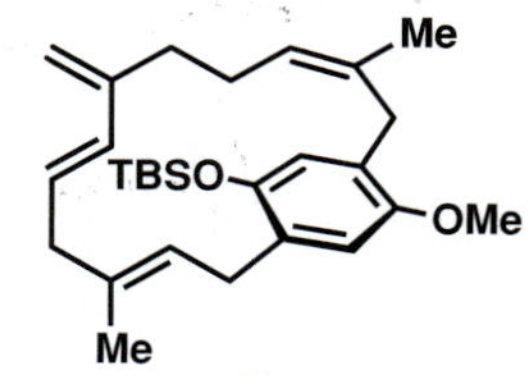

As one would expect, the initial stages of this retrosynthetic analysis mirror the presumed biogenesis of longithorone A (**1**, Scheme 2) with the exception of viewing the two critical Diels–Alder reactions as separate synthetic operations rather than as a series of events to be conducted in a single cascade. Accordingly, following what would hopefully be the smooth formation of **7** from [12]-paracyclophanes **8** and **9**, both aromatic systems in this [4+2]-product would then need to be oxidized to their quinone counterparts in order to induce the final transannular Diels–Alder reaction that would lead to longithorone A (**1**).

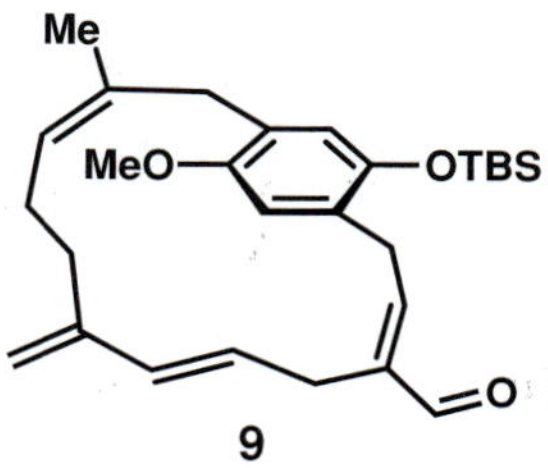

As with all biogenetically inspired synthetic strategies, however, the odds of achieving success with this plan in the laboratory are predicated less on the general feasibility of the hypothesis itself than on the ability to execute the same operations as Nature without the enzymes she employs to accomplish both substrate activation and stereochemical control. For example, in the proposed Diels–Alder union between **8** and **9**, one could be almost certain that successful cycloaddition would proceed with *endo* selectivity due to the aldehyde functionality within **9**. Less obvious was whether or not this event would occur with facial selectivity, as

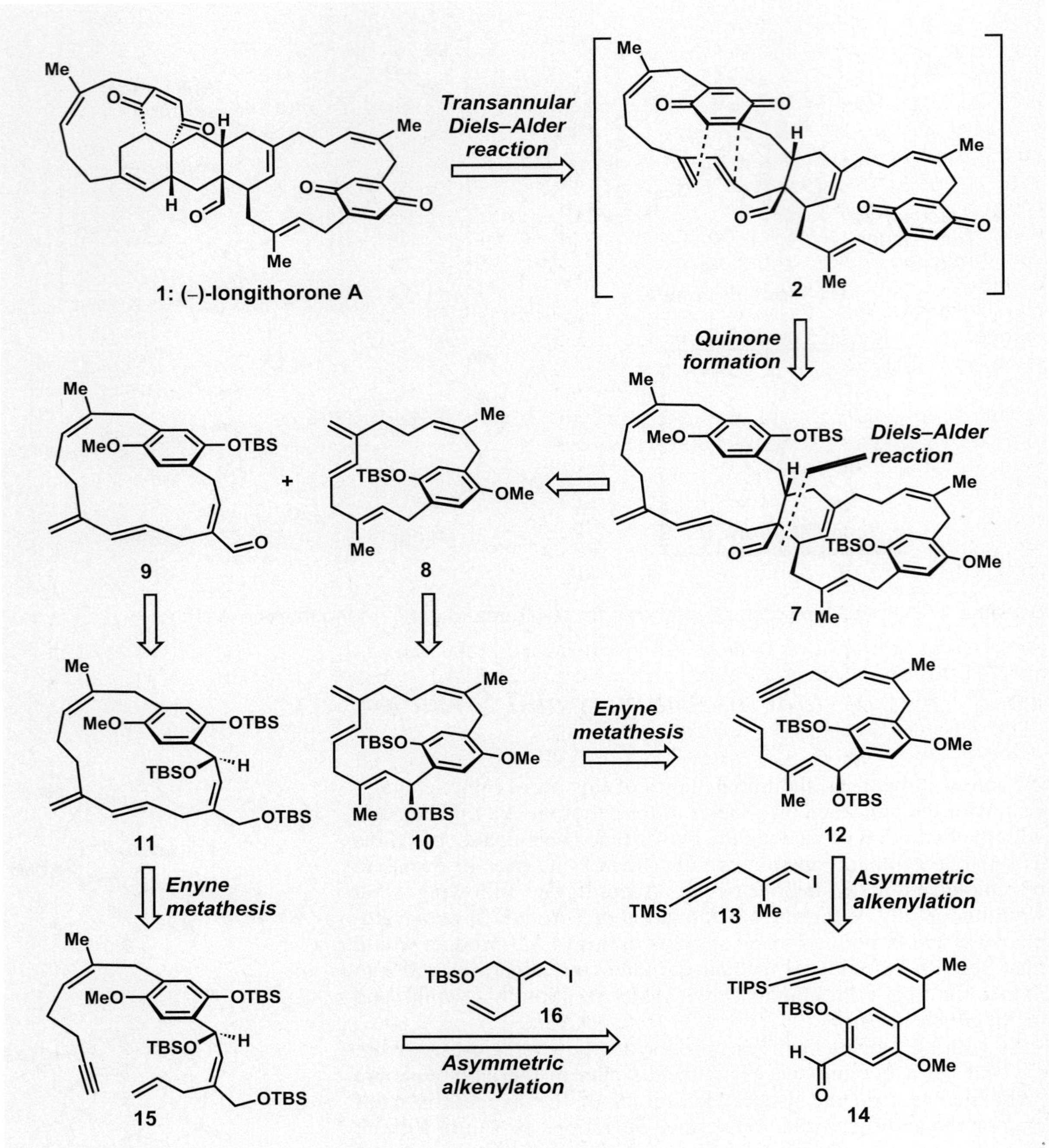

Scheme 2. Shair's retrosynthetic analysis of (–)-longithorone A (**1**): initial stages.

there are no extenuating stereochemical features in either **8** or **9** suggestive of the notion that their union would proceed in an enantioselective manner. Thus, the fact that only a single isomeric form of longithorone A (**1**) has been isolated from Nature (at least for the present) seems to implicate the existence of either a Diels–Alderase or some other enzyme that forces the two reacting [12]-cyclophanes into a specific conformation prior to cycloaddition, creating a homochiral environment.[7] If true, then the invocation of enzymatic encapsulation might also give the impression that this Diels–Alder reaction would not likely proceed spontaneously under ambient conditions, but would require activation to initiate. While the first of these general concerns could only be probed empirically, for the second the Shair group could enlist either thermal activation or one of a wide variety of Lewis acid catalysts to achieve the same general effect.

Leaving both these issues to be negotiated during the actual synthesis, longithorone A (**1**) has thus been reduced to the assembly of **8** and **9**, two structurally related 16-membered cyclophanes. Since both these subtargets exist in a specific atropisomeric configuration, effective synthetic strategies to access them must consider not only how to form their macrocycles, but also how to do so atropselectively. While the problem is simply stated, achieving both of these objectives in a single operation is far from pedestrian as demonstrated in the syntheses of the vancomycin aglycon presented in Chapter 9. As we shall see, though, both the Evans and Nicolaou groups left behind several clues during their extensive studies towards this target which ultimately inspired the design that was adopted in the case at hand to attempt the synthesis of **8** and **9** as single atropisomers.

Before this insightful strategy reached maturity, the Shair group first had to decide upon a method to effect a macrocyclic ring closure for both targeted intermediates. While **8** and **9** superficially possess the same ring size, the more important feature of their structural homology is an olefin substitution pattern in the form of a 1,3-disubstituted 1,3-diene, since it was this motif that led to the recognition that both macrocycles could potentially arise from ring-closing enyne metathesis reactions.[8] The intriguing aspect of this analysis, however, is all precedents for this transformation since its discovery by Thomas J. Katz in 1985[9] indicated this arrangement of unsaturation could not be obtained from an intramolecular reaction.[10,11] As shown in Scheme 3, such events had always provided products with a 1,2-disubstituted 1,3-diene (**22**), a result that can be rationalized by the regioselective insertion of a catalyst such as the Grubbs' ruthenium alkylidene **18**[12] into the electron-rich alkyne to form the more substituted metal carbene (**20**), followed by its addition to the neighboring C–C double bond. This final addition is the critical feature for the ultimate stereochemistry of the product in that, although metal carbenes always prefer to add to double or triple bonds such that the more-substituted product results,[13] the short atom tether constrains the olefin in **20** from adding in this

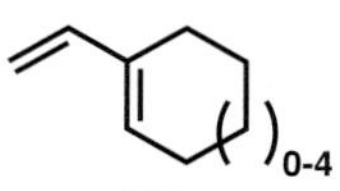

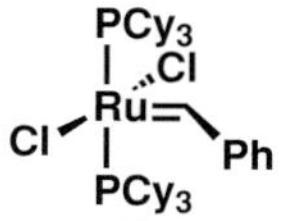

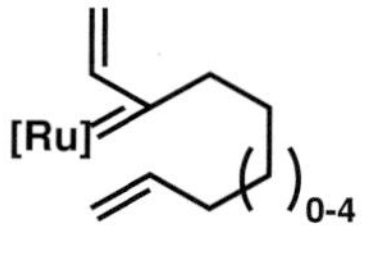

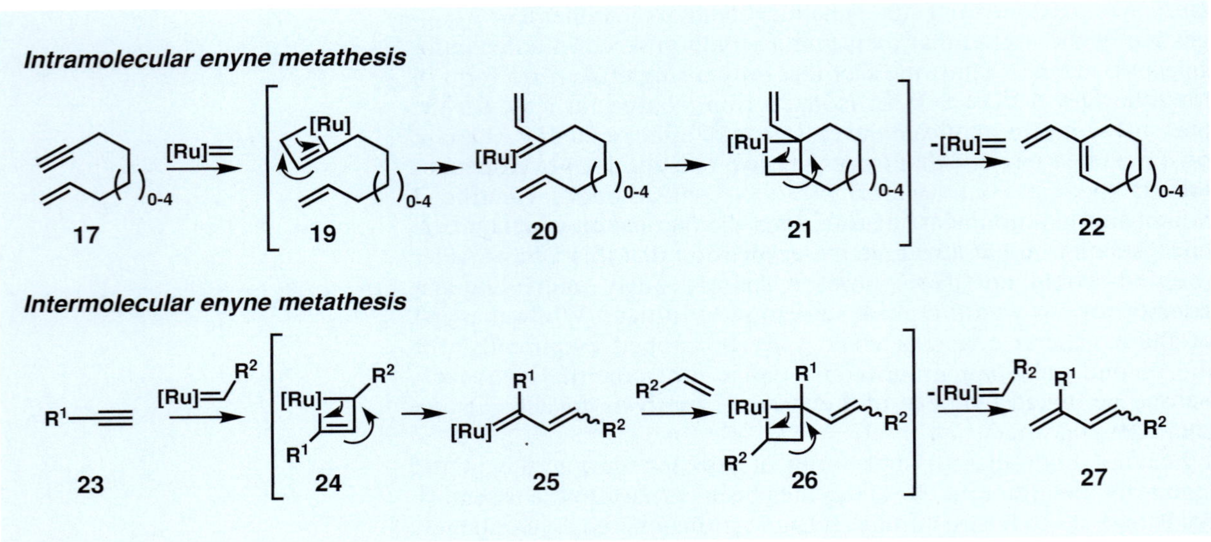

Intramolecular enyne metathesis

Intermolecular enyne metathesis

Scheme 3. Observed trend for diene-substitution patterns in intra- and intermolecular enyne metathesis reactions.

direction. As such, the less-substituted metallacyclobutane intermediate (**21**) results. Of course, when the olefin and alkyne reactants are not part of the same molecule, each addition proceeds with the expected formation of the more substituted carbene, thus leading to enyne metathesis products with a 1,3-disubstituted 1,3-diene system (as in **27**).*

While this picture would seem to indicate that intramolecular enyne metathesis of substrates such as **12** and **15** could not afford the desired olefinic substitution pattern needed for longithorone A (**1**), the Shair group took note of the fact that only five- to eight-membered rings had ever been constructed with the process,[10] systems in which the olefinic portion was highly constrained.**

* We should note that both mechanistic pictures shown in Scheme 3 serve merely to explain the observed set of results; evidence for their accuracy is far from extensive. For example, in intramolecular enyne metathesis the insertion of the alkylidene catalyst into the olefin of **17** rather than into its alkyne would afford the same product (**22**). The important feature is the regioselectivity of carbene addition, which is not in question.

** As an aside, intramolecular enyne metathesis constitutes one of the most powerful techniques to form challenging eight-membered rings (entities which can be found in natural products such as Taxol™, a molecule discussed in Chapter 34 of *Classics I*). One should note, however, that while related metathesis ring closures to access such products (such as ring-closing olefin metathesis or alkyne metathesis) are driven by the loss of an olefin-containing by-product, enyne metathesis cannot benefit from this contributing feature to the ΔS term of the reaction since the event is wholly atom-economic. Instead, the accomplishment of conjugation in the final product drives the reaction, and thus ensures that once either **22** or **27** has been formed the process is no longer reversible.

In a macrocyclic context, however, perhaps the greater mobility of the olefin could enable this pattern to be broken if it reacted with the intermediate carbene (i. e. **20**, Scheme 3) to form the more-substituted metallacyclobutane, thus leading to a 1,3-disubstituted product (i. e. a 16-membered ring) through an intermolecular-type enyne-metathesis mechanism. Thus, this argument affords a conceivable rationale for how this unprecedented stereochemical outcome, which was required in the case of longithorone A, might occur. Irrespective of its ultimate result, though, the proposed strategy would unquestionably pave new ground in the metathesis field as the first explorations in a macrocyclic context for this class of ring-closing reactions.

Assuming that these reactions would proceed in the desired sense, the lingering question was how to induce this enyne macrocyclization to occur atropselectively. The team's response was to append an extraneous silyl-protected benzylic alcohol to both cyclization precursors (**12** and **15**, Scheme 2) with the assumption that their presence could potentially govern which transition state would be adopted during the enyne-metathesis reaction. As shown in Scheme 4, if these benzylic groups existed as single enantiomers, they could then provide serious $A_{1,3}$ steric interactions for

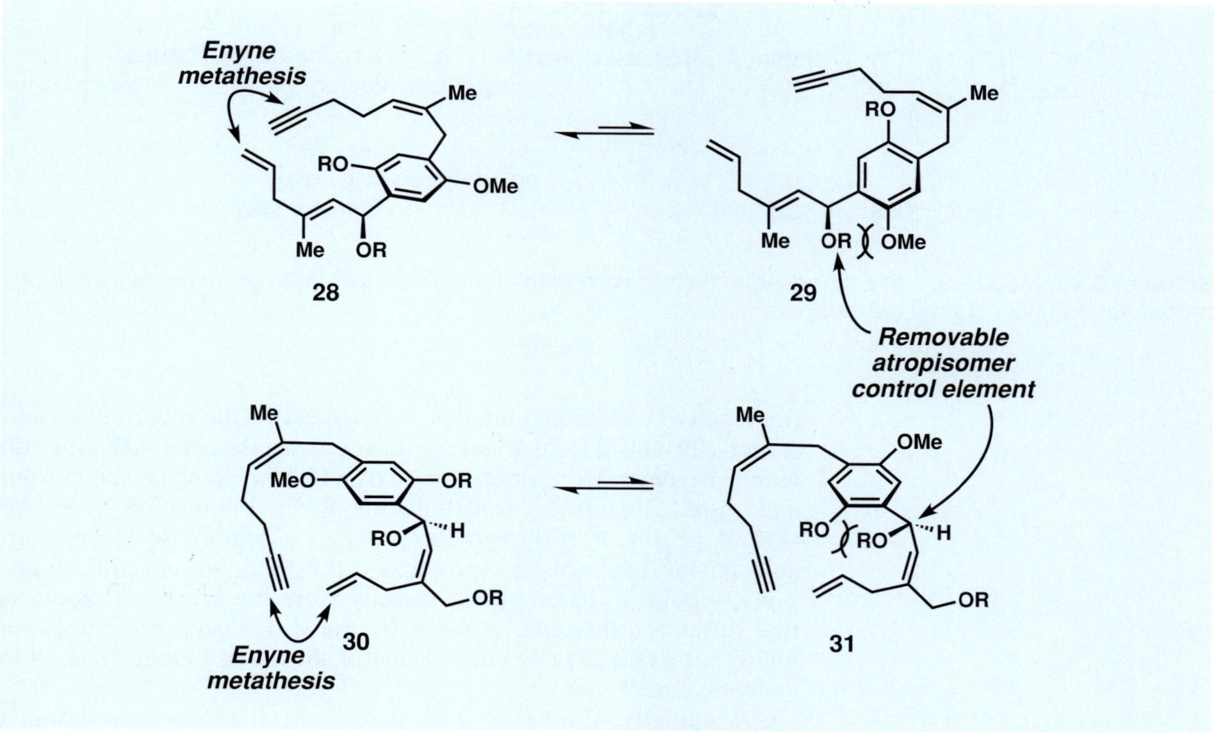

Scheme 4. Anticipated paracyclophane synthesis using enyne metathesis macrocyclizations based on a removable atropisomer control element.

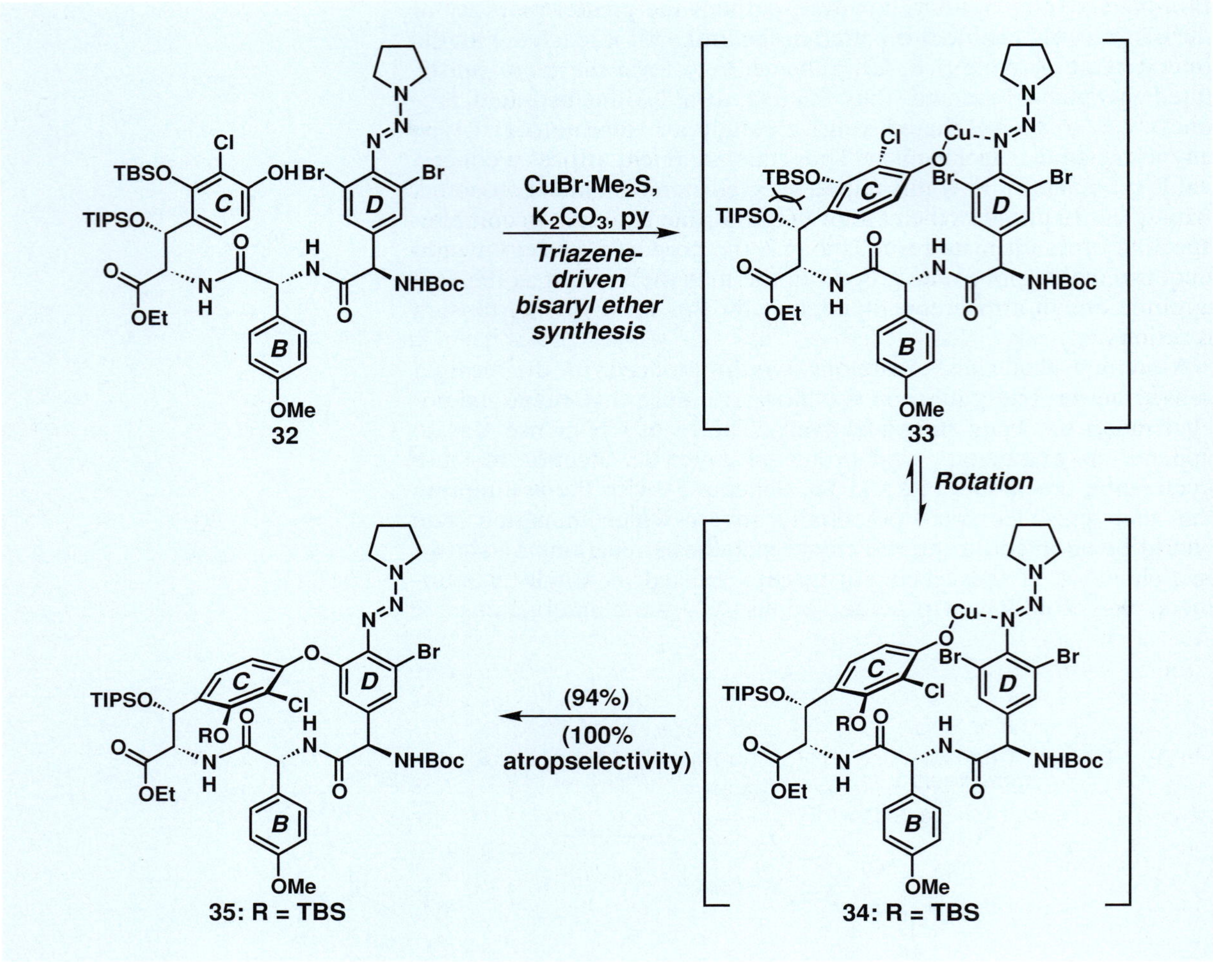

Scheme 5. Nicolaou's use of a disposable directing group to effect an atropselective macrocyclization of a model vancomycin bisaryl macrocycle.

the reactive conformations that would lead to the incorrect atropisomer (**29** and **31**), while their desired counterparts (**28** and **30**) would be devoid (or at least would suffer far less) of such strain and should, therefore, constitute the favored transition states by several kcal/mol. Although this strategy certainly seems reasonable on the basis of first principles, the Shair group could additionally point to its empirical validity from the successful application of such a directing element by the Evans group to achieve a highly atropselective closure of one of the bisaryl ether systems in vancomycin.[14a]

Conceptually similar to this precedent, the Nicolaou group's placement of a silyl ether directly on the C-ring of **32** (Scheme 5) in a model study towards the same antibiotic also accomplished a completely atropselective synthesis of a macrocyclic bisaryl ether (**35**) by virtue of the difference in the strain of the likely transition states (**33** and **34**).[14bc] Significantly, in the absence of this controlling group, the same closure was non-selective. Thus, based on the levels of success displayed by these examples, application of the same general approach for longithorone A certainly seemed promising, though a method would then be needed to excise these superfluous benzylic motifs following macrocyclization to complete the paracyclophane targets (**8** and **9**, Scheme 2). Since there are several means to execute such a requirement, whether leaving group formation followed by hydride displacement or simply acid-mediated lysis, it was forecasted that this objective would not prove especially challenging to execute during the synthesis.

With these strategic decisions taken, the remaining retrosynthetic disconnections from **12** and **15** (Scheme 2) to simple starting materials relied on a series of transformations with far greater precedent than those already discussed. First, since the upper domain of each of these pieces is completely homologous, it was anticipated that their lower portions possessing the essential benzylic directing groups could arise from an asymmetric alkenylation between aldehyde **14** and a nucleophile formed from either vinyl iodide **16** or **13**, governed by an external chiral auxiliary. With several techniques available to achieve this general transformation, following the minor adjustment of **14** to **36** (Scheme 6), it was then assumed that this essential starting material could be derived from a palladium-mediated Negishi-type cross-coupling between iodide **38** and an organozinc substrate derived from **37**.[15] On the whole, the complete strategy developed to synthesize longithorone A (**1**) not only displayed a great degree of convergence, but also a host of challenging operations whose success would unquestionably extend the frontiers of several important reactions into some of the most complex contexts yet explored.

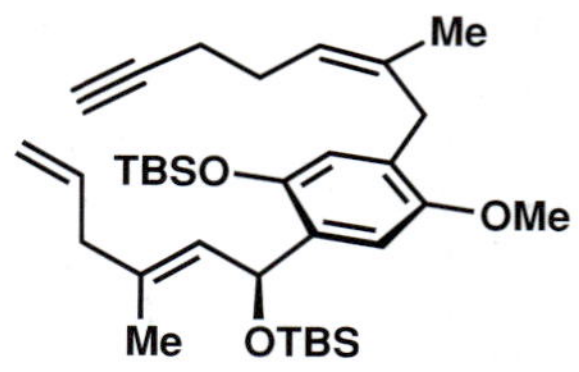

Me TBSO OMe Me OTBS **12**

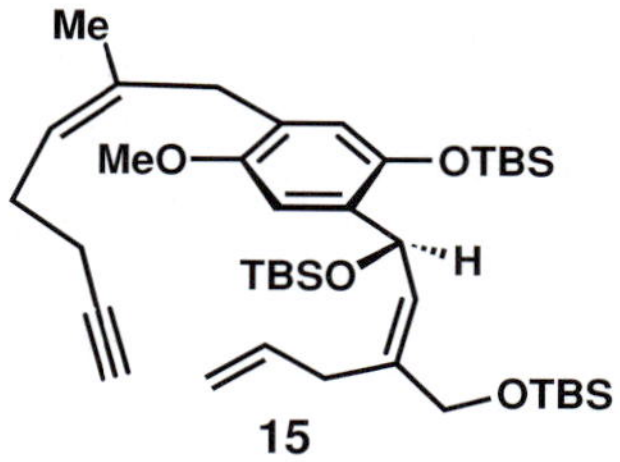

Me MeO OTBS TBSO H OTBS **15**

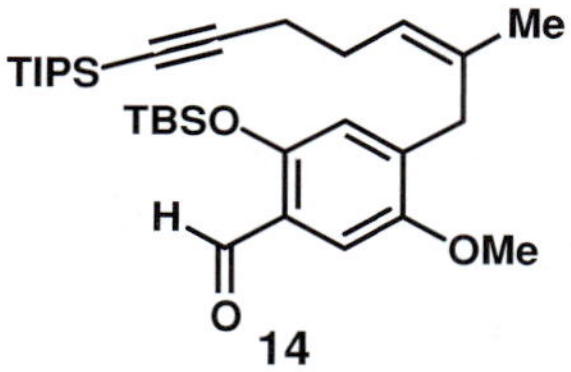

TIPS Me TBSO H O OMe **14**

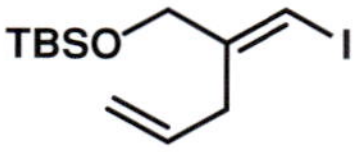

TBSO I **16**

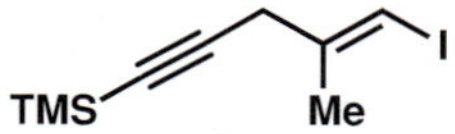

TMS I Me **13**

TIPS Me TBSO H O OMe **14** ⟹ TIPS Me MeO Br OMe **36** ⟹ *Pd-mediated cross-coupling* MeO Br Br OMe **37** TIPS Me I **38**

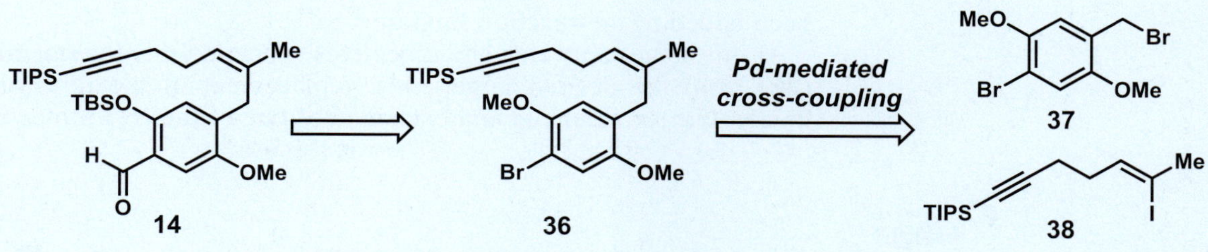

Scheme 6. Shair's retrosynthetic analysis of longithorone A (**1**): final stages.

16.3 Total Synthesis

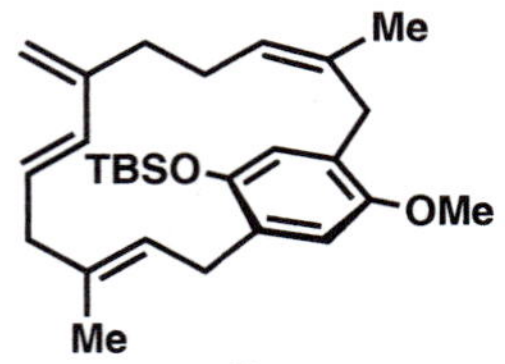

Exploration of the envisioned route towards longithorone A (**1**) began with attempts to construct the vinyl iodide **38** which would ultimately serve as the tethered alkyne in both enyne metathesis macrocyclization precursors. As shown in Scheme 7, this fragment was readily accessed in just a few operations from the commercially available 4-pentyn-1-ol (**39**). First, the terminal alkyne was "protected" as a TIPS acetylide (TIPSCl) following the initial generation of an alkynyl Grignard species by refluxing the starting material with EtMgCl in THF for 12 hours. Having effected this operation in 94 % yield, subsequent oxidation of the primary alcohol in **40** with Dess–Martin periodinane (CH_2Cl_2, 25 °C, 30 min) then set the stage for a *Z* selective Wittig homologation with the nonstabilized ylide derived from 2-iodoethyltriphenylphosphonium iodide, which completed the target (**38**).

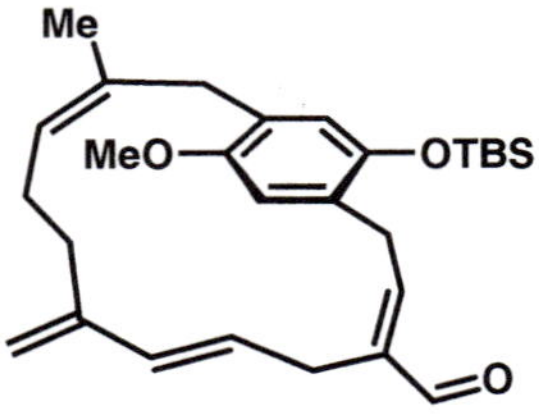

With this fragment in hand, efforts could now commence towards fashioning benzaldehyde **14**, the critical starting point for the divergent aspects of the synthetic sequence that would hopefully lead to cyclophanes **8** and **9**. The material needed to begin these studies was benzylic bromide **37**, which was ultimately accessed from 1,4-dibromo-2,5-dimethoxybenzene (**41**) in just three steps as shown in Scheme 8: 1) replacement of one of its halides with an aldehyde through a sequence involving aryl lithiation followed by a quench with DMF (a well-known source of formyl groups), 2) reduction with $NaBH_4$, and 3) exchange of the resultant benzylic alcohol for a bromine residue using PBr_3. These operations proceeded without incident in 77 % overall yield, permitting experiments to begin aimed at the merger of this building block (**37**) with the vinyl iodide **38** to reach intermediate **36**. Fortunately, this requirement was smoothly met in near quantitative yield through the initial conversion of **37** into benzylic zinc bromide **43** with freshly activated zinc[16] over the course of 30 minutes in THF at 0 °C, followed by a room temperature, palladium-mediated Negishi cross-coupling once both **38** and catalytic $Pd(PPh_3)_4$ had been added to the reaction mixture.[15]

At this stage, only two key objectives separated this intermediate (**36**) from the desired target (**14**): replacement of the remaining halide with an aldehyde and exchange of one of the aryl protecting

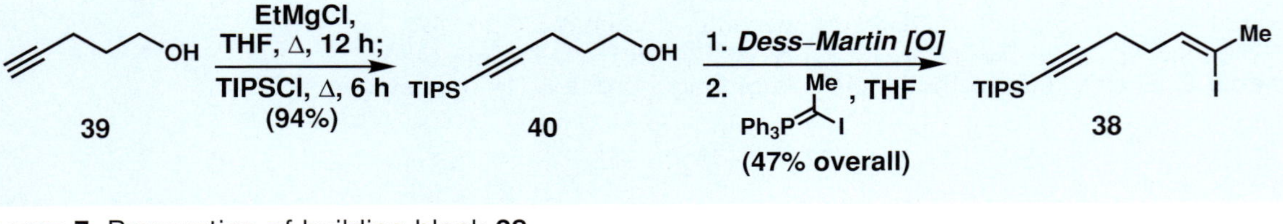

Me
TBSO
OMe
Me
8

Me
MeO
OTBS
O
9

39
EtMgCl, THF, Δ, 12 h; TIPSCl, Δ, 6 h (94%)
TIPS
OH
40
1. Dess–Martin [O]
2. Me, Ph$_3$P, I, THF
(47% overall)
TIPS
Me
I
38

Scheme 7. Preparation of building block **38**.

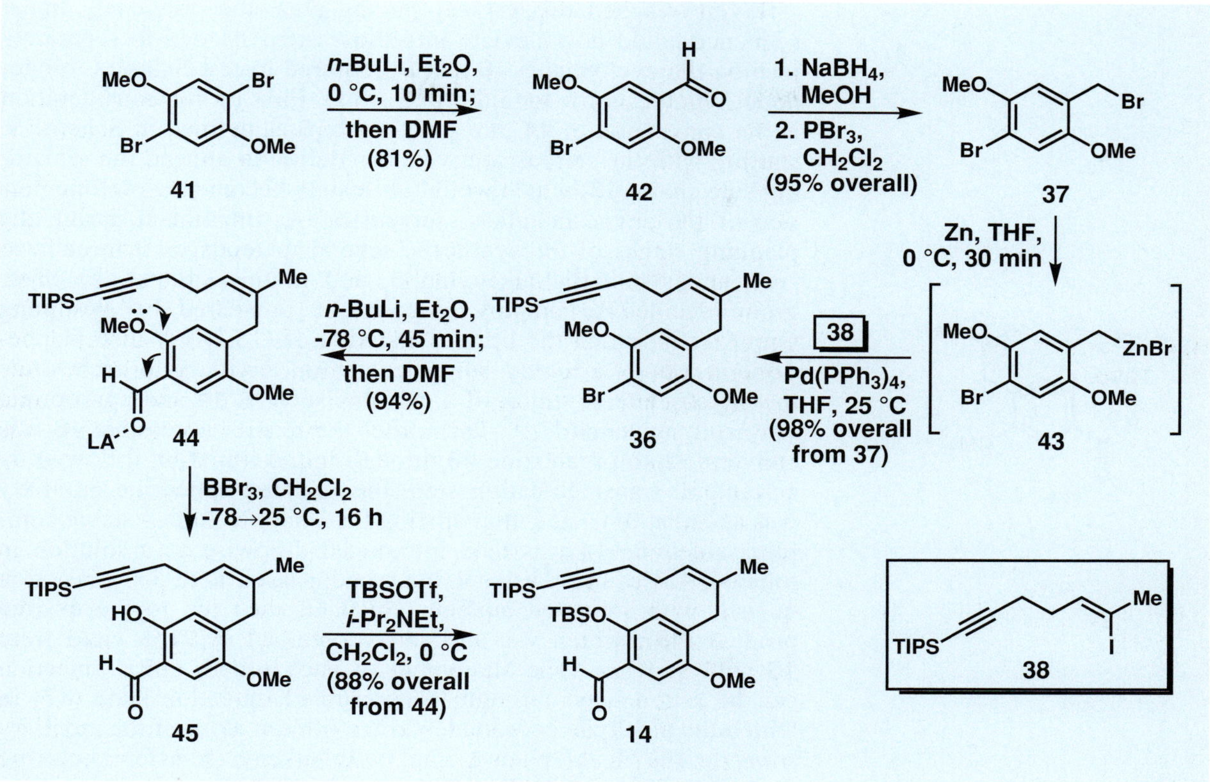

Scheme 8. Synthesis of advanced intermediate **14**.

groups for a bulkier one to hopefully control the anticipated enyne metathesis reactions in regard to their atropselectivity. The first of these tasks was handled using the same transformation that opened Scheme 8, namely aryl lithiation with *n*-BuLi in Et_2O at −78 °C followed by a DMF quench to afford an aldehyde (**44**), leaving only the need to differentiate the two aryl methoxy groups in this compound. Although achieving a chemoselective deprotection of only one of several such ethers is typically next to impossible, it was anticipated that this mission would be uniquely easy to accomplish on **44** since the aldehyde that had just been incorporated could coordinate with a Lewis acid (LA) catalyst, directing it towards the adjacent methyl ether, activating it for preferential cleavage.[17] Moreover, a strong electronic component should favor this reaction as well since only this methyl ether could delocalize into the adjacent aldehyde (as shown by the indicated arrows). Accordingly, upon exposure of **44** to a solution of BBr_3 in CH_2Cl_2, this operation was effected as anticipated, and a final reprotection of the newly unveiled phenol (**45**) using TBSOTf and Hünig's base in CH_2Cl_2 at 0 °C then completed the assembly of **14** in 88 % yield from **44**.

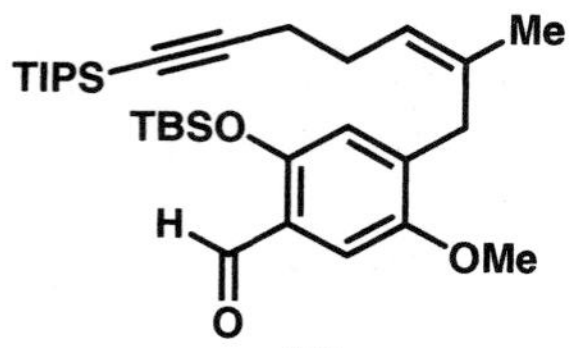

Having reached this critical staging area, the previously linear sequence could now deviate into those steps needed to separately fashion the cyclophanes (**8** and **9**) required to test Schmitz' biogenetic hypothesis for longithorone A (**1**). First up for consideration is the conversion of **14** into macrocycle **8** as defined in Scheme 9, starting with the stereoselective alkenylation to append the remaining side chain (**13**) which would ultimately become the olefinic portion of the enyne metathesis precursor. As mentioned during the planning stages of the synthesis, several methods exist to achieve such an asymmetric alkenylation, and in this context the Shair group decided to employ a technique pioneered by Wolfgang Oppolzer in which the lithium alkoxide of (1*S*,2*R*)-*N*-methylephedrine serves as a highly competent chiral auxiliary to orchestrate the stereoselective union of aldehydes such as **14** and appropriate vinylzinc nucleophiles.[18] Thus, after the readily accessible **13** was converted into organozinc **46** through initial lithiation followed by a standard transmetallation step, the lithiated ephedrine auxiliary was added at 0 °C, and after stirring for 1 hour to form a stable complex, aldehyde **14** was then introduced dropwise as a solution in toluene. Once an additional hour had passed at 0 °C, a reaction quench with saturated aqueous $NaHCO_3$ then led to the desired product (**48**), which was ultimately obtained in 91 % yield from **13** with 95 % *ee*. The attainment of such superb chiral selection can be rationalized through the presumed transition state (**47**) in which the aldehyde coordinates to the lithium atom of the auxiliary *trans* to the distal phenyl ring, with alkenyl transfer occurring through a six-membered transition state. Apart from such excellent stereoselectivity, equally important from a process standpoint was the recovery of the auxiliary unscathed at the end of the reaction through a simple extraction. Moreover, the reaction could be successfully driven to completion using equimolar amounts of both the bromozinc reagent and the aldehyde component, thereby ensuring optimum material economy.

With this key union effected, only a few operations separated **48** from the substrate needed to test enyne metathesis (i. e. **12**, Scheme 9). First, the controlled exposure of this compound (**48**) to 2 equivalents of TBAF in THF at 0 °C effected the lysis of both the phenolic silyl ether and the TMS group appended onto the terminal position of the alkyne, but, importantly, not the TIPS group on the other acetylene group. As such, in the next operation partial reduction with Lindlar's catalyst (Pd on $BaSO_4$ poisoned with quinoline) was accomplished selectively on only one alkyne to provide the needed terminal olefin. Finally, cleavage of the alkynyl TIPS moiety under more forcing conditions (TBAF, THF, 25 °C), followed by silylation of both the allylic hydroxy and the phenolic groups (TBSCl, imid, DMF), then completed the assembly of enyne metathesis precursor **12** in 62 % overall yield from **48**.

Pressing forward to test the operation at the heart of the entire proposed synthesis, the Shair group pleasantly discovered that the designed enyne macrocyclization reaction could be effected using

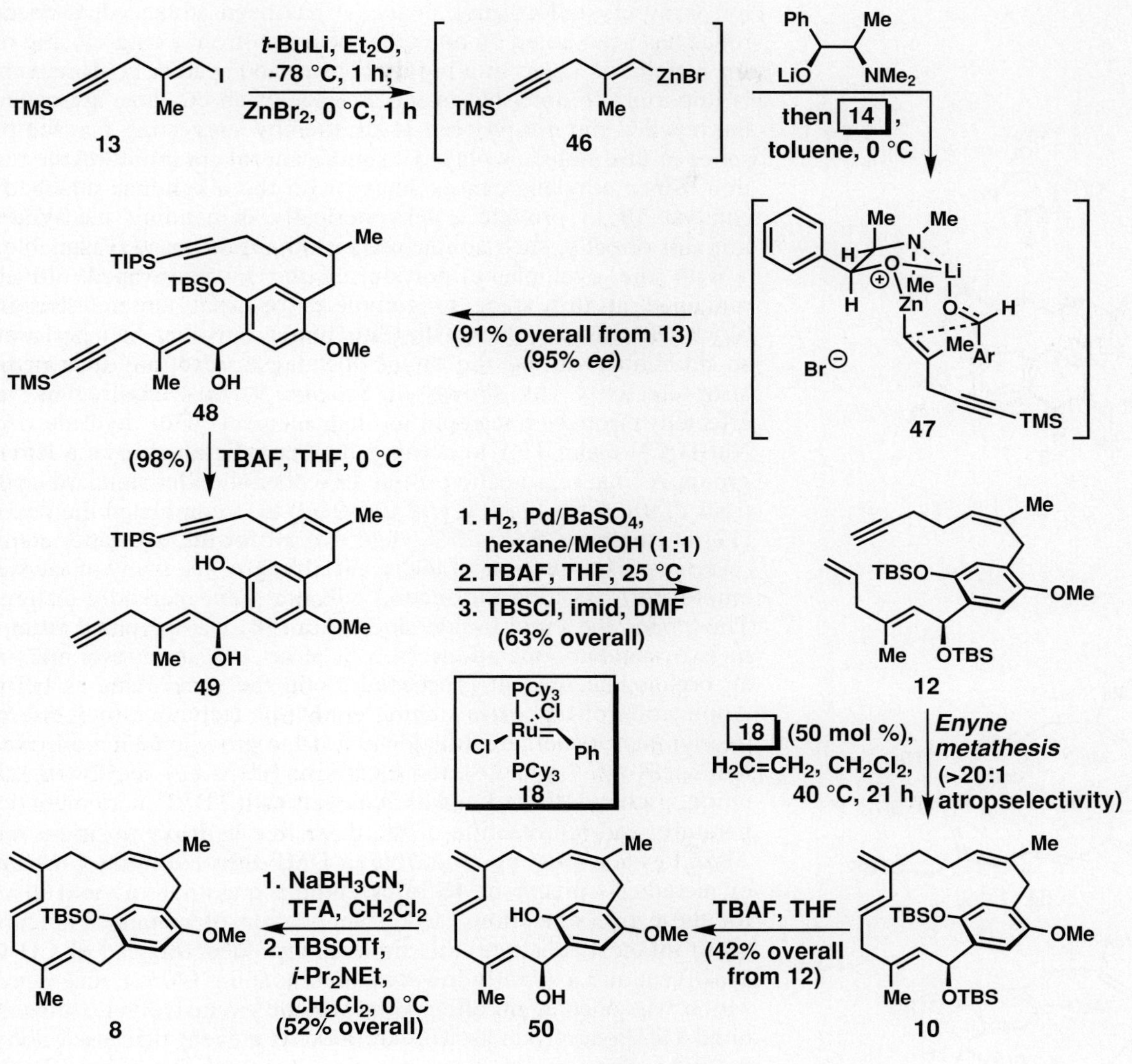

Scheme 9. Synthesis of advanced paracyclophane 8.

a 50 mol % loading of Grubbs' ruthenium catalyst **18** in CH_2Cl_2 at 40 °C under an ethylene atmosphere. Indeed, upon subsequent reaction of the crude products with TBAF in THF, cyclophane **50** was isolated in 42 % yield with complete selectivity in terms of both olefin geometry and atropisomerism. Although this material throughput may seem low, it reflects the formation of a major by-product (**51**, see column figure) in which one carbon atom was lost, presumably as part of a molecule of propene formed with the carbene of the

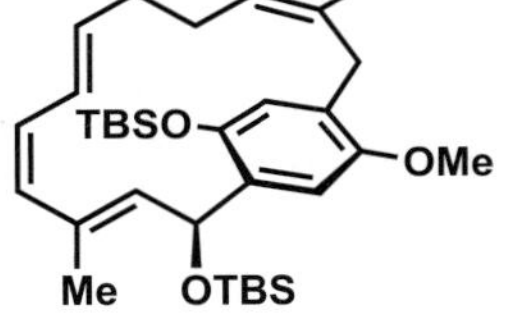

51

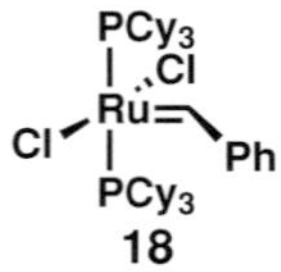

catalyst. While no mechanistic rationale for this result (confirmed by X-ray crystallographic analysis) has been advanced, its occurrence has been noted on one other occasion from a ring-closing olefin metathesis event in a highly constrained system.[19] Moreover, it is important to note that in the absence of an ethylene atmosphere the reaction did not proceed at all, thereby suggesting that the presence of this molecule played a significant role in initiating the reaction. Since ethylene can exchange with the alkylidene attached to catalyst **18** to provide a less sterically demanding methylidene initiator directly, such conjecture would appear to be reasonable.[10f]

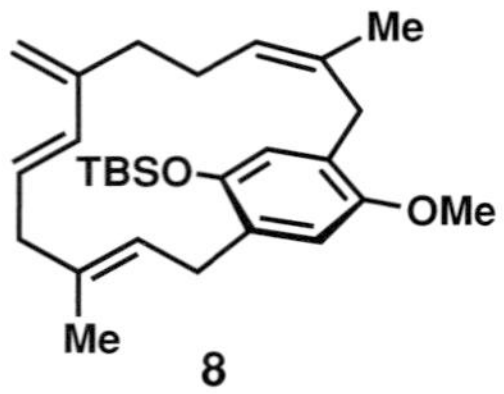

With the cyclophane portion of the target formed, all that remained at this stage to complete the final targeted building block (**8**) was to remove the directing group that had performed so admirably during the enyne metathesis reaction in providing atropselectivity. As shown in Scheme 9, this requirement was effected through nucleophilic displacement with hydride from $NaBH_3CN$ using TFA to activate the benzylic alcohol as a leaving group. A final reprotection of the free phenol under standard conditions (TBSOTf, *i*-Pr_2NEt, CH_2Cl_2, 0 °C) then completed the desired [12]-cyclophane (**8**) in 52 % yield overall for the two latter steps.

Having discussed one of the required fragments, many of the steps employed to fashion the second will not seem markedly different. Thus, once the vinyl iodide side chain (**16**) was prepared from **52** in two standard operations (see Scheme 10), its conversion into an organozinc reagent proceeded along the same lines as before. Application of Oppolzer's chiral ephedrine technique then effected its asymmetric merger with aldehyde **14** to provide **53** in both excellent yield (97 %) and stereoselectivity (90 % *ee*). Following this union, global desilylation was achieved with TBAF at ambient temperature, and reprotection of the three free hydroxy positions with TBSCl as activated by imidazole in DMF then led to the formation of metathesis precursor **15** in essentially quantitative yield (99 %) for these two operations. Upon application of metathesis catalyst **18** to this test substrate, this time using toluene instead of CH_2Cl_2 as solvent and a slightly lower catalyst loading (40 %), macrocyclization was once again effected to provide exclusively 1,3-disubstituted 1,3-diene products. Unlike the enyne event discussed above, however, this second enyne reaction was both less atropselective (2.8:1) and failed to completely control endocyclic olefin geometry (3.9:1 *E*/*Z*) as **54**, **55**, and **56** (see column figures) were obtained in addition to the desired product (**11**) in the yields indicated in the adjoining column. Since this event proceeded with less selectivity overall, this result would seem to suggest that there was less steric differentiation between the two possible atropisomeric transition states than in the macrocyclization leading to **10** (cf. Scheme 9). Although this outcome was probably slightly disappointing from the standpoint of yield, the overall atropselectivity accomplished was more than adequate since the required compound predominated. Moreover, the exclusive formation of 1,3-disubstituted 1,3-diene products in an additional test substrate lends further credence

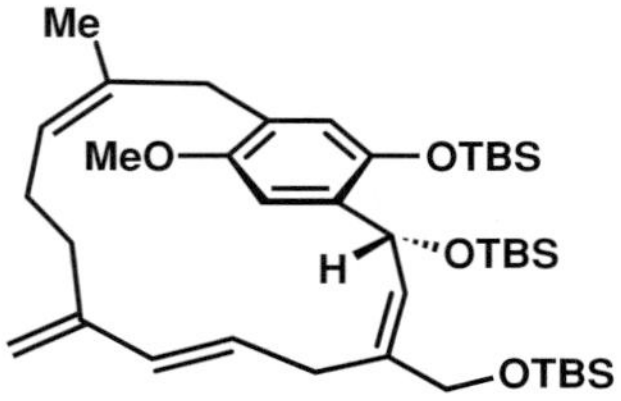

18

8

54 (11%)

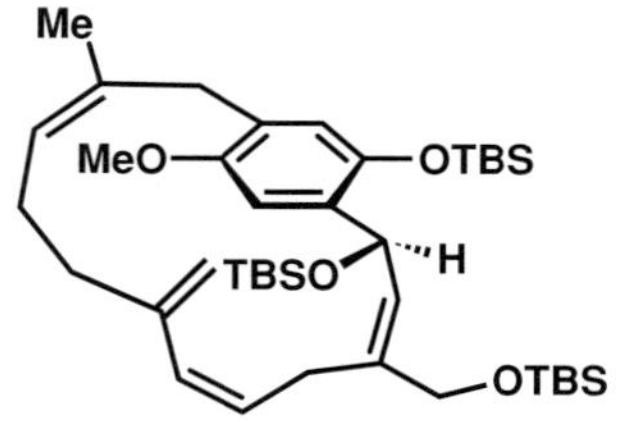

55 (8%)

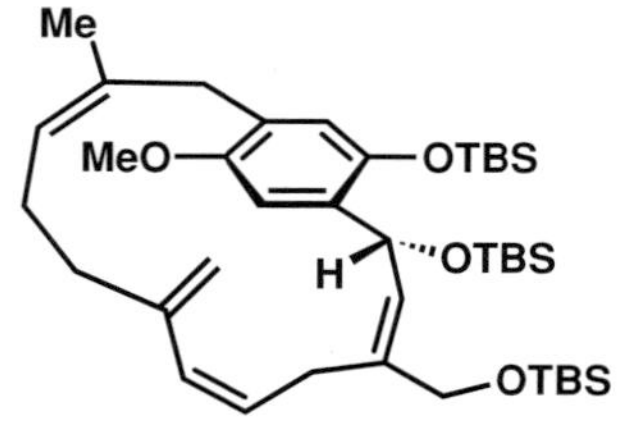

56 (1%)

to the possibility that this outcome might be general for the construction of other macrocycles through enyne metathesis.

From this key intermediate (**11**), the protected benzylic directing group was then removed chemoselectively through an ionic-type reduction using Et_3SiH and TFA at ambient temperature. Subsequent treatment with PPTS selectively excised the TBS group from the primary alcohol to provide **57** in 46 % overall yield. A final oxidation using Dess–Martin periodinane then smoothly completed the assembly of dienophile **9** in 99 % yield. Before moving to the final biomimetic stages of the synthesis, a few comments about both cyclophanes **8** and **9** and the routes employed to access them are in order. First, the selective nature of several of the silyl deprotections effected en route to both these fragments merits careful attention since they illustrate that similar protecting groups can be independently manipulated through an appreciation of the inherent topology and innate reactivity of the substrates. As such, the numer-

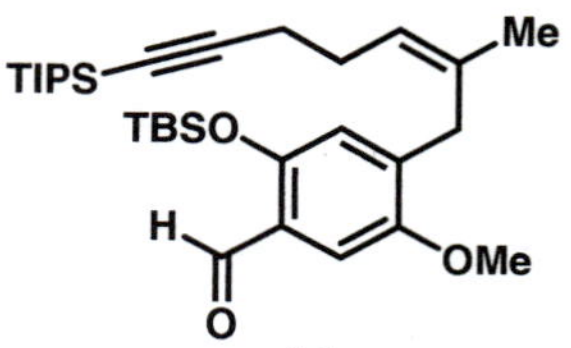

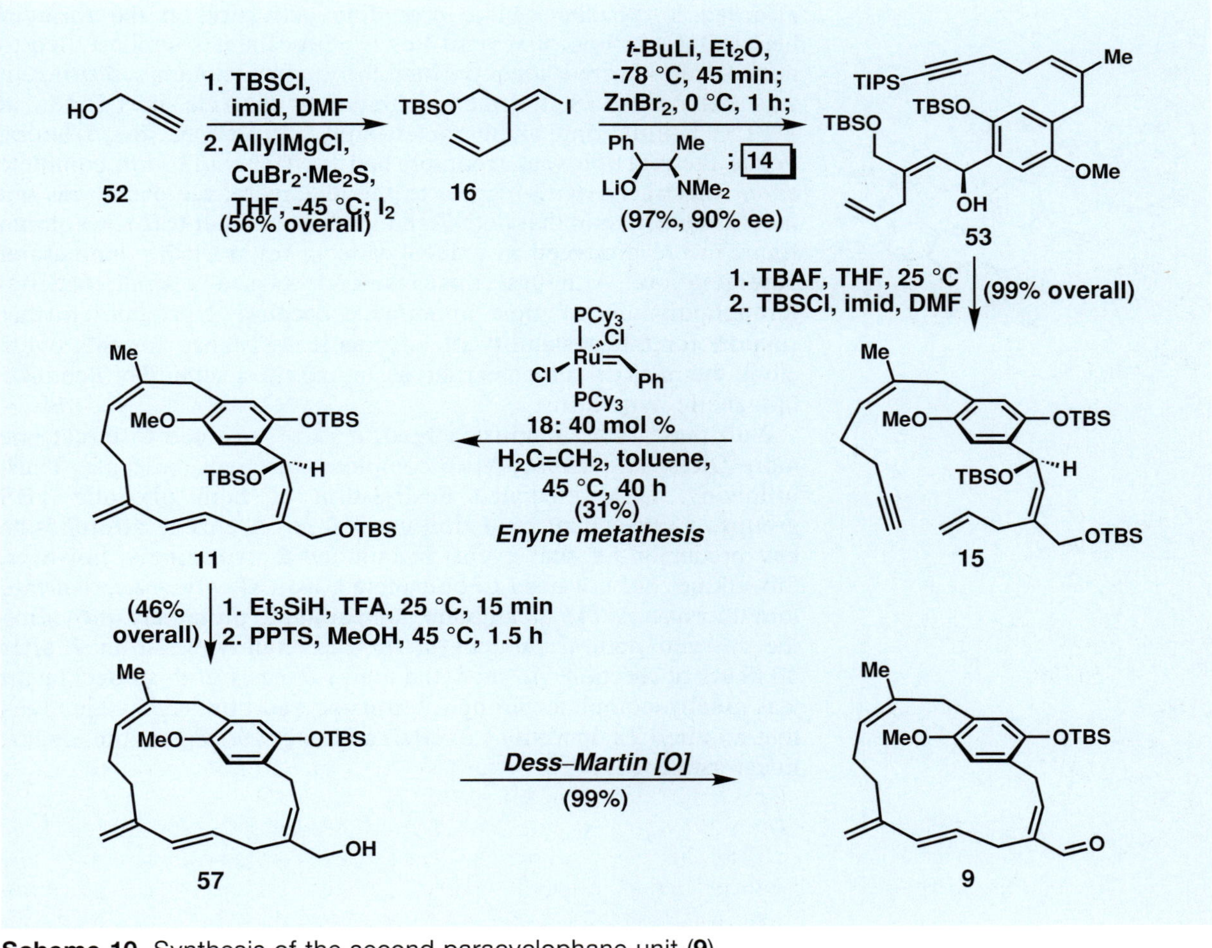

Scheme 10. Synthesis of the second paracyclophane unit (**9**).

ous operations needed to incorporate a fully orthogonal array of protecting groups may not always be needed for a projected total synthesis. Equally interesting from an experimental standpoint is the absence of any conjugated products in the operations that excised the benzylic directing groups, particularly under the strongly acidic conditions employed. Finally, one should note that both **8** and **9** demonstrated no ability to undergo thermal equilibration to their alternate atropisomeric forms at temperatures up to 100 °C. While this result is significant in its own right since it revealed the truly constrained nature of these substrates, it also afforded important reconnaissance for the next operation in the sequence, the biomimetic Diels–Alder reaction, by establishing an upper window for the levels of thermal activation which could potentially be employed to effect their union.

Unfortunately, despite the beauty and ease of the Diels–Alder reaction between **8** and **9** to form **7** on paper (see Scheme 11), this merger met only with failure under initial empirical investigation. For example, 15 hours of reaction at ambient temperature afforded no product, while providing activation in the form of heat (80 °C) or one of several Lewis acid catalysts similarly led to none of the desired adduct. Thankfully, after enough screening, it was determined that dimethylaluminum chloride in CH_2Cl_2 at −20 °C could bring about the desired outcome in just 5 hours. While the reaction was accomplished in 70 % yield with complete *endo* selectively (with respect to the aldehyde), the event was not facially selective in that both **7** and its diastereomer (**58**, see column figure) were produced in a 1:1.4 ratio in favor of the non-natural configuration. Although this result decreased overall material throughput, it is of great importance because it provides further support for the possibility of enzymatic assistance for this event along the lines discussed earlier, assuming the accuracy of Schmitz' biogenetic hypothesis.

With these two subunits merged, it now remained to effect one more Diels–Alder reaction to complete the target molecule. Thus, following TBAF-mediated desilylation of both phenolic TBS groups, a subsequent oxidation using iodosylbenzene afforded the key precursor for that event, bisquinone **2**. Amazingly, however, this adduct did not need to be isolated, as it slowly converted into longithorone A (**1**) at ambient temperature, ultimately providing the coveted natural product in 90 % overall yield from **7** after 40 hours of reaction. As such, the total synthesis of this novel target was finally complete through a concise and convergent approach that required 32 operations overall, and only 19 steps in the longest linear sequence.

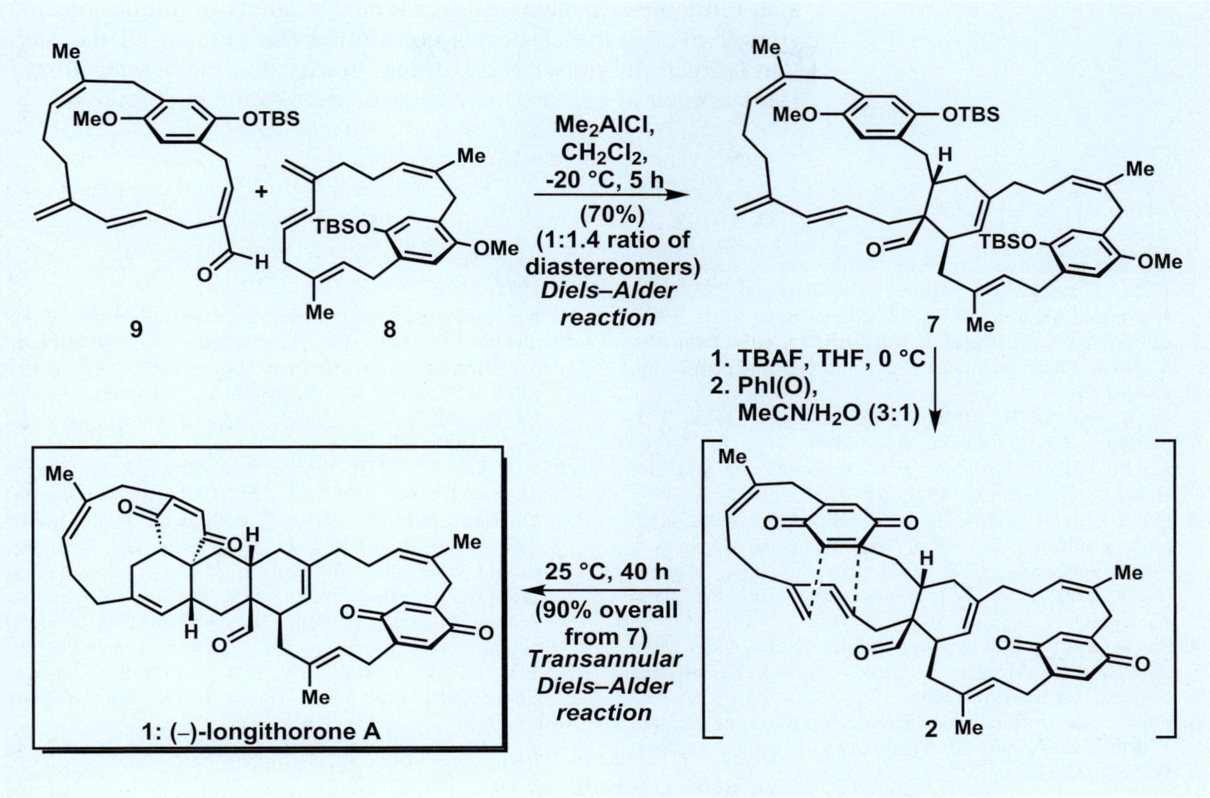

Scheme 11. Final stages and completion of the total synthesis of (–)-longithorone A (**1**).

16.4 Conclusion

Although the Shair group's total synthesis of longithorone A (**1**) is noteworthy for providing further verification of Schmitz' biogenetic hypothesis, its more lasting and significant accomplishment resides in the developed synthetic strategy which included the first example of the enyne metathesis reaction to form macrocycles. Not only did the resultant olefinic stereochemistry of these [12]-paracyclophane products go against conventional wisdom, thereby establishing new precedent for the reaction in complex molecule synthesis, but the insightful design of the test substrates enabled these macrocyclizations to proceed with admirable levels of atropselectivity. As such, this work has gained membership, and perhaps even a leadership position, in a very exclusive club of syntheses that have handled the issue of atropisomerism with efficacy. Beyond these fundamental contributions, the developed sequence also reinforces the value of organozinc reagents to build molecular complexity, as these species constituted the glue in all the steps that merged the prescribed building blocks into the advanced scaffolds needed to explore the critical aspects of the strategy.

References

1. For representative discussion of this topic, see: a) *The Total Synthesis of Natural Products, Vol 2* (Ed.: J. ApSimon), John Wiley & Sons, New York, **1973**, p. 754; b) K. G. Torssell, *Natural Product Chemistry*, Swedish Pharmaceutical Press, Stockholm, **1997**, pp. 251–312.
2. a) X. Fu, M. B. Hossain, D. van der Helm, F. J. Schmitz, *J. Am. Chem. Soc.* **1994**, *116*, 12125; b) X. Fu, M. B. Hossain, F. J. Schmitz, D. van der Helm, *J. Org. Chem.* **1997**, *62*, 3810.
3. For a general review on the Diels–Alder reaction in total synthesis, see: K. C. Nicolaou, S. A. Snyder, T. Montagnon, G. E. Vassilikogiannakis, *Angew. Chem.* **2002**, *114*, 1742; *Angew. Chem. Int. Ed.* **2002**, *41*, 1668.
4. For a recent review of transannular Diels–Alder reactions, see: E. Marsault, A. Toró, P. Nowak, P. Deslongchamps, *Tetrahedron* **2001**, *57*, 4243.
5. For a total synthesis of racemic longithorone B, see: T. Kato, K. Nagae, M. Hoshikawa, *Tetrahedron Lett.* **1999**, *40*, 1941.
6. M. E. Layton, C. A. Morales, M. D. Shair, *J. Am. Chem. Soc.* **2002**, *124*, 773.
7. For persuasive reports on the isolation of true Diels–Alderases, see: a) T. Ose, K. Watanabe, T. Mie, M. Honma, H. Watanabe, M. Yao, H. Oikawa, I. Tanaka, *Nature* **2003**, *422*, 185; b) K. Auclair, A. Sutherland, J. Kennedy, D. J. Witter, J. P. Van den Heever, C. R. Hutchinson, J. C. Vederas, *J. Am. Chem. Soc.* **2000**, *122*, 11519. For commentary on their existence, see: Ref. 3 and c) G. Pohnert, *ChemBioChem* **2001**, *2*, 873.
8. For a discussion of enyne metathesis, see: M. Mori in *Topics in Organometallic Chemistry, Vol. 1* (Ed.: A. Fürstner), Springer-Verlag, Berlin, **1998**, pp. 133–154.
9. T. J. Katz, T. M. Sivavec, *J. Am. Chem. Soc.* **1985**, *107*, 737.
10. For selected examples of intramolecular enyne metathesis, see: a) M. Mori, T. Kitamura, Y. Sato, *Synthesis* **2001**, 654; b) A. Rückert, D. Eisele, S. Blechert, *Tetrahedron Lett.* **2001**, *42*, 5245; c) M. Mori, T. Kitamura, N. Sakakibara, Y. Sato, *Org. Lett.* **2000**, *2*, 543; d) T. R. Hoye, S. M. Donaldson, T. J. Vos, *Org. Lett.* **1999**, *1*, 277; e) M. Mori, N. Sakakibara, A. Kinoshita, *J. Org. Chem.* **1998**, *63*, 6082; f) A. Kinoshita, N. Sakakibara, M. Mori, *J. Am. Chem. Soc.* **1997**, *119*, 12388; g) A. Kinoshita, M. Mori, *J. Org. Chem.* **1996**, *61*, 8356; h) A. Kinoshita, M. Mori, *Synlett* **1994**, 1020.
11. For selected examples of intermolecular enyne metathesis, see: a) R. Stragies, U. Voigtmann, S. Blechert, *Tetrahedron Lett.* **2000**, *41*, 5465; b) R. Stragies, M. Schuster, S. Blechert, *Angew. Chem.* **1997**, *109*, 2629; *Angew. Chem. Int. Ed. Engl.* **1997**, *36*, 2518.
12. a) P. Schwab, M. B. France, J. W. Ziller, R. H. Grubbs, *Angew. Chem.* **1995**, *107*, 2179; *Angew. Chem. Int. Ed. Engl.* **1995**, *34*, 2039; b) P. Schwab, R. H. Grubbs, J. W. Ziller, *J. Am. Chem. Soc.* **1996**, *118*, 100. One should note that although this class of ruthenium catalysts is often drawn with trigonal bipyramidal geometry, recent evidence has shown that a square planar orientation is the correct representation: T. M. Trnka, M. W. Day, R. H. Grubbs, *Angew. Chem.* **2001**, *113*, 3549; *Angew. Chem. Int. Ed.* **2001**, *40*, 3441.
13. a) J. McGinnis, T. J. Katz, S. Hurwitz, *J. Am. Chem. Soc.* **1976**, *98*, 605; b) T. J. Katz, J. McGinnis, C. Altus, *J. Am. Chem. Soc.* **1976**, *98*, 606; c) C. P. Casey, H. E. Tuinstra, M. C. Saeman, *J. Am. Chem. Soc.* **1976**, *98*, 608; d) W. D. Wulff, P.-C. Tang, J. S. McCallum, *J. Am. Chem. Soc.* **1981**, *103*, 7677.
14. a) D. A. Evans, C. J. Dinsmore, P. S. Watson, M. R. Wood, T. I. Richardson, B. W. Trotter, J. L. Katz, *Angew. Chem.* **1998**, *110*, 2868; *Angew. Chem. Int. Ed.* **1998**, *37*, 2704. b) K. C. Nicolaou, C. N. C. Boddy, S. Bräse, N. Winssinger, *Angew. Chem.* **1999**, *111*, 2230; *Angew. Chem. Int. Ed.* **1999**, *38*, 2096; c) K. C. Nicolaou, C. N. C. Boddy, *J. Am. Chem. Soc.* **2002**, *124*, 10451.
15. E. Negishi, H. Matsushita, N. Okukado, *Tetrahedron Lett.* **1981**, *29*, 2715.
16. C. Jubert, P. Knochel, *J. Org. Chem.* **1992**, *57*, 5425.
17. For examples of this deprotection in chemical synthesis, see: a) K. A. Parker, J. A. Petraitis, *Tetrahedron Lett.* **1981**, *22*, 397; b) H. Nagaoka, G. Schmid, H. Iio, Y. Kishi, *Tetrahedron Lett.* **1981**, *22*, 899. For mechanistic studies, see: c) Y. Kawamura, H. Takatsuki, F. Torii, T. Horie, *Bull. Chem. Soc. Jpn.* **1994**, *67*, 511.
18. W. Oppolzer, R. N. Radinov, *Tetrahedron Lett.* **1991**, *32*, 5777.
19. D. Joe, L. E. Overman, *Tetrahedron Lett.* **1997**, *38*, 8635.

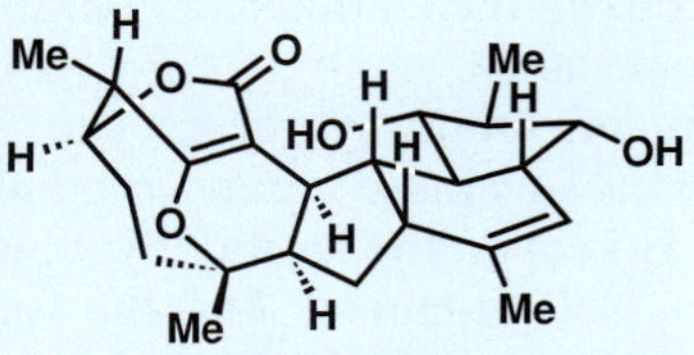

1: (–)-FR182877

17

E. J. Sorensen (2002)

(–)-FR182877

17.1 Introduction

There is little doubt that the Diels–Alder reaction is one of the most popular tools for the laboratory construction of complex molecules.[1] What is perhaps more interesting to ponder is if Nature is equally enamored with this powerful ring-forming reaction as are synthetic chemists.[2] This question has long been contemplated the chemical community, and present evidence does not permit it to be answered with complete certainty.

For instance, during the past few years several investigators have identified definitive examples of "Diels–Alderases," enzymes that catalyze and stereoselectively orchestrate the merger of biologically generated dienes and dienophiles.[3] These discoveries prove that Nature does indeed employ the Diels–Alder reaction in her syntheses of certain natural products. The lingering mystery, however, is just how frequently and with what specificity Nature uses this reaction, considering that only a minuscule number of these enzymes have been uncovered in the course of nearly a quarter century of searching.

Importantly, though, Nature does not have to rely on enzymatic assistance in order to utilize the full range of synthetic power afforded by the Diels–Alder reaction since she can also construct compounds whose unsaturated motifs are disposed to participate spontaneously in this pericyclic reaction on their own accord. For instance, in the previous chapter on longithorone A, we chronicled how a transannular Diels–Alder reaction was initiated under ambient conditions the instant that a quinone-based dienophile was unveiled in a macrocyclic ring bearing a neighboring diene system.[4] Similarly, as described in Chapter 17 of *Classics I*, heating a small

Key concepts:

- **The Diels–Alder reaction in Nature**
- **Transannular Diels–Alder reactions**
- **Molecular self-assembly**

collection of unsaturated acyclic starting materials was all that was required to induce cascades of pericyclic reactions (including one Diels–Alder reaction) which led directly to the complete and highly complex architectures of the endiandric acids.[5]

Neither of these impressive biomimetic total syntheses can, of course, definitively prove that Nature creates these natural products using similar Diels–Alder reactions. However, the facility by which their starting substrates were able to "self-organize" into the final target molecules in the laboratory certainly is a strong indication that Nature would attempt to accomplish similar levels of efficiency in her synthesis of these natural products. In fact, because of the dozens of examples of such Diels–Alder self-assemblies in the chemical literature, it would seem highly probable that Nature frequently employs the Diels–Alder reaction in her biosynthetic pathways, but in ways that are often subtle and difficult to detect since they do not require direct enzymatic participation.[6]

To help underscore this point, we have elected to begin this chapter by briefly delineating a few biomimetic total syntheses in which the participation of a [4+2] cycloaddition seems virtually inevitable for the formation of the observed natural product. We start with Kishi's elegant total synthesis of the marine-derived calcium-channel activator pinnatoxin A (**5**, Scheme 1) in which a late-stage Diels–Alder reaction was employed to generate its central 19-membered macrocycle concurrently with its cyclohexene ring (**2**→**4**).[7] Although the accomplishment of this cycloaddition is impressive, as Diels–Alder based macrocyclizations are often difficult to achieve, what is even more amazing is its stereoselectivity. Indeed, compound **4**, bearing the desired *exo* geometry of the natural product, was the major product (in 34 % yield) when any of the other seven possible intramolecular Diels–Alder adducts could have been generated preferentially. Thus, the conclusion that can be drawn from this fortuitous result is that the unique structural features of the starting material must have been predisposed to enable the reaction to occur with the desired selectivity, as no external catalyst was employed to orchestrate a specific reactive conformation. Or, stated more simply, this experiment provides strong circumstantial evidence of a Diels–Alder-based "self-construction."

The Nicolaou group's total synthesis of 1-*O*-methylforbesione (**9**, Scheme 2) in 2001 provides an equally impressive example of this general theme.[8] In this instance, following the synthesis of a tri-reverse-prenylated phenol (**7**), heating in DMF at 120 °C for only 20 minutes brought about a cascade sequence which directly provided **9** in 63 % yield, presumably through the indicated Claisen rearrangements and terminating Diels–Alder reaction. The dramatic and smooth increase in molecular complexity clearly makes this biomimetic route (which was first proposed by Quillinan and Scheinmann)[9] an attractive hypothesis for how this and several other related natural products may be formed in Nature.

It is important to recognize, however, that Diels–Alder reactions do not have to be involved in intramolecular contexts for them to

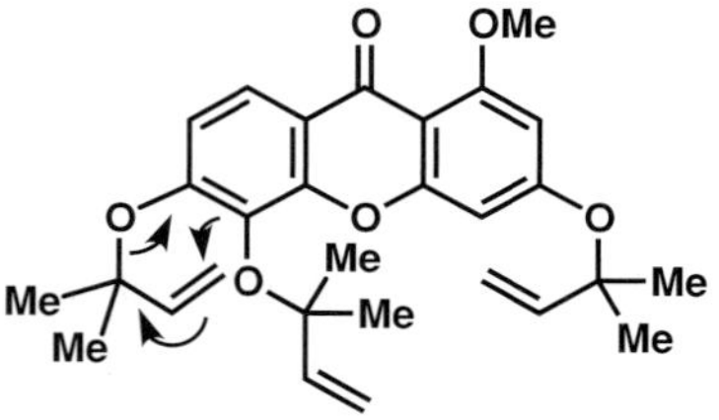

7

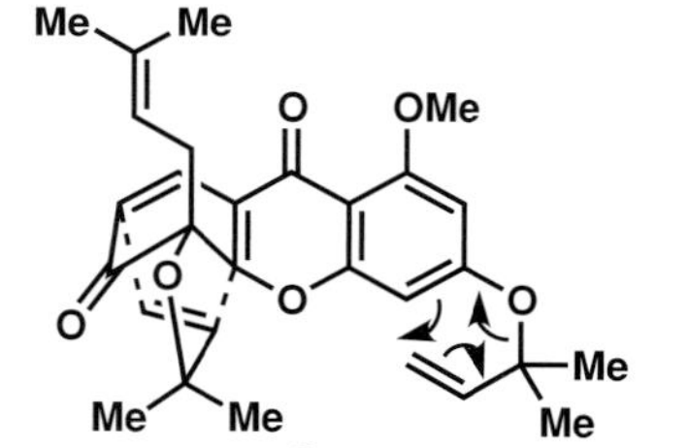

8

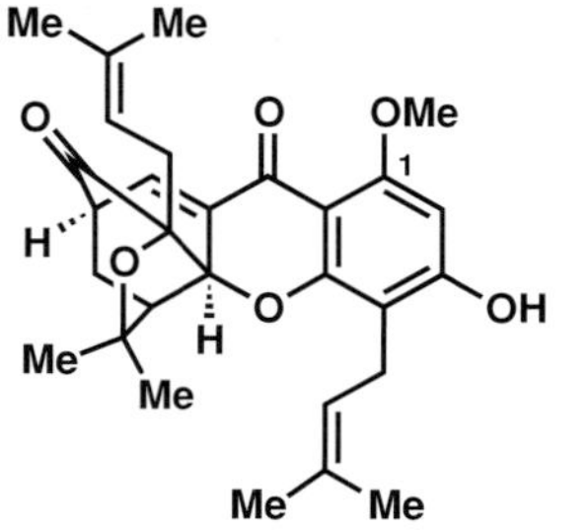

9: 1-*O*-methylforbesione

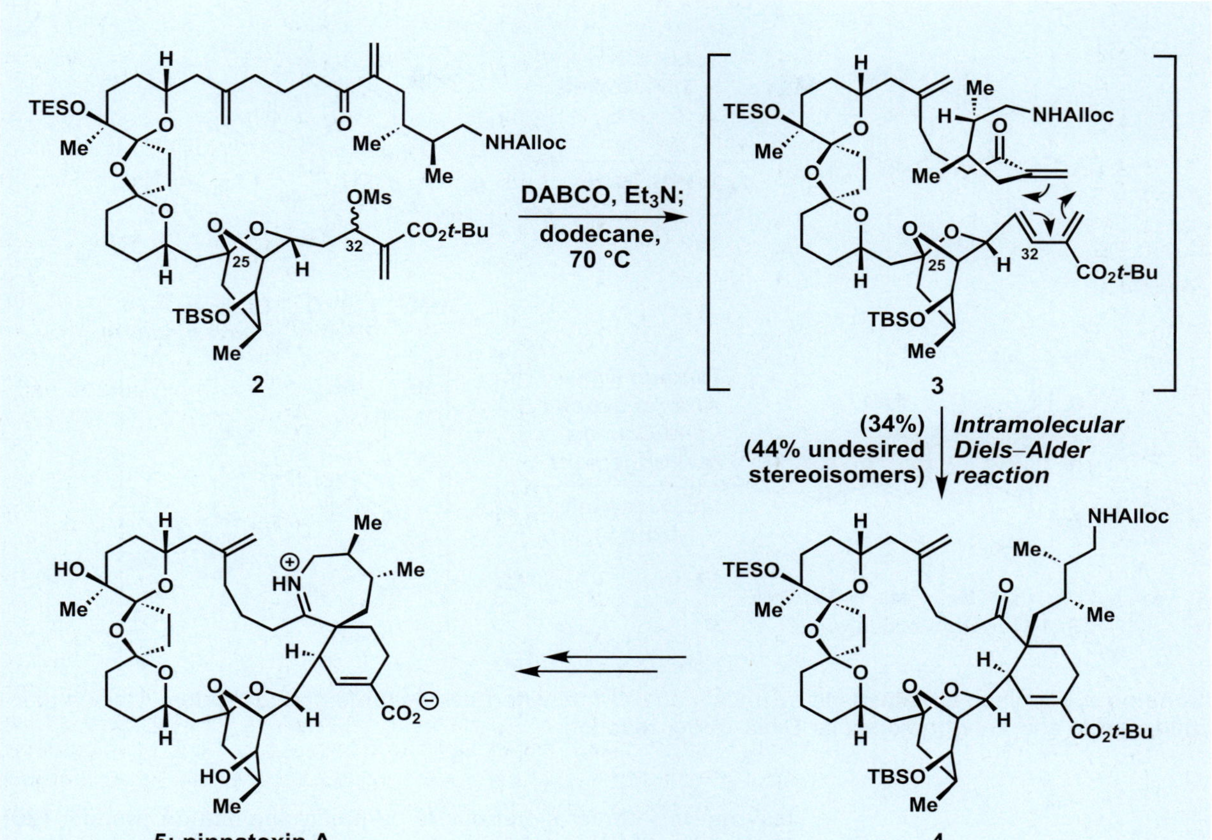

Scheme 1. Kishi's biomimetic synthesis of pinnatoxin A (**5**) based on an intramolecular Diels–Alder reaction.

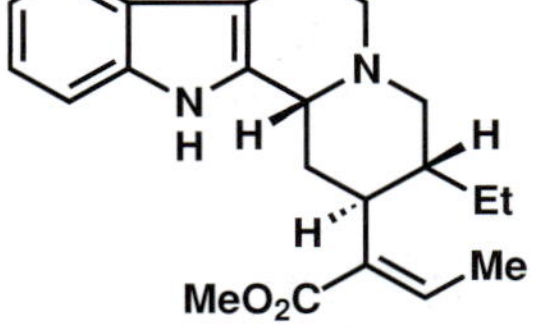

appear as likely candidates for a given natural product's biosynthesis. Intermolecular couplings are just as conceivable, and, to illustrate this alternate pathway, we have delineated an especially inventive biomimetic total synthesis of the alkaloid hirsutine (**20**) from the Tietze group in Scheme 3.[10] The entire sequence revolves around one key domino sequence of reactions induced by the sonication of a solution of tetrahydro-γ-carboline **10**, Meldrum's acid (**11**), enol ether **13** (1:1 mixture of *E*/*Z* isomers), and a few crystals of ethylenediamine diacetate (EDDA) in benzene for 12 hours. During this time, these conditions promoted an intermolecular Knoevenagel condensation to generate a highly reactive 1-oxa-1,3-butadiene (**12**), which immediately participated in a subsequent stereo- and regioselective hetero-Diels–Alder union with **13** (governed by the stereogenic center initially present in **10**) to afford **14**. *In situ* thermolysis of the unstable acetal motif in this cycloadduct then afforded dihydropyran **15** directly in 84 % yield,

20: hirsutine

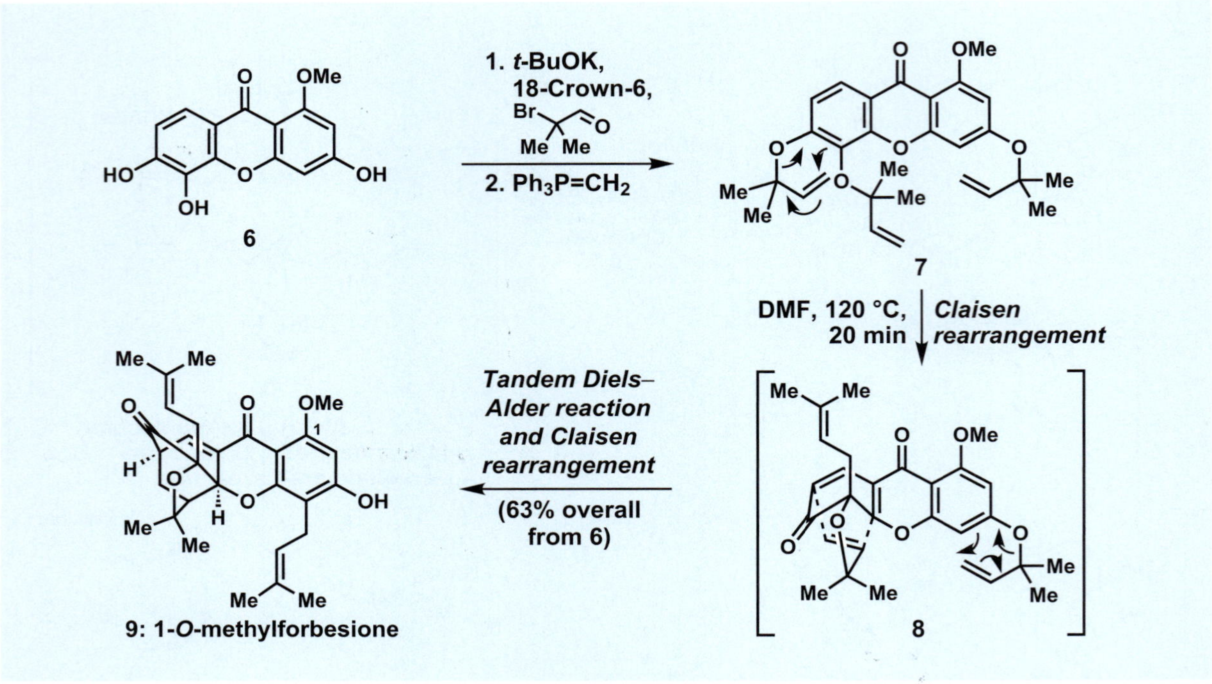

Scheme 2. Nicolaou's biomimetic synthesis of 1-*O*-methylforbesione (**9**) featuring tandem Claisen rearrangements and an intramolecular Diels–Alder reaction.

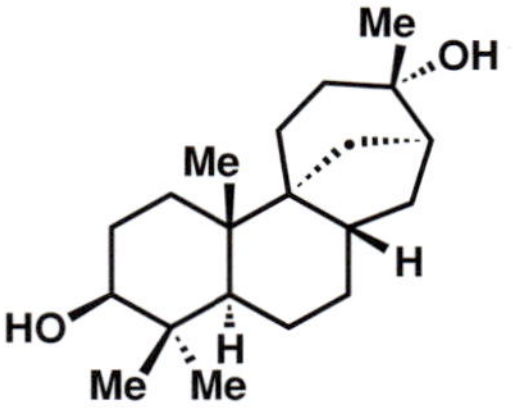

leaving only three operations to complete the natural product (**20**). This sequence certainly provides an expedient entry into diverse classes of alkaloid natural products,[11] and although it is quite challenging to verify if Nature similarly employs such a set of operations in her construction of these compounds, its synthetic efficiency is quite appealing.

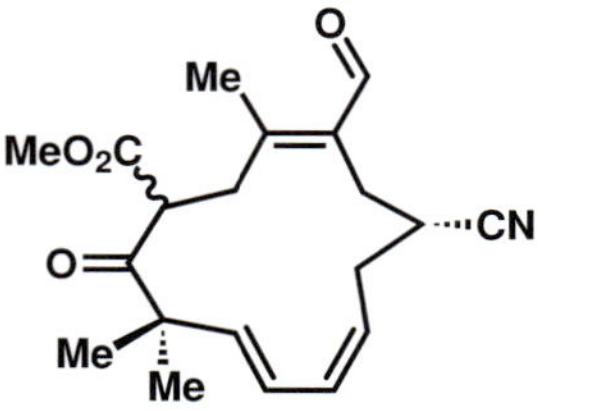

The one type of Diels–Alder reaction that is hard to dispute as one of Nature's tools is the transannular [4+2] cycloaddition.[12] We already mentioned one example of this ever-expanding class of pericyclic reactions above (longithorone A, Chapter 16), and you will find a second example in the synthesis of tri-*O*-methyldynemicin A methyl ester by Schreiber and co-workers discussed in Chapter 4. At this juncture, we thought that we would briefly highlight a recent example from one of the true pioneers of this type of cycloaddition process in target-oriented synthesis, namely Pierre Deslongchamps and his group at the University of Sherbrooke in Quebec, Canada. Their total synthesis of (+)-maritimol (**26**, Scheme 4) serves as a good vehicle through which to discuss the virtues of this likely biogenetic process.[13] As indicated, the key Diels–Alder step involved heating the 13-membered macrocyclic triene **21** in a solvent mixture of DMSO and H_2O at 155 °C, conditions that coerced the substrate to adopt the reactive conformation (**22**) required for a productive transannular

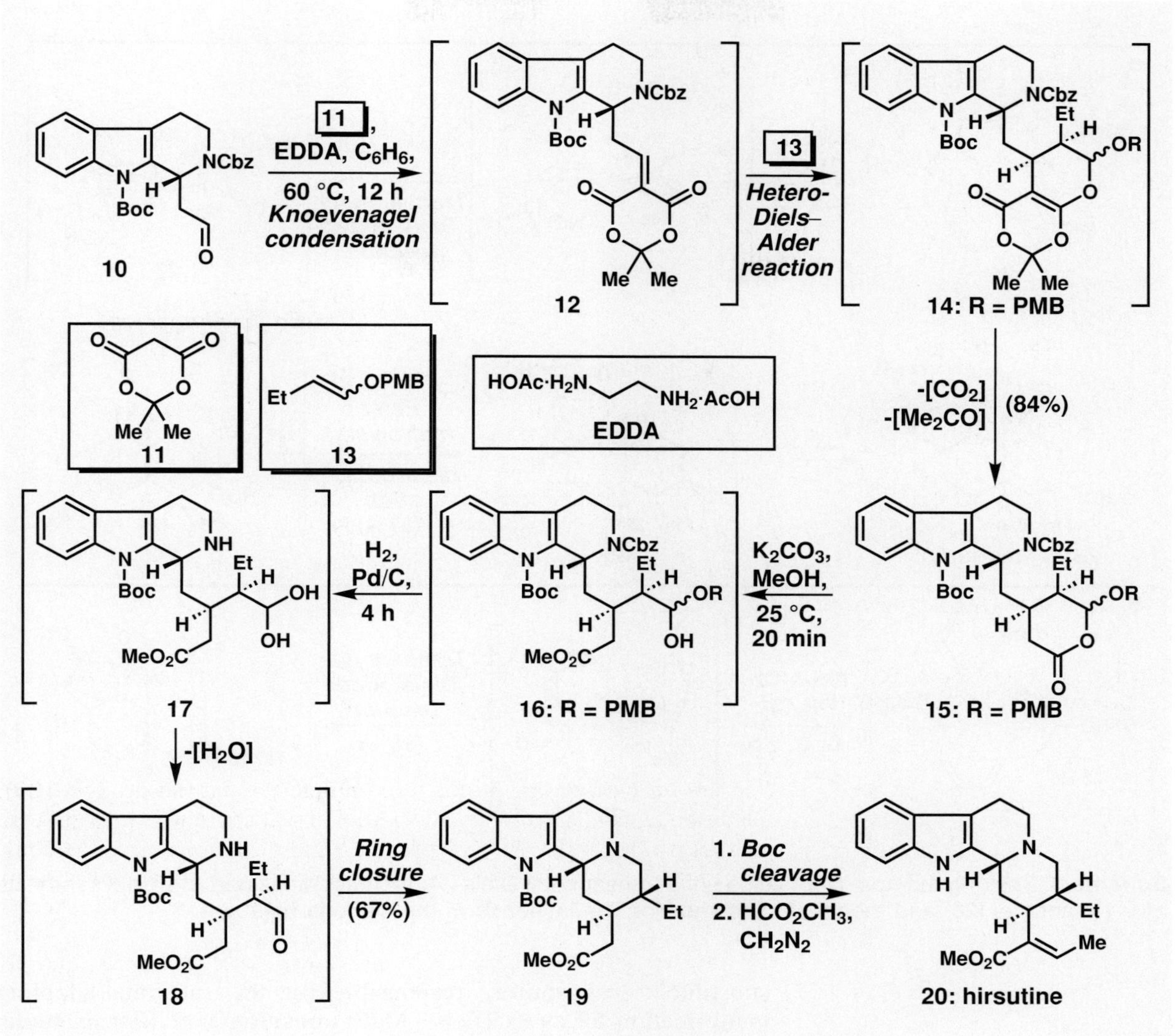

Scheme 3. Tietze's biomimetic synthesis of the alkaloid hirsutine (**20**) through a domino Knoevenagel/hetero-Diels–Alder sequence.

cycloaddition leading to **23**. Subsequent enolization, methyl ester hydrolysis, and decarboxylation, also promoted by these reaction conditions, then directly converted this intermediate into the isolated tricyclic product **25** in 87 % overall yield. Apart from its facility in enabling the completion of the target molecule (**26**) by concurrently installing all three of its rings and four of its stereocenters, the important lesson from this transannular Diels–Alder example is the way in which subtle stereochemical features of the substrate governed the stereoselectivity of the process. In fact, the stereochemical orientation of the relatively remote nitrile

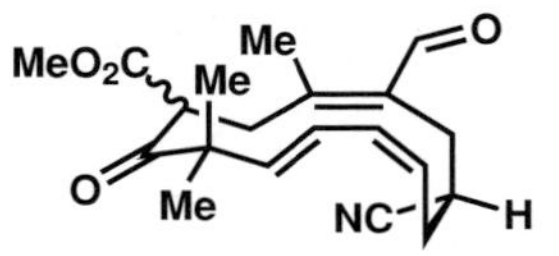

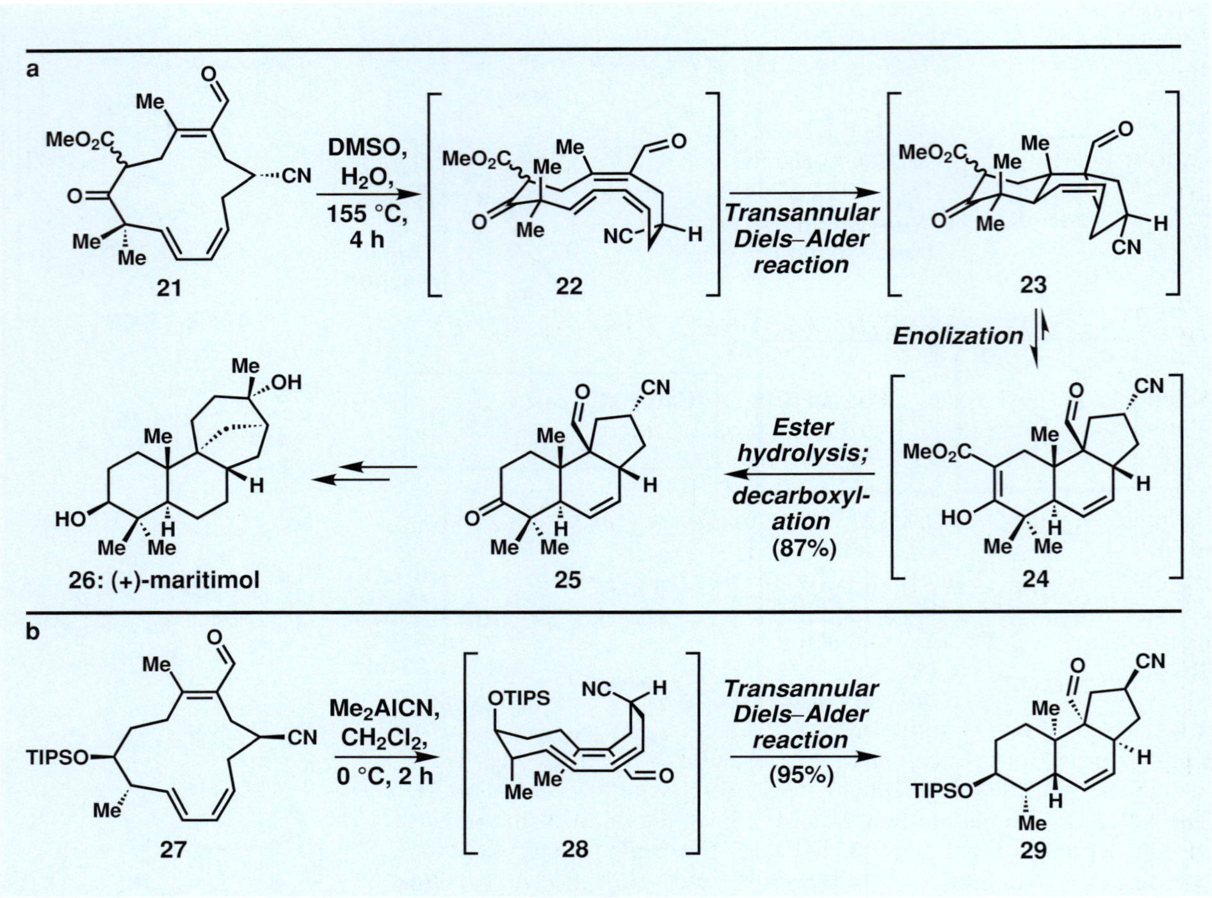

Scheme 4. Deslongchamps' creative use of a transannular Diels–Alder reaction as part of a total synthesis of (+)-maritimol (**26**) and model studies showing the influence of the nitrile substituent.

substituent was entirely responsible for the substrate adopting conformation **22** as its Diels–Alder transition state. For example, during early model studies these researchers discovered that the opposite stereochemistry at the cyanide-bearing carbon center led to structural isomer **29** in 95 % yield, presumably as a consequence of an entirely different transition state (**28**).* As such, this example provides a powerful demonstration of how Nature could similarly accomplish Diels–Alder stereocontrol not through enzymatic presentation, but through the appropriate construction of homochiral starting materials.

---

* Although the other structural alterations and reaction conditions in this model system would appear to prevent an "apples-to-apples" comparison with the example in part a of Scheme 4, several other studies beyond the scope of our presentation here verify that only the nitrile-bearing stereocenter is responsible for the stereochemical outcome of the transannular Diels–Alder reaction.

In the remainder of this chapter, we will present another total synthesis in which transannular Diels–Alder reactions would appear to play an essential role in Nature's construction of a natural product. That work is the 2001 total synthesis of (−)-FR182877 (**1**) in which the Sorensen group at The Scripps Research Institute executed not one, but two transannular Diels–Alder reactions in a single cascade to establish five rings and seven stereogenic centers from a 19-membered macrocyclic pentaene precursor.[14]

17.2 Retrosynthetic Analysis and Strategy

Scientists at the Fujisawa Pharmaceutical Company in Japan first obtained (−)-FR182877 (**1**) from the fermentation broth of *Streptomyces* sp. No9885 in 1998 as just one of several cytotoxic agents identified during a search for new cell-cycle inhibitors.[15] While the pharmaceutical potential of most of these compounds evaporated following subsequent batteries of *in vitro* and *in vivo* tests, (−)-FR182877 (**1**) persisted as a particularly strong clinical candidate since these examinations revealed that it could bind and stabilize microtubules with a level of efficacy commensurate with that of Taxol™ and the epothilones (see Chapter 7). Equally exciting, (−)-FR182877 (**1**) possessed an especially intricate, compact, and novel molecular structure comprised of a contiguous array of six rings adorned by twelve stereocenters and a strained anti-Bredt bridgehead olefin (see Chapter 13 for a discussion on such motifs). The latter feature has, in fact, been implicated as part of the potent biological activity of (−)-FR182877 since amines and alcohols can smoothly undergo conjugate addition to the double bond under physiological conditions, suggesting that it could be a site for covalent attachment to tubulin if that is its mode of cytotoxicity.

With little question, the conglomeration of all these features into a single molecule renders (−)-FR182877 (**1**) an enticing target for a research program in total synthesis. There was an additional attribute, however, that piqued the Sorensen group's interest, perhaps one of more importance than any of the others discussed above: could the complex polycyclic architecture of **1** be yet another example of a spontaneous molecular self-reorganization using the Diels–Alder reaction?

This idea is presented more formally in Scheme 5 with the thought that five of the six rings of the target molecule and seven of its stereocenters (as expressed in **34**) could arise from a polyunsaturated precursor (**30**) through a series of two cycloaddition reactions and a macrocyclization event. For instance, if **30** initially participated in a Type I intramolecular Diels–Alder reaction with *endo* selectivity to afford **32**, a subsequent intramolecular condensation of the Knoevenagel type could then provide **33**, a substrate poised for an intramolecular transannular hetero-Diels–Alder reaction between its electron-poor enone system and the proximal trisubstituted double

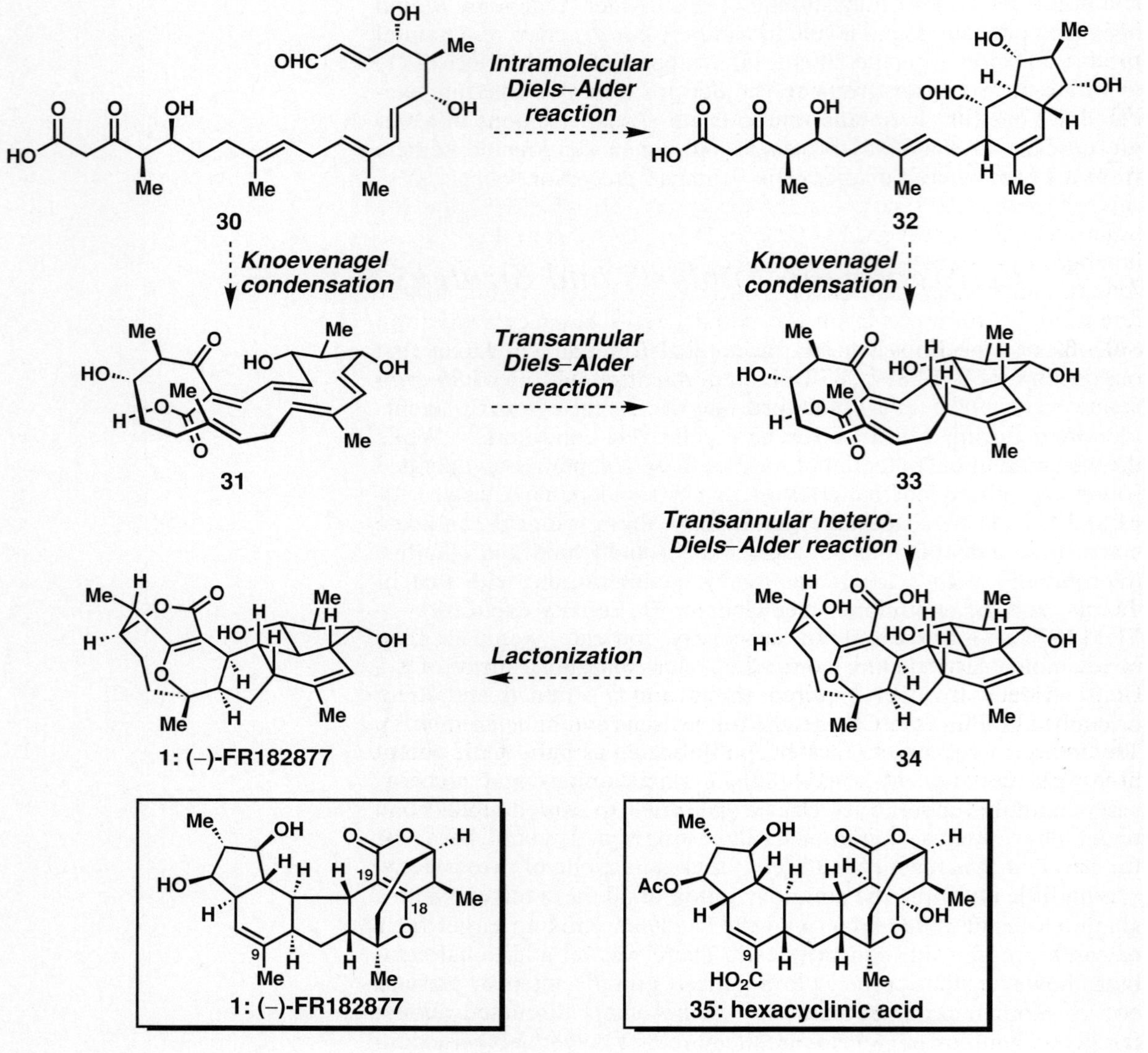

Scheme 5. Potential biogenetic sequence for the synthesis of (–)-FR182877 (**1**).

bond. Alternatively, if **30** first participated in a Knoevenagel condensation reaction to afford the 19-membered macrocyclic pentaene **31**, then a carbocycle-based transannular Diels–Alder reaction leading to **33** could be followed by the same transannular hetero-Diels–Alder reaction just discussed to provide **34**. In either scenario, a final lactonization would complete the target molecule (**1**).

Although the above hypothesis certainly provides an appealing rationalization for the biogenesis of (–)-FR182877 (**1**) on its own merits, further substantiation for the involvement of a Diels–Alder reaction derives from the structural features of hexacyclinic acid (**35**), a natural product obtained by Zeeck and co-workers in 2000 from a different strain of *Streptomyces* bacteria.[16] As indicated by a comparison of its structure to the redrawn version of (–)-FR182877 (**1**) in the neighboring box in Scheme 5, these two natural products have identical carbon skeletons, differing only with respect to the orientation of two stereocenters on their cyclohexene rings, the oxidation states of the carbon attached to C-9, and the functionalization at the C18–C19 junction. The first of these subtle incongruities is really the only important one as far as biogenetic schemes are concerned, since it could simply be the product of different modes of approach in the initial Diels–Alder cycloaddition (**30**→**32** or **31**→**33**) with *endo* attack affording an architecture corresponding to (–)-FR182877 (**1**) and *exo* approach providing the connectivities of hexacyclinic acid (**35**).

In order to convert this biogenetic scheme into a full blown retrosynthetic blueprint, the Sorensen group would have to determine which of the two alternatives in Scheme 5 to pursue in the laboratory. This choice was neither easy nor obvious, though literature precedent indicated that the first route discussed above (**30**→**32**→**33**→**34**) was probably more conservative, considering the success of other Knoevenagel/hetero-Diels–Alder cascades like the one in Tietze's synthesis of hirsutine presented earlier. In contrast, no precedent existed for the series of iterative transannular Diels–Alder reactions required to accomplish the other route. Accordingly, the Sorensen group began their synthetic studies by attempting to follow this "safer" pathway. Pleasingly, preliminary investigations soon revealed that compounds such as **32** could, indeed, be generated through a Type I Diels–Alder reaction of structures resembling **30**. All attempts to accomplish Knoevenagel-based macrocyclizations from these staging areas to provide compounds such as **33**, unfortunately, met with failure.[17] As a result, the Sorensen group ultimately elected to pursue the other pathway in Scheme 5 keeping with the double transannular Diels–Alder sequence. However, they based their retrosynthetic analysis for its precursor on an approach that projected macrocycle formation not through a Knoevenagel condensation,[18] but through a synthetic equivalent. Their final plan is presented in Scheme 6 in retrosynthetic format.

As indicated, these researchers expected that compound **37** might serve as a reasonable synthetic analog of the hypothetical acyclic polyunsaturated building block **30** discussed above. The expectation was that if initial macrocyclization could be accomplished through a Tsuji–Trost reaction (see Chapter 14) to establish the C1–C19 bond of **36**, then the subsequent introduction of unsaturation at the same site would properly activate the resultant compound to undergo the desired series of transannular Diels–Alder reactions. A final lactonization could then complete (–)-FR182877 (**1**). This sequence of operations was reasonable given the power of the Tsu-

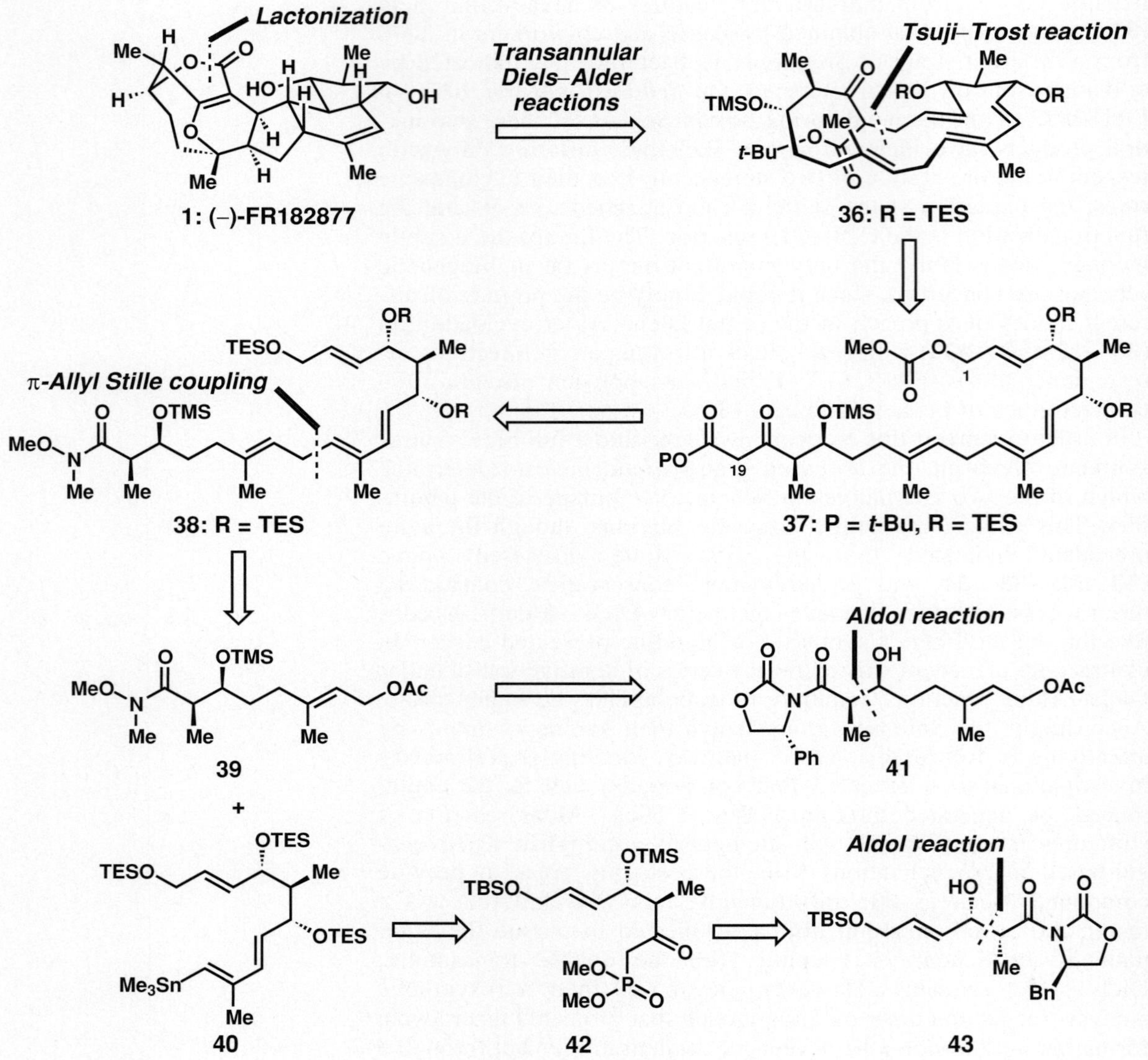

Scheme 6. Sorensen's retrosynthetic analysis of (–)-FR182877 (1).

ji–Trost reaction in complex contexts[19] and the variety of protocols which could be enlisted to introduce the required unsaturation, such as a selenation–elimination sequence.[20] Equally enticing, this idea would be relatively easy to test. For example, if **37** was modified to **38**, this new subgoal structure could be broken to reveal two compounds of approximately equal size and stereochemical complexity (allylic acetate **39** and vinyl stannane **40**) by projecting another palladium-based reaction, a π-allyl Stille reaction,[21] to accomplish their merger. In turn, because each of these new fragments possess a 1,2-*syn* arrangement of stereocenters adjacent to a site that either is (in the case of **39**), or could be (if **40** was modified to **42**), a carbonyl group, their chirality could arise through substrate-controlled Evans aldol reactions (see Chapter 3).[22] In fact, both final targets identified through this analysis (**41** and **43**) are quite similar to products formed by the Evans group during their total synthesis of cytovaricin (see Chapter 28 of *Classics I*), suggesting that their formation here would likely not prove problematic.*

17.3 Total Synthesis

We begin our discussion of how this intriguing plan was put into action with the operations that were executed to prepare the two building blocks essential to testing the biomimetic part of the proposed sequence, namely allylic acetate **39** and vinyl stannane **40**. Fortunately, neither proved too challenging to synthesize as intimated at above, with both routes actually mirroring each other in their key operations.

The optimized sequence developed by the Sorensen group for the first of these fragments is shown in Scheme 7, starting with a stereoselective Evans aldol reaction to establish its two adjacent stereocenters. As indicated, admixing (*E*)-6-acetoxy-4-methyl-4-hexenal (**44**) with the preformed *Z* boron enolate **45** at −78 °C in CH_2Cl_2 for 1 hour, followed by stirring at −25 °C for 18 hours, provided the desired aldol product (**41**) in quantitative yield once the reaction had been quenched with MeOH at 0 °C. As discussed in Chapter 3, the high level of diastereocontrol achieved in this event can be rationalized with a transition state such as **46** in which the propensity to minimize dipole interactions orients the carbonyl group of the auxiliary in the opposite direction to the C−O bond of its adjacent enolate, leading to substrate-controlled aldehyde approach dictated by the stereocenter of the auxiliary.[23] With this initial merger out of the way, the now-extraneous oxazolidinone was then converted into a Weinreb amide upon reaction with *N*,*O*-dimethylhydroxylamine hydrochloride and Me_3Al in THF at −15 °C,[24] and, finally, the free secondary alcohol function of the resultant product was protected as a TMS ether. These final two steps completed the synthesis of **39** in 87 % overall yield from **44**.

As shown in Scheme 8, the same three operations also opened the sequence leading to vinyl stannane **40**, providing in this case the

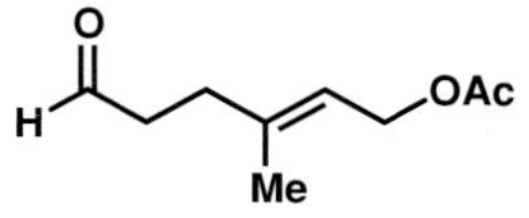

O H OAc Me **44**

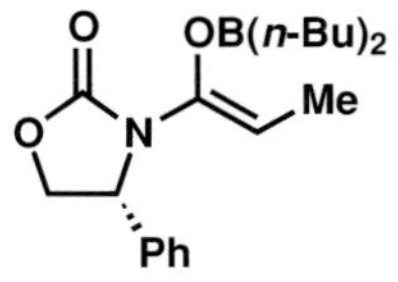

O OB(*n*-Bu)$_2$ O N Me Ph **45**

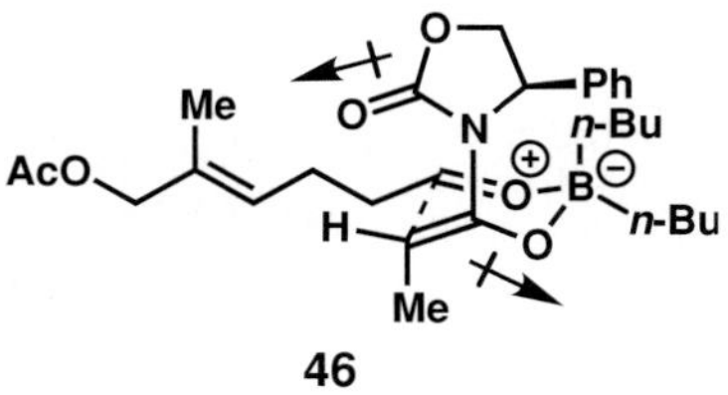

O Ph Me O N *n*-Bu AcO ⊕ O B ⊖ H O *n*-Bu Me **46**

---

* We should note that the Fujisawa scientists originally assigned **1** with the opposite absolute configuration to that shown in these pages, an assignment that was later corrected.[15e] The first total synthesis of **1** by the Sorensen group was actually that of the non-natural antipode;[14a] in this chapter, however, we present their optimized sequence to the correct, naturally occurring enantiomer.[14b]

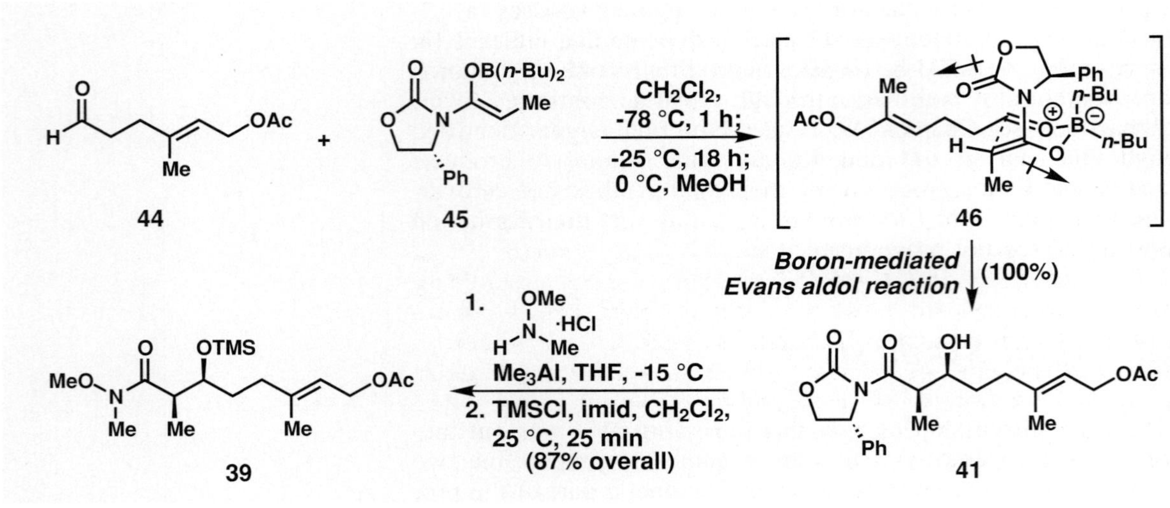

Scheme 7. Synthesis of Weinreb amide building block **39**.

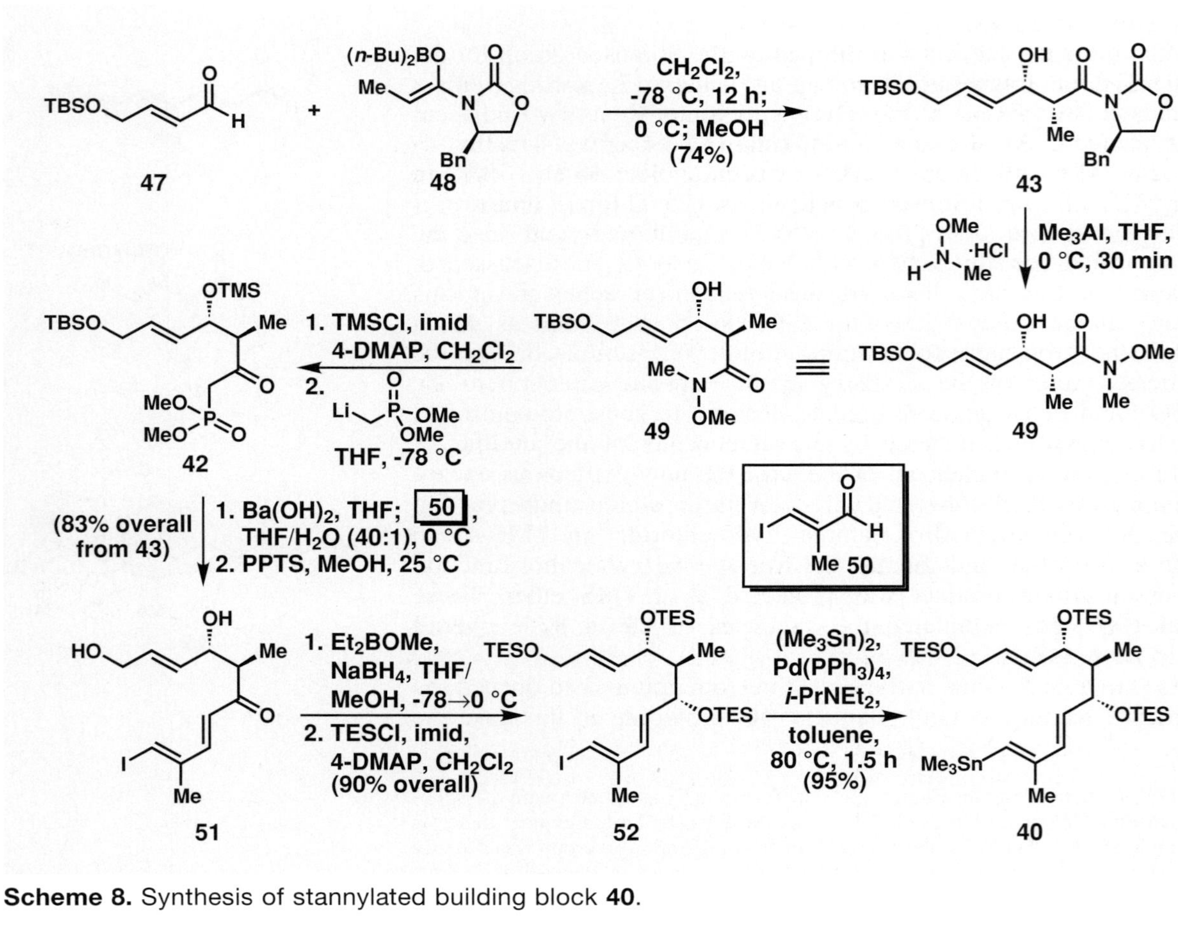

Scheme 8. Synthesis of stannylated building block **40**.

TBS-protected variant of **49** in high yield and with complete stereocontrol starting from aldehyde **47** and *Z* boron enolate **48**. Subsequent reaction of this intermediate with excess dimethyl lithiomethylphosphonate then provided **42**, setting the stage for a $Ba(OH)_2$-promoted Horner–Wadsworth–Emmons coupling with aldehyde **50** to install the diene motif of the target molecule. The final stereocenter was then set by cleaving the TMS ether in this new product with PPTS to provide an alcohol handle that could enable a subsequent diastereoselective 1,3-*syn* reduction with Et_2BOMe and $NaBH_4$ in THF/MeOH.[25] Finally, the desired building block (**40**) was completed in two more steps through tandem protection of all alcohol groups as TES ethers (TESCl, imid, 4-DMAP, CH_2Cl_2) and exchange of the vinyl iodide for a trimethyltin group under typical conditions.[26] Overall, the nine operations delineated in Scheme 8 proceeded in a combined yield of 53 %, an outcome that corresponds to an impressive average yield of 92 % per step. If it seems that we have gone through these operations rather quickly, it is not due to their lack of importance, but due to our eagerness to discuss the pending and most challenging steps in the projected sequence.

The first of these operations was the union of these two advanced building blocks (**39** and **40**) through a π-allyl Stille reaction, and, as shown in Scheme 9, this objective was accomplished quite smoothly. In the event, the allylic acetate (**39**) was first mixed with excess LiCl (4 equiv) and a catalytic amount of Pd_2dba_3 in *N*-methyl-2-pyrrolidinone (NMP). A solution of stannane (**40**) and *i*-Pr_2NEt in the same solvent was then added, and the resultant mixture was warmed to 40 °C for 3 hours during which time the materials slowly funneled into **38** in 91 % yield. Key to the consistent success of this reaction, especially on a large scale, was the addition of the palladium catalyst in batches and at a relatively high loading (10–20 %), as some residual Ph_3P from the final step needed to prepare **39** always inhibited a certain portion of the catalyst. Equally important, the reaction had to be conducted at a temperature no greater than 40 °C to maintain the stereochemical integrity of the reactive π-allyl palladium intermediate (**53**). Otherwise not only **38**, but also material with *Z* geometry at the newly generated double bond, was obtained.

With ample supplies of **38** provided through this protocol, the Sorensen group could next attempt to attach the atoms needed to prepare **37**, the projected intermediate for a second reaction based on π-allyl palladium complexes (a Tsuji–Trost reaction) that would hopefully lead to the 19-membered macrocycle **36**. In essence, this requirement boiled down to only two key synthetic objectives: generating a ketoester moiety from the Weinreb amide, and converting the allylic TES-protected alcohol function at C1 into a methyl carbonate. Neither of these tasks ultimately proved to be overly challenging to carry out, with the first accomplished by treating **38** with excess quantities of the lithium enolate of *t*-butyl acetate to provide **54**,[27] and the second requiring three rela-

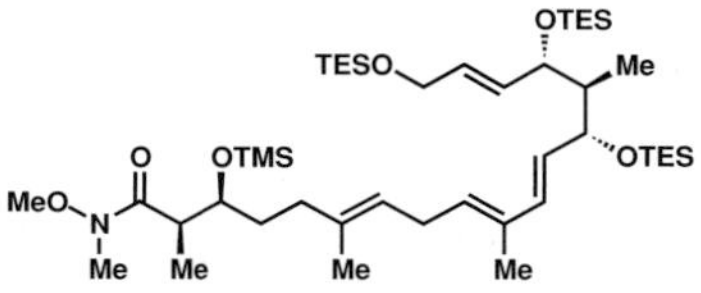

38

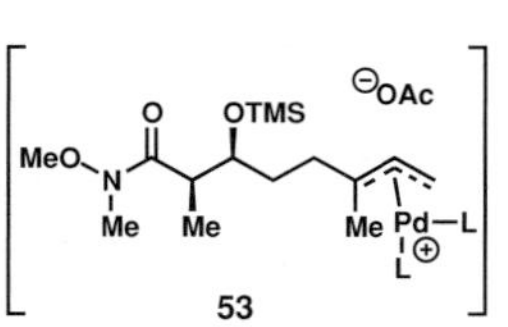

53

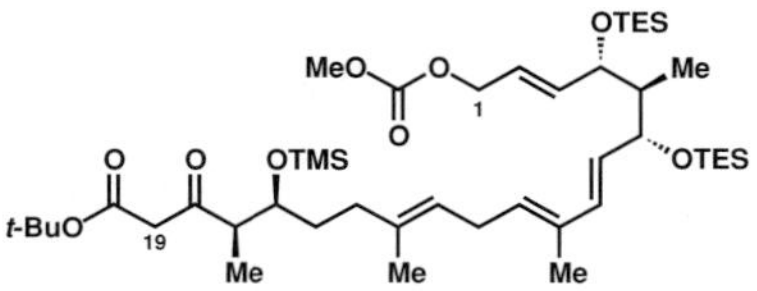

37

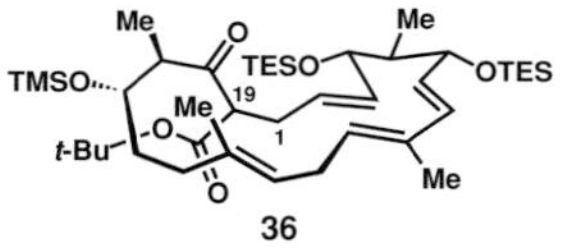

36

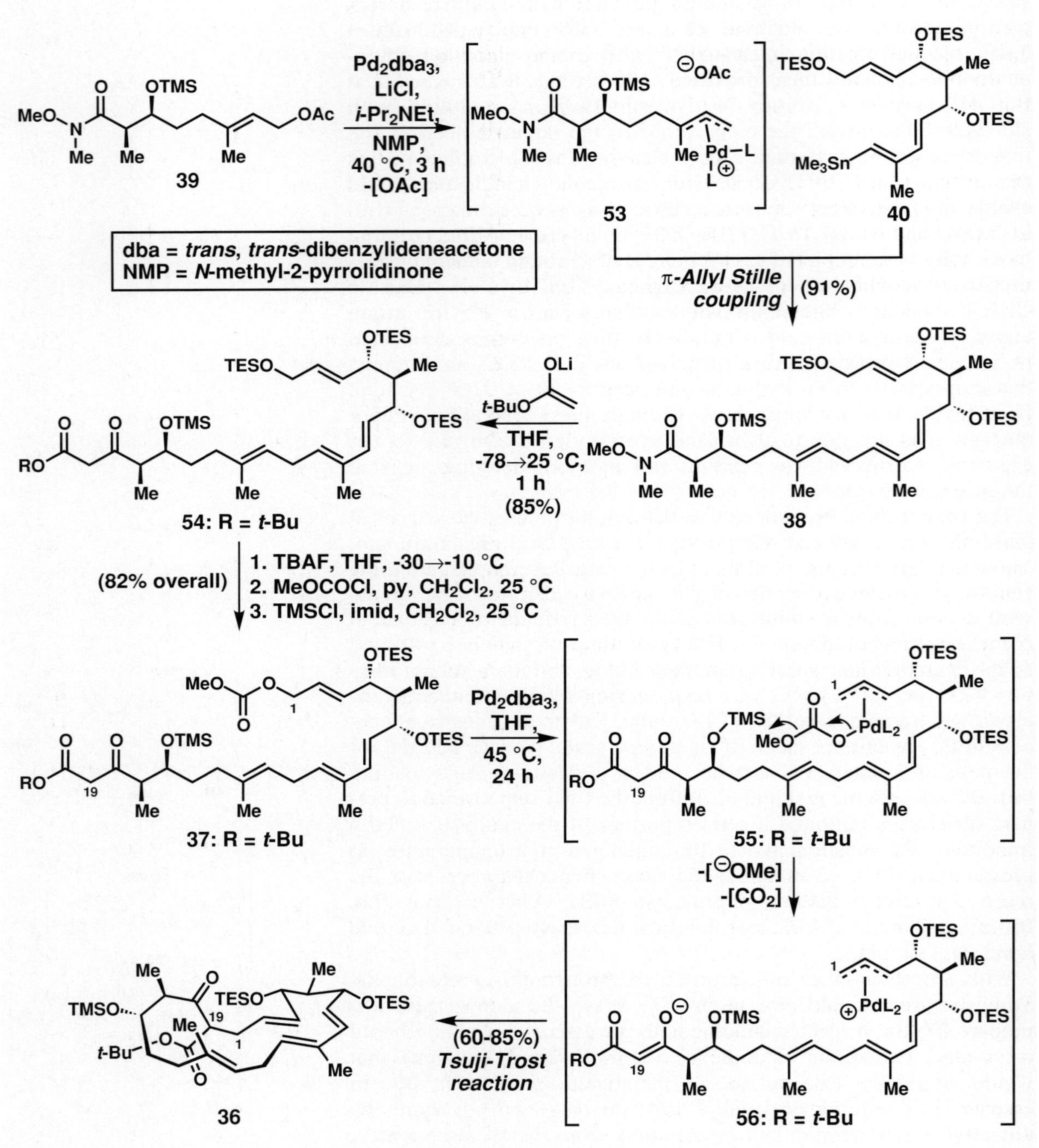

dba = *trans, trans*-dibenzylideneacetone
NMP = *N*-methyl-2-pyrrolidinone

Scheme 9. Synthesis of advanced intermediate **36**.

tively routine steps from this new intermediate: 1) selective deprotection of the primary TES and secondary TMS ethers with TBAF in THF at low temperature (−30→10 °C), 2) chemoselective methoxycarbonylation of the primary allylic hydroxy group, and 3) resilylation of the remaining secondary alcohol function as a TMS ether (TMSCl, imid, CH_2Cl_2).

With these three steps affording **37** in 63 % overall yield from **38**, the stage was now set for the Tsuji–Trost macrocyclization. Most gratifyingly, this reaction was a resounding success. Exposure of a dilute solution of **37** in THF (0.05 M, to prevent dimerization) to a catalytic amount (10 mol %) of Pd_2dba_3 at 45 °C for 24 hours ultimately provided **36** in 60–85 % yield depending on the reaction scale. Its smooth prosecution is undoubtedly due to the selection of a methyl carbonate as the allylic activating group, since, as mentioned in Chapter 14, these groups enable irreversible palladium insertion under mild conditions promoted by the loss of CO_2 and an alkoxide from the initial reactive intermediate (i. e. **55**). In fact, it was this alkoxide by-product that ensured full enolization of the ketoester to provide **56** and enable the ring closure to occur. Before moving forward, it is worth noting that the success of both π-allyl coupling reactions in Scheme 9 indicates the true utility and versatility of these transformations in complex contexts. For instance, if one considers that most of the steps that stitch together building blocks or form macrocyclic rings usually proceed in modest yield, their accomplishment with better than 80 % efficiency here makes them remarkable indeed. Yet, if **36** could not ultimately participate in the final double transannular Diels–Alder sequence to complete (−)-FR182877, then the progress achieved thus far with the sequence would be a hollow victory, at least as far as this total synthesis was concerned.

So, pressing forward without any further delay, the Sorensen group now attempted to install the final unit of unsaturation at C1–C19 to create the projected macrocyclic pentaene for this key event. As preliminary forays soon revealed, however, the challenge was not installing that C–C double bond onto **36**, but generating it with *E* selectivity. Scheme 10 provides the optimized protocol, with initial enolate formation using NaHMDS in Et_2O at 25 °C followed by a PhSeBr quench delivering **57** as a 10:1 mixture of diastereomeric C-19 phenylselenides (absolute stereochemistry not assigned). Brief exposure to *m*CPBA in CH_2Cl_2 at −78 °C then induced the oxidation-elimination sequence, leading to the desired *E*-disposed intermediate (**58**) and its *Z*-disposed counterpart (**59**, see column figure) in a 2.2:1 ratio. Although it might seem surprising that the opening ratio of C-19 phenylselenides did not correlate to the final *E*/*Z* ratio of the elimination products, it is important to remember that the formation of a selenoxide introduces an additional element of chirality. Thus, this simple feature could certainly play a major part in determining the overall distribution of **58** and **59** obtained through the terminating *syn*-elimination.

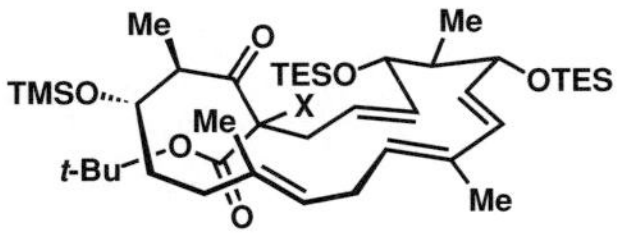

57: X = SePh

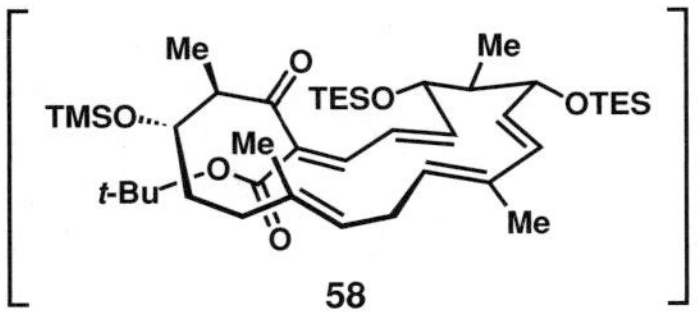

58

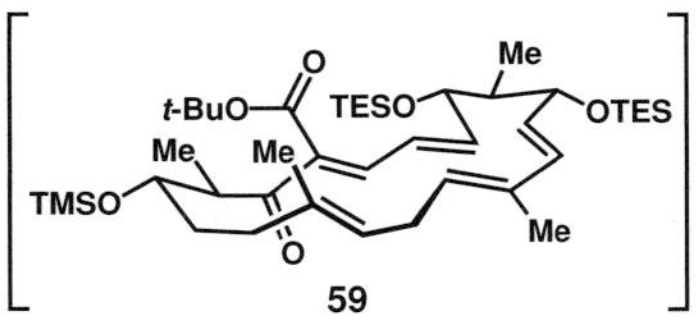

59

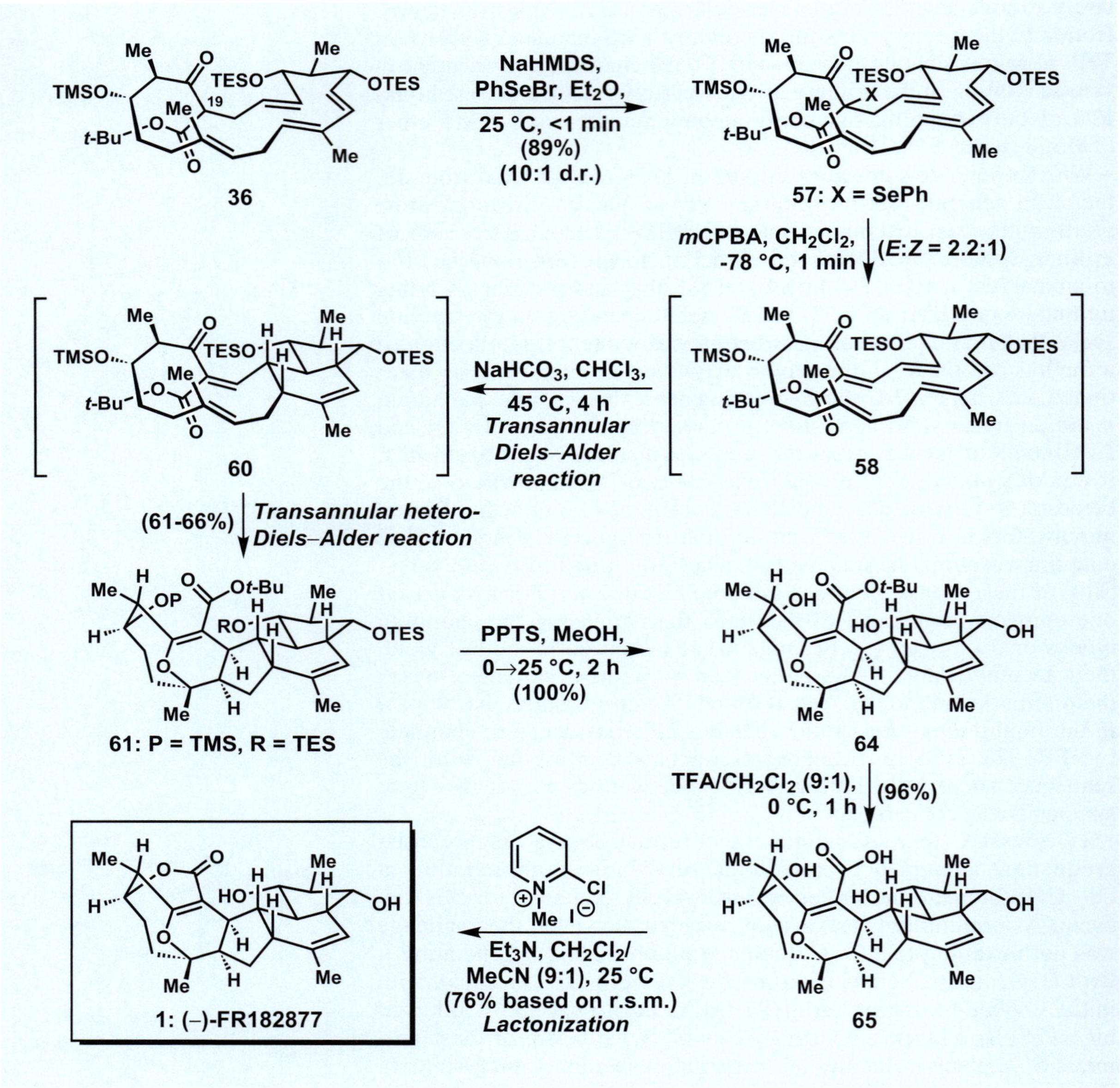

Scheme 10. Final stages and completion of Sorensen's total synthesis of (–)-FR182877 (**1**).

Nevertheless, with a means to obtain **58** as the major diastereomer, following the extractive workup needed to remove the extraneous *m*CPBA by-products, the mixture of **58** and **59** was immediately dissolved in $NaHCO_3$-buffered $CHCl_3$ and then heated at 45 °C for 4 hours. Amazingly, at the end of this time the desired double transannular Diels–Alder adduct was obtained in 61–66% yield (depending on reaction scale) along with minor amounts of **62** and **63** (both formed in 8–15 % yield, see column figures) arising from Diels–Alder reactions involving the pentaene system of **59**.* Considering that **58** accounted for only 70 % of the original mixture of starting materials, this throughput is quite striking. Moreover, with the event installing five new rings and seven stereogenic centers with complete diastereocontrol, the only major remaining operation to complete the target molecule was a lactonization to generate the one still-missing ring system. In preparation for this key step, **61** was exposed to PPTS in MeOH for 2 hours to dismantle all three of its silyl ethers, and then exposed to a 9:1 mixture of TFA and CH_2Cl_2 to hydrolyze its *t*-butyl ester, providing hydroxy acid **65** in 96 % overall yield. The synthesis of (–)-FR182877 (**1**) was then completed using Mukaiyama's reagent (*N*-methyl-2-chloropyridinium iodide)[28] and Et_3N in a 9:1 mixture of CH_2Cl_2 and MeCN to form the final ring.** Overall, this elegant sequence required a total of 23 steps (20 in its longest linear sequence), and proceeded with such a high level of efficiency that it could provide multiple grams of the final natural product and, thereby, fuel the advanced biological evaluations needed to determine whether (–)-FR182877 (**1**) truly has potential as a drug candidate for cancer chemotherapy.

59

$NaHCO_3$, $CHCl_3$, 45 °C, 4 h

62: R = *t*-Bu

+

63: R = *t*-Bu

17.4 Conclusion

Even though the Diels–Alder reaction has arguably been explored more deeply than any other transformation over the course of the past 75 years, this inventive total synthesis has managed to pave new ground for its utility in chemical synthesis by providing the first example of a successful double transannular Diels–Alder sequence. In the process, it has also provided yet another powerful demonstration of how properly formatted polyunsaturated precursors can be coaxed to contract spontaneously into stereochemically rich polycyclic products. While syntheses based on biogenetic hypotheses cannot prove definitively if Nature creates a particular natural product through an analogous set of reactions, when they are as efficient as this one, it is hard to imagine that the master chemical artisan with millenia of "bench experience" would employ a more circuitous route.

---

* It is interesting to note that **63**, the product of only one transannular Diels–Alder reaction, was formed with *exo* selectivity to afford the stereochemical arrangements of hexacyclinic acid (**35**, cf. Scheme 5).

** As an aside, an exhaustive screening of virtually every major lactonization protocol was attempted in this step, but none worked as well as the indicated conditions. Equally important, even with the given protocol the use of MeCN was critical for a high yield of **1** since both Mukaiyama's reagent and the triol carboxylic acid (**65**) are poorly soluble in neat CH_2Cl_2.

17.5 Evans' Total Synthesis of (–)-FR182877

As matters would transpire, the Sorensen group would not be the only team of researchers enticed by the intriguing structure of (–)-FR182877 (**1**) and the potential for a cascade sequence of transannular Diels–Alder reactions to generate its central polycycle stereoselectively from a macrocyclic precursor. Early in 2002, the Evans group at Harvard University reported their successful completion of this formidable target (**1**) along the lines of the same synthetic blueprint as the work presented above, but executed through a set of different operations.[29]

The key elements of this second total synthesis of (–)-FR182877 (**1**) are depicted in Scheme 11, starting with the steps required to generate the macrocyclic precursor for the projected transannular Diels–Alder cascade sequence (i.e. **70**). As indicated, the first of these events was the merger of building blocks **66** and **67** through a palladium-mediated Suzuki coupling reaction to form advanced intermediate **68** in 84 % yield. This fragment is, of course, reminiscent of compound **38** (cf. Scheme 9) generated by the Sorensen group through a π-allyl-type Stille reaction as discussed earlier. However, there is one subtle, but important, difference, namely the incorporation of a bromine substituent at C-9 rather than the corresponding methyl group of the target molecule. Although this choice might seem odd in light of the successful sequence discussed above, at the time this decision was made it was unclear whether or not the transannular Diels–Alder sequence would proceed with *endo* or *exo* selectivity. As a result, the Evans group designed a substrate that could be advanced to either (–)-FR182877 (**1**) or hexacyclinic acid (**35**) irrespective of that outcome.

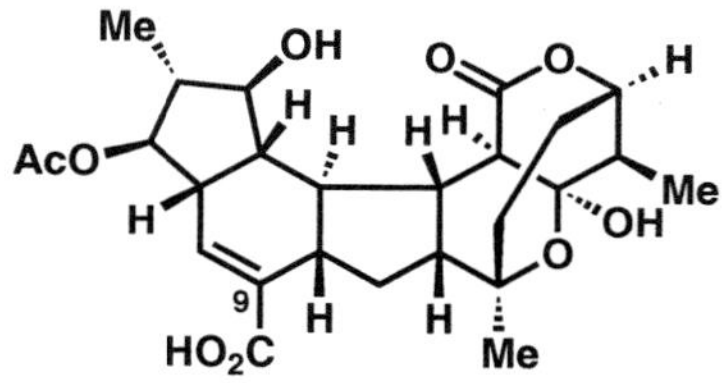

35: hexacyclinic acid

Before moving on, a few additional comments about this opening Suzuki reaction are in order. First, although either of the two bromine substituents in **67** could have participated in this event in principle, extensive precedent has always indicated that the *E*-disposed bromine is far more reactive towards palladium insertion, which explains why **68** was formed stereoselectively.[30] It is important to note that the smooth nature of this reaction and the survival of the bromine atom in the product would not have occurred in the absence of the thallium base, since this species accelerated the reaction to the degree that the initial oxidative addition was the rate-determining step. Thus, as soon as insertion occurred, coupling with **66** ensued. The Kishi group at Harvard was the first to pioneer this type of base (they used TlOH) to accelerate a Suzuki reaction as part of their elegant total synthesis of palytoxin (see Chapter 36 of *Classics I*).[31]

Returning to the work at hand, with the two building blocks merged together, **68** was advanced to **69** through a series of four separate synthetic operations, and then the complete 19-membered

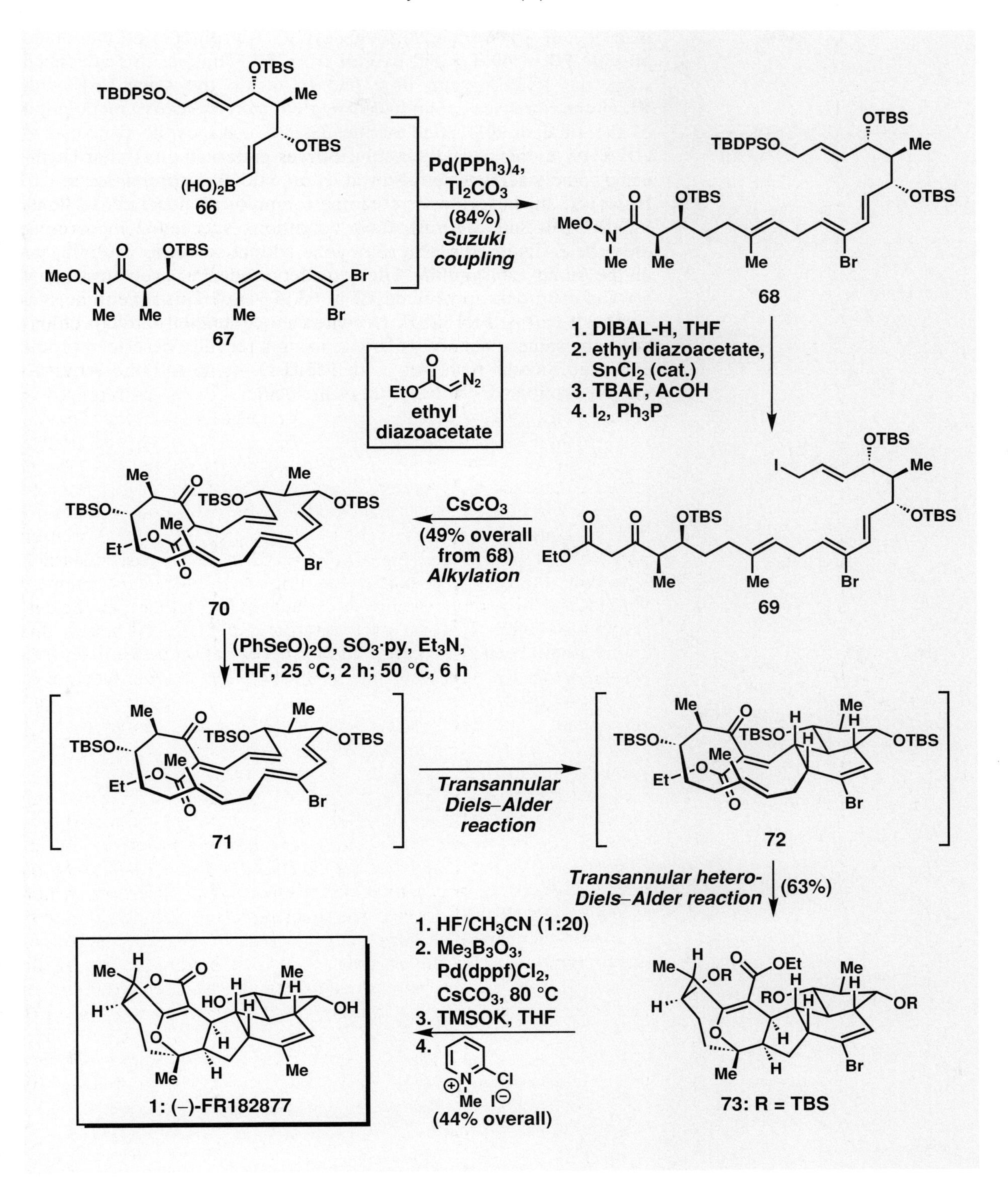

Scheme 11. Evans' total synthesis of (–)-FR182877 (**1**).

macrocycle was formed through a $CsCO_3$-promoted alkylation to provide **70** in 49 % yield overall from **68**. Thus, at this advanced stage the Evans group now had to tackle the same task with which the Sorensen group had struggled: stereoselective installation of the final double bond of the desired macrocyclic pentaene at C1−C19. In this case, their solution was patterned on similar chemistry, namely selenation followed by an oxidative elimination event. However, because these researchers employed different reagents [$(PhSeO)_2O$ and SO_3·py], their conditions succeeded in forming only the *E*-disposed olefin (**71**). This adduct smoothly participated in the same transannular Diels−Alder sequence upon heating at 50 °C for 6 hours to provide **73** in 63 % yield. This inventive total synthesis of (−)-FR182877 (**1**) was then completed through essentially the same sequence as above, using a recently described protocol[32] for Suzuki reactions with $Me_3B_3O_3$ to install the requisite methyl group at C-9 of the target molecule.

References

1. For a review on the Diels–Alder reaction in total synthesis, see: K. C. Nicolaou, S. A. Snyder, T. Montagnon, G. Vassilikogiannakis, *Angew. Chem.* **2002**, *114*, 1742; *Angew. Chem. Int. Ed.* **2002**, *41*, 1668.
2. For insightful reviews on this subject, see: a) G. Pohnert, *ChemBioChem* **2001**, *2*, 873; b) S. Laschat, *Angew. Chem.* **1996**, *108*, 313; *Angew. Chem. Int. Ed. Engl.* **1996**, *35*, 289.
3. a) T. Ose, K. Watanabe, T. Mie, M. Honma, H. Watanabe, M. Yao, H. Oikawa, I. Tanaka, *Nature* **2003**, *422*, 185; b) K. Auclair, A. Sutherland, J. Kennedy, D. J. Witter, J. P. Van den Heever, C. R. Hutchinson, J. C. Vederas, *J. Am. Chem. Soc.* **2000**, *122*, 11519; c) H. Oikawa, T. Kobayashi, K. Katayama, Y. Suzuki, A. Ichihara, *J. Org. Chem.* **1998**, *63*, 8748.
4. M. E. Layton, C. A. Morales, M. D. Shair, *J. Am. Chem. Soc.* **2002**, *124*, 773.
5. a) K. C. Nicolaou, N. A. Petasis, R. E. Zipkin, J. Uenishi, *J. Am. Chem. Soc.* **1982**, *104*, 5555; b) K. C. Nicolaou, N. A. Petasis, J. Uenishi, R. E. Zipkin, *J. Am. Chem. Soc.* **1982**, *104*, 5557; c) K. C. Nicolaou, R. E. Zipkin, N. A. Petasis, *J. Am. Chem. Soc.* **1982**, *104*, 5558; d) K. C. Nicolaou, N. A. Petasis, R. E. Zipkin, *J. Am. Chem. Soc.* **1982**, *104*, 5560.
6. For a review on Diels–Alder reactions in the biosynthesis of natural products, see: A. Ichihara, H. Oikawa in *Comprehensive Natural Products Chemistry, Vol. 5* (Eds.: D. H. R. Barton, K. Nakanishi, O. Meth-Cohn), Elsevier, New York, **1999**, pp. 367–408. For other examples in this book showing biomimetic Diels–Alder reactions, see Chapter 8 on manzamine A, Chapter 11 on the bisorbicillinoids, and Chapter 18 on vinblastine.
7. J. A. McCauley, K. Nagasawa, P. A. Lander, S. G. Mischke, M. A. Semones, Y. Kishi, *J. Am. Chem. Soc.* **1998**, *120*, 7647.
8. K. C. Nicolaou, J. Li, *Angew. Chem.* **2001**, *113*, 4394; *Angew. Chem. Int. Ed.* **2001**, *40*, 4264.
9. A. J. Quillinan, F. Scheinmann, *J. Chem. Soc. (D)* **1971**, 966.
10. L. F. Tietze, Y. Zhou, *Angew. Chem.* **1999**, *111*, 2076; *Angew. Chem. Int. Ed.* **1999**, *38*, 2045.
11. For reviews and other examples of this general sequence, see: a) L. F. Tietze, J. Bachmann, J. Wichmann, Y. Zhou, T. Raschke, *Liebigs Ann.* **1997**, 881; b) L. F. Tietze, *Chem. Rev.* **1996**, *96*, 115; c) L. F. Tietze, J. Bachmann, J. Wichmann, O. Burkhardt, *Synthesis* **1994**, 1185.
12. For reviews, see: a) E. Marsault, A. Toró, P. Nowak, P. Deslongchamps, *Tetrahedron* **2001**, *57*, 4243; b) P. Deslongchamps, *Pure & Appl. Chem.* **1992**, *64*, 1831.
13. a) A. Toró, P. Nowak, P. Deslongchamps, *J. Am. Chem. Soc.* **2000**, *122*, 4526; b) A. Toró, C.-A. Lemelin, P. Préville, G. Bélanger, P. Deslongchamps, *Tetrahedron* **1999**, *55*, 4655.
14. a) D. A. Vosburg, C. D. Vanderwal, E. J. Sorensen, *J. Am. Chem. Soc.* **2002**, *124*, 4552; b) C. D. Vanderwal, D. A. Vosburg, S. Weiler, E. J. Sorensen, *J. Am. Chem. Soc.* **2003**, *125*, 5393.
15. a) H. Muramatsu, M. Miyauchi, B. Sato, S. Yoshimura, *40th Symposium on the Chemistry of Natural Products*, Fukuoka, Japan, **1998**, pp. 487–492; b) B. Sato, H. Muramatsu, M. Miyauchi, Y. Hori, S. Takase, M. Hino, S. Hashimoto, H. Terano, *J. Antibiot.* **2000**, *53*, 123; c) B. Sato, H. Nakajima, Y. Hori, M. Hino, S. Hashimoto, H. Terano, *J. Antibiot.* **2000**, *53*, 204; d) S. Yoshimura, B. Sato, T. Kinoshita, S. Takase, H. Terano, *J. Antibiot.* **2000**, *53*, 615; e) S. Yoshimura, B. Sato, T. Kinoshita, S. Takase, H. Terano, *J. Antibiot.* **2002**, *55*, C1.
16. R. Höfs, M. Walker, A. Zeeck, *Angew. Chem.* **2000**, *112*, 3400; *Angew. Chem. Int. Ed.* **2000**, *39*, 3258.
17. a) C. D. Vanderwal, D. A. Vosburg, S. Weiler, E. J. Sorensen, *Org. Lett.* **1999**, *1*, 645; b) C. D. Vanderwal, D. A. Vosburg, E. J. Sorensen, *Org. Lett.* **2001**, *3*, 4307.
18. Macrocyclizations through Knoevenagel condensations are actually quite rare, and may only be limited to the formation of cyclic oligomers. For an example, see: Y. Zhang, T. Wada, H. Sasabe, *Chem. Commun.* **1996**, 621.
19. For reviews, see: a) B. M. Trost, *Angew. Chem.* **1989**, *101*, 1199; *Angew. Chem. Int. Ed. Engl.* **1989**, *28*, 1173; b) J. Tsuji, *Palladium Reagents and Catalysts*, John Wiley & Sons, Chichester, **1995**, pp. 290–422. For an early example of a macrocyclization with this reaction, see: c) B. M. Trost, S. J. Brickner, *J. Am. Chem. Soc.* **1983**, *105*, 568.
20. a) H. J. Reich, I. L. Reich, J. M. Renga, *J. Am. Chem. Soc.* **1973**, *95*, 5813; b) K. B. Sharpless, R. F. Lauer, *J. Am. Chem. Soc.* **1973**, *95*, 2697.
21. For a review, see: V. Farina, V. Krishnamurthy, W. J. Scott in *Organic Reactions*, *Vol. 50* (Eds.: L. A. Paquette, et al.), John Wiley & Sons, New York, **1997**, pp. 1–652.
22. D. A. Evans, S. W. Kaldor, T. K. Jones, J. Clardy, T. J. Stout, *J. Am. Chem. Soc.* **1990**, *112*, 7001.
23. For a general review on boron-mediated aldol reactions, see: C. J. Cowden, I. Paterson in *Organic Reactions, Vol. 51* (Eds.: L. A. Paquette, et al.), John Wiley & Sons, New York, **1997**, pp. 1–200.
24. A. Basha, M. Lipton, S. M. Weinreb, *Tetrahedron Lett.* **1977**, *18*, 4171.
25. K.-M. Chen, G. E. Hardtmann, K. Prasad, O. Repic, M. J. Shapiro, *Tetrahedron Lett.* **1987**, *28*, 155.
26. a) W. D. Wulff, G. A. Peterson, W. E. Bauta, K.-S. Chan, K. L. Faron, S. R. Gilbertson, R. W. Kaesler, D. C. Yang, C. K. Murray, *J. Org. Chem.* **1986**, *51*, 277; b) W. J. Scott, J. K. Stille, *J. Am. Chem. Soc.* **1986**, *108*, 3033.

27. J. A. Turner, W. S. Jacks, *J. Org. Chem.* **1989**, *54*, 4229.
28. T. Mukaiyama, M. Usui, K. Saigo, *Chem. Lett.* **1976**, 49.
29. D. A. Evans, J. T. Starr, *Angew. Chem.* **2002**, *114*, 1865; *Angew. Chem. Int. Ed.* **2002**, *41*, 1787.
30. a) R. Rossi, A. Carpita, *Tetrahedron Lett.* **1986**, *27*, 2529; b) W. R. Roush, R. Riva, *J. Org. Chem.* **1988**, *53*, 710.
31. J. Uenishi, J.-M. Beau, R. W. Armstrong, Y. Kishi, *J. Am. Chem. Soc.* **1987**, *109*, 4756. For the first use of Tl_2CO_3 to accelerate a Suzuki reaction, see: I. E. Markó, F. Murphy, S. Dolan, *Tetrahedron Lett.* **1996**, *37*, 2507.
32. M. Gray, I. P. Andrews, D. F. Hook, J. Kitteringham, M. Voyle, *Tetrahedron Lett.* **2000**, *41*, 6237.

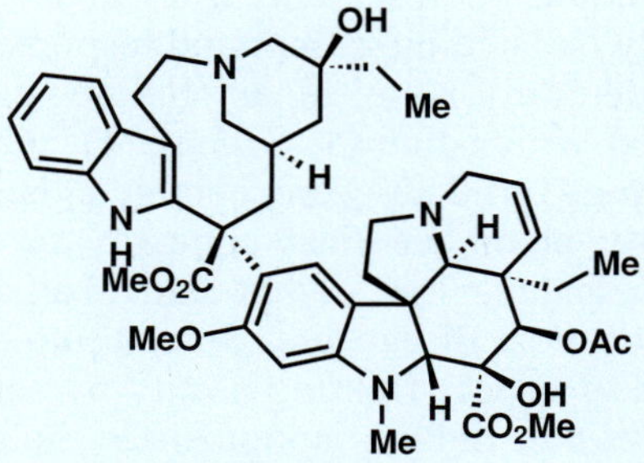

1: vinblastine

18

T. Fukuyama (2002)

Vinblastine

18.1 Introduction

Key concepts:

- **Oxidative couplings**
- **Skraup quinoline synthesis**
- **Fukuyama indole synthesis**
- **Reaction optimization**

For several decades, cultures from distant regions of the globe have used medicines procured from the leaves of the Madagascar periwinkle plant (*Cantharanthus roseus*) to treat a host of ailments such as the common cold, insect stings, and even eye infections. For the most part, however, these therapeutics remained unscrutinized by Western medical science until 1952 when Dr. Clark Noble, a Canadian doctor, came upon a unique elixir derived from this plant during his travels in Jamaica in the form of a tea given to diabetic patients when supplies of insulin ran out at the neighborhood clinics. Since this alternative remedy appeared to be efficacious, Noble believed that he might have uncovered a new life-saving treatment for a major medical disease. As a result, he decided to pass along his observations and an ample supply of periwinkle plant leaves to his younger brother at the University of Western Ontario, Dr. Robert Noble, who had the resources to probe his finding empirically. Unfortunately, the lead proved to be false as the junior Dr. Noble found no indication that any chemical entity from this natural source could induce the biochemical events necessary to treat diabetes successfully. Intriguingly, when he attempted to examine the role of the leaves' extracts in modulating blood sugar levels, he observed that the white blood cell count of his samples diminished over time, thereby suggesting the leaves contained an agent that perturbed the normal replication of these cells. Since the aberrant synthesis of white blood cells is a telltale sign of leukemia, Noble hypothesized that the Madagascar periwinkle plant might still hold medical promise, not as a treatment for diabetes, but as the source of a new anticancer drug.[1]

Following several years of intensive research, in 1958 Robert Noble and his co-workers finally isolated the compound responsible for this intriguing effect, a structurally novel binary indole–indoline alkaloid, which they named vinblastine (**1**, Scheme 1).[2] Once sufficient quantities of the natural product were obtained, Noble joined forces with the Eli Lilly pharmaceutical company to test the agent clinically. As anticipated, the compound displayed impressive cytotoxicity against a host of cancers. Serendipitously, it also possessed a novel mode of cancer-fighting ability by selective binding to tubulin, an event that induces apoptosis by halting spindle formation during mitosis. Unsurprisingly, vinblastine (**1**) was immediately adopted in cancer therapy, and today, after almost forty years of clinical use, it continues to serve as one of the most important anticancer drugs on the market.[3]

Although its value to society is clear, vinblastine constitutes one of the most expensive cancer chemotherapeutics available due to the dual problems of its low natural abundance within *Cantharanthus roseus* and the labor intensive separation procedures required to isolate it from other alkaloids. As such, from the moment of its complete structural characterization, the synthetic community has devoted significant attention and financial resources to achieving a laboratory synthesis of this agent. Fortunately, these endeavors have proven quite fruitful in that four separate syntheses of vinblastine (**1**) were achieved in the three decades following the initial disclosure of its structure.[4] Apart from leading to an increased supply of this potent compound, these successful routes have also enabled access to a host of vinblastine analogues bearing deep-seated structural modifications, yielding much insight into the structure–activity profile of this natural product. Arguably, however, none of these endeavors can formally be classified under the heading of "total" synthesis because they all employed another natural product from *Cantharanthus roseus*, (–)-vindoline (**2**, Scheme 1),[5] as the source of the "bottom" half of vinblastine (**1**). Nevertheless, such a decision can hardly be criticized since adopting the use of this starting material engendered both simplicity and cost effectiveness, and also reflected vinblastine's postulated biogenesis.

As shown in Scheme 1a, one can formally consider vinblastine as arising from the nucleophilic addition of (–)-vindoline (**2**) to an activated form of the secondary metabolite catharanthine (**3**), a natural product also obtained from *Cantharanthus roseus*. If this union was attended by bond cleavage at the indicated site within **3**, then a subsequent stereoselective oxidation process would complete the assembly of (+)-vinblastine (**1**). While several assays using enzymatic extracts have verified this general order of biosynthetic events,[6] the first synthetic evidence supporting its viability came from the work of Pierre Potier and members of his group at the Institut de Chimie des Substances Naturelles in France.[3a,4a] In their pioneering investigations, these researchers discovered that following the initial reaction of catharanthine *N*-oxide

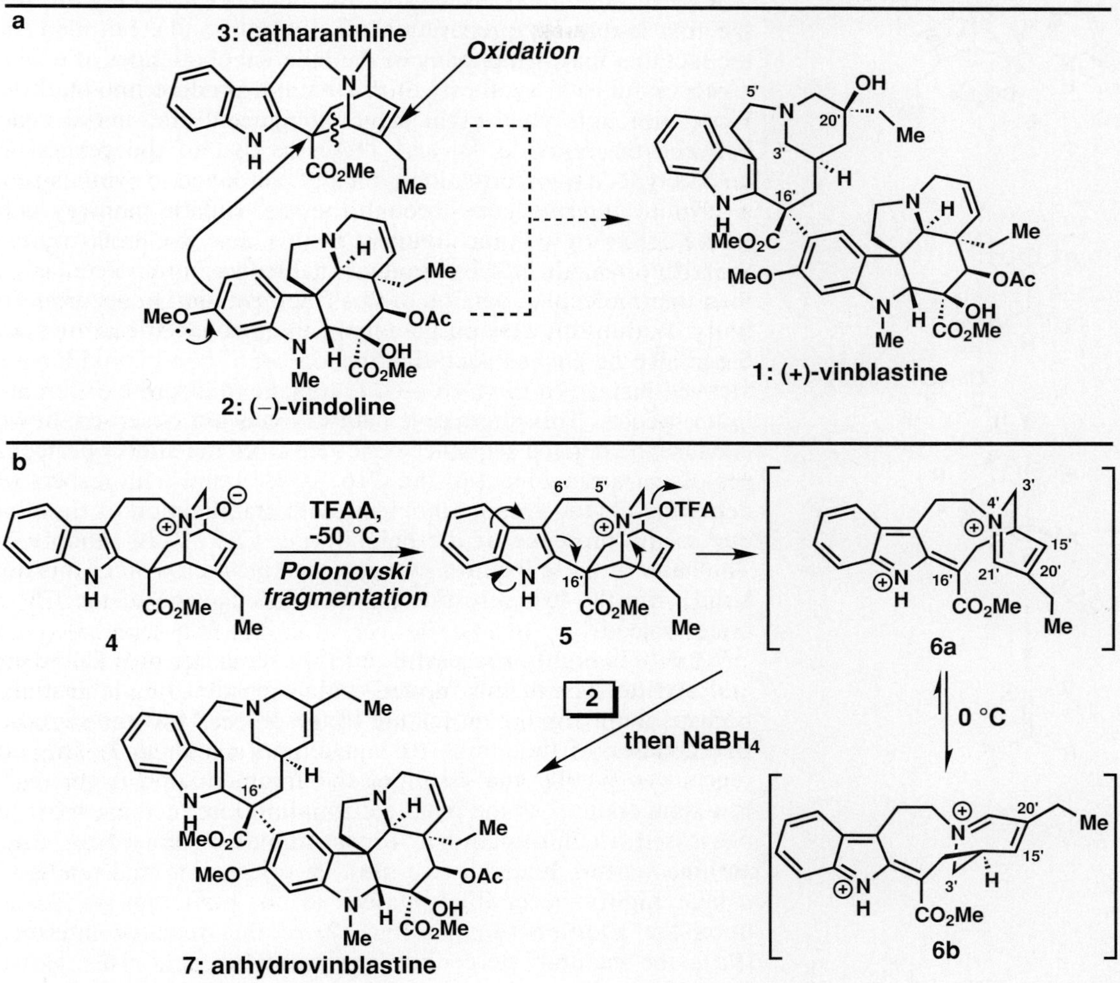

Scheme 1. General biosynthetic pathway (a) for binary indole–indoline alkaloids such as vinblastine (**1**) and Potier's synthetic approach (b) based on that hypothesis.

(**4**, Scheme 1b) with trifluoroacetic anhydride (TFAA) to form intermediate **5**, intramolecular fragmentation through the indicated mechanistic pathway afforded iminium system **6a**, an intermediate suitably activated for nucleophilic interception at C-16′ by vindoline (**2**). Indeed, when this sequence was performed at low temperature (−50 °C) in the presence of vindoline (**2**), anhydrovinblastine (**7**) was formed in approximately 50 % yield following the addition of $NaBH_4$ (to reduce the C21′–N4′ iminium species originally formed within **6a**). This new product (**7**) could then be converted into **1** through a short sequence of reactions.

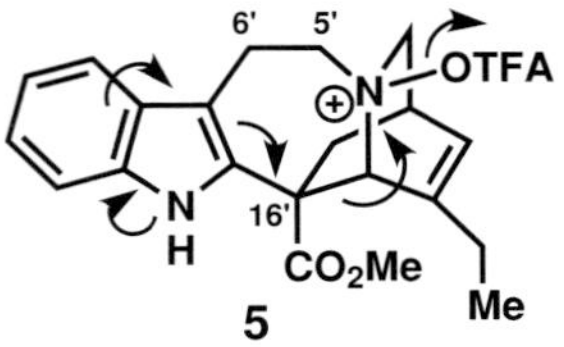

Several aspects of this instructive biomimetic synthesis are worthy of additional discussion, not only because they are part of the first laboratory preparation of vinblastine (**1**), but also due to the fact that they bring many of the inherent challenges of executing a successful total synthesis of this natural product into stark relief. First, although you might have recognized the initial reaction between the *N*-oxide (**4**) and TFAA as that of the general Polonovski type, a transformation typically employed to synthesize reactive imine intermediates through the mechanistic pathway defined in the adjoining column figures,[7] in this case the weak basicity of the trifluoroacetic acid by-product enabled an intramolecular, rather than intermolecular, version of this fragmentation to occur preferentially. Within this general paradigm, the mechanistic arrows within **5** can also be pushed such that the C5′–C6′ bond could have been cleaved instead of the C16′–C21′ linkage to afford a different iminium species. This alternative pathway was not observed, however, because the desired sequence benefited from the almost perfect antiperiplanar arrangement of the C16′–C21′ bond with respect to the departing OTFA group, combined with stabilization of the product due to the presence of the neighboring C15′–C20′ double bond. The latter of these features was critical, since removal of this double bond from the test substrate resulted in cleavage of the C5′–C6′ bond instead.[8]

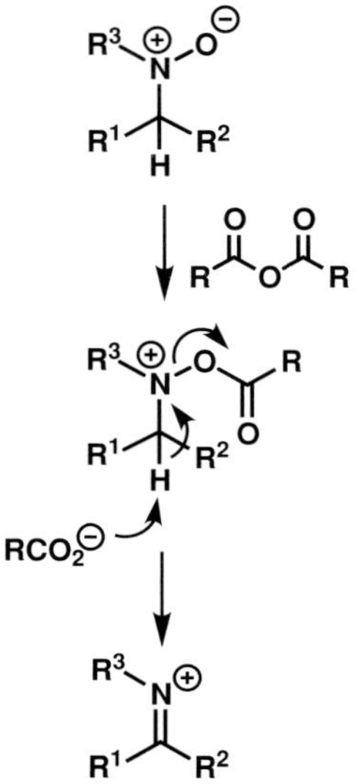

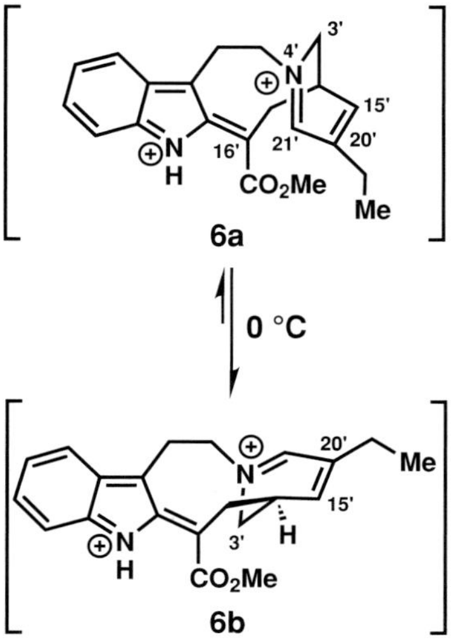

While this highly successful reaction sequence provided the general architecture of the "upper" indole portion of vinblastine, the perhaps more significant feature of the process was the stereoselective creation of the new C-16′ quaternary center in **7**. This occurrence was wholly the result of the fragmentation occurring at a low temperature, as the resultant iminium intermediate most likely possessed a conformational format similar to that of the catharanthine starting material (i. e. **6a**), an orientation that rendered the α-face highly accessible relative to its β-alternative. As such, upon the addition of vindoline (**2**) to this reactive intermediate (**6a**), the natural stereochemistry resulted at C-16′. However, when the reaction was merely warmed to 0 °C before the introduction of **2**, anhydrovinblastine was generated exclusively as its C-16′ epimer, a result that can be rationalized by assuming thermal equilibration of **6a** to its more stable conformer (**6b**) in which the apical C-3′ methylene effectively blocks any approach to the α-face. This general issue in forming the C-16′ quaternary center of vinblastine stereoselectively through an oxidative coupling perhaps constitutes the most critical challenge incurred by adopting such a biomimetic strategy; in this regard, Potier's solution should be noted as relatively unique in both its efficiency and selectivity. For example, the formation of a similar reactive intermediate (**10**, Scheme 2) from **8** in the presence of an electrophilic chlorine source such as *t*-BuOCl afforded only *epi*-C-16′-anhydrovinblastine (**11**).[9]

Although these biomimetic studies by Potier (as well as by many others whose accomplishments are reviewed elsewhere)[3] in the early 1980s established many of the essential aspects of what

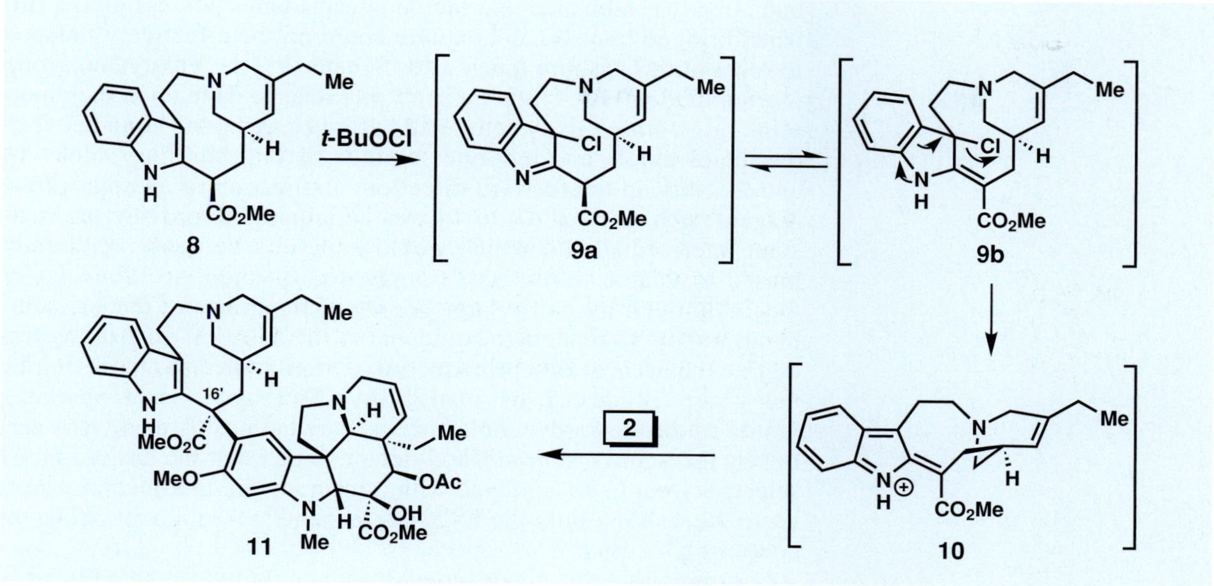

Scheme 2. The challenges of establishing the C-16′ stereocenter of vinblastine: the "chloroindoline" approach.

could constitute the final stages of an entirely synthetic enterprise towards vinblastine (**1**), the first such route would not be achieved until 2002 by Tohru Fukuyama and members of his group at the University of Tokyo.[10] This work is the subject of the remainder of this chapter. As we shall see, not only did they implement an effective strategy to form the C-16′ center of **1** stereoselectively based on an oxidative coupling, but they also blended a number of carefully conceived modifications of standard synthetic transformations with a series of inventive synthetic strategies to overcome several other challenging features within the molecular structure of vinblastine.

18.2 Retrosynthetic Analysis and Strategy

Because of the highly successful nature of the chemistry described in Scheme 1, the Fukuyama group was certainly enticed by the prospects of effecting an oxidative coupling event between vindoline (**2**) and some advanced intermediate corresponding to the indole portion of vinblastine during the late stages of their synthetic drive towards **1**. As we have already seen, however, the major issue associated with adopting such a convergent strategy is selecting a building block with an appropriate spatial orientation so that vindoline (**2**) can add to its correct face and establish the natural stereochemistry of vinblastine at C-16′. Since ample precedent had already established that indole intermediates possessing the finished bridged bicyclic architecture could not be effectively induced to react in this fashion (such as **8**, Scheme 2), the Fukuyama group elected in their retrosynthetic analysis (Scheme 3) to open the piperidine ring within this system to **12** before cleaving the bond between the major indole and indoline subunits of **1** to building blocks **14** and **2**. Thus, in the forward direction, application of an appropriate reagent such as *t*-BuOCl to **14** was anticipated to provide the transient intermediate **13**, which would hopefully be conformationally biased to enable the selective approach of vindoline (**2**) to its α-face. Although the fairly large degree of flexibility of the 11-membered ring in **13** renders the outcome of this reaction difficult to predict on the basis of first principles, extensive molecular modeling by the group suggested its plausibility. Moreover, some synthetic model studies towards vinblastine executed almost fifteen years earlier by the Schill group in Basel demonstrated that the desired facial selectivity could be achieved with a macrocyclic intermediate similar to **13** (lacking only the C-20′ stereocenter) through an oxidative coupling process.[11]

Assuming that this paper-based design could indeed be reduced to practice, then this initial set of simplifications has broken vinblastine down into two advanced subtargets, vindoline (**2**) and compound **14**, each endowed with its own set of unique synthetic challenges. Although we will get to vindoline (**2**) in due course, for now the remainder of Scheme 3 provides a possible retrosynthetic analysis of the second of these critical subgoal structures (i. e. **14**). As you might have already noted, a somewhat unique protecting group has been appended onto the nitrogen atom within the macrocycle of **14** and advanced intermediate **12**, namely a 2-nitrobenzenesulfonamide (also known as a nosyl group, Ns). Its selection, however, was far from arbitrary in that the Fukuyama group had already established that amines bearing this protecting group could readily participate in Mitsunobu-type alkylations as well as macrocyclizations (with the added benefit of facile deprotection upon exposure to a sulfide, vide infra).[12] Accordingly, while this functionality could thus be envisioned to facilitate the operations needed to convert **12** into vinblastine (**1**) in the final stages of the synthesis, its continued presence within **14** signaled the possibility that the macrocyclic ring of this intermediate could be derived from the nucleophilic addition of an Ns-protected amine to an electrophilic acceptor such as the epoxide in **15**. In turn, this epoxide could arise from a diol through a displacement reaction and the nosyl-protected amine could be the product of a Mitsunobu addition to a primary alcohol precursor (**16**).[13] As such, this analysis has productively removed one ring, but left two major synthetic challenges for further consideration in its wake: finding a method to fashion a 2,3-substituted indole and establishing the two stereogenic centers on the carbon chain appended to the 2-position of that indole.

Fortunately, the Fukuyama group had already developed a novel methodology that would hopefully prove capable of handling the

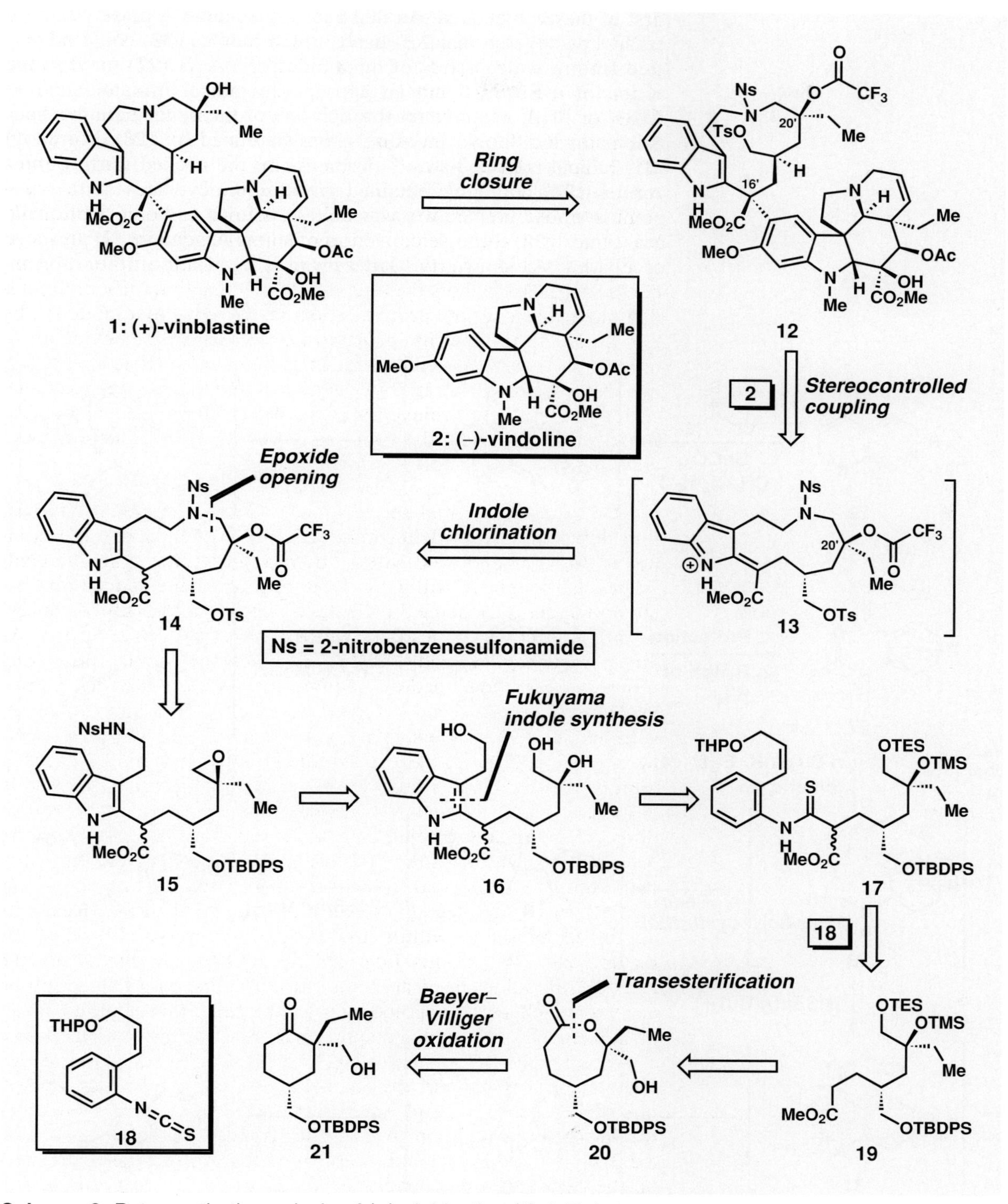

Scheme 3. Retrosynthetic analysis of (+)-vinblastine (**1**): initial stages.

first of these objectives. As illustrated in Scheme 4, these researchers had discovered that 2,3-disubstituted indoles (**32**) could be created from a wide variety of thioamide precursors (**27**) through the action of n-Bu_3SnH and an appropriate radical initiator such as AIBN or Et_3B, presumably through one of the degenerate mechanistic pathways shown invoking either stabilized sp^3 (**28**) or imidoyl (**31**) radical intermediates.[14] Furthermore, the needed starting thioamides (**27**) could be obtained from isothiocyanates (**26**), compounds whose preparation was already known to be possible from quinolines (**22**) through the action of thiophosgene in the presence of a base.[15] As shown, this latter event is the result of initial forma-

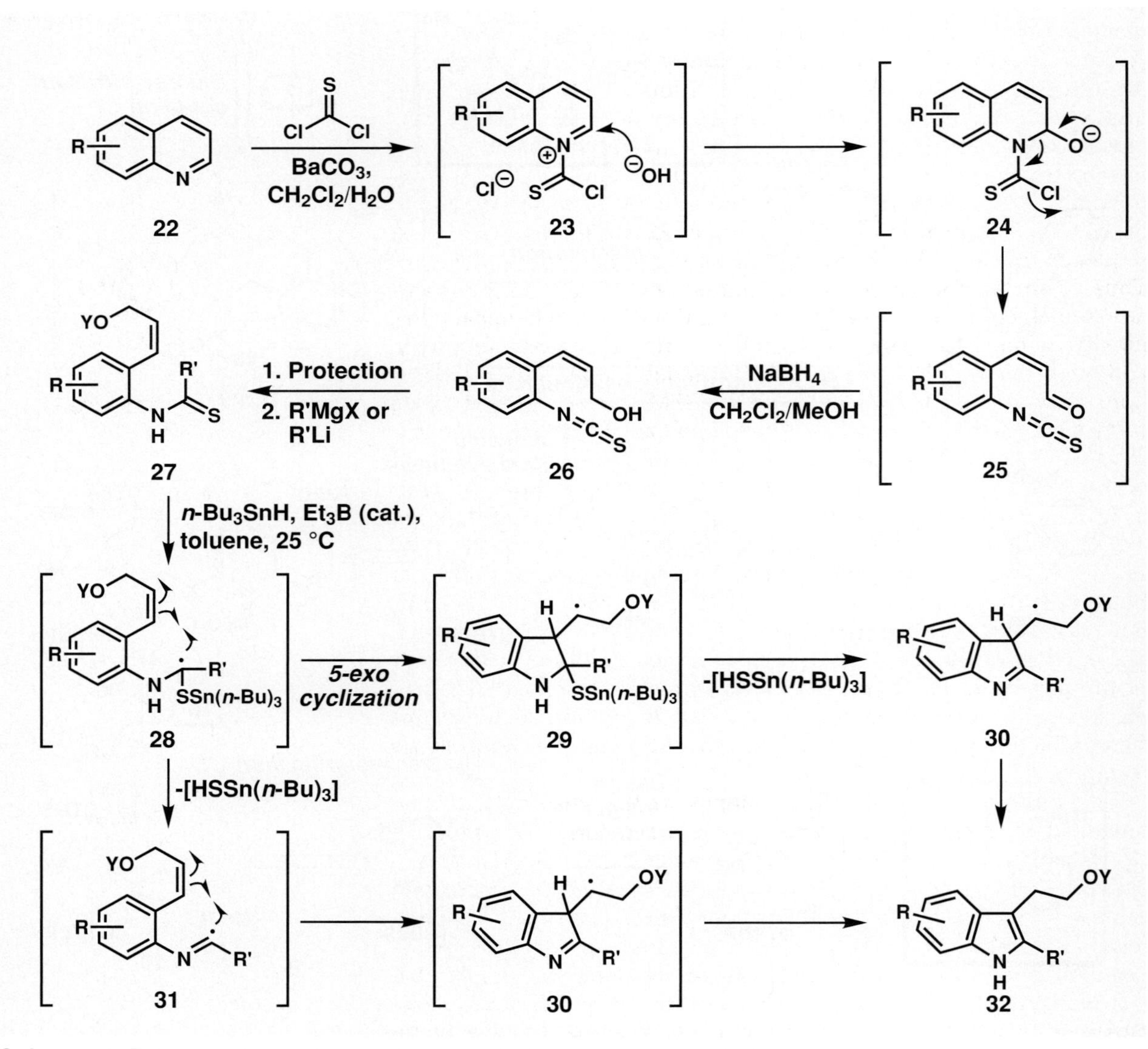

Scheme 4. Fukuyama's synthesis of indoles (**32**) from quinolines (**22**).

tion of **23**, followed by hydroxide attack at the indicated carbon atom to afford **24**. Since this intermediate is suitably disposed for an intramolecular ring fragmentation, **24** smoothly rearranges *in situ* to aldehyde **25**, a compound that can then be converted into **26** by a standard reducing agent (such as $NaBH_4$ in CH_2Cl_2/MeOH). Although this isothiocyanate-formation process constitutes well-established chemistry, it is instructive to note that the ring opening of quinolines with this protocol is unique to thiocarbonyl species such as **23**. Had phosgene been used instead of thiophosgene to afford a Reissert-type intermediate (C=O instead of C=S), then, on the basis of matching hard acids with hard bases, the hard hydroxide species would have preferentially attacked the carbonyl group and reformed the quinoline starting material rather than lead to an isocyanate product.

Expecting that this general methodology would be applicable to the formation of the indole subunit within **16** (Scheme 3), then this heterocycle could be retrosynthetically adjusted to thioamide precursor **17**, an intermediate amenable to further disconnection into methyl ester **19** and isothiocyanate **18** assuming a nucleophilic-addition-based transform. The latter of these materials could, of course, be obtained from quinoline **22** through the reaction sequence already discussed within the confines of Scheme 4. Thus, with the formation of the indole ring system addressed, it now remained to determine how the additional stereochemical complexity within **19** could be installed with its proper topology. Although this problem might seem formidable upon initial inspection, if both the methyl ester and the TMS group were excised from **19** to reveal a free carboxylic acid and an alcohol, then the disconnections needed to meet this objective should be relatively clear in that this variant of **19** could likely be derived from the opening of a seven-membered lactone precursor such as **20**. The fact that the oxygen atom within this newly adopted ring is adjacent to a fully substituted carbon atom implicates its possible insertion through a Baeyer–Villiger oxidation of a predecessor such as **21**. Thus, with the revelation of a six-membered cyclic template through this final disconnection, the means to generate both stereocenters of this new target structure should no longer seem problematic in that a wealth of literature precedent exists to fashion such systems stereoselectively through cyclic stereocontrol and/or asymmetric reactions.

With one fragment reduced to simple starting materials, it now remained to determine a viable synthetic sequence for vindoline (**2**, Scheme 5). Although this target possesses far greater stereochemical complexity than **14** with its six contiguous stereogenic centers (three of which are quaternary) and five rings, in many respects it is an easier molecular architecture to disassemble retrosynthetically in that the construction of its stereochemically rich six-membered C-ring could confidently be assigned to an intramolecular Diels–Alder reaction. Indeed, if **2** was traced to **34** through the operations indicated in Scheme 5, then application of

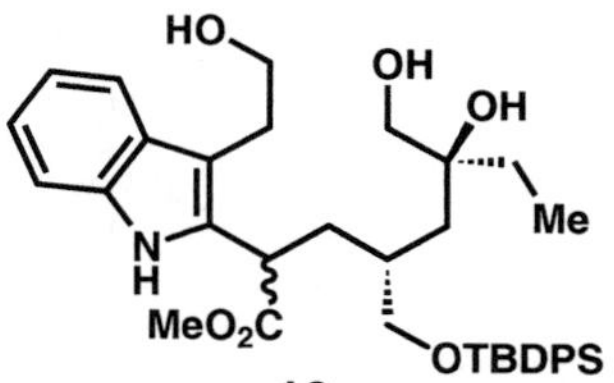

16

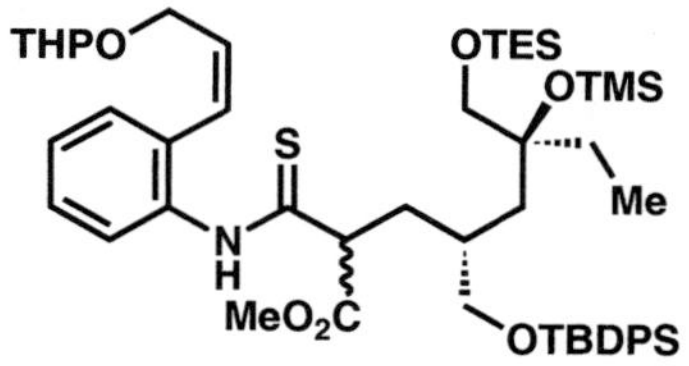

17

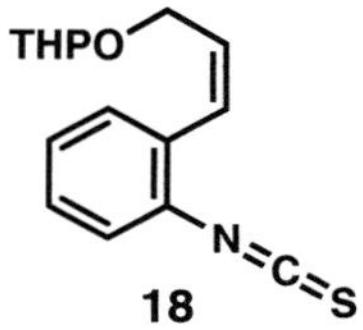

18

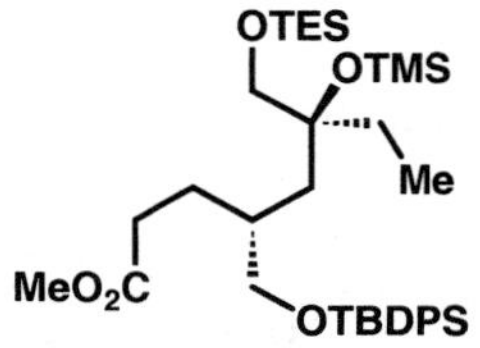

19

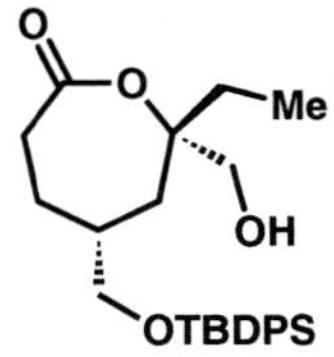

20

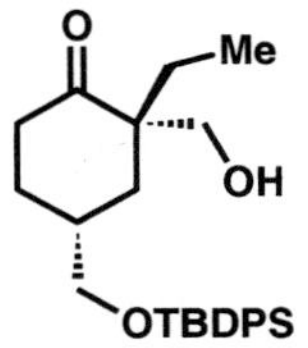

21

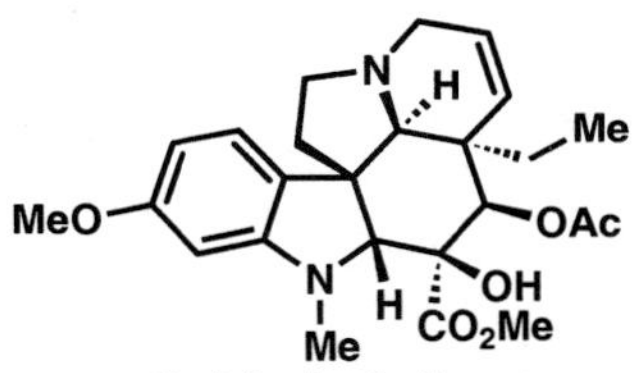

2: (–)-vindoline

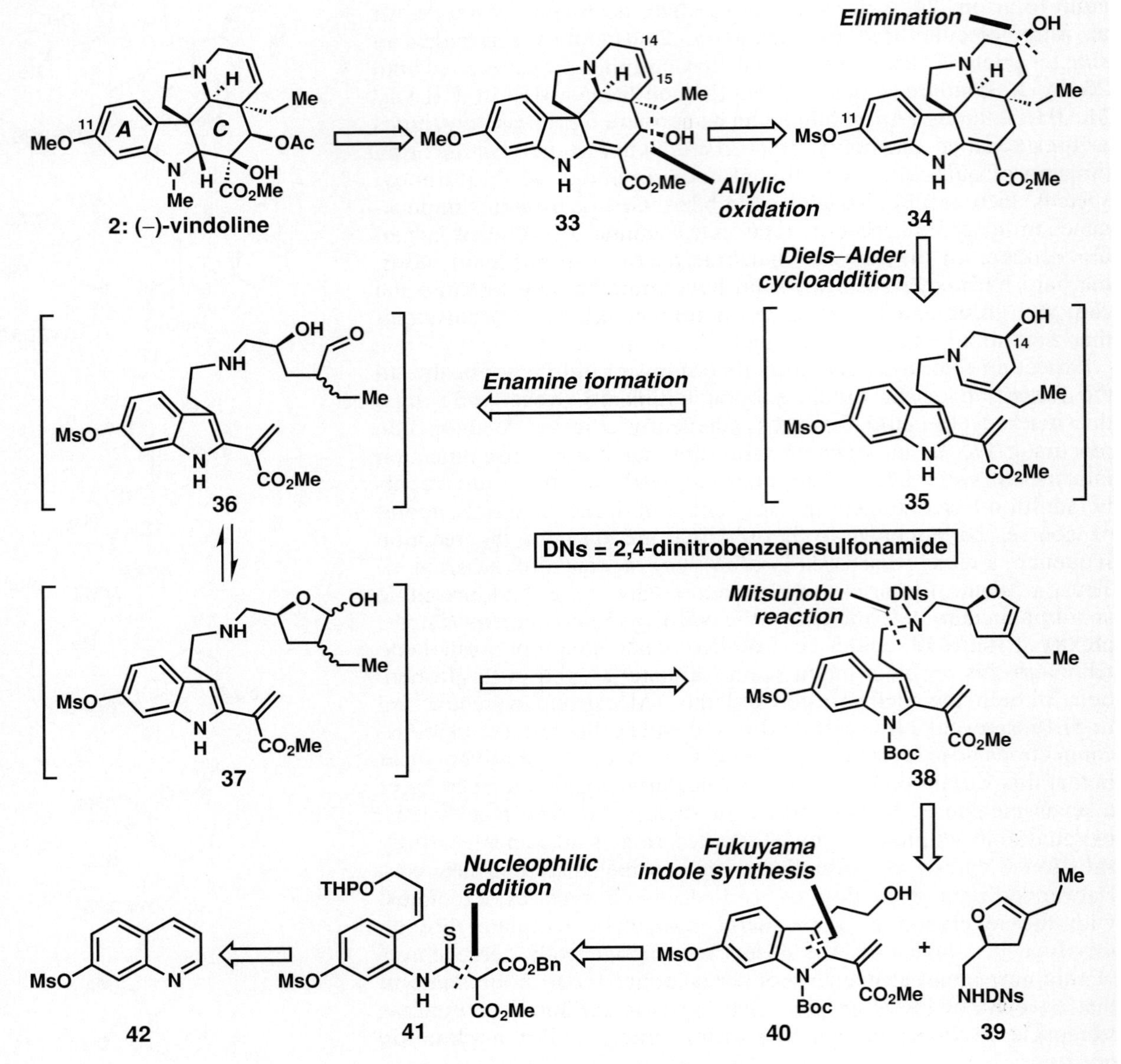

Scheme 5. Retrosynthetic analysis of (+)-vinblastine (1): final stages.

a Diels–Alder transform to this new subgoal structure would lead back to **35** as a potential retrosynthetic precursor. Rather than try to target this specific intermediate for synthesis, the Fukuyama group reasoned that the enamine comprising its dienophile could potentially arise in the same pot through a condensation between the aldehyde and secondary amine units of **36**. However, with an alcohol in close proximity to this aldehyde, **36** should not constitute a distinct (or isolable) species either, but instead should exist in equilibrium with its lactol congener (**37**), a compound accessible through hydration of a cyclic enol such as **38**. Thus, in a single cascade operation, application of acidic conditions during the forward synthesis could conceivably drive the equilibria needed to convert **38** directly into **34**, with the opportunity to apply an additional driving force (heat or Lewis acid catalysis) to induce the final intramolecular Diels–Alder reaction if necessary.

While the execution of Diels–Alder reactions to establish molecular complexity constitutes a relatively common occurrence in natural product total synthesis,[16] two features of the proposed transformation in this context are far from conventional. First, the lone and somewhat distal C-14 stereocenter in **35** would have to facially bias the presentation of the diene and dienophile components for the correct stereochemical outcome to occur, and, second, the orbital coefficients of these participants would need to be appropriately matched for a facile inverse-electron-demand Diels–Alder reaction. Precedent in several contexts, such as the intramolecular Diels–Alder reaction executed in Martin's synthesis of manzamine A (see Chapter 8), indicated that the first of these objectives should not prove problematic. The second was far less certain. Although the electron-withdrawing ester appended to the diene and the electron-rich nitrogen atom adjacent to the dienophile should facilitate a $LUMO_{diene}$-controlled cycloaddition of **35**, the ability of the indole nitrogen atom to donate electron density into the diene system could potentially obviate any benefits afforded by the ester group. As such, the Diels–Alder reaction could fall under a neutral case and thus prove challenging (as discussed in Chapter 2). Following careful analysis of this problem, the Fukuyama group envisioned that they could increase their chances of success by placing an electron-withdrawing substituent on the neighboring aromatic ring, as its inductive and mesomeric properties could render the entire indole ring electron-poor and thus create a better matched case for an inverse-electron-demand Diels–Alder reaction. Accordingly, although vindoline (**2**) possesses a methoxy group at C-11, **34** and its synthetic precursors were modified to bear a methanesulfonate instead. Before continuing, however, we should note that Diels–Alder-based approaches to indole alkaloids bearing the gross structure of vindoline (**2**) are legion, particularly since this reaction is part of their presumed biosynthesis. As an especially inventive example among these precedents, we have delineated the critical aspects of Grieco's total synthesis of the related natural product pseudotabersonine (**49**) in Scheme 6.[17]

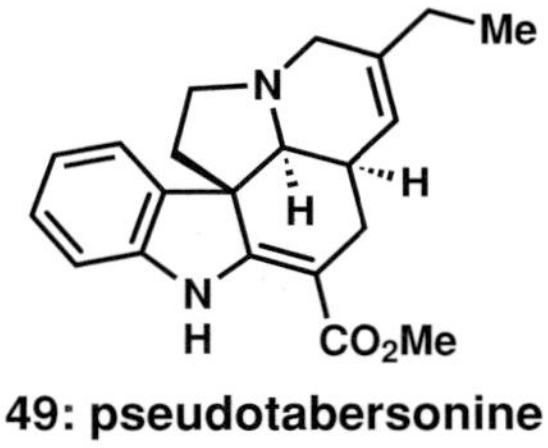

With the assembly of vindoline (**2**) now reduced to **38** (Scheme 5), the remaining retrosynthetic disconnections mirror those applied to the upper domain of vinblastine. First, by appending a protecting group related to the nosyl group on **38**, a 2,4-dinitrobenzenesulfonamide (DNs), this intermediate could then be separated into fragments **40** and **39**, assuming that a Mitsunobu reaction could merge these

49: pseudotabersonine

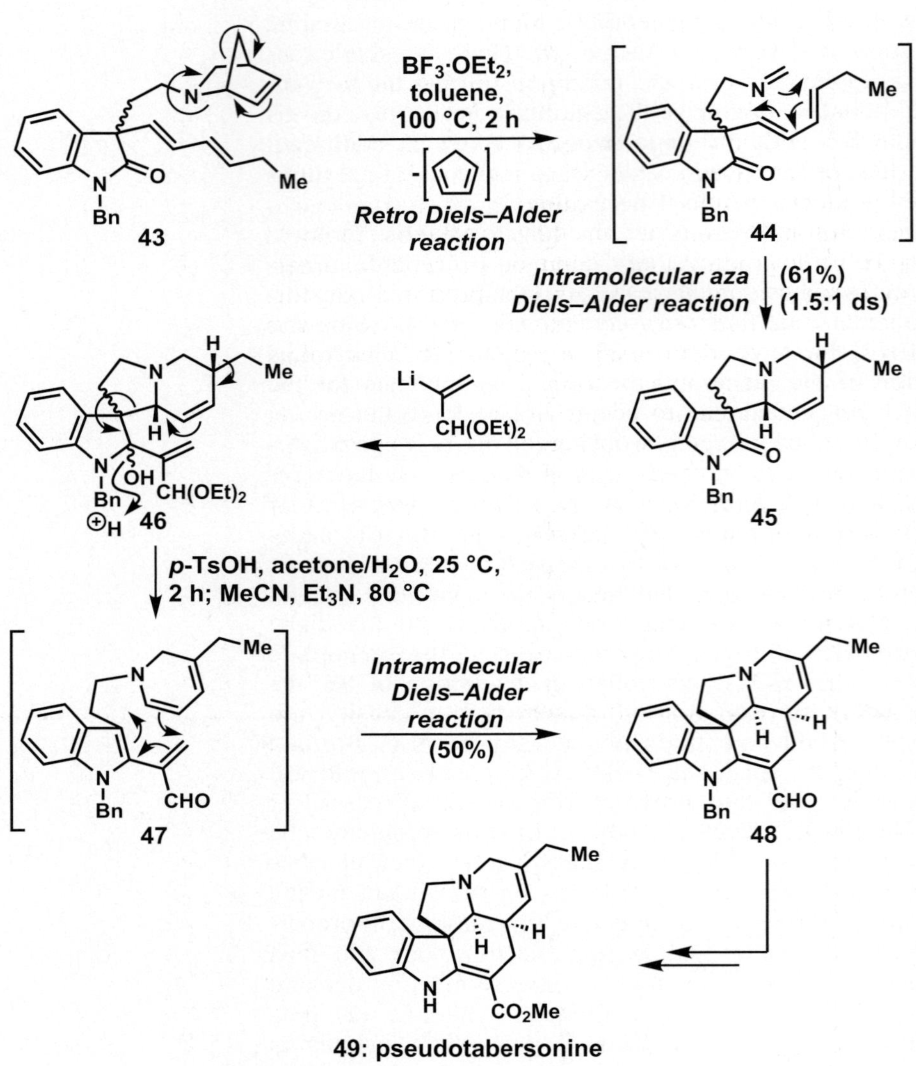

Scheme 6. Grieco's total synthesis of the alkaloid pseudotabersonine (**49**) based on a Diels–Alder design (1993).[17]

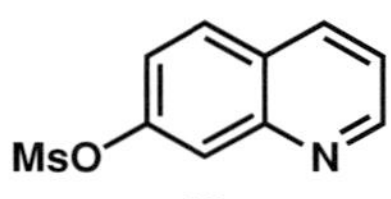

pieces together during the actual synthesis. Next, with a simplified 2,3-disubstituted indole unveiled in the form of **40**, a second application of the Fukuyama indole synthesis protocol would then engender disassembly to thioamide **41**, an intermediate which, in turn, could be derived from an isothiocyanate obtainable from quinoline **42** and commercially available benzyl methyl malonate.

Before closing this section, a few additional remarks about quinoline **42** are in order. Most typically, heterocycles of this type are prepared by the Skraup method, which, as shown in

Scheme 7a, involves the initial addition of an aniline derivative in a Michael fashion to acrolein (**51**), followed by cyclization, elimination of water, and oxidation.[18] While the strongly acidic conditions needed to accomplish this event are arguably harsh, most substrates work well under this protocol. However, this general approach is less than ideal when aniline starting materials bearing substituents *meta* to its amine are employed, as a mixture of difficult to separate regioisomeric products results. Accordingly, since the *meta*-oriented methanesulfonate group within **42** (cf. Scheme 5) could thus conceivably be a product that would be difficult to obtain regioselectively using the standard Skraup method, the Fukuyama group developed a modified form of this reaction to overcome this potential problem.[19] As delineated in Scheme 7b, they found that by appending an electron-withdrawing group onto the aniline nitrogen atom, the *para*-directing power of the R substituent in **56** was markedly enhanced relative to intermediates lacking such nitrogen protection (i. e. **52**), providing the desired regioisomer (**57**) in typically better than 8:1 selectivity. Following this cyclization, treatment with base then completed the assembly of the final quinoline (**58**) through a process that can be rationalized as either direct hydrolysis of the sulfonamide group followed by air oxidation

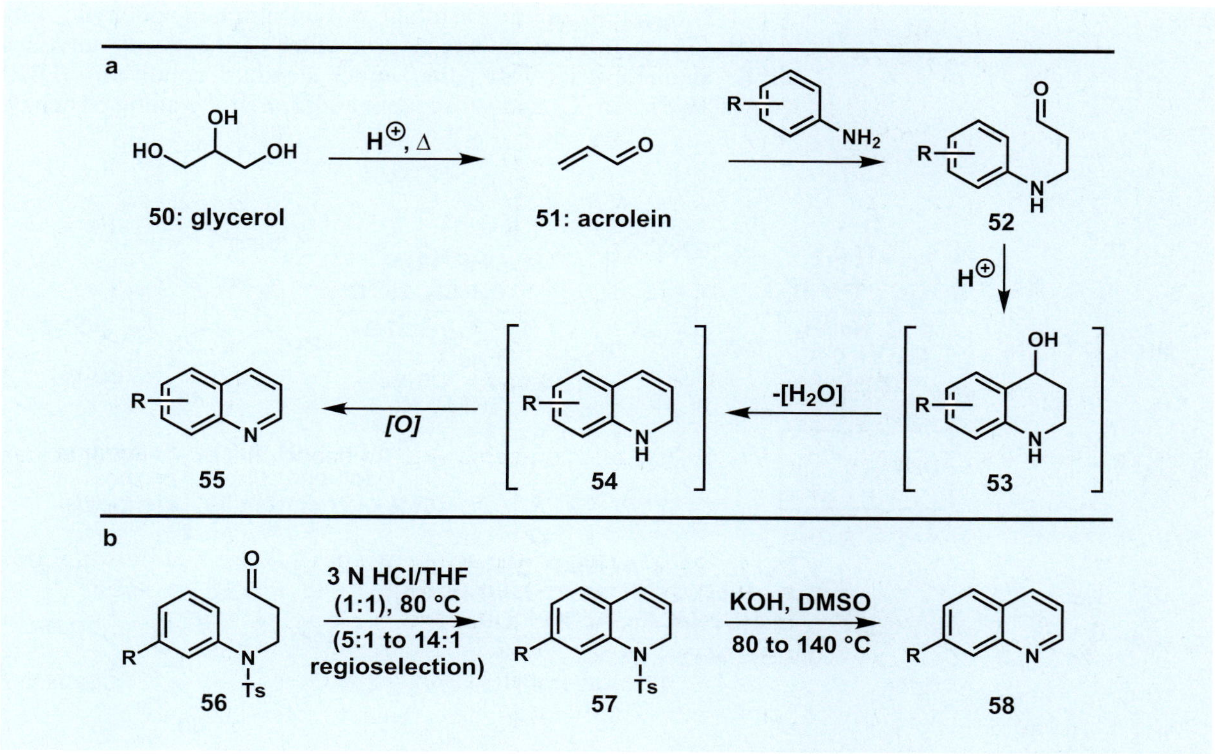

Scheme 7. The classical Skraup quinoline synthesis (a) and Fukuyama's modified approach (b) using *N*-aryl-*N*-sulfonylaminopropionaldehydes (**56**).

2: (–)-vindoline

or β-elimination of the sulfonyl group. As such, this four-step sequence could provide a useful alternative for the selective formation of quinoline **42** in the likely scenario that the more concise Skraup method should afford no such fidelity.

18.3 Total Synthesis

We begin our discussion of this total synthesis with efforts directed towards the assembly of vindoline (**2**), the natural product that, as mentioned above, could hopefully be merged with the other half of vinblastine through an oxidative coupling at the end of the campaign. In order to gain entry into the challenging aspects of their proposed synthesis of this essential piece, the Fukuyama group first had to prepare building blocks **40** and **39**. The successful route developed to access the larger of these entities, indole **40**, is outlined in Scheme 8. Following the large-scale preparation of quinoline **42** through the unique methodology discussed in Scheme 7b (as the Skraup synthesis intriguingly failed to deliver any quinoline product whatsoever), its conversion into isothiocyanate **59** proceeded smoothly upon initial ring opening, effected with thiophosgene and Na_2CO_3 in a mixture THF/H_2O at 0 °C, followed by selective 1,2-reduction of the resultant α,β-unsaturated aldehyde with $NaBH_4$. From this intermediate, protection of the newly unveiled allylic alcohol as its THP ether under standard conditions (DHP, CSA, CH_2Cl_2, 25 °C) and subsequent addition of the anion of benzyl

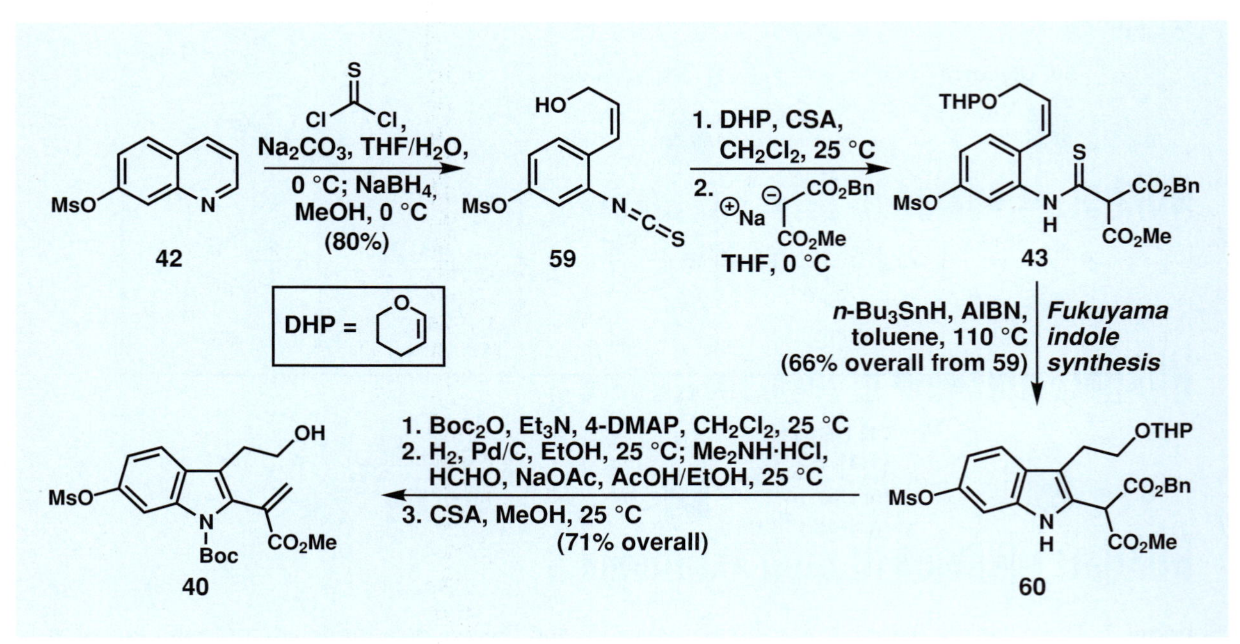

Scheme 8. Synthesis of indole building block **40**.

methyl malonate completed the assembly of thioamide **43** in 82 % overall yield from **59**. As such, these operations afforded the key precursor needed for the Fukuyama indole synthesis protocol (cf. Scheme 4) based on radical cyclization.[14] Pleasingly, this methodology rose to the occasion as treatment of **43** with n-Bu_3SnH and catalytic AIBN in toluene at 110 °C gave rise to the desired 2,3-disubstituted indole product (**60**) in 80 % yield. With this operation effected, only two main objectives remained to complete the assembly of **40**, namely formation of a vinyl ester from the malonate portion and alteration of the resident protecting group ensemble to enable its merger with amine **39** once its synthesis was complete. As shown, these requirements were accomplished in just three steps through protection of the indole nitrogen atom as a Boc carbamate, hydrogenolytic removal of the benzyl ester followed by a decarboxylative Mannich reaction with formaldehyde, and cleavage of the THP ether by CSA. These operations proceeded in a combined yield of 71 %.

Having realized a synthesis of indole **40**, attention could now be focused on generating its eventual coupling partner, amine **39** (Scheme 9). The critical challenge posed by this building block was finding a means to obtain it with complete enantiopurity, as its lone stereogenic center would eventually be enlisted to control

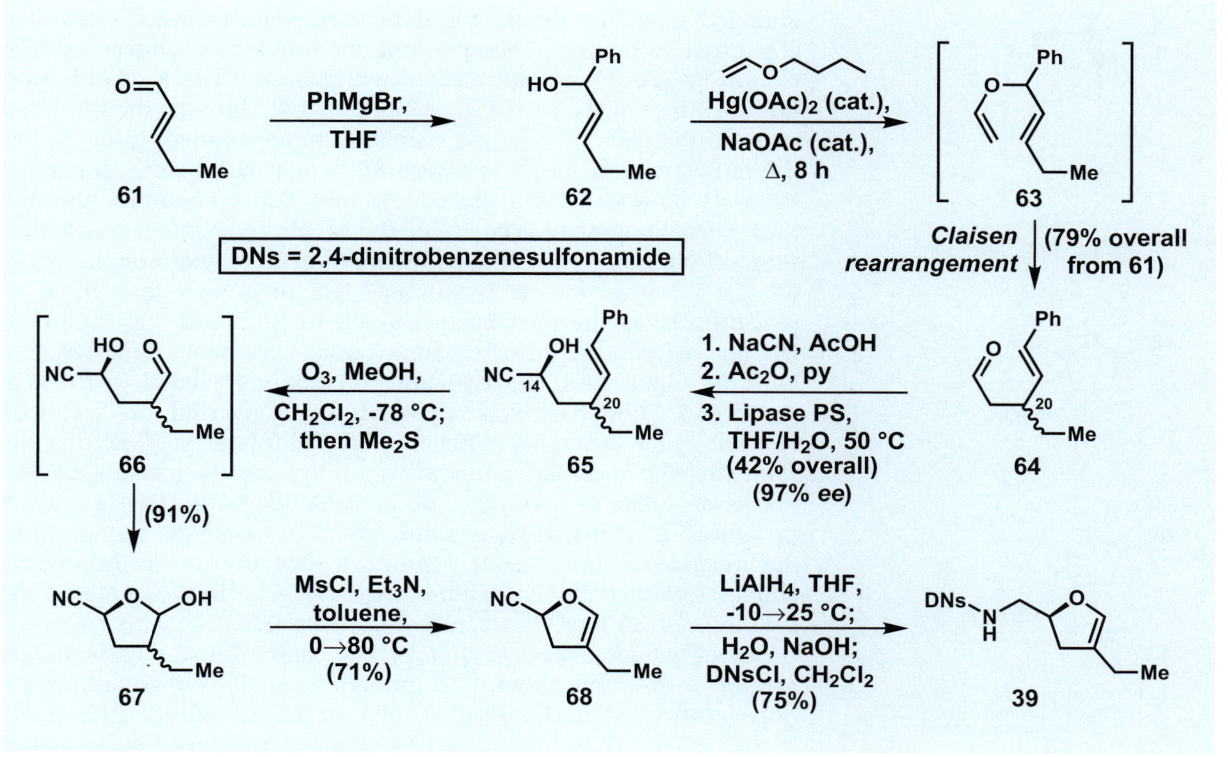

Scheme 9. Synthesis of amine building block **39**.

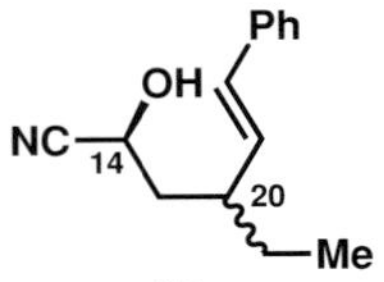

the facial selectivity of the intramolecular Diels–Alder reaction at the heart of the entire synthetic strategy for vindoline (**2**). Rather than employ a chiral starting material or an asymmetric reaction to meet this requirement, however, the Fukuyama group envisioned that an enzymatic resolution of a cyanohydrin such as **65** might be more effective since this motif could eventually afford a means to generate both the five-membered ring of **39** and its appended amine. In order to test this proposition, they first needed access to an intermediate such as γ,δ-unsaturated aldehyde **64**, a precursor whose structure clearly implicated the possibility of employing a Claisen rearrangement to effect its construction. Indeed, following the formation of allylic alcohol **62** through a Grignard reaction between phenylmagnesium bromide and aldehyde **61**, subsequent treatment with *n*-butyl vinyl ether in the presence of catalytic $Hg(OAc)_2$ and NaOAc smoothly initiated a Claisen rearrangement to afford **64** in 79 % overall yield from **61**. Although this latter conversion is relatively standard chemistry, one should note that the selection of *n*-butyl vinyl ether from the variety of reagents capable of effecting vinyl ether formation stemmed from the fact that excess amounts of this reactant can readily be removed by concentration on a rotary evaporator.[20]

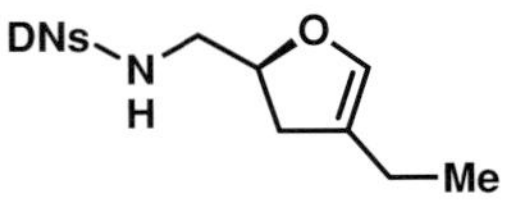

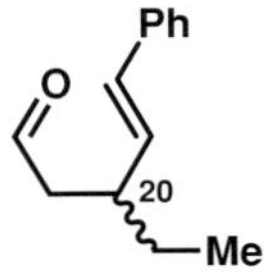

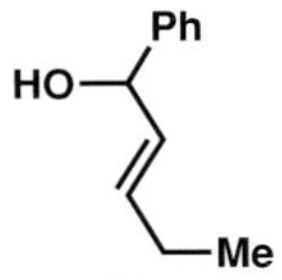

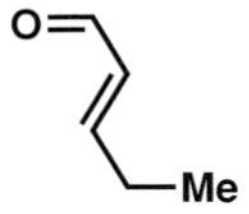

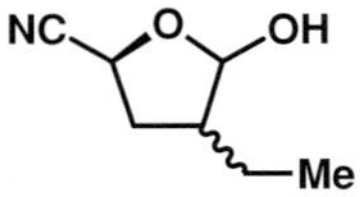

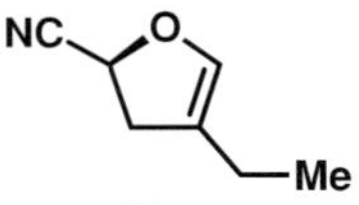

With this material in hand, the aldehyde within **64** was then successfully converted into a cyanohydrin upon reaction with NaCN in the presence of acid. To resolve the two isomeric C-14 products obtained from this transformation, these intermediates were then converted into their corresponding cyanoacetates under standard conditions (Ac_2O, py) and treated with Lipase PS in a solvent mixture of THF and H_2O at 50 °C. As anticipated, this enzyme smoothly effected the lysis of only the acetate group appended to the *S*-configured starting material to afford **65**, which was readily separated from its untouched *R*-disposed cyanoacetate counterpart through flash chromatography. The effectiveness of this resolution is underscored by the fact that **65** was obtained in 42 % yield overall from **64** (50 % would be the maximum) and in greater than 97 % *ee* (*ee*, not *de*, as the previously formed C-20 center was racemic). Having secured the critical stereochemical element of the targeted building block (**39**), the remaining operations needed to finish it proceeded without incident. First, **65** was converted into lactol **67** in 91 % yield through a conventional ozonolysis step, setting the stage for the eventual elimination of this newly formed alcohol via its mesylate to enol ether **68** as achieved with MsCl and Et_3N in toluene at elevated temperature (80 °C). Complete reduction of the cyanide within the new product (**68**) to the corresponding amine, followed by its capture with DNsCl, then completed the synthesis of **39** in 53 % overall yield from **67**.

Having gained access to both pieces, amine **39** was immediately merged with indole subunit **40** through a standard Mitsunobu reaction protocol (DEAD, Ph_3P, C_6H_6) to afford **38** in 79 % yield (Scheme 10). With this adduct in hand, an advanced staging area had now been reached from which to probe the viability of the proposed cascade that would hopefully culminate in an intramolecular Diels–Alder reaction, and thereby complete all of the ring systems and three of the six stereocenters of vindoline. Although a fair amount of reaction scouting and optimization were required to execute the designed sequence from **38** to pentacycle **34**, the process was eventually realized in an impressive overall yield of 73 %. In the event, treatment of **38** with an excess of TFA and Me_2S in CH_2Cl_2 at 25 °C over the course of 1 hour presumably gave rise to **37** by way of amine deprotection, Boc cleavage, and hydration of the enol ether. Following a standard aqueous workup, the crude organic extracts from this step were then suspended in a 5:1 mixture of $MeOH/CH_3CN$ and exposed to 5 equivalents of pyrrolidine at 0 °C. Upon warming this solution to 50 °C and stirring for an additional 3 hours, the assembly of **34** was then presumably effected by recourse to intermediates **36** and **35**. As desired, **34** was formed as a single stereoisomer, most likely due to the governing

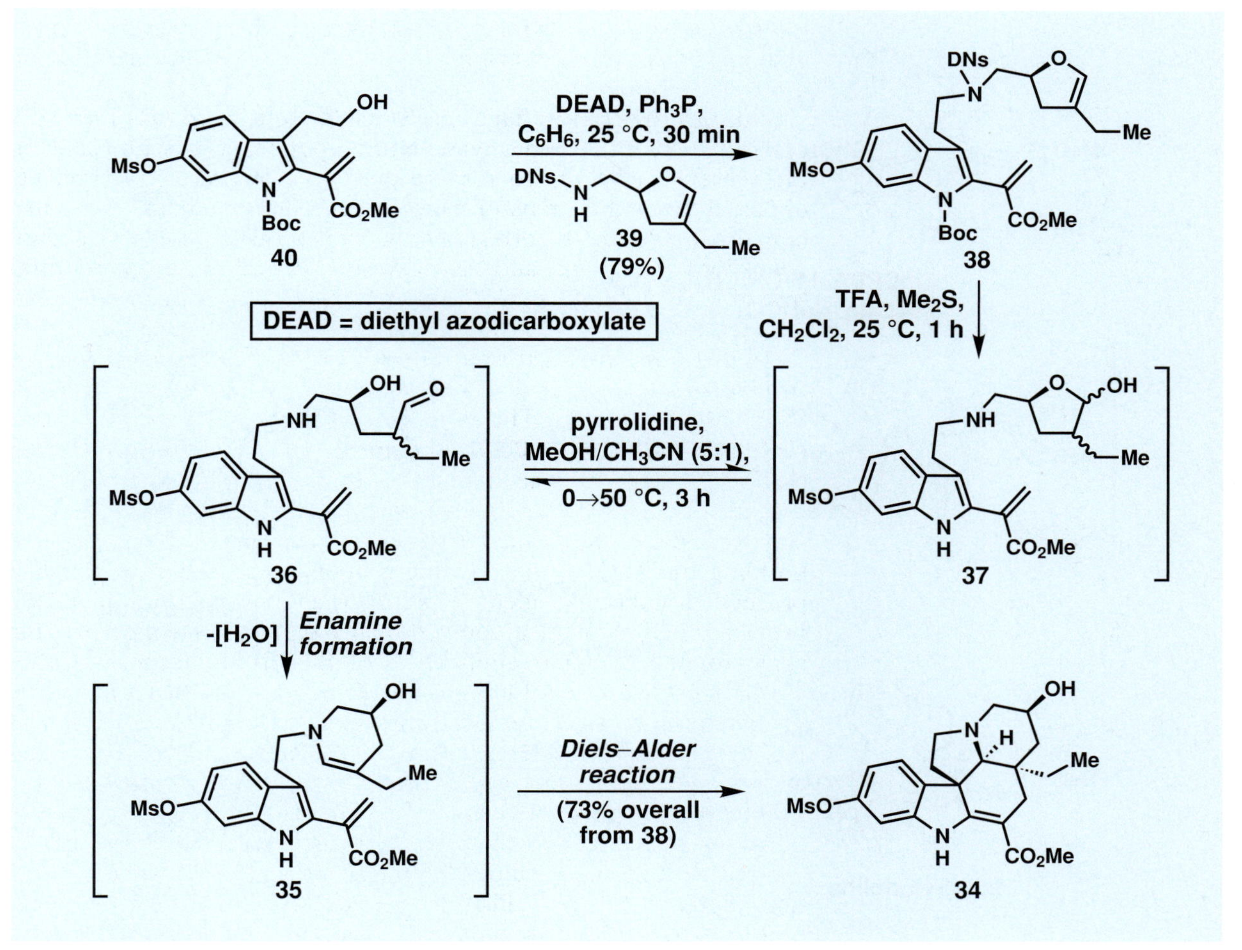

Scheme 10. Synthesis of advanced intermediate **34**.

influence of the lone stereogenic center within **35** for the final [4+2] cycloaddition. In line with expectations, the presence of the methanesulfonate group at C-11 was essential for the overall material throughput of this sequence as evidenced by the decreased yield (58 %) observed when this same protocol was applied to a starting

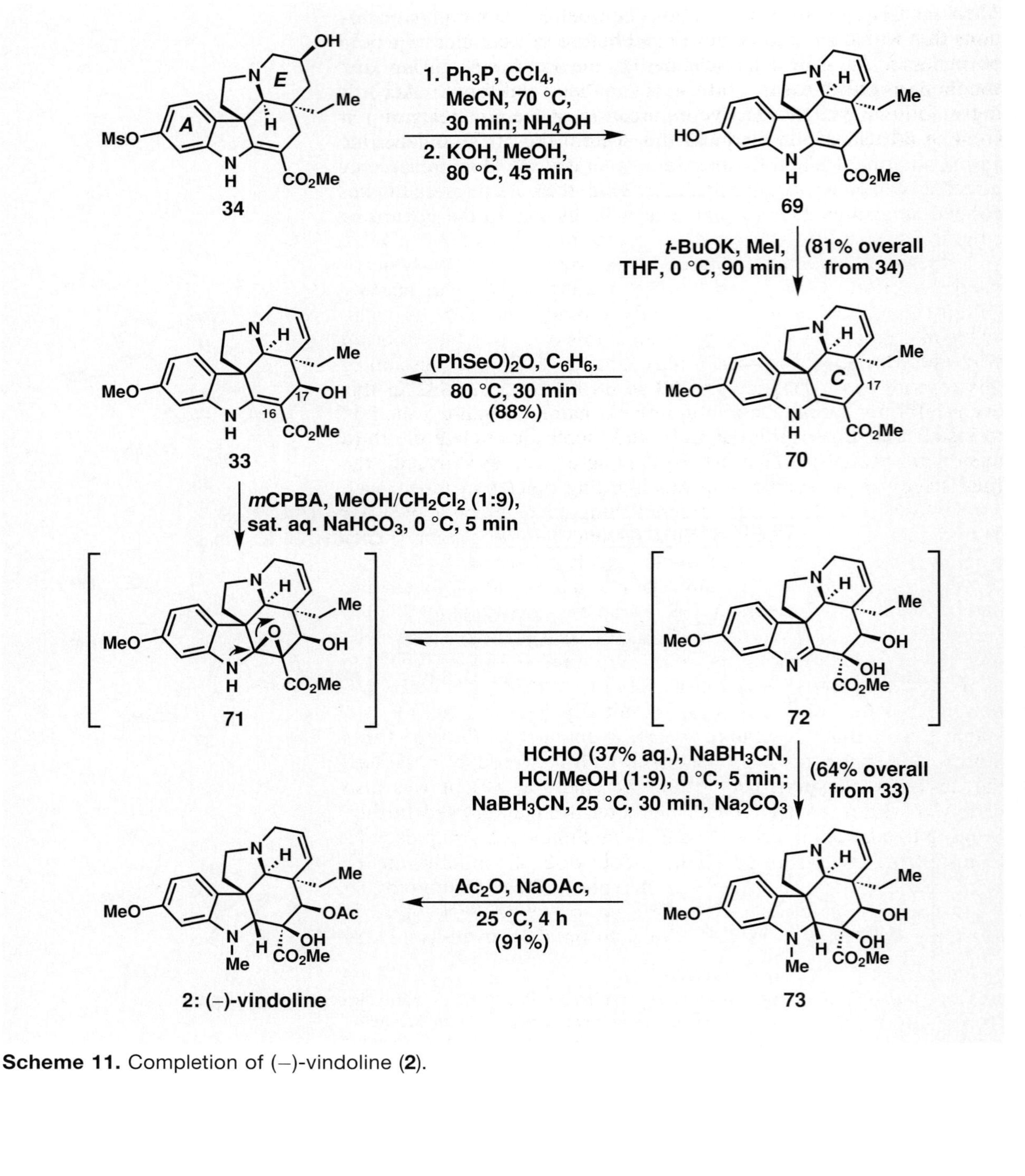

Scheme 11. Completion of (–)-vindoline (**2**).

material bearing a hydrogen at this site.[21] As such, the electron-withdrawing properties of the methanesulfonate substituent clearly enhanced the overall orbital matching, enabling a more facile cycloaddition to occur.

At this stage, with the majority of the architecture of vindoline (**2**) expressed by **34**, only a few functional group adjustments were required before a synthesis of this target would be complete. Their seemingly simple nature, however, belied the creative operations that would have to be developed before this compound would be in hand. As shown in Scheme 11, the Fukuyama group first sought to establish the A- and E-rings in their final format. Accordingly, following regioselective elimination of the alcohol group at C-14 in **34** using CCl_4 and Ph_3P in acetonitrile at 70 °C, alkaline hydrolysis of the phenolic mesylate generated **69**. Since this compound tended to decompose upon standing, the newly unveiled phenol was immediately methylated at 0 °C using a combination of *t*-BuOK and MeI in THF, leading to the formation of **70** in 81 % yield for these three steps. With both these rings now properly decorated, all efforts concentrated on completing the C-ring, starting with a chemo- and stereoselective allylic oxidation of **70** to install an alcohol group at C-17. As delineated, this operation was effected with phenylselenenic anhydride (see Chapter 6 for a discussion of this reaction) to afford alcohol **33** in 88 % yield. Success in this event left only a single stumbling block, namely finding a method to install a hydroxy group at C-16 and thereby complete the third quaternary center of vindoline. By having elected to wait until the final stages of the synthesis of this building block to address such a transformation, however, accomplishing this objective was far from trivial in that it required the approach of some oxygen source from the more hindered face of **33**. While most reagents cannot overcome such substrate bias alone, since an adjacent allylic alcohol had just been installed these researchers envisioned that they could enlist this motif to direct the approach of such a reagent along the less favored trajectory. This conjecture indeed proved accurate as treatment of **33** with 2 equivalents of *m*CPBA in a $NaHCO_3$-buffered mixture of $MeOH/CH_2Cl_2$ initially gave rise to **71** as a single stereoisomer, a compound that eventually isomerized to its more stable imine tautomer (**72**). Although this product possessed the desired C-16 quaternary center of vindoline (**2**), it was also endowed with a reactive imine that could be manipulated further. Seeking to take advantage of this fact, the Fukuyama group decided to install the methyl group in the indole ring of vindoline in the same pot through a reductive amination reaction employing HCHO and $NaBH_3CN$, an event that led to the synthesis of **73** in 64 % overall yield from **33**.* Critical to obtaining this yield was

---

* Just as the allylic alcohol group at C-17 in **33** directed *m*CPBA to the more hindered face of the molecule, in this step the alcohol group at C-16 played a similar role as it ensured the initial delivery of hydride onto the more sterically shielded side of **72** at C-15.

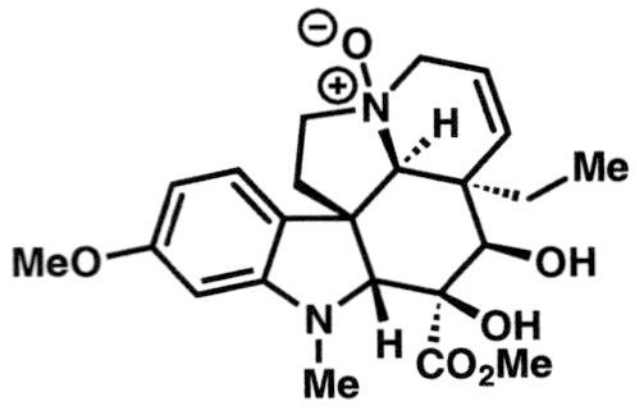

the brief exposure of the reaction products to sodium hydrogensulfite during the final workup, as a small amount of the desired adduct existed as its *N*-oxidized congener (**74**, see column figure) due to the excess *m*CPBA employed in the initial step; this operation served to remove this unwanted oxygen atom. Having successfully forged **73**, selective acetylation of its more accessible secondary alcohol at C-17 then completed the assembly of vindoline (**2**) in 91 % yield. In total, only 16 linear steps were required to synthesize this piece, and their overall efficiency enabled the preparation of hundreds of milligrams of this natural product for use in the final stages of the campaign.

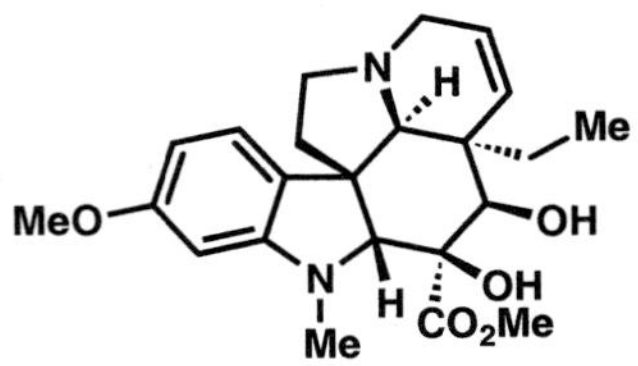

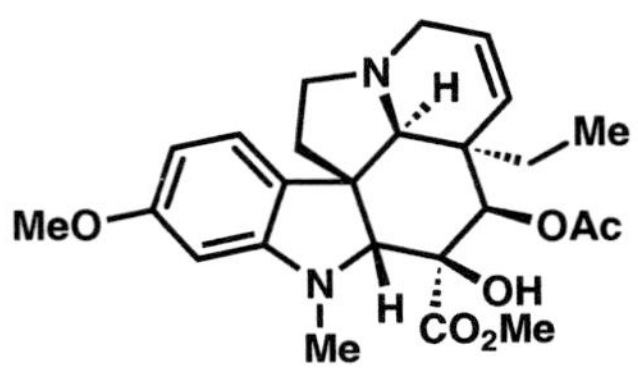

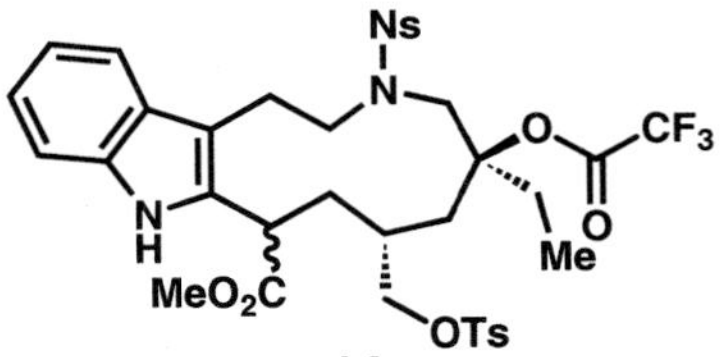

Before we can discuss these transformations, however, the synthesis of the other requisite building block, indole **14**, must be addressed. As mentioned above, the proposed preparation of this fragment rested, at least in its opening stages, upon the homochiral assembly of a six-membered cyclic product such as **21**. As shown in Scheme 12, the Fukuyama group was able to develop a unique synthesis of this compound through a route that featured both standard asymmetric chemistry as well as a facially selective intramolecular 1,3-dipolar cycloaddition reaction. The opening operations involved the installation of a single stereocenter using a member of the highly powerful family of oxazolidinone chiral auxiliaries pioneered by David Evans at Harvard University.[22] As indicated, after forming the pivaloate mixed anhydride of 4-ethylpent-4-enoic acid (**75**), the addition of the deprotonated form of (4*R*)-4-benzyloxazolidin-2-one to this reactive intermediate smoothly afforded **76**, a compound primed for the selective incorporation of a new stereocenter. Upon chelation of its two carbonyl groups with titanium [using (*i*-PrO)$TiCl_3$ generated *in situ* through the initial reaction of $TiCl_4$ with Ti(O*i*-Pr)$_4$], stereoselective cyanoethylation was then achieved with acrylonitrile and Hünig's base in CH_2Cl_2 at 0 °C.[23] With these steps affording **77** in 73 % yield from **75**, the chiral auxiliary within this product was no longer needed, and thus it was excised with $NaBH_4$ in a solvent mixture of THF and H_2O to afford its primary alcohol congener.[24] Protection of this newly unveiled group with TBDPSCl under standard conditions (imidazole in DMF) then completed the synthesis of **78**. Before continuing with the synthesis of this piece, one should note that although there are several alternative hydride sources that could have been used to cleave the oxazolidinone from **77** (many of which were discussed in the cytovaricin chapter in *Classics I*), only $NaBH_4$ in the solvent mixture employed enabled its excision without any epimerization of the newly installed stereocenter.

From this key intermediate (**78**), the steps required to effect its subsequent conversion into **21** proceeded smoothly. Thus, partial reduction of the cyanide to its aldehyde counterpart in CH_2Cl_2 at −78 °C was followed by reaction with $H_2NOH{\cdot}HCl$ to afford oxime intermediate **80**. With this new functional group appended in proper relation to the disubstituted C–C double bond, the stage was set to attempt ring formation through a nitrile oxide-based

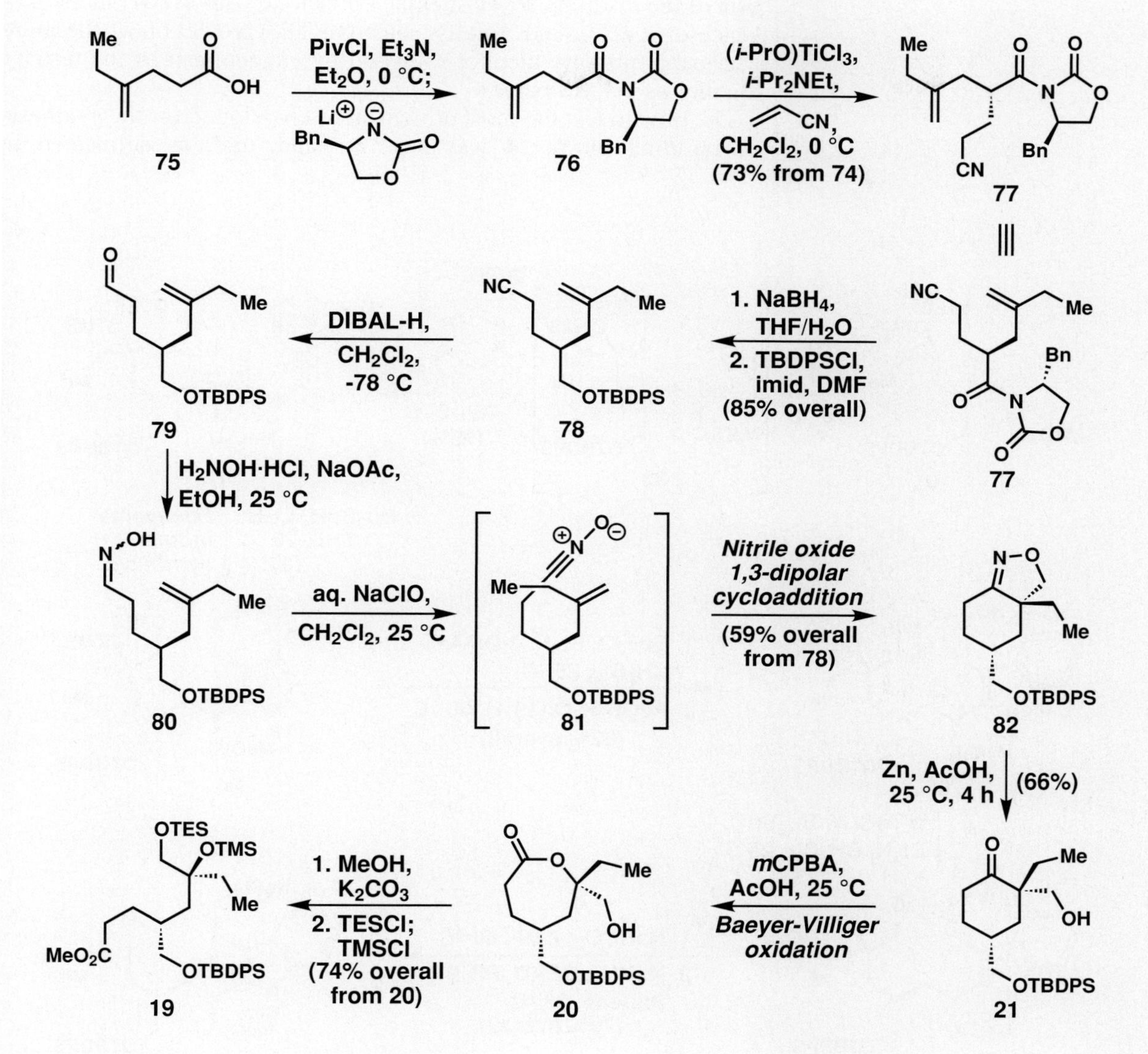

Scheme 12. Synthesis of building block **19**.

1,3-dipolar cycloaddition. Most gratifyingly, this event proceeded without incident as isoxazoline **82** was acquired as a single stereoisomer following the formation of the requisite nitrile oxide dipole (**81**) using aqueous sodium hypochlorite.[25] Reductive cleavage of the N-O bond within this product then completed the construction of hydroxy ketone **21** in 66 % yield. From this cyclic scaffold, the remaining steps to the final building block (**19**) proceeded in 74 % overall yield exactly as discussed earlier through initial Baeyer–Villiger oxidation using *m*CPBA in AcOH, transesterification of the resulting seven-membered lactone with MeOH in the presence of K_2CO_3, and, finally, selective TES protection of the more accessible primary alcohol followed by engagement of the tertiary alcohol as a TMS ether.

The time to test the most critical steps envisioned for the synthesis of building block **14** was now at hand, and, as delineated in

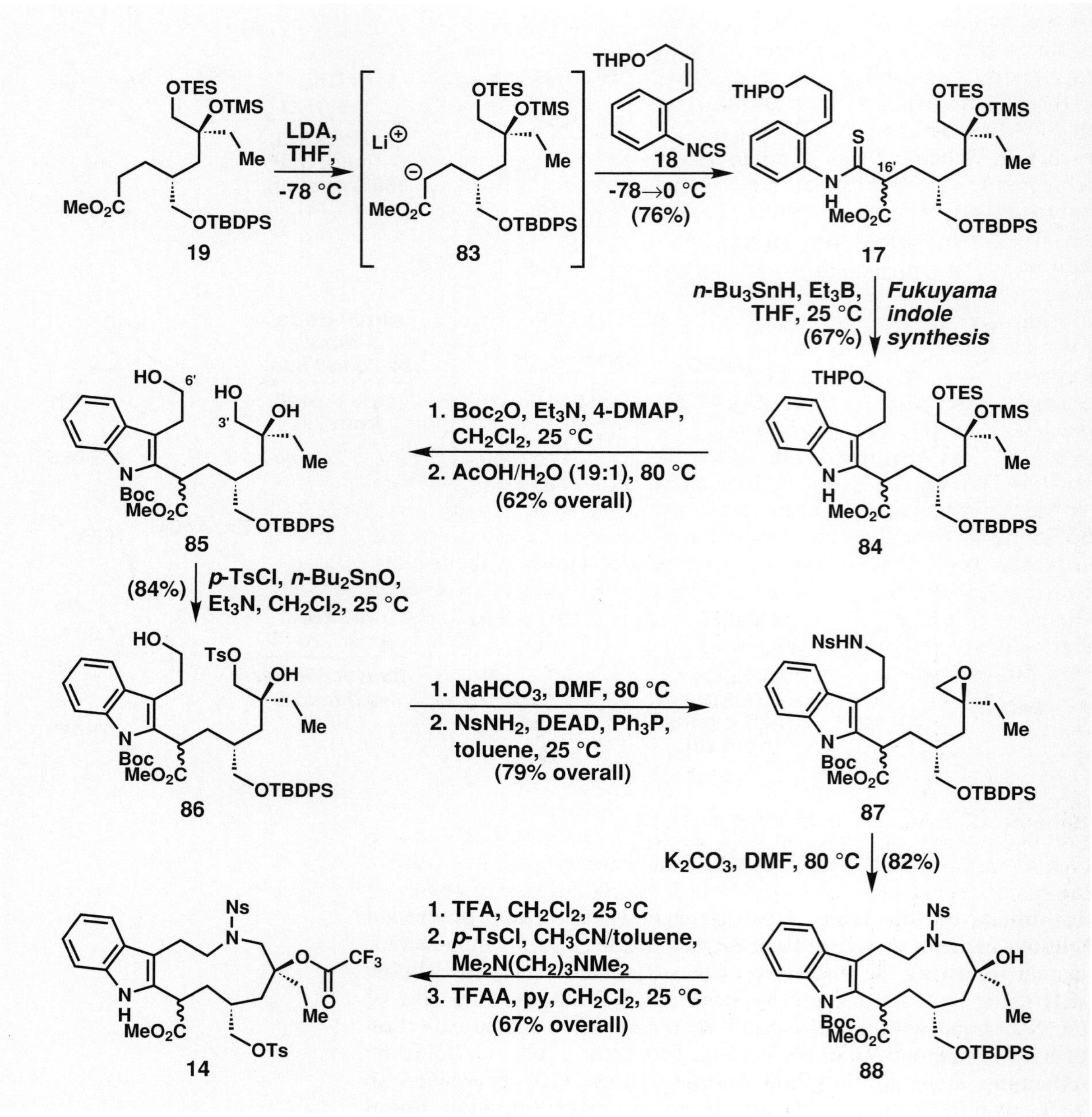

Scheme 13. Synthesis of advanced intermediate **14**.

Scheme 13, the first set of these operations proceeded as originally designed. Thus, following reaction of **19** with LDA in THF at −78 °C to form a lithiated nucleophile (**83**), the dropwise addition of isothiocyanate **18**[14] to this reactant afforded thioamide **17** in 76 % yield. Although the stereochemistry of the C-16′ center was not controlled in this reaction, the result was ultimately of no consequence as this position would become part of an element of unsaturation during the critical oxidative coupling step at the end of the synthesis. Having now generated a thioamide in proximity to an alkene group, the proper array of functionality was present to attempt indole formation through the Fukuyama conditions of *n*-Bu_3SnH and a radical initiator. As expected, this reaction succeeded in providing **84** in 67 % yield using Et_3B rather than AIBN (as used earlier in Scheme 8) as it could initiate the reaction at a lower temperature (25 °C). In fact, borane-based radical initiators are the only reagents known that can initiate radical reactions at low temperature (even as low as −78 °C); by contrast, AIBN decomposes into its active radical form with useful efficiency only when heated at approximately 80 °C.

With a 2,3-disubstituted indole now in hand (**84**), it was nearly time to attempt the critical macrocyclization step envisioned for the creation of the 11-membered ring of **14**, namely an epoxide opening with an amine nucleophile. To access the key intermediate needed to test this hypothesis (i. e. **87**), the indole nitrogen was first protected as a Boc carbamate under standard conditions (Boc_2O, Et_3N, 4-DMAP), and then both the TES and TMS ethers as well as the THP protecting group were cleaved with AcOH to afford **85** in 62 % overall yield. Since the remainder of the proposed sequence to **87** required the selective differentiation of these three hydroxy groups, their concomitant deprotection in this step could appear to afford a tactical disadvantage at first glance, in that several reprotections might be necessary to achieve success. However, their different substitution patterns ensured that such travail was not needed. Seeking first to form the epoxide needed for macrocyclization, the Fukuyama group found that they could selectively convert the central primary alcohol at C-3′ into a tosylate (**86**) in 84 % yield with *p*-TsCl in the presence of catalytic *n*-Bu_2SnO and Et_3N.[26] Those of you who have read the everninomicin chapter should recognize this reaction as a variant of the selective opening of five-membered tin-acetals at their less hindered position. Importantly, products corresponding to a C-6′ tosylate were not observed from this event by virtue of the fact that the activating tin species benefited from superior bidentate chelation with a diol system versus that provided by a solitary hydroxyl group. Thus, the alcohol group at C-3′ was more rapidly converted into a tosylate than was its C-6′ competitor, enabling the chemoselective formation of **86** as long as reaction progress was carefully monitored.

Upon simple exposure of this product to $NaHCO_3$ in hot DMF, the tertiary alcohol within **86** displaced the neighboring leaving group to form the epoxide acceptor, and the remaining primary alcohol was then converted into a nosyl-protected amine under Mitsunobu conditions to complete the assembly of **87** in 79% overall yield. With these motifs now installed into the growing scaffold, macrocyclization was pursued immediately. Pleasingly, upon treatment of **87** with K_2CO_3 in DMF at 80 °C, this critical step forming **88** was accomplished in 82% yield. At this juncture, one might ask why the earlier reaction of **86** with $NaHCO_3$ under essentially the same conditions used to convert **87** into **88** did not afford a cyclized product, as base would have been present following epoxide formation. Due to the outcome shown, $NaHCO_3$ did not provide a strong enough conjugate base to enable this event to occur even though alcohols can, of course, serve as competent nucleophiles. Had the much stronger (by several orders of magnitude) K_2CO_3 been employed instead, as used in the final operation leading to **88**, then this side reaction would perhaps have been observed. Apart from this feature of judicious reagent selection, one should also realize that although carrying an epoxide for several synthetic operations constitutes a relatively rare (and sometimes risky) event in complex molecule synthesis, the installation of the amine in this sequence after epoxide formation was not negotiable. Indeed, it would have been impossible to effect a selective Mitsunobu displacement of only one of the three alcohols in **85**.

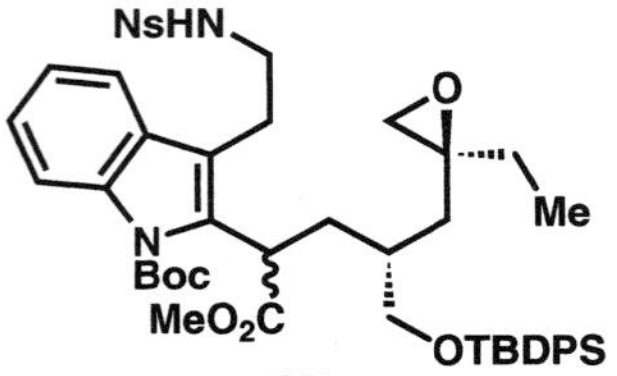

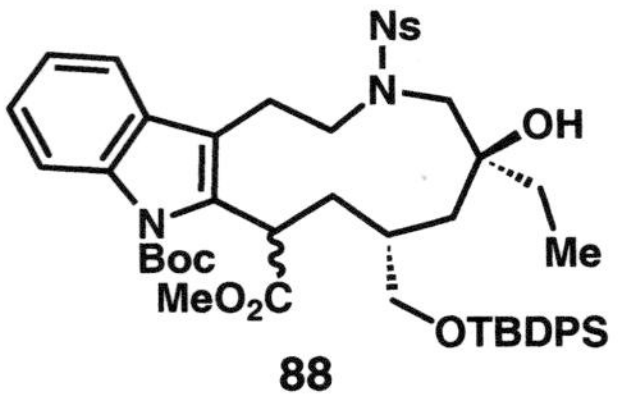

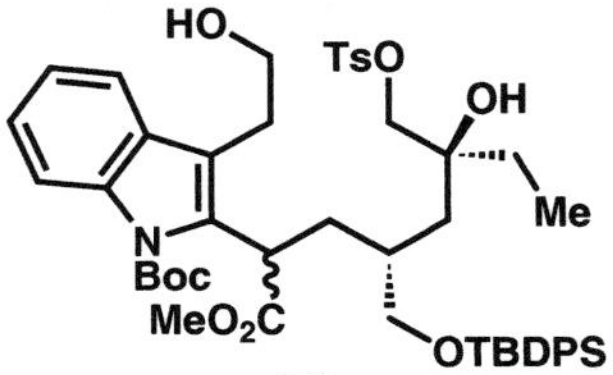

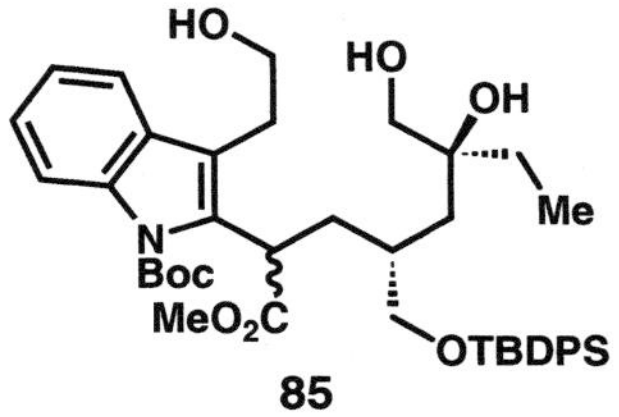

At this juncture, with an effective route established to advanced macrocycle **88**, only a few minor protecting group adjustments were required to complete building block **14**. As shown, these events were readily executed through initial treatment of **88** with TFA to effect both Boc deprotection and TBDPS cleavage, followed by engagement of the resultant free primary alcohol as a tosylate and protection of the tertiary alcohol as a trifluoroacetate. Before moving forward into the final stages of the total synthesis of vinblastine (**1**), one should note that while the base employed to facilitate the penultimate of these operations might seem esoteric, namely $Me_2N(CH_2)_3NMe_2$, its use was crucial in preventing the possible displacement of the desired tosylate by chloride, a side reaction that can occur under more standard tosylation conditions employing *p*-TsCl and pyridine since the py·HCl by-product is an effective nucleophile. This eventuality was circumvented in this case as the HCl salt of this alternative base has extremely poor nucleophilicity.[27]

Once ample quantities of both **14** and vindoline (**2**) were in hand using the developed sequences, the final operations required to complete the synthesis of vinblastine (**1**) were pursued with vigor. The first of these, their stereoselective coupling, was by far the most critical. Seeking to explore this union, the Fukuyama group observed that upon reaction of **14** (Scheme 14) with *t*-BuOCl in CH_2Cl_2 at 0 °C, they could quantitatively obtain chloroindolenine **90**. When this intermediate (**90**) was exposed to TFA in the presence of vindoline (**2**), oxidative coupling via **13** afforded the desired adduct (**12**) bearing the requisite stereochemistry of the final natural product at C-16′ in 97% yield! From this advanced compound, only

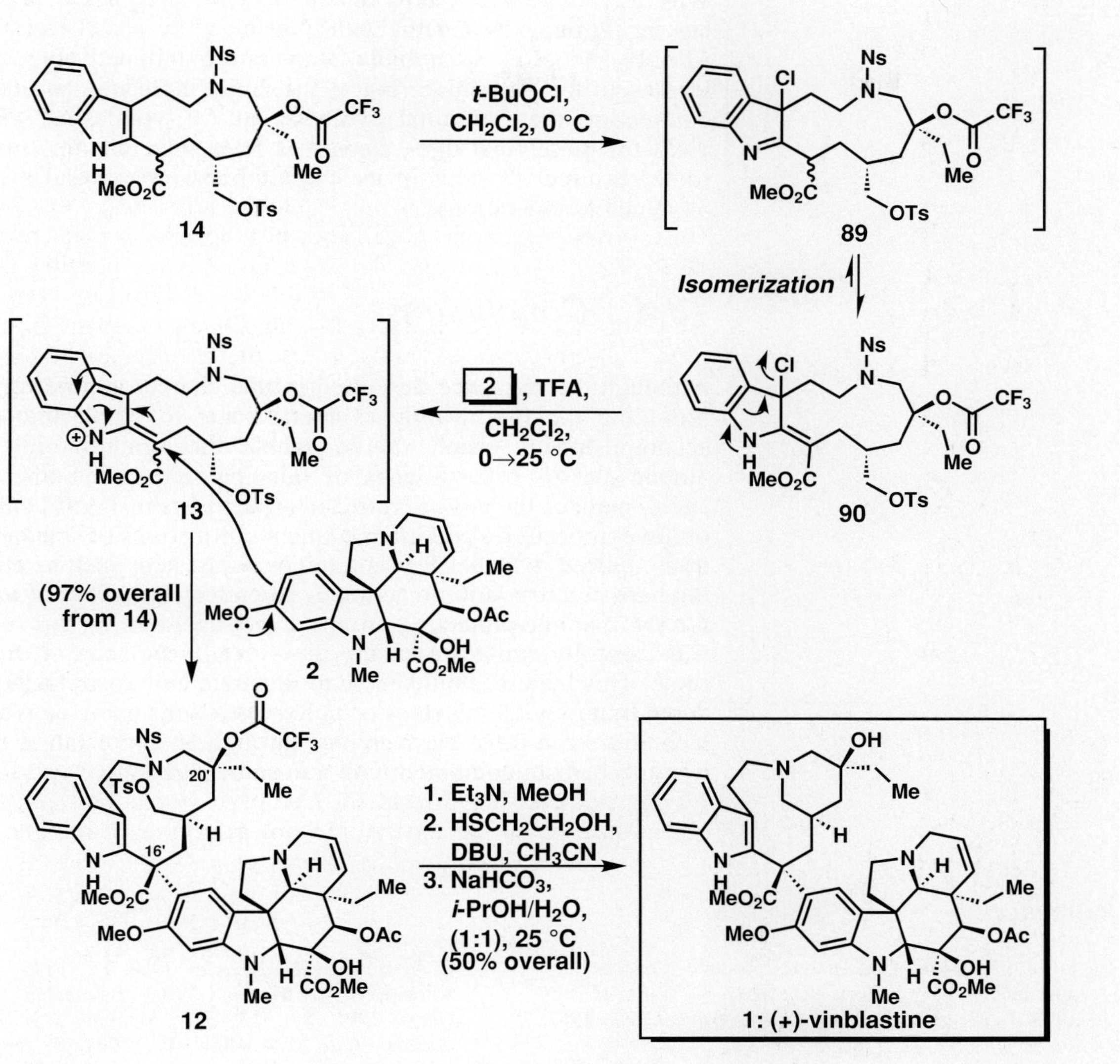

Scheme 14. Final stages and completion of the total synthesis of (+)-vinblastine (1).

the completion of the final piperidine ring system in the upper portion remained to be effected along with two deprotections. The latter of these objectives was tackled first with treatment of **12** with Et_3N in MeOH accomplishing the chemoselective lysis of the tertiary trifluoroacetate protecting group in the presence of the acetate at C-17. Next, the nosyl group was removed under mild conditions upon exposure to $HSCH_2CH_2OH$. This deprotection can be rationalized as proceeding through a Meisenheimer intermediate (**91**, see column figure) in which the better leaving group (i.e. the sulfonate) is expelled with the loss of SO_2 to afford the desired free amine.

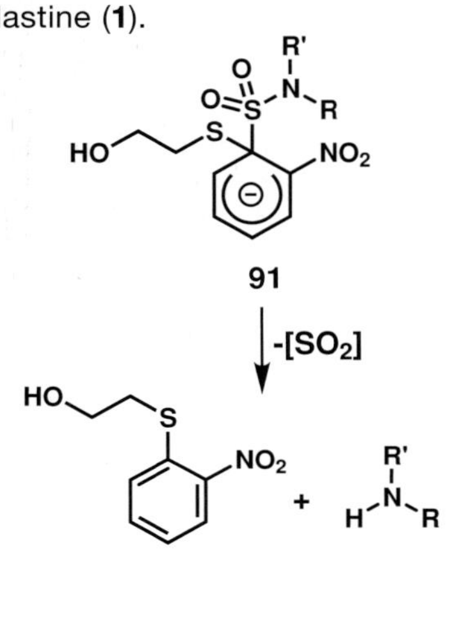

With a reactive amine now unveiled in the presence of a tosylate leaving group, treatment with $NaHCO_3$ in a 1:1 mixture of *i*-PrOH and H_2O at ambient temperature initiated ring closure, leading to the piperidine ring of the target molecule, and, thereby, completing the first total synthesis of (+)-vinblastine (**1**). The yield for these final three steps was 50%. Overall, this inventive route required 27 steps in its longest linear chain, and a total of 34 synthetic operations.

18.4 Conclusion

Although the sequence described in this chapter is noteworthy for providing the first wholly synthetic route to vinblastine (**1**), an accomplishment which should enable the synthesis of several unique classes of analogues of this important chemotherapeutic agent, perhaps the more significant aspect is its insightful collection of novel procedures and often subtle modifications of standard reactions. Indeed, whether it was a minor adjustment such as changing the base in a tosylation reaction or extending the utility of a unique family of amine protecting groups, every operation in this sequence was carefully crafted to ensure the overall efficiency of the entire route. This feature should serve to illustrate how roadblocks engendered by low-yielding steps or failed operations using conventional techniques can often be overcome through an appreciation of reaction mechanism combined with a dose of ingenuity.

References

1. For a discussion of this natural product in cancer therapy and its early history, see: a) H.-K. Wang, *IDrugs* **1998**, *1*, 92; b) K. Folkers, *Pure & Appl. Chem.* **1967**, *14*, 1.
2. R.L. Noble, C.T. Beer, J.H. Cutts, *Ann. N.Y. Acad. Sci.* **1958**, *76*, 882. Independent isolation of vinblastine was also accomplished a year later: G.H. Svoboda, N. Neuss, M. Gorman, *J. Am. Pharm. Assoc.* **1959**, *48*, 659. Complete structural elucidation was accomplished through X-ray analysis: J.W. Moncrief, W.N. Lipscomb, *J. Am. Chem. Soc.* **1965**, *87*, 4963.
3. For reviews, see: a) P. Potier, *J. Nat. Prod.* **1980**, *43*, 72; b) M.E. Kuehne, I. Markó in *The Alkaloids, Vol. 37* (Eds.: A. Brossi, M. Suffness), Academic Press, San Diego, **1990**, pp. 77–131.
4. a) P. Mangeney, R.Z. Andriamialisoa, N. Langlois, Y. Langlois, P. Potier, *J. Am. Chem. Soc.* **1979**, *101*, 2243; b) J.P. Kutney, L.S.L. Choi, J. Nakano, H. Tsukamoto, *Heterocycles* **1988**, *27*, 1837; c) J.P. Kutney, L.S.L. Choi, J. Nakano, H. Tsukamoto, M. McHugh, C.A. Boulet, *Heterocycles* **1988**, *27*, 1845; d) M.E. Kuehne, P.A. Matson, W.G. Bornmann, *J. Org. Chem.* **1991**, *56*, 513; e) P. Magnus, J.S. Mendoza, A. Stamford, M. Ladlow, P. Willis, *J. Am. Chem. Soc.* **1992**, *114*, 10232.
5. To date, seven total syntheses of vindoline have been achieved, although the material obtained from these routes has never been used to prepare vinblastine: a) M. Ando, G. Büchi, T. Ohnuma, *J. Am. Chem. Soc.* **1975**, *97*, 6880; b) Y. Ban, Y. Sekine, T. Oishi, *Tetrahedron Lett.* **1978**, 151; c) J.P. Kutney, U. Bunzli-Trepp, K.K. Chan, J.P. de Souza, Y. Fujise, T. Honda, J. Katsube, F.K. Klein, A. Leutwiler, S. Morehead, M. Rohr, B.R. Worth, *J. Am. Chem. Soc.* **1978**, *100*, 4220; d) B. Danieli, G. Lesma, G. Palmisano, R. Riva, *J. Chem. Soc., Chem. Commun.* **1984**, 909; e) P.L. Feldman, H. Rapoport, *J. Am. Chem. Soc.* **1987**, *109*, 1603; f) M.E. Kuehne, D.E. Podhorez, T. Mulamba, W.G. Bornmann, *J. Org. Chem.* **1987**, *52*, 347.

6. For representative biosynthetic studies, see Ref. 4b and the following: a) J. P. Kutney, *Pure & Appl. Chem.* **1989**, *61*, 449; b) J. P. Kutney, L. S. L. Choi, T. Honda, N. G. Lewis, T. Sato, K. L. Stuart, B. R. Worth, *Helv. Chim. Acta* **1982**, *65*, 2088; c) F. Gueritte, V. B. Nguyen, Y. Langlois, P. Potier, *J. Chem. Soc., Chem. Commun.* **1980**, 452.
7. For reviews, see: a) D. Grierson, *Organic Reactions, Vol. 39* (Eds.: L. Paquette, et al.), John Wiley & Sons, New York, **1990**, pp. 85–295; b) M. Lounasmaa, A. Koskinen, *Heterocycles* **1984**, *22*, 1591.
8. a) A. Husson, Y. Langlois, C. Riche, H.-P. Husson, P. Potier, *Tetrahedron* **1973**, *29*, 3095; b) A. I. Scott, C.-L. Yeh, D. Greenslade, *J. Chem. Soc., Chem. Commun.* **1978**, 947.
9. a) N. Neuss, M. Gorman, N. J. Cone, L. L. Huckstep, *Tetrahedron Lett.* **1968**, *9*, 783; b) J. P. Kutney, J. Beck, F. Bylsma, J. Cook, W. J. Cretney, K. Fuji, R. Imhof, A. M. Treasurywala, *Helv. Chim. Acta* **1975**, *58*, 1690; c) A. Rahman, A. Basha, M. Ghazala, *Tetrahedron Lett.* **1976**, *17*, 2351.
10. S. Yokoshima, T. Uedo, S. Kobayashi, A. Sato, T. Kuboyama, H. Tokuyama, T. Fukuyama, *J. Am. Chem. Soc.* **2002**, *124*, 2137. For earlier studies which culminated in a synthesis of vindoline, see: S. Kobayashi, T. Ueda, T. Fukuyama, *Synlett* **2000**, 883.
11. G. Schill, C. U. Priester, U. F. Windhövel, H. Fritz, *Tetrahedron* **1987**, *43*, 3765.
12. a) A. Fujiwara, T. Kan, T. Fukuyama, *Synlett* **2000**, 1667; b) T. Fukuyama, M. Cheung, C.-K. Jow, Y. Hidai, T. Kan, *Tetrahedron Lett.* **1997**, *38*, 5831; c) T. Fukuyama, C.-K. Jow, M. Cheung, *Tetrahedron Lett.* **1995**, *36*, 6373.
13. For a review of the Mitsunobu reaction, see: O. Mitsunobu, *Synthesis* **1981**, 1.
14. H. Tokuyama, T. Yamashita, M. T. Reding, Y. Kaburagi, T. Fukuyama, *J. Am. Chem. Soc.* **1999**, *121*, 3791.
15. R. Hull, *J. Chem. Soc. (C)* **1968**, 1777.
16. K. C. Nicolaou, S. A. Snyder, T. Montagnon, G. Vassilikogiannakis, *Angew. Chem.* **2002**, *114*, 1742; *Angew. Chem. Int. Ed.* **2002**, *41*, 1668.
17. a) W. A. Carroll, P. A. Grieco, *J. Am. Chem. Soc.* **1993**, *115*, 1164; b) P. A. Grieco, A. Bahsas, *J. Org. Chem.* **1987**, *52*, 5746. For related expressions of this biogenetic concept, see: c) E. Wenkert, *J. Am. Chem. Soc.* **1962**, *84*, 98; d) A. I. Scott, *Acc. Chem. Res.* **1970**, *3*, 51.
18. For a review of the Skraup quinoline synthesis, see: R. H. F. Manske, M. Kulka, *Organic Reactions, Vol. 7* (Eds.: R. Adams, et al.), John Wiley & Sons, London, **1953**, pp. 59–98.
19. H. Tokuyama, M. Sato, T. Ueda, T. Fukuyama, *Heterocycles* **2001**, *54*, 105.
20. H. Tokuyama, T. Makido, T. Uedo, T. Fukuyama, *Synth. Commun.* **2002**, *32*, 869.
21. S. Kobayashi, G. Peng, T. Fukuyama, *Tetrahedron Lett.* **1999**, *40*, 1519.
22. For an early discussion of the Evans oxazolidinone chiral auxiliaries, see: D. A. Evans, J. V. Nelson, T. R. Taber, *Top. Stereochem.* **1982**, *13*, 1.
23. a) D. A. Evans, J. R. Gage, J. L. Leighton, *J. Am. Chem. Soc.* **1992**, *114*, 9434; b) D. A. Evans, M. T. Bilodeau, T. C. Somers, J. Clardy, D. Cherry, Y. Kato, *J. Org. Chem.* **1991**, *56*, 5750.
24. M. Prashad, D. Har, H.-Y. Kim, O. Repic, *Tetrahedron Lett.* **1998**, *39*, 7067.
25. For a recent review, see: S. Karlsson, H.-E. Högberg, *Org. Prep. Proc. Int.* **2001**, *33*, 103. For an excellent treatment of several classes of 1,3-dipolar cycloadditions, see: W. Carruthers, *Cycloaddition Reactions in Organic Synthesis*, Pergamon, Oxford, **1990**, pp. 269–314.
26. M. J. Martinelli, N. K. Nayyar, E. D. Moher, U. P. Dhokte, J. M. Pawlak, R. Vaidyanathan, *Org. Lett.* **1999**, *1*, 447.
27. Y. Yoshida, K. Shimonishi, Y. Sakakura, S. Okada, N. Aso, Y. Tanabe, *Synthesis* **1999**, 1633.

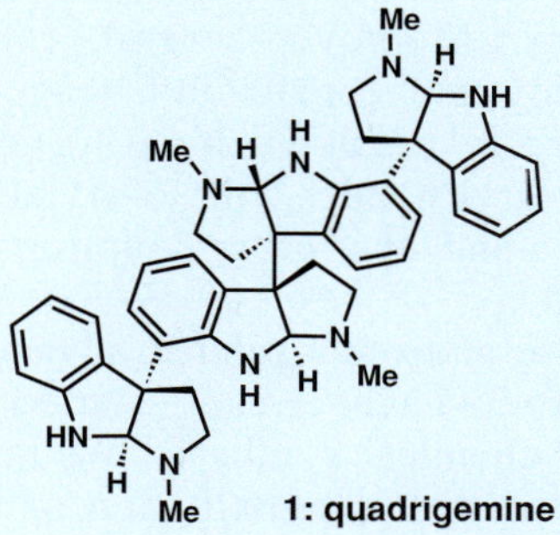

1: quadrigemine C

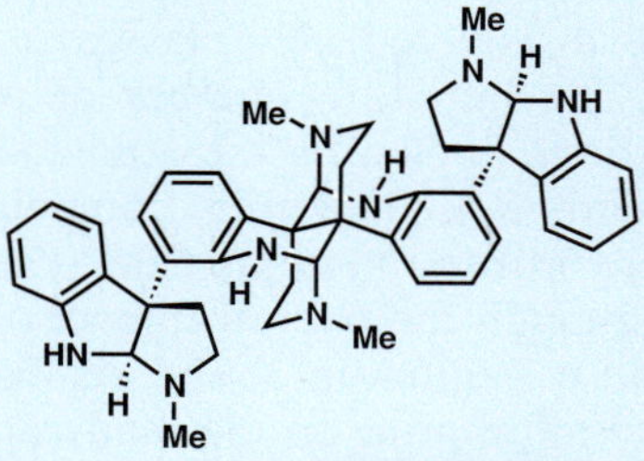

2: psycholeine

19

L. E. Overman (2002)

Quadrigemine C and Psycholeine

19.1 Introduction

If you ever wanted to evaluate just how much the practice of total synthesis has evolved during the past one or two decades, spend some time comparing the syntheses in this book to those in *Classics I.* What you will find are some truly profound differences, all of which derive from the power, selectivity, and types of reactions available in each era. For instance, fifteen years ago a synthetic chemist would not have dreamed of using ring-closing metathesis to construct an olefin-containing ring on a functionally complex molecule because appropriate carbene initiators did not exist. Today, of course, a number of methodological developments in catalyst design and synthesis have changed this situation, transforming this reaction into a standard synthetic tool for diverse applications and thereby changing the manner in which chemists think about constructing such motifs.

The area in chemical synthesis where some of the most perceptible changes have occurred recently involves the collection of C–C bond-forming reactions initiated by palladium reagents and catalysts, such as those of the Suzuki, Stille, Heck, Negishi, and Sonogashira type.[1] Although these reactions have long been employed in chemical synthesis, fundamental mechanistic research during the past few years has enabled these processes to be sharpened into tools with constructive capabilities far beyond what they possessed even in the early 1990s. For example, we now know how to generate catalyst systems that are sufficiently active to execute Suzuki reactions at ambient temperature with many

Key concepts:

- **Two-directional synthesis**
- **Catalytic asymmetric Heck cyclizations**

aryl halides, and, in some instances, induce unactivated aryl fluorides to serve as coupling partners.[2] Moreover, several groups have discovered new classes of highly active palladium reagents that can initiate a number of previously unknown bond-forming reactions, including the amination of aryl halides,[3] the α-arylation of carbonyl-containing compounds,[4] and the cross-coupling of sp^3-hybridized systems (see Scheme 1).[5]

Although the overall impact of these and other related advances is only beginning to be realized, they have already started to leave an imprint on the landscape of chemical synthesis. We hope to demonstrate this fact in this chapter with a presentation of the Overman group's highly elegant and concise total syntheses of quadrigemine C (**1**) and psycholeine (**2**) published in 2002.[6] As we shall see, these researchers beautifully orchestrated a sequence of three key palladium-mediated reactions to complete the most formidable elements of these target molecules with levels of synthetic efficiency and stereocontrol that would not have been achievable with the existing variants of these same reactions prior to the late 1990s.

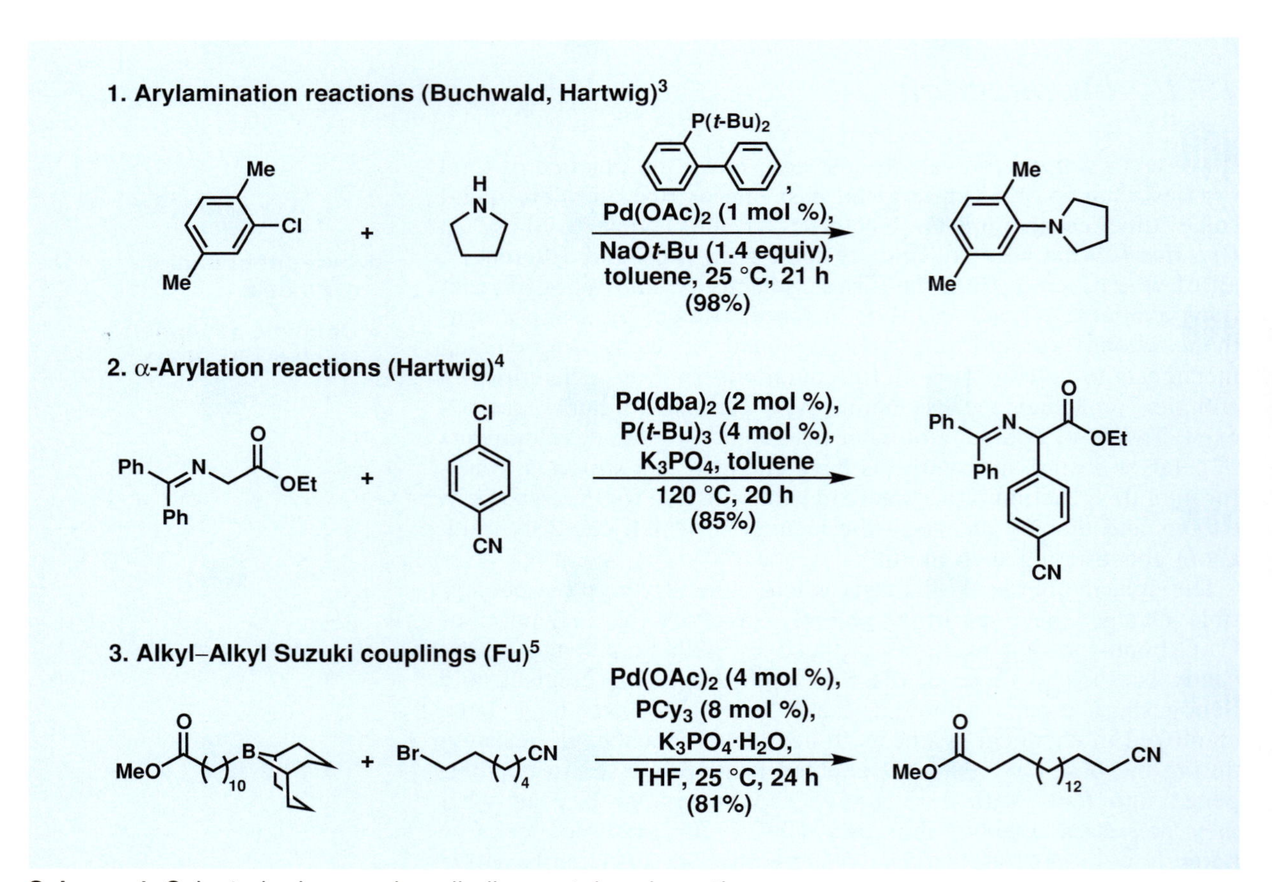

Scheme 1. Selected advances in palladium-catalyzed reactions.

19.2 Retrosynthetic Analysis and Strategy

Over the course of the past forty years or so, chemists have isolated a collection of structurally related tryptophan-derived alkaloids from a diverse group of higher order plants and fungi, a few of which, including quadrigemine C (**1**), are shown in Figure 1.[7,8] As you can see, these natural products incorporate varying numbers of pyrrolidinoindoline (cyclotryptamine) subunits attached to the benzylic position of a bispyrrolidinoindoline core possessing either *meso* (as in **3**) or C_2-symmetry (as in **4**). In fact, although only com-

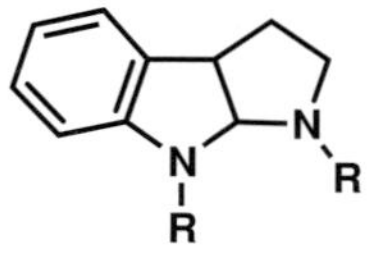

cyclotryptamine

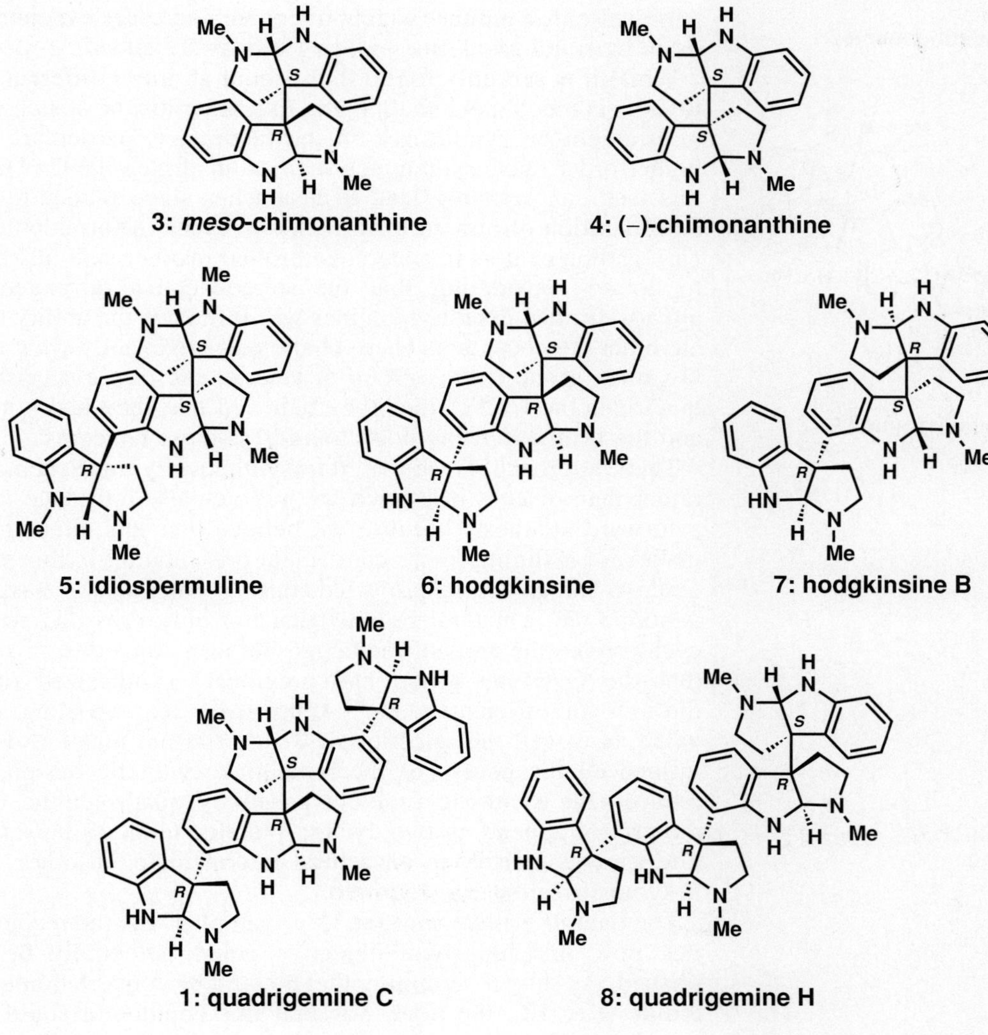

3: ***meso*-chimonanthine**

4: (–)-chimonanthine

5: idiospermuline

6: hodgkinsine

7: hodgkinsine B

1: quadrigemine C

8: quadrigemine H

Figure 1. Structures of selected polypyrrolidinoindoline (cyclotryptamine) alkaloids.

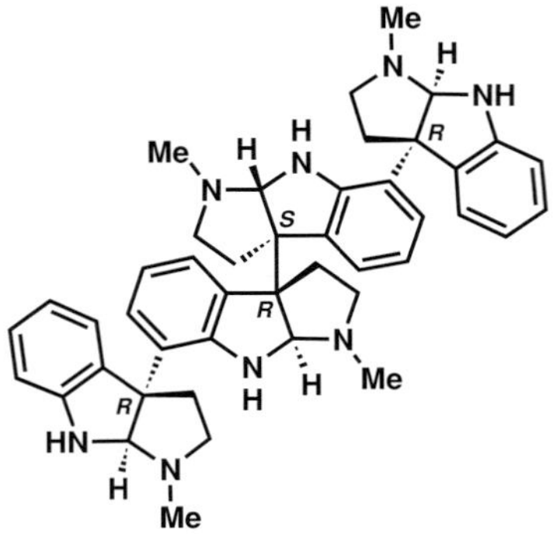

1: quadrigemine C

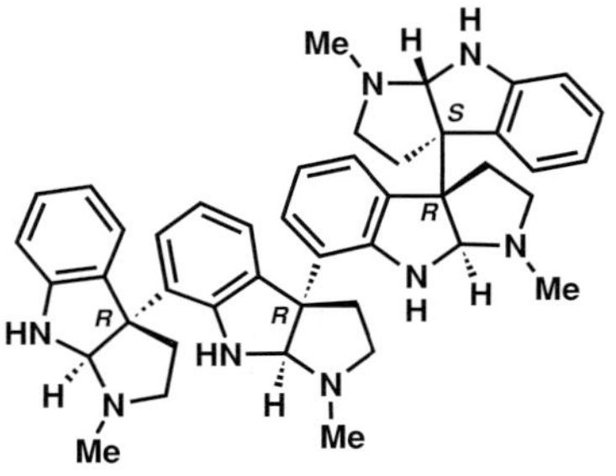

8: quadrigemine H

pounds with up to four cyclotryptamines are shown in this figure, several natural products that contain five, six, seven, and even eight of these subunits have also been isolated from natural sources.

What is perhaps more amazing than the number of these building blocks that Nature can amalgamate together is the molecular diversity that she can achieve with these synthons merely by altering the stereogenicity of their quaternary carbon centers and/or their linking pattern. For instance, both tetrapyrrolidinoindoline alkaloids shown in Figure 1 (**1** and **8**) are made up of the same set of four building blocks in terms of their absolute stereochemistry, but since they are attached in different ways, these compounds have completely unique three-dimensional structures and biological activity profiles. Over twenty additional variations of just this group of cyclotryptamine architectures are possible based on these two general structural permutations, a number which, of course, increases exponentially as more pyrrolidinoindoline building blocks are added.

While it is certainly fun to think about all these different possibilities, it is much more challenging to ponder just how such architectures might be synthesized in the laboratory, particularly for the higher-order cyclotryptamines such as quadrigemine C (**1**). Indeed, this task can certainly lead to headaches since it requires careful consideration of how to form their two types of formidable quaternary carbon centers in a stereocontrolled manner, and in what order to do so considering that the stereochemical arrangements of already-formed cyclotryptamines will influence the ability to generate other stereocenters. These challenges are exactly what drew the Overman group to this group of targets, and, more specifically, to quadrigemine C (**1**), since they believed that they had both a plan and the synthetic technology to address these concerns.

Their approach is shown retrosynthetically in Scheme 2, but rather than discuss it as such we will actually introduce it here as a forward synthesis because we believe that this alternate format makes for a slightly more convenient presentation in this particular case. As indicated, the projected starting material was *meso*-chimonanthine (**3**), a natural product that not only provided two of the cyclotryptamine units of the target, but also constituted a substance that the Overman group had previously synthesized (in 2000) through an efficient route.[9] Apart from its appealing practical value, however, the selection of **3** as a starting material decisively influenced the course of the remaining synthetic design since it reduced the synthetic problem posed by quadrigemine C (**1**) to the task of appending two cyclotryptamine units of the same absolute stereochemistry to an achiral intermediate. In other words, a desymmetrization was required.

The overall brilliance of the Overman plan was the recognition of just how this nontrivial objective could potentially be accomplished. As shown, assuming that **3** could be converted into a diaryl iodide (i. e. **12**), the hope was that this compound could then be coupled to 2 equivalents of stannane **11** through Stille couplings to provide **10**. The next operation, however, would be the key

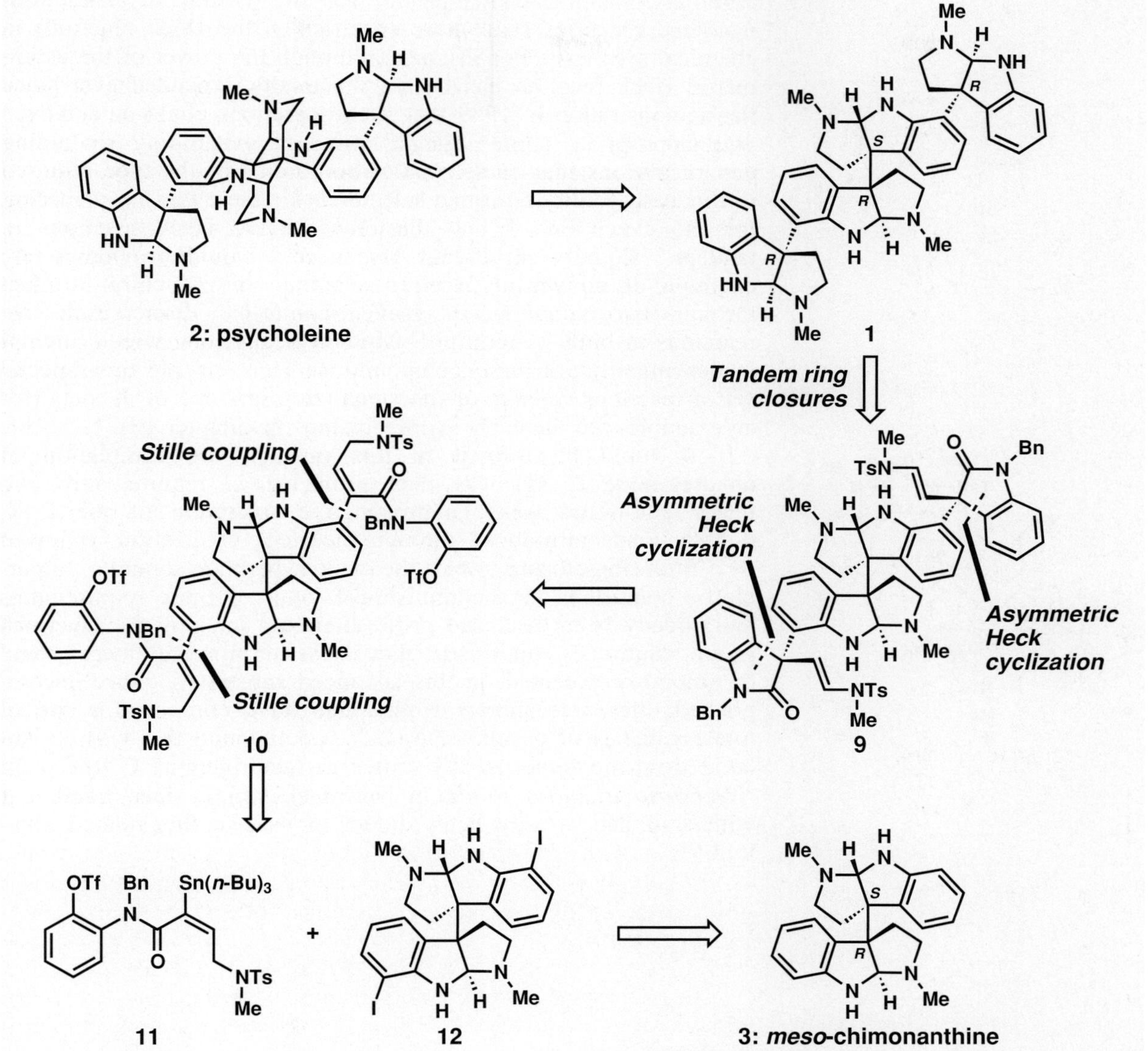

Scheme 2. Overman's retrosynthetic analysis of quadrigemine C (**1**) and psycholeine (**2**).

step of the projected sequence, but also the most risky: inducing this new *meso* material (**10**) to participate in a set of tandem, reagent-controlled, asymmetric Heck reactions leading to the desymmetrized intermediate **9** bearing two new oxindole rings with *R*-disposed quaternary carbon centers.[10]

With little question, if this event could be accomplished, it would definitely establish new ground for the virtues of palladium-mediated reactions (and more specifically, the Heck reaction) in chemical synthesis. For instance, although the power of the asymmetric Heck reaction has been dramatically expanded ever since the demonstration in 1989 that such operations could be achieved with appropriate chiral ligands bound to palladium[11] (including demonstrations that quaternary carbon atoms of the type required in this case could be formed selectively),[12] there was no precedent for the execution of two diastereoselective Heck reactions in tandem.* Equally significant, its success would provide a rare example of converting a *meso* substrate into a chiral product through two simultaneous reagent-controlled diastereoselective reactions at both its termini.[13] More typically, one would attempt to desymmetrize a *meso* compound with an enzyme or a special set of reactions capable of functionalizing just one of its sites (for an example, see the early synthetic steps in Chapter 13).[14]

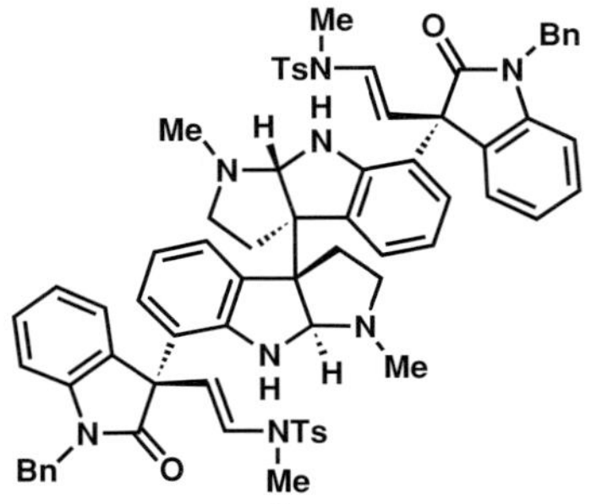

If **9** could be formed in this manner, the completion of quadregimine C (**1**) was then projected to require only two major additional tasks: hydrogenative saturation of the C–C double bonds introduced through the Heck cyclization, followed by a final ring closure to cast the cyclotryptamine subunits. Importantly, conditions to accomplish the latter of these requirements had already been disclosed in the literature, making the Overman group cautiously optimistic that these terminating steps could be smoothly executed on this advanced substrate.[15] Once accomplished, these researchers would also have completed a formal total synthesis of psycholeine (**2**),[8b] a compound that was co-isolated from the same natural source as quadrigemine C (the plant *Psychotria oleoides* found in New Caledonia), since treating **1** with acid had already been shown to lead to this related alkaloid.**

Me TsN H Me N H N O Bn N N H H N Me Bn N O NTs Me

9

* Although we will discuss the parameters that influence successful asymmetric Heck reactions in more detail during the synthesis, at this juncture it is important to note a couple of well-established trends that are reflected in the design of the test substrate (**10**). First, *Z* alkenes consistently provide higher levels of enantioselectivity than their *E*-disposed counterparts. Second, the presence of functional groups near the reactive site can greatly influence the reactive conformation and prevent reagent control in accomplishing asymmetry, which is why in this case the indoline nitrogen atoms in the central core have been left without protection.

** Because psycholeine (**2**) can be formed through an acid-induced rearrangement of **1**, it may not actually constitute a natural product produced by *Psychotria oleoides*, but rather an artifact arising from the isolation procedures that were used to obtain **1**.

19.3 Total Synthesis

As a prelude to discussing the implementation of this bold plan towards **1**, we thought that we would start our discussions by mentioning a few aspects of the route that the Overman group had developed previously to access their key starting material, *meso*-chimonanthine (**3**).[9] Scheme 3 provides the steps of this creative total synthesis in which the most important reaction was unquestionably the conversion of **17** into **20**, since this operation was responsible for generating the two adjacent quaternary centers of the target molecule.

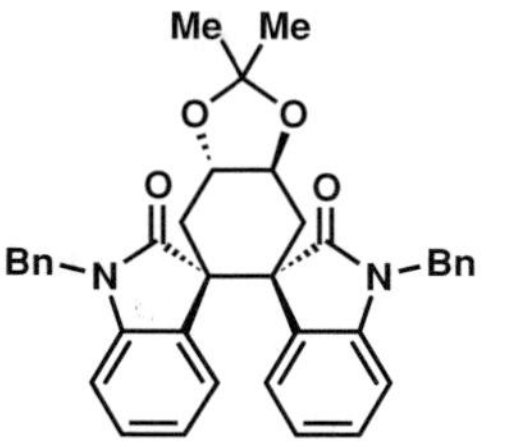

In this event, the addition of 2.1 equivalents of NaHMDS in THF at −78 °C converted **17** into a prostereogenic cyclic enolate which then attacked the tartrate-derived dielectrophile **18**[16] with a high level of facial selectivity to generate an intermediate of type **19**. Subsequent cyclization was then biased to provide the desired pseudo *meso* product **20**, a compound that was ultimately obtained in 92% yield. In contrast, when the conditions were changed to prevent chelation by using LiHMDS instead of NaHMDS in a solvent mixture of THF and HMPA (hexamethylphosphoramide), the C_2-symmetric variant of this product (**27**, see column figures) was formed in 58 % yield. This alternate outcome presumably reflects the preliminary formation of an acyclic lithium dienolate which led to a monoalkylated product corresponding to **26**.*

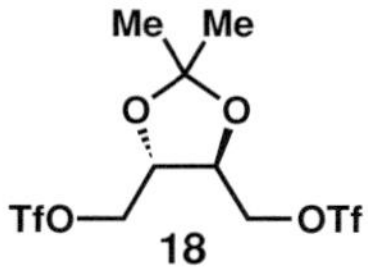

Thus, what these two subtly different protocols provide is a means to use the properties of the dienolate (either chelated or nonchelated) to initially control the relative configuration, with the chirality of the dielectrophile then providing absolute stereocontrol to afford selectively either a pseudo *meso* or a C_2-symmetric product. With little question, this general alkylation approach is worth remembering as it provides a rare example of high diastereoselectivity arising from the combination of a prostereogenic enolate with a chiral electrophile containing sp^3 carbon centers. Moreover, its efficiency is quite remarkable considering that vicinal quaternary carbon atoms of the type present in **20** are more typically tackled sequentially through multistep protocols.[17]

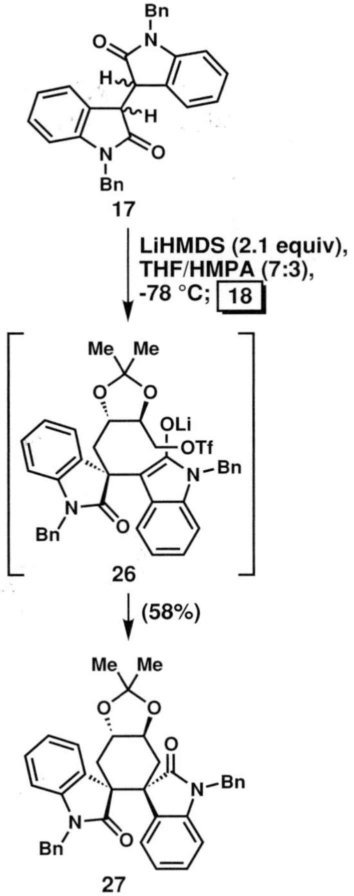

With these two stereogenic centers set beautifully through this creative approach, *meso*-chimonanthine (**3**) was then obtained in a few additional operations from **20** to ultimately provide a reliable route capable of delivering the natural product in an overall yield of 34 %. As an additional option for obtaining supplies of this compound, the Overman group could also turn to a 3-step

Me Me O O O O Bn–N N–Bn **20**

Me Me O O TfO OTf **18**

Bn O N H H N O Bn **17**

LiHMDS (2.1 equiv), THF/HMPA (7:3), -78 °C; **18**

Me Me O O OLi OTf N–Bn N Bn O **26**

(58%)

Me Me O O O N–Bn N Bn O **27**

* As more-detailed studies of this reaction have since indicated, chelation is favored by larger metal cations, with both sodium and potassium providing markedly superior formation of **20** than protocols in which lithium is used.

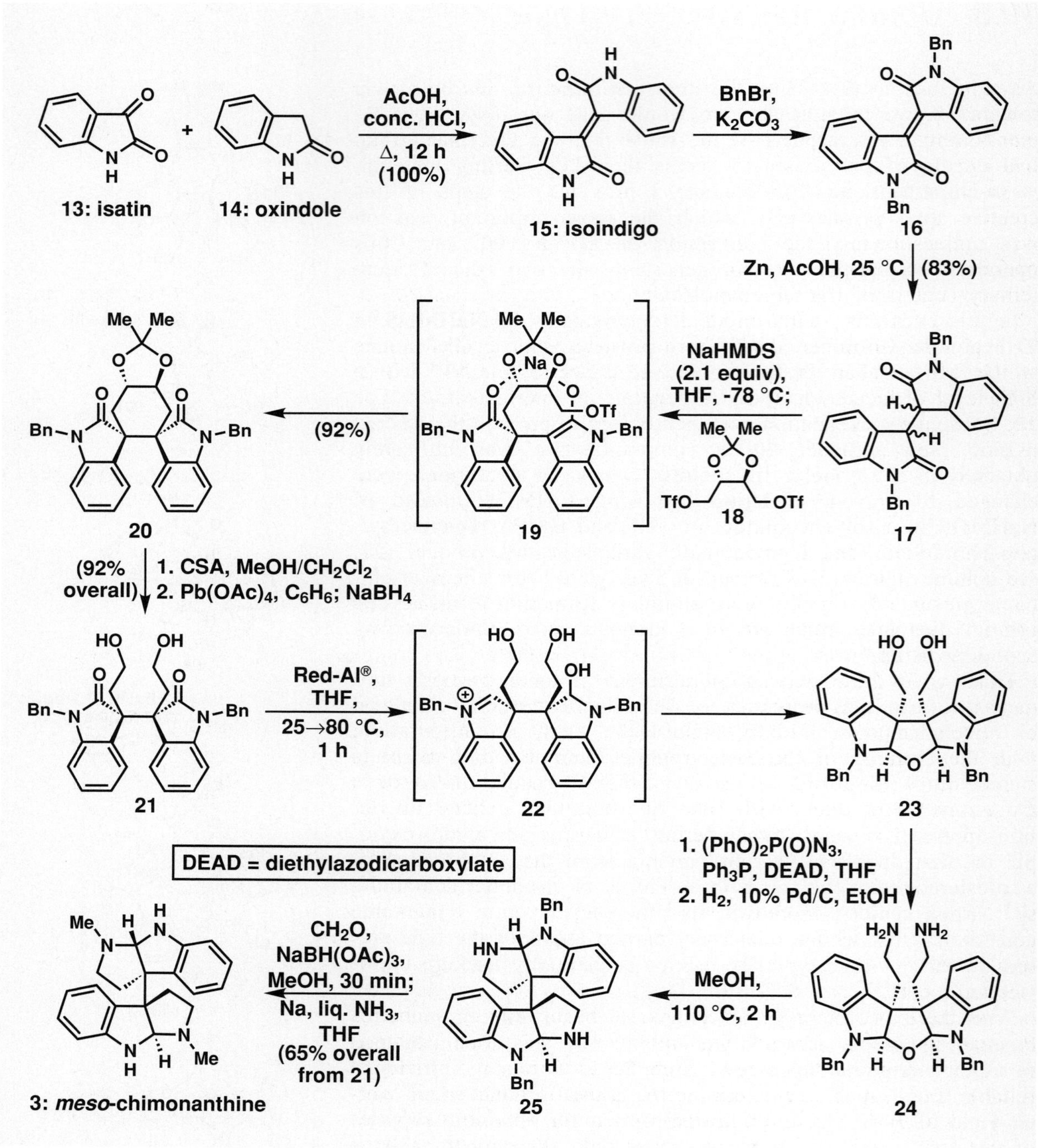

13: isatin + 14: oxindole

AcOH, conc. HCl, Δ, 12 h (100%)

15: isoindigo

BnBr, K_2CO_3

16

Zn, AcOH, 25 °C (83%)

17

NaHMDS (2.1 equiv), THF, -78 °C; 18

19

(92%)

20

1. CSA, MeOH/CH_2Cl_2
2. $Pb(OAc)_4$, C_6H_6; $NaBH_4$
(92% overall)

21

Red-Al®, THF, 25→80 °C, 1 h

22

23

1. $(PhO)_2P(O)N_3$, Ph_3P, DEAD, THF
2. H_2, 10% Pd/C, EtOH

24

MeOH, 110 °C, 2 h

25

CH_2O, $NaBH(OAc)_3$, MeOH, 30 min; Na, liq. NH_3, THF (65% overall from 21)

3: *meso*-chimonanthine

DEAD = diethylazodicarboxylate

Scheme 3. Overman's total synthesis of *meso*-chimonanthine (**3**).

preparation of **3** published by the Takayama group in 2002 which started from tryptamine (**28**, see column figures) and employed a hypervalent iodine-mediated radical dimerization to ultimately generate both racemic and *meso*-chimonanthine.[18] While this alternate approach lacked stereocontrol, it could provide pure *meso*-chimonanthine (**3**) in 30% yield following column chromatography.

Having presented two different options for how these researchers could obtain adequate supplies of *meso*-chimonanthine (**3**), we can now address the main topic of this chapter: how they converted this material into quadrigemine C (**1**) and psycholeine (**2**). Based on the retrosynthetic analysis presented in Scheme 2, the first task was to properly outfit this natural product for an eventual Stille coupling by appending an iodine substituent onto each of its aromatic rings at the site adjacent to the free nitrogen atom. As indicated in Scheme 4, this objective was accomplished through a sequence exploiting the well-documented power of Boc protecting groups to direct the *ortho*-lithiation of anilines.[19] Thus, following the initial formation of **32** from **3** using NaHMDS and Boc_2O, subsequent exposure of this new compound to excess *s*-BuLi (5 equiv) in Et_2O at −78 °C for 45 minutes led to the formation of dilithio species **33**. This reactive intermediate was quenched with a solution of 1,2-diiodoethane in THF to provide diiodide **34** in 88% yield,[20] leaving only the excision of the now-extraneous Boc directing groups to complete the synthesis of **12**. As shown, this concluding task was accomplished in near quantitative yield under mild conditions using TMSOTf (see Chapter 8 for a discussion of this technique). Before moving on with our commentary, we should remark on a couple of important features of the *ortho*-lithiation sequence used to assemble this fragment. First, although a number of carbamates can direct lithium to sites adjacent to typical aniline nitrogen atoms, only the Boc group has proven competent in this task when those nitrogen atoms are part of indoline systems as in **32**.[19] Second, had the indoline been an indole instead, then lithiation would not have occurred at C-7, but instead at its adjacent C-2 position (see column figures), thereby further reflecting how substrate properties are important determinants for the regioselectivity of these events.

With one fragment discussed, we can now focus our attention on the reactions that were required to access the coupling partner of this compound, stannane **11**. The Overman group's efficient synthesis of this target is delineated in Scheme 5, starting with benzylation of the known oxazolidinone **35** using standard conditions (BnCl, K_2CO_3).[21] While this opening operation was conventional, it served an extremely important purpose as it set the stage for a cascade that delivered most of the molecular complexity of the target in the next step. In the event, a solution of **36** in THF at −78 °C was treated with lithium acetylide **37** over the course of 3 hours with slow warming to −5 °C, conditions that led to the rupture of its cyclic carbamate and the formation of an intermediate phenolate

28
(100%) $ClCO_2Me$, NaOH/ H_2O, CH_2Cl_2, 0 °C
29
$PhI(OCOCF_3)_2$, CF_3CH_2OH, -30 °C
30
Biomimetic coupling
31
(30%) Red-Al®, toluene, Δ
3: *meso*-chimonanthine

s-BuLi- -O Ot-Bu 7
t-BuO O- -*s*-BuLi 2

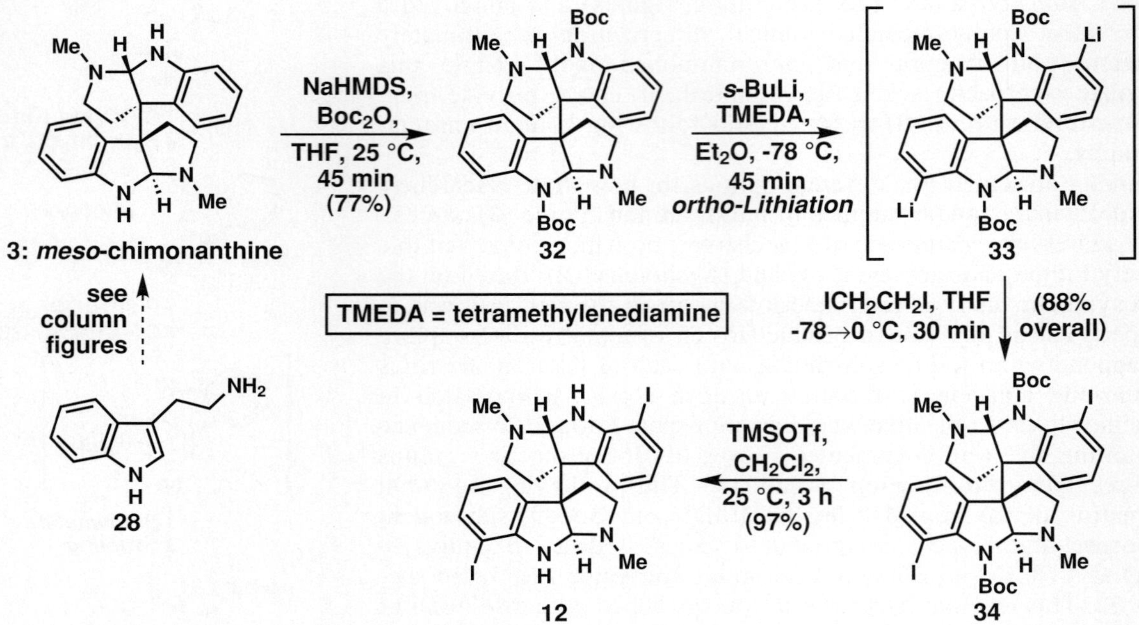

3: *meso*-chimonanthine

NaHMDS, Boc_2O, THF, 25 °C, 45 min (77%)

32

s-BuLi, TMEDA, Et_2O, -78 °C, 45 min *ortho-Lithiation*

33

see column figures

28

TMEDA = tetramethylenediamine

ICH_2CH_2I, THF -78→0 °C, 30 min (88% overall)

34

TMSOTf, CH_2Cl_2, 25 °C, 3 h (97%)

12

Scheme 4. Synthesis of advanced diaryliodide **12**.

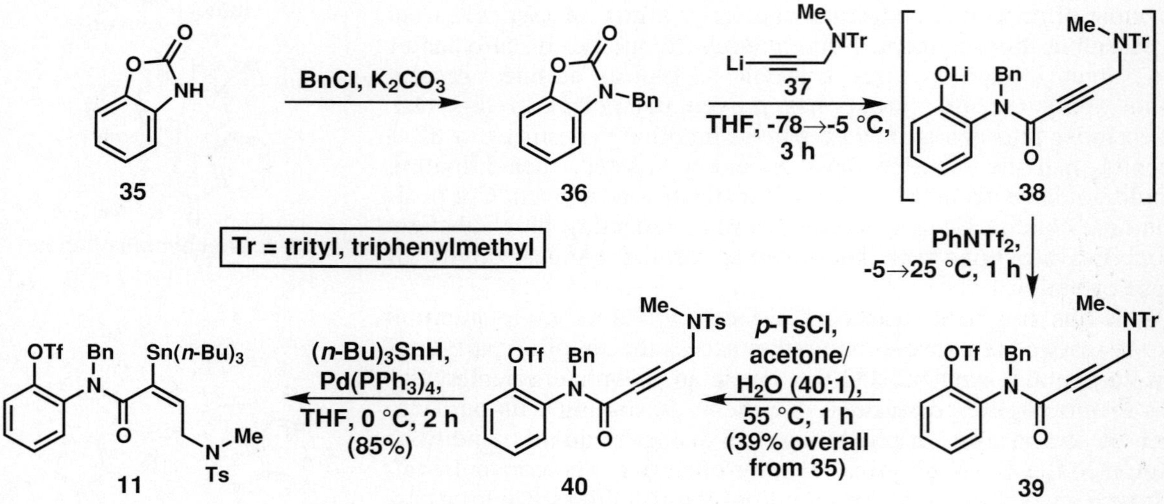

35

BnCl, K_2CO_3

36

37

THF, -78→-5 °C, 3 h

38

Tr = trityl, triphenylmethyl

$PhNTf_2$, -5→25 °C, 1 h

39

p-TsCl, acetone/H_2O (40:1), 55 °C, 1 h (39% overall from 35)

40

$(n\text{-Bu})_3SnH$, $Pd(PPh_3)_4$, THF, 0 °C, 2 h (85%)

11

Scheme 5. Preparation of stannane building block **11**.

(**38**). A terminating quench with phenyl triflimide (the McMurry-Hendrickson reagent)[22] then provided aryl triflate **39** bearing all the atoms of the target molecule (**11**) except for its tin motif. Since this new compound (**39**) was not particularly stable, its trityl protecting group was exchanged for a tosylate by heating **39** with *p*-TsCl in a mixture of acetone and H_2O (40:1) at 55 °C for 1 hour. The resultant, more-robust adduct (**40**) was then converted into the desired building block (**11**) in 85 % yield upon reaction with *n*-Bu_3SnH and catalytic $Pd(PPh_3)_4$ in THF at 0 °C, with the addition of the tri-*n*-butylstannane group occurring regioselectively as directed by the neighboring heteroatom in the starting alkyne.

With these two building blocks in hand, the critical palladium-mediated steps of the projected sequence could now begin, starting, of course, with their merger through a Stille reaction. As shown in Scheme 6, this objective was accomplished quite smoothly using a catalyst system consisting of $Pd_2(dba)_3{\cdot}CHCl_3$, tri-2-furylphosphine, and CuI to give the desired product (*meso*-**10**) in 71 % yield following 36 hours of stirring at ambient temperature in *N*-methyl-2-pyrrolidinone. Several features of this reaction are worthy of commentary. First and foremost, although this catalyst combination might seem esoteric, each component in this mixture served a specific purpose in facilitating the smooth prosecution of this step. For instance, Pd_2dba_3 was selected as the source of palladium among several other alternatives because it more rapidly exchanges its ligands with phosphines to create the active catalyst system (PdL_2) *in situ*.[23] Here, that phosphine was tri-2-furylphosphine, an electron-rich species whose use also remarkably accelerates the rate-determinig transmetallation step (see Chapter 31 in *Classics I* for the mechanism of this reaction) compared to more typical triarylphosphine ligands such as Ph_3P.[24] The addition of CuI also served to increase the rate of this particular step in the catalytic cycle.[25] Finally, although you might expect that the aryl triflates within the starting stannane (**11**) could also participate in this Stille reaction, these motifs are actually significantly less susceptible to palladium insertion than are aryl iodides,[26] especially under the mild conditions achieved in this case through the use of a highly active catalyst system. As such, they were merely idle observers in this transformation.

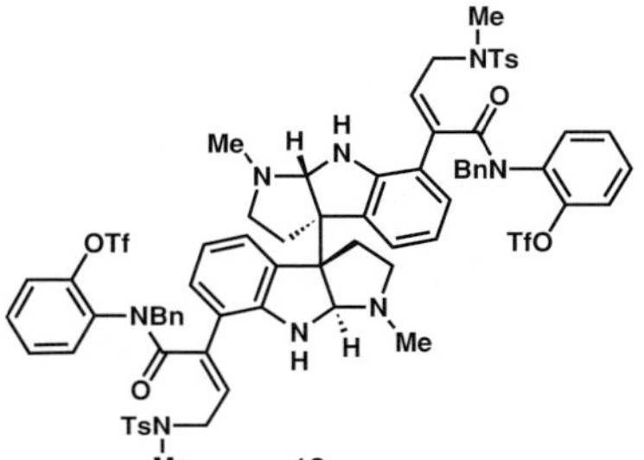

With this step accomplished as desired, probes of the key step of the sequence, the tandem asymmetric Heck reaction, could now commence in earnest. Based on several years of intensive research in developing appropriate systems to accomplish asymmetric Heck reactions with the ability to form all-carbon quaternary carbon centers,[12] the Overman group expected that the exposure of **10** to a 1:2 mixture of $Pd(OAc)_2$ and (*R*)-BINAP* would lead to the desired

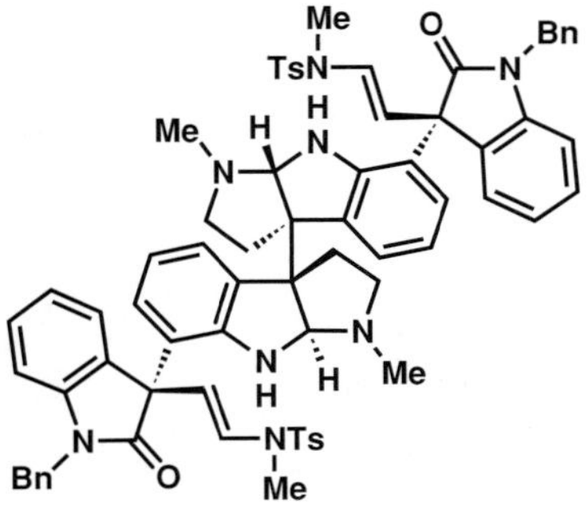

Me NTs O BnN TfO Me H H N N OTf NBn N H H N Me O TsN Me

10

Me TsN H O N Bn Me H N N N H H N Me N Bn O NTs Me

9

* Two equivalents of the phosphine ligand are required with the use of $Pd(OAc)_2$ since 1 equivalent of this phosphine is consumed during the reduction of Pd(II) to Pd(0).

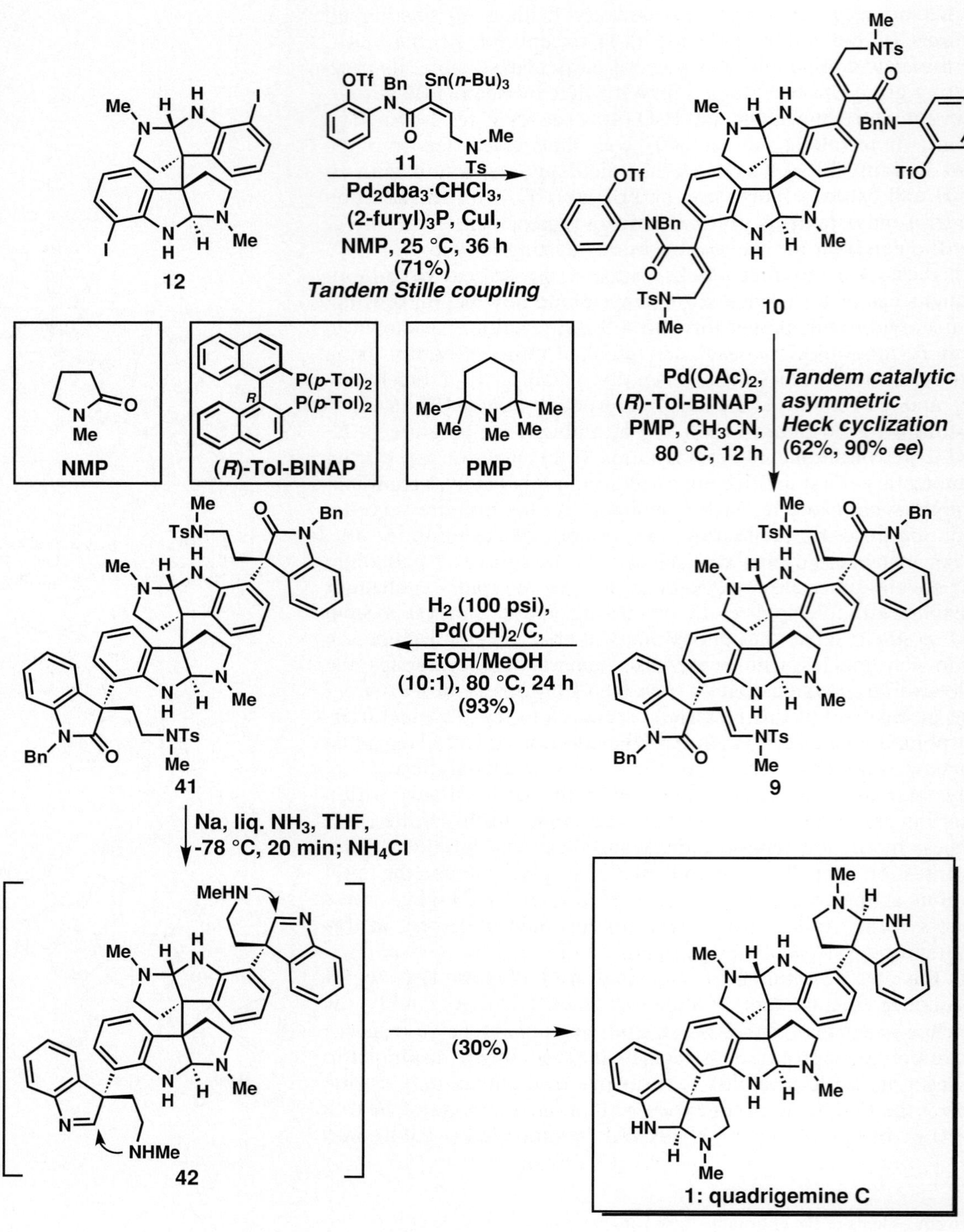

Scheme 6. Final stages and completion of Overman's total synthesis of quadrigemine C (**1**).

product (**9**). As matters transpired, this conjecture indeed proved to be true, although in early attempts the reaction proceeded in modest yield and in 65 % *ee*. Careful optimization, however, led to an ideal protocol which ultimately provided the desired chiral adduct (**9**) bearing its two oxindole rings and properly formed diaryl quaternary carbon centers in 62 % overall yield and 90 % *ee*. As shown, these conditions involved exchanging BINAP for a structurally-related ligand, (*R*)-Tol-BINAP, and reacting the substrate in CH_3CN at 80 °C for 12 hours in the presence of PMP (to scavenge the triflic acid by-product formed through successful coupling).*

Needless to say, there are several elements of this signature reaction (i. e. **10**⟶**9**) of the total synthesis that we should mention. First, as intimated above, the absence of any indoline protection on the central core was critical to accomplishing its high enantioselectivity. For instance, if an alternate substrate such as **43** with benzyl groups on these sites was reacted under related conditions, no enantioselection was observed whatsoever; only *meso* products such as **44** (see column figures) were obtained.[27] Since the ability of the substrate to place its unsaturated sites in a coplanar alignment is critical to achieving high levels of asymmetry, the absence of any reagent control in this experiment presumably indicates that non-bonding interactions imposed by the benzyl groups prevented such a conformation from being adopted.[12a] It is important to note that even with an appropriately designed substrate many other factors can also influence the overall enantioselectivity of the process. For example, the choice of solvent can have a dramatic effect, with polar solvents such as MeCN or NMP usually providing superior results. Equally crucial is the overall electron density on the appended aromatic rings since overly electron-rich systems lead to decreased *ee* values due to their faster rate of oxidative addition/reductive elimination.[12] Finally, we should mention that, in general, intramolecular Heck reactions proceed with far higher levels of enantioselectivity than their intermolecular counterparts, an outcome that primarily reflects the difficulty in controlling the regioselectivity of alkene insertion in the absence of a connecting tether. Furthermore, intramolecular asymmetric Heck reactions are also more versatile in terms of substrate scope since they tolerate both electron-neutral and electron-poor olefins, while their intermolecular cousins typically only succeed when electron-rich olefin substrates are used.

With this operation behind us, we can now present the concluding steps of this elegant sequence, which only involved forming the final rings needed to complete the peripheral cyclotryptamines of the target molecules. Two steps were required to accomplish this task. First, hydrogenation in a 10:1 solvent mixture of EtOH and MeOH in the presence of $Pd(OH)_2/C$ (Pearlman's catalyst) at rela-

43

$Pd(OAc)_2$, (*R*)-BINAP, PMP, DMA, 100 °C, 12 h — *Attempted tandem catalytic asymmetric Heck cyclization*

44: *meso*

* Two other *meso* dioxindole products of unassigned configuration were also formed from this step in yields of 14 and 7 %, respectively.

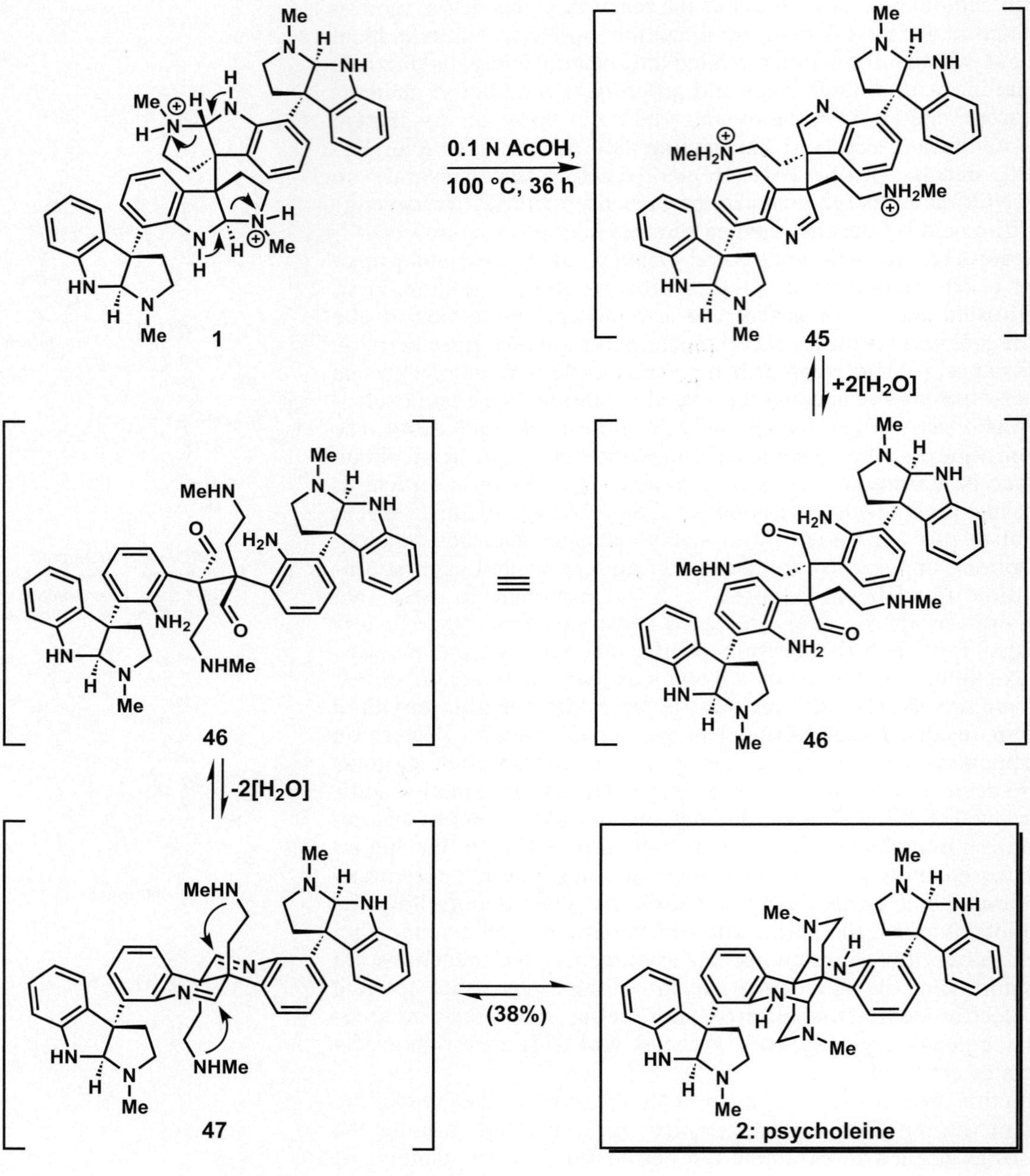

Scheme 7. Acid-catalyzed conversion of quadrigemine C (**1**) into psycholeine (**2**).

tively high pressure (100 psi) and temperature (80 °C) saturated the olefins introduced by the successful Heck reaction in the previous operation to afford **41**. Then, following precedent established in 1935 by Julian and Pikl,[15] this new compound (**41**) was exposed to a large excess of sodium in a mixture of THF and liquid NH_3, reductive conditions that cleaved both benzyl ethers, both tosyl groups, and permitted the final cyclizations to provide quadrigemine C (**1**) in 30 % yield! Overall, only 10 steps were required to complete this natural product from commercially available materials [based on Takayama's sequence to prepare *meso*-chimonanthine (**3**)]. Four of these operations involved some type of palladium catalyst.

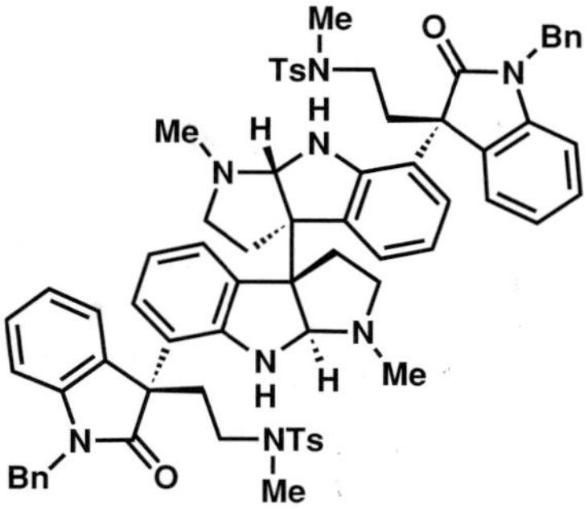

As a final exercise, the Overman group exposed their synthetic sample of quadrigemine C (**1**) to a solution of AcOH (0.1 N) at 100 °C for 36 hours in order to convert it into psycholeine (**2**). In line with literature precedent,[8b] these conditions provided this intriguing alkaloid (**2**) in 38 % yield. Scheme 7 provides a mechanism that can account for this impressive rearrangement based simply on a series of equilibria in which the six-membered rings of the central core of psycholeine (**2**) provide a thermodynamic sink.[28]

19.4 Conclusion

It is hard to predict the exciting directions that palladium-based reagents and catalysts will take chemical synthesis in the future, but this total synthesis of quadrigemine C (**1**) and psycholeine (**2**) should persist, at least for a while, as a premier example of some of their capabilities. Indeed, the way in which the Overman group creatively used these facilitators to orchestrate a series of three consecutive reactions terminating in a tandem asymmetric Heck reaction provided a highly effective solution to the central challenge posed by the target molecules that is both admirable and exemplary. In the absence of these C–C bond-forming reactions and the recent methodological developments that turned them into sharp tools, it would be exceedingly hard to envision how these natural products could be synthesized in the laboratory with such impressive efficiency.[29]

References

1. For leading monographs, see: a) J. Tsuji, *Palladium Reagents and Catalysts: Innovations in Organic Synthesis*, John Wiley & Sons, Chichester, **1995**, p. 560; b) *Metal-catalyzed Cross-coupling Reactions*, (Eds.: F. Diederich, P.J. Stang), Wiley-VCH, Weinheim, **1998**, p. 517.
2. For leading examples, see: a) S.-Y. Liu, M.J. Choi, G.C. Fu, *Chem. Commun.* **2001**, 2408; b) J.P. Wolfe, R.A. Singer, B.H. Yang, S.L. Buchwald, *J. Am. Chem. Soc.* **1999**, *121*, 9550.
3. a) J.P. Wolfe, H. Tomori, J.P. Sadighi, J. Yin, S.L. Buchwald, *J. Org. Chem.* **2000**, *65*, 1158; b) J.F. Hartwig, M. Kawatsura, S.I. Hauck, K.H. Shaughnessy, L.M. Alcazar-Roman, *J. Org. Chem.* **1999**, *64*, 5575, and references cited in each.
4. S. Lee, N.A. Beare, J.F. Hartwig, *J. Am. Chem. Soc.* **2001**, *123*, 8410.
5. M.R. Netherton, C. Dai, K. Neuschütz, G.C. Fu, *J. Am. Chem. Soc.* **2001**, *123*, 10099.
6. A.D. Lebsack, J.T. Link, L.E. Overman, B.A. Stearns, *J. Am. Chem. Soc.* **2002**, *124*, 9008.
7. For selected papers on the isolation of these compounds, see: a) H.F. Hodson, B. Robinson, G.F. Smith, *Proc. Chem. Soc. London* **1961**, 465; b) T. Tokuyama, J.W. Daly, *Tetrahedron* **1983**, *39*, 41; c) Y. Adjibadé, B. Weniger, J.C. Quirion, B. Kuballa, P. Cabalion, R. Anton, *Phytochemistry* **1992**, *31*, 317; d) R.K. Duke, R.D. Allan, G.A.R. Johnston, K.N. Mewett, A.D. Mitrovic, C.C. Duke, T.W. Hambley, *J. Nat. Prod.* **1995**, *58*, 1200. For a recent review on these natural products, see: U. Anthoni, C. Christophersen, P.H. Nielsen in *Alkaloids: Chemical and Biological Perspectives*, *Vol. 13* (Ed.: S.W. Pelletier), Pergamon, New York, **1999**, pp. 163–236.
8. For the isolation and structural characterization of quadrigemine C and psycholeine, see: a) F. Libot, C. Miet, N. Kunesch, J.E. Poisson, J. Pusset, T. Sévenet, *J. Nat. Prod.* **1987**, *50*, 468; b) F. Guéritte-Voegelein, T. Sévenet, J. Pusset, M.-T. Adeline, B. Gillet, J.-C. Beloeil, D. Guénard, P. Potier, R. Rasolonjanahary, C. Kordon, *J. Nat. Prod.* **1992**, *55*, 923; c) L. Verotta, T. Pilati, M. Tatò, E. Elisabetsky, T.A. Amador, D.S. Nunes, *J. Nat. Prod.* **1998**, *61*, 392; d) V. Jannic, F. Guéritte, O. Laprévote, L. Serani, M.-T. Martin, T. Sévenet, P. Potier, *J. Nat. Prod.* **1999**, *62*, 838.
9. a) L.E. Overman, J.F. Larrow, B.A. Stearns, J.M. Vance, *Angew. Chem.* **2000**, *112*, 219; *Angew. Chem. Int. Ed.* **2000**, *39*, 213; b) L.E. Overman, D.V. Paone, B.A. Stearns, *J. Am. Chem. Soc.* **1999**, *121*, 7702; c) S.B. Hoyt, L.E. Overman, *Org. Lett.* **2000**, *2*, 3241. For an earlier total synthesis using a different sequence, see: d) J.T. Link, L.E. Overman, *J. Am. Chem. Soc.* **1996**, *118*, 8166.
10. For reviews on the asymmetric Heck reaction, see: a) J.T. Link in *Organic Reactions*, *Vol. 60* (Eds.: L.E. Overman, et al.), John Wiley & Sons, New York, **2002**, pp. 157–534; b) Y. Donde, L.E. Overman in *Catalytic Asymmetric Synthesis*, 2nd *Ed.* (Ed.: I. Ojima), Wiley-VCH, Weinheim, **2000**, pp. 675–697; c) M. Shibasaki, C.D.J. Boden, A. Kojima, *Tetrahedron* **1997**, *53*, 7371. For an earlier review covering contributions by the Overman group, see: d) L.E. Overman, *Pure & Appl. Chem.* **1994**, *66*, 1423.
11. a) N.E. Carpenter, D.J. Kucera, L.E. Overman, *J. Org. Chem.* **1989**, *54*, 5846; b) A. Ashimori, L.E. Overman, *J. Org. Chem.* **1992**, *57*, 4571. For the first demonstration of an asymmetric Heck reaction in chemical synthesis, see: Y. Sato, M. Sodeoka, M. Shibasaki, *J. Org. Chem.* **1989**, *54*, 4738.
12. a) A.B. Dounay, K. Hatanaka, J.J. Kodanko, M. Oestreich, L.E. Overman, L.A. Pfeifer, M.M. Weiss, *J. Am. Chem. Soc.* **2003**, *125*, 6261; b) A. Ashimori, B. Bachand, L.E. Overman, D.J. Poon, *J. Am. Chem. Soc.* **1998**, *120*, 6477; c) A. Ashimori, B. Bachand, M.A. Calter, S.P. Govek, L.E. Overman, D.J. Poon, *J. Am. Chem. Soc.* **1998**, *120*, 6488.
13. For the first successful example of this approach, see: Z. Wang, D. Deschênes, *J. Am. Chem. Soc.* **1992**, *114*, 1090.
14. For reviews on these approaches, see: a) C.S. Poss, S.L. Schreiber, *Acc. Chem. Res.* **1994**, *27*, 9; b) S.R. Magnuson, *Tetrahedron* **1995**, *51*, 2167. For some of the difficulties in preparing *meso* compounds, see: c) R.W. Hoffmann, *Angew. Chem.* **2003**, *115*, 1120; *Angew. Chem. Int. Ed.* **2003**, *42*, 1096.
15. P.L. Julian, J. Pikl, *J. Am. Chem. Soc.* **1935**, *57*, 539.
16. E.A. Mash, K.A. Nelson, S. Van Deusen, S.B. Hemperly in *Organic Syntheses*, *Vol. 68* (Eds.: J.D. White, et al.), John Wiley & Sons, New York, **1990**, pp. 92–103.
17. For one example, see: R.M. Lemieux, A.I. Meyers, *J. Am. Chem. Soc.* **1998**, *120*, 5453. For a review on the asymmetric synthesis of quaternary centers, see: E.J. Corey, A. Guzman-Perez, *Angew. Chem.* **1998**, *110*, 402; *Angew. Chem. Int. Ed.* **1998**, *37*, 388.
18. H. Ishikawa, H. Takayama, N. Aimi, *Tetrahedron Lett.* **2002**, *43*, 5637.
19. a) J.M. Muchowski, M.C. Venuti, *J. Org. Chem.* **1980**, *45*, 4798; b) P. Stanetty, H. Koller, M. Mihovilovic, *J. Org. Chem.* **1992**, *57*, 6833; c) P. Beak, W.K. Lee, *J. Org. Chem.* **1993**, *58*, 1109. For a review on directed *ortho*-lithation, see: V. Snieckus, *Chem. Rev.* **1990**, *90*, 879.
20. a) M. Iwao, T. Kuraishi in *Organic Syntheses*, *Vol. 73* (Eds.: R.K. Boeckman, et al.), John Wiley & Sons, New York, **1996**, pp. 85–93; b) M. Iwao, T. Kuraishi, *Heterocycles* **1992**, *34*, 1031.
21. H. Ucar, K. Van derpoorten, S. Cacciaguerra, S. Spampinato, J.P. Stables, P. Depovere, M. Isa, B. Masereel,

J. Delarge, J.H. Poupaert, *J. Med. Chem.* **1998**, *41*, 1138.

22. a) J.B. Hendrickson, R. Bergeron, *Tetrahedron Lett.* **1973**, *14*, 4607; b) J.E. McMurry, W.J. Scott, *Tetrahedron Lett.* **1983**, *24*, 979. For a review, see: W.J. Scott, J.E. McMurry, *Acc. Chem. Res.* **1988**, *21*, 47.
23. C.G. Pierpont, M.C. Mazza, *Inorg. Chem.* **1974**, *13*, 1891.
24. a) V. Farina, S.R. Baker, D.A. Benigni, C. Sapino, *Tetrahedron Lett.* **1988**, *29*, 5739; b) V. Farina, S.R. Baker, C. Sapino, *Tetrahedron Lett.* **1988**, *29*, 6043; c) V. Farina, B. Krishnan, *J. Am. Chem. Soc.* **1991**, *113*, 9585.
25. a) L.S. Liebeskind, R.W. Fengl, *J. Org. Chem.* **1990**, *55*, 5359; b) V. Farina, S. Kapadia, B. Krishnan, C. Wang, L.S. Liebeskind, *J. Org. Chem.* **1994**, *59*, 5905; c) X. Hun, B.M. Stoltz, E.J. Corey, *J. Am. Chem. Soc.* **1999**, *121*, 7600.
26. L.M. Alcazar-Roman, J.F. Hartwig, *Organometallics* **2002**, *21*, 491.
27. Personal communication with Professor Larry E. Overman.
28. For related mechanisms, see: a) J.B. Hendrickson, R. Göschke, R. Rees, *Tetrahedron* **1964**, *20*, 565; b) E.S. Hall, F. McCapra, A.I. Scott, *Tetrahedron* **1967**, *23*, 4131.
29. For recent applications of related strategies in the syntheses of some of the other natural products shown in Figure 1, see: a) J.J. Kodanko, L.E. Overman, *Angew. Chem.* **2003**, *115*, 2632; *Angew. Chem. Int. Ed.* **2003**, *42*, 2528; b) L.E. Overman, E.A. Peterson, *Angew. Chem.* **2003**, *115*, 2629; *Angew. Chem. Int. Ed.* **2003**, *42*, 2525.

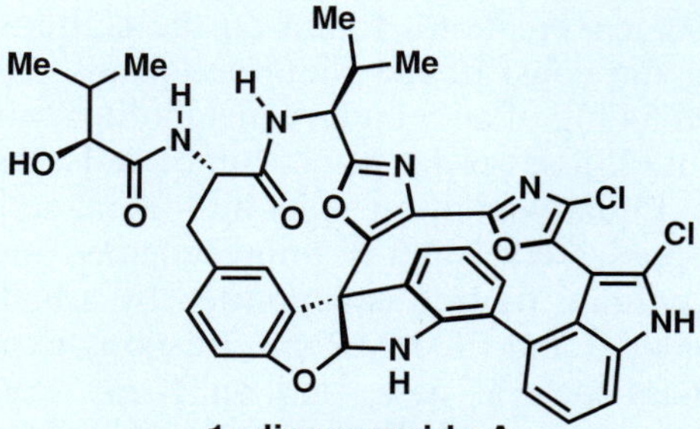

1: diazonamide A

20

K. C. Nicolaou (2002; 2003)

Diazonamide A

20.1 Introduction

Key concepts:

- **Structure elucidation**
- **Robinson–Gabriel cyclodehydration**
- **Challenges in macrocycle synthesis**
- **SmI_2 in organic synthesis**

Throughout the early part of the 20th century, chemists often performed a total synthesis merely to verify the molecular structure of a natural product, expecting that if their synthetic material bearing the proposed architecture had physical properties identical to those of the natural sample, they could establish unambiguous proof of its connectivities. Although these endeavors required painstaking attention to detail, years of effort, and the assumption that reactions followed prescribed pathways, they succeeded in unraveling several profoundly complex molecular architectures with an amazingly high degree of accuracy. Today, however, with the advent of modern analytical methods such as NMR spectroscopy, this role for total synthesis has all but disappeared, with such endeavors serving only to establish absolute or relative stereochemical configurations in those rare instances in which X-ray crystallographic data are unavailable. Yet, despite all our present advantages for determining the architectural disposition of complex molecules in the absence of chemical synthesis, mistakes are still possible even when the most advanced techniques have been applied. Such was the case with the natural product that is the subject of this chapter.

Our story begins in the late 1980s, when Professor William Fenical and his student Niels Lindquist of the Scripps Institution of Oceanography in La Jolla, California, collected various colonial ascidians from remote areas throughout the Pacific, hoping that some of them would contain unusual secondary metabolites endowed with potent and useful biological activity. Their efforts were handsomely rewarded, as one of these organisms possessed just such a compound. That species was *Diazona angulata*, a beautifully colored ascidian (see cover photo) found on the ceilings of several small caves along the coast of the Siquijor Islands. From this organism was obtained 54 mg of a secondary metabolite, coined diazonamide A, which could kill several types of tumor cells at low nanomolar concentrations. Equally exciting as its biological action, this natural product also possessed a totally unprecedented molecular architecture (**2**, see column figure) as supported by a battery of one- and two-dimensional ^{1}H and ^{13}C NMR spectroscopy experiments, high-resolution mass-spectral data, and an X-ray crystal structure obtained in collaboration with Professor Jon Clardy and Gregory Van Duyne at Cornell University.[1]

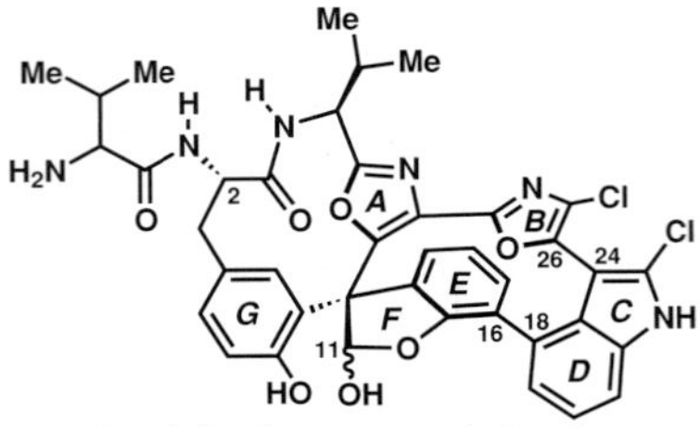

2: originally proposed structure of diazonamide A

Superficially, it is relatively easy to apprehend most of the unique elements of this amazing architecture in which Nature has somehow blended two distinctly different 12-membered macrocyclic rings, one composed of a peptide backbone and the other of heteroaromatic rings, into a single architectural unit. However, what is far more difficult to ascertain from a two-dimensional representation is its overall strain, compactness, and rigidity. In fact, these features restrict rotational freedom to the degree that both the C16–C18 and C24–C26 biaryl axes are trapped as single atropisomers (see Chapter 9 for a discussion of this topic). And, although you might expect diazonamide A to possess a rather unique UV signature as it possesses several units of unsaturation adjacent to each other, the strict requirements of its structural elements twist the constituent aromatic rings so dramatically that they cannot achieve coplanarity, leading to almost no detectable UV activity. Its lone quaternary center not withstanding, the heterocycle-based macrocycle of diazonamide A constitutes as near a Möbius strip as yet encountered in Nature.

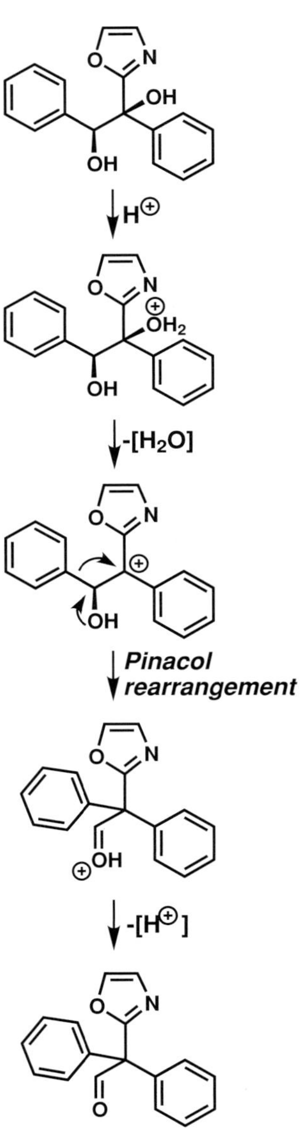

Of course, it would be impossible for a natural product gifted with such impressive structural complexity and promising biological activity to escape the attention of the synthetic community, and, indeed, following the initial report of its structure in 1991, several research groups sought to accomplish the total synthesis of diazonamide A.[2] These efforts finally reached fruition at the end of 2001 when Professor Patrick Harran and his colleagues at the Southwestern Medical Center in Dallas, Texas, chartered the first pathway to **2**.[3] The key elements of their insightful solution to this formidable synthetic puzzle are shown in Scheme 1; two highly powerful reactions were creatively enlisted to forge the major macrocyclic subunits. The first of these critical steps was an acid-catalyzed pinacol rearrangement of chiral diol **3** (see column figures for a mechanism on a simplified substrate), which served to contract the 13-membered ring of this intermediate, concurrently forming the 12-membered AG macrocycle and the daunting C-10 quaternary center of the target structure.[4] With this process stereospecifically affording **4** in 54 % yield following a protection step, a few additional operations led to **6**, a precursor suitably outfitted for the generation of the second 12-membered ring through photocyclization.

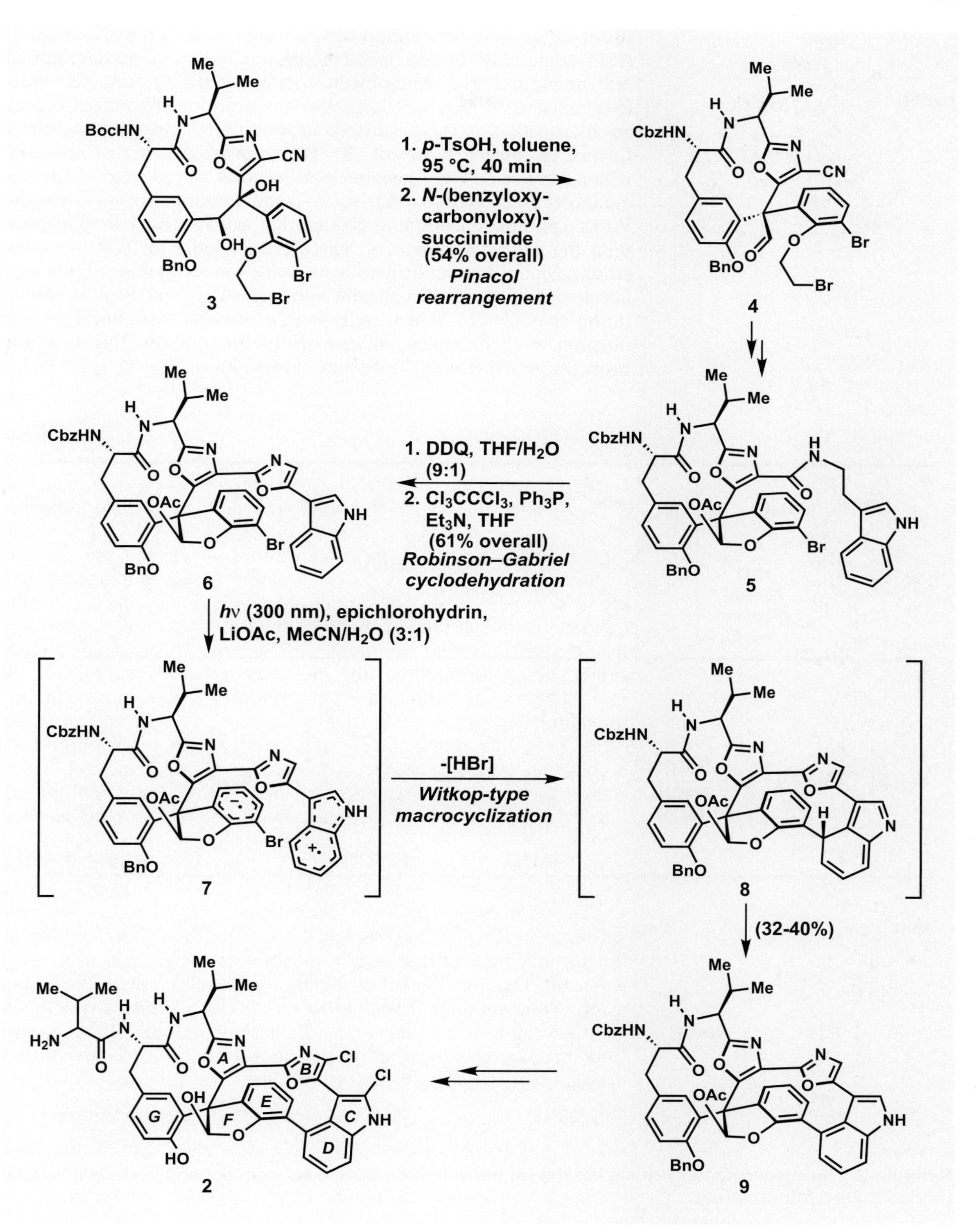

Scheme 1. Harran's synthesis of the originally proposed structure of diazonamide A (**2**).

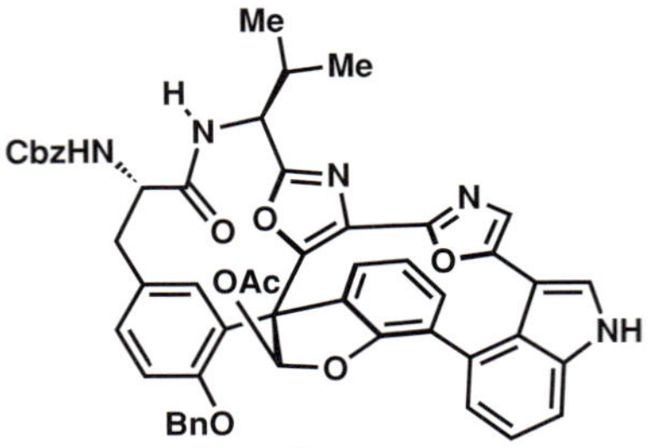

In the event, photoexcitation of a solution of **6** in MeCN and H_2O (3:1) containing LiOAc and epichlorohydrin at a wavelength of 300 nm initiated a single-electron transfer (SET) from the electron-rich indole ring to its adjacent electron-poor E-ring, affording an intermediate (**7**) of suitable reactivity to form the hindered C16–C18 biaryl bond of **9**. This remarkable transformation, which atropselectively generated the second macrocycle of the target molecule in yields of 32–40 %, is more commonly known as the Witkop photocyclization, a reaction that was first described in 1966 with the example shown in Scheme 2a.[5] Since then, it has been employed only sporadically in natural product synthesis, perhaps because in many contexts there is a lack of regiocontrol as shown in the examples in Schemes 2b and 2c.[6] In this case, however, the reaction worked beautifully, and with **9** in hand the Harran group soon converted it into the desired target (**2**).

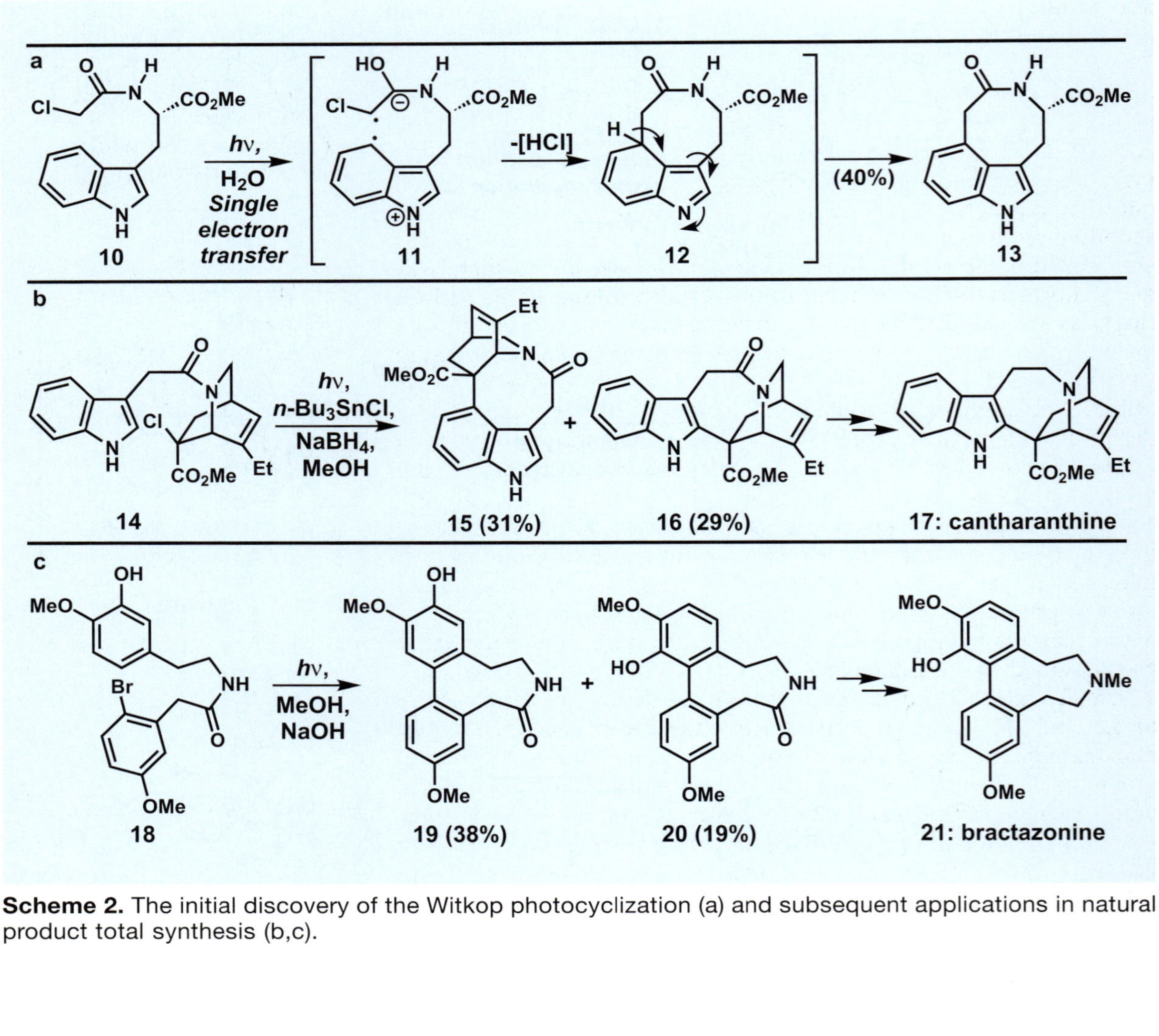

Scheme 2. The initial discovery of the Witkop photocyclization (a) and subsequent applications in natural product total synthesis (b,c).

Unfortunately, these researchers could not declare a total synthesis of diazonamide A at this point, as both the chemical stability as well as the spectral data of their synthetic material (**2**) differed drastically from those reported for the natural product. Since the original structure assignment was based on X-ray crystallographic data, one might be tempted to believe that the synthetic chemists had made a mistake somewhere along their route and had actually synthesized something other than what they intended. Amazingly, though, the problem did not reside with their developed sequence, but, instead, with the proposed structure (**2**) of diazonamide A due to a series of seemingly logical, but ultimately misleading assumptions made during its characterization.

During the early stages of their structure elucidation, the Fenical and Clardy groups wished to obtain a crystal structure for diazonamide A in order to support their initial NMR assignments. Unfortunately, no crystals of suitable quality could be grown. Yet, hope still remained that they could employ this powerful method to confirm the structure because *Diazona angulata* provided several other natural products, including diazonamide B (**22**, see column figure), which appeared to be structurally similar to diazonamide A on the basis of NMR spectroscopic analysis.[1] If one of these compounds, or a derivative thereof, could be crystallized, then at least parts of the structure of diazonamide A could be confirmed. Pleasingly, the *p*-bromobenzamide derivative of diazonamide B turned out to be a beautifully crystalline solid, revealing a structure corresponding to **23**, which, for the most part, verified much of what was anticipated based on the initial NMR data set. However, there was one major surprise in the form of an acetal bridging its F- and G-rings, as the NMR data for the diazonamides seemed to indicate the existence of an open hemiacetal instead (as indicated in structure **2**). Somehow this one element of structural incongruity between the different techniques had to be reconciled. The answer proposed by the Fenical and Clardy groups was that the closed acetal observed in crystal structure **23** was, in fact, an artifact resulting from the conditions employed to attach the *p*-bromobenzamide group onto **22**. Thus, if a hemiacetal was accepted for the F-ring of diazonamide A, then the one element that the crystal structure of **23** could not determine (the amino acid tethered from C-2) must be a valine group in order to provide a compound that could match a mass spectral signature corresponding to $[M+H-H_2O]^+$. As such, the structure **2** was assigned to diazonamide A.

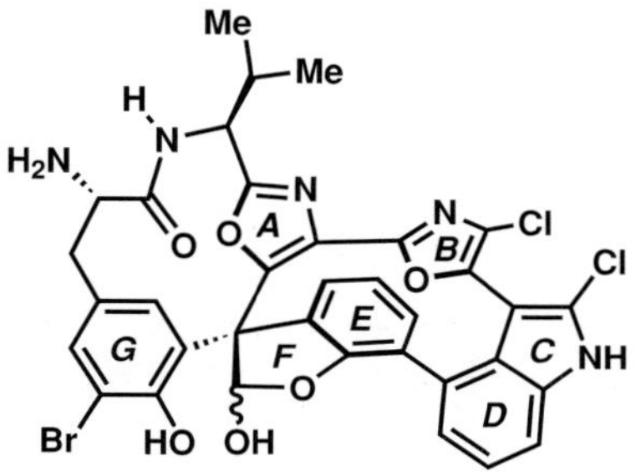

Armed with the knowledge that both the X-ray crystal structure of **23** and the formula derived from the mass spectral data of diazonamide A involved the loss of a molecule of water, the Harran group speculated that perhaps the correct structure for the natural product simply differed from **2** by a closed acetal. Indeed, this alternate compound would match all of the critical elements of structure **23** and would have a mass signature corresponding directly to $[M+H]^+$. So, using their sequence, these investigators succeeded in generating this new structure. But, amazingly, once again the

22: putative structure of diazonamide B

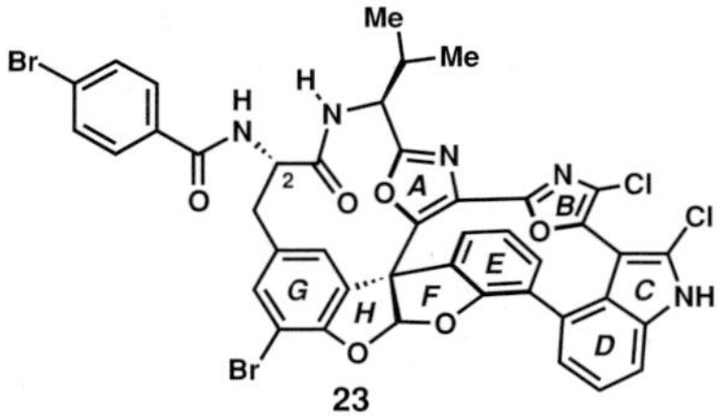

23

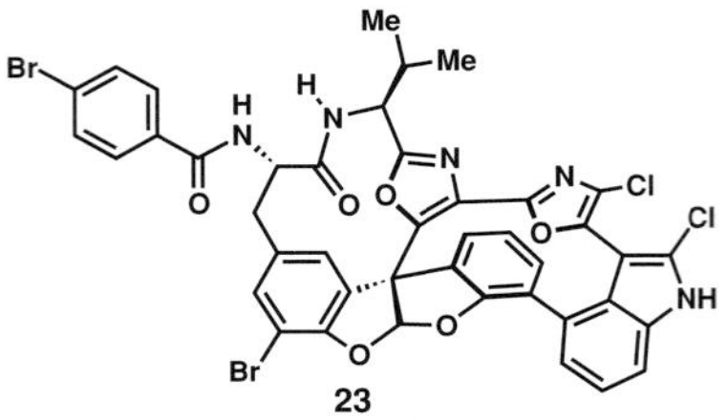

23

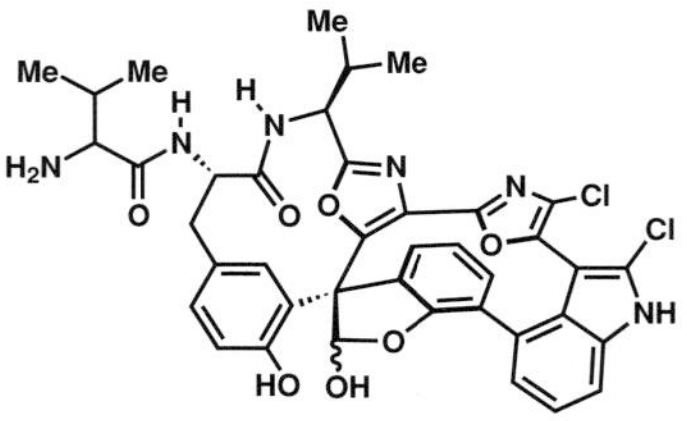

2: originally proposed structure of diazonamide A

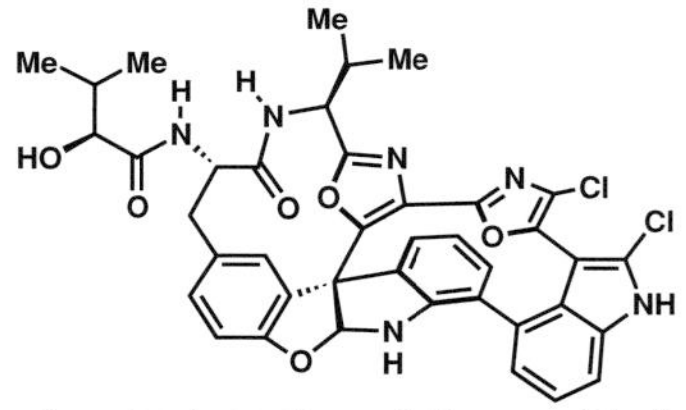

1: revised structure of diazonamide A

physical properties of their synthetic material did not correlate to data obtained from the natural sample of diazonamide A. Where, then, was the discrepancy?

The answer appeared to reside within the original crystal structure of **23**. Although X-ray crystallography is often viewed as infallible, its one weakness is its inability to detect hydrogen atoms. As a result, it is sometimes difficult to distinguish whether an atom is an oxygen devoid of any hydrogen atoms or a nitrogen atom whose bound hydrogen atom(s) cannot be revealed by X-ray crystallographic analysis. The Harran group logically hypothesized that this shortcoming might be the origin of the problem in this case, with the oxygen atom in the F-ring of **23** (and thus in **2**) misassigned when it should really be an NH group within an aminal system as in revised structure **1**.[3] From a biosynthetic standpoint, this change would appear to provide an easier structure for Nature to synthesize since the aminal motif could be formed directly from a tryptophan rather than the unnatural collection of amino acids that would be required to generate the E-ring system of **2**. Accepting this alteration, a second modification somewhere else in the molecule would be needed to account for the mass profile of diazonamide A. The obvious candidate was the terminal amino acid attached to the C-2 amine. If this fragment was a non-natural 2-hydroxyisovaleric acid as shown in **1**, then all the previously incongruent data would appear to be reconciled.

Although we opened this chapter by stating that total synthesis played a minimal role in modern structure elucidation, at this point it now constituted the only tool that could finally prove if the structure of diazonamide A was indeed **1**. As we shall see, it took just a few months to put this problem to rest, as the Nicolaou group at The Scripps Research Institute was ultimately able to develop not one, but two, completely divergent routes to this revised structure (**1**) which verified its accuracy.[7,8] These endeavors are the focus of the remainder of this chapter.

20.2 Retrosynthetic Analysis

Since the revised architecture of diazonamide A (**1**) was based on a major constitutional change from **2** at its epicenter, any synthetic plan that had been developed for the originally proposed structure of diazonamide A would need to be significantly redesigned in order to allow access to building blocks with appropriate functionality. Considerable scouting would likely be required before suitable routes to such intermediates could be defined. Once these obstacles were overcome, however, the total synthesis was not necessarily near or guaranteed. Experience garnered during campaigns towards **2** consistently demonstrated that although it is relatively easy to create fragments bearing the disparate elements of the diazonamide structure, merging these pieces into structures containing both of its highly strained and rigid 12-membered macrocyclic systems is incredibly difficult.[2] For example, few of the creative strategies developed towards **2** managed to identify a sequence that could generate just *one* of these 12-membered systems in any measurable yield.[9] Thus, from the standpoint of synthetic design, this information meant that one should seek answers to the questions of how and in what order to form these macrocyclic rings before addressing any other element of the target molecule (**1**).

The latter question is, of course, the simpler of the two, as there are only two possible choices for the ordering of macrocycle construction. Upon initial inspection, the 12-membered AG-ring system of **1** would appear to be the easier macrocycle to construct since an obvious site to effect its closure is through its amide bond, a motif that can, in principle, be forged by several reagents. In contrast, the highly complex heteroaromatic core has no single definitive site where macrocyclization could be achieved, and the requirement to generate its biaryl axes in an atropselective manner meant that it would likely be far harder to construct in any case. As such, one might be tempted to go for the more challenging system first, and then fashion the "easier" AG system last. Unfortunately, assessing the challenges posed by individual macrocycles does not always lead to the best order needed to create the final polycycle, because when these systems share several atoms the conformational requirements of one can dramatically impact the ease with which the other is formed. Faced with this uncertainty, the Nicolaou group decided to maximize their chances of success by pursuing both options for the order of macrocycle closure. Thus, they developed two unique and distinct retrosynthetic analyses of the target (**1**), which, as we shall see, had entirely different answers to the "how" question for macrocycle construction.

The first of these plans is shown retrosynthetically in Scheme 3, starting with three minor, but necessary, modifications to prepare the ground for disconnections seeking to open the macrocyclic systems of diazonamide A (**1**).[7] The first of these alterations was the removal of the hydroxyisovaleric acid side chain (**24**) from C-2, a disconnection that was implemented early in the analysis because the stereogenic center of this fragment (C-37) could not be assigned with the available data. Thus, by removing this element at the outset, both enantiomers of **24** could be attached to a suitable amine precursor at the end of the sequence in order to assign the stereochemistry at C-37 with minimum effort. The other two modifications leading to **25** from **1**, the excision of the two aryl chlorine substituents and opening of the FH aminal ring, served to remove functionality that would likely not survive the arduous synthetic journey and diverse reaction conditions needed to create the other elements of the architecture of diazonamide A. In the forward direction, it was expected that these motifs could be generated in either order, with the chlorine atoms installed through a site-selective electrophi-

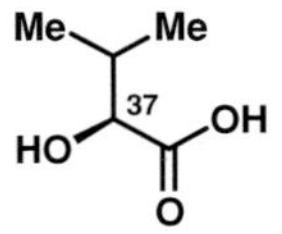

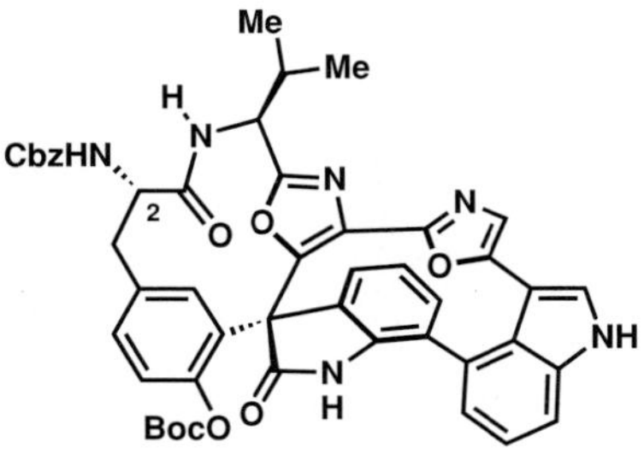

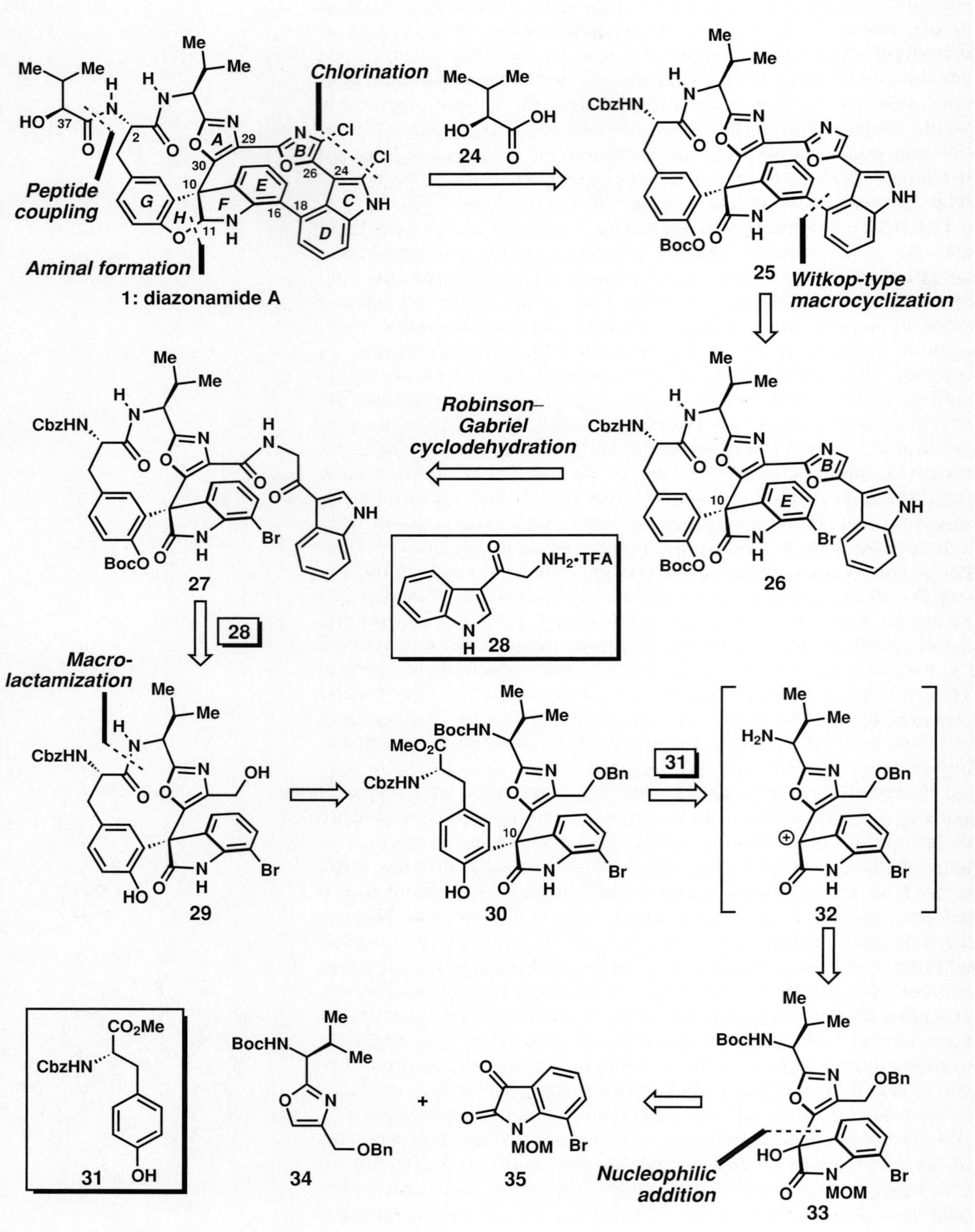

Scheme 3. Retrosynthetic analysis of diazonamide A (**1**): the photocyclization route.

lic aromatic substitution reaction and the aminal ring formed upon the addition of the G-ring phenolic group to a reactive imine species derived from the F-ring lactam.

Accepting these initial adjustments, a new subgoal structure had been identified (**25**) from which either macrocyclic system could now be ruptured. For this particular retrosynthetic analysis, the Nicolaou group elected to unfurl the heterocycle-based ring system first at its C-16/C-18 biaryl linkage through a Witkop-type photocyclization transform,[5] the same approach that the Harran group had so beautifully employed to form this macrocyclic system in **2**.[3] As such, the chances that **26** could successfully be converted into **25** during the actual synthesis were relatively high, although the presence of new functionality in the E-ring of **26** (cf. **5**→**9**) meant that some modification of the conditions would likely be needed to best promote the single-electron transfer step in this case. Equally important from the standpoint of design, successful cyclization should occur atropselectively as governed by the chirality of the C-10 quaternary center and the likely strong driving force to orient the B- and E-rings in a stacked arrangement (an orientation that only the correct atropisomer can deliver).

Having elected to pursue this method for macrocyclization, the heterocyclic chain of this new target (**26**) could then be simplified to **29** by unraveling its B-ring oxazole to a ketoamide precursor (**27**) and breaking apart the newly unveiled amide linkage to remove the CD-indole portion as amine **28**. Neither of these steps (oxazole and amide bond formation) was expected to pose any particular difficulties in the synthesis as they both constituted relatively conventional transformations. However, before pressing on with this retrosynthetic analysis, a few additional words about the projected conversion of **27** into **26** are in order since this step involves a reaction that we have not yet discussed, the Robinson–Gabriel cyclodehydration. Traditionally, this named reaction encompasses all those transformations in which the exposure of an appropriately functionalized 1,4-dicarbonyl compound to a mineral acid (such as H_2SO_4) serves to effect a dehydrative ring closure leading to a heteroaromatic system.[10] This process can be applied to the synthesis of oxazoles, thiazoles, and furans, and, for the sake of illustration, a mechanism accounting for oxazole formation is delineated in the adjoining column.[11] While a powerful technique to form isolated heterocycles, this reaction was not applied widely in natural products synthesis for most of the 20th century as the strength of conditions were often too harsh for the heavily functionalized substrates typical of such endeavors. However, since numerous natural products contain these heterocyclic systems, especially those of marine origin like diazonamide A (**1**), several investigators have devoted significant attention to resolving this problem. Fortunately, their efforts have provided several reagent combinations that can now accomplish this reaction under essentially neutral conditions.[12] Rather than exhaustively cover these formulations here, we shall leave their discussion for the synthesis, in which the unique struc-

$-[H_3O^{\oplus}]$

oxazole

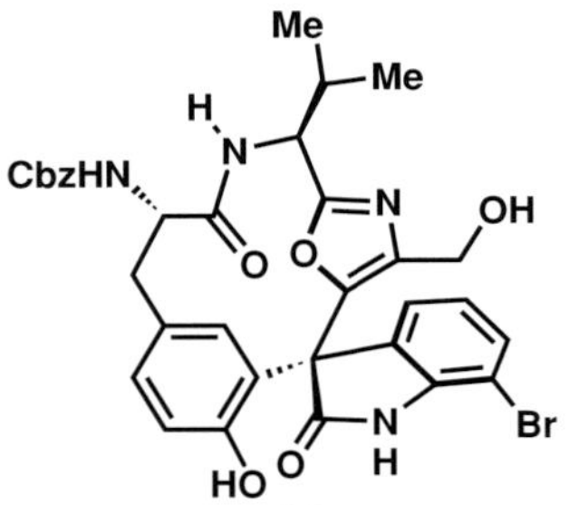

tural features of diazonamide A would provide a severe test of their power.

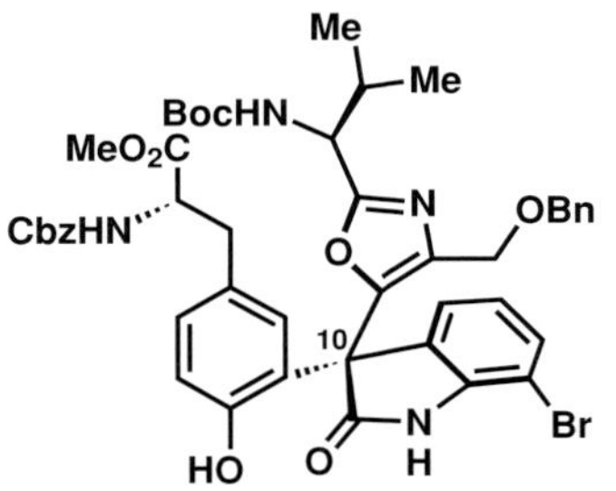

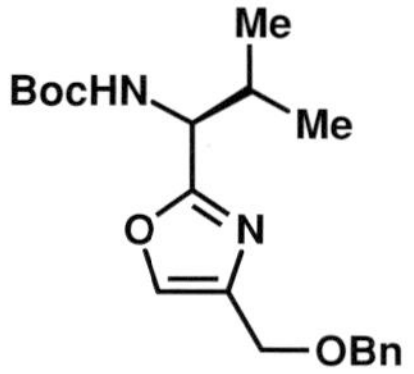

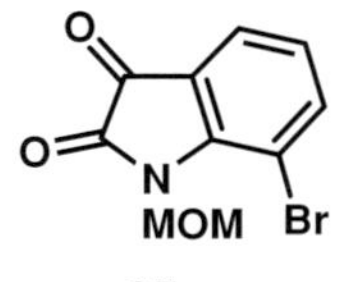

With simplification of the target molecule to **29**, an entity lacking the formidable heterocycle-based macrocycle, the Nicolaou group found themselves in a position to disassemble the remaining 12-membered ring of diazonamide A (**1**). Thus, the retrosynthetic sword cut this ring at its most logical site, its amide, to provide an amino acid precursor (**30**) modified by the addition of a few protecting groups. While this choice was an easy and obvious one, the operations that could, in turn, convert this new target (**30**) into simple building blocks were far from self-evident since that task would entail the dissection of three disparate aromatic systems from the quaternary C-10 center at the heart of diazonamide A (**1**). Indeed, literature precedent for how to form such systems was relatively sparse.[13,14] Little as it was, the available information pointed the Nicolaou group in the direction of a possible solution.

The initial seeds of inspiration came from the work of the famous German chemist Adolf von Baeyer (1905 recipient of the Nobel Prize in Chemistry), who was exploring the reactivity of the brilliantly red-colored dye isatin (**36**, Scheme 4) in the hopes of identifying new synthetic colorants at the tail end of the 19th century. One experiment that he performed along these lines with his student M. J. Lazarus was the reaction of **36** with sulfuric acid in the presence of anisole. As shown in Scheme 4a, this event gave rise to compound **41**, an outcome that can be rationalized as the product of two iterative electrophilic aromatic substitution reactions proceeding by way of intermediates **37**–**40**.[13] Thus, this reaction revealed how a quaternary carbon center bearing aromatic ring systems could be formed, though it did not indicate how two different aromatic systems might be appended to a starting isatin, as required for diazonamide A. During the next hundred years, only one report (from G. A. Olah, the 1994 recipient of the Nobel Prize in Chemistry) suggested, albeit in random fashion, how this goal might be accomplished. This accomplishment, which is shown in Scheme 4b, involved the same reaction process, but instead used two aromatic reactants simultaneously to obtain a statistical mixture of all three possible products in which **42** was favored.[14]

Although a good start, these reports underscored the fact that attaching two disparate aromatics to isatin in a single step based on electrophilic aromatic substitution would be quite challenging to accomplish in a reasonable yield. Moreover, since an oxazole and a phenol have far different reactivities, this approach might not work at all. How, then, should one proceed on the basis of this information? The answer that these researchers devised was based on the hypothesis that if they could form a compound such as **38** first, then perhaps the second aromatic moiety could be introduced through this same acid-catalyzed eletrophilic aromatic substitution process that Baeyer and Olah had used so productively. As shown in Scheme 3, this idea translated into the initial nucleophilic addition of a dianion derived from oxazole **34** to the more reactive

29

30

34

35

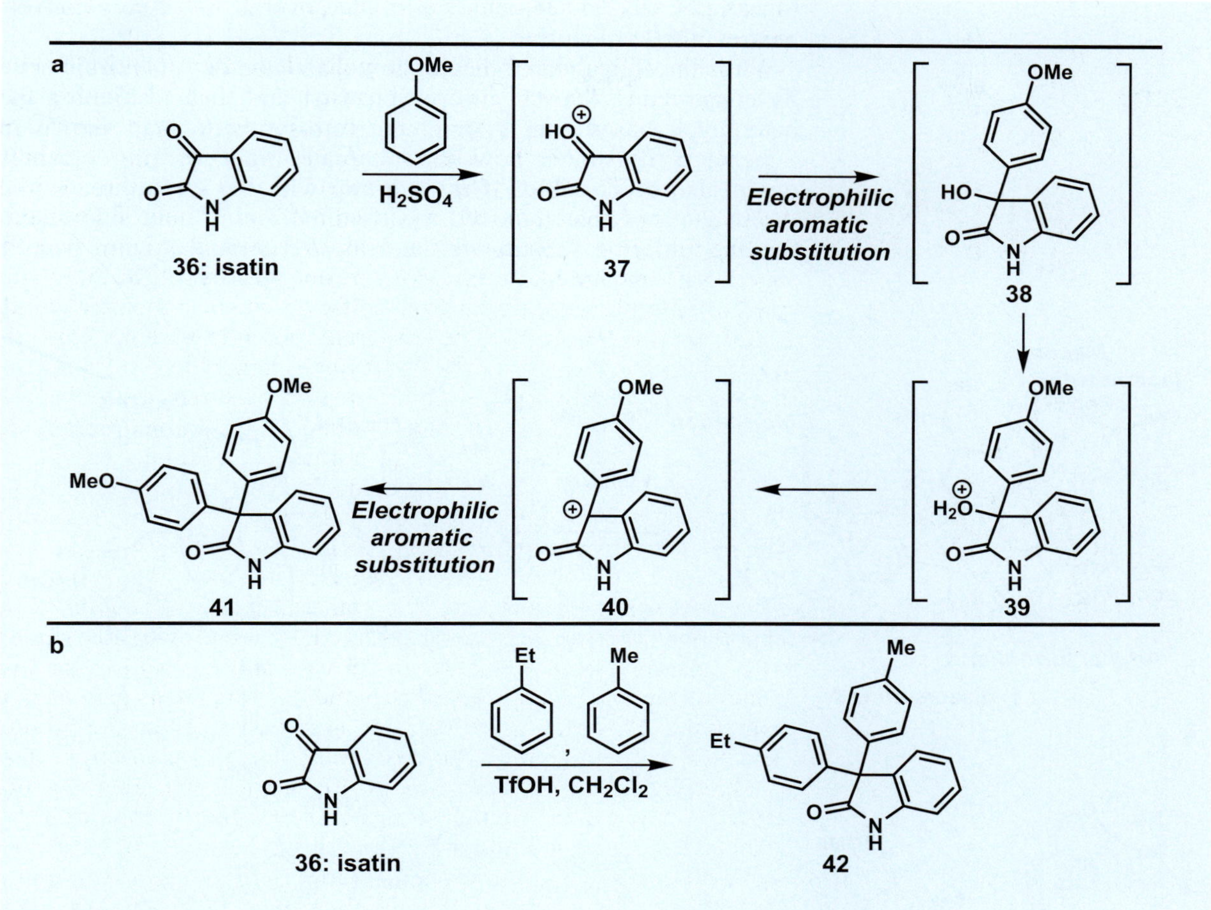

Scheme 4. Precedent for the reaction of aromatics with isatins under acidic conditions.

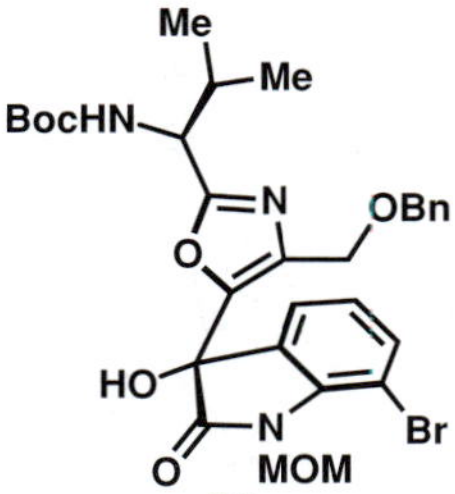

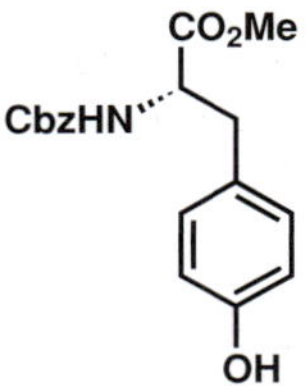

C-3 carbonyl group of isatin **35**, a process that would lead to the requisite test compound (**33**) bearing a tertiary alcohol function. Subsequent exposure of this product (**33**) to acid in the presence of the tyrosine-derived building block **31** could then give rise to **30**. On paper, this approach seems reasonable, and, moreover, if this strategy could be reduced to practice, then it would afford an attractively expedient entry into the diazonamide skeleton. While enticing, this idea also came with an inherent weakness: it could never lead to the stereocontrolled synthesis of the C-10 center in the absence of additional tricks. However, because it is challenging to envision a more expeditious synthesis of such a fragment, the loss of a certain portion of material due to shortcomings imposed by its conciseness was anticipated to provide a level of material throughput that could match, or, surpass a longer route featuring stereocontrol. As a result of this analysis, the formidable architecture of diazonamide A (**1**) has now been reduced to five simple building blocks (**24**, **28**, **31**, **34**, and **35**) in what overall is a highly convergent synthetic blueprint.

While the above plan indicates how diazonamide A (**1**) could arise by constructing the AG macrocycle first and then fashioning the heterocycle-based ring system, the retrosynthetic map shown in Scheme 5 illustrates how the alternate order of ring closures might also be possible.[8] Thus, starting with the same three initial disconnections that removed the terminal amino acid side chain, the aryl chlorine substituents, and the FH aminal system from **1**

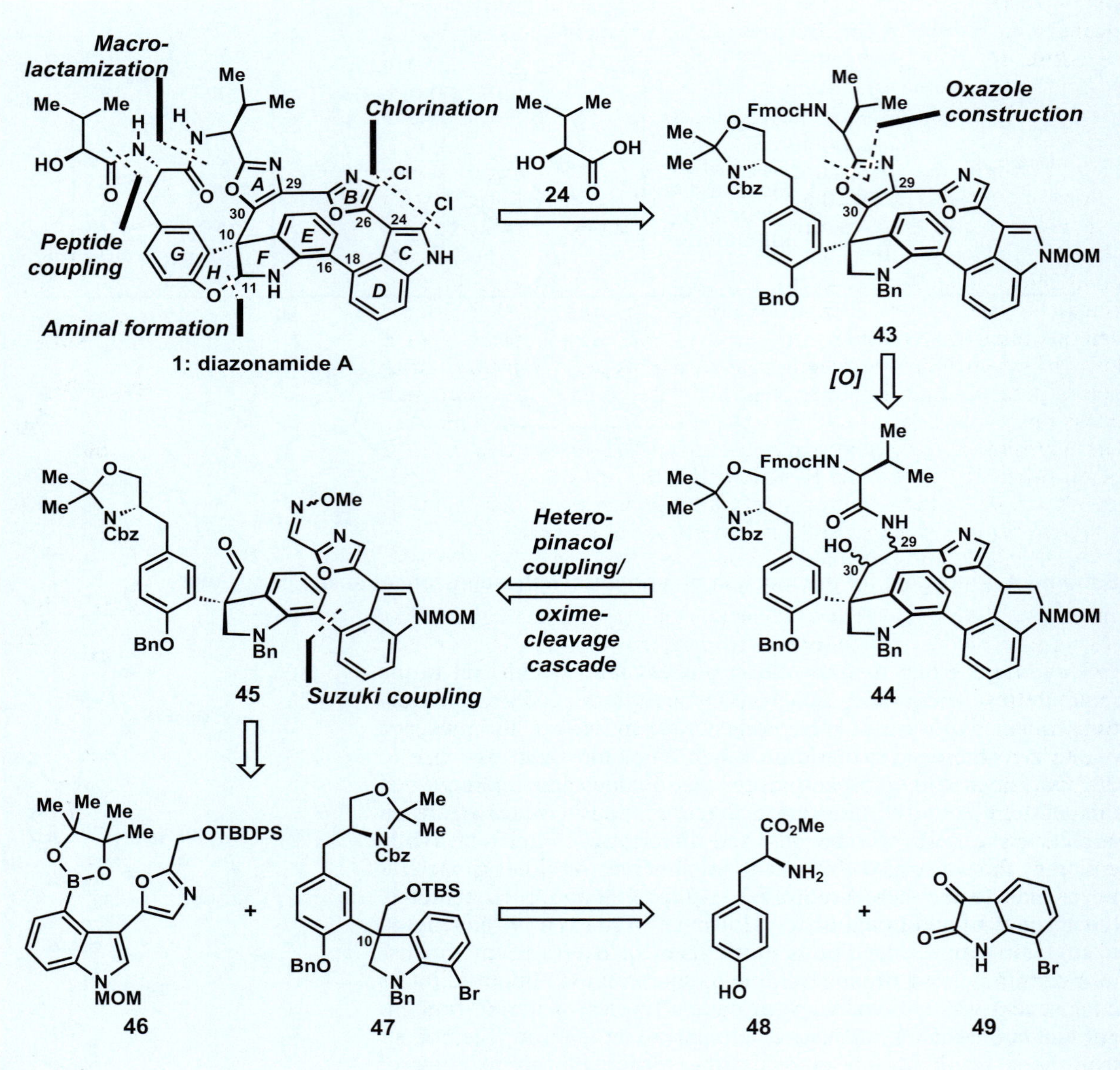

Scheme 5. Alternate retrosynthetic analysis of diazonamide A (**1**): the hetero-pinacol route.

as discussed earlier, rupture of the amide bond served to open the macrolactam ring of the target, leading to a protected amino acid precursor (**43**). With these relatively straightforward retrosynthetic transformations implemented, the critical issue now facing the Nicolaou group was the manner in which the heterocycle-based macrocycle would be retrosynthetically unlocked, as this operation would dictate virtually every additional modification in the analysis. The chosen strategy entailed disconnection of the C29–C30 linkage, an idea that became apparent once the A-ring oxazole of **43** had been simplified to a ketoamide precursor through a Robinson–Gabriel transform. Indeed, if the C-30 ketone of this new compound was retrosynthetically reduced to an alcohol as shown in structure **44**, then perhaps its C29–C30 bond could be formed with the concurrent installation of both its alcohol and amine groups through an intramolecular hetero-pinacol cyclization of an open chain precursor such as aldehyde-oxime **45**.

Even though we have not seen this particular transformation before, if you have read Chapter 34 on Taxol™ in *Classics I*, then you might recognize that this reaction is simply a variant of the classical pinacol cyclization in which ketyl radicals derived from dialdehyde precursors react to generate cyclic 1,2-diols. In fact, since the late 1970s when the Corey, Hart, and Bartlett groups demonstrated that oximes can serve as radical acceptors with a level of competence commensurate to aldehydes,[15] intramolecular hetero-pinacol coupling reactions have been used on numerous occasions to fashion a diverse range of rings.[16] For example, the Naito group at the Kobe Pharmaceutical University employed a hetero-pinacol cyclization to efficiently convert **50** into a seven-membered ring (**51**) appropriately functionalized to complete a total synthesis of balanol (**52**) as shown in Scheme 6a.[17]

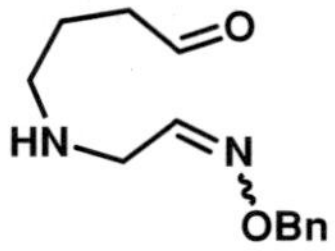

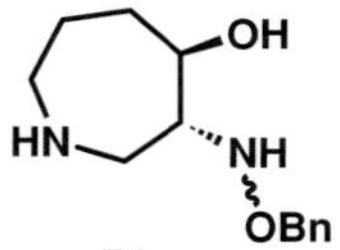

Significantly, however, hetero-pinacol coupling reactions possess two key differences from their pinacol counterparts. First, there is a much more limited arsenal of reagents capable of initiating the hetero variant, with the example in Scheme 6a showing the optimal system identified to date: SmI_2 complexed with 4 equivalents of an activating ligand (such as HMPA) to maximize its radical-generating power.[18,19] Second, while pinacol coupling reactions have proven to be competent in affording both medium-sized as well as macrocyclic systems, hetero-pinacol reactions had never been successfully employed in the construction of rings possessing more than seven atoms. All attempts to form medium-sized rings had met with failure until 2002. As such, this information might lead one to question why the Nicolaou group would even attempt to convert **45** into **44** through this method since it involved the formation of a 12-membered ring. The answer is that the diazonamide context seemed to provide a unique case, as molecular modeling of **45** indicated that highly favorable π–π stacking between the B- and E-rings would bring the aldehyde and oxime motifs quite close to each other, significantly reducing their rotational freedom. In fact, their proximity appeared more akin to a six- or seven-membered

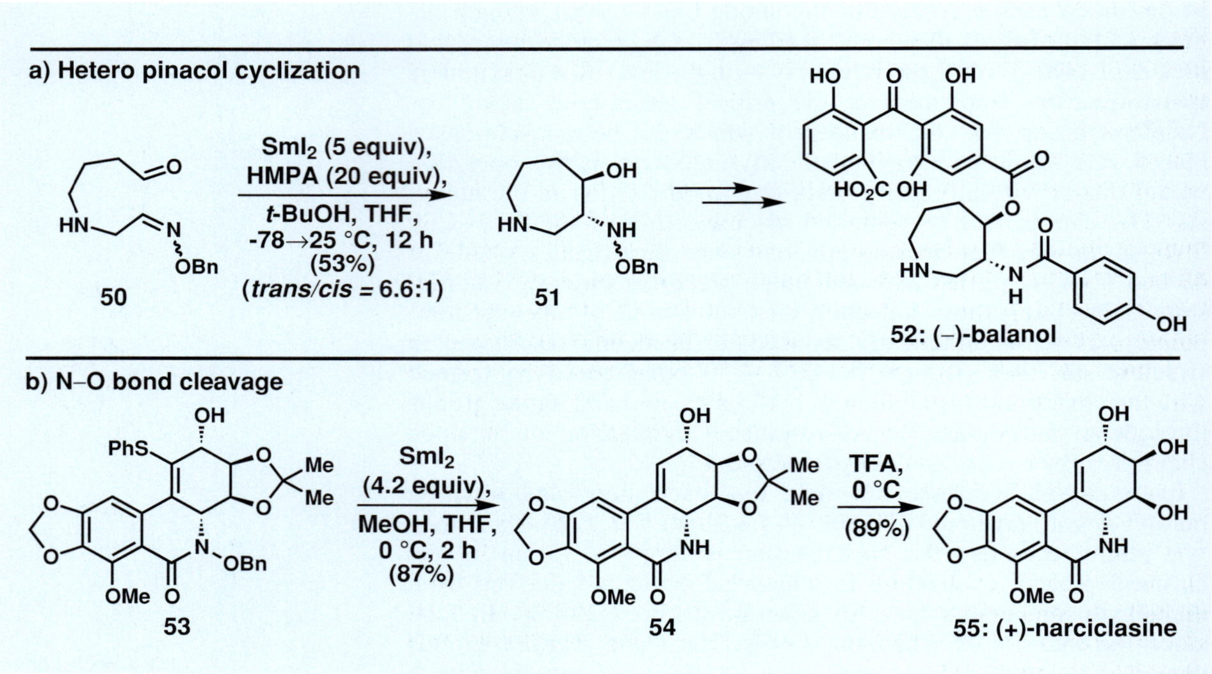

Scheme 6. Selected uses of SmI_2 in organic synthesis as pertinent precedents to the Nicolaou retro-synthetic analysis of diazonamide A (**1**).

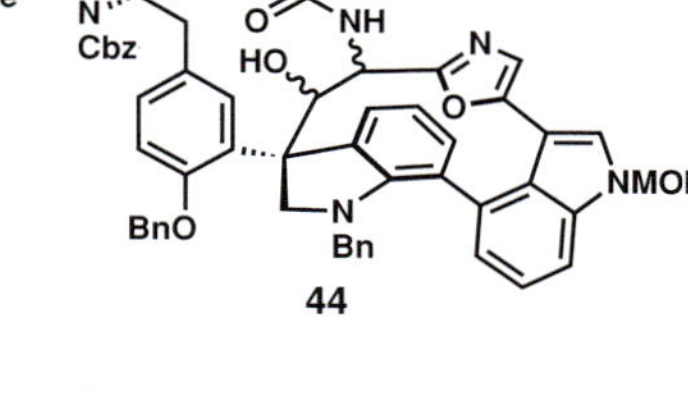

chain than to a standard, entropically more flexible, 12-membered system. Thus, the reaction appeared to have at least some chance for success, although conventional wisdom seemed to dictate otherwise.

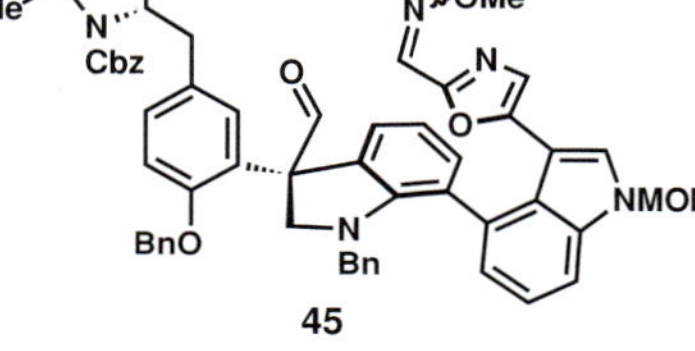

Despite the inherent challenge and risk for failure of this reaction, the Nicolaou group heightened the bar for its implementation even further by seeking to accomplish two additional operations in the same pot following cyclization. As shown in Scheme 6b, SmI_2 also possesses the power to break apart N–O bonds as revealed by the conversion of **53** into amide **54** during the synthesis of (+)-narciclasine (**55**) by the Keck group.[20] Intriguingly, though, while these conditions are essentially the same as those employed in Scheme 6a to effect a hetero-pinacol cyclization, in that example only material bearing an intact N–O linkage (**51**) had been isolated. This outcome appeared to be due solely to the amount of SmI_2 employed, in that if 4–5 equivalents of this lanthanide reagent were sufficient to accomplish "one reaction," then at least 8–10 equivalents would probably be required to accomplish two SmI_2-induced operations. Accordingly, if **45** were exposed to such a large excess of SmI_2, then perhaps both cyclization and N–O cleavage could be accomplished in the same pot to afford a 1,2-aminoalcohol product upon workup, a compound that could then be coupled with a protected form of L-valine to form **44** directly from **45**. Thus, if this sequence could be realized, not only would it constitute the first example of a hetero-pinacol reaction leading to a medium- or large-sized ring, but it also would productively combine this transformation into a new reaction cascade that could have potentially wider applications.

Although this novel strategy was certainly daring, part of its appeal resided in the fact that its plausibility would be relatively easy to ascertain as **45** was not expected to be that hard to synthesize. If its C16–C18 biaryl linkage was severed through a palladium-mediated coupling reaction transform, such as the venerable Suzuki reaction,[21] then this intermediate could be dissected into two fragments of commensurate size, indole-oxazole **46** and EFG building block **47**. The first of these appeared to be straightforward to synthesize, and the second (**47**) could be traced to an initial reaction between tyrosine methyl ester (**48**) and 7-bromoisatin (**49**) through a carefully controlled acid-catalyzed merger of the type discussed earlier in the context of the chemistry depicted in Scheme 4. As such, a second potential strategy towards diazonamide A (**1**) has been reduced to the construction of simple building blocks.

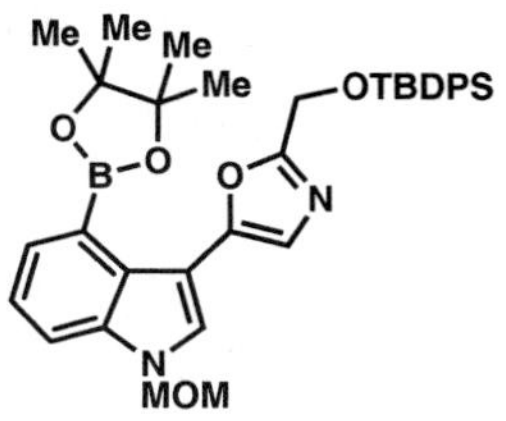

Having analyzed the Nicolaou group's approaches to diazonamide A at some length on paper, it is now time to see whether or not these strategies could lead to the revised structure of diazonamide A (**1**) in practice, and, if so, whether it actually was the correct assignment.

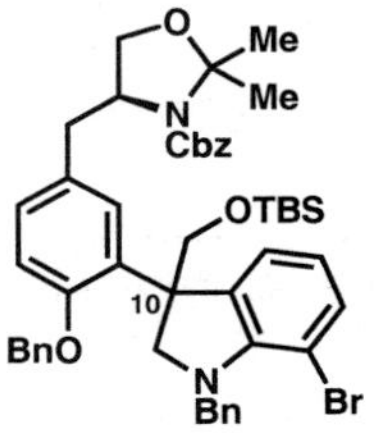

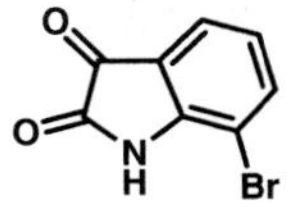

20.3 Total Synthesis

20.3.1 The First Total Synthesis of Diazonamide A (1)

We begin our discussion with those operations seeking to explore the approach delineated in Scheme 3, the route based on the initial formation of the AG macrocycle followed by the eventual construction of the heterocyclic core through a Witkop-type photocyclization. The first objective for this campaign was the creation of an appropriately formatted quaternary carbon center bearing three aromatic rings, a task that required the following building blocks: Cbz-protected L-tyrosine methyl ester (**31**), MOM-protected 7-bromoisatin (**35**), and oxazole **34** (see Scheme 3). Of these, only the last demanded any route development since the first two were compounds accessible in just a few steps from known or commercially available materials.[22]

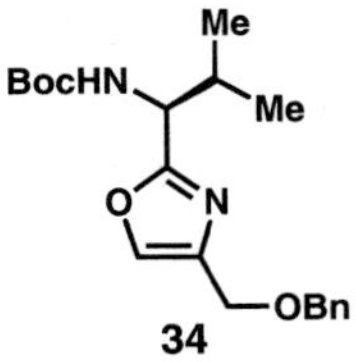

The construction of oxazole fragment **34** is shown in Scheme 7, starting with the merger of Boc-protected L-valine (**56**) and D,L-serine methyl ester (**57**) into **58** through a standard peptide-coupling reaction induced by EDC and HOBt.[23] Although this was a routine reaction, it served to provide a product (**58**) bearing all the structural motifs required to cast the oxazole ring of the target. Indeed, if one wanted to employ a Robinson–Gabriel cyclodehydration to create this heterocycle, **58** only needed to be oxidized to its aldehyde congener in order to provide the appropriate 1,4-dicarbonyl substrate.

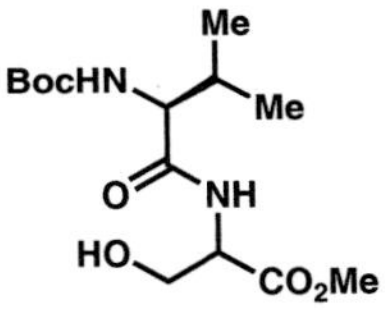

Me Me Me Me O B O OTBDPS O N N MOM
46

O Me Me N Cbz OTBS BnO 10 N Bn Br
47

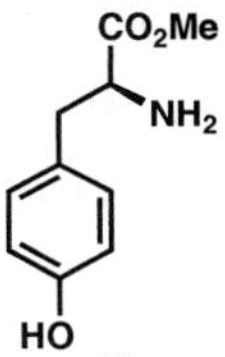

CO_2Me NH_2 HO
48

O O N H Br
49

Me BocHN Me O N OBn
34

Me BocHN Me O NH HO CO_2Me
58

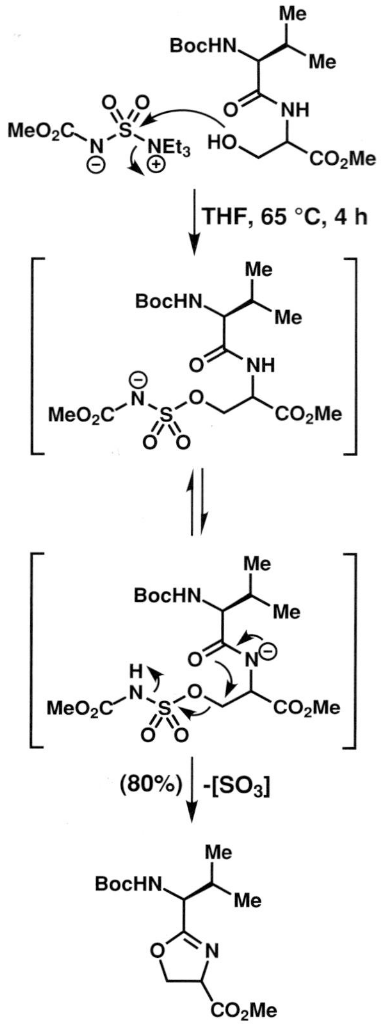

However, since investigations along these lines afforded the desired oxazole product (**60**) in poor yields, the Nicolaou group elected to generate this ring by a different route involving a two-step sequence which provided better material throughput. The first of these operations involved the conversion of **58** into an oxazoline (**59**) through the use of the Burgess reagent in refluxing THF, a step in which this zwitterionic substance transformed the primary alcohol into an activated leaving group (in the vein of a tosylate) which could then be internally displaced (see column figure).[24] With this cyclization accomplished in 80 % yield, the newly formed ring system was then aromatized to the desired oxazole (**60**) in excellent yield (90 %) by the action of $BrCCl_3$ in CH_2Cl_2.[25] In this event, the halogenated reagent served to deliver a bromine atom to **59** at what constituted the C-29 position of diazonamide A, presumably through a radical mechanism, which led to **60** through the DBU-promoted loss of HBr. The desired building block (**34**) was then completed in 83 % overall yield through the conventional reduction of the methyl ester with $LiBH_4$ in THF, followed by capture of the resultant primary alcohol as a benzyl ether (LiHMDS, TBAI, BnBr, THF).

Having accomplished this preparative work to afford **34** along with the trivial syntheses of **31** and **35**, attempts to merge these pieces into the diazonamide skeleton could now begin in earnest, starting with the operations needed to effect the union of oxazole **34** and isatin **35**. Pleasingly, these fragments could be smoothly merged into **33** (Scheme 8) simply by treating **34** with 2 equivalents of *n*-BuLi in THF at −78 °C for 20 minutes to generate an oxazole nucleophile (**61**), and then adding **35** to the reaction flask. With this operation unveiling a tertiary hydroxy group at the carbon atom corresponding to C-10 of the target, the compound was perfectly outfitted to attempt the incorporation of the final aromatic ring onto this site in the next step. As mentioned during the planning stages, this task anticipated that the exposure of **33** to an appropriate acid would lead to the reactive cation **32**, a species that could then undergo an electrophilic aromatic substitution reaction with the L-tyrosine derivative (**31**) to afford a product such as **62**. Following several weeks of scouting, this process was indeed accomplished in an optimized yield of 33 % yield by refluxing a solution of **33**, **31**, and *p*-TsOH in 1,2-dichloroethane for 15 minutes. Although the material return for this reaction was modest, this shortcoming was not a measure of the efficiency of the reaction itself, but rather reflected the difficulty in isolating the final product (**62**) due to the presence of the free amine group which had been unveiled unintentionally by virtue of the acidic conditions. This unavoidable outcome was dealt with in the next operation by reprotecting the amine group, now as a *t*-butyl carbamate (Boc), to provide **30** as a mixture of chromatographically separable diastereomers (as the C-10 center of **62** had been formed in an unselective manner) in 76 % yield. While this separation was fortunate in that it led to stereochemically pure isomers, NMR spectroscopic analysis could not provide sufficient information to unambiguously assign their

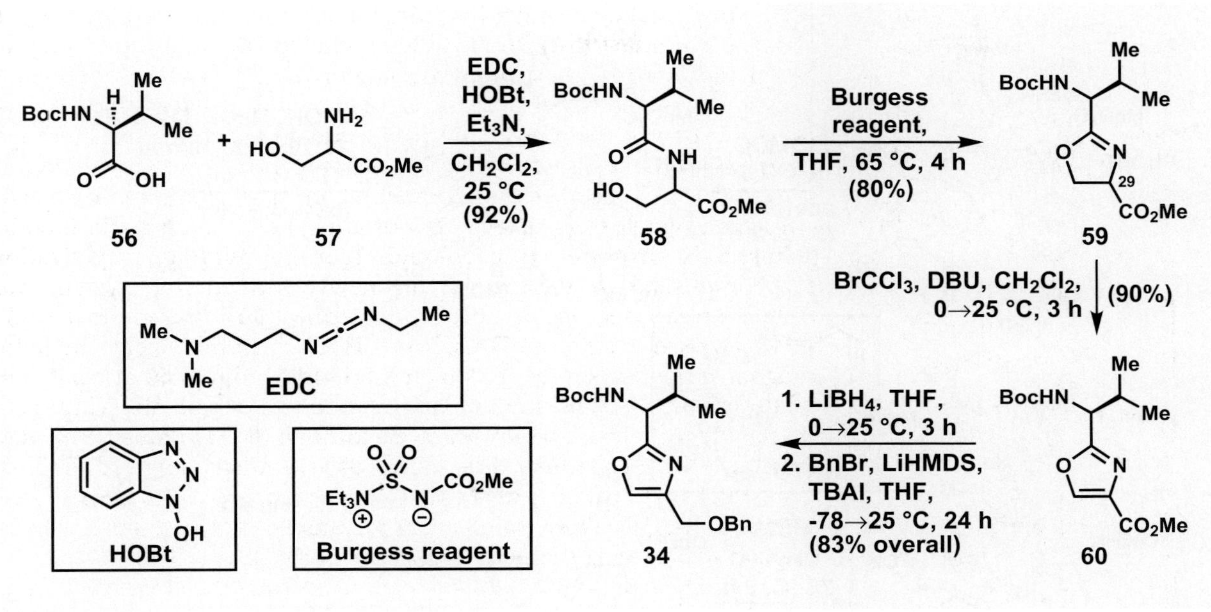

Scheme 7. Synthesis of oxazole building block **34**.

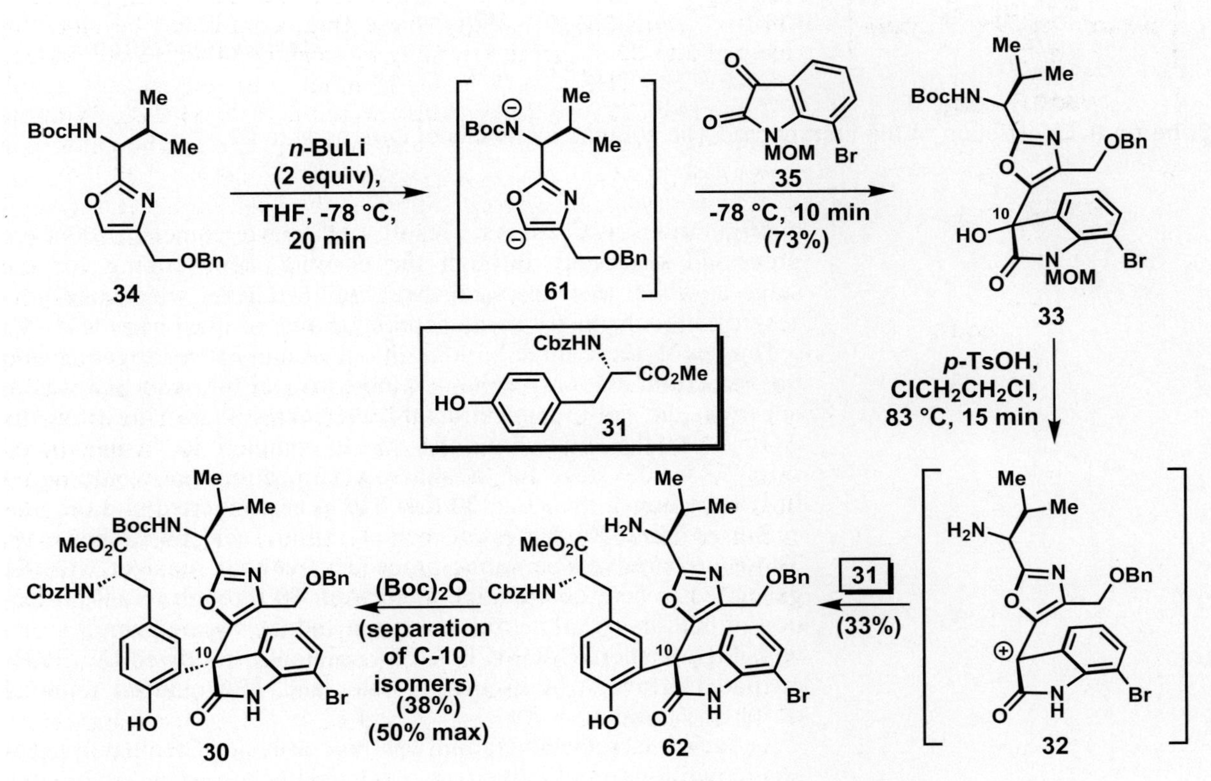

Scheme 8. Construction of key intermediate **30** bearing the critical C-10 quaternary center of diazonamide A.

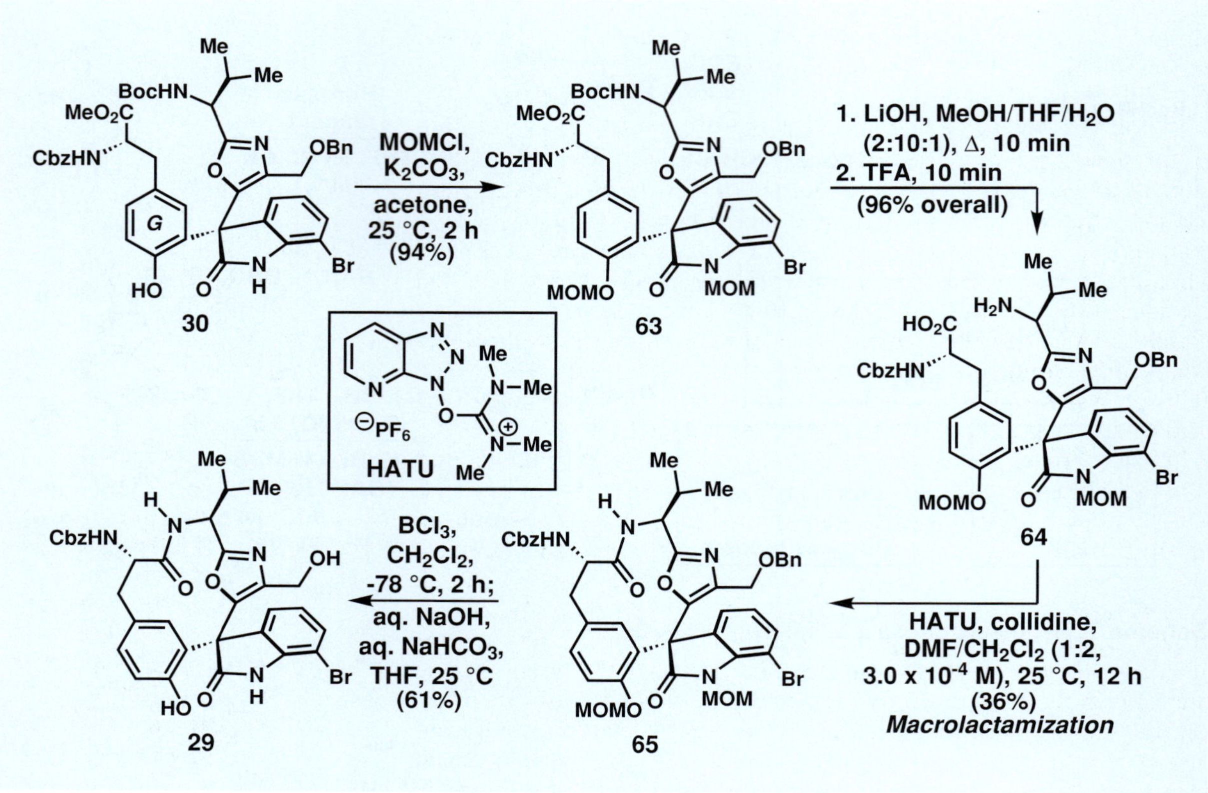

Scheme 9. Completion of the first macrocyclic subunit: synthesis of intermediate **29**.

configurations at C-10. As a result, both diastereomers of **30** were processed separately through the ensuing steps hoping for the stage at which their physical data would indicate which one possessed stereochemistry corresponding to that of diazonamide A (**1**).

Thus, with these three building blocks finally fused together into one structural entourage, the next objective of the sequence was to construct the only major structural element still missing from the AG system, the amide bond that would complete its 12-membered ring. To set the stage for the macrolactamization that would hopefully accomplish this goal, **30** first had to be converted into the precursor required for the event, namely amino acid **64** (Scheme 9). These preparative operations proceeded without incident with **64** generated in near quantitative yield from **30** through initial protection of both its oxindole nitrogen atom and the G-ring phenol group with MOM ethers (MOMCl, K_2CO_3, acetone), followed by LiOH-mediated cleavage of its methyl ester and TFA-induced removal of its Boc group.

As such, macrolactamization was now at hand. Fortunately, several conditions could overcome the challenges imposed by this ring closure, with the use of HATU and collidine in a 1:2 mixture of DMF and CH_2Cl_2 at a final concentration of 3.0×10^{-4} M constituting the best protocol, delivering **65** in 36 % yield.[26] As you might expect, the use of high dilution conditions was required to minimize the formation of dimers and other oligomers. However, the result that you might not have anticipated, and which the Nicolaou group did not predict either, was that only one of the two C-10 epimers of **64** successfully underwent macrocyclization, with the other leading primarily to dimerized material irrespective of concentration. Thus, the question raised by this interesting outcome was whether or not it was the correct diastereomer of **64** that had cyclized to **65** as drawn in Scheme 9, because if the wrong epimer of **65** had been formed, then this approach was effectively dead. Thankfully, some positive evidence that the correct epimer had cyclized was provided once the benzyl ether and two MOM groups of **65** had been removed through the action of BCl_3 in CH_2Cl_2 at −78 °C to afford **29**, as the ^{1}H NMR data of this new compound were highly reminiscent of those exhibited by the natural product. While not conclusive proof, it certainly provided enough of an impetus to press forward and attempt the construction of the remaining macrocyclic unit. Before we describe these endeavors, we should note in passing that although the presence of the MOM groups throughout the sequence shown in Scheme 9 may appear needless, they served to considerably improve the yield of the ring closure leading to **65** by facilitating the isolation of this product. In addition, other, more daring strategies for macrocycle formation were also attempted using elements of this same sequence, such as a ring closure through an acid-catalyzed electrophilic aromatic substitution of compounds like **66** (see column figure). Unfortunately, the one bond needed to complete the macrocycle from these intermediates could not be formed despite an exhaustive screening of both Brønsted and Lewis acidic reagents.

Nevertheless, with a route to **29** identified and faith in its stereochemical disposition, work could now be directed towards the synthesis of the heterocyclic core. Thus, as shown in Scheme 10, these efforts commenced with selective Boc protection of the H-ring phenol group with $(Boc)_2O$ in a 1:2 solvent mixture of aqueous $NaHCO_3$ and 1,4-dioxane over the course of 24 hours, and then conversion of the free primary alcohol into a carboxylic acid through two iterative oxidations (IBX; $NaClO_2$) to afford **69** in 63 % overall yield from **29**. This new motif was then enlisted to complete the remaining structural elements of the target, namely the B-, C-, and D-rings, through an initial peptide coupling with **28** to afford ketoamide **27**, followed by a Robinson–Gabriel cyclodehydration with pyridine-buffered $POCl_3$ at ambient temperature (a set of conditions developed by the Nicolaou group to form oxazoles in challenging settings).[9b] With these latter two operations executed in 34 % overall yield, a substrate had been synthesized that could now be subjected to Witkop-type photocyclization conditions. Fortunately, little alteration of the protocol developed by the Harran group to accomplish this ring closure in their synthesis of **2** (cf.

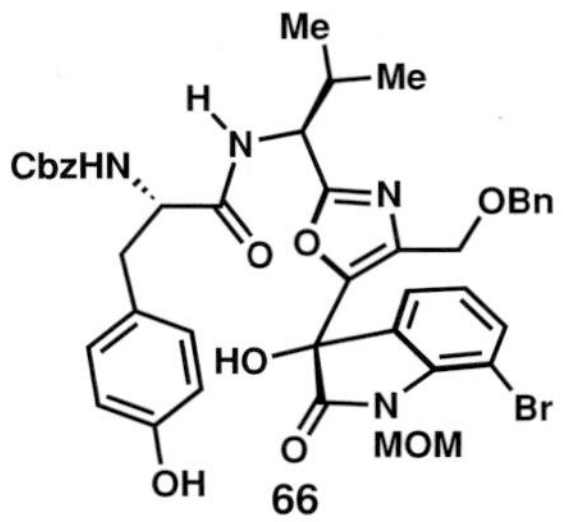

66

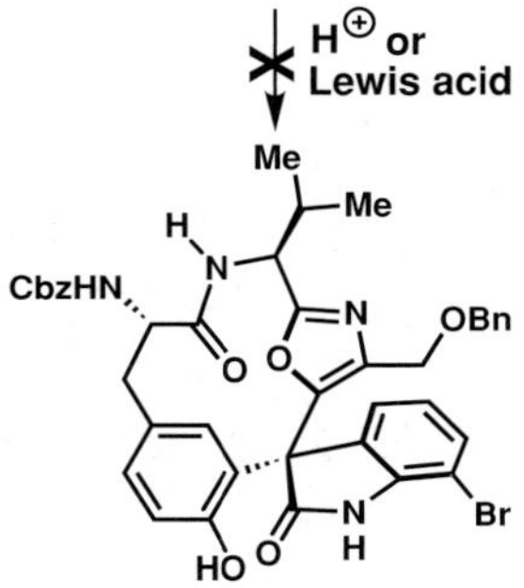

$H^{\oplus}$ or Lewis acid

67

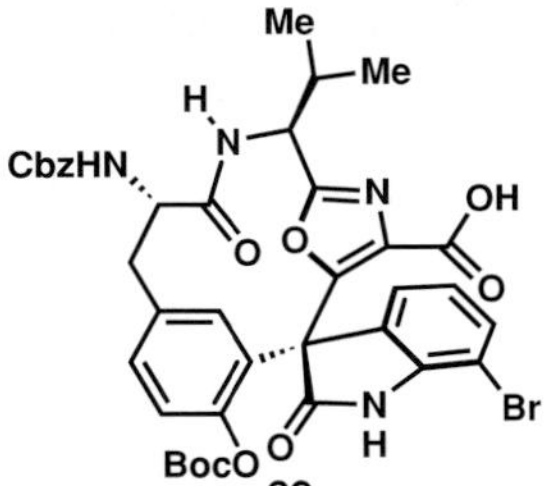

69

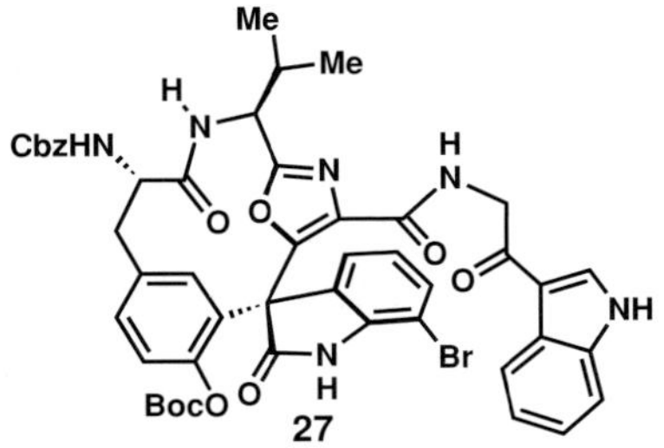

27

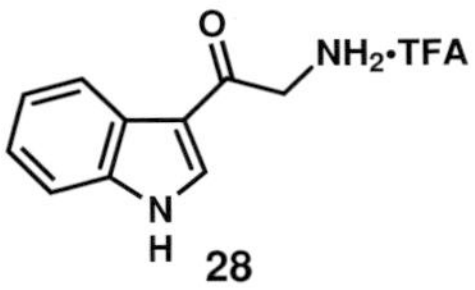

28

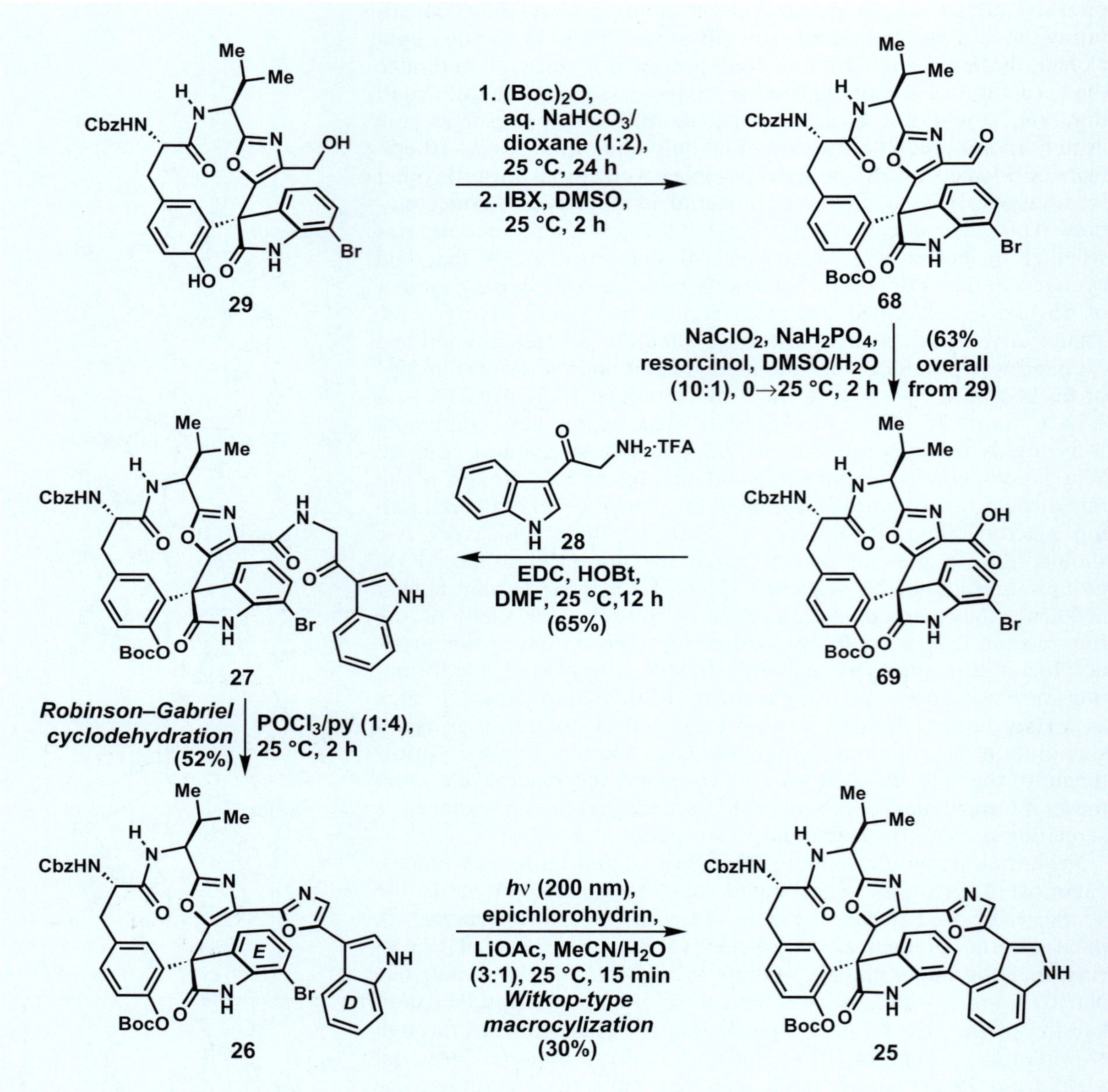

Scheme 10. Completion of the second macrocyclic subunit of diazonamide A: synthesis of advanced intermediate **25**.

Scheme 1)[3] was required here, with excitation at 200 nm serving to afford **25** in a reproducible yield of 30 %. As expected, the reaction proceeded with complete atropselectivity for both the C16–C18 and C24–C26 biaryl axes, and the remaining material balance was mostly recovered starting material that could be recycled to enhance the overall throughput of material. As an aside, the final cyclization could also be accomplished under radical conditions by treating **26** with Ph_3SnH and catalytic amounts of AIBN in refluxing benzene, though this alternate event proceeded in only 10% yield.

With both macrocycles successfully generated through the formation of **25**, the completion of the target now required only a few finishing touches, the most notable being the installation of the two missing aryl chlorine substituents and the formation of the FH aminal system of diazonamide A (see Scheme 11). The first of

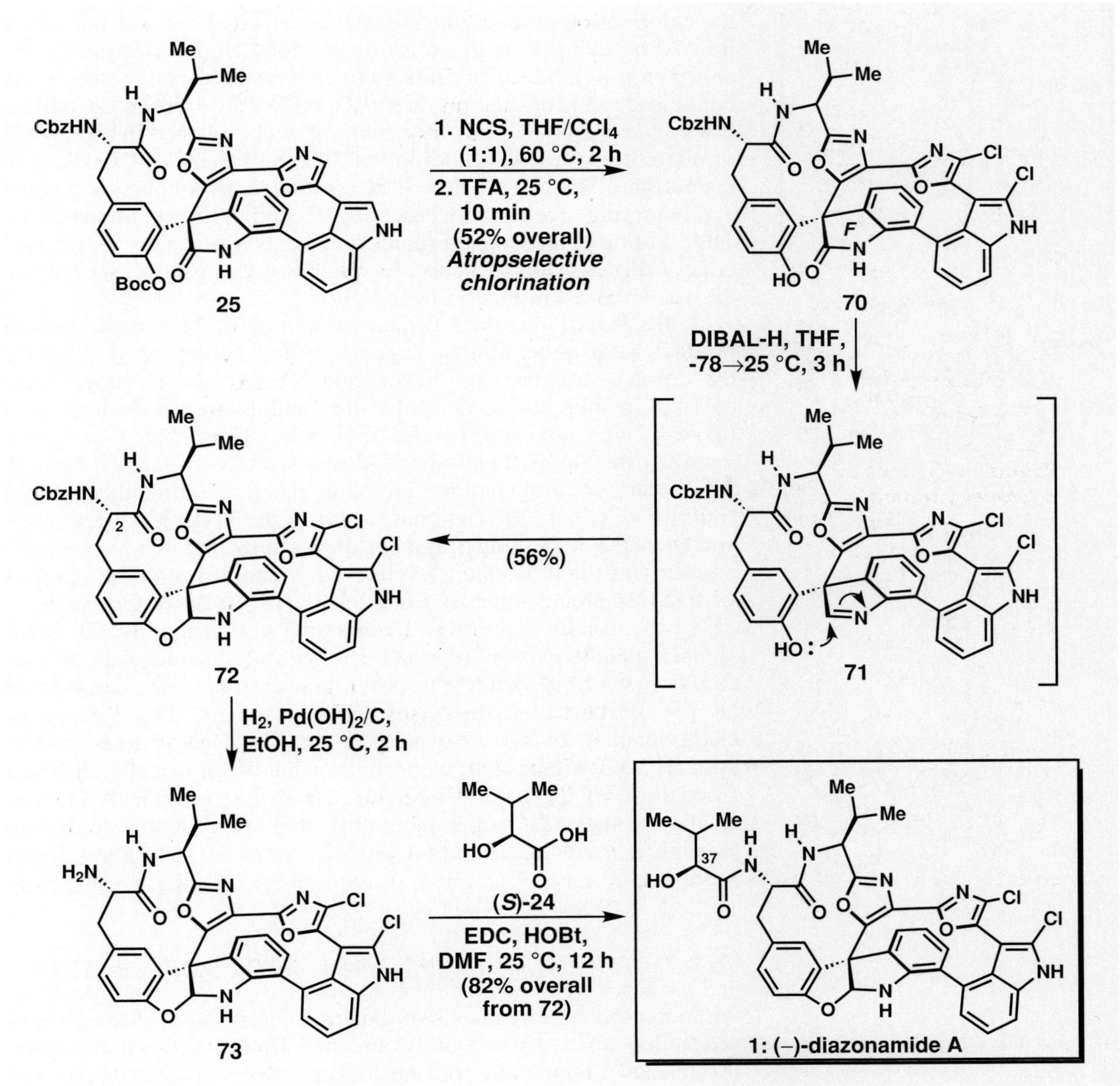

1. NCS, THF/CCl_4 (1:1), 60 °C, 2 h
2. TFA, 25 °C, 10 min
(52% overall)
Atropselective chlorination

DIBAL-H, THF, -78→25 °C, 3 h

(56%)

H_2, $Pd(OH)_2$/C, EtOH, 25 °C, 2 h

(*S*)-24
EDC, HOBt, DMF, 25 °C, 12 h
(82% overall from 72)

25, **70**, **71**, **72**, **73**, **1**: (–)-diazonamide A

Scheme 11. Final stages and completion of the first total synthesis of diazonamide A (**1**).

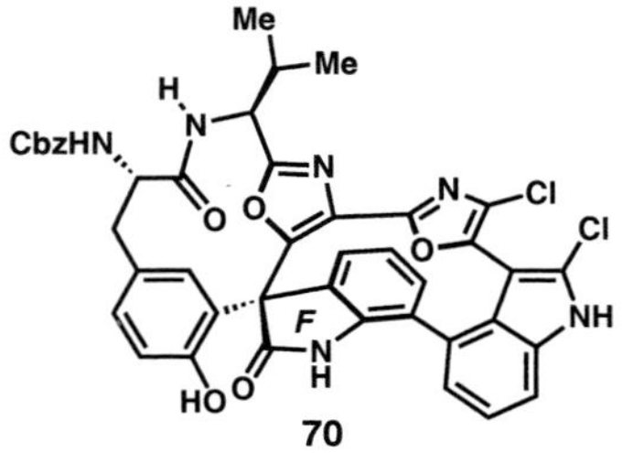

these objectives was tackled in a single step by treating **25** with NCS at 60 °C for 2 hours in a 1:1 mixture of THF and CCl_4 (which was required to help solubilize the NCS), conditions which led to the site- and atropselective incorporation of the two requisite chlorine atoms. Seeking now to form the FH aminal system, the Nicolaou group expected that if they could convert the F-ring oxindole into an imine or iminium species (such as **71**) through a reaction with an appropriate hydride source, then a free C-7 phenol could engage that reactive intermediate and thus generate this motif.[27] As a result, the chlorination product obtained as described above was converted into **70** by using TFA to excise its phenolic Boc protecting group, and then a number of hydride sources were screened to see if any could lead to the desired product (**72**). Following extensive scouting, it was discovered that this task could be accomplished in 56 % yield through the portionwise addition of 100 equivalents of DIBAL-H to a solution of **70** in THF at −78 °C, followed by 3 hours of cooling and warming cycles between −78 °C and ambient temperature. The addition of the hydride source in batches was critical for the success of this cascade sequence, as otherwise the reaction stalled and **72** was obtained in much lower yields.

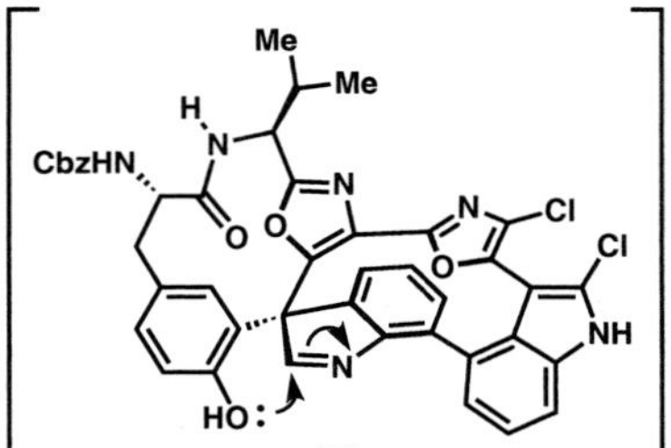

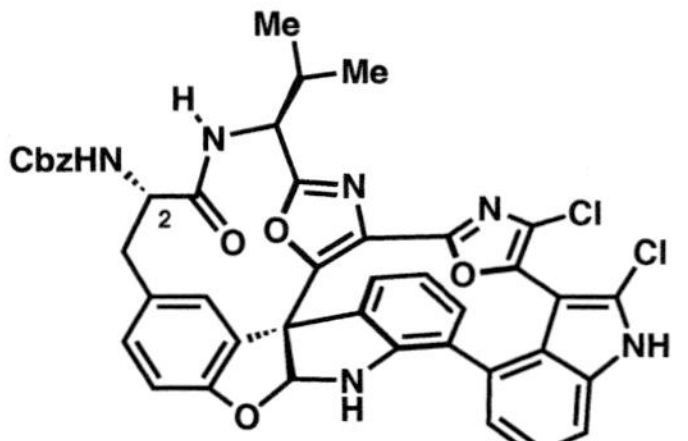

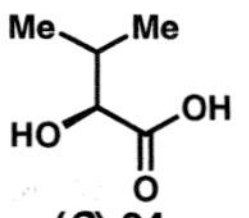

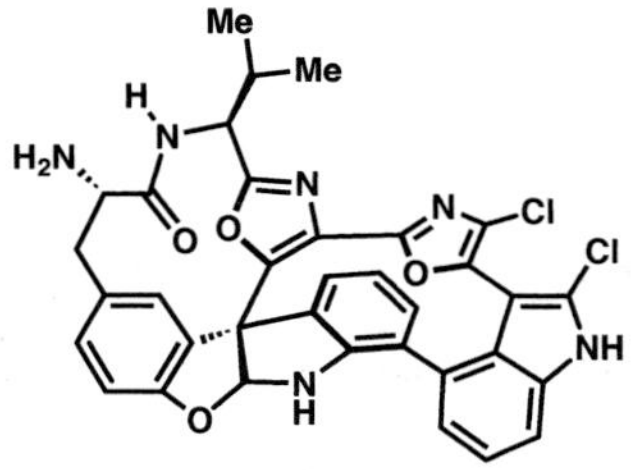

At this stage, the spectral data for synthetic **72** and the natural product were quite similar, suggesting that perhaps **1** was indeed the correct structure for diazonamide A. However, the moment of truth would not occur until the final hydroxyisovaleric acid [(*S*)-**24**] was attached to the molecule. Thus, the Cbz group guarding the amine function at C-2 was selectively excised through a standard hydrogenation protocol using Pearlman's catalyst [$Pd(OH)_2/C$] in EtOH (without cleaving the aryl chlorine atoms), and then the remaining fragment of diazonamide A was installed onto the resultant product (**73**) through a peptide-coupling reaction with (*S*)-**24** orchestrated by EDC and HOBt. As such, the synthesis of **1** was finally complete. Even more gratifying, its ^{1}H NMR signals matched those reported for natural diazonamide A perfectly.[1] As a final exercise to prove that diazonamide A possessed an *S*-configured hydroxyisovaleric residue, the *R*-disposed enantiomer of **24** was also merged onto **73**, leading to a product with ^{1}H NMR signals that were slightly, but diagnostically, different from those of the natural material. Thus, diazonamide A (**1**) was finally synthesized in the laboratory and its structure confirmed through a route that required only 21 steps in its longest linear sequence.

20.3.2 The Second Total Synthesis of Diazonamide A (1)

Although the total synthesis of diazonamide A (**1**) discussed above proceeded relatively smoothly and had finally verified the long-mysterious architecture of the target, the Nicolaou group still wanted to see whether their alternate approach to **1** was viable, as it was predicated on entirely different chemistry. Of its projected steps, the most important, as well as the most challenging, was a SmI_2-promoted hetero-pinacol reaction to forge the final C–C bond of the heterocyclic core of the molecule. Thus, as a prelude to testing this daring hypothesis, the Nicolaou group embarked on their second total synthesis of diazonamide A by developing syntheses of appropriately functionalized building blocks corresponding to the BCD- and EFG-ring systems of diazonamide A (**46** and **47**, cf. Scheme 5).

The successful route identified for the first of these fragments, indole-oxazole **46**, is shown in Scheme 12 starting from the known 4-bromoindole (**74**),[28] a compound that already contained the C- and D-rings of the target. As such, the use of this material meant that only a functionalized oxazole had to be appended onto it in order to complete the desired building block, a task that would rely upon the familiar Robinson–Gabriel cyclodehydration reaction. Thus, the first operations sought to install some of the functionality needed to create the ketoamide precursor for that event (i. e. compound **77**), starting with a TFA-promoted condensation of 4-bromoindole (**74**) with dimethylaminonitroethylene.[29] This operation served to add a three-atom nitroethylene tether to the indole fragment through the mechanism shown in the adjoining col-

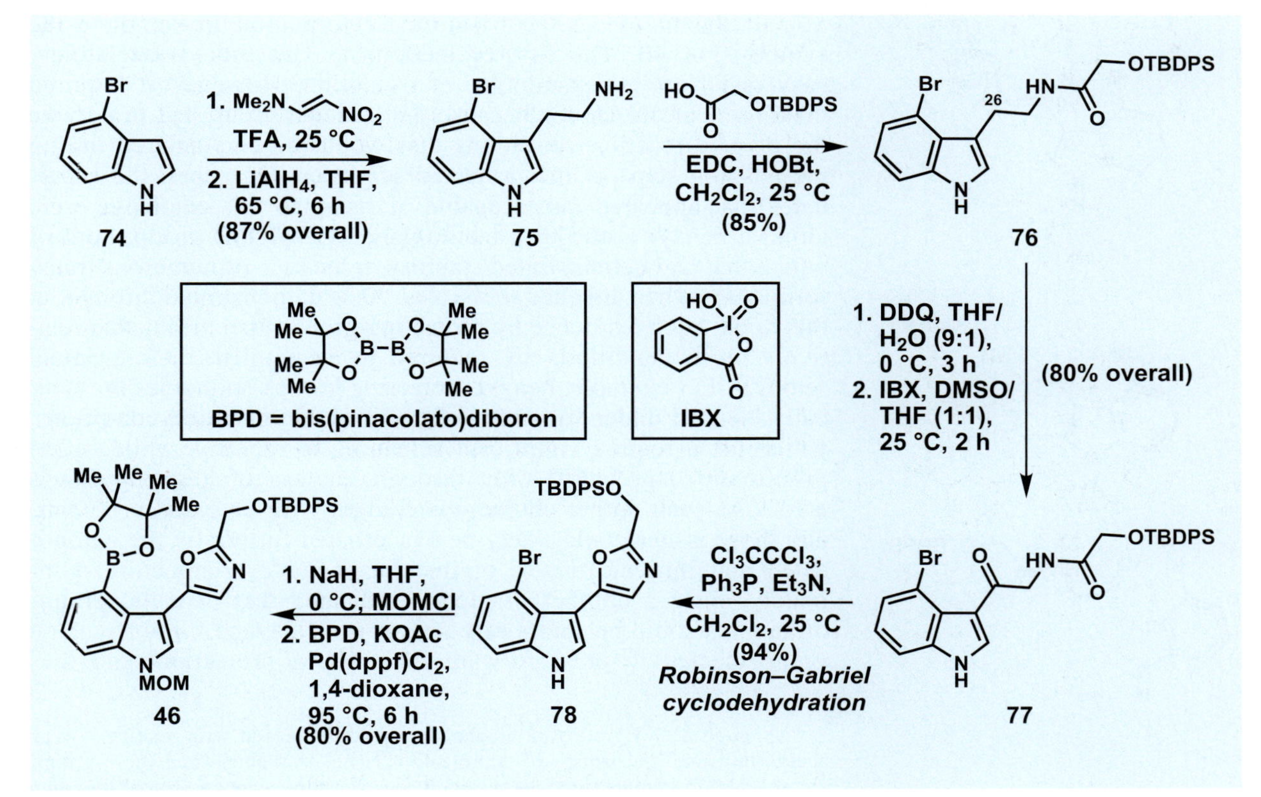

Scheme 12. Synthesis of indole-oxazole intermediate **46**.

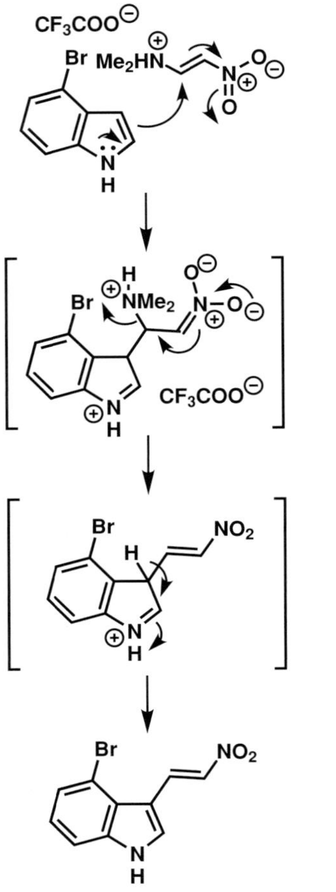

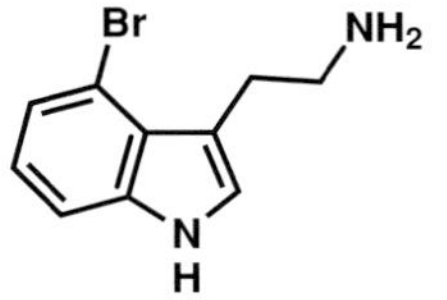

umn, enabling the subsequent formation of 4-bromotryptamine (**75**) in 87 % yield upon concomitant reduction of the newly-installed olefin and its nitro group with $LiAlH_4$ in refluxing THF.

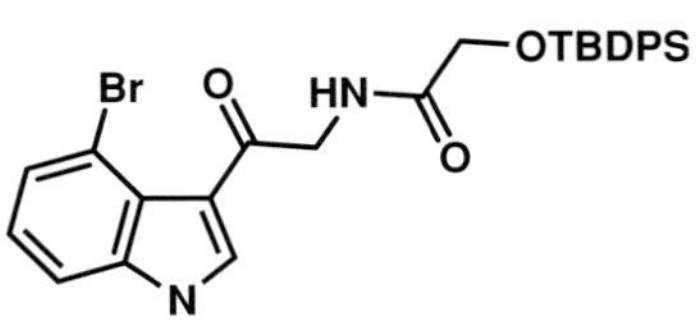

With a reactive handle now unveiled on **75** in the form of a free amine, a subsequent peptide-coupling step with TBDPS-protected glycolic acid led to the smooth synthesis of **76** in 85 % yield, a compound separated from the desired ketoamide substrate (**77**) only by the absence of a C-26 carbonyl group. This deficiency was rectified over the course of two steps in 80 % yield through the initial installation of a hydroxy group as accomplished by the action of DDQ in aqueous THF at 0 °C, followed by oxidation of the resulting alcohol with IBX (the synthetic precursor for the Dess–Martin periodinane). Interestingly, although aqueous DDQ is known to convert materials such as **76** directly into **77** through the putative mechanism shown in the column figure on the next page,[30] the second oxidation step could not be induced with this particular substrate. As revealed by several model studies, the bromine substituent was the likely culprit, as application of the same conditions to the debrominated variant of **76** led directly to a ketone.

Irrespective of this interesting observation on chemical reactivity, with a synthesis of **77** accomplished, use of Wipf's mild conditions for the Robinson–Gabriel reaction (Cl_3CCCl_3, Ph_3P, Et_3N, CH_2Cl_2, 25 °C)[12b] served to form the oxazole ring (**78**), leaving only protection of the indole and boronate ester formation to complete the synthesis of **46**. The first of these tasks constituted a relatively easy one from the standpoint of execution. However, it required making a careful choice because the selected group had to survive the diverse reaction conditions that would be encountered during the ensuing steps of the projected sequence. On paper, the candidates that appeared most capable of rising to this challenge were either a benzyl group or a methoxymethyl (MOM) group, both of which that had demonstrated stability in the face of numerous transformations when attached to indoles. As is demonstrated throughout this book, such resilience by protecting groups also means that relatively harsh conditions are required to accomplish their eventual removal. For example, benzyl protecting groups on indoles are typically cleaved under substrate-dependent hydrogenation conditions, which often require significant screening to identify, while MOM groups are deprotected only through the use of a strong Lewis acid.* As such, either choice was fraught with potential problems, and those issues could likely be exacerbated further by the complicated and intricate nature of the diazonamide architecture. Ultimately, since a number of reports indicated that benzylic groups on indoles could be stubbornly resilient to cleavage,[32] the Nicolaou group elected to press forward with MOM protection, and thus

CF3COO⊖ Br Me2HN⊕ N⊕ O⊖ O N H

Br ⊕NMe2 H N⊕ O⊖ O⊖ CF3COO⊖ ⊕N H

Br H NO2 ⊕N H

Br NO2 N H

Br NH2 N H **75**

Br O HN OTBDPS O N H **77**

* Although MOM groups on alcohols are readily cleaved with relatively weak acids, the same mechanism of deprotection is not available when these groups are attached to indoles since the nitrogen center cannot accept a proton, thus rendering the group far more resilient.[31]

attached this group onto **78** through a standard protocol (NaH, THF, 0 °C; MOMCl).

With this step accomplished, a palladium-mediated reaction at 95 °C with bis(pinacolato)diboron and KOAc in dioxane then completed the assembly of **46** by exchanging the aryl halide for a boronic ester in 85 % yield. This powerful transformation, which was first described by Miyaura and co-workers in 1995,[33] proceeds through the standard catalytic cycle invoked for the palladium-based Suzuki reaction in which the KOAc serves to accelerate the generally slow transmetallation step. As an aside, although the corresponding boronic acid could also be formed from this substrate by first converting it into an aryl lithium, the ester (**46**) was far easier to handle. From a process standpoint, the boronic ester was also the simpler of the two to synthesize since it only required the admixing of several reagents in an oxygen-free atmosphere. Indeed, the facile execution and mild nature of this method has rendered it a highly popular technique to form boron-containing species for Suzuki reactions. In fact, for compounds possessing esters, ketones, and nitriles, groups incompatible with the aryl lithium or aryl Grignard reagents required for the standard aryl boronic acid synthesis protocol, this methodology affords the only direct method to outfit them for coupling.

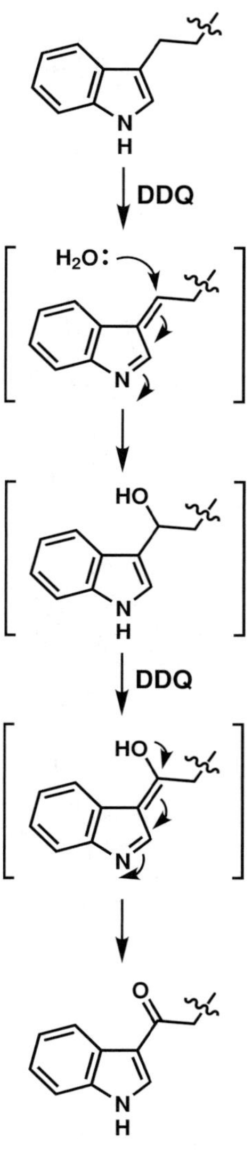

Having discussed the final step in the synthesis of this piece, we can now turn our attention to those steps needed to prepare the other key building block, EFG fragment **47**. As indicated in Scheme 13, its synthesis commenced with the protection of commercially available L-tyrosine methyl ester (**48**) as its corresponding benzyloxycarbamate (Cbz) derivative, a step that proceeded in 94 % yield. While a conventional beginning, it served a very important purpose as it set the stage for a subsequent $TiCl_4$-mediated merger with 7-bromoisatin (**49**), a reaction that gave rise to a product (**80**) bearing the majority of the architecture of the targeted fragment in 58 % yield (70 % based on recovered starting material). The only major structural element separating **80** from **47** was appropriate functionalization of its C-10 center. Thus, in preparation for the sequence that would eventually fix this deficiency, the newly formed tertiary alcohol in **80** was first exchanged for a hydrogen (**81**) in 76 % yield over two steps by way of an intermediate chloride formed upon several hours of exposure of **80** to $SOCl_2$ at 25 °C, followed by its reduction with $NaCNBH_3$. Next, the upper tyrosine domain was transformed into an acetonide (**82**) in near quantitative yield through reduction of the methyl ester ($LiBH_4$, THF, 0→25 °C), followed by reaction with 2,2-dimethoxypropane catalyzed by *p*-TsOH, thereby providing a substrate (**82**) onto which the required hydroxymethyl function could now be attached at C-10. This task was accomplished by enlisting a two-step procedure developed by the Padwa group at Emory University in which the initial formation of a silyl enol ether (**83**) served to create a latent nucleophile that could be unleashed upon a subsequent reaction with an appropriate Lewis acid.[34] When that species was $Yb(OTf)_3$, and formaldehyde was present in solution, the desired product (**84**) was obtained as a 1:1 mixture

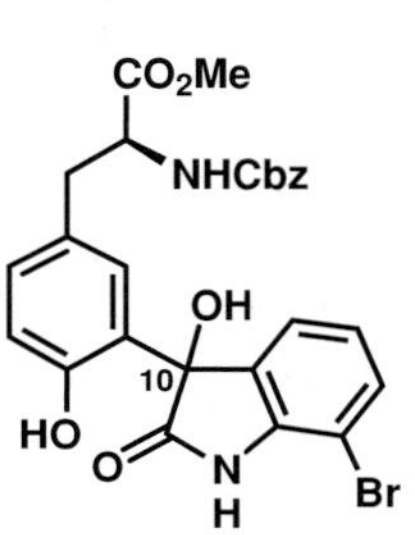

DDQ

H_2O:

HO

DDQ

HO

O

CO_2Me

NHCbz

OH

10

HO

O

N

H

Br

80

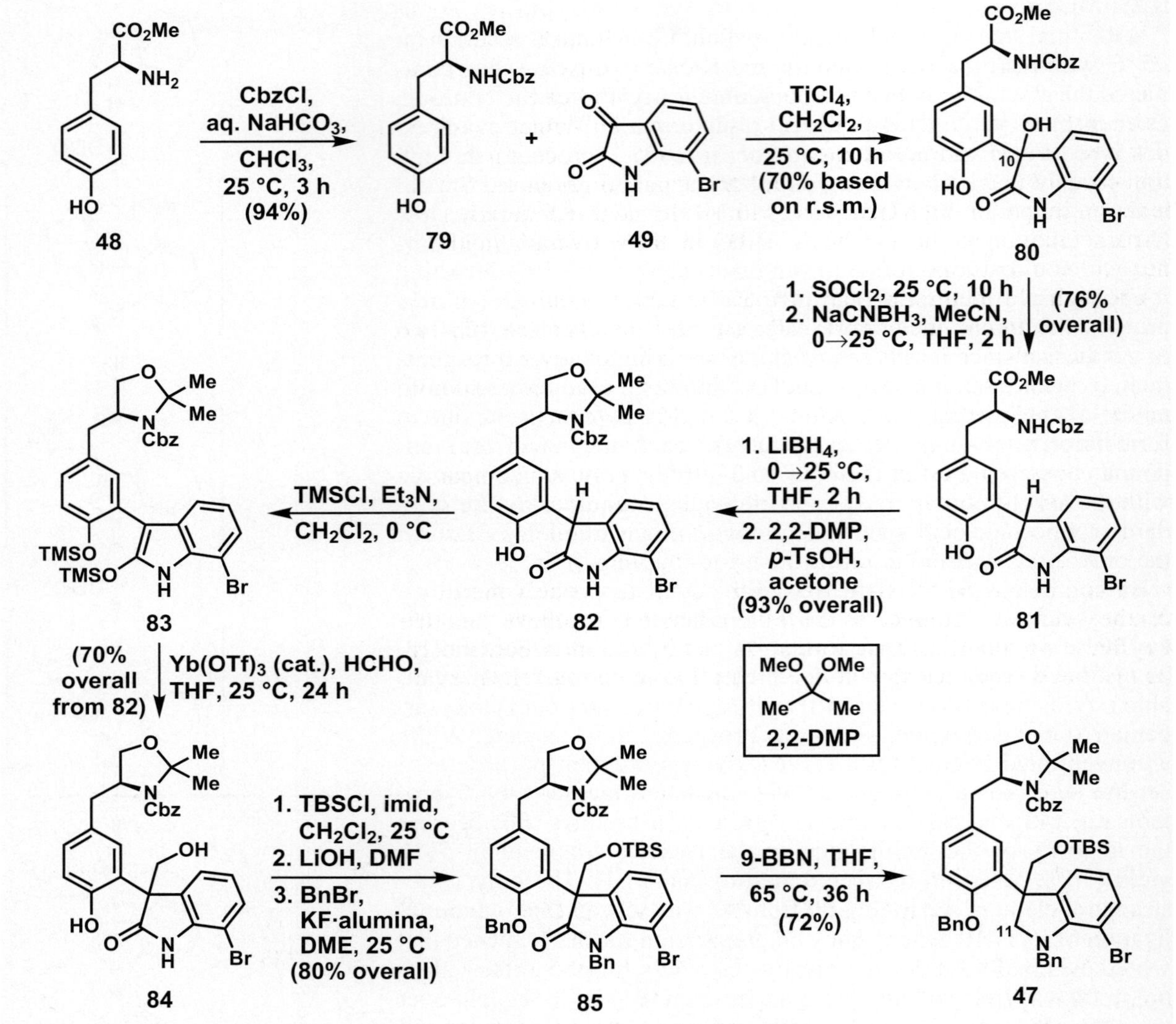

Scheme 13. Synthesis of EFG intermediate **47**.

of both C-10 stereoisomers in 70 % overall yield from **82**. Intriguingly, these diastereomeric compounds could not be separated at this stage either through flash-column chromatography, or selective crystallization. As such, they were carried forward together, hoping that the incorporation of an appropriate functional group would eventually confer sufficient differences in their physical properties to allow their separation.

With the complete architectural framework of the EFG fragment now established as expressed in **84**, only a few protecting and functional group manipulations remained before the substrate would be suitably outfitted to attempt Suzuki coupling with **46**. These adjustments were accomplished in just a few steps, starting with silylation of the free primary alcohol within **84** with TBSCl and imidazole in CH_2Cl_2, conditions that also led to protection of the free A-ring phenol. Since the latter event was unintended, a subsequent reaction with LiOH served to excise this superfluous group. With the phenol now free, the desired protecting group array was then completed by protecting both the phenol as well as the lactam nitrogen atom with benzyl groups through reaction with BnBr and KF•alumina in DME,[35] leading to the synthesis of **85** in 80% overall yield for these three operations.

Although we mentioned earlier that the targeted building block was **47**, the intermediate that had just been synthesized (**85**) in fact possessed all the motifs that the projected strategy required in order to press forward. Of particular importance was the carbonyl group of the E-ring, as the first total synthesis described above had productively employed this motif to eventually construct the FH aminal system of diazonamide A.[7] Unfortunately, as experimentation seeking to advance this material (**85**) soon revealed, its presence instigated several problems, most notably its ability to induce the hydroxymethylene group at C-10 to fragment. For example, when compound **85** was treated with TBAF to probe conditions for the selective removal of its TBS ether, the resultant product was **88**, a compound whose formation can be rationalized through the indicated sequence shown in Scheme 14a invoking the interme-

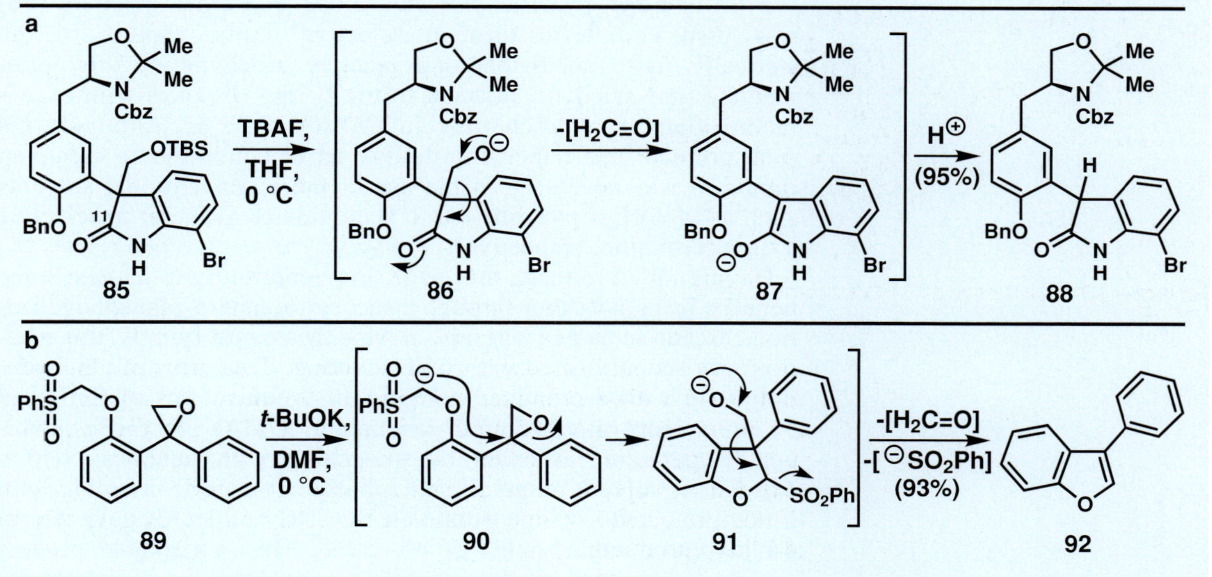

Scheme 14. Problems with the C-11 carbonyl: deformylation through a novel cascade (a) observed earlier during attempts to create the originally proposed structure of diazonamide A (b).

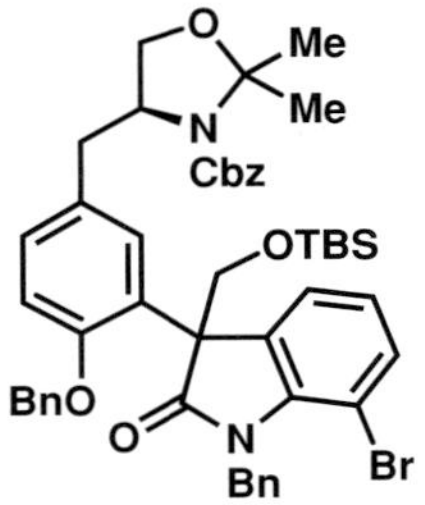

diacy of a fully aromatic enolate (**87**). Interestingly, this decomposition pathway closely parallels a cascade observed during studies conducted by the Nicolaou group seeking to create the C-10 quaternary center of the original structure of diazonamide A (**2**) through a 5-*exo*-tet cyclization as shown in Scheme 14b.[15a] In that event, treatment of substrates such as **89** with a strong base (such as *t*-BuOK) led to the generation of **91**, an alkoxide intermediate that could similarly lose a molecule of formaldehyde as well as a phenylsulfinate anion to afford 3-phenylbenzofuran (**92**) if not quenched immediately with acid.[36]

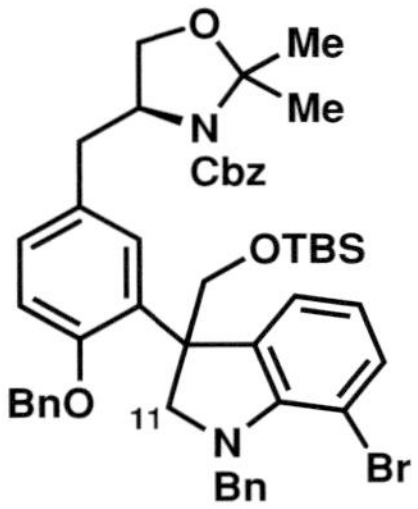

Thus, in light of these unforeseen challenges with **85**, its oxindole system was fully reduced to an indoline (see Scheme 13) through the action of excess 9-BBN in refluxing THF over the course of 36 hours to provide an alternate EFG system (**47**) that enabled the synthesis to proceed forward (at least for the time being). As a consequence of this alteration, the successful completion of the target molecule was now predicated upon finding a suitable method to accomplish its eventual reintroduction. Before these researchers could worry about this likely nontrivial operation, however, the heterocyclic core had to be constructed, and, as shown in Scheme 15, the initial steps leading to the macrocyclization precursor proceeded without incident. First, the aryl halide (**47**) and aryl boronate (**46**) building blocks were smoothly amalgamated into **93** through a Suzuki reaction employing catalytic Pd(dppf)Cl_2 and K_2CO_3 in refluxing DME, an event that proceeded in 78% yield after 12 hours of reaction. Next, **93** was converted into a dialdehyde intermediate through a tandem deprotection/oxidation sequence employing TBAF at a slightly elevated temperature in THF (45 °C) and SO_3·py. The projected aldehyde-oxime substrate (**45**) was then completed through selective oxime capture of the sterically more accessible and reactive aldehyde of this intermediate (the aldehyde adjacent to the B-ring oxazole) with excess methoxylamine hydrochloride in DMSO at 25 °C. Although this final protocol is a rather unorthodox set of conditions to create an oxime, it was necessary as the use of more conventional solvents such as MeOH or pyridine led to acetonide cleavage or unselective oxime formation, respectively.

Having toiled to reach this stage, the opportunity to harvest some benefits from that labor through a successful hetero-pinacol cyclization cascade sequence was now at hand. Most gratifyingly, this reaction was accomplished with little scouting. Treatment of aldehyde-oxime **45** with a premixed complex of 9 equivalents of SmI_2 and 36 equivalents of *N*,*N*-dimethylacetamide (DMA)[37] in THF at ambient temperature, followed by quenching with aqueous NH_4Cl, extraction, solvent removal, and subsequent peptide coupling with Fmoc-protected L-valine promoted by EDC and HOBt gave rise to **44** in reproducible yields of 45–50%. This remarkable process can be rationalized mechanistically by invoking the initial formation of a diradical intermediate (**94**) upon exposure of **45** to SmI_2/DMA. Once this intermediate cyclized to provide **95**, the presence

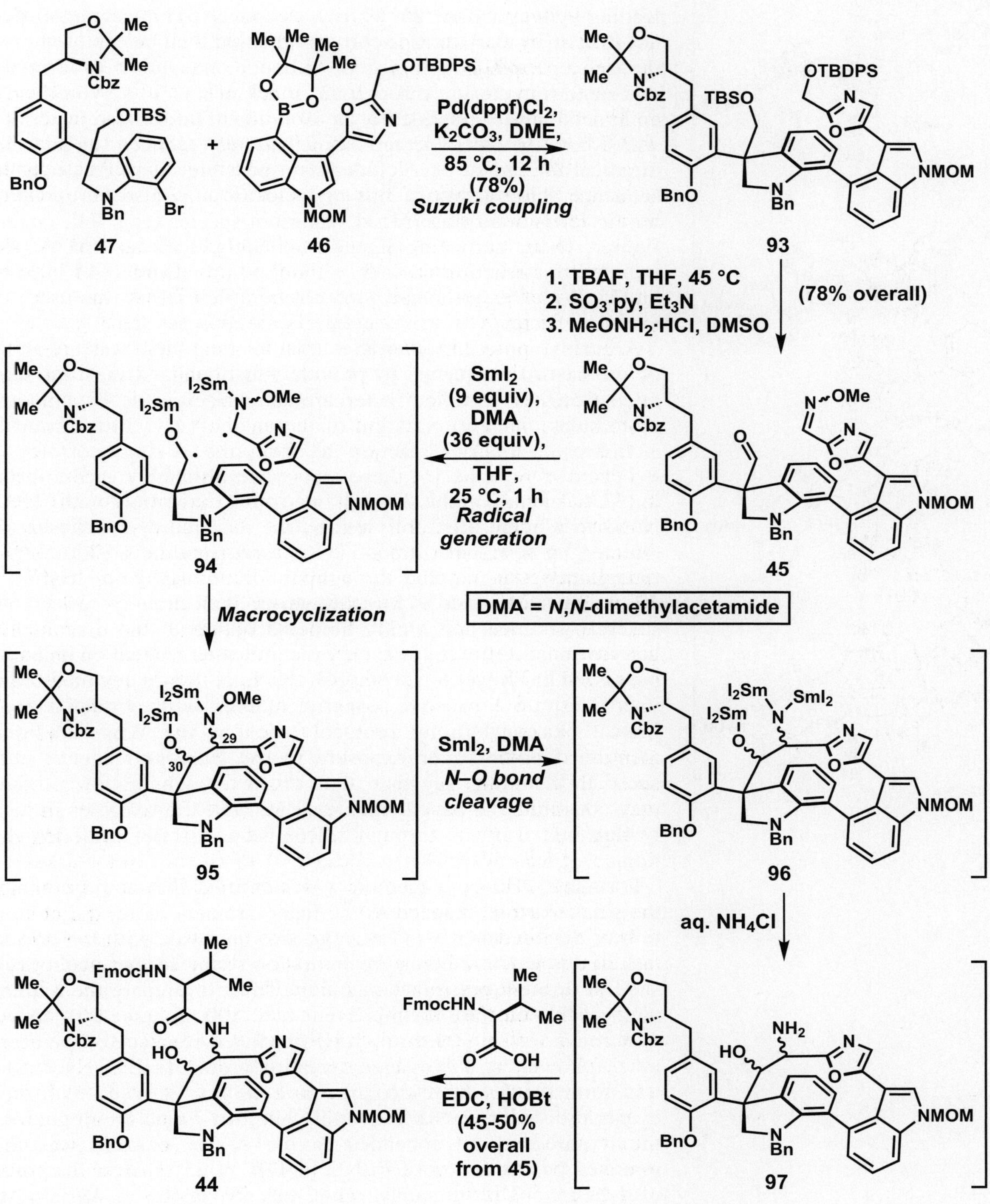

Scheme 15. Completion of the heterocyclic subunit of diazonamide A through a hetero-pinacol macrocyclization sequence: synthesis of intermediate **44**.

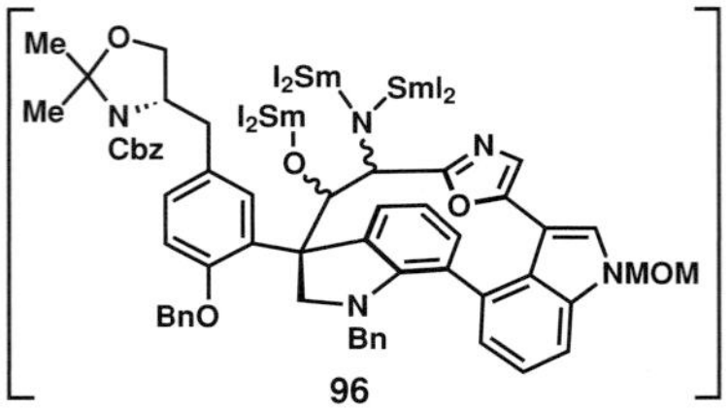

of excess SmI_2 complexed with DMA then effected N–O cleavage, leading to intermediate **96**, which was converted into an amino alcohol (**97**) upon workup; this compound could then be captured as **44** through a peptide-coupling step. Although one could also formulate this same ring closure through the attack of a C-30 ketyl radical at an intact oxime to provide **95**, the isolation of uncyclized material in which both the aldehyde and oxime had been reduced suggests that diradical **94** cannot be excluded as a possible intermediate in this sequence. The facility of this ring closure may also be explained by the formation of a bridged samarium species (e. g. **94′**, column figure). Other activating ligands for SmI_2, such as HMPA, also initiated this reaction cascade, although they afforded **44** in lower yields. With the optimized protocol in which DMA was used, the average efficiency of this sequence was 79 % per step.

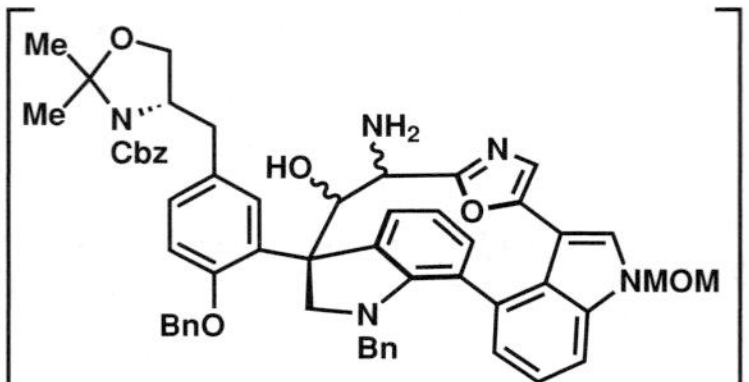

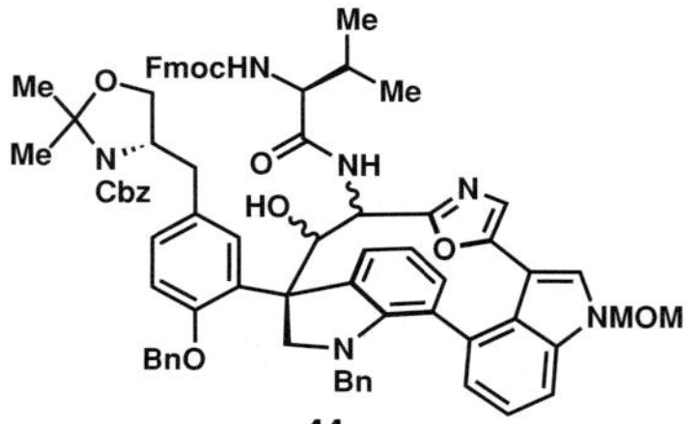

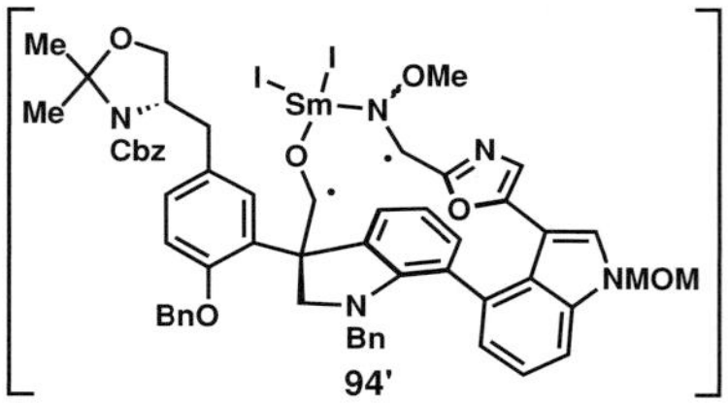

Arguably, more important than its numerical yield was the ability of the cascade sequence to provide functionalized material separated from the complete heteroaromatic macrocycle by only two more steps, namely oxidation of the alcohol at C-30 followed by a Robinson–Gabriel reaction to form the A-ring oxazole. As indicated in Scheme 16, these tasks were ultimately accomplished in 33 % overall yield through an initial oxidation with TPAP/NMO to provide **98**,[38] followed by the required cyclodehydration, effected by a 1:2 mixture of $POCl_3$ and pyridine at 70° C.[9b] At first glance, the material throughput for this sequence leading to **43** might seem unduly low. However, it accurately reflects the severely strained and highly hindered nature of the diazonamide heteroaromatic core. In fact, only the indicated reagent combination possessed the power to accomplish this final task in any measurable yield despite exhaustive experimentation with virtually every known oxazole-forming protocol (such as the Wipf conditions mentioned earlier[12b] or exposure to the Burgess reagent[12c]). As such, this finding suggests that this cyclodehydration protocol may constitute the best available system to form oxazoles in highly hindered contexts through a Robinson–Gabriel cyclodehydration.

Fortunately, since a method was identified that could complete the final oxazole attached to the heteroaromatic core, the completion of diazonamide A (**1**) was not that far away, with the primary task at this juncture being the formation of the second macrocyclic subunit through macrolactamization. Thus, to prepare the required amino acid substrate for this event (i. e. **100**, Scheme 16), **43** was first converted into **99** through HF-mediated cleavage of the acetonide, followed by a two-stage oxidation protocol (IBX; $NaClO_2$) to transform the resultant alcohol into a carboxylic acid. With these events proceeding in 85 % overall yield, the Fmoc group guarding the required amine appended to the A-ring oxazole was then removed by the action of Et_2NH in THF at 25° C over the course of 4 hours.[39] Unfortunately, while the conversion of **43** into **100** proceeded relatively smoothly, attempts to accomplish macrolactam formation from this advanced compound met with significant

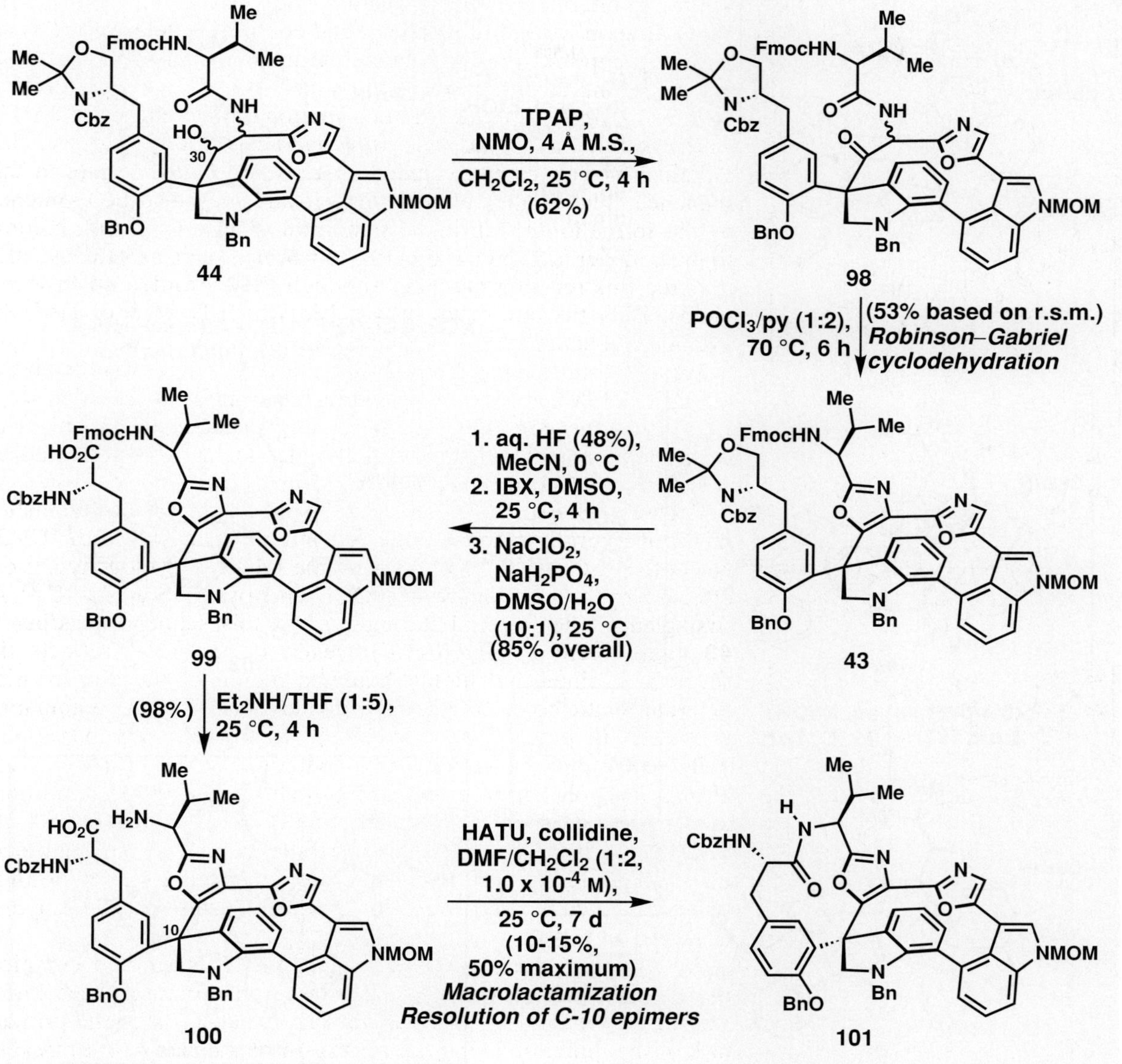

Scheme 16. Completion of the second macrocyclic subunit of diazonamide A (**1**): synthesis of advanced intermediate **101**.

resistance despite an almost exhaustive screening of peptide-forming reagents, leading in all cases either to decomposition or dimerization (even when the reaction was run at exceedingly low concentrations). This result was unexpected, as molecular models suggested that these motifs were well-disposed for a productive reaction. Accordingly, some aspect of the existing motifs must have prevented the smooth formation of this system, as it was far

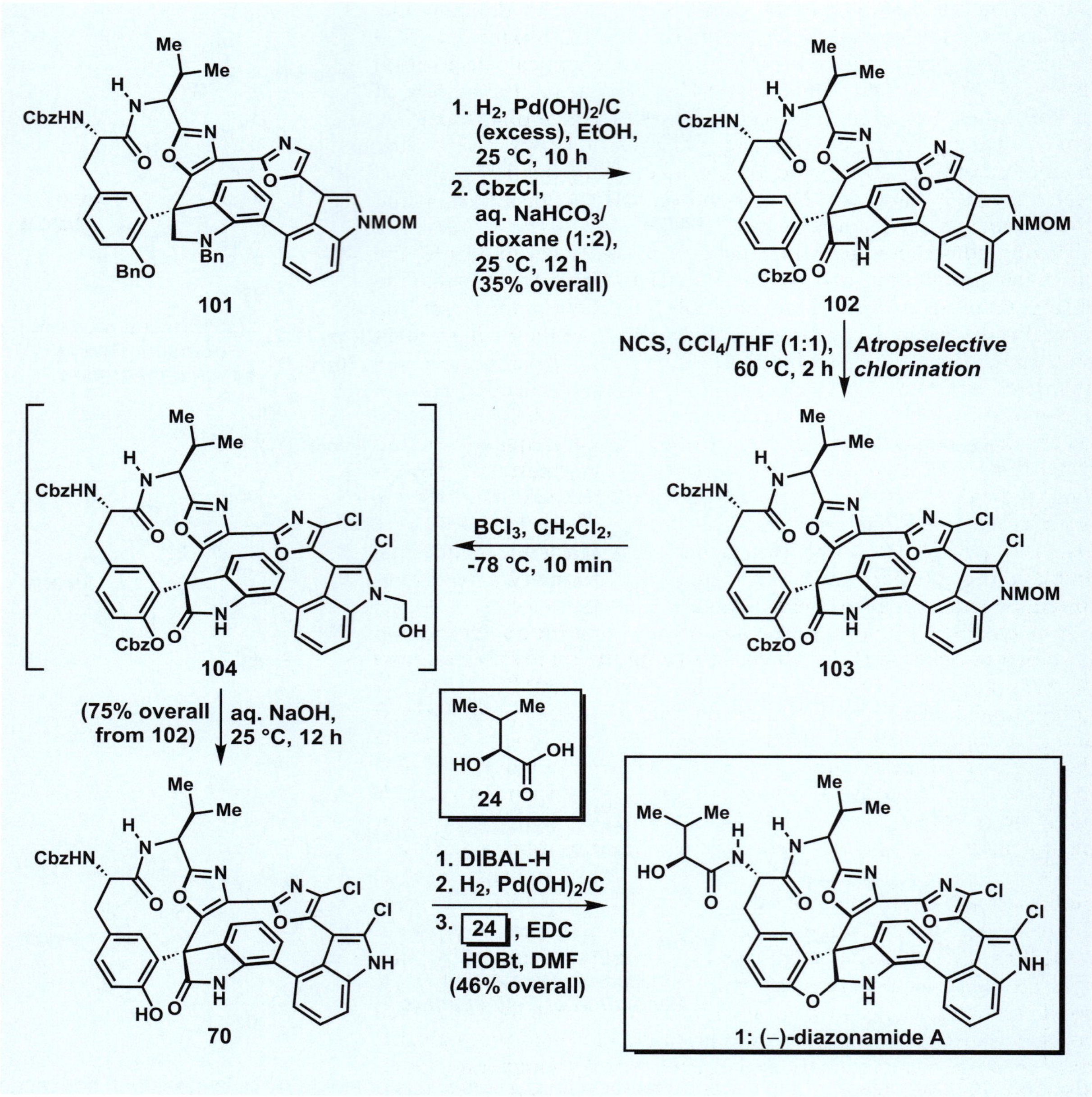

Scheme 17. Final steps and completion of the second total synthesis of diazonamide A (1).

more readily synthesized in the absence of the heteroaromatic core. With enough experimentation, a solution can usually be found, and, in this case, the desired product (**101**) could be obtained in 10–15 % yield when **100** was treated with HATU and collidine for several days at ambient temperature in a 1:2 mixture of DMF and CH_2Cl_2 at a final concentration of 1.0×10^{-4} M. Importantly, apart from completing the second highly strained macrocyclic domain, this step also served to resolve the previously formed mixture of C-10 epimers, as only material with the stereochemical disposition drawn in **101** resulted. Thus, just as in the first total synthesis of diazonamide A, the success or failure of this macrolactamization was intricately linked to the stereochemistry of its intervening chain. Accordingly, the yield of the macrolactamization step was really 20–30 % based on the one and only C-10 diastereomer that could undergo cyclization.

Having constructed both macrocycles of the target molecule, the Nicolaou group now had to face the challenge of oxidizing the E-ring indoline in **101** to an oxindole in order to introduce the aminal ring system. Expecting that the cleavage of the F-ring benzyl protecting group would facilitate this task, these researchers attempted to remove both benzyl groups of the molecule through a standard hydrogenation reaction facilitated by excess $Pd(OH)_2/C$ (Pearlman's catalyst). Amazingly, following subsequent reaction of the intermediate product with benzyl chloroformate, all spectral data indicated that while the debenzylations had occurred as intended, the desired conversion of indoline into oxindole had also taken place to afford **102** directly! As such, the conditions employed in this hydrogenation reaction had formally served both to reduce and to oxidize the substrate.

A mechanistic proposal to account for this unusual transformation is shown in Scheme 18 in which, following the removal of the two benzyl groups to afford **105**, one could envision the $Pd(OH)_2$ undergoing ligand exchange by inserting into both the phenolic and amino groups of **105** to afford an intermediate of type **106**. While the insertion of a palladium species into an amine would not normally lead to a subsequent β-hydride elimination, in this case it might be reasonable to presume that coordination with the adjacent phenol altered its typical geometry so that this operation could occur, thereby leading to **107**. Migratory insertion of the phenol onto the C-11 position would then afford **108**, in which the palladium species is now bound exclusively to the nitrogen atom. A second β-hydride elimination to **109**, followed by the loss of Pd^0 and hydrogen gas, would afford an imine whose capture by hydroxide could account for the formation of **110**, an intermediate that then could collapse to the observed product (**112**) through the indicated ring opening and tautomerization sequence. Although direct evidence for this chain of events has not yet been obtained, this conjecture appears reasonable due to its close similarity to the mechanistic underpinnings of the Wacker oxidation.[40] Moreover, some additional explorations of this reaction have revealed that it cannot be accomplished without an excess of $Pd(OH)_2/C$ and that it is substrate-specific. For example, application of the same conditions to **47** (cf. Scheme 13) served only to excise the benzyl groups. Both these observations are reflected by the delineated mechanism in which the rigidity of the substrate is the crucial feature leading to the formation of **106**.

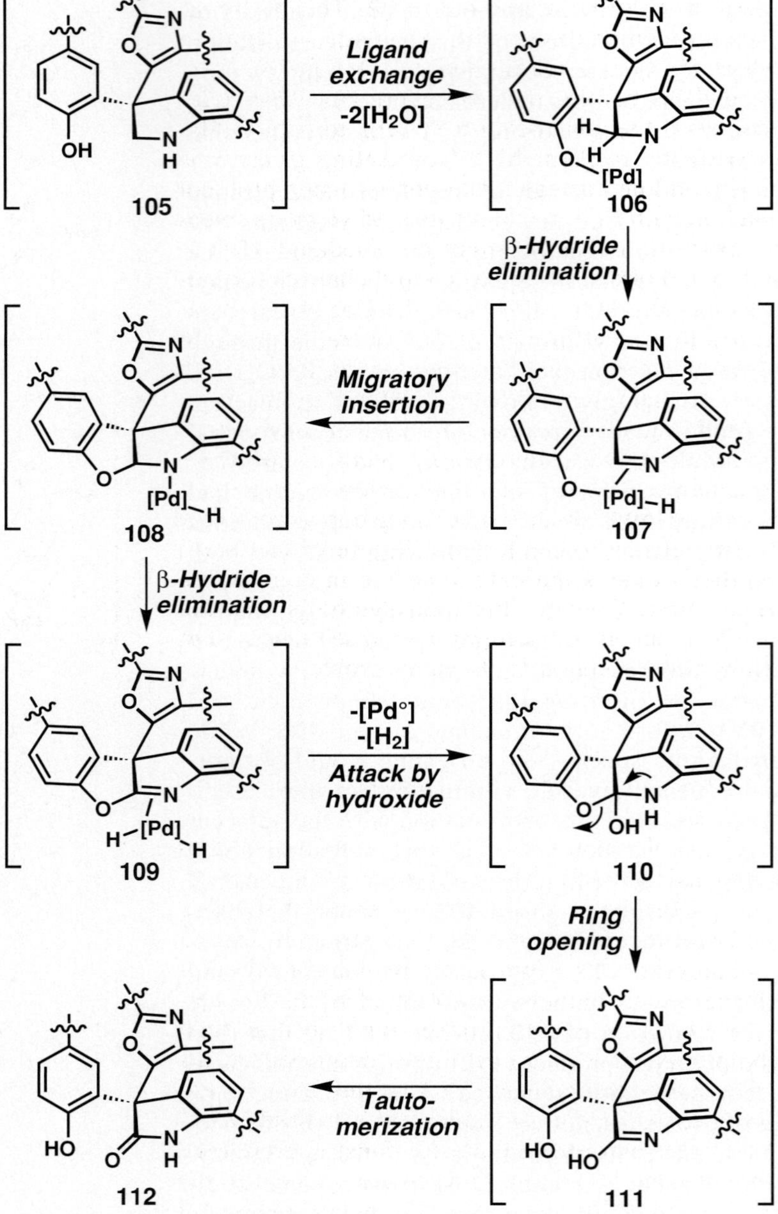

Scheme 18. Proposed mechanism to account for concomitant reduction/oxidation of intermediate **105** to lactam **112** in the presence of excess $Pd(OH)_2/C$.

Irrespective of how this event actually transpired, with this fortuitous transformation accomplished, the completion of diazonamide A (**1**) required only a few finishing touches, starting with the installation of the two requisite chlorine atoms onto **102**. This task was readily accomplished through an electrophilic aromatic substitution reaction using NCS at 60 °C in a 1:1 mixture of THF and CCl_4, a reaction that proceeded, as before, with complete atropselectivity due to the constraints of the macrocyclic system to afford **103** (see Scheme 17). Next, the resilient MOM protecting group was excised from the C-ring indole through a one-pot, two-step protocol involving initial reaction with BCl_3 to cleave its methyl group selectively, followed by exposure of the resultant intermediate (**104**) to NaOH in order to expel the residual hydroxymethyl chain as formaldehyde. These conditions also served to cleave the Cbz group residing on the phenol, thereby generating **70** in 75 % overall yield from **102**. As you might remember, this intermediate had also been encountered during the final stages of the first total synthesis of diazonamide A.[7] Gratifyingly, all its spectral data matched those of the compound obtained previously, thus verifying the integrity of the developed sequence. As a result, application of the same three steps previously employed by the Nicolaou group to complete diazonamide A (**1**) from this advanced intermediate served to finish the second total synthesis of this intricate molecule in comparable yields to those of the first. Overall, this total synthesis was the longer of the two (31 steps in its longest linear sequence), but afforded a completely novel solution to its many problems.

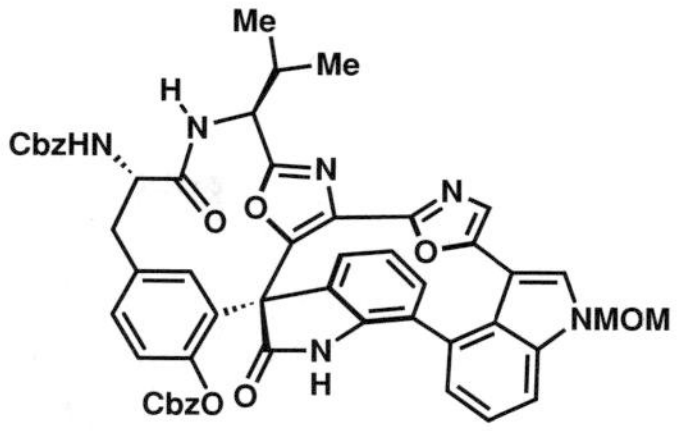

Me Me H N CbzHN N O O N O NMOM N H CbzO O
102

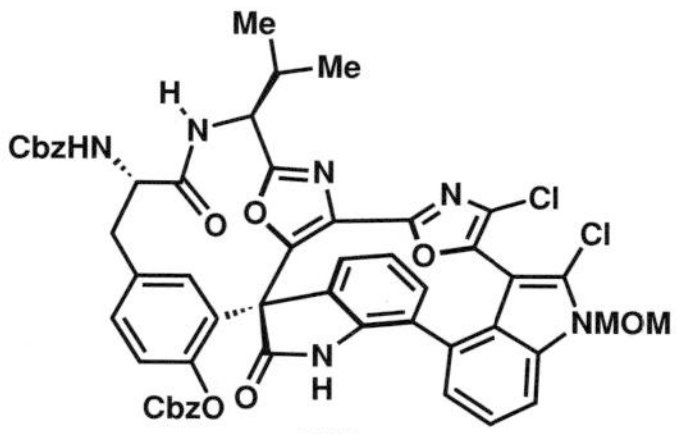

Me Me H N CbzHN N O O N O Cl Cl NMOM N H CbzO O
103

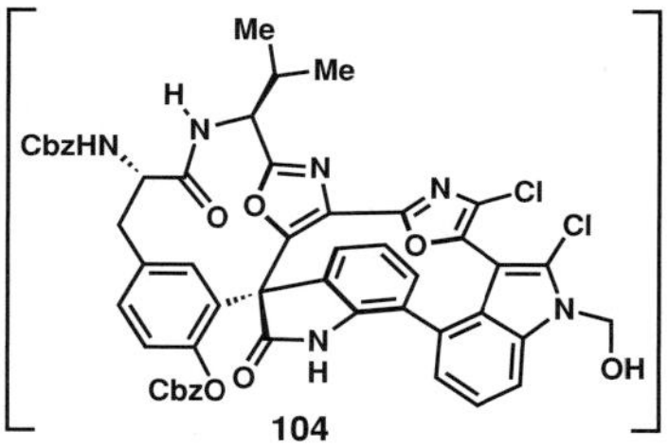

Me Me H N CbzHN N O O N O Cl Cl N OH N H CbzO O
104

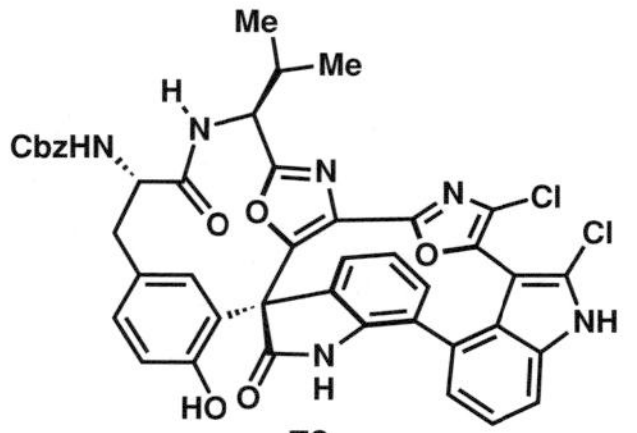

Me Me H N CbzHN N O O N O Cl Cl NH N H HO O
70

20.4 Conclusion

Unique and challenging molecular motifs within secondary metabolites have long presented synthetic chemists with golden opportunities for discovery, whether as a source of inspiration leading to creative strategies and tactics or, instead, as a stringent testing ground revealing weaknesses in the power of available methodology to effectively fashion such complexity worthy of repair. For certain, diazonamide A (**1**) is a natural product with several such domains, as the completion of the two syntheses described in this chapter could not have been accomplished without the development of several unique synthetic strategies and complexity-building reaction cascades. Most notable among these are the construction of the quaternary carbon center and its adjoining aromatic systems in the first synthesis of diazonamide A, and the development of a novel SmI_2-promoted hetero-pinacol cyclization sequence to create the heteroaromatic core in the second. In addition to the challenges overcome by these inventive tactics, both syntheses also had to confront a series of steps that were nearly impossible to execute due to the extreme rigidity of the substrate and the steric shielding imposed by its constituent ring systems. Only by the development of creative reagent combinations and new reaction protocols were some of these obstacles ultimately overcome. In some instances, the complexity of the architecture of diazonamide A even led to some unexpected examples of chemical reactivity, such as the fortuitous oxidation of an indoline to an oxindole during a hydrogenation step.

Beyond the importance of these contributions to the field of chemical synthesis, diazonamide A (**1**) also holds considerable promise as a potential agent for the treatment of cancer. Indeed, preliminary biological evaluation of synthetic diazonamide A has provided some exciting results, including the observation of single digit nanomolar cytotoxicity values against a variety of human cancer cell lines of distinct origin, including lines resistant to powerful drugs such as Taxol™. Its mechanism of action, though postulated to involve stabilization of tubulin, still requires further investigation. As such, the developed sequences should now render sufficient supplies of this scarce natural substance available for the sophisticated assays needed to resolve this pressing question. Indeed, their value in this regard is underscored by the fact that efforts to acquire additional diazonamide A from Nature have met with failure due to an inability to locate its host. Equally significant as generating stockpiles of the natural product, these routes can also be used to construct simplified structural analogues of either of its macrocyclic domains since the pathways towards the target started from entirely different directions. Hopefully the synthesis and biological testing of such agents will reveal both a clear structure–activity profile for the diazonamide class as well as simplified compounds that are as potent as the parent natural product.

References

1. N. Lindquist, W. Fenical, G.D. Van Duyne, J. Clardy, *J. Am. Chem. Soc.* **1991**, *113*, 2303.
2. For highlights of synthetic studies towards diazonamide A, see: a) V. Wittmann, *Nachrichten aus der Chemie* **2002**, *50*, 477; b) T. Ritter, E.M. Carreira, *Angew. Chem.* **2002**, *114*, 2601; *Angew. Chem. Int. Ed.* **2002**, *41*, 2489.
3. a) J. Li, S. Jeong, L. Esser, P.G. Harran, *Angew. Chem.* **2001**, *113*, 4901; *Angew. Chem. Int. Ed.* **2001**, *40*, 4765; b) J. Li, A.W.G. Burgett, L. Esser, C. Amezcua, P.G. Harran, *Angew. Chem.* **2001**, *113*, 4906; *Angew. Chem. Int. Ed.* **2001**, *40*, 4770. For earlier studies by the Harran group, see: c) J. Li, X. Chen, A.W.G. Burgett, P.G. Harran, *Angew. Chem.* **2001**, *113*, 2754; *Angew. Chem. Int. Ed.* **2001**, *40*, 2682; d) X. Chen, L. Esser, P.G. Harran, *Angew. Chem.* **2000**, *112*, 967; *Angew. Chem. Int. Ed.* **2000**, *39*, 937; e) S. Jeong, X. Chen, P.G. Harran, *J. Org. Chem.* **1998**, *63*, 8640.
4. For a discussion of this rearrangement, see: Y. Pocker in *Molecular Rearrangements* (Ed.: P. de Mayo), Wiley-Interscience, New York, **1963**, pp. 15–25.
5. O. Yonemitsu, P. Cerutti, B. Witkop, *J. Am. Chem. Soc.* **1966**, *88*, 3941.
6. a) H.G. Theuns, H.B.M. Lenting, C.A. Salemink, H. Tanaka, M. Shibata, K. Ito, R.J.J. Ch. Lousberg, *Heterocycles* **1984**, *22*, 2007; b) C. Szántay, H. Bolcskei, E. Gács-Baitz, *Tetrahedron* **1990**, *46*, 1711.
7. K.C. Nicolaou, M. Bella, D.Y.-K. Chen, X. Huang, T. Ling, S.A. Snyder, *Angew. Chem.* **2002**, *114*, 3645; *Angew. Chem. Int. Ed.* **2002**, *41*, 3495.
8. K.C. Nicolaou, P. Bheema Rao, J. Hao, M.V. Reddy, G. Rassias, X. Huang, D.Y.-K. Chen, S.A. Snyder, *Angew. Chem.* **2003**, *115*, 1795; *Angew. Chem. Int. Ed.* **2003**, *42*, 1753.
9. For examples that were successful, see: a) K.C. Nicolaou, S.A. Snyder, K.B. Simonsen, A.E. Koumbis, *Angew. Chem.* **2000**, *112*, 3615; *Angew. Chem. Int. Ed.* **2000**, *39*, 3473; b) K.C. Nicolaou, X. Huang, N. Giuseppone, P. Bheema Rao, M. Bella, M.V. Reddy, S.A. Snyder, *Angew. Chem.* **2001**, *113*, 4841; *Angew. Chem. Int. Ed.* **2001**, *40*, 4705; c) E. Vedejs, M A. Zajac, *Org. Lett.* **2001**, *3*, 2451.
10. a) R. Robinson, *J. Chem. Soc.* **1909**, *95*, 2167; b) S. Gabriel, *Chem. Ber.* **1910**, *43*, 1283. For a review, see: I.J. Turchi, *The Chemistry of Heterocyclic Compounds, Vol. 45*, John Wiley and Sons, New York, **1986**, pp. 1–342.
11. H.H. Wasserman, F.J. Vinick, *J. Org. Chem.* **1973**, *38*, 2407.
12. For selected examples of these protocols, see: a) P. Wipf, C.P. Miller, *J. Org. Chem.* **1993**, *58*, 3604; b) P. Wipf, F. Yokokawa, *Tetrahedron Lett.* **1998**, *39*, 2223; c) C.T. Brain, J.M. Paul, *Synlett* **1999**, 1642; d) R.L. Dow, *J. Org. Chem.* **1990**, *55*, 386.
13. A. Baeyer, M.J. Lazarus, *Chem. Ber.* **1885**, *18*, 2642.
14. D.A. Klumpp, K.Y. Yeung, G.K. Surya Prakash, G.A. Olah, *J. Org. Chem.* **1998**, *63*, 4481.
15. a) E.J. Corey, S.G. Pyne, *Tetrahedron Lett.* **1983**, *24*, 2821; b) D.J. Hart, F.L. Seely, *J. Am. Chem. Soc.* **1988**, *110*, 1631; c) P.A. Bartlett, K.L. McLaren, P.C. Ting, *J. Am. Chem. Soc.* **1988**, *110*, 1633.
16. For selected intramolecular examples, see: a) D. Riber, R. Hazell, T. Skrydstrup, *J. Org. Chem.* **2000**, *65*, 5382; b) J. Tormo, D.S. Hays, G.C. Fu, *J. Org. Chem.* **1998**, *63*, 201; c) P. Camps, M. Font-Bardia, D. Munoz-Torrero, X. Solans, *Liebigs Ann.* **1995**, 523; d) T. Shono, N. Kise, T. Fujimoto, A. Yamanami, R. Nomura, *J. Org. Chem.* **1994**, *59*, 1730; e) J. Marco-Contelles, P. Gallego, M. Rodríguez-Fernández, N. Khiar, C. Destabel, M. Berbabé, A. Martínez-Grau, J.L. Chiara, *J. Org. Chem.* **1997**, *62*, 7397.
17. H. Miyabe, M. Torieda, K. Inoue, K. Tajiri, T. Kiguchi, T. Naito, *J. Org. Chem.* **1998**, *63*, 4397.
18. This reagent was pioneered by H. Kagan as described in the following accounts: a) J.L. Namy, P. Girard, H.B. Kagan, *Nouv. J. Chem.* **1977**, *1*, 5; b) P. Girard, J.L. Namy, H.B. Kagan, *J. Am. Chem. Soc.* **1980**, *102*, 2693.
19. For recent review articles on the application of SmI_2 in organic synthesis, see: a) A. Krief, A.-M. Laval, *Chem. Rev.* **1999**, *99*, 745; b) G.A. Molander, C.R. Harris, *Tetrahedron* **1998**, *54*, 3321; c) G.A. Molander, C.R. Harris, *Chem. Rev.* **1996**, *96*, 307; d) T. Skrydstrup, *Angew. Chem.* **1997**, *110*, 355; *Angew. Chem. Int. Ed. Engl.* **1997**, *36*, 345.
20. G.E. Keck, T.T. Wager, J.F.D. Rodriguez, *J. Am. Chem. Soc.* **1999**, *121*, 5176. For related methodology, see: G.E. Keck, T.T. Wager, S.F. McHardy, *Tetrahedron* **1999**, *55*, 11755.
21. For an excellent review on this reaction, see: A. Suzuki in *Metal-catalyzed Cross-coupling Reactions* (Eds.: F. Diederich, P.J. Stang), Wiley-VCH, Weinheim, **1998**, pp. 49–98.
22. For the synthesis of isatin fragments, see: V. Lisowski, M. Robba, S. Rault, *J. Org. Chem.* **2000**, *65*, 4193.
23. For a related synthesis of this fragment, see: S.V. Downing E. Aguilar, A.I. Meyers, *J. Org. Chem.* **1999**, *64*, 826.
24. For reviews on the chemistry of the Burgess reagent, see: a) P. Taibe, S. Mobashery in *Encyclopedia of Reagents for Organic Synthesis, Vol. 5* (Ed.: L.A. Paquette), John Wiley & Sons, Chichester, **1995**, pp. 3345–3347; b) S. Burckhardt, *Synlett* **2000**, 559. For the initial report of this reagent, see: c) G.M. Atkins, E.M. Burgess, *J. Am. Chem. Soc.* **1968**, *90*, 4744; d) G.M. Atkins, E.M. Burgess, *J. Am. Chem. Soc.* **1972**, *94*, 6135; e) E.M. Burgess, H.R. Penton, E.A. Taylor, *J. Org. Chem.* **1973**, *38*, 26.
25. D.R. Williams, P.D. Lowder, Y.-G. Gu, D.A. Brooks, *Tetrahedron Lett.* **1997**, *38*, 331.

26. For other syntheses that benefited from this reagent combination, see: a) K. C. Nicolaou, A. E. Koumbis, M. Takayanagi, S. Natarajan, N. F. Jain, T. Bando, H. Li, R. Hughes, *Chem. Eur. J.* **1999**, *5*, 2622; b) T. Hu, J. S. Panek, *J. Org. Chem.* **1999**, *64*, 3000; c) D. A. Evans, M. R. Wood, B. W. Trotter, T. I. Richardson, J. C. Barrow, J. L. Katz, *Angew. Chem.* **1998**, *110*, 2864; *Angew. Chem. Int. Ed.* **1998**, *37*, 2700.
27. For examples of other related iminium closures, see: a) K. Shishido, E. Shitara, H. Komatsu, K. Hiroya, K. Fukumoto, T. Kametani, *J. Org. Chem.* **1986**, *51*, 3007; b) K. Shishido, K. Hiroya, H. Komatsu, K. Fukumoto, T. Kametani, *J. Chem. Soc, Perkin Trans. 1* **1987**, 2491; c) J. Ezquerra, C. Pedregal, B. Yruretagoyena, A. Rubio, M. C. Carreño, A. Escribano, J. L. G. Ruano, *J. Org. Chem.* **1995**, *60*, 2925; d) I. Collado, J. Ezquerra, A. I. Mateo, A. Rubio, *J. Org. Chem.* **1998**, *63*, 1995; e) R. Downham, F. W. Ng, L. E. Overman, *J. Org. Chem.* **1998**, *63*, 8096.
28. P. J. Harrington, L. S. Hegedus, *J. Org. Chem.* **1984**, *49*, 2657.
29. T. Severin, H.-J. Böhme, *Chem. Ber.* **1968**, *101*, 2925.
30. Y. Oikawa, T. Toshioka, K. Mohri, O. Yonemitsu, *Heterocycles* **1979**, *12*, 1457.
31. A. I. Meyers, T. K. Highsmith, P. T. Buonora, *J. Org. Chem.* **1991**, *56*, 2960.
32. For some discussion on this topic, see: a) P. J. Kocienski, *Protecting Groups*, Georg Thieme Verlag, Stuttgart, **1994**, pp. 220–227; b) T. Watanabe, A. Kobayashi, M. Nishiura, H. Takahashi, T. Usui, I. Kamiyama, N. Mochizuki, K. Noritake, Y. Yokoyama, Y. Murakami, *Chem. Pharm. Bull.* **1991**, *39*, 1152.
33. T. Ishiyama, M. Murata, N. Miyaura, *J. Org. Chem.* **1995**, *60*, 7508. For a later preparation of aryl boronic esters, see: M. Murata, S. Watanabe, Y. Masuda, *J. Org. Chem.* **1997**, *62*, 6458.
34. A. Padwa, D. Dehm, T. Oine, G. A. Lee, *J. Am. Chem. Soc.* **1975**, *97*, 1837.
35. For a related protection using this reagent combination, see: a) T. Ando, J. Yamawaki, T. Kawate, S. Sumi, T. Hanafusa, *Bull. Chem. Soc. Jpn.* **1982**, *55*, 2504; b) J. Yamawaki, T. Ando, T. Hanafusa, *Chem. Lett.* **1981**, 1143.
36. For a productive use of this pathway, see: K. C. Nicolaou, S. A. Snyder, A. Bigot, J. A. Pfefferkorn, *Angew. Chem.* **2000**, *112*, 1135; *Angew. Chem. Int. Ed.* **2000**, *39*, 1093.
37. M. Nishiura, K. Katagiri, T. Imamoto, *Bull. Chem. Soc. Jpn.* **2001**, *74*, 1417.
38. For an excellent review on this oxidant, see: S. V. Ley, J. Norman, W. P. Griffith, S. P. Marsden, *Synthesis* **1994**, 639. For its initial report, see: W. P. Griffith, S. V. Ley, G. P. Whitcombe, A. D. White, *J. Chem. Soc., Chem. Commun.* **1987**, 1625.
39. For another example of this set of conditions to remove an Fmoc group, see the first total synthesis of calicheamicin γ_1^I: K. C. Nicolaou, C. W. Hummel, M. Nakada, K. Shibayama, E. N. Pitsinos, H. Saimoto, Y. Mizuno, K.-U. Baldenius, A. L. Smith, *J. Am. Chem. Soc.* **1993**, *115*, 7625.
40. For a review, see: J. M. Takacs, X. Jiang, *Curr. Org. Chem.* **2003**, *7*, 369.

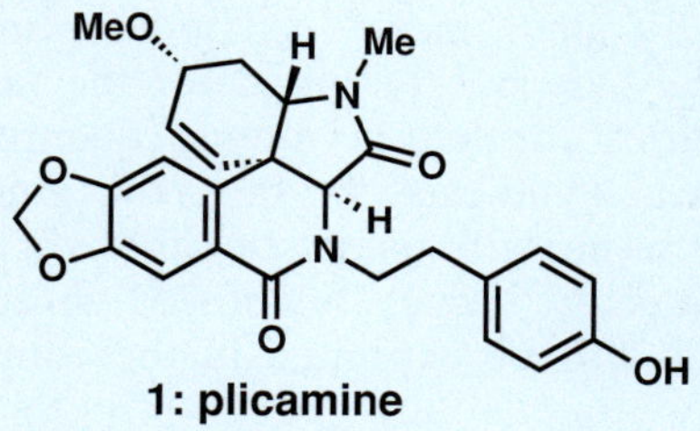

1: plicamine

21

S. V. Ley (2002)

Plicamine

21.1 Introduction

Key concepts:

- **Purification techniques**
- **Solid-supported reagents**
- **Phenolic oxidative couplings**
- **Solid-phase synthesis**

Up to this point, our discussions of individual syntheses have focused almost exclusively on the manner in which given reactions or sequences of transformations can convert synthetic intermediates into the target molecule. In truth, however, these tell only half the story, because the ability to purify and isolate individual compounds effectively from unwanted material is just as intimately tied to the overall effectiveness and success of a synthesis as the power of reagents to effect specific conversions. After all, what reaction is useful if it proceeds quantitatively in the flask, but affords the desired product in mediocre yield following isolation? Nevertheless, synthetic chemists typically discuss most, if not all, elements of product separation and purification within the buried confines of experimental sections in their articles, and skip them completely in book chapters.

Why, then, is so little attention devoted to a component of synthesis that clearly has a high degree of importance? The answer most likely lies in the fact that the basic set of tools for separation and purification (chromatography, distillation, extraction, and crystallization) has fundamentally remained the same for the past quarter of a century. Therefore, in the absence of a new purification method that is truly different from those known, or considered, to be "standard" techniques, chemists are far more likely to tout the development of a novel synthetic reaction or strategy in their manuscripts.

In order to meet the demands that will be placed on organic synthesis in the future, however, there is little question that synthetic practitioners will need to develop new purification methods with a range of capabilities that greatly exceed those presently in use.[1] For example, current techniques do not meet all the objectives of environmentally benign chemistry since many lead to inordinate amounts of waste. Moreover, they are poorly suited for the rapid purification of compound libraries formed in minute quantities through parallel or combinatorial synthesis. In fact, the ability of chemists to solve the latter issue will unquestionably dictate the degree to which new therapeutic targets identified through genomic and proteomic methods will be exploited in the coming decades.

In this chapter, we will address an emerging synthetic technology that may hold the key to addressing both these concerns: the application of solid-supported reagents in multistep synthesis.[2] As we shall see, the creative use of several such immobilized reactants enabled Professor Steven V. Ley and members of his group to prosecute an inventive strategy towards the complex alkaloid natural product (+)-plicamine (**1**) in an expeditious and high-yielding manner without a single chromatographic, distillation, or crystallization step.[3]

21.1.1 Solid-Supported Reagents

If you have had the opportunity to read our discussion on the solid-phase total synthesis of epothilone A in Chapter 7, then you are already aware of the primary synthetic benefits afforded by immobilizing chemical entities onto solid supports.[4] Just to review, these advantages are the ability: 1) to drive reactions to completion by using reagents in excess, and 2) to accomplish facile purification because resin-bound materials can be separated from those in solution simply through filtration. However, while these general advantages are active irrespective of whether it is the substrate or a reagent that is immobilized, experience has shown that the former is far less general and practical. For instance, loading a substrate onto a solid support requires a reactive handle, something that not every compound possesses. Moreover, once that intermediate is attached, it is impossible to monitor the progress of the reaction directly through techniques such as thin-layer chromatography (TLC). Equally problematic, in the absence of a selective method to cleave the final product from the resin at the end of a sequence, any partially or improperly reacted intermediates will be released into solution at the same time as the desired compound. Thus, even though one can rapidly generate libraries through substrate-supported synthesis, the final products may not be obtained in sufficient purity to enable their use in the desired application.

Fortunately, recent work has shown that supported reagents can overcome all these disadvantages and provide some additional perks as well.[2] To illustrate this point, we will use the reactants shown in Figures 1 and 2 to discuss the numerous ways in which these entities can be productively employed in chemical synthesis, starting with how they can facilitate the sometimes onerous task of purification. For the sake of argument, imagine that you successfully managed to cleave a 1,3-dioxane protecting group with a protic acid to form a ketone (see column figures). As a result, both excess acid and ethylene glycol would have to be separated from the product in order to obtain spectroscopically pure material. One could, of course, apply conventional techniques such as extraction or chromatography to achieve this objective. In this case, solid-supported reagents can accomplish the same task far more quickly and with dramatically less waste. For example, any of the basic resins listed in Figure 1 could be used to scavenge the residual acid, and the electrophilic aldehyde reagent could trap the ethylene glycol by providing a more reactive carbonyl handle than a ketone for acetal formation. Simple filtration to remove these immobilized reagents and their attached by-products, followed by evaporation of the residual solvent, would then afford the pure ketone.

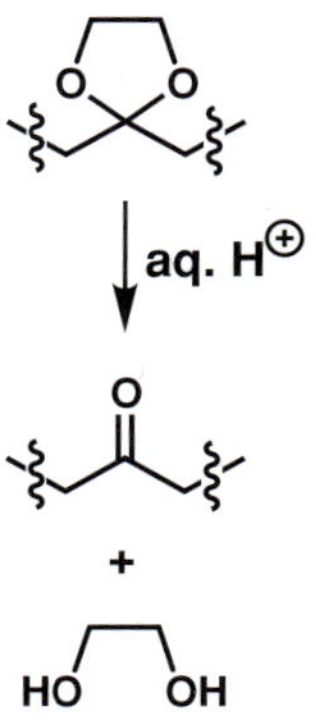

The same general approach can also be applied to sequester products. For example, if one wanted to purify a ketone, the simple diol resin shown in the neighboring column figure can be used to capture that material through acetal formation. After filtration to remove any undesired material still in solution, transfer of the resin to a different flask and subsequent exposure to appropriate conditions can then break that acetal apart, releasing the ketone into solution free from any of the initial impurities. This and related techniques, more commonly termed "catch-and-release" purifications, are particularly well-suited for reactions that fail to proceed to completion. For instance, had this ketone been the product of the oxidation of an alcohol, the diol resin would have left any residual starting material in solution. Normally, one would need column chromatography for such a separation, assuming that the physical properties of the starting material and product are sufficiently different for this method to work.

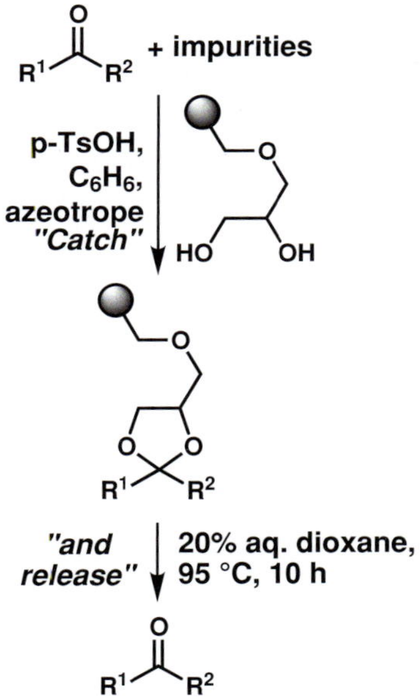

aq. $H^{\oplus}$

HO OH

R^1 R^2 + impurities

p-TsOH, C_6H_6, azeotrope *"Catch"*

HO OH

"and release" 20% aq. dioxane, 95 °C, 10 h

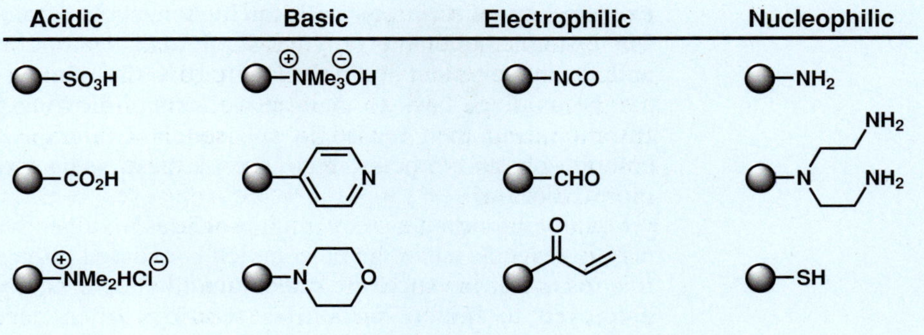

Acidic	Basic	Electrophilic	Nucleophilic
$-SO_3H$	$-\overset{\oplus}{N}Me_3\overset{\ominus}{O}H$	–NCO	$-NH_2$
$-CO_2H$	4-pyridyl	–CHO	$-N(CH_2CH_2NH_2)_2$
$-\overset{\oplus}{N}Me_2HCl^{\ominus}$	N-morpholinyl	acryloyl	–SH

Figure 1. Selected examples of scavenger resins.

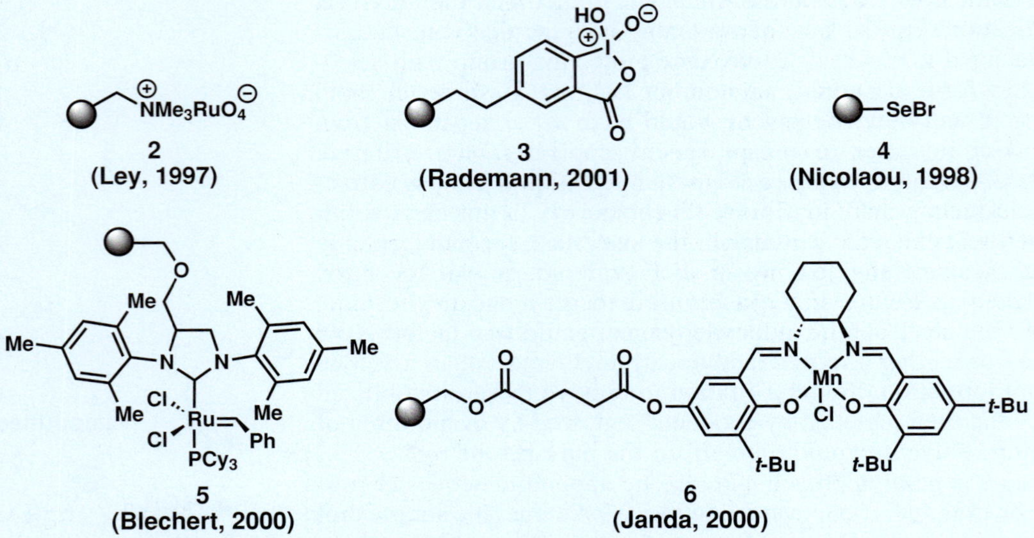

Figure 2. Selected examples of solid-supported reagents and catalysts.

While the utility of solid-supported reagents in purification cannot be underestimated, their value as stoichiometric or catalytic reagents is, perhaps, even more important, especially since they enable synthetic chemists to accomplish several tasks that conventional solution-phase synthesis cannot. For instance, because polymer-bound reagents are isolated (hidden) within the support matrix, accessible only to dissolved material and not to other immobilized reagents, one can concurrently employ two different solid-supported reagents that would be mutually incompatible if they were both in solution, such as an oxidant and a reductant. Consequently, this property provides synthetic practitioners with the ability to execute entirely novel reaction cascades which would otherwise be impossible to perform. Apart from new reaction sequences, supported reagents also provide an opportunity to execute known reactions with far higher levels of atom economy. For example, both the polymer-supported versions of TPAP (**2**)[5] and the hypervalent iodine oxidant IBX (**3**)[6] shown in Figure 2 can be oxidized back to their active forms following post-reaction filtration, and then reused in subsequent oxidations. In standard solution-phase synthesis, recovery of these spent oxidants is far more laborious.

Equally important as recycling, problems involved with the use of certain reagents can often be completely avoided if they are attached to a resin. For instance, the ruthenium alkylidene catalysts typically employed to initiate metathesis reactions often decompose into darkly colored, metal-containing by-products, unwanted materials that sometimes require special protocols to remove and which are capable of causing olefin isomerization within the desired products if left behind.[7] Neither of these problems are encountered if one enlists the services of a ruthenium alkylidene attached to a solid-support, such as catalyst **5** developed by the Blechert group, as all the metal by-products remain on-bead.[8,9] Similarly, the selenenyl bromide resin (**4**) developed by the Nicolaou group in 1998 possesses all the same synthetic powers as phenylselenenyl bromide, but without its noxious odor or high toxicity.[10] The short total synthesis of peracetyl macrophylloside D (**13**) shown in Scheme 1 serves to illustrate some of the virtues of this polymer-supported reagent.[11] As can be seen, the loading and cleavage steps involving this polymer-supported reagent enabled an expedient means by which to generate the 2,2-dimethylbenzopyran ring system of the target

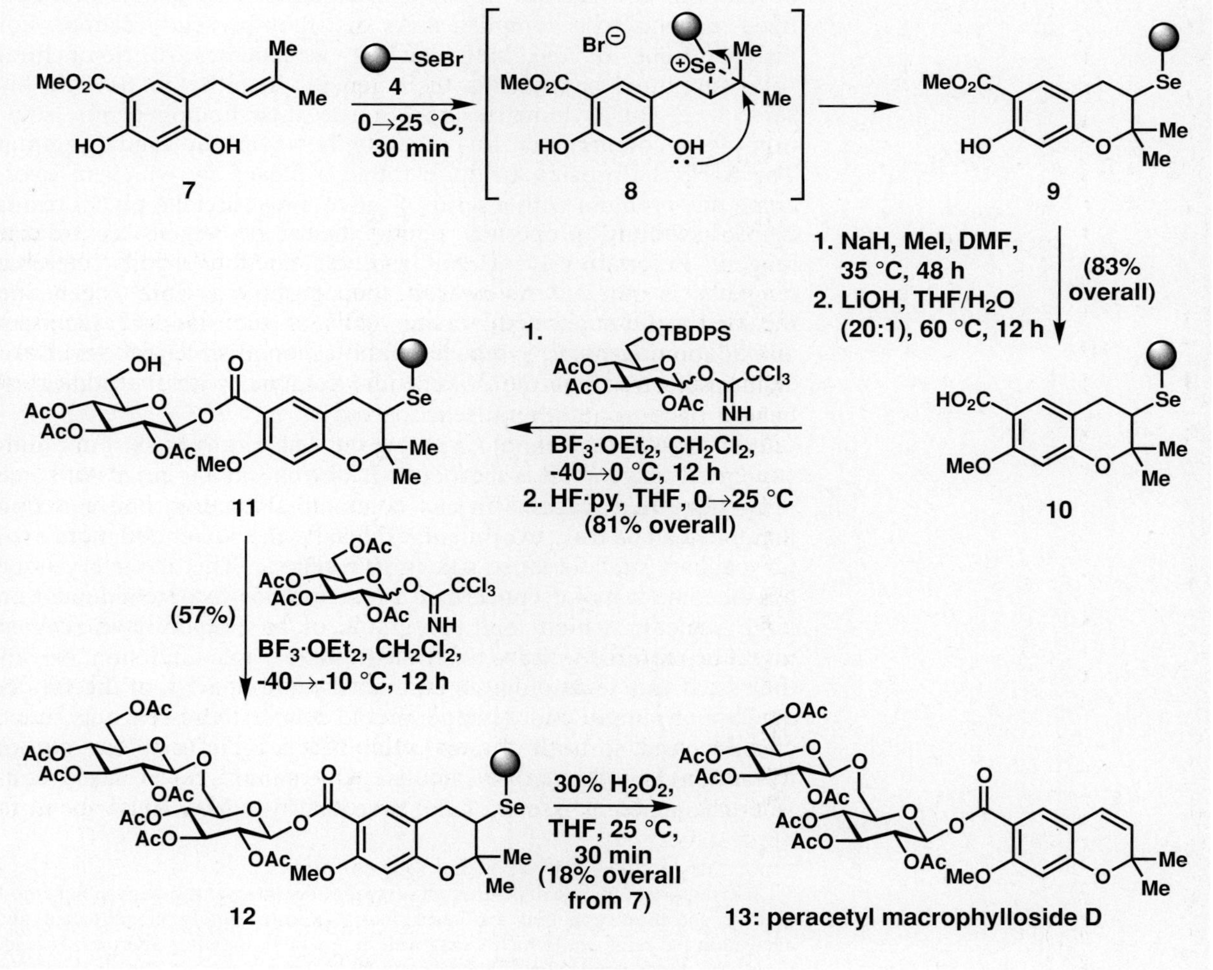

Scheme 1. Use of the polystyrene-based selenenyl bromide resin in the solid-phase total synthesis of peracetyl macrophylloside D (**13**). (Nicolaou, 2000)[11]

molecule from a prenylated phenol starting material (**7**).* Amazingly, no significant trace of selenium was detected in solution following the elimination step which excised the final product (**13**) from the support! As such, this compound, as well as 10,000 additional benzopyran-containing products which the Nicolaou group synthesized in a similar manner,[12] could be used directly in high throughput biological assays without fear of any false positives arising from residual selenium.[13]

Finally, because a variety of materials with far different physical properties can be used to immobilize chemicals (such as glass balls, clay particulates, or polystyrene), it is possible for chemists to tailor both the kinetics of a given reaction and the reactivity profile of the appended reagent. Thus, a nonselective reagent in standard solution-phase reactions could conceivably demonstrate higher fidelity once attached to a support. This concept also means that developing new classes of support materials may enable immobilized reagents to accomplish tasks that they presently cannot. For instance, one of the long-standing weaknesses of homochiral solid-supported catalysts is their general inability to provide the same levels of asymmetric induction as their homogeneous, solution-phase counterparts. Just recently, however, the Janda group at The Scripps Research Institute found a means by which to overcome this problem with a series of novel polyethylene glycol resins whose swelling properties permit them to behave as soluble reagents in certain solvents, but insoluble, and thus readily filterable reagents, in others.[14] As a result, their group was able to generate the first solid-supported version (**6**)[15] of the standard Jacobsen epoxidation catalyst,[16] which could accomplish equal levels of enantioselectivity as the commercial catalyst, with the additional benefit of reuse in several reaction cycles.

Every synthetic technology, however, has a drawback. For solid-supported reagents, it is their cost. If one checks the prices of some resin-supported reagents in any commercial catalog, one will find that they can be truly exorbitant, especially the newer and more exotic reagents such as those shown in Figure 2. This monetary issue can become a major concern if a reaction has to be conducted on large scale in which multiple grams of an immobilized reagent might be required to drive that transformation to completion. Nevertheless, if this technology is embraced more heavily in the future, the law of supply and demand should render these reagents much cheaper. And, in truth, the real price of a reagent (or better put, of a reaction) is quite hard to calculate. One should, for example, consider the inherent issues of the time that a worker must spend to

---

* It is important to note that although several of the intervening steps in Scheme 1 involve substrate-supported synthesis, had the resin simply been loaded and cleaved in the same pot (which works well in practice), then this approach is readily placed under the heading of reagent-supported synthesis. As shown, **4** reflects the virtues of both a safety-catch linker (see Chapter 7 for a discussion of this concept) and an immobilized reagent.

purify a compound as well as the costs involved in disposing the wastes generated by that method, features which solid-supported reagents handle beautifully.

Rather than continue to discuss these reagents along general lines, it is time to illustrate their virtues in the most stringent testing ground possible: a complicated natural product whose architecture dictates a bold synthetic plan of attack if it is to be produced in the laboratory. That secondary metabolite is (+)-plicamine (**1**).

21.2 Retrosynthetic Analysis and Strategy

(+)-Plicamine (**1**) was first isolated in 1999 from the plant *Galanthus plicatus* subspecies *byzantinus* native to northwestern Turkey, and, although initial screens have not revealed any interesting biological activity, it is certainly endowed with a highly complicated polycyclic architecture.[17] Indeed, even a cursory examination of its structure reveals the formidable test that its molecular connectivities pose to the power of modern organic synthesis. As such, most synthetic chemists would be highly satisfied if they could establish an efficient approach to accomplish its total synthesis. From the Ley group's perspective, however, a successful synthesis of (+)-plicamine (**1**) would not only mean developing an inventive sequence, but also finding a way to use solid-supported reagents to render that route efficient enough that it could be performed without any conventional purification steps.

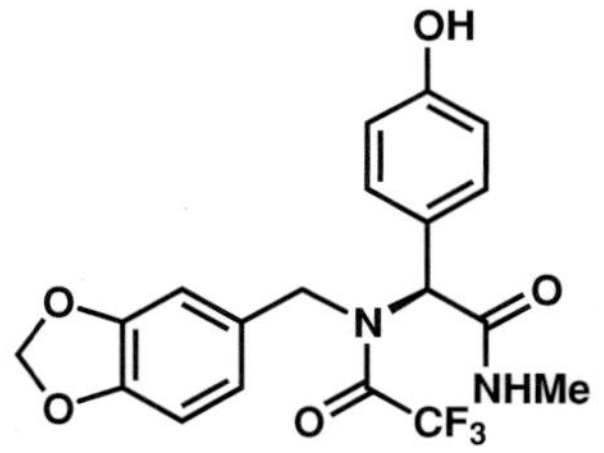

Their creative retrosynthetic analysis is shown in Scheme 2, with the critical insight being the idea that an intramolecular phenolic oxidative coupling could deliver an intermediate appropriately functionalized to enable the facile completion of the target. More specifically, if an appropriate resin-supported oxidant could oxidatively convert an intermediate such as **17** into **16**,[18] then there would be an opportunity to attempt the construction of the remaining ring systems through a substrate-controlled conjugate addition of the amide at C-7 onto the proximal Michael system. If these steps leading to **15** could be accomplished, accessing (+)-plicamine (**1**) would then require a stereocontrolled 1,2-reduction of its α,β-unsaturated ketone, sequential alkylations to attach the remaining functionality, and an oxidation step to generate the C-8 carbonyl. Overall, these bold operations appear highly practical and efficient on paper, and, equally pleasing, quite feasible based on an extensive body of literature precedent.[18] As such, the only remaining design element that the Ley group had to consider in advance of laboratory experiments was how to obtain **17** as a single stereoisomer. Their solution was to rupture **17** at the junction indicated in Scheme 2, expecting that a reductive amination of aldehyde **18** with hydroxyphenylglycine derivative **19** could accomplish the construction of this subtarget during the synthesis. In turn, the

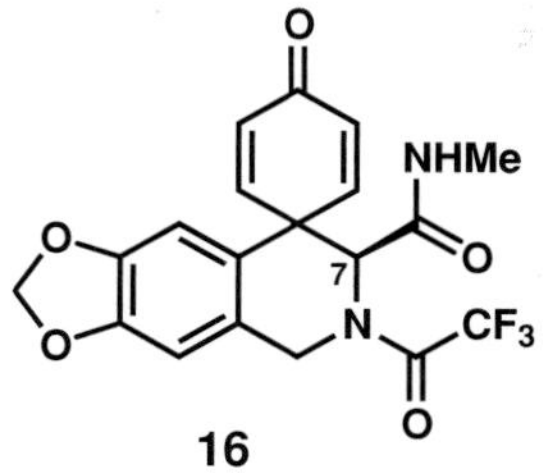

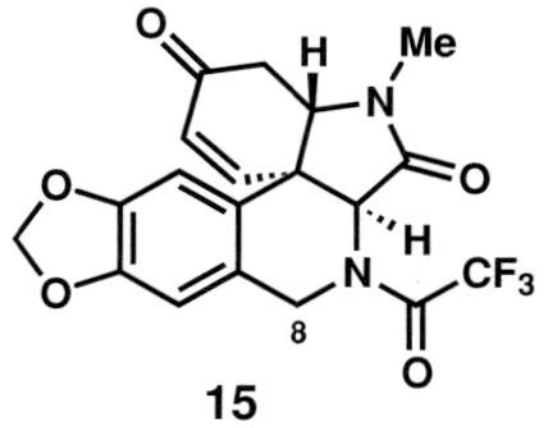

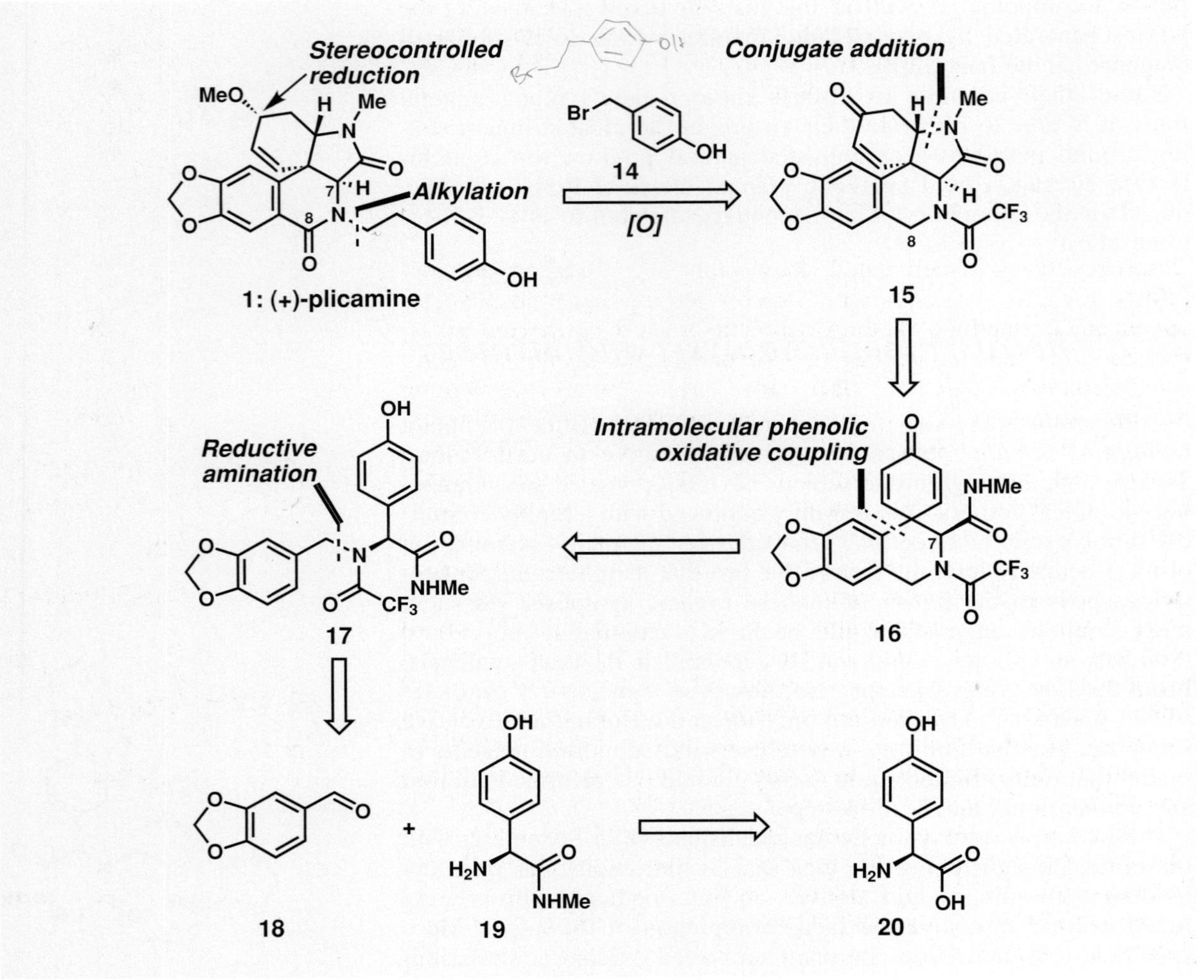

Scheme 2. Ley's retrosynthetic analysis of (+)-plicamine (**1**).

structural homology of **19** and the commercially available **20** suggested that the latter compound, a member of the chiral pool, could serve as a plausible point from which to begin synthetic operations.

Before we analyze how this expedient and creative strategy was put into practice, we should comment on why the plan delineated in Scheme 2 sought to install the C-8 carbonyl at the end of the sequence rather than at the beginning, when it would certainly be easier to incorporate. The reason is one of viability because phenolic couplings of the type sought in this case depend heavily on the oxidation potential of the substrate (which we shall talk about in more detail shortly). For example, the incorporation of an electron-withdrawing carbonyl group adjacent to that of the reacting rings would likely provide a compound unsuitable for the desired reaction.[18] In fact, preliminary model studies bore out this conjecture, as the exposure of compounds such as **21** (see column figures) to suitable oxidants led only to aziridine products (**23**), not the desired 6,6-spirocyclic system (**22**).[3a]

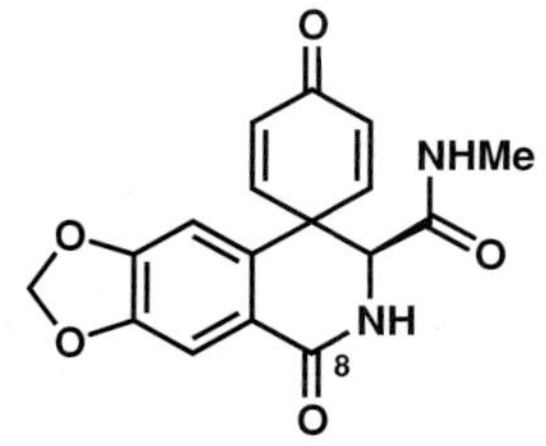

21.3 Total Synthesis

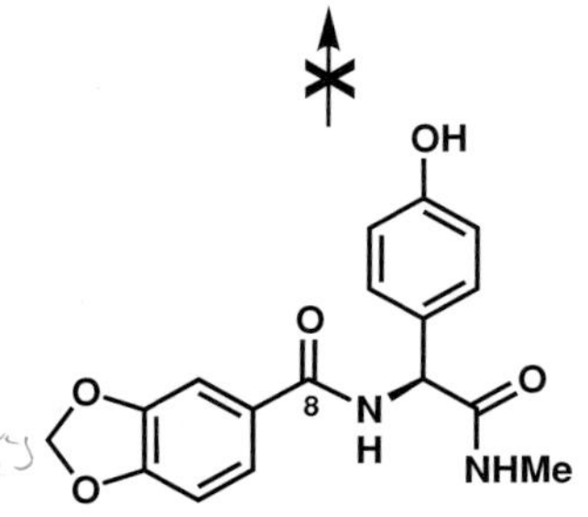

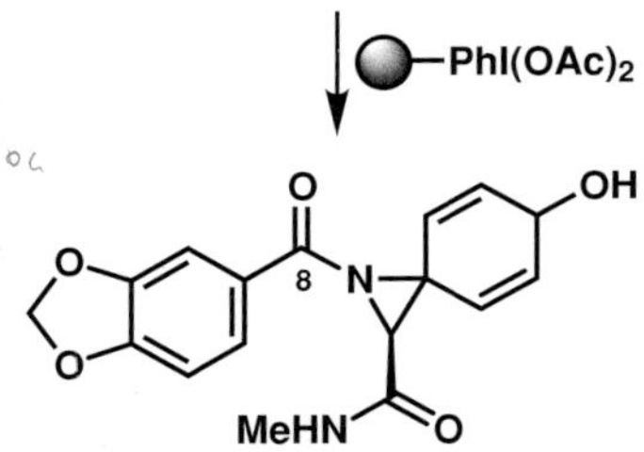

Based on the plan delineated above, the Ley group began their efforts towards (+)-plicamine (**1**) by attempting to synthesize advanced intermediate **17**, the key precursor for the projected oxidative coupling step. Thus, they first needed to find a way to convert L-4-hydroxyphenylglycine (**20**) into amide **19**. As shown in Scheme 3, this task was smoothly accomplished by treating **20** with MeOH in the presence of a Lewis acid catalyst (TMSCl) to generate an intermediate methyl ester, mopping up the resultant HCl by-product with the indicated resin, and then adding excess methylamine to the reaction mixture to form the amide. Following removal of the solvent, these operations afforded **19** in near quantitative yield (>95 % purity, determined by LC–MS analysis).

With the desired amide synthesized, its merger with aldehyde **18** was then achieved through a typical reductive amination protocol using polymer-supported borohydride rather than a more conventional reagent to execute the reduction step that completed the assembly of **24**. As a result, instead of performing an extraction or column chromatography to obtain clean material, simple filtration of the immobilized reductant, followed by evaporation of the solvent, provided **24** in sufficient purity to attempt the final operation required to complete **17**, namely trifluoroacetylation of the secondary amine of **24**. In general, such a protection requires the action of Tf_2O and a bulky base like 2,6-di-*tert*-butylpyridine.[19] However, because chromatography is often required to remove these bases, the Ley group sought to develop an alternate protocol using solid-supported bases to accomplish the same reaction and avoid this type of purification. To their delight, they discovered that a combination of polyvinylpyridine (PVP) and polymer-supported 4-DMAP (which served as base and catalyst, respectively) was more than sufficient to enable the synthesis of **17** upon their addition to a solution of **24** and Tf_2O in CH_2Cl_2 at 0 °C. Following filtration of the polymer-supported bases and removal of the solvent, **17** was obtained in 91 % overall yield from **20** (>97 % purity, based on HPLC analysis). Equally impressive as a demonstration of the power of immobilized reagents to conduct reactions and achieve purifications, all six operations leading to **17** from **20** could be performed with equal facility on both milli- and multigram scales.

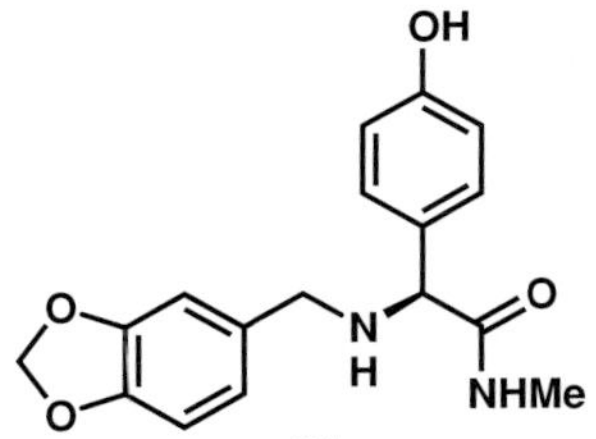

With a highly efficient route identified for the synthesis of advanced intermediate **17**, these researchers were now in a position to determine whether this adduct could undergo the intramolecular

22

21

PhI(OAc)2

23

24

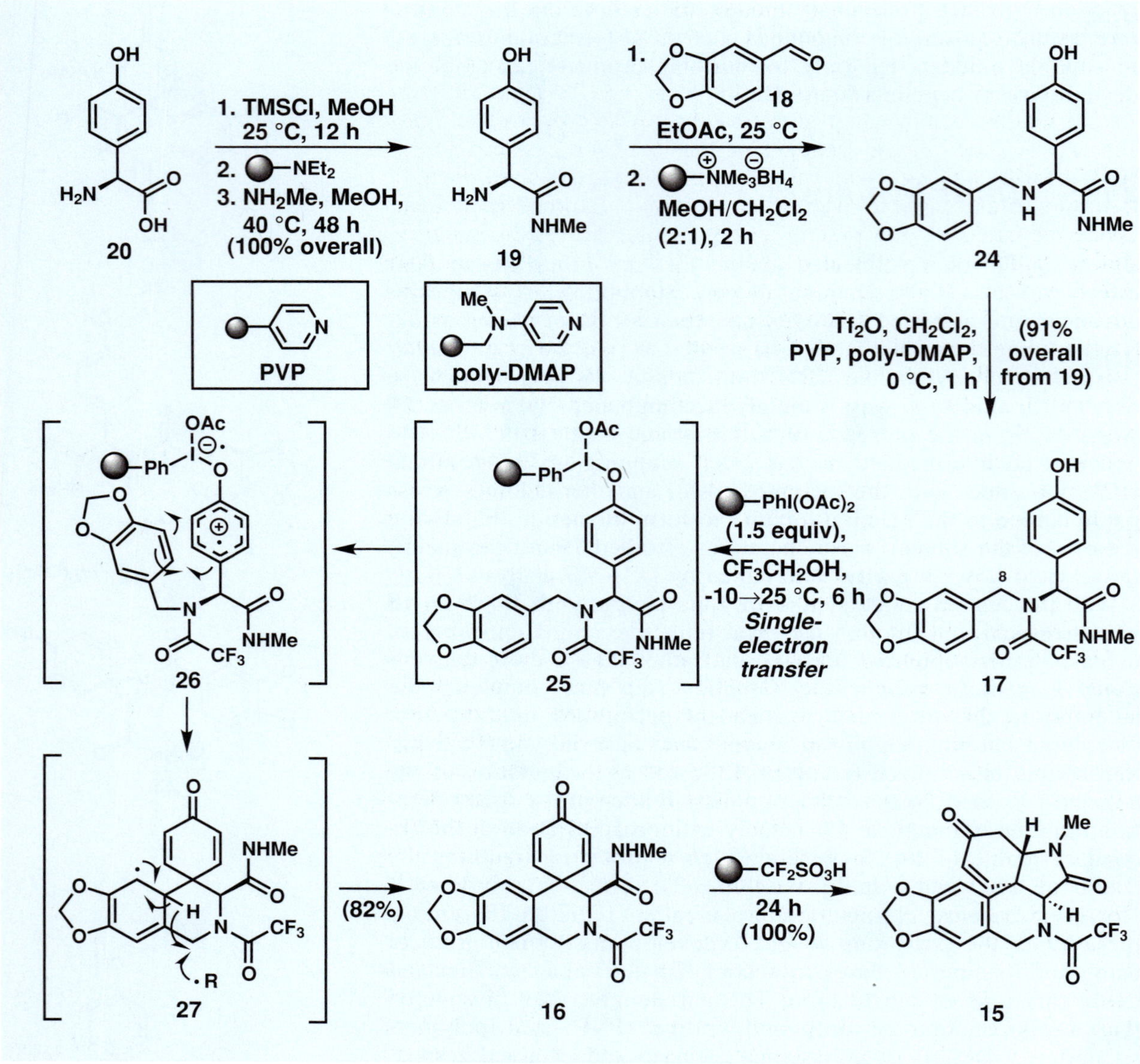

Scheme 3. Synthesis of advanced intermediate **15** through a phenolic oxidative coupling.

phenolic coupling required to access the 6,6-spirocyclic product (**16**) essential to the entire sequence. Fortunately, **17** was a willing participant when exposed to the shown solid-supported iodonium diacetate reagent developed by Togo and co-workers,[20] with 6 hours of reaction in 2,2,2-trifluoroethanol leading to **16**. As indicated by the structures of the postulated intermediates shown in Scheme 3, the mechanism for this powerful annulation reaction is relatively complicated, with the initial step constituting the hypervalent-iodine-mediated conversion of the free phenol in **17** into an electron-deficient radical cation (see **26**). Once formed, this reactive intermediate is then engaged at its *para* position (in terms of the starting phenol) by the most electron-rich position of the neighboring catechol system to accomplish ring formation (**27**), with a terminating rearomatization step affording the final product (**16**). Thus, this mechanistic picture accounts for why starting materials similar to **17** bearing a C-8 carbonyl failed to undergo the desired phenolic coupling as discussed above, as that carbonyl would severely deactivate the catechol ring system and retard its ability to engage the radical cation of **26**.* It also explains why 2,2,2-trifluoroethanol was employed as the reaction solvent, since only a highly polar reaction media could appropriately stabilize the single-electron transfer complex and thereby accelerate the overall reaction rate to synthetically useful levels. Perhaps what is equally amazing as the realization of this step was that the final compound was uncontaminated by any other side products even though its material return was not quantitative (**16** was isolated in 82 % yield). Although this outcome would not normally be expected for a reaction that failed to go to completion, it is reasonable in this case because the putative radical intermediates encountered en route to **16** (i. e. **26** and **27**) could have reacted with the aromatic rings of the polymeric support or its appended functionality.[21] As such, their covalent attachment to the resin enabled their facile removal.

Since the formation of **16** was achieved cleanly, the Ley group could now explore the ability of its methyl amide to add to the neighboring dienone system in a Michael fashion and thereby complete the remaining ring of the target in one fell swoop. Gratifyingly, exposure of **16** to a solid-supported version of triflic acid (Nafion-H)[22] was all that was required. And, with this process creating **15** as a single diastereomer in quantitative yield, a subsequent substrate-controlled reduction using polymer-supported borohydride then served to install the final stereocenter of the target as expressed in **28** (see Scheme 4) with equal efficiency.

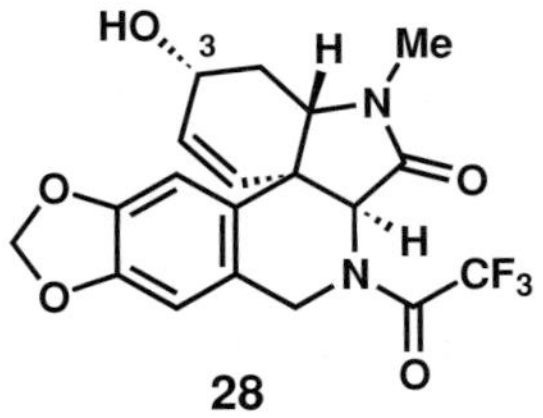

Having beautifully executed these bold operations, all that remained to complete the target molecule (**1**) from this advanced staging area (**28**) were two alkylations and oxidation of the amine group to form an amide. In principle, these tasks could be tackled in either order. However, since the previous step had served to unveil a nucleophile that could be used in one of the alkylation steps, it seemed logical to start with those operations first. As a result, the Ley group set out to identify conditions for appending a methyl group onto the allylic alcohol of **28** without inducing its epimerization or elimination. Following a brief period of screening, they found that the use of trimethylsilyldiazomethane and a macroporous sulfonic acid exchange resin in a mixture of MeOH and CH_2Cl_2 (3:2) could smoothly accomplish that task, delivering **29**

HO 3 H Me N O H N CF_3 O

28

* Another feature that should also be considered is the ability of that tether to adopt the required transition state for productive cyclization.

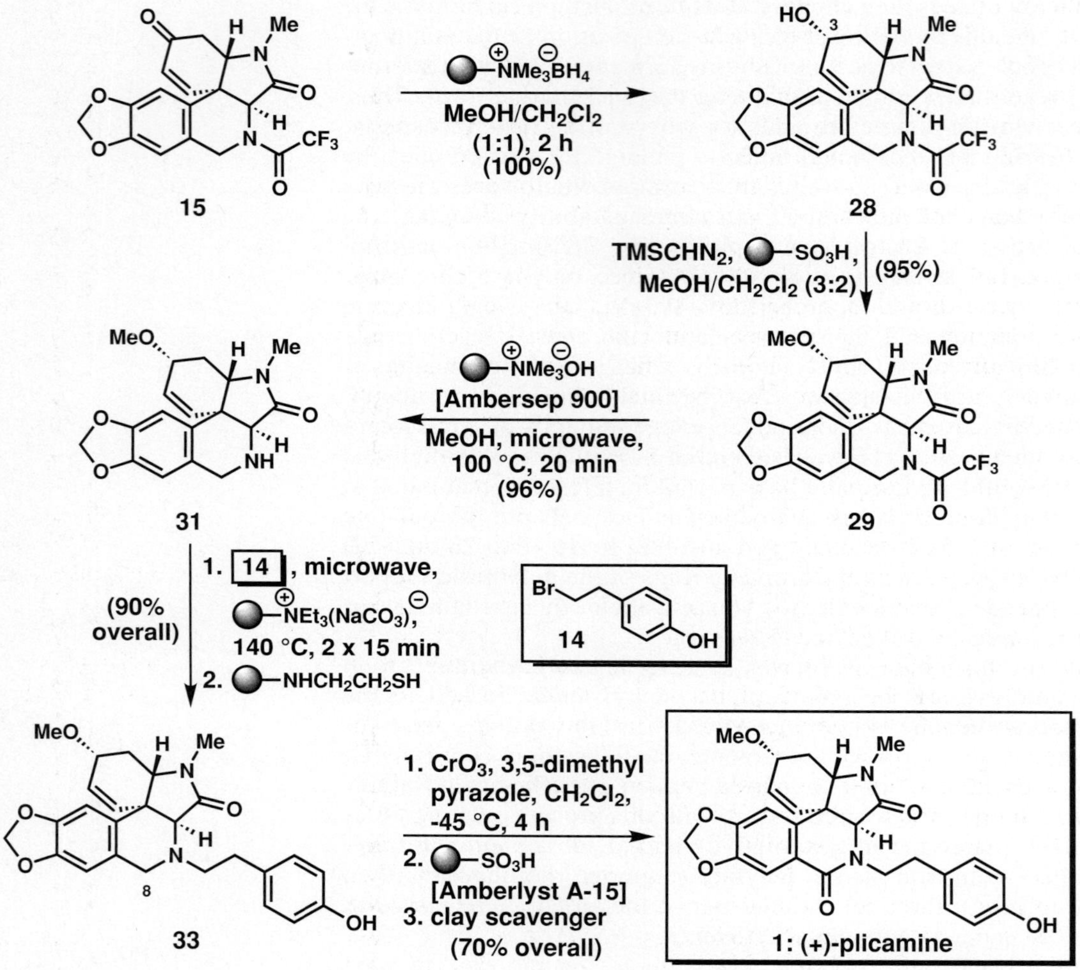

Scheme 4. Final stages and completion of Ley's total synthesis of (+)-plicamine (1).

in 95 % yield. Unlike the other steps shown thus far, this particular operation did not prove as amenable to scaling, with decreased throughput observed whenever more than 2 grams of **28** were subjected to these conditions. Fortunately, an alternate protocol using MeOTf, a polymer-supported version of 2,6-di-*t*-butylpyridine, and thermal activation in the form of microwave irradiation proved to be an equally effective method for the clean preparation of **29** when larger-scale reactions had to be performed.[23] As an important aside, one should note that whenever a reaction involving solid-supported reagents or substrates has to be conducted above ambient temperature, sharp exposure to heat supplied by microwave irradiation is almost always superior to heating with a more typical source.

Not only does prolonged and focused heating often lead to the decomposition of most solid-supports, it also typically affords numerous side products.[2] For instance, when the reaction with MeOTf just described was conducted at elevated temperature using an oil bath rather than microwave irradiation, significant amounts of the elimination product **30** (see column figure) were always observed in addition to **29**.

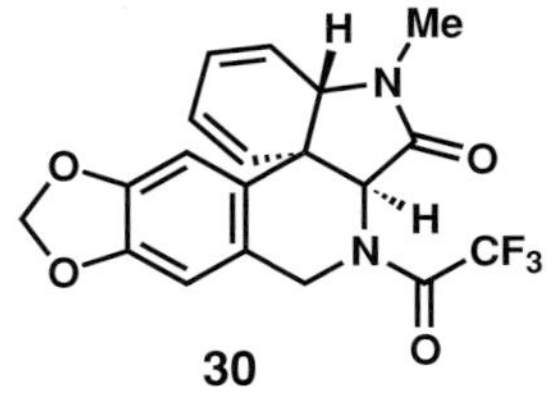

Having effected the first alkylation step, efforts towards the second were then begun in earnest by attempting to excise the trifluoroacetate group guarding the amine function of **29**. As shown in Scheme 4, this obligatory task was accomplished using the basic form of a resin known as Ambersep 900 in methanol, with 20 minutes of irradiation at 100 °C in a sealed tube microwave reactor leading to the formation of **31** in 96 % yield. Alkylation was then achieved by treating this amine intermediate with excess bromide **14** (synthesized from **32** using a polymer-supported version of Ph_3P as shown in the adjacent column) in the presence of the indicated immobilized carbonate base at 140 °C. Filtration of this resin, followed by a final purification step employing a mercaptoaminoethyl resin to sequester the residual traces of **14** still present in solution, served to deliver **33** in 90 % yield (>95 % purity). Before moving on, one should note that attempts to execute this alkylation with any other resin-supported base led to markedly inferior conversion; stronger bases induced the bromide residue within **14** to preferentially eliminate to 4-vinylphenol, and milder bases proved too weak to enable efficient conversion of the starting material into **33**.

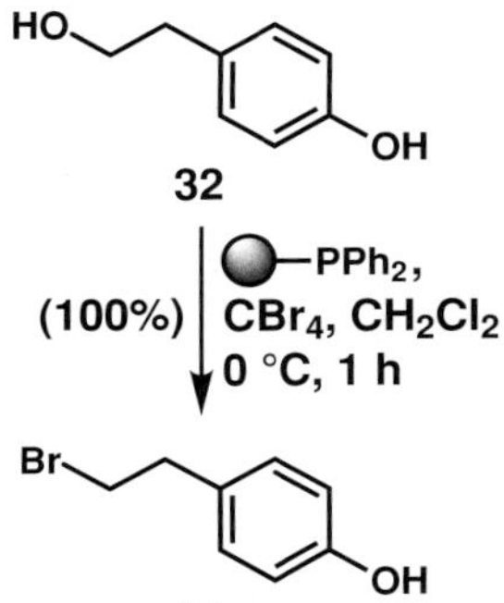

With a means identified to attach this final tether in high yield, only a carbonyl group had to be installed at the benzylic C-8 position of **33** in order to complete (+)-plicamine (**1**). Despite the simply stated nature of this task, this concluding oxidation provided the most stringent purification test for solid-supported reagents in the entire sequence because no screened oxidant led to a quantitative conversion, and many led to a series of by-products. Fortunately, a set of conditions was eventually identified that could enable success. Thus, as the operation in Scheme 4 indicates, that procedure began with the exposure of a solution of **33** in CH_2Cl_2 to equimolar amounts of CrO_3 and 3,5-dimethylpyrazole (a reagent combination that we discussed at some length in Chapter 8).[24] Following 4 hours of reaction at −45 °C, approximately 70 % of the starting material was successfully converted into **1** in the absence of any additional side products. As such, it was now up to solid-supported reagents to perform their final task in this synthesis, namely the removal of the unoxidized amine starting material, the residual chromium salts, and the 3,5-dimethylpyrazole from **1**. This chore was accomplished using an Amberlyst A-15 resin to sequester both amine-containing by-products with a quick passage through a small bed of Chem Elut CE1005 packing material combined with montmorillonite K 10 clay to excise the metal-containing salts. Combined, these operations afforded **1** with a purity greater than

H Me N O H N CF_3 O O O

30

HO OH

32

(100%) ●–PPh_2, CBr_4, CH_2Cl_2 0 °C, 1 h

Br OH

14

90 % (HPLC). An analytically pure sample of (+)-plicamine (**1**) was finally obtained simply by filtering the remaining materials through a short plug of silica. As such, this concluding operation has brought to a close a highly efficient and creative total synthesis, one in which a total of twelve different immobilized reagents enabled the assembly of a highly complicated natural product in an overall yield of ~40 % without having to employ a single conventional purification!

21.4 Conclusion

Although predicting the future is always a risky business, it is probably safe to say that solid-supported reagents will play an increasingly important role in chemical synthesis during the coming decades. Indeed, as this inventive total synthesis has elegantly demonstrated, their ability to drive reactions to completion, sequester unwanted by-products, and expedite the general process of purification with the production of minimal waste affords several tactical advantages over conventional solution-phase synthesis and techniques more commonly employed to isolate synthetic intermediates. In their absence, it is unlikely that this insightful and quite synthetically challenging route would have been executed with such amazing efficiency. As such, this synthesis soundly reinforces the notion that only by approaching total synthesis in novel ways, from the standpoint of both executing reactions and accomplishing purifications, can the power of synthetic chemistry reach new heights.

References

1. For some examples of potential solutions, see: D. P. Curran, *Angew. Chem.* **1998**, *110*, 1230; *Angew. Chem. Int. Ed.* **1998**, *37*, 1174.
2. For excellent and comprehensive reviews, see: a) S. V. Ley, I. R. Baxendale, R. N. Bream, P. S. Jackson, A. G. Leach, D. A. Longbottom, M. Nesi, J. S. Scott, R. I. Storer, S. J. Taylor, *J. Chem. Soc., Perkin Trans 1* **2000**, 3815; b) S. V. Ley, I. R. Baxendale, *Nature Reviews: Drug Discovery* **2002**, *1*, 573.
3. a) I. R. Baxendale, S. V. Ley, M. Nessi, C. Piutti, *Tetrahedron* **2002**, *58*, 6285; b) I. R. Baxendale, S. V. Ley, C. Piutti, *Angew. Chem.* **2002**, *114*, 2298; *Angew. Chem. Int. Ed.* **2002**, *41*, 2194.
4. For selected discussions on solid-phase organic synthesis, see the *Handbook of Combinational Chemistry, Vols. 1 & 2* (Eds.: K. C. Nicolaou, R. Hanko, W. Hartwig), Wiley-VCH, Weinheim, **2002**, p. 1114.
5. B. Hinzen, S. V. Ley, *J. Chem. Soc., Perkin Trans. 1* **1997**, 1907. For other uses of this oxidant, see: a) B. Hinzen, S. V. Ley, *J. Chem. Soc., Perkin Trans. 1* **1998**, 1; b) B. Hinzen, R. Lenz, S. V. Ley, *Synthesis* **1998**, 977.
6. G. Sorg, A. Mengel, G. Jung, J. Rademann, *Angew. Chem.* **2001**, *113*, 4532; *Angew. Chem. Int. Ed.* **2001**, *40*, 4395. For another example of IBX on solid support, see: M. Mülbaier, A. Giannis, *Angew. Chem.* **2001**, *113*, 4530; *Angew. Chem. Int. Ed.* **2001**, *40*, 4393.
7. a) L. A. Paquette, J. D. Schloss, I. Efremov, F. Fabris, F. Gallou, J. Méndez-Andino, J. Yang, *Org. Lett.* **2000**, *2*, 1259; b) H. D. Maynard, R. H. Grubbs, *Tetrahedron Lett.* **1999**, *40*, 4137.
8. S. C. Schürer, S. Gessler, N. Buschmann, S. Blechert, *Angew. Chem.* **2000**, *112*, 4062; *Angew. Chem. Int. Ed.* **2000**, *39*, 3898.
9. For selected examples of solid-supported ruthenium alkylidenes active for metathesis, see: a) S. T. Nguyen, R. H. Grubbs, *J. Organomet. Chem.* **1995**, *497*, 195; b) Q. Yao, *Angew. Chem.* **2000**, *112*, 4060; *Angew. Chem. Int. Ed.* **2000**, *39*, 3896; c) J. S. Kingsbury, S. B. Garber, J. M. Giftos, B. L. Gray, M. M. Okamoto, R. A. Farrer, J. T. Fourkas, A. H. Hoveyda, *Angew. Chem.* **2001**, *113*, 4381; *Angew. Chem. Int. Ed.* **2001**, *40*, 4251; d) P. Nieczypor, W. Buchowicz, W. J. N. Meester, F. P. J. T. Rutjes, J. C. Mol, *Tetrahedron Lett.* **2001**, *42*, 7103; e) L. Jafarpour, M.-P. Heck, C. Baylon, H. M. Lee, C. Mioskowski, S. P. Nolan, *Organometallics* **2002**, *21*, 671; f) R. Akiyama, S. Kobayashi, *Angew. Chem.* **2002**, *114*, 2714; *Angew. Chem. Int. Ed.* **2002**, *41*, 2602; g) S. J. Connon, A. M. Dunne, S. Blechert, *Angew. Chem.* **2002**, *114*, 3989; *Angew. Chem. Int. Ed.* **2002**, *41*, 3835.
10. K. C. Nicolaou, J. Pastor, S. Barluenga, N. Winssinger, *Chem. Commun.* **1998**, 1947. For related selenium-based resins, see: a) T. Ruhland, K. Andersen, H. Pedersen, *J. Org. Chem.* **1998**, *63*, 9204; b) K. Fujita, K. Watanabe, A. Oishi, Y. Ikeda, Y. Taguchi, *Synlett* **1999**, 1760.
11. a) K. C. Nicolaou, J. A. Pfefferkorn, G.-Q. Cao, *Angew. Chem.* **2000**, *112*, 750; *Angew. Chem. Int. Ed.* **2000**, *39*, 734; b) K. C. Nicolaou, G.-Q. Cao, J. A. Pfefferkorn, *Angew. Chem.* **2000**, *112*, 755; *Angew. Chem. Int. Ed.* **2000**, *39*, 739.
12. a) K. C. Nicolaou, J. A. Pfefferkorn, A. J. Roecker, G.-Q. Cao, S. Barluenga, H. J. Mitchell, *J. Am. Chem. Soc.* **2000**, *122*, 9939; b) K. C. Nicolaou, J. A. Pfefferkorn, H. J. Mitchell, A. J. Roecker, S. Barluenga, G.-Q. Cao, R. L. Affleck, J. E. Lillig, *J. Am. Chem. Soc.* **2000**, *122*, 9954; c) K. C. Nicolaou, J. A. Pfefferkorn, S. Barluenga, H. J. Mitchell, A. J. Roecker, G.-Q. Cao, *J. Am. Chem. Soc.* **2000**, *122*, 9968.
13. For other applications of this resin in chemical synthesis, see: a) K. C. Nicolaou, A. J. Roecker, J. A. Pfefferkorn, G.-Q. Cao, *J. Am. Chem. Soc.* **2000**, *122*, 2966; b) K. C. Nicolaou, N. Winssinger, R. Hughes, C. Smethurst, S. Y. Cho, *Angew. Chem.* **2000**, *112*, 1126; *Angew. Chem. Int. Ed.* **2000**, *39*, 1084.
14. For a review on these resins, see: D. J. Gravert, K. D. Janda, *Chem. Rev.* **1997**, *97*, 489.
15. T. S. Reger, K. D. Janda, *J. Am. Chem. Soc.* **2000**, *122*, 6929.
16. W. Zhang, J. L. Loebach, S. R. Wilson, E. N. Jacobsen, *J. Am. Chem. Soc.* **1990**, *112*, 2801.
17. a) N. Ünver, T. Gözler, N. Walch, B. Gözler, M. Hesse, *Phytochemistry* **1999**, *50*, 1255; b) N. Ünver, S. Noyan, B. Gözler, T. Gözler, C. Werner, M. Hesse, *Heterocycles* **2001**, *55*, 641.
18. For related precedents, see: a) Y. Kita, M. Arisawa, M. Gyoten, M. Nakajima, R. Hamada, H. Tohma, T. Takada, *J. Org. Chem.* **1998**, *63*, 6625; b) S. V. Ley, O. Schucht, A. W. Thomas, P. J. Murray, *J. Chem. Soc., Perkin Trans. 1* **1999**, 1251; c) M. Arisawa, N. G. Ramesh, M. Nakajima, H. Tohma, Y. Kita, *J. Org. Chem.* **2001**, *66*, 59.
19. A. G. Martínez, L. R. Subramanian, M. Hanack in *Encyclopedia of Reagents for Organic Synthesis, Vol. 7* (Ed.: L. A. Paquette), John Wiley & Sons, Chichester, **1995**, pp. 5146–5152.
20. H. Togo, G. Nogami, M. Yokoyama, *Synlett* **1998**, 534. For other uses of this oxidant, see: S. V. Ley, A. W. Thomas, H. Finch, *J. Chem. Soc., Perkin Trans. 1* **1999**, 669.
21. For a general and mechanistic discussion of phenolic oxidative couplings, see: D. A. Whiting in *Comprehensive Organic Synthesis, Vol. 3* (Eds.: B. M. Trost, I. Fleming), Pergamon, Oxford, **1991**, pp. 659–703.
22. G. A. Olah, T. Yamato, P. S. Iyer, G. K. Surya Prakash, *J. Org. Chem.* **1986**, *51*, 2826.
23. For related solution-phase precedents for these methylation reactions, see: a) S. A. Hartsel, W. S. Marshall, *Bioorg. Med. Chem. Lett.* **1996**, *6*, 2993; b) R. M. Burk, T. S. Gac, M. B. Roof, *Tetrahedron Lett.* **1994**, *35*, 8111.
24. W. G. Salmond, M. A. Barta, J. L. Havens, *J. Org. Chem.* **1978**, *43*, 2057.

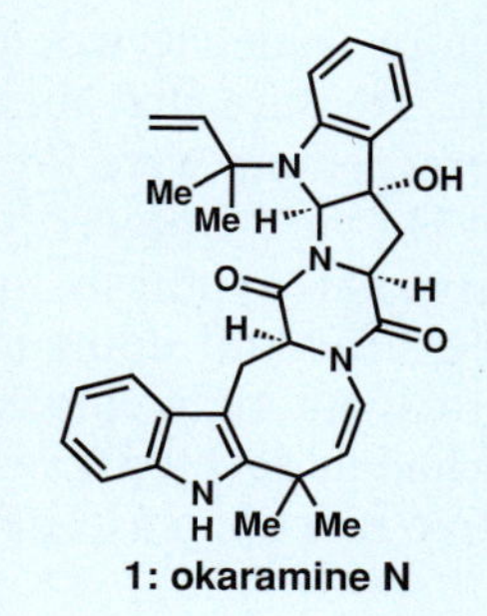
1: okaramine N

22

E. J. Corey (2003)

Okaramine N

22.1 Introduction

Certain natural products seem destined to become hotly pursued targets from the very moment their structures are published, with their rise to fame assured by their novel molecular architectures and/or their unprecedented types and levels of biological activity. Virtually every molecule that we have discussed in this book can be classified within this special category. In this chapter, however, we will present the story of a group of natural products whose ascent to synthetic prominence took a far longer and much less certain course.

These compounds are the okaramines, a modestly sized family of biologically active indole alkaloids isolated over the course of the past few years from various strains of *Penicillium simplicissum* by Professor Hideo Hiyashi and his collaborators at the Osaka Prefecture University.[1,2] With little question, their most important property is their distinctive structure, which, as indicated by the drawings of the selected okaramines shown in the adjoining column, includes a common heptacyclic core dominated by the presence of a daunting eight-membered dihydroindoloazocine ring. Indeed, ever since the first members of this family were identified in 1988 (i.e. **2** and **3**),[1] no other group of natural products has been isolated with the same assortment of structural motifs. Curiously, the architectural uniqueness of the okaramines drew little, if any, attention from the synthetic community. In fact, prior to 2003, there was no indication that synthetic chemists were even aware of their existence, as the primary literature was devoid of any examples of synthetic studies being directed towards them. Equally surprising, major compilations of natural products did not include entries for them.[3]

Key concepts:

- **Cascade reactions**
- **Indole protections**
- **Singlet oxygen chemistry**

2: okaramine A

3: okaramine B

With practitioners always sifting the literature for novel synthetic targets, it was really only a matter of time before someone would notice the okaramines and collect the synthetic treasure that their structures seemed to promise. The successful prospectors were Professor E. J. Corey and two of his co-workers at Harvard University, who, in early 2003, accomplished the first total synthesis of okaramine N (**1**).[4] As we shall see, the unique structural domains of this target afforded a stringent test of modern synthetic methods, an examination so taxing that these researchers had to develop a series of entirely novel transformations and creative reaction cascades in order to orchestrate their assembly.

22.2 Retrosynthetic Analysis and Strategy

Before we begin our analysis of the synthetic plan developed by the Corey group to attack the structural complexity of okaramine N (**1**), we feel compelled to reiterate that our discussions in this section always focus on the key elements of only the final synthetic route. As mentioned in the introductory chapters to both this book and *Classics I*, there are two main reasons for this approach. First, we feel that it provides the student with a better appreciation for the inherent risks and challenges incurred by adopting a particular strategy than would a more global and broad discussion in the absence of specifics. Second, since we can never truly ascertain the full retrosynthetic thoughts of the original authors, as these musings are often kept private and are typically based on failed experiments that may also not be mentioned, this format might be the only reconstruction possible by a third party. Our expectation is that the didactic value of all our retrosynthetic sections will supercede their potential inaccuracies in terms of how the synthesis was really approached. We mention this issue here, rather than elsewhere, because the retrosynthetic analysis that we are about to present certainly does not reflect the original strategy pursued by the Corey group towards this target molecule. Rather, it is the final product of a number of failed strategies that eventually evolved into a successful solution.* So, keeping this "post-synthesis" nature of all of our retrosynthetic analyses in this book in mind, we shall now press forward with the discussion of the molecule at hand.

If you have read the chapters on vancomycin and everninomicin 13,384-1 in this book, then you already know that most polycyclic natural products are quite difficult to analyze retrosynthetically because they provide an almost limitless number of options for how, and in what order, to construct their various domains. Superficially, okaramine N (**1**) appears to be essentially no different. However, it does possess one characteristic that, if recognized,

---

* Personal communication with Professor E. J. Corey and Dr. Phil S. Baran.

can dramatically limit the size of its retrosynthetic tree. That feature is the prominent location of most of its nitrogen atoms at, or near, the junctions of its central ring systems, because it suggests one could develop a retrosynthetic plan for this natural product using only these atoms to instigate their assembly. As the next few paragraphs will reveal, it was exactly with this general thought in mind that the Corey group executed their initial dissections of okaramine N (**1**) to convert it into simpler subgoal structures.

As even a cursory examination of the general structure of this target reveals, there are many points where one could begin the process of setting such a strategy into motion. Breaking apart either of the amide bonds that hold the D-ring diketopiperazine together would seem to be an obvious choice of disconnection, as would some type of transform that could appropriately unlock the 8-membered C-ring. However, as indicated by the initial set of alterations in Scheme 1, the Corey group elected instead to focus their attention on the stereochemically rich five-membered E-ring, opening it to reveal intermediate **6** as a potential synthetic precursor. In the synthetic direction, if singlet oxygen[5] could add with facial selectivity to the electron-rich C–C double bond of the *tert*-prenylated indole moiety in **6** to afford a reactive epoxide-like electrophile (**5**), then the nitrogen atom in the neighboring ring could likely engage it as shown to form **4**. Subsequent lysis of the resultant hydroperoxide, a task easily accomplished with Me_2S,[6] would then complete the assembly of okaramine N (**1**).

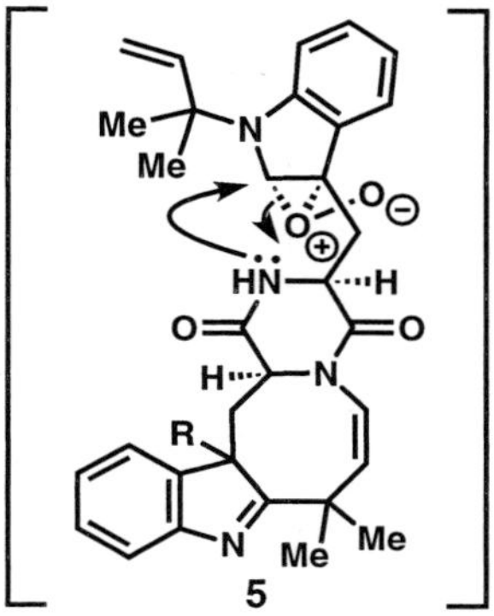

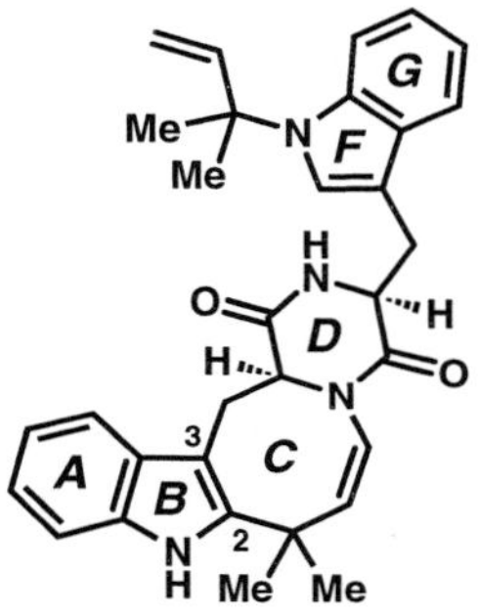

At first glance, this proposed photooxidation sequence seems reasonable as singlet oxygen is well-precedented to react with olefins to form perepoxides[7] such as **5**,* and the existing stereogenic center in **6** bridging the D- and F-rings should govern the approach of that reagent with the correct facial selectivity. Unfortunately, there is one major question for its viability: could singlet oxygen be counted upon to react only with the isolated C–C double bond of the FG indole and not its counterpart in the AB indole system? As a matter of fact, because the reverse prenyl group electronically deactivates the FG indole as an enophile, the C-2/C-3 double bond of the AB indole is the more likely target for singlet oxygen.** So, in order

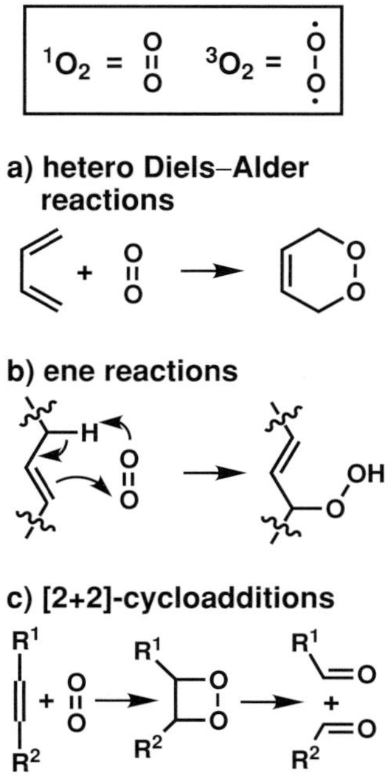

$^1O_2$ = O=O $^3O_2$ = •O–O•

a) hetero Diels–Alder reactions

b) ene reactions

c) [2+2]-cycloadditions

Figure 1. Singlet oxygen in organic synthesis.

* As an aside, singlet oxygen is a highly valuable reagent for diverse applications in organic synthesis since it can engage in three completely different, but equally useful, types of reactions with alkenes: 1) it can serve as a dienophile in hetero-Diels–Alder reactions with appropriate dienes as discussed in some detail in Chapter 2 on isochrysohermidin; 2) it can add to olefins bearing an allylic proton through an ene reaction to form hydroperoxides; and 3) it can react with electron-rich or strained olefins in [2+2]-cycloadditions to afford disparate types of final products. These modes of reactivity are shown graphically in Figure 1, and we refer you to several review articles if you wish to learn more about this reagent.[5] The proposed reaction here is a variant of this final mode of reactivity, with the perepoxide constituting a likely intermediate en route to the final product of a stepwise [2+2]-cycloaddition.

** Although compound **6** bears two other electron-rich C–C double bonds, they cannot participate in ene reactions because they lack allylic hydrogen atoms.

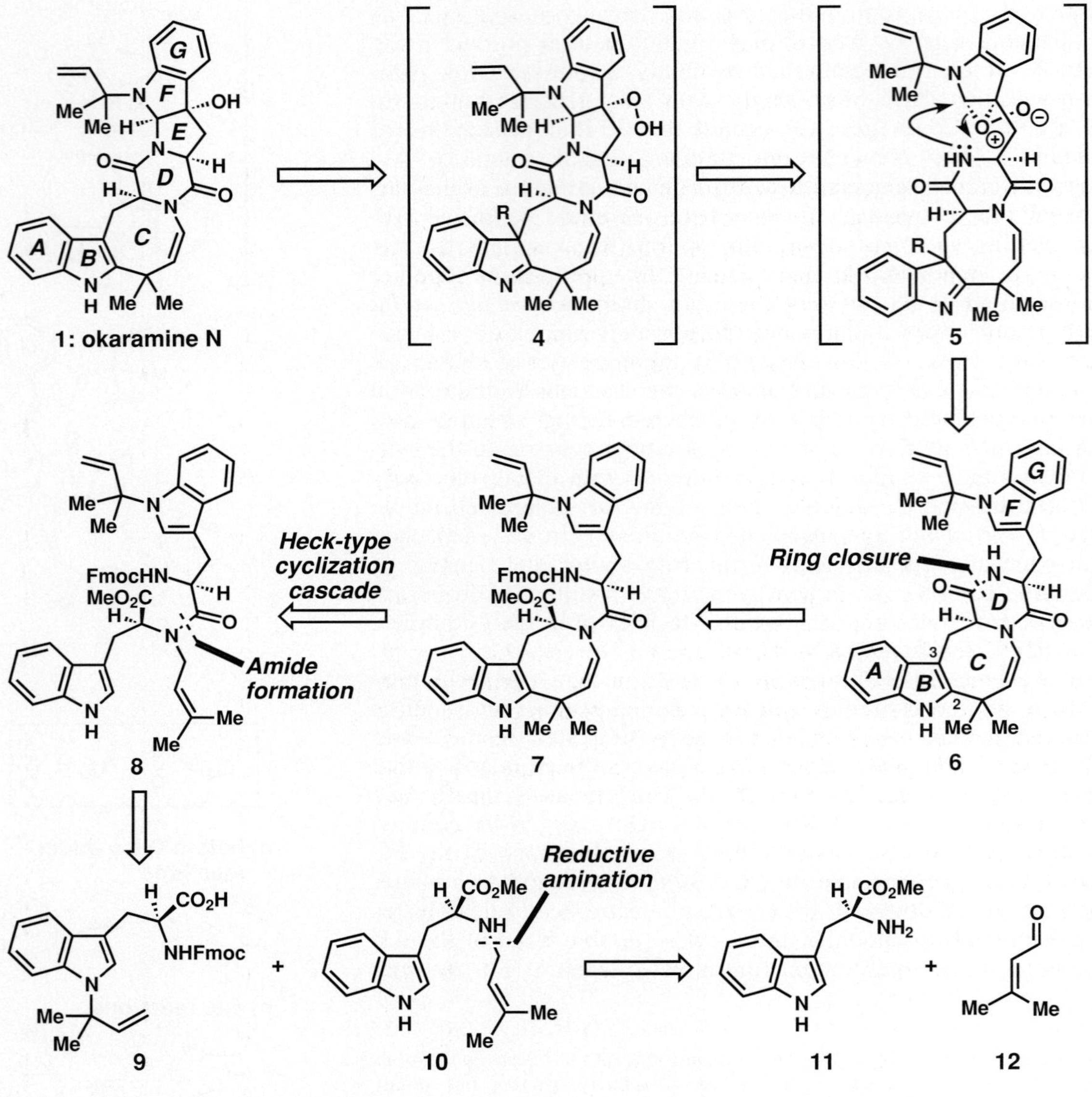

Scheme 1. Corey's retrosynthetic analysis of okaramine N (1).

for this bold strategy to have any realistic chance for success in the laboratory, the Corey group had to identify a way to distinguish these two indole systems. Their solution was conventional: attaching a protecting group onto the AB indole of **6** in order to shield it from attack by the singlet oxygen. The specific method they had in mind to implement this tactic, however, was quite novel. Namely, rather than append a protective device directly to the free indole of **6**, a task that would be difficult in light of the steric bulk of the adjacent C-ring methyl groups, they sought to accomplish the required differentiation by protecting its C2–C3 olefin instead.

This general idea is expressed in Scheme 1 by the presence of the generic R groups attached to intermediates **4** and **5**, motifs that could be installed through the well-established addition of electrophiles to the C-3 position of indoles.[8] Just what that R group would be, however, was an important detail, since despite the ability of many different electrophiles to react in the desired sense, none were known to be subsequently removable. Fortunately, such intellectual challenges often inspire interesting ideas, and, here, the Corey group hypothesized that perhaps an appropriate R group could be found if the electrophile was added to the indole not through a nucleophilic alkylation, but through an ene reaction instead. Since these reactions proceed through a pericyclic mechanism, perhaps the addition could be reversed at the appropriate juncture in the synthesis by heating the substrate to promote a retro-ene reaction.[9]

As shown in Scheme 2, initial probes of this idea using a commercially available ene reagent, *N*-methyltriazolinedione (**14**, MTAD),[10] were quite promising. Simply mixing a model indole-azocine system (**13**) with this reagent (**14**) in CH_2Cl_2 at 0 °C afforded **15** quantitatively in just 1 minute through the ene mechanism drawn in the inset box. Subsequent heating of this urazole product at 150 °C for 1 minute in the absence of solvent then accomplished the reverse ene reaction to re-form the starting indole (**13**). Thus, this model study not only established a new method for how indoles could be protected in general,[11] but it also afforded more than sufficient proof-of-principle to suggest that the initial disconnection of okaramine N (**1**) to **6** was worth pursuing in the forward synthesis.

Moving ahead with this analysis, the retrosynthetic sword next cut the D-ring diketopiperazine ring of **6** (see Scheme 1) at the indicated amide linkage to reveal **7** as a protected amino acid precursor. Although the selection of a methyl ester and a 9-fluorenylmethyl carbamate (Fmoc) group to protect these new functionalities

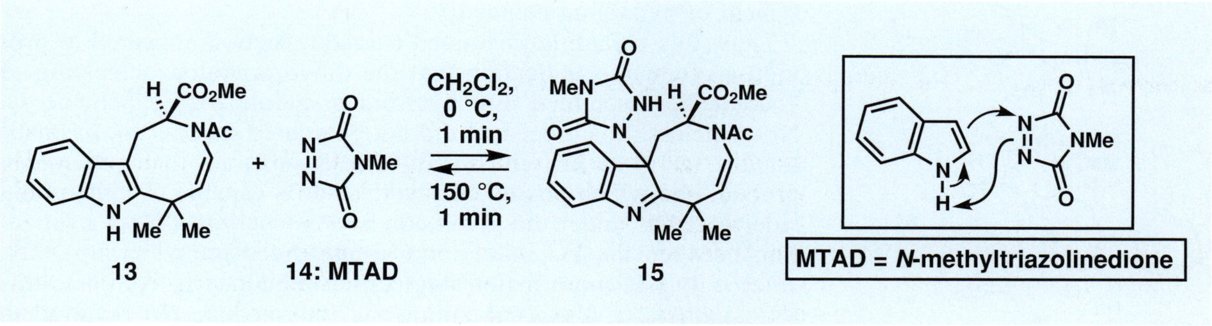

13 + 14: MTAD → (CH_2Cl_2, 0 °C, 1 min) 15 ; 15 → (150 °C, 1 min) 13

MTAD = *N*-methyltriazolinedione

Scheme 2. Reversible ene reaction of a dihydroindoloazocine with MTAD.

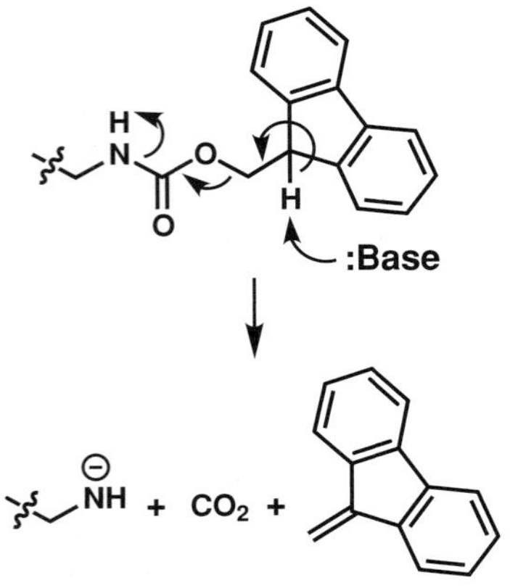

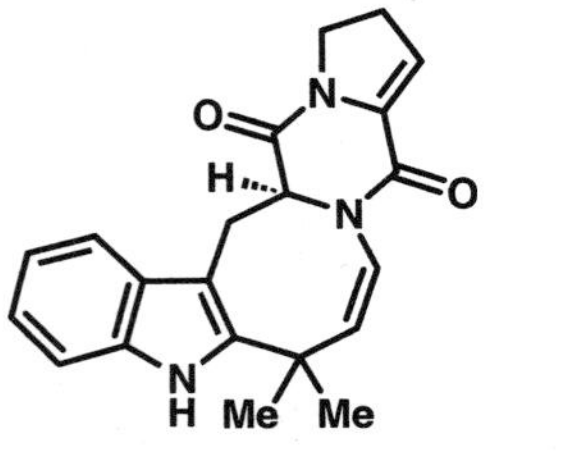

might seem arbitrary at first glance, they were critical to a plan to convert **7** into **6** in a single operation during the synthesis. For example, if **7** was treated with a suitable base to cleave the Fmoc protecting group guarding its amine (see column figures), then that operation should transform the innocuous nitrogen atom into a powerful nucleophile capable of forming the diketopiperazine ring directly by attacking its neighboring methyl ester.

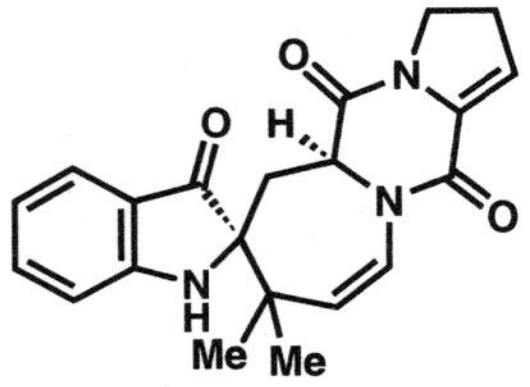

As such, the two general retrosynthetic operations outlined thus far have managed to reduce the complexity of okaramine N (**1**) by a total of two rings and two stereocenters, leaving the eight-membered dihydroindoloazocine ring as the one obvious synthetic challenge to be addressed. The next disconnection sought to tackle this lingering problem by projecting that this ring could arise in a single step using a novel palladium-mediated reaction sequence that had been developed in-house to generate the similar systems of the austamides (**16** and **17**, see column figures).[12] Scheme 3 outlines the details of this powerful and highly efficient cascade in more depth using model system **18** to illustrate its mechanistic underpinnings. As shown, treatment of **18** with stoichiometric amounts of $Pd(OAc)_2$ at ambient temperature leads to the initial insertion of that palladium species at the C-2 position of its indole ring in order to generate a reactive intermediate (**19**).[13] This species can then engage the pendant C–C double bond through a 7-*exo*-trig Heck cyclization to form a seven-membered ring (**20**). The alternate 8-*endo*-trig Heck reaction would not be expected to occur because it would require the attack of the metallated species (**19**) at the more substituted position of the double bond through a much more highly strained transition state.[14] From this point forward, however, what happens to intermediate **20** depends exclusively on the nature of the solvent. When that mixture is AcOH and H_2O, the major reaction product is an eight-membered dihydroindoloazocine (**23**), a compound whose formation can be rationalized as occurring through heterolysis of the C–Pd bond in **20** to generate a cationic intermediate (**21**) followed by an energetically favored ring-expansion/elimination sequence. In contrast to this succession of events, when **20** is formed in anhydrous AcOH, then only a seven-membered product (**27**) is observed. As shown, this compound is the likely product of a more conventional, β-elimination-type sequence typical of palladium chemistry.

Thus, this palladium-mediated cascade reaction appeared to provide a promising indication that the dihydroindoloazocine ring of **7** could be assembled directly from bisindole **8** (see Scheme 1). Nevertheless, this precedent did not guarantee success, as **7** constituted a far more heavily functionalized substrate than previously probed, one which possessed several motifs capable of undergoing side reactions under the given conditions (such as a Heck cyclization between the FG indole and its appended prenyl group). Yet, since only one more major surgical cut remained to complete this retrosynthetic analysis, obtaining the intermediate (**8**) required to test this potentially problematic, but highly appealing, reaction

16: (+)-deoxyisoaustamide

17: (+)-austamide

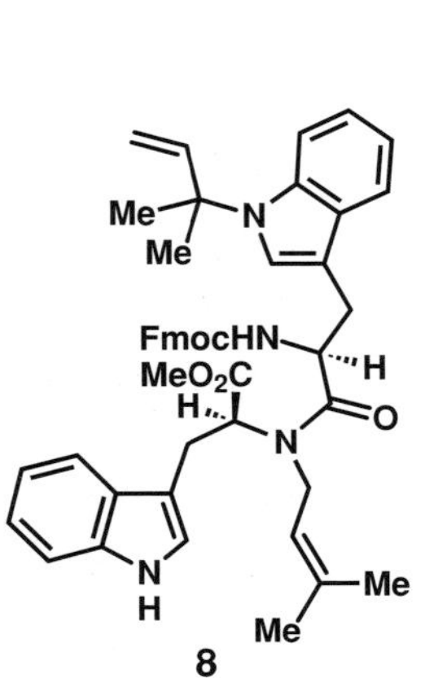

8

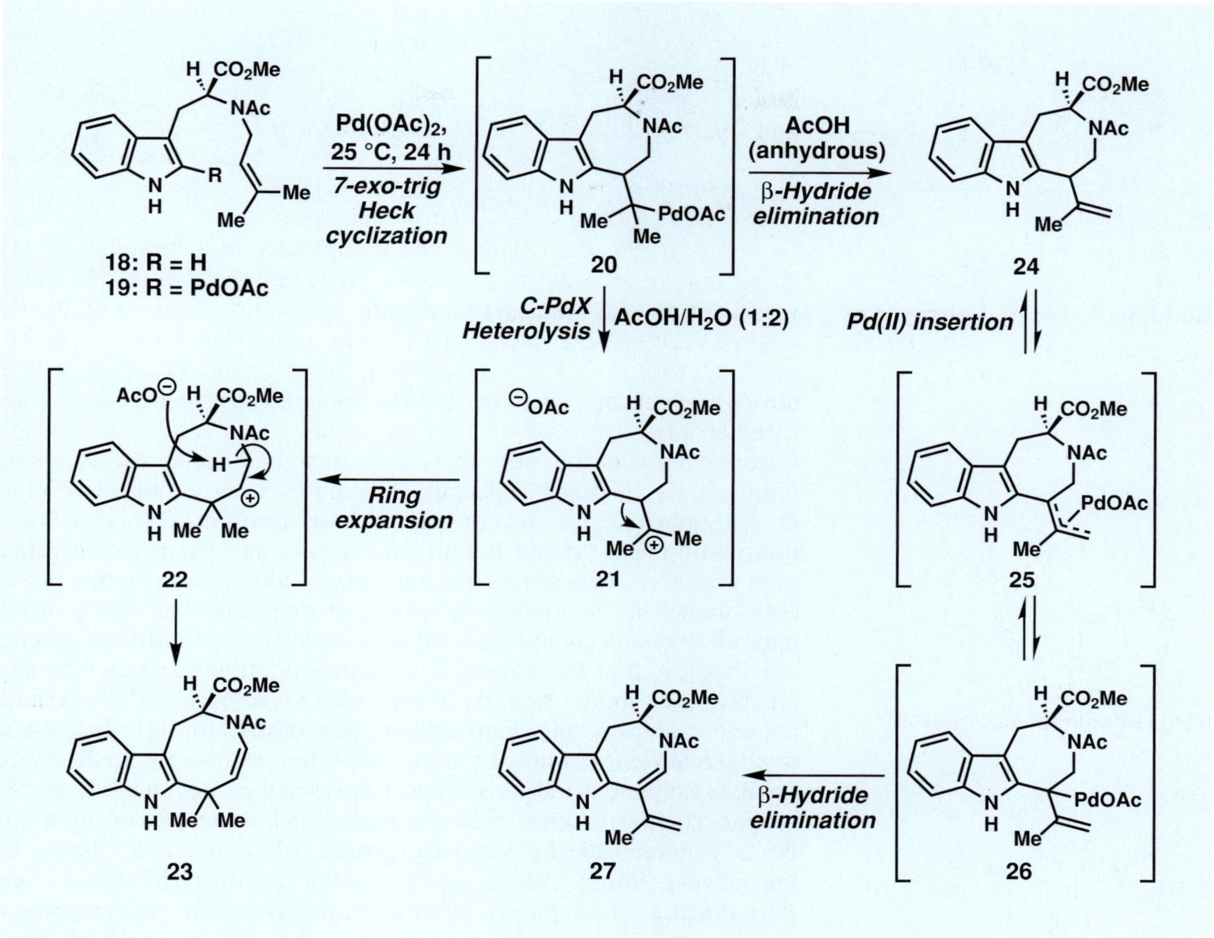

Scheme 3. Formation of dihydroindoloazocine **23** and dihydroindoloazepine **27** through a novel Pd-mediated cyclization.

was not expected to be an overly laborious task. That final retrosynthetic operation was the dissection of the amide at the heart of **8** to reveal carboxylic acid **9** and amine **10**, two equally sized precursors whose stereogenic centers and indole rings could be traced to L-tryptophan. Indeed, removal of the prenyl side chain from **10** as aldehyde **12** indicated that commercially available L-tryptophan methyl ester (**11**) would be a reasonable starting material, while indole **9** is obviously a Fmoc-protected version of the same amino acid with a reverse prenyl group attached to its indole core. As an aside, although the identification of tryptophan-derived building blocks to start this synthesis would appear to be a relatively obvious design, it is important to note that this decision was, in fact, counter to most other synthetic approaches toward related alkaloid natural products. A more typical strategy would be to start with an indole and then build the remainder of the tryptophan-type fragment

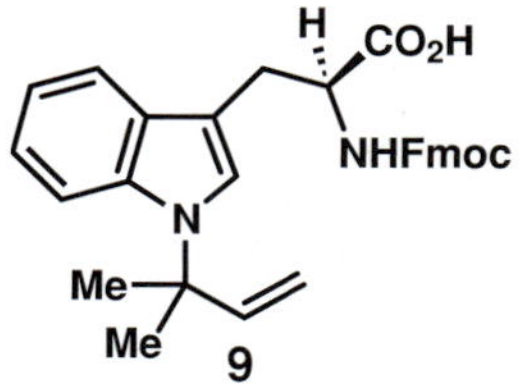

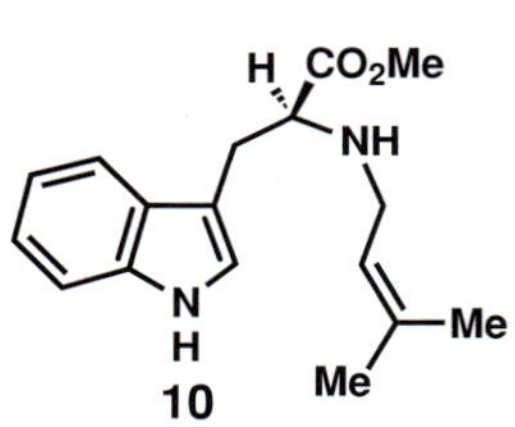

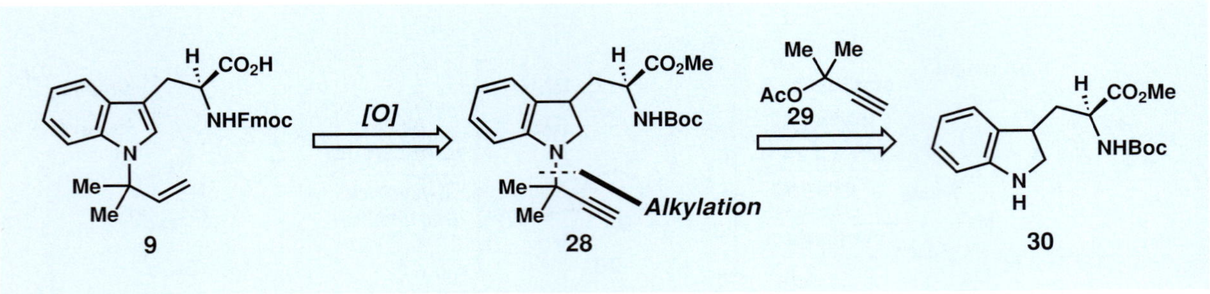

Scheme 4. Corey's retrosynthetic analysis of reverse prenylated indole **9**.

through reactions such as a Vilsmeier–Haack formylation (see Chapter 12).

Before concluding this analysis, a few additional words about indole **9** are in order, especially in regards to its attached reverse prenyl group. In line with the fact that we have not really discussed this motif in any depth before, this functional group is not often seen directly attached to a nitrogen atom, although there are probably numerous instances in which its presence has likely been masked by subsequent biosynthetic modifications (for example, see structure **3** at the start of this chapter).[15] Reverse prenyl groups are, however, more often found appended to phenols (see Chapter 17 for one example), and there are several methods for attaching them to either oxygen or nitrogen atoms. The short retrosynthetic analysis for **9** in Scheme 4 indicates one of these techniques, namely an *N*-propargylation reaction between propargyl acetate **29** and indoline **30** to generate **28**, followed by partial reduction of its alkyne to the corresponding alkene under Lindlar conditions.[16] While we shall discuss this sequence in far more detail shortly, it is important to note that indole itself is not a powerful enough nucleophile to accomplish the alkylation, which is why the proposed starting material in Scheme 4 is the indoline variant (**30**) of L-tryptophan rather than L-tryptophan itself. Consequently, the adoption of this general strategy meant that an oxidation step would have to be executed following the synthesis of **28** in order to finalize the structural format of the desired building block (**9**).

22.3 Total Synthesis

Although the overall plan that we have just delineated is both bold and creative, in order for it to have more than just intellectual appeal, it had to translate into a successful laboratory synthesis of okaramine N (**1**). Accordingly, the Corey group began their investigations of its feasibility by attempting to construct the reverse prenylated indole building block (**9**) along the lines of the generalized sequence shown in Scheme 4, a strategy which required the initial

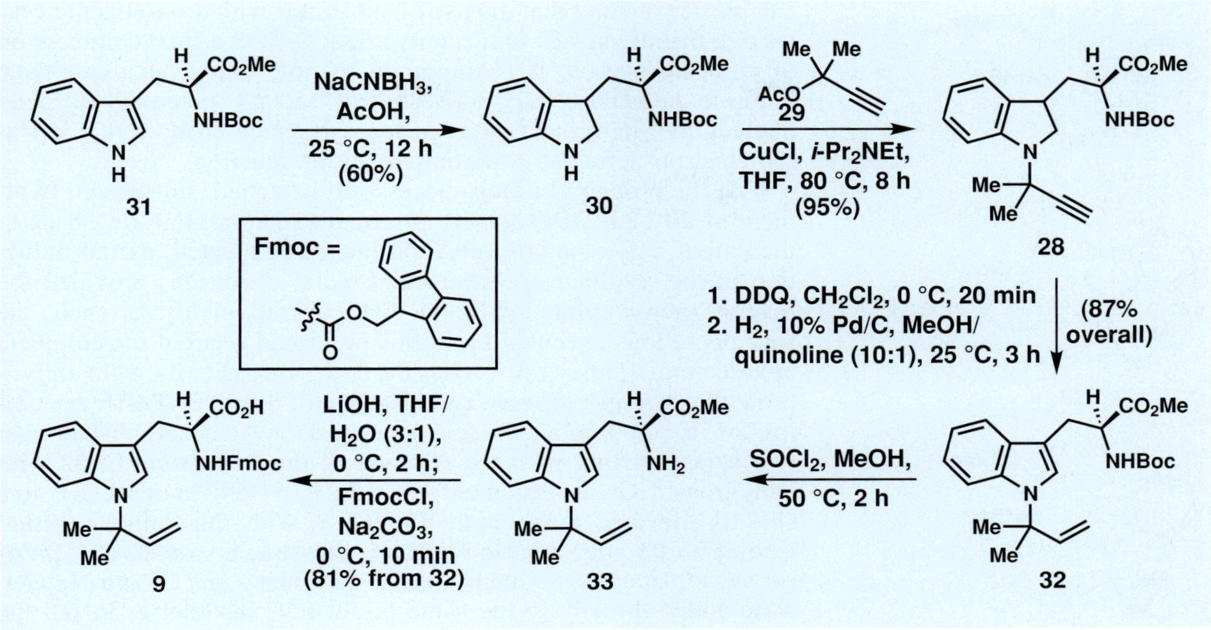

Scheme 5. Synthesis of reverse-prenylated indole **9**.

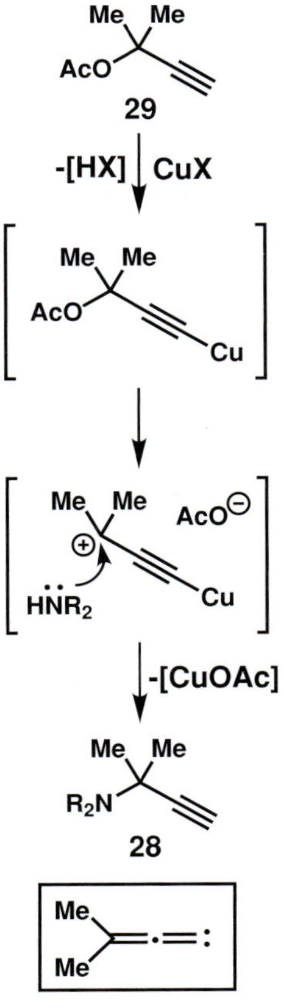

construction of indoline **30**. As indicated in Scheme 5, this introductory task was only of marginal difficulty, as it could be accomplished in a single step by treating the commercially available *N*-Boc-protected L-tryptophan methyl ester (**31**) with $NaCNBH_3$ in AcOH at ambient temperature.[17] This step delivered **30** in 60 % yield and provided an opportunity to attempt the key reaction of the projected route to **9**, namely its *N*-propargylation.

Fortunately, conditions pioneered by the Murahashi and Yokokawa groups for this specific purpose worked quite smoothly here.[16] In the event, a slight excess of 2-acetoxy-2-methyl-3-butyne (**29**) and a catalytic amount of CuCl in THF were added to a solution of **30** in THF, and the reaction mixture was then heated at reflux (80 °C) for a total of 8 hours. During this prolonged period of heating, the starting material was slowly converted into the desired propargylated product (**28**) in near quantitative yield (95 %), presumably as a consequence of the mechanism delineated in the neighboring column. As shown, this sequence commences with the insertion of the copper (I) reagent into the terminal position of the alkyne reactant (**29**), an operation that eventually leads to the generation of a highly reactive tertiary carbocation intermediate through the solvolysis of its acetate group. Direct attack of the indoline nitrogen atom (**30**) at this electrophile or a carbene derived from it (see the boxed structure) then completes the assembly of **28**. Before moving forward, we should emphasize that the mechanism we have just provided for this reaction is only a reasonable explanation based on empirical observations, such as the fact that only terminal alkynes can participate and that the use of substrates with a stereogenic center α to the alkyne afford racemic products.[16] As a final comment on this transformation, it is important to note that even though the indoline substrate (**30**) possessed a second potential nitrogen nucleophile, its protection as a Boc carbamate ensured that it was too electron-deficient to participate in the reaction.

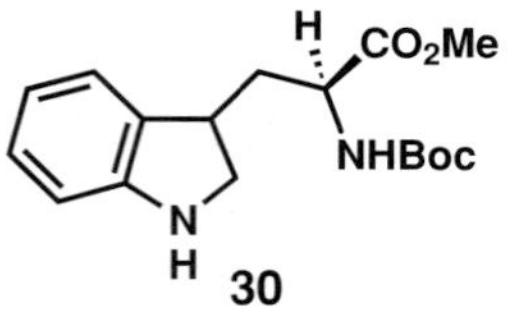

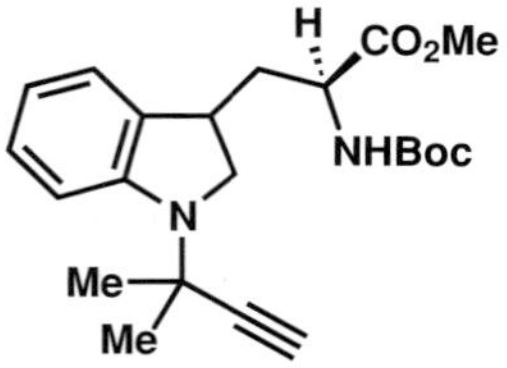

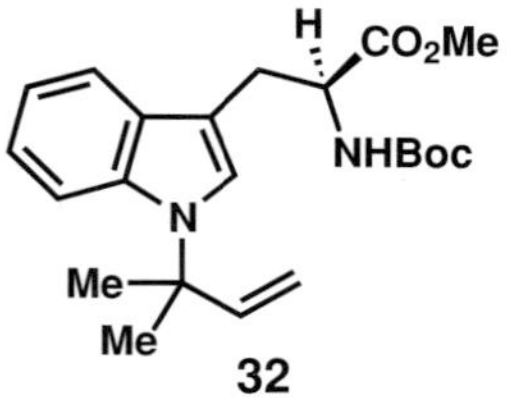

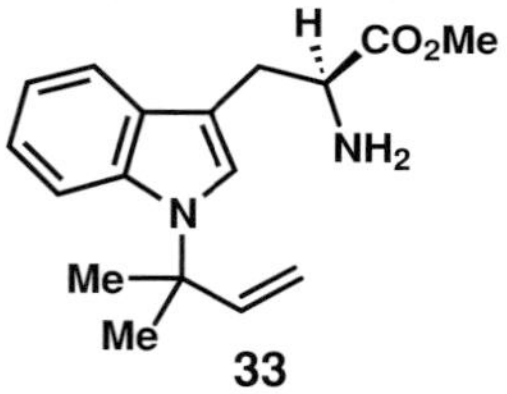

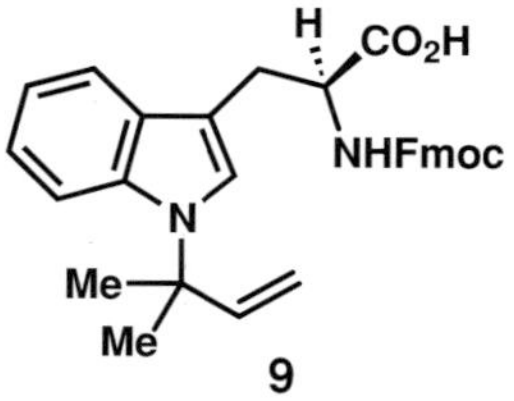

With the propargyl group successfully attached, subsequent treatment of **28** with DDQ served to reoxidize its indoline ring back to the parent indole heterocycle, and then, as expected, partial reduction of the resultant alkyne under Lindlar conditions provided the reverse prenyl group of **32** in 87 % overall yield. As such, the four operations executed up to this point had secured the complete architectural framework of the desired subtarget (**9**), with only a protecting group exchange and cleavage of the methyl ester remaining for its completion. These cursory tasks were accomplished over two steps, starting with the removal of the Boc group in **32** with anhydrous HCl as generated by mixing $SOCl_2$ and MeOH at a slightly elevated temperature (50 °C).[18] With this initial reaction leading to **33** after a basic work-up, subsequent exposure to LiOH served to saponify its methyl ester, and then FmocCl and Na_2CO_3 were added directly to the same pot at 0 °C in order to install the desired Fmoc protecting group and complete the assembly of **9**.* And, perhaps even more exciting than finalizing this fragment (**9**) in 81 % yield, the smooth completion of these steps meant that efforts could now be directed toward executing the challenging reactions that would hopefully construct the rest of okaramine N (**1**).

As discussed during the planning stages, the next major target was bisindole **8**, a compound whose synthesis was projected to require an initial reductive amination step between aldehyde **12** and commercially available L-tryptophan **11** to generate **10**, followed by its coupling with **9** (cf. Scheme 1). In practice, these steps did lead to the synthesis of **8**, but what is perhaps more amazing is their execution as a single-pot cascade sequence. The successful protocol is shown in Scheme 6, starting with the mixing of L-tryptophan methyl ester with aldehyde **12** in CH_2Cl_2 at ambient temperature for 3 hours to generate Schiff base **34**. Following removal of the solvent, all the residual adducts were resuspended in MeOH and treated with $NaBH_4$ at 0 °C to complete the reductive amination sequence leading to **10**. A second solvent exchange back to CH_2Cl_2, followed by the addition of the reverse prenylated indole (**9**) and the peptide coupling reagent BOP-Cl, then accomplished the synthesis of amide **8** in 70 % yield from **11**. As such, this overall material return correlated to an average success rate of 89 % for

* Even though one could envision starting this sequence with the Fmoc-protected version of **31** in order to obviate the need for this final protecting-group exchange, the conditions of the propargylation step are too strongly basic for this protecting group to survive, making this option unviable.

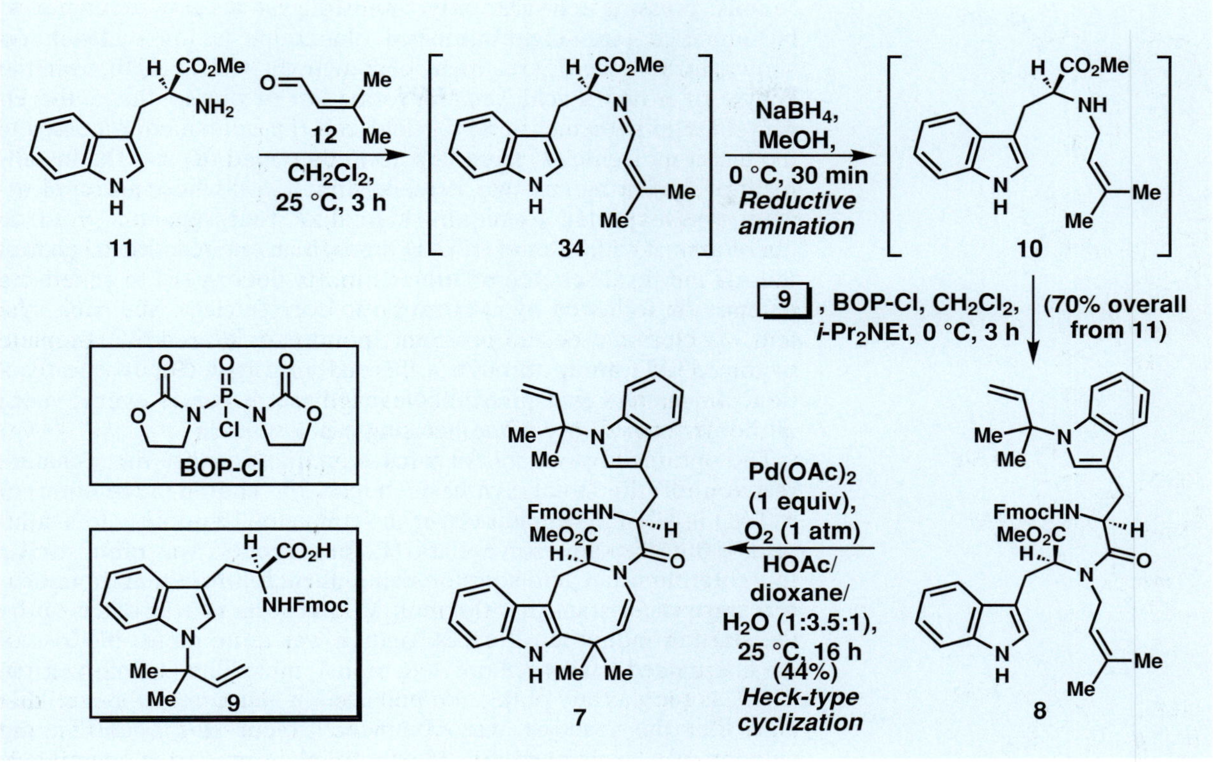

Scheme 6. Synthesis of advanced intermediate **7**.

each of the three separate steps accomplished during this one-flask operation.

With the synthesis of okaramine N (**1**) well on its way, the stage was now set to examine whether the proposed Pd-mediated Heck-type cascade sequence discussed earlier[12] could deliver the eight-membered dihydroindoloazocine C-ring. Pleasingly, the potential side reactions that could have occurred were not observed, and the synthesis of **7** was accomplished in 44 % yield based on recovered starting material (38 % yield of isolated **7**) using stoichiometric amounts of freshly prepared $Pd(OAc)_2$ in a 1:3.5:1 solvent mixture of HOAc, 1,4-dioxane, and water (the dioxane aids in the solubilization of **8**).[19] In line with our previous discussions of this multistep transformation, almost none of the alternate seven-membered ring products were observed at its completion, and, in light of the number of events accomplished during this reaction, the formation of the dihydroindoloazocine ring of **7** in any yield is particularly impressive. Equally important, this palladium-mediated annulation was robust enough that it could be performed on a multigram scale with no loss in efficiency, thereby ensuring that sufficient material supplies were available to probe the final steps of the projected campaign.

Thus, pressing forward, only two ring systems now remained to be engraved onto the established okaramine N (**1**) scaffold. As shown in Scheme 7, treatment of **7** with Et_2NH in THF over the course of 6 hours achieved the formation of one of these, the D-ring diketopiperazine, in 95 % yield. All that remained to complete the target molecule (**1**) from this new compound (**6**) was the installment of the E-ring and two stereocenters, a task whose accomplishment was expected to require a total of four separate synthetic operations: 1) addition of MTAD through an ene reaction to protect the AB indole, 2) photooxidation using singlet oxygen to generate a perepoxide followed by cyclization to complete the final ring system, 3) cleavage of the resultant hydroperoxide, and 4) excision of the MTAD group through a thermally-promoted retro-ene reaction. As matters transpired, however, these disparate events could all be performed in yet another single-pot cascade!

The optimized protocol for what is unquestionably the signature reaction of this total synthesis began by adding a solution of MTAD in CH_2Cl_2 dropwise over the course of 10 minutes to a solution of **6** in the same solvent at 0 °C. This process was quite similar to a titration of an acid solution using phenolphthalein as an indicator, since the instant that the pink MTAD reagent was taken up by the starting indole to afford **36** (which was quite rapid) the loss of its conjugated chromophore led to its immediate decolorization. Thus, as soon as any pink color persisted in solution, it was a telltale sign that the reaction was complete.* Once this occurred, the solvent was exchanged for MeOH, a photosensitizer (methylene blue) was added, and then oxygen was bubbled through the solution for several minutes. With the reaction media now primed for the generation of singlet oxygen, the solution was cooled to −28 °C and irradiated with a sunlamp for 7.5 hours, during which time sufficient quantities of this short-lived reactant were produced to convert most of **36** into hydroperoxide **39**. Excess Me_2S was then added directly to this flask at −28 °C to excise the superfluous oxygen atom installed by this cyclization sequence, and once sufficient time had passed to enable the completion of this essential task leading to **40**, the volatile materials were removed by rotary evaporation. The remaining residue was heated in the absence of solvent at 110 °C for 30 minutes. These conditions smoothly accomplished the retro-ene reaction, expelling the MTAD-based urazole protecting group, thereby furnishing okaramine N (**1**). Overall, the final natural product was isolated in 36 % yield from this sequence, with most of the remaining material balance constituting recovered **6**; if a yield was calculated based on that fact, this final cascade

---

* If more than one equivalent of MTAD was added to this flask, then the *tert*-prenylated FG indole system would begin to react with it as well, thereby preventing its participation in the required singlet oxygen chemistry. Interestingly, as subsequent studies have shown,[11] the MTAD addition is best promoted in MeOH. Thus, the use of CH_2Cl_2 as an alternative solvent here helped to ensure its chemoselective addition to the AB indole by tempering its reactivity.

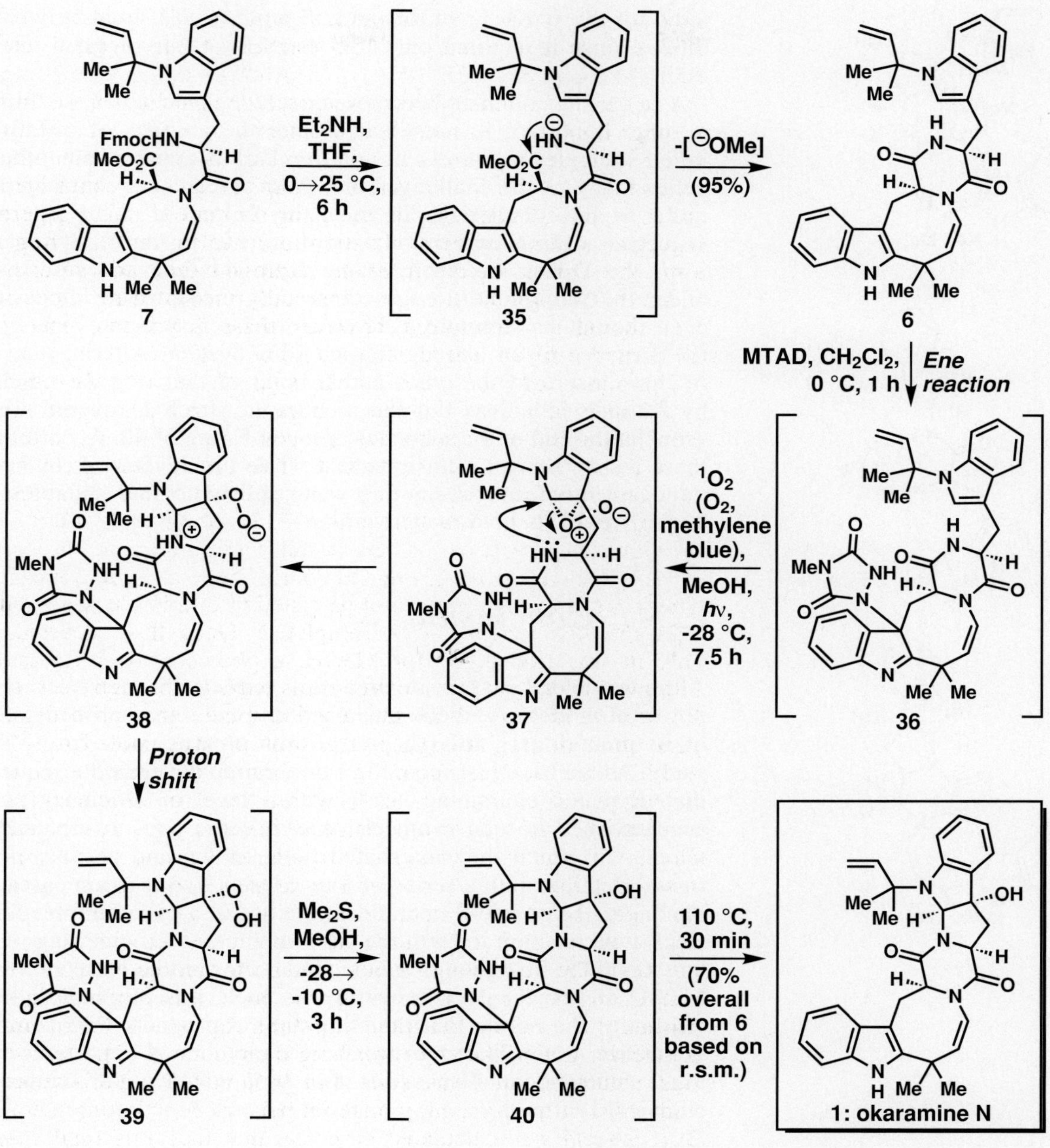

Scheme 7. Final stages and completion of Corey's total synthesis of okaramine N (**1**).

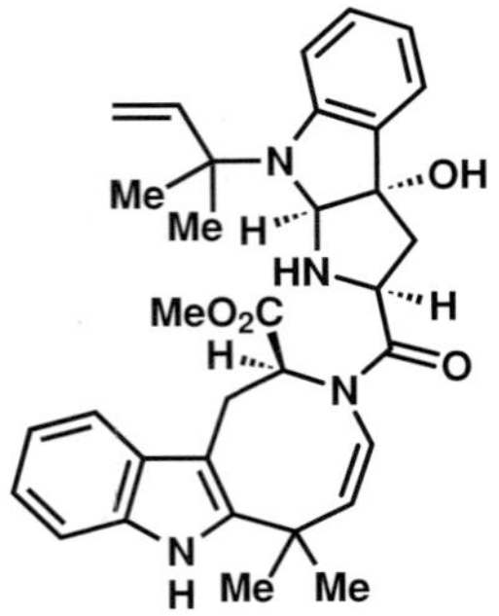

reaction was prosecuted in 70 % overall yield! In truth, though, not only was this operation quite efficient, but the entire pathway chartered for the synthesis of this natural product was stunningly expeditious since it required only four steps from L-tryptophan methyl ester (**11**).

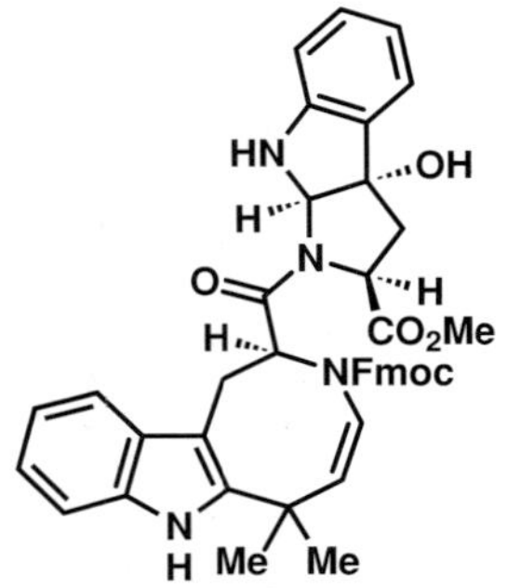

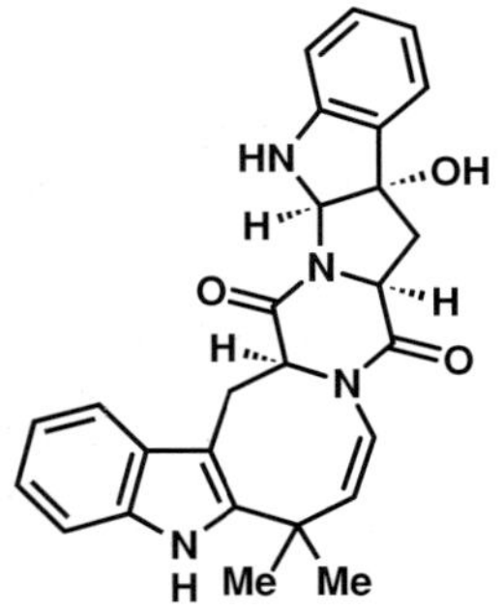

As a parting comment before we conclude this chapter, we think it is quite important to note that despite the absence of pitfalls or failed strategies in our discussion thus far, its apparent smoothness belies many of the challenges that these researchers confronted en route to its completion. In fact, the failure of many alternate sequences defied conventional wisdom. For instance, attempts to form the D-ring diketopiperazine from **41** (see column figure) under the same conditions as those described proved impossible, even though its structure differed from **7** (cf. Scheme 7) only by the presence of an already formed EFG system. Altering the site of that closure to the other amide bond of that ring as indicated by **42** also failed, as did the attempt to attach a reverse prenyl group at the end of the sequence onto the F-ring of **43**. Accordingly, these results serve to illustrate that while the developed route was quite inventive, much scouting was required to find suitable substrates for all its bold maneuvers.

22.4 *Conclusion*

Although every chapter in this book reflects the presently sharp power of modern synthetic chemistry and the remarkable ingenuity of its practitioners, this chapter affords a particularly strong case study. As we have just seen, the Corey group developed a sequence that delivered okaramine N (**1**) with a level of efficiency rarely accomplished in total synthesis, due, in large part, to the creative manner in which they marshaled both known and entirely novel transformations into a series of four consecutive reaction cascades. Among these powerful operations, the use of a palladium-mediated Heck-type reaction to form the eight-membered azocine ring of the target and the invention of a new method for indole protection with MTAD are especially noteworthy. As such, this elegant synthesis reinforces the notion that targeting natural products with complex or strange motifs is certain to afford a plethora of opportunities to make fundamental discoveries that will advance the science of chemical synthesis.

References

1. For the isolation of the first members of the okaramines, see: a) S. Murao, H. Hayashi, K. Takiuchi, M. Arai, *Agric. Biol. Chem.* **1988**, *52*, 885; b) H. Hayashi, K. Takiuchi, S. Murao, M. Arai, *Agric. Biol. Chem.* **1988**, *52*, 2131; c) H. Hayashi, K. Takiuchi, S. Murao, M. Arai, *Agric. Biol. Chem.* **1989**, *53*, 461.
2. For the isolation of other family members, including okaramine N, see: a) H. Hayashi, T. Fujiwara, S. Murao, M. Arai, *Agric. Biol. Chem.* **1991**, *55*, 3143; b) H. Hayashi, Y. Asabu, S. Murao, M. Arai, *Biosci. Biotechnol. Biochem.* **1995**, *59*, 246; c) H. Hayashi, A. Sakaguchi, *Biosci. Biotechnol. Biochem.* **1998**, *62*, 804; d) Y. Shiono, K. Akiyama, H. Hayashi, *Biosci. Biotechnol. Biochem.* **1999**, *63*, 1910; e) H. Hiyashi, K. Furutsuka, Y. Shiono, *J. Nat. Prod.* **1999**, *62*, 315; f) Y. Shiono, K. Akiyama, H. Hayashi, *Biosci. Biotechnol. Biochem.* **2000**, *64*, 103; g) Y. Shiono, K. Akiyama, H. Hayashi, *Biosci. Biotechnol. Biochem.* **2000**, *64*, 1519.
3. For selected examples, see: a) *Rompp Encyclopedia Natural Products* (Eds.: W. Steglich, B. Fugmann, S. Lang-Fugmann), Georg Thieme Verlag, Stuttgart, **2000**, p. 748; b) *The Merck Index, 13th Ed.*, Merck & Co., Inc., Whitehouse Station, **2001**, p. 1818.
4. P. S. Baran, C. A. Guerrero, E. J. Corey, *J. Am. Chem. Soc.* **2003**, *125*, 5628.
5. For selected review articles on the chemistry of singlet oxygen, see: a) A. G. Leach, K. N. Houk, *Chem. Commun.* **2002**, 1243; b) M. Stratakis, M. Orfanopoulos, *Tetrahedron* **2000**, *56*, 1595; c) H. H. Wasserman, R. W. DeSimone in *Encyclopedia of Reagents for Organic Synthesis*, *Vol. 6* (Ed.: L. A. Paquette), John Wiley & Sons, Chichester, **1995**, pp. 4478–4484.
6. H. H. Wasserman, I. Saito, *J. Am. Chem. Soc.* **1975**, *97*, 905.
7. For lead references, see: a) N. M. Hasty, D. R. Kearns, *J. Am. Chem. Soc.* **1973**, *95*, 3380; b) A. P. Schaap, G. R. Faler, *J. Am. Chem. Soc.* **1973**, *95*, 3381; c) S. Inagaki, K. Fukui, *J. Am. Chem. Soc.* **1975**, *97*, 7480. For a review, see: d) H. H. Wasserman, J. L. Ives, *Tetrahedron* **1981**, *37*, 1825.
8. For representative examples and discussions, see: T. L. Gilchrist, *Heterocyclic Chemistry, 3rd Ed.*, Addison Wesley Longman, Harlow, **1997**, pp. 237–241.
9. For selected reviews on both ene and retro-ene reactions, see: a) J.-L. Ripoll, Y. Vallée, *Synthesis* **1993**, 659; b) K. Mikami, M. Shimizu, *Chem. Rev.* **1992**, *92*, 1021; c) H. M. R. Hoffmann, *Angew. Chem.* **1969**, *81*, 597; *Angew. Chem. Int. Ed. Engl.* **1969**, *8*, 556.
10. For selected examples of the use of MTAD in ene reactions, see: a) W. Adam, A. Pastor, T. Wirth, *Org. Lett.* **2000**, *2*, 1295; b) W. Adam, T. Wirth, A. Pastor, K. Peters, *Eur. J. Org. Chem.* **1998**, 501; c) G. Vassilikogiannakis, Y. Elemes, M. Orfanopoulos, *J. Am. Chem. Soc.* **2000**, *122*, 9540; d) G. Vassilikogiannakis, M. Stratakis, M. Orfanopoulos, C. S. Foote, *J. Org. Chem.* **1999**, *64*, 4130.
11. P. S. Baran, C. A. Guerrero, E. J. Corey, *Org. Lett.* **2003**, *5*, 1999.
12. P. S. Baran, E. J. Corey, *J. Am. Chem. Soc.* **2002**, *124*, 7904. For another total synthesis of some members of the austamide family, see: A. J. Hutchison, Y. Kishi, *J. Am. Chem. Soc.* **1979**, *101*, 6786.
13. For an example of a related insertion using excess silver salts and stoichiometric Pd(0), see: B. M. Trost, S. A. Godleski, J. P. Genêt, *J. Am. Chem. Soc.* **1978**, *100*, 3930.
14. For a discussion on this mechanistic pathway as well as a general review of the Heck reaction, see: I. P. Beletskaya, A. V. Cheprakov, *Chem. Rev.* **2000**, *100*, 3009.
15. For relevant discussions and lead references, see: a) E. M. Stocking, R. M. Williams, J. F. Sanz-Cervera, *J. Am. Chem. Soc.* **2000**, *122*, 9089; b) T. D. Cushing, J. F. Sanz-Cervera, R. M. Williams, *J. Am. Chem. Soc.* **1996**, *118*, 557.
16. a) H. Sugiyama, F. Yokokawa, T. Aoyama, T. Shioiri, *Tetrahedron Lett.* **2001**, *42*, 7277; b) Y. Imada, M. Yuasa, I. Nakamura, S.-I. Murahashi, *J. Org. Chem.* **1994**, *59*, 2282. The fact that copper salts could promote such amination reactions was known for some time prior to the above reports: G. F. Hennion, R. S. Hanzel, *J. Am. Chem. Soc.* **1960**, *82*, 4908.
17. For an alternate synthesis of this compound, see: T. D. Dinh, D. L. Van Vranken, *J. Peptide Res.* **1999**, *53*, 465.
18. For a related preparation of anhydrous HCl, see: A. Nudelman, Y. Bechor, E. Falb, B. Fischer, B. A. Wexler, A. Nudelman, *Synth. Commun.* **1998**, *28*, 471.
19. T. A. Stephenson, S. M. Morehouse, A. R. Powell, J. P. Heffer, G. Wilkinson, *J. Chem. Soc.* **1965**, 3632.

Author Index

Subject Index

B

C

E

F

N

Q

S